Electrical

John Bird

In Memory of Elizabeth

Electrical Circuit Theory and Technology

Second Edition

John Bird, BSc(Hons), CEng, MIEE, FIEIE, CMath, FIMA, FCollP

OXFORD AUCKLAND BOSTON JOHANNESBURG MELBOURNE NEW DELHI

Newnes
An imprint of Butterworth-Heinemann
Linacre House, Jordan Hill, Oxford OX2 8DP
225 Wildwood Avenue, Woburn, MA 01801-2041
A division of Reed Educational and Professional Publishing Ltd

 A member of the Reed Elsevier plc group

First published 1997
Second edition 2001

British Library Cataloguing in Publication Data

A catalogue record for this book is available from the British Library

ISBN 0 7506 4989 5

Typeset by Laser Words, Madras, India
Printed and bound in Great Britain

Contents

Preface

'Electrical Circuit Theory and Technology, Second Edition' provides coverage for a wide range of courses that contain electrical principles, circuit theory and technology in their syllabuses, from introductory to degree level. In this second edition, material has been added on semiconductor diodes, transistors and operational amplifiers. Page extent restraints have meant that Chapter 42 from the first edition — 'Transients and Laplace transforms' has been removed from this edition; however, for those who need this chapter please refer to www.newnespress.com

The text is set out in four parts as follows:

PART 1, involving chapters 1 to 12, contains **'Basic Electrical Engineering Principles'** which any student wishing to progress in electrical engineering would need to know. An introduction to electrical circuits, resistance variation, chemical effects of electricity, series and parallel circuits, capacitors and capacitance, magnetic circuits, electromagnetism, electromagnetic induction, semiconductor diodes, transistors and electrical measuring instruments and measurements are all included in this section.

PART 2, involving chapters 13 to 22, contains **'Electrical Principles and Technology'** suitable for Advanced GNVQ, National Certificate, National Diploma and City and Guilds courses in electrical and electronic engineering. D.c. circuit theory, alternating voltages and currents, single-phase series and parallel circuits, d.c. transients, operational amplifiers, three-phase systems, transformers, d.c. machines and three-phase induction motors are all included in this section.

PART 3, involving chapters 23 to 44, contains **'Advanced Circuit Theory and Technology'** suitable for Degree, Higher National Certificate/Diploma and City and Guilds courses in electrical and electronic/telecommunications engineering. The two earlier sections of the book will provide a valuable reference/revision for students at this level.

Complex numbers and their application to series and parallel networks, power in a.c. circuits, a.c. bridges, series and parallel resonance and Q-factor, network analysis involving Kirchhoff's laws, mesh and nodal analysis, the superposition theorem, Thévenin's and Norton's theorems, delta-star and star-delta transforms, maximum power transfer theorems and impedance matching, complex waveforms, harmonic analysis, magnetic materials, dielectrics and dielectric loss, field theory, attenuators, filter networks, magnetically coupled circuits and transmission line theory are all included in this section.

PART 4 provides a short, **'General Reference'** for standard electrical quantities - their symbols and units, the Greek alphabet, common prefixes and resistor colour coding and ohmic values.

At the beginning of each of the 44 chapters **learning objectives** are listed.

At the end of each of the first three parts of the text is a handy reference of the **main formulae** used.

It is not possible to acquire a thorough understanding of electrical principles, circuit theory and technology without working through a good number of numerical problems. It is for this reason that *'Electrical Circuit Theory and Technology, Second Edition'* contains some **700 detailed worked problems,** together with over **1050 further problems,** all with answers in brackets immediately following each question. Over **1050 line diagrams** further enhance the understanding of the theory.

Fourteen Assignments have been included, interspersed within the text every few chapters. For example, Assignment 1 tests understanding of chapters 1 to 4, Assignment 2 tests understanding of chapters 5 to 7, Assignment 3 tests understanding of chapters 8 to 12, and so on.

(A lecturers' resource guide is available, containing the solutions to the Assignments).

'Learning by Example' is at the heart of *'Electrical Circuit Theory and Technology, Second Edition'*.

JOHN BIRD
University of Portsmouth

Part 1 Basic Electrical Engineering Principles

1 Units associated with basic electrical quantities

At the end of this chapter you should be able to:

- state the basic SI units
- recognize derived SI units
- understand prefixes denoting multiplication and division
- state the units of charge, force, work and power and perform simple calculations involving these units
- state the units of electrical potential, e.m.f., resistance, conductance, power and energy and perform simple calculations involving these units

1.1 SI units

The system of units used in engineering and science is the Système Internationale d'Unités (International system of units), usually abbreviated to SI units, and is based on the metric system. This was introduced in 1960 and is now adopted by the majority of countries as the official system of measurement.

The basic units in the SI system are listed with their symbols, in Table 1.1.

TABLE 1.1 *Basic SI Units*

Quantity	*Unit*
length	metre, m
mass	kilogram, kg
time	second, s
electric current	ampere, A
thermodynamic temperature	kelvin, K
luminous intensity	candela, cd
amount of substance	mole, mol

Derived SI units use combinations of basic units and there are many of them. Two examples are:

- Velocity — metres per second (m/s)
- Acceleration — metres per second squared (m/s^2)

SI units may be made larger or smaller by using prefixes which denote multiplication or division by a particular amount. The six most common multiples, with their meaning, are listed in Table 1.2.

TABLE 1.2

Prefix	*Name*	*Meaning*	
M	mega	multiply by 1 000 000	(i.e. $\times 10^6$)
k	kilo	multiply by 1000	(i.e. $\times 10^3$)
m	milli	divide by 1000	(i.e. $\times 10^{-3}$)
μ	micro	divide by 1 000 000	(i.e. $\times 10^{-6}$)
n	nano	divide by 1 000 000 000	(i.e. $\times 10^{-9}$)
p	pico	divide by 1 000 000 000 000	(i.e. $\times 10^{-12}$)

1.2 Charge

The **unit of charge** is the coulomb (C) where one coulomb is one ampere second. (1 coulomb $= 6.24 \times 10^{18}$ electrons). The coulomb is defined as the quantity of electricity which flows past a given point in an electric circuit when a current of one ampere is maintained for one second. Thus,

charge, in coulombs $\boldsymbol{Q = It}$

where I is the current in amperes and t is the time in seconds.

Problem 1. If a current of 5 A flows for 2 minutes, find the quantity of electricity transferred.

Quantity of electricity $Q = It$ coulombs

$I = 5$ A, $t = 2 \times 60 = 120$ s

Hence $Q = 5 \times 120 =$ **600 C**

1.3 Force

The **unit of force** is the newton (N) where one newton is one kilogram metre per second squared. The newton is defined as the force which, when applied to a mass of one kilogram, gives it an acceleration of one metre per second squared. Thus,

force, in newtons $\boldsymbol{F = ma}$

where m is the mass in kilograms and a is the acceleration in metres per second squared. Gravitational force, or weight, is mg, where $g = 9.81$ m/s^2

Problem 2. A mass of 5000 g is accelerated at 2 m/s^2 by a force. Determine the force needed.

Force = mass × acceleration

$$= 5\text{ kg} \times 2\text{ m/s}^2 = 10\frac{\text{kg m}}{\text{s}^2} = \mathbf{10\ N}$$

Problem 3. Find the force acting vertically downwards on a mass of 200 g attached to a wire.

Mass = 200 g = 0.2 kg and acceleration due to gravity, $g = 9.81\text{ m/s}^2$

Force acting downwards = weight = mass × acceleration

$= 0.2\text{ kg} \times 9.81\text{ m/s}^2$

$= \mathbf{1.962\ N}$

1.4 Work

The **unit of work or energy** is the **joule (J)** where one joule is one newton metre. The joule is defined as the work done or energy transferred when a force of one newton is exerted through a distance of one metre in the direction of the force. Thus

work done on a body, in joules $\boxed{\boldsymbol{W = Fs}}$

where F is the force in newtons and s is the distance in metres moved by the body in the direction of the force. Energy is the capacity for doing work.

1.5 Power

The **unit of power** is the watt (W) where one watt is one joule per second. Power is defined as the rate of doing work or transferring energy. Thus,

power in watts, $\boxed{\boldsymbol{P = \frac{W}{t}}}$

where W is the work done or energy transferred in joules and t is the time in seconds. Thus

energy, in joules, $\boxed{\boldsymbol{W = Pt}}$

Problem 4. A portable machine requires a force of 200 N to move it. How much work is done if the machine is moved 20 m and what average power is utilized if the movement takes 25 s?

Work done = force × distance = 200 N × 20 m = **4000 Nm or 4 kJ**

$$\text{Power} = \frac{\text{work done}}{\text{time taken}} = \frac{4000\ \text{J}}{25\ \text{s}} = \mathbf{160\ J/s = 160\ W}$$

Problem 5. A mass of 1000 kg is raised through a height of 10 m in 20 s. What is (a) the work done and (b) the power developed?

(a) Work done = force × distance and force = mass × acceleration

Hence, work done $= (1000\ \text{kg} \times 9.81\ \text{m/s}^2) \times (10\ \text{m})$

$= 98\,100\ \text{Nm} =$ **98.1 kNm or 98.1 kJ**

(b) $$\text{Power} = \frac{\text{work done}}{\text{time taken}} = \frac{98100\ \text{J}}{20\ \text{s}} = 4905\ \text{J/s}$$

$= $ **4905 W or 4.905 kW**

1.6 Electrical potential and e.m.f.

The **unit of electric potential** is the volt (V) where one volt is one joule per coulomb. One volt is defined as the difference in potential between two points in a conductor which, when carrying a current of one ampere, dissipates a power of one watt, i.e.

$$\text{volts} = \frac{\text{watts}}{\text{amperes}} = \frac{\text{joules/second}}{\text{amperes}} = \frac{\text{joules}}{\text{ampere seconds}} = \frac{\text{joules}}{\text{coulombs}}$$

A change in electric potential between two points in an electric circuit is called a **potential difference**. The **electromotive force (e.m.f.)** provided by a source of energy such as a battery or a generator is measured in volts.

1.7 Resistance and conductance

The **unit of electric resistance** is the **ohm (Ω)** where one ohm is one volt per ampere. It is defined as the resistance between two points in a conductor when a constant electric potential of one volt applied at the two points produces a current flow of one ampere in the conductor. Thus,

resistance, in ohms $$\boldsymbol{R = \frac{V}{I}}$$

where V is the potential difference across the two points in volts and I is the current flowing between the two points in amperes.

The reciprocal of resistance is called **conductance** and is measured in siemens (S). Thus,

conductance, in siemens $\boldsymbol{G = \frac{1}{R}}$

where R is the resistance in ohms.

Problem 6. Find the conductance of a conductor of resistance (a) 10 Ω, (b) 5 kΩ and (c) 100 mΩ

(a) Conductance $G = \frac{1}{R} = \frac{1}{10}$ siemen = **0.1 s**

(b) $G = \frac{1}{R} = \frac{1}{5 \times 10^3}\ \text{S} = 0.2 \times 10^{-3}\ \text{S} = \mathbf{0.2\ mS}$

(c) $G = \frac{1}{R} = \frac{1}{100 \times 10^{-3}}\ \text{S} = \frac{10^3}{100}\ \text{S} = \mathbf{10\ S}$

1.8 Electrical power and energy

When a direct current of I amperes is flowing in an electric circuit and the voltage across the circuit is V volts, then

power, in watts $\boldsymbol{P = VI}$

Electrical energy = Power × time

= ***VIt* Joules**

Although the unit of energy is the joule, when dealing with large amounts of energy, the unit used is the **kilowatt hour (kWh)** where

$$1\ \text{kWh} = 1000\ \text{watt hour}$$
$$= 1000 \times 3600\ \text{watt seconds or joules}$$
$$= 3\,600\,000\ \text{J}$$

Problem 7. A source e.m.f. of 5 V supplies a current of 3 A for 10 minutes. How much energy is provided in this time?

Energy = power × time and power = voltage × current. Hence

Energy $= VIt = 5 \times 3 \times (10 \times 60) = 9000$ Ws or J

$= \mathbf{9\ kJ}$

Problem 8. An electric heater consumes 1.8 MJ when connected to a 250 V supply for 30 minutes. Find the power rating of the heater and the current taken from the supply.

i.e. **Power rating of heater = 1 kW**

Power $P = VI$, thus $I = \dfrac{P}{V} = \dfrac{1000}{250} = 4$ A

Hence the current taken from the supply is 4 A

1.9 Summary of terms, units and their symbols

Quantity	Quantity Symbol	Unit	Unit symbol
Length	l	metre	m
Mass	m	kilogram	kg
Time	t	second	s
Velocity	v	metres per second	m/s or m s^{-1}
Acceleration	a	metres per second squared	m/s^2 or m s^{-2}
Force	F	newton	N
Electrical charge or quantity	Q	coulomb	C
Electric current	I	ampere	A
Resistance	R	ohm	Ω
Conductance	G	siemen	S
Electromotive force	E	volt	V
Potential difference	V	volt	V
Work	W	joule	J
Energy	E (or W)	joule	J
Power	P	watt	W

As progress is made through *Electrical Circuit Theory and Technology* many more terms will be met. A full list of electrical quantities, together with their symbols and units are given in Part 4, page 911.

1.10 Further problems on units associated with basic electrical quantities

(Take $g = 9.81$ m/s^2 where appropriate)

1. What force is required to give a mass of 20 kg an acceleration of 30 m/s^2? [600 N]
2. Find the accelerating force when a car having a mass of 1.7 Mg increases its speed with a constant acceleration of 3 m/s^2 [5.1 kN]
3. A force of 40 N accelerates a mass at 5 m/s^2. Determine the mass. [8 kg]
4. Determine the force acting downwards on a mass of 1500 g suspended on a string. [14.72 N]
5. A force of 4 N moves an object 200 cm in the direction of the force. What amount of work is done? [8 J]
6. A force of 2.5 kN is required to lift a load. How much work is done if the load is lifted through 500 cm? [12.5 kJ]
7. An electromagnet exerts a force of 12 N and moves a soft iron armature through a distance of 1.5 cm in 40 ms. Find the power consumed. [4.5 W]
8. A mass of 500 kg is raised to a height of 6 m in 30 s. Find (a) the work done and (b) the power developed. [(a) 29.43 kNm (b) 981 W]
9. What quantity of electricity is carried by 6.24×10^{21} electrons? [1000 C]
10. In what time would a current of 1 A transfer a charge of 30 C? [30 s]
11. A current of 3 A flows for 5 minutes. What charge is transferred? [900 C]
12. How long must a current of 0.1 A flow so as to transfer a charge of 30 C? [5 minutes]
13. Find the conductance of a resistor of resistance (a) 10Ω (b) 2 kΩ (c) 2 mΩ [(a) 0.1 S (b) 0.5 mS (c) 500 S]
14. A conductor has a conductance of 50 μS. What is its resistance? [20 kΩ]
15. An e.m.f. of 250 V is connected across a resistance and the current flowing through the resistance is 4 A. What is the power developed? [1 kW]
16. 450 J of energy are converted into heat in 1 minute. What power is dissipated? [7.5 W]
17. A current of 10 A flows through a conductor and 10 W is dissipated. What p.d. exists across the ends of the conductor? [1 V]
18. A battery of e.m.f. 12 V supplies a current of 5 A for 2 minutes. How much energy is supplied in this time? [7.2 kJ]
19. A dc electric motor consumes 36 MJ when connected to a 250 V supply for 1 hour. Find the power rating of the motor and the current taken from the supply. [10 kW, 40 A]

2 An introduction to electric circuits

At the end of this chapter you should be able to:

- recognize common electrical circuit diagram symbols
- understand that electric current is the rate of movement of charge and is measured in amperes
- appreciate that the unit of charge is the coulomb
- calculate charge or quantity of electricity Q from $Q = It$
- understand that a potential difference between two points in a circuit is required for current to flow
- appreciate that the unit of p.d. is the volt
- understand that resistance opposes current flow and is measured in ohms
- appreciate what an ammeter, a voltmeter, an ohmmeter, a multimeter and a C.R.O. measure
- distinguish between linear and non-linear devices
- state Ohm's law as $V = IR$ or $I = \frac{V}{R}$ or $R = \frac{V}{I}$
- use Ohm's law in calculations, including multiples and sub-multiples of units
- describe a conductor and an insulator, giving examples of each
- appreciate that electrical power P is given by

  $$P = VI = I^2R = \frac{V^2}{R} \text{ watts}$$

- calculate electrical power
- define electrical energy and state its unit
- calculate electrical energy
- state the three main effects of an electric current, giving practical examples of each
- explain the importance of fuses in electrical circuits

2.1 Standard symbols for electrical components

Symbols are used for components in electrical circuit diagrams and some of the more common ones are shown in Figure 2.1.

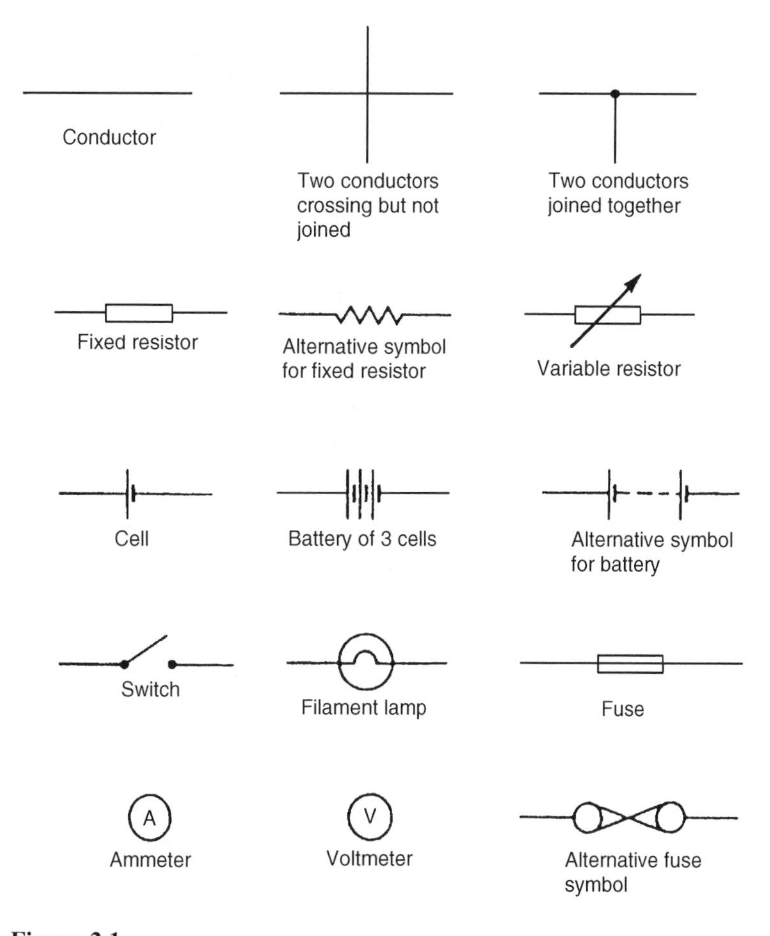

Figure 2.1

2.2 Electric current and quantity of electricity

All **atoms** consist of **protons, neutrons** and **electrons**. The protons, which have positive electrical charges, and the neutrons, which have no electrical charge, are contained within the **nucleus**. Removed from the nucleus are minute negatively charged particles called electrons. Atoms of different materials differ from one another by having different numbers of protons, neutrons and electrons. An equal number of protons and electrons exist within an atom and it is said to be electrically balanced, as the positive and negative charges cancel each other out. When there are more than two electrons in an atom the electrons are arranged into **shells** at various distances from the nucleus.

All atoms are bound together by powerful forces of attraction existing between the nucleus and its electrons. Electrons in the outer shell of an atom, however, are attracted to their nucleus less powerfully than are electrons whose shells are nearer the nucleus.

It is possible for an atom to lose an electron; the atom, which is now called an **ion**, is not now electrically balanced, but is positively charged and is thus able to attract an electron to itself from another atom. Electrons that move from one atom to another are called free electrons and such random motion can continue indefinitely. However, if an electric pressure or **voltage** is applied across any material there is a tendency for electrons to move in a particular direction. This movement of free electrons, known as **drift**, constitutes an electric current flow. **Thus current is the rate of movement of charge**.

Conductors are materials that contain electrons that are loosely connected to the nucleus and can easily move through the material from one atom to another.

Insulators are materials whose electrons are held firmly to their nucleus.

The unit used to measure the **quantity of electrical charge Q** is called the **coulomb C** (where 1 coulomb $= 6.24 \times 10^{18}$ electrons)

If the drift of electrons in a conductor takes place at the rate of one coulomb per second the resulting current is said to be a current of one ampere.

Thus, 1 ampere = 1 coulomb per second or 1 A = 1 C/s

Hence, 1 coulomb = 1 ampere second or 1 C = 1 As

Generally, if I is the current in amperes and t the time in seconds during which the current flows, then $I \times t$ represents the quantity of electrical charge in coulombs, i.e.

quantity of electrical charge transferred, $\boldsymbol{Q = I \times t}$ **coulombs**

Problem 1. What current must flow if 0.24 coulombs is to be transferred in 15 ms?

Since the quantity of electricity, $Q = It$, then

$$I = \frac{Q}{t} = \frac{0.24}{15 \times 10^{-3}} = \frac{0.24 \times 10^{3}}{15} = \frac{240}{15} = \mathbf{16\ A}$$

Problem 2. If a current of 10 A flows for four minutes, find the quantity of electricity transferred.

Quantity of electricity, $Q = It$ coulombs

$I = 10$ A; $t = 4 \times 60 = 240$ s

Hence $Q = 10 \times 240 = \mathbf{2400\ C}$

Further problems on $Q = I \times t$ may be found in Section 2.12, problems 1 to 3, page 21.

2.3 Potential difference and resistance

For a continuous current to flow between two points in a circuit a **potential difference (p.d.)** or **voltage**, V, is required between them; a complete conducting path is necessary to and from the source of electrical energy. The unit of p.d. is the **volt, V**

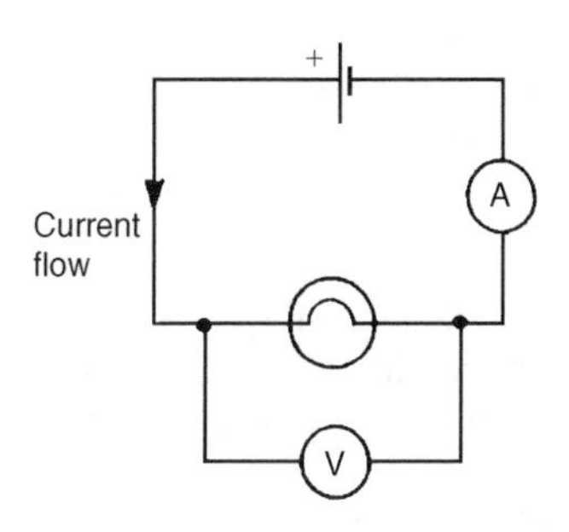

Figure 2.2

Figure 2.2 shows a cell connected across a filament lamp. Current flow, by convention, is considered as flowing from the positive terminal of the cell, around the circuit to the negative terminal.

The flow of electric current is subject to friction. This friction, or opposition, is called **resistance *R*** and is the property of a conductor that limits current. The unit of resistance is the **ohm**; 1 ohm is defined as the resistance which will have a current of 1 ampere flowing through it when 1 volt is connected across it, i.e.

$$\textbf{resistance } R = \frac{\textbf{potential difference}}{\textbf{current}}$$

2.4 Basic electrical measuring instruments

An **ammeter** is an instrument used to measure current and must be connected **in series** with the circuit. Figure 2.2 shows an ammeter connected in series with the lamp to measure the current flowing through it. Since all the current in the circuit passes through the ammeter it must have a very **low resistance**.

A **voltmeter** is an instrument used to measure p.d. and must be connected **in parallel** with the part of the circuit whose p.d. is required. In Figure 2.2, a voltmeter is connected in parallel with the lamp to measure the p.d. across it. To avoid a significant current flowing through it a voltmeter must have a very **high resistance**.

An **ohmmeter** is an instrument for measuring resistance.

A **multimeter**, or universal instrument, may be used to measure voltage, current and resistance. An 'Avometer' is a typical example.

The **cathode ray oscilloscope (CRO)** may be used to observe waveforms and to measure voltages and currents. The display of a CRO involves a spot of light moving across a screen. The amount by which the spot is deflected from its initial position depends on the p.d. applied to the terminals of the CRO and the range selected. The displacement is calibrated in 'volts per cm'. For example, if the spot is deflected 3 cm and the volts/cm switch is on 10 V/cm then the magnitude of the p.d. is 3 cm × 10 V/cm, i.e. 30 V (See Chapter 10 for more detail about electrical measuring instruments and measurements.)

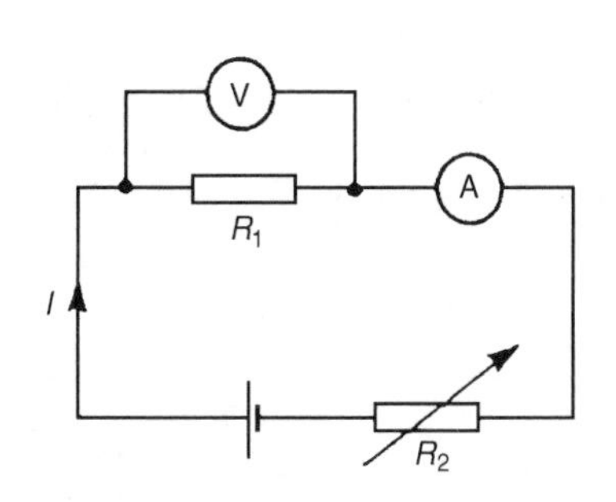

Figure 2.3

2.5 Linear and non-linear devices

Figure 2.3 shows a circuit in which current I can be varied by the variable resistor R_2. For various settings of R_2, the current flowing in resistor R_1, displayed on the ammeter, and the p.d. across R_1, displayed on the voltmeter, are noted and a graph is plotted of p.d. against current. The result is shown in Figure 2.4(a) where the straight line graph passing through the origin indicates that current is directly proportional to the p.d. Since the gradient i.e. (p.d./current) is constant, resistance R_1 is constant. A resistor is thus an example of a **linear device**.

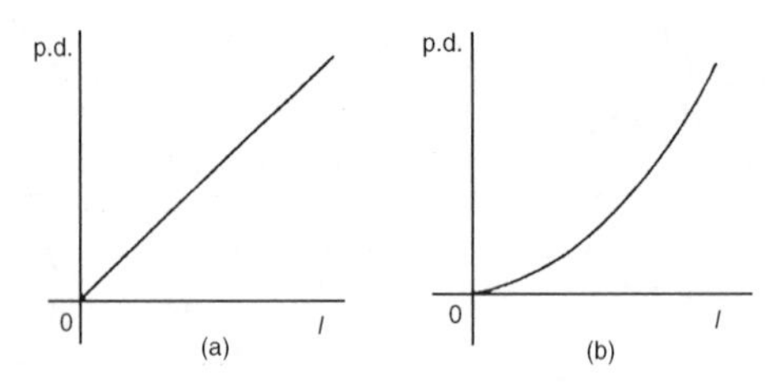

Figure 2.4

If the resistor R_1 in Figure 2.3 is replaced by a component such as a lamp then the graph shown in Figure 2.4(b) results when values of p.d. are noted for various current readings. Since the gradient is changing, the lamp is an example of a **non-linear device**.

2.6 Ohm's law

Ohm's law states that the current I flowing in a circuit is directly proportional to the applied voltage V and inversely proportional to the resistance R, provided the temperature remains constant. Thus,

$$\boldsymbol{I = \frac{V}{R} \text{ or } V = IR \text{ or } R = \frac{V}{I}}$$

Problem 3. The current flowing through a resistor is 0.8 A when a p.d. of 20 V is applied. Determine the value of the resistance.

From Ohm's law, resistance $R = \frac{V}{I} = \frac{20}{0.8} = \frac{200}{8} = \mathbf{25\ \Omega}$

2.7 Multiples and sub-multiples

Currents, voltages and resistances can often be very large or very small. Thus multiples and sub-multiples of units are often used, as stated in chapter 1. The most common ones, with an example of each, are listed in Table 2.1

TABLE 2.1

Prefix	*Name*	*Meaning*	*Example*
M	mega	multiply by 1 000 000 (i.e., $\times 10^6$)	$2\ \text{M}\Omega = 2\,000\,000$ ohms
k	kilo	multiply by 1000 (i.e., $\times 10^3$)	$10\ \text{kV} = 10\,000$ volts
m	milli	divide by 1000 (i.e., $\times 10^{-3}$)	$25\ \text{mA} = \frac{25}{1000}\ \text{A} = 0.025$ amperes
μ	micro	divide by 1 000 000 (i.e., $\times 10^{-6}$)	$50\ \mu\text{V} = \frac{50}{1\,000\,000}\ \text{V} = 0.000\,05$ volts

A more extensive list of common prefixes are given on page 915.

Problem 4. Determine the p.d. which must be applied to a 2 kΩ resistor in order that a current of 10 mA may flow.

Resistance $R = 2\ \text{k}\Omega = 2 \times 10^3 = 2000\ \Omega$

Current $I = 10\ \text{mA} = 10 \times 10^{-3}\ \text{A}$ or $\dfrac{10}{10^3}$ or $\dfrac{10}{1000}\ \text{A} = 0.01\ \text{A}$

From Ohm's law, potential difference, $V = IR = (0.01)(2000) = \mathbf{20\ V}$

Problem 5. A coil has a current of 50 mA flowing through it when the applied voltage is 12 V. What is the resistance of the coil?

$$\text{Resistance, } R = \frac{V}{I} = \frac{12}{50 \times 10^{-3}} = \frac{12 \times 10^3}{50} = \frac{12\,000}{50} = \mathbf{240\ \Omega}$$

Problem 6. A 100 V battery is connected across a resistor and causes a current of 5 mA to flow. Determine the resistance of the resistor. If the voltage is now reduced to 25 V, what will be the new value of the current flowing?

$$\text{Resistance } R = \frac{V}{I} = \frac{100}{5 \times 10^{-3}} = \frac{100 \times 10^3}{5} = 20 \times 10^3 = \mathbf{20\ k\Omega}$$

Current when voltage is reduced to 25 V,

$$I = \frac{V}{R} = \frac{25}{20 \times 10^3} = \frac{25}{20} \times 10^{-3} = \mathbf{1.25\ mA}$$

Problem 7. What is the resistance of a coil which draws a current of (a) 50 mA and (b) 200 μA from a 120 V supply?

(a) Resistance $R = \dfrac{V}{I} = \dfrac{120}{50 \times 10^{-3}}$

$$= \frac{120}{0.05} = \frac{12\,000}{5} = \mathbf{2\,400\ \Omega} \text{ or } \mathbf{2.4\ k\Omega}$$

(b) Resistance $R = \dfrac{120}{200 \times 10^{-6}} = \dfrac{120}{0.0002}$

$$= \frac{1200\,000}{2} = \mathbf{600\,000\ \Omega} \text{ or } \mathbf{600\ k\Omega} \text{ or } \mathbf{0.6\ M\Omega}$$

Further problems on Ohm's law may be found in Section 2.12, problems 4 to 7, page 21.

2.8 Conductors and insulators

A **conductor** is a material having a low resistance which allows electric current to flow in it. All metals are conductors and some examples include copper, aluminium, brass, platinum, silver, gold and carbon.

An **insulator** is a material having a high resistance which does not allow electric current to flow in it. Some examples of insulators include plastic, rubber, glass, porcelain, air, paper, cork, mica, ceramics and certain oils.

2.9 Electrical power and energy

Electrical power

Power P in an electrical circuit is given by the product of potential difference V and current I, as stated in Chapter 1. The unit of power is the **watt, W**. Hence

$$\boxed{P = V \times I \text{ watts}} \qquad (2.1)$$

From Ohm's law, $V = IR$
Substituting for V in equation (2.1) gives:

$$P = (IR) \times I$$

i.e. $$\boxed{P = I^2R \text{ watts}}$$

Also, from Ohm's law, $I = \dfrac{V}{R}$
Substituting for I in equation (2.1) gives:

$$P = V \times \frac{V}{R}$$

i.e. $$\boxed{P = \frac{V^2}{R} \text{ watts}}$$

There are thus three possible formulae which may be used for calculating power.

> Problem 8. A 100 W electric light bulb is connected to a 250 V supply. Determine (a) the current flowing in the bulb, and (b) the resistance of the bulb.

Power $P = V \times I$, from which, current $I = \dfrac{P}{V}$

(a) Current $I = \dfrac{100}{250} = \dfrac{10}{25} = \dfrac{2}{5} = \mathbf{0.4\ A}$

(b) Resistance $R = \dfrac{V}{I} = \dfrac{250}{0.4} = \dfrac{2500}{4} = \mathbf{625\ \Omega}$

Problem 9. Calculate the power dissipated when a current of 4 mA flows through a resistance of 5 kΩ

Power $P = I^2R = (4 \times 10^{-3})^2(5 \times 10^3)$

$= 16 \times 10^{-6} \times 5 \times 10^3 = 80 \times 10^{-3}$

$= \mathbf{0.08\ W}$ or $\mathbf{80\ mW}$

Alternatively, since $I = 4 \times 10^{-3}$ and $R = 5 \times 10^3$ then from Ohm's law, voltage $V = IR = 4 \times 10^{-3} \times 5 \times 10^{-3} = 20$ V
Hence, power $P = V \times I = 20 \times 4 \times 10^{-3} = \mathbf{80\ mW}$

Problem 10. An electric kettle has a resistance of 30 Ω. What current will flow when it is connected to a 240 V supply? Find also the power rating of the kettle.

Current, $I = \frac{V}{R} = \frac{240}{30} = \mathbf{8\ A}$

Power, $P = VI = 240 \times 8 = 1920\ \text{W} = \mathbf{1.92\ kW}$

= power rating of kettle

Problem 11. A current of 5 A flows in the winding of an electric motor, the resistance of the winding being 100 Ω. Determine (a) the p.d. across the winding, and (b) the power dissipated by the coil.

(a) Potential difference across winding, $V = IR = 5 \times 100 = \mathbf{500\ V}$

(b) Power dissipated by coil, $P = I^2R = 5^2 \times 100$

$= \mathbf{2500\ W\ or\ 2.5\ kW}$

(Alternatively, $P = V \times I = 500 \times 5 = \mathbf{2500\ W\ or\ 2.5\ kW}$)

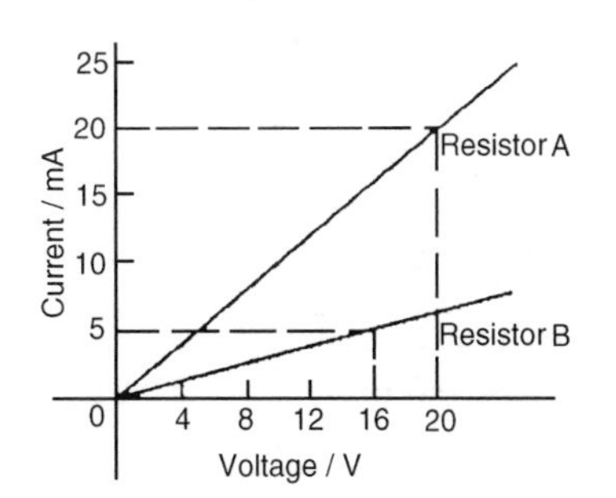

Figure 2.5

Problem 12. The current/voltage relationship for two resistors A and B is as shown in Figure 2.5. Determine the value of the resistance of each resistor.

For resistor A, $R = \frac{V}{I} = \frac{20\ \text{A}}{20\ \text{mA}} = \frac{20}{0.02} = \frac{2\,000}{2} = \mathbf{1\,000\ \Omega\ or\ 1\ k\Omega}$

For resistor B, $R = \frac{V}{I} = \frac{16\ \text{V}}{5\ \text{mA}} = \frac{16}{0.005} = \frac{16\,000}{5} = \mathbf{3\,200\ \Omega\ or\ 3.2\ k\Omega}$

Problem 13. The hot resistance of a 240 V filament lamp is 960 Ω. Find the current taken by the lamp and its power rating.

From Ohm's law, current $I = \frac{V}{R} = \frac{240}{960} = \frac{24}{96} = \frac{1}{4}$ **A or 0.25 A**

Power rating $P = VI = (240)\left(\frac{1}{4}\right) = \mathbf{60\ W}$

Electrical energy

Electrical energy = power × time

If the power is measured in watts and the time in seconds then the unit of energy is watt-seconds or **joules**. If the power is measured in kilowatts and the time in hours then the unit of energy is **kilowatt-hours**, often called the '**unit of electricity**'. The 'electricity meter' in the home records the number of kilowatt-hours used and is thus an energy meter.

Problem 14. A 12 V battery is connected across a load having a resistance of 40 Ω. Determine the current flowing in the load, the power consumed and the energy dissipated in 2 minutes.

Current $I = \frac{V}{R} = \frac{12}{40} = \mathbf{0.3\ A}$

Power consumed, $P = VI = (12)(0.3) = \mathbf{3.6\ W}$

Energy dissipated = power × time = (3.6 W)(2 × 60 s) = **432 J**

(since 1 J = 1 Ws)

Problem 15. A source of e.m.f. of 15 V supplies a current of 2 A for six minutes. How much energy is provided in this time?

Energy = power × time, and power = voltage × current

Hence energy = $VIt = 15 \times 2 \times (6 \times 60) = 10\,800$ Ws or J = **10.8 kJ**

Problem 16. Electrical equipment in an office takes a current of 13 A from a 240 V supply. Estimate the cost per week of electricity if the equipment is used for 30 hours each week and 1 kWh of energy costs 7p

Power = VI watts = 240 × 13 = 3120 W = 3.12 kW

Energy used per week = power × time = (3.12 kW) × (30 h)

= 93.6 kWh

Cost at 7p per kWh = 93.6 × 7 = 655.2 p

Hence **weekly cost of electricity = £6.55**

> Problem 17. An electric heater consumes 3.6 MJ when connected to a 250 V supply for 40 minutes. Find the power rating of the heater and the current taken from the supply.

$$\text{Power} = \frac{\text{energy}}{\text{time}} = \frac{3.6 \times 10^6}{40 \times 60} \frac{\text{J}}{\text{s}} (\text{or W}) = 1500 \text{ W}$$

i.e. Power rating of heater = **1.5 kW**

$$\text{Power } P = VI, \text{ thus } I = \frac{P}{V} = \frac{1500}{250} = 6 \text{ A}$$

Hence the current taken from the supply is **6 A**

> Problem 18. Determine the power dissipated by the element of an electric fire of resistance 20 Ω when a current of 10 A flows through it. If the fire is on for 6 hours determine the energy used and the cost if 1 unit of electricity costs 7p.

Power$P = I^2R = 10^2 \times 20 = 100 \times 20 =$ **2 000 W or 2 kW**
(Alternatively, from Ohm's law, $V = IR = 10 \times 20 = 200$ V, hence power $P = V \times I = 200 \times 10 = 2000$ W = 2 kW)
Energy used in 6 hours = power × time = 2 kW × 6 h = **12 kWh**

1 unit of electricity = 1 kWh
Hence the number of units used is 12
Cost of energy = 12 × 7 = **84p**

> Problem 19. A business uses two 3 kW fires for an average of 20 hours each per week, and six 150 W lights for 30 hours each per week. If the cost of electricity is 7p per unit, determine the weekly cost of electricity to the business.

Energy = power × time

Energy used by one 3 kW fire in 20 hours = 3 kW × 20 h = 60 kWh

Hence weekly energy used by two 3 kW fires = 2 × 60 = 120 kWh

Energy used by one 150 W light for 30 hours = 150 W × 30 h

= 4500 Wh = 4.5 kWh

Hence weekly energy used by six 150 W lamps = 6 × 4.5 = 27 kWh
Total energy used per week = 120 + 27 = 147 kWh

1 unit of electricity = 1 kWh of energy

Thus weekly cost of energy at 7p per kWh = 7 × 147 = 1029p

= **£10.29**

Further problems on power and energy may be found in Section 2.12, problems 8 to 17, page 21.

2.10 Main effects of electric current

The three main effects of an electric current are:

(a) magnetic effect
(b) chemical effect
(c) heating effect

Some practical applications of the effects of an electric current include:

Magnetic effect: bells, relays, motors, generators, transformers, telephones, car-ignition and lifting magnets

Chemical effect: primary and secondary cells and electroplating

Heating effect: cookers, water heaters, electric fires, irons, furnaces, kettles and soldering irons

2.11 Fuses

A **fuse** is used to prevent overloading of electrical circuits. The fuse, which is made of material having a low melting point, utilizes the heating effect of an electric current. A fuse is placed in an electrical circuit and if the current becomes too large the fuse wire melts and so breaks the circuit. A circuit diagram symbol for a fuse is shown in Figure 2.1, on page 11.

Problem 20. If 5 A, 10 A and 13 A fuses are available, state which is most appropriate for the following appliances which are both connected to a 240 V supply (a) Electric toaster having a power rating of 1 kW (b) Electric fire having a power rating of 3 kW

Power $P = VI$, from which, current $I = \dfrac{P}{V}$

(a) For the toaster, current $I = \dfrac{P}{V} = \dfrac{1000}{240} = \dfrac{100}{24} = 4\dfrac{1}{6}$ A

Hence a **5 A fuse** is most appropriate

(b) For the fire, current $I = \dfrac{P}{V} = \dfrac{3000}{240} = \dfrac{300}{24} = 12\dfrac{1}{2}$ A

Hence a **13 A fuse** is most appropriate

A further problem on fuses may be found in Section 2.12 following, problem 18, page 22.

2.12 Further problems on the introduction to electric circuits

$Q = I \times t$

1 In what time would a current of 10 A transfer a charge of 50 C? [5 s]

2 A current of 6 A flows for 10 minutes. What charge is transferred? [3600 C]

3 How long must a current of 100 mA flow so as to transfer a charge of 80 C? [13 min 20 s]

Ohm's law

4 The current flowing through a heating element is 5 A when a p.d. of 35 V is applied across it. Find the resistance of the element. [7 Ω]

5 A 60 W electric light bulb is connected to a 240 V supply. Determine (a) the current flowing in the bulb and (b) the resistance of the bulb. [(a) 0.25 A (b) 960 Ω]

6 Graphs of current against voltage for two resistors P and Q are shown in Figure 2.6. Determine the value of each resistor. [2 mΩ, 5 mΩ]

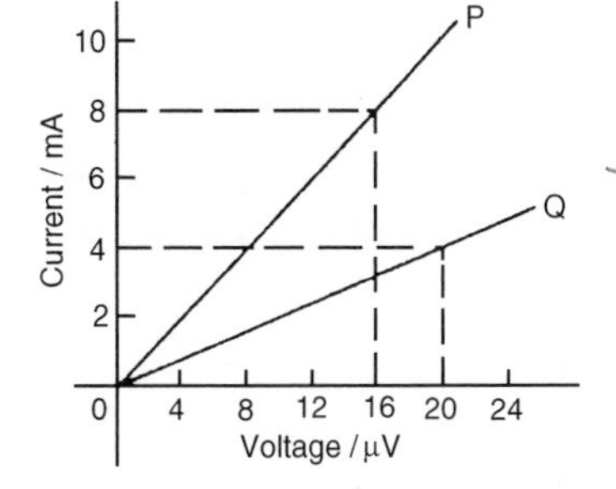

Figure 2.6

7 Determine the p.d. which must be applied to a 5 kΩ resistor such that a current of 6 mA may flow. [30 V]

Power and energy

8 The hot resistance of a 250 V filament lamp is 625 Ω. Determine the current taken by the lamp and its power rating. [0.4 A, 100 W]

9 Determine the resistance of a coil connected to a 150 V supply when a current of (a) 75 mA (b) 300 μA flows through it. [(a) 2 kΩ (b) 0.5 MΩ]

10 Determine the resistance of an electric fire which takes a current of 12 A from a 240 V supply. Find also the power rating of the fire and the energy used in 20 h. [20 Ω, 2.88 kW, 57.6 kWh]

11 Determine the power dissipated when a current of 10 mA flows through an appliance having a resistance of 8 kΩ. [0.8 W]

12 85.5 J of energy are converted into heat in nine seconds. What power is dissipated? [9.5 W]

13 A current of 4 A flows through a conductor and 10 W is dissipated. What p.d. exists across the ends of the conductor? [2.5 V]

14 Find the power dissipated when:
(a) a current of 5 mA flows through a resistance of 20 kΩ
(b) a voltage of 400 V is applied across a 120 kΩ resistor
(c) a voltage applied to a resistor is 10 kV and the current flow is 4 mA. [(a) 0.5 W (b) $1\frac{1}{3}$ W (c) 40 W]

15 A battery of e.m.f. 15 V supplies a current of 2 A for 5 min. How much energy is supplied in this time? [9 kJ]

16 In a household during a particular week three 2 kW fires are used on average 25 h each and eight 100 W light bulbs are used on average 35 h each. Determine the cost of electricity for the week if 1 unit of electricity costs 7p. [£12.46]

17 Calculate the power dissipated by the element of an electric fire of resistance 30 Ω when a current of 10 A flows in it. If the fire is on for 30 hours in a week determine the energy used. Determine also the weekly cost of energy if electricity costs 7.2p per unit. [3 kW, 90 kWh, £6.48]

Fuses

18 A television set having a power rating of 120 W and electric lawnmower of power rating 1 kW are both connected to a 240 V supply. If 3 A, 5 A and 10 A fuses are available state which is the most appropriate for each appliance. [3 A, 5 A]

3 Resistance variation

At the end of this chapter you should be able to:

- appreciate that electrical resistance depends on four factors
- appreciate that resistance $R = \dfrac{\rho l}{a}$, where ρ is the resistivity
- recognize typical values of resistivity and its unit
- perform calculations using $R = \dfrac{\rho l}{a}$
- define the temperature coefficient of resistance, α
- recognize typical values for α
- perform calculations using $R_\theta = R_0(1 + \alpha\theta)$

3.1 Resistance and resistivity

The resistance of an electrical conductor depends on 4 factors, these being: (a) the length of the conductor, (b) the cross-sectional area of the conductor, (c) the type of material and (d) the temperature of the material.

Resistance, R, is directly proportional to length, l, of a conductor, i.e. $R \propto l$. Thus, for example, if the length of a piece of wire is doubled, then the resistance is doubled.

Resistance, R, is inversely proportional to cross-sectional area, a, of a conductor, i.e. $R \propto 1/a$. Thus, for example, if the cross-sectional area of a piece of wire is doubled then the resistance is halved.

Since $R \propto l$ and $R \propto 1/a$ then $R \propto l/a$. By inserting a constant of proportionality into this relationship the type of material used may be taken into account. The constant of proportionality is known as the **resistivity** of the material and is given the symbol ρ (Greek rho). Thus,

$$\text{resistance } \boxed{\mathbf{R = \frac{\rho l}{a} \text{ ohms}}}$$

ρ is measured in ohm metres (Ωm)

The value of the resistivity is that resistance of a unit cube of the material measured between opposite faces of the cube.

Resistivity varies with temperature and some typical values of resistivities measured at about room temperature are given below:

Copper 1.7×10^{-8} Ωm (or 0.017 $\mu\Omega$m)

Aluminium 2.6×10^{-8} Ωm (or 0.026 $\mu\Omega$m)

Carbon (graphite) 10×10^{-8} Ωm (or 0.10 $\mu\Omega$m)

Glass 1×10^{10} Ωm (or 10^4 μΩm)

Mica 1×10^{13} Ωm (or 10^7 μΩm)

Note that good conductors of electricity have a low value of resistivity and good insulators have a high value of resistivity.

> Problem 1. The resistance of a 5 m length of wire is 600 Ω. Determine (a) the resistance of an 8 m length of the same wire, and (b) the length of the same wire when the resistance is 420 Ω.

(a) Resistance, R, is directly proportional to length, l, i.e. $R \propto l$

Hence, 600 Ω ∝ 5 m or $600 = (k)(5)$, where k is the coefficient of proportionality. Hence,

$$k = \frac{600}{5} = 120$$

When the length l is 8 m, then resistance

$$R = kl = (120)(8) = \mathbf{960\ \Omega}$$

(b) When the resistance is 420 Ω, $420 = kl$, from which,

$$\text{length } l = \frac{420}{k} = \frac{420}{120} = \mathbf{3.5\ m}$$

> Problem 2. A piece of wire of cross-sectional area 2 mm² has a resistance of 300 Ω. Find (a) the resistance of a wire of the same length and material if the cross-sectional area is 5 mm², (b) the cross-sectional area of a wire of the same length and material of resistance 750 Ω

Resistance R is inversely proportional to cross-sectional area, a, i.e. $R \propto \frac{1}{a}$

$$\text{Hence } 300\ \Omega \propto \frac{1}{2\ \text{mm}^2} \text{ or } 300 = (k)\left(\frac{1}{2}\right),$$

from which, the coefficient of proportionality, $k = 300 \times 2 = 600$

(a) When the cross-sectional area $a = 5$ mm² then $R = (k)\left(\frac{1}{5}\right)$

$$= (600)\left(\tfrac{1}{5}\right) = \mathbf{120\ \Omega}$$

(Note that resistance has decreased as the cross-sectional is increased.)

(b) When the resistance is 750 Ω then $750 = (k)(1/a)$, from which

$$\text{cross-sectional area, } a = \frac{k}{750} = \frac{600}{750} = \mathbf{0.8\ mm^2}$$

Problem 3. A wire of length 8 m and cross-sectional area 3 mm^2 has a resistance of 0.16 Ω. If the wire is drawn out until its cross-sectional area is 1 mm^2, determine the resistance of the wire.

Resistance R is directly proportional to length l, and inversely proportional to the cross-sectional area, a, i.e.,

i.e., $R \propto \frac{l}{a}$ or $R = k\left(\frac{l}{a}\right)$, where k is the coefficient of proportionality.

Since $R = 0.16$, $l = 8$ and $a = 3$, then $0.16 = (k)\left(\frac{8}{3}\right)$, from which

$k = 0.16 \times \frac{3}{8} = 0.06$

If the cross-sectional area is reduced to $\frac{1}{3}$ of its original area then the length must be tripled to 3×8, i.e., 24 m

New resistance $\quad R = k\left(\frac{l}{a}\right) = 0.06\left(\frac{24}{1}\right) = \mathbf{1.44\ \Omega}$

Problem 4. Calculate the resistance of a 2 km length of aluminium overhead power cable if the cross-sectional area of the cable is 100 mm^2. Take the resistivity of aluminium to be 0.03×10^{-6} Ωm

Length $l = 2$ km $= 2000$ m; area, $a = 100\ mm^2 = 100 \times 10^{-6} m^2$; resistivity $\rho = 0.03 \times 10^{-6}$ Ωm

$$\text{Resistance} \quad R = \frac{\rho l}{a} = \frac{(0.03 \times 10^{-6}\ \Omega m)(2000\ m)}{(100 \times 10^{-6}\ m^2)} = \frac{0.03 \times 2000}{100}\ \Omega$$

$$= \mathbf{0.6\ \Omega}$$

Problem 5. Calculate the cross-sectional area, in mm^2, of a piece of copper wire, 40 m in length and having a resistance of 0.25 Ω. Take the resistivity of copper as 0.02×10^{-6} Ωm

Resistance $R = \frac{\rho l}{a}$ hence cross-sectional area $a = \frac{\rho l}{R}$

$$= \frac{(0.02 \times 10^{-6}\ \Omega m)(40\ m)}{0.25\ \Omega} = 3.2 \times 10^{-6}\ m^2$$

$$= (3.2 \times 10^{-6}) \times 10^6\ mm^2 = \mathbf{3.2\ mm^2}$$

Problem 6. The resistance of 1.5 km of wire of cross-sectional area 0.17 mm^2 is 150 Ω. Determine the resistivity of the wire.

Resistance, $R = \frac{\rho l}{a}$

hence, resistivity $\rho = \frac{Ra}{l} = \frac{(150\ \Omega)(0.17 \times 10^{-6}\ \text{m}^2)}{(1500\ \text{m})}$

$= \mathbf{0.017 \times 10^{-6}\ \Omega m\ or\ 0.017\ \mu\Omega m}$

Problem 7. Determine the resistance of 1200 m of copper cable having a diameter of 12 mm if the resistivity of copper is $1.7 \times 10^{-8}\ \Omega$m

Cross-sectional area of cable, $a = \pi r^2 = \pi \left(\frac{12}{2}\right)^2$

$= 36\pi\ \text{mm}^2 = 36\pi \times 10^{-6}\ \text{m}^2$

Resistance $R = \frac{\rho l}{a} = \frac{(1.7 \times 10^{-8}\ \Omega\text{m})\ (1200\ \text{m})}{(36\pi \times 10^{-6}\ \text{m}^2)}$

$= \frac{1.7 \times 1200 \times 10^6}{10^8 \times 36\pi}\Omega = \frac{1.7 \times 12}{36\pi}\Omega$

$= \mathbf{0.180\ \Omega}$

Further problems on resistance and resistivity may be found in Section 3.3, problems 1 to 7, page 29.

3.2 Temperature coefficient of resistance

In general, as the temperature of a material increases, most conductors increase in resistance, insulators decrease in resistance, whilst the resistance of some special alloys remain almost constant.

The **temperature coefficient of resistance** of a material is the increase in the resistance of a 1 Ω resistor of that material when it is subjected to a rise of temperature of 1°C. The symbol used for the temperature coefficient of resistance is α (Greek alpha). Thus, if some copper wire of resistance 1 Ω is heated through 1°C and its resistance is then measured as 1.0043 Ω then $\alpha = 0.0043\ \Omega/\Omega°\text{C}$ for copper. The units are usually expressed only as 'per °C', i.e., $\alpha = 0.0043/°\text{C}$ for copper. If the 1 Ω resistor of copper is heated through 100°C then the resistance at 100°C would be $1 + 100 \times 0.0043 = 1.43\Omega$

Some typical values of temperature coefficient of resistance measured at 0°C are given below:

Copper	0.0043/°C	Aluminium	0.0038/°C
Nickel	0.0062/°C	Carbon	−0.000 48/°C
Constantan	0	Eureka	0.000 01/°C

(Note that the negative sign for carbon indicates that its resistance falls with increase of temperature.)

If the resistance of a material at 0°C is known the resistance at any other temperature can be determined from:

$$\boldsymbol{R_\theta = R_0(1 + \alpha_0\theta)}$$

where R_0 = resistance at 0°C

R_θ = resistance at temperature θ°C

α_0 = temperature coefficient of resistance at 0°C

Problem 8. A coil of copper wire has a resistance of 100 Ω when its temperature is 0°C. Determine its resistance at 70°C if the temperature coefficient of resistance of copper at 0°C is 0.0043/°C

Resistance $R_\theta = R_0(1 + \alpha_0\theta)$

Hence resistance at 70°C, $R_{70} = 100[1 + (0.0043)(70)]$

$= 100[1 + 0.301] = 100(1.301)$

$= \mathbf{130.1\ \Omega}$

Problem 9. An aluminium cable has a resistance of 27 Ω at a temperature of 35°C. Determine its resistance at 0°C. Take the temperature coefficient of resistance at 0°C to be 0.0038/°C

Resistance at θ°C, $R_\theta = R_0(1 + \alpha_0\theta)$

Hence resistance at 0°C, $R_0 = \dfrac{R_\theta}{(1 + \alpha_0\theta)} = \dfrac{27}{[1 + (0.0038)(35)]}$

$= \dfrac{27}{1 + 0.133} = \dfrac{27}{1.133} = \mathbf{23.83\ \Omega}$

Problem 10. A carbon resistor has a resistance of 1 kΩ at 0°C. Determine its resistance at 80°C. Assume that the temperature coefficient of resistance for carbon at 0°C is −0.0005/°C

Resistance at temperature θ°C, $R_\theta = R_0(1 + \alpha_0\ \theta)$

i.e., $R_\theta = 1000[1 + (-0.0005)(80)]$

$= 1000[1 - 0.040] = 1000(0.96) = \mathbf{960\ \Omega}$

If the resistance of a material at room temperature (approximately 20°C), R_{20}, and the temperature coefficient of resistance at 20°C, α_{20}, are known then the resistance R_θ at temperature θ°C is given by:

$$\boldsymbol{R_\theta = R_{20}[1 + \alpha_{20}(\theta - 20)]}$$

Problem 11. A coil of copper wire has a resistance of 10 Ω at 20°C. If the temperature coefficient of resistance of copper at 20°C is 0.004/°C determine the resistance of the coil when the temperature rises to 100°C

Resistance at θ°C, $R = R_{20}[1 + \alpha_{20}(\theta - 20)]$

Hence resistance at 100°C,
$$R_{100} = 10[1 + (0.004)(100 - 20)]$$
$$= 10[1 + (0.004)(80)]$$
$$= 10[1 + 0.32]$$
$$= 10(1.32) = \mathbf{13.2\ \Omega}$$

Problem 12. The resistance of a coil of aluminium wire at 18°C is 200 Ω. The temperature of the wire is increased and the resistance rises to 240 Ω. If the temperature coefficient of resistance of aluminium is 0.0039/°C at 18°C determine the temperature to which the coil has risen.

Let the temperature rise to θ°

Resistance at θ°C, $R_\theta = R_{18}[1 + \alpha_{18}(\theta - 18)]$

i.e.
$$240 = 200[1 + (0.0039)(\theta - 18)]$$
$$240 = 200 + (200)(0.0039)(\theta - 18)$$
$$240 - 200 = 0.78(\theta - 18)$$
$$40 = 0.78(\theta - 18)$$
$$\frac{40}{0.78} = \theta - 18$$
$$51.28 = \theta - 18, \text{ from which, } \theta = 51.28 + 18 = 69.28°\text{C}$$

Hence the temperature of the coil increases to 69.28°C

If the resistance at 0°C is not known, but is known at some other temperature θ_1, then the resistance at any temperature can be found as follows:

$$R_1 = R_0(1 + \alpha_0\theta_1) \text{ and } R_2 = R_0(1 + \alpha_0\theta_2)$$

Dividing one equation by the other gives:

$$\boldsymbol{\frac{R_1}{R_2} = \frac{1+\alpha_0\theta_1}{1+\alpha_0\theta_2}}$$

where R_2 = resistance at temperature θ_2

Problem 13. Some copper wire has a resistance of 200 Ω at 20°C. A current is passed through the wire and the temperature rises to 90°C. Determine the resistance of the wire at 90°C, correct to the nearest ohm, assuming that the temperature coefficient of resistance is 0.004/°C at 0°C

$R_{20} = 200\ \Omega$, $\alpha_0 = 0.004/°\text{C}$

$$\frac{R_{20}}{R_{90}} = \frac{[1+\alpha_0(20)]}{[1+\alpha_0(90)]}$$

Hence $$R_{90} = \frac{R_{20}[1+90\alpha_0]}{[1+20\alpha_0]} = \frac{200[1+90(0.004)]}{[1+20(0.004)]} = \frac{200[1+0.36]}{[1+0.08]}$$

$$= \frac{200(1.36)}{(1.08)} = \mathbf{251.85\ \Omega}$$

i.e., the resistance of the wire at 90°C is 252 Ω

Further problems on temperature coefficient of resistance may be found in Section 3.3, following, problems 8 to 14, page 30.

3.3 Further problems on resistance variation

Resistance and resistivity

1 The resistance of a 2 m length of cable is 2.5 Ω. Determine (a) the resistance of a 7 m length of the same cable and (b) the length of the same wire when the resistance is 6.25 Ω. [(a) 8.75 Ω (b) 5 m]

2 Some wire of cross-sectional area 1 mm^2 has a resistance of 20 Ω. Determine (a) the resistance of a wire of the same length and material if the cross-sectional area is 4 mm^2, and (b) the cross-sectional area of a wire of the same length and material if the resistance is 32 Ω. [(a) 5 Ω (b) 0.625 mm^2]

3 Some wire of length 5 m and cross-sectional area 2 mm^2 has a resistance of 0.08 Ω. If the wire is drawn out until its cross-sectional area is 1 mm^2, determine the resistance of the wire. [0.32 Ω]

4 Find the resistance of 800 m of copper cable of cross-sectional area 20 mm^2. Take the resistivity of copper as 0.02 μΩm. [0.8 Ω]

5 Calculate the cross-sectional area, in mm^2, of a piece of aluminium wire 100 m long and having a resistance of 2 Ω. Take the resistivity of aluminium as 0.03×10^{-6} Ωm. [1.5 mm^2]

6 (a) What does the resistivity of a material mean?

(b) The resistance of 500 m of wire of cross-sectional area 2.6 mm^2 is 5 Ω. Determine the resistivity of the wire in μΩm. [0.026 μΩm]

7 Find the resistance of 1 km of copper cable having a diameter of 10 mm if the resistivity of copper is 0.017×10^{-6} Ωm. [0.216 Ω]

Temperature coefficient of resistance

8 A coil of aluminium wire has a resistance of 50 Ω when its temperature is 0°C. Determine its resistance at 100°C if the temperature coefficient of resistance of aluminium at 0°C is 0.0038/°C. [69 Ω]

9 A copper cable has a resistance of 30 Ω at a temperature of 50°C. Determine its resistance at 0°C. Take the temperature coefficient of resistance of copper at 0°C as 0.0043/°C. [24.69 Ω]

10 The temperature coefficient of resistance for carbon at 0°C is −0.00048/°C. What is the significance of the minus sign? A carbon resistor has a resistance of 500 Ω at 0°C. Determine its resistance at 50°C. [488 Ω]

11 A coil of copper wire has a resistance of 20 Ω at 18°C. If the temperature coefficient of resistance of copper at 18°C is 0.004/°C, determine the resistance of the coil when the temperature rises to 98°C [26.4 Ω]

12 The resistance of a coil of nickel wire at 20°C is 100 Ω. The temperature of the wire is increased and the resistance rises to 130 Ω. If the temperature coefficient of resistance of nickel is 0.006/°C at 20°C, determine the temperature to which the coil has risen. [70°C]

13 Some aluminium wire has a resistance of 50 Ω at 20°C. The wire is heated to a temperature of 100°C. Determine the resistance of the wire at 100°C, assuming that the temperature coefficient of resistance at 0°C is 0.004/°C [64.8 Ω]

14 A copper cable is 1.2 km long and has a cross-sectional area of 5 mm^2. Find its resistance at 80°C if at 20°C the resistivity of copper is 0.02×10^{-6} Ωm and its temperature coefficient of resistance is 0.004/°C [5.952 Ω]

4 Chemical effects of electricity

At the end of this chapter you should be able to:

- understand electrolysis and its applications, including electroplating
- appreciate the purpose and construction of a simple cell
- explain polarization and local action
- explain corrosion and its effects
- define the terms e.m.f., E, and internal resistance, r, of a cell
- perform calculations using $V = E - Ir$
- determine the total e.m.f. and total internal resistance for cells connected in series and in parallel
- distinguish between primary and secondary cells
- explain the construction and practical applications of the Leclanché, mercury, lead-acid and alkaline cells
- list the advantages and disadvantages of alkaline cells over lead-acid cells
- understand the term 'cell capacity' and state its unit

4.1 Introduction

A material must contain **charged particles** to be able to conduct electric current. In **solids**, the current is carried by **electrons**. Copper, lead, aluminium, iron and carbon are some examples of solid conductors. In **liquids and gases**, the current is carried by the part of a molecule which has acquired an electric charge, called **ions**. These can possess a positive or negative charge, and examples include hydrogen ion H^+, copper ion Cu^{++} and hydroxyl ion OH^-. Distilled water contains no ions and is a poor conductor of electricity whereas salt water contains ions and is a fairly good conductor of electricity.

4.2 Electrolysis

Electrolysis is the decomposition of a liquid compound by the passage of electric current through it. Practical applications of electrolysis include the electroplating of metals (see Section 4.3), the refining of copper and the extraction of aluminium from its ore.

An **electrolyte** is a compound which will undergo electrolysis. Examples include salt water, copper sulphate and sulphuric acid.

The **electrodes** are the two conductors carrying current to the electrolyte. The positive-connected electrode is called the **anode** and the negative-connected electrode the **cathode**.

When two copper wires connected to a battery are placed in a beaker containing a salt water solution, current will flow through the solution. Air bubbles appear around the wires as the water is changed into hydrogen and oxygen by electrolysis.

4.3 Electroplating

Electroplating uses the principle of electrolysis to apply a thin coat of one metal to another metal. Some practical applications include the tin-plating of steel, silver-plating of nickel alloys and chromium-plating of steel. If two copper electrodes connected to a battery are placed in a beaker containing copper sulphate as the electrolyte it is found that the cathode (i.e. the electrode connected to the negative terminal of the battery) gains copper whilst the anode loses copper.

4.4 The simple cell

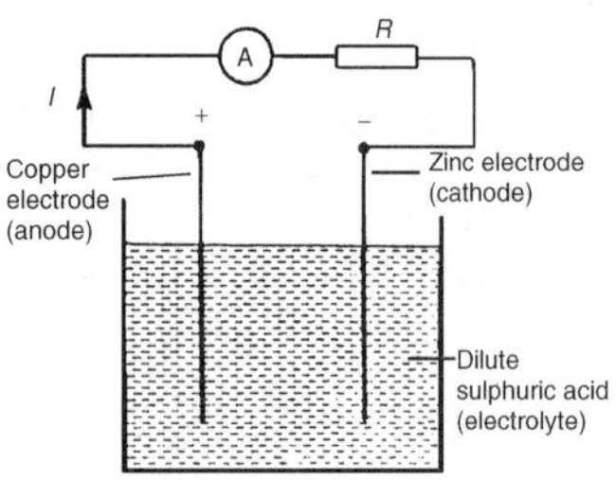

The purpose of an **electric cell** is to convert chemical energy into electrical energy.

A **simple cell** comprises two dissimilar conductors (electrodes) in an electrolyte. Such a cell is shown in Figure 4.1, comprising copper and zinc electrodes. An electric current is found to flow between the electrodes. Other possible electrode pairs exist, including zinc-lead and zinc-iron. The electrode potential (i.e. the p.d. measured between the electrodes) varies for each pair of metals. By knowing the e.m.f. of each metal with respect to some standard electrode the e.m.f. of any pair of metals may be determined. The standard used is the hydrogen electrode. The **electrochemical series** is a way of listing elements in order of electrical potential, and Table 4.1 shows a number of elements in such a series.

Figure 4.1

TABLE 4.1 *Part of the electrochemical series*

Potassium
sodium
aluminium
zinc
iron
lead
hydrogen
copper
silver
carbon

In a simple cell two faults exist — those due to **polarization** and **local action**.

Polarization

If the simple cell shown in Figure 4.1 is left connected for some time, the current *I* decreases fairly rapidly. This is because of the formation of a film of hydrogen bubbles on the copper anode. This effect is known as the polarization of the cell. The hydrogen prevents full contact between the copper electrode and the electrolyte and this increases the internal resistance of the cell. The effect can be overcome by using a chemical depolarizing agent or depolarizer, such as potassium dichromate which removes the hydrogen bubbles as they form. This allows the cell to deliver a steady current.

Local action

When commercial zinc is placed in dilute sulphuric acid, hydrogen gas is liberated from it and the zinc dissolves. The reason for this is that impurities, such as traces of iron, are present in the zinc which set up small primary cells with the zinc. These small cells are short-circuited by the electrolyte, with the result that localized currents flow causing corrosion. This action is known as local action of the cell. This may be prevented by rubbing a small amount of mercury on the zinc surface, which forms a protective layer on the surface of the electrode.

When two metals are used in a simple cell the electrochemical series may be used to predict the behaviour of the cell:

(i) The metal that is higher in the series acts as the negative electrode, and vice-versa. For example, the zinc electrode in the cell shown in Figure 4.1 is negative and the copper electrode is positive.
(ii) The greater the separation in the series between the two metals the greater is the e.m.f. produced by the cell.

The electrochemical series is representative of the order of reactivity of the metals and their compounds:

(i) The higher metals in the series react more readily with oxygen and vice-versa.
(ii) When two metal electrodes are used in a simple cell the one that is higher in the series tends to dissolve in the electrolyte.

4.5 Corrosion

Corrosion is the gradual destruction of a metal in a damp atmosphere by means of simple cell action. In addition to the presence of moisture and air required for rusting, an electrolyte, an anode and a cathode are required for corrosion. Thus, if metals widely spaced in the electrochemical series, are used in contact with each other in the presence of an electrolyte, corrosion will occur. For example, if a brass valve is fitted to a heating system made of steel, corrosion will occur.

The **effects of corrosion** include the weakening of structures, the reduction of the life of components and materials, the wastage of materials and the expense of replacement.

Corrosion may be **prevented** by coating with paint, grease, plastic coatings and enamels, or by plating with tin or chromium. Also, iron may be galvanized, i.e., plated with zinc, the layer of zinc helping to prevent the iron from corroding.

4.6 E.m.f. and internal resistance of a cell

The **electromotive force (e.m.f.), E**, of a cell is the p.d. between its terminals when it is not connected to a load (i.e. the cell is on 'no load').

The e.m.f. of a cell is measured by using a **high resistance voltmeter** connected in parallel with the cell. The voltmeter must have a high resistance otherwise it will pass current and the cell will not be on no-load. For example, if the resistance of a cell is 1 Ω and that of a voltmeter 1 MΩ then the equivalent resistance of the circuit is 1 MΩ + 1Ω, i.e. approximately 1 MΩ, hence no current flows and the cell is not loaded.

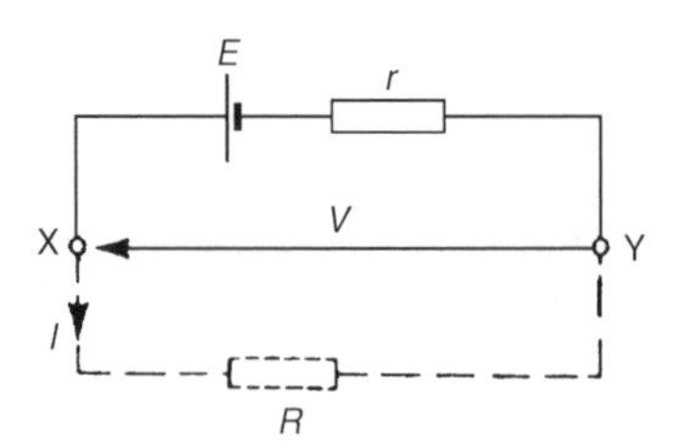

Figure 4.2

The voltage available at the terminals of a cell falls when a load is connected. This is caused by the **internal resistance** of the cell which is the opposition of the material of the cell to the flow of current. The internal resistance acts in series with other resistances in the circuit. Figure 4.2 shows a cell of e.m.f. E volts and internal resistance, r, and XY represents the terminals of the cell.

When a load (shown as resistance R) is not connected, no current flows and the terminal p.d., $V = E$. When R is connected a current I flows which causes a voltage drop in the cell, given by Ir. The p.d. available at the cell terminals is less than the e.m.f. of the cell and is given by:

$$\boxed{V = E - Ir}$$

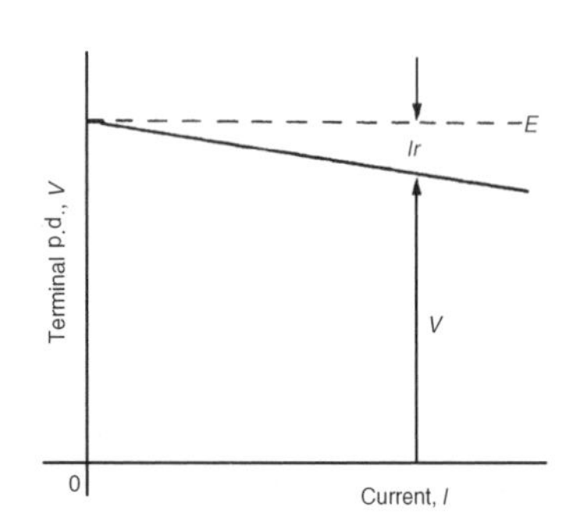

Figure 4.3

Thus if a battery of e.m.f. 12 volts and internal resistance 0.01 Ω delivers a current of 100 A, the terminal p.d.,

$$V = 12 - (100)(0.01)$$
$$= 12 - 1 = 11 \text{ V}$$

When different values of potential difference V, across a cell or power supply are measured for different values of current I, a graph may be plotted as shown in Figure 4.3. Since the e.m.f. E of the cell or power supply is the p.d. across its terminals on no load (i.e. when $I = 0$), then E is as shown by the broken line.

Since $V = E - Ir$ then the internal resistance may be calculated from

$$\boxed{r = \frac{E - V}{I}}$$

When a current is flowing in the direction shown in Figure 4.2 the cell is said to be **discharging** ($E > V$)

When a current flows in the opposite direction to that shown in Figure 4.2 the cell is said to be **charging** ($V > E$)

A **battery** is a combination of more than one cell. The cells in a battery may be connected in series or in parallel.

(i) **For cells connected in series:**
Total e.m.f. = sum of cell's e.m.f.'s
Total internal resistance = sum of cell's internal resistances

(ii) **For cells connected in parallel:**
If each cell has the same e.m.f. and internal resistance:
Total e.m.f. = e.m.f. of one cell
Total internal resistance of n cells

$$= \frac{1}{n} \times \text{internal resistance of one cell}$$

Problem 1. Eight cells, each with an internal resistance of 0.2 Ω and an e.m.f. of 2.2 V are connected (a) in series, (b) in parallel. Determine the e.m.f. and the internal resistance of the batteries so formed.

(a) When connected in series, total e.m.f. = sum of cell's e.m.f.

$$= 2.2 \times 8 = \mathbf{17.6\ V}$$

Total internal resistance = sum of cell's internal resistance

$$= 0.2 \times 8 = \mathbf{1.6\ \Omega}$$

(b) When connected in parallel, total e.m.f. = e.m.f. of one cell

$$= \mathbf{2.2\ V}$$

Total internal resistance of 8 cells

$$= \tfrac{1}{8} \times \text{internal resistance of one cell}$$

$$= \tfrac{1}{8} \times 0.2 = \mathbf{0.025\ \Omega}$$

Problem 2. A cell has an internal resistance of 0.02 Ω and an e.m.f. of 2.0 V. Calculate its terminal p.d. if it delivers (a) 5 A, (b) 50 A

(a) Terminal p.d., $V = E - Ir$ where E = e.m.f. of cell, I = current flowing and r = internal resistance of cell

$$E = 2.0\ \text{V},\ I = 5\ \text{A and } r = 0.02\ \Omega$$

Hence $V = 2.0 - (5)(0.02) = 2.0 - 0.1 = \mathbf{1.9\ V}$

(b) When the current is 50 A, terminal p.d.,

$$V = E - Ir = 2.0 - 50(0.02)$$

i.e., $V = 2.0 - 1.0 = \mathbf{1.0\ V}$

Thus the terminal p.d. decreases as the current drawn increases.

Problem 3. The p.d. at the terminals of a battery is 25 V when no load is connected and 24 V when a load taking 10 A is connected. Determine the internal resistance of the battery.

When no load is connected the e.m.f. of the battery, E, is equal to the terminal p.d., V, i.e., $E = 25$ V

When current $I = 10$ A and terminal p.d. $V = 24$ V, then $V = E - Ir$ i.e., $24 = 25 - (10)r$

Hence, rearranging, gives $10r = 25 - 24 = 1$ and the internal resistance, $r = \frac{1}{10} = \mathbf{0.1\ \Omega}$

Problem 4. Ten 1.5 V cells, each having an internal resistance of 0.2 Ω, are connected in series to a load of 58 Ω. Determine (a) the current flowing in the circuit and (b) the p.d. at the battery terminals.

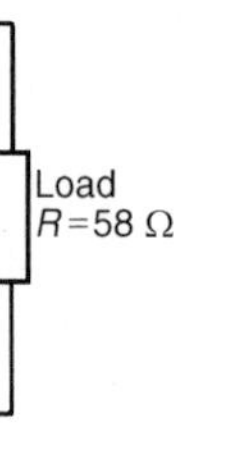

Figure 4.4

(a) For ten cells, battery e.m.f., $E = 10 \times 1.5 = 15$ V, and the total internal resistance, $r = 10 \times 0.2 = 2\ \Omega$ When connected to a 58 Ω load the circuit is as shown in Figure 4.4.

$$\text{Current I} = \frac{\text{e.m.f.}}{\text{total resistance}} = \frac{15}{58+2} = \frac{15}{60} = \mathbf{0.25\ A}$$

(b) P.d. to battery terminals, $V = E - Ir$

i.e. $V = 15 - (0.25)(2) = \mathbf{14.5\ V}$

4.7 Primary cells

Primary cells cannot be recharged, that is, the conversion of chemical energy to electrical energy is irreversible and the cell cannot be used once the chemicals are exhausted. Examples of primary cells include the Leclanché cell and the mercury cell.

Lechlanché cell

A typical dry Lechlanché cell is shown in Figure 4.5. Such a cell has an e.m.f. of about 1.5 V when new, but this falls rapidly if in continuous use due to polarization. The hydrogen film on the carbon electrode forms faster than can be dissipated by the depolarizer. The Lechlanché cell is suitable only for intermittent use, applications including torches, transistor radios, bells, indicator circuits, gas lighters, controlling switch-gear, and so on. The cell is the most commonly used of primary cells, is cheap, requires little maintenance and has a shelf life of about 2 years.

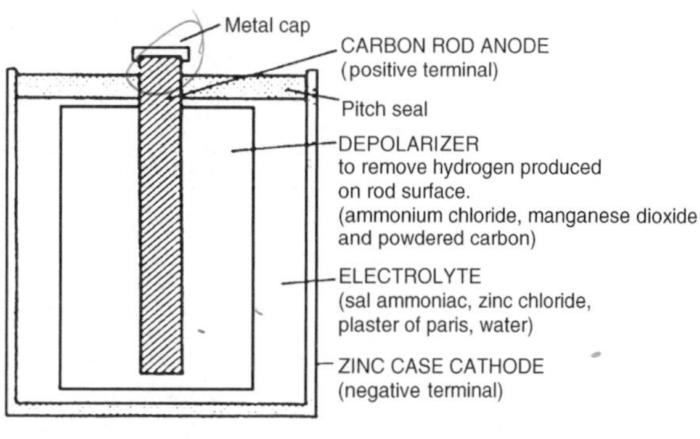

Figure 4.5

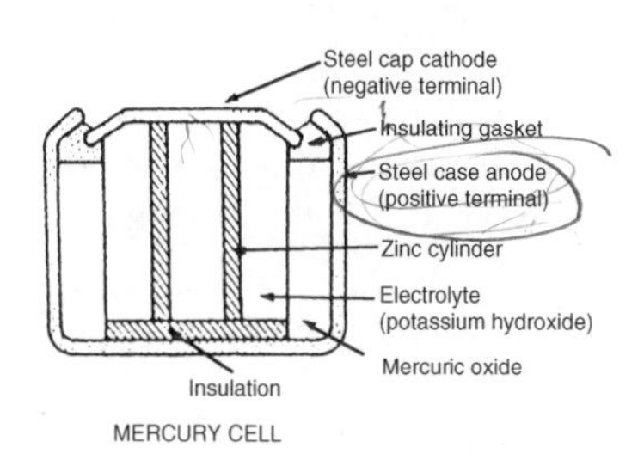

Figure 4.6

Mercury cell

A typical mercury cell is shown in Figure 4.6. Such a cell has an e.m.f. of about 1.3 V which remains constant for a relatively long time. Its main advantages over the Lechlanché cell is its smaller size and its long shelf life. Typical practical applications include hearing aids, medical electronics, cameras and for guided missiles.

4.8 Secondary cells

Secondary cells can be recharged after use, that is, the conversion of chemical energy to electrical energy is reversible and the cell may be used many times. Examples of secondary cells include the lead-acid cell and the alkaline cell. Practical applications of such cells include car batteries, telephone circuits and for traction purposes — such as milk delivery vans and fork lift trucks.

Lead-acid cell

A typical lead-acid cell is constructed of:

(i) A container made of glass, ebonite or plastic.

(ii) **Lead plates**

(a) the negative plate (cathode) consists of spongy lead
(b) the positive plate (anode) is formed by pressing lead peroxide into the lead grid.

The plates are interleaved as shown in the plan view of Figure 4.7 to increase their effective cross-sectional area and to minimize internal resistance.

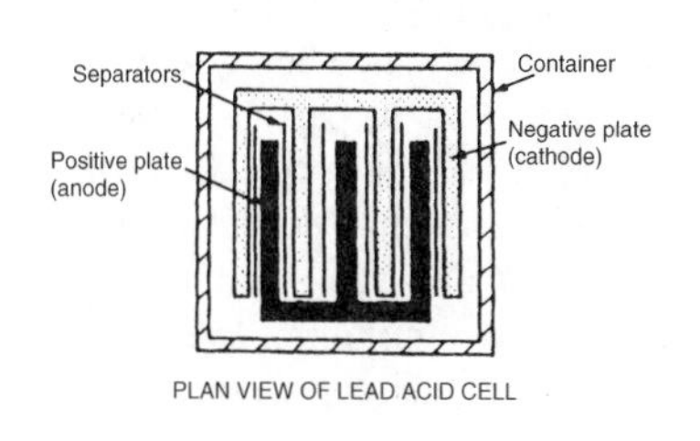

Figure 4.7

(iii) **Separators** made of glass, celluloid or wood.

(iv) An **electrolyte** which is a mixture of sulphuric acid and distilled water.

The relative density (or specific gravity) of a lead-acid cell, which may be measured using a hydrometer, varies between about 1.26 when the cell is fully charged to about 1.19 when discharged. The terminal p.d. of a lead-acid cell is about 2 V.

When a cell supplies current to a load it is said to be **discharging**. During discharge:

(i) the lead peroxide (positive plate) and the spongy lead (negative plate) are converted into lead sulphate, and
(ii) the oxygen in the lead peroxide combines with hydrogen in the electrolyte to form water. The electrolyte is therefore weakened and the relative density falls.

The terminal p.d. of a lead-acid cell when fully discharged is about 1.8 V.

A cell is **charged** by connecting a d.c. supply to its terminals, the positive terminal of the cell being connected to the positive terminal of the supply. The charging current flows in the reverse direction to the discharge current and the chemical action is reversed. During charging:

(i) the lead sulphate on the positive and negative plates is converted back to lead peroxide and lead respectively, and
(ii) the water content of the electrolyte decreases as the oxygen released from the electrolyte combines with the lead of the positive plate. The relative density of the electrolyte thus increases.

The colour of the positive plate when fully charged is dark brown and when discharged is light brown. The colour of the negative plate when fully charged is grey and when discharged is light grey.

Alkaline cell

There are two main types of alkaline cell — the nickel-iron cell and the nickel-cadmium cell. In both types the positive plate is made of nickel hydroxide enclosed in finely perforated steel tubes, the resistance being reduced by the addition of pure nickel or graphite. The tubes are assembled into nickel-steel plates.

In the nickel-iron cell, (sometimes called the Edison cell or nife cell), the negative plate is made of iron oxide, with the resistance being reduced by a little mercuric oxide, the whole being enclosed in perforated steel tubes and assembled in steel plates. In the nickel-cadmium cell the negative plate is made of cadmium. The electrolyte in each type of cell is a solution of potassium hydroxide which does not undergo any chemical change and thus the quantity can be reduced to a minimum. The plates are separated by insulating rods and assembled in steel containers which are then enclosed in a non-metallic crate to insulate the cells from one another. The average discharge p.d. of an alkaline cell is about 1.2 V.

Advantages of an alkaline cell (for example, a nickel-cadmium cell or a nickel-iron cell) over a lead-acid cell include:

(i) More robust construction
(ii) Capable of withstanding heavy charging and discharging currents without damage
(iii) Has a longer life
(iv) For a given capacity is lighter in weight
(v) Can be left indefinitely in any state of charge or discharge without damage
(vi) Is not self-discharging

Disadvantages of an alkaline cell over a lead-acid cell include:

(i) Is relatively more expensive
(ii) Requires more cells for a given e.m.f.
(iii) Has a higher internal resistance
(iv) Must be kept sealed
(v) Has a lower efficiency

Alkaline cells may be used in extremes of temperature, in conditions where vibration is experienced or where duties require long idle periods or heavy discharge currents. Practical examples include traction and marine work, lighting in railway carriages, military portable radios and for starting diesel and petrol engines.

However, the lead-acid cell is the most common one in practical use.

4.9 Cell capacity

The **capacity** of a cell is measured in ampere-hours (Ah). A fully charged 50 Ah battery rated for 10 h discharge can be discharged at a steady current of 5 A for 10 h, but if the load current is increased to 10 A then the battery is discharged in 3-4 h, since the higher the discharge current, the lower is the effective capacity of the battery. Typical discharge characteristics for a lead-acid cell are shown in Figure 4.8.

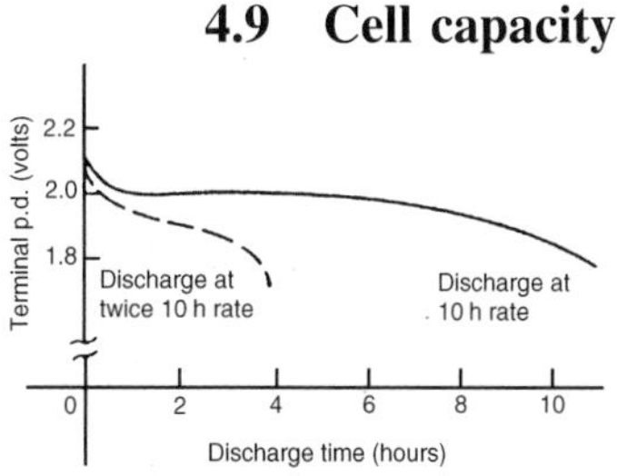

Figure 4.8

4.10 Further problems on the chemical effects of electricity

1 Twelve cells, each with an internal resistance of 0.24 Ω and an e.m.f. of 1.5 V are connected (a) in series, (b) in parallel. Determine the e.m.f. and internal resistance of the batteries so formed.
[(a) 18 V, 2.88 Ω (b) 1.5 V, 0.02 Ω]

2 A cell has an internal resistance of 0.03 Ω and an e.m.f. of 2.2 V. Calculate its terminal p.d. if it delivers (a) 1 A, (b) 20 A, (c) 50 A
[(a) 2.17 V (b) 1.6 V (c) 0.7 V]

3 The p.d. at the terminals of a battery is 16 V when no load is connected and 14 V when a load taking 8 A is connected. Determine the internal resistance of the battery. [0.25 Ω]

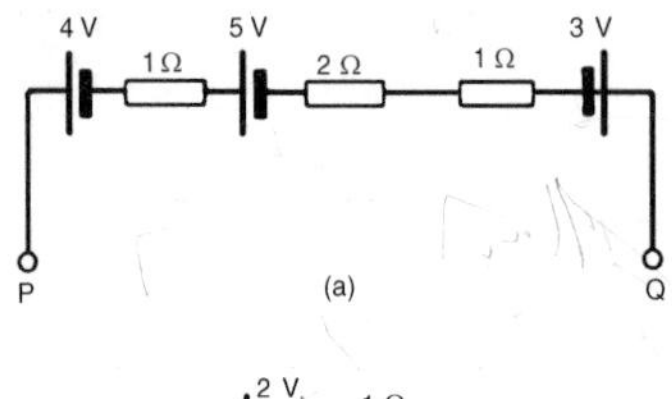

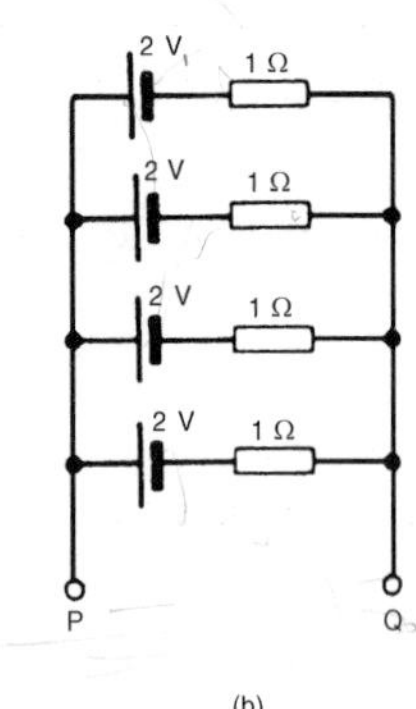

Figure 4.9

4 A battery of e.m.f. 20 V and internal resistance 0.2 Ω supplies a load taking 10 A. Determine the p.d. at the battery terminals and the resistance of the load. [18 V, 1.8 Ω]

5 Ten 2.2 V cells, each having an internal resistance of 0.1 Ω are connected in series to a load of 21 Ω. Determine (a) the current flowing in the circuit, and (b) the p.d. at the battery terminals. [(a) 1 A (b) 21 V]

6 For the circuits shown in Figure 4.9 the resistors represent the internal resistance of the batteries. Find, in each case: (a) the total e.m.f. across PQ (b) the total equivalent internal resistances of the batteries. [(a)(i) 6 V (ii) 2 V (b)(i) 4 Ω (ii) 0.25 Ω]

7 The voltage at the terminals of a battery is 52 V when no load is connected and 48.8 V when a load taking 80 A is connected. Find the internal resistance of the battery. What would be the terminal voltage when a load taking 20 A is connected? [0.04 Ω, 51.2 V]

Assignment 1

This assignment covers the material contained in chapters 1 to 4.

The marks for each question are shown in brackets at the end of each question.

1 An electromagnet exerts a force of 15 N and moves a soft iron armature through a distance of 12 mm in 50 ms. Determine the power consumed. (5)

2 A d.c. motor consumes 47.25 MJ when connected to a 250 V supply for 1 hour 45 minutes. Determine the power rating of the motor and the current taken from the supply. (5)

3 A 100 W electric light bulb is connected to a 200 V supply. Calculate (a) the current flowing in the bulb, and (b) the resistance of the bulb. (4)

4 Determine the charge transferred when a current of 5 mA flows for 10 minutes. (4)

5 A current of 12 A flows in the element of an electric fire of resistance 25 Ω. Determine the power dissipated by the element. If the fire is on for 5 hours every day, calculate for a one week period (a) the energy used, and (b) cost of using the fire if electricity cost 7p per unit. (6)

6 Calculate the resistance of 1200 m of copper cable of cross-sectional area 15 mm^2. Take the resistivity of copper as 0.02 μΩm. (5)

7 At a temperature of 40°C, an aluminium cable has a resistance of 25 Ω. If the temperature coefficient of resistance at 0°C is 0.0038/°C, calculate it's resistance at 0°C. (5)

8 (a) State six typical applications of primary cells.
(b) State six typical applications of secondary cells. (6)

9 Four cells, each with an internal resistance of 0.40 Ω and an e.m.f. of 2.5 V are connected in series to a load of 38.40 Ω. (a) Determine the current flowing in the circuit and the p.d. at the battery terminals. (b) If the cells are connected in parallel instead of in series, determine the current flowing and the p.d. at the battery terminals. (10)

5 Series and parallel networks

At the end of this chapter you should be able to:

- calculate unknown voltages, current and resistances in a series circuit
- understand voltage division in a series circuit
- calculate unknown voltages, currents and resistances in a parallel network
- calculate unknown voltages, currents and resistances in series-parallel networks
- understand current division in a two-branch parallel network
- describe the advantages and disadvantages of series and parallel connection of lamps

5.1 Series circuits

Figure 5.1 shows three resistors R_1, R_2 and R_3 connected end to end, i.e., in series, with a battery source of V volts. Since the circuit is closed a current I will flow and the p.d. across each resistor may be determined from the voltmeter readings V_1, V_2 and V_3

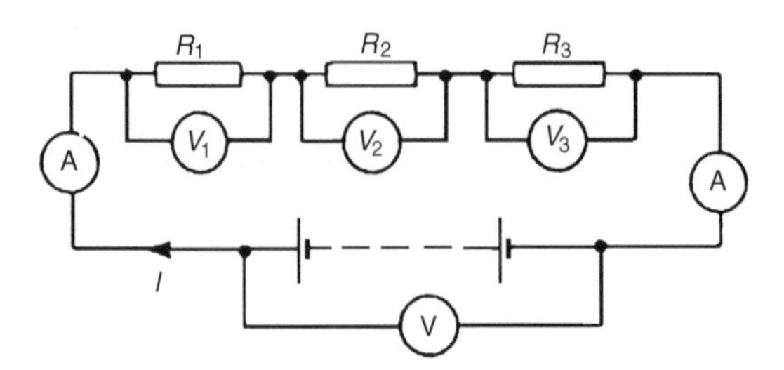

Figure 5.1

In a series circuit

(a) the current I is the same in all parts of the circuit and hence the same reading is found on each of the two ammeters shown, and
(b) the sum of the voltages V_1, V_2 and V_3 is equal to the total applied voltage, V, i.e.

$$\boldsymbol{V = V_1 + V_2 + V_3}$$

From Ohm's law:

$$V_1 = IR_1, V_2 = IR_2, V_3 = IR_3 \text{ and } V = IR$$

where R is the total circuit resistance.
Since $V = V_1 + V_2 + V_3$
then $IR = IR_1 + IR_2 + IR_3$
Dividing throughout by I gives

$$\boldsymbol{R = R_1 + R_2 + R_3}$$

Thus for a series circuit, the total resistance is obtained by adding together the values of the separate resistances.

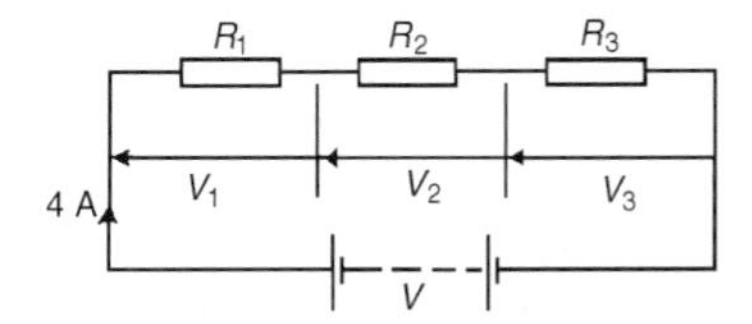

Figure 5.2

Problem 1. For the circuit shown in Figure 5.2, determine (a) the battery voltage V, (b) the total resistance of the circuit, and (c) the values of resistance of resistors R_1, R_2 and R_3, given that the p.d.'s across R_1, R_2 and R_3 are 5 V, 2 V and 6 V respectively.

(a) Battery voltage $V = V_1 + V_2 + V_3$

$$= 5 + 2 + 6 = \mathbf{13\ V}$$

(b) Total circuit resistance $R = \dfrac{V}{I} = \dfrac{13}{4} = \mathbf{3.25\ \Omega}$

(c) Resistance $R_1 = \dfrac{V_1}{I} = \dfrac{5}{4} = \mathbf{1.25\ \Omega}$

Resistance $R_2 = \dfrac{V_2}{I} = \dfrac{2}{4} = \mathbf{0.5\ \Omega}$

Resistance $R_3 = \dfrac{V_3}{I} = \dfrac{6}{4} = \mathbf{1.5\ \Omega}$

(Check: $R_1 + R_2 + R_3 = 1.25 + 0.5 + 1.5 = 3.25\ \Omega = R$)

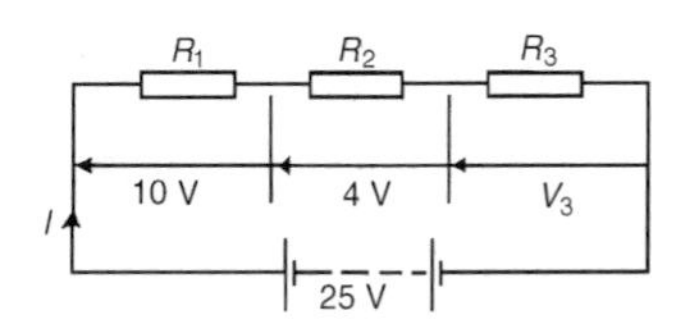

Figure 5.3

Problem 2. For the circuit shown in Figure 5.3, determine the p.d. across resistor R_3. If the total resistance of the circuit is 100 Ω, determine the current flowing through resistor R_1. Find also the value of resistor R_2

P.d. across R_3, $V_3 = 25 - 10 - 4 = \mathbf{11\ V}$

Current $I = \dfrac{V}{R} = \dfrac{25}{100} = \mathbf{0.25\ A}$, which is the current flowing in each resistor

Resistance $R_2 = \dfrac{V_2}{I} = \dfrac{4}{0.25} = \mathbf{16\ \Omega}$

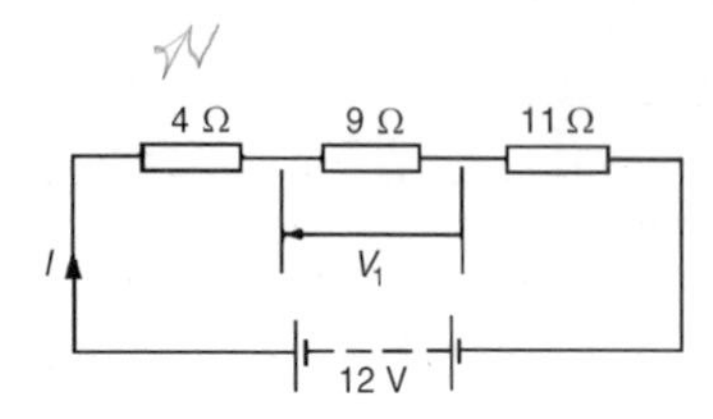

Figure 5.4

Problem 3. A 12 V battery is connected in a circuit having three series-connected resistors having resistances of 4 Ω, 9 Ω and 11 Ω. Determine the current flowing through, and the p.d. across the 9 Ω resistor. Find also the power dissipated in the 11 Ω resistor.

The circuit diagram is shown in Figure 5.4.
Total resistance $R = 4 + 9 + 11 = 24\ \Omega$

Current $I = \dfrac{V}{R} = \dfrac{12}{24} = \mathbf{0.5\ A}$, which is the current in the 9 Ω resistor.

P.d. across the 9 Ω resistor, $V_1 = I \times 9 = 0.5 \times 9 = \mathbf{4.5\ V}$

Power dissipated in the 11 Ω resistor, $P = I^2R = 0.5^2(11)$

$$= (0.25)(11) = \mathbf{2.75\ W}$$

5.2 Potential divider

The voltage distribution for the circuit shown in Figure 5.5(a) is given by:

$$\boldsymbol{V_1 = \left(\frac{R_1}{R_1 + R_2}\right) V}$$

$$\boldsymbol{V_2 = \left(\frac{R_2}{R_1 + R_2}\right) V}$$

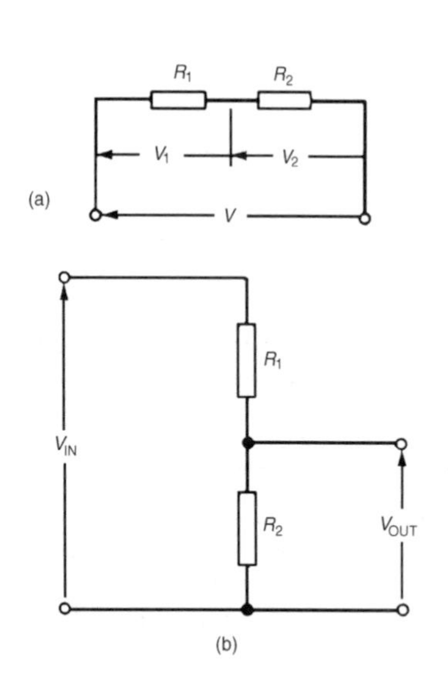

Figure 5.5

The circuit shown in Figure 5.5(b) is often referred to as a **potential divider** circuit. Such a circuit can consist of a number of similar elements in series connected across a voltage source, voltages being taken from connections between the elements. Frequently the divider consists of two resistors as shown in Figure 5.5(b), where

$$\boldsymbol{V_{OUT} = \left(\frac{R_2}{R_1 + R_2}\right) V_{IN}}$$

Problem 4. Determine the value of voltage V shown in Figure 5.6.

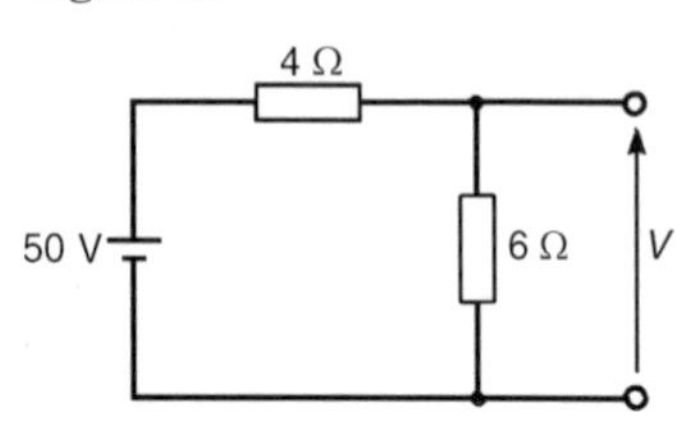

Figure 5.6

Figure 5.6 may be redrawn as shown in Figure 5.7, and voltage

$$V = \left(\frac{6}{6 + 4}\right)(50) = \mathbf{30\ V}$$

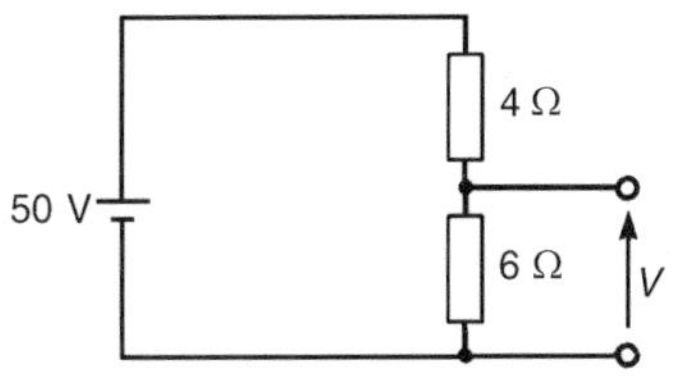

Figure 5.7

Problem 5. Two resistors are connected in series across a 24 V supply and a current of 3 A flows in the circuit. If one of the resistors has a resistance of 2 Ω determine (a) the value of the other resistor, and (b) the p.d. across the 2 Ω resistor. If the circuit is connected for 50 hours, how much energy is used?

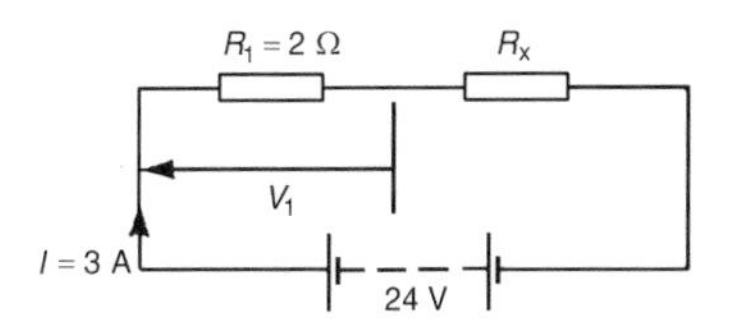

Figure 5.8

The circuit diagram is shown in Figure 5.8

(a) Total circuit resistance $R = \dfrac{V}{I} = \dfrac{24}{3} = 8\ \Omega$

Value of unknown resistance, $R_x = 8 - 2 = \mathbf{6\ \Omega}$

(b) P.d. across 2 Ω resistor, $V_1 = IR_1 = 3 \times 2 = \mathbf{6\ V}$
Alternatively, from above,

$$V_1 = \left(\frac{R_1}{R_1 + R_x}\right) \text{V} = \left(\frac{2}{2+6}\right)(24) = 6 \text{ V}$$

$$\begin{aligned} \text{Energy used} &= \text{power} \times \text{time} \\ &= V \times I \times t \\ &= (24 \times 3 \text{ W})(50 \text{ h}) \\ &= 3600 \text{ Wh} = \mathbf{3.6\ kWh} \end{aligned}$$

5.3 Parallel networks

Figure 5.9 shows three resistors, R_1, R_2 and R_3 connected across each other, i.e., in parallel, across a battery source of V volts.

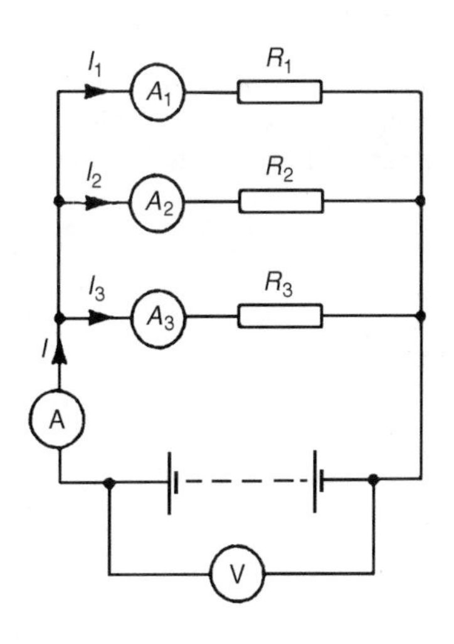

Figure 5.9

In a parallel circuit:

(a) the sum of the currents I_1, I_2 and I_3 is equal to the total circuit current, I, i.e. $\boldsymbol{I = I_1 + I_2 + I_3}$, and
(b) the source p.d., V volts, is the same across each of the resistors.

From Ohm's law:

$$I_1 = \frac{V}{R_1},\ I_2 = \frac{V}{R_2},\ I_3 = \frac{V}{R_3} \text{ and } I = \frac{V}{R}$$

where R is the total circuit resistance.

Since $I = I_1 + I_2 + I_3$

then, $\dfrac{V}{R} = \dfrac{V}{R_1} + \dfrac{V}{R_2} + \dfrac{V}{R_3}$

Dividing throughout by V gives:

$$\boldsymbol{\frac{1}{R} = \frac{1}{R_1} + \frac{1}{R_2} + \frac{1}{R_3}}$$

This equation must be used when finding the total resistance R of a parallel circuit. For the special case of **two resistors in parallel**

$$\frac{1}{R} = \frac{1}{R_1} + \frac{1}{R_2} = \frac{R_2 + R_1}{R_1 R_2}$$

Hence $\boldsymbol{R = \frac{R_1 R_2}{R_1 + R_2}}$ $\left(\text{i.e. } \frac{\text{product}}{\text{sum}}\right)$

Problem 6. For the circuit shown in Figure 5.10, determine (a) the reading on the ammeter, and (b) the value of resistor R_2

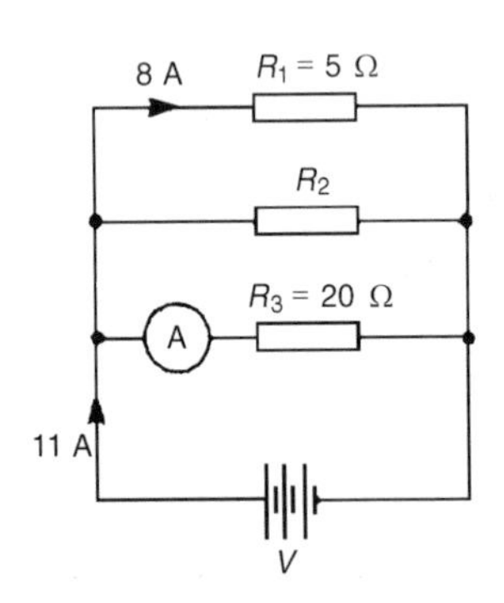

Figure 5.10

P.d. across R_1 is the same as the supply voltage V.

Hence supply voltage, $V = 8 \times 5 = 40$ V

(a) Reading on ammeter, $I = \frac{V}{R_3} = \frac{40}{20} = \mathbf{2\ A}$

(b) Current flowing through $R_2 = 11 - 8 - 2 = 1$ A

Hence, $R_2 = \frac{V}{I_2} = \frac{40}{1} = \mathbf{40\ \Omega}$

Problem 7. Two resistors, of resistance 3 Ω and 6 Ω, are connected in parallel across a battery having a voltage of 12 V. Determine (a) the total circuit resistance and (b) the current flowing in the 3 Ω resistor.

The circuit diagram is shown in Figure 5.11.

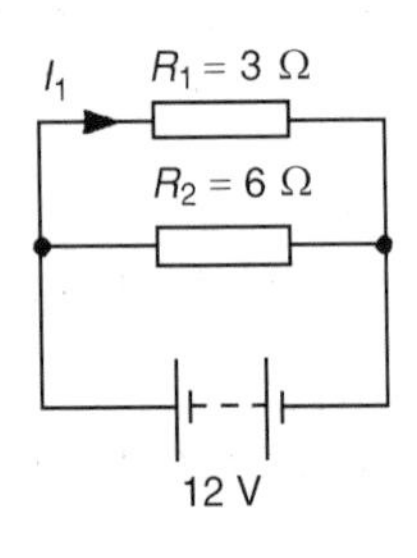

Figure 5.11

(a) The total circuit resistance R is given by

$$\frac{1}{R} = \frac{1}{R_1} + \frac{1}{R_2} = \frac{1}{3} + \frac{1}{6}$$

$$\frac{1}{R} = \frac{2+1}{6} = \frac{3}{6}$$

Hence, $R = \frac{6}{3} = \mathbf{2\ \Omega}$

$$\left(\text{Alternatively, } R = \frac{R_1 R_2}{R_1 + R_2} = \frac{3 \times 6}{3 + 6} = \frac{18}{9} = \mathbf{2\ \Omega}\right)$$

(b) Current in the 3 Ω resistance, $I_1 = \dfrac{V}{R_1} = \dfrac{12}{3} = \mathbf{4\ A}$

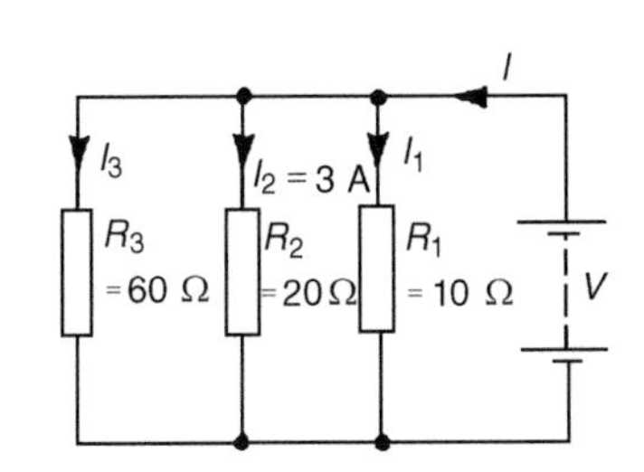

Figure 5.12

Problem 8. For the circuit shown in Figure 5.12, find (a) the value of the supply voltage V and (b) the value of current I.

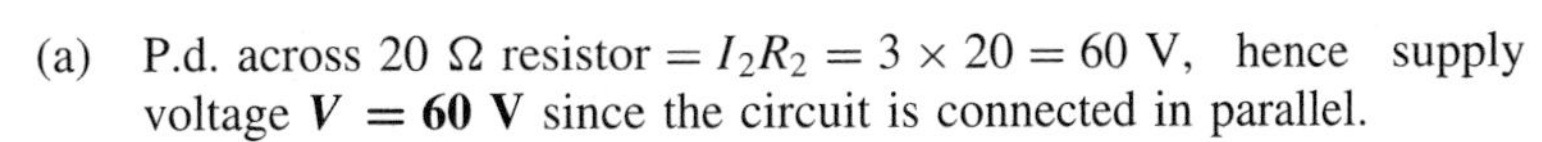

(a) P.d. across 20 Ω resistor $= I_2 R_2 = 3 \times 20 = 60$ V, hence supply voltage $\mathbf{V = 60\ V}$ since the circuit is connected in parallel.

(b) Current $I_1 = \dfrac{V}{R_1} = \dfrac{60}{10} = 6$ A; $I_2 = 3$ A

$$I_3 = \frac{V}{R_3} = \frac{60}{60} = 1 \text{ A}$$

Current $I = I_1 + I_2 + I_3$ and hence $I = 6 + 3 + 1 = \mathbf{10\ A}$

$$\text{Alternatively, } \frac{1}{R} = \frac{1}{60} + \frac{1}{20} + \frac{1}{10} = \frac{1 + 3 + 6}{60} = \frac{10}{60}$$

$$\text{Hence total resistance } R = \frac{60}{10} = 6\ \Omega$$

$$\text{Current } I = \frac{V}{R} = \frac{60}{6} = \mathbf{10\ A}$$

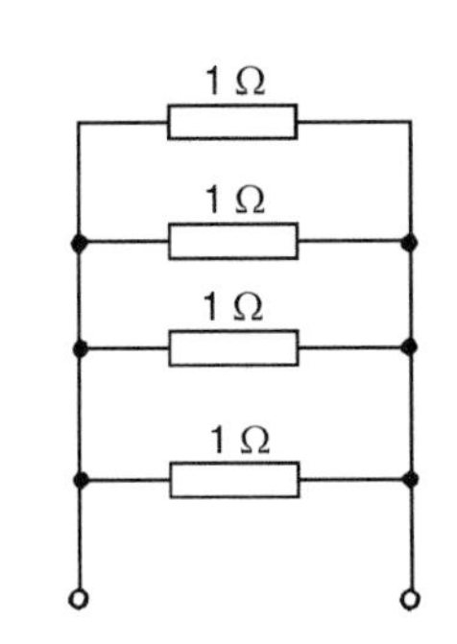

Figure 5.13

Problem 9. Given four 1 Ω resistors, state how they must be connected to give an overall resistance of (a) $\frac{1}{4}\Omega$ (b) 1 Ω (c) $1\frac{1}{3}$ Ω (d) $2\frac{1}{2}$ Ω, all four resistors being connected in each case.

(a) **All four in parallel** (see Figure 5.13),

$$\text{since } \frac{1}{R} = \frac{1}{1} + \frac{1}{1} + \frac{1}{1} + \frac{1}{1} = \frac{4}{1}, \text{ i.e., } R = \frac{1}{4}\Omega$$

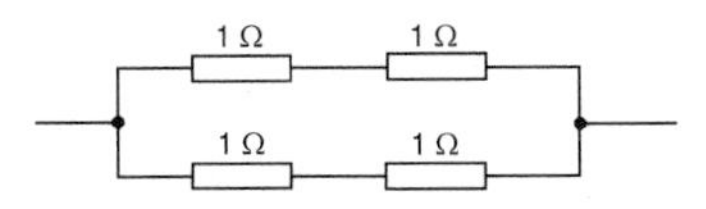

Figure 5.14

(b) **Two in series, in parallel with another two in series** (see Figure 5.14), since 1 Ω and 1 Ω in series gives 2 Ω, and 2 Ω in parallel with 2 Ω gives: $\dfrac{2 \times 2}{2 + 2} = \dfrac{4}{4} = 1\ \Omega$

(c) **Three in parallel, in series with one** (see Figure 5.15), since for the three in parallel,

$$\frac{1}{R} = \frac{1}{1} + \frac{1}{1} + \frac{1}{1} = \frac{3}{1}, \text{ i.e., } R = \frac{1}{3}\ \Omega \text{ and } \frac{1}{3}\ \Omega \text{ in series}$$

with 1 Ω gives $1\frac{1}{3}$ Ω

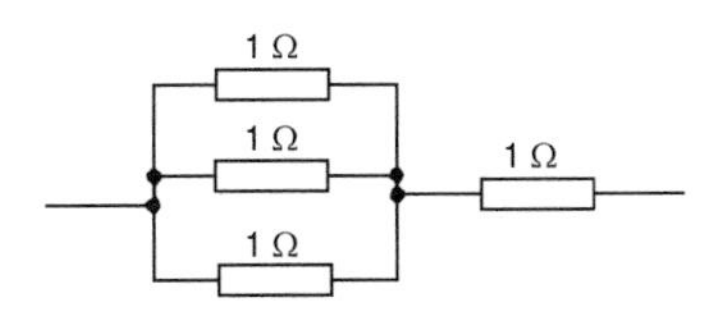

Figure 5.15

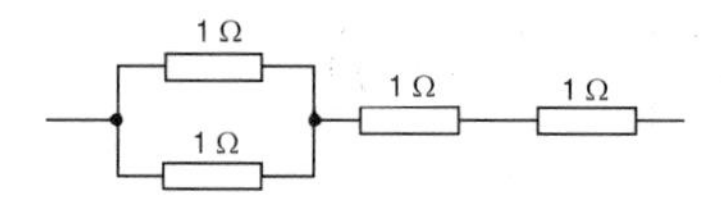

Figure 5.16

(d) **Two in parallel, in series with two in series** (see Figure 5.16), since for the two in parallel

$$R = \frac{1 \times 1}{1 + 1} = \frac{1}{2}\ \Omega, \text{ and } \frac{1}{2}\ \Omega,\ 1\ \Omega \text{ and } 1\ \Omega \text{ in series gives } 2\frac{1}{2}\ \Omega$$

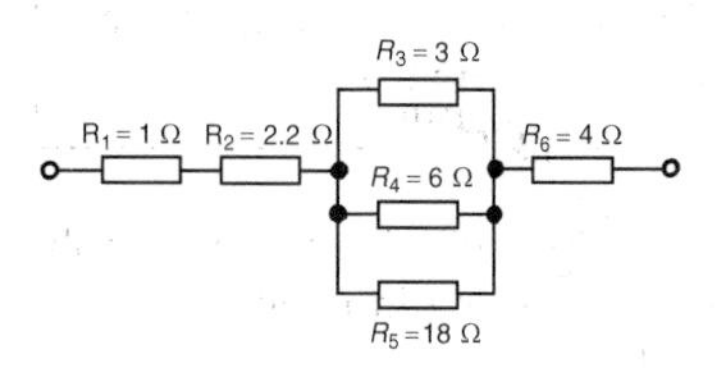

Figure 5.17

Problem 10. Find the equivalent resistance for the circuit shown in Figure 5.17.

R_3, R_4 and R_5 are connected in parallel and their equivalent resistance R is given by:

$$\frac{1}{R} = \frac{1}{3} + \frac{1}{6} + \frac{1}{18} = \frac{6 + 3 + 1}{18} = \frac{10}{18}$$

Hence $R = \dfrac{18}{10} = 1.8\ \Omega$

The circuit is now equivalent to four resistors in series and the equivalent circuit resistance $= 1 + 2.2 + 1.8 + 4 = \mathbf{9\ \Omega}$

5.4 Current division

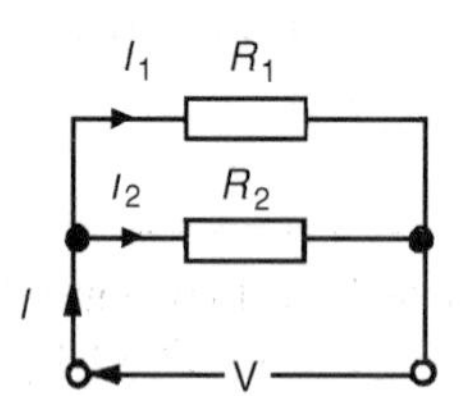

Figure 5.18

For the circuit shown in Figure 5.18, the total circuit resistance, R_T is given by:

$$R_T = \frac{R_1 R_2}{R_1 + R_2}$$

and $V = IR_T = I\left(\dfrac{R_1 R_2}{R_1 + R_2}\right)$

$$\text{Current } I_1 = \frac{V}{R_1} = \frac{I}{R_1}\left(\frac{R_1 R_2}{R_1 + R_2}\right) = \left(\frac{R_2}{R_1 + R_2}\right)(I)$$

Similarly,

$$\text{current } I_2 = \frac{V}{R_2} = \frac{I}{R_2}\left(\frac{R_1 R_2}{R_1 + R_2}\right) = \left(\frac{R_1}{R_1 + R_2}\right)(I)$$

Summarizing, with reference to Figure 5.18

$$\boldsymbol{I_1 = \left(\frac{R_2}{R_1 + R_2}\right)(I)} \quad \textbf{and} \quad \boldsymbol{I_2 = \left(\frac{R_1}{R_1 + R_2}\right)(I)}$$

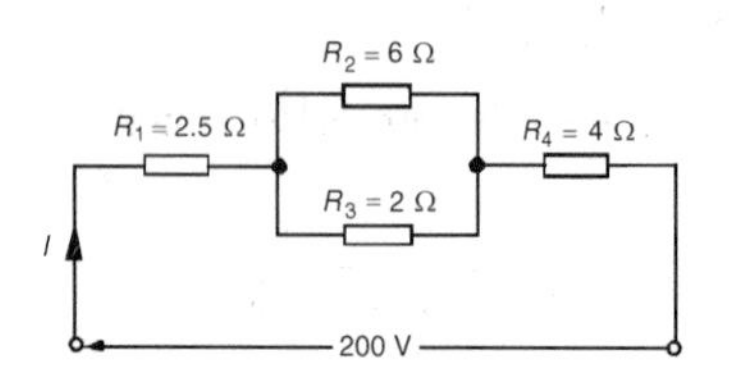

Figure 5.19

Problem 11. For the series-parallel arrangement shown in Figure 5.19, find (a) the supply current, (b) the current flowing through each resistor and (c) the p.d. across each resistor.

(a) The equivalent resistance R_x of R_2 and R_3 in parallel is:

$$R_x = \frac{6 \times 2}{6+2} = \frac{12}{8} = 1.5\ \Omega$$

The equivalent resistance R_T of R_1, R_x and R_4 in series is:

$$R_T = 2.5 + 1.5 + 4 = 8\ \Omega$$

$$\text{Supply current } I = \frac{V}{R_T} = \frac{200}{8} = \mathbf{25\ A}$$

(b) The current flowing through R_1 and R_4 is 25 A

$$\text{The current flowing through } R_2 = \left(\frac{R_3}{R_2+R_3}\right) I = \left(\frac{2}{6+2}\right) 25$$

$$= \mathbf{6.25\ A}$$

$$\text{The current flowing through } R_3 = \left(\frac{R_2}{R_2+R_3}\right) I = \left(\frac{6}{6+2}\right) 25$$

$$= \mathbf{18.75\ A}$$

(Note that the currents flowing through R_2 and R_3 must add up to the total current flowing into the parallel arrangement, i.e. 25 A)

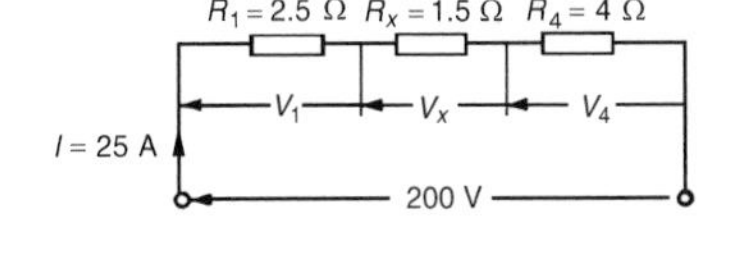

Figure 5.20

(c) The equivalent circuit of Figure 5.19 is shown in Figure 5.20.

p.d. across R_1, i.e., $V_1 = IR_1 = (25)(2.5) = \mathbf{62.5\ V}$

p.d. across R_x, i.e., $V_x = IR_x = (25)(1.5) = \mathbf{37.5\ V}$

p.d. across R_4, i.e., $V_4 = IR_4 = (25)(4) = \mathbf{100\ V}$

Hence the p.d. across $R_2 =$ p.d. across $R_3 = \mathbf{37.5\ V}$

Problem 12. For the circuit shown in Figure 5.21 calculate (a) the value of resistor R_x such that the total power dissipated in the circuit is 2.5 kW, and (b) the current flowing in each of the four resistors.

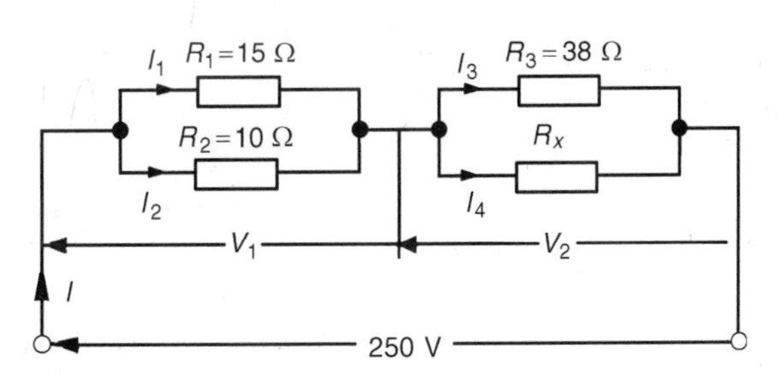

Figure 5.21

(a) Power dissipated $P = VI$ watts, hence $2500 = (250)(I)$

$$I = \frac{2500}{250} = 10 \text{ A}$$

From Ohm's law, $R_T = \frac{V}{I} = \frac{250}{10} = 25\ \Omega$, where R_T is the equivalent circuit resistance.

The equivalent resistance of R_1 and R_2 in parallel is

$$\frac{15 \times 10}{15 + 10} = \frac{150}{25} = 6\ \Omega$$

The equivalent resistance of resistors R_3 and R_x in parallel is equal to $25\ \Omega - 6\ \Omega$, i.e., $19\ \Omega$

There are three methods whereby R_x can be determined.

Method 1

The voltage $V_1 = IR$, where R is $6\ \Omega$, from above,

i.e. $V_1 = (10)(6) = 60$ V

Hence $V_2 = 250\text{ V} - 60\text{ V} = 190\text{ V} = \text{p.d. across } R_3$
$= \text{p.d. across } R_x$

$I_3 = \frac{V_2}{R_3} = \frac{190}{38} = 5$ A. Thus $I_4 = 5$ A also, since $I = 10$ A

Thus $R_x = \frac{V_2}{I_4} = \frac{190}{5} = 38\ \Omega$

Method 2

Since the equivalent resistance of R_3 and R_x in parallel is $19\ \Omega$,

$19 = \frac{38R_x}{38 + R_x}$ $\left(\text{i.e. } \frac{\text{product}}{\text{sum}}\right)$ Hence

$$19(38 + R_x) = 38R_x$$

$$722 + 19R_x = 38R_x$$

$$722 = 38R_x - 19R_x = 19R_x$$

Thus $R_x = \frac{722}{19} = \mathbf{38\ \Omega}$

Method 3

When two resistors having the same value are connected in parallel the equivalent resistance is always half the value of one of the resistors. Thus, in this case, since $R_T = 19\ \Omega$ and $R_3 = 38\ \Omega$, then $R_x = 38\ \Omega$ could have been deduced on sight.

(b) Current $I_1 = \left(\frac{R_2}{R_1 + R_2}\right) I = \left(\frac{10}{15 + 10}\right)(10)$

$= \left(\frac{2}{5}\right)(10) = \mathbf{4\ A}$

Current $I_2 = \left(\frac{R_1}{R_1 + R_2}\right) I = \left(\frac{15}{15 + 10}\right)(10)$

$= \left(\frac{3}{5}\right)(10) = \mathbf{6\ A}$

From part (a), method 1, $\mathbf{I_3 = I_4 = 5\ A}$

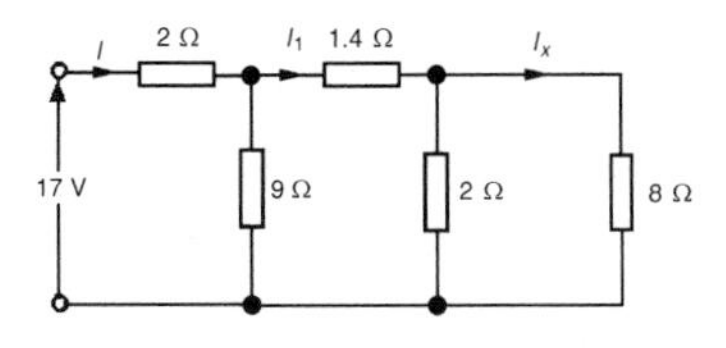

Figure 5.22

Problem 13. For the arrangement shown in Figure 5.22, find the current I_x

Commencing at the right-hand side of the arrangement shown in Figure 5.22, the circuit is gradually reduced in stages as shown in Figure 5.23(a)–(d).

From Figure 5.23(d) $I = \frac{17}{4.25} = 4$ A

From Figure 5.23(b) $I_1 = \left(\frac{9}{9+3}\right)(I) = \left(\frac{9}{12}\right)(4) = 3$ A

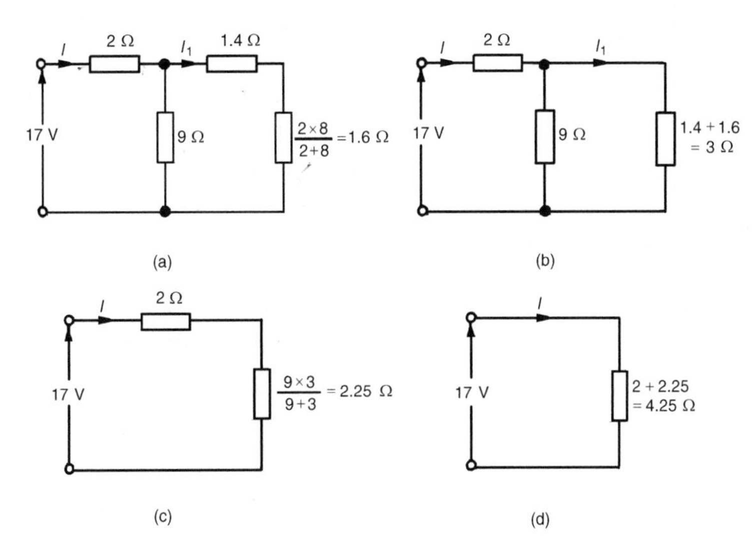

Figure 5.23

From Figure 5.22 $I_x = \left(\frac{2}{2+8}\right)(I_1) = \left(\frac{2}{10}\right)(3) = \mathbf{0.6\ A}$

5.5 Wiring lamps in series and in parallel

Series connection

Figure 5.24 shows three lamps, each rated at 240 V, connected in series across a 240 V supply.

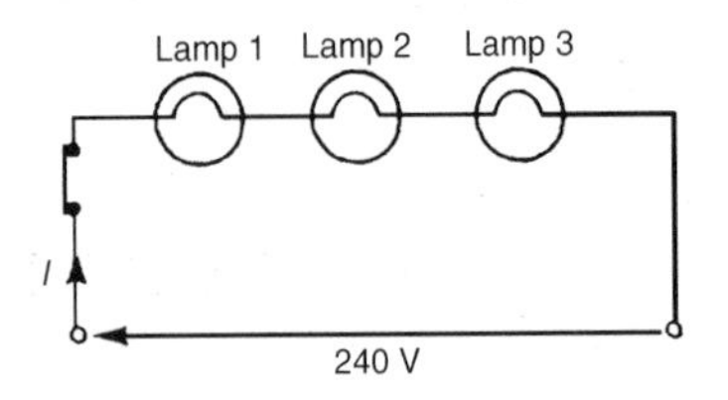

Figure 5.24

(i) Each lamp has only $\frac{240}{3}$ V, i.e., 80 V across it and thus each lamp glows dimly.

(ii) If another lamp of similar rating is added in series with the other three lamps then each lamp now has $\frac{240}{4}$ V, i.e., 60 V across it and each now glows even more dimly.

(iii) If a lamp is removed from the circuit or if a lamp develops a fault (i.e. an open circuit) or if the switch is opened then the circuit is broken, no current flows, and the remaining lamps will not light up.

(iv) Less cable is required for a series connection than for a parallel one.

The series connection of lamps is usually limited to decorative lighting such as for Christmas tree lights.

Parallel connection

Figure 5.25 shows three similar lamps, each rated at 240 V, connected in parallel across a 240 V supply.

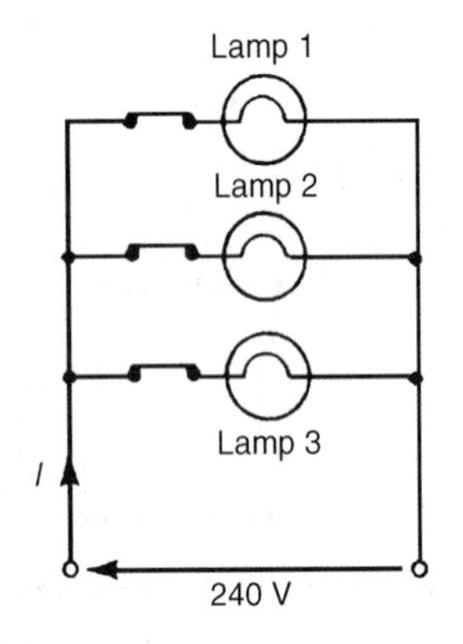

Figure 5.25

(i) Each lamp has 240 V across it and thus each will glow brilliantly at their rated voltage.

(ii) If any lamp is removed from the circuit or develops a fault (open circuit) or a switch is opened, the remaining lamps are unaffected.

(iii) The addition of further similar lamps in parallel does not affect the brightness of the other lamps.

(iv) More cable is required for parallel connection than for a series one.

The parallel connection of lamps is the most widely used in electrical installations.

Problem 14. If three identical lamps are connected in parallel and the combined resistance is 150 Ω, find the resistance of one lamp.

Let the resistance of one lamp be R, then,

$$\frac{1}{150} = \frac{1}{R} + \frac{1}{R} + \frac{1}{R} = \frac{3}{R}, \text{ from which, } R = 3 \times 150 = \mathbf{450\ \Omega}$$

Problem 15. Three identical lamps A, B and C are connected in series across a 150 V supply. State (a) the voltage across each lamp, and (b) the effect of lamp C failing.

(a) Since each lamp is identical and they are connected in series there is $\frac{150}{3}$ V, i.e. 50 V across each.

(b) If lamp C fails, i.e., open circuits, no current will flow and **lamps A and B will not operate**.

5.6 Further problems on series and parallel networks

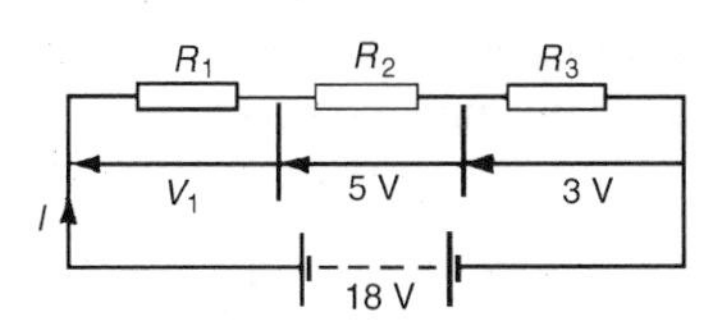

Figure 5.26

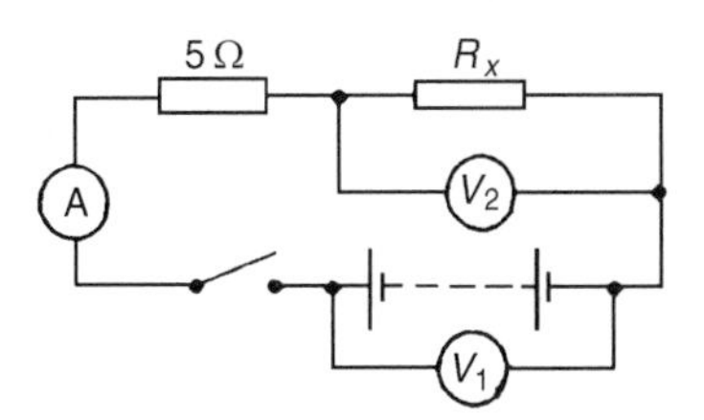

Figure 5.27

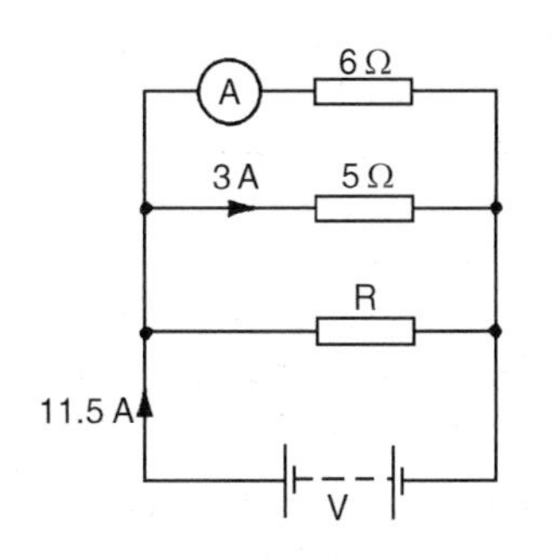

Figure 5.28

1 The p.d's measured across three resistors connected in series are 5 V, 7 V and 10 V, and the supply current is 2 A. Determine (a) the supply voltage, (b) the total circuit resistance and (c) the values of the three resistors. [(a) 22 V (b) 11 Ω (c) 2.5 Ω, 3.5 Ω, 5 Ω]

2 For the circuit shown in Figure 5.26, determine the value of V_1. If the total circuit resistance is 36 Ω determine the supply current and the value of resistors R_1, R_2 and R_3. [10 V, 0.5 A, 20 Ω, 10 Ω, 6 Ω]

3 When the switch in the circuit in Figure 5.27 is closed the reading on voltmeter 1 is 30 V and that on voltmeter 2 is 10 V. Determine the reading on the ammeter and the value of resistor R_x [4 A, 2.5 Ω]

4 Two resistors are connected in series across an 18 V supply and a current of 5 A flows. If one of the resistors has a value of 2.4 Ω determine (a) the value of the other resistor and (b) the p.d. across the 2.4 Ω resistor. [(a) 1.2 Ω (b) 12 V]

5 Resistances of 4 Ω and 12 Ω are connected in parallel across a 9 V battery. Determine (a) the equivalent circuit resistance, (b) the supply current, and (c) the current in each resistor. [(a) 3 Ω (b) 3 A (c) 2.25 A, 0.75 A]

6 For the circuit shown in Figure 5.28 determine (a) the reading on the ammeter, and (b) the value of resistor R. [2.5 A, 2.5 Ω]

7 Find the equivalent resistance when the following resistances are connected (a) in series, (b) in parallel
(i) 3 Ω and 2 Ω (ii) 20 kΩ and 40 kΩ
(iii) 4 Ω, 8 Ω and 16 Ω (iv) 800 Ω, 4 kΩ and 1500 Ω
[(a) (i) 5 Ω (ii) 60 kΩ (iii) 28 Ω (iv) 6.3 kΩ
(b) (i) 1.2 Ω (ii) $13\frac{1}{3}$ kΩ (iii) $2\frac{2}{7}$ Ω (iv) 461.5 kΩ]

8 Find the total resistance between terminals A and B of the circuit shown in Figure 5.29(a) [8 Ω]

9 Find the equivalent resistance between terminals C and D of the circuit shown in Figure 5.29(b) [27.5 Ω]

10 Resistors of 20 Ω, 20 Ω and 30 Ω are connected in parallel. What resistance must be added in series with the combination to obtain a total resistance of 10 Ω. If the complete circuit expends a power of 0.36 kW, find the total current flowing. [2.5 Ω, 6 A]

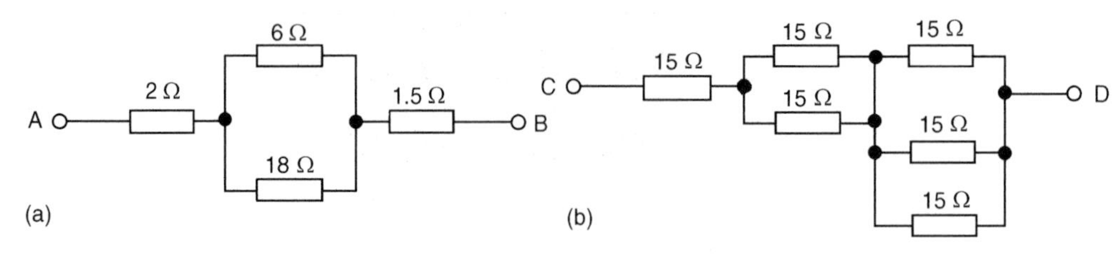

Figure 5.29

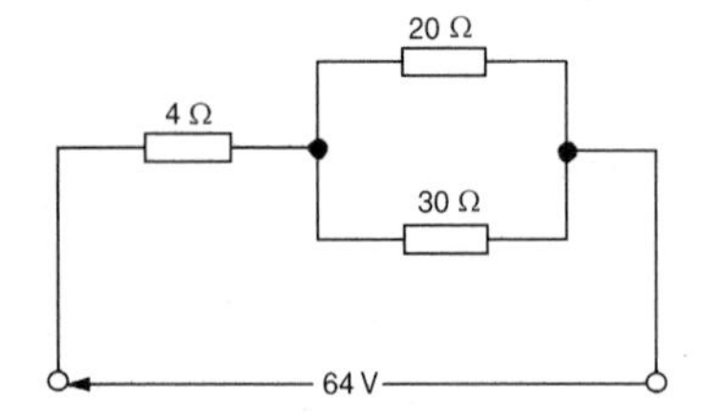

Figure 5.30

11 (a) Calculate the current flowing in the 30 Ω resistor shown in Figure 5.30
(b) What additional value of resistance would have to be placed in parallel with the 20 Ω and 30 Ω resistors to change the supply current to 8 A, the supply voltage remaining constant.
[(a) 1.6 A (b) 6 Ω]

12 Determine the currents and voltages indicated in the circuit shown in Figure 5.31.
[$I_1 = 5$ A, $I_2 = 2.5$ A, $I_3 = 1\frac{2}{3}$ A, $I_4 = \frac{5}{6}$ A
$I_5 = 3$ A, $I_6 = 2$ A, $V_1 = 20$ V, $V_2 = 5$ V, $V_3 = 6$ V]

13 Find the current I in Figure 5.32. [1.8 A]

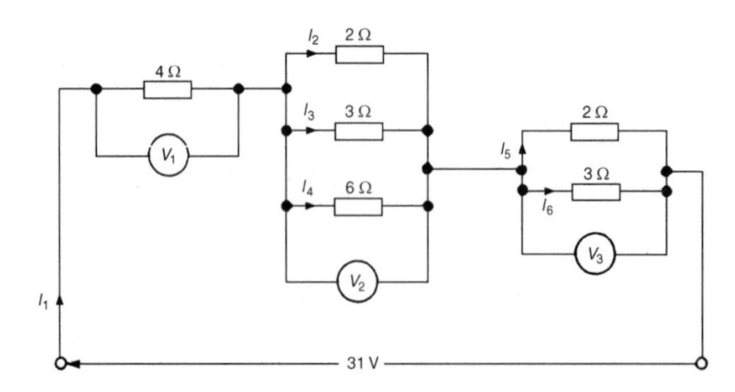

Figure 5.31

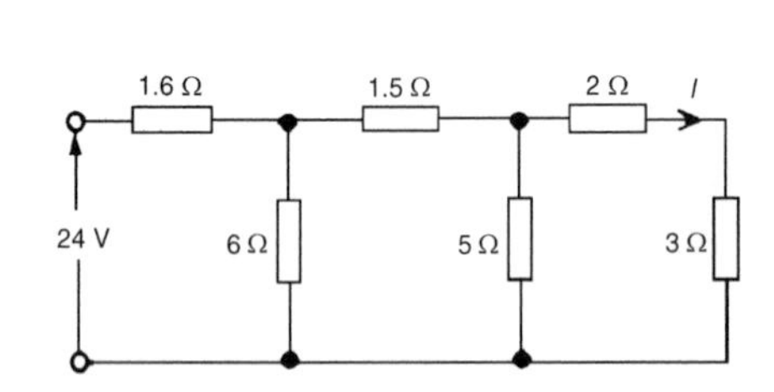

Figure 5.32

14 If four identical lamps are connected in parallel and the combined resistance is 100 Ω, find the resistance of one lamp. [400 Ω]

15 Three identical filament lamps are connected (a) in series, (b) in parallel across a 210 V supply. State for each connection the p.d. across each lamp. [(a) 70 V (b) 210 V]

6 Capacitors and capacitance

At the end of this chapter you should be able to:

- describe an electrostatic field
- define electric field strength E and state its unit
- define capacitance and state its unit
- describe a capacitor and draw the circuit diagram symbol
- perform simple calculations involving $C = \frac{Q}{V}$ and $Q = It$
- define electric flux density D and state its unit
- define permittivity, distinguishing between ε_0, ε_r and ε
- perform simple calculations involving $D = \frac{Q}{A}$, $E = \frac{V}{D}$ and $\frac{D}{E} = \varepsilon_0\varepsilon_r$
- understand that for a parallel plate capacitor, $C = \frac{\varepsilon_0\varepsilon_r A(n-1)}{d}$
- perform calculations involving capacitors connected in parallel and in series
- define dielectric strength and state its unit
- state that the energy stored in a capacitor is given by $W = \frac{1}{2}\,CV^2$ joules
- describe practical types of capacitor
- understand the precautions needed when discharging capacitors

6.1 Electrostatic field

Figure 6.1 represents two parallel metal plates, A and B, charged to different potentials. If an electron that has a negative charge is placed between the plates, a force will act on the electron tending to push it away from the negative plate B towards the positive plate, A. Similarly, a positive charge would be acted on by a force tending to move it toward the negative plate. Any region such as that shown between the plates in Figure 1, in which an electric charge experiences a force, is called an **electrostatic field**. The direction of the field is defined as that of the force acting on a positive charge placed in the field. In Figure 6.1, the direction of the force is from the positive plate to the negative plate.

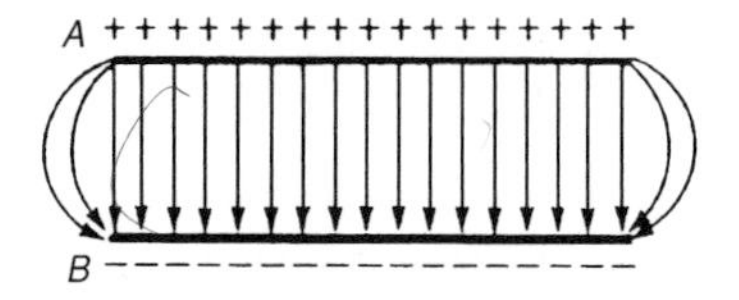

Figure 6.1 *Electrostatic field*

Such a field may be represented in magnitude and direction by **lines of electric force** drawn between the charged surfaces. The closeness of the lines is an indication of the field strength. Whenever a p.d. is established between two points, an electric field will always exist. Figure 6.2(a) shows a typical field pattern for an isolated point charge, and Figure 6.2(b) shows the field pattern for adjacent charges of opposite polarity. Electric lines of force (often called electric flux lines) are continuous and start and finish on point charges. Also, the lines cannot cross each other. When a charged body is placed close to an uncharged body, an induced charge of opposite sign appears on the surface of the uncharged body. This is because lines of force from the charged body terminate on its surface.

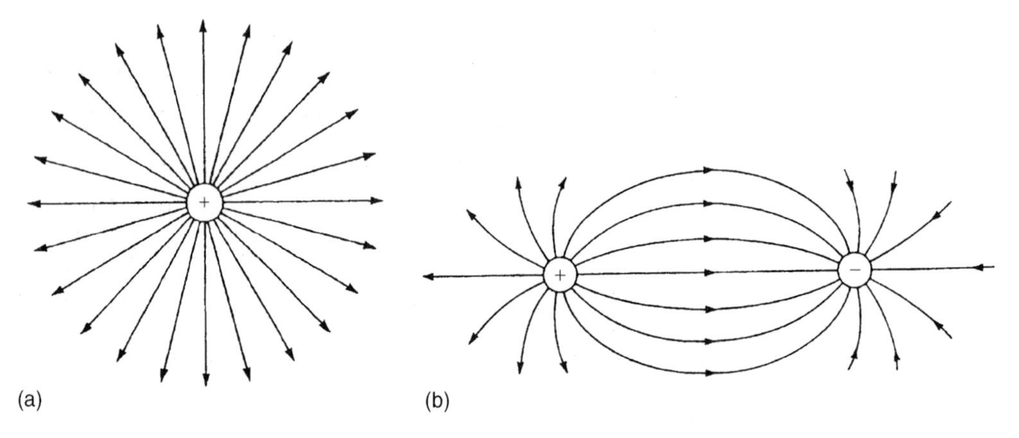

Figure 6.2 *(a) Isolated point charge; (b) adjacent charges of opposite polarity*

The concept of field lines or lines of force is used to illustrate the properties of an electric field. However, it should be remembered that they are only aids to the imagination.

The **force of attraction or repulsion** between two electrically charged bodies is proportional to the magnitude of their charges and inversely proportional to the square of the distance separating them,

i.e. force $\propto \dfrac{q_1 q_2}{d^2}$ or $\boxed{\text{force} = k\dfrac{q_1 q_2}{d^2}}$ where constant $k \approx 9 \times 10^9$ in air

This is known as **Coulomb's law**.

Hence the force between two charged spheres in air with their centres 16 mm apart and each carrying a charge of $+1.6\ \mu C$ is given by:

$$\text{force} = k\frac{q_1 q_2}{d^2} \approx (9 \times 10^9)\frac{(1.6 \times 10^{-6})^2}{(16 \times 10^{-3})^2} = \textbf{90 newtons}$$

6.2 Electric field strength

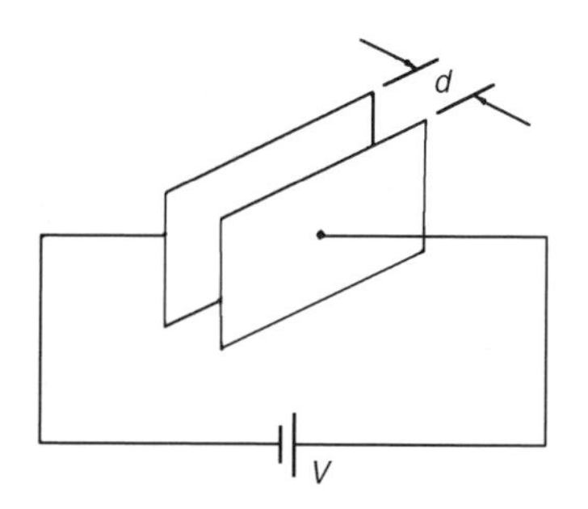

Figure 6.3

Figure 6.3 shows two parallel conducting plates separated from each other by air. They are connected to opposite terminals of a battery of voltage V volts.

There is therefore an electric field in the space between the plates. If the plates are close together, the electric lines of force will be straight and parallel and equally spaced, except near the edge where fringing will occur (see Figure 6.1). Over the area in which there is negligible fringing,

$$\textbf{Electric field strength, } E = \frac{V}{d} \textbf{ volts/metre}$$

where d is the distance between the plates. Electric field strength is also called **potential gradient**.

6.3 Capacitance

Static electric fields arise from electric charges, electric field lines beginning and ending on electric charges. Thus the presence of the field indicates the presence of equal positive and negative electric charges on the two plates of Figure 6.3. Let the charge be $+Q$ coulombs on one plate and $-Q$ coulombs on the other. The property of this pair of plates which determines how much charge corresponds to a given p.d. between the plates is called their capacitance:

$$\textbf{capacitance } C = \frac{Q}{V}$$

The **unit of capacitance** is the **farad F** (or more usually $\mu F = 10^{-6}$ F or pF $= 10^{-12}$ F), which is defined as the capacitance when a pd of one volt appears across the plates when charged with one coulomb.

6.4 Capacitors

Every system of electrical conductors possesses capacitance. For example, there is capacitance between the conductors of overhead transmission lines and also between the wires of a telephone cable. In these examples the capacitance is undesirable but has to be accepted, minimized or compensated for. There are other situations where capacitance is a desirable property.

Devices specially constructed to possess capacitance are called **capacitors** (or condensers, as they used to be called). In its simplest form a capacitor consists of two plates which are separated by an insulating material known as a **dielectric**. A capacitor has the ability to store a quantity of static electricity.

The symbols for a fixed capacitor and a variable capacitor used in electrical circuit diagrams are shown in Figure 6.4.

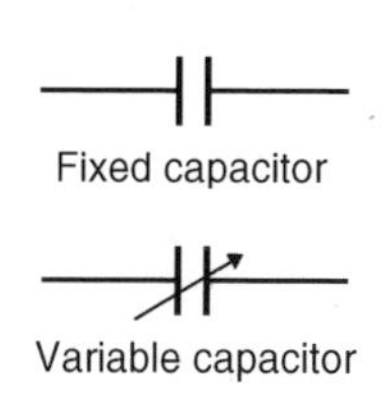

Figure 6.4

The **charge Q** stored in a capacitor is given by:

$$\boldsymbol{Q = I \times t} \textbf{ coulombs,}$$

where I is the current in amperes and t the time in seconds.

Problem 1. (a) Determine the p.d. across a 4 μF capacitor when charged with 5 mC.

(b) Find the charge on a 50 pF capacitor when the voltage applied to it is 2 kV.

(a) $C = 4\ \mu F = 4 \times 10^{-6}$ F; $Q = 5$ mC $= 5 \times 10^{-3}$ C

Since $C = \dfrac{Q}{V}$ then $V = \dfrac{Q}{C} = \dfrac{5 \times 10^{-3}}{4 \times 10^{-6}} = \dfrac{5 \times 10^{6}}{4 \times 10^{3}} = \dfrac{5000}{4}$

Hence p.d. = 1250 V or 1.25 kV

(b) $C = 50$ pF $= 50 \times 10^{-12}$ F; $V = 2$ kV $= 2000$ V

$$Q = CV = 50 \times 10^{-12} \times 2000 = \frac{5 \times 2}{10^8} = 0.1 \times 10^{-6}$$

Hence charge = 0.1 μC

Problem 2. A direct current of 4 A flows into a previously uncharged 20 μF capacitor for 3 ms. Determine the pd between the plates.

$I = 4$ A; $C = 20\ \mu F = 20 \times 10^{-6}$ F; $t = 3$ ms $= 3 \times 10^{-3}$ s

$Q = It = 4 \times 3 \times 10^{-3}$ C

$$V = \frac{Q}{C} = \frac{4 \times 3 \times 10^{-3}}{20 \times 10^{-6}} = \frac{12 \times 10^{6}}{20 \times 10^{3}} = 0.6 \times 10^{3} = 600 \text{ V}$$

Hence, the pd between the plates is 600 V

Problem 3. A 5 μF capacitor is charged so that the pd between its plates is 800 V. Calculate how long the capacitor can provide an average discharge current of 2 mA.

$C = 5\ \mu F = 5 \times 10^{-6}$ F; $V = 800$ V; $I = 2$ mA $= 2 \times 10^{-3}$ A

$Q = CV = 5 \times 10^{-6} \times 800 = 4 \times 10^{-3}$ C

Also, $Q = It$. Thus, $t = \dfrac{Q}{I} = \dfrac{4 \times 10^{-3}}{2 \times 10^{-3}} = 2$ s

Hence the capacitor can provide an average discharge current of 2 mA for 2 s

Further problems on charge and capacitance may be found in Section 6.13, problems 1 to 5, page 70.

6.5 Electric flux density

Unit flux is defined as emanating from a positive charge of 1 coulomb. Thus electric flux Ψ is measured in coulombs, and for a charge of Q coulombs, the flux $\Psi = Q$ coulombs.

Electric flux density D is the amount of flux passing through a defined area A that is perpendicular to the direction of the flux:

$$\textbf{electric flux density, } D = \frac{Q}{A} \textbf{ coulombs/metre}^2$$

Electric flux density is also called **charge density,** σ

6.6 Permittivity

At any point in an electric field, the electric field strength E maintains the electric flux and produces a particular value of electric flux density D at that point. For a field established in **vacuum** (or for practical purposes in air), the ratio D/E is a constant ε_0, i.e.

$$\frac{D}{E} = \varepsilon_0$$

where ε_0 is called the **permittivity of free space** or the free space constant. The value of ε_0 is 8.85×10^{-12} F/m.

When an insulating medium, such as mica, paper, plastic or ceramic, is introduced into the region of an electric field the ratio of D/E is modified:

$$\frac{D}{E} = \varepsilon_0\varepsilon_r$$

where ε_r, the **relative permittivity** of the insulating material, indicates its insulating power compared with that of vacuum:

$$\text{relative permittivity } \varepsilon_r = \frac{\text{flux density in material}}{\text{flux density in vacuum}}$$

ε_r has no unit. Typical values of ε_r include air, 1.00; polythene, 2.3; mica, 3–7; glass, 5–10; water, 80; ceramics, 6–1000.

The product $\varepsilon_0\varepsilon_r$ is called the **absolute permittivity**, ε, i.e.,

$$\varepsilon = \varepsilon_0\varepsilon_r$$

The insulating medium separating charged surfaces is called a **dielectric.** Compared with conductors, dielectric materials have very high resistivities. They are therefore used to separate conductors at different potentials, such as capacitor plates or electric power lines.

Problem 4. Two parallel rectangular plates measuring 20 cm by 40 cm carry an electric charge of 0.2 μC. Calculate the electric flux density. If the plates are spaced 5 mm apart and the voltage between them is 0.25 kV determine the electric field strength.

Charge $Q = 0.2\ \mu\text{C} = 0.2 \times 10^{-6} C$;

Area $A = 20\ \text{cm} \times 40\ \text{cm} = 800\ \text{cm}^2 = 800 \times 10^{-4}\ \text{m}^2$

Electric flux density $\boldsymbol{D} = \dfrac{Q}{A} = \dfrac{0.2 \times 10^{-6}}{800 \times 10^{-4}} = \dfrac{0.2 \times 10^4}{800 \times 10^6}$

$= \dfrac{2000}{800} \times 10^{-6} = \mathbf{2.5\ \mu C/m^2}$

Voltage $V = 0.25\ \text{kV} = 250\ \text{V}$; Plate spacing, $d = 5\ \text{mm} = 5 \times 10^{-3}\ \text{m}$

Electric field strength $\boldsymbol{E} = \dfrac{V}{d} = \dfrac{250}{5 \times 10^{-3}} = \mathbf{50\ kV/m}$

Problem 5. The flux density between two plates separated by mica of relative permittivity 5 is 2 μC/m². Find the voltage gradient between the plates.

Flux density $D = 2\ \mu\text{C/m}^2 = 2 \times 10^{-6}\ \text{C/m}^2$;

$\varepsilon_0 = 8.85 \times 10^{-12}$ F/m; $\varepsilon_r = 5$.

$\dfrac{D}{E} = \varepsilon_0\varepsilon_r$, hence **voltage gradient** $\boldsymbol{E} = \dfrac{D}{\varepsilon_0\varepsilon_r}$

$= \dfrac{2 \times 10^{-6}}{8.85 \times 10^{-12} \times 5}$ V/m

$= \mathbf{45.2\ kV/m}$

Problem 6. Two parallel plates having a pd of 200 V between them are spaced 0.8 mm apart. What is the electric field strength? Find also the flux density when the dielectric between the plates is (a) air, and (b) polythene of relative permittivity 2.3

Electric field strength $\boldsymbol{E} = \dfrac{V}{D} = \dfrac{200}{0.8 \times 10^{-3}} = \mathbf{250\ kV/m}$

(a) For air: $\varepsilon_r = 1$

$\dfrac{D}{E} = \varepsilon_0\varepsilon_r$. Hence

electric flux density $D = E\varepsilon_0\varepsilon_r$

$= (250 \times 10^3 \times 8.85 \times 10^{-12} \times 1)\ \text{C/m}^2$

$= \mathbf{2.213\ \mu C/m^2}$

(b) For polythene, $\varepsilon_r = 2.3$

Electric flux density $D = E\varepsilon_0\varepsilon_r$

$= (250 \times 10^3 \times 8.85 \times 10^{-12} \times 2.3)\ \text{C/m}^2$

$= \mathbf{5.089\ \mu C/m^2}$

Further problems on electric field strength, electric flux density and permittivity may be found in Section 6.13, problems 6 to 10, page 71.

6.7 The parallel plate capacitor

For a parallel-plate capacitor, as shown in Figure 6.5(a), experiments show that capacitance C is proportional to the area A of a plate, inversely proportional to the plate spacing d (i.e., the dielectric thickness) and depends on the nature of the dielectric:

$$\textbf{Capacitance, } C = \frac{\varepsilon_0\varepsilon_r A}{d} \textbf{ farads}$$

where $\varepsilon_0 = 8.85 \times 10^{-12}$ F/m (constant)
ε_r = relative permittivity
A = area of one of the plates, in m^2, and
d = thickness of dielectric in m

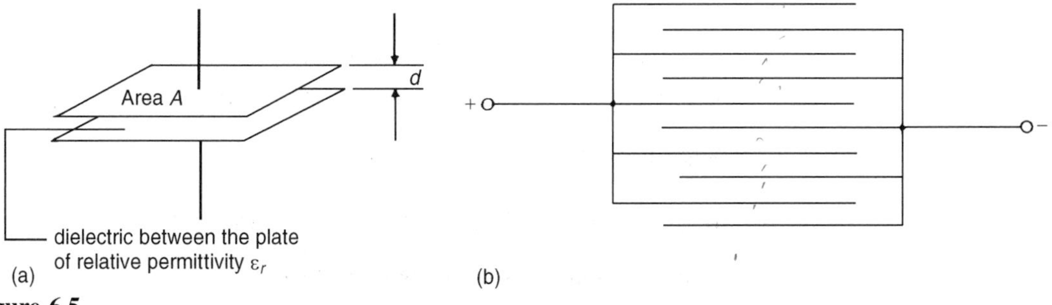

Figure 6.5

Another method used to increase the capacitance is to interleave several plates as shown in Figure 6.5(b). Ten plates are shown, forming nine capacitors with a capacitance nine times that of one pair of plates.

If such an arrangement has n plates then capacitance $C \propto (n - 1)$.

Thus capacitance $\boxed{C = \dfrac{\varepsilon_0 \varepsilon_r A(n-1)}{d} \text{ farads}}$

> Problem 7. (a) A ceramic capacitor has an effective plate area of 4 cm^2 separated by 0.1 mm of ceramic of relative permittivity 100. Calculate the capacitance of the capacitor in picofarads. (b) If the capacitor in part (a) is given a charge of 1.2 μC what will be the pd between the plates?

(a) Area A $= 4 \text{ cm}^2 = 4 \times 10^{-4} \text{ m}^2$; $d = 0.1 \text{ mm} = 0.1 \times 10^{-3} \text{ m}$;

$\varepsilon_0 = 8.85 \times 10^{-12}$ F/m; $\varepsilon_r = 100$

$$\textbf{Capacitance } C = \frac{\varepsilon_0\, \varepsilon_r\, A}{d} \text{ farads}$$

$$= \frac{8.85 \times 10^{-12} \times 100 \times 4 \times 10^{-4}}{0.1 \times 10^{-3}} \text{ F}$$

$$= \frac{8.85 \times 4}{10^{10}} \text{ F} = \frac{8.85 \times 4 \times 10^{12}}{10^{10}} \text{ pF}$$

$$= \mathbf{3540\ pF}$$

(b) $Q = CV$ thus $V = \dfrac{Q}{C} = \dfrac{1.2 \times 10^{-6}}{3540 \times 10^{-12}}$ $V = \mathbf{339\ V}$

> Problem 8. A waxed paper capacitor has two parallel plates, each of effective area 800 cm^2. If the capacitance of the capacitor is 4425 pF determine the effective thickness of the paper if its relative permittivity is 2.5

$A = 800 \text{ cm}^2 = 800 \times 10^{-4} \text{ m}^2 = 0.08 \text{ m}^2$;

$C = 4425 \text{ pF} = 4425 \times 10^{-12}$ F; $\varepsilon_0 = 8.85 \times 10^{-12}$ F/m; $\varepsilon_r = 2.5$

Since $C = \dfrac{\varepsilon_0 \varepsilon_r A}{d}$ then $d = \dfrac{\varepsilon_0 \varepsilon_r A}{C}$

Hence, $d = \dfrac{8.85 \times 10^{-12} \times 2.5 \times 0.08}{4425 \times 10^{-12}} = 0.0004$ m

Hence the thickness of the paper is 0.4 mm

> Problem 9. A parallel plate capacitor has nineteen interleaved plates each 75 mm by 75 mm separated by mica sheets 0.2 mm thick. Assuming the relative permittivity of the mica is 5, calculate the capacitance of the capacitor.

$n = 19;\ n - 1 = 18;\ A = 75 \times 75 = 5625\ \text{mm}^2 = 5625 \times 10^{-6}\ \text{m}^2;$
$\varepsilon_r = 5;\ \varepsilon_0 = 8.85 \times 10^{-12}\ \text{F/m};\ d = 0.2\ \text{mm} = 0.2 \times 10^{-3}\ \text{m}$

$$\text{Capacitance } C = \frac{\varepsilon_0\ \varepsilon_r A(n-1)}{d}$$

$$= \frac{8.85 \times 10^{-12} \times 5 \times 5625 \times 10^{-6} \times 18}{0.2 \times 10^{-3}}\ \text{F}$$

$$= \mathbf{0.0224\ \mu F\ or\ 22.4\ nF}$$

Further problems on parallel plate capacitors may be found in Section 6.13, problems 11 to 17, page 71.

6.8 Capacitors connected in parallel and series

(a) Capacitors connected in parallel

Figure 6.6 shows three capacitors, C_1, C_2 and C_3, connected in parallel with a supply voltage V applied across the arrangement.

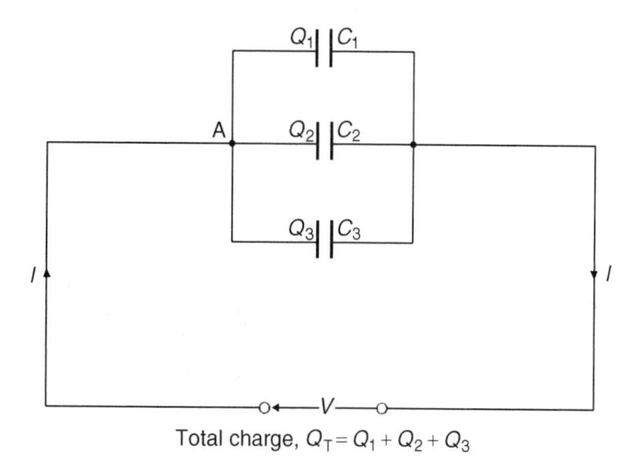

Figure 6.6

When the charging current I reaches point A it divides, some flowing into C_1, some flowing into C_2 and some into C_3. Hence the total charge $Q_T (= I \times t)$ is divided between the three capacitors. The capacitors each store a charge and these are shown as Q_1, Q_2 and Q_3 respectively. Hence

$$Q_T = Q_1 + Q_2 + Q_3$$

But $Q_T = CV$, $Q_1 = C_1V$, $Q_2 = C_2V$ and $Q_3 = C_3V$
Therefore $CV = C_1V + C_2V + C_3V$ where C is the total equivalent circuit capacitance,

i.e. $\mathbf{C = C_1 + C_2 + C_3}$

It follows that for n parallel-connected capacitors,

$$C = C_1 + C_2 + C_3 \ldots + C_n,$$

i.e. the equivalent capacitance of a group of parallel-connected capacitors is the sum of the capacitances of the individual capacitors. (Note that this formula is similar to that used for **resistors** connected in **series**)

(b) Capacitors connected in series

Figure 6.7 shows three capacitors, C_1, C_2 and C_3, connected in series across a supply voltage V. Let the p.d. across the individual capacitors be V_1, V_2 and V_3 respectively as shown.

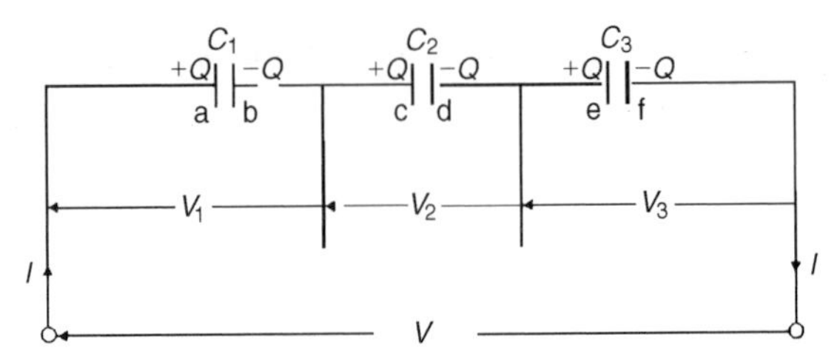

Figure 6.7

Let the charge on plate 'a' of capacitor C_1 be $+Q$ coulombs. This induces an equal but opposite charge of $-Q$ coulombs on plate 'b'. The conductor between plates 'b' and 'c' is electrically isolated from the rest of the circuit so that an equal but opposite charge of $+Q$ coulombs must appear on plate 'c', which, in turn, induces an equal and opposite charge of $-Q$ coulombs on plate 'd', and so on.
Hence when capacitors are connected in series the charge on each is the same.

In a series circuit: $V = V_1 + V_2 + V_3$

Since $V = \frac{Q}{C}$ then $\frac{Q}{C} = \frac{Q}{C_1} + \frac{Q}{C_2} + \frac{Q}{C_3}$

where C is the total equivalent circuit capacitance,

i.e. $\frac{1}{C} = \frac{1}{C_1} + \frac{1}{C_2} + \frac{1}{C_3}$

It follows that for n series-connected capacitors:

$$\frac{1}{C} = \frac{1}{C_1} + \frac{1}{C_2} + \frac{1}{C_3} + \ldots + \frac{1}{C_n},$$

i.e. for series-connected capacitors, the reciprocal of the equivalent capacitance is equal to the sum of the reciprocals of the individual capacitances. (Note that this formula is similar to that used for **resistors** connected in **parallel**)

For the special case of **two capacitors in series:**

$$\frac{1}{C} = \frac{1}{C_1} + \frac{1}{C_2} = \frac{C_2 + C_1}{C_1 C_2}$$

Hence $$\boxed{C = \frac{C_1 C_2}{C_1 + C_2}} \quad \left(\text{i.e. } \frac{\text{product}}{\text{sum}}\right)$$

> Problem 10. Calculate the equivalent capacitance of two capacitors of 6 μF and 4 μF connected (a) in parallel and (b) in series

(a) In parallel, equivalent capacitance $\boldsymbol{C} = C_1 + C_2 = 6\ \mu\text{F} + 4\ \mu\text{F} =$ **10 μF**

(b) In series, equivalent capacitance C is given by: $C = \dfrac{C_1 C_2}{C_1 + C_2}$

This formula is used for the special case of **two** capacitors in series.

Thus $\boldsymbol{C} = \dfrac{6 \times 4}{6 + 4} = \dfrac{24}{10} = \mathbf{2.4\ \mu F}$

> Problem 11. What capacitance must be connected in series with a 30 μF capacitor for the equivalent capacitance to be 12 μF?

Let $C = 12\ \mu\text{F}$ (the equivalent capacitance), $C_1 = 30\ \mu\text{F}$ and C_2 be the unknown capacitance.

For two capacitors in series $\dfrac{1}{C} = \dfrac{1}{C_1} + \dfrac{1}{C_2}$

Hence $\dfrac{1}{C_2} = \dfrac{1}{C} - \dfrac{1}{C_1} = \dfrac{C_1 - C}{C C_1}$

and $\boldsymbol{C_2} = \dfrac{C C_1}{C_1 - C} = \dfrac{12 \times 30}{30 - 12} = \dfrac{360}{18} = \mathbf{20\ \mu F}$

> Problem 12. Capacitances of 1 μF, 3 μF, 5 μF and 6 μF are connected in parallel to a direct voltage supply of 100 V. Determine (a) the equivalent circuit capacitance, (b) the total charge and (c) the charge on each capacitor.

(a) The equivalent capacitance C for four capacitors in parallel is given by:

$$C = C_1 + C_2 + C_3 + C_4$$

i.e. $\boldsymbol{C = 1 + 3 + 5 + 6 = 15\ \mu F}$

(b) Total charge $Q_T = CV$ where C is the equivalent circuit capacitance i.e. $\boldsymbol{Q_T} = 15 \times 10^{-6} \times 100 = 1.5 \times 10^{-3}\ C = \mathbf{1.5\ mC}$

(c) The charge on the 1 μF capacitor $\boldsymbol{Q_1} = C_1V = 1 \times 10^{-6} \times 100$
$= \mathbf{0.1\ mC}$

The charge on the 3 μF capacitor $\boldsymbol{Q_2} = C_2V = 3 \times 10^{-6} \times 100$
$= \mathbf{0.3\ mC}$

The charge on the 5 μF capacitor $\boldsymbol{Q_3} = C_3V = 5 \times 10^{-6} \times 100$
$= \mathbf{0.5\ mC}$

The charge on the 6 μF capacitor $\boldsymbol{Q_4} = C_4V = 6 \times 10^{-6} \times 100$
$= \mathbf{0.6\ mC}$

[Check: In a parallel circuit $Q_T = Q_1 + Q_2 + Q_3 + Q_4$
$Q_1 + Q_2 + Q_3 + Q_4 = 0.1 + 0.3 + 0.5 + 0.6 = 1.5\text{ mC} = Q_T$]

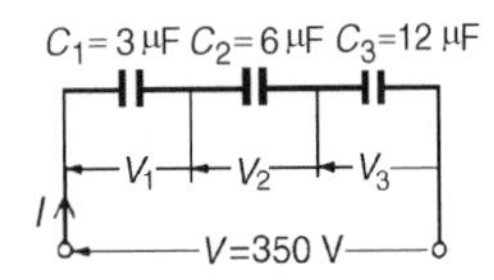

Figure 6.8

Problem 13. Capacitances of 3 μF, 6 μF and 12 μF are connected in series across a 350 V supply. Calculate (a) the equivalent circuit capacitance, (b) the charge on each capacitor and (c) the pd across each capacitor.

The circuit diagram is shown in Figure 6.8.

(a) The equivalent circuit capacitance C for three capacitors in series is given by:

$$\frac{1}{C} = \frac{1}{C_1} + \frac{1}{C_2} + \frac{1}{C_3}$$

i.e. $$\frac{1}{C} = \frac{1}{3} + \frac{1}{6} + \frac{1}{12} = \frac{4+2+1}{12} = \frac{7}{12}$$

Hence the equivalent circuit capacitance $\boldsymbol{C = \frac{12}{7} = 1\frac{5}{7}\ \mu F}$

(b) Total charge $Q_T = CV$,

hence $$Q_T = \frac{12}{7} \times 10^{-6} \times 350 = 600\ \mu\text{C or } 0.6\text{ mC}$$

Since the capacitors are connected in series 0.6 mC is the charge on each of them.

(c) The voltage across the 3 μF capacitor, $V_1 = \dfrac{Q}{C_1} = \dfrac{0.6 \times 10^{-3}}{3 \times 10^{-6}}$

$= \mathbf{200\ V}$

The voltage across the 6 μF capacitor, $V_2 = \frac{Q}{C_2} = \frac{0.6 \times 10^{-3}}{6 \times 10^{-6}}$

$= \mathbf{100\ V}$

The voltage across the 12 μF capacitor, $V_3 = \frac{Q}{C_3} = \frac{0.6 \times 10^{-3}}{12 \times 10^{-6}}$

$= \mathbf{50\ V}$

[Check: In a series circuit $V = V_1 + V_2 + V_3$

$V_1 + V_2 + V_3 = 200 + 100 + 50 = 350\ V =$ supply voltage.]

In practice, capacitors are rarely connected in series unless they are of the same capacitance. The reason for this can be seen from the above problem where the lowest valued capacitor (i.e. 3 μF) has the highest pd across it (i.e. 200 V) which means that if all the capacitors have an identical construction they must all be rated at the highest voltage.

Further problems on capacitors in parallel and series may be found in *Section 6.13, problems 18 to 25, page 72.*

6.9 Dielectric strength

The maximum amount of field strength that a dielectric can withstand is called the dielectric strength of the material.

Dielectric strength, $\boxed{E_m = \frac{V_m}{d}}$

> Problem 14. A capacitor is to be constructed so that its capacitance is 0.2 μF and to take a p.d. of 1.25 kV across its terminals. The dielectric is to be mica which, after allowing a safety factor of 2, has a dielectric strength of 50 MV/m. Find (a) the thickness of the mica needed, and (b) the area of a plate assuming a two-plate construction. (Assume ε_r for mica to be 6)

(a) Dielectric strength, $E = \frac{V}{d}$, i.e. $d = \frac{V}{E} = \frac{1.25 \times 10^3}{50 \times 10^6}$ m

$= \mathbf{0.025\ mm}$

(b) Capacitance, $C = \frac{\varepsilon_0 \varepsilon_r A}{d}$, hence area

$$A = \frac{Cd}{\varepsilon_0 \varepsilon_r} = \frac{0.2 \times 10^{-6} \times 0.025 \times 10^{-3}}{8.85 \times 10^{-12} \times 6}\ \text{m}^2$$

$= 0.09416\ \text{m}^2 = \mathbf{941.6\ cm^2}$

6.10 Energy stored

The energy, W, stored by a capacitor is given by

$$W = \frac{1}{2}CV^2 \text{ joules}$$

Problem 15. (a) Determine the energy stored in a 3 μF capacitor when charged to 400 V. (b) Find also the average power developed if this energy is dissipated in a time of 10 μs

(a) **Energy stored** $W = \frac{1}{2} C V^2$ joules

$$= \frac{1}{2} \times 3 \times 10^{-6} \times 400^2 = \frac{3}{2} \times 16 \times 10^{-2}$$

$$= \mathbf{0.24\ J}$$

(b) **Power** $= \frac{\text{Energy}}{\text{time}} = \frac{0.24}{10 \times 10^{-6}}$ W $=$ **24 kW**

Problem 16. A 12 μF capacitor is required to store 4 J of energy. Find the pd to which the capacitor must be charged.

Energy stored $W = \frac{1}{2} CV^2$ hence $V^2 = \frac{2W}{C}$

and $V = \sqrt{\left(\frac{2W}{C}\right)} = \sqrt{\left(\frac{2 \times 4}{12 \times 10^{-6}}\right)} = \sqrt{\left(\frac{2 \times 10^6}{3}\right)} =$ **816.5 V**

Problem 17. A capacitor is charged with 10 mC. If the energy stored is 1.2 J find (a) the voltage and (b) the capacitance.

Energy stored $W = \frac{1}{2} CV^2$ and $C = \frac{Q}{V}$

Hence $W = \frac{1}{2}\left(\frac{Q}{V}\right) V^2 = \frac{1}{2} QV$

from which $V = \frac{2W}{Q}$

$Q = 10$ mc $= 10 \times 10^{-3}$ C and $W = 1.2$ J

(a) Voltage $V = \frac{2W}{Q} = \frac{2 \times 1.2}{10 \times 10^{-3}} =$ **0.24 kV or 240 V**

(b) Capacitance $C = \frac{Q}{V} = \frac{10 \times 10^{-3}}{240}$ F $= \frac{10 \times 10^6}{240 \times 10^3}$ μF $=$ **41.67 μF**

Further problems on energy stored may be found in Section 6.13, problems 26 to 30, page 73.

6.11 Practical types of capacitor

Practical types of capacitor are characterized by the material used for their dielectric. The main types include: variable air, mica, paper, ceramic, plastic, titanium oxide and electrolytic.

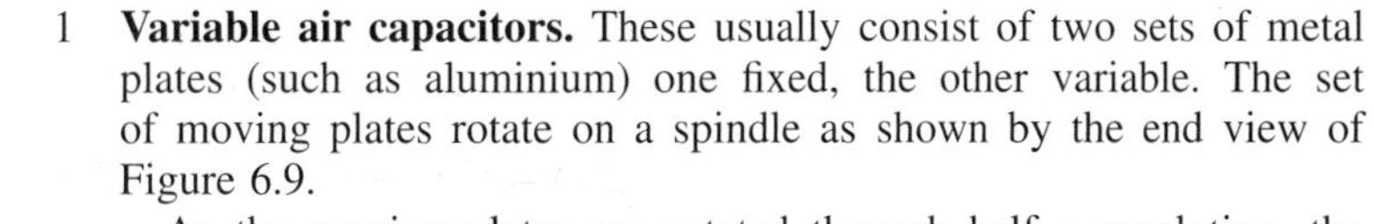

1 **Variable air capacitors.** These usually consist of two sets of metal plates (such as aluminium) one fixed, the other variable. The set of moving plates rotate on a spindle as shown by the end view of Figure 6.9.

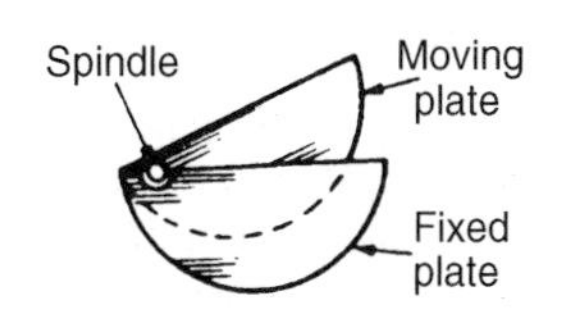

Figure 6.9

As the moving plates are rotated through half a revolution, the meshing, and therefore the capacitance, varies from a minimum to a maximum value. Variable air capacitors are used in radio and electronic circuits where very low losses are required, or where a variable capacitance is needed. The maximum value of such capacitors is between 500 pF and 1000 pF.

2 **Mica capacitors.** A typical older type construction is shown in Figure 6.10.

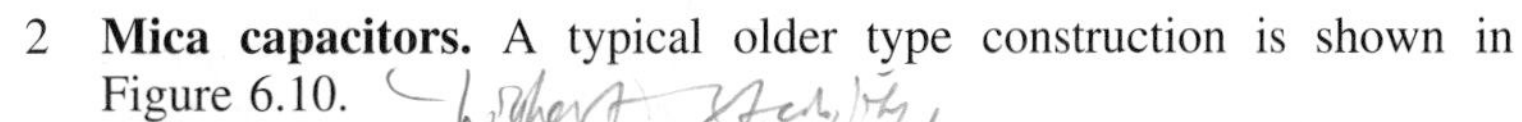

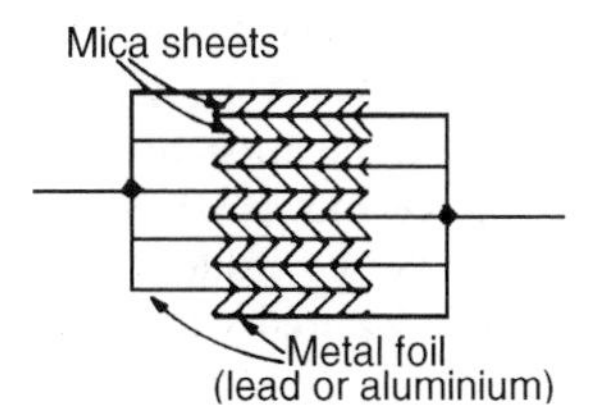

Figure 6.10

Usually the whole capacitor is impregnated with wax and placed in a bakelite case. Mica is easily obtained in thin sheets and is a good insulator. However, mica is expensive and is not used in capacitors above about 0.2 μF. A modified form of mica capacitor is the silvered mica type. The mica is coated on both sides with a thin layer of silver which forms the plates. Capacitance is stable and less likely to change with age. Such capacitors have a constant capacitance with change of temperature, a high working voltage rating and a long service life and are used in high frequency circuits with fixed values of capacitance up to about 1000 pF.

3 **Paper capacitors.** A typical paper capacitor is shown in Figure 6.11 where the length of the roll corresponds to the capacitance required. The whole is usually impregnated with oil or wax to exclude moisture, and then placed in a plastic or aluminium container for protection. Paper capacitors are made in various working voltages up to about 150 kV and are used where loss is not very important. The maximum value of this type of capacitor is between 500 pF and 10 μF. Disadvantages of paper capacitors include variation in capacitance with temperature change and a shorter service life than most other types of capacitor.

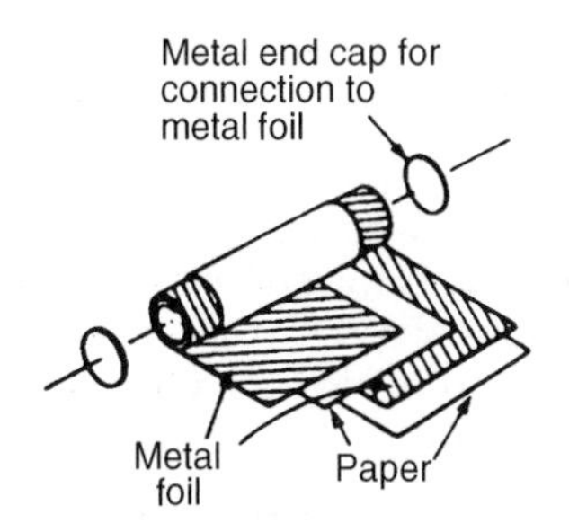

Figure 6.11

4 **Ceramic capacitors.** These are made in various forms, each type of construction depending on the value of capacitance required. For high values, a tube of ceramic material is used as shown in the cross section of Figure 6.12. For smaller values the cup construction is used as shown in Figure 6.13, and for still smaller values the disc construction shown in Figure 6.14 is used. Certain ceramic materials have a very high permittivity and this enables capacitors of high capacitance to be made which are of small physical size with a high working voltage

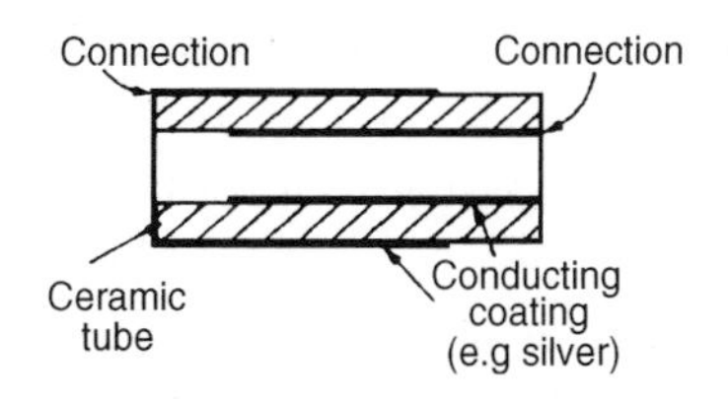

Figure 6.12

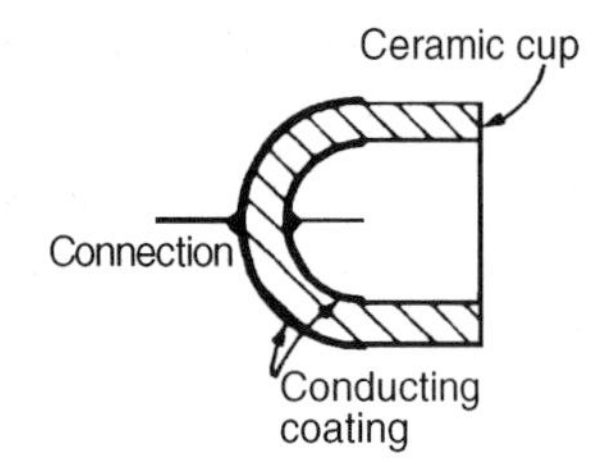

Figure 6.13

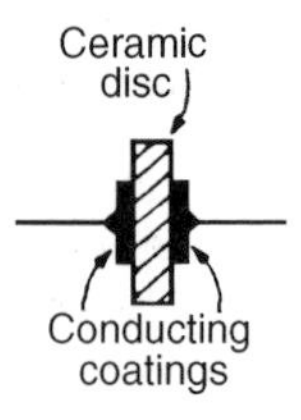

Figure 6.14

rating. Ceramic capacitors are available in the range 1 pF to 0.1 μF and may be used in high frequency electronic circuits subject to a wide range of temperatures.

5 **Plastic capacitors.** Some plastic materials such as polystyrene and Teflon can be used as dielectrics. Construction is similar to the paper capacitor but using a plastic film instead of paper. Plastic capacitors operate well under conditions of high temperature, provide a precise value of capacitance, a very long service life and high reliability.

6 **Titanium oxide capacitors** have a very high capacitance with a small physical size when used at a low temperature.

7 **Electrolytic capacitors.** Construction is similar to the paper capacitor with aluminium foil used for the plates and with a thick absorbent material, such as paper, impregnated with an electrolyte (ammonium borate), separating the plates. The finished capacitor is usually assembled in an aluminium container and hermetically sealed. Its operation depends on the formation of a thin aluminium oxide layer on the positive plate by electrolytic action when a suitable direct potential is maintained between the plates. This oxide layer is very thin and forms the dielectric. (The absorbent paper between the plates is a conductor and does not act as a dielectric.) Such capacitors **must always be used on dc** and must be connected with the correct polarity; if this is not done the capacitor will be destroyed since the oxide layer will be destroyed. Electrolytic capacitors are manufactured with working voltage from 6 V to 600 V, although accuracy is generally not very high. These capacitors possess a much larger capacitance than other types of capacitors of similar dimensions due to the oxide film being only a few microns thick. The fact that they can be used only on dc supplies limit their usefulness.

6.12 Discharging capacitors

When a capacitor has been disconnected from the supply it may still be charged and it may retain this charge for some considerable time. Thus precautions must be taken to ensure that the capacitor is automatically discharged after the supply is switched off. This is done by connecting a high value resistor across the capacitor terminals.

6.13 Further problems on capacitors and capacitance

(Where appropriate take ε_0 as 8.85×10^{-12} F/m)

Charge and capacitance

1 Find the charge on a 10 μF capacitor when the applied voltage is 250 V. [2.5 mC]

2 Determine the voltage across a 1000 pF capacitor to charge it with 2 μC. [2 kV]

3 The charge on the plates of a capacitor is 6 mC when the potential between them is 2.4 kV. Determine the capacitance of the capacitor. [2.5 μF]

4 For how long must a charging current of 2 A be fed to a 5 μF capacitor to raise the pd between its plates by 500 V. [1.25 ms]

5 A steady current of 10 A flows into a previously uncharged capacitor for 1.5 ms when the pd between the plates is 2 kV. Find the capacitance of the capacitor. [7.5 μF]

Electric field strength, electric flux density and permittivity

6 A capacitor uses a dielectric 0.04 mm thick and operates at 30 V. What is the electric field strength across the dielectric at this voltage? [750 kV/m]

7 A two-plate capacitor has a charge of 25 C. If the effective area of each plate is 5 cm^2 find the electric flux density of the electric field. [50 kC/m^2]

8 A charge of 1.5 μC is carried on two parallel rectangular plates each measuring 60 mm by 80 mm. Calculate the electric flux density. If the plates are spaced 10 mm apart and the voltage between them is 0.5 kV determine the electric field strength. [312.5 $\mu C/m^2$, 50 kV/m]

9 The electric flux density between two plates separated by polystyrene of relative permittivity 2.5 is 5 $\mu C/m^2$. Find the voltage gradient between the plates. [226 kV/m]

10 Two parallel plates having a pd of 250 V between them are spaced 1 mm apart. Determine the electric field strength. Find also the electric flux density when the dielectric between the plates is (a) air and (b) mica of relative permittivity 5. [250 kV/m (a) 2.213 $\mu C/m^2$ (b) 11.063 $\mu C/m^2$]

Parallel plate capacitor

11 A capacitor consists of two parallel plates each of area 0.01 m^2, spaced 0.1 mm in air. Calculate the capacitance in picofarads. [885 pF]

12 A waxed paper capacitor has two parallel plates, each of effective area 0.2 m^2. If the capacitance is 4000 pF determine the effective thickness of the paper if its relative permittivity is 2. [0.885 mm]

13 Calculate the capacitance of a parallel plate capacitor having 5 plates, each 30 mm by 20 mm and separated by a dielectric 0.75 mm thick having a relative permittivity of 2.3 [65.14 pF]

14 How many plates has a parallel plate capacitor having a capacitance of 5 nF, if each plate is 40 mm by 40 mm and each dielectric is 0.102 mm thick with a relative permittivity of 6. [7]

15 A parallel plate capacitor is made from 25 plates, each 70 mm by 120 mm interleaved with mica of relative permittivity 5. If the capacitance of the capacitor is 3000 pF determine the thickness of the mica sheet. [2.97 mm]

16 The capacitance of a parallel plate capacitor is 1000 pF. It has 19 plates, each 50 mm by 30 mm separated by a dielectric of thickness 0.40 mm. Determine the relative permittivity of the dielectric. [1.67]

17 A capacitor is to be constructed so that its capacitance is 4250 pF and to operate at a pd of 100 V across its terminals. The dielectric is to be polythene ($\varepsilon_r = 2.3$) which, after allowing a safety factor, has a dielectric strength of 20 MV/m. Find (a) the thickness of the polythene needed, and (b) the area of a plate. [(a) 0.005 mm (b) 10.44 cm^2]

Capacitors in parallel and series

18 Capacitors of 2 μF and 6 μF are connected (a) in parallel and (b) in series. Determine the equivalent capacitance in each case. [(a) 8 μF (b) 1.5 μF]

19 Find the capacitance to be connected in series with a 10 μF capacitor for the equivalent capacitance to be 6 μF [15 μF]

20 Two 6 μF capacitors are connected in series with one having a capacitance of 12 μF. Find the total equivalent circuit capacitance. What capacitance must be added in series to obtain a capacitance of 1.2 μF? [2.4 μF, 2.4 μF]

21 Determine the equivalent capacitance when the following capacitors are connected (a) in parallel and (b) in series:
(i) 2 μF, 4 μF and 8 μF
(ii) 0.02 μF, 0.05 μF and 0.10 μF
(iii) 50 pF and 450 pF
(iv) 0.01 μF and 200 pF
[(a) (i) 14 μF (ii) 0.17 μF (iii) 500 pF (iv) 0.0102 μF
(b) (i) $1\frac{1}{7}$ μF (ii) 0.0125 μF (iii) 45 pF (iv) 196.1 pF]

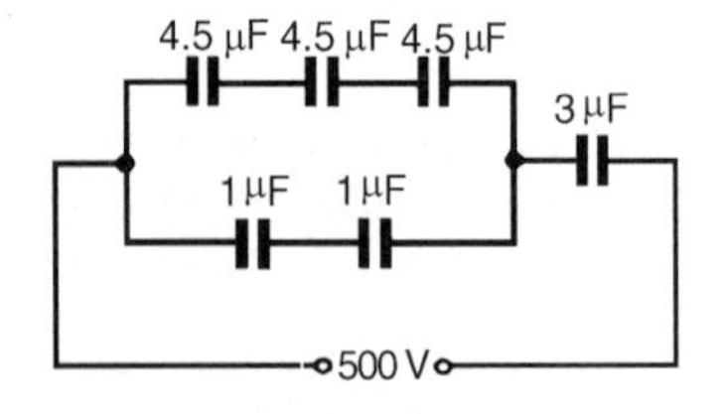

Figure 6.15

22 For the arrangement shown in Figure 6.15 find (a) the equivalent circuit capacitance and (b) the voltage across a 4.5 μF capacitor. [(a) 1.2 μF (b) 100 V]

23 Three 12 μF capacitors are connected in series across a 750 V supply. Calculate (a) the equivalent capacitance, (b) the charge on each capacitor and (c) the pd across each capacitor. [(a) 4 μF (b) 3 mC (c) 250 V]

24 If two capacitors having capacitances of 3 μF and 5 μF respectively are connected in series across a 240 V supply, determine (a) the p.d. across each capacitor and (b) the charge on each capacitor. [(a) 150 V, 90 V (b) 0.45 mC on each]

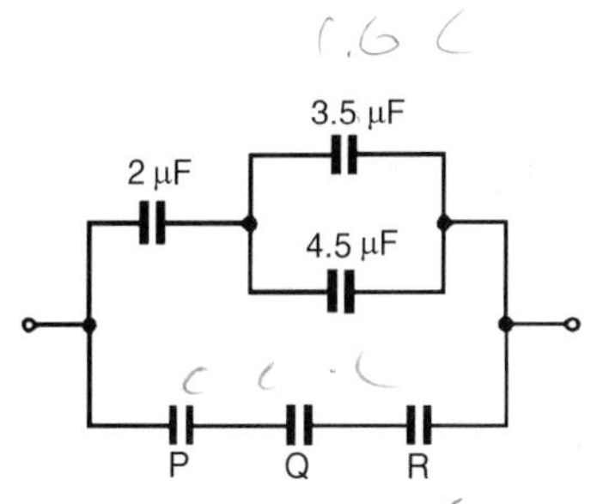

Figure 6.16

25 In Figure 6.16 capacitors P, Q and R are identical and the total equivalent capacitance of the circuit is 3 μF. Determine the values of P, Q and R [4.2 μF each]

Energy stored

26 When a capacitor is connected across a 200 V supply the charge is 4 μC. Find (a) the capacitance and (b) the energy stored. [(a) 0.02 μF (b) 0.4 mJ]

27 Find the energy stored in a 10 μF capacitor when charged to 2 kV. [20 J]

28 A 3300 pF capacitor is required to store 0.5 mJ of energy. Find the pd to which the capacitor must be charged. [550 V]

29 A capacitor, consisting of two metal plates each of area 50 cm^2 and spaced 0.2 mm apart in air, is connected across a 120 V supply. Calculate (a) the energy stored, (b) the electric flux density and (c) the potential gradient [(a) 1.593 μJ (b) 5.31 $\mu C/m^2$ (c) 600 kV/m]

30 A bakelite capacitor is to be constructed to have a capacitance of 0.04 μF and to have a steady working potential of 1 kV maximum. Allowing a safe value of field stress of 25 MV/m find (a) the thickness of bakelite required, (b) the area of plate required if the relative permittivity of bakelite is 5, (c) the maximum energy stored by the capacitor and (d) the average power developed if this energy is dissipated in a time of 20 μs. [(a) 0.04 mm (b) 361.6 cm^2 (c) 0.02 J (d) 1 kW]

7 Magnetic circuits

At the end of this chapter you should be able to:

- describe the magnetic field around a permanent magnet
- state the laws of magnetic attraction and repulsion for two magnets in close proximity
- define magnetic flux, Φ, and magnetic flux density, B, and state their units
- perform simple calculations involving $B = \dfrac{\Phi}{A}$
- define magnetomotive force, F_m, and magnetic field strength, H, and state their units
- perform simple calculations involving $F_m = NI$ and $H = \dfrac{NI}{l}$
- define permeability, distinguishing between μ_0, μ_r and μ
- understand the B–H curves for different magnetic materials
- appreciate typical values of μ_r
- perform calculations involving $B = \mu_0 \mu_r H$
- define reluctance, S, and state its units
- perform calculations involving $S = \dfrac{\text{mmf}}{\Phi} = \dfrac{l}{\mu_0 \mu_r A}$
- perform calculations on composite series magnetic circuits
- compare electrical and magnetic quantities
- appreciate how a hysteresis loop is obtained and that hysteresis loss is proportional to its area

7.1 Magnetic fields

A **permanent magnet** is a piece of ferromagnetic material (such as iron, nickel or cobalt) which has properties of attracting other pieces of these materials. A permanent magnet will position itself in a north and south direction when freely suspended. The north-seeking end of the magnet is called the **north pole**, ***N***, and the south-seeking end the **south pole**, ***S***.

The area around a magnet is called the **magnetic field** and it is in this area that the effects of the **magnetic force** produced by the magnet can be detected. A magnetic field cannot be seen, felt, smelt or heard and therefore is difficult to represent. Michael Faraday suggested that the magnetic field could be represented pictorially, by imagining the field to consist of **lines of magnetic flux**, which enables investigation of the distribution and density of the field to be carried out.

The distribution of a magnetic field can be investigated by using some iron filings. A bar magnet is placed on a flat surface covered by, say,

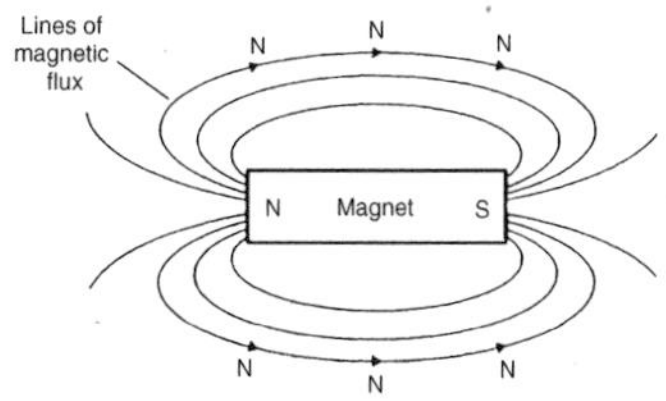

Figure 7.1

cardboard, upon which is sprinkled some iron filings. If the cardboard is gently tapped the filings will assume a pattern similar to that shown in Figure 7.1. If a number of magnets of different strength are used, it is found that the stronger the field the closer are the lines of magnetic flux and vice versa. Thus a magnetic field has the property of exerting a force, demonstrated in this case by causing the iron filings to move into the pattern shown. The strength of the magnetic field decreases as we move away from the magnet. It should be realized, of course, that the magnetic field is three dimensional in its effect, and not acting in one plane as appears to be the case in this experiment.

If a compass is placed in the magnetic field in various positions, the direction of the lines of flux may be determined by noting the direction of the compass pointer. The direction of a magnetic field at any point is taken as that in which the north-seeking pole of a compass needle points when suspended in the field. The direction of a line of flux is from the north pole to the south pole on the outside of the magnet and is then assumed to continue through the magnet back to the point at which it emerged at the north pole. Thus such lines of flux always form complete closed loops or paths, they never intersect and always have a definite direction. The laws of magnetic attraction and repulsion can be demonstrated by using two bar magnets. In Figure 7.2(a), **with unlike poles adjacent, attraction takes place**. Lines of flux are imagined to contract and the magnets try to pull together. The magnetic field is strongest in between the two magnets, shown by the lines of flux being close together. In Figure 7.2(b), **with similar poles adjacent (i.e. two north poles), repulsion occurs**, i.e. the two north poles try to push each other apart, since magnetic flux lines running side by side in the same direction repel.

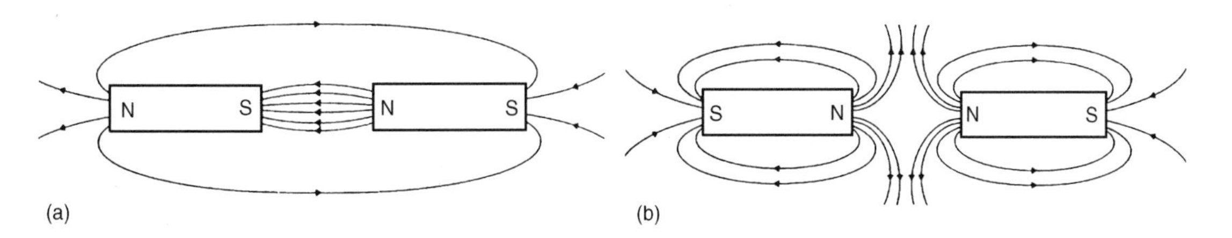

Figure 7.2

7.2 Magnetic flux and flux density

Magnetic flux is the amount of magnetic field (or the number of lines of force) produced by a magnetic source. The symbol for magnetic flux is Φ (Greek letter 'phi'). The unit of magnetic flux is the **weber, Wb**

Magnetic flux density is the amount of flux passing through a defined area that is perpendicular to the direction of the flux:

$$\textbf{Magnetic flux density} = \frac{\textbf{magnetic flux}}{\textbf{area}}$$

The symbol for magnetic flux density is B. The unit of magnetic flux density is the **tesla, T**, where $1\ \mathrm{T} = 1\ \mathrm{Wb/m^2}$ Hence

$$\boxed{B = \frac{\Phi}{A} \textbf{ tesla}}, \text{ where } A(\text{m}^2) \text{ is the area}$$

Problem 1. A magnetic pole face has a rectangular section having dimensions 200 mm by 100 mm. If the total flux emerging from the pole is 150 μWb, calculate the flux density.

Flux $\Phi = 150\ \mu\text{Wb} = 150 \times 10^{-6}$ Wb

Cross sectional area $A = 200 \times 100 = 20000\ \text{mm}^2$
$= 20000 \times 10^{-6}\ \text{m}^2$

$$\text{Flux density } B = \frac{\Phi}{A} = \frac{150 \times 10^{-6}}{20000 \times 10^{-6}}$$
$= \mathbf{0.0075\ T\ or\ 7.5\ mT}$

Problem 2. The maximum working flux density of a lifting electromagnet is 1.8 T and the effective area of a pole face is circular in cross-section. If the total magnetic flux produced is 353 mWb, determine the radius of the pole face.

Flux density $B = 1.8$ T; flux $\Phi = 353\ \text{mWb} = 353 \times 10^{-3}$ Wb

$$\text{Since } B = \frac{\Phi}{A}, \text{ cross-sectional area } A = \frac{\Phi}{B} = \frac{353 \times 10^{-3}}{1.8}\ \text{m}^2$$
$= 0.1961\ \text{m}^2$

The pole face is circular, hence area $= \pi r^2$, where r is the radius.

Hence $\pi r^2 = 0.1961$

from which $r^2 = \dfrac{0.1961}{\pi}$ and radius $r = \sqrt{\left(\dfrac{0.1961}{\pi}\right)} = 0.250$ m

i.e. **the radius of the pole face is 250 mm**

7.3 Magnetomotive force and magnetic field strength

Magnetomotive force (mmf) is the cause of the existence of a magnetic flux in a magnetic circuit,

$$\boxed{\textbf{mmf}, F_m = NI \textbf{ amperes}}$$

where N is the number of conductors (or turns) and I is the current in amperes. The unit of mmf is sometimes expressed as 'ampere-turns'. However since 'turns' have no dimensions, the SI unit of mmf is the ampere. **Magnetic field strength** (or **magnetizing force**),

$$\mathbf{H = NI/l} \text{ ampere per metre,}$$

where l is the mean length of the flux path in metres.

Thus **mmf = NI = Hl amperes**.

> Problem 3. A magnetizing force of 8000 A/m is applied to a circular magnetic circuit of mean diameter 30 cm by passing a current through a coil wound on the circuit. If the coil is uniformly wound around the circuit and has 750 turns, find the current in the coil.

$H = 8000$ A/m; $l = \pi d = \pi \times 30 \times 10^{-2}$ m; $N = 750$ turns

Since $H = \frac{NI}{l}$ then, $I = \frac{Hl}{N} = \frac{8000 \times \pi \times 30 \times 10^{-2}}{750}$

Thus, **current I = 10.05 A**

7.4 Permeability and B–H curves

For air, or any non-magnetic medium, the ratio of magnetic flux density to magnetizing force is a constant, i.e. B/H = a constant. This constant is μ_0, the **permeability of free space** (or the magnetic space constant) and is equal to $4\pi \times 10^{-7}$ H/m, i.e., **for air, or any non-magnetic medium,** the ratio $\boxed{B/H = \mu_0}$ (Although all non-magnetic materials, including air, exhibit slight magnetic properties, these can effectively be neglected.)

For all media other than free space, $\boxed{B/H = \mu_0\mu_r}$

where u_r is the relative permeability, and is defined as

$$\mu_r = \frac{\textbf{flux density in material}}{\textbf{flux density in a vacuum}}$$

μ_r varies with the type of magnetic material and, since it is a ratio of flux densities, it has no unit. From its definition, μ_r for a vacuum is 1.

$\mu_0\mu_r = \mu$, called the **absolute permeability**

By plotting measured values of flux density B against magnetic field strength H, a **magnetization curve** (or **B–H curve**) is produced. For non-magnetic materials this is a straight line. Typical curves for four magnetic materials are shown in Figure 7.3.

The **relative permeability** of a ferromagnetic material is proportional to the slope of the B–H curve and thus varies with the magnetic field strength. The approximate range of values of relative permeability μ_r for some common magnetic materials are:

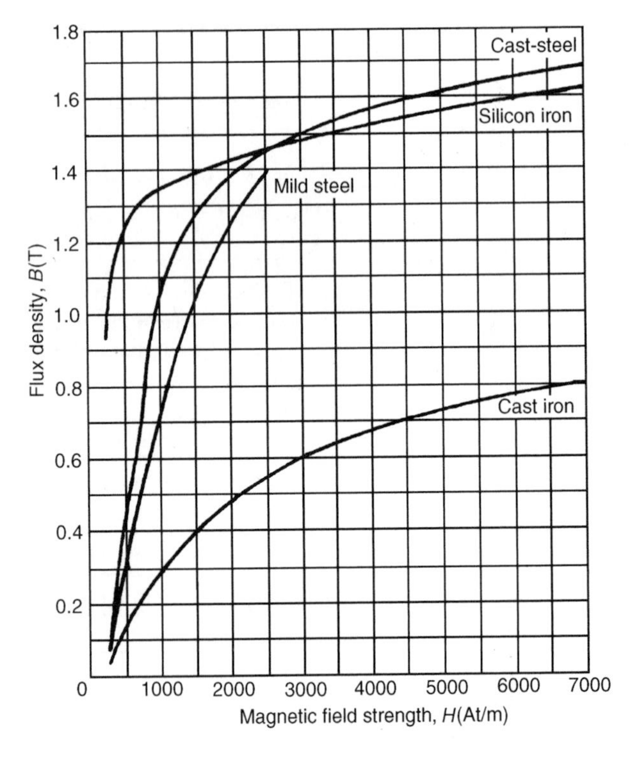

Figure 7.3 *B–H curves for four materials*

Cast iron	$\mu_r = 100–250$	Mild steel	$\mu_r = 200–800$
Silicon iron	$\mu_r = 1000–5000$	Cast steel	$\mu_r = 300–900$
Mumetal	$\mu_r = 200–5000$	Stalloy	$\mu_r = 500–6000$

Problem 4. A flux density of 1.2 T is produced in a piece of cast steel by a magnetizing force of 1250 A/m. Find the relative permeability of the steel under these conditions.

For a magnetic material:

$$B = \mu_0 \mu_r H$$

i.e. $$u_r = \frac{B}{\mu_0 H} = \frac{1.2}{(4\pi \times 10^{-7})(1250)} = \mathbf{764}$$

Problem 5. Determine the magnetic field strength and the mmf required to produce a flux density of 0.25 T in an air gap of length 12 mm.

For air: $B = \mu_0 H$ (since $\mu_r = 1$)

$$\textbf{Magnetic field strength } H = \frac{B}{\mu_0} = \frac{0.25}{4\pi \times 10^{-7}} = \mathbf{198\,940\ A/m}$$

$$\textbf{mmf} = Hl = 198\,940 \times 12 \times 10^{-3} = \mathbf{2387\ A}$$

> Problem 6. A coil of 300 turns is wound uniformly on a ring of non-magnetic material. The ring has a mean circumference of 40 cm and a uniform cross sectional area of 4 cm^2. If the current in the coil is 5 A, calculate (a) the magnetic field strength, (b) the flux density and (c) the total magnetic flux in the ring.

(a) Magnetic field strength $H = \frac{NI}{l} = \frac{300 \times 5}{40 \times 10^{-2}} = \mathbf{3750\ A/m}$

(b) For a non-magnetic material $\mu_r = 1$, thus flux density $B = \mu_0 H$

i.e. $\mathbf{B} = 4\pi \times 10^{-7} \times 3750 = \mathbf{4.712\ mT}$

(c) Flux $\Phi = BA = (4.712 \times 10^{-3})(4 \times 10^{-4}) = \mathbf{1.885\ \mu Wb}$

> Problem 7. An iron ring of mean diameter 10 cm is uniformly wound with 2000 turns of wire. When a current of 0.25 A is passed through the coil a flux density of 0.4 T is set up in the iron. Find (a) the magnetizing force and (b) the relative permeability of the iron under these conditions.

$l = \pi d = \pi \times 10$ cm $= \pi \times 10 \times 10^{-2}$ m; $N = 2000$ turns; $I = 0.25$ A; $B = 0.4$ T

(a) $\mathbf{H} = \frac{NI}{l} = \frac{2000 \times 0.25}{\pi \times 10 \times 10^{-2}} = \frac{5000}{\pi} = \mathbf{1592\ A/m}$

(b) $B = \mu_0 \mu_r H$, hence $\boldsymbol{\mu_r} = \frac{B}{\mu_0 H} = \frac{0.4}{(4\pi \times 10^{-7})(1592)} = \mathbf{200}$

> Problem 8. A uniform ring of cast iron has a cross-sectional area of 10 cm^2 and a mean circumference of 20 cm. Determine the mmf necessary to produce a flux of 0.3 mWb in the ring. The magnetization curve for cast iron is shown on page 78.

$A = 10\ \text{cm}^2 = 10 \times 10^{-4}\ \text{m}^2$; $l = 20$ cm $= 0.2$ m; $\Phi = 0.3 \times 10^{-3}$ Wb

Flux density $B = \frac{\Phi}{A} = \frac{0.3 \times 10^{-3}}{10 \times 10^{-4}} = 0.3$ T

From the magnetization curve for cast iron on page 78, when $B = 0.3$ T, $H = 1000$ A/m, hence mmf $= Hl = 1000 \times 0.2 = \mathbf{200\ A}$

A tabular method could have been used in this problem. Such a solution is shown below.

Part of circuit	Material	Φ (Wb)	A (m^2)	$B = \frac{\Phi}{A}$ (T)	H from graph	l (m)	mmf = Hl (A)
Ring	Cast iron	0.3×10^{-3}	10×10^{-4}	0.3	1000	0.2	200

7.5 Reluctance

Reluctance S (or R_M) is the 'magnetic resistance' of a magnetic circuit to the presence of magnetic flux.

$$\textbf{Reluctance } S = \frac{F_M}{\Phi} = \frac{NI}{\Phi} = \frac{Hl}{BA} = \frac{l}{(B/H)A} = \frac{\boldsymbol{l}}{\boldsymbol{\mu_0 \mu_r A}}$$

The unit of reluctance is 1/H (or H^{-1}) or A/Wb

Ferromagnetic materials have a low reluctance and can be used as **magnetic screens** to prevent magnetic fields affecting materials within the screen.

Problem 9. Determine the reluctance of a piece of mumetal of length 150 mm and cross-sectional area 1800 mm^2 when the relative permeability is 4000. Find also the absolute permeability of the mumetal.

$$\textbf{Reluctance } S = \frac{l}{\mu_0 \mu_r A} = \frac{150 \times 10^{-3}}{(4\pi \times 10^{-7})(4000)(1800 \times 10^{-6})}$$
$$= \mathbf{16\ 580/H}$$

$$\textbf{Absolute permeability, } \mu = \mu_0 \mu_r = (4\pi \times 10^{-7})(4000)$$
$$= \mathbf{5.027 \times 10^{-3}\ H/m}$$

Problem 10. A mild steel ring has a radius of 50 mm and a cross-sectional area of 400 mm^2. A current of 0.5 A flows in a coil wound uniformly around the ring and the flux produced is 0.1 mWb. If the relative permeability at this value of current is 200 find (a) the reluctance of the mild steel and (b) the number of turns on the coil.

$l = 2\pi r = 2 \times \pi \times 50 \times 10^{-3}$ m; $A = 400 \times 10^{-6}$ m^2; $I = 0.5$ A; $\Phi = 0.1 \times 10^{-3}$ Wb; $\mu_r = 200$

(a) $$\textbf{Reluctance } S = \frac{l}{\mu_0 \mu_r A} = \frac{2 \times \pi \times 50 \times 10^{-3}}{(4\pi \times 10^{-7})(200)(400 \times 10^{-6})}$$
$$= \mathbf{3.125 \times 10^6/H}$$

(b) $S = \frac{\text{mmf}}{\Phi}$ i.e. $\text{mmf} = S\Phi$

so that $NI = S\Phi$ and

hence $N = \frac{S\Phi}{I} = \frac{3.125 \times 10^6 \times 0.1 \times 10^{-3}}{0.5}$ = **625 turns**

Further problems on magnetic circuit quantities may be found in Section 7.9, problems 1 to 14, page 85.

7.6 Composite series magnetic circuits

For a series magnetic circuit having n parts, the **total reluctance S** is given by:

$$\mathbf{S = S_1 + S_2 + \ldots + S_n}$$

(This is similar to resistors connected in series in an electrical circuit.)

> Problem 11. A closed magnetic circuit of cast steel contains a 6 cm long path of cross-sectional area 1 cm^2 and a 2 cm path of cross-sectional area 0.5 cm^2. A coil of 200 turns is wound around the 6 cm length of the circuit and a current of 0.4 A flows. Determine the flux density in the 2 cm path, if the relative permeability of the cast steel is 750.

For the 6 cm long path:

$$\text{Reluctance } S_1 = \frac{l_1}{\mu_0 \mu_r A_1} = \frac{6 \times 10^{-2}}{(4\pi \times 10^{-7})(750)(1 \times 10^{-4})}$$

$$= 6.366 \times 10^5/\text{H}$$

For the 2 cm long path:

$$\text{Reluctance } S_2 = \frac{l_2}{\mu_0 \mu_r A_2} = \frac{2 \times 10^{-2}}{(4\pi \times 10^{-7})(750)(0.5 \times 10^{-4})}$$

$$= 4.244 \times 10^5/\text{H}$$

$$\text{Total circuit reluctance } S = S_1 + S_2 = (6.366 + 4.244) \times 10^5$$

$$= 10.61 \times 10^5/\text{H}$$

$$S = \frac{\text{mmf}}{\Phi}, \text{ i.e. } \Phi = \frac{\text{mmf}}{S} = \frac{NI}{S} = \frac{200 \times 0.4}{10.61 \times 10^5} = 7.54 \times 10^{-5} \text{ Wb}$$

Flux density in the 2 cm path, $B = \frac{\Phi}{A} = \frac{7.54 \times 10^{-5}}{0.5 \times 10^{-4}}$ = **1.51 T**

Problem 12. A silicon iron ring of cross-sectional area 5 cm^2 has a radial air gap of 2 mm cut into it. If the mean length of the silicon iron path is 40 cm, calculate the magnetomotive force to produce a flux of 0.7 mWb. The magnetization curve for silicon is shown on page 78.

There are two parts to the circuit — the silicon iron and the air gap. The total mmf will be the sum of the mmf's of each part.

For the silicon iron: $B = \dfrac{\Phi}{A} = \dfrac{0.7 \times 10^{-3}}{5 \times 10^{-4}} = 1.4$ T

From the B–H curve for silicon iron on page 78, when $B = 1.4$ T, $H = 1650$ At/m.

Hence the mmf for the iron path $= Hl = 1650 \times 0.4 = 660$ A

For the air gap:

The flux density will be the same in the air gap as in the iron, i.e. 1.4 T. (This assumes no leakage or fringing occurring.)

For air, $H = \dfrac{B}{\mu_0} = \dfrac{1.4}{4\pi \times 10^{-7}}$

$= 1\,114\,000$ A/m

Hence the mmf for the air gap $= Hl = 1\,114\,000 \times 2 \times 10^{-3}$

$= 2228$ A

Total mmf to produce a flux of 0.7 mWb $= 660 + 2228$

$= \mathbf{2888\ A}$

A tabular method could have been used as shown below.

Part of circuit	Material	Φ (Wb)	A (m^2)	B (T)	H (A/m)	l (m)	mmf = Hl (A)
Ring	Silicon iron	0.7×10^{-3}	5×10^{-4}	1.4	1650 (from graph)	0.4	660
Air-gap	Air	0.7×10^{-3}	5×10^{-4}	1.4	$\dfrac{1.4}{4\pi \times 10^{-7}}$ $= 1\,114\,000$	2×10^{-3}	2228
						Total:	**2888 A**

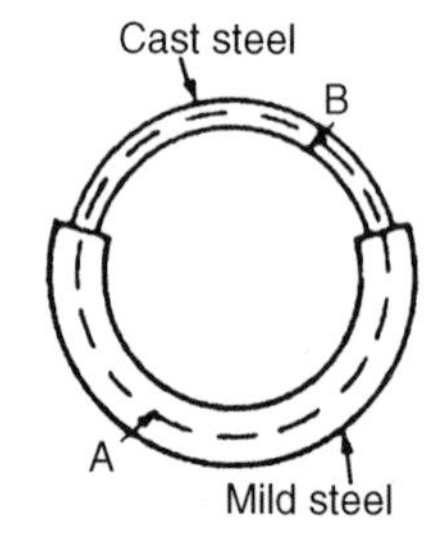

Figure 7.4

Problem 13. Figure 7.4 shows a ring formed with two different materials — cast steel and mild steel. The dimensions are:

	mean length	**cross-sectional area**
Mild steel	400 mm	500 mm^2
Cast steel	300 mm	312.5 mm^2

Find the total mmf required to cause a flux of 500 μWb in the magnetic circuit. Determine also the total circuit reluctance.

A tabular solution is shown below.

Part of circuit	Material	Φ (Wb)	A (m²)	B (T) (= Φ/A)	H (A/m) (from graphs p 78)	l (m)	mmf= Hl (A)
A	Mild steel	500×10^{-6}	500×10^{-6}	1.0	1400	400×10^{-3}	560
B	Cast steel	500×10^{-6}	312.5×10^{-6}	1.6	4800	300×10^{-3}	1440
							Total: **2000 A**

Total circuit reluctance $S = \frac{\text{mmf}}{\Phi} = \frac{2000}{500 \times 10^{-6}}$

$= \mathbf{4 \times 10^6/H}$

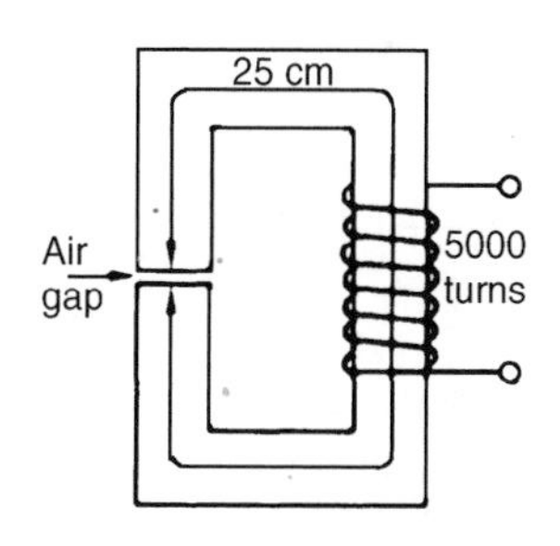

Figure 7.5

Problem 14. A section through a magnetic circuit of uniform cross-sectional area 2 cm² is shown in Figure 7.5. The cast steel core has a mean length of 25 cm. The air gap is 1 mm wide and the coil has 5000 turns. The B–H curve for cast steel is shown on page 78. Determine the current in the coil to produce a flux density of 0.80 T in the air gap, assuming that all the flux passes through both parts of the magnetic circuit.

For the cast steel core, when $B = 0.80$ T, $H = 750$ A/m (from page 78)

Reluctance of core $S_1 = \frac{l_1}{\mu_0 \mu_r A_1}$ and since $B = \mu_0 \mu_r H$,

then $\mu_r = \frac{B}{\mu_0 H}$. Thus $S_1 = \frac{l_1}{\mu_0 \left(\frac{B}{\mu_0 H}\right) A} = \frac{l_1 H}{BA} = \frac{(25 \times 10^{-2})(750)}{(0.8)(2 \times 10^{-4})}$

$= 1\,172\,000/\text{H}$

For the air gap: Reluctance, $S_2 = \frac{l_2}{\mu_0 \mu_r A_2} = \frac{l_2}{\mu_0 A_2}$

(since $\mu_r = 1$ for air)

$= \frac{1 \times 10^{-3}}{(4\pi \times 10^{-7})(2 \times 10^{-4})}$

$= 3\,979\,000/\text{H}$

Total circuit reluctance $S = S_1 + S_2 = 1\,172\,000 + 3\,979\,000$

$= 5\,151\,000/\text{H}$

Flux $\Phi = BA = 0.80 \times 2 \times 10^{-4} = 1.6 \times 10^{-4}$ Wb

$$S = \frac{\text{mmf}}{\Phi}, \text{ thus mmf} = S\Phi$$

Hence $NI = S\Phi$

and current $I = \frac{S\Phi}{N} = \frac{(5\,151\,000)(1.6 \times 10^{-4})}{5000} = \mathbf{0.165\ A}$

Further problems on composite series magnetic circuits may be found in Section 7.9, problems 15 to 19, page 86.

7.7 Comparison between electrical and magnetic quantities

Electrical circuit	Magnetic circuit
e.m.f. E (V) current I (A) resistance R (Ω) $I = \frac{E}{R}$ $R = \frac{\rho l}{A}$	mmf F_m (A) flux Φ (Wb) reluctance S (H^{-1}) $\Phi = \frac{\text{mmf}}{S}$ $S = \frac{l}{\mu_0 \mu_r A}$

7.8 Hysteresis and hysteresis loss

Hysteresis is the 'lagging' effect of flux density B whenever there are changes in the magnetic field strength H. When an initially unmagnetized ferromagnetic material is subjected to a varying magnetic field strength H, the flux density B produced in the material varies as shown in Figure 7.6, the arrows indicating the direction of the cycle. Figure 7.6 is known as a **hysteresis loop**.

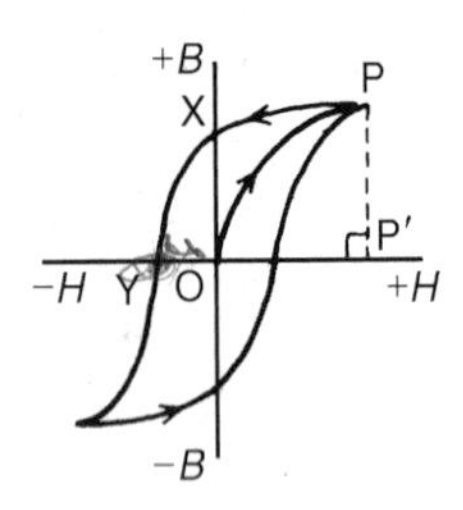

Figure 7.6

From Figure 7.6, distance OX indicates the **residual flux density** or **remanence**, OY indicates the **coercive force**, and PP' is the **saturation flux density**.

Hysteresis results in a dissipation of energy which appears as a heating of the magnetic material. **The energy loss associated with hysteresis is proportional to the area of the hysteresis loop**.

The production of the hysteresis loop and hysteresis loss are explained in greater detail in Chapter 38, Section 3, page 692.

The area of a hysteresis loop varies with the type of material. The area, and thus the energy loss, is much greater for hard materials than for soft materials.

For AC-excited devices the hysteresis loop is repeated every cycle of alternating current. Thus a hysteresis loop with a large area (as with hard steel) is often unsuitable since the energy loss would be considerable. Silicon steel has a narrow hysteresis loop, and thus small hysteresis loss, and is suitable for transformer cores and rotating machine armatures.

7.9 Further problems on magnetic circuits

(Where appropriate, assume $\mu_0 = 4\pi \times 10^{-7}$ H/m)

Magnetic circuit quantities

1 What is the flux density in a magnetic field of cross-sectional area 20 cm^2 having a flux of 3 mWb? [1.5 T]

2 Determine the total flux emerging from a magnetic pole face having dimensions 5 cm by 6 cm, if the flux density is 0.9 T. [2.7 mWb]

3 The maximum working flux density of a lifting electromagnet is 1.9 T and the effective area of a pole face is circular in cross-section. If the total magnetic flux produced is 611 mWb determine the radius of the pole face. [32 cm]

4 Find the magnetic field strength and the magnetomotive force needed to produce a flux density of 0.33 T in an air-gap of length 15 mm. [(a) 262 600 A/m (b) 3939 A]

5 An air-gap between two pole pieces is 20 mm in length and the area of the flux path across the gap is 5 cm^2. If the flux required in the air-gap is 0.75 mWb find the mmf necessary. [23 870 A]

6 Find the magnetic field strength applied to a magnetic circuit of mean length 50 cm when a coil of 400 turns is applied to it carrying a current of 1.2 A. [960 A/m]

7 A solenoid 20 cm long is wound with 500 turns of wire. Find the current required to establish a magnetizing force of 2500 A/m inside the solenoid. [1 A]

8 A magnetic field strength of 5000 A/m is applied to a circular magnetic circuit of mean diameter 250 mm. If the coil has 500 turns find the current in the coil. [7.85 A]

9 Find the relative permeability of a piece of silicon iron if a flux density of 1.3 T is produced by a magnetic field strength of 700 A/m [1478]

10 Part of a magnetic circuit is made from steel of length 120 mm, cross-sectional area 15 cm^2 and relative permeability 800. Calculate (a) the reluctance and (b) the absolute permeability of the steel. [(a) 79 580 /H (b) 1 mH/m]

11 A steel ring of mean diameter 120 mm is uniformly wound with 1500 turns of wire. When a current of 0.30 A is passed through the coil a flux density of 1.5 T is set up in the steel. Find the relative permeability of the steel under these conditions. [1000]

12 A mild steel closed magnetic circuit has a mean length of 75 mm and a cross-sectional area of 320.2 mm^2. A current of 0.40 A flows in a coil wound uniformly around the circuit and the flux produced is 200 μWb. If the relative permeability of the steel at this value of current is 400 find (a) the reluctance of the material and (b) the number of turns of the coil. [(a) 466 000/H (b) 233]

13 A uniform ring of cast steel has a cross-sectional area of 5 cm^2 and a mean circumference of 15 cm. Find the current required in a coil

of 1200 turns wound on the ring to produce a flux of 0.8 mWb. (Use the magnetization curve for cast steel shown on page 78.) [0.60 A]

14 (a) A uniform mild steel ring has a diameter of 50 mm and a cross-sectional area of 1 cm^2. Determine the mmf necessary to produce a flux of 50 μWb in the ring. (Use the *B–H* curve for mild steel shown on page 78.)

(b) If a coil of 440 turns is wound uniformly around the ring in part (a) what current would be required to produce the flux? [(a) 110 A (b) 0.25 A]

Composite series magnetic circuits

15 A magnetic circuit of cross-sectional area 0.4 cm^2 consists of one part 3 cm long, of material having relative permeability 1200, and a second part 2 cm long of material having relative permeability 750. With a 100 turn coil carrying 2 A, find the value of flux existing in the circuit. [0.195 mWb]

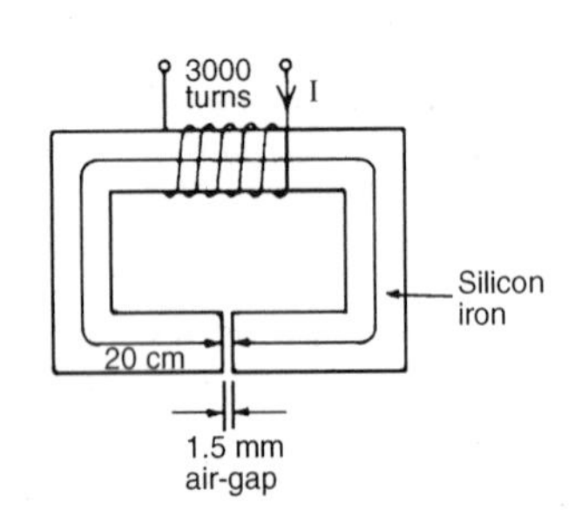

Figure 7.7

16 (a) A cast steel ring has a cross-sectional area of 600 mm^2 and a radius of 25 mm. Determine the mmf necessary to establish a flux of 0.8 mWb in the ring. Use the *B–H* curve for cast steel shown on page 78.

(b) If a radial air gap 1.5 mm wide is cut in the ring of part (a) find the mmf now necessary to maintain the same flux in the ring. [(a) 270 A (b) 1860 A]

17 For the magnetic circuit shown in Figure 7.7 find the current I in the coil needed to produce a flux of 0.45 mWb in the air-gap. The silicon iron magnetic circuit has a uniform cross-sectional area of 3 cm^2 and its magnetization curve is as shown on page 78. [0.83 A]

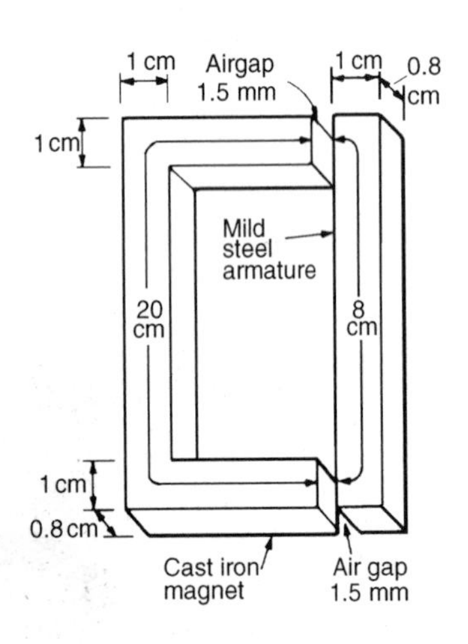

Figure 7.8

18 A ring forming a magnetic circuit is made from two materials; one part is mild steel of mean length 25 cm and cross-sectional area 4 cm^2, and the remainder is cast iron of mean length 20 cm and cross-sectional area 7.5 cm^2. Use a tabular approach to determine the total mmf required to cause a flux of 0.30 mWb in the magnetic circuit. Find also the total reluctance of the circuit. Use the magnetization curves shown on page 78. [550 A, 18.3×10^5/H]

19 Figure 7.8 shows the magnetic circuit of a relay. When each of the air gaps are 1.5 mm wide find the mmf required to produce a flux density of 0.75 T in the air gaps. Use the *B–H* curves shown on page 78. [2970 A]

Assignment 2

This assignment covers the material contained in chapters 5 to 7.

The marks for each question are shown in brackets at the end of each question.

1 Resistance's of 5 Ω, 7 Ω, and 8 Ω are connected in series. If a 10 V supply voltage is connected across the arrangement determine the current flowing through and the p.d. across the 7 Ω resistor. Calculate also the power dissipated in the 8 Ω resistor. (6)

2 For the series-parallel network shown in Figure A2.1, find (a) the supply current, (b) the current flowing through each resistor, (c) the p.d. across each resistor, (d) the total power dissipated in the circuit, (e) the cost of energy if the circuit is connected for 80 hours. Assume electrical energy costs 7.2 p per unit. (15)

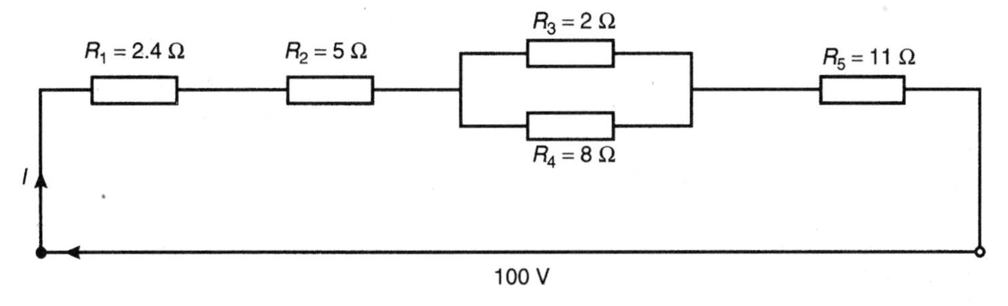

Figure A2.1

3 The charge on the plates of a capacitor is 8 mC when the potential between them is 4 kV. Determine the capacitance of the capacitor. (2)

4 Two parallel rectangular plates measuring 80 mm by 120 mm are separated by 4 mm of mica and carry an electric charge of 0.48 μC. The voltage between the plates is 500 V. Calculate (a) the electric flux density (b) the electric field strength, and (c) the capacitance of the capacitor, in picofarads, if the relative permittivity of mica is 5. (7)

5 A 4 μF capacitor is connected in parallel with a 6 μF capacitor. This arrangement is then connected in series with a 10 μF capacitor. A supply p.d. of 250 V is connected across the circuit. Find (a) the equivalent capacitance of the circuit, (b) the voltage across the 10 μF capacitor, and (c) the charge on each capacitor. (7)

6 A coil of 600 turns is wound uniformly on a ring of non-magnetic material. The ring has a uniform cross-sectional area of $200\,mm^2$ and a mean circumference of 500 mm. If the current in the coil is 4 A, determine (a) the magnetic field strength, (b) the flux density, and (c) the total magnetic flux in the ring. (5)

7 A mild steel ring of cross-sectional area $4\,cm^2$ has a radial air-gap of 3 mm cut into it. If the mean length of the mild steel path is 300 mm, calculate the magnetomotive force to produce a flux of 0.48 mWb. (Use the B-H curve on page 78) (8)

8 Electromagnetism

At the end of this chapter you should be able to:

- understand that magnetic fields are produced by electric currents
- apply the screw rule to determine direction of magnetic field
- recognize that the magnetic field around a solenoid is similar to a magnet
- apply the screw rule or grip rule to a solenoid to determine magnetic field direction
- recognize and describe practical applications of an electromagnet, i.e. electric bell, relay, lifting magnet, telephone receiver
- appreciate factors upon which the force F on a current-carrying conductor depends
- perform calculations using $F = BIl$ and $F = BIl \sin\theta$
- recognize that a loudspeaker is a practical application of force F
- use Fleming's left-hand rule to pre-determine direction of force in a current-carrying conductor
- describe the principle of operation of a simple d.c. motor
- describe the principle of operation and construction of a moving coil instrument
- appreciate the force F on a charge in a magnetic field is given by $F = QvB$
- perform calculations using $F = QvB$

8.1 Magnetic field due to an electric current

Magnetic fields can be set up not only by permanent magnets, as shown in Chapter 7, but also by electric currents.

Let a piece of wire be arranged to pass vertically through a horizontal sheet of cardboard, on which is placed some iron filings, as shown in Figure 8.1(a).

If a current is now passed through the wire, then the iron filings will form a definite circular field pattern with the wire at the centre, when the cardboard is gently tapped. By placing a compass in different positions the lines of flux are seen to have a definite direction as shown in Figure 8.1(b). If the current direction is reversed, the direction of the lines of flux is also reversed. The effect on both the iron filings and the compass needle disappears when the current is switched off. The magnetic field is thus

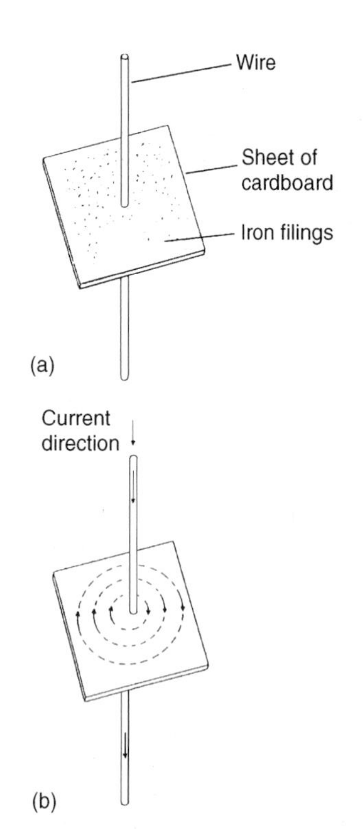

Figure 8.1

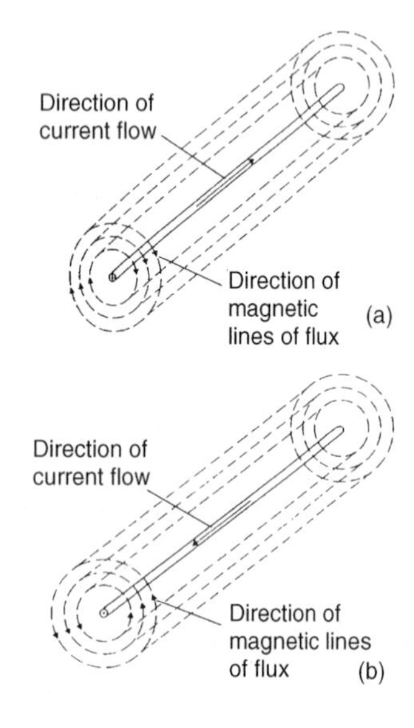

Figure 8.2

produced by the electric current. The magnetic flux produced has the same properties as the flux produced by a permanent magnet. If the current is increased the strength of the field increases and, as for the permanent magnet, the field strength decreases as we move away from the current-carrying conductor.

In Figure 8.1, the effect of only a small part of the magnetic field is shown.

If the whole length of the conductor is similarly investigated it is found that the magnetic field around a straight conductor is in the form of concentric cylinders as shown in Figure 8.2, the field direction depending on the direction of the current flow.

When dealing with magnetic fields formed by electric current it is usual to portray the effect as shown in Figure 8.3. The convention adopted is:

(i) Current flowing away from the viewer, i.e. into the paper, is indicated by ⊕. This may be thought of as the feathered end of the shaft of an arrow. See Figure 8.3(a).

(ii) Current flowing towards the viewer, i.e. out of the paper, is indicated by ⊙. This may be thought of as the point of an arrow. See Figure 8.3(b).

The direction of the magnetic lines of flux is best remembered by the **screw rule**. This states that:

'If a normal right-hand thread screw is screwed along the conductor in the direction of the current, the direction of rotation of the screw is in the direction of the magnetic field.'

For example, with current flowing away from the viewer (Figure 8.3(a)) a right-hand thread screw driven into the paper has to be rotated clockwise. Hence the direction of the magnetic field is clockwise.

A magnetic field set up by a long coil, or **solenoid**, is shown in Figure 8.4(a) and is seen to be similar to that of a bar magnet. If the solenoid is wound on an iron bar, as shown in Figure 8.4(b), an even stronger magnetic field is produced, the iron becoming magnetized and behaving like a permanent magnet.

The direction of the magnetic field produced by the current I in the solenoid may be found by either of two methods, i.e. the screw rule or the grip rule.

(a) **The screw rule** states that if a normal right-hand thread screw is placed along the axis of the solenoid and is screwed in the direction of the current it moves in the direction of the magnetic field **inside** the solenoid. The direction of the magnetic field **inside** the solenoid is from south to north. Thus in Figures 8.4(a) and (b) the north pole is to the right.

(b) **The grip rule** states that if the coil is gripped with the **right** hand, with the fingers pointing in the direction of the current, then the thumb, outstretched parallel to the axis of the solenoid, points in the direction of the magnetic field **inside** the solenoid.

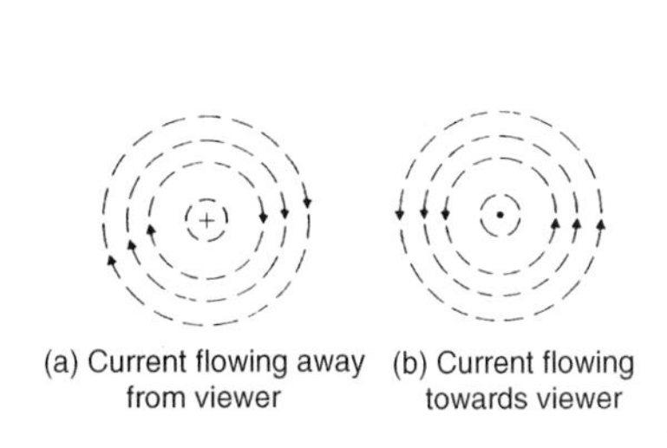

Figure 8.3

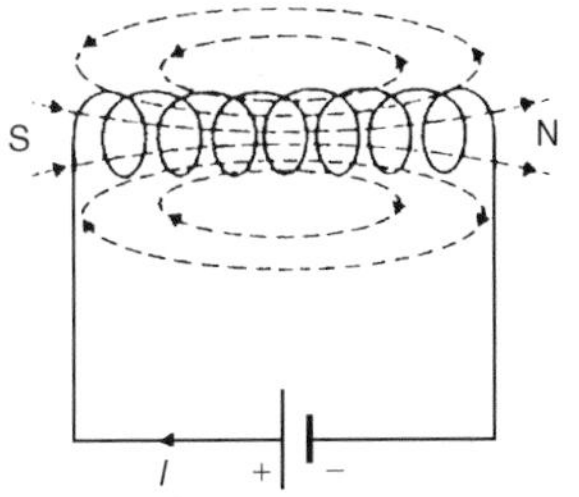

S
N
I
(a) Magnetic field of a solenoid

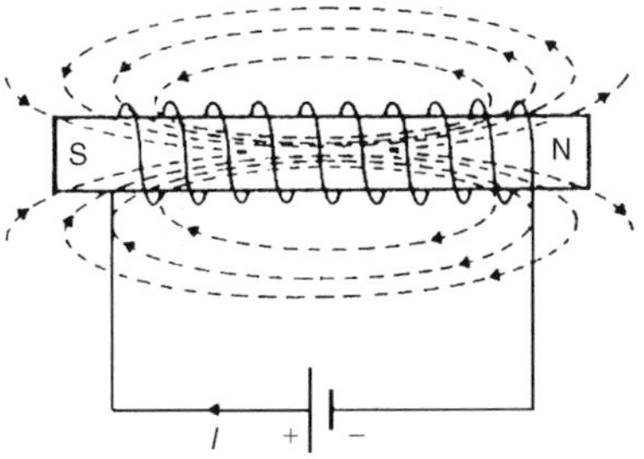

S
N
I
(b) Magnetic field of an iron cored solenoid

Figure 8.4

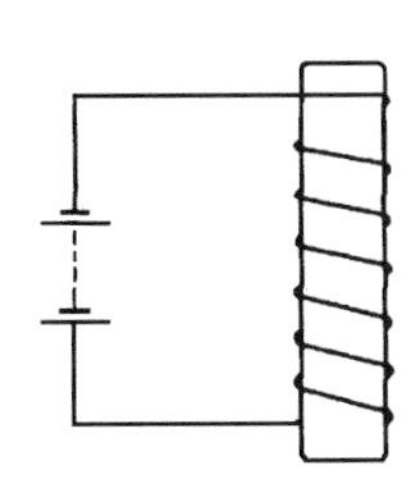

Figure 8.5

Problem 1. Figure 8.5 shows a coil of wire wound on an iron core connected to a battery. Sketch the magnetic field pattern associated with the current carrying coil and determine the polarity of the field.

The magnetic field associated with the solenoid in Figure 8.5 is similar to the field associated with a bar magnet and is as shown in Figure 8.6. The polarity of the field is determined either by the screw rule or by the grip rule. Thus the north pole is at the bottom and the south pole at the top.

8.2 Electromagnets

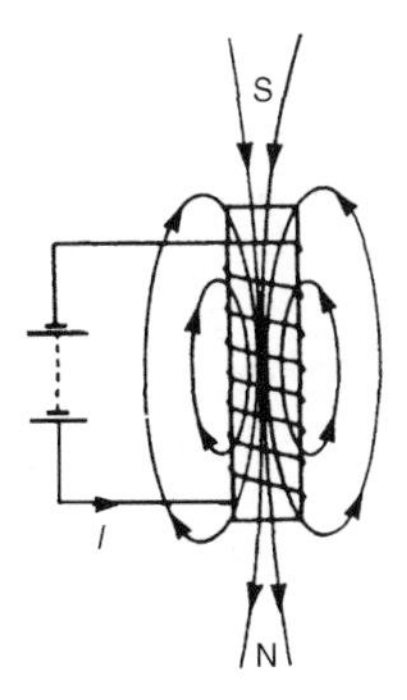

Figure 8.6

The solenoid is very important in electromagnetic theory since the magnetic field inside the solenoid is practically uniform for a particular current, and is also versatile, inasmuch that a variation of the current can alter the strength of the magnetic field. An electromagnet, based on the solenoid, provides the basis of many items of electrical equipment, examples of which include electric bells, relays, lifting magnets and telephone receivers.

(i) Electric bell

There are various types of electric bell, including the single-stroke bell, the trembler bell, the buzzer and a continuously ringing bell, but all depend on the attraction exerted by an electromagnet on a soft iron armature. A typical single stroke bell circuit is shown in Figure 8.7. When the push button is operated a current passes through the coil. Since the iron-cored coil is energized the soft iron armature is attracted to the electromagnet. The armature also carries a striker which hits the gong. When the circuit is broken the coil becomes demagnetized and the spring steel strip pulls the armature back to its original position. The striker will only operate when the push is operated.

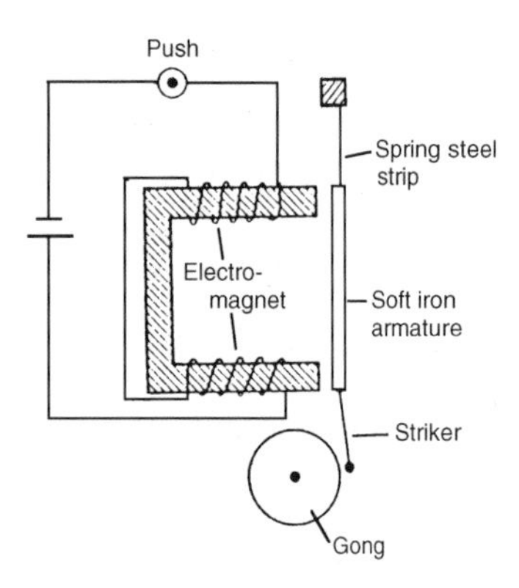

Figure 8.7

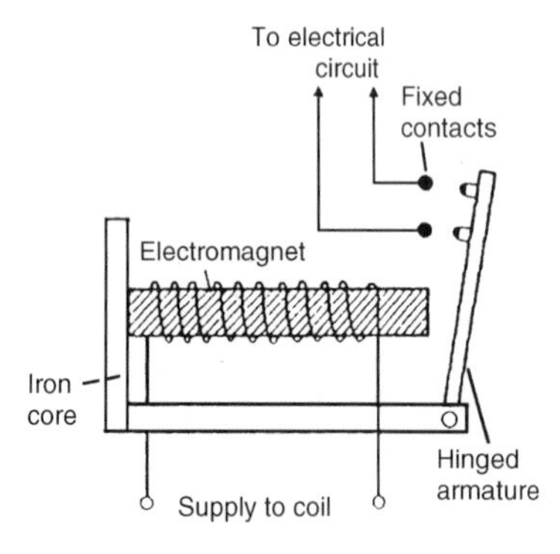

Figure 8.8

(ii) Relay

A relay is similar to an electric bell except that contacts are opened or closed by operation instead of a gong being struck. A typical simple relay is shown in Figure 8.8, which consists of a coil wound on a soft iron core. When the coil is energized the hinged soft iron armature is attracted to the electromagnet and pushes against two fixed contacts so that they are connected together, thus closing some other electrical circuit.

(iii) Lifting magnet

Lifting magnets, incorporating large electromagnets, are used in iron and steel works for lifting scrap metal. A typical robust lifting magnet, capable of exerting large attractive forces, is shown in the elevation and plan view of Figure 8.9 where a coil, C, is wound round a central core, P, of the iron casting. Over the face of the electromagnet is placed a protective non-magnetic sheet of material, R. The load, Q, which must be of magnetic material is lifted when the coils are energized, the magnetic flux paths, M, being shown by the broken lines.

(iv) Telephone receiver

Whereas a transmitter or microphone changes sound waves into corresponding electrical signals, a telephone receiver converts the electrical waves back into sound waves. A typical telephone receiver is shown in Figure 8.10 and consists of a permanent magnet with coils wound on its poles. A thin, flexible diaphragm of magnetic material is held in position near to the magnetic poles but not touching them. Variation in current from the transmitter varies the magnetic field and the diaphragm consequently vibrates. The vibration produces sound variations corresponding to those transmitted.

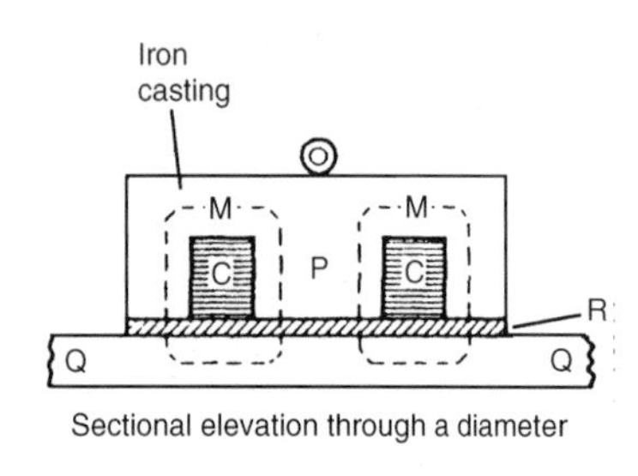

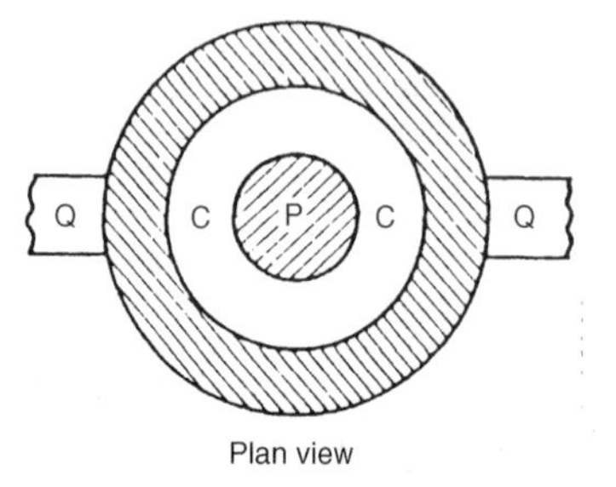

Figure 8.9

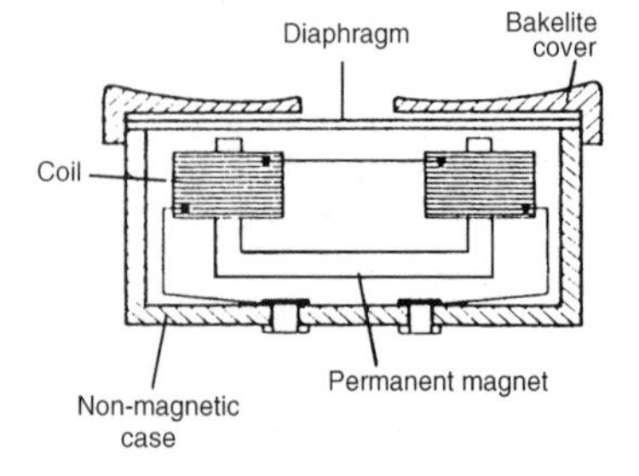

Figure 8.10

8.3 Force on a current-carrying conductor

If a current-carrying conductor is placed in a magnetic field produced by permanent magnets, then the fields due to the current-carrying conductor and the permanent magnets interact and cause a force to be exerted on the conductor. The force on the current-carrying conductor in a magnetic field depends upon:

(a) the flux density of the field, B teslas
(b) the strength of the current, I amperes,
(c) the length of the conductor perpendicular to the magnetic field, l metres, and
(d) the directions of the field and the current.

When the magnetic field, the current and the conductor are mutually at right angles then:

Force $F = BIl$ newtons

When the conductor and the field are at an angle $\theta°$ to each other then:

Force $F = BIl \sin\theta$ newtons

Since when the magnetic field, current and conductor are mutually at right angles, $F = BIl$, the magnetic flux density B may be defined by $B = F/Il$, i.e. the flux density is 1 T if the force exerted on 1 m of a conductor when the conductor carries a current of 1 A is 1 N.

Loudspeaker

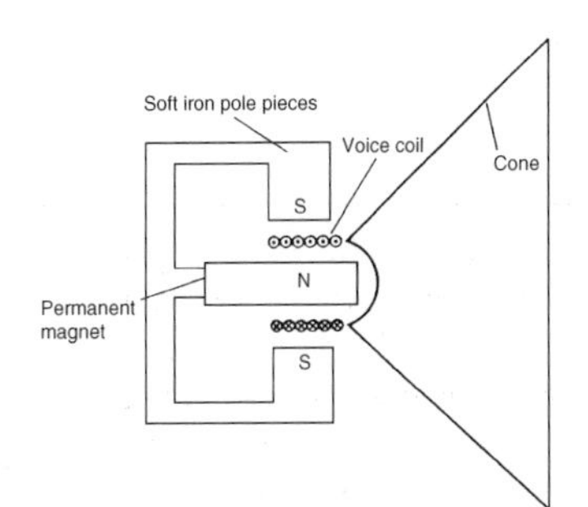

Figure 8.11

A simple application of the above force is the moving coil loudspeaker. The loudspeaker is used to convert electrical signals into sound waves.

Figure 8.11 shows a typical loudspeaker having a magnetic circuit comprising a permanent magnet and soft iron pole pieces so that a strong magnetic field is available in the short cylindrical airgap. A moving coil, called the voice or speech coil, is suspended from the end of a paper or plastic cone so that it lies in the gap. When an electric current flows through the coil it produces a force which tends to move the cone backwards and forwards according to the direction of the current. The cone acts as a piston, transferring this force to the air, and producing the required sound waves.

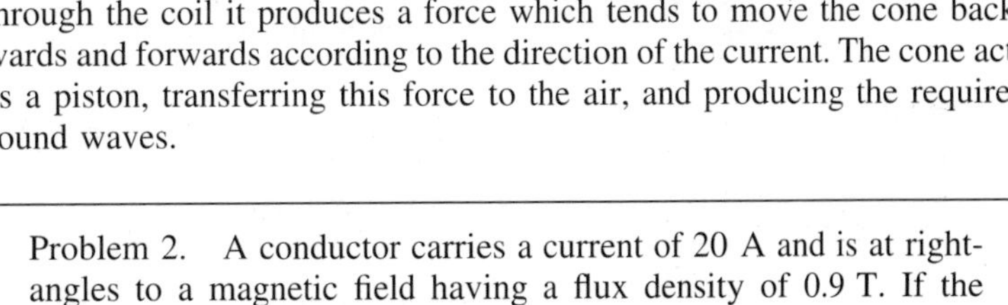
Problem 2. A conductor carries a current of 20 A and is at right-angles to a magnetic field having a flux density of 0.9 T. If the length of the conductor in the field is 30 cm, calculate the force acting on the conductor.
Determine also the value of the force if the conductor is inclined at an angle of 30° to the direction of the field.

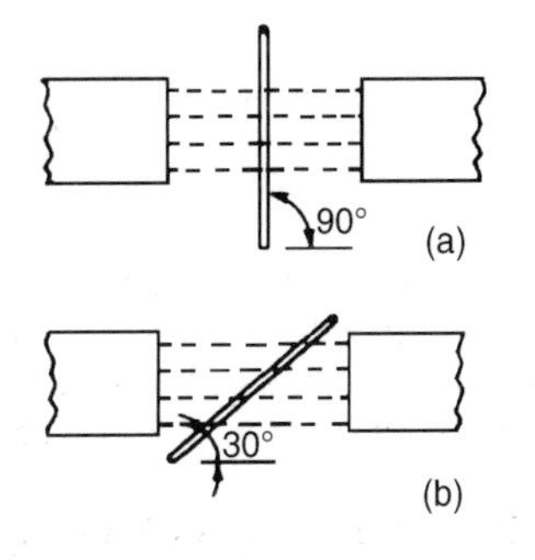

Figure 8.12

$B = 0.9$ T; $I = 20$ A; $l = 30$ cm $= 0.30$ m

Force $F = BIl = (0.9)(20)(0.30)$ newtons when the conductor is at right-angles to the field, as shown in Figure 8.12(a), i.e. ***F* = 5.4 N**

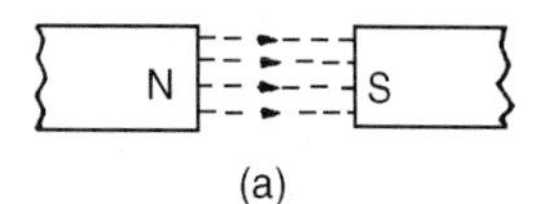

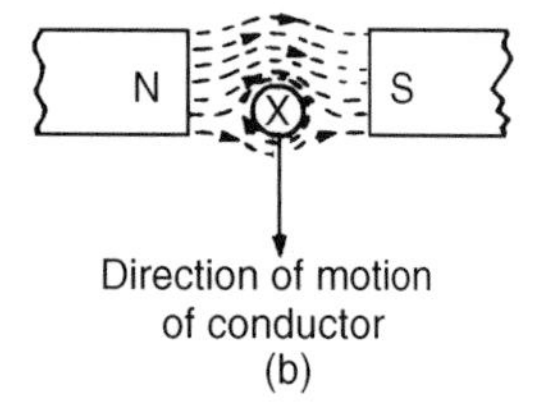

Figure 8.13

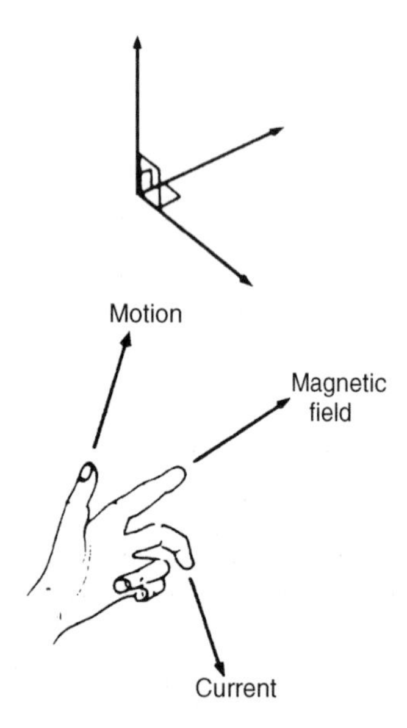

Figure 8.14

When the conductor is inclined at 30° to the field, as shown in Figure 8.12(b), then force $F = BIl \sin\theta$

$$= (0.9)(20)(0.30)\sin 30°$$

i.e. $\boldsymbol{F = 2.7\ \text{N}}$

If the current-carrying conductor shown in Figure 8.3(a) is placed in the magnetic field shown in Figure 8.13(a), then the two fields interact and cause a force to be exerted on the conductor as shown in Figure 8.13(b). The field is strengthened above the conductor and weakened below, thus tending to move the conductor downwards. This is the basic principle of operation of the electric motor (see Section 8.4) and the moving-coil instrument (see Section 8.5).

The direction of the force exerted on a conductor can be pre-determined by using **Fleming's left-hand rule** (often called the motor rule) which states:

Let the thumb, first finger and second finger of the left hand be extended such that they are all at right-angles to each other, (as shown in Figure 8.14). If the first finger points in the direction of the magnetic field, the second finger points in the direction of the current, then the thumb will point in the direction of the motion of the conductor.

Summarizing:

First finger - Field

SeCond finger - Current

ThuMb - Motion

> Problem 3. Determine the current required in a 400 mm length of conductor of an electric motor, when the conductor is situated at right-angles to a magnetic field of flux density 1.2 T, if a force of 1.92 N is to be exerted on the conductor. If the conductor is vertical, the current flowing downwards and the direction of the magnetic field is from left to right, what is the direction of the force?

Force = 1.92 N; $l = 400$ mm $= 0.40$ m; $B = 1.2$ T

Since $F = BIl$, then $I = \dfrac{F}{Bl}$

hence current $I = \dfrac{1.92}{(1.2)(0.4)} = \mathbf{4\ A}$

If the current flows downwards, the direction of its magnetic field due to the current alone will be clockwise when viewed from above. The lines of flux will reinforce (i.e. strengthen) the main magnetic field at the back of the conductor and will be in opposition in the front (i.e. weaken the field).

Hence the force on the conductor will be from back to front (i.e. toward the viewer). This direction may also have been deduced using Fleming's left-hand rule.

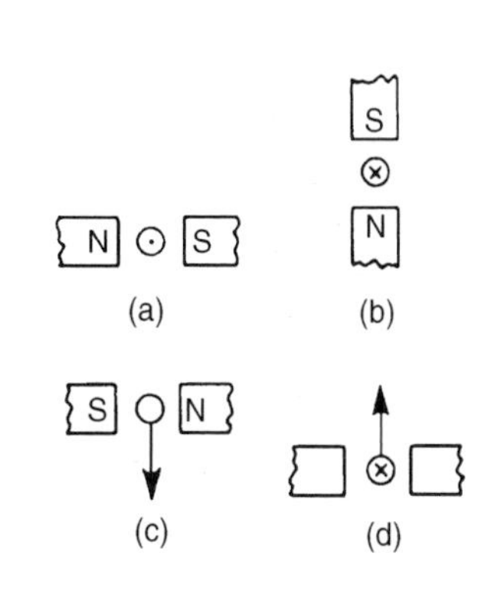

Figure 8.15

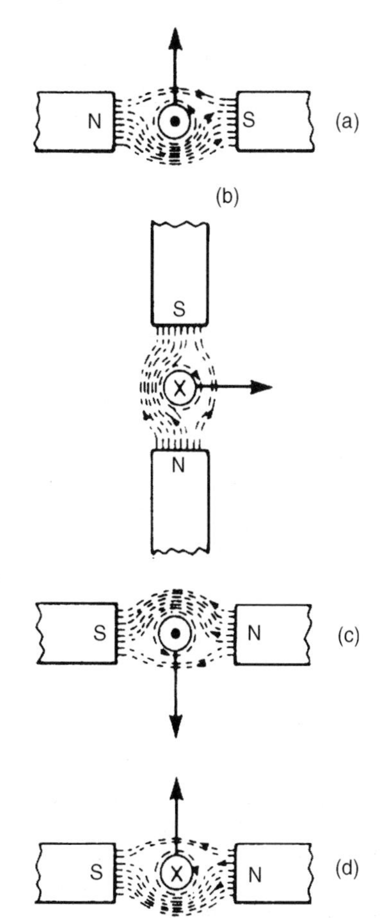

Figure 8.16

> Problem 4. A conductor 350 mm long carries a current of 10 A and is at right-angles to a magnetic field lying between two circular pole faces each of radius 60 mm. If the total flux between the pole faces is 0.5 mWb, calculate the magnitude of the force exerted on the conductor.

$l = 350$ mm $= 0.35$ m; $I = 10$ A;

Area of pole face $A = \pi r^2 = \pi(0.06)^2$ m^2;

$\Phi = 0.5$ mWb $= 0.5 \times 10^{-3}$ Wb

Force $F = BIl$, and $B = \dfrac{\Phi}{A}$

hence force $F = \left(\dfrac{\Phi}{A}\right) Il = \dfrac{(0.5 \times 10^{-3})}{\pi(0.06)^2}(10)(0.35)$ newtons

i.e. force = 0.155 N

> Problem 5. With reference to Figure 8.15 determine (a) the direction of the force on the conductor in Figure 8.15(a), (b) the direction of the force on the conductor in Figure 8.15(b), (c) the direction of the current in Figure 8.15(c), (d) the polarity of the magnetic system in Figure 8.15(d).

(a) The direction of the main magnetic field is from north to south, i.e. left to right. The current is flowing towards the viewer, and using the screw rule, the direction of the field is anticlockwise. Hence either by Fleming's left-hand rule, or by sketching the interacting magnetic field as shown in Figure 8.16(a), the direction of the force on the conductor is seen to be upward.

(b) Using a similar method to part (a) it is seen that the force on the conductor is to the right — see Figure 8.16(b).

(c) Using Fleming's left-hand rule, or by sketching as in Figure 8.16(c), it is seen that the current is toward the viewer, i.e. out of the paper.

(d) Similar to part (c), the polarity of the magnetic system is as shown in Figure 8.16(d).

> Problem 6. A coil is wound on a rectangular former of width 24 mm and length 30 mm. The former is pivoted about an axis passing through the middle of the two shorter sides and is placed in a uniform magnetic field of flux density 0.8 T, the axis being perpendicular to the field. If the coil carries a current of 50 mA, determine the force on each coil side (a) for a single-turn coil, (b) for a coil wound with 300 turns.

(a) Flux density $B = 0.8$ T; length of conductor lying at right-angles to field $l = 30$ mm $= 30 \times 10^{-3}$ m; current $I = 50$ mA $= 50 \times 10^{-3}$ A

For a single-turn coil, force on each coil side

$$F = BIl = 0.8 \times 50 \times 10^{-3} \times 30 \times 10^{-3}$$

$$= \mathbf{1.2 \times 10^{-3}\ N}, \text{ or } \mathbf{0.0012\ N}$$

(b) When there are 300 turns on the coil there are effectively 300 parallel conductors each carrying a current of 50 mA. Thus the total force produced by the current is 300 times that for a single-turn coil. Hence force on coil side $F = 300\ BIl = 300 \times 0.0012 = \mathbf{0.36\ N}$

Further problems on the force on a current-carrying conductor may be found in Section 8.7, problems 1 to 6, page 98.

8.4 Principle of operation of a simple d.c. motor

A rectangular coil which is free to rotate about a fixed axis is shown placed inside a magnetic field produced by permanent magnets in Figure 8.17. A direct current is fed into the coil via carbon brushes bearing on a commutator, which consists of a metal ring split into two halves separated by insulation.

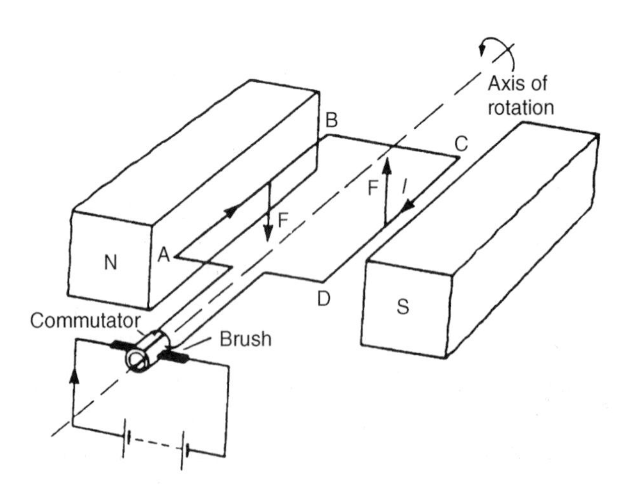

Figure 8.17

When current flows in the coil a magnetic field is set up around the coil which interacts with the magnetic field produced by the magnets. This causes a force F to be exerted on the current-carrying conductor which, by Fleming's left-hand rule, is downwards between points A and B and upward between C and D for the current direction shown. This causes a torque and the coil rotates anticlockwise. When the coil has turned through 90° from the position shown in Figure 8.17 the brushes connected to the positive and negative terminals of the supply make contact with different halves of the commutator ring, thus reversing the direction of the current flow in the conductor. If the current is not reversed and the coil rotates past this position the forces acting on it change direction and it rotates in the opposite direction thus never making more than half a revolution. The current direction is reversed every time the coil swings through the vertical position and thus the coil rotates anti-clockwise for as long as the current flows. This is the principle of operation of a d.c. motor which is thus a device that takes in electrical energy and converts it into mechanical energy.

8.5 Principle of operation of a moving coil instrument

A moving-coil instrument operates on the motor principle. When a conductor carrying current is placed in a magnetic field, a force F is exerted on the conductor, given by $F = BIl$. If the flux density B is made constant (by using permanent magnets) and the conductor is a fixed length (say, a coil) then the force will depend only on the current flowing in the conductor.

In a moving-coil instrument a coil is placed centrally in the gap between shaped pole pieces as shown by the front elevation in Figure 8.18(a). (The airgap is kept as small as possible, although for clarity it is shown exaggerated in Figure 8.18). The coil is supported by steel pivots, resting in jewel bearings, on a cylindrical iron core. Current is led into and out of the coil by two phosphor bronze spiral hairsprings which are wound in opposite directions to minimize the effect of temperature change and to limit the coil swing (i.e. to **control** the movement) and return the movement to zero position when no current flows. Current flowing in the coil produces forces as shown in Fig 8.18(b), the directions being obtained by Fleming's left-hand rule. The two forces, F_A and F_B, produce a torque which will move the coil in a clockwise direction, i.e. move the pointer from left to right. Since force is proportional to current the scale is linear.

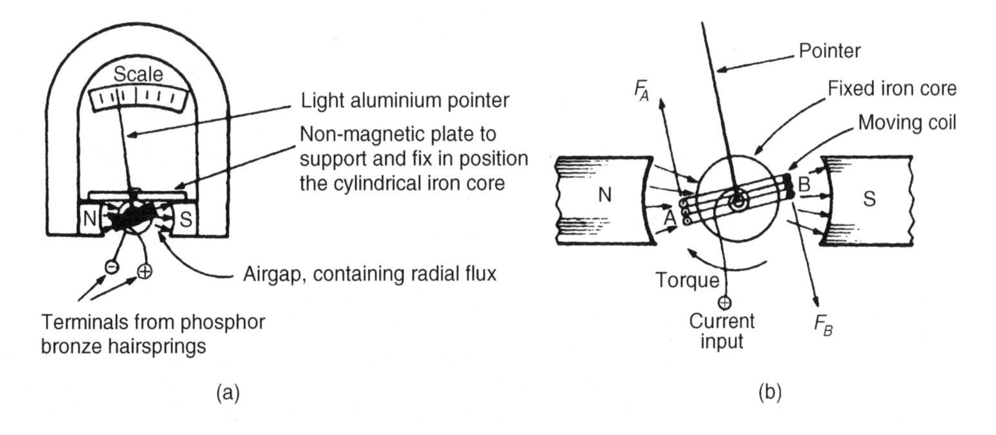

Figure 8.18

When the aluminium frame, on which the coil is wound, is rotated between the poles of the magnet, small currents (called eddy currents) are induced into the frame, and this provides automatically the necessary **damping** of the system due to the reluctance of the former to move within the magnetic field. The moving-coil instrument will measure only direct current or voltage and the terminals are marked positive and negative to ensure that the current passes through the coil in the correct direction to deflect the pointer 'up the scale'.

The range of this sensitive instrument is extended by using shunts and multipliers (see Chapter 10).

8.6 Force on a charge

When a charge of Q coulombs is moving at a velocity of v m/s in a magnetic field of flux density B teslas, the charge moving perpendicular to the field, then the magnitude of the force F exerted on the charge is given by:

$$\mathbf{F = QvB \text{ newtons}}$$

Problem 17. An electron in a television tube has a charge of 1.6×10^{-19} coulombs and travels at 3×10^7 m/s perpendicular to a field of flux density 18.5 μT. Determine the force exerted on the electron in the field.

From above, force $F = QvB$ newtons, where

Q = charge in coulombs $= 1.6 \times 10^{-19}$ C;

v = velocity of charge $= 3 \times 10^7$ m/s;

and B = flux density $= 18.5 \times 10^{-6}$ T

Hence force on electron $F = 1.6 \times 10^{-19} \times 3 \times 10^7 \times 18.5 \times 10^{-6}$

$= 1.6 \times 3 \times 18.5 \times 10^{-18}$

$= 88.8 \times 10^{-18} = \mathbf{8.88 \times 10^{-17}\ N}$

Further problems on the force on a charge may be found in Section 8.7 following, problems 7 and 8, page 99.

8.7 Further problems on electromagnetism

Force on a current-carrying conductor

1 A conductor carries a current of 70 A at right-angles to a magnetic field having a flux density of 1.5 T. If the length of the conductor in the field is 200 mm calculate the force acting on the conductor. What is the force when the conductor and field are at an angle of 45°?

[21.0 N, 14.8 N]

2 Calculate the current required in a 240 mm length of conductor of a d.c. motor when the conductor is situated at right-angles to the magnetic field of flux density 1.25 T, if a force of 1.20 N is to be exerted on the conductor. [4.0 A]

3 A conductor 30 cm long is situated at right-angles to a magnetic field. Calculate the strength of the magnetic field if a current of 15 A in the conductor produces a force on it of 3.6 N. [0.80 T]

4 A conductor 300 mm long carries a current of 13 A and is at right-angles to a magnetic field between two circular pole faces, each of diameter 80 mm. If the total flux between the pole faces is 0.75 mWb calculate the force exerted on the conductor. [0.582 N]

5 (a) A 400 mm length of conductor carrying a current of 25 A is situated at right-angles to a magnetic field between two poles of an electric motor. The poles have a circular cross-section. If the force exerted on the conductor is 80 N and the total flux between the pole faces is 1.27 mWb, determine the diameter of a pole face.

(b) If the conductor in part (a) is vertical, the current flowing downwards and the direction of the magnetic field is from left to right, what is the direction of the 80 N force?

[(a) 14.2 mm (b) towards the viewer]

6 A coil is wound uniformly on a former having a width of 18 mm and a length of 25 mm. The former is pivoted about an axis passing through the middle of the two shorter sides and is placed in a uniform magnetic field of flux density 0.75 T, the axis being perpendicular to the field. If the coil carries a current of 120 mA, determine the force exerted on each coil side, (a) for a single-turn coil, (b) for a coil wound with 400 turns. [(a) 2.25×10^{-3} N (b) 0.9 N]

Force on a charge

7 Calculate the force exerted on a charge of 2×10^{-18} C travelling at 2×10^{6} m/s perpendicular to a field of density 2×10^{-7} T.

[8×10^{-19} N]

8 Determine the speed of a 10^{-19} C charge travelling perpendicular to a field of flux density 10^{-7} T, if the force on the charge is 10^{-20} N.

[10^{6} m/s]

9 Electromagnetic induction

At the end of this chapter you should be able to:

- understand how an e.m.f. may be induced in a conductor
- state Faraday's laws of electromagnetic induction
- state Lenz's law
- use Fleming's right-hand rule for relative directions
- appreciate that the induced e.m.f., $E = Blv$ or $E = Blv\sin\theta$
- calculate induced e.m.f. given B, l, v and θ and determine relative directions
- define inductance L and state its unit
- define mutual inductance
- appreciate that e.m.f. $E = -N\dfrac{d\Phi}{dt} = -L\dfrac{dI}{dt}$
- calculate induced e.m.f. given N, t, L, change of flux or change of current
- appreciate factors which affect the inductance of an inductor
- draw the circuit diagram symbols for inductors
- calculate the energy stored in an inductor using $W = \frac{1}{2}LI^2$ joules
- calculate inductance L of a coil, given $L = \dfrac{N\Phi}{I}$
- calculate mutual inductance using $E_2 = -M\dfrac{dI_1}{dt}$

9.1 Introduction to electromagnetic induction

When a conductor is moved across a magnetic field so as to cut through the lines of force (or flux), an electromotive force (e.m.f.) is produced in the conductor. If the conductor forms part of a closed circuit then the e.m.f. produced causes an electric current to flow round the circuit. Hence an e.m.f. (and thus current) is 'induced' in the conductor as a result of its movement across the magnetic field. This effect is known as '**electromagnetic induction**'.

Figure 9.1(a) shows a coil of wire connected to a centre-zero galvanometer, which is a sensitive ammeter with the zero-current position in the centre of the scale.

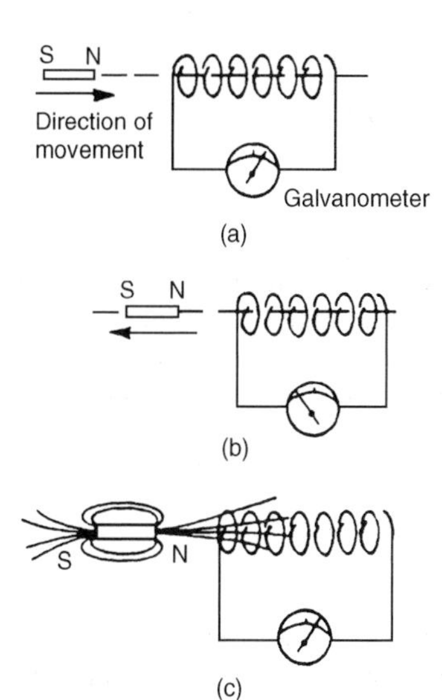

Figure 9.1

(a) When the magnet is moved at constant speed towards the coil (Figure 9.1(a)), a deflection is noted on the galvanometer showing that a current has been produced in the coil.
(b) When the magnet is moved at the same speed as in (a) but away from the coil the same deflection is noted but is in the opposite direction (see Figure 9.1(b))
(c) When the magnet is held stationary, even within the coil, no deflection is recorded.
(d) When the coil is moved at the same speed as in (a) and the magnet held stationary the same galvanometer deflection is noted.
(e) When the relative speed is, say, doubled, the galvanometer deflection is doubled.
(f) When a stronger magnet is used, a greater galvanometer deflection is noted.
(g) When the number of turns of wire of the coil is increased, a greater galvanometer deflection is noted.

Figure 9.1(c) shows the magnetic field associated with the magnet. As the magnet is moved towards the coil, the magnetic flux of the magnet moves across, or cuts, the coil. **It is the relative movement of the magnetic flux and the coil that causes an e.m.f. and thus current, to be induced in the coil**. This effect is known as electromagnetic induction. The laws of electromagnetic induction stated in Section 9.2 evolved from experiments such as those described above.

9.2 Laws of electromagnetic induction

Faraday's laws of electromagnetic induction state:

(i) *'An induced e.m.f. is set up whenever the magnetic field linking that circuit changes.'*
(ii) *'The magnitude of the induced e.m.f. in any circuit is proportional to the rate of change of the magnetic flux linking the circuit.'*

Lenz's law states:

'The direction of an induced e.m.f. is always such that it tends to set up a current opposing the motion or the change of flux responsible for inducing that e.m.f.'.

An alternative method to Lenz's law of determining relative directions is given by **Fleming's Right-hand rule** (often called the gene*R*ator rule) which states:

Let the thumb, first finger and second finger of the right hand be extended such that they are all at right angles to each other (as shown in Figure 9.2). If the first finger points in the direction of the magnetic field, the thumb points in the direction of motion of the conductor relative to the magnetic field, then the second finger will point in the direction of the induced e.m.f.

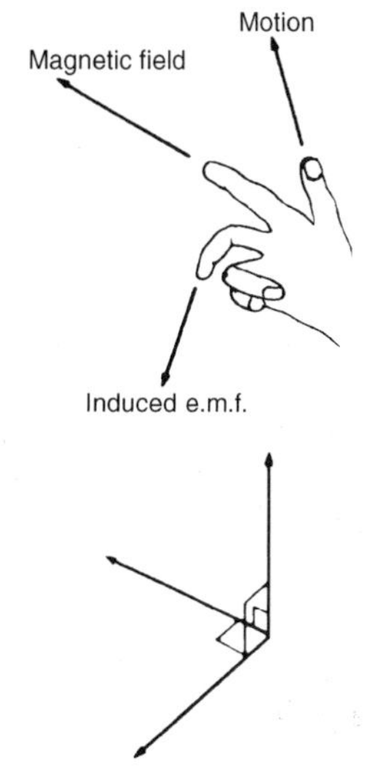

Figure 9.2

Summarizing:

First finger — Field

ThuMb — Motion

SEcond finger — E.m.f.

In a generator, conductors forming an electric circuit are made to move through a magnetic field. By Faraday's law an e.m.f. is induced in the conductors and thus a source of e.m.f. is created. A generator converts mechanical energy into electrical energy. (The action of a simple a.c. generator is described in Chapter 14.) The induced e.m.f. E set up between the ends of the conductor shown in Figure 9.3 is given by:

$$\boldsymbol{E = Blv} \textbf{ volts,}$$

where B, the flux density, is measured in teslas, l, the length of conductor in the magnetic field, is measured in metres, and v, the conductor velocity, is measured in metres per second.

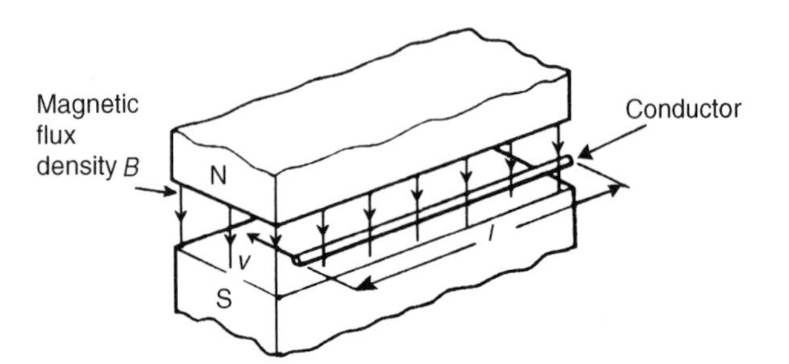

Figure 9.3

If the conductor moves at an angle θ° to the magnetic field (instead of at 90° as assumed above) then

$$\boldsymbol{E = Blv \sin\theta} \textbf{ volts}$$

Problem 1. A conductor 300 mm long moves at a uniform speed of 4 m/s at right-angles to a uniform magnetic field of flux density 1.25 T. Determine the current flowing in the conductor when (a) its ends are open-circuited, (b) its ends are connected to a load of 20 Ω resistance.

When a conductor moves in a magnetic field it will have an e.m.f. induced in it but this e.m.f. can only produce a current if there is a closed circuit.

Induced e.m.f. $E = Blv = (1.25)\left(\frac{300}{1000}\right)(4) = 1.5$ V

(a) If the ends of the conductor are open circuited no current will flow even though 1.5 V has been induced.

(b) From Ohm's law, $I = \frac{E}{R} = \frac{1.5}{20} = \mathbf{0.075\ A}$ or $\mathbf{75\ mA}$

Problem 2. At what velocity must a conductor 75 mm long cut a magnetic field of flux density 0.6 T if an e.m.f. of 9 V is to be induced in it? Assume the conductor, the field and the direction of motion are mutually perpendicular.

Induced e.m.f. $E = Blv$, hence velocity $v = \frac{E}{Bl}$

$$\text{Hence } v = \frac{9}{(0.6)(75 \times 10^{-3})} = \frac{9 \times 10^3}{0.6 \times 75} = \mathbf{200\ m/s}$$

Problem 3. A conductor moves with a velocity of 15 m/s at an angle of (a) 90°, (b) 60° and (c) 30° to a magnetic field produced between two square-faced poles of side length 2 cm. If the flux leaving a pole face is 5 μWb, find the magnitude of the induced e.m.f. in each case.

$v = 15$ m/s; length of conductor in magnetic field, $l = 2$ cm $= 0.02$ m; $A = 2 \times 2\ \text{cm}^2 = 4 \times 10^{-4}\ \text{m}^2$, $\Phi = 5 \times 10^{-6}$ Wb

(a) $$E_{90} = Blv\sin 90^\circ = \left(\frac{\Phi}{A}\right) lv\sin 90^\circ = \frac{(5 \times 10^{-6})}{(4 \times 10^{-4})}(0.02)(15)(1) = \mathbf{3.75\ mV}$$

(b) $E_{60} = Blv\sin 60^\circ = E_{90}\sin 60^\circ = 3.75\sin 60^\circ = \mathbf{3.25\ mV}$

(c) $E_{30} = Blv\sin 30^\circ = E_{90}\sin 30^\circ = 3.75\sin 30^\circ = \mathbf{1.875\ mV}$

Problem 4. The wing span of a metal aeroplane is 36 m. If the aeroplane is flying at 400 km/h, determine the e.m.f. induced between its wing tips. Assume the vertical component of the earth's magnetic field is 40 μT

Induced e.m.f. across wing tips, $E = Blv$

$B = 40\ \mu\text{T} = 40 \times 10^{-6}$ T; $l = 36$ m

$$v = 400\frac{\text{km}}{\text{h}} \times 1000\frac{\text{m}}{\text{km}} \times \frac{1\ \text{h}}{60 \times 60\ \text{s}} = \frac{(400)(1000)}{3600} = \frac{4000}{36}\ \text{m/s}$$

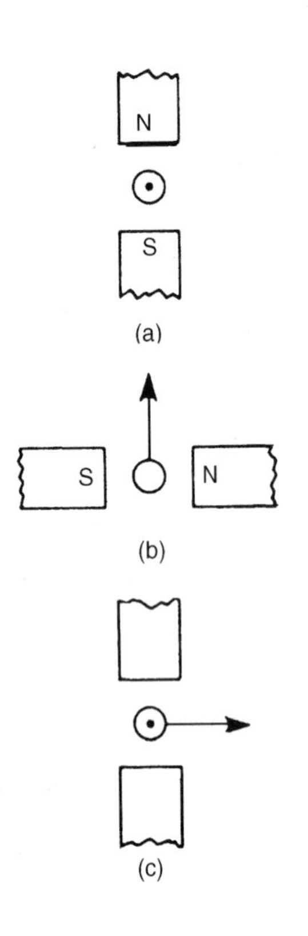

Figure 9.4

$$\text{Hence } E = Blv = (40 \times 10^{-6})(36)\left(\frac{4000}{36}\right)$$
$$= \mathbf{0.16\ V}$$

> Problem 5. The diagram shown in Figure 9.4 represents the generation of e.m.f's. Determine (i) the direction in which the conductor has to be moved in Figure 9.4(a), (ii) the direction of the induced e.m.f. in Figure 9.4(b), (iii) the polarity of the magnetic system in Figure 9.4(c).

The direction of the e.m.f., and thus the current due to the e.m.f. may be obtained by either Lenz's law or Fleming's Right-hand rule (i.e. GeneRator rule).

(i) Using Lenz's law: The field due to the magnet and the field due to the current-carrying conductor are shown in Figure 9.5(a) and are seen to reinforce to the left of the conductor. Hence the force on the conductor is to the right. However Lenz's law states that the direction of the induced e.m.f. is always such as to oppose the effect producing it. **Thus the conductor will have to be moved to the left.**

(ii) Using Fleming's right-hand rule:

First finger — Field, i.e. $N \rightarrow S$, or right to left;

ThuMb — Motion, i.e. upwards;

SEcond finger — E.m.f., i.e. **towards the viewer or out of the paper**, as shown in Figure 9.5(b)

(iii) The polarity of the magnetic system of Figure 9.4(c) is shown in Figure 9.5(c) and is obtained using Fleming's right-hand rule.

Further problems on induced e.m.f.'s may be found in Section 9.8, problems 1 to 5, page 109.

9.3 Inductance

Inductance is the name given to the property of a circuit whereby there is an e.m.f. induced into the circuit by the change of flux linkages produced by a current change.

When the e.m.f. is induced in the same circuit as that in which the current is changing, the property is called **self inductance, *L***

When the e.m.f. is induced in a circuit by a change of flux due to current changing in an adjacent circuit, the property is called **mutual inductance, *M***.

The unit of inductance is the **henry, H**.

'A circuit has an inductance of one henry when an e.m.f. of one volt is induced in it by a current changing at the rate of one ampere per second.'

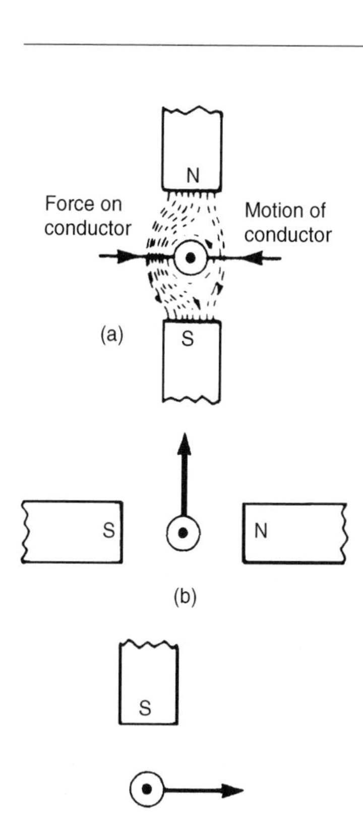

Figure 9.5

Induced e.m.f. in a coil of N turns,

$$\boldsymbol{E = -N\frac{d\Phi}{dt}} \textbf{ volts,}$$

where $d\Phi$ is the change in flux in Webers, and dt is the time taken for the flux to change in seconds(i.e., $d\Phi/dt$ is the rate of change of flux).

Induced e.m.f. in a coil of inductance L henrys,

$$\boldsymbol{E = -L\frac{dI}{dt}} \textbf{ volts,}$$

where dI is the change in current in amperes and dt is the time taken for the current to change in seconds(i.e., dI/dt is the rate of change of current). The minus sign in each of the above two equations remind us of its direction (given by Lenz's law).

Problem 6. Determine the e.m.f. induced in a coil of 200 turns when there is a change of flux of 25 mWb linking with it in 50 ms

$$\text{Induced e.m.f. } E = -N\frac{d\Phi}{dt} = -(200)\left(\frac{25 \times 10^{-3}}{50 \times 10^{-3}}\right) = \mathbf{-100\ volts}$$

Problem 7. A flux of 400 μWb passing through a 150-turn coil is reversed in 40 ms. Find the average e.m.f. induced.

Since the flux reverses, the flux changes from +400 μWb to −400 μWb, a total change of flux of 800 μWb

$$\text{Induced e.m.f. } E = -N\frac{d\Phi}{dt} = -(150)\left(\frac{800 \times 10^{-6}}{40 \times 10^{-3}}\right)$$

$$= -\left(\frac{150 \times 800 \times 10^{3}}{40 \times 10^{6}}\right)$$

Hence **the average e.m.f. induced** $\boldsymbol{E = -3}$ **volts**

Problem 8. Calculate the e.m.f. induced in a coil of inductance 12 H by a current changing at the rate of 4 A/s

$$\text{Induced e.m.f. } E = -L\frac{dI}{dt} = -(12)(4) = \mathbf{-48\ volts}$$

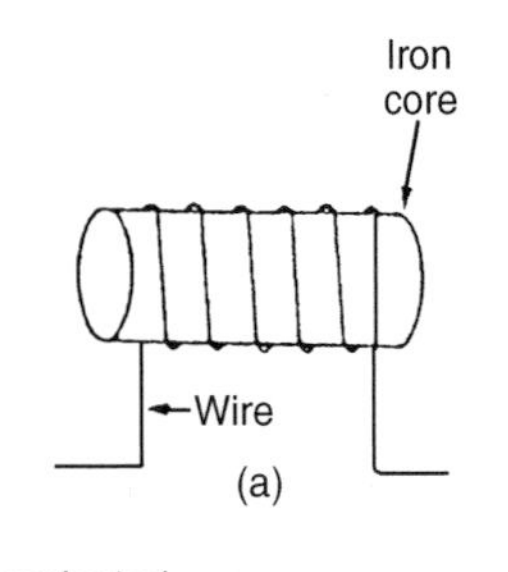

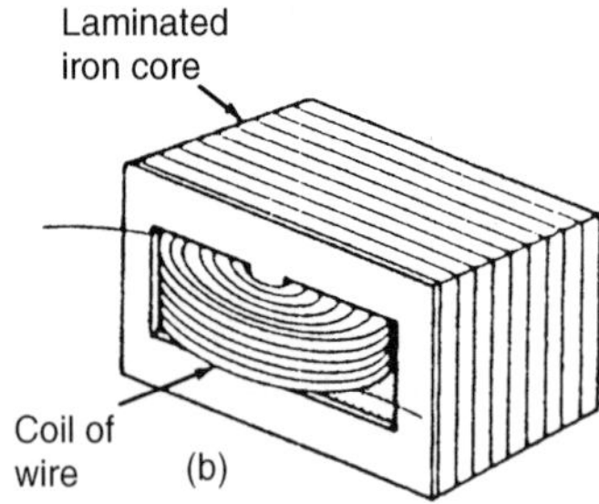

Figure 9.6

> Problem 9. An e.m.f. of 1.5 kV is induced in a coil when a current of 4 A collapses uniformly to zero in 8 ms. Determine the inductance of the coil.

Change in current, $dI = (4 - 0) = 4$ A; $dt = 8$ ms $= 8 \times 10^{-3}$ s;

$$\frac{dI}{dt} = \frac{4}{8 \times 10^{-3}} = \frac{4000}{8} = 500 \text{ A/s}; \; E = 1.5 \text{ kV} = 1500 \text{ V}$$

$$\text{Since } |E| = L\left(\frac{dI}{dt}\right), \text{ inductance, } L = \frac{|E|}{(dI/dt)} = \frac{1500}{500} = \mathbf{3\ H}$$

(Note that $|E|$ means the 'magnitude of E', which disregards the minus sign)

Further problems on inductance may be found in Section 9.8, problems 6 to 9, page 110.

9.4 Inductors

A component called an inductor is used when the property of inductance is required in a circuit. The basic form of an inductor is simply a coil of wire.

Factors which affect the inductance of an inductor include:

(i) the number of turns of wire — the more turns the higher the inductance

(ii) the cross-sectional area of the coil of wire — the greater the cross-sectional area the higher the inductance

(iii) the presence of a magnetic core — when the coil is wound on an iron core the same current sets up a more concentrated magnetic field and the inductance is increased

(iv) the way the turns are arranged — a short thick coil of wire has a higher inductance than a long thin one.

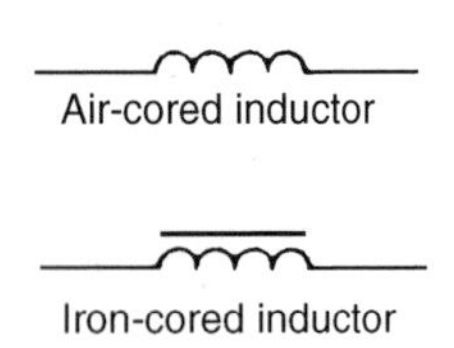

Figure 9.7

Two examples of practical inductors are shown in Figure 9.6, and the standard electrical circuit diagram symbols for air-cored and iron-cored inductors are shown in Figure 9.7.

An iron-cored inductor is often called a **choke** since, when used in a.c. circuits, it has a choking effect, limiting the current flowing through it. Inductance is often undesirable in a circuit. To reduce inductance to a minimum the wire may be bent back on itself, as shown in Figure 9.8, so that the magnetizing effect of one conductor is neutralized by that of the adjacent conductor. The wire may be coiled around an insulator, as shown, without increasing the inductance. Standard resistors may be non-inductively wound in this manner.

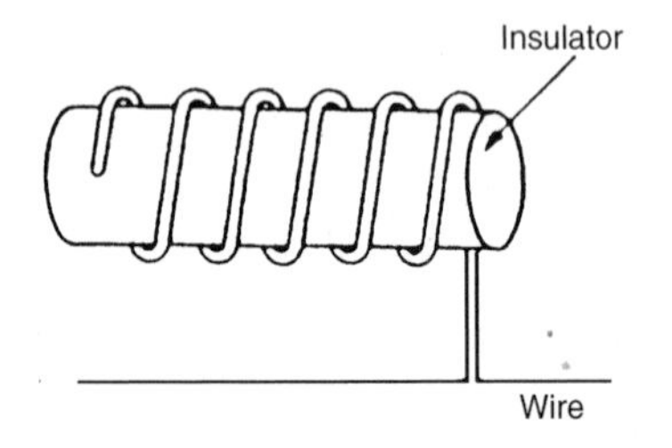

Figure 9.8

9.5 Energy stored

An inductor possesses an ability to store energy. The energy stored, W, in the magnetic field of an inductor is given by:

$$W = \tfrac{1}{2}LI^2 \text{ joules}$$

> Problem 10. An 8 H inductor has a current of 3 A flowing through it. How much energy is stored in the magnetic field of the inductor?

Energy stored, $W = \frac{1}{2}LI^2 = \frac{1}{2}(8)(3)^2 = \mathbf{36\ joules}$

Further problems on energy stored may be found in Section 9.8, problems 10 and 11, page 110.

9.6 Inductance of a coil

If a current changing from 0 to I amperes, produces a flux change from 0 to Φ Webers, then $dI = I$ and $d\Phi = \Phi$. Then, from Section 9.3, induced e.m.f. $E = N\Phi/t = LI/t$, from which

inductance of coil, $$L = \frac{N\Phi}{I} \text{ henrys}$$

> Problem 11. Calculate the coil inductance when a current of 4 A in a coil of 800 turns produces a flux of 5 mWb linking with the coil.

For a coil, inductance $L = \dfrac{N\Phi}{I} = \dfrac{(800)(5 \times 10^{-3})}{4} = \mathbf{1\ H}$

> Problem 12. A flux of 25 mWb links with a 1500 turn coil when a current of 3 A passes through the coil. Calculate (a) the inductance of the coil, (b) the energy stored in the magnetic field, and (c) the average e.m.f. induced if the current falls to zero in 150 ms.

(a) **Inductance,** $L = \dfrac{N\Phi}{I} = \dfrac{(1500)(25 \times 10^{-3})}{3} = \mathbf{12.5\ H}$

(b) **Energy stored,** $W = \frac{1}{2}LI^2 = \frac{1}{2}(12.5)(3)^2 = \mathbf{56.25\ J}$

(c) **Induced e.m.f.,** $E = -L\dfrac{dI}{dt} = -(12.5)\left(\dfrac{3-0}{150 \times 10^{-3}}\right) = \mathbf{-250\ V}$

(Alternatively, $E = -N\left(\frac{d\Phi}{dt}\right) = -(1500)\left(\frac{25 \times 10^{-3}}{150 \times 10^{-3}}\right)$

$= \mathbf{-250\ V}$

since if the current falls to zero so does the flux)

Problem 13. A 750 turn coil of inductance 3 H carries a current of 2 A. Calculate the flux linking the coil and the e.m.f. induced in the coil when the current collapses to zero in 20 ms

Coil inductance, $L = \frac{N\Phi}{I}$ from which, flux $\Phi = \frac{LI}{N}$

$= \frac{(3)(2)}{750} = 8 \times 10^{-3} = \mathbf{8\ mWb}$

Induced e.m.f. $E = -L\left(\frac{dI}{dt}\right) = -3\left(\frac{2-0}{20 \times 10^{-3}}\right) = \mathbf{-300\ V}$

(Alternatively, $E = -N\frac{d\Phi}{dt} = -(750)\left(\frac{8 \times 10^{-3}}{20 \times 10^{-3}}\right) = \mathbf{-300\ V}$)

Further problems on the inductance of a coil may be found in Section 9.8, problems 12 to 18, page 110.

9.7 Mutual inductance

Mutually induced e.m.f. in the second coil,

$$\boldsymbol{E_2 = -M\frac{dI_1}{dt}}\ \textbf{volts,}$$

where M is the **mutual inductance** between two coils, in henrys, and dI_1/dt is the rate of change of current in the first coil.

The phenomenon of mutual inductance is used in **transformers** (see Chapter 20, page 315). Mutual inductance is developed further in Chapter 43 on magnetically coupled circuits (see page 841).

Problem 14. Calculate the mutual inductance between two coils when a current changing at 200 A/s in one coil induces an e.m.f. of 1.5 V in the other.

Induced e.m.f. $|E_2| = M\frac{dI_1}{dt}$, i.e., $1.5 = M(200)$

Thus **mutual inductance**, $\boldsymbol{M} = \frac{1.5}{200} = \mathbf{0.0075\ H\ or\ 7.5\ mH}$

Problem 15. The mutual inductance between two coils is 18 mH. Calculate the steady rate of change of current in one coil to induce an e.m.f. of 0.72 V in the other.

Induced e.m.f., $|E_2| = M\dfrac{dI_1}{dt}$

Hence rate of change of current, $\dfrac{dI_1}{dt} = \dfrac{|E_2|}{M} = \dfrac{0.72}{0.018} = \mathbf{40\ A/s}$

Problem 16. Two coils have a mutual inductance of 0.2 H. If the current in one coil is changed from 10 A to 4 A in 10 ms, calculate (a) the average induced e.m.f. in the second coil, (b) the change of flux linked with the second coil if it is wound with 500 turns.

(a) Induced e.m.f. $E_2 = -M\dfrac{dI_1}{dt} = -(0.2)\left(\dfrac{10-4}{10\times 10^{-3}}\right) = \mathbf{-120\ V}$

(b) Induced e.m.f. $|E_2| = N\dfrac{d\Phi}{dt}$, hence $d\Phi = \dfrac{|E_2|dt}{N}$

Thus the change of flux, $d\Phi = \dfrac{120(10\times 10^{-3})}{500} = \mathbf{2.4\ mWb}$

Further problems on mutual inductance may be found in Section 9.8 following, problems 19 to 22, page 111.

9.8 Further problems on electromagnetic induction

Induced e.m.f.

1 A conductor of length 15 cm is moved at 750 mm/s at right-angles to a uniform flux density of 1.2 T. Determine the e.m.f. induced in the conductor. [0.135 V]

2 Find the speed that a conductor of length 120 mm must be moved at right angles to a magnetic field of flux density 0.6 T to induce in it an e.m.f. of 1.8 V. [25 m/s]

3 A 25 cm long conductor moves at a uniform speed of 8 m/s through a uniform magnetic field of flux density 1.2 T. Determine the current flowing in the conductor when (a) its ends are open-circuited, (b) its ends are connected to a load of 15 ohms resistance. [(a) 0 (b) 0.16 A]

4 A car is travelling at 80 km/h. Assuming the back axle of the car is 1.76 m in length and the vertical component of the earth's magnetic field is 40 μT, find the e.m.f. generated in the axle due to motion. [1.56 mV]

5 A conductor moves with a velocity of 20 m/s at an angle of (a) 90° (b) 45° (c) 30°, to a magnetic field produced between two square-faced poles of side length 2.5 cm. If the flux on the pole face is 60 mWb, find the magnitude of the induced e.m.f. in each case. [(a) 48 V (b) 33.9 V (c) 24 V]

Inductance

6 Find the e.m.f. induced in a coil of 200 turns when there is a change of flux of 30 mWb linking with it in 40 ms. [−150 V]

7 An e.m.f. of 25 V is induced in a coil of 300 turns when the flux linking with it changes by 12 mWb. Find the time, in milliseconds, in which the flux makes the change. [144 ms]

8 An ignition coil having 10 000 turns has an e.m.f. of 8 kV induced in it. What rate of change of flux is required for this to happen? [0.8 Wb/s]

9 A flux of 0.35 mWb passing through a 125-turn coil is reversed in 25 ms. Find the magnitude of the average e.m.f. induced. [3.5 V]

Energy stored

10 Calculate the value of the energy stored when a current of 30 mA is flowing in a coil of inductance 400 mH. [0.18 mJ]

11 The energy stored in the magnetic field of an inductor is 80 J when the current flowing in the inductor is 2 A. Calculate the inductance of the coil. [40 H]

Inductance of a coil

12 A flux of 30 mWb links with a 1200 turn coil when a current of 5 A is passing through the coil. Calculate (a) the inductance of the coil, (b) the energy stored in the magnetic field, and (c) the average e.m.f. induced if the current is reduced to zero in 0.20 s. [(a) 7.2 H (b) 90 J (c) 180 V]

13 An e.m.f. of 2 kV is induced in a coil when a current of 5 A collapses uniformly to zero in 10 ms. Determine the inductance of the coil. [4 H]

14 An average e.m.f. of 60 V is induced in a coil of inductance 160 mH when a current of 7.5 A is reversed. Calculate the time taken for the current to reverse. [40 ms]

15 A coil of 2500 turns has a flux of 10 mWb linking with it when carrying a current of 2 A. Calculate the coil inductance and the e.m.f. induced in the coil when the current collapses to zero in 20 ms. [12.5 H, 1.25 kV]

16 A coil is wound with 600 turns and has a self inductance of 2.5 H. What current must flow to set up a flux of 20 mWb?
[4.8 A]

17 When a current of 2 A flows in a coil, the flux linking with the coil is 80 μWb. If the coil inductance is 0.5 H, calculate the number of turns of the coil. [12 500]

18 A steady current of 5 A when flowing in a coil of 1000 turns produces a magnetic flux of 500 μWb. Calculate the inductance of the coil. The current of 5 A is then reversed in 12.5 ms. Calculate the e.m.f. induced in the coil. [0.1 H, 80 V]

Mutual inductance

19 The mutual inductance between two coils is 150 mH. Find the magnitude of the e.m.f. induced in one coil when the current in the other is increasing at a rate of 30 A/s. [4.5 V]

20 Determine the mutual inductance between two coils when a current changing at 50 A/s in one coil induces an e.m.f. of 80 mV in the other. [1.6 mH]

21 Two coils have a mutual inductance of 0.75 H. Calculate the magnitude of the e.m.f. induced in one coil when a current of 2.5 A in the other coil is reversed in 15 ms. [250 V]

22 The mutual inductance between two coils is 240 mH. If the current in one coil changes from 15 A to 6 A in 12 ms, calculate (a) the average e.m.f. induced in the other coil, (b) the change of flux linked with the other coil if it is wound with 400 turns.
[(a) −180 V (b) 5.4 mWb]

10 Electrical measuring instruments and measurements

At the end of this chapter you should be able to:

- recognize the importance of testing and measurements in electric circuits
- appreciate the essential devices comprising an analogue instrument
- explain the operation of an attraction and a repulsion type of moving-iron instrument
- explain the operation of a moving coil rectifier instrument
- compare moving coil, moving iron and moving coil rectifier instruments
- calculate values of shunts for ammeters and multipliers for voltmeters
- understand the advantages of electronic instruments
- understand the operation of an ohmmeter/megger
- appreciate the operation of multimeters/Avometers
- understand the operation of a wattmeter
- appreciate instrument 'loading' effect
- understand the operation of a C.R.O. for d.c. and a.c. measurements
- calculate periodic time, frequency, peak to peak values from waveforms on a C.R.O.
- recognize harmonics present in complex waveforms
- determine ratios of powers, currents and voltages in decibels
- understand null methods of measurement for a Wheatstone bridge and d.c. potentiometer
- understand the operation of a.c. bridges
- understand the operation of a Q-meter
- appreciate the most likely source of errors in measurements
- appreciate calibration accuracy of instruments

10.1 Introduction

Tests and measurements are important in designing, evaluating, maintaining and servicing electrical circuits and equipment. In order to detect electrical quantities such as current, voltage, resistance or power, it is necessary to transform an electrical quantity or condition into a visible indication. This is done with the aid of instruments (or meters) that indicate the magnitude of quantities either by the position of a pointer moving over a graduated scale (called an analogue instrument) or in the form of a decimal number (called a digital instrument).

10.2 Analogue instruments

All analogue electrical indicating instruments require three essential devices:

(a) A **deflecting or operating device**. A mechanical force is produced by the current or voltage which causes the pointer to deflect from its zero position.

(b) **A controlling device**. The controlling force acts in opposition to the deflecting force and ensures that the deflection shown on the meter is always the same for a given measured quantity. It also prevents the pointer always going to the maximum deflection. There are two main types of controlling device — spring control and gravity control.

(c) **A damping device**. The damping force ensures that the pointer comes to rest in its final position quickly and without undue oscillation. There are three main types of damping used — eddy-current damping, air-friction damping and fluid-friction damping.

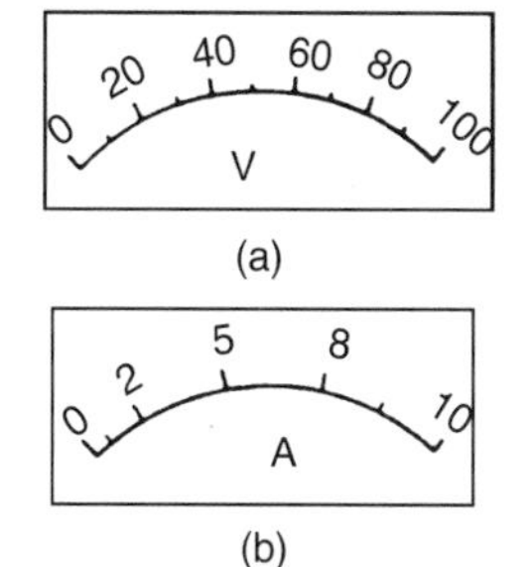

Figure 10.1

There are basically two types of scale — linear and non-linear.

A **linear scale** is shown in Figure 10.1(a), where the divisions or graduations are evenly spaced. The voltmeter shown has a range 0–100 V, i.e. a full-scale deflection (f.s.d.) of 100 V. A **non-linear scale** is shown in Figure 10.1(b). The scale is cramped at the beginning and the graduations are uneven throughout the range. The ammeter shown has a f.s.d. of 10 A.

10.3 Moving-iron instrument

(a) An **attraction type** of moving-iron instrument is shown diagrammatically in Figure 10.2(a). When current flows in the solenoid, a pivoted soft-iron disc is attracted towards the solenoid and the movement causes a pointer to move across a scale.

(b) In the **repulsion type** moving-iron instrument shown diagrammatically in Figure 10.2(b), two pieces of iron are placed inside the solenoid, one being fixed, and the other attached to the spindle carrying the pointer. When current passes through the solenoid, the two pieces of iron are magnetized in the same direction and therefore repel each other. The pointer thus moves across the scale. The force moving the pointer is, in each type, proportional to I^2. Because

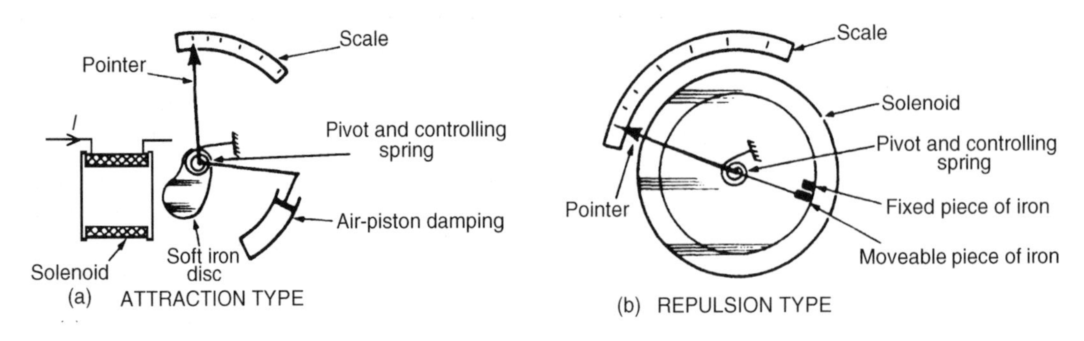

Figure 10.2

of this the direction of current does not matter and the moving-iron instrument can be used on d.c. or a.c. The scale, however, is non-linear.

10.4 The moving-coil rectifier instrument

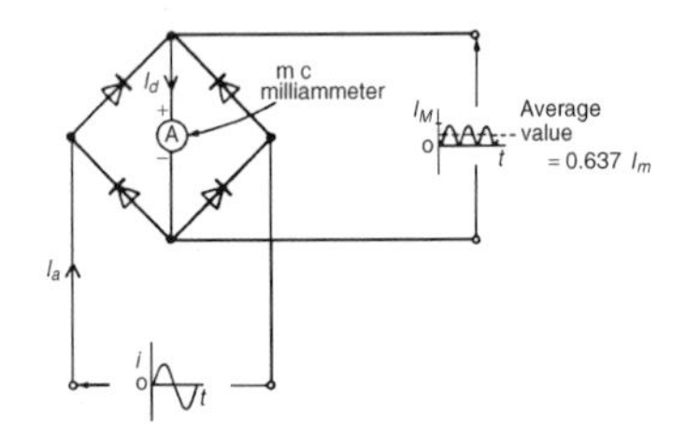

Figure 10.3

A moving-coil instrument, which measures only d.c., may be used in conjunction with a bridge rectifier circuit as shown in Figure 10.3 to provide an indication of alternating currents and voltages (see Chapter 14). The average value of the full wave rectified current is $0.637\,I_m$. However, a meter being used to measure a.c. is usually calibrated in r.m.s. values. For sinusoidal quantities the indication is $(0.707\,I_m)/(0.637\,I_m)$ i.e. 1.11 times the mean value. Rectifier instruments have scales calibrated in r.m.s. quantities and it is assumed by the manufacturer that the a.c. is sinusoidal.

10.5 Comparison of moving-coil, moving-iron and moving-coil rectifier instruments

Type of instrument	*Moving-coil*	*Moving-iron*	*Moving-coil rectifier*
Suitable for measuring	Direct current and voltage	Direct and alternating currents and voltage (reading in rms value)	Alternating current and voltage (reads average value but scale is adjusted to give rms value for sinusoidal waveforms)
Scale	Linear	Non-linear	Linear
Method of control	Hairsprings	Hairsprings	Hairsprings
Method of damping	Eddy current	Air	Eddy current

Type of instrument	*Moving-coil*	*Moving-iron*	*Moving-coil rectifier*
Frequency limits	—	20–200 Hz	20–100 kHz
Advantages	1 Linear scale 2 High sensitivity 3 Well shielded from stray magnetic fields 4 Lower power consumption	1 Robust construction 2 Relatively cheap 3 Measures dc and ac 4 In frequency range 20–100 Hz reads rms correctly regardless of supply wave-form	1 Linear scale 2 High sensitivity 3 Well shielded from stray magnetic fields 4 Low power consumption 5 Good frequency range
Disadvantages	1 Only suitable for dc 2 More expensive than moving iron type 3 Easily damaged	1 Non-linear scale 2 Affected by stray magnetic fields 3 Hysteresis errors in dc circuits 4 Liable to temperature errors 5 Due to the inductance of the solenoid, readings can be affected by variation of frequency	1 More expensive than moving iron type 2 Errors caused when supply is non-sinusoidal

(For the principle of operation of a moving-coil instrument, see Chapter 8, page 97).

10.6 Shunts and multipliers

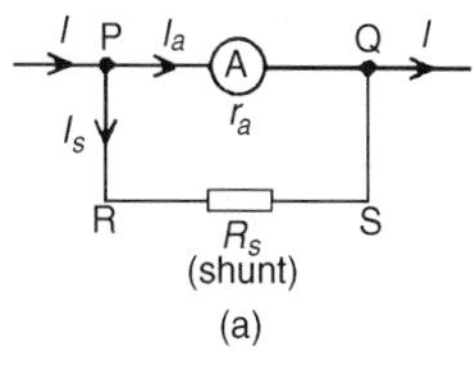

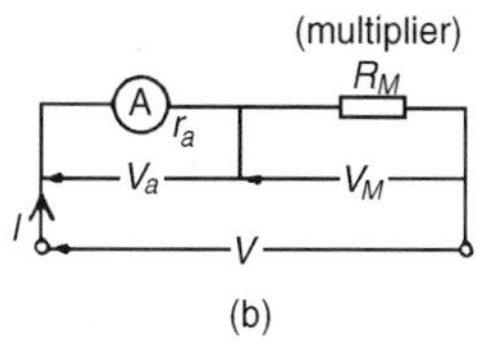

Figure 10.4

An **ammeter**, which measures current, has a low resistance (ideally zero) and must be connected in series with the circuit.

A **voltmeter**, which measures p.d., has a high resistance (ideally infinite) and must be connected in parallel with the part of the circuit whose p.d. is required.

There is no difference between the basic instrument used to measure current and voltage since both use a milliammeter as their basic part. This is a sensitive instrument which gives f.s.d. for currents of only a few milliamperes. When an ammeter is required to measure currents of larger magnitude, a proportion of the current is diverted through a low-value resistance connected in parallel with the meter. Such a diverting resistor is called a **shunt**.

From Figure 10.4(a), $V_{PQ} = V_{RS}$. Hence $I_a r_a = I_S R_S$

Thus the value of the shunt, $\boldsymbol{R_s = \frac{I_a r_a}{I_s}}$ **ohms**

The milliammeter is converted into a voltmeter by connecting a high value resistance (called a **multiplier**) in series with it as shown in Figure 10.4(b). From Figure 10.4(b), $V = V_a + V_M = Ir_a + IR_M$

Thus the value of the multiplier, $\boldsymbol{R_M = \dfrac{V - Ir_a}{I}}$ **ohms**

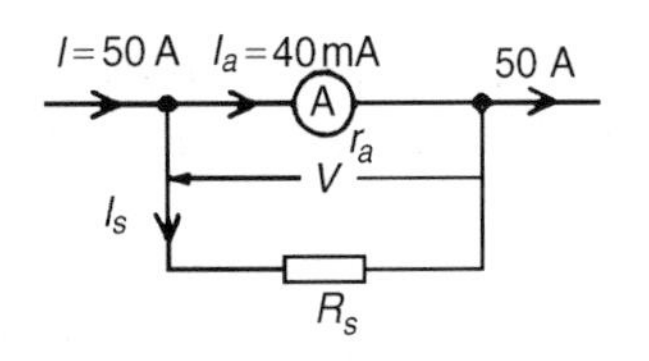

Figure 10.5

Problem 1. A moving-coil instrument gives a f.s.d. when the current is 40 mA and its resistance is 25 Ω. Calculate the value of the shunt to be connected in parallel with the meter to enable it to be used as an ammeter for measuring currents up to 50 A.

The circuit diagram is shown in Figure 10.5,

where r_a = resistance of instrument = 25 Ω,

R_s = resistance of shunt,

I_a = maximum permissible current flowing in instrument

= 40 mA = 0.04 A,

I_s = current flowing in shunt,

I = total circuit current required to give f.s.d. = 50 A

Since $I = I_a + I_s$ then $I_s = I - I_a = 50 - 0.04 = 49.96$ A

$V = I_a r_a = I_s R_s$

Hence $R_s = \dfrac{I_a r_a}{I_s} = \dfrac{(0.04)(25)}{49.96} = 0.02002\ \Omega =$ **20.02 mΩ**

Thus for the moving-coil instrument to be used as an ammeter with a range 0–50 A, a resistance of value 20.02 mΩ needs to be connected in parallel with the instrument.

Problem 2. A moving-coil instrument having a resistance of 10 Ω, gives a f.s.d. when the current is 8 mA. Calculate the value of the multiplier to be connected in series with the instrument so that it can be used as a voltmeter for measuring p.d.s. up to 100 V.

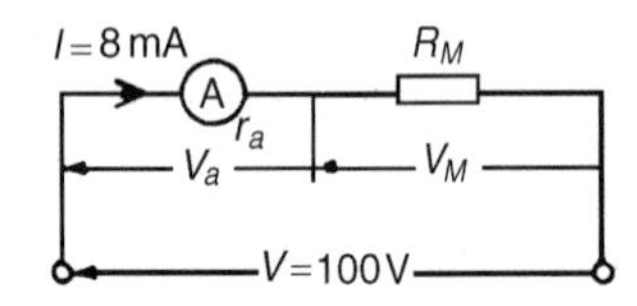

Figure 10.6

The circuit diagram is shown in Figure 10.6,

where r_a = resistance of instrument = 10 Ω,

R_M = resistance of multiplier,

$I =$ total permissible instrument current $= 8$ mA $= 0.008$ A,

$V =$ total p.d. required to give f.s.d. $= 100$ V

$V = V_a + V_M = Ir_a + IR_M$

i.e. $100 = (0.008)(10) + (0.008)\ R_M$, or $100 - 0.08 = 0.008\ R_M$

thus $R_M = \dfrac{99.92}{0.008} = 12\,490\ \Omega =$ **12.49 kΩ**

Hence for the moving-coil instrument to be used as a voltmeter with a range 0–100 V, a resistance of value 12.49 kΩ needs to be connected in series with the instrument.

Further problems on shunts and multipliers may be found in Section 10.20, problems 1 to 4, page 133.

10.7 Electronic instruments

Electronic measuring instruments have advantages over instruments such as the moving-iron or moving-coil meters, in that they have a much higher input resistance (some as high as 1000 MΩ) and can handle a much wider range of frequency (from d.c. up to MHz).

The digital voltmeter (DVM) is one which provides a digital display of the voltage being measured. Advantages of a DVM over analogue instruments include higher accuracy and resolution, no observational or parallex errors (see Section 10.20) and a very high input resistance, constant on all ranges.

A digital multimeter is a DVM with additional circuitry which makes it capable of measuring a.c. voltage, d.c. and a.c. current and resistance.

Instruments for a.c. measurements are generally calibrated with a sinusoidal alternating waveform to indicate r.m.s. values when a sinusoidal signal is applied to the instrument. Some instruments, such as the moving-iron and electro-dynamic instruments, give a true r.m.s. indication. With other instruments the indication is either scaled up from the mean value (such as with the rectifier moving-coil instrument) or scaled down from the peak value.

Sometimes quantities to be measured have complex waveforms (see Section 10.13), and whenever a quantity is non-sinusoidal, errors in instrument readings can occur if the instrument has been calibrated for sine waves only.

Such waveform errors can be largely eliminated by using electronic instruments.

10.8 The ohmmeter

An **ohmmeter** is an instrument for measuring electrical resistance.

A simple ohmmeter circuit is shown in Figure 10.7(a). Unlike the ammeter or voltmeter, the ohmmeter circuit does not receive the energy

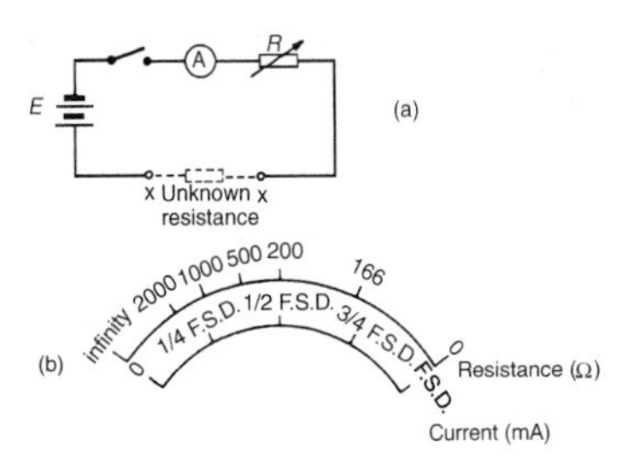

Figure 10.7

necessary for its operation from the circuit under test. In the ohmmeter this energy is supplied by a self-contained source of voltage, such as a battery. Initially, terminals XX are short-circuited and R adjusted to give f.s.d. on the milliammeter. If current I is at a maximum value and voltage E is constant, then resistance $R = E/I$ is at a minimum value. Thus f.s.d. on the milliammeter is made zero on the resistance scale. When terminals XX are open circuited no current flows and $R(= E/O)$ is infinity, ∞

The milliammeter can thus be calibrated directly in ohms. A cramped (non-linear) scale results and is 'back to front', as shown in Figure 10.7(b). When calibrated, an unknown resistance is placed between terminals XX and its value determined from the position of the pointer on the scale. An ohmmeter designed for measuring low values of resistance is called a **continuity tester**. An ohmmeter designed for measuring high values of resistance (i.e. megohms) is called an **insulation resistance tester** (e.g. **'Megger'**).

10.9 Multimeters

Instruments are manufactured that combine a moving-coil meter with a number of shunts and series multipliers, to provide a range of readings on a single scale graduated to read current and voltage. If a battery is incorporated then resistance can also be measured. Such instruments are called **multimeters** or **universal instruments** or **multirange instruments.** An 'Avometer' is a typical example. A particular range may be selected either by the use of separate terminals or by a selector switch. Only one measurement can be performed at a time. Often such instruments can be used in a.c. as well as d.c. circuits when a rectifier is incorporated in the instrument.

10.10 Wattmeters

A **wattmeter** is an instrument for measuring electrical power in a circuit. Figure 10.8 shows typical connections of a wattmeter used for measuring power supplied to a load. The instrument has two coils:

(i) a current coil, which is connected in series with the load, like an ammeter, and

(ii) a voltage coil, which is connected in parallel with the load, like a voltmeter.

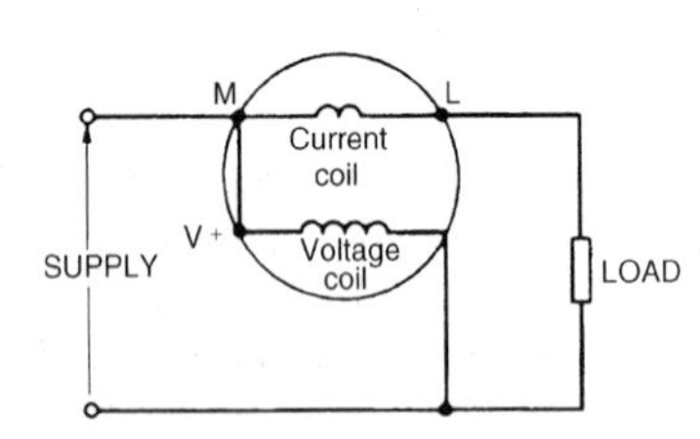

Figure 10.8

10.11 Instrument 'loading' effect

Some measuring instruments depend for their operation on power taken from the circuit in which measurements are being made. Depending on the 'loading' effect of the instrument (i.e. the current taken to enable it to operate), the prevailing circuit conditions may change.

The resistance of voltmeters may be calculated since each have a stated sensitivity (or 'figure of merit'), often stated in 'kΩ per volt' of f.s.d. A voltmeter should have as high a resistance as possible (— ideally infinite).

In a.c. circuits the impedance of the instrument varies with frequency and thus the loading effect of the instrument can change.

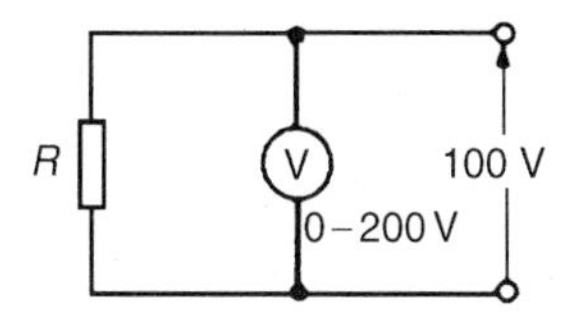

Figure 10.9

Problem 3. Calculate the power dissipated by the voltmeter and by resistor R in Figure 10.9 when (a) $R = 250\ \Omega$ (b) $R = 2\ \text{M}\Omega$. Assume that the voltmeter sensitivity (sometimes called figure of merit) is $10\ \text{k}\Omega/\text{V}$.

(a) Resistance of voltmeter, R_v = sensitivity × f.s.d.

Hence, $R_v = (10\ \text{k}\Omega/\text{V}) \times (200\ \text{V}) = 2000\ \text{k}\Omega = 2\ \text{M}\Omega$

Current flowing in voltmeter, $I_v = \dfrac{V}{R_v} = \dfrac{100}{2 \times 10^6} = 50 \times 10^{-6}\text{A}$

Power dissipated by voltmeter $= VI_v = (100)(50 \times 10^{-6}) = \mathbf{5\ mW}$

When $R = 250\ \Omega$, current in resistor, $I_R = \dfrac{V}{R} = \dfrac{100}{250} = \mathbf{0.4\ A}$

Power dissipated in load resistor $R = VI_R = (100)(0.4) = \mathbf{40\ W}$

Thus the power dissipated in the voltmeter is insignificant in comparison with the power dissipated in the load.

(b) When $R = 2\ \text{M}\Omega$, current in resistor, $I_R = \dfrac{V}{R} = \dfrac{100}{2 \times 10^6}$
$= 50 \times 10^{-6}\text{A}$

Power dissipated in load resistor $R = VI_R = 100 \times 50 \times 10^{-6}$
$= \mathbf{5\ mW}$

In this case the higher load resistance reduced the power dissipated such that the voltmeter is using as much power as the load.

Problem 4. An ammeter has a f.s.d. of 100 mA and a resistance of 50 Ω. The ammeter is used to measure the current in a load of resistance 500 Ω when the supply voltage is 10 V. Calculate (a) the ammeter reading expected (neglecting its resistance), (b) the actual current in the circuit, (c) the power dissipated in the ammeter, and (d) the power dissipated in the load.

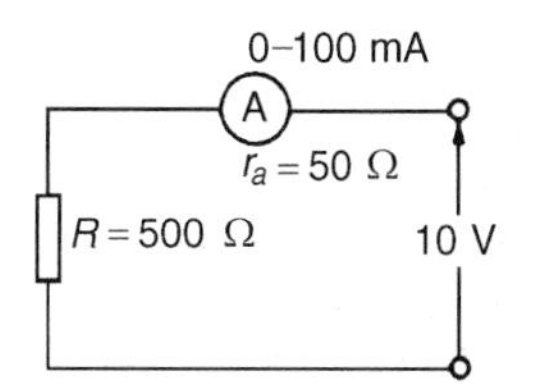

Figure 10.10

From Figure 10.10,

(a) expected ammeter reading $= \dfrac{V}{R} = \dfrac{10}{500} = \mathbf{20\ mA}$

(b) Actual ammeter reading $= \dfrac{V}{R + r_a} = \dfrac{10}{500 + 50} = \mathbf{18.18\ mA}$

Thus the ammeter itself has caused the circuit conditions to change from 20 mA to 18.18 mA

(c) Power dissipated in the ammeter $= I^2 r_a = (18.18 \times 10^{-3})^2(50)$

$= \mathbf{16.53\ mW}$

(d) Power dissipated in the load resistor $= I^2 R = (18.18 \times 10^{-3})^2(500)$

$= \mathbf{165.3\ mW}$

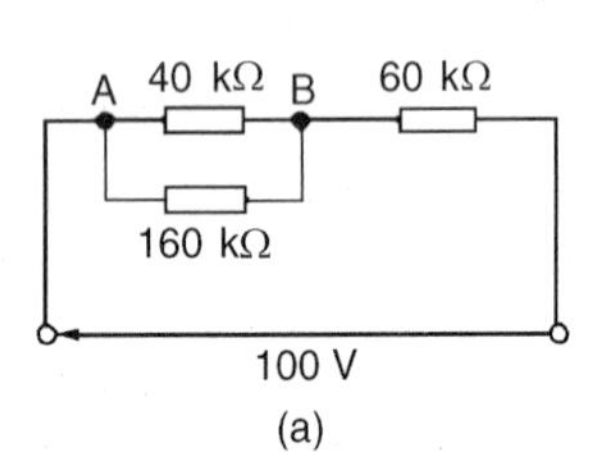

Figure 10.11

Problem 5. A voltmeter having a f.s.d. of 100 V and a sensitivity of 1.6 kΩ/V is used to measure voltage V_1 in the circuit of Figure 10.11. Determine (a) the value of voltage V_1 with the voltmeter not connected, and (b) the voltage indicated by the voltmeter when connected between A and B.

(a) By voltage division, $V_1 = \left(\dfrac{40}{40+60}\right)100 = \mathbf{40\ V}$

(b) The resistance of a voltmeter having a 100 V f.s.d. and sensitivity $1.6\ k\Omega/V$ is $100\ V \times 1.6\ k\Omega/V = 160\ k\Omega$.

When the voltmeter is connected across the 40 kΩ resistor the circuit is as shown in Figure 10.12(a) and the equivalent resistance of the parallel network is given by

$$\left(\frac{40 \times 160}{40+160}\right) \text{k}\Omega \text{ i.e. } \left(\frac{40 \times 160}{200}\right) \text{k}\Omega = 32\ \text{k}\Omega$$

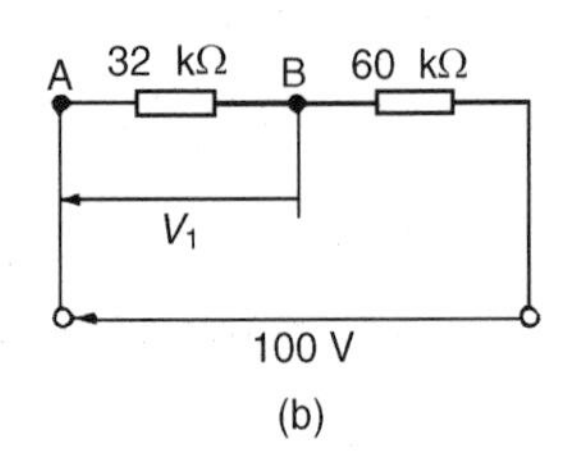

Figure 10.12

The circuit is now effectively as shown in Figure 10.12(b).

Thus the voltage indicated on the voltmeter is

$$\left(\frac{32}{32+60}\right) 100\ \text{V} = \mathbf{34.78\ V}$$

A considerable error is thus caused by the loading effect of the voltmeter on the circuit. The error is reduced by using a voltmeter with a higher sensitivity.

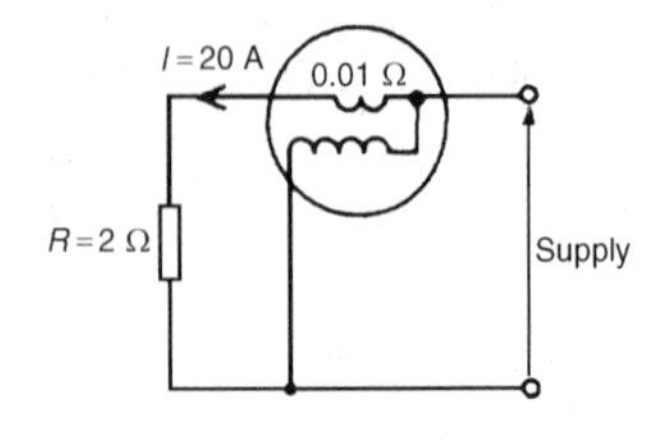

Figure 10.13

Problem 6. (a) A current of 20 A flows through a load having a resistance of 2 Ω. Determine the power dissipated in the load. (b) A wattmeter, whose current coil has a resistance of 0.01 Ω is connected as shown in Figure 10.13. Determine the wattmeter reading.

(a) Power dissipated in the load, $P = I^2 R = (20)^2(2) = \mathbf{800\ W}$

(b) With the wattmeter connected in the circuit the total resistance R_T is $2 + 0.01 = 2.01\ \Omega$

The wattmeter reading is thus $I^2R_T = (20)^2(2.01) = \mathbf{804\ W}$

Further problems on instrument 'loading' effects may be found in Section 10.20, problems 5 to 7, page 134.

10.12 The cathode ray oscilloscope

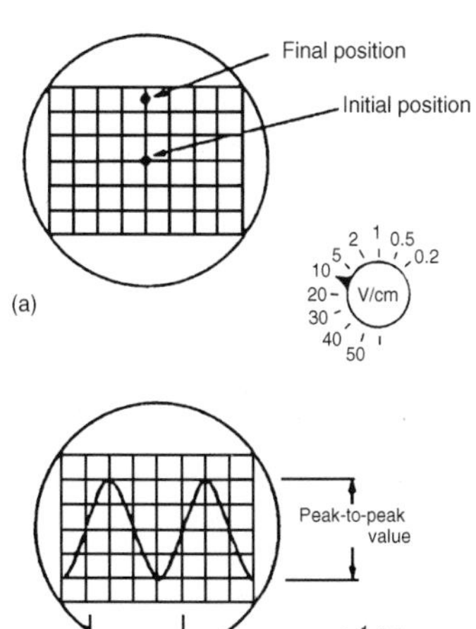

Figure 10.14

The **cathode ray oscilloscope** (c.r.o.) may be used in the observation of waveforms and for the measurement of voltage, current, frequency, phase and periodic time. For examining periodic waveforms the electron beam is deflected horizontally (i.e. in the X direction) by a sawtooth generator acting as a timebase. The signal to be examined is applied to the vertical deflection system (Y direction) usually after amplification.

Oscilloscopes normally have a transparent grid of 10 mm by 10 mm squares in front of the screen, called a graticule. Among the timebase controls is a 'variable' switch which gives the sweep speed as time per centimetre. This may be in s/cm, ms/cm or μs/cm, a large number of switch positions being available. Also on the front panel of a c.r.o. is a Y amplifier switch marked in volts per centimetre, with a large number of available switch positions.

(i) With **direct voltage measurements,** only the Y amplifier 'volts/cm' switch on the c.r.o. is used. With no voltage applied to the Y plates the position of the spot trace on the screen is noted. When a direct voltage is applied to the Y plates the new position of the spot trace is an indication of the magnitude of the voltage. For example, in Figure 10.14(a), with no voltage applied to the Y plates, the spot trace is in the centre of the screen (initial position) and then the spot trace moves 2.5 cm to the final position shown, on application of a d.c. voltage. With the 'volts/cm' switch on 10 volts/cm the magnitude of the direct voltage is 2.5 cm × 10 volts/cm, i.e. 25 volts.

(ii) With **alternating voltage measurements**, let a sinusoidal waveform be displayed on a c.r.o. screen as shown in Figure 10.14(b). If the time/cm switch is on, say, 5 ms/cm then the **periodic time *T*** of the sinewave is 5 ms/cm × 4 cm, i.e. **20 ms or 0.02 s**

$$\text{Since frequency } f = \frac{1}{T}, \ \textbf{frequency} = \frac{\mathbf{1}}{\mathbf{0.02}} = \mathbf{50\ Hz}$$

If the 'volts/cm' switch is on, say, 20 volts/cm then the **amplitude** or **peak value** of the sinewave shown is 20 volts/cm × 2 cm, i.e. 40 V.

$$\text{Since r.m.s. Voltage} = \frac{\text{peak voltage}}{\sqrt{2}}, \ \text{(see Chapter 14)},$$

$$\textbf{r.m.s. voltage} = \frac{40}{\sqrt{2}} = \mathbf{28.28\ volts}$$

Double beam oscilloscopes are useful whenever two signals are to be compared simultaneously.

The c.r.o. demands reasonable skill in adjustment and use. However its greatest advantage is in observing the shape of a waveform — a feature not possessed by other measuring instruments.

Problem 7. Describe how a simple c.r.o. is adjusted to give (a) a spot trace, (b) a continuous horizontal trace on the screen, explaining the functions of the various controls.

(a) To obtain a spot trace on a typical c.r.o. screen:

(i) Switch on the c.r.o.

(ii) Switch the timebase control to off. This control is calibrated in time per centimetres — for example, 5 ms/cm or 100 μs/cm. Turning it to zero ensures no signal is applied to the X-plates. The Y-plate input is left open-circuited.

(iii) Set the intensity, X-shift and Y-shift controls to about the mid-range positions.

(iv) A spot trace should now be observed on the screen. If not, adjust either or both of the X and Y-shift controls. The X-shift control varies the position of the spot trace in a horizontal direction whilst the Y-shift control varies its vertical position.

(v) Use the X and Y-shift controls to bring the spot to the centre of the screen and use the focus control to focus the electron beam into a small circular spot.

(b) To obtain a continuous horizontal trace on the screen the same procedure as in (a) is initially adopted. Then the timebase control is switched to a suitable position, initially the millisecond timebase range, to ensure that the repetition rate of the sawtooth is sufficient for the persistence of the vision time of the screen phosphor to hold a given trace.

Problem 8. For the c.r.o. square voltage waveform shown in Figure 10.15 determine (a) the periodic time, (b) the frequency and (c) the peak-to-peak voltage. The 'time/cm' (or timebase control) switch is on 100 μs/cm and the 'volts/cm' (or signal amplitude control) switch is on 20 V/cm.

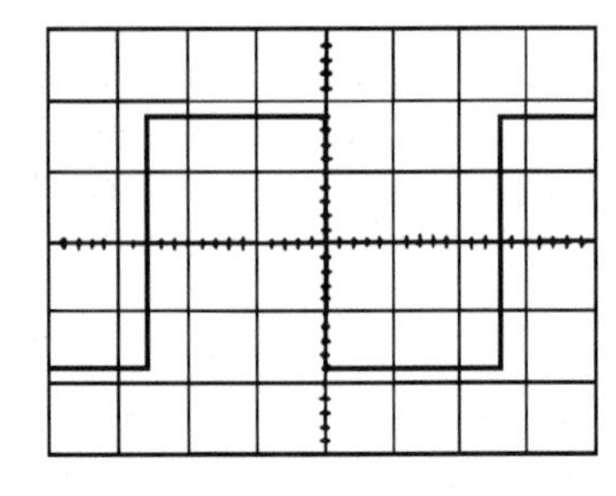

Figure 10.15

(In Figures 10.15 to 10.18 assume that the squares shown are 1 cm by 1 cm)

(a) The width of one complete cycle is 5.2 cm

Hence the periodic time, $T = 5.2\text{ cm} \times 100 \times 10^{-6}\text{ s/cm} = \mathbf{0.52\ ms}$

(b) Frequency, $f = \dfrac{1}{T} = \dfrac{1}{0.52 \times 10^{-3}} = \mathbf{1.92\ kHz}$

(c) The peak-to-peak height of the display is 3.6 cm, hence the peak-to-peak voltage $= 3.6\ \text{cm} \times 20\ \text{V/cm} = \mathbf{72\ V}$

Figure 10.16

Problem 9. For the c.r.o. display of a pulse waveform shown in Figure 10.16 the 'time/cm' switch is on 50 ms/cm and the 'volts/cm' switch is on 0.2 V/cm. Determine (a) the periodic time, (b) the frequency, (c) the magnitude of the pulse voltage.

(a) The width of one complete cycle is 3.5 cm

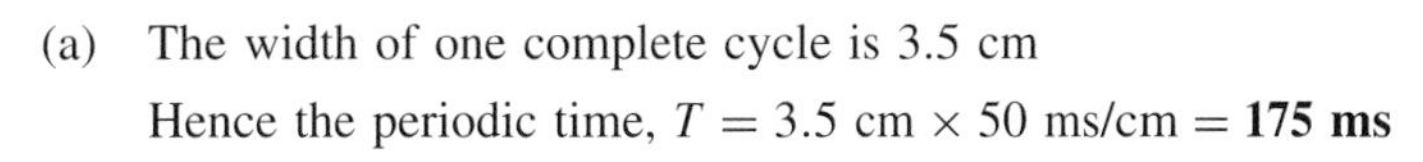

Hence the periodic time, $T = 3.5\ \text{cm} \times 50\ \text{ms/cm} = \mathbf{175\ ms}$

(b) Frequency, $f = \dfrac{1}{T} = \dfrac{1}{0.175} = \mathbf{5.71\ Hz}$

(c) The height of a pulse is 3.4 cm hence the magnitude of the pulse voltage $= 3.4\ \text{cm} \times 0.2\ \text{V/cm} = \mathbf{0.68\ V}$

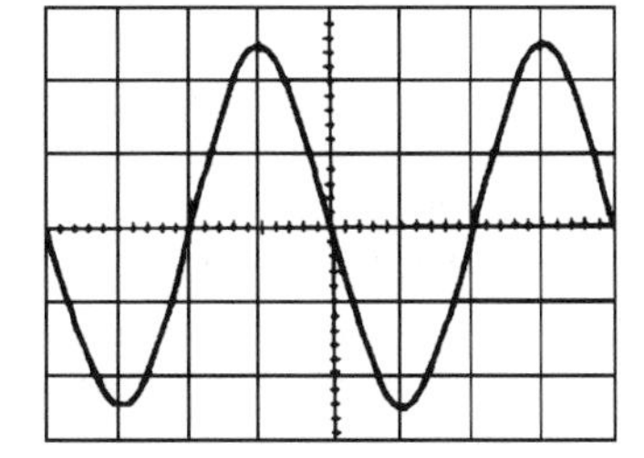

Figure 10.17

Problem 10. A sinusoidal voltage trace displayed by a c.r.o. is shown in Figure 10.17. If the 'time/cm' switch is on 500 μs/cm and the 'volts/cm' switch is on 5 V/cm, find, for the waveform, (a) the frequency, (b) the peak-to-peak voltage, (c) the amplitude, (d) the r.m.s. value.

(a) The width of one complete cycle is 4 cm. Hence the periodic time, T is 4 cm × 500 μs/cm, i.e. 2 ms

Frequency, $f = \dfrac{1}{T} = \dfrac{1}{2 \times 10^{-3}} = \mathbf{500\ Hz}$

(b) The peak-to-peak height of the waveform is 5 cm. Hence the peak-to-peak voltage $= 5\ \text{cm} \times 5\ \text{V/cm} = \mathbf{25\ V}$

(c) Amplitude $\frac{1}{2} \times 25\ V = \mathbf{12.5\ V}$

(d) The peak value of voltage is the amplitude, i.e. 12.5 V.

$$\text{r.m.s voltage} = \frac{\text{peak voltage}}{\sqrt{2}} = \frac{12.5}{\sqrt{2}} = \mathbf{8.84\ V}$$

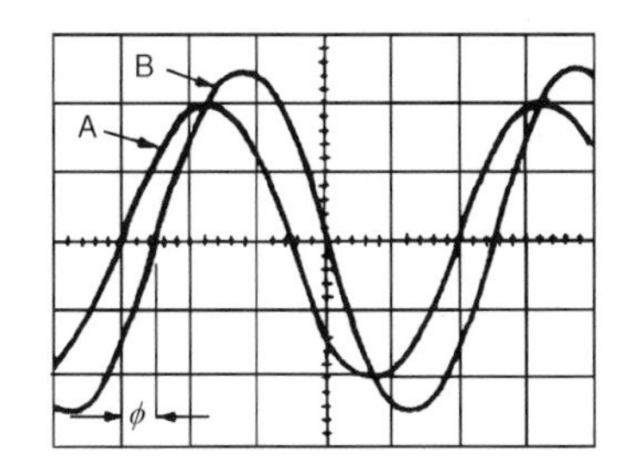

Figure 10.18

Problem 11. For the double-beam oscilloscope displays shown in Figure 10.18 determine (a) their frequency, (b) their r.m.s. values, (c) their phase difference. The 'time/cm' switch is on 100 μs/cm and the 'volts/cm' switch on 2 V/cm.

(a) The width of each complete cycle is 5 cm for both waveforms.

Hence the periodic time, T, of each waveform is 5 cm × 100 μs/cm, i.e. 0.5 ms.

Frequency of each waveform, $f = \frac{1}{T} = \frac{1}{0.5 \times 10^{-3}} = \mathbf{2\ kHz}$

(b) The peak value of waveform A is 2 cm × 2 V/cm = **4 V**,

hence the r.m.s. value of waveform A $= \frac{4}{\sqrt{2}} = \mathbf{2.83\ V}$

The peak value of waveform B is 2.5 cm × 2 V/cm = 5 V,

hence the r.m.s. value of waveform B $= \frac{5}{\sqrt{2}} = \mathbf{3.54\ V}$

(c) Since 5 cm represents 1 cycle, then 5 cm represents 360°,

i.e. 1 cm represents $\frac{360}{5} = 72°$.

The phase angle $\phi = 0.5$ cm $= 0.5$ cm $\times$ 72°/cm = 36°

Hence waveform A leads waveform B by 36°

Further problems on the c.r.o. may be found in Section 10.20, problems 8 to 10, page 134.

10.13 Waveform harmonics

(i) Let an instantaneous voltage v be represented by $v = V_m \sin 2\pi f t$ volts. This is a waveform which varies sinusoidally with time t, has a frequency f, and a maximum value V_m. Alternating voltages are usually assumed to have wave-shapes which are sinusoidal where only one frequency is present. If the waveform is not sinusoidal it is called a **complex wave**, and, whatever its shape, it may be split up mathematically into components called the **fundamental** and a number of **harmonics**. This process is called harmonic analysis. The fundamental (or first harmonic) is sinusoidal and has the supply frequency, f; the other harmonics are also sine waves having frequencies which are integer multiples of f. Thus, if the supply frequency is 50 Hz, then the third harmonic frequency is 150 Hz, the fifth 250 Hz, and so on.

(ii) A complex waveform comprising the sum of the fundamental and a third harmonic of about half the amplitude of the fundamental is shown in Figure 10.19(a), both waveforms being initially in phase with each other. If further odd harmonic waveforms of the appropriate amplitudes are added, a good approximation to a square wave results. In Figure 10.19(b), the third harmonic is shown having an initial phase displacement from the fundamental. The positive and negative half cycles of each of the complex waveforms shown in Figures 10.19(a) and (b) are identical in shape, and this is a feature of waveforms containing the fundamental and only odd harmonics.

(iii) A complex waveform comprising the sum of the fundamental and a second harmonic of about half the amplitude of the fundamental is shown in Figure 10.19(c), each waveform being initially in phase

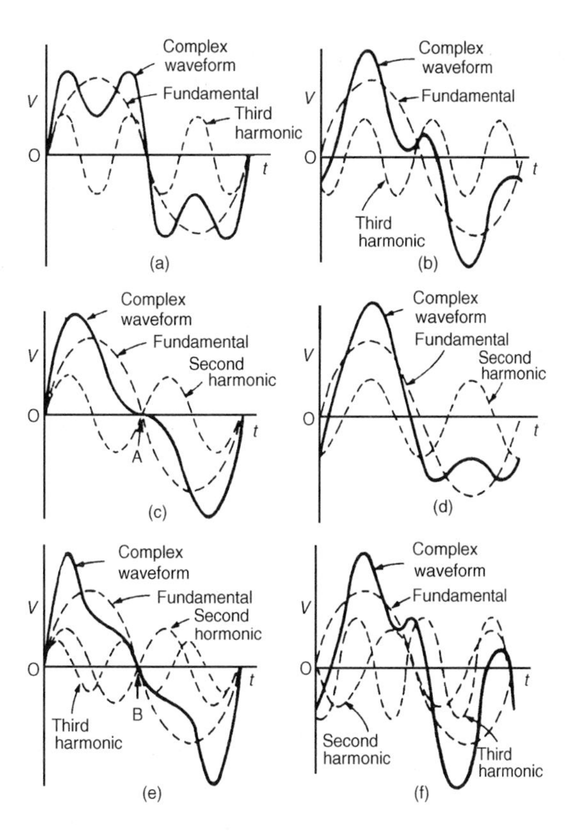

Figure 10.19

with each other. If further even harmonics of appropriate amplitudes are added a good approximation to a triangular wave results. In Figure 10.19(c) the negative cycle appears as a mirror image of the positive cycle about point A. In Figure 10.19(d) the second harmonic is shown with an initial phase displacement from the fundamental and the positive and negative half cycles are dissimilar.

(iv) A complex waveform comprising the sum of the fundamental, a second harmonic and a third harmonic is shown in Figure 10.19(e), each waveform being initially 'in-phase'. The negative half cycle appears as a mirror image of the positive cycle about point B. In Figure 10.19(f), a complex waveform comprising the sum of the fundamental, a second harmonic and a third harmonic are shown with initial phase displacement. The positive and negative half cycles are seen to be dissimilar.

The features mentioned relative to Figures 10.19(a) to (f) make it possible to recognize the harmonics present in a complex waveform displayed on a CRO.

More on complex waveforms may be found in Chapter 36, page 631.

10.14 Logarithmic ratios

In electronic systems, the ratio of two similar quantities measured at different points in the system, are often expressed in logarithmic units. By definition, if the ratio of two powers P_1 and P_2 is to be expressed in **decibel (dB) units** then the number of decibels, X, is given by:

$$\boxed{X = 10\ \lg\left(\frac{P_2}{P_1}\right)\ \text{dB}} \qquad (10.1)$$

Thus, when the power ratio, $\frac{P_2}{P_1} = 1$ then the decibel power ratio

$= 10\ \lg 1 = 0$

when the power ratio, $\frac{P_2}{P_1} = 100$ then the decibel power ratio

$= 10\ \lg 100 = +20$

(i.e. a power gain),

and when the power ratio, $\frac{P_2}{P_1} = \frac{1}{100}$ then the decibel power ratio

$= 10\ \lg \frac{1}{100} = -20$

(i.e. a power loss or attenuation).

Logarithmic units may also be used for voltage and current ratios.

Power, P, is given by $P = I^2R$ or $P = V^2/R$

Substituting in equation (10.1) gives:

$$X = 10\ \lg\left(\frac{I_2^2 R_2}{I_1^2 R_1}\right)\ \text{dB} \quad \text{or } X = 10\ \lg\left(\frac{V_2^2/R_2}{V_1^2/R_1}\right)\ \text{dB}$$

If $R_1 = R_2$ then $X = 10\ \lg\left(\frac{I_2^2}{I_1^2}\right)$ dB or $X = 10\ \lg\left(\frac{V_2^2}{V_1^2}\right)$ dB

i.e. $\boxed{X = 20\ \lg\left(\frac{I_2}{I_1}\right)\ \text{dB}}$ **or** $\boxed{X = 20\ \lg\left(\frac{V_2}{V_1}\right)\ \text{dB}}$

(from the laws of logarithms).

From equation (10.1), X decibels is a logarithmic ratio of two similar quantities and is not an absolute unit of measurement. It is therefore necessary to state a **reference level** to measure a number of decibels above or below that reference. The most widely used reference level for power is 1 mW, and when power levels are expressed in decibels, above or below the 1 mW reference level, the unit given to the new power level is dBm.

A voltmeter can be re-scaled to indicate the power level directly in decibels. The scale is generally calibrated by taking a reference level of 0 dB when a power of 1 mW is dissipated in a 600 Ω resistor (this being the natural impedance of a simple transmission line). The reference voltage V is then obtained from

$$P = \frac{V^2}{R}, \text{ i.e. } 1 \times 10^{-3} = \frac{V^2}{600} \text{ from which, } V = 0.775 \text{ volts.}$$

In general, the number of dBm, $X = 20 \lg\left(\frac{V}{0.775}\right)$

Thus $V = 0.20$ V corresponds to $20 \lg\left(\frac{0.20}{0.775}\right) = -11.77$ dBm and

$V = 0.90$ V corresponds to $20 \lg\left(\frac{0.90}{0.775}\right) = +1.3$ dBm, and so on.

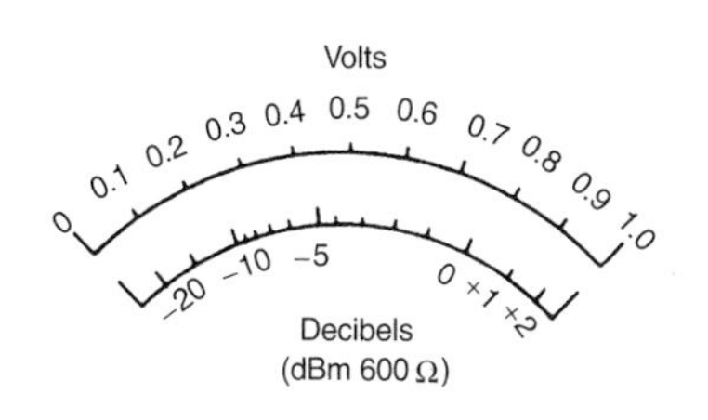

Figure 10.20

A typical **decibelmeter**, or **dB meter**, scale is shown in Figure 10.20. Errors are introduced with dB meters when the circuit impedance is not 600 Ω.

Problem 12. The ratio of two powers is (a) 3 (b) 20 (c) 400 (d) $\frac{1}{20}$ Determine the decibel power ratio in each case.

From above, the power ratio in decibels, X, is given by: $X = 10 \lg\left(\frac{P_2}{P_1}\right)$

(a) When $\frac{P_2}{P_1} = 3, X = 10 \lg(3) = 10(0.477) = \mathbf{4.77\ dB}$

(b) When $\frac{P_2}{P_1} = 20, X = 10 \lg(20) = 10(1.30) = \mathbf{13.0\ dB}$

(c) When $\frac{P_2}{P_1} = 400, X = 10 \lg(400) = 10(2.60) = \mathbf{26.0\ dB}$

(d) When $\frac{P_2}{P_1} = \frac{1}{20} = 0.05, X = 10 \lg(0.05) = 10(-1.30)$

$$= \mathbf{-13.0\ dB}$$

(a), (b) and (c) represent power gains and (d) represents a power loss or attenuation.

Problem 13. The current input to a system is 5 mA and the current output is 20 mA. Find the decibel current ratio assuming the input and load resistances of the system are equal.

From above, the decibel current ratio is $20 \lg\left(\frac{I_2}{I_1}\right) = 20 \lg\left(\frac{20}{5}\right)$

$$= 20 \lg 4$$

$$= 20(0.60)$$

$$= \mathbf{12\ dB\ gain}$$

Problem 14. 6% of the power supplied to a cable appears at the output terminals. Determine the power loss in decibels.

If P_1 = input power and P_2 = output power then $\dfrac{P_2}{P_1} = \dfrac{6}{100} = 0.06$

$$\text{Decibel power ratio} = 10 \lg\left(\frac{P_2}{P_1}\right) = 10 \lg(0.06) = 10(-1.222) = -12.22 \text{ dB}$$

Hence the decibel power loss, or attenuation, is 12.22 dB

Problem 15. An amplifier has a gain of 14 dB. Its input power is 8 mW. Find its output power.

$$\text{Decibel power ratio} = 10 \lg\left(\frac{P_2}{P_1}\right)$$ where P_1 = input power = 8 mW, and P_2 = output power

Hence $14 = 10 \lg\left(\dfrac{P_2}{P_1}\right)$

$$1.4 = \lg\left(\frac{P_2}{P_1}\right)$$

and $10^{1.4} = \dfrac{P_2}{P_1}$ from the definition of a logarithm

i.e. $25.12 = \dfrac{P_2}{P_1}$

Output power, $P_2 = 25.12P_1 = (25.12)(8) =$ **201 mW or 0.201 W**

Problem 16. The output voltage from an amplifier is 4 V. If the voltage gain is 27 dB, calculate the value of the input voltage assuming that the amplifier input resistance and load resistance are equal.

$$\text{Voltage gain in decibels} = 27 = 20 \lg\left(\frac{V_2}{V_1}\right) = 20 \lg\left(\frac{4}{V_1}\right)$$

Hence $\dfrac{27}{20} = \lg\left(\dfrac{4}{V_1}\right)$

$$1.35 = \lg\left(\frac{4}{V_1}\right)$$

$10^{1.35} = \dfrac{4}{V_1}$, from which $V_1 = \dfrac{4}{10^{1.35}} = \dfrac{4}{22.39} = 0.179 \text{ V}$

Hence the input voltage V_1 is 0.179 V

Further problems on logarithmic ratios may be found in Section 10.20, problems 11 to 17, page 134.

10.15 Null method of measurement

A **null method of measurement** is a simple, accurate and widely used method which depends on an instrument reading being adjusted to read zero current only. The method assumes:

(i) if there is any deflection at all, then some current is flowing;

(ii) if there is no deflection, then no current flows (i.e. a null condition).

Hence it is unnecessary for a meter sensing current flow to be calibrated when used in this way. A sensitive milliammeter or microammeter with centre zero position setting is called a **galvanometer**. Examples where the method is used are in the Wheatstone bridge (see Section 10.16), in the d.c. potentiometer (see Section 10.17) and with a.c. bridges (see Section 10.18).

10.16 Wheatstone bridge

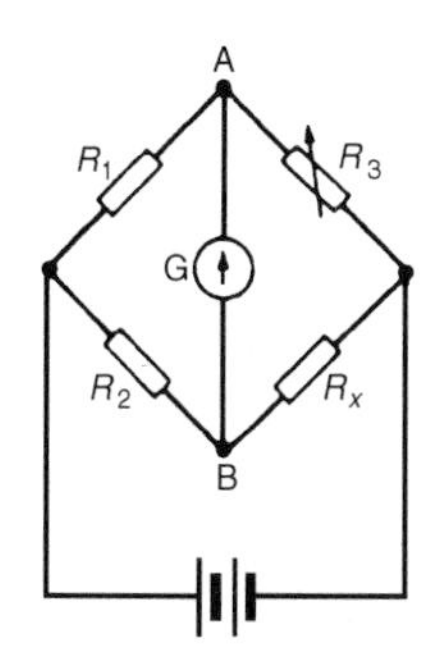

Figure 10.21

Figure 10.21 shows a Wheatstone bridge circuit which compares an unknown resistance R_x with others of known values, i.e. R_1 and R_2, which have fixed values, and R_3, which is variable. R_3 is varied until zero deflection is obtained on the galvanometer G. No current then flows through the meter, $V_A = V_B$, and the bridge is said to be 'balanced'.

At balance, $R_1R_x = R_2R_3$, i.e. $\boxed{\mathbf{R_x = \frac{R_2R_3}{R_1}\ \text{ohms}}}$

Problem 17. In a Wheatstone bridge ABCD, a galvanometer is connected between A and C, and a battery between B and D. A resistor of unknown value is connected between A and B. When the bridge is balanced, the resistance between B and C is 100 Ω, that between C and D is 10 Ω and that between D and A is 400 Ω. Calculate the value of the unknown resistance.

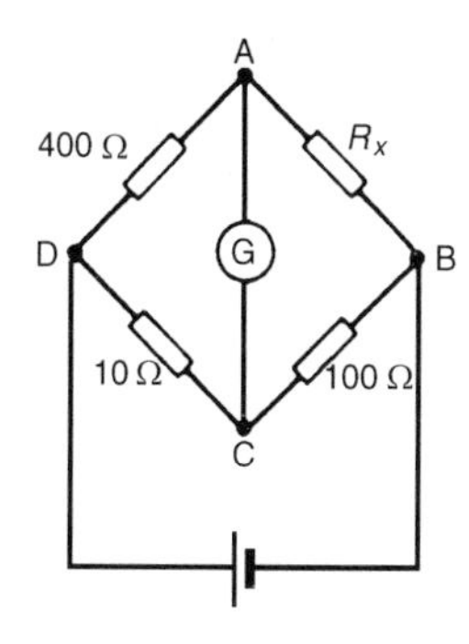

Figure 10.22

The Wheatstone bridge is shown in Figure 10.22 where R_x is the unknown resistance. At balance, equating the products of opposite ratio arms, gives:

$(R_x)(10) = (100)(400)$

$$\text{and } R_x = \frac{(100)(400)}{10} = 4000\ \Omega$$

Hence the unknown resistance, R_x = 4 kΩ

10.17 D.c. potentiometer

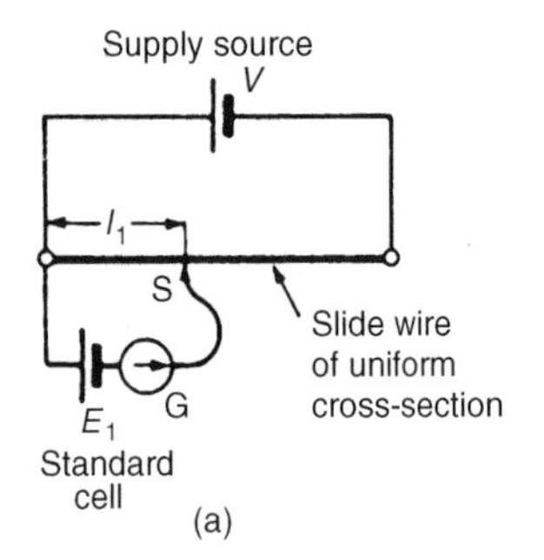

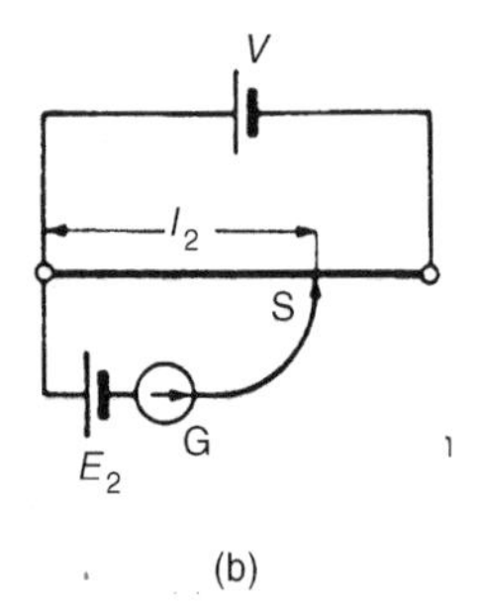

Figure 10.23

The **d.c. potentiometer** is a null-balance instrument used for determining values of e.m.f.'s and p.d.s. by comparison with a known e.m.f. or p.d. In Figure 10.23(a), using a standard cell of known e.m.f. E_1, the slider S is moved along the slide wire until balance is obtained (i.e. the galvanometer deflection is zero), shown as length l_1.

The standard cell is now replaced by a cell of unknown e.m.f. E_2 (see Figure 10.23(b)) and again balance is obtained (shown as l_2).

Since $E_1 \alpha l_1$ and $E_2 \alpha l_2$ then $\frac{E_1}{E_2} = \frac{l_1}{l_2}$ and $\boxed{\mathbf{E_2 = E_1\left(\frac{l_2}{l_1}\right)\ volts}}$

A potentiometer may be arranged as a resistive two-element potential divider in which the division ratio is adjustable to give a simple variable d.c. supply. Such devices may be constructed in the form of a resistive element carrying a sliding contact which is adjusted by a rotary or linear movement of the control knob.

> Problem 18. In a d.c. potentiometer, balance is obtained at a length of 400 mm when using a standard cell of 1.0186 volts. Determine the e.m.f. of a dry cell if balance is obtained with a length of 650 mm

$E_1 = 1.0186$ V, $l_1 = 400$ mm, $l_2 = 650$ mm

With reference to Figure 10.23, $\frac{E_1}{E_2} = \frac{l_1}{l_2}$

from which, $E_2 = E_1\left(\frac{l_2}{l_1}\right) = (1.0186)\left(\frac{650}{400}\right) = \mathbf{1.655\ volts}$

Further problems on the Wheatstone bridge and d.c. potentiometer may be found in Section 10.20, problems 18 to 20, page 135.

10.18 A.c. bridges

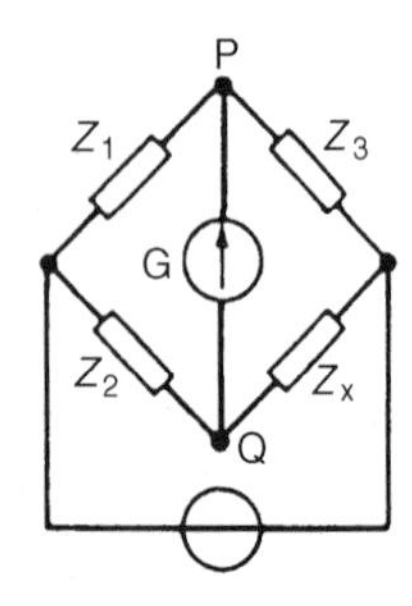

Figure 10.24

A Wheatstone bridge type circuit, shown in Figure 10.24, may be used in a.c. circuits to determine unknown values of inductance and capacitance, as well as resistance.

When the potential differences across Z_3 and Z_x (or across Z_1 and Z_2) are equal in magnitude and phase, then the current flowing through the galvanometer, G, is zero.

At balance, $Z_1 Z_x = Z_2 Z_3$, from which, $\boxed{\mathbf{Z_x = \frac{Z_2 Z_3}{Z_1}\ \Omega}}$

There are many forms of a.c. bridge, and these include: the Maxwell, Hay, Owen and Heaviside bridges for measuring inductance, and the De Sauty,

Schering and Wien bridges for measuring capacitance. A **commercial or universal bridge** is one which can be used to measure resistance, inductance or capacitance.

A.c. bridges require a knowledge of complex numbers, as explained in Chapter 23 and such bridges are discussed in detail in Chapter 27.

10.19 Measurement errors

Errors are always introduced when using instruments to measure electrical quantities. The errors most likely to occur in measurements are those due to:

(i) the limitations of the instrument
(ii) the operator
(iii) the instrument disturbing the circuit

(i) Errors in the limitations of the instrument

The **calibration accuracy** of an instrument depends on the precision with which it is constructed. Every instrument has a margin of error which is expressed as a percentage of the instruments full scale deflection.

For example, industrial grade instruments have an accuracy of ±2% of f.s.d. Thus if a voltmeter has a f.s.d. of 100 V and it indicates 40 V say, then the actual voltage may be anywhere between 40 ± (2% of 100), or 40 ± 2, i.e. between 38 V and 42 V.

When an instrument is calibrated, it is compared against a standard instrument and a graph is drawn of 'error' against 'meter deflection'.

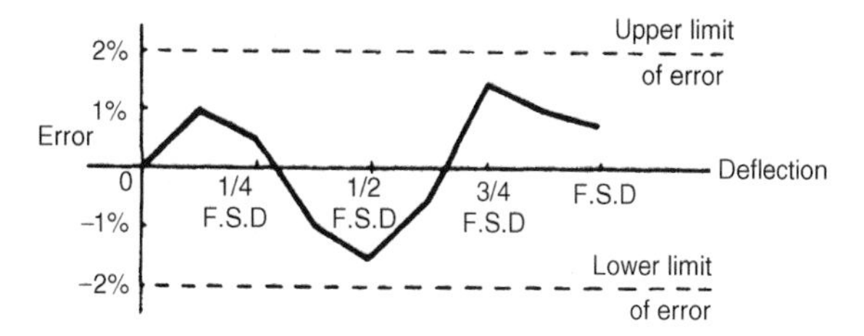

Figure 10.25

A typical graph is shown in Figure 10.25 where it is seen that the accuracy varies over the scale length. Thus a meter with a ±2% f.s.d. accuracy would tend to have an accuracy which is much better than ±2% f.s.d. over much of the range.

(ii) Errors by the operator

It is easy for an operator to misread an instrument. With linear scales the values of the sub-divisions are reasonably easy to determine; non-linear scale graduations are more difficult to estimate. Also, scales differ from instrument to instrument and some meters have more than one scale (as with multimeters) and mistakes in reading

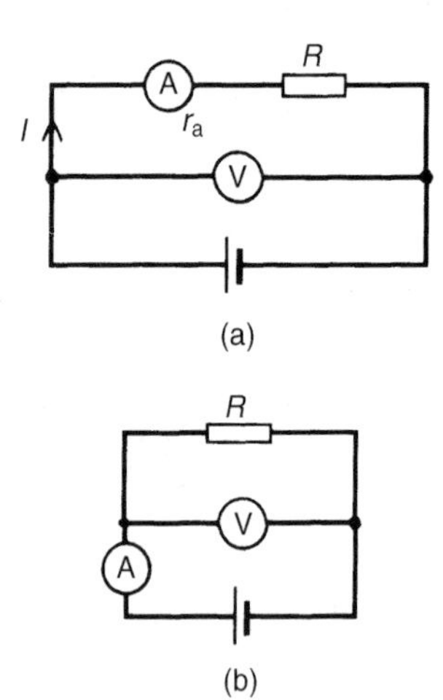

Figure 10.26

indications are easily made. When reading a meter scale it should be viewed from an angle perpendicular to the surface of the scale at the location of the pointer; a meter scale should not be viewed 'at an angle'.

(iii) **Errors due to the instrument disturbing the circuit**
Any instrument connected into a circuit will affect that circuit to some extent. Meters require some power to operate, but provided this power is small compared with the power in the measured circuit, then little error will result. Incorrect positioning of instruments in a circuit can be a source of errors. For example, let a resistance be measured by the voltmeter-ammeter method as shown in Figure 10.26. Assuming 'perfect instruments, the resistance should be given by the voltmeter reading divided by the ammeter reading (i.e. $R = V/I$).

However, in Figure 10.26(a), $V/I = R + r_a$ and in Figure 10.26(b) the current through the ammeter is that through the resistor plus that through the voltmeter. Hence the voltmeter reading divided by the ammeter reading will not give the true value of the resistance R for either method of connection.

Problem 19. The current flowing through a resistor of 5 k$\Omega \pm$ 0.4% is measured as 2.5 mA with an accuracy of measurement of $\pm$0.5%. Determine the nominal value of the voltage across the resistor and its accuracy.

Voltage, $V = IR = (2.5 \times 10^{-3})(5 \times 10^3) = 12.5$ V.

The maximum possible error is $0.4\% + 0.5\% = 0.9\%$

Hence the voltage, $V = 12.5\text{ V} \pm 0.9\%$ of $12.5\text{ V} = 0.9/100 \times 12.5 = 0.1125\text{ V} = 0.11$ V correct to 2 significant figures. Hence the voltage V may also be expressed as **12.5 ± 0.11 volts** (i.e. a voltage lying between 12.39 V and 12.61 V).

Problem 20. The current I flowing in a resistor R is measured by a 0–10 A ammeter which gives an indication of 6.25 A. The voltage V across the resistor is measured by a 0–50 V voltmeter, which gives an indication of 36.5 V. Determine the resistance of the resistor, and its accuracy of measurement if both instruments have a limit of error of 2% of f.s.d. Neglect any loading effects of the instruments.

Resistance, $R = \dfrac{V}{I} = \dfrac{36.5}{6.25} = 5.84\ \Omega$

Voltage error is $\pm 2\%$ of 50 $V = \pm 1.0$ V and expressed as a percentage of the voltmeter reading gives $\dfrac{\pm 1}{36.5} \times 100\% = \pm 2.74\%$

Current error is $\pm 2\%$ of 10 A $= \pm 0.2$ A and expressed as a percentage of the ammeter reading gives $\dfrac{\pm 0.2}{6.25} \times 100\% = \pm 3.2\%$

Maximum relative error = sum of errors = $2.74\% + 3.2\% = \pm 5.94\%$,

5.94% of 5.84 $\Omega = 0.347\ \Omega$

Hence the resistance of the resistor may be expressed as:

5.84 Ω ± 5.94%, or **5.84 ± 0.35 Ω** (rounding off)

Problem 21. The arms of a Wheatstone bridge ABCD have the following resistances: AB: $R_1 = 1000\ \Omega \pm 1.0\%$; BC: $R_2 = 100\ \Omega \pm 0.5\%$; CD: unknown resistance R_x; DA: $R_3 = 432.5\ \Omega \pm 0.2\%$. Determine the value of the unknown resistance and its accuracy of measurement.

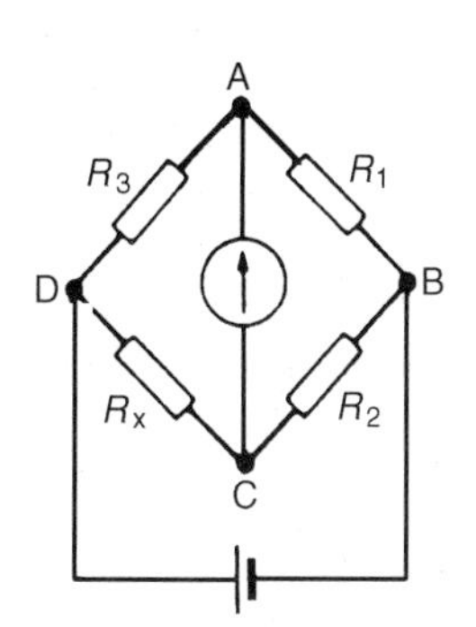

Figure 10.27

The Wheatstone bridge network is shown in Figure 10.27 and at balance:

$$R_1 R_x = R_2 R_3, \quad \text{i.e. } R_x = \frac{R_2 R_3}{R_1} = \frac{(100)(432.5)}{1000} = 43.25\ \Omega$$

The maximum relative error of R_x is given by the sum of the three individual errors, i.e. $1.0\% + 0.5\% + 0.2\% = 1.7\%$

Hence $R_x = 43.25\ \Omega \pm 1.7\%$

1.7% of 43.25 $\Omega = 0.74\ \Omega$ (rounding off).

Thus R_x may also be expressed as $\boldsymbol{R_x = 43.25 \pm 0.74\ \Omega}$

Further problems on measurement errors may be found in Section 10.20 following, problems 21 to 23, page 135.

10.20 Further problems on electrical measuring instruments and measurements

Shunts and multipliers

1 A moving-coil instrument gives f.s.d. for a current of 10 mA. Neglecting the resistance of the instrument, calculate the approximate value of series resistance needed to enable the instrument to measure up to (a) 20 V (b) 100 V (c) 250 V.

[(a) 2 kΩ (b) 10 kΩ (c) 25 kΩ]

2 A meter of resistance 50 Ω has a f.s.d. of 4 mA. Determine the value of shunt resistance required in order that f.s.d. should be (a) 15 mA (b) 20 A (c) 100 A.

[(a) 18.18 Ω (b) 10.00 mΩ (c) 2.00 mΩ]

3 A moving-coil instrument having a resistance of 20 Ω, gives a f.s.d. when the current is 5 mA. Calculate the value of the multiplier to be

connected in series with the instrument so that it can be used as a voltmeter for measuring p.d.s up to 200 V. [39.98 kΩ]

4 A moving-coil instrument has a f.s.d. of 20 mA and a resistance of 25 Ω. Calculate the values of resistance required to enable the instrument to be used (a) as a 0–10 A ammeter, and (b) as a 0–100 V voltmeter. State the mode of resistance connection in each case. [(a) 50.10 mΩ in parallel (b) 4.975 kΩ in series]

Instrument 'loading' effects

5 A 0–1 A ammeter having a resistance of 50 Ω is used to measure the current flowing in a 1 kΩ resistor when the supply voltage is 250 V. Calculate: (a) the approximate value of current (neglecting the ammeter resistance), (b) the actual current in the circuit, (c) the power dissipated in the ammeter, (d) the power dissipated in the 1 kΩ resistor. [(a) 0.250 A (b) 0.238 A (c) 2.832 W (d) 56.64 W]

6 (a) A current of 15 A flows through a load having a resistance of 4 Ω. Determine the power dissipated in the load. (b) A wattmeter, whose current coil has a resistance of 0.02 Ω is connected (as shown in Figure 10.13) to measure the power in the load. Determine the wattmeter reading assuming the current in the load is still 15 A. [(a) 900 W (b) 904.5 W]

7 A voltage of 240 V is applied to a circuit consisting of an 800 Ω resistor in series with a 1.6 kΩ resistor. What is the voltage across the 1.6 kΩ resistor? The p.d. across the 1.6 kΩ resistor is measured by a voltmeter of f.s.d. 250 V and sensitivity 100 Ω/V. Determine the voltage indicated. [160 V; 156.7 V]

Cathode ray oscilloscope

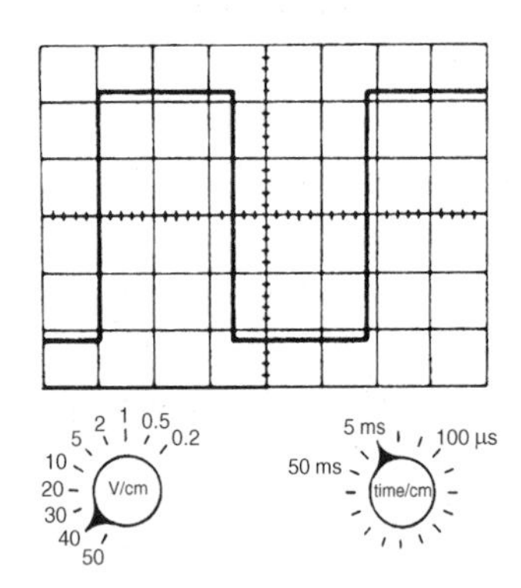

Figure 10.28

8 For the square voltage waveform displayed on a c.r.o. shown in Figure 10.28, find (a) its frequency, (b) its peak-to-peak voltage. [(a) 41.7 Hz (b) 176 V]

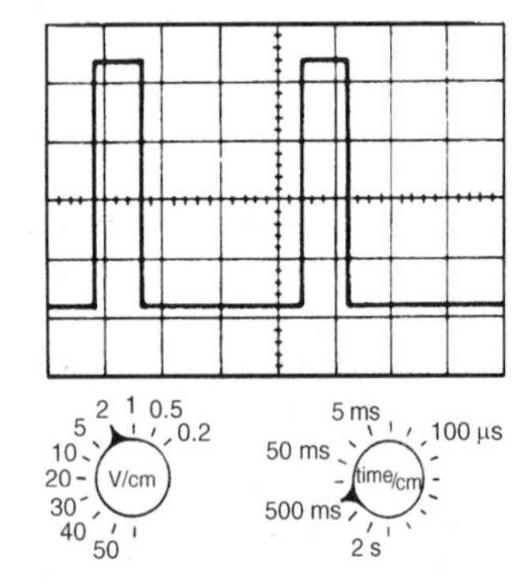

Figure 10.29

9 For the pulse waveform shown in Figure 10.29, find (a) its frequency, (b) the magnitude of the pulse voltage. [(a) 0.56 Hz (b) 8.4 V]

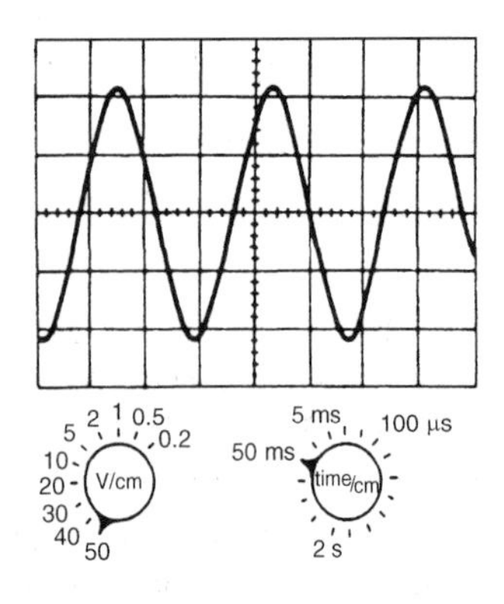

Figure 10.30

10 For the sinusoidal waveform shown in Figure 10.30, determine (a) its frequency, (b) the peak-to-peak voltage, (c) the r.m.s. voltage. [(a) 7.14 Hz (b) 220 V (c) 77.78 V]

Logarithmic ratios

11 The ratio of two powers is (a) 3 (b) 10 (c) 20 (d) 10000. Determine the decibel power ratio for each. [(a) 4.77 dB (b) 10 dB (c) 13 dB (d) 40 dB]

12 The ratio of two powers is (a) $\frac{1}{10}$ (b) $\frac{1}{3}$ (c) $\frac{1}{40}$ (d) $\frac{1}{100}$

Determine the decibel power ratio for each.
[(a) −10 dB (b) −4.77 dB (c) −16.02 dB (d) −20 dB]

13 The input and output currents of a system are 2 mA and 10 mA respectively. Determine the decibel current ratio of output to input current assuming input and output resistances of the system are equal. [13.98 dB]

14 5% of the power supplied to a cable appears at the output terminals. Determine the power loss in decibels. [13 dB]

15 An amplifier has a gain of 24 dB. Its input power is 10 mW. Find its output power. [2.51 W]

16 The output voltage from an amplifier is 7 mV. If the voltage gain is 25 dB calculate the value of the input voltage assuming that the amplifier input resistance and load resistance are equal. [0.39 mV]

17 The scale of a voltmeter has a decibel scale added to it, which is calibrated by taking a reference level of 0 dB when a power of 1 mW is dissipated in a 600 Ω resistor. Determine the voltage at (a) 0 dB (b) 1.5 dB and (c) −15 dB (d) What decibel reading corresponds to 0.5 V? [(a) 0.775 V (b) 0.921 V (c) 0.138 V (d) −3.807 dB]

Wheatstone bridge and d.c. potentiometer

18 In a Wheatstone bridge PQRS, a galvanometer is connected between Q and S and a voltage source between P and R. An unknown resistor R_x is connected between P and Q. When the bridge is balanced, the resistance between Q and R is 200 Ω, that between R and S is 10 Ω and that between S and P is 150 Ω. Calculate the value of R_x [3 kΩ]

19 Balance is obtained in a d.c. potentiometer at a length of 31.2 cm when using a standard cell of 1.0186 volts. Calculate the e.m.f. of a dry cell if balance is obtained with a length of 46.7 cm. [1.525 V]

20 A Wheatstone bridge PQRS has the following arm resistances:

PQ, 1 kΩ ± 2%; QR, 100 Ω ± 0.5%; RS, unknown resistance; SP, 273.6 Ω ± 0.1%. Determine the value of the unknown resistance, and its accuracy of measurement.
[27.36 Ω ± 2.6% or 27.36 Ω ± 0.71 Ω]

Measurement errors

21 The p.d. across a resistor is measured as 37.5 V with an accuracy of ±0.5%. The value of the resistor is 6 kΩ ± 0.8%. Determine the current flowing in the resistor and its accuracy of measurement.
[6.25 mA ±1.3% or 6.25± 0.08 mA]

22 The voltage across a resistor is measured by a 75 V f.s.d. voltmeter which gives an indication of 52 V. The current flowing in the resistor is measured by a 20 A f.s.d. ammeter which gives an indication of

12.5 A. Determine the resistance of the resistor and its accuracy if both instruments have an accuracy of $\pm 2\%$ of f.s.d.

[4.16 Ω ± 6.08% or 4.16 ± 0.25 Ω]

23 A 240 V supply is connected across a load resistance R. Also connected across R is a voltmeter having a f.s.d. of 300 V and a figure of merit (i.e. sensitivity) of 8 kΩ/V. Calculate the power dissipated by the voltmeter and by the load resistance if (a) $R = 100\ \Omega$ (b) $R = 1\ \text{M}\Omega$. Comment on the results obtained.

[(a) 24 mW, 576 W (b) 24 mW, 57.6 mW]

11 Semiconductor diodes

At the end of this chapter you should be able to:

- classify materials as conductors, semiconductors or insulators
- appreciate the importance of silicon and germanium
- understand n-type and p-type materials
- understand the p-n junction
- appreciate forward and reverse bias of p-n junctions
- draw the circuit diagram symbol for a semiconductor diode

11.1 Types of materials

Materials may be classified as **conductors, semiconductors** or **insulators**. The classification depends on the value of resistivity of the material. Good conductors are usually metals and have resistivities in the order of 10^{-7} to 10^{-8} Ωm, semiconductors have resistivities in the order of 10^{-3} to 3×10^{3} Ωm and the resistivities of insulators are in the order of 10^{4} to 10^{14} Ωm. Some typical approximate values at normal room temperatures are:

Conductors:

Aluminium	2.7×10^{-8} Ωm
Brass (70 Cu/30 Zn)	8×10^{-8} Ωm
Copper (pure annealed)	1.7×10^{-8} Ωm
Steel (mild)	15×10^{-8} Ωm

Semiconductors:

Silicon	2.3×10^{3} Ωm	at 27°C
Germanium	0.45 Ωm	at 27°C

Insulators:

Glass $\geq 10^{10}$ Ωm
Mica $\geq 10^{11}$ Ωm
PVC $\geq 10^{13}$ Ωm
Rubber (pure) 10^{12} to 10^{14} Ωm

In general, over a limited range of temperatures, the resistance of a conductor increases with temperature increase, the resistance of insulators remains approximately constant with variation of temperature and

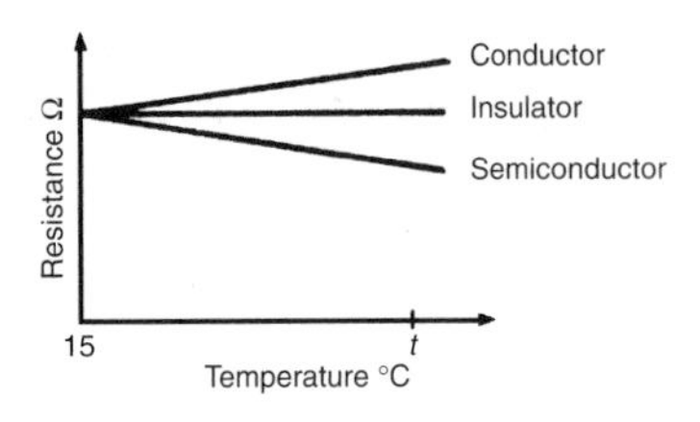

Figure 11.1

the resistance of semiconductor materials decreases as the temperature increases. For a specimen of each of these materials, having the same resistance (and thus completely different dimensions), at say, 15°C, the variation for a small increase in temperature to t °C is as shown in Figure 11.1.

11.2 Silicon and germanium

The most important semiconductors used in the electronics industry are silicon and germanium. As the temperature of these materials is raised above room temperature, the resistivity is reduced and ultimately a point is reached where they effectively become conductors. For this reason, silicon should not operate at a working temperature in excess of 150°C to 200°C, depending on its purity, and germanium should not operate at a working temperature in excess of 75°C to 90°C, depending on its purity. As the temperature of a semiconductor is reduced below normal room temperature, the resistivity increases until, at very low temperatures the semiconductor becomes an insulator.

11.3 n-type and p-type materials

Adding extremely small amounts of impurities to pure semiconductors in a controlled manner is called **doping**. Antimony, arsenic and phosphorus are called n-type impurities and form an **n-type material** when any of these impurities are added to silicon or germanium. The amount of impurity added usually varies from 1 part impurity in 10^5 parts semiconductor material to 1 part impurity to 10^8 parts semiconductor material, depending on the resistivity required. Indium, aluminium and boron are called p-type impurities and form a **p-type material** when any of these impurities are added to a semiconductor.

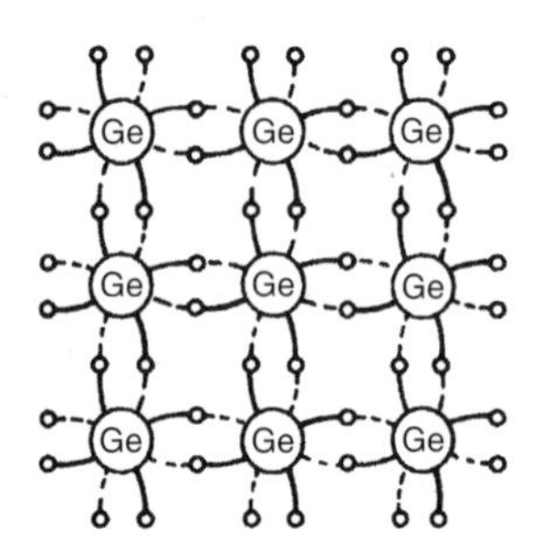

Figure 11.2

In semiconductor materials, there are very few charge carriers per unit volume free to conduct. This is because the 'four electron structure' in the outer shell of the atoms (called valency electrons), form strong covalent bonds with neighbouring atoms, resulting in a tetrahedral structure with the electrons held fairly rigidly in place. A two-dimensional diagram depicting this is shown for germanium in Figure 11.2.

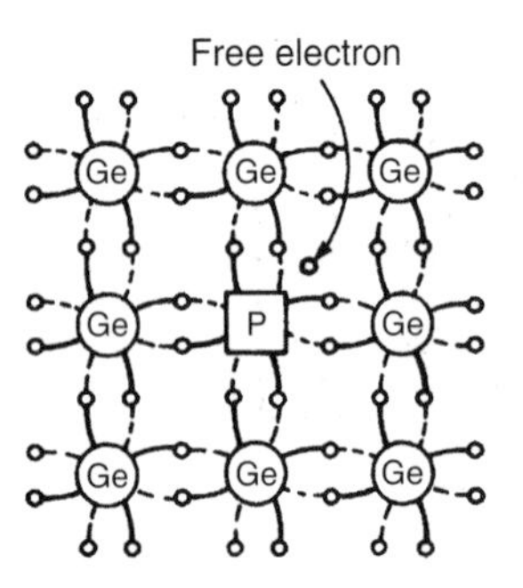

Figure 11.3

Arsenic, antimony and phosphorus have five valency electrons and when a semiconductor is doped with one of these substances, some impurity atoms are incorporated in the tetrahedral structure. The 'fifth' valency electron is not rigidly bonded and is free to conduct, the impurity atom donating a charge carrier. A two-dimensional diagram depicting this is shown in Figure 11.3, in which a phosphorus atom has replaced one of the germanium atoms. The resulting material is called n-type material, and contains free electrons.

Indium, aluminium and boron have three valency electrons and when a semiconductor is doped with one of these substances, some of the semiconductor atoms are replaced by impurity atoms. One of the four bonds associated with the semiconductor material is deficient by one electron and this deficiency is called a **hole**. Holes give rise to conduction when a potential difference exists across the semiconductor material due to movement of electrons from one hole to another, as shown in Figure 11.4. In this figure, an electron moves from A to B, giving the appearance that the hole moves from B to A. Then electron C moves to A, giving the appearance that the hole moves to C, and so on. The resulting material is p-type material containing holes.

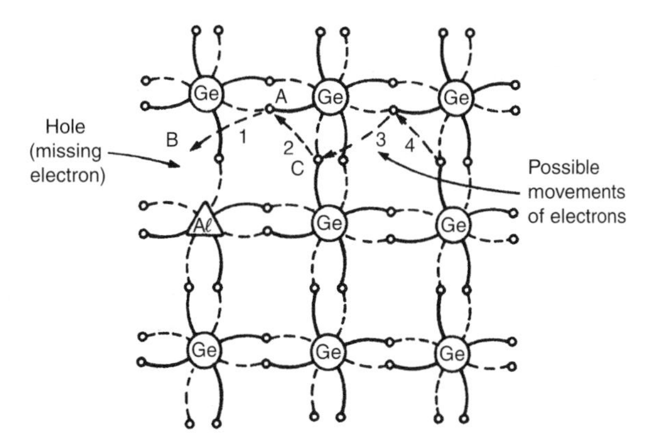

Figure 11.4

11.4 The p-n junction

A **p-n junction** is piece of semiconductor material in which part of the material is p-type and part is n-type. In order to examine the charge situation, assume that separate blocks of p-type and n-type materials are pushed together. Also assume that a hole is a positive charge carrier and that an electron is a negative charge carrier.

At the junction, the donated electrons in the n-type material, called **majority carriers**, diffuse into the p-type material (diffusion is from an

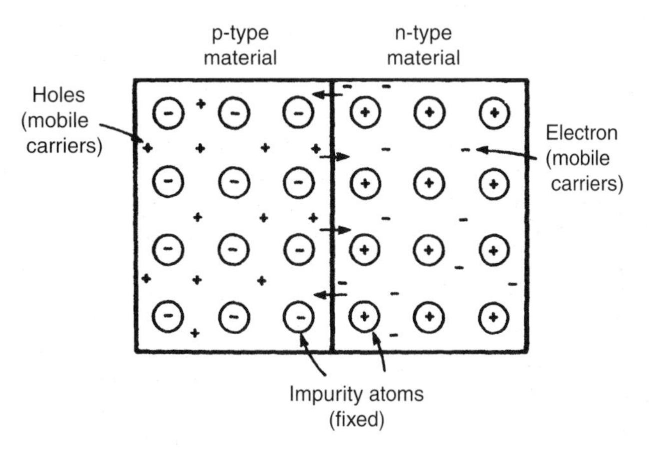

Figure 11.5

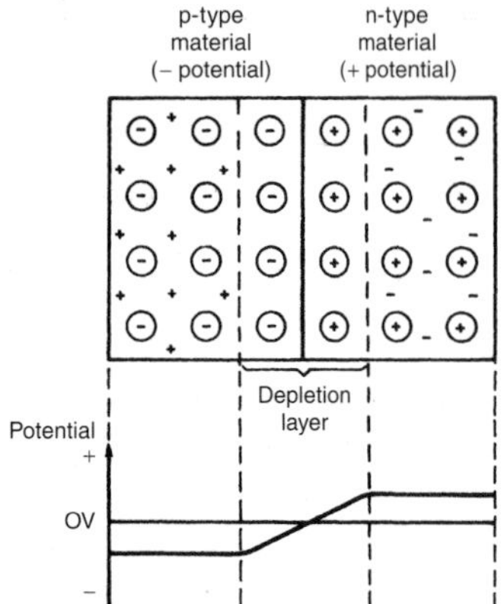

Figure 11.6

area of high density to an area of lower density) and the acceptor holes in the p-type material diffuse into the n-type material as shown by the arrows in Figure 11.5. Because the n-type material has lost electrons, it acquires a positive potential with respect to the p-type material and thus tends to prevent further movement of electrons. The p-type material has gained electrons and becomes negatively charged with respect to the n-type material and hence tends to retain holes. Thus after a short while, the movement of electrons and holes stops due to the potential difference across the junction, called the **contact potential**. The area in the region of the junction becomes depleted of holes and electrons due to electron-hole recombinations, and is called a **depletion layer**, as shown in Figure 11.6.

11.5 Forward and reverse bias

When an external voltage is applied to a p-n junction making the p-type material positive with respect to the n-type material, as shown in Figure 11.7, the p-n junction is **forward biased**. The applied voltage opposes the contact potential, and, in effect, closes the depletion layer. Holes and electrons can now cross the junction and a current flows.

An increase in the applied voltage above that required to narrow the depletion layer (about 0.2 V for germanium and 0.6 V for silicon), results in a rapid rise in the current flow. Graphs depicting the current-voltage relationship for forward biased p-n junctions, for both germanium and silicon, called the **forward characteristics**, are shown in Figure 11.8.

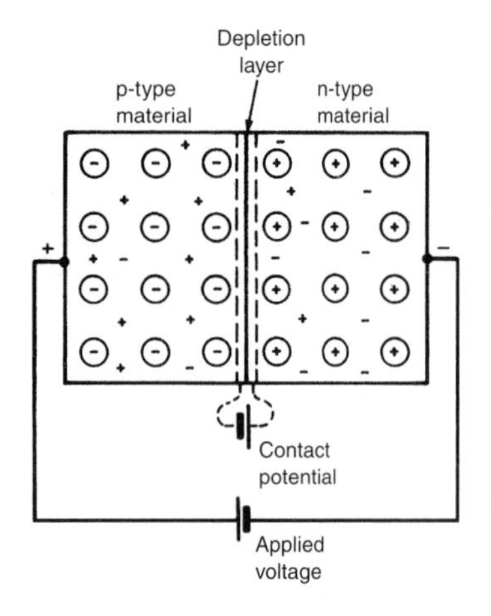

Figure 11.7

When an external voltage is applied to a p-n junction making the p-type material negative with respect to the n-type material as in shown in Figure 11.9, the p-n junction is **reverse biased**. The applied voltage is now in the same sense as the contact potential and opposes the movement of holes and electrons due to opening up the depletion layer. Thus, in theory, no current flows. However at normal room temperature certain electrons in the covalent bond lattice acquire sufficient energy from the heat available to leave the lattice, generating mobile electrons and holes. This process is called electron-hole generation by thermal excitation.

The electrons in the p-type material and holes in the n-type material caused by thermal excitation, are called **minority carriers** and these will be attracted by the applied voltage. Thus, in practice, a small current of a few microamperes for germanium and less than one microampere for silicon, at normal room temperature, flows under reverse bias conditions. Typical **reverse characteristics** are shown in Figure 11.10 for both germanium and silicon.

11.6 Semiconductor diodes

A semiconductor diode is a device having a p-n junction mounted in a container, suitable for conducting and dissipating the heat generated in operation, and having connecting leads. Its operating characteristics are as shown in Figures 11.8 and 11.10. Two circuit diagram symbols for semiconductor diodes are in common use and are as shown in Figure 11.11.

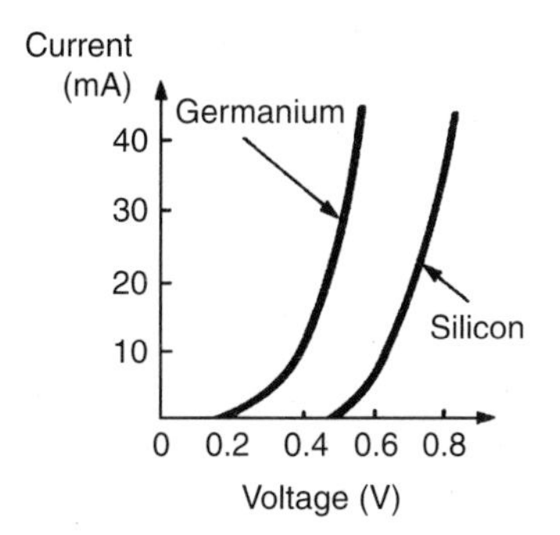

Figure 11.8

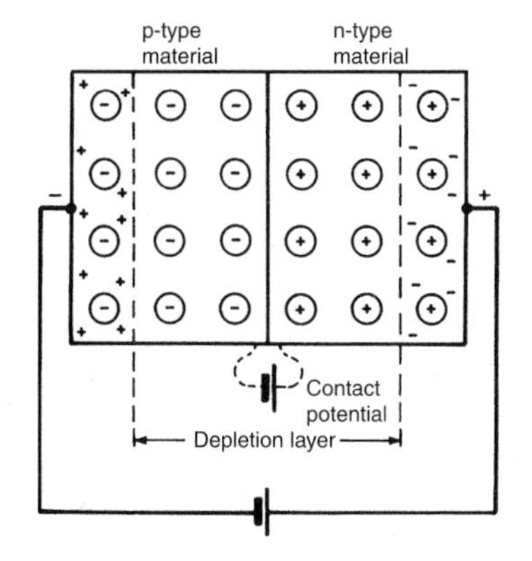

Figure 11.9

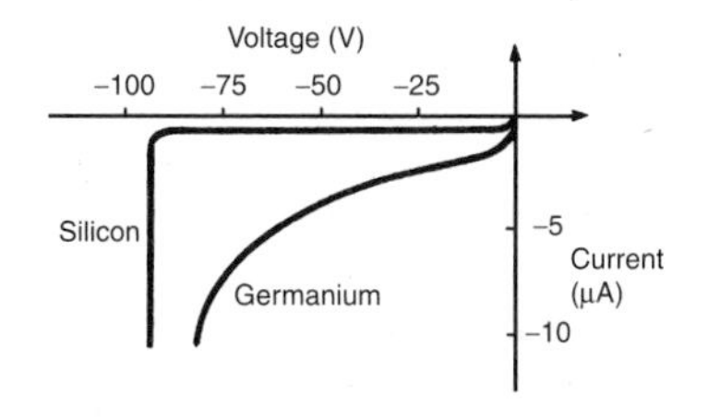

Figure 11.10

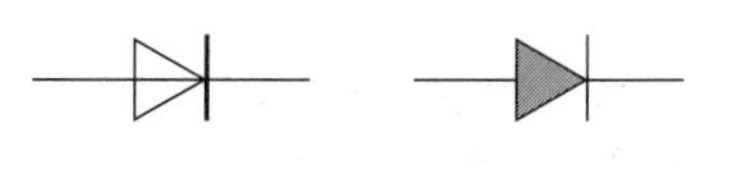

Figure 11.11

Sometimes the symbols are encircled as shown in Figures 14.14–14.16 on pages 208 and 209.

> Problem 1. Explain briefly the terms given below when they are associated with a p-n junction: (a) conduction in intrinsic semiconductors (b) majority and minority carriers, and (c) diffusion

(a) Silicon or germanium with no doping atoms added are called intrinsic semiconductors. At room temperature, some of the electrons acquire sufficient energy for them to break the covalent bond between atoms and become free mobile electrons. This is called thermal generation of electron-hole pairs. Electrons generated thermally create a gap in the crystal structure called a hole, the atom associated with the hole being positively charged, since it has lost an electron. This positive charge may attract another electron released from another atom, creating a hole elsewhere.
When a potential is applied across the semiconductor material, holes drift towards the negative terminal (unlike charges attract), and electrons towards the positive terminal, and hence a small current flows.

(b) When additional mobile electrons are introduced by doping a semiconductor material with pentavalent atoms (atoms having five valency electrons), these mobile electrons are called majority carriers. The relatively few holes in the n-type material produced by intrinsic action are called minority carriers.
For p-type materials, the additional holes are introduced by doping with trivalent atoms (atoms having three valency electrons). The holes are apparently positive mobile charges and are majority carriers in the p-type material. The relatively few mobile electrons in the p-type material produced by intrinsic action are called minority carriers.

(c) Mobile holes and electrons wander freely within the crystal lattice of a semiconductor material. There are more free electrons in n-type material than holes and more holes in p-type material than electrons. Thus, in their random wanderings, on average, holes pass into the n-type material and electrons into the p-type material. This process is called diffusion.

> Problem 2. Explain briefly why a junction between p-type and n-type materials creates a contact potential.

Intrinsic semiconductors have resistive properties, in that when an applied voltage across the material is reversed in polarity, a current of the same magnitude flows in the opposite direction. When a p-n junction is formed, the resistive property is replaced by a rectifying property, that is, current passes more easily in one direction than the other.

An n-type material can be considered to be a stationary crystal matrix of fixed positive charges together with a number of mobile negative charge carriers (electrons). The total number of positive and negative charges are equal. A p-type material can be considered to be a number of stationary negative charges together with mobile positive charge carriers (holes).

Again, the total number of positive and negative charges are equal and the material is neither positively nor negatively charged. When the materials are brought together, some of the mobile electrons in the n-type material diffuse into the p-type material. Also, some of the mobile holes in the p-type material diffuse into the n-type material.

Many of the majority carriers in the region of the junction combine with the opposite carriers to complete covalent bonds and create a region on either side of the junction with very few carriers. This region, called the depletion layer, acts as an insulator and is in the order of 0.5 μm thick. Since the n-type material has lost electrons, it becomes positively charged. Also, the p-type material has lost holes and becomes negatively charged, creating a potential across the junction, called the barrier or contact potential.

Problem 3. Sketch the forward and reverse characteristics of a silicon p-n junction diode and describe the shapes of the characteristics drawn.

A typical characteristic for a silicon p-n junction having a forward bias is shown in Figure 11.8 and having a reverse bias in Figure 11.10. When the positive terminal of the battery is connected to the p-type material and the negative terminal to the n-type material, the diode is forward biased. Due to like charges repelling, the holes in the p-type material drift towards the junction. Similarly the electrons in the n-type material are repelled by the negative bias voltage and also drift towards the junction. The width of the depletion layer and size of the contact potential are reduced. For applied voltages from 0 to about 0.6 V, very little current flows. At about 0.6 V, majority carriers begin to cross the junction in large numbers and current starts to flow. As the applied voltage is raised above 0.6 V, the current increases exponentially (see Figure 11.8).

When the negative terminal of the battery is connected to the p-type material and the positive terminal to the n-type material the diode is reverse biased. The holes in the p-type material are attracted towards the negative terminal and the electrons in the n-type material are attracted towards the positive terminal (unlike charges attract). This drift increases the magnitude of both the contact potential and the thickness of the depletion layer, so that only very few majority carriers have sufficient energy to surmount the junction.

The thermally excited minority carriers, however, can cross the junction since it is, in effect, forward biased for these carriers. The movement of minority carriers results in a small constant current flowing. As the magnitude of the reverse voltage is increased a point will be reached where a large current suddenly starts to flow. The voltage at

Figure 11.12

which this occurs is called the breakdown voltage. This current is due to two effects:

(i) the **zener effect**, resulting from the applied voltage being sufficient to break some of the covalent bonds, and
(ii) the **avalanche effect**, resulting from the charge carriers moving at sufficient speed to break covalent bonds by collision.

A **zener diode** is used for voltage reference purposes or for voltage stabilisation. Two common circuit diagram symbols for a zener diode are shown in Figure 11.12.

11.7 Rectification

The process of obtaining unidirectional currents and voltages from alternating currents and voltages is called **rectification**. Automatic switching in circuits is carried out by diodes. For methods of half-wave and full-wave rectification, see Section 14.7, page 208.

11.8 Further problems on semiconductor diodes

1. Explain what you understand by the term intrinsic semiconductor and how an intrinsic semiconductor is turned into either a p-type or an n-type material.
2. Explain what is meant by minority and majority carriers in an n-type material and state whether the numbers of each of these carriers are affected by temperature.
3. A piece of pure silicon is doped with (a) pentavalent impurity and (b) trivalent impurity. Explain the effect these impurities have on the form of conduction in silicon.
4. With the aid of simple sketches, explain how pure germanium can be treated in such a way that conduction is predominantly due to (a) electrons and (b) holes.
5. Explain the terms given below when used in semiconductor terminology: (a) covalent bond (b) trivalent impurity (c) pentavalent impurity (d) electron-hole pair generation.
6. Explain briefly why although both p-type and n-type materials have resistive properties when separate, they have rectifying properties when a junction between them exists.
7. The application of an external voltage to a junction diode can influence the drift of holes and electrons. With the aid of diagrams explain this statement and also how the direction and magnitude of the applied voltage affects the depletion layer.
8. State briefly what you understand by the terms:

   (a) reverse bias (b) forward bias (c) contact potential (d) diffusion (e) minority carrier conduction.

9. Explain briefly the action of a p-n junction diode:

   (a) on open-circuit, (b) when provided with a forward bias, and (c) when provided with a reverse bias. Sketch the characteristic curves for both forward and reverse bias conditions.

10. Draw a diagram illustrating the charge situation for an unbiased p-n junction. Explain the change in the charge situation when compared with that in isolated p-type and n-type materials. Mark on the diagram the depletion layer and the majority carriers in each region.

12 Transistors

At the end of this chapter you should be able to:

- understand the structure of a bipolar junction transistor
- understand transistor action for p-n-p and n-p-n types
- draw the circuit diagram symbols for p-n-p and n-p-n transistors
- appreciate common-base, common-emitter and common-collector transistor connections
- interpret transistor characteristics
- appreciate how the transistor is used as an amplifier
- determine the load line on transistor characteristics
- estimate current, voltage and power gains from transistor characteristics
- understand thermal runaway in a transistor

12.1 The bipolar junction transistor

The bipolar junction transistor consists of three regions of semiconductor material. One type is called a p-n-p transistor, in which two regions of p-type material sandwich a very thin layer of n-type material. A second type is called an n-p-n transistor, in which two regions of n-type material sandwich a very thin layer of p-type material. Both of these types of transistors consist of two p-n junctions placed very close to one another in a back-to-back arrangement on a single piece of semiconductor material. Diagrams depicting these two types of transistors are shown in Figure 12.1.

The two p-type material regions of the p-n-p transistor are called the **emitter** and **collector** and the n-type material is called the **base**. Similarly, the two n-type material regions of the n-p-n transistor are called the emitter and collector and the p-type material region is called the base, as shown in Figure 12.1.

Transistors have three connecting leads and in operation an electrical input to one pair of connections, say the emitter and base connections can control the output from another pair, say the collector and emitter connections. This type of operation is achieved by appropriately biasing the two internal p-n junctions. When batteries and resistors are connected to a p-n-p transistor, as shown in Figure 12.2(a), the base-emitter junction is **forward biased** and the base-collector junction is **reverse biased**.

Similarly, an n-p-n transistor has its base-emitter junction forward biased and its base-collector junction reverse biased when the batteries are connected as shown in Figure 12.2(b).

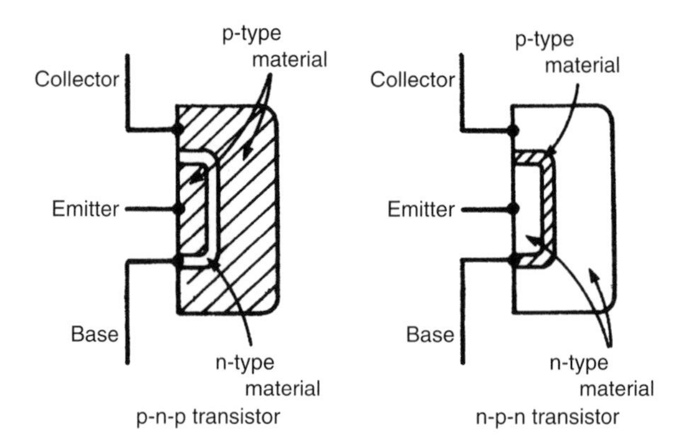

Figure 12.1

For a silicon p-n-p transistor, biased as shown in Figure 12.2(a), if the base-emitter junction is considered on its own, it is forward biased and a current flows. This is depicted in Figure 12.3(a). For example, if R_E is 1000 Ω, the battery is 4.5 V and the voltage drop across the junction is taken as 0.7 V, the current flowing is given by $(4.5 - 0.7)/1000 =$ 3.8 mA.

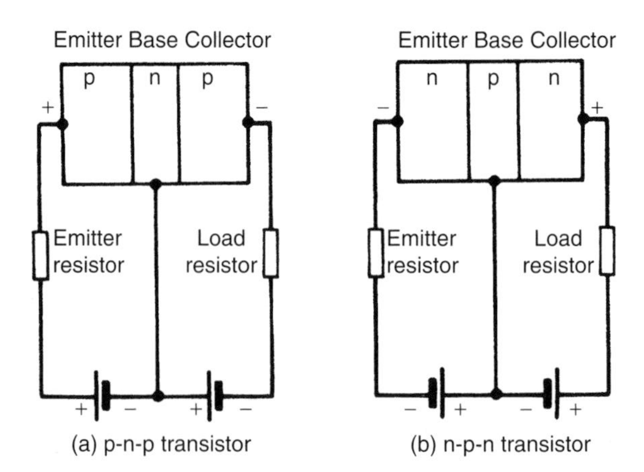

Figure 12.2

When the base-collector junction is considered on its own, as shown in Figure 12.3(b), it is reverse biased and the collector current is something less than 1 μA.

However, when both external circuits are connected to the transistor, most of the 3.8 mA of current flowing in the emitter, which previously flowed from the base connection, now flows out through the collector connection due to transistor action.

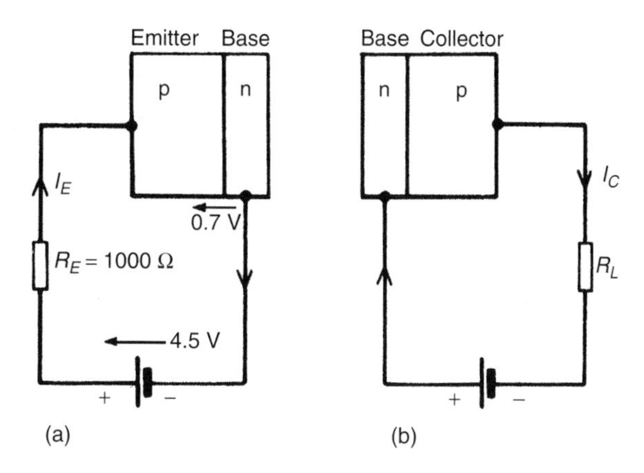

Figure 12.3

12.2 Transistor action

In a **p-n-p transistor**, connected as shown in Figure 12.2(a), transistor action is accounted for as follows:

(a) The majority carriers in the emitter p-type material are holes
(b) The base-emitter junction is forward biased to the majority carriers and the holes cross the junction and appear in the base region
(c) The base region is very thin and is only lightly doped with electrons so although some electron-hole pairs are formed, many holes are left in the base region
(d) The base-collector junction is reverse biased to electrons in the base region and holes in the collector region, but forward biased to holes in the base region; these holes are attracted by the negative potential at the collector terminal
(e) A large proportion of the holes in the base region cross the base-collector junction into the collector region, creating a collector current; conventional current flow is in the direction of hole movement.

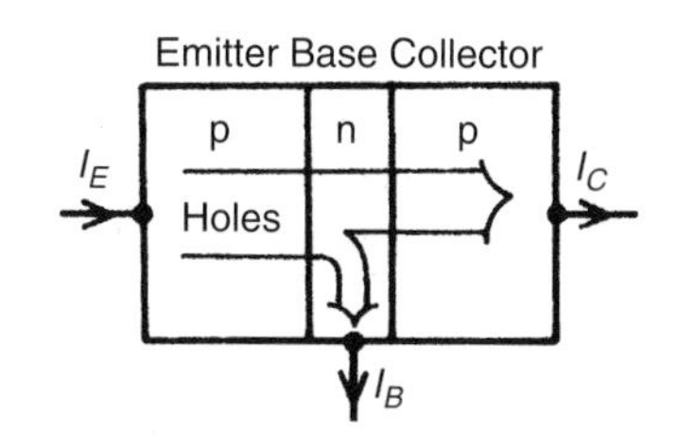

Figure 12.4

The transistor action is shown diagrammatically in Figure 12.4. For transistors having very thin base regions, up to 99.5% of the holes leaving the emitter cross the base collector junction.

In an **n-p-n transistor**, connected as shown in Figure 12.2(b), transistor action is accounted for as follows:

(a) The majority carriers in the n-type emitter material are electrons
(b) The base-emitter junction is forward biased to these majority carriers and electrons cross the junction and appear in the base region
(c) The base region is very thin and only lightly doped with holes, so some recombination with holes occurs but many electrons are left in the base region
(d) The base-collector junction is reverse biased to holes in the base region and electrons in the collector region, but is forward biased

to electrons in the base region; these electrons are attracted by the positive potential at the collector terminal

(e) A large proportion of the electrons in the base region cross the base collector junction into the collector region, creating a collector current.

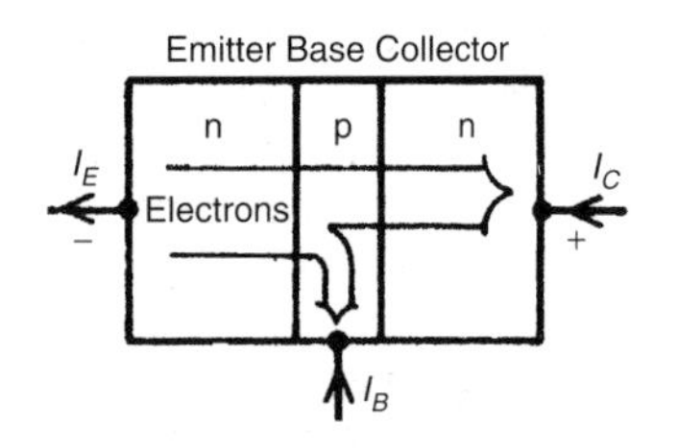

Figure 12.5

The transistor action is shown diagrammatically in Figure 12.5. As stated in Section 12.1, conventional current flow is taken to be in the direction of hole flow, that is, in the opposite direction to electron flow, hence the directions of the conventional current flow are as shown in Figure 12.5.

For a p-n-p transistor, the base-collector junction is reverse biased for majority carriers. However, a small leakage current, I_{CBO} flows from the base to the collector due to thermally generated minority carriers (electrons in the collector and holes in the base), being present.

The base-collector junction is forward biased to these minority carriers. If a proportion, α, (having a value of up to 0.995 in modern transistors), of the holes passing into the base from the emitter, pass through the base-collector junction, then the various currents flowing in a p-n-p transistor are as shown in Figure 12.6(a).

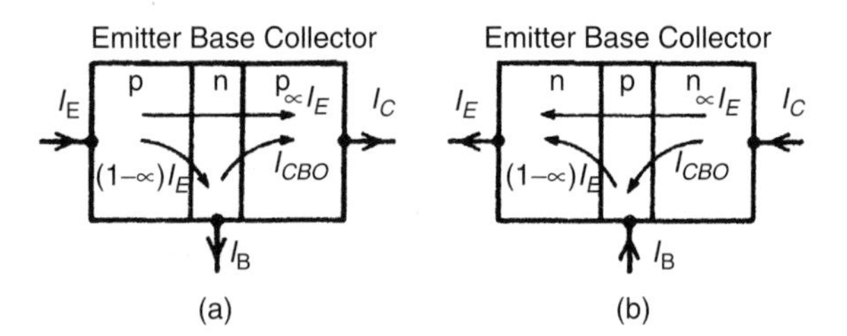

Figure 12.6

Similarly, for an n-p-n transistor, the base-collector junction is reversed biased for majority carriers, but a small leakage current, I_{CBO} flows from the collector to the base due to thermally generated minority carriers (holes in the collector and electrons in the base), being present. The base-collector junction is forward biased to these minority carriers. If a proportion, α, of the electrons passing through the base-emitter junction also pass through the base-collector junction then the currents flowing in an n-p-n transistor are as shown in Figure 12.6(b).

> Problem 1. With reference to a p-n-p transistor, explain briefly what is meant by the term transistor action and why a bipolar junction transistor is so named.

For the transistor as depicted in Figure 12.4, the emitter is relatively heavily doped with acceptor atoms (holes). When the emitter terminal is made sufficiently positive with respect to the base, the base-emitter junction is forward biased to the majority carriers. The majority carriers are holes in the emitter and these drift from the emitter to the base. The

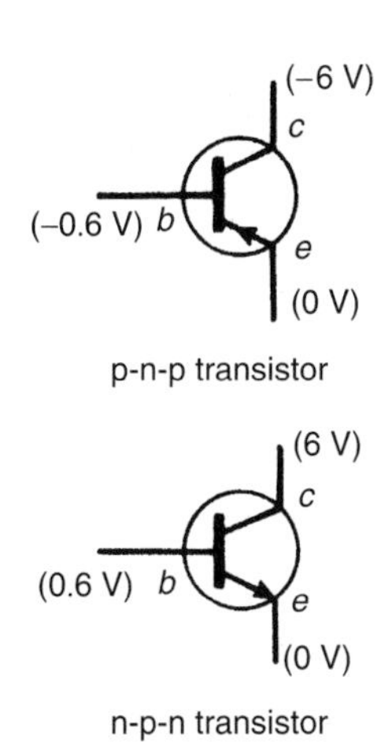

Figure 12.7

base region is relatively lightly doped with donor atoms (electrons) and although some electron-hole recombination's take place, perhaps 0.5%, most of the holes entering the base, do not combine with electrons.

The base-collector junction is reverse biased to electrons in the base region, but forward biased to holes in the base region. Since the base is very thin and now is packed with holes, these holes pass the base-emitter junction towards the negative potential of the collector terminal. The control of current from emitter to collector is largely independent of the collector-base voltage and almost wholly governed by the emitter-base voltage. The essence of transistor action is this current control by means of the base-emitter voltage.

In a p-n-p transistor, holes in the emitter and collector regions are majority carriers, but are minority carriers when in the base region. Also thermally generated electrons in the emitter and collector regions are minority carriers as are holes the base region. However, both majority and minority carriers contribute towards the total current flow (see Figure 12.6(a)). It is because a transistor makes use of both types of charge carriers (holes and electrons) that they are called bipolar. The transistor also comprises two p-n junctions and for this reason it is a junction transistor. Hence the name — bipolar junction transistor.

12.3 Transistor symbols

Symbols are used to represent p-n-p and n-p-n transistors in circuit diagrams and are as shown in Figure 12.7. The arrowhead drawn on the emitter of the symbol is in the direction of conventional emitter current (hole flow). The potentials marked at the collector, base and emitter are typical values for a silicon transistor having a potential difference of 6 V between its collector and its emitter.

The voltage of 0.6 V across the base and emitter is that required to reduce the potential barrier and if it is raised slightly to, say, 0.62 V, it is likely that the collector current will double to about 2 mA. Thus a small change of voltage between the emitter and the base can give a relatively large change of current in the emitter circuit; because of this, transistors can be used as amplifiers.

12.4 Transistor connections

There are three ways of connecting a transistor, depending on the use to which it is being put. The ways are classified by the electrode that is common to both the input and the output. They are called:

(a) common-base configuration, shown in Figure 12.8(a)
(b) common-emitter configuration, shown in Figure 12.8(b)
(c) common-collector configuration, shown in Figure 12.8(c)

These configurations are for an n-p-n transistor. The current flows shown are all reversed for a p-n-p transistor.

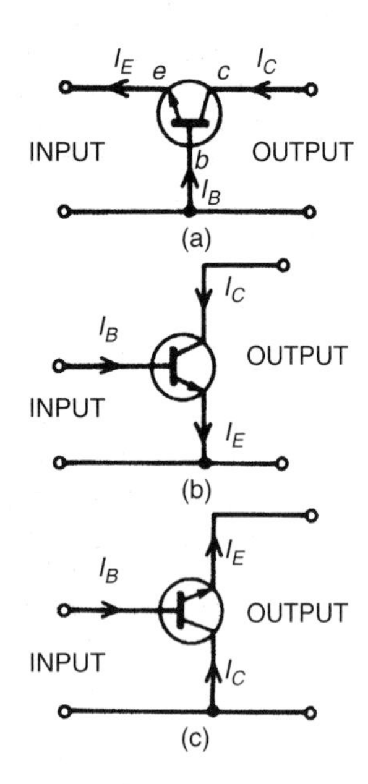

Figure 12.8

Problem 2. The basic construction of an n-p-n transistor makes it appear that the emitter and collector can be interchanged. Explain why this is not usually done.

In principle, a bipolar junction transistor will work equally well with either the emitter or collector acting as the emitter. However, the conventional emitter current largely flows from the collector through the base to the emitter, hence the emitter region is far more heavily doped with donor atoms (electrons) than the base is with acceptor atoms (holes). Also, the base-collector junction is normally reverse biased and in general, doping density increases the electric field in the junction and so lowers the breakdown voltage. Thus, to achieve a high breakdown voltage, the collector region is relatively lightly doped.

In addition, in most transistors, the method of production is to diffuse acceptor and donor atoms onto the n-type semiconductor material, one after the other, so that one overrides the other. When this is done, the doping density in the base region is not uniform but decreases from emitter to collector. This results in increasing the effectiveness of the transistor. Thus, because of the doping densities in the three regions and the non-uniform density in the base, the collector and emitter terminals of a transistor should not be interchanged when making transistor connections.

12.5 Transistor characteristics

The effect of changing one or more of the various voltages and currents associated with a transistor circuit can be shown graphically and these graphs are called the **characteristics** of the transistor. As there are five variables (collector, base and emitter currents and voltages across the collector and base and emitter and base) and also three configurations, many characteristics are possible. Some of the possible characteristics are given below.

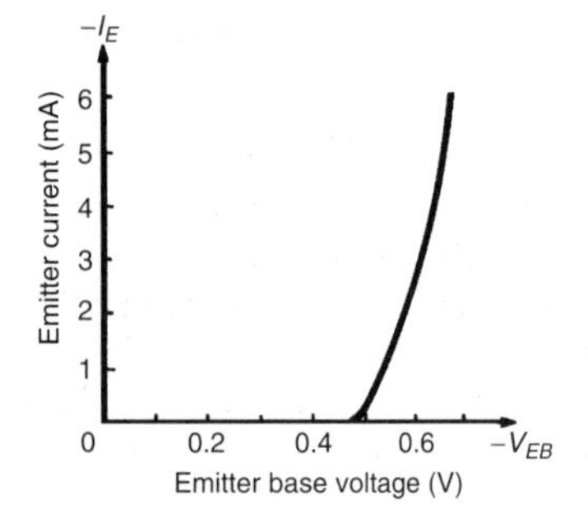

Figure 12.9

(a) Common-base configuration

(i) **Input characteristic**. With reference to Figure 12.8(a), the input to a common-base transistor is the emitter current, I_E, and can be varied by altering the base emitter voltage V_{EB}. The base-emitter junction is essentially a forward biased junction diode, so as V_{EB} is varied, the current flowing is similar to that for a junction diode, as shown in Figure 12.9 for a silicon transistor. Figure 12.9 is called the input characteristic for an n-p-n transistor having common-base configuration. The variation of the collector-base voltage V_{CB} has little effect on the characteristic. A similar characteristic can be obtained for a p-n-p transistor, these having reversed polarities.

(ii) **Output characteristics**. The value of the collector current I_C is very largely determined by the emitter current, I_E. For a given value of I_E the collector-base voltage, V_{CB}, can be varied and has little effect on the value of I_C. If V_{CB} is made slightly negative, the collector no longer

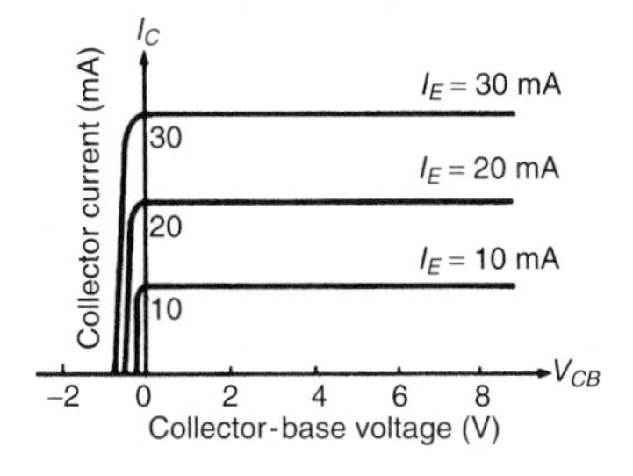

Figure 12.10

attracts the majority carriers leaving the emitter and I_C falls rapidly to zero. A family of curves for various values of I_E are possible and some of these are shown in Figure 12.10. Figure 12.10 is called the output characteristics for an n-p-n transistor having common-base configuration. Similar characteristics can be obtained for a p-n-p transistor, these having reversed polarities.

(b) Common-emitter configuration

(i) **Input characteristic**. In a common-emitter configuration (see Figure 12.8(b)), the base current is now the input current. As V_{EB} is varied, the characteristic obtained is similar in shape to the input characteristic for a common-base configuration shown in Figure 12.9, but the values of current are far less. With reference to Figure 12.6(a), as long as the junctions are biased as described, the three currents I_E, I_C and I_B keep the ratio $1 : \alpha : (1 - \alpha)$, whichever configuration is adopted. Thus the base current changes are much smaller than the corresponding emitter current changes and the input characteristic for an n-p-n transistor is as shown in Figure 12.11. A similar characteristic can be obtained for a p-n-p transistor, these having reversed polarities.

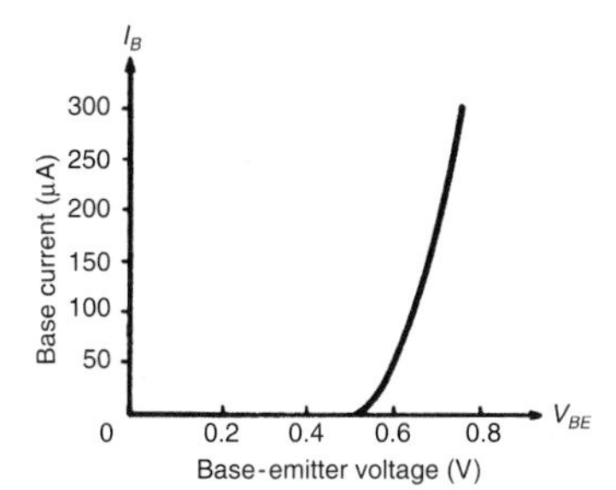

Figure 12.11

(ii) **Output characteristics**. A family of curves can be obtained, depending on the value of base current I_B and some of these for an n-p-n transistor are shown in Figure 12.12. A similar set of characteristics can be obtained for a p-n-p transistor, these having reversed polarities. These characteristics differ from the common base output characteristics in two ways:

the collector current reduces to zero without having to reverse the collector voltage, and

the characteristics slope upwards indicating a lower output resistance (usually kilohms for a common-emitter configuration compared with megohms for a common-base configuration).

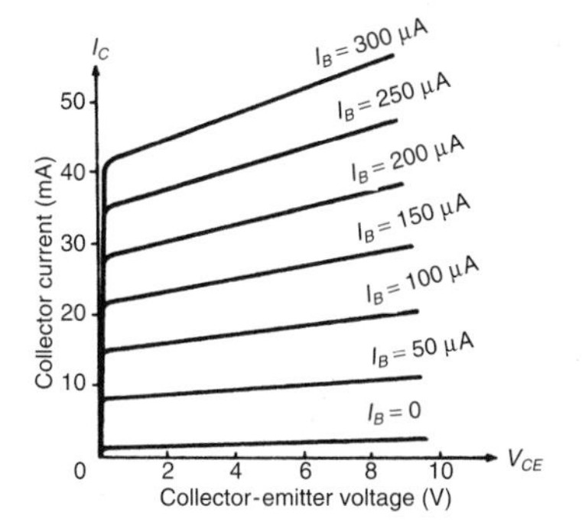

Figure 12.12

> Problem 3. With the aid of a circuit diagram, explain how the input and output characteristics of an n-p-n transistor having a common-base configuration can be obtained.

A circuit diagram for obtaining the input and output characteristics for an n-p-n transistor connected in common-base configuration is shown in Figure 12.13. The input characteristic can be obtained by varying R_1, which varies V_{EB}, and noting the corresponding values of I_E. This is repeated for various values of V_{CB}. It will be found that the input characteristic is almost independent of V_{CB} and it is usual to give only one characteristic, as shown in Figure 12.9.

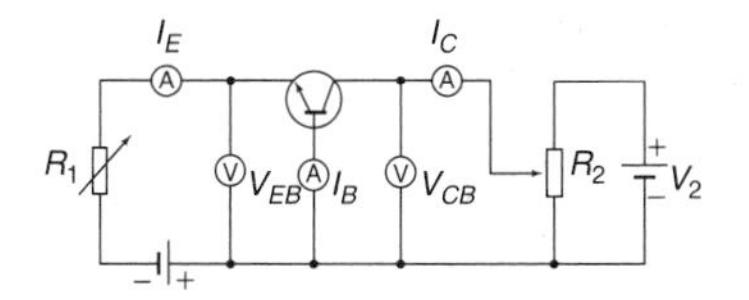

Figure 12.13

To obtain the output characteristics, as shown in Figure 12.10, I_E is set to a suitable value by adjusting R_1. For various values of V_{CB}, set by adjusting R_2, I_C is noted. This procedure is repeated for various values of I_E. To obtain the full characteristics, the polarity of battery V_2 has to be reversed to reduce I_C to zero. This must be done very carefully or else values of I_C will rapidly increase in the reverse direction and burn out the transistor.

12.6 The transistor as an amplifier

The amplifying properties of a transistor depend upon the fact that current flowing in a low-resistance circuit is transferred to a high-resistance circuit with negligible change in magnitude. If the current then flows through a load resistance, a voltage is developed. This voltage can be many times greater than the input voltage which caused the original current flow.

(a) Common-base amplifier

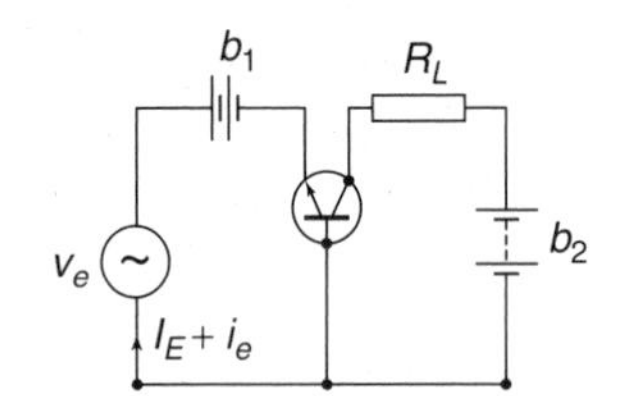

Figure 12.14

The basic circuit for a transistor is shown in Figure 12.14 where an n-p-n transistor is biased with batteries b_1 and b_2. A sinusoidal alternating input signal, v_e, is placed in series with the input bias voltage, and a load resistor, R_L, is placed in series with the collector bias voltage. The input signal is therefore the sinusoidal current i_e resulting from the application of the sinusoidal voltage v_e **superimposed** on the direct current I_E established by the base-emitter voltage V_{BE}.

Let the signal voltage v_e be 100 mV and the base-emitter circuit resistance be 50 Ω. Then the emitter signal current will be $100/50 = 2$ mA. Let the load resistance $R_L = 2.5$ kΩ. About 0.99 of the emitter current will flow in R_L. Hence the collector signal current will be about $0.99 \times 2 = 1.98$ mA and the signal voltage across the load will be $2500 \times 1.98 \times 10^{-3} = 4.95$ V. Thus a signal voltage of 100 mV at the emitter has produced a voltage of 4950 mV across the load. The voltage amplification or gain is therefore $4950/100 = 49.5$ times. This example illustrates the action of a common-base amplifier where the input signal is applied between emitter and base and the output is taken from between collector and base.

(b) Common-emitter amplifier

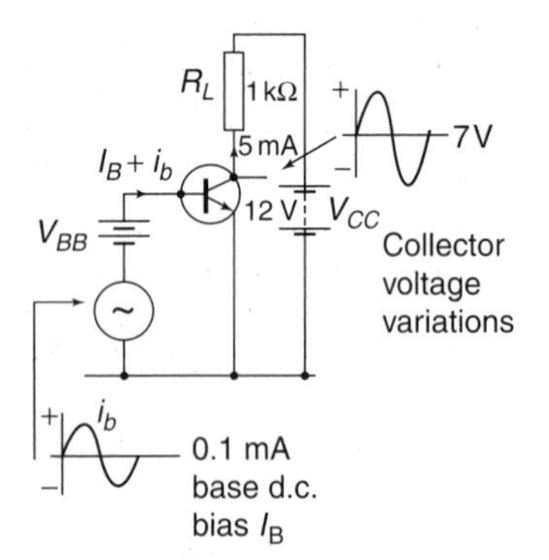

Figure 12.15

The basic circuit arrangement of a common-emitter amplifier is shown in Figure 12.15. Although two batteries are shown, it is more usual to employ only one to supply all the necessary bias. The input signal is applied between base and emitter, and the load resistor R_L is connected between collector and emitter. Let the base bias battery provide a voltage which causes a base current I_B of 0.1 mA to flow. This value of base current determines the mean d.c. level upon which the a.c. input signal will be superimposed. This is the **d.c. base current operating point**.

Let the static current gain of the transistor, α_E, be 50. Since 0.1 mA is the steady base current, the collector current I_C will be $\alpha_E \times I_B = 50 \times 0.1 = 5$ mA. This current will flow through the load resistor R_L (= 1 kΩ), and there will be a steady voltage drop across R_L given by $I_C R_L = 5 \times 10^{-3} \times 1000 = 5$ V. The voltage at the collector, V_{CE}, will therefore be $V_{CC} - I_C R_L = 12 - 5 = 7$ V. This value of V_{CE} is the mean (or quiescent) level about which the output signal voltage will swing alternately positive and negative. This is the **collector voltage d.c. operating**

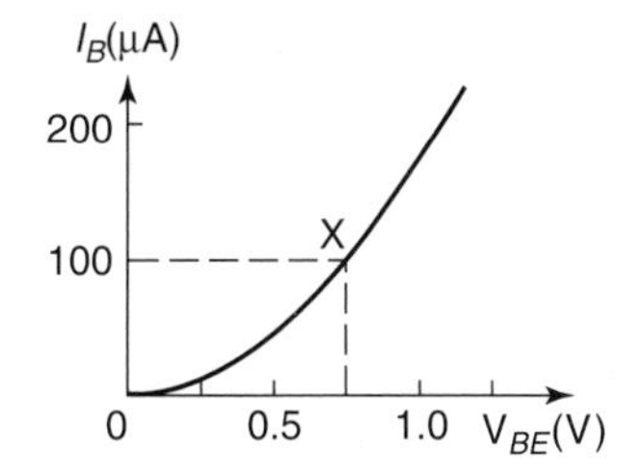

Figure 12.16

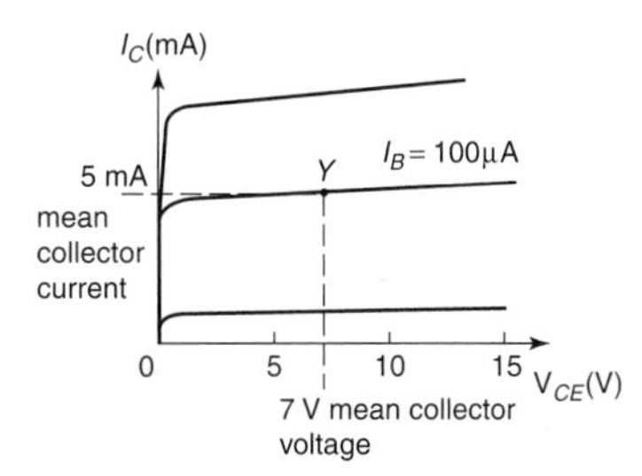

Figure 12.17

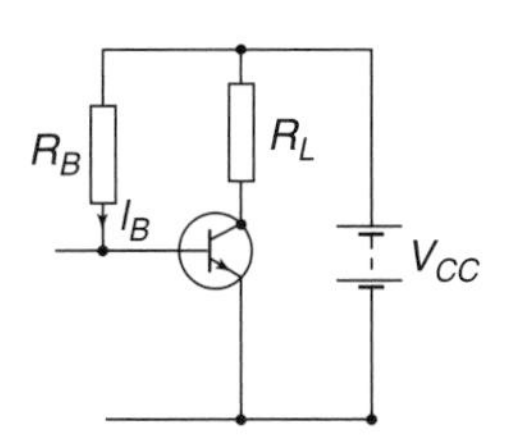

Figure 12.18

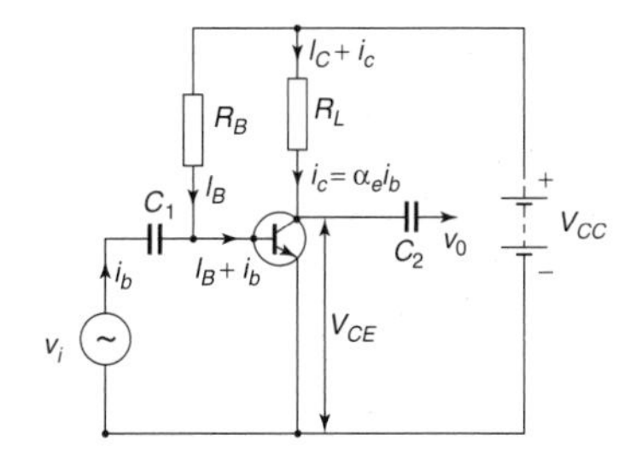

Figure 12.19

point. Both of these d.c. operating points can be pin-pointed on the input and output characteristics of the transistor. Figure 12.16 shows the I_B/V_{BE} characteristic with the operating point X positioned at $I_B = 100\,\mu A$, $V_{BE} = 0.75\,V$, say.

Figure 12.17 shows the I_C/V_{CE} characteristics, with the operating point Y positioned at $I_C = 5\,mA$, $V_{CE} = 7V$. It is usual to choose the operating point Y somewhere near the centre of the graph.

It is possible to remove the bias battery V_{BB} and obtain base bias from the collector supply battery V_{CC} instead. The simplest way to do this is to connect a bias resistor R_B between the positive terminal of the V_{CC} supply and the base as shown in Figure 12.18. The resistor must be of such a value that it allows $100\,\mu A$ to flow in the base-emitter diode.

For a silicon transistor, the voltage drop across the junction for forward bias conditions is about 0.6 V. The voltage across R_B must then be $12 - 0.6 = 11.4\,V$. Hence, the value of R_B must be such that $I_B \times R_B = 11.4\,V$, i.e.

$$R_B = \frac{11.4}{I_B} = \frac{11.4}{100 \times 10^{-6}} = 114\,k\Omega.$$

With the inclusion of the $1\,k\Omega$ load resistor, R_L, a steady 5 mA collector current, and a collector-emitter voltage of 7 V, the d.c. conditions are established.

An alternating input signal (v_i) can now be applied. In order not to disturb the bias condition established at the base, the input must be fed to the base by way of a capacitor C_1. This will permit the alternating signal to pass to the base but will prevent the passage of direct current. The reactance of this capacitor must be such that it is very small compared with the input resistance of the transistor. The circuit of the amplifier is now as shown in Figure 12.19. The a.c. conditions can now be determined.

When an alternating signal voltage v_1 is applied to the base via capacitor C_1 the base current i_b varies. When the input signal swings positive, the base current increases; when the signal swings negative, the base current decreases. The base current consists of two components: I_B, the static base bias established by R_B, and i_b, the signal current. The current variation i_b will in turn vary the collector current, i_c. The relationship between i_c and i_b is given by $i_c = \alpha_e i_b$, where α_e is the **dynamic current gain** of the transistor and is not quite the same as the static current gain α_E; the difference is usually small enough to be insignificant.

The current through the load resistor R_L also consists of two components: I_C, the static collector current, and i_c, the signal current. As i_b increases, so does i_c and so does the voltage drop across R_L. Hence, from the circuit:

$$V_{CE} = V_{CC} - (I_C + i_c)R_L$$

The d.c. components of this equation, though necessary for the amplifier to operate at all, need not be considered when the a.c. signal conditions are being examined. Hence, the signal voltage variation relationship is:

$$v_{ce} = -\alpha_e \times i_b \times R_L = i_c R_L$$

the negative sign being added because v_{ce} decreases when i_b increases and vice versa. The signal output and input voltages are of opposite polarity, i.e. a phase shift of 180° has occurred. So that the collector d.c. potential is not passed on to the following stage, a second capacitor, C_2, is added as shown in Figure 12.19. This removes the direct component but permits the signal voltage $v_o = i_c R_L$ to pass to the output terminals.

12.7 The load line

The relationship between the collector-emitter voltage (V_{CE}) and collector current (I_C) is given by the equation: $V_{CE} = V_{CC} - I_C R_L$ in terms of the **d.c. conditions**. Since V_{CC} and R_L are constant in any given circuit, this represents the equation of a straight line which can be written in the $y = mx + c$ form. Transposing $V_{CE} = V_{CC} - I_C R_L$ for I_C gives:

$$I_C = \frac{V_{CC} - V_{CE}}{R_L} = \frac{V_{CC}}{R_L} - \frac{V_{CE}}{R_L} = -\left(\frac{1}{R_L}\right) V_{CE} + \frac{V_{CC}}{R_L}$$

i.e. $$I_C = -\left(\frac{1}{R_L}\right) V_{CE} + \frac{V_{CC}}{R_L}$$

which is of the straight line form $y = mx + c$; hence if I_c is plotted vertically and V_{CE} horizontally, then the gradient is given by $-(1/R_L)$ and the vertical axis intercept is V_{CC}/R_L.

A family of collector static characteristics drawn on such axes is shown in Figure 12.12 on page 151, and so the line may be superimposed on these as shown in Figure 12.20.

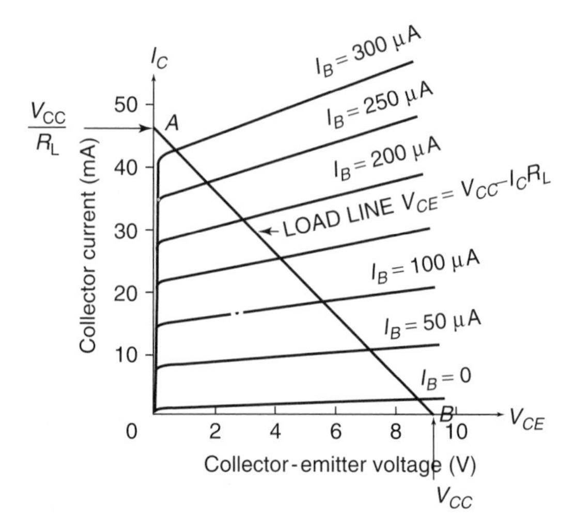

Figure 12.20

The reason why this line is necessary is because the static curves relate I_C to V_{CE} for a series of fixed values of I_B. When a signal is applied to the base of the transistor, the base current varies and can instantaneously take any of the values between the extremes shown. Only two points are necessary to draw the line and these can be found conveniently by considering extreme conditions. From the equation:

$$V_{CE} = V_{CC} - I_C R_L$$

(i) when $I_C = 0, V_{CE} = V_{CC}$ (ii) when $V_{CE} = 0, I_C = V_{CC}/R_L$

Thus the points A and B respectively are located on the axes of the I_C/V_{CE} characteristics. This line is called the **load line** and it is dependent for its position upon the value of V_{CC} and for its gradient upon R_L. As the gradient is given by $-(1/R_L)$, the slope of the line is negative.

For every value assigned to R_L in a particular circuit there will be a corresponding (and different) load line. If V_{CC} is maintained constant, all the possible lines will start at the same point (B) but will cut the I_C axis at different points A. Increasing R_L will reduce the gradient of the line and vice-versa. Quite clearly the collector voltage can never exceed V_{CC} (point B) and equally the collector current can never be greater than that value which would make V_{CE} zero (point A).
Using the circuit example of Figure 12.15, we have

$$V_{CE} = V_{CC} = 12\,\text{V}, \quad \text{when } I_C = 0$$

$$I_C = \frac{V_{CC}}{R_L} = \frac{12}{1000} = 12\,\text{mA}, \quad \text{when } V_{CE} = 0$$

The load line is drawn on the characteristics shown in Figure 12.21, which we assume are characteristics for the transistor used in the circuit of Figure 12.15 earlier. Notice that the load line passes through the operating point X, as it should, since every position on the line represents a relationship between V_{CE} and I_C for the particular values of V_{CC} and R_L given. Suppose that the base current is caused to vary ± 0.1 mA about the d.c. base bias of 0.1 mA. The result is I_B changes from 0 mA to 0.2 mA and back again to 0 mA during the course of each input cycle. Hence the operating point moves up and down the load line in phase with the input current and hence the input voltage. A sinusoidal input cycle is shown on Figure 12.21.

12.8 Current and voltage gains

The output signal voltage (v_{ce}) and current (i_c) can be obtained by projecting vertically from the load line on to V_{CE} and I_C axes respectively. When the input current i_b varies sinusoidally as shown in Figure 12.21, then v_{ce} varies sinusoidally if the points E and F at the extremities of the input variations are equally spaced on either side of X.

The peak to peak output voltage is seen to be 8.5 V, giving an r.m.s. value of 3 V (i.e. $0.707 \times 8.5/2$). The peak to peak output current is 8.75 mA, giving an r.m.s. value of 3.1 mA. From these figures the voltage and current amplifications can be obtained.

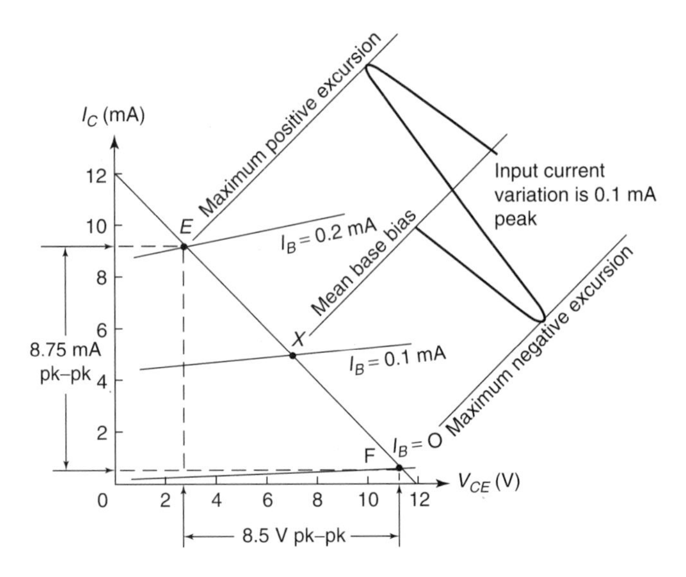

Figure 12.21

The **dynamic current gain** A_i ($= \alpha_e$ as opposed to the static gain α_E), is given by:

$$A_i = \frac{\textbf{change in collector current}}{\textbf{change in base current}}$$

This always leads to a different figure from that obtained by the direct division of I_C/I_B which assumes that the collector load resistor is zero. From Figure 12.21 the peak input current is 0.1 mA and the peak output current is 4.375 mA. Hence

$$A_i = \frac{4.375 \times 10^{-3}}{0.1 \times 10^{-3}} = 43.75$$

The **voltage gain** A_v is given by:

$$A_v = \frac{\textbf{change in collector voltage}}{\textbf{change in base voltage}}$$

This cannot be calculated from the data available, but if we assume that the base current flows in the input resistance, then the base voltage can be determined. The input resistance can be determined from an input characteristic such as was shown earlier.

Then $\quad R_i = \dfrac{\text{change in } V_{BC}}{\text{change in } I_B} \quad$ and $\quad v_i = i_b R_C$ and $v_o = i_c R_L$

and $$A_v = \frac{i_c R_L}{I_b R_i} = \alpha_e \frac{R_L}{R_i}$$

For a resistive load, **power gain, A_p**, is given by

$$\mathbf{A_p = A_v \times A_i}$$

Problem 4. An n-p-n transistor has the following characteristics, which may be assumed to be linear between the values of collector voltage stated.

Base current (μA)	Collector current (mA) for collector voltages of	
	1 V	5 V
30	1.4	1.6
50	3.0	3.5
70	4.6	5.2

The transistor is used as a common-emitter amplifier with load resistor $R_L = 1.2\,\text{k}\Omega$ and a collector supply of 7 V. The signal input resistance is 1 kΩ. Estimate the voltage gain A_v, the current gain A_i and the power gain A_p when an input current of 20 μA peak varies sinusoidally about a mean bias of 50 μA.

The characteristics are drawn in Figure 12.22. The load line equation is $V_{CC} = V_{CE} - I_C R_L$ which enables the extreme points of the line to be calculated.

When $I_C = 0, V_{CE} = V_C = 7.0\,\text{V}$

and when $V_{CE} = 0, |I_C| = \dfrac{V_{CC}}{R_L} = \dfrac{7}{1200} = 5.83\,\text{mA}$

The load line is shown superimposed on the characteristic curves with the operating point marked X at the intersection of the line and the 50 μA characteristic.

From the diagram, the output voltage swing is 3.6 V peak to peak. The input voltage swing is $i_b R_i$ where i_b is the base current swing and R_i is the input resistance.

Therefore $v_i = (70 - 30) \times 10^{-6} \times 1 \times 10^3 = 40\,\text{mV}$ peak to peak

Hence, **voltage gain, A_v** $= \dfrac{\text{output volts}}{\text{input volts}} = \dfrac{3.6}{40 \times 10^{-3}} = \mathbf{90}$

Note that peak to peak values are taken at both input and output. There is no need to convert to r.m.s. as only ratios are involved.

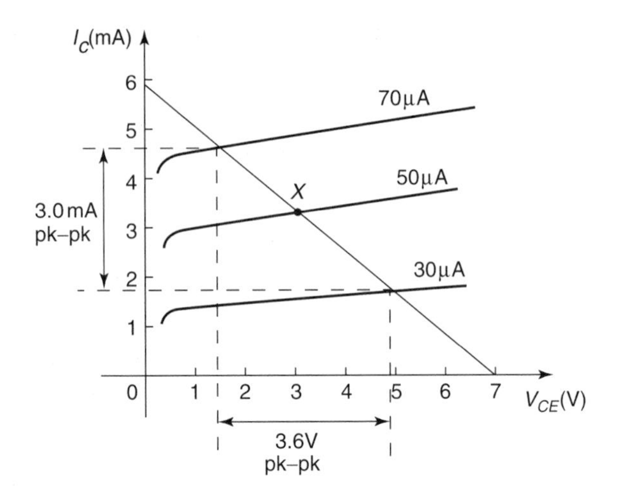

Figure 12.22

From the diagram, the output current swing is 3.0 mA peak to peak. The input base current swing is 40 μA peak to peak.

Hence, **current gain,** $A_i = \dfrac{\text{output current}}{\text{input current}} = \dfrac{3 \times 10^{-3}}{40 \times 10^{-6}} = \mathbf{75}$

For a resistive load R_L the **power gain**, $\mathbf{A_p}$, is given by:

$\mathbf{A_p}$ = voltage gain × current gain = $\mathbf{A_v} \times \mathbf{A_i} = 90 \times 75 = \mathbf{6750}$

12.9 Thermal runaway

When a transistor is used as an amplifier it is necessary to ensure that it does not overheat. Overheating can arise from causes outside of the transistor itself, such as the proximity of radiators or hot resistors, or within the transistor as the result of dissipation by the passage of current through it. Power dissipated within the transistor, which is given approximately by the product $I_C V_{CE}$, is wasted power; it contributes nothing to the signal output power and merely raises the temperature of the transistor. Such overheating can lead to very undesirable results.

The increase in the temperature of a transistor will give rise to the production of hole electron pairs, hence an increase in leakage current represented by the additional minority carriers. In turn, this leakage current leads to an increase in collector current and this increases the product $I_C V_{CE}$. The whole effect thus becomes self-perpetuating and results in **thermal runaway**. This rapidly leads to the destruction of the transistor.

Problem 5. Explain how thermal runaway might be prevented in a transistor.

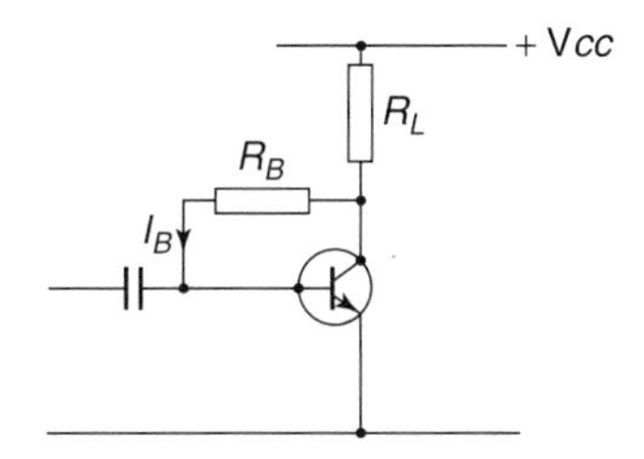

Figure 12.23

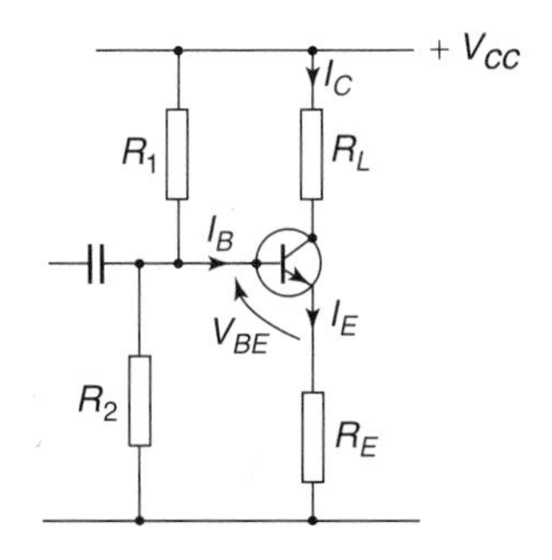

Figure 12.24

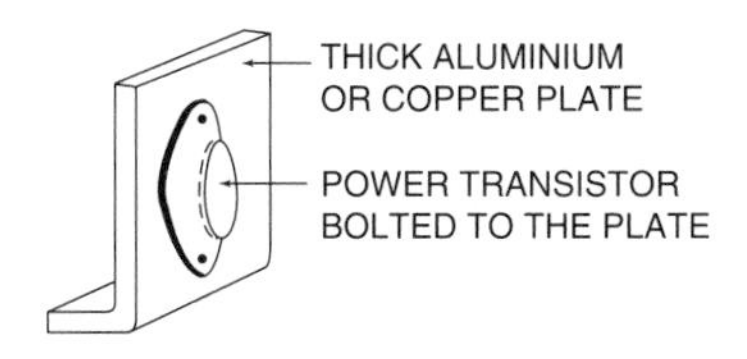

Figure 12.25

Two basic methods are available and either or both may be used in a particular application.

Method 1 is in the circuit design itself. The use of a single biasing resistor R_B as shown earlier in Figure 12.18 is not particularly good practice. If the temperature of the transistor increases, the leakage current also increases. The collector current, collector voltage and base current are thereby changed, the base current decreasing as I_C increases. An alternative is shown in Figure 12.23. Here the resistor R_B is returned, not to the V_{CC} line, but to the collector itself.

If the collector current increases for any reason, the collector voltage V_{CE} will fall. Therefore, the d.c. base current I_B will fall, since $I_B = V_{CE}/R_B$. Hence the collector current $I_C = \alpha_E I_B$ will also fall and compensate for the original increase.

A commonly used bias arrangement is shown in Figure 12.24. If the total resistance value of resistors R_1 and R_2 is such that the current flowing through the divider is large compared with the d.c. bias current I_B, then the base voltage V_{BE} will remain substantially constant regardless of variations in collector current. The emitter resistor R_E in turn determines the value of emitter current which flows for a given base voltage at the junction of R_1 and R_2. Any increase in I_C produces an increase in I_E and a corresponding increase in the voltage drop across R_E. This reduces the forward bias voltage V_{BE} and leads to a compensating reduction in I_C.

Method 2 concerns some means of keeping the transistor temperature down by external cooling. For this purpose, a heat sink is employed, as shown in Figure 12.25. If the transistor is clipped or bolted to a large conducting area of aluminium or copper plate (which may have cooling fins), cooling is achieved by convection and radiation.

Heat sinks are usually blackened to assist radiation and are normally used where large power dissipation's are involved. With small transistors, heat sinks are unnecessary. Silicon transistors particularly have such small leakage currents that thermal problems rarely arise.

12.10 Further problems on transistors

1 Explain with the aid of sketches, the operation of an n-p-n transistor and also explain why the collector current is very nearly equal to the emitter current.

2 Explain what is meant by the term 'transistor action'.

3 Describe the basic principle of operation of a bipolar junction transistor including why majority carriers crossing into the base from the emitter pass to the collector and why the collector current is almost unaffected by the collector potential.

4 For a transistor connected in common-emitter configuration, sketch the output characteristics relating collector current and the collector-emitter voltage, for various values of base current. Explain the shape of the characteristics.

5 Sketch the input characteristic relating emitter current and the emitter-base voltage for a transistor connected in common-base configuration, and explain its shape.

6 With the aid of a circuit diagram, explain how the output characteristics of an n-p-n transistor having common-base configuration may be obtained and any special precautions which should be taken.

7 Draw sketches to show the direction of the flow of leakage current in both n-p-n and p-n-p transistors. Explain the effect of leakage current on a transistor connected in common-base configuration.

8 Using the circuit symbols for transistors show how (a) common-base, and (b) common-emitter configuration can be achieved. Mark on the symbols the inputs, the outputs, polarities under normal operating conditions to give correct biasing and current directions.

9 Draw a diagram showing how a transistor can be used in common emitter configuration. Mark on the sketch the input and output polarities under normal operating conditions, and current directions.

10 Sketch the circuit symbols for (a) a p-n-p and (b) an n-p-n transistor. Mark on the emitter electrodes the direction of conventional current flow and explain why the current flows in the direction indicated.

11 State whether the following statements are true or false:

(a) The purpose of a transistor amplifier is to increase the frequency of an input signal

(b) The gain of an amplifier is the ratio of the output signal amplitude to the input signal amplitude

(c) The output characteristics of a transistor relate the collector current to the base voltage.

(d) The equation of the load line is $V_{CE} = V_{CC} - I_C R_L$

(e) If the load resistor value is increased the load line gradient is reduced

(f) In a common-emitter amplifier, the output voltage is shifted through 180° with reference to the input voltage

(g) In a common-emitter amplifier, the input and output currents are in phase

(h) If the temperature of a transistor increases, V_{BE}, I_C and α_E all increase

(i) A heat sink operates by artificially increasing the surface area of a transistor

(j) The dynamic current gain of a transistor is always greater than the static current.

[(a) false (b) true (c) false (d) true (e) true (f) true (g) true (h) false (V_{BE} decreases) (i) true (j) true]

12 An amplifier has $A_i = 40$ and $A_v = 30$. What is the power gain ?

[1200]

13 What will be the gradient of a load line for a load resistor of value 4 kΩ? What unit is the gradient measured in? [−1/4000 siemen]

14 A transistor amplifier, supplied from a 9 V battery, requires a d.c. bias current of 100 μA. What value of bias resistor would be connected from base to the V_{CC} line (a) if V_{CE} is ignored (b) if V_{CE} is 0.6 V?
[(a) 90 kΩ (b) 84 kΩ]

15 The output characteristics of a transistor in common-emitter configuration can be regarded as straight lines connecting the following points

	$I_B = 20\,\mu A$		50 μA		80 μA	
V_{CE} (V)	1.0	8.0	1.0	8.0	1.0	8.0
I_C (mA)	1.2	1.4	3.4	4.2	6.1	8.1

Plot the characteristics and superimpose the load line for a 1 kΩ load, given that the supply voltage is 9 V and the d.c. base bias is 50 μA. The signal input resistance is 800 Ω. When a peak input current of 30 μA varies sinusoidally about a mean bias of 50 μA, determine (a) the quiescent collector current (b) the current gain (c) the voltage gain (d) the power gain

[(a) 4 mA (b) 104 (c) 83 (d) 8632]

16 Explain briefly what is meant by 'thermal runaway'.

Assignment 3

This assignment covers the material contained in chapters 8 to 12.

The marks for each question are shown in brackets at the end of each question.

1 A conductor, 25 cm long, is situated at right angles to a magnetic field. Determine the strength of the magnetic field if a current of 12 A in the conductor produces a force on it of 4.5 N. (3)

2 An electron in a television tube has a charge of 1.5×10^{-19} C and travels at 3×10^{7} m/s perpendicular to a field of flux density 20 µT. Calculate the force exerted on the electron in the field. (3)

3 A lorry is travelling at 100 km/h. Assuming the vertical component of the earth's magnetic field is 40 µT and the back axle of the lorry is 1.98 m, find the e.m.f. generated in the axle due to motion. (5)

4 An e.m.f. of 2.5 kV is induced in a coil when a current of 2 A collapses to zero in 5 ms. Calculate the inductance of the coil. (4)

5 Two coils, P and Q, have a mutual inductance of 100 mH. If a current of 3 A in coil P is reversed in 20 ms, determine (a) the average e.m.f. induced in coil Q, and (b) the flux change linked with coil Q if it wound with 200 turns. (5)

6 A moving coil instrument gives a f.s.d. when the current is 50 mA and has a resistance of 40 Ω. Determine the value of resistance required to enable the instrument to be used (a) as a 0–5 A ammeter, and (b) as a 0–200 V voltmeter. State the mode of connection in each case. (6)

7 An amplifier has a gain of 20 dB. It's input power is 5 mW. Calculate it's output power. (3)

8 A sinusoidal voltage trace displayed on a c.r.o. is shown in Figure A3.1; the 'time/cm' switch is on 50 ms and the 'volts/cm' switch is on 2 V/cm. Determine for the waveform (a) the frequency (b) the peak-to-peak voltage (c) the amplitude (d) the r.m.s. value. (7)

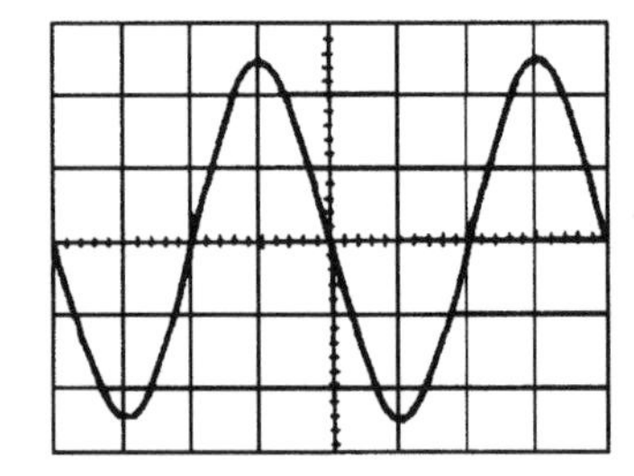

Figure A3.1

9 With reference to a p-n junction, briefly explain the terms:

(a) majority carriers (b) contact potential (c) depletion layer (d) forward bias (e) reverse bias. (5)

10 The output characteristics of a common-emitter transistor amplifier are given below. Assume that the characteristics are linear between

the values of collector voltage stated.

	$I_B = 10\,\mu A$		$40\,\mu A$		$70\,\mu A$	
V_{CE} (V)	1.0	7.0	1.0	7.0	1.0	7.0
I_C (mA)	0.6	0.7	2.5	2.9	4.6	5.35

Plot the characteristics and superimpose the load line for a $1.5\,k\Omega$ load resistor and collector supply voltage of 8 V. The signal input resistance is $1.2\,k\Omega$. Determine (a) the voltage gain, (b) the current gain, and (c) the power gain when an input current of $30\,\mu A$ peak varies sinusoidally about a mean bias of $40\,\mu A$ (9)

Main formulae for Part 1

General

Charge $Q = It$ Force $F = ma$ Work $W = Fs$ Power $P = \frac{W}{t}$

Energy $W = Pt$

Ohm's law $V = IR$ or $I = \frac{V}{R}$ or $R = \frac{V}{I}$ Conductance $G = \frac{1}{R}$

Power $P = VI = I^2R = \frac{V^2}{R}$ Resistance $R = \frac{\rho l}{a}$

Resistance at $\theta°C$, $R_\theta = R_0(1 + \alpha_0\theta)$

Terminal p.d. of source, $V = E - Ir$

Series circuit $R = R_1 + R_2 + R_3 + \cdots$

Parallel network $\frac{1}{R} = \frac{1}{R_1} + \frac{1}{R_2} + \frac{1}{R_3} + \cdots$

Capacitors and capacitance

$E = \frac{V}{d}$ $C = \frac{Q}{V}$ $Q = It$ $D = \frac{Q}{A}$ $\frac{D}{E} = \varepsilon_0\varepsilon_r$

$C = \frac{\varepsilon_0\varepsilon_r A(n-1)}{d}$

Capacitors in parallel $C = C_1 + C_2 + C_3 + \cdots$

Capacitors in series $\frac{1}{C} = \frac{1}{C_1} + \frac{1}{C_2} + \frac{1}{C_3} + \cdots$ $W = \frac{1}{2}CV^2$

Magnetic circuits

$B = \frac{\Phi}{A}$ $F_m = NI$ $H = \frac{NI}{l}$ $\frac{B}{H} = \mu_0\mu_r$

$S = \frac{\text{mmf}}{\Phi} = \frac{1}{\mu_0\mu_r A}$

Electromagnetism

$F = Bil\sin\theta$ $F = QvB$

Electromagnetic induction

$E = Blv\sin\theta$ $E = -N\frac{d\Phi}{dt} = -L\frac{dI}{dt}$ $W = \frac{1}{2}LI^2$ $L = \frac{N\Phi}{I}$

$E_2 = -M\frac{dI_1}{dt}$

Measurements

Shunt $R_s = \frac{I_a r_a}{I_s}$ Multiplier $R_M = \frac{V - Ir_a}{\text{I}}$

Power in decibels $= 10\ \log\frac{P_2}{P_1} = 20\ \log\frac{I_2}{I_1} = 20\ \log\frac{V_2}{V_1}$

Wheatstone bridge $R_x = \frac{R_2R_3}{R_1}$ Potentiometer $E_2 = E_1\left(\frac{l_2}{l_1}\right)$

Part 2 Electrical Principles and Technology

13 D.c. circuit theory

At the end of this chapter you should be able to:

- state and use Kirchhoff's laws to determine unknown currents and voltages in d.c. circuits
- understand the superposition theorem and apply it to find currents in d.c. circuits
- understand general d.c. circuit theory
- understand Thévenin's theorem and apply a procedure to determine unknown currents in d.c. circuits
- recognize the circuit diagram symbols for ideal voltage and current sources
- understand Norton's theorem and apply a procedure to determine unknown currents in d.c. circuits
- appreciate and use the equivalence of the Thévenin and Norton equivalent networks
- state the maximum power transfer theorem and use it to determine maximum power in a d.c. circuit

13.1 Introduction

The laws which determine the currents and voltage drops in d.c. networks are: (a) Ohm's law (see Chapter 2), (b) the laws for resistors in series and in parallel (see Chapter 5), and (c) Kirchhoff's laws (see Section 13.2 following). In addition, there are a number of circuit theorems which have been developed for solving problems in electrical networks. These include:

(i) the superposition theorem (see Section 13.3),
(ii) Thévenin's theorem (see Section 13.5),
(iii) Norton's theorem (see Section 13.7), and
(iv) the maximum power transfer theorem (see Section 13.8).

13.2 Kirchhoff's laws

Kirchhoff's laws state:

(a) **Current Law.** *At any junction in an electric circuit the total current flowing towards that junction is equal to the total current flowing away from the junction*, i.e. $\Sigma I = 0$
Thus, referring to Figure 13.1:

$$I_1 + I_2 = I_3 + I_4 + I_5 \quad \text{or} \quad I_1 + I_2 - I_3 - I_4 - I_5 = 0$$

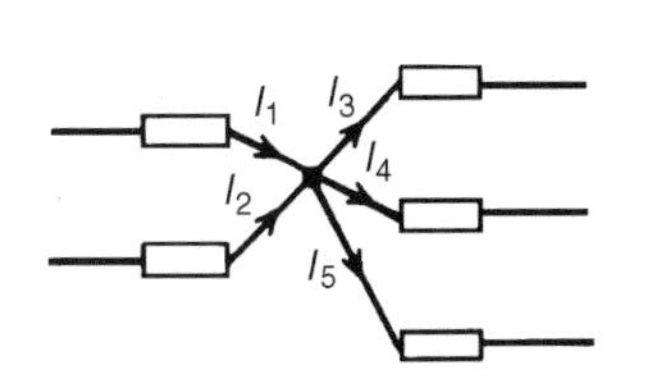

Figure 13.1

(b) **Voltage Law.** *In any closed loop in a network, the algebraic sum of the voltage drops (i.e. products of current and resistance) taken* *around the loop is equal to the resultant e.m.f. acting in that loop.*

Thus, referring to Figure 13.2: $E_1 - E_2 = IR_1 + IR_2 + IR_3$

(Note that if current flows away from the positive terminal of a source, that source is considered by convention to be positive. Thus moving anticlockwise around the loop of Figure 13.2, E_1 is positive and E_2 is negative.)

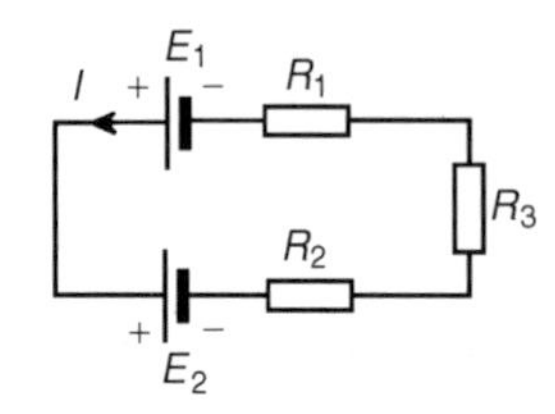

Figure 13.2

Problem 1. (a) Find the unknown currents marked in Figure 13.3(a). (b) Determine the value of e.m.f. E in Figure 13.3(b).

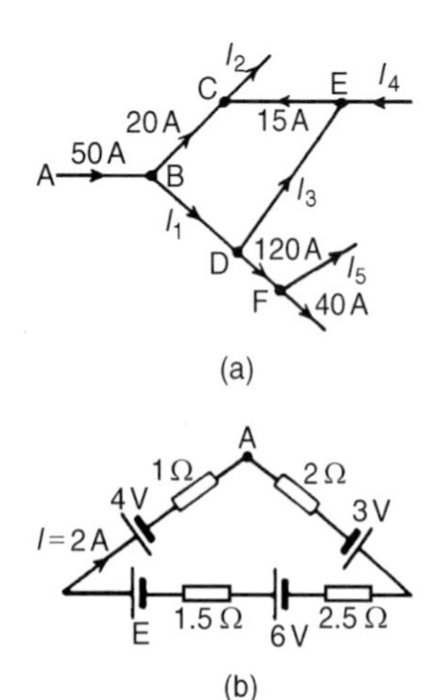

Figure 13.3

(a) Applying Kirchhoff's current law:

For junction B: $50 = 20 + I_1$. Hence $\boldsymbol{I_1 = 30}$ **A**

For junction C: $20 + 15 = I_2$. Hence $\boldsymbol{I_2 = 35}$ **A**

For junction D: $I_1 = I_3 + 120$

i.e. $30 = I_3 + 120$. Hence $\boldsymbol{I_3 = -90}$ **A**

(i.e. in the opposite direction to that shown in Figure 13.3(a))

For junction E: $I_4 + I_3 = 15$

i.e. $I_4 = 15 - (-90)$. Hence $\boldsymbol{I_4 = 105}$ **A**

For junction F: $120 = I_5 + 40$. Hence $\boldsymbol{I_5 = 80}$ **A**

(b) Applying Kirchhoff's voltage law and moving clockwise around the loop of Figure 13.3(b) starting at point A:

$$3 + 6 + E - 4 = (I)(2) + (I)(2.5) + (I)(1.5) + (I)(1)$$

$$= I(2 + 2.5 + 1.5 + 1)$$

i.e. $5 + E = 2(7)$, since $I = 2$ A

Hence $E = 14 - 5 = \mathbf{9\ V}$

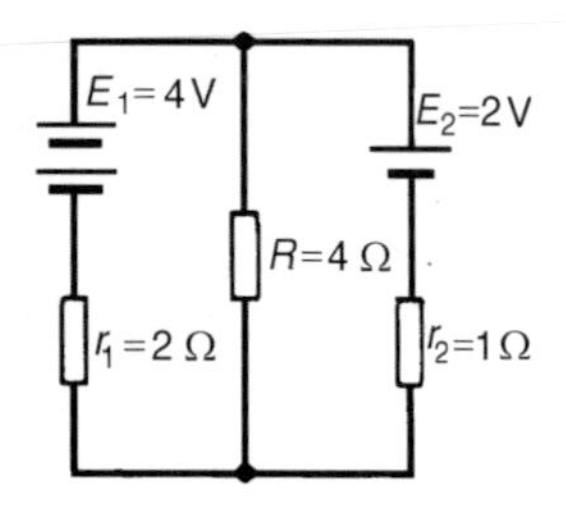

Figure 13.4

Problem 2. Use Kirchhoff's laws to determine the currents flowing in each branch of the network shown in Figure 13.4.

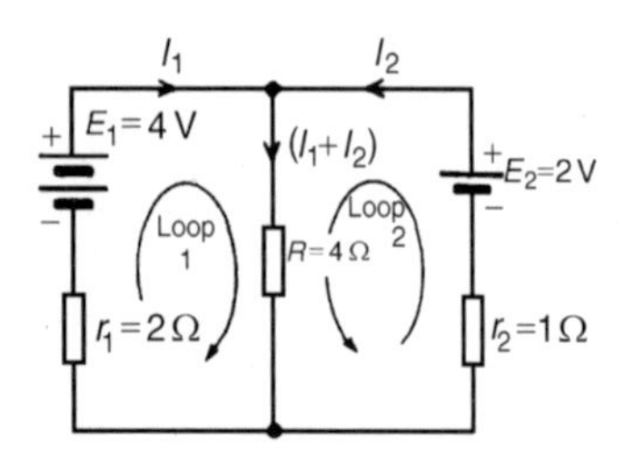

Figure 13.5

Procedure

1 Use Kirchhoff's current law and label current directions on the original circuit diagram. The directions chosen are arbitrary, but it is usual, as a starting point, to assume that current flows from the positive terminals of the batteries. This is shown in Figure 13.5 where the three branch currents are expressed in terms of I_1 and I_2 only, since the current through R is $I_1 + I_2$.

2 Divide the circuit into two loops and apply Kirchhoff's voltage law to each. From loop 1 of Figure 13.5, and moving in a clockwise direction

as indicated (the direction chosen does not matter), gives

$$E_1 = I_1 r_1 + (I_1 + I_2)R, \text{ i.e. } 4 = 2I_1 + 4(I_1 + I_2),$$

$$\text{i.e. } 6I_1 + 4I_2 = 4 \qquad (1)$$

From loop 2 of Figure 13.5, and moving in an anticlockwise direction as indicated (once again, the choice of direction does not matter; it does not have to be in the same direction as that chosen for the first loop), gives:

$$E_2 = I_2 r_2 + (I_1 + I_2)R, \text{ i.e. } 2 = I_2 + 4(I_1 + I_2),$$

$$\text{i.e. } 4I_1 + 5I_2 = 2 \qquad (2)$$

3 Solve equations (1) and (2) for I_1 and I_2.

$2 \times (1)$ gives: $12I_1 + 8I_2 = 8$ (3)

$3 \times (2)$ gives: $12I_1 + 15I_2 = 6$ (4)

$(3) - (4)$ gives: $-7I_2 = 2$ hence $I_2 = -\frac{2}{7} = \mathbf{-0.286\ A}$

(i.e. I_2 is flowing in the opposite direction to that shown in Figure 13.5.)

From (1) $6I_1 + 4(-0.286) = 4$

$6I_1 = 4 + 1.144$

Hence $I_1 = \frac{5.144}{6} = \mathbf{0.857\ A}$

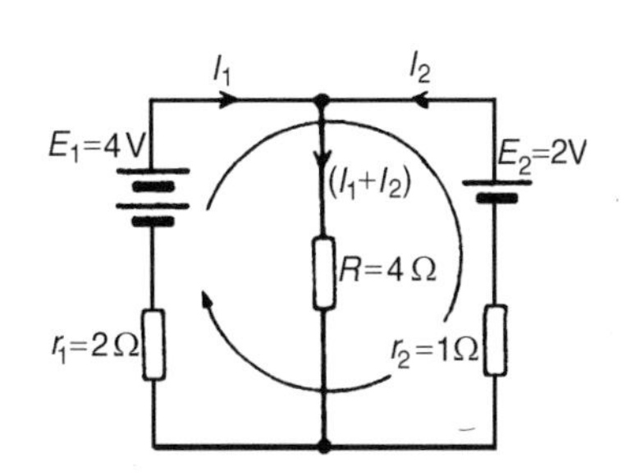

Figure 13.6

Current flowing through resistance R is

$$I_1 + I_2 = 0.857 + (-0.286) = \mathbf{0.571\ A}$$

Note that a third loop is possible, as shown in Figure 13.6, giving a third equation which can be used as a check:

$$E_1 - E_2 = I_1 r_1 - I_2 r_2$$

$$4 - 2 = 2I_1 - I_2$$

$$2 = 2I_1 - I_2$$

[Check: $2I_1 - I_2 = 2(0.857) - (-0.286) = 2$]

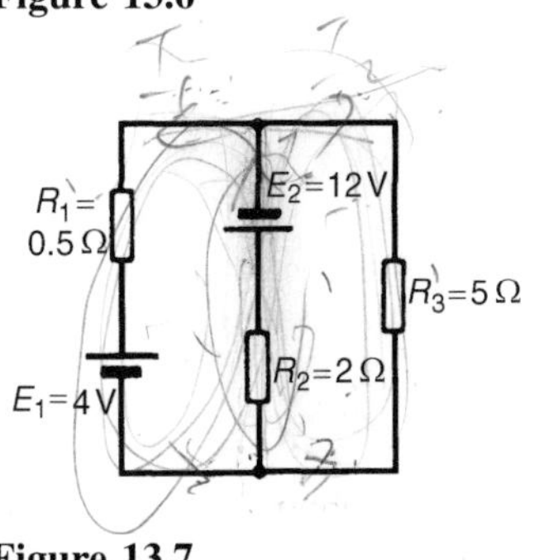

Figure 13.7

Problem 3. Determine, using Kirchhoff's laws, each branch current for the network shown in Figure 13.7.

1 Currents, and their directions are shown labelled in Figure 13.8 following Kirchhoff's current law. It is usual, although not essential,

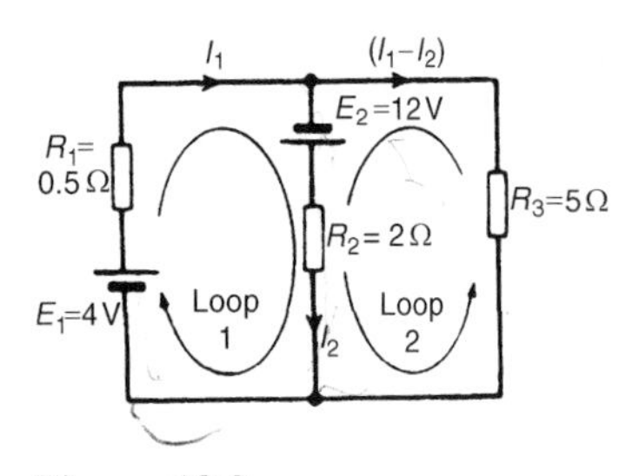

Figure 13.8

to follow conventional current flow with current flowing from the positive terminal of the source.

2 The network is divided into two loops as shown in Figure 13.8. Applying Kirchhoff's voltage law gives:

For loop 1:

$$E_1 + E_2 = I_1R_1 + I_2R_2$$

i.e. $\quad 16 = 0.5I_1 + 2I_2 \qquad (1)$

For loop 2:

$$E_2 = I_2R_2 - (I_1 - I_2)R_3$$

Note that since loop 2 is in the opposite direction to current$(I_1 - I_2)$, the volt drop across R_3 (i.e. $(I_1 - I_2)(R_3)$ is by convention negative).

Thus $\;12 = 2I_2 - 5(I_1 - I_2)$ i.e. $12 = -5I_1 + 7I_2 \qquad (2)$

3 Solving equations (1) and (2) to find I_1 and I_2:

$10 \times (1)$ gives $160 = 5I_1 + 20I_2 \qquad (3)$

$(2) + (3)$ gives $172 = 27I_2$ hence $I_2 = \dfrac{172}{27} = \mathbf{6.37\ A}$

From (1): $\;16 = 0.5I_1 + 2(6.37)$

$$I_1 = \frac{16 - 2(6.37)}{0.5} = \mathbf{6.52\ A}$$

Current flowing in R_3 $= I_1 - I_2 = 6.52 - 6.37 = \mathbf{0.15\ A}$

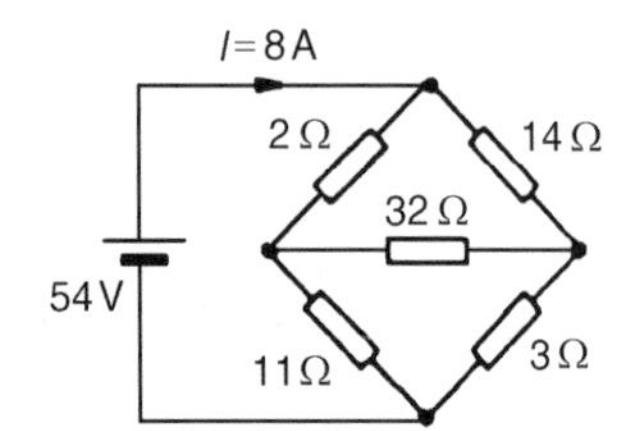

Figure 13.9

Problem 4. For the bridge network shown in Figure 13.9 determine the currents in each of the resistors.

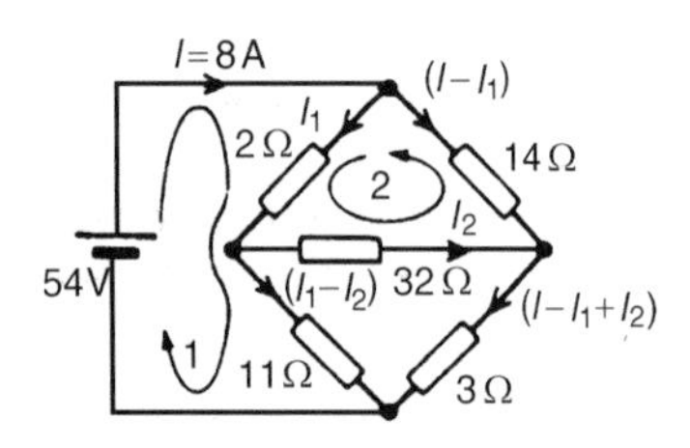

Figure 13.10

Let the current in the 2 Ω resistor be I_1, then by Kirchhoff's current law, the current in the 14 Ω resistor is $(I - I_1)$. Let the current in the 32 Ω resistor be I_2 as shown in Figure 13.10. Then the current in the 11 Ω resistor is $(I_1 - I_2)$ and that in the 3 Ω resistor is $(I - I_1 + I_2)$. Applying Kirchhoff's voltage law to loop 1 and moving in a clockwise direction as shown in Figure 13.10 gives:

$$54 = 2I_1 + 11(I_1 - I_2)$$

i.e. $\quad 13I_1 - 11I_2 = 54 \qquad (1)$

Applying Kirchhoff's voltage law to loop 2 and moving in an anticlockwise direction as shown in Figure 13.10 gives:

$$0 = 2I_1 + 32I_2 - 14(I - I_1)$$

However $\quad I = 8$ A

Hence $\quad 0 = 2I_1 + 32I_2 - 14(8 - I_1)$

i.e. $\quad 16I_1 + 32I_2 = 112 \qquad (2)$

Equations (1) and (2) are simultaneous equations with two unknowns, I_1 and I_2.

$16 \times (1)$ gives: $\quad 208I_1 - 176I_2 = 864 \qquad (3)$

$13 \times (2)$ gives: $\quad 208I_1 + 416I_2 = 1456 \qquad (4)$

$(4) - (3)$ gives: $\quad 592I_2 = 592$

$$I_2 = 1 \text{ A}$$

Substituting for I_2 in (1) gives:

$$13I_1 - 11 = 54$$

$$I_1 = \frac{65}{13} = 5\ A$$

Hence,

the current flowing in the 2 Ω resistor $= I_1 =$ **5 A**

the current flowing in the 14 Ω resistor $= I - I_1 = 8 - 5 =$ **3 A**

the current flowing in the 32 Ω resistor $= I_2 =$ **1 A**

the current flowing in the 11 Ω resistor $= I_1 - I_2 = 5 - 1 =$ **4 A** and

the current flowing in the 3 Ω resistor $= I - I_1 + I_2 = 8 - 5 + 1$

$=$ **4 A**

Further problems on Kirchhoff's laws may be found in Section 13.10, problems 1 to 6, page 189.

13.3 The superposition theorem

The **superposition theorem** states:

'In any network made up of linear resistances and containing more than one source of e.m.f., the resultant current flowing in any branch is the algebraic sum of the currents that would flow in that branch if each source was considered separately, all other sources being replaced at that time by their respective internal resistances.'

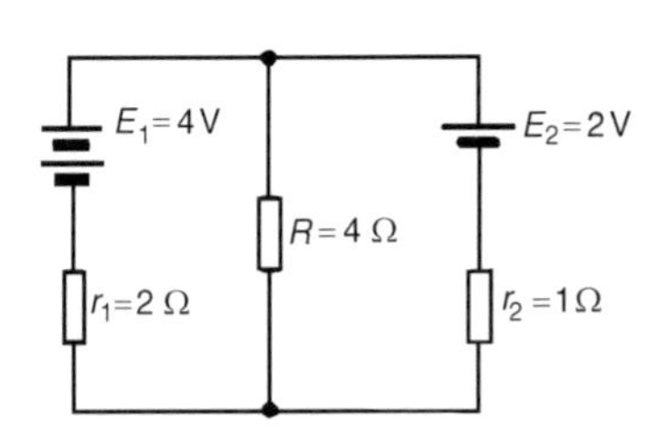

Figure 13.11

Problem 5. Figure 13.11 shows a circuit containing two sources of e.m.f., each with their internal resistance. Determine the current in each branch of the network by using the superposition theorem.

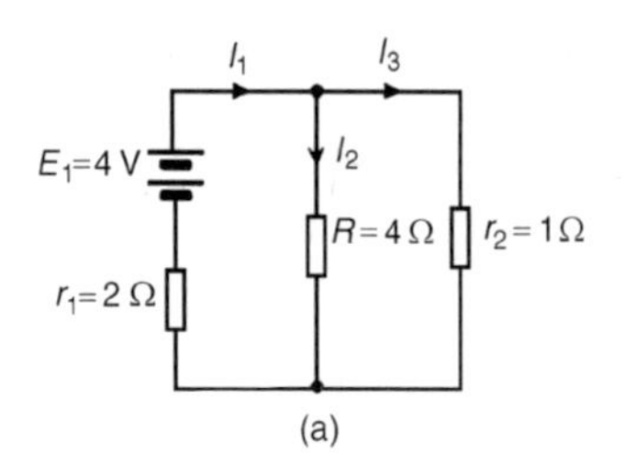

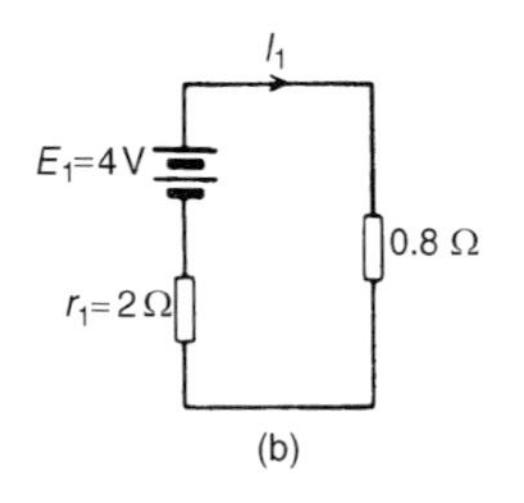

Figure 13.12

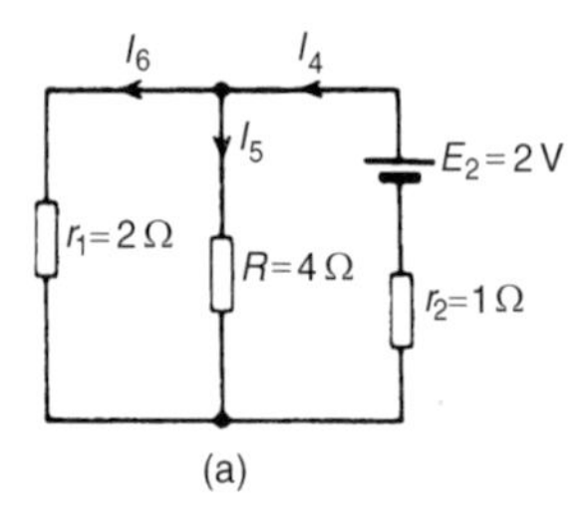

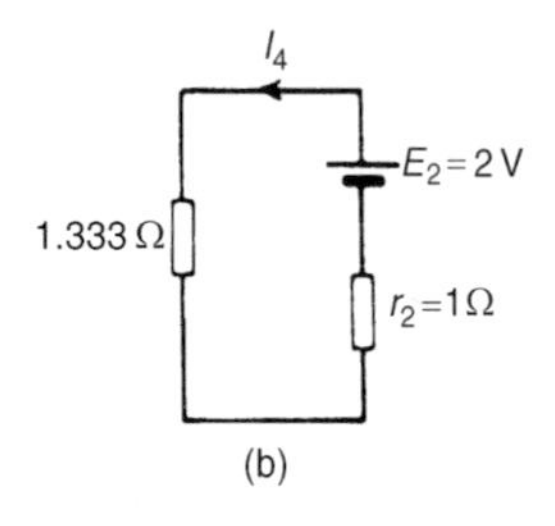

Figure 13.13

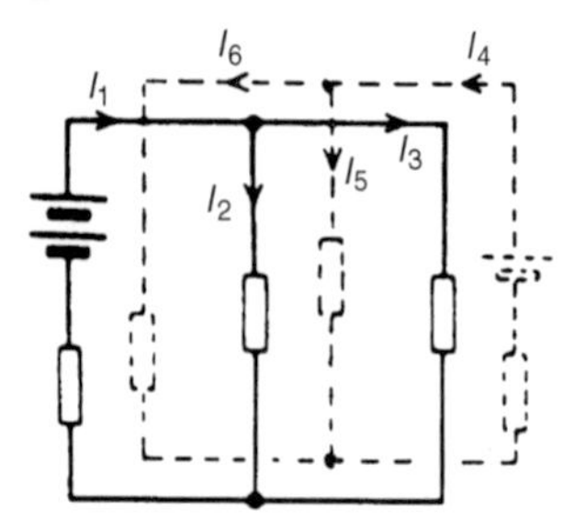

Figure 13.14

Procedure:

1 Redraw the original circuit with source E_2 removed, being replaced by r_2 only, as shown in Figure 13.12(a).

2 Label the currents in each branch and their directions as shown in Figure 13.12(a) and determine their values. (Note that the choice of current directions depends on the battery polarity, which, by convention is taken as flowing from the positive battery terminal as shown.) R in parallel with r_2 gives an equivalent resistance of:

$$\frac{4 \times 1}{4 + 1} = 0.8\ \Omega$$

From the equivalent circuit of Figure 13.12(b)

$$I_1 = \frac{E_1}{r_1 + 0.8} = \frac{4}{2 + 0.8} = 1.429\ \text{A}$$

From Figure 13.12(a)

$$I_2 = \left(\frac{1}{4 + 1}\right) I_1 = \frac{1}{5}(1.429) = 0.286\ \text{A}$$

and

$$I_3 = \left(\frac{4}{4 + 1}\right) I_1 = \frac{4}{5}(1.429) = 1.143\ \text{A by current division}$$

3 Redraw the original circuit with source E_1 removed, being replaced by r_1 only, as shown in Figure 13.13(a).

4 Label the currents in each branch and their directions as shown in Figure 13.13(a) and determine their values.
r_1 in parallel with R gives an equivalent resistance of:

$$\frac{2 \times 4}{2 + 4} = \frac{8}{6} = 1.333\ \Omega$$

From the equivalent circuit of Figure 13.13(b)

$$I_4 = \frac{E_2}{1.333 + r_2} = \frac{2}{1.333 + 1} = 0.857\ \text{A}$$

From Figure 13.13(a)

$$I_5 = \left(\frac{2}{2 + 4}\right) I_4 = \frac{2}{6}(0.857) = 0.286\ \text{A}$$

$$I_6 = \left(\frac{4}{2 + 4}\right) I_4 = \frac{4}{6}(0.857) = 0.571\ \text{A}$$

5 Superimpose Figure 13.13(a) on to Figure 13.12(a) as shown in Figure 13.14.

6 Determine the algebraic sum of the currents flowing in each branch. Resultant current flowing through source 1, i.e.

$$I_1 - I_6 = 1.429 - 0.571$$
$$= \mathbf{0.858\ A\ (discharging)}$$

Resultant current flowing through source 2, i.e.

$$I_4 - I_3 = 0.857 - 1.143$$
$$= \mathbf{-0.286\ A\ (charging)}$$

Resultant current flowing through resistor R, i.e.

$$I_2 + I_5 = 0.286 + 0.286$$
$$= \mathbf{0.572\ A}$$

Figure 13.15

The resultant currents with their directions are shown in Figure 13.15.

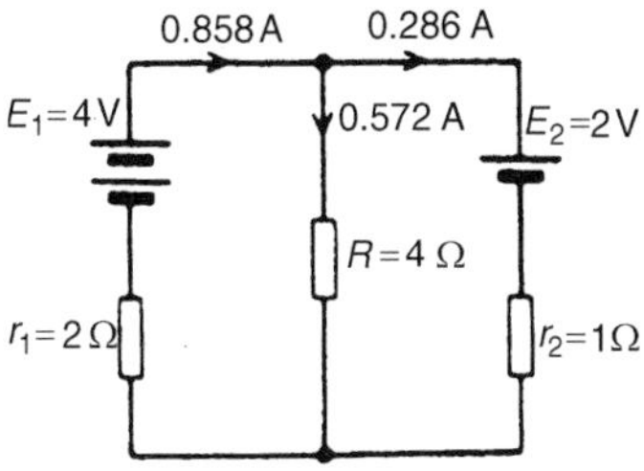

Figure 13.16

Problem 6. For the circuit shown in Figure 13.16, find, using the superposition theorem, (a) the current flowing in and the pd across the 18 Ω resistor, (b) the current in the 8 V battery and (c) the current in the 3 V battery.

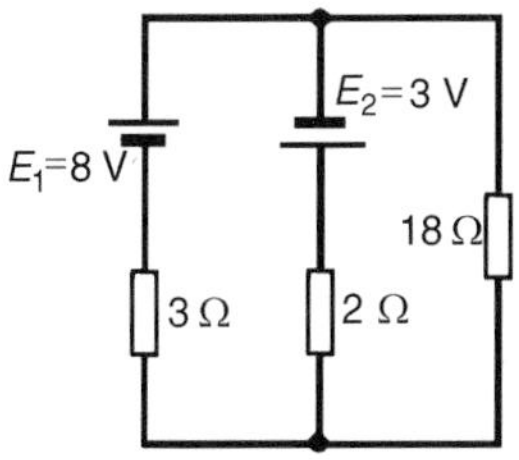

1 Removing source E_2 gives the circuit of Figure 13.17(a).

2 The current directions are labelled as shown in Figure 13.17(a), I_1 flowing from the positive terminal of E_1.

From Figure 13.17(b), $I_1 = \dfrac{E_1}{3 + 1.8} = \dfrac{8}{4.8} = 1.667$ A

From Figure 13.17(a), $I_2 = \left(\dfrac{18}{2 + 18}\right) I_1 = \dfrac{18}{20}(1.667) = 1.500$ A

and $I_3 = \left(\dfrac{2}{2 + 18}\right) I_1 = \dfrac{2}{20}(1.667) = 0.167$ A

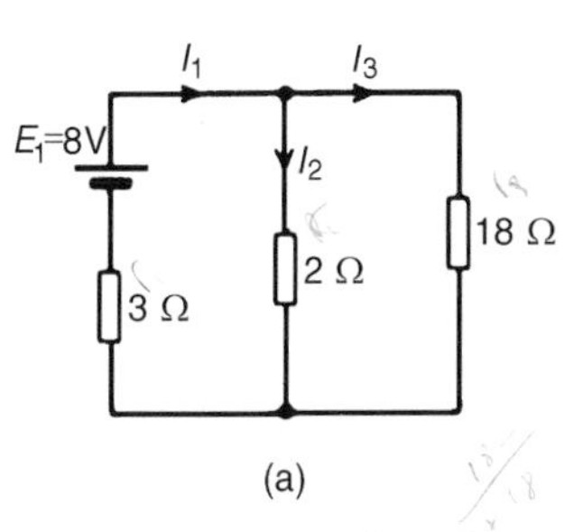

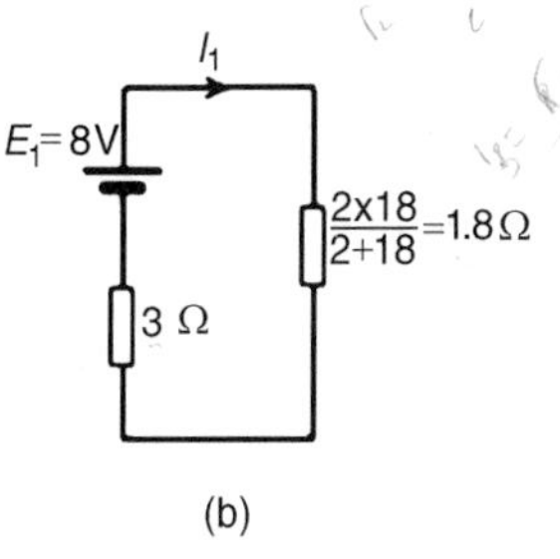

Figure 13.17

3 Removing source E_1 gives the circuit of Figure 13.18(a) (which is the same as Figure 13.18(b)).

4 The current directions are labelled as shown in Figures 13.18(a) and 13.18(b), I_4 flowing from the positive terminal of E_2

From Figure 13.18(c), $I_4 = \dfrac{E_2}{2 + 2.571} = \dfrac{3}{4.571} = 0.656$ A

From Figure 13.18(b), $I_5 = \left(\dfrac{18}{3 + 18}\right) I_4 = \dfrac{18}{21}(0.656) = 0.562$ A

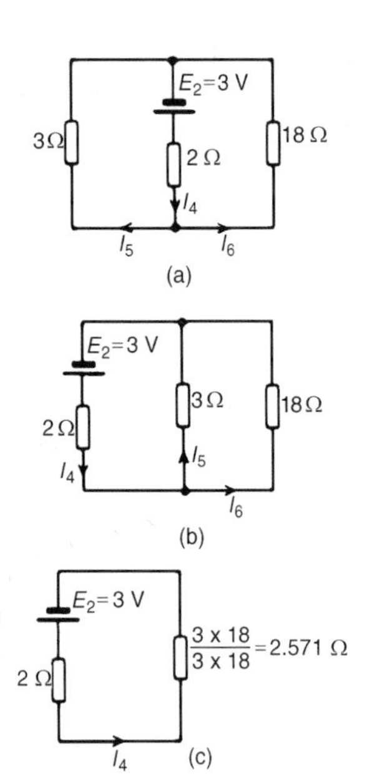

Figure 13.18

$$I_6 = \left(\frac{3}{3+18}\right) I_4 = \frac{3}{21}(0.656) = 0.094 \text{ A}$$

5 Superimposing Figure 13.18(a) on to Figure 13.17(a) gives the circuit in Figure 13.19.

6 (a) Resultant current in the 18 Ω resistor $= I_3 - I_6$

$$= 0.167 - 0.094$$

$$= \mathbf{0.073\ A}$$

P.d. across the 18 Ω resistor $= 0.073 \times 18 =$ **1.314 V**

(b) Resultant current in the 8 V battery $= I_1 + I_5 = 1.667 + 0.562$

$$= \mathbf{2.229\ A\ (discharging)}$$

(c) Resultant current in the 3 V battery $= I_2 + I_4 = 1.500 + 0.656$

$$= \mathbf{2.156\ A\ (discharging)}$$

Further problems on the superposition theorem may be found in Section 13.10, problems 7 to 10, page 190.

13.4 General d.c. circuit theory

The following points involving d.c. circuit analysis need to be appreciated before proceeding with problems using Thévenin's and Norton's theorems:

(i) The open-circuit voltage, E, across terminals AB in Figure 13.20 is equal to 10 V, since no current flows through the 2 Ω resistor and hence no voltage drop occurs.

(ii) The open-circuit voltage, E, across terminals AB in Figure 13.21(a) is the same as the voltage across the 6 Ω resistor. The circuit may be redrawn as shown in Figure 13.21(b).

$$E = \left(\frac{6}{6+4}\right)(50)$$

by voltage division in a series circuit, i.e. $\mathbf{E = 30\ V}$

(iii) For the circuit shown in Figure 13.22(a) representing a practical source supplying energy, $V = E - Ir$, where E is the battery e.m.f., V is the battery terminal voltage and r is the internal resistance of the battery (as shown in Section 4.6). For the circuit shown in Figure 13.22(b), $V = E - (-I)r$, i.e. $V = E + Ir$

(iv) The resistance 'looking-in' at terminals AB in Figure 13.23(a) is obtained by reducing the circuit in stages as shown in Figures 13.23(b) to (d). Hence the equivalent resistance across AB is 7 Ω

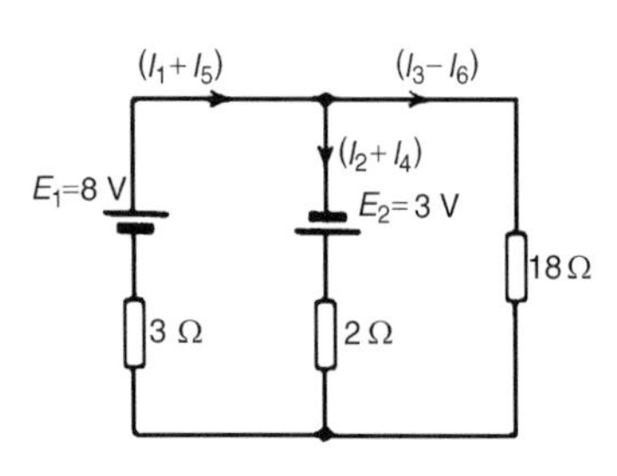

Figure 13.19

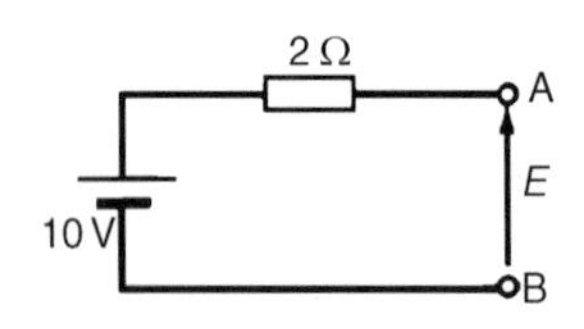

Figure 13.20

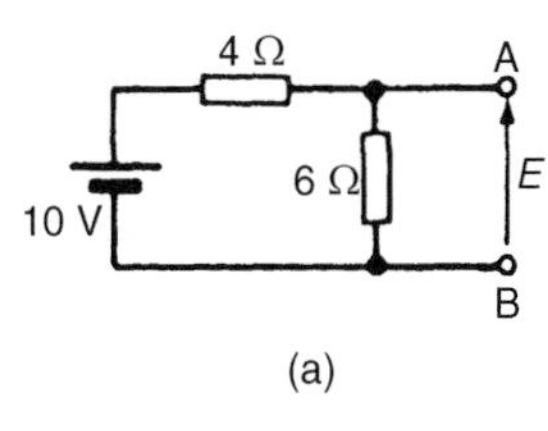

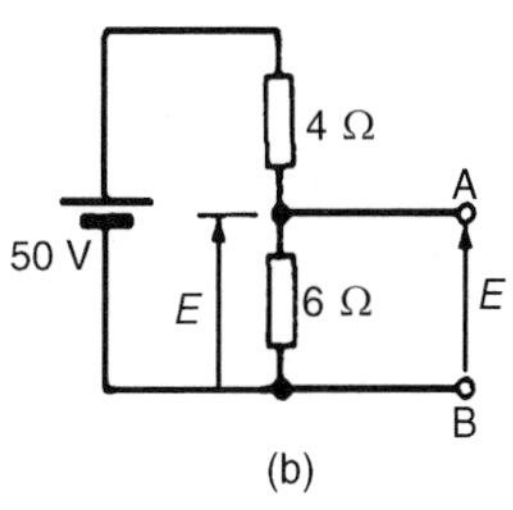

Figure 13.21

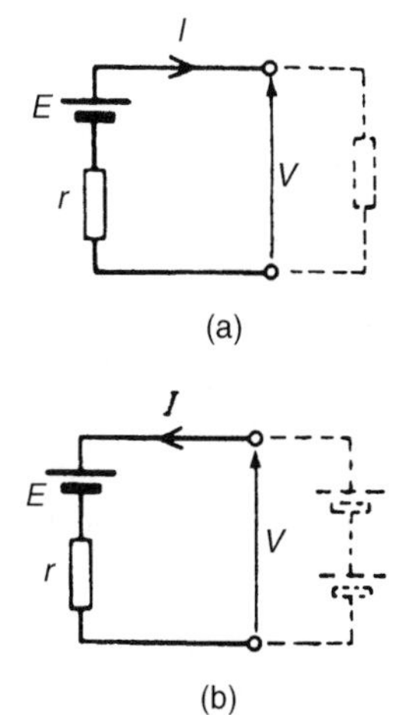

Figure 13.22

(v) For the circuit shown in Figure 13.24(a), the 3 Ω resistor carries no current and the p.d. across the 20 Ω resistor is 10 V. Redrawing the circuit gives Figure 13.24(b), from which

$$E = \left(\frac{4}{4+6}\right) \times 10 = \mathbf{4\ V}$$

(vi) If the 10 V battery in Figure 13.24(a) is removed and replaced by a short-circuit, as shown in Figure 13.24(c), then the 20 Ω resistor may be removed. The reason for this is that a short-circuit has zero resistance, and 20 Ω in parallel with zero ohms gives an equivalent resistance of: $(20 \times 0/20 + 0)$, i.e. 0 Ω. The circuit is then as shown in Figure 13.24(d), which is redrawn in Figure 13.24(e). From Figure 13.24(e), the equivalent resistance across AB,

$$r = \frac{6 \times 4}{6+4} + 3 = 2.4 + 3 = \mathbf{5.4\ \Omega}$$

(vii) To find the voltage across AB in Figure 13.25:
Since the 20 V supply is across the 5 Ω and 15 Ω resistors in series then, by voltage division, the voltage drop across AC,

$$V_{AC} = \left(\frac{5}{5+15}\right)(20) = 5\ \text{V}$$

Similarly, $V_{CB} = \left(\frac{12}{12+3}\right)(20) = 16$ V.

V_C is at a potential of +20 V.

$V_A = V_C - V_{AC} = +20 - 5 = 15\ V$ and

$V_B = V_C - V_{BC} = +20 - 16 = 4\ V$.

Hence the voltage between AB is $V_A - V_B = 15 - 4 = 11\ V$ and current would flow from A to B since A has a higher potential than B.

(viii) In Figure 13.26(a), to find the equivalent resistance across AB the circuit may be redrawn as in Figures 13.26(b) and (c). From Figure 13.26(c), the equivalent resistance across

$$\text{AB} = \frac{5 \times 15}{5+15} + \frac{12 \times 3}{12+3} = 3.75 + 2.4 = \mathbf{6.15\ \Omega}$$

(ix) In the worked problems in Sections 13.5 and 13.7 following, it may be considered that Thévenin's and Norton's theorems have

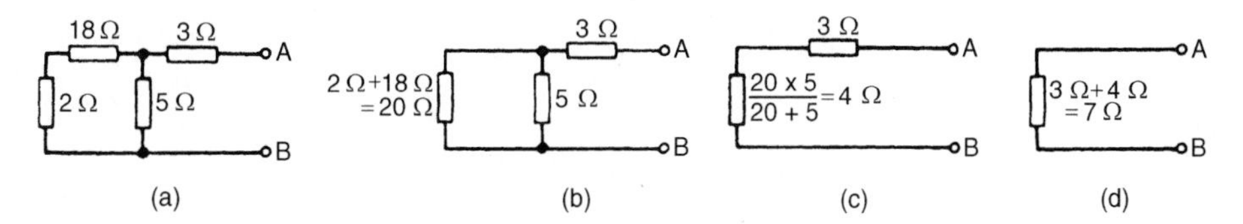

Figure 13.23

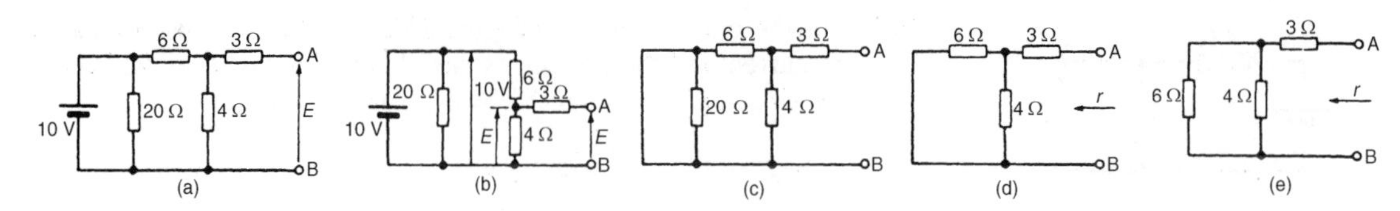

Figure 13.24

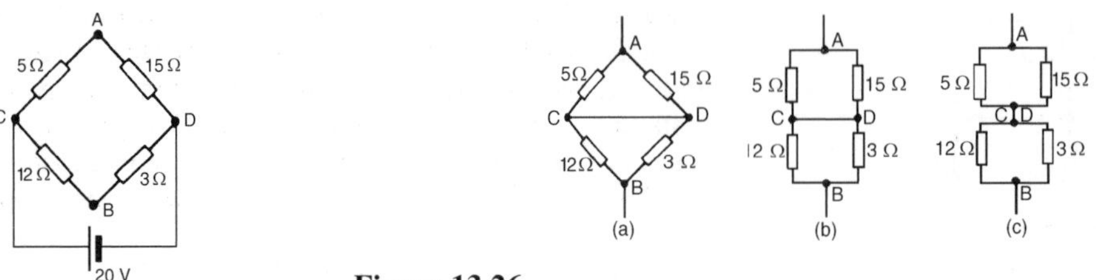

Figure 13.25

Figure 13.26

no obvious advantages compared with, say, Kirchhoff's laws. However, these theorems can be used to analyse part of a circuit and in much more complicated networks the principle of replacing the supply by a constant voltage source in series with a resistance (or impedance) is very useful.

13.5 Thévenin's theorem

Thévenin's theorem states:

'The current in any branch of a network is that which would result if an e.m.f. equal to the p.d. across a break made in the branch, were introduced into the branch, all other e.m.f.'s being removed and represented by the internal resistances of the sources.'

The procedure adopted when using Thévenin's theorem is summarized below. To determine the current in any branch of an active network (i.e. one containing a source of e.m.f.):

(i) remove the resistance R from that branch,

(ii) determine the open-circuit voltage, E, across the break,

(iii) remove each source of e.m.f. and replace them by their internal resistances and then determine the resistance, r, 'looking-in' at the break,

(iv) determine the value of the current from the equivalent circuit shown in Figure 13.27, i.e. $\boldsymbol{I = \dfrac{E}{R + r}}$

> Problem 7. Use Thévenin's theorem to find the current flowing in the 10 Ω resistor for the circuit shown in Figure 13.28(a).

Following the above procedure:

(i) The 10 Ω resistance is removed from the circuit as shown in Figure 13.28(b)

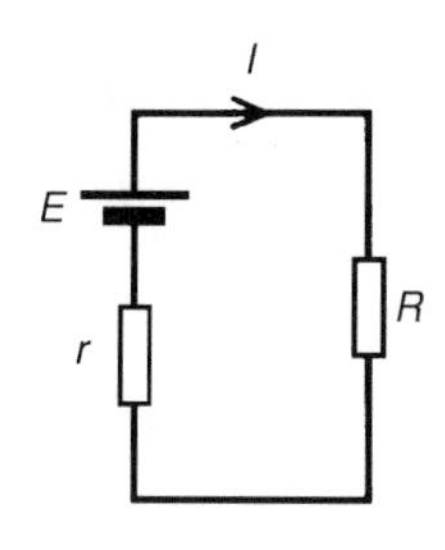

Figure 13.27

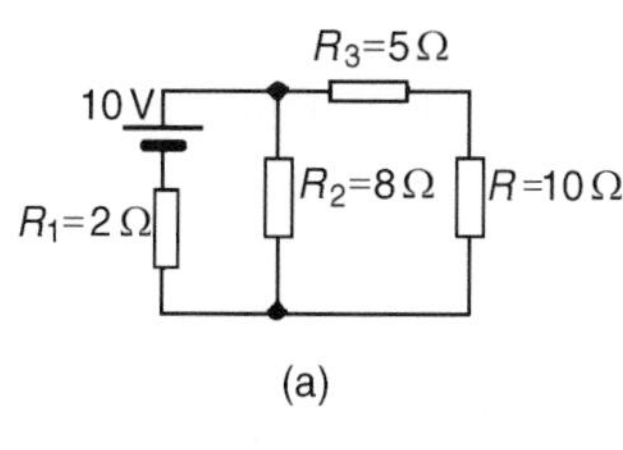

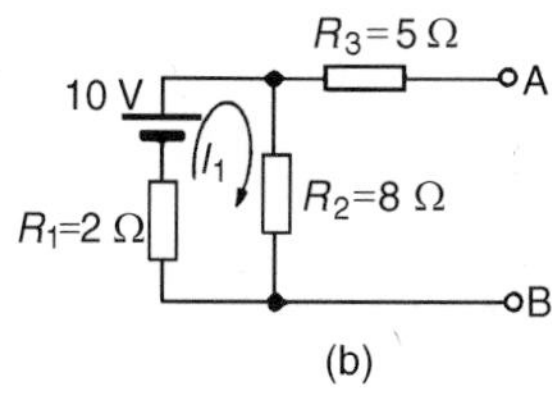

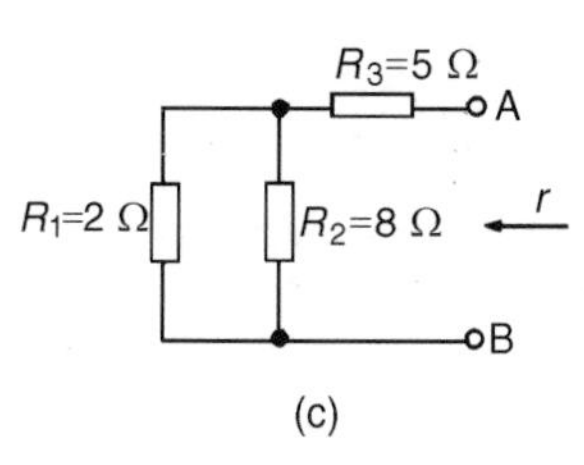

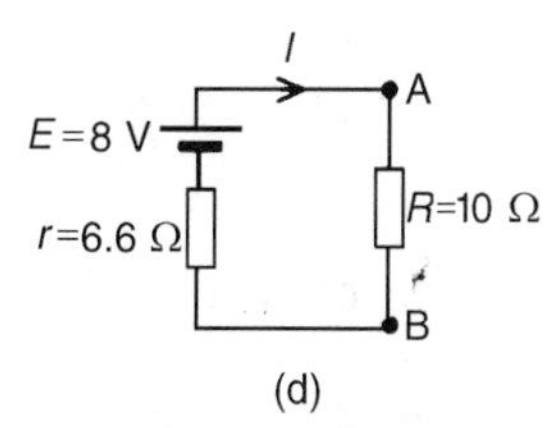

Figure 13.28

(ii) There is no current flowing in the 5 Ω resistor and current I_1 is given by:

$$I_1 = \frac{10}{R_1 + R_2} = \frac{10}{2+8} = 1 \text{ A}$$

P.d. across $R_2 = I_1 R_2 = 1 \times 8 = 8$ V

Hence p.d. across AB, i.e. the open-circuit voltage across the break, $E = 8$ V.

(iii) Removing the source of e.m.f. gives the circuit of Figure 13.28(c).

$$\text{Resistance, } r = R_3 + \frac{R_1 R_2}{R_1 + R_2} = 5 + \frac{2 \times 8}{2+8}$$
$$= 5 + 1.6 = 6.6 \ \Omega$$

(iv) The equivalent Thévenin's circuit is shown in Figure 13.28(d).

$$\text{Current } I = \frac{E}{R+r} = \frac{8}{10+6.6} = \frac{8}{16.6} = 0.482 \text{ A}$$

Hence the current flowing in the 10 Ω resistor of Figure 28(a) is **0.482 A**

Problem 8. For the network shown in Figure 13.29(a) determine the current in the 0.8 Ω resistor using Thévenin's theorem.

Following the procedure:

(i) The 0.8 Ω resistor is removed from the circuit as shown in Figure 13.29(b).

(ii) $$\text{Current } I_1 = \frac{12}{1+5+4} = \frac{12}{10} = 1.2 \text{ A}$$

P.d. across 4 Ω resistor $= 4I_1 = (4)(1.2) = 4.8$ V

Hence p.d. across AB, i.e. the open-circuit voltage across AB, $E = 4.8$ V

(iii) Removing the source of e.m.f. gives the circuit shown in Figure 13.29(c). The equivalent circuit of Figure 13.29(c) is shown in Figure 13.29(d), from which,

$$\text{resistance } r = \frac{4 \times 6}{4+6} = \frac{24}{10} = 2.4 \ \Omega$$

(iv) The equivalent Thévenin's circuit is shown in Figure 13.29(e), from which,

$$\text{current } I = \frac{E}{r+R} = \frac{4.8}{2.4+0.8} + \frac{4.8}{3.2}$$

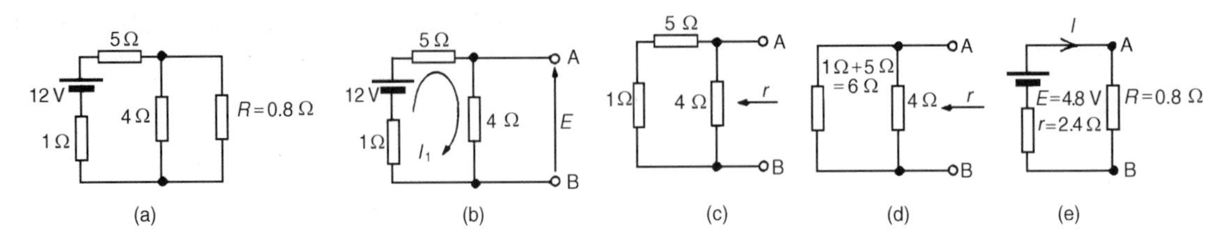

Figure 13.29

$$I = 1.5 \text{ A} = \textbf{current in the 0.8 } \boldsymbol{\Omega} \textbf{ resistor}$$

Problem 9. Use Thévenin's theorem to determine the current I flowing in the 4 Ω resistor shown in Figure 13.30(a). Find also the power dissipated in the 4 Ω resistor.

Following the procedure:

(i) The 4 Ω resistor is removed from the circuit as shown in Figure 13.30(b).

(ii) Current $I_1 = \dfrac{E_1 - E_2}{r_1 + r_2} = \dfrac{4-2}{2+1} = \dfrac{2}{3}$ A

P.d. across AB, $E = E_1 - I_1 r_1 = 4 - \left(\frac{2}{3}\right)(2) = 2\frac{2}{3}$ V

(see Section 13.4(iii))

(Alternatively, p.d. across AB, $E = E_2 - I_1 r_2$

$= 2 - -\left(\frac{2}{3}\right)(1) = 2\frac{2}{3}$ V)

(iii) Removing the sources of e.m.f. gives the circuit shown in Figure 13.30(c), from which resistance

$$r = \frac{2 \times 1}{2+1} = \frac{2}{3}\ \Omega$$

(iv) The equivalent Thévenin's circuit is shown in Figure 13.30(d), from which,

$$\text{current, } I = \frac{E}{r+R} = \frac{2\frac{2}{3}}{\frac{2}{3}+4} = \frac{8/3}{14/3} = \frac{8}{14} = \mathbf{0.571\ A}$$

$$= \textbf{current in the 4 } \boldsymbol{\Omega} \textbf{ resistor}$$

Power dissipated in 4 Ω resistor, $P = I^2 R = (0.571)^2(4) = \mathbf{1.304\ W}$

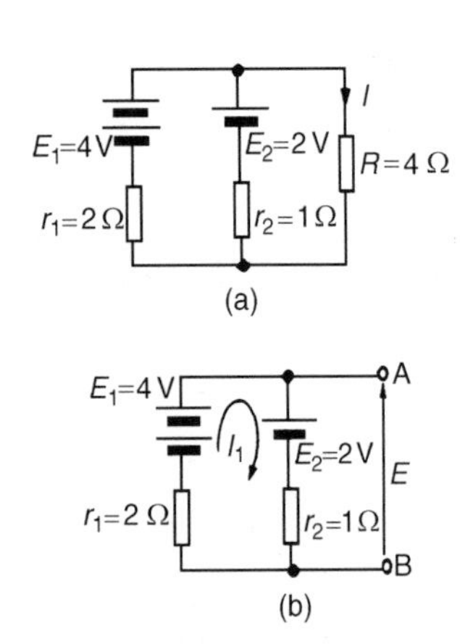

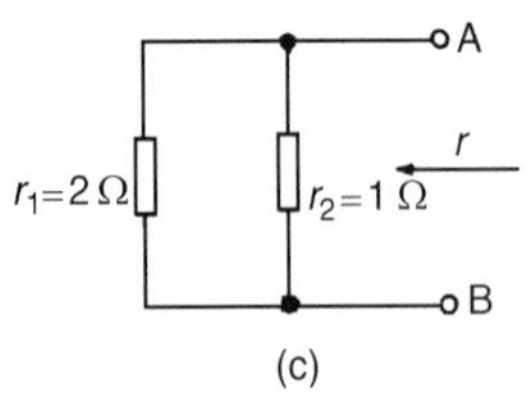

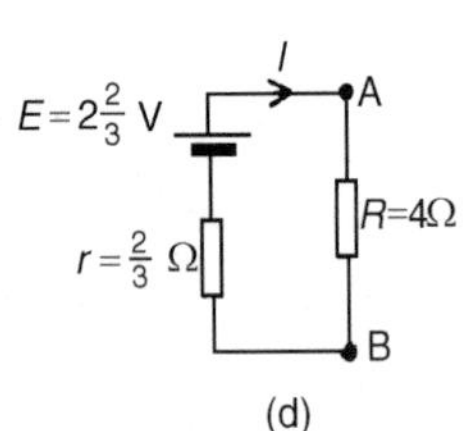

Figure 13.30

Problem 10. Use Thévenin's theorem to determine the current flowing in the 3 Ω resistance of the network shown in Figure 13.31(a). The voltage source has negligible internal resistance.

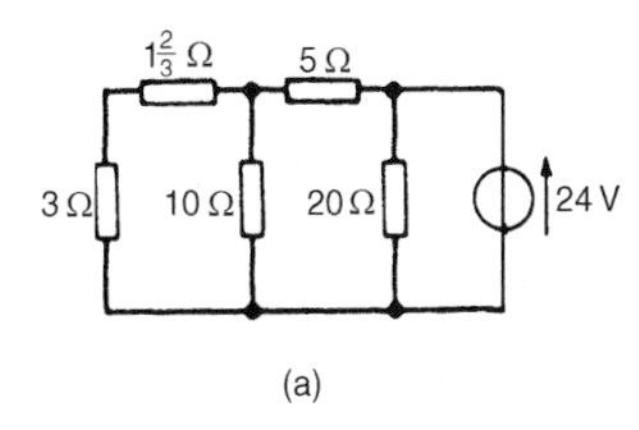

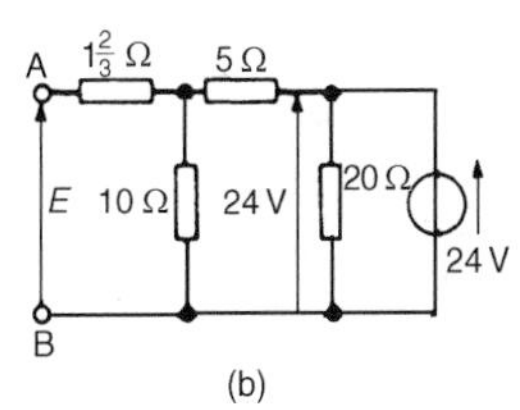

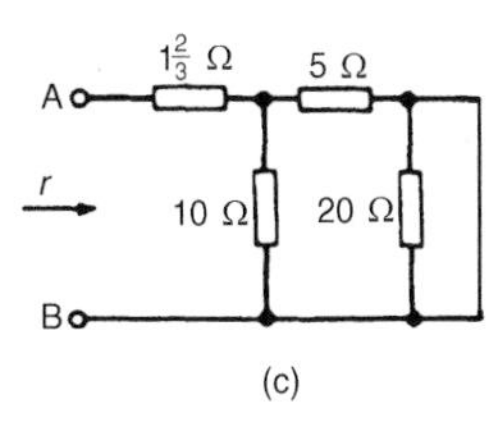

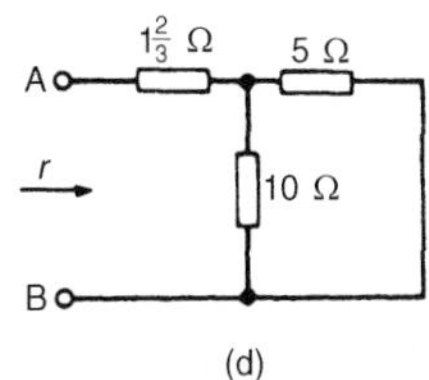

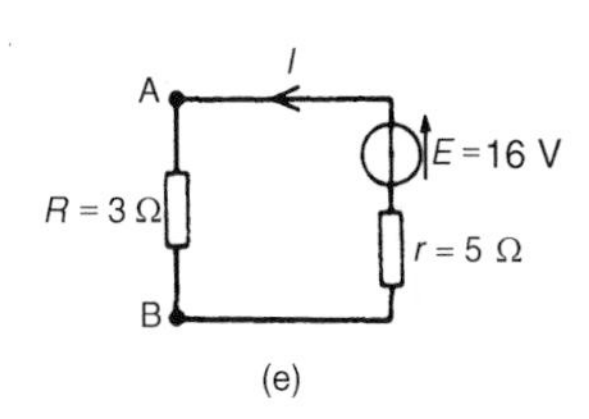

Figure 13.31

(Note the symbol for an ideal voltage source in Figure 13.31(a) which may be used as an alternative to the battery symbol.)

Following the procedure

(i) The 3 Ω resistance is removed from the circuit as shown in Figure 13.31(b).

(ii) The $1\frac{2}{3}\Omega$ resistance now carries no current.

$$\text{P.d. across } 10\ \Omega \text{ resistor} = \left(\frac{10}{10+5}\right)(24)$$

$$= \mathbf{16\ V} \text{ (see Section 13.4(v)).}$$

Hence p.d. across AB, $E = 16$ V

(iii) Removing the source of e.m.f. and replacing it by its internal resistance means that the 20 Ω resistance is short-circuited as shown in Figure 13.31(c) since its internal resistance is zero. The 20 Ω resistance may thus be removed as shown in Figure 13.31(d) (see Section 13.4 (vi)).

$$\text{From Figure 13.31(d), resistance, } r = 1\frac{2}{3} + \frac{10 \times 5}{10+5}$$

$$= 1\frac{2}{3} + \frac{50}{15} = 5\ \Omega$$

(iv) The equivalent Thévenin's circuit is shown in Figure 13.31(e), from which

$$\text{current, } I = \frac{E}{r+R} = \frac{16}{3+5} = \frac{16}{8} = \mathbf{2\ A}$$

$$= \textbf{current in the 3 } \boldsymbol{\Omega} \textbf{ resistance}$$

> Problem 11. A Wheatstone Bridge network is shown in Figure 13.32(a). Calculate the current flowing in the 32 Ω resistor, and its direction, using Thévenin's theorem. Assume the source of e.m.f. to have negligible resistance.

Following the procedure:

(i) The 32 Ω resistor is removed from the circuit as shown in Figure 13.32(b)

(ii) The p.d. between A and C, $V_{AC} = \left(\frac{R_1}{R_1 + R_4}\right)(E)$

$$= \left(\frac{2}{2+11}\right)(54) = 8.31\ \text{V}$$

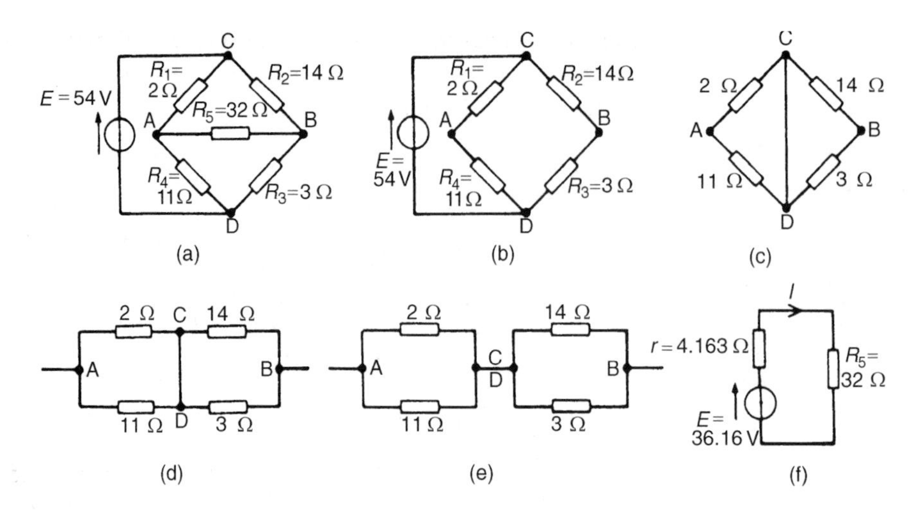

Figure 13.32

The p.d. between B and C, $V_{BC} = \left(\frac{R_2}{R_2 + R_3}\right)(E)$

$= \left(\frac{14}{14 + 3}\right)(54) = 44.47$ V

Hence the p.d. between A and B $= 44.47 - 8.31 =$ **36.16 V**
Point C is at a potential of + 54 V. Between C and A is a voltage drop of 8.31 V. Hence the voltage at point A is $54 - 8.31 = 45.69$ V. Between C and B is a voltage drop of 44.47 V. Hence the voltage at point B is $54 - 44.47 = 9.53$ V. Since the voltage at A is greater than at B, current must flow in the direction A to B. (See Section 13.4 (vii)).

(iii) Replacing the source of e.m.f. with a short-circuit (i.e. zero internal resistance) gives the circuit shown in Figure 13.32(c). The circuit is redrawn and simplified as shown in Figure 13.32(d) and (e), from which the resistance between terminals A and B,

$$r = \frac{2 \times 11}{2 + 11} + \frac{14 \times 3}{14 + 3} = \frac{22}{13} + \frac{42}{17} = 1.692 + 2.471 = \mathbf{4.163\ \Omega}$$

(iv) The equivalent Thévenin's circuit is shown in Figure 13.32(f), from which,

$$\text{current } I = \frac{E}{r + R_5} = \frac{36.16}{4.163 + 32} = 1 \text{ A}$$

Hence the current in the 32 Ω resistor of Figure 13.32(a) is 1 A, flowing from A to B

Further problems on Thévenin's theorem may be found in Section 13.10, problems 11 to 15, page 190.

13.6 Constant-current source

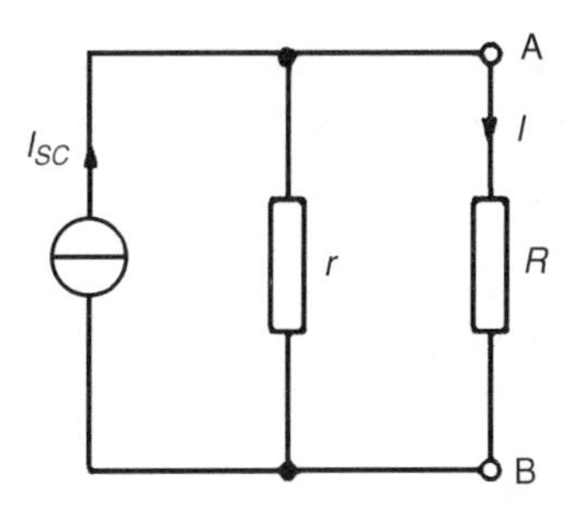

Figure 13.33

A source of electrical energy can be represented by a source of e.m.f. in series with a resistance. In Section 13.5, the Thévenin constant-voltage source consisted of a constant e.m.f. E in series with an internal resistance r. However this is not the only form of representation. A source of electrical energy can also be represented by a constant-current source in parallel with a resistance. It may be shown that the two forms are equivalent. An **ideal constant-voltage generator** is one with zero internal resistance so that it supplies the same voltage to all loads. An **ideal constant-current generator** is one with infinite internal resistance so that it supplies the same current to all loads.

Note the symbol for an ideal current source (BS 3939, 1985), shown in Figure 13.33.

13.7 Norton's theorem

Norton's theorem states:

'The current that flows in any branch of a network is the same as that which would flow in the branch if it were connected across a source of electrical energy, the short-circuit current of which is equal to the current that would flow in a short-circuit across the branch, and the internal resistance of which is equal to the resistance which appears across the open-circuited branch terminals.'

The procedure adopted when using Norton's theorem is summarized below.

To determine the current flowing in a resistance R of a branch AB of an active network:

(i) short-circuit branch AB

(ii) determine the short-circuit current I_{SC} flowing in the branch

(iii) remove all sources of e.m.f. and replace them by their internal resistance (or, if a current source exists, replace with an open-circuit), then determine the resistance r, 'looking-in' at a break made between A and B

(iv) determine the current I flowing in resistance R from the Norton equivalent network shown in Figure 13.33, i.e.

$$I = \left(\frac{r}{r+R}\right) I_{SC}$$

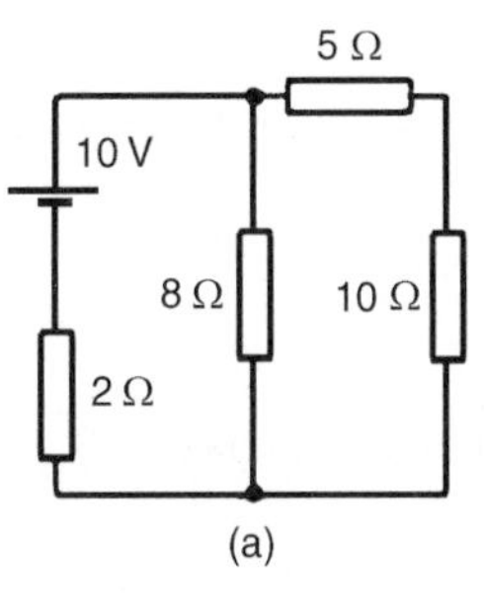

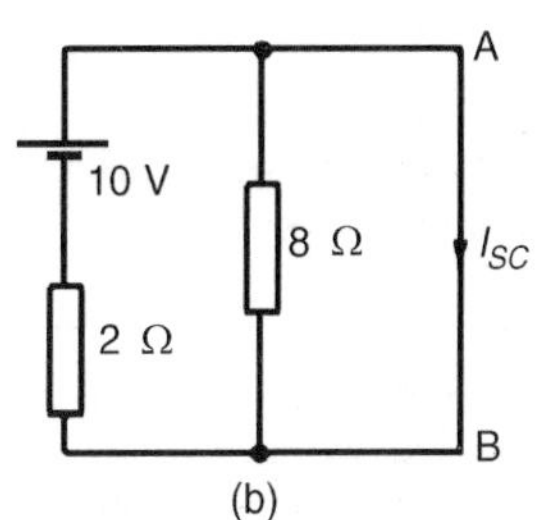

Figure 13.34

Problem 12. Use Norton's theorem to determine the current flowing in the 10 Ω resistance for the circuit shown in Figure 13.34(a).

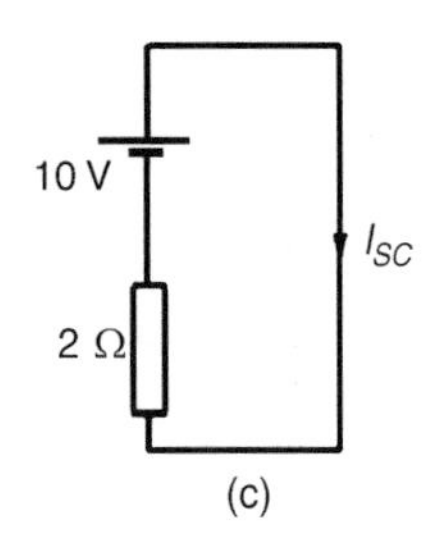

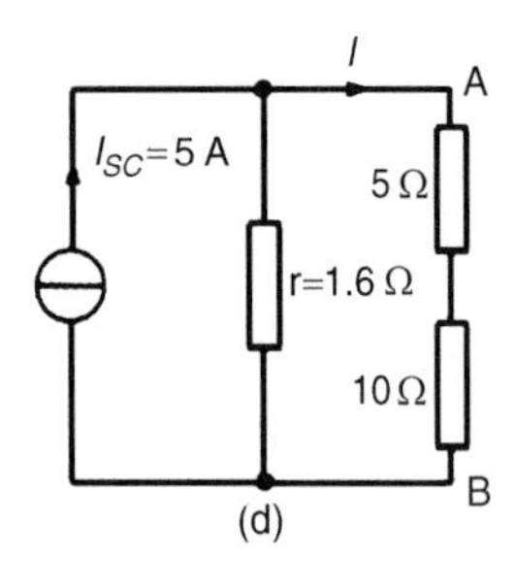

Figure 13.34 *continued*

Following the above procedure:

(i) The branch containing the 10 Ω resistance is short-circuited as shown in Figure 13.34(b).

(ii) Figure 13.34(c) is equivalent to Figure 13.34(b). Hence

$$I_{SC} = \tfrac{10}{2} = 5 \text{ A}$$

(iii) If the 10 V source of e.m.f. is removed from Figure 13.34(b) the resistance 'looking-in' at a break made between A and B is given by:

$$r = \frac{2 \times 8}{2 + 8} = 1.6 \ \Omega$$

(iv) From the Norton equivalent network shown in Figure 13.34(d) the current in the 10 Ω resistance, by current division, is given by:

$$I = \left(\frac{1.6}{1.6 + 5 + 10}\right)(5) = \mathbf{0.482 \ A}$$

as obtained previously in problem 7 using Thévenin's theorem.

> Problem 13. Use Norton's theorem to determine the current I flowing in the 4 Ω resistance shown in Figure 13.35(a).

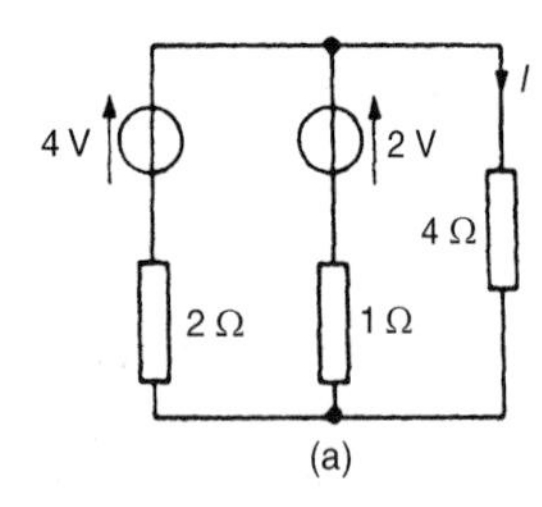

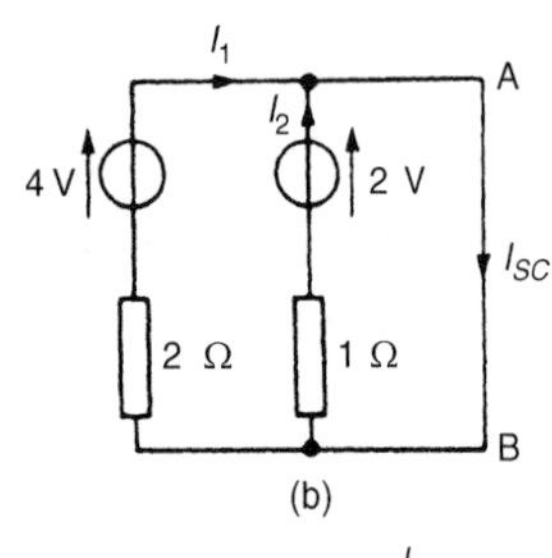

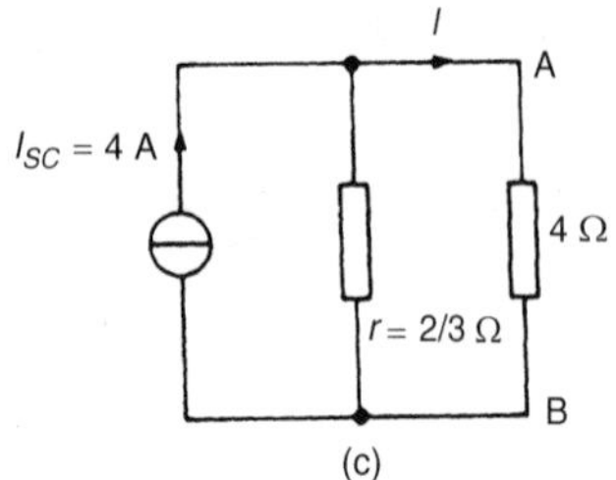

Figure 13.35

Following the procedure:

(i) The 4 Ω branch is short-circuited as shown in Figure 13.35(b).

(ii) From Figure 13.35(b), $I_{SC} = I_1 + I_2 = \tfrac{4}{2} + \tfrac{2}{1} = 4$ A

(iii) If the sources of e.m.f. are removed the resistance 'looking-in' at a break made between A and B is given by:

$$r = \frac{2 \times 1}{2 + 1} = \frac{2}{3} \ \Omega$$

(iv) From the Norton equivalent network shown in Figure 13.35(c) the current in the 4 Ω resistance is given by:

$$I = \left[\frac{2/3}{(2/3) + 4}\right](4) = \mathbf{0.571 \ A},$$

as obtained previously in problems 2, 5 and 9 using Kirchhoff's laws and the theorems of superposition and Thévenin.

> Problem 14. Use Norton's theorem to determine the current flowing in the 3 Ω resistance of the network shown in Figure 13.36(a). The voltage source has negligible internal resistance.

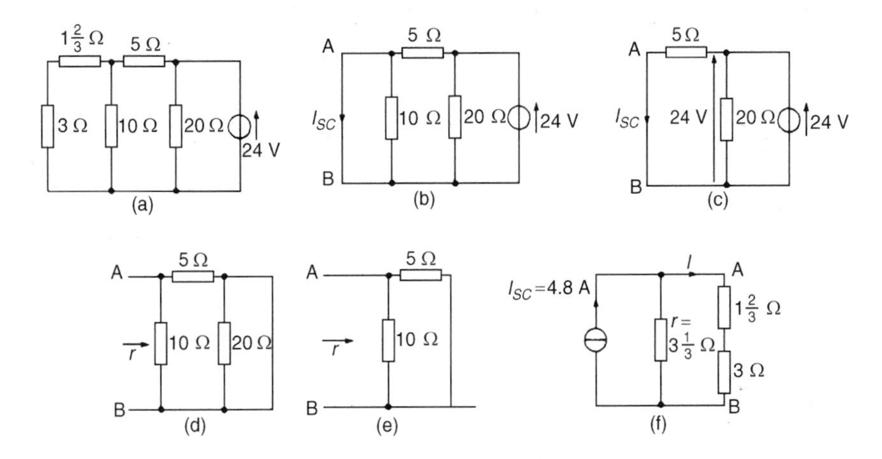

Figure 13.36

Following the procedure:

(i) The branch containing the 3 Ω resistance is short-circuited as shown in Figure 13.36(b).

(ii) From the equivalent circuit shown in Figure 13.36(c),

$$I_{SC} = \frac{24}{5} = 4.8 \text{ A}$$

(iii) If the 24 V source of e.m.f. is removed the resistance 'looking-in' at a break made between A and B is obtained from Figure 13.36(d) and its equivalent circuit shown in Figure 13.36(e) and is given by:

$$r = \frac{10 \times 5}{10 + 5} = \frac{50}{15} = 3\frac{1}{3}\,\Omega$$

(iv) From the Norton equivalent network shown in Figure 13.36(f) the current in the 3 Ω resistance is given by:

$$I = \left[\frac{3\frac{1}{3}}{3\frac{1}{3} + 1\frac{2}{3} + 3}\right](4.8) = \mathbf{2\ A,}$$

as obtained previously in problem 10 using Thévenin's theorem.

Problem 15. Determine the current flowing in the 2 Ω resistance in the network shown in Figure 13.37(a).

Following the procedure:

(i) The 2 Ω resistance branch is short-circuited as shown in Figure 13.37(b).

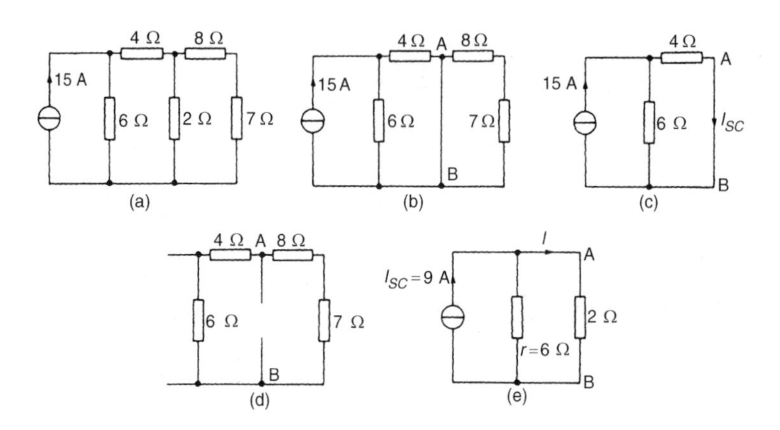

Figure 13.37

(ii) Figure 13.37(c) is equivalent to Figure 13.37(b).

Hence $I_{SC} = \left(\frac{6}{6+4}\right)(15) = \mathbf{9\ A}$ by current division.

(iii) If the 15 A current source is replaced by an open-circuit then from Figure 13.37(d) the resistance 'looking-in' at a break made between A and B is given by (6 + 4) Ω in parallel with (8 + 7) Ω, i.e.

$$r = \frac{(10)(15)}{10+15} = \frac{150}{25} = 6\ \Omega$$

(iv) From the Norton equivalent network shown in Figure 13.37(e) the current in the 2 Ω resistance is given by:

$$I = \left(\frac{6}{6+2}\right)(9) = \mathbf{6.75\ A}$$

13.8 Thévenin and Norton equivalent networks

The Thévenin and Norton networks shown in Figure 13.38 are equivalent to each other. The resistance 'looking-in' at terminals AB is the same in each of the networks, i.e. r.

If terminals AB in Figure 13.38(a) are short-circuited, the short-circuit current is given by E/r. If terminals AB in Figure 13.38(b) are short-circuited, the short-circuit current is I_{SC}. For the circuit shown in Figure 13.38(a) to be equivalent to the circuit in Figure 13.38(b) the same short-circuit current must flow. Thus $I_{SC} = E/r$.

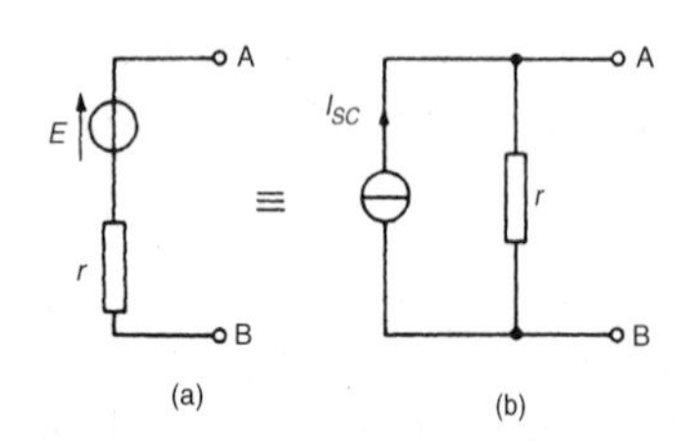

Figure 13.38

Figure 13.39 shows a source of e.m.f. E in series with a resistance r feeding a load resistance R.

From Figure 13.39, $I = \frac{E}{r+R} = \frac{E/r}{(r+R)/r} = \left(\frac{r}{r+R}\right)\frac{E}{r}$

$$\text{i.e. } I = \left(\frac{r}{r+R}\right) I_{SC}$$

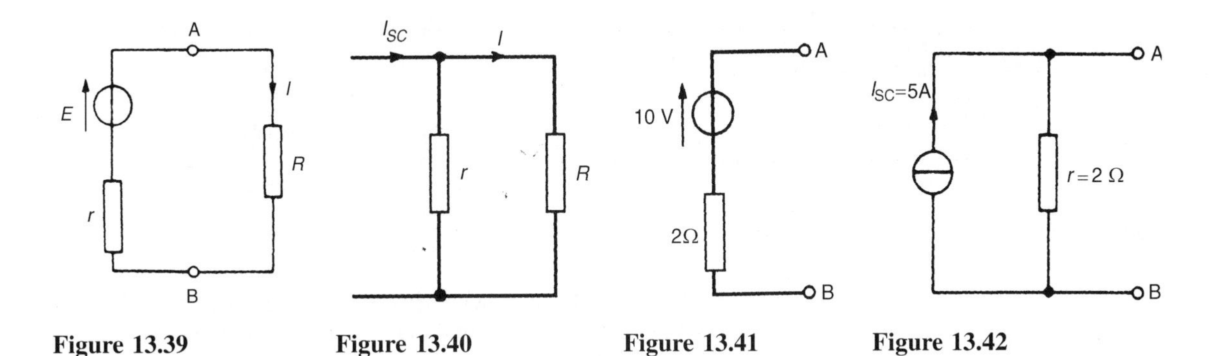

Figure 13.39 **Figure 13.40** **Figure 13.41** **Figure 13.42**

From Figure 13.40 it can be seen that, when viewed from the load, the source appears as a source of current I_{SC} which is divided between r and R connected in parallel.

Thus the two representations shown in Figure 13.38 are equivalent.

Problem 16. Convert the circuit shown in Figure 13.41 to an equivalent Norton network.

If terminals AB in Figure 13.41 are short-circuited, the short-circuit current $I_{SC} = \frac{10}{2} = 5$ A

The resistance 'looking-in' at terminals AB is 2 Ω. Hence the equivalent Norton network is as shown in Figure 13.42.

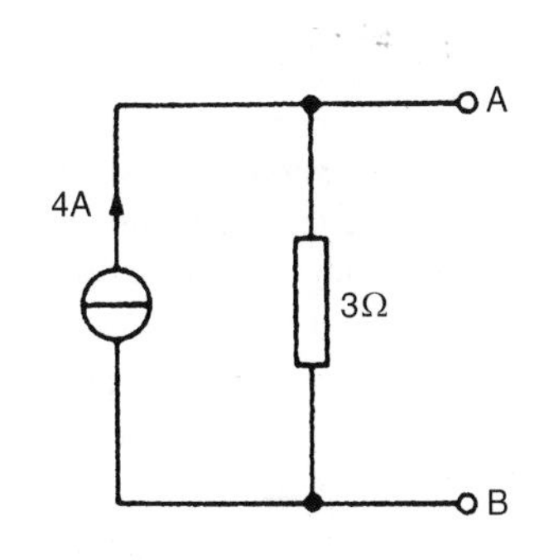

Figure 13.43

Problem 17. Convert the network shown in Figure 13.43 to an equivalent Thévenin circuit.

The open-circuit voltage E across terminals AB in Figure 13.43 is given by: $E = (I_{SC})(r) = (4)(3) = 12$ V.

The resistance 'looking-in' at terminals AB is 3 Ω. Hence the equivalent Thévenin circuit is as shown in Figure 13.44.

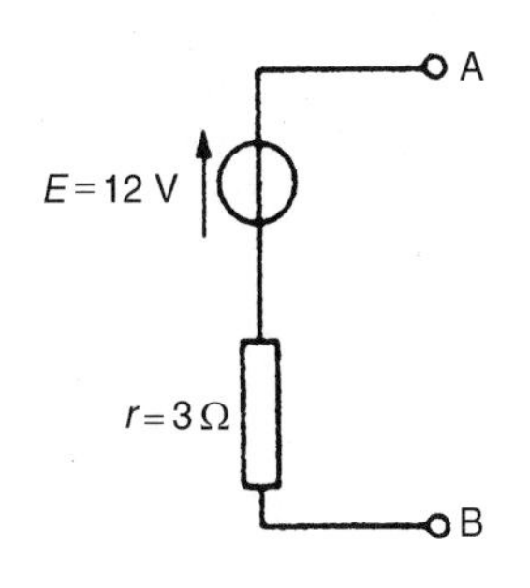

Figure 13.44

Problem 18. (a) Convert the circuit to the left of terminals AB in Figure 13.45(a) to an equivalent Thévenin circuit by initially converting to a Norton equivalent circuit. (b) Determine the current flowing in the 1.8 Ω resistor.

(a) For the branch containing the 12 V source, converting to a Norton equivalent circuit gives $I_{SC} = 12/3 = 4$ A and $r_1 = 3\Omega$. For the branch containing the 24 V source, converting to a Norton equivalent circuit gives $I_{SC2} = 24/2 = 12$ A and $r_2 = 2\ \Omega$.

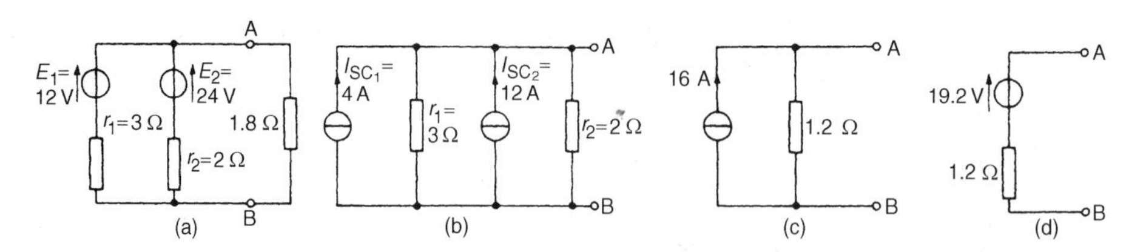

Figure 13.45

Thus Figure 13.45(b) shows a network equivalent to Figure 13.45(a). From Figure 13.45(b) the total short-circuit current is $4 + 12 = 16$ A

and the total resistance is given by: $\dfrac{3 \times 2}{3 + 2} = 1.2\ \Omega$

Thus Figure 13.45(b) simplifies to Figure 13.45(c).
The open-circuit voltage across AB of Figure 13.45(c),

$E = (16)(1.2) = 19.2$ V, and the resistance 'looking-in' at AB is 1.2 Ω. Hence the Thévenin equivalent circuit is as shown in Figure 13.45(d).

(b) When the 1.8 Ω resistance is connected between terminals A and B of Figure 13.45(d) the current I flowing is given by:

$$I = \frac{19.2}{1.2 + 1.8} = \mathbf{6.4\ A}$$

Problem 19. Determine by successive conversions between Thévenin and Norton equivalent networks a Thévenin equivalent circuit for terminals AB of Figure 13.46(a). Hence determine the current flowing in the 200 Ω resistance.

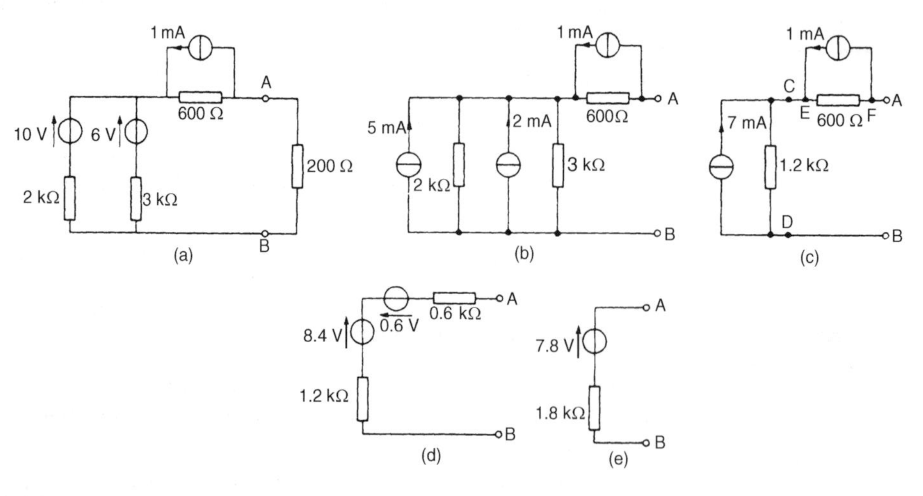

Figure 13.46

For the branch containing the 10 V source, converting to a Norton equivalent network gives

$$I_{SC} = \frac{10}{2000} = 5 \text{ mA and } r_1 = 2 \text{ k}\Omega.$$

For the branch containing the 6 V source, converting to a Norton equivalent network gives

$$I_{SC} = \frac{6}{3000} = 2 \text{ mA and } r_2 = 3 \text{ k}\Omega.$$

Thus the network of Figure 13.46(a) converts to Figure 13.46(b).

Combining the 5 mA and 2 mA current sources gives the equivalent network of Figure 13.46(c) where the short-circuit current for the original two branches considered is 7 mA and the resistance is

$$\frac{2 \times 3}{2 + 3} = 1.2 \text{ k}\Omega.$$

Both of the Norton equivalent networks shown in Figure 13.46(c) may be converted to Thévenin equivalent circuits. The open-circuit voltage across CD is $(7 \times 10^{-3})(1.2 \times 10^3) = 8.4\ V$ and the resistance 'looking-in' at CD is 1.2 kΩ.

The open-circuit voltage across EF is $(1 \times 10^{-3})(600) = 0.6\ V$ and the resistance 'looking-in' at EF is 0.6 kΩ. Thus Figure 13.46(c) converts to Figure 13.46(d). Combining the two Thévenin circuits gives

$E = 8.4 - 0.6 = \mathbf{7.8\ V}$ and the resistance

$r = (1.2 + 0.6)\text{ k}\Omega = \mathbf{1.8\ k\Omega}.$

Thus the Thévenin equivalent circuit for terminals AB of Figure 13.46(a) is as shown in Figure 13.46(e).

Hence the current I flowing in a 200 Ω resistance connected between A and B is given by:

$$I = \frac{7.8}{1800 + 200} = \frac{7.8}{2000} = \mathbf{3.9\ mA}$$

Further problems on Norton's theorem may be found in Section 13.10, problems 16 to 21, page 191.

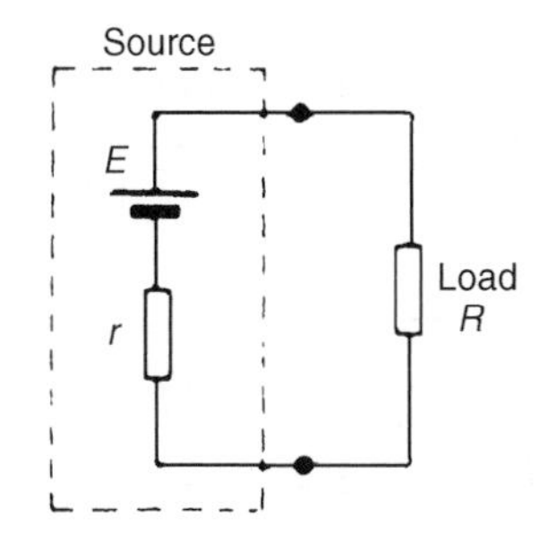

Figure 13.47

13.9 Maximum power transfer theorem

The **maximum power transfer theorem** states:

'The power transferred from a supply source to a load is at its maximum when the resistance of the load is equal to the internal resistance of the source.'

Hence, in Figure 13.47, when $R = r$ the power transferred from the source to the load is a maximum.

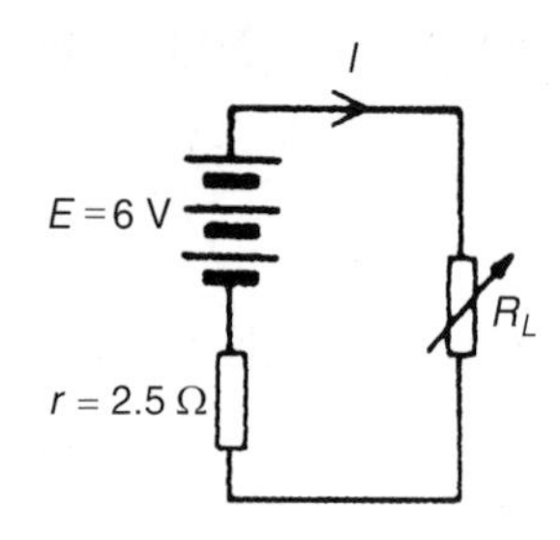

Figure 13.48

Problem 20. The circuit diagram of Figure 13.48 shows dry cells of source e.m.f. 6 V, and internal resistance 2.5Ω. If the load resistance R_L is varied from 0 to 5 Ω in 0.5 Ω steps, calculate the power dissipated by the load in each case. Plot a graph of R_L (horizontally) against power (vertically) and determine the maximum power dissipated.

When $R_L = 0$, current $I = \dfrac{E}{r + R_L} = \dfrac{6}{2.5} = 2.4\ A$ and power dissipated in R_L, $P = I^2 R_L$, i.e. $P = (2.4)^2(0) = 0$ W

When $R_L = 0.5\ \Omega$, current $I = \dfrac{E}{r + R_L} = \dfrac{6}{2.5 + 0.5} = 2$ A

and $\quad P = I^2 R_L = (2)^2(0.5) = 2$ W

When $R_L = 1.0\Omega$, current $I = \dfrac{6}{2.5 + 1.0} = 1.714$ A

and $\quad P = (1.714)^2(1.0) = 2.94$ W

With similar calculations the following table is produced:

$R_L(\Omega)$	0	0.5	1.0	1.5	2.0	2.5	3.0	3.5	4.0	4.5	5.0
$I = \dfrac{E}{r + R_L}$	2.4	2.0	1.714	1.5	1.333	1.2	1.091	1.0	0.923	0.857	0.8
$P = I^2 R_L(W)$	0	2.00	2.94	3.38	3.56	3.60	3.57	3.50	3.41	3.31	3.20

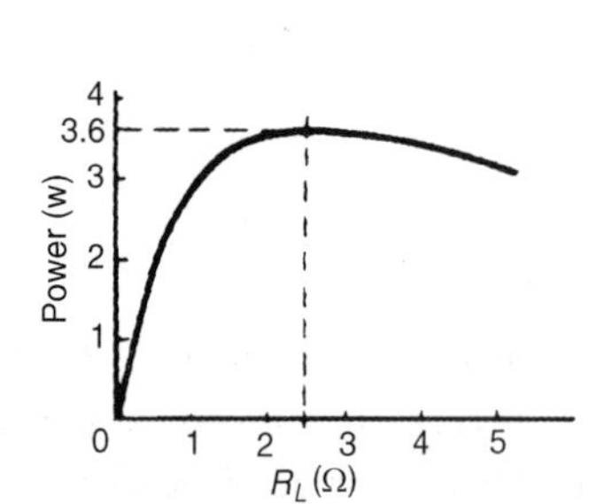

Figure 13.49

A graph of R_L against P is shown in Figure 13.49. **The maximum value of power is 3.60 W** which occurs when R_L is 2.5 Ω, i.e. **maximum power occurs when $R_L = r$**, which is what the maximum power transfer theorem states.

Problem 21. A d.c. source has an open-circuit voltage of 30 V and an internal resistance of 1.5 Ω. State the value of load resistance that gives maximum power dissipation and determine the value of this power.

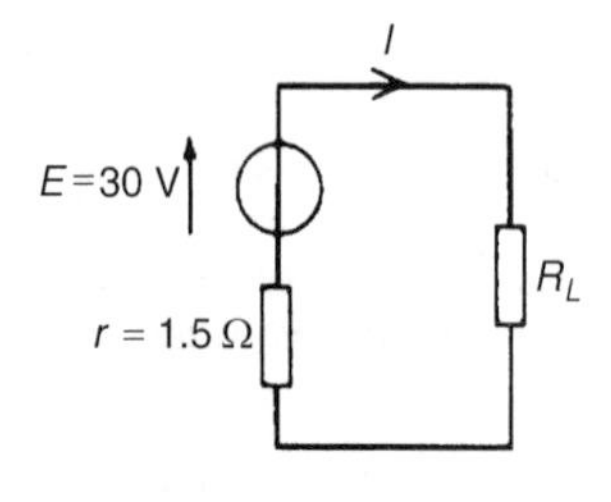

Figure 13.50

The circuit diagram is shown in Figure 13.50. From the maximum power transfer theorem, for maximum power dissipation,

$\mathbf{R_L = r = 1.5\ \Omega}$

From Figure 13.50, current $I = \dfrac{E}{r + R_L} = \dfrac{30}{1.5 + 1.5} = 10$ A

Power $P = I^2 R_L = (10)^2(1.5) =$ **150 W** = maximum power dissipated

Problem 22. Find the value of the load resistor R_L shown in Figure 13.51(a) that gives maximum power dissipation and determine the value of this power.

Using the procedure for Thévenin's theorem:

(i) Resistance R_L is removed from the circuit as shown in Figure 13.51(b).

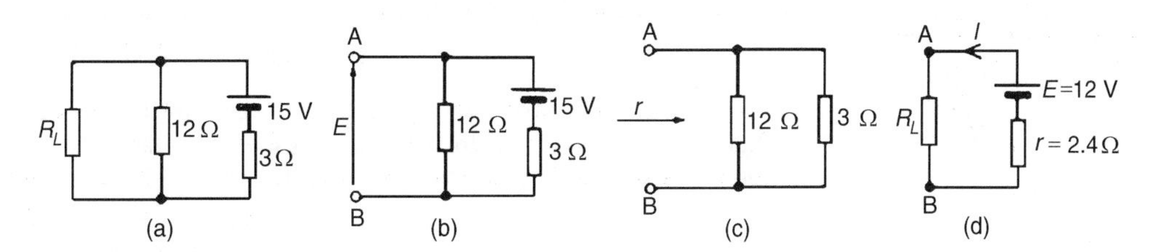

Figure 13.51

(ii) The p.d. across AB is the same as the p.d. across the 12 Ω resistor.

Hence $E = \left(\frac{12}{12+3}\right)(15) = 12$ V

(iii) Removing the source of e.m.f. gives the circuit of Figure 13.51(c),

from which resistance, $r = \frac{12 \times 3}{12+3} = \frac{36}{15} = 2.4\ \Omega$

(iv) The equivalent Thévenin's circuit supplying terminals AB is shown in Figure 13.51(d), from which, current, $I = E/(r + R_L)$

For maximum power, $R_L = r =$ **2.4 Ω**. Thus current,

$$I = \frac{12}{2.4 + 2.4} = 2.5 \text{ A.}$$

Power, P, dissipated in load R_L, $P = I^2 R_L = (2.5)^2(2.4) =$ **15 W**

Further problems on the maximum power transfer theorem may be found in Section 13.10 following, problems 22 and 23, page 192.

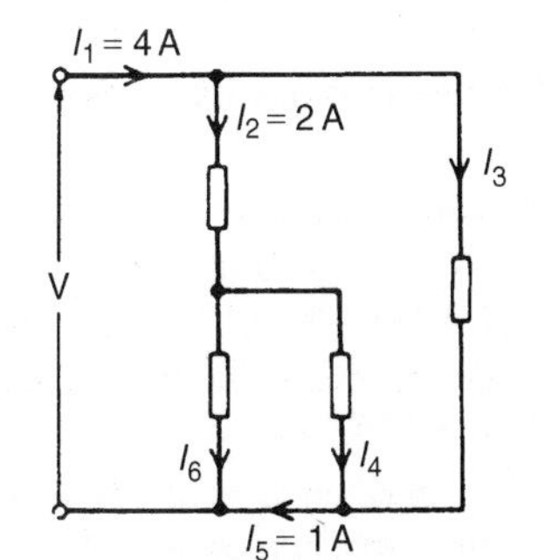

Figure 13.52

13.10 Further problems on d.c. circuit theory

Kirchhoff's laws

1 Find currents I_3, I_4 and I_6 in Figure 13.52

[$I_3 = 2$ A; $I_4 = -1$ A; $I_6 = 3$ A]

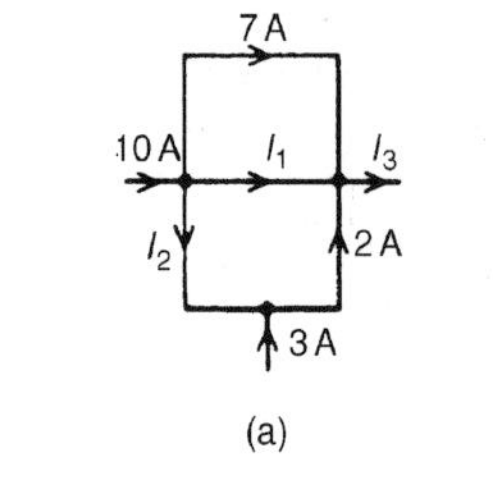

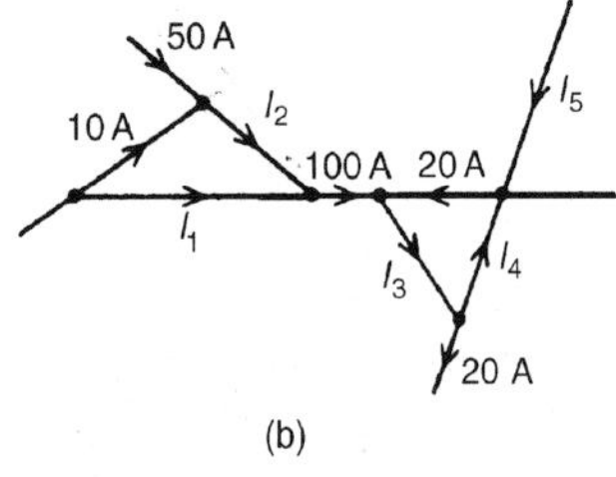

Figure 13.53

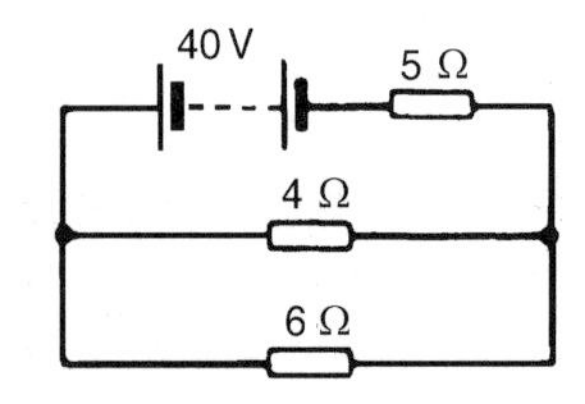

Figure 13.54

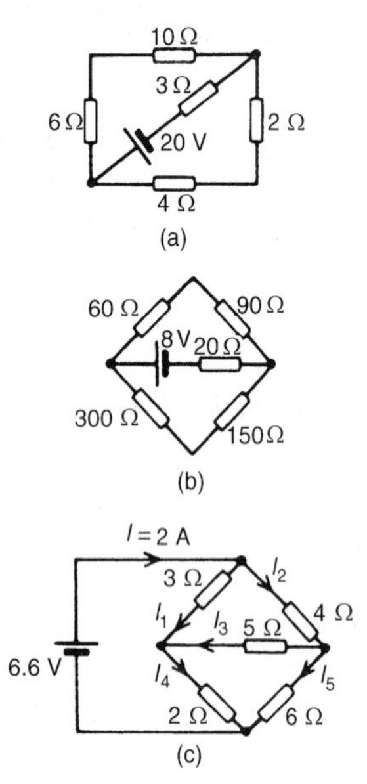

Figure 13.55

2 For the networks shown in Figure 13.53, find the values of the currents marked.

[(a) $I_1 = 4$ A, $I_2 = -1$ A, $I_3 = 13$ A
(b) $I_1 = 40$ A, $I_2 = 60$ A, $I_3 = 120$ A,
$I_4 = 100$ A, $I_5 = -80$ A]

3 Use Kirchhoff's laws to find the current flowing in the 6 Ω resistor of Figure 13.54 and the power dissipated in the 4 Ω resistor.

[2.162 A, 42.07 W]

4 Find the current flowing in the 3 Ω resistor for the network shown in Figure 13.55(a). Find also the p.d. across the 10 Ω and 2 Ω resistors.

[2.715 A, 7.410 V, 3.948 V]

5 For the networks shown in Figure 13.55(b) find: (a) the current in the battery, (b) the current in the 300 Ω resistor, (c) the current in the 90 Ω resistor, and (d) the power dissipated in the 150 Ω resistor.

[(a) 60.38 mA (b) 15.10 mA
(c) 45.28 mA (d) 34.20 mW]

6 For the bridge network shown in Figure 13.55(c), find the currents I_1 to I_5.

[$I_1 = 1.26$ A, $I_2 = 0.74$ A, $I_3 = 0.16$ A
$I_4 = 1.42$ A, $I_5 = 0.59$ A]

Superposition theorem

7 Use the superposition theorem to find currents I_1, I_2 and I_3 of Figure 13.56(a). [$I_1 = 2$ A, $I_2 = 3$ A, $I_3 = 5$ A]

8 Use the superposition theorem to find the current in the 8 Ω resistor of Figure 13.56(b). [0.385 A]

9 Use the superposition theorem to find the current in each branch of the network shown in Figure 13.56(c).

[10 V battery discharges at 1.429 A
4 V battery charges at 0.857 A
Current through 10 Ω resistor is 0.572 A]

10 Use the superposition theorem to determine the current in each branch of the arrangement shown in Figure 13.56(d).

[24 V battery charges at 1.664 A
52 V battery discharges at 3.280 A
Current in 20 Ω resistor is 1.616 A]

Thévenin's theorem

11 Use Thévenin's theorem to find the current flowing in the 14 Ω resistor of the network shown in Figure 13.57. Find also the power dissipated in the 14 Ω resistor. [0.434 A, 2.64 W]

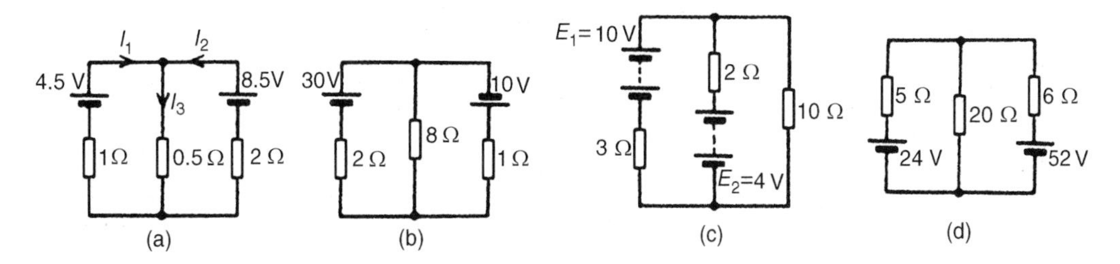

Figure 13.56

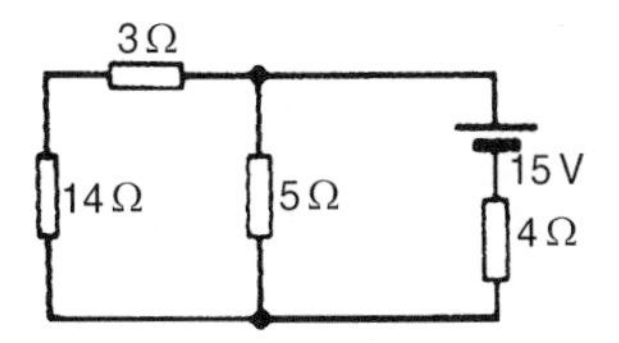

Figure 13.57

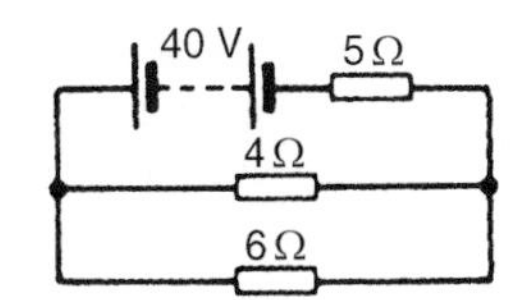

Figure 13.58

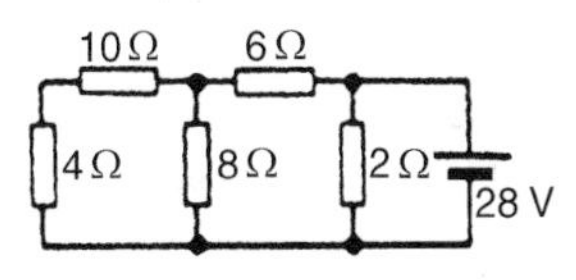

Figure 13.59

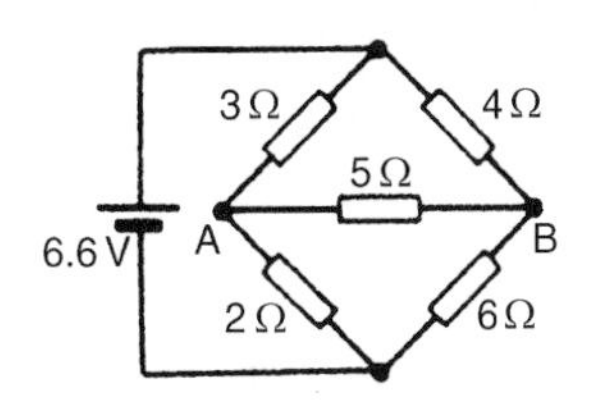

Figure 13.60

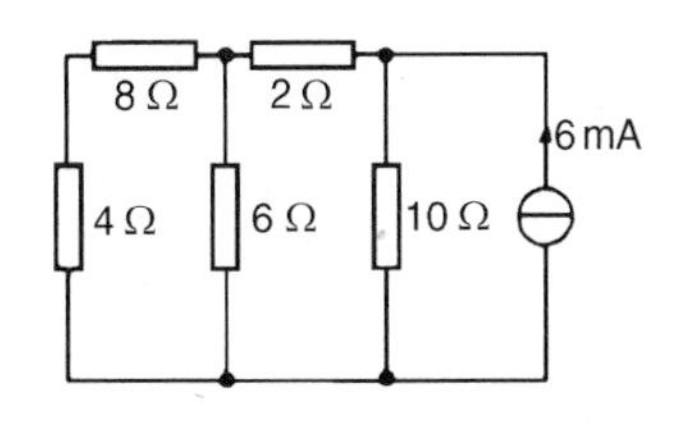

Figure 13.61

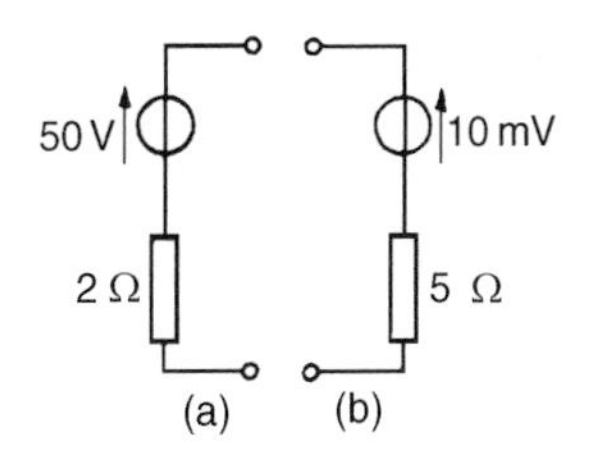

Figure 13.62

12 Use Thévenin's theorem to find the current flowing in the 6 Ω resistor shown in Figure 13.58 and the power dissipated in the 4 Ω resistor. [2.162 A, 42.07 W]

13 Repeat problems 7–10 using Thévenin's theorem.

14 In the network shown in Figure 13.59, the battery has negligible internal resistance. Find, using Thévenin's theorem, the current flowing in the 4 Ω resistor. [0.918 A]

15 For the bridge network shown in Figure 13.60, find the current in the 5 Ω resistor, and its direction, by using Thévenin's theorem. [0.153 A from B to A]

Norton's theorem

16 Repeat problems 7–12, 14 and 15 using Norton's theorem.

17 Determine the current flowing in the 6 Ω resistance of the network shown in Figure 13.61 by using Norton's theorem. [2.5 mA]

18 Convert the circuits shown in Figure 13.62 to Norton equivalent networks.
[(a) $I_{SC} = 25$ A, $r = 2$ Ω
(b) $I_{SC} = 2$ mA, $r = 5$ Ω]

19 Convert the networks shown in Figure 13.63 to Thévenin equivalent circuits.
[(a) $E = 20$ V, $r = 4$ Ω
(b) $E = 12$ mV, $r = 3$ Ω]

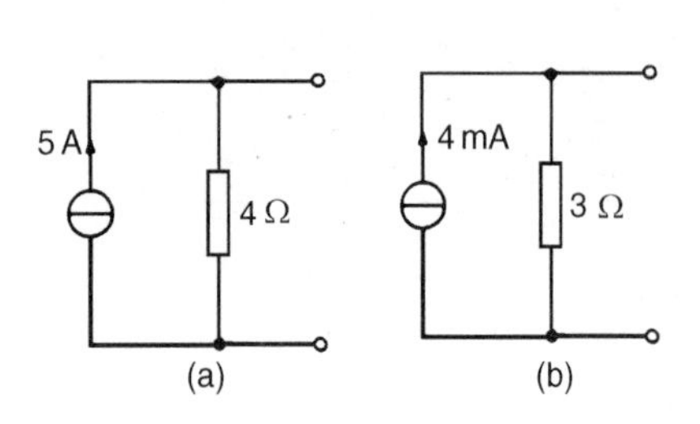

Figure 13.63

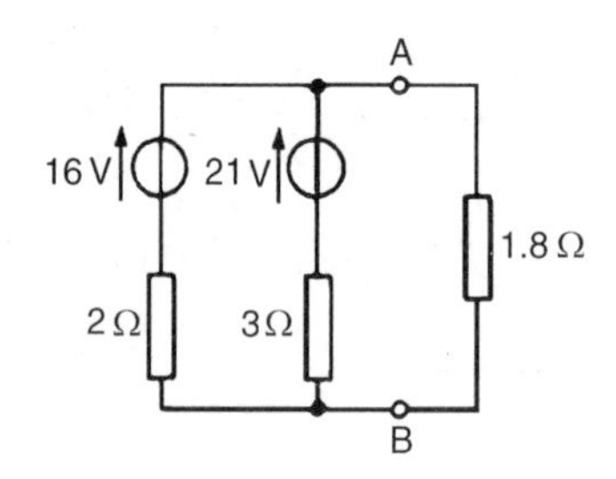

Figure 13.64

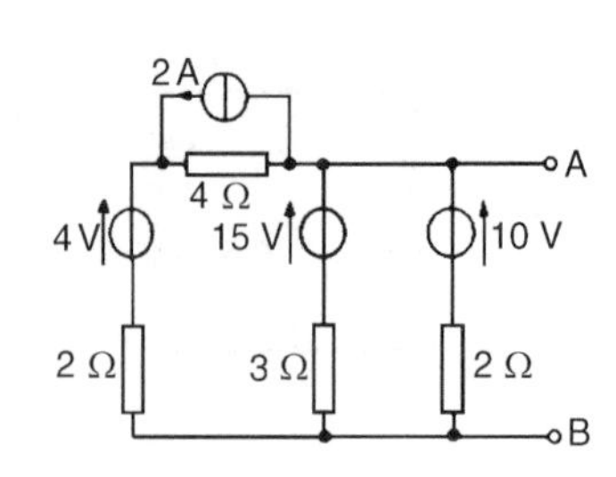

Figure 13.65

20 (a) Convert the network to the left of terminals AB in Figure 13.64 to an equivalent Thévenin circuit by initially converting to a Norton equivalent network.
(b) Determine the current flowing in the 1.8 Ω resistance connected between A and B in Figure 13.64.

[(a) $E = 18$ V, $r = 1.2$ Ω (b) 6 A]

21 Determine, by successive conversions between Thévenin and Norton equivalent networks, a Thévenin equivalent circuit for terminals AB of Figure 13.65. Hence determine the current flowing in a 6 Ω resistor connected between A and B.

[$E = 9\frac{1}{3}$ V, $r = 1$ Ω, $1\frac{1}{3}$ A]

Maximum power transfer theorem

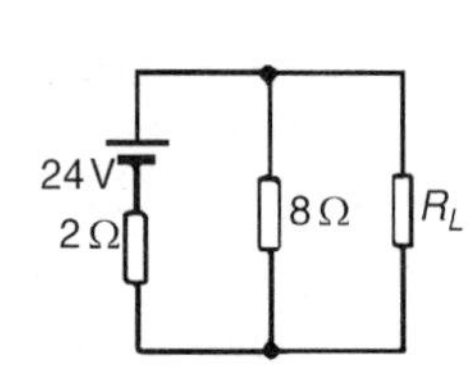

Figure 13.66

22 A d.c. source has an open-circuit voltage of 20 V and an internal resistance of 2 Ω. Determine the value of the load resistance that gives maximum power dissipation. Find the value of this power.

[2 Ω, 50 W]

23 Determine the value of the load resistance R_L shown in Figure 13.66 that gives maximum power dissipation and find the value of the power. [$R_L = 1.6$ Ω, $P = 57.6$ W]

14 Alternating voltages and currents

At the end of this chapter you should be able to:

- appreciate why a.c. is used in preference to d.c.
- describe the principle of operation of an a.c. generator
- distinguish between unidirectional and alternating waveforms
- define cycle, period or periodic time T and frequency f of a waveform
- perform calculations involving $T = \frac{1}{f}$
- define instantaneous, peak, mean and rms values, and form and peak factors for a sine wave
- calculate mean and rms values and form and peak factors for given waveforms
- understand and perform calculations on the general sinusoidal equation $v = V_m \sin(\omega t \pm \phi)$
- understand lagging and leading angles
- combine two sinusoidal waveforms (a) by plotting graphically, (b) by drawing phasors to scale and (c) by calculation
- understand rectification, and describe methods of obtaining half-wave and full-wave rectification

14.1 Introduction

Electricity is produced by generators at power stations and then distributed by a vast network of transmission lines (called the National Grid system) to industry and for domestic use. It is easier and cheaper to generate alternating current (a.c.) than direct current (d.c.) and a.c. is more conveniently distributed than d.c. since its voltage can be readily altered using transformers. Whenever d.c. is needed in preference to a.c., devices called rectifiers are used for conversion (see Section 14.7).

14.2 The a.c. generator

Let a single turn coil be free to rotate at constant angular velocity symmetrically between the poles of a magnet system as shown in Figure 14.1.

An e.m.f. is generated in the coil (from Faraday's Laws) which varies in magnitude and reverses its direction at regular intervals. The reason for this is shown in Figure 14.2. In positions (a), (e) and (i) the conductors

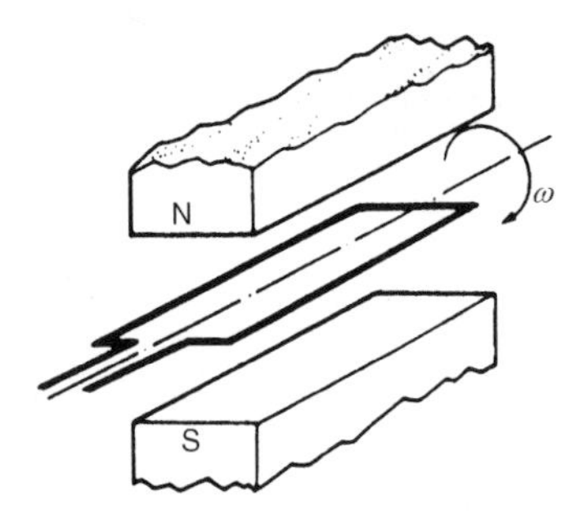

Figure 14.1

of the loop are effectively moving along the magnetic field, no flux is cut and hence no e.m.f. is induced. In position (c) maximum flux is cut and hence maximum e.m.f. is induced. In position (g), maximum flux is cut and hence maximum e.m.f. is again induced. However, using Fleming's right-hand rule, the induced e.m.f. is in the opposite direction to that in position (c) and is thus shown as $-E$. In positions (b), (d), (f) and (h) some flux is cut and hence some e.m.f. is induced. If all such positions of the coil are considered, in one revolution of the coil, one cycle of alternating e.m.f. is produced as shown. This is the principle of operation of the ac generator (i.e. the alternator).

14.3 Waveforms

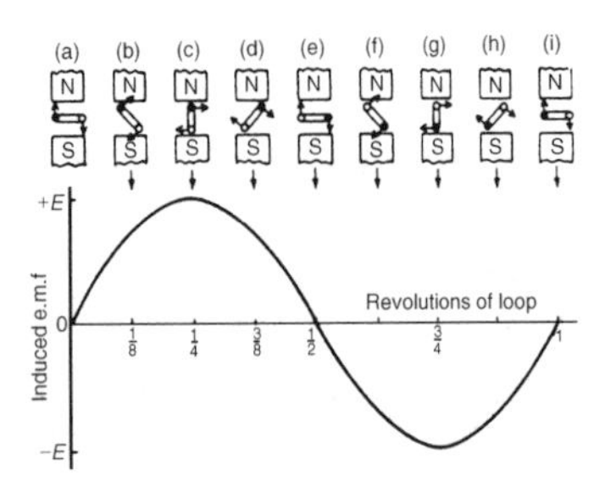

Figure 14.2

If values of quantities which vary with time t are plotted to a base of time, the resulting graph is called a **waveform**. Some typical waveforms are shown in Figure 14.3. Waveforms (a) and (b) are **unidirectional waveforms**, for, although they vary considerably with time, they flow in one direction only (i.e. they do not cross the time axis and become negative). Waveforms (c) to (g) are called **alternating waveforms** since their quantities are continually changing in direction (i.e. alternately positive and negative).

A waveform of the type shown in Figure 14.3(g) is called a **sine wave**. It is the shape of the waveform of e.m.f. produced by an alternator and thus the mains electricity supply is of 'sinusoidal' form.

One complete series of values is called a **cycle** (i.e. from O to P in Figure 14.3(g)).

The time taken for an alternating quantity to complete one cycle is called the **period** or the **periodic time, *T***, of the waveform.

The number of cycles completed in one second is called the **frequency**, f, of the supply and is measured in **hertz, Hz**. The standard frequency of the electricity supply in Great Britain is 50 Hz.

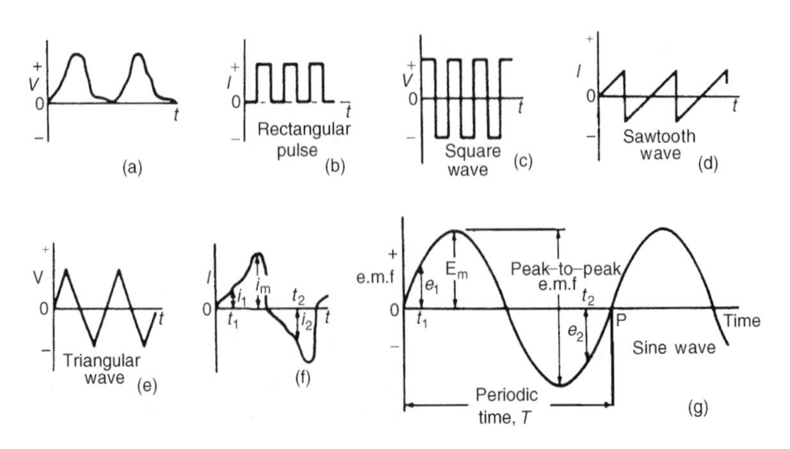

Figure 14.3

$$T = \frac{1}{f} \text{ or } f = \frac{1}{T}$$

Problem 1. Determine the periodic time for frequencies of (a) 50 Hz and (b) 20 kHz

(a) Periodic time $T = \frac{1}{f} = \frac{1}{50} = \mathbf{0.02\ s\ or\ 20\ ms}$

(b) Periodic time $T = \frac{1}{f} = \frac{1}{20\,000} = \mathbf{0.000\,05\ s\ or\ 50\ \mu s}$

Problem 2. Determine the frequencies for periodic times of (a) 4 ms, (b) 4 μs

(a) Frequency $f = \frac{1}{T} = \frac{1}{4 \times 10^{-3}} = \frac{1000}{4} = \mathbf{250\ Hz}$

(b) Frequency $f = \frac{1}{T} = \frac{1}{4 \times 10^{-6}} = \frac{1\,000\,000}{4}$

$= \mathbf{250\,000\ Hz}$ or $\mathbf{250\ kHz}$ or $\mathbf{0.25\ MHz}$

Problem 3. An alternating current completes 5 cycles in 8 ms. What is its frequency?

Time for 1 cycle $= \frac{8}{5}$ ms $= 1.6$ ms $=$ periodic time T

Frequency $f = \frac{1}{T} = \frac{1}{1.6 \times 10^{-3}} = \frac{1000}{1.6} = \frac{10000}{16} = \mathbf{625\ Hz}$

Further problems on frequency and periodic time may be found in Section 14.8, problems 1 to 3, page 209.

14.4 A.c. values

Instantaneous values are the values of the alternating quantities at any instant of time. They are represented by small letters, i, v, e etc., (see Figures 14.3(f) and (g)).

The largest value reached in a half cycle is called the **peak value** or the **maximum value** or the **crest value** or the **amplitude** of the waveform. Such values are represented by V_m, I_m, etc. (see Figures 14.3(f) and (g)). A **peak-to-peak** value of e.m.f. is shown in Figure 14.3(g) and is the difference between the maximum and minimum values in a cycle.

The **average** or **mean value** of a symmetrical alternating quantity, (such as a sine wave), is the average value measured over a half cycle, (since over a complete cycle the average value is zero).

$$\textbf{Average or mean value} = \frac{\textbf{area under the curve}}{\textbf{length of base}}$$

The area under the curve is found by approximate methods such as the trapezoidal rule, the mid-ordinate rule or Simpson's rule. Average values are represented by V_{AV}, I_{AV}, etc.

For a sine wave, average value = 0.637 × maximum value (i.e. $2/\pi$ × maximum value)

The **effective value** of an alternating current is that current which will produce the same heating effect as an equivalent direct current. The effective value is called the **root mean square (rms) value** and whenever an alternating quantity is given, it is assumed to be the rms value. For example, the domestic mains supply in Great Britain is 240 V and is assumed to mean '240 V rms'. The symbols used for rms values are I, V, E, etc. For a non-sinusoidal waveform as shown in Figure 14.4 the rms value is given by:

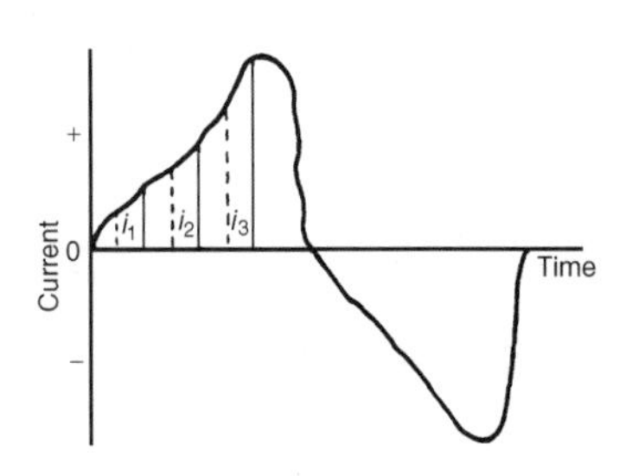

Figure 14.4

$$I = \sqrt{\left(\frac{i_1^2 + i_2^2 + \ldots + i_n^2}{n}\right)}$$

where n is the number of intervals used.

For a sine wave, rms value = 0.707 × maximum value (i.e. $1/\sqrt{2}$ × maximum value)

$$\textbf{Form factor} = \frac{\textbf{rms value}}{\textbf{average value}}$$

For a sine wave, form factor = 1.11

$$\textbf{Peak factor} = \frac{\textbf{maximum value}}{\textbf{rms value}}$$

For a sine wave, peak factor = 1.41

The values of form and peak factors give an indication of the shape of waveforms.

Problem 4. For the periodic waveforms shown in Figure 14.5 determine for each: (i) frequency (ii) average value over half a cycle (iii) rms value (iv) form factor and (v) peak factor

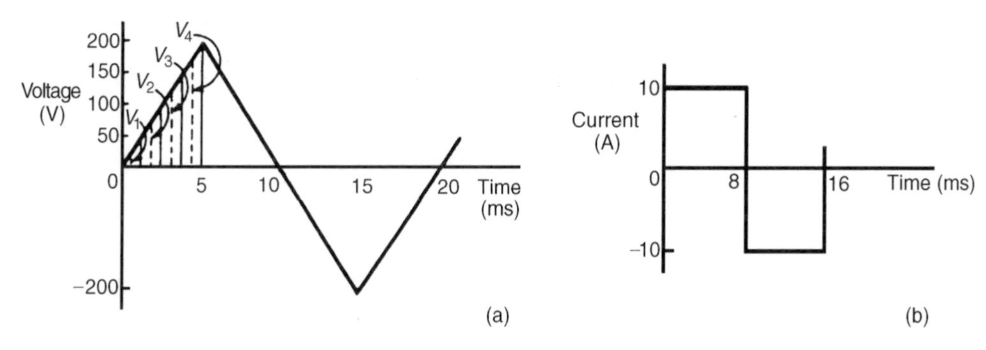

Figure 14.5

(a) **Triangular waveform (Figure 14.5(a))**

(i) Time for 1 complete cycle = 20 ms = periodic time, T

$$\text{Hence frequency } f = \frac{1}{T} = \frac{1}{20 \times 10^{-3}} = \frac{1000}{20} = \mathbf{50\ Hz}$$

(ii) Area under the triangular waveform for a half cycle

$$= \tfrac{1}{2} \times \text{base} \times \text{height} = \tfrac{1}{2} \times (10 \times 10^{-3}) \times 200$$

$$= 1 \text{ volt second}$$

Average value of waveform

$$= \frac{\text{area under curve}}{\text{length of base}} = \frac{1 \text{ volt second}}{10 \times 10^{-3} \text{ second}} = \frac{1000}{10} = \mathbf{100\ V}$$

(iii) In Figure 14.5(a), the first 1/4 cycle is divided into 4 intervals.

$$\text{Thus rms value} = \sqrt{\left(\frac{v_1^2 + v_2^2 + v_3^2 + v_4^2}{4}\right)}$$

$$= \sqrt{\left(\frac{25^2 + 75^2 + 125^2 + 175^2}{4}\right)}$$

$$= \mathbf{114.6\ V}$$

(Note that the greater the number of intervals chosen, the greater the accuracy of the result. For example, if twice the number of ordinates as that chosen above are used, the rms value is found to be 115.6 V)

(iv) $$\text{Form factor} = \frac{\text{rms value}}{\text{average value}} = \frac{114.6}{100} = \mathbf{1.15}$$

(v) $$\text{Peak factor} = \frac{\text{maximum value}}{\text{rms value}} = \frac{200}{114.6} = \mathbf{1.75}$$

(b) **Rectangular waveform (Figure 14.5(b))**

(i) Time for 1 complete cycle = 16 ms = periodic time, T

$$\text{Hence frequency, } f = \frac{1}{T} = \frac{1}{16 \times 10^{-3}} = \frac{1000}{16}$$

$$= \mathbf{62.5\ Hz}$$

(ii) $$\text{Average value over half a cycle} = \frac{\text{area under curve}}{\text{length of base}}$$

$$= \frac{10 \times (8 \times 10^{-3})}{8 \times 10^{-3}} = \mathbf{10\ A}$$

(iii) $$\text{The rms value} = \sqrt{\left(\frac{i_1^2 + i_2^2 + \ldots + i_n^2}{n}\right)} = \mathbf{10\ A}$$

however many intervals are chosen, since the waveform is rectangular.

(iv) $$\text{Form factor} = \frac{\text{rms value}}{\text{average value}} = \frac{10}{10} = \mathbf{1}$$

(v) $$\text{Peak factor} = \frac{\text{maximum value}}{\text{rms value}} = \frac{10}{10} = \mathbf{1}$$

Problem 5. The following table gives the corresponding values of current and time for a half cycle of alternating current.

time t (ms)	0	0.5	1.0	1.5	2.0	2.5	3.0	3.5	4.0	4.5	5.0
current i (A)	0	7	14	23	40	56	68	76	60	5	0

Assuming the negative half cycle is identical in shape to the positive half cycle, plot the waveform and find (a) the frequency of the supply, (b) the instantaneous values of current after 1.25 ms and 3.8 ms, (c) the peak or maximum value, (d) the mean or average value, and (e) the rms value of the waveform.

The half cycle of alternating current is shown plotted in Figure 14.6

(a) Time for a half cycle = 5 ms. Hence the time for 1 cycle, i.e. the periodic time, T = 10 ms or 0.01 s

$$\text{Frequency, } f = \frac{1}{T} = \frac{1}{0.01} = \mathbf{100\ Hz}$$

(b) Instantaneous value of current after 1.25 ms is **19 A**, from Figure 14.6

Instantaneous value of current after 3.8 ms is **70 A**, from Figure 14.6

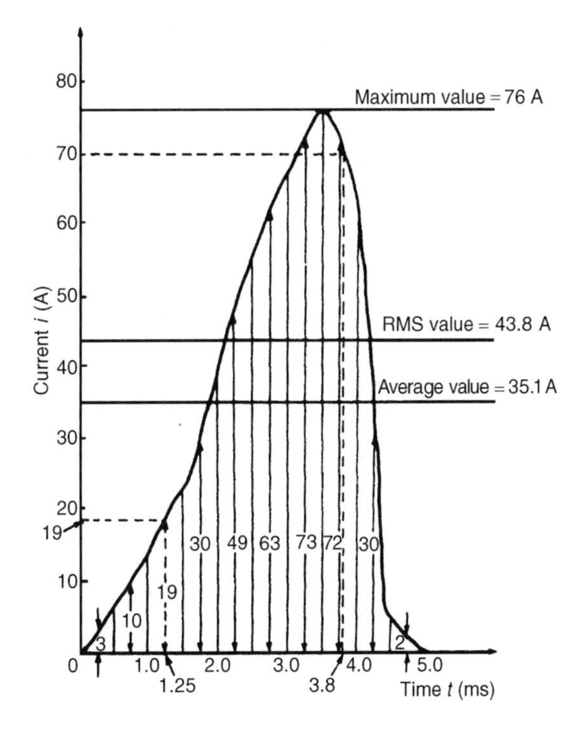

Figure 14.6

(c) Peak or maximum value = **76 A**

(d) Mean or average value $= \dfrac{\text{area under curve}}{\text{length of base}}$

Using the mid-ordinate rule with 10 intervals, each of width 0.5 ms gives:

$$\text{area under curve} = (0.5 \times 10^{-3})[3 + 10 + 19 + 30 + 49 + 63 + 73 + 72 + 30 + 2] \text{ (see Figure 14.6)}$$

$$= (0.5 \times 10^{-3})(351)$$

$$\text{Hence mean or average value} = \frac{(0.5 \times 10^{-3})(351)}{5 \times 10^{-3}} = \mathbf{35.1\ A}$$

(e) rms value

$$= \sqrt{\left(\frac{3^2 + 10^2 + 19^2 + 30^2 + 49^2 + 63^2 + 73^2 + 72^2 + 30^2 + 2^2}{10}\right)}$$

$$= \sqrt{\left(\frac{19\,157}{10}\right)} = \mathbf{43.8\ A}$$

Problem 6. Calculate the rms value of a sinusoidal current of maximum value 20 A

For a sine wave, rms value $= 0.707 \times$ maximum value

$$= 0.707 \times 20 = \mathbf{14.14\ A}$$

Problem 7. Determine the peak and mean values for a 240 V mains supply.

For a sine wave, rms value of voltage $V = 0.707 \times V_m$
A 240 V mains supply means that 240 V is the rms value, hence

$$V_m = \frac{V}{0.707} = \frac{240}{0.707} = \mathbf{339.5\ V = peak\ value}$$

Mean value $V_{AV} = 0.637V_m = 0.637 \times 339.5 = \mathbf{216.3\ V}$

Problem 8. A supply voltage has a mean value of 150 V. Determine its maximum value and its rms value

For a sine wave, mean value $= 0.637 \times$ maximum value

Hence maximum value $= \dfrac{\text{mean value}}{0.637} = \dfrac{150}{0.637} = \mathbf{235.5\ V}$

rms value $= 0.707 \times$ maximum value $= 0.707 \times 235.5 = \mathbf{166.5\ V}$

Further problems on a.c. values of waveforms may be found in Section 14.8, problems 4 to 10, page 209.

14.5 The equation of a sinusoidal waveform

In Fig 14.7, OA represents a vector that is free to rotate anticlockwise about 0 at an angular velocity of ω rad/s. A rotating vector is known as a **phasor.**

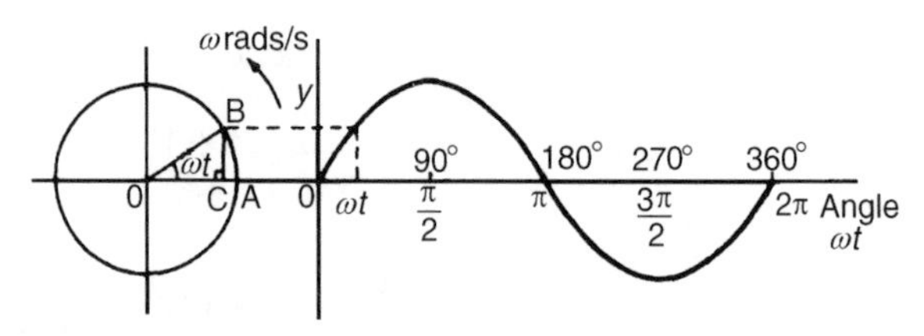

Figure 14.7

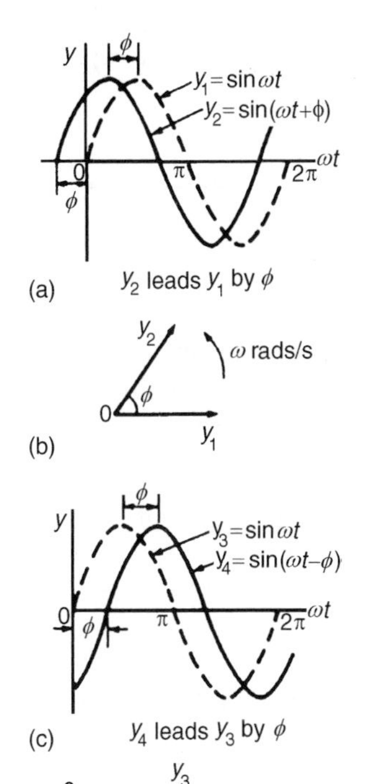

Figure 14.8

After time t seconds the vector OA has turned through an angle ωt. If the line BC is constructed perpendicular to OA as shown, then

$$\sin \omega t = \frac{\text{BC}}{\text{OB}} \quad \text{i.e. BC} = \text{OB} \ \sin \omega t$$

If all such vertical components are projected on to a graph of y against angle ωt (in radians), a sine curve results of maximum value OA. Any quantity which varies sinusoidally can thus be represented as a phasor.

A sine curve may not always start at 0°. To show this a periodic function is represented by $y = \sin(\omega t \pm \phi)$, where ϕ is the phase (or angle) difference compared with $y = \sin \omega t$. In Figure 14.8(a), $y_2 = \sin(\omega t + \phi)$ starts ϕ radians earlier than $y_1 = \sin \omega t$ and is thus said to **lead** y_1 by ϕ radians. Phasors y_1 and y_2 are shown in Figure 14.8(b) at the time when $t = 0$.

In Figure 14.8(c), $y_4 = \sin(\omega t - \phi)$ starts ϕ radians later than $y_3 = \sin \omega t$ and is thus said to **lag** y_3 by ϕ radians. Phasors y_3 and y_4 are shown in Figure 14.8(d) at the time when $t = 0$.

Given the general sinusoidal voltage, $\boldsymbol{v = V_m \sin(\omega t \pm \phi)}$, then

(i) Amplitude or maximum value $= V_m$
(ii) Peak to peak value $= 2V_m$
(iii) Angular velocity $= \omega$ rad/s
(iv) Periodic time, $T = 2\pi/\omega$ seconds
(v) Frequency, $f = \omega/2\pi$ Hz (since $\omega = 2\pi$ f)
(vi) $\phi =$ angle of lag or lead (compared with $v = V_m \sin \omega t$)

Problem 9. An alternating voltage is given by $v = 282.8 \sin 314t$ volts. Find (a) the rms voltage, (b) the frequency and (c) the instantaneous value of voltage when $t = 4$ ms

(a) The general expression for an alternating voltage is

$$v = V_m \sin(\omega t \pm \phi).$$

Comparing $v = 282.8 \sin 314t$ with this general expression gives the peak voltage as 282.8 V

Hence the rms voltage $= 0.707 \times$ maximum value

$$= 0.707 \times 282.8 = \mathbf{200\ V}$$

(b) Angular velocity, $\omega = 314$ rad/s, i.e. $2\pi f = 314$

Hence frequency, $f = \dfrac{314}{2\pi} = \mathbf{50\ Hz}$

(c) When $t = 4$ ms, $v = 282.8 \sin(314 \times 4 \times 10^{-3})$

$$= 282.8 \sin(1.256) = \mathbf{268.9\ V}$$

(Note that 1.256 radians $= \left[1.256 \times \frac{180}{\pi}\right]^\circ = 71.96^\circ = 71^\circ 58'$

Hence $v = 282.8 \sin 71^\circ 58' =$ **268.9 V**)

Problem 10. An alternating voltage is given by

$v = 75\sin(200\pi t - 0.25)$ volts.

Find (a) the amplitude, (b) the peak-to-peak value, (c) the rms value, (d) the periodic time, (e) the frequency, and (f) the phase angle (in degrees and minutes) relative to $75\sin 200\pi t$

Comparing $v = 75\sin(200\pi t - 0.25)$ with the general expression $v = V_m \sin(\omega t \pm \phi)$ gives:

(a) Amplitude, or peak value = **75 V**

(b) Peak-to-peak value = 2 × 75 = **150 V**

(c) The rms value = 0.707 × maximum value = 0.707 × 75 = **53 V**

(d) Angular velocity, $\omega = 200\pi$ rad/s

Hence periodic time, $T = \frac{2\pi}{\omega} = \frac{2\pi}{200\pi} = \frac{1}{100} =$ **0.01 s or 10 ms**

(e) Frequency, $f = \frac{1}{T} = \frac{1}{0.01} =$ **100 Hz**

(f) Phase angle, $\phi = 0.25$ radians lagging $75 \sin 200\pi t$

$0.25 \text{ rads} = \left(0.25 \times \frac{180}{\pi}\right)^\circ = 14.32^\circ = 14^\circ 19'$

Hence phase angle = **14°19′ lagging**

Problem 11. An alternating voltage, v, has a periodic time of 0.01 s and a peak value of 40 V. When time t is zero, $v = -20$ V. Express the instantaneous voltage in the form $v = V_m \sin(\omega t \pm \phi)$

Amplitude, $V_m = 40$ V

Periodic time $T = \frac{2\pi}{\omega}$ hence angular velocity,

$$\omega = \frac{2\pi}{T} = \frac{2\pi}{0.01} = 200\pi \text{ rad/s}$$

$v = V_m \sin(\omega t + \phi)$ thus becomes $v = 40\sin(200\pi t + \phi)$ V

When time $t = 0$, $v = -20$ V

i.e. $-20 = 40 \sin \phi$

so that $\sin \phi = \dfrac{-20}{40} = -0.5$

Hence $\phi = \arcsin(-0.5) = -30° = \left(-30 \times \dfrac{\pi}{180}\right)$ rads $= -\dfrac{\pi}{6}$ rads

Thus $\boldsymbol{v = 40 \sin\left(200\pi t - \dfrac{\pi}{6}\right)}$ **V**

Problem 12. The current in an a.c. circuit at any time t seconds is given by: $i = 120 \sin(100\pi t + 0.36)$ amperes. Find:

(a) the peak value, the periodic time, the frequency and phase angle relative to $120 \sin 100\pi t$

(b) the value of the current when $t = 0$

(c) the value of the current when $t = 8$ ms

(d) the time when the current first reaches 60 A, and

(e) the time when the current is first a maximum

(a) Peak value = **120 A**

Periodic time $T = \dfrac{2\pi}{\omega} = \dfrac{2\pi}{100\pi}$ (since $\omega = 100\pi$)

$= \dfrac{1}{50} =$ **0.02 s or 20 ms**

Frequency, $f = \dfrac{1}{T} = \dfrac{1}{0.02} =$ **50 Hz**

Phase angle $= 0.36$ rads $= \left(0.36 \times \dfrac{180}{\pi}\right)^{\circ} =$ **20°38′ leading**

(b) When $t = 0$, $i = 120 \sin(0 + 0.36) = 120 \sin 20°38' =$ **49.3 A**

(c) When $t = 8$ ms, $i = 120 \sin\left[100\pi\left(\dfrac{8}{10^3}\right) + 0.36\right]$

$= 120 \sin 2.8733 (= 120 \sin 164°38') =$ **31.8 A**

(d) When $i = 60$ A, $60 = 120 \sin(100\pi t + 0.36)$

thus $\dfrac{60}{120} = \sin(100\pi t + 0.36)$

so that $(100\pi t + 0.36) = \arcsin 0.5 = 30° = \dfrac{\pi}{6}$ rads $= 0.5236$ rads

Hence time, $t = \dfrac{0.5236 - 0.36}{100\pi} =$ **0.521 ms**

(e) When the current is a maximum, $i = 120$ A

Thus $120 = 120\sin(100\pi t + 0.36)$

$1 = \sin(100\pi t + 0.36)$

$(100\pi t + 0.36) = \arcsin 1 = 90^\circ = \frac{\pi}{2}$ rads $= 1.5708$ rads

Hence time, $t = \frac{1.5708 - 0.36}{100\pi} = \mathbf{3.85\ ms}$

Further problems on $v = V_m\sin(\omega t \pm \phi)$ may be found in Section 14.8, problems 11 to 15, page 210.

14.6 Combination of waveforms

The resultant of the addition (or subtraction) of two sinusoidal quantities may be determined either:

(a) by plotting the periodic functions graphically (see worked Problems 13 and 16), or

(b) by resolution of phasors by drawing or calculation (see worked Problems 14 and 15).

Problem 13. The instantaneous values of two alternating currents are given by $i_1 = 20\sin\omega t$ amperes and $i_2 = 10\sin(\omega t + \pi/3)$ amperes. By plotting i_1 and i_2 on the same axes, using the same scale, over one cycle, and adding ordinates at intervals, obtain a sinusoidal expression for $i_1 + i_2$

$i_1 = 20\sin\omega t$ and $i_2 = 10\sin\left(\omega t + \frac{\pi}{3}\right)$ are shown plotted in Figure 14.9

Ordinates of i_1 and i_2 are added at, say, 15° intervals (a pair of dividers are useful for this).

For example,

at 30°, $i_1 + i_2 = 10 + 10 = 20$ A
at 60°, $i_1 + i_2 = 8.7 + 17.3 = 26$ A
at 150°, $i_1 + i_2 = 10 + (-5) = 5$ A, and so on.

The resultant waveform for $i_1 + i_2$ is shown by the broken line in Figure 14.9. It has the same period, and hence frequency, as i_1 and i_2. The amplitude or peak value is 26.5 A.

The resultant waveform leads the curve $i_1 = 20\sin\omega t$ by 19°

i.e. $\left(19 \times \frac{\pi}{180}\right)$ rads $= 0.332$ rads

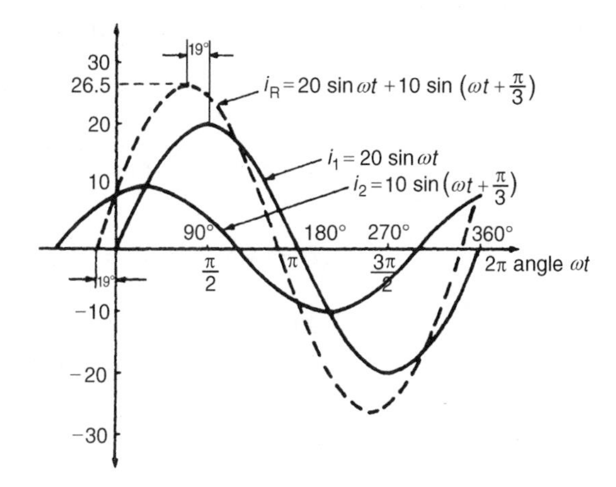

Figure 14.9

Hence the sinusoidal expression for the resultant $i_1 + i_2$ is given by:

$$\boldsymbol{i_R = i_1 + i_2 = 26.5 \sin(\omega t + 0.332)\ \mathrm{A}}$$

> Problem 14. Two alternating voltages are represented by $v_1 = 50 \sin \omega t$ volts and $v_2 = 100 \sin(\omega t - \pi/6)$ V. Draw the phasor diagram and find, by calculation, a sinusoidal expression to represent $v_1 + v_2$

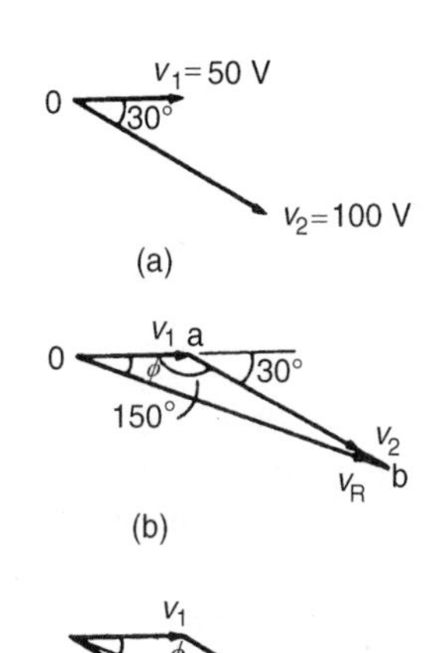

Figure 14.10

Phasors are usually drawn at the instant when time $t = 0$. Thus v_1 is drawn horizontally 50 units long and v_2 is drawn 100 units long lagging v_1 by $\pi/6$ rads, i.e. 30°. This is shown in Figure 14.10(a) where 0 is the point of rotation of the phasors.

Procedure to draw phasor diagram to represent $v_1 + v_2$:

(i) Draw v_1 horizontal 50 units long, i.e. oa of Figure 14.10(b)

(ii) Join v_2 to the end of v_1 at the appropriate angle, i.e. ab of Figure 14.10(b)

(iii) The resultant $v_R = v_1 + v_2$ is given by the length ob and its phase angle may be measured with respect to v_1

Alternatively, when two phasors are being added the resultant is always the diagonal of the parallelogram, as shown in Figure 14.10(c).

From the drawing, by measurement, $v_R = 145$ V and angle $\phi = 20°$ lagging v_1.

A more accurate solution is obtained by calculation, using the cosine and sine rules. Using the cosine rule on triangle oab of Figure 14.10(b) gives:

$$v_R^2 = v_1^2 + v_2^2 - 2v_1v_2 \cos 150°$$

$$= 50^2 + 100^2 - 2(50)(100) \cos 150°$$

$$= 2500 + 10000 - (-8660)$$

$$v_R = \sqrt{(21\,160)} = 145.5 \text{ V}$$

Using the sine rule, $\dfrac{100}{\sin\phi} = \dfrac{145.5}{\sin 150°}$

from which $\sin\phi = \dfrac{100 \sin 150°}{145.5} = 0.3436$

and $\phi = \arcsin 0.3436 = 20°6' = 0.35$ radians, and lags v_1

Hence $v_R = v_1 + v_2 =$ **$145.5 \sin(\omega t - 0.35)$ V**

> Problem 15. Find a sinusoidal expression for $(i_1 + i_2)$ of Problem 13, (a) by drawing phasors, (b) by calculation.

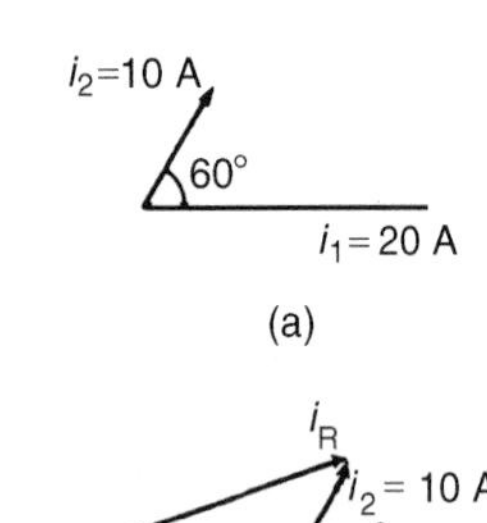

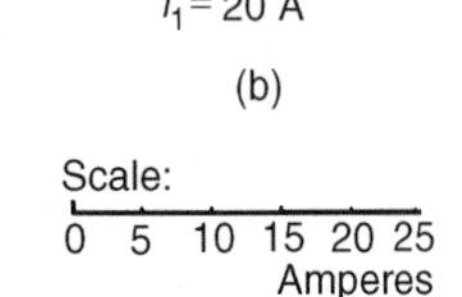

Figure 14.11

(a) The relative positions of i_1 and i_2 at time $t = 0$ are shown as phasors in Figure 14.11(a). The phasor diagram in Figure 14.11(b) shows the resultant i_R, and i_R is measured as 26 A and angle ϕ as 19° or 0.33 rads leading i_1.

Hence, by drawing, $i_R = 26 \sin(\omega t + 0.33)$ A

(b) From Figure 14.11(b), by the cosine rule:

$$i_R^2 = 20^2 + 10^2 - 2(20)(10)(\cos 120°)$$

from which $i_R =$ **26.46 A**

By the sine rule: $\dfrac{10}{\sin\phi} = \dfrac{26.46}{\sin 120°}$

from which $\phi = 19.10°$ (i.e. 0.333 rads)

Hence, by calculation $i_R = 26.46 \sin(\omega t + 0.333)$ A

> Problem 16. Two alternating voltages are given by $v_1 = 120 \sin \omega t$ volts and $v_2 = 200 \sin(\omega t - \pi/4)$ volts. Obtain sinusoidal expressions for $v_1 - v_2$ (a) by plotting waveforms, and (b) by resolution of phasors.

(a) $v_1 = 120 \sin \omega t$ and $v_2 = 200 \sin(\omega t - \pi/4)$ are shown plotted in Figure 14.12. Care must be taken when subtracting values of ordinates especially when at least one of the ordinates is negative. For example

at 30°, $v_1 - v_2 = 60 - (-52) = 112$ V

at 60°, $v_1 - v_2 = 104 - 52 = 52$ V

at 150°, $v_1 - v_2 = 60 - 193 = -133$ V and so on

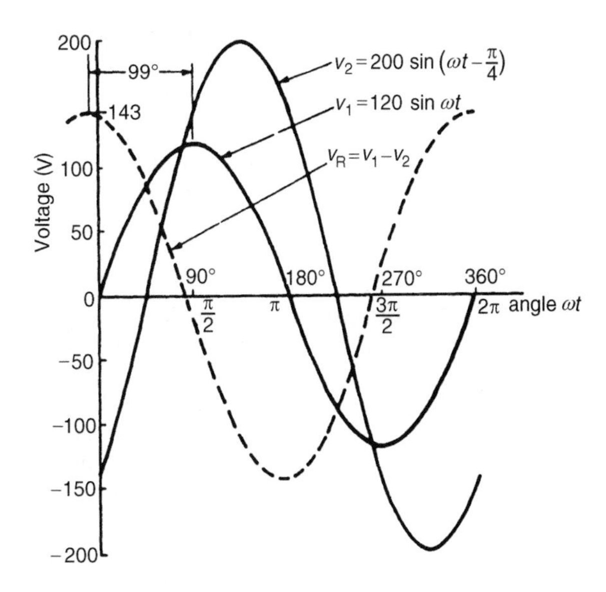

Figure 14.12

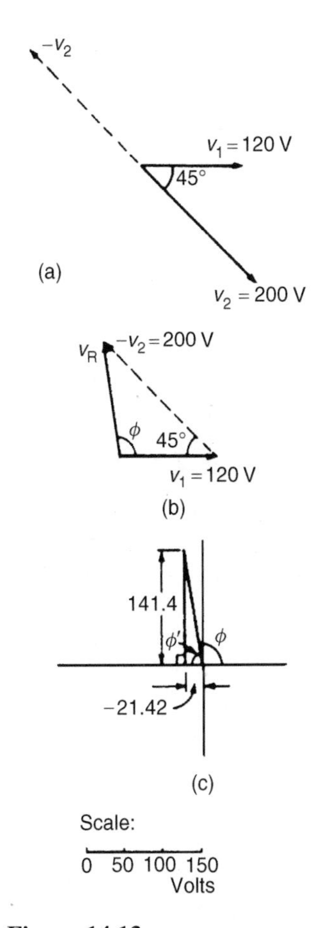

Figure 14.13

The resultant waveform, $v_R = v_1 - v_2$, is shown by the broken line in Figure 14.12. The maximum value of v_R is 143 V and the waveform is seen to lead v_1 by 99° (i.e. 1.73 radians)

Hence, by drawing, $v_R = v_1 - v_2 = 143 \sin(\omega t + 1.73)$ volts

(b) The relative positions of v_1 and v_2 are shown at time $t = 0$ as phasors in Figure 14.13(a). Since the resultant of $v_1 - v_2$ is required, $-v_2$ is drawn in the opposite direction to $+v_2$ and is shown by the broken line in Figure 14.13(a). The phasor diagram with the resultant is shown in Figure 14.13(b) where $-v_2$ is added phasorially to v_1
By resolution:

Sum of horizontal components of v_1 and v_2

$$= 120 \cos 0° + 200 \cos 135° = -21.42$$

Sum of vertical components of v_1 and v_2

$$= 120 \sin 0° + 200 \sin 135° = 141.4$$

From Figure 14.13(c), resultant

$$v_R = \sqrt{[(-21.42)^2 + (141.4)^2]} = 143.0,$$

and $\tan \phi' = \dfrac{141.4}{21.42} = \tan 6.6013$, from which

$$\phi' = \arctan 6.6013 = 81°23' \text{ and}$$

$$\phi = 98°37' \text{ or } 1.721 \text{ radians}$$

Hence, by resolution of phasors,

$$v_R = v_1 - v_2 = \mathbf{143.0\sin(\omega t + 1.721)\ volts}$$

Further problems on the combination of periodic functions may be found in Section 14.8, problems 16 to 19, page 211.

14.7 Rectification

The process of obtaining unidirectional currents and voltages from alternating currents and voltages is called **rectification**. Automatic switching in circuits is carried out by devices called diodes.

Using a single diode, as shown in Figure 14.14, **half-wave rectification** is obtained. When P is sufficiently positive with respect to Q, diode D is switched on and current i flows. When P is negative with respect to Q, diode D is switched off. Transformer T isolates the equipment from direct connection with the mains supply and enables the mains voltage to be changed.

Two diodes may be used as shown in Figure 14.15 to obtain **full wave rectification**. A centre-tapped transformer T is used. When P is sufficiently positive with respect to Q, diode D_1 conducts and current flows (shown by the broken line in Figure 14.15). When S is positive with respect to Q, diode D_2 conducts and current flows (shown by the continuous line in Figure 14.15). The current flowing in R is in the same direction for both half cycles of the input. The output waveform is thus as shown in Figure 14.15.

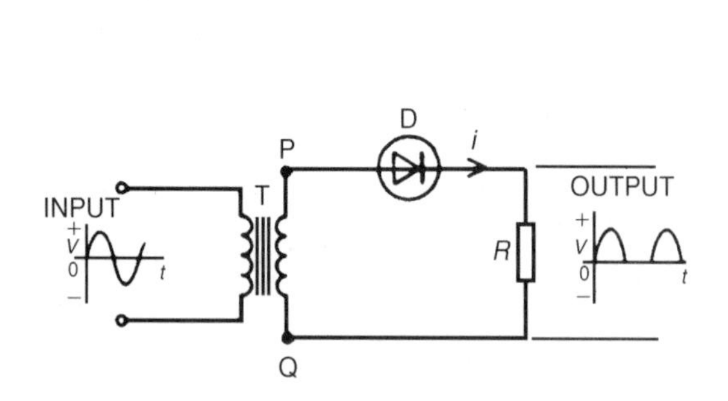

Figure 14.14

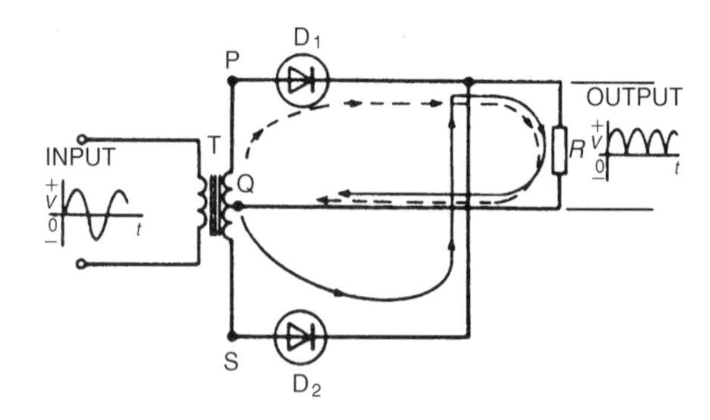

Figure 14.15

Four diodes may be used in a **bridge rectifier** circuit, as shown in Figure 14.16 to obtain **full wave rectification**. As for the rectifier shown in Figure 14.15, the current flowing in R is in the same direction for both half cycles of the input giving the output waveform shown.

To smooth the output of the rectifiers described above, capacitors having a large capacitance may be connected across the load resistor R. The effect of this is shown on the output in Figure 14.17.

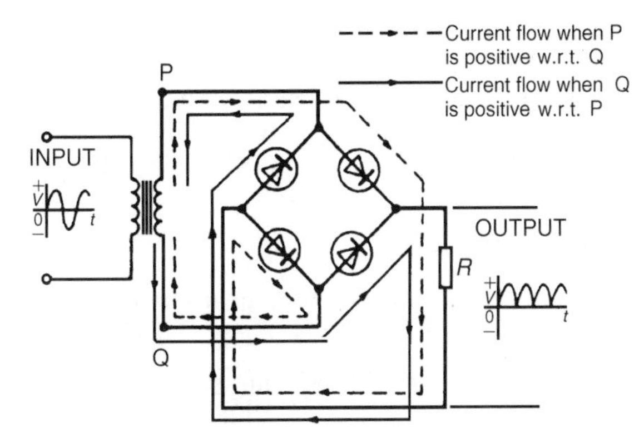

Figure 14.16

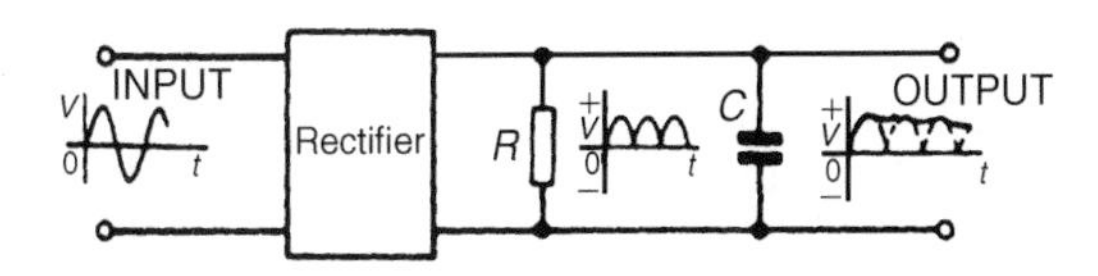

Figure 14.17

14.8 Further problems on alternating voltages and currents

Frequency and periodic time

1 Determine the periodic time for the following frequencies:
(a) 2.5 Hz (b) 100 Hz (c) 40 kHz
[(a) 0.4 s (b) 10 ms (c) 25 μs]

2 Calculate the frequency for the following periodic times:
(a) 5 ms (b) 50 μs (c) 0.2 s
[(a) 0.2 kHz (b) 20 kHz (c) 5 Hz]

3 An alternating current completes 4 cycles in 5 ms. What is its frequency? [800 Hz]

A.c. values of non-sinusoidal waveforms

4 An alternating current varies with time over half a cycle as follows:

Current (A)	0	0.7	2.0	4.2	8.4	8.2	2.5	1.0	0.4	0.2	0
time (ms)	0	1	2	3	4	5	6	7	8	9	10

The negative half cycle is similar. Plot the curve and determine: (a) the frequency (b) the instantaneous values at 3.4 ms and 5.8 ms (c) its mean value and (d) its rms value
[(a) 50 Hz (b) 5.5 A, 3.4 A (c) 2.8 A (d) 4.0 A]

5 For the waveforms shown in Figure 14.18 determine for each (i) the frequency (ii) the average value over half a cycle (iii) the rms value (iv) the form factor (v) the peak factor.
[(a) (i) 100 Hz (ii) 2.50 A (iii) 2.88 A (iv) 1.15 (v) 1.74
(b) (i) 250 Hz (ii) 20 V (iii) 20 V (iv) 1.0 (v) 1.0
(c) (i) 125 Hz (ii) 18 A (iii) 19.56 A (iv) 1.09 (v) 1.23
(d) (i) 250 Hz (ii) 25 V (iii) 50 V (iv) 2.0 (v) 2.0]

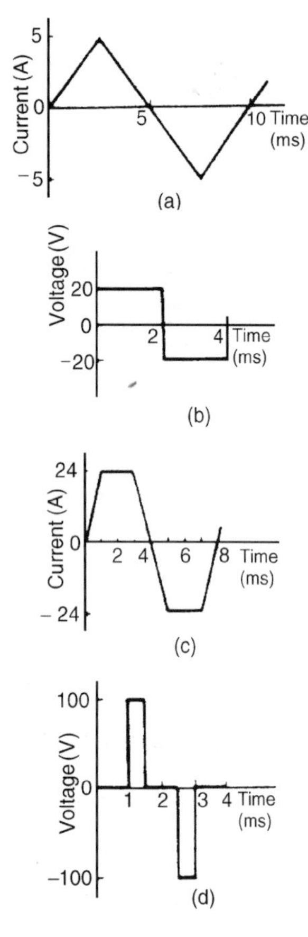

Figure 14.18

6 An alternating voltage is triangular in shape, rising at a constant rate to a maximum of 300 V in 8 ms and then falling to zero at a constant rate in 4 ms. The negative half cycle is identical in shape to the positive half cycle. Calculate (a) the mean voltage over half a cycle, and (b) the rms voltage [(a) 150 V (b) 170 V]

A.c. values of sinusoidal waveforms

7 Calculate the rms value of a sinusoidal curve of maximum value 300 V [212.1 V]

8 Find the peak and mean values for a 200 V mains supply [282.9 V, 180.2 V]

9 A sinusoidal voltage has a maximum value of 120 V. Calculate its rms and average values. [84.8 V, 76.4 V]

10 A sinusoidal current has a mean value of 15.0 A. Determine its maximum and rms values. [23.55 A, 16.65 A]

$v = V_m \sin(\omega t \pm \phi)$

11 An alternating voltage is represented by $v = 20 \sin 157.1\, t$ volts. Find (a) the maximum value (b) the frequency (c) the periodic time. (d) What is the angular velocity of the phasor representing this waveform? [(a) 20 V (b) 25 Hz (c) 0.04 s (d) 157.1 rads/s]

12 Find the peak value, the rms value, the periodic time, the frequency and the phase angle (in degrees and minutes) of the following alternating quantities:

(a) $v = 90 \sin 400\pi t$ volts
[90 V, 63.63 V, 5 ms, 200 Hz, 0°]

(b) $i = 50 \sin(100\pi t + 0.30)$ amperes
[50 A, 35.35 A, 0.02 s, 50 Hz, 17°11′ lead]

(c) $e = 200 \sin(628.4t - 0.41)$ volts
[200 V, 141.4 V, 0.01 s, 100 Hz, 23°29′ lag]

13 A sinusoidal current has a peak value of 30 A and a frequency of 60 Hz. At time $t = 0$, the current is zero. Express the instantaneous current i in the form $i = I_m \sin \omega t$ [$i = 30 \sin 120\pi t$]

14 An alternating voltage v has a periodic time of 20 ms and a maximum value of 200 V. When time $t = 0$, $v = -75$ volts. Deduce a sinusoidal expression for v and sketch one cycle of the voltage showing important points. [$v = 200 \sin(100\pi t - 0.384)$]

15 The instantaneous value of voltage in an a.c. circuit at any time t seconds is given by $v = 100 \sin(50\pi t - 0.523)$ V. Find:

(a) the peak-to-peak voltage, the periodic time, the frequency and the phase angle

(b) the voltage when $t = 0$
(c) the voltage when $t = 8$ ms
(d) the times in the first cycle when the voltage is 60 V
(e) the times in the first cycle when the voltage is −40 V, and
(f) the first time when the voltage is a maximum.

Sketch the curve for one cycle showing relevant points.

[(a) 200 V, 0.04 s, 25 Hz, 29°58′ lagging
(b) −49.95 V (c) 66.96 V (d) 7.426 ms, 19.23 ms
(e) 25.95 ms, 40.71 ms (f) 13.33 ms]

Combination of periodic functions

16 The instantaneous values of two alternating voltages are given by $v_1 = 5 \sin \omega t$ and $v_2 = 8 \sin (\omega t - \pi/6)$. By plotting v_1 and v_2 on the same axes, using the same scale, over one cycle, obtain expressions for (a) $v_1 + v_2$ and (b) $v_1 - v_2$

[(a) $v_1 + v_2 = 12.58 \sin(\omega t - 0.325)$ V
(b) $v_1 - v_2 = 4.44 \sin(\omega t + 2.02)$ V]

17 Repeat Problem 16 by resolution of phasors.

18 Construct a phasor diagram to represent $i_1 + i_2$ where $i_1 = 12 \sin \omega t$ and $i_2 = 15 \sin(\omega t + \pi/3)$. By measurement, or by calculation, find a sinusoidal expression to represent $i_1 + i_2$

[(23.43 $\sin(\omega t + 0.588)$]

19 Determine, either by plotting graphs and adding ordinates at intervals, or by calculation, the following periodic functions in the form $v = V_m \sin(\omega t \pm \phi)$

(a) $10 \sin \omega t + 4 \sin(\omega t + \pi/4)$ [$13.14 \sin(\omega t + 0.217)$]
(b) $80 \sin(\omega t + \pi/3) + 50 \sin(\omega t - \pi/6)$
[$94.34 \sin(\omega t + 0.489)$]
(c) $100 \sin(\omega t - 70 \sin(\omega t - \pi/3)$ [$88.88 \sin(\omega t + 0.751)$]

Assignment 4

This assignment covers the material contained in chapters 13 and 14.

The marks for each question are shown in brackets at the end of each question.

1 Find the current flowing in the 5 Ω resistor of the circuit shown in Figure A4.1 using (a) Kirchhoff's laws, (b) the Superposition theorem, (c) Thévenin's theorem, (d) Norton's theorem. Demonstrate that the same answer results from each method. Find also the current flowing in each of the other two branches of the circuit. (27)

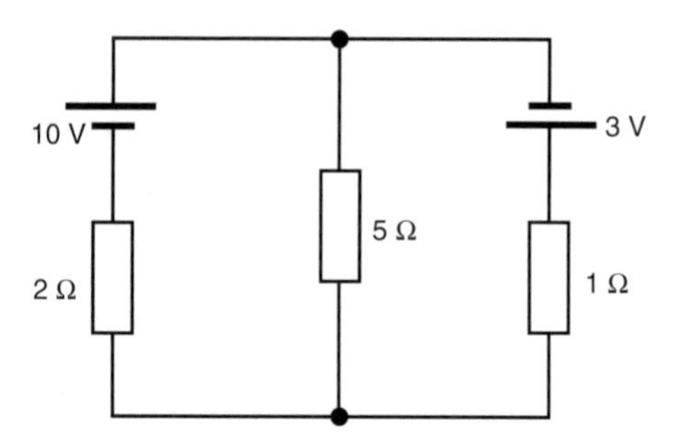

Figure A4.1

2 A d.c. voltage source has an internal resistance of 2 Ω and an open circuit voltage of 24 V. State the value of load resistance that gives maximum power dissipation and determine the value of this power. (5)

3 A sinusoidal voltage has a mean value of 3.0 A. Determine it's maximum and r.m.s. values. (4)

4 The instantaneous value of current in an a.c. circuit at any time t seconds is given by: $i = 50\sin(100\pi t - 0.45)$ mA. Determine

(a) the peak to peak current, the periodic time, the frequency and the phase angle (in degrees and minutes)
(b) the current when $t = 0$
(c) the current when $t = 8$ ms
(d) the first time when the current is a maximum.

Sketch the current for one cycle showing relevant points. (14)

15 Single-phase series a.c. circuits

At the end of this chapter you should be able to:

- draw phasor diagrams and current and voltage waveforms for (a) purely resistive (b) purely inductive and (c) purely capacitive a.c. circuits
- perform calculations involving $X_L = 2\pi f L$ and $X_C = \dfrac{1}{2\pi f C}$
- draw circuit diagrams, phasor diagrams and voltage and impedance triangles for R–L, R–C and R–L–C series a.c. circuits and perform calculations using Pythagoras' theorem, trigonometric ratios and $Z = \dfrac{V}{I}$
- understand resonance
- derive the formula for resonant frequency and use it in calculations
- understand Q-factor and perform calculations using $\dfrac{V_L(\text{or } V_C)}{V}$ or $\dfrac{\omega_r L}{R}$ or $\dfrac{1}{\omega_r CR}$ or $\dfrac{1}{R}\sqrt{\left(\dfrac{L}{C}\right)}$
- understand bandwidth and half-power points
- perform calculations involving $(f_2 - f_1) = \dfrac{f_r}{Q}$
- understand selectivity and typical values of Q-factor
- appreciate that power P in an a.c. circuit is given by $P = VI\cos\phi$ or $I_R^2 R$ and perform calculations using these formulae
- understand true, apparent and reactive power and power factor and perform calculations involving these quantities

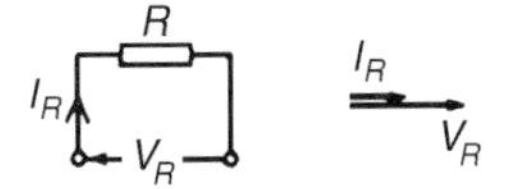

CIRCUIT DIAGRAM   PHASOR DIAGRAM

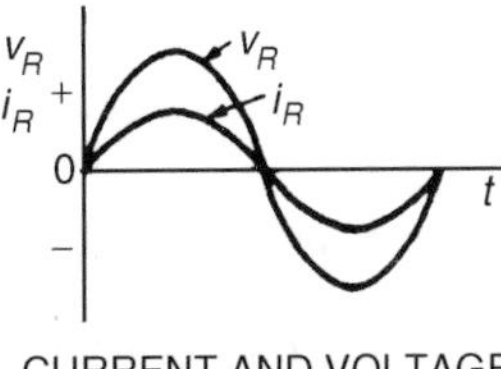

CURRENT AND VOLTAGE WAVEFORMS

Figure 15.1

15.1 Purely resistive a.c. circuit

In a purely resistive a.c. circuit, the current I_R and applied voltage V_R are in phase. See Figure 15.1.

15.2 Purely inductive a.c. circuit

In a purely inductive a.c. circuit, the current I_L **lags** the applied voltage V_L by 90° (i.e. $\pi/2$ rads). See Figure 15.2.

In a purely inductive circuit the opposition to the flow of alternating current is called the **inductive reactance, X_L**

$$X_L = \frac{V_L}{I_L} = 2\pi f L\ \Omega$$

where f is the supply frequency, in hertz, and L is the inductance, in henry's. X_L is proportional to f as shown in Figure 15.3.

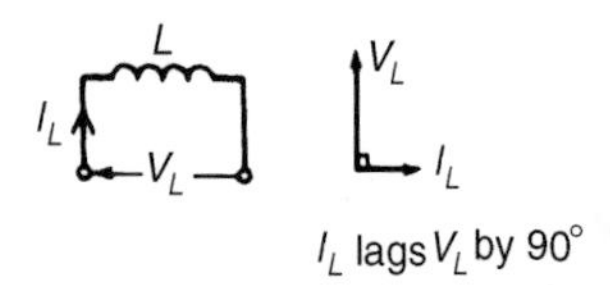

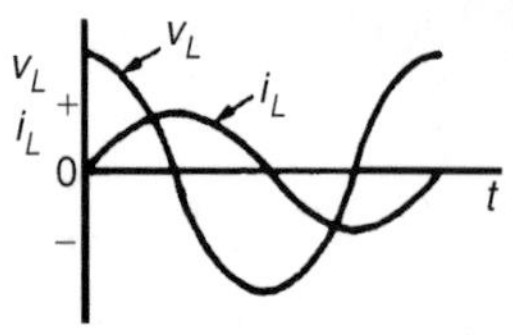

Figure 15.2

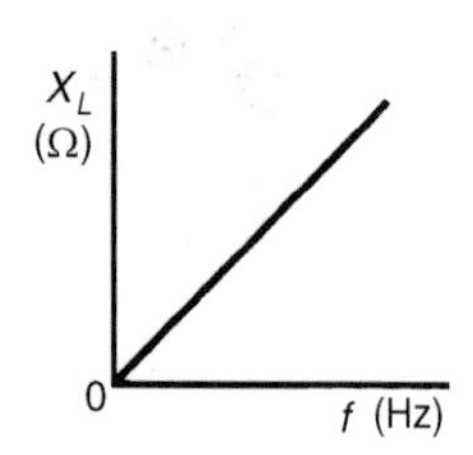

Figure 15.3

Problem 1. (a) Calculate the reactance of a coil of inductance 0.32 H when it is connected to a 50 Hz supply. (b) A coil has a reactance of 124 Ω in a circuit with a supply of frequency 5 kHz. Determine the inductance of the coil.

(a) Inductive reactance, $X_L = 2\pi f L = 2\pi(50)(0.32) = \mathbf{100.5\ \Omega}$

(b) Since $X_L = 2\pi f L$, inductance $L = \dfrac{X_L}{2\pi f} = \dfrac{124}{2\pi(5000)}\text{H} = \mathbf{3.95\ mH}$

Problem 2. A coil has an inductance of 40 mH and negligible resistance. Calculate its inductive reactance and the resulting current if connected to (a) a 240 V, 50 Hz supply, and (b) a 100 V, 1 kHz supply.

(a) Inductive reactance, $X_L = 2\pi f L = 2\pi(50)(40 \times 10^{-3}) = \mathbf{12.57\ \Omega}$

$$\text{Current, } I = \frac{V}{X_L} = \frac{240}{12.57} = \mathbf{19.09\ A}$$

(b) Inductive reactance, $X_L = 2\pi(1000)(40 \times 10^{-3}) = \mathbf{251.3\ \Omega}$

$$\text{Current, } I = \frac{V}{X_L} = \frac{100}{251.3} = \mathbf{0.398\ A}$$

15.3 Purely capacitive a.c. circuit

In a purely capacitive a.c. circuit, the current I_C **leads** the applied voltage V_C by 90° (i.e. $\pi/2$ rads). See Figure 15.4.

In a purely capacitive circuit the opposition to the flow of alternating current is called the **capacitive reactance, X_C**

$$X_C = \frac{V_C}{I_C} = \frac{1}{2\pi f C}\ \Omega$$

where C is the capacitance in farads.

X_C varies with frequency f as shown in Figure 15.5.

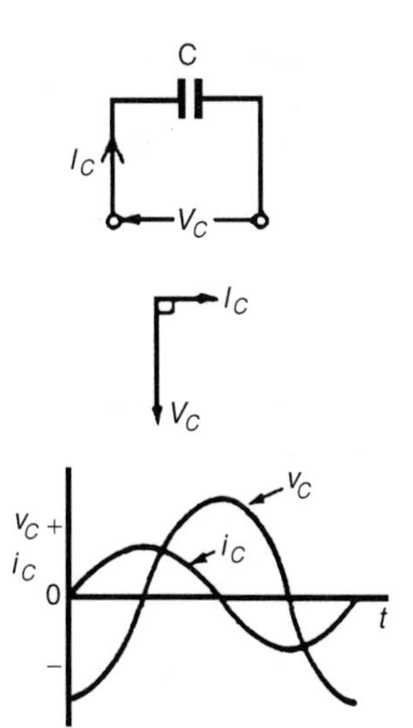

Figure 15.4

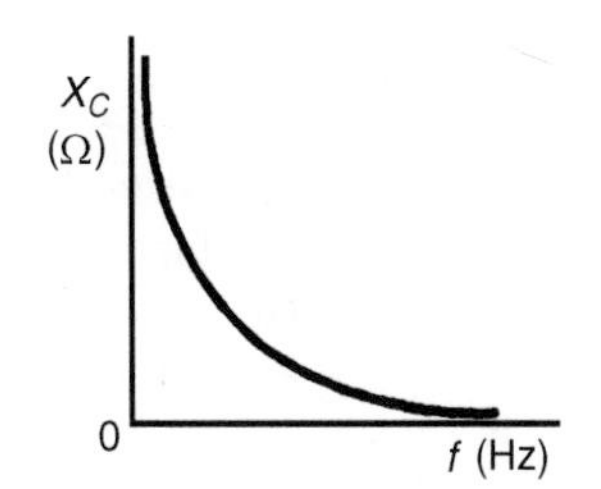

Figure 15.5

> Problem 3. Determine the capacitive reactance of a capacitor of 10 μF when connected to a circuit of frequency (a) 50 Hz (b) 20 kHz

(a) Capacitive reactance $X_C = \dfrac{1}{2\pi fC} = \dfrac{1}{2\pi(50)(10 \times 10^{-6})}$

$$= \frac{10^6}{2\pi(50)(10)}$$

$$= \mathbf{318.3\ \Omega}$$

(b) $X_C = \dfrac{1}{2\pi fC} = \dfrac{1}{2\pi(20 \times 10^3)(10 \times 10^{-6})} = \dfrac{10^6}{2\pi(20 \times 10^3)(10)}$

$$= \mathbf{0.796\ \Omega}$$

Hence as the frequency is increased from 50 Hz to 20 kHz, X_C decreases from 318.3 Ω to 0.796 Ω (see Figure 15.5).

> Problem 4. A capacitor has a reactance of 40 Ω when operated on a 50 Hz supply. Determine the value of its capacitance.

Since $X_C = \dfrac{1}{2\pi fC}$, capacitance $C = \dfrac{1}{2\pi fX_C} = \dfrac{1}{2\pi(50)(40)}$ F

$$= \frac{10^6}{2\pi(50)(40)}\ \mu\text{F} = \mathbf{79.58\ \mu F}$$

> Problem 5. Calculate the current taken by a 23 μF capacitor when connected to a 240 V, 50 Hz supply.

Current $I = \dfrac{V}{X_C} = \dfrac{V}{\left(\dfrac{1}{2\pi fC}\right)} = 2\pi fCV = 2\pi(50)(23 \times 10^{-6})(240)$

$$= \mathbf{1.73\ A}$$

Further problems on purely inductive and capacitive a.c. circuits may be found in Section 15.12, problems 1 to 8, page 234.

15.4 *R–L* series a.c. circuit

In an a.c. circuit containing inductance L and resistance R, the applied voltage V is the phasor sum of V_R and V_L (see Figure 15.6), and thus the current I lags the applied voltage V by an angle lying between 0° and 90° (depending on the values of V_R and V_L), shown as angle ϕ. In any

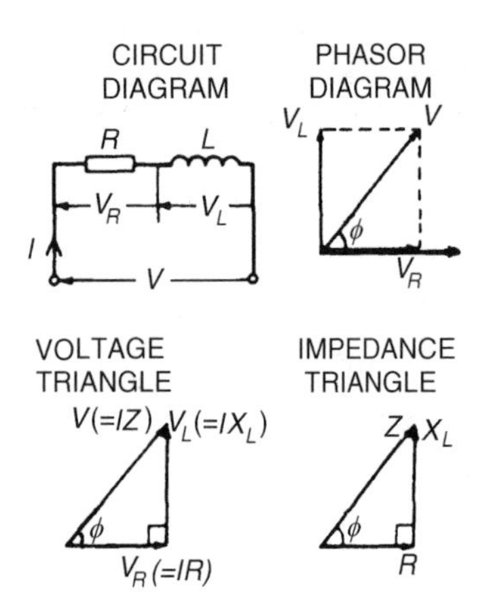

Figure 15.6

a.c. series circuit the current is common to each component and is thus taken as the reference phasor.

From the phasor diagram of Figure 15.6, the **'voltage triangle'** is derived.

For the R–L circuit: $V = \sqrt{(V_R^2 + V_L^2)}$ (by Pythagoras' theorem)

and $\tan\phi = \dfrac{V_L}{V_R}$ (by trigonometric ratios)

In an a.c. circuit, the ratio $\dfrac{\text{applied voltage } V}{\text{current } I}$ is called the **impedance Z**, i.e.

$$\boldsymbol{Z = \frac{V}{I}\ \Omega}$$

If each side of the voltage triangle in Figure 15.6 is divided by current I then the **'impedance triangle'** is derived.

For the R–L circuit: $Z = \sqrt{(R^2 + X_L^2)}$

$\tan\phi = \dfrac{X_L}{R}$, $\sin\phi = \dfrac{X_L}{Z}$ and $\cos\phi = \dfrac{R}{Z}$

> Problem 6. In a series R–L circuit the p.d. across the resistance R is 12 V and the p.d. across the inductance L is 5 V. Find the supply voltage and the phase angle between current and voltage.

From the voltage triangle of Figure 15.6,

supply voltage $V = \sqrt{(12^2 + 5^2)}$ i.e. $\boldsymbol{V = 13\ \text{V}}$

(Note that in a.c. circuits, the supply voltage is **not** the arithmetic sum of the p.d's across components. It is, in fact, the **phasor sum**.)

$\tan\phi = \dfrac{V_L}{V_R} = \dfrac{5}{12}$, from which $\phi = \arctan\left(\dfrac{5}{12}\right) = 22.62°$

$= \mathbf{22°37'}$ **lagging**

('Lagging' infers that the current is 'behind' the voltage, since phasors revolve anticlockwise.)

> Problem 7. A coil has a resistance of 4 Ω and an inductance of 9.55 mH. Calculate (a) the reactance, (b) the impedance, and (c) the current taken from a 240 V, 50 Hz supply. Determine also the phase angle between the supply voltage and current.

$R = 4\ \Omega$; $L = 9.55\ \text{mH} = 9.55 \times 10^{-3}\ \text{H}$; $f = 50\ \text{Hz}$; $V = 240\ \text{V}$

(a) Inductive reactance, $X_L = 2\pi fL = 2\pi(50)(9.55 \times 10^{-3}) = \mathbf{3\ \Omega}$

(b) Impedance, $Z = \sqrt{(R^2 + X_L^2)} = \sqrt{(4^2 + 3^2)} = \mathbf{5\ \Omega}$

(c) Current, $I = \frac{V}{Z} = \frac{240}{5} = \mathbf{48\ A}$

The circuit and phasor diagrams and the voltage and impedance triangles are as shown in Figure 15.6.

$$\text{Since } \tan\phi = \frac{X_L}{R},\ \phi = \arctan\frac{X_L}{R} = \arctan\frac{3}{4} = 36.87^\circ$$
$$= \mathbf{36^\circ 52' \text{ lagging}}$$

Problem 8. A coil takes a current of 2 A from a 12 V d.c. supply. When connected to a 240 V, 50 Hz supply the current is 20 A. Calculate the resistance, impedance, inductive reactance and inductance of the coil.

$$\text{Resistance } R = \frac{\text{d.c. voltage}}{\text{d.c. current}} = \frac{12}{2} = 6\ \Omega$$

$$\text{Impedance } Z = \frac{\text{a.c. voltage}}{\text{a.c. current}} = \frac{240}{20} = 12\ \Omega$$

$$\text{Since } Z = \sqrt{(R^2 + X_L^2)},\ \text{inductive reactance, } X_L = \sqrt{(Z^2 - R^2)}$$
$$= \sqrt{(12^2 - 6^2)}$$
$$= 10.39\ \Omega$$

$$\text{Since } X_L = 2\pi fL, \text{ inductance } L = \frac{X_L}{2\pi f} = \frac{10.39}{2\pi(50)} = \mathbf{33.1\ mH}$$

This problem indicates a simple method for finding the inductance of a coil, i.e. firstly to measure the current when the coil is connected to a d.c. supply of known voltage, and then to repeat the process with an a.c. supply.

Problem 9. A coil of inductance 318.3 mH and negligible resistance is connected in series with a 200 Ω resistor to a 240 V, 50 Hz supply. Calculate (a) the inductive reactance of the coil, (b) the impedance of the circuit, (c) the current in the circuit, (d) the p.d. across each component, and (e) the circuit phase angle.

$L = 318.3$ mH $= 0.3183$ H; $R = 200\ \Omega$; $V = 240$ V; $f = 50$ Hz

The circuit diagram is as shown in Figure 15.6.

(a) Inductive reactance $X_L = 2\pi fL = 2\pi(50)(0.3183) = \mathbf{100\ \Omega}$

(b) Impedance $Z = \sqrt{(R^2 + X_L^2)} = \sqrt{[(200)^2 + (100)^2]} = \mathbf{223.6\ \Omega}$

(c) Current $I = \frac{V}{Z} = \frac{240}{223.6} = \mathbf{1.073\ A}$

(d) The p.d. across the coil, $V_L = IX_L = 1.073 \times 100 = \mathbf{107.3\ V}$

The p.d. across the resistor, $V_R = IR = 1.073 \times 200 = \mathbf{214.6\ V}$

[Check: $\sqrt{(V_R^2 + V_L^2)} = \sqrt{[(214.6)^2 + (107.3)^2]} = 240$ V, the supply voltage]

(e) From the impedance triangle, angle $\phi = \arctan \frac{X_L}{R} = \arctan\left(\frac{100}{200}\right)$

Hence the phase angle $\phi = 26.57° = 26°34'$ lagging

Problem 10. A coil consists of a resistance of 100 Ω and an inductance of 200 mH. If an alternating voltage, v, given by $v = 200 \sin 500t$ volts is applied across the coil, calculate (a) the circuit impedance, (b) the current flowing, (c) the p.d. across the resistance, (d) the p.d. across the inductance and (e) the phase angle between voltage and current.

Since $v = 200 \sin 500t$ volts then $V_m = 200$ V and $\omega = 2\pi f = 500$ rad/s

Hence rms voltage $V = 0.707 \times 200 = 141.4$ V

Inductive reactance, $X_L = 2\pi f L = \omega L = 500 \times 200 \times 10^{-3} = 100\ \Omega$

(a) Impedance $Z = \sqrt{(R^2 + X_L^2)} = \sqrt{(100^2 + 100^2)} = \mathbf{141.4\ \Omega}$

(b) Current $I = \frac{V}{Z} = \frac{141.4}{141.4} = \mathbf{1\ A}$

(c) p.d. across the resistance $V_R = IR = 1 \times 100 = \mathbf{100\ V}$

p.d. across the inductance $V_L = IX_L = 1 \times 100 = \mathbf{100\ V}$

(e) Phase angle between voltage and current is given by: $\tan\phi = \left(\frac{X_L}{R}\right)$

from which, $\phi = \arctan(100/100)$, hence $\boldsymbol{\phi = 45°}$ **or** $\frac{\pi}{4}$ **rads**

Problem 11. A pure inductance of 1.273 mH is connected in series with a pure resistance of 30 Ω. If the frequency of the sinusoidal supply is 5 kHz and the p.d. across the 30 Ω resistor is 6 V, determine the value of the supply voltage and the voltage across the 1.273 mH inductance. Draw the phasor diagram.

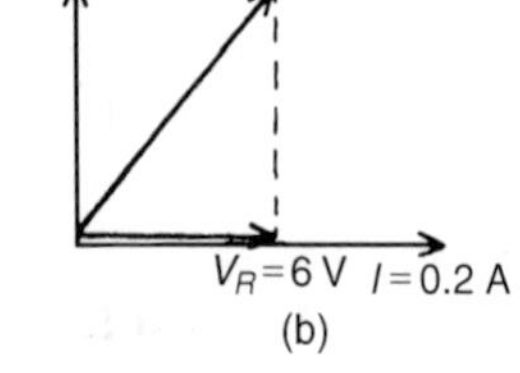

The circuit is shown in Figure 15.7(a).

Supply voltage, $V = IZ$

Figure 15.7

Current $I = \frac{V_R}{R} = \frac{6}{30} = 0.20$ A

Inductive reactance $X_L = 2\pi f L = 2\pi(5 \times 10^3)(1.273 \times 10^{-3})$
$= 40\ \Omega$

Impedance, $Z = \sqrt{(R^2 + X_L^2)} = \sqrt{(30^2 + 40^2)} = 50\ \Omega$

Supply voltage $V = IZ = (0.20)(50) = \mathbf{10\ V}$

Voltage across the 1.273 mH inductance, $V_L = IX_L = (0.2)(40) = \mathbf{8\ V}$

The phasor diagram is shown in Figure 15.7(b).

(Note that in a.c. circuits, the supply voltage is **not** the arithmetic sum of the p.d.'s across components but the **phasor sum**)

> Problem 12. A coil of inductance 159.2 mH and resistance 20 Ω is connected in series with a 60 Ω resistor to a 240 V, 50 Hz supply. Determine (a) the impedance of the circuit, (b) the current in the circuit, (c) the circuit phase angle, (d) the p.d. across the 60 Ω resistor and (e) the p.d. across the coil. (f) Draw the circuit phasor diagram showing all voltages.

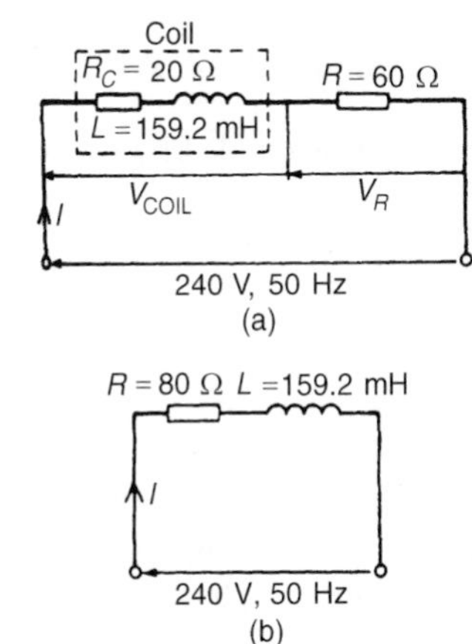

Figure 15.8

The circuit diagram is shown in Figure 15.8(a). When impedances are connected in series the individual resistances may be added to give the total circuit resistance. The equivalent circuit is thus shown in Figure 15.8(b).

Inductive reactance $X_L = 2\pi f L = 2\pi(50)(159.2 \times 10^{-3}) = 50\ \Omega$

(a) Circuit impedance, $Z = \sqrt{(R_T^2 + X_L^2)} = \sqrt{(80^2 + 50^2)} = 94.34\ \Omega$

(b) Circuit current, $I = \frac{V}{Z} = \frac{240}{94.34} = \mathbf{2.544\ A}$

(c) Circuit phase angle $\phi = \arctan\left(\frac{X_L}{R}\right) = \arctan(50/80)$
$= \mathbf{32^\circ\ lagging}$

From Figure 15.8(a):

(d) $V_R = IR = (2.544)(60) = \mathbf{152.6\ V}$

(e) $V_{COIL} = IZ_{COIL}$, where $Z_{COIL} = \sqrt{(R_C^2 + X_L^2)} = \sqrt{(20^2 + 50^2)}$
$= 53.85\ \Omega$

Hence $V_{COIL} = (2.544)(53.85) = \mathbf{137.0\ V}$

(f) For the phasor diagram, shown in Figure 15.9,

$V_L = IX_L = (2.544)(50) = 127.2$ V

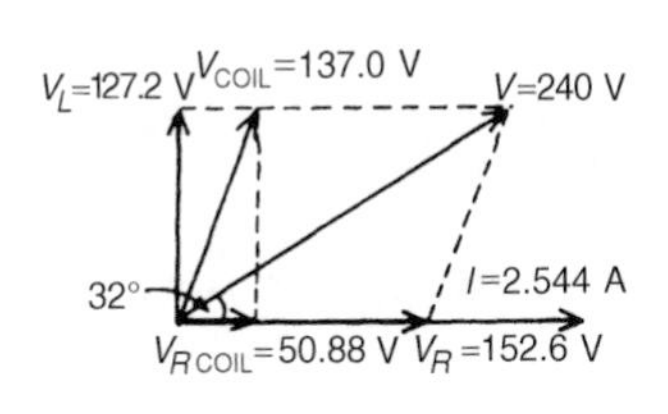

Figure 15.9

$V_{R\ \text{COIL}} = IR_C = (2.544)(20) = 50.88$ V

The 240 V supply voltage is the phasor sum of V_{COIL} and V_R

Further problems on R–L a.c. series circuits may be found in Section 15.12, problems 9 to 13, page 234.

15.5 R–C series a.c. circuit

In an a.c. series circuit containing capacitance C and resistance R, the applied voltage V is the phasor sum of V_R and V_C (see Figure 15.10) and thus the current I leads the applied voltage V by an angle lying between 0° and 90° (depending on the values of V_R and V_C), shown as angle α.

From the phasor diagram of Figure 15.10, the **'voltage triangle'** is derived. For the R–C circuit:

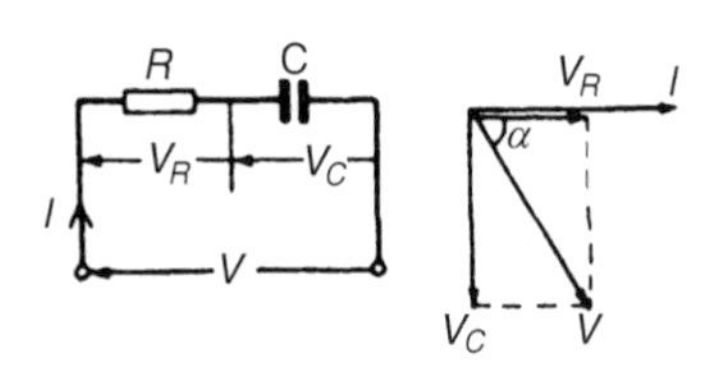

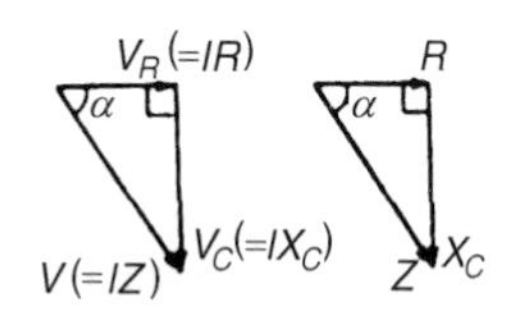

Figure 15.10

$$V = \sqrt{(V_R^2 + V_C^2)} \text{ (by Pythagoras' theorem)}$$

$$\text{and } \tan\alpha = \frac{V_C}{V_R} \text{ (by trigonometric ratios)}$$

As stated in Section 15.4, in an a.c. circuit, the ratio

(applied voltage V)/(current I) is called the **impedance Z**, i.e. $Z = \dfrac{V}{I}\ \Omega$

If each side of the voltage triangle in Figure 15.10 is divided by current I then the **'impedance triangle'** is derived.

For the R–C circuit: $Z = \sqrt{(R^2 + X_C^2)}$

$$\tan\alpha = \frac{X_C}{R},\ \sin\alpha = \frac{X_C}{Z} \text{ and } \cos\alpha = \frac{R}{Z}$$

> Problem 13. A resistor of 25 Ω is connected in series with a capacitor of 45 μF. Calculate (a) the impedance, and (b) the current taken from a 240 V, 50 Hz supply. Find also the phase angle between the supply voltage and the current.

$R = 25\ \Omega$; $C = 45\ \mu\text{F} = 45 \times 10^{-6}$ F; $V = 240$ V; $f = 50$ Hz

The circuit diagram is as shown in Figure 15.10

$$\text{Capacitive reactance, } X_C = \frac{1}{2\pi fC} = \frac{1}{2\pi(50)(45 \times 10^{-6})} = 70.74\ \Omega$$

(a) Impedance $Z = \sqrt{(R^2 + X_C^2)} = \sqrt{[(25)^2 + (70.74)^2]} = \mathbf{75.03\ \Omega}$

(b) Current $I = \dfrac{V}{Z} = \dfrac{240}{75.03} = \mathbf{3.20\ A}$

Phase angle between the supply voltage and current, $\alpha = \arctan\left(\frac{X_C}{R}\right)$

hence $\alpha = \arctan\left(\frac{70.74}{25}\right) = 70.54° = \mathbf{70°32' \text{ leading}}$

('Leading' infers that the current is 'ahead' of the voltage, since phasors revolve anticlockwise.)

Problem 14. A capacitor C is connected in series with a 40 Ω resistor across a supply of frequency 60 Hz. A current of 3 A flows and the circuit impedance is 50 Ω. Calculate: (a) the value of capacitance, C, (b) the supply voltage, (c) the phase angle between the supply voltage and current, (d) the p.d. across the resistor, and (e) the p.d. across the capacitor. Draw the phasor diagram.

(a) Impedance $Z = \sqrt{(R^2 + X_C^2)}$

Hence $X_C = \sqrt{(Z^2 - R^2)} = \sqrt{(50^2 - 40^2)} = 30\ \Omega$

$$X_C = \frac{1}{2\pi f C} \text{ hence } C = \frac{1}{2\pi f X_C} = \frac{1}{2\pi(60)30} \text{ F}$$

$$= \mathbf{88.42\ \mu F}$$

(b) Since $Z = \frac{V}{I}$ then $V = IZ = (3)(50) = \mathbf{150\ V}$

(c) Phase angle, $\alpha = \arctan\frac{X_C}{R} = \arctan\left(\frac{30}{40}\right) = 36.87°$

$$= \mathbf{36°52' \text{ leading}}$$

(d) P.d. across resistor, $V_R = IR = (3)(40) = \mathbf{120\ V}$

(e) P.d. across capacitor, $V_C = IX_C = (3)(30) = \mathbf{90\ V}$

The phasor diagram is shown in Figure 15.11, where the supply voltage V is the phasor sum of V_R and V_C.

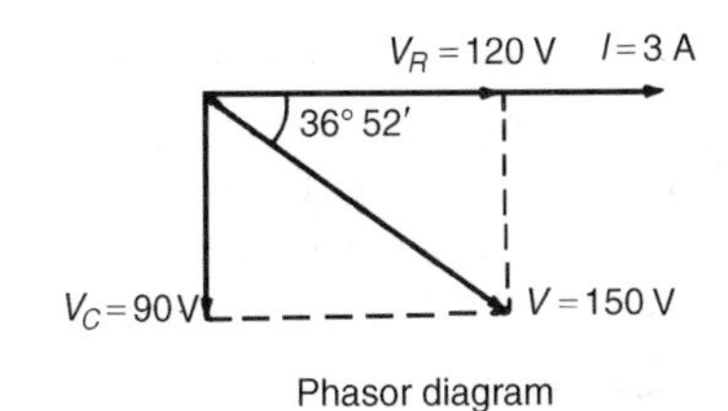

Figure 15.11

Further problems on R–C a.c. circuits may be found in Section 15.12, problems 14 to 17, page 235.

15.6 *R–L–C* series a.c. circuit

In an a.c. series circuit containing resistance R, inductance L and capacitance C, the applied voltage V is the phasor sum of V_R, V_L and V_C (see Figure 15.12). V_L and V_C are anti-phase, i.e. displaced by 180°, and there are three phasor diagrams possible — each depending on the relative values of V_L and V_C

When $X_L > X_C$ (Figure 15.12(b)): $Z = \sqrt{[R^2 + (X_L - X_C)^2]}$

and $\tan\phi = \frac{(X_L - X_C)}{R}$

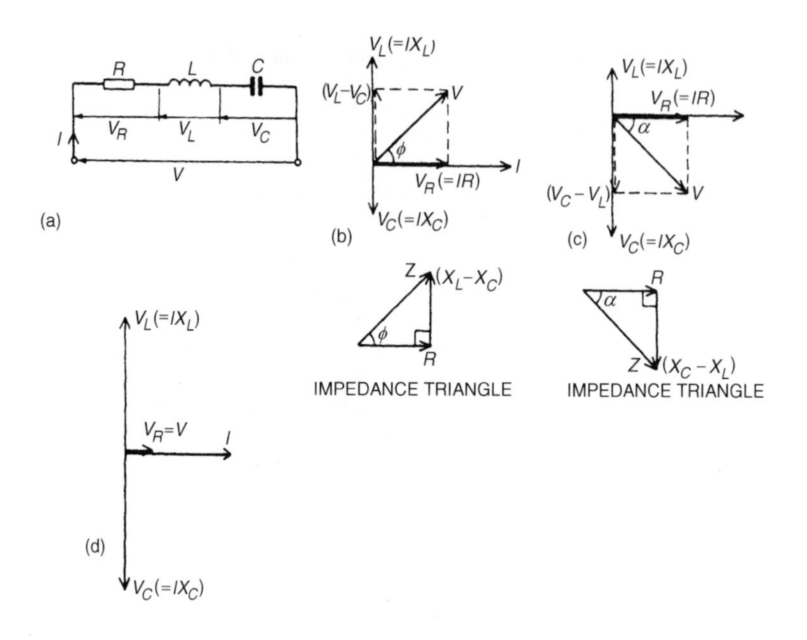

Figure 15.12

When $X_C > X_L$ (Figure 15.12(c)): $Z = \sqrt{[R^2 + (X_C - X_L)^2]}$

$$\text{and } \tan\alpha = \frac{(X_C - X_L)}{R}$$

When $X_L = X_C$ (Figure 15.12(d)), the applied voltage V and the current I are in phase. This effect is called **series resonance** (see Section 15.7)

Problem 15. A coil of resistance 5 Ω and inductance 120 mH in series with a 100 μF capacitor, is connected to a 300 V, 50 Hz supply. Calculate (a) the current flowing, (b) the phase difference between the supply voltage and current, (c) the voltage across the coil and (d) the voltage across the capacitor.

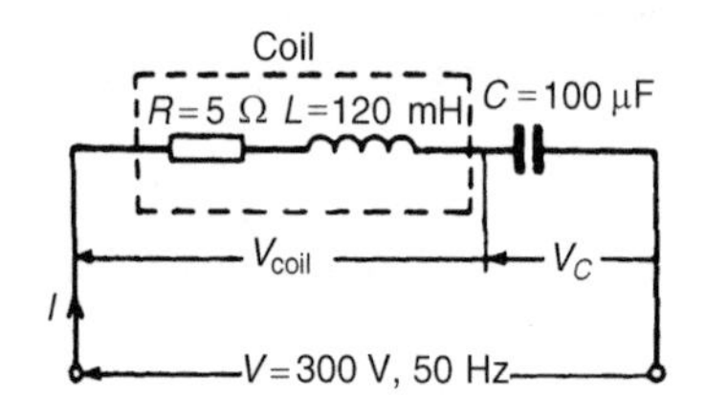

Figure 15.13

The circuit diagram is shown in Figure 15.13

$$X_L = 2\pi fL = 2\pi(50)(120 \times 10^{-3}) = \mathbf{37.70\ \Omega}$$

$$X_C = \frac{1}{2\pi fC} = \frac{1}{2\pi(50)(100 \times 10^{-6})} = \mathbf{31.83\ \Omega}$$

Since X_L is greater than X_C the circuit is inductive.
$X_L - X_C = 37.70 - 31.83 = 5.87\ \Omega$

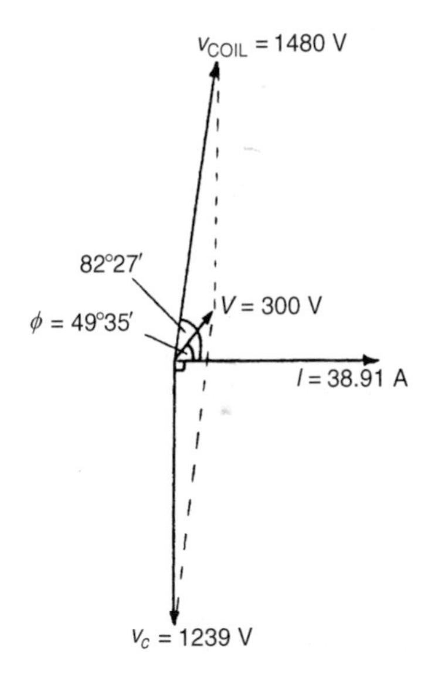

Figure 15.14

Impedance $Z = \sqrt{[R^2 + (X_L - X_C)^2]} = \sqrt{[(5)^2 + (5.87)^2]} = 7.71\ \Omega$

(a) Current $I = \dfrac{V}{Z} = \dfrac{300}{7.71} = \mathbf{38.91\ A}$

(b) Phase angle $\phi = \arctan\left(\dfrac{X_L - X_C}{R}\right) = \arctan\dfrac{5.87}{5} = 49.58° = \mathbf{49°35'}$

(c) Impedance of coil $Z_{COIL} = \sqrt{(R^2 + X_L^2)} = \sqrt{[(5)^2 + (37.70)^2]} = 38.03\ \Omega$

Voltage across coil $V_{COIL} = IZ_{COIL} = (38.91)(38.03) = \mathbf{1480\ V}$

Phase angle of coil $= \arctan\dfrac{X_L}{R} = \arctan\left(\dfrac{37.70}{5}\right) = 82.45° = 82°27'$ lagging

(d) Voltage across capacitor $V_C = IX_C = (38.91)(31.83) = \mathbf{1239\ V}$

The phasor diagram is shown in Figure 15.14. The supply voltage V is the phasor sum of V_{COIL} and V_C

Series connected impedances

For series-connected impedances the total circuit impedance can be represented as a single L–C–R circuit by combining all values of resistance together, all values of inductance together and all values of capacitance together,

(remembering that for series connected capacitors $\dfrac{1}{C} = \dfrac{1}{C_1} + \dfrac{1}{C_2} + \ldots$).

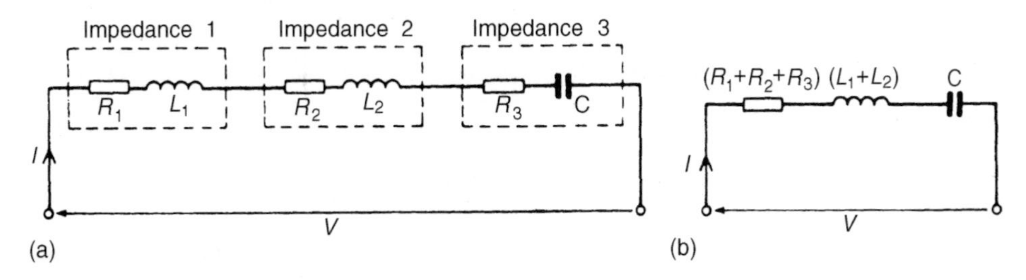

Figure 15.15

For example, the circuit of Figure 15.15(a) showing three impedances has an equivalent circuit of Figure 15.15(b).

Problem 16. The following three impedances are connected in series across a 40 V, 20 kHz supply: (i) a resistance of 8 Ω, (ii) a coil of inductance 130 μH and 5 Ω resistance, and (iii) a 10 Ω resistor in series with a 0.25 μF capacitor. Calculate (a) the circuit current, (b) the circuit phase angle and (c) the voltage drop across each impedance.

The circuit diagram is shown in Figure 15.16(a). Since the total circuit resistance is $8 + 5 + 10$, i.e. 23 Ω, an equivalent circuit diagram may be drawn as shown in Figure 15.16(b)

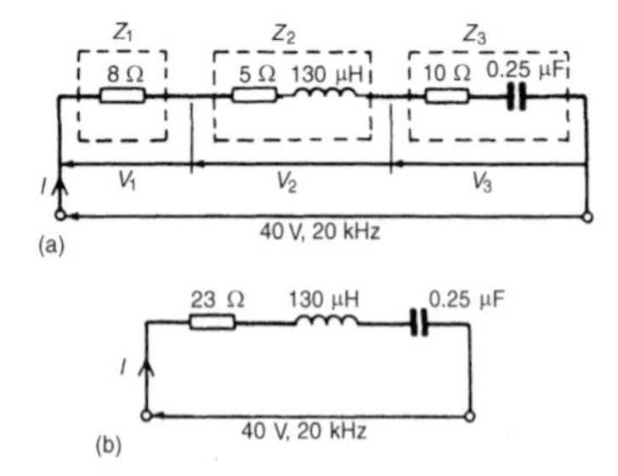

Figure 15.16

Inductive reactance, $X_L = 2\pi fL = 2\pi(20 \times 10^3)(130 \times 10^{-6})$

$$= 16.34\ \Omega$$

Capacitive reactance, $X_C = \dfrac{1}{2\pi fC} = \dfrac{1}{2\pi(20 \times 10^3)(0.25 \times 10^{-6})}$

$$= 31.83\ \Omega$$

Since $X_C > X_L$, the circuit is capacitive (see phasor diagram in Figure 15.12(c)). $X_C - X_L = 31.83 - 16.34 = 15.49\ \Omega$

(a) Circuit impedance, $Z = \sqrt{[R^2 + (X_C - X_L)^2]} = \sqrt{[23^2 + 15.49^2]}$

$$= 27.73\ \Omega$$

Circuit current, $I = \dfrac{V}{Z} = \dfrac{40}{27.73} = \mathbf{1.442\ A}$

From Figure 15.12(c), circuit phase angle $\phi = \arctan\left(\dfrac{X_C - X_L}{R}\right)$

i.e, $\phi = \arctan\left(\dfrac{15.49}{23}\right) = 33.96° = \mathbf{33°58'\ leading}$

(b) From Figure 15.16(a), $V_1 = IR_1 = (1.442)(8) = \mathbf{11.54\ V}$

$V_2 = IZ_2 = I\sqrt{(5^2 + 16.34^2)} = (1.442)(17.09) = \mathbf{24.64\ V}$

$V_3 = IZ_3 = I\sqrt{(10^2 + 31.83^2)} = (1.442)(33.36) = \mathbf{48.11\ V}$

The 40 V supply voltage is the phasor sum of V_1, V_2 and V_3

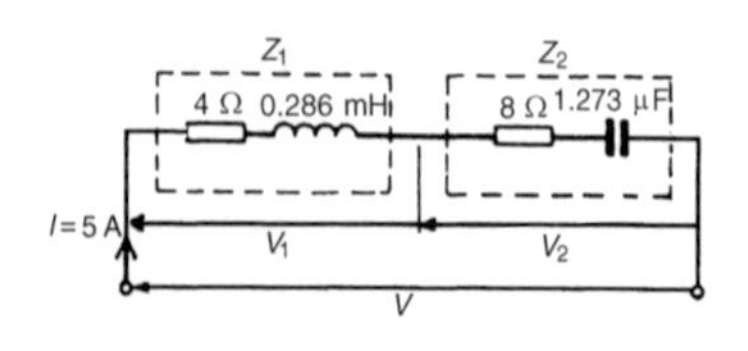

Figure 15.17

Problem 17. Determine the p.d.'s V_1 and V_2 for the circuit shown in Figure 15.17 if the frequency of the supply is 5 kHz. Draw the phasor diagram and hence determine the supply voltage V and the circuit phase angle.

For impedance Z_1:

$$R_1 = 4\ \Omega \text{ and } X_L = 2\pi fL = 2\pi(5\times10^3)(0.286\times10^{-3}) = 8.985\ \Omega$$

$$V_1 = IZ_1 = I\sqrt{(R^2+X_L^2)} = 5\sqrt{(4^2+8.985^2)} = 49.18\ \text{V}$$

$$\text{Phase angle } \phi_1 = \arctan\left(\frac{X_L}{R}\right) = \arctan\left(\frac{8.985}{4}\right) = \mathbf{66°0'\ lagging}$$

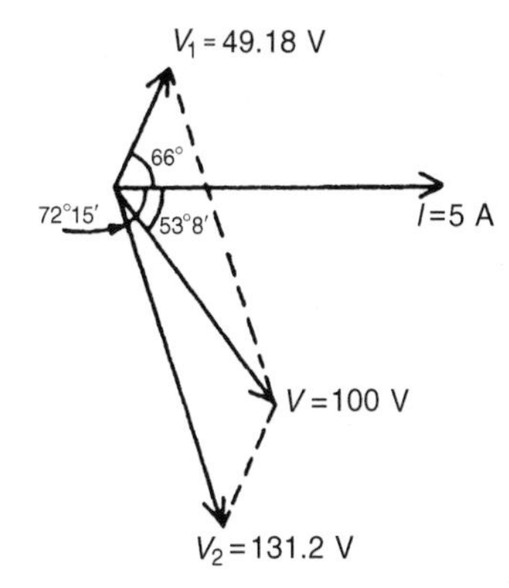

Figure 15.18

For impedance Z_2:

$$R_2 = 8\ \Omega \text{ and } X_C = \frac{1}{2\pi fC} = \frac{1}{2\pi(5\times10^3)(1.273\times10^{-6})} = 25.0\ \Omega$$

$$V_2 = IZ_2 = I\sqrt{(R^2+X_C^2)} = 5\sqrt{(8^2+25.0^2)} = 131.2\ \text{V}$$

$$\text{Phase angle } \phi_2 = \arctan\left(\frac{X_C}{R}\right) = \arctan\left(\frac{25.0}{8}\right) = 72°15'\ \text{leading}$$

The phasor diagram is shown in Figure 15.18.

The phasor sum of V_1 and V_2 gives the supply voltage V of 100 V at a phase angle of **53°8′ leading**. These values may be determined by drawing or by calculation — either by resolving into horizontal and vertical components or by the cosine and sine rules.

Further problems on R–L–C a.c. circuits may be found in Section 15.12, problems 18 to 20, page 235.

15.7 Series resonance

As stated in Section 15.6, for an R–L–C series circuit, when $X_L = X_C$ (Figure 15.12(d)), the applied voltage V and the current I are in phase. This effect is called **series resonance**. At resonance:

(i) $V_L = V_C$

(ii) $Z = R$ (i.e. the minimum circuit impedance possible in an L–C–R circuit)

(iii) $I = \dfrac{V}{R}$ (i.e. the maximum current possible in an L–C–R circuit)

(iv) Since $X_L = X_C$, then $2\pi f_r L = \dfrac{1}{2\pi f_r C}$

from which, $f_r^2 = \dfrac{1}{(2\pi)^2 LC}$

and, $$\boxed{f_r = \frac{1}{2\pi\sqrt{(LC)}}\ \text{Hz},}$$

where f_r is the resonant frequency.

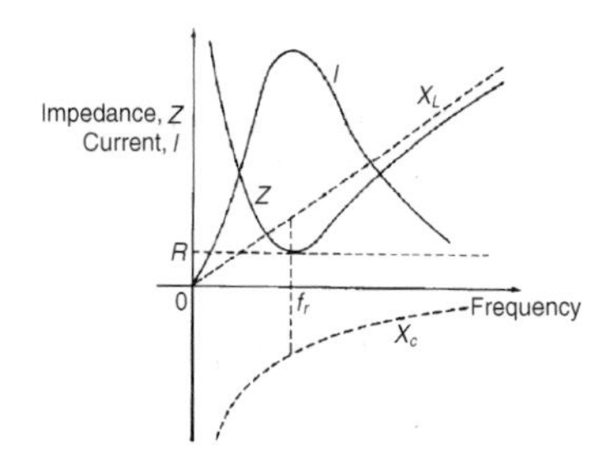

Figure 15.19

(v) The series resonant circuit is often described as an **acceptor circuit** since it has its minimum impedance, and thus maximum current, at the resonant frequency.

(vi) Typical graphs of current I and impedance Z against frequency are shown in Figure 15.19.

Problem 18. A coil having a resistance of 10 Ω and an inductance of 125 mH is connected in series with a 60 μF capacitor across a 120 V supply. At what frequency does resonance occur? Find the current flowing at the resonant frequency.

$$\text{Resonant frequency, } f_r = \frac{1}{2\pi\sqrt{(LC)}}\ \text{Hz}$$

$$= \frac{1}{2\pi\sqrt{\left[\left(\frac{125}{10^3}\right)\left(\frac{60}{10^6}\right)\right]}}\ \text{Hz}$$

$$= \frac{1}{2\pi\sqrt{\left(\frac{125\times 6}{10^8}\right)}} = \frac{1}{2\pi\frac{\sqrt{[(125)(6)]}}{10^4}}$$

$$= \frac{10^4}{2\pi\sqrt{[(125)(6)]}} = \mathbf{58.12\ Hz}$$

At resonance, $X_L = X_C$ and impedance $Z = R$

$$\text{Hence current, } I = \frac{V}{R} = \frac{120}{10} = \mathbf{12\ A}$$

Problem 19. The current at resonance in a series L–C–R circuit is 100 μA. If the applied voltage is 2 mV at a frequency of 200 kHz, and the circuit inductance is 50 μH, find (a) the circuit resistance, and (b) the circuit capacitance.

(a) $I = 100\ \mu\text{A} = 100 \times 10^{-6}$ A; $V = 2$ mV $= 2 \times 10^{-3}$ V

At resonance, impedance $Z =$ resistance R

$$\text{Hence } R = \frac{V}{I} = \frac{20\times 10^{-3}}{100\times 10^{-6}} = \frac{2\times 10^6}{100\times 10^3} = \mathbf{20\ \Omega}$$

(b) At resonance $X_L = X_C$

$$\text{i.e. } 2\pi f L = \frac{1}{2\pi f C}$$

$$\text{Hence capacitance } C = \frac{1}{(2\pi f)^2 L}$$

$$= \frac{1}{(2\pi \times 200 \times 10^3)^2 (50 \times 10^{-6})} \text{F}$$

$$= \frac{(10^6)(10^6)}{(4\pi)^2 (10^{10})(50)} \mu\text{F}$$

$$= \mathbf{0.0127\ \mu F\ or\ 12.7\ nF}$$

15.8 Q-factor

At resonance, if R is small compared with X_L and X_C, it is possible for V_L and V_C to have voltages many times greater than the supply voltage (see Figure 15.12(d)).

$$\textbf{Voltage magnification at resonance} = \frac{\textbf{voltage across } L \textbf{ (or } C\textbf{)}}{\textbf{supply voltage } V}$$

This ratio is a measure of the quality of a circuit (as a resonator or tuning device) and is called the **Q-factor**.

$$\text{Hence Q-factor} = \frac{V_L}{V} = \frac{IX_L}{IR} = \frac{X_L}{R} = \frac{\mathbf{2\pi f_r L}}{\mathbf{R}}$$

$$\text{Alternatively, Q-factor} = \frac{V_C}{V} = \frac{IX_C}{IR} = \frac{X_C}{R} = \frac{\mathbf{1}}{\mathbf{2\pi f_r CR}}$$

$$\text{At resonance } f_r = \frac{1}{2\pi\sqrt{(LC)}} \text{ i.e. } 2\pi f_r = \frac{1}{\sqrt{(LC)}}$$

$$\text{Hence } Q\text{-factor} = \frac{2\pi f_r L}{R} = \frac{1}{\sqrt{(LC)}}\left(\frac{L}{R}\right) = \frac{\mathbf{1}}{\mathbf{R}}\sqrt{\left(\frac{\mathbf{L}}{\mathbf{C}}\right)}$$

(Q-factor is explained more fully in Chapter 28, page 495)

Problem 20. A coil of inductance 80 mH and negligible resistance is connected in series with a capacitance of 0.25 μF and a resistor of resistance 12.5 Ω across a 100 V, variable frequency supply. Determine (a) the resonant frequency, and (b) the current at resonance. How many times greater than the supply voltage is the voltage across the reactances at resonance?

(a) Resonant frequency f_r

$$= \frac{1}{2\pi\sqrt{\left[\left(\frac{80}{10^3}\right)\left(\frac{0.25}{10^6}\right)\right]}} = \frac{1}{2\pi\sqrt{\left[\frac{(8)(0.25)}{10^8}\right]}}$$

$$= \frac{10^4}{2\pi\sqrt{2}}$$

$$= \mathbf{1125.4\ Hz} = \mathbf{1.1254\ kHz}$$

(b) Current at resonance $I = \frac{V}{R} = \frac{100}{12.5} = \mathbf{8\ A}$

Voltage across inductance, at resonance,

$$V_L = IX_L = (I)(2\pi f L)$$
$$= (8)(2\pi)(1125.4)(80 \times 10^{-3})$$
$$= 4525.5\ \mathrm{V}$$

(Also, voltage across capacitor,

$$V_C = IX_C = \frac{I}{2\pi f C} = \frac{8}{2\pi(1125.4)(0.25 \times 10^{-6})} = 4525.5\ V)$$

Voltage magnification at resonance $= \frac{V_L}{V}$ or $\frac{V_c}{V} = \frac{4525.5}{100}$

$$= \mathbf{45.255\ V}$$

i.e. at resonance, the voltage across the reactances are 45.255 times greater than the supply voltage. Hence Q-factor of circuit is 45.255.

Problem 21. A series circuit comprises a coil of resistance 2 Ω and inductance 60 mH, and a 30 μF capacitor. Determine the Q-factor of the circuit at resonance.

At resonance, Q-factor $= \frac{1}{R}\sqrt{\left(\frac{L}{C}\right)} = \frac{1}{2}\sqrt{\left(\frac{60 \times 10^{-3}}{30 \times 10^{-6}}\right)}$

$$= \frac{1}{2}\sqrt{\left(\frac{60 \times 10^{6}}{30 \times 10^{3}}\right)}$$

$$= \frac{1}{2}\sqrt{(2000)} = \mathbf{22.36}$$

Problem 22. A coil of negligible resistance and inductance 100 mH is connected in series with a capacitance of 2 μF and a resistance of 10 Ω across a 50 V, variable frequency supply. Determine (a) the resonant frequency, (b) the current at resonance, (c) the voltages across the coil and the capacitor at resonance, and (d) the Q-factor of the circuit.

(a) Resonant frequency, $f_r = \frac{1}{2\pi\sqrt{(LC)}} = \frac{1}{2\pi\sqrt{\left[\left(\frac{100}{10^3}\right)\left(\frac{2}{10^6}\right)\right]}}$

$$= \frac{1}{2\pi\sqrt{\left(\frac{20}{10^8}\right)}} = \frac{1}{\left(\frac{2\pi\sqrt{20}}{10^4}\right)} = \frac{10^4}{2\pi\sqrt{20}}$$

$$= \mathbf{355.9\ Hz}$$

(b) Current at resonance $I = \frac{V}{R} = \frac{50}{10} = \mathbf{5\ A}$

(c) Voltage across coil at resonance,

$$V_L = IX_L = I(2\pi f_r L)$$
$$= (5)(2\pi \times 355.9 \times 100 \times 10^{-3})$$
$$= \mathbf{1118\ V}$$

Voltage across capacitance at resonance,

$$V_C = IX_C = \frac{I}{2\pi f_r C}$$
$$= \frac{5}{2\pi(355.9)(2 \times 10^{-6})}$$
$$= \mathbf{1118\ V}$$

(d) Q-factor (i.e. voltage magnification at resonance) $= \frac{V_L}{V}$ or $\frac{V_C}{V}$

$$= \frac{1118}{50} = \mathbf{22.36}$$

Q-factor may also have been determined by $\frac{2\pi f_r L}{R}$ or $\frac{1}{2\pi f_r CR}$ or $\frac{1}{R}\sqrt{\left(\frac{L}{C}\right)}$

Further problems on series resonance and Q-factor may be found in Section 15.12, problems 21 to 25, page 236.

15.9 Bandwidth and selectivity

Figure 15.20 shows how current I varies with frequency in an R–L–C series circuit. At the resonant frequency f_r, current is a maximum value, shown as I_r. Also shown are the points A and B where the current is 0.707 of the maximum value at frequencies f_1 and f_2. The power delivered to

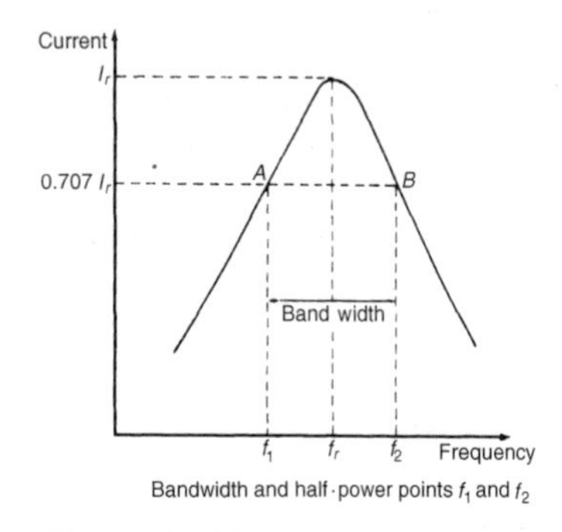

Figure 15.20

the circuit is I^2R. At $I = 0.707I_r$, the power is $(0.707I_r)^2R = 0.5I_r^2R$, i.e., half the power that occurs at frequency f_r. The points corresponding to f_1 and f_2 are called the **half-power points**. The distance between these points, i.e. $(f_2 - f_1)$, is called the **bandwidth**.

It may be shown that

$$\boxed{Q = \frac{f_r}{f_2 - f_1}} \quad \text{or} \quad \boxed{(f_2 - f_1) = \frac{f_r}{Q}}$$

(This formula is proved in Chapter 28, page 495)

Problem 23. A filter in the form of a series L–R–C circuit is designed to operate at a resonant frequency of 5 kHz. Included within the filter is a 20 mH inductance and 10 Ω resistance. Determine the bandwidth of the filter.

Q-factor at resonance is given by

$$Q_r = \frac{\omega_r L}{R} = \frac{(2\pi 5000)(20 \times 10^{-3})}{10} = 62.83$$

Since $Q_r = f_r/(f_2 - f_1)$

$$\textbf{bandwidth}, (f_2 - f_1) = \frac{f_r}{Q_r} = \frac{5000}{62.83} = \mathbf{79.6\ Hz}$$

Selectivity is the ability of a circuit to respond more readily to signals of a particular frequency to which it is tuned than to signals of other frequencies. The response becomes progressively weaker as the frequency departs from the resonant frequency. The higher the Q-factor, the narrower the bandwidth and the more selective is the circuit. Circuits having high Q-factors (say, in the order of 100 to 300) are therefore useful in communications engineering. A high Q-factor in a series power circuit has disadvantages in that it can lead to dangerously high voltages across the insulation and may result in electrical breakdown.

(For more on bandwidth and selectivity see Chapter 28, page 504)

15.10 Power in a.c. circuits

In Figures 15.21(a)–(c), the value of power at any instant is given by the product of the voltage and current at that instant, i.e. the instantaneous power, $p = vi$, as shown by the broken lines.

(a) For a purely resistive a.c. circuit, the average power dissipated, P, is given by:

$$\boldsymbol{P = VI = I^2R = \frac{V^2}{R}} \textbf{ watts } (V \text{ and } I \text{ being rms values}).$$

See Figure 15.21(a).

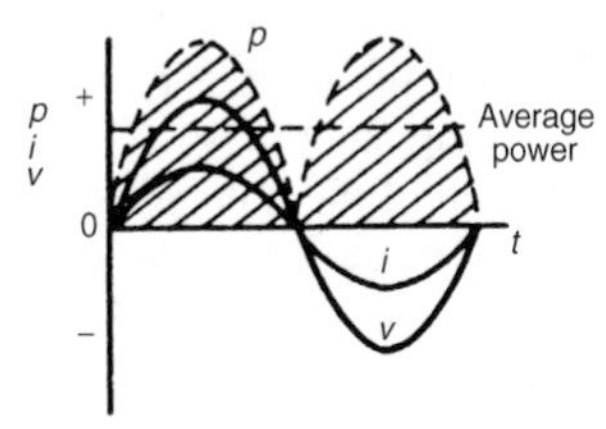

PURE RESISTANCE-AVERAGE POWER = VI

(a)

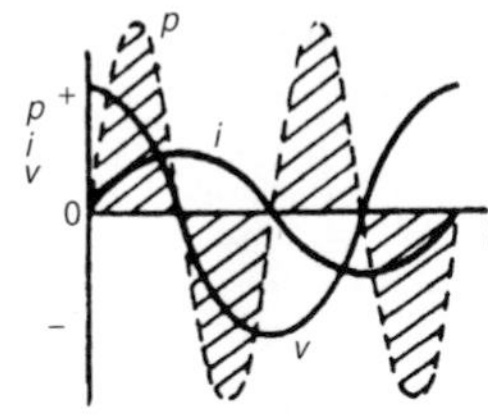

PURE INDUCTANCE-AVERAGE POWER = 0

(b)

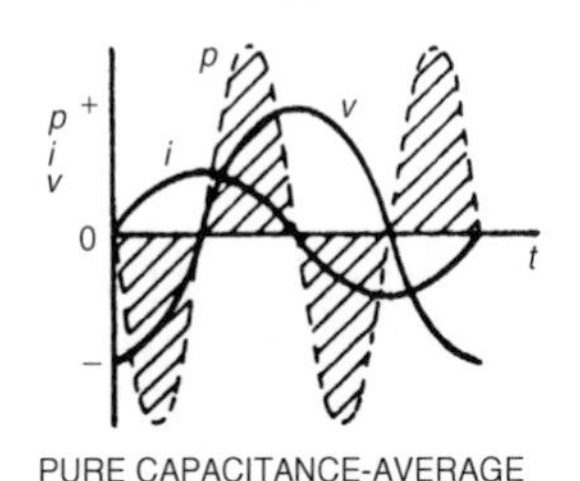

PURE CAPACITANCE-AVERAGE POWER = 0

(c)

Figure 15.21

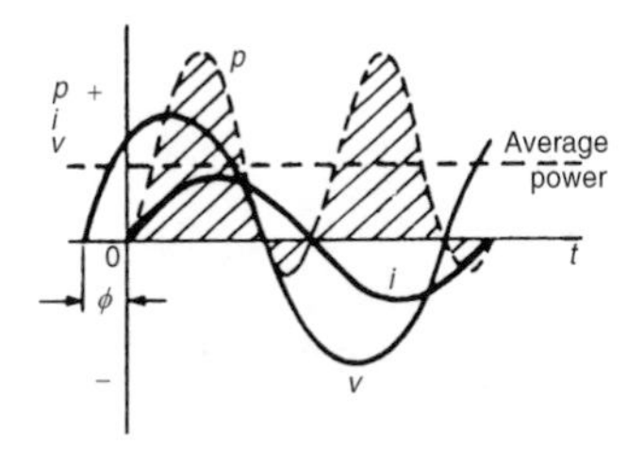

Figure 15.22

(b) For a purely inductive a.c. circuit, the average power is zero. See Figure 15.21(b).

(c) For a purely capacitive a.c. circuit, the average power is zero. See Figure 15.21(c).

Figure 15.22 shows current and voltage waveforms for an R–L circuit where the current lags the voltage by angle ϕ. The waveform for power (where $p = vi$) is shown by the broken line, and its shape, and hence average power, depends on the value of angle ϕ.

For an R–L, R–C or R–L–C series a.c. circuit, the average power P is given by:

$$\boldsymbol{P = VI\cos\phi \text{ watts}}$$

or

$$\boldsymbol{P = I^2R \text{ watts}} \quad (V \text{ and } I \text{ being rms values})$$

The formulae for power are proved in Chapter 26, page 459.

Problem 24. An instantaneous current, $i = 250\sin\omega t$ mA flows through a pure resistance of 5 kΩ. Find the power dissipated in the resistor.

Power dissipated, $P = I^2R$ where I is the rms value of current.

If $i = 250\sin\omega t$ mA, then $I_m = 0.250$ A and rms current, $I = (0.707 \times 0.250)$ A

Hence power $P = (0.707 \times 0.250)^2(5000) = $ **156.2 watts**

Problem 25. A series circuit of resistance 60 Ω and inductance 75 mH is connected to a 110 V, 60 Hz supply. Calculate the power dissipated.

Inductive reactance, $X_L = 2\pi fL = 2\pi(60)(75 \times 10^{-3}) = 28.27\ \Omega$

Impedance, $Z = \sqrt{(R^2 + X_L^2)} = \sqrt{[(60)^2 + (28.27)^2]} = 66.33\ \Omega$

Current, $I = \dfrac{V}{Z} = \dfrac{110}{66.33} = 1.658$ A

To calculate power dissipation in an a.c. circuit two formulae may be used:

(i) $P = I^2R = (1.658)^2(60) = \mathbf{165\ W}$

or (ii) $P = VI\cos\phi$ where $\cos\phi = \dfrac{R}{Z} = \dfrac{60}{66.33} = 0.9046$

Hence $P = (110)(1.658)(0.9046) = \mathbf{165\ W}$

15.11 Power triangle and power factor

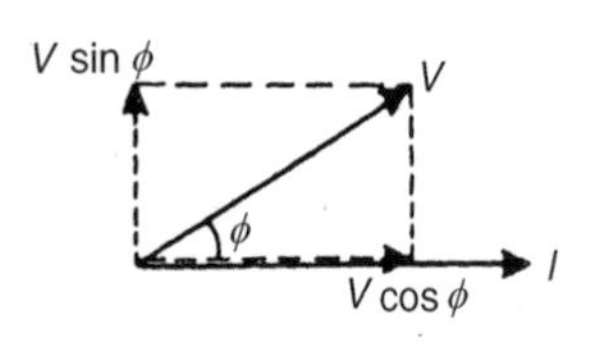

(a) PHASOR DIAGRAM

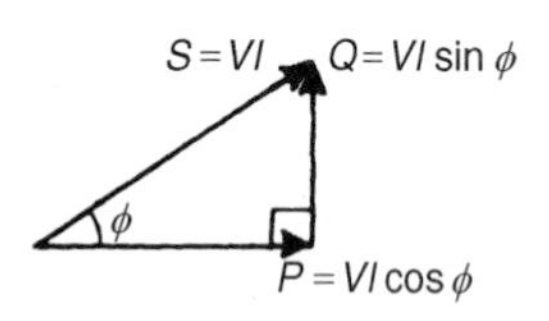

(b) POWER TRIANGLE

Figure 15.23

Figure 15.23(a) shows a phasor diagram in which the current I lags the applied voltage V by angle ϕ. The horizontal component of V is $V\cos\phi$ and the vertical component of V is $V\sin\phi$. If each of the voltage phasors is multiplied by I, Figure 15.23(b) is obtained and is known as the **'power triangle'**.

Apparent power, $S = VI$ **voltamperes (VA)**
True or active power, $P = VI\cos\phi$ **watts (W)**
Reactive power, $Q = VI\sin\phi$ **reactive voltamperes (var)**

$$\textbf{Power factor} = \frac{\textbf{True power } P}{\textbf{Apparent power } S}$$

For sinusoidal voltages and currents, power factor $= \dfrac{P}{S} = \dfrac{VI\cos\phi}{VI}$, i.e.

$$\textbf{p.f.} = \cos\phi = \frac{R}{Z}$$ (from Figure 15.6)

The relationships stated above are also true when current I leads voltage V. More on the power triangle and power factor is contained in Chapter 26, page 464.

Problem 26. A pure inductance is connected to a 150 V, 50 Hz supply, and the apparent power of the circuit is 300 VA. Find the value of the inductance.

Apparent power $S = VI$

Hence current $I = \dfrac{S}{V} = \dfrac{300}{150} = 2$ A

Inductive reactance $X_L = \dfrac{V}{I} = \dfrac{150}{2} = 75\ \Omega$

Since $X_L = 2\pi fL$, inductance $L = \dfrac{X_L}{2\pi f} = \dfrac{75}{2\pi(50)} = \mathbf{0.239\ H}$

Problem 27. A transformer has a rated output of 200 kVA at a power factor of 0.8. Determine the rated power output and the corresponding reactive power.

$VI = 200$ kVA $= 200 \times 10^3$; p.f. $= 0.8 = \cos\phi$

Power output, $P = VI\cos\phi = (200 \times 10^3)(0.8) =$ **160 kW**

Reactive power, $Q = VI \sin\phi$

If $\cos\phi = 0.8$, then $\phi = \arccos 0.8 = 36.87^\circ = 36^\circ 52'$

Hence $\sin\phi = \sin 36.87^\circ = 0.6$

Hence reactive power, $Q = (200 \times 10^3)(0.6) = \mathbf{120\ kvar}$

> Problem 28. The power taken by an inductive circuit when connected to a 120 V, 50 Hz supply is 400 W and the current is 8 A. Calculate (a) the resistance, (b) the impedance, (c) the reactance, (d) the power factor, and (e) the phase angle between voltage and current.

(a) Power $P = I^2R$. Hence $R = \dfrac{P}{I^2} = \dfrac{400}{(8)^2} = \mathbf{6.25\ \Omega}$

(b) Impedance $Z = \dfrac{V}{I} = \dfrac{120}{8} = \mathbf{15\ \Omega}$

(c) Since $Z = \sqrt{(R^2 + X_L^2)}$, then $X_L = \sqrt{(Z^2 - R^2)}$

$$= \sqrt{[(15)^2 - (6.25)^2]}$$

$$= \mathbf{13.64\ \Omega}$$

(d) Power factor $= \dfrac{\text{true power}}{\text{apparent power}} = \dfrac{VI\cos\phi}{VI} = \dfrac{400}{(120)(8)} = \mathbf{0.4167}$

(e) p.f. $= \cos\phi = 0.4167$.Hence phase angle $\phi = \arccos 0.4167$

$$= 65.37^\circ$$

$$= \mathbf{65^\circ 22'\ lagging}$$

> Problem 29. A circuit consisting of a resistor in series with a capacitor takes 100 watts at a power factor of 0.5 from a 100 V, 60 Hz supply. Find (a) the current flowing, (b) the phase angle, (c) the resistance, (d) the impedance, and (e) the capacitance.

(a) Power factor $= \dfrac{\text{true power}}{\text{apparent power}}$

i.e. $0.5 = \dfrac{100}{(100)(I)}$. Hence $I = \dfrac{100}{(0.5)(100)} = \mathbf{2\ A}$

(b) Power factor $= 0.5 = \cos\phi$. Hence phase angle $\phi = \arccos 0.5$

$$= \mathbf{60^\circ\ leading}$$

(c) Power $P = I^2R$. Hence resistance $R = \dfrac{P}{I^2} = \dfrac{100}{(2)^2} = \mathbf{25\ \Omega}$

(d) Impedance $Z = \dfrac{V}{I} = \dfrac{100}{2} = \mathbf{50\ \Omega}$

(e) Capacitive reactance, $X_C = \sqrt{(Z^2 - R^2)} = \sqrt{(50^2 - 25^2)}$

$= 43.30\ \Omega$

$$X_C = \frac{1}{2\pi f C} \text{ hence capacitance } C = \frac{1}{2\pi f X_c} = \frac{1}{2\pi(60)(43.30)} \text{ F}$$

$= \mathbf{61.26\ \mu F}$

Further problems on power in a.c. circuits may be found in Section 15.12 following, problems 26 to 36, page 237.

15.12 Further problems on single-phase series a.c. circuits

A.c. circuits containing pure inductance and pure capacitance

1 Calculate the reactance of a coil of inductance 0.2 H when it is connected to (a) a 50 Hz, (b) a 600 Hz and (c) a 40 kHz supply. [(a) 62.83 Ω (b) 754 Ω (c) 50.27 kΩ]

2 A coil has a reactance of 120 Ω in a circuit with a supply frequency of 4 kHz. Calculate the inductance of the coil. [4.77 mH]

3 A supply of 240 V, 50 Hz is connected across a pure inductance and the resulting current is 1.2 A. Calculate the inductance of the coil. [0.637 H]

4 An e.m.f. of 200 V at a frequency of 2 kHz is applied to a coil of pure inductance 50 mH. Determine (a) the reactance of the coil, and (b) the current flowing in the coil. [(a) 628 Ω (b) 0.318 A]

5 Calculate the capacitive reactance of a capacitor of 20 μF when connected to an a.c. circuit of frequency (a) 20 Hz, (b) 500 Hz, (c) 4 kHz [(a) 397.9 Ω (b) 15.92 Ω (c) 1.989 Ω]

6 A capacitor has a reactance of 80 Ω when connected to a 50 Hz supply. Calculate the value of its capacitance. [39.79 μF]

7 A capacitor has a capacitive reactance of 400 Ω when connected to a 100 V, 25 Hz supply. Determine its capacitance and the current taken from the supply. [15.92 μF, 0.25 A]

8 Two similar capacitors are connected in parallel to a 200 V, 1 kHz supply. Find the value of each capacitor if the circuit current is 0.628 A. [0.25 μF]

***R–L* a.c. circuits**

9 Determine the impedance of a coil which has a resistance of 12 Ω and a reactance of 16 Ω [20 Ω]

10 A coil of inductance 80 mH and resistance 60 Ω is connected to a 200 V, 100 Hz supply. Calculate the circuit impedance and the

current taken from the supply. Find also the phase angle between the current and the supply voltage.
[78.27 Ω, 2.555 A, 39°57′ lagging]

11 An alternating voltage given by $v = 100 \sin 240t$ volts is applied across a coil of resistance 32 Ω and inductance 100 mH. Determine (a) the circuit impedance, (b) the current flowing, (c) the p.d. across the resistance, and (d) the p.d. across the inductance.
[(a) 40 Ω (b) 1.77 A (c) 56.64 V (d) 42.48 V]

12 A coil takes a current of 5 A from a 20 V d.c. supply. When connected to a 200 V, 50 Hz a.c. supply the current is 25 A. Calculate the (a) resistance, (b) impedance and (c) inductance of the coil. [(a) 4 Ω (b) 8 Ω (c) 22.05 mH]

13 A coil of inductance 636.6 mH and negligible resistance is connected in series with a 100 Ω resistor to a 250 V, 50 Hz supply. Calculate (a) the inductive reactance of the coil, (b) the impedance of the circuit, (c) the current in the circuit, (d) the p.d. across each component, and (e) the circuit phase angle.
[(a) 200 Ω (b) 223.6 Ω (c)1.118 A
(d) 223.6 V, 111.8 V (e) 63°26′ lagging]

R–C a.c. circuits

14 A voltage of 35 V is applied across a *C–R* series circuit. If the voltage across the resistor is 21 V, find the voltage across the capacitor. [28 V]

15 A resistance of 50 Ω is connected in series with a capacitance of 20 μF. If a supply of 200 V, 100 Hz is connected across the arrangement find (a) the circuit impedance, (b) the current flowing, and (c) the phase angle between voltage and current.
[(a) 93.98 Ω (b) 2.128 A (c) 57°51′ leading]

16 An alternating voltage $v = 250 \sin 800$ t volts is applied across a series circuit containing a 30 Ω resistor and 50 μF capacitor. Calculate (a) the circuit impedance, (b) the current flowing, (c) the p.d. across the resistor, (d) the p.d. across the capacitor, and (e) the phase angle between voltage and current
[(a) 39.05 Ω (b) 4.527 A (c) 135.8 V
(d) 113.2 V (e) 39°48′]

17 A 400 Ω resistor is connected in series with a 2358 pF capacitor across a 12 V a.c. supply. Determine the supply frequency if the current flowing in the circuit is 24 mA. [225 kHz]

R–L–C a.c. circuits

18 A 40 μF capacitor in series with a coil of resistance 8 Ω and inductance 80 mH is connected to a 200 V, 100 Hz supply. Calculate (a) the circuit impedance, (b) the current flowing, (c) the phase angle

between voltage and current, (d) the voltage across the coil, and (e) the voltage across the capacitor.
[(a) 13.18 Ω (b) 15.17 A (c) 52°38′
(d) 772.1 V (e) 603.6 V]

19 Three impedances are connected in series across a 100 V, 2 kHz supply. The impedances comprise:

(i) an inductance of 0.45 mH and 2 Ω resistance,
(ii) an inductance of 570 μH and 5 Ω resistance, and
(iii) a capacitor of capacitance 10 μF and resistance 3 Ω.

Assuming no mutual inductive effects between the two inductances calculate (a) the circuit impedance, (b) the circuit current, (c) the circuit phase angle and (d) the voltage across each impedance. Draw the phasor diagram.
[(a) 11.12 Ω (b) 8.99 A (c) 25°55′ lagging
(d) 53.92 V, 78.53 V, 76.46 V]

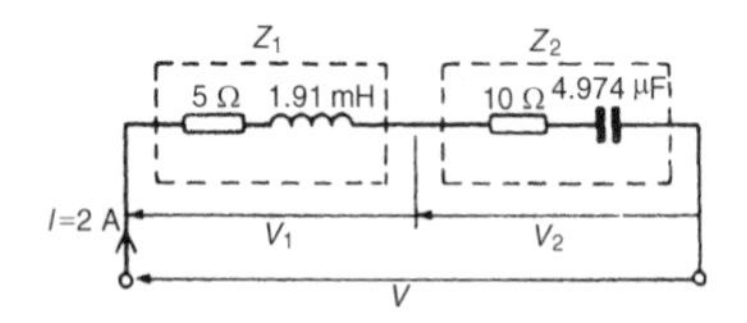

Figure 15.24

20 For the circuit shown in Figure 15.24 determine the voltages V_1 and V_2 if the supply frequency is 1 kHz. Draw the phasor diagram and hence determine the supply voltage V and the circuit phase angle.
[$V_1 = 26.0$ V, $V_2 = 67.05$ V,
$V = 50$ V, 53°8′ leading]

Series resonance and Q-factor

21 Find the resonant frequency of a series a.c. circuit consisting of a coil of resistance 10 Ω and inductance 50 mH and capacitance 0.05 μF. Find also the current flowing at resonance if the supply voltage is 100 V. [3.183 kHz, 10 A]

22 The current at resonance in a series L–C–R circuit is 0.2 mA. If the applied voltage is 250 mV at a frequency of 100 kHz and the circuit capacitance is 0.04 μF, find the circuit resistance and inductance.
[1.25 kΩ, 63.3 μH]

23 A coil of resistance 25 Ω and inductance 100 mH is connected in series with a capacitance of 0.12 μF across a 200 V, variable frequency supply. Calculate (a) the resonant frequency, (b) the current at resonance and (c) the factor by which the voltage across the reactance is greater than the supply voltage.
[(a) 1.453 kHz (b) 8 A (c) 36.52]

24 Calculate the inductance which must be connected in series with a 1000 pF capacitor to give a resonant frequency of 400 kHz.
[0.158 mH]

25 A series circuit comprises a coil of resistance 20 Ω and inductance 2 mH and a 500 pF capacitor. Determine the Q-factor of the circuit at resonance. If the supply voltage is 1.5 V, what is the voltage across the capacitor? [100, 150 V]

Power in a.c. circuits

26 A voltage $v = 200 \sin \omega t$ volts is applied across a pure resistance of 1.5 kΩ. Find the power dissipated in the resistor.

[13.33 W]

27 A 50 μF capacitor is connected to a 100 V, 200 Hz supply. Determine the true power and the apparent power.

[0, 628.3 VA]

28 A motor takes a current of 10 A when supplied from a 250 V a.c. supply. Assuming a power factor of 0.75 lagging find the power consumed. Find also the cost of running the motor for 1 week continuously if 1 kWh of electricity costs 7.20 p.

[1875 W, £22.68]

29 A motor takes a current of 12 A when supplied from a 240 V a.c. supply. Assuming a power factor of 0.75 lagging, find the power consumed. [2.16 kW]

30 A substation is supplying 200 kVA and 150 kvar. Calculate the corresponding power and power factor. [132 kW, 0.66]

31 A load takes 50 kW at a power factor of 0.8 lagging. Calculate the apparent power and the reactive power.

[62.5 kVA, 37.5 kvar]

32 A coil of resistance 400 Ω and inductance 0.20 H is connected to a 75 V, 400 Hz supply. Calculate the power dissipated in the coil.

[5.452 W]

33 An 80 Ω resistor and a 6 μF capacitor are connected in series across a 150 V, 200 Hz supply. Calculate (a) the circuit impedance, (b) the current flowing and (c) the power dissipated in the circuit.

[(a) 154.9 Ω (b) 0.968 A (c) 75 W]

34 The power taken by a series circuit containing resistance and inductance is 240 W when connected to a 200 V, 50 Hz supply. If the current flowing is 2 A find the values of the resistance and inductance. [60 Ω, 255 mH]

35 A circuit consisting of a resistor in series with an inductance takes 210 W at a power factor of 0.6 from a 50 V, 100 Hz supply. Find (a) the current flowing, (b) the circuit phase angle, (c) the resistance, (d) the impedance and (e) the inductance.

[(a) 7 A (b) 53°8′ lagging (c) 4.286 Ω
(d) 7.143 Ω (e) 9.095 mH]

36 A 200 V, 60 Hz supply is applied to a capacitive circuit. The current flowing is 2 A and the power dissipated is 150 W. Calculate the values of the resistance and capacitance.

[37.5 Ω, 28.61 μF]

16 Single-phase parallel a.c. circuits

At the end of this chapter you should be able to:

- calculate unknown currents, impedances and circuit phase angle from phasor diagrams for (a) R–L (b) R–C (c) L–C (d) LR–C parallel a.c. circuits
- state the condition for parallel resonance in an LR–C circuit
- derive the resonant frequency equation for an LR–C parallel a.c. circuit
- determine the current and dynamic resistance at resonance in an LR–C parallel circuit
- understand and calculate Q-factor in an LR–C parallel circuit
- understand how power factor may be improved

16.1 Introduction

In parallel circuits, such as those shown in Figures 16.1 and 16.2, the voltage is common to each branch of the network and is thus taken as the reference phasor when drawing phasor diagrams.
For any parallel a.c. circuit:

True or active power, $P = VI\cos\phi$ watts (W)

or $P = I_R{}^2 R$ watts

Apparent power, S = VI voltamperes (VA)

Reactive power, $Q = VI\sin\phi$ reactive voltamperes (var)

$$\text{Power factor} = \frac{\text{true power}}{\text{apparent power}} = \frac{P}{S} = \cos\phi$$

(These formulae are the same as for series a.c. circuits as used in Chapter 15.)

CIRCUIT DIAGRAM

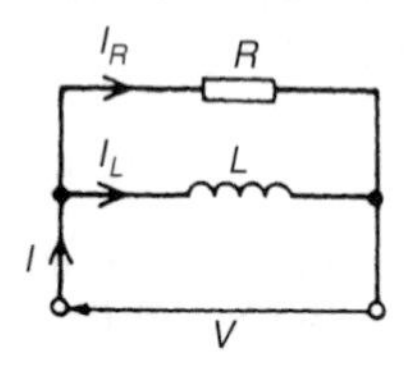

PHASOR DIAGRAM

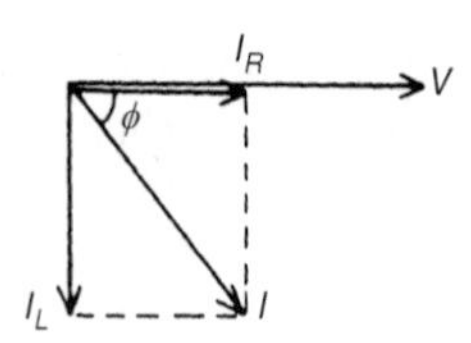

Figure 16.1

16.2 R–L parallel a.c. circuit

In the two branch parallel circuit containing resistance R and inductance L shown in Figure 16.1, the current flowing in the resistance, I_R, is in-phase with the supply voltage V and the current flowing in the inductance, I_L, lags the supply voltage by 90°. The supply current I is the phasor sum of I_R and I_L and thus the current I lags the applied voltage V by an angle lying between 0° and 90° (depending on the values of I_R and I_L), shown as angle ϕ in the phasor diagram.

From the phasor diagram:

$$I = \sqrt{(I_R^2 + I_L^2)}, \text{ (by Pythagoras' theorem)}$$

where $I_R = \dfrac{V}{R}$ and $I_L = \dfrac{V}{X_L}$

$\tan\phi = \dfrac{I_L}{I_R}$, $\sin\phi = \dfrac{I_L}{I}$ and $\cos\phi = \dfrac{I_R}{I}$ (by trigonometric ratios)

Circuit impedance, $Z = \dfrac{V}{I}$

> Problem 1. A 20 Ω resistor is connected in parallel with an inductance of 2.387 mH across a 60 V, 1 kHz supply. Calculate (a) the current in each branch, (b) the supply current, (c) the circuit phase angle, (d) the circuit impedance, and (e) the power consumed.

The circuit and phasor diagrams are as shown in Figure 16.1.

(a) Current flowing in the resistor $I_R = \dfrac{V}{R} = \dfrac{60}{20} = \mathbf{3\ A}$

Current flowing in the inductance $I_L = \dfrac{V}{X_L} = \dfrac{V}{2\pi f L}$

$$= \frac{60}{2\pi(1000)(2.387 \times 10^{-3})}$$

$$= \mathbf{4\ A}$$

(b) From the phasor diagram, supply current, $I = \sqrt{(I_R{}^2 + I_L^2)}$

$$= \sqrt{(3^2 + 4^2)}$$

$$= \mathbf{5\ A}$$

(c) Circuit phase angle, $\phi = \arctan\dfrac{I_L}{I_R} = \arctan\left(\dfrac{4}{3}\right) = 53.13°$

$$= \mathbf{53°8' \text{ lagging}}$$

(d) Circuit impedance, $Z = \dfrac{V}{I} = \dfrac{60}{5} = \mathbf{12\ \Omega}$

(e) Power consumed $P = VI\cos\phi = (60)(5)(\cos 53°8') = \mathbf{180\ W}$
(Alternatively, power consumed $P = I_R{}^2R = (3)^2(20) = \mathbf{180\ W}$)

Further problems on R–L parallel a.c. circuits may be found in Section 16.8, problems 1 and 2, page 256.

16.3 R–C parallel a.c. circuit

In the two branch parallel circuit containing resistance R and capacitance C shown in Figure 16.2, I_R is in-phase with the supply voltage V and the current flowing in the capacitor, I_C, leads V by 90°. The supply current I is the phasor sum of I_R and I_C and thus the current I leads the applied voltage V by an angle lying between 0° and 90° (depending on the values of I_R and I_C), shown as angle α in the phasor diagram.

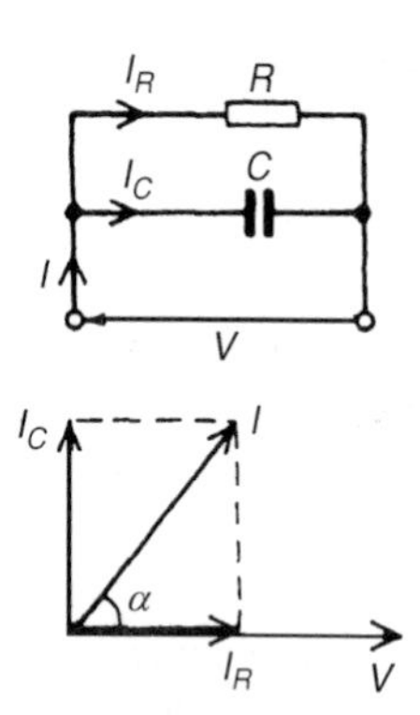

Figure 16.2

From the phasor diagram:

$$I = \sqrt{(I_R^2 + I_C^2)}, \text{ (by Pythagoras' theorem)}$$

where $I_R = \dfrac{V}{R}$ and $I_C = \dfrac{V}{X_C}$

$$\tan\alpha = \frac{I_C}{I_R}, \sin\alpha = \frac{I_C}{I} \text{ and } \cos\alpha = \frac{I_R}{\mathrm{I}} \text{ (by trigonometric ratios)}$$

Circuit impedance $Z = \dfrac{V}{I}$

> Problem 2. A 30 μF capacitor is connected in parallel with an 80 Ω resistor across a 240 V, 50 Hz supply. Calculate (a) the current in each branch, (b) the supply current, (c) the circuit phase angle, (d) the circuit impedance, (e) the power dissipated, and (f) the apparent power.

The circuit and phasor diagrams are as shown in Figure 16.2.

(a) Current in resistor, $I_R = \dfrac{V}{R} = \dfrac{240}{80} = \mathbf{3\ A}$

Current in capacitor, $I_C = \dfrac{V}{X_C} = \dfrac{V}{\left(\dfrac{1}{2\pi fC}\right)}$

$= 2\pi\mathrm{fCV}$

$= 2\pi(50)(30 \times 10^6)(240)$

$= \mathbf{2.262\ A}$

(b) Supply current, $I = \sqrt{({I_R}^2 + {I_C}^2)} = \sqrt{(3^2 + 2.262^2)}$

$= \mathbf{3.757\ A}$

(c) Circuit phase angle, $\alpha = \arctan\dfrac{I_C}{I_R} = \arctan\left(\dfrac{2.262}{3}\right)$

$= \mathbf{37°1'\ leading}$

(d) Circuit impedance, $Z = \dfrac{V}{I} = \dfrac{240}{3.757} = \mathbf{63.88\ \Omega}$

(e) True or active power dissipated, $P = VI\cos\alpha$

$$= 240(3.757)\cos 37°1'$$

$$= \mathbf{720\ W}$$

(Alternatively, true power $P = I_R{}^2\,R = (3)^2(80) = 720$ W)

(f) Apparent power, $S = VI = (240)(3.757) = \mathbf{901.7\ VA}$

Problem 3. A capacitor C is connected in parallel with a resistor R across a 120 V, 200 Hz supply. The supply current is 2 A at a power factor of 0.6 leading. Determine the values of C and R.

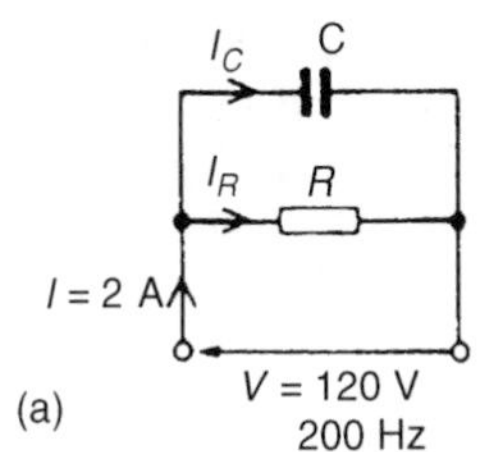

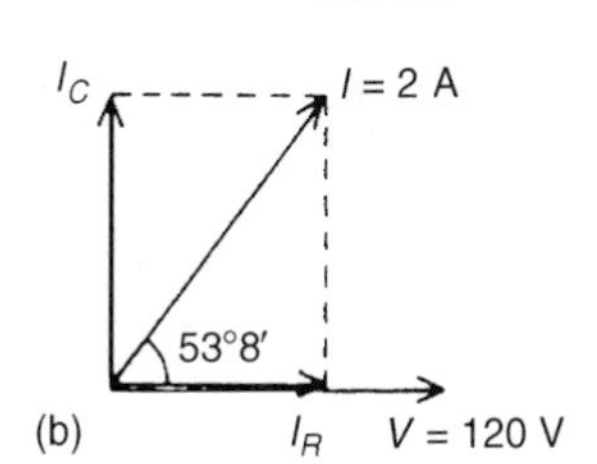

Figure 16.3

The circuit diagram is shown in Figure 16.3(a).

Power factor $= \cos\phi = 0.6$ leading, hence $\phi = \arccos 0.6 = 53.13°$ leading.

From the phasor diagram shown in Figure 16.3(b),

$$I_R = I\cos 53.13° = (2)(0.6)$$

$$= \mathbf{1.2\ A}$$

and $I_C = I\sin 53.13° = (2)(0.8)$

$$= \mathbf{1.6\ A}$$

(Alternatively, I_R and I_C can be measured from the scaled phasor diagram.)

From the circuit diagram,

$$I_R = \frac{V}{R} \text{ from which } R = \frac{V}{I_R} = \frac{120}{1.2} = \mathbf{100\ \Omega}$$

and $I_C = \dfrac{V}{X_C} = 2\pi fCV$, from which, $C = \dfrac{I_C}{2\pi fV}$

$$= \frac{1.6}{2\pi(200)(120)}$$

$$= \mathbf{10.61\ \mu F}$$

Further problems on R–C parallel a.c. circuits may be found in Section 16.8, problems 3 and 4, page 256.

16.4 *L–C* parallel a.c. circuit

In the two branch parallel circuit containing inductance L and capacitance C shown in Figure 16.4, I_L lags V by 90° and I_C leads V by 90°.

Theoretically there are three phasor diagrams possible — each depending on the relative values of I_L and I_C:

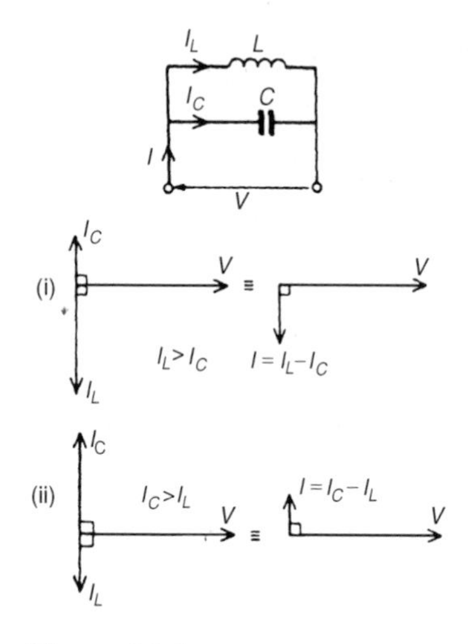

Figure 16.4

(i) $I_L > I_C$ (giving a supply current, $I = I_L - I_C$ lagging V by 90°)

(ii) $I_C > I_L$ (giving a supply current, $I = I_C - I_L$ leading V by 90°)

(iii) $I_L = I_C$ (giving a supply current, I = 0).

The latter condition is not possible in practice due to circuit resistance inevitably being present (as in the circuit described in Section 16.5).

For the L–C parallel circuit, $I_L = \dfrac{V}{X_L}$, $I_C = \dfrac{V}{X_C}$

I = phasor difference between I_L and I_C, and $Z = \dfrac{V}{I}$

Problem 4. A pure inductance of 120 mH is connected in parallel with a 25 μF capacitor and the network is connected to a 100 V, 50 Hz supply. Determine (a) the branch currents, (b) the supply current and its phase angle, (c) the circuit impedance, and (d) the power consumed.

The circuit and phasor diagrams are as shown in Figure 16.4.

(a) Inductive reactance, $X_L = 2\pi fL = 2\pi(50)(120 \times 10^{-3})$

$$= 37.70\ \Omega$$

Capacitive reactance, $X_C = \dfrac{1}{2\pi fC} = \dfrac{1}{2\pi(50)(25 \times 10^{-6})}$

$$= 127.3\ \Omega$$

Current flowing in inductance, $I_L = \dfrac{V}{X_L} = \dfrac{100}{37.70} = \mathbf{2.653\ A}$

Current flowing in capacitor, $I_C = \dfrac{V}{X_C} = \dfrac{100}{127.3} = \mathbf{0.786\ A}$

(b) I_L and I_C are anti-phase. Hence supply current,

$I = I_L - I_C = 2.653 - 0.786 =$ **1.867 A and the current lags the supply voltage V by 90°** (see Figure 16.4(i))

(c) Circuit impedance, $Z = \dfrac{V}{I} = \dfrac{100}{1.867} = \mathbf{53.56\ \Omega}$

(d) Power consumed, $P = VI\cos\phi = (100)(1.867)(\cos 90°)$

$$= \mathbf{0\ W}$$

Problem 5. Repeat Problem 4 for the condition when the frequency is changed to 150 Hz.

(a) Inductive reactance, $X_L = 2\pi(150)(120 \times 10^{-3}) = 113.1\ \Omega$

Capacitive reactance, $X_C = \dfrac{1}{2\pi(150)(25 \times 10^{-6})} = 42.44\ \Omega$

Current flowing in inductance, $I_L = \dfrac{V}{X_L} = \dfrac{100}{113.1} = \mathbf{0.884\ A}$

Current flowing in capacitor, $I_C = \dfrac{V}{X_C} = \dfrac{100}{42.44} = \mathbf{2.356\ A}$

(b) Supply current, $I = I_C - I_L = 2.356 - 0.884 =$ **1.472 A leading V by 90°** (see Figure 4(ii))

(c) Circuit impedance, $Z = \dfrac{V}{I} = \dfrac{100}{1.472} = \mathbf{67.93\ \Omega}$

(d) Power consumed, $P = VI\cos\phi =$ **0 W** (since $\phi = 90°$)

From Problems 4 and 5:

(i) When $X_L < X_C$ then $I_L > I_C$ and I lags V by 90°

(ii) When $X_L > X_C$ then $I_L < I_C$ and I leads V by 90°

(iii) In a parallel circuit containing no resistance the power consumed is zero

---

Further problems on L–C parallel a.c. circuits may be found in Section 16.8, problems 5 and 6, page 256.

16.5 *LR–C* parallel a.c. circuit

In the two branch circuit containing capacitance C in parallel with inductance L and resistance R in series (such as a coil) shown in Figure 16.5(a), the phasor diagram for the LR branch alone is shown in Figure 16.5(b) and the phasor diagram for the C branch is shown alone in Figure 16.5(c). Rotating each and superimposing on one another gives the complete phasor diagram shown in Figure 16.5(d).

The current I_{LR} of Figure 16.5(d) may be resolved into horizontal and vertical components. The horizontal component, shown as op is $I_{LR}\cos\phi_1$ and the vertical component, shown as pq is $I_{LR}\sin\phi_1$. There are three possible conditions for this circuit:

(i) $I_C > I_{LR}\sin\phi_1$ (giving a supply current I leading V by angle ϕ—as shown in Figure 16.5(e))

(ii) $I_{LR}\sin\phi_1 > I_C$ (giving I lagging V by angle ϕ—as shown in Figure 16.5(f))

(iii) $I_C = I_{LR}\sin\phi_1$ (this is called parallel resonance, see Section 16.6).

There are two methods of finding the phasor sum of currents I_{LR} and I_C in Figures 16.5(e) and (f). These are: (i) by a scaled phasor diagram,

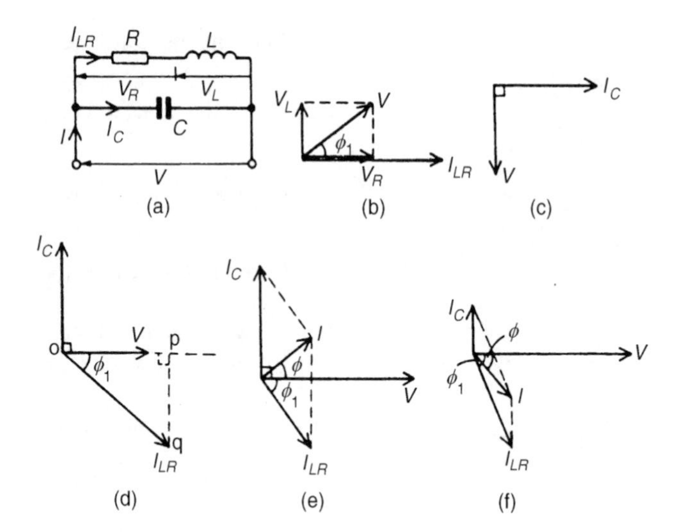

Figure 16.5

or (ii) by resolving each current into their '**in-phase**' (i.e. horizontal) and '**quadrature**' (i.e. vertical) **components,** as demonstrated in problems 6 and 7. With reference to the phasor diagrams of Figure 16.5:

Impedance of LR branch, $Z_{LR} = \sqrt{(R^2 + X_L{}^2)}$

Current, $I_{LR} = \dfrac{V}{Z_{LR}}$ and $I_C = \dfrac{V}{X_C}$

Supply current I = phasor sum of I_{LR} and I_C (by drawing)

$= \sqrt{\{(I_{LR}\cos\phi_1)^2 + (I_{LR}\sin\phi_1 \sim I_C)^2\}}$ (by calculation)

where $\sim$ means 'the difference between'.

Circuit impedance $Z = \dfrac{V}{I}$

$$\tan\phi_1 = \frac{V_L}{V_R} = \frac{X_L}{R}, \ \sin\phi_1 = \frac{X_L}{Z_{LR}} \text{ and } \cos\phi_1 = \frac{R}{Z_{LR}}$$

$$\tan\phi = \frac{I_{LR}\sin\phi_1 \sim I_C}{I_{LR}\cos\phi_1} \text{ and } \cos\phi = \frac{I_{LR}\cos\phi_1}{I}$$

Problem 6. A coil of inductance 159.2 mH and resistance 40 Ω is connected in parallel with a 30 μF capacitor across a 240 V, 50 Hz supply. Calculate (a) the current in the coil and its phase angle, (b) the current in the capacitor and its phase angle, (c) the supply current and its phase angle,(d) the circuit impedance, (e) the power consumed, (f) the apparent power, and (g) the reactive power. Draw the phasor diagram.

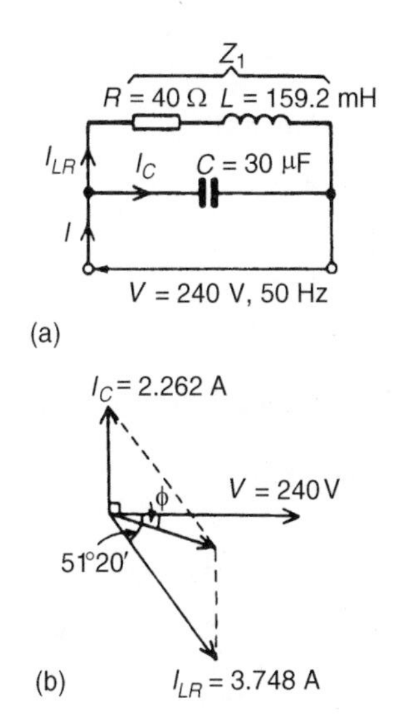

Figure 16.6

The circuit diagram is shown in Figure 16.6(a).

(a) For the coil, inductive reactance $X_L = 2\pi f L$

$$= 2\pi(50)(159.2 \times 10^{-3})$$

$$= 50\ \Omega$$

Impedance $Z_1 = \sqrt{(R^2 + X_L^2)} = \sqrt{(40^2 + 50^2)} = 64.03\ \Omega$

Current in coil, $I_{LR} = \dfrac{V}{Z_1} = \dfrac{240}{64.03} = \mathbf{3.748\ A}$

Branch phase angle $\phi_1 = \arctan\dfrac{X_L}{R} = \arctan\left(\dfrac{50}{40}\right) = \arctan 1.25$

$$= 51.34^\circ = \mathbf{51^\circ 20' \ lagging}$$

(see phasor diagram in Figure 16.6(b))

(b) Capacitive reactance, $X_C = \dfrac{1}{2\pi f C} = \dfrac{1}{2\pi(50)(30 \times 10^{-6})}$

$$= 106.1\ \Omega$$

Current in capacitor, $I_C = \dfrac{V}{X_C} = \dfrac{240}{106.1}$

= 2.262 A leading the supply voltage by 90°

(see phasor diagram of Figure 16.6(b)).

(c) The supply current I is the phasor sum of I_{LR} and I_C This may be obtained by drawing the phasor diagram to scale and measuring the current I and its phase angle relative to V. (Current I will always be the diagonal of the parallelogram formed as in Figure 16.6(b)).

Alternatively the current I_{LR} and I_C may be resolved into their horizontal (or 'in-phase') and vertical (or 'quadrant') components. The horizontal component of I_{LR} is

$$I_{LR}\cos(51^\circ 20') = 3.748\cos 51^\circ 20' = 2.342\ \text{A}$$

The horizontal component of I_C is $I_C \cos 90^\circ = 0$

Thus the total horizontal component, $I_H = \mathbf{2.342\ A}$

The vertical component of $I_{LR} = -I_{LR}\sin(51^\circ 20')$

$$= -3.748 \sin 51^\circ 20'$$

$$= -2.926\ \text{A}$$

The vertical component of $I_C = I_C \sin 90^\circ$

$$= 2.262 \sin 90^\circ = 2.262\ \text{A}$$

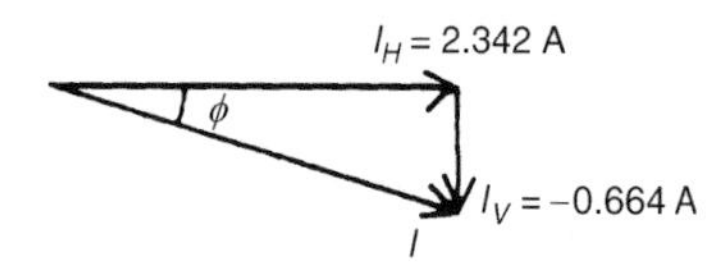

Figure 16.7

Thus the total vertical component, $I_V = -2.926 + 2.262$

$$= \mathbf{-0.664\ A}$$

I_H and I_V are shown in Figure 16.7, from which,

$$I = \sqrt{[(2.342)^2 + (-0.664)^2]} = 2.434 \text{ A}$$

$$\text{Angle } \phi = \arctan\left(\frac{0.664}{2.342}\right) = 15.83° = 15°50' \text{ lagging}$$

Hence the supply current I = 2.434 A lagging V by 15°50′.

(d) Circuit impedance, $Z = \dfrac{V}{I} = \dfrac{240}{2.434} = \mathbf{98.60\ \Omega}$

(e) Power consumed, $P = VI\cos\phi = (240)(2.434)\cos 15°50'$

$$= \mathbf{562\ W}$$

(Alternatively, $P = I_R{}^2R = I_{LR}{}^2R$ (in this case)

$$= (3.748)^2(40) = \mathbf{562\ W})$$

(f) Apparent power, $S = VI = (240)(2.434) = \mathbf{584.2\ VA}$

(g) Reactive power, $Q = VI\sin\phi = (240)(2.434)(\sin 15°50')$

$$= \mathbf{159.4\ var}$$

Problem 7. A coil of inductance 0.12 H and resistance 3 kΩ is connected in parallel with a 0.02 μF capacitor and is supplied at 40 V at a frequency of 5 kHz. Determine (a) the current in the coil, and (b) the current in the capacitor. (c) Draw to scale the phasor diagram and measure the supply current and its phase angle; check the answer by calculation. Determine (d) the circuit impedance and (e) the power consumed.

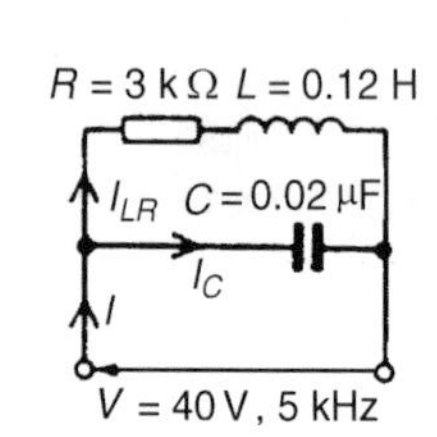

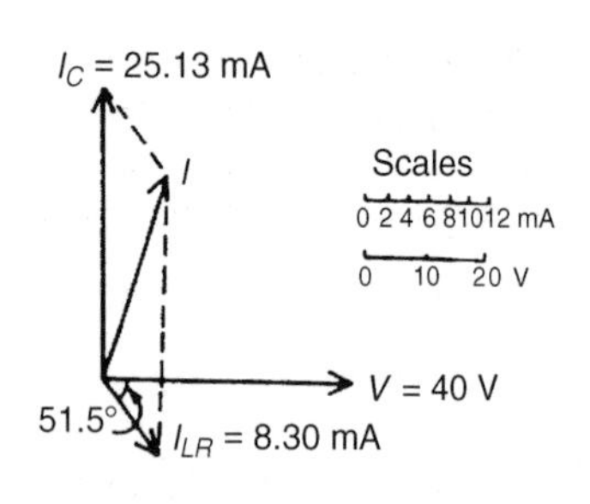

Figure 16.8

The circuit diagram is shown in Figure 16.8(a).

(a) Inductive reactance, $X_L = 2\pi fL = 2\pi(5000)(0.12) = 3770\ \Omega$

$$\text{Impedance of coil, } Z_1 = \sqrt{(R^2 + X_L{}^2)} = \sqrt{[(3000)^2 + (3770)^2]}$$

$$= 4818\ \Omega$$

$$\text{Current in coil, } I_{LR} = \frac{V}{Z_1} = \frac{40}{4818} = \mathbf{8.30\ mA}$$

$$\text{Branch phase angle } \phi = \arctan\frac{X_L}{R} = \arctan\frac{3770}{3000}$$

$$= \mathbf{51.5°\ lagging}$$

(b) Capacitive reactance, $X_C = \dfrac{1}{2\pi fC} = \dfrac{1}{2\pi(5000)(0.02 \times 10^{-6})}$

$= 1592\ \Omega$

Capacitor current, $I_C = \dfrac{V}{X_C} = \dfrac{40}{1592}$

$=$ **25.13 mA leading V by 90°**

(c) Currents I_{LR} and I_C are shown in the phasor diagram of Figure 16.8(b). The parallelogram is completed as shown and the supply current is given by the diagonal of the parallelogram. The current I is measured as **19.3 mA** leading voltage V by **74.5°**

By calculation, $I = \sqrt{[(I_{LR}\cos 51.5°)^2 + (I_C - I_{LR}\sin 51.5°)^2]}$

$= 19.34$ mA

and $\phi = \arctan\left(\dfrac{I_C - I_{LR}\sin 51.5°}{I_{LR}\cos 51.5°}\right) = 74.50°$

(d) Circuit impedance, $Z = \dfrac{V}{I} = \dfrac{40}{19.34 \times 10^{-3}} =$ **2.068 kΩ**

(e) Power consumed, $P = VI\cos\phi = (40)(19.34 \times 10^{-3})(\cos 74.50°)$

$=$ **206.7 mW**

(Alternatively, $P = I_R{}^2R = I_{LR}{}^2R = (8.30 \times 10^{-3})^2(3000)$

$= 206.7$ mW)

Further problems on the LR–C parallel a.c. circuit may be found in Section 16.8, problems 7 and 8, page 256.

16.6 Parallel resonance and Q-factor

Parallel resonance

Resonance occurs in the two branch network containing capacitance C in parallel with inductance L and resistance R in series (see Figure 16.5(a)) when the quadrature (i.e. vertical) component of current I_{LR} is equal to I_C. At this condition the supply current I is in-phase with the supply voltage V.

Resonant frequency

When the quadrature component of I_{LR} is equal to I_C then: $I_C = I_{LR}\sin\phi_1$ (see Figure 16.9)

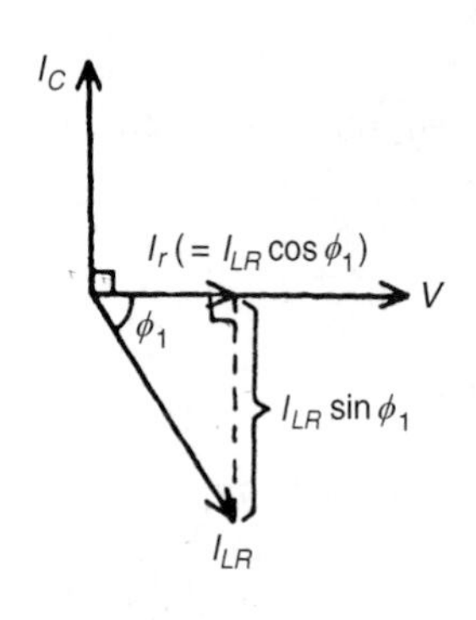

Figure 16.9

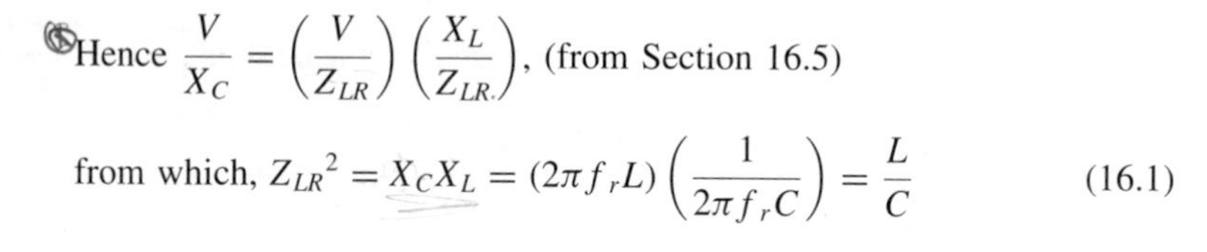

Hence $\dfrac{V}{X_C} = \left(\dfrac{V}{Z_{LR}}\right)\left(\dfrac{X_L}{Z_{LR}}\right)$, (from Section 16.5)

from which, $Z_{LR}{}^2 = X_C X_L = (2\pi f_r L)\left(\dfrac{1}{2\pi f_r C}\right) = \dfrac{L}{C}$ (16.1)

Hence $[\sqrt{(R^2 + X_L{}^2)}]^2 = \dfrac{L}{C}$ and $R^2 + X_L{}^2 = \dfrac{L}{C}$

Thus $(2\pi f_r L)^2 = \dfrac{L}{C} - R^2$ and $2\pi f_r L = \sqrt{\left(\dfrac{L}{C} - R^2\right)}$

and $$f_r = \frac{1}{2\pi L}\sqrt{\left(\frac{L}{C} - R^2\right)} = \frac{1}{2\pi}\sqrt{\left(\frac{L}{L^2 C} - \frac{R^2}{L^2}\right)}$$

i.e. parallel resonant frequency, $$\boldsymbol{f_r = \frac{1}{2\pi}\sqrt{\left(\frac{1}{LC} - \frac{R^2}{L^2}\right)}}\ \textbf{Hz}$$

(When R is negligible, then $f_r = \dfrac{1}{2\pi\sqrt{(LC)}}$, which is the same as for series resonance.)

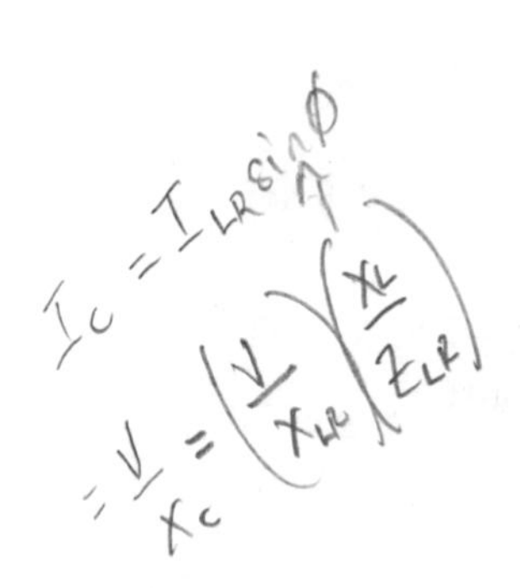

Current at resonance

Current at resonance, $I_r = I_{LR}\cos\phi_1$ (from Figure 16.9)

$$= \left(\frac{V}{Z_{LR}}\right)\left(\frac{R}{Z_{LR}}\right) \quad \text{(from Section 16.5)}$$

$$= \frac{VR}{Z_{LR}^2}$$

However from equation (16.1), $Z_{LR}^2 = \dfrac{L}{C}$

hence $\boldsymbol{I_r = \dfrac{VR}{\frac{L}{C}} = \dfrac{VRC}{L}}$ (16.2)

The current is at a **minimum** at resonance.

Dynamic resistance

Since the current at resonance is in-phase with the voltage the impedance of the circuit acts as a resistance. This resistance is known as the **dynamic resistance, R_D** (or sometimes, the dynamic impedance).

From equation (16.2), impedance at resonance $= \dfrac{V}{I_r} = \dfrac{V}{\left(\dfrac{VRC}{L}\right)} = \dfrac{L}{RC}$

i.e. dynamic resistance, $\boldsymbol{R_D = \dfrac{L}{RC}}$ **ohms**

Rejector circuit

The parallel resonant circuit is often described as a **rejector** circuit since it presents its maximum impedance at the resonant frequency and the resultant current is a minimum.

Q-factor

Currents higher than the supply current can circulate within the parallel branches of a parallel resonant circuit, the current leaving the capacitor and establishing the magnetic field of the inductor, this then collapsing and recharging the capacitor, and so on. The **Q-factor** of a parallel resonant circuit is the ratio of the current circulating in the parallel branches of the circuit to the supply current, i.e. the current magnification.

$$\text{Q-factor at resonance} = \text{current magnification} = \frac{\text{circulating current}}{\text{supply current}}$$

$$= \frac{I_C}{I_r} = \frac{I_{LR}\sin\phi_1}{I_r}$$

$$= \frac{I_{LR}\sin\phi_1}{I_{LR}\cos\phi_1} = \frac{\sin\phi_1}{\cos\phi_1}$$

$$= \tan\phi_1 = \frac{X_L}{R}$$

i.e. **Q-factor at resonance** $\boldsymbol{= \dfrac{2\pi f_r L}{R}}$

(which is the same as for a series circuit)

Note that in a **parallel** circuit the Q-factor is a measure of **current magnification**, whereas in a **series** circuit it is a measure of **voltage magnification**.

At mains frequencies the Q-factor of a parallel circuit is usually low, typically less than 10, but in radio-frequency circuits the Q-factor can be very high.

Problem 8. A pure inductance of 150 mH is connected in parallel with a 40 μF capacitor across a 50 V, variable frequency supply. Determine (a) the resonant frequency of the circuit and (b) the current circulating in the capacitor and inductance at resonance.

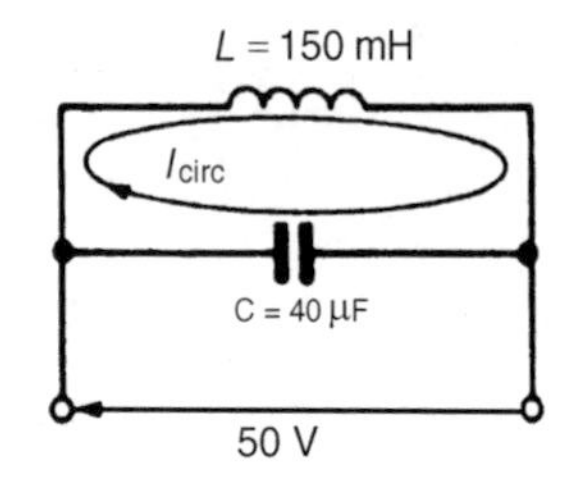

Figure 16.10

The circuit diagram is shown in Figure 16.10.

(a) Parallel resonant frequency, $f_r = \frac{1}{2\pi}\sqrt{\left(\frac{1}{LC} - \frac{R^2}{L^2}\right)}$

However, resistance $R = 0$. Hence,

$$f_r = \frac{1}{2\pi}\sqrt{\left(\frac{1}{LC}\right)} = \frac{1}{2\pi}\sqrt{\left[\frac{1}{(150 \times 10^{-3})(40 \times 10^{-6})}\right]}$$

$$= \frac{1}{2\pi}\sqrt{\left(\frac{10^7}{(15)(4)}\right)}$$

$$= \frac{10^3}{2\pi}\sqrt{\left(\frac{1}{6}\right)} = \mathbf{64.97\ Hz}$$

(b) Current circulating in L and C at resonance,

$$I_{CIRC} = \frac{V}{X_C} = \frac{V}{\left(\frac{1}{2\pi f_r C}\right)} = 2\pi f_r CV$$

Hence $I_{CIRC} = 2\pi(64.97)(40 \times 10^{-6})(50) = \mathbf{0.816\ A}$

(Alternatively, $I_{CIRC} = \frac{V}{X_L} = \frac{V}{2\pi f_r L} = \frac{50}{2\pi(64.97)(0.15)}$
$= \mathbf{0.817\ A}$)

> Problem 9. A coil of inductance 0.20 H and resistance 60 Ω is connected in parallel with a 20 μF capacitor across a 20 V, variable frequency supply. Calculate (a) the resonant frequency, (b) the dynamic resistance, (c) the current at resonance and (d) the circuit Q-factor at resonance.

(a) Parallel resonant frequency,

$$f_r = \frac{1}{2\pi}\sqrt{\left(\frac{1}{LC} - \frac{R^2}{L^2}\right)}$$

$$= \frac{1}{2\pi}\sqrt{\left(\frac{1}{(0.20)(20 \times 10^{-6})} - \frac{(60)^2}{(0.2)^2}\right)}$$

$$= \frac{1}{2\pi}\sqrt{(250\,000 - 90\,000)}$$

$$= \frac{1}{2\pi}\sqrt{(160\,000)} = \frac{1}{2\pi}(400)$$

$$= \mathbf{63.66\ Hz}$$

(b) Dynamic resistance, $R_D = \frac{L}{RC} = \frac{0.20}{(60)(20 \times 10^{-6})} = \mathbf{166.7\ \Omega}$

(c) Current at resonance, $I_r = \frac{V}{R_D} = \frac{20}{166.7} = \mathbf{0.12\ A}$

(d) Circuit Q-factor at resonance $= \frac{2\pi f_r L}{R} = \frac{2\pi(63.66)(0.2)}{60} = \mathbf{1.33}$

Alternatively, Q-factor at resonance = current magnification (for a parallel circuit) $= I_c/I_r$

$$I_c = \frac{V}{X_c} = \frac{V}{\left(\frac{1}{2\pi f_r C}\right)} = 2\pi f_r C V = 2\pi(63.66)(20 \times 10^{-6})(20)$$

$$= 0.16\ A$$

Hence Q-factor $= \frac{I_c}{I_r} = \frac{0.16}{0.12} = \mathbf{1.33}$, as obtained above

Problem 10. A coil of inductance 100 mH and resistance 800 Ω is connected in parallel with a variable capacitor across a 12 V, 5 kHz supply. Determine for the condition when the supply current is a minimum: (a) the capacitance of the capacitor, (b) the dynamic resistance, (c) the supply current, and (d) the Q-factor.

(a) The supply current is a minimum when the parallel circuit is at resonance.

Resonant frequency, $f_r = \frac{1}{2\pi}\sqrt{\left(\frac{1}{LC} - \frac{R^2}{L^2}\right)}$

Transposing for C gives: $(2\pi f_r)^2 = \frac{1}{LC} - \frac{R^2}{L^2}$

$$(2\pi f_r)^2 + \frac{R^2}{L^2} = \frac{1}{LC}$$

$$C = \frac{1}{L\left\{(2\pi f_r)^2 + \frac{R^2}{L^2}\right\}}$$

When $L = 100$ mH, $R = 800\ \Omega$ and $f_r = 5000$ Hz,

$$C = \frac{1}{100 \times 10^{-3}\left\{2\pi(5000)^2 + \frac{800^2}{(100 \times 10^{-3})^2}\right\}}$$

$$= \frac{1}{0.1[\pi^2 10^8 + (0.64)10^8]} F$$

$$= \frac{10^6}{0.1(10.51 \times 10^8)} \mu F = \mathbf{0.009515\ \mu F\ or\ 9.515\ nF}$$

(b) Dynamic resistance, $R_D = \frac{L}{CR} = \frac{100 \times 10^{-3}}{(9.515 \times 10^{-9})(800)}$

$$= \mathbf{13.14\ k\Omega}$$

(c) Supply current at resonance, $I_r = \frac{V}{R_D} = \frac{12}{13.14 \times 10^3} = \mathbf{0.913\ mA}$

(d) Q-factor at resonance $= \frac{2\pi f_r L}{R} = \frac{2\pi(5000)(100 \times 10^{-3})}{800} = \mathbf{3.93}$

Alternatively, Q-factor at resonance $= \frac{I_c}{I_r} = \frac{V/X_c}{I_r} = \frac{2\pi f_r CV}{I_r}$

$$= \frac{2\pi(5000)(9.515 \times 10^{-9})(12)}{0.913 \times 10^{-3}}$$

$$= \mathbf{3.93}$$

Further problems on parallel resonance and Q-factor may be found in Section 16.8, problems 9 to 12, page 257.

16.7 Power factor improvement

For a particular power supplied, a high power factor reduces the current flowing in a supply system and therefore reduces the cost of cables, switch-gear, transformers and generators. Supply authorities use tariffs which encourage electricity consumers to operate at a reasonably high power factor. Industrial loads such as a.c. motors are essentially inductive (R–L) and may have a low power factor. One method of improving (or correcting) the power factor of an inductive load is to connect a static capacitor C in parallel with the load (see Figure 16.11(a)). The supply current is reduced from I_{LR} to I, the phasor sum of I_{LR} and I_C, and the circuit power factor improves from $\cos\phi_1$ to $\cos\phi_2$ (see Figure 16.11(b)).

Problem 11. A single-phase motor takes 50 A at a power factor of 0.6 lagging from a 240 V, 50 Hz supply. Determine (a) the current taken by a capacitor connected in parallel with the motor to correct the power factor to unity, and (b) the value of the supply current after power factor correction.

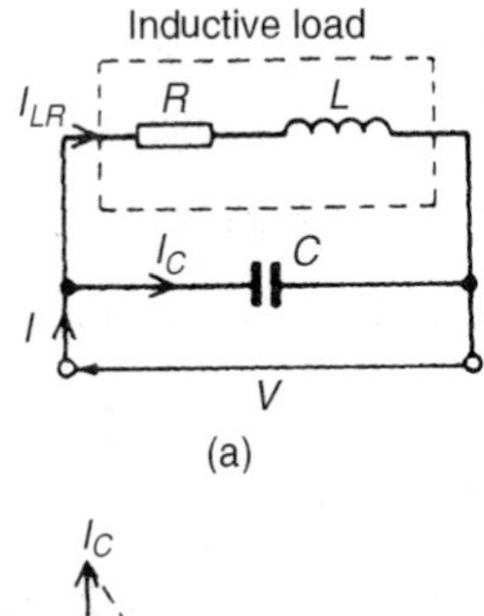

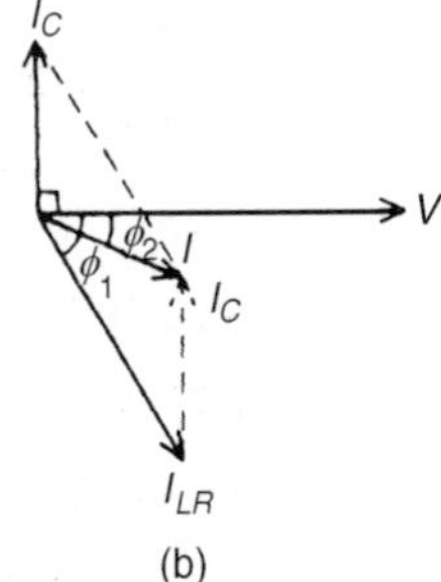

Figure 16.11

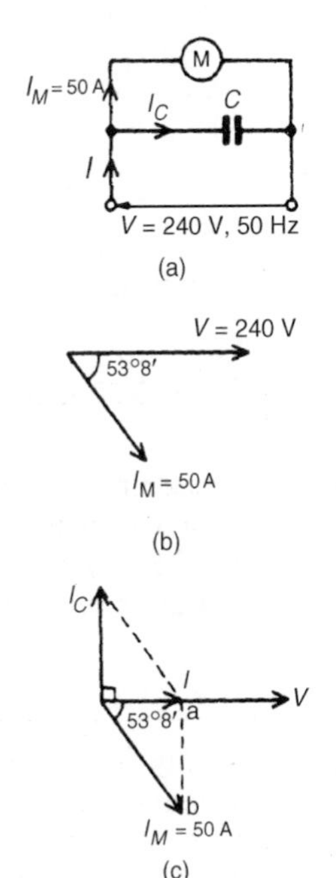

Figure 16.12

The circuit diagram is shown in Figure 16.12(a).

(a) A power factor of 0.6 lagging means that $\cos\phi = 0.6$ i.e. $\phi = \arccos 0.6 = 53°8'$

Hence I_M lags V by $53°8'$ as shown in Figure 16.12(b).

If the power factor is to be improved to unity then the phase difference between supply current I and voltage V is 0°, i.e. I is in phase with V as shown in Figure 16.12(c). For this to be so, I_C must equal the length ab, such that the phasor sum of I_M and I_C is I. $ab = I_M \sin 53°8' = 50(0.8) = 40$ A

Hence the capacitor current I_c must be 40 A for the power factor to be unity.

(b) Supply current $I = I_M \cos 53°8' = 50(0.6) =$ **30 A**

> Problem 12. A motor has an output of 4.8 kW, an efficiency of 80% and a power factor of 0.625 lagging when operated from a 240 V, 50 Hz supply. It is required to improve the power factor to 0.95 lagging by connecting a capacitor in parallel with the motor. Determine (a) the current taken by the motor, (b) the supply current after power factor correction, (c) the current taken by the capacitor, (d) the capacitance of the capacitor, and (e) the kvar rating of the capacitor.

(a) Efficiency $= \dfrac{\text{power output}}{\text{power input}}$ hence $\dfrac{80}{100} = \dfrac{4800}{\text{power input}}$

Power input $= \dfrac{4800}{0.8} = 6000$ W

Hence, $6000 = VI_M \cos\phi = (240)(I_M)(0.625)$,

since $\cos\phi = \text{p.f.} = 0.625$

Thus current taken by the motor, $I_M = \dfrac{6000}{(240)(0.625)} =$ **40 A**

The circuit diagram is shown in Figure 16.13(a).

The phase angle between I_M and V is given by:

$\phi = \arccos 0.625 = 51.32° = 51°19'$, hence the phasor diagram is as shown in Figure 16.13(b).

(b) When a capacitor C is connected in parallel with the motor a current I_C flows which leads V by 90°. The phasor sum of I_M and I_C gives the supply current I, and has to be such as to change the circuit power factor to 0.95 lagging, i.e. a phase angle of arccos 0.95 or $18°12'$ lagging, as shown in Figure 16.13(c).

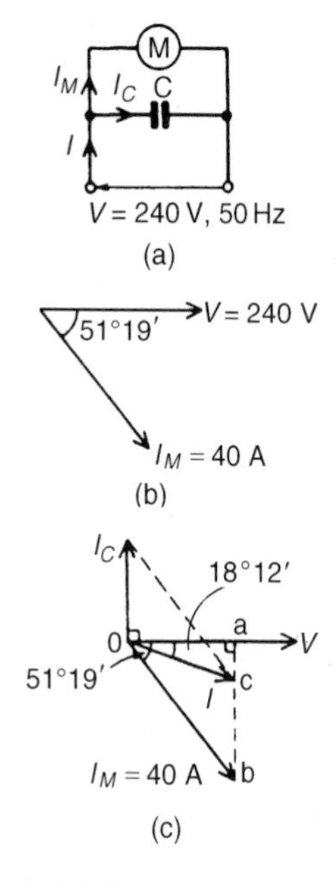

Figure 16.13

The horizontal component of I_M (shown as oa) $= I_M \cos 51°19'$

$$= 40 \cos 51°19'$$
$$= 25 \text{ A}$$

The horizontal component of I (also given by oa) $= I \cos 18°12'$

$$= 0.95\, I$$

Equating the horizontal components gives: $25 = 0.95$ I

Hence the supply current after p.f. correction, $I = \dfrac{25}{0.95} = \mathbf{26.32\ A}$

(c) The vertical component of I_M (shown as ab) $= I_M \sin 51°19'$

$$= 40 \sin 51°19'$$
$$= 31.22\ A$$

The vertical component of I (shown as ac) $= I \sin 18°12'$

$$= 26.32 \sin 18°12'$$
$$= 8.22\ A$$

The magnitude of the capacitor current I_C (shown as bc) is given by $ab - ac$, i.e. $31.22 - 8.22 =$ **23 A**

(d) Current $I_C = \dfrac{V}{X_c} = \dfrac{V}{\left(\dfrac{1}{2\pi f C}\right)} = 2\pi f C V,$

from which, $C = \dfrac{I_C}{2\pi f V} = \dfrac{23}{2\pi(50)(240)}$ F $=$ **305 μF**

(e) kvar rating of the capacitor $= \dfrac{VI_c}{1000} = \dfrac{(240)(23)}{1000} =$ **5.52 kvar**

In this problem the supply current has been reduced from 40 A to 26.32 A without altering the current or power taken by the motor. This means that the size of generating plant and the cross-sectional area of conductors supplying both the factory and the motor can be less — with an obvious saving in cost.

Problem 13. A 250 V, 50 Hz single-phase supply feeds the following loads (i) incandescent lamps taking a current of 10 A at unity power factor, (ii) fluorescent lamps taking 8 A at a power factor of 0.7 lagging, (iii) a 3 kVA motor operating at full load and at a power factor of 0.8 lagging and (iv) a static capacitor. Determine, for the lamps and motor, (a) the total current, (b) the overall power factor and (c) the total power. (d) Find the value of the static capacitor to improve the overall power factor to 0.975 lagging.

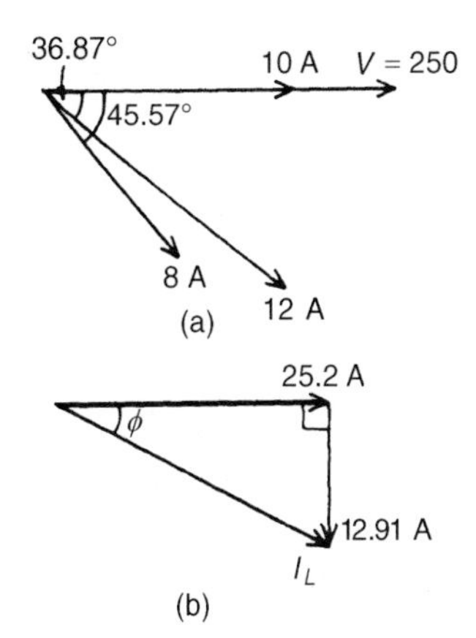

Figure 16.14

A phasor diagram is constructed as shown in Figure 16.14(a), where 8 A is lagging voltage V by arccos 0.7, i.e. 45.57°, and the motor current is 3000/250, i.e. 12 A lagging V by arccos 0.8, i.e. 36.87°

(a) The horizontal component of the currents

$$= 10\cos 0^\circ + 12\cos 36.87^\circ + 8\cos 45.57^\circ$$

$$= 10 + 9.6 + 5.6 = 25.2\ A$$

The vertical component of the currents

$$= 10\sin 0^\circ - 12\sin 36.87^\circ - 8\sin 45.57^\circ$$

$$= 0 - 7.2 - 5.713 = -12.91\ \text{A}$$

From Figure 16.14(b), total current, $I_L = \sqrt{[(25.2)^2 + (12.91)^2]}$
$= \mathbf{28.31\ A}$

at a phase angle of $\phi = \arctan\left(\dfrac{12.91}{25.2}\right)$, i.e. 27.13° lagging

(b) Power factor $= \cos\phi = \cos 27.13^\circ =$ **0.890 lagging**

(c) Total power, $P = VI_L\cos\phi = (250)(28.31)(0.890) =$ **6.3 kW**

(d) To improve the power factor, a capacitor is connected in parallel with the loads. The capacitor takes a current I_C such that the supply current falls from 28.31 A to I, lagging V by arccos 0.975, i.e. 12.84°. The phasor diagram is shown in Figure 16.15.

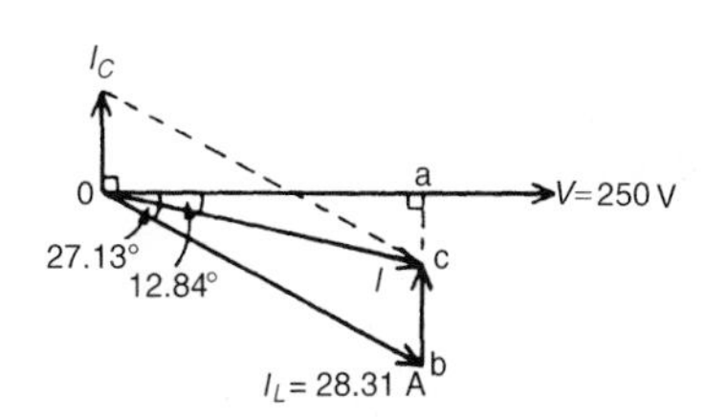

Figure 16.15

$$\text{oa} = 28.31\cos 27.13^\circ = I\cos 12.84^\circ$$

$$\text{Hence } I = \frac{28.31\cos 27.13^\circ}{\cos 12.84^\circ} = 25.84\ \text{A}$$

$$\text{Current } I_C = bc = (ab - ac)$$

$$= 28.31\sin 27.13^\circ - 25.84\sin 12.84^\circ$$

$$= 12.91 - 5.742$$

$$= 7.168\ \text{A}$$

$$I_c = \frac{V}{X_c} = \frac{V}{\left(\dfrac{1}{2\pi fC}\right)} = 2\pi fCV$$

$$\text{Hence capacitance } C = \frac{I_c}{2\pi fV} = \frac{7.168}{2\pi(50)(250)}\ \mathbf{F}$$

$$= \mathbf{91.27\ \mu F}$$

Thus to improve the power factor from 0.890 to 0.975 lagging a 91.27 μF capacitor is connected in parallel with the loads.

Further problems on power factor improvement may be found in Section 16.8 following, problems 13 to 16, page 257.

16.8 Further problems on single-phase parallel a.c. circuits

R–L parallel a.c. circuit

1 A 30 Ω resistor is connected in parallel with a pure inductance of 3 mH across a 110 V, 2 kHz supply. Calculate (a) the current in each branch, (b) the circuit current, (c) the circuit phase angle, (d) the circuit impedance, (e) the power consumed, and (f) the circuit power factor.
[(a) $I_R = 3.67$ A, $I_L = 2.92$ A (b) 4.69 A (c) 38°30′ lagging (d) 23.45 Ω (e) 404 W (f) 0.783 lagging]

2 A 40 Ω resistance is connected in parallel with a coil of inductance L and negligible resistance across a 200 V, 50 Hz supply and the supply current is found to be 8 A. Draw a phasor diagram to scale and determine the inductance of the coil. [102 mH]

R–C parallel a.c. circuit

3 A 1500 nF capacitor is connected in parallel with a 16 Ω resistor across a 10 V, 10 kHz supply. Calculate (a) the current in each branch, (b) the supply current, (c) the circuit phase angle, (d) the circuit impedance, (e) the power consumed, (f) the apparent power, and (g) the circuit power factor. Draw the phasor diagram.
[(a) $I_R = 0.625$ A, $I_C = 0.943$ A (b) 1.13 A (c) 56°28′ leading (d) 8.85 Ω (e) 6.25 W (f) 11.3 VA (g) 0.55 leading]

4 A capacitor C is connected in parallel with a resistance R across a 60 V, 100 Hz supply. The supply current is 0.6 A at a power factor of 0.8 leading. Calculate the value of R and C.
[$R = 125$ Ω, $C = 9.55$ μF]

L–C parallel a.c. circuit

5 An inductance of 80 mH is connected in parallel with a capacitance of 10 μF across a 60 V, 100 Hz supply. Determine (a) the branch currents, (b) the supply current, (c) the circuit phase angle, (d) the circuit impedance and (e) the power consumed.
[(a) $I_C = 0.377$ A, $I_L = 1.194$ A (b) 0.817 A (c) 90° lagging (d) 73.44 Ω (e) 0 W]

6 Repeat problem 5 for a supply frequency of 200 Hz.
[(a) $I_C = 0.754$ A, $I_L = 0.597$ A (b) 0.157 A (c) 90° leading (d) 382.2 Ω (e) 0 W]

LR–C parallel a.c. circuit

7 A coil of resistance 60 Ω and inductance 318.4 mH is connected in parallel with a 15 μF capacitor across a 200 V, 50 Hz supply.

Calculate (a) the current in the coil, (b) the current in the capacitor, (c) the supply current and its phase angle, (d) the circuit impedance, (e) the power consumed, (f) the apparent power and (g) the reactive power. Draw the phasor diagram.

[(a) 1.715 A (b) 0.943 A (c) 1.028 A at 30°54′ lagging (d) 194.6 Ω (e) 176.5 W (f) 205.6 VA (g) 105.6 var]

8 A 25 nF capacitor is connected in parallel with a coil of resistance 2 kΩ and inductance 0.20 H across a 100 V, 4 kHz supply. Determine (a) the current in the coil, (b) the current in the capacitor, (c) the supply current and its phase angle (by drawing a phasor diagram to scale, and also by calculation), (d) the circuit impedance, and (e) the power consumed.

[(a) 18.48 mA (b) 62.83 mA (c) 46.17 mA at 81°29′ leading (d) 2.166 kΩ (e) 0.683 W]

Parallel resonance and Q-factor

9 A 0.15 μF capacitor and a pure inductance of 0.01 H are connected in parallel across a 10 V, variable frequency supply. Determine (a) the resonant frequency of the circuit, and (b) the current circulating in the capacitor and inductance.

[(a) 4.11 kHz (b) 38.73 mA]

10 A 30 μF capacitor is connected in parallel with a coil of inductance 50 mH and unknown resistance R across a 120 V, 50 Hz supply. If the circuit has an overall power factor of 1 find (a) the value of R, (b) the current in the coil, and (c) the supply current.

[(a) 37.7 Ω (b) 2.94 A (c) 2.714 A]

11 A coil of resistance 25 Ω and inductance 150 mH is connected in parallel with a 10 μF capacitor across a 60 V, variable frequency supply. Calculate (a) the resonant frequency, (b) the dynamic resistance, (c) the current at resonance and (d) the Q-factor at resonance.

[(a) 127.2 Hz (b) 600 Ω (c) 0.10 A (d) 4.80]

12 A coil of resistance 1.5 kΩ and 0.25 H inductance is connected in parallel with a variable capacitance across a 10 V, 8 kHz supply. Calculate (a) the capacitance of the capacitor when the supply current is a minimum, (b) the dynamic resistance, and (c) the supply current.

[(a) 1561 pF (b) 106.8 kΩ (c) 93.66 μA]

Power factor improvement

13 A 415 V alternator is supplying a load of 55 kW at a power factor of 0.65 lagging. Calculate (a) the kVA loading and (b) the current taken from the alternator. (c) If the power factor is now raised to unity find the new kVA loading.

[(a) 84.6 kVA (b) 203.9 A (c) 84.6 kVA]

14 A single phase motor takes 30 A at a power factor of 0.65 lagging from a 240 V, 50 Hz supply. Determine (a) the current taken by the

capacitor connected in parallel to correct the power factor to unity, and (b) the value of the supply current after power factor correction.
[(a) 22.80 A (b) 19.5 A]

15 A motor has an output of 6 kW, an efficiency of 75% and a power factor of 0.64 lagging when operated from a 250 V, 60 Hz supply. It is required to raise the power factor to 0.925 lagging by connecting a capacitor in parallel with the motor. Determine (a) the current taken by the motor, (b) the supply current after power factor correction, (c) the current taken by the capacitor, (d) the capacitance of the capacitor and (e) the kvar rating of the capacitor.
[(a) 50 A (b) 34.59 A (c) 25.28 A (d) 268.2 μF (e) 6.32 kvar]

16 A 200 V, 50 Hz single-phase supply feeds the following loads: (i) fluorescent lamps taking a current of 8 A at a power factor of 0.9 leading, (ii) incandescent lamps taking a current of 6 A at unity power factor, (iii) a motor taking a current of 12 A at a power factor of 0.65 lagging. Determine the total current taken from the supply and the overall power factor. Find also the value of a static capacitor connected in parallel with the loads to improve the overall power factor to 0.98 lagging. [21.74 A, 0.966 lagging, 21.68 μF]

17 D.c. transients

At the end of this chapter you should be able to:

- understand the term 'transient'
- describe the transient response of capacitor and resistor voltages, and current in a series C–R d.c. circuit
- define the term 'time constant'
- calculate time constant in a C–R circuit
- draw transient growth and decay curves for a C–R circuit
- use equations $v_C = V(1 - e^{-t/\tau})$, $v_R = Ve^{-t/\tau}$ and $i = Ie^{-t/\tau}$ for a CR circuit
- describe the transient response when discharging a capacitor
- describe the transient response of inductor and resistor voltages, and current in a series L–R d.c. circuit
- calculate time constant in an L–R circuit
- draw transient growth and decay curves for an LR circuit
- use equations $v_L = Ve^{-t/\tau}$, $v_R = V(1 - e^{-t/\tau})$ and $i = I(1 - e^{-t/\tau})$
- describe the transient response for current decay in an LR circuit
- understand the switching of inductive circuits
- describe the effects of time constant on a rectangular waveform via integrator and differentiator circuits

17.1 Introduction

When a d.c. voltage is applied to a capacitor C, and resistor R connected in series, there is a short period of time immediately after the voltage is connected, during which the current flowing in the circuit and voltages across C and R are changing.

Similarly, when a d.c. voltage is connected to a circuit having inductance L connected in series with resistance R, there is a short period of time immediately after the voltage is connected, during which the current flowing in the circuit and the voltages across L and R are changing.

These changing values are called **transients**.

17.2 Charging a capacitor

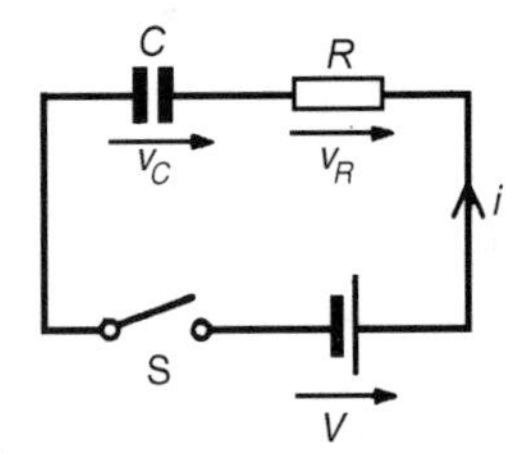

Figure 17.1

(a) The circuit diagram for a series connected C–R circuit is shown in Figure 17.1. When switch S is closed then by Kirchhoff's voltage law:

$$V = v_C + v_R \tag{17.1}$$

(b) The battery voltage V is constant. The capacitor voltage v_C is given by q/C, where q is the charge on the capacitor. The voltage drop across R is given by iR, where i is the current flowing in the circuit. Hence at all times:

$$V = \frac{q}{C} + iR \tag{17.2}$$

At the instant of closing S, (initial circuit condition), assuming there is no initial charge on the capacitor, q_0 is zero, hence v_{Co} is zero. Thus from equation (17.1), $V = 0 + v_{Ro}$, i.e. $v_{Ro} = V$. This shows that the resistance to current is solely due to R, and the initial current flowing, $i_o = I = V/R$.

(c) A short time later at time t_1 seconds after closing S, the capacitor is partly charged to, say, q_1 coulombs because current has been flowing. The voltage v_{C1} is now q_1/C volts. If the current flowing is i_1 amperes, then the voltage drop across R has fallen to i_1R volts. Thus, equation (17.2) is now $V = (q_1/C) + i_1R$.

(d) A short time later still, say at time t_2 seconds after closing the switch, the charge has increased to q_2 coulombs and v_C has increased to q_2/C volts. Since $V = v_C + v_R$ and V is a constant, then v_R decreases to i_2R, Thus v_C is increasing and i and v_R are decreasing as time increases.

(e) Ultimately, a few seconds after closing S, (i.e. at the final or **steady state** condition), the capacitor is fully charged to, say, Q coulombs, current no longer flows, i.e. $i = 0$, and hence $v_R = iR = 0$. It follows from equation (17.1) that $v_C = V$.

(f) Curves showing the changes in v_C, v_R and i with time are shown in Figure 17.2.

The curve showing the variation of v_C with time is called an **exponential growth curve** and the graph is called the 'capacitor voltage/time' characteristic. The curves showing the variation of v_R and i with time are called **exponential decay curves**, and the graphs are called 'resistor voltage/time' and 'current/time' characteristics respectively. (The name 'exponential' shows that the shape can be expressed mathematically by an exponential mathematical equation, as shown in Section 17.4).

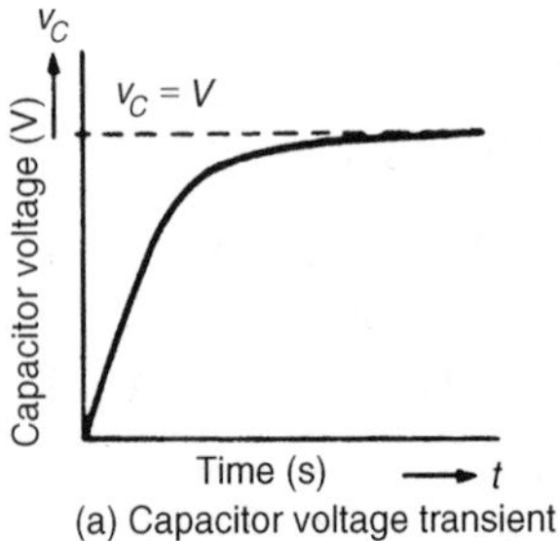

(a) Capacitor voltage transient

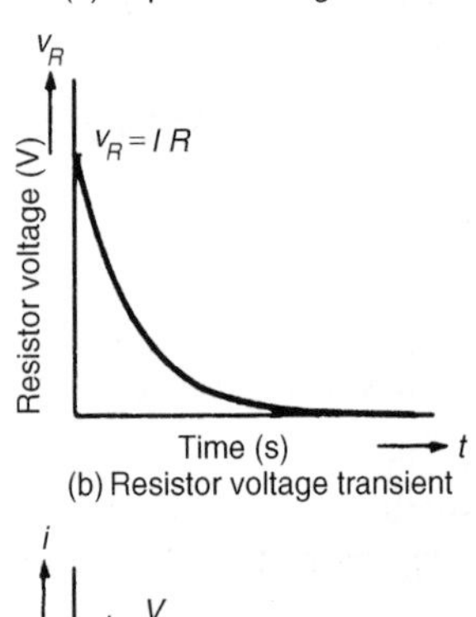

(b) Resistor voltage transient

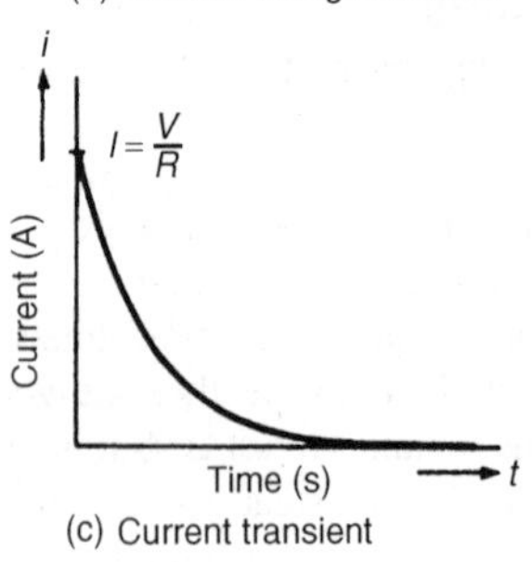

(c) Current transient

Figure 17.2

17.3 Time constant for a C–R circuit

(a) If a constant d.c. voltage is applied to a series connected C–R circuit, a transient curve of capacitor voltage v_C is as shown in Figure 17.2(a).

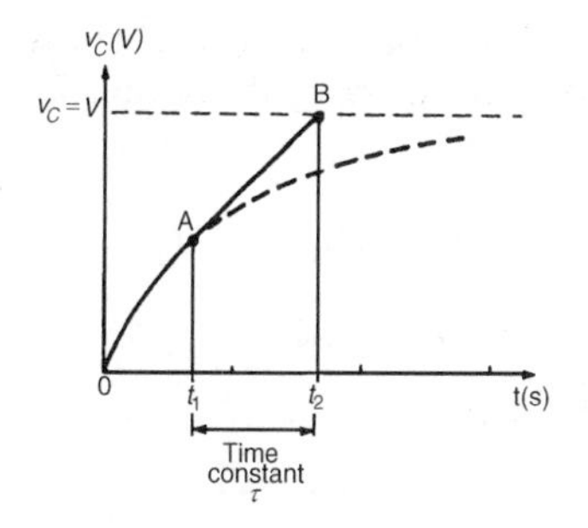

Figure 17.3

(b) With reference to Figure 17.3, let the constant voltage supply be replaced by a variable voltage supply at time t_1 seconds. Let the voltage be varied so that the **current** flowing in the circuit is **constant**.

(c) Since the current flowing is a constant, the curve will follow a tangent, AB, drawn to the curve at point A.

(d) Let the capacitor voltage v_C reach its final value of V at time t_2 seconds.

(e) The time corresponding to $(t_2 - t_1)$ seconds is called the **time constant** of the circuit, denoted by the Greek letter 'tau', τ. The value of the time constant is CR seconds, i.e., for a series connected C–R circuit,

time constant $\tau = CR$ seconds

Since the variable voltage mentioned in para (b) above can be applied to any instant during the transient change, it may be applied at $t = 0$, i.e., at the instant of connecting the circuit to the supply. If this is done, then the time constant of the circuit may be defined as:

'*the time taken for a transient to reach its final state if the initial rate of change is maintained*'.

17.4 Transient curves for a C–R circuit

There are two main methods of drawing transient curves graphically, these being:

(a) the **tangent method** — this method is shown in Problem 1 below and

(b) the **initial slope and three point method**, which is shown in Problem 2, and is based on the following properties of a transient exponential curve:

(i) for a growth curve, the value of a transient at a time equal to one time constant is 0.632 of its steady state value (usually taken as 63% of the steady state value), at a time equal to two and a half time constants is 0.918 if its steady state value (usually taken as 92% of its steady state value) and at a time equal to five time constants is equal to its steady state value,

(ii) for a decay curve, the value of a transient at a time equal to one time constant is 0.368 of its initial value (usually taken as 37% of its initial value), at a time equal to two and a half time constants is 0.082 of its initial value (usually taken as 8% of its initial value) and at a time equal to five time constants is equal to zero.

The transient curves shown in Figure 17.2 have mathematical equations, obtained by solving the differential equations representing the circuit. The equations of the curves are:

growth of capacitor voltage, $v_C = V(1 - e^{-t/CR}) = V(1 - e^{-t/\tau})$

decay of resistor voltage, $v_R = Ve^{-t/CR} = Ve^{-t/\tau}$ and

decay of current flowing, $i = Ie^{-t/CR} = Ie^{-t/\tau}$

Problem 1. A 15 μF uncharged capacitor is connected in series with a 47 kΩ resistor across a 120 V, d.c. supply. Use the tangential graphical method to draw the capacitor voltage/time characteristic of the circuit. From the characteristic, determine the capacitor voltage at a time equal to one time constant after being connected to the supply, and also two seconds after being connected to the supply. Also, find the time for the capacitor voltage to reach one half of its steady state value.

To construct an exponential curve, the time constant of the circuit and steady state value need to be determined.

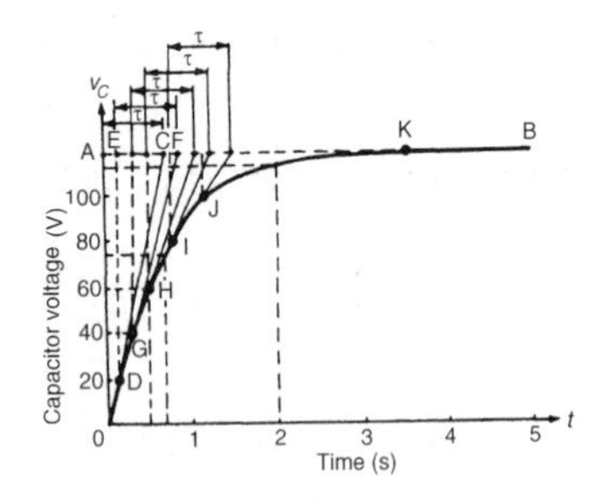

Figure 17.4

$$\text{Time constant} = CR = 15\ \mu\text{F} \times 47\ \text{k}\Omega = 15 \times 10^{-6} \times 47 \times 10^3$$
$$= 0.705\ \text{s}$$

Steady state value of $v_C = V$, i.e. $v_C = 120$ V.

With reference to Figure 17.4, the scale of the horizontal axis is drawn so that it spans at least five time constants, i.e. 5×0.705 or about 3.5 seconds. The scale of the vertical axis spans the change in the capacitor voltage, that is, from 0 to 120 V. A broken line AB is drawn corresponding to the final value of v_C.

Point C is measured along AB so that AC is equal to 1τ, i.e., AC = 0.705 s. Straight line OC is drawn. Assuming that about five intermediate points are needed to draw the curve accurately, a point D is selected on OC corresponding to a v_C value of about 20 V. DE is drawn vertically. EF is made to correspond to 1τ, i.e. EF = 0.705 s. A straight line is drawn joining DF. This procedure of

(a) drawing a vertical line through point selected,

(b) at the steady-state value, drawing a horizontal line corresponding to 1τ, and

(c) joining the first and last points,

is repeated for v_C values of 40, 60, 80 and 100 V, giving points G, H, I and J.

The capacitor voltage effectively reaches its steady-state value of 120 V after a time equal to five time constants, shown as point K. Drawing a smooth curve through points O, D, G, H, I, J and K gives the exponential growth curve of capacitor voltage.

From the graph, the value of capacitor voltage at a time equal to the time constant is about **75 V**. It is a characteristic of all exponential growth curves, that after a time equal to one time constant, the value of the transient is 0.632 of its steady-state value. In this problem, $0.632 \times 120 = 75.84$ V. Also from the graph, when t is two seconds, v_C is about **115 Volts**. [This value may be checked using the equation $v_C(1 - e^{-t/\tau})$, where $V = 120$ V, $\tau = 0.705$ s and $t = 2$ s. This calculation gives $v_C = 112.97$ V.]

The time for v_C to rise to one half of its final value, i.e. 60 V, can be determined from the graph and is about **0.5 s**. [This value may be checked using $v_C = V(1 - e^{-t/\tau})$ where $V = 120$ V, $v_C = 60$ V and $\tau = 0.705$ s, giving $t = 0.489$ s.]

Problem 2. A 4 μF capacitor is charged to 24 V and then discharged through a 220 kΩ resistor. Use the 'initial slope and three point' method to draw: (a) the capacitor voltage/time characteristic, (b) the resistor voltage/time characteristic and (c) the current/time characteristic, for the transients which occur. From the characteristics determine the value of capacitor voltage, resistor voltage and current one and a half seconds after discharge has started.

To draw the transient curves, the time constant of the circuit and steady state values are needed.

Time constant, $\tau = CR = 4 \times 10^{-6} \times 220 \times 10^3 = 0.88$ s

Initially, capacitor voltage $v_C = v_R = 24$ V, $i = \frac{V}{R} = \frac{24}{220 \times 10^3}$

$= 0.109$ mA

Finally, $v_C = v_R = i = 0$

(a) The exponential decay of capacitor voltage is from 24 V to 0 V in a time equal to five time constants, i.e., $5 \times 0.88 = 4.4$ s. With reference to Figure 17.5, to construct the decay curve:

(i) the horizontal scale is made so that it spans at least five time constants, i.e. 4.4 s,

(ii) the vertical scale is made to span the change in capacitor voltage, i.e., 0 to 24 V,

(iii) point A corresponds to the initial capacitor voltage, i.e, 24 V,

(iv) OB is made equal to one time constant and line AB is drawn. This gives the initial slope of the transient,

(v) the value of the transient after a time equal to one time constant is 0.368 of the initial value, i.e. $0.368 \times 24 =$

8.83 V; a vertical line is drawn through B and distance BC is made equal to 8.83 V,

(vi) the value of the transient after a time equal to two and a half time constants is 0.082 of the initial value, i.e., $0.082 \times 24 = 1.97$ V, shown as point D in Figure 17.5,

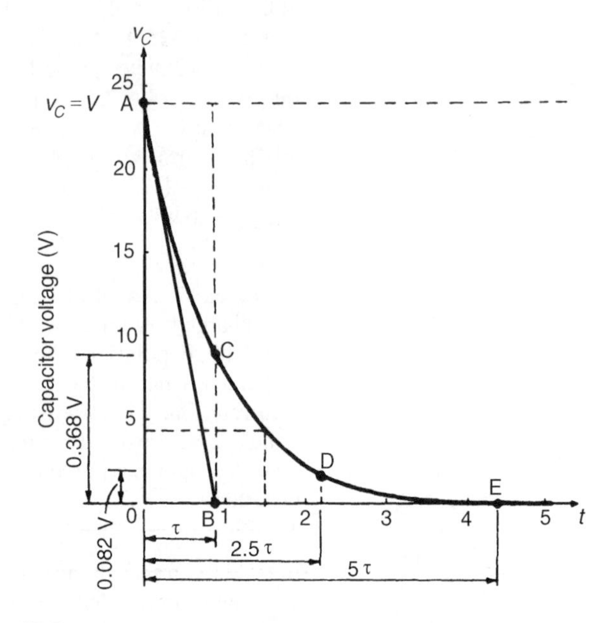

Figure 17.5

(vii) the transient effectively dies away to zero after a time equal to five time constants, i.e., 4.4 s, giving point E.

The smooth curve drawn through points A, C, D and E represents the decay transient. At $1\frac{1}{2}$ s after decay has started, $v_C \approx$ **4.4 V**. [This may be checked using $v_C = Ve^{-t/\tau}$, where $V = 24$, $t = 1\frac{1}{2}$ and $\tau = 0.88$, giving $v_C = 4.36$ V]

(b) The voltage drop across the resistor is equal to the capacitor voltage when a capacitor is discharging through a resistor, thus the resistor voltage/time characteristic is identical to that shown in Figure 17.5.

Since $v_R = v_C$, then at $1\frac{1}{2}$ seconds after decay has started, $v_R \approx$ **4.4 V** (see (vii) above).

(c) The current/time characteristic is constructed in the same way as the capacitor voltage/time characteristic, shown in part (a) of this problem, and is as shown in Figure 17.6. The values are:

point A: initial value of current = 0.109 mA

point C: at 1τ, $i = 0.368 \times 0.109 = 0.040$ mA

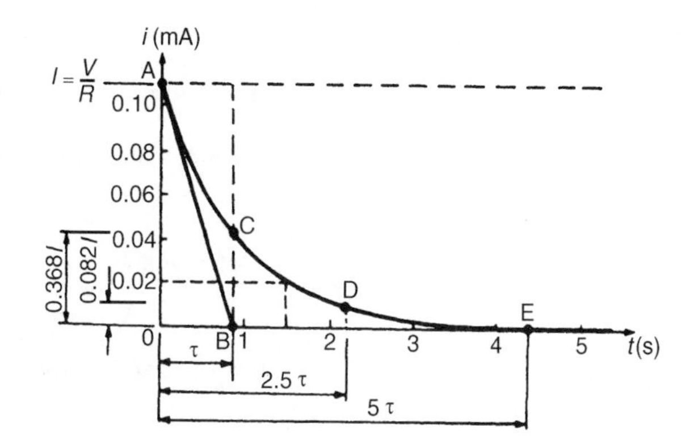

Figure 17.6

point D: at 2.5τ, $i = 0.082 \times 0.109 = 0.009$ mA

point E: at 5τ, $i = 0$

Hence the current transient is as shown. At a time of $1\frac{1}{2}$ seconds, the value of current, from the characteristic is **0.02 mA**. [This may be checked using $i = Ie^{(-t/\tau)}$ where $I = 0.109$, $t = 1\frac{1}{2}$ and $\tau = 0.88$, giving $i = 0.0198$ mA or 19.8 μA]

Problem 3. A 20 μF capacitor is connected in series with a 50 kΩ resistor and the circuit is connected to a 20 V, d.c. supply. Determine

(a) the initial value of the current flowing,
(b) the time constant of the circuit,
(c) the value of the current one second after connection,
(d) the value of the capacitor voltage two seconds after connection, and
(e) the time after connection when the resistor voltage is 15 V

Parts (c), (d) and (e) may be determined graphically, as shown in Problems 1 and 2 or by calculation as shown below.

$V = 20$ V, $C = 20\ \mu\text{F} = 20 \times 10^{-6}$ F, $R = 50\ \text{k}\Omega = 50 \times 10^{3}$ V

(a) The initial value of the current flowing is

$$I = \frac{V}{R}, \text{ i.e. } \frac{20}{50 \times 10^3} = \mathbf{0.4\ mA}$$

(b) From Section 17.3 the time constant,

$$\tau = CR = (20 \times 10^{-6}) \times (50 \times 10^3) = \mathbf{1\ s}$$

(c) Current, $i = Ie^{-t/\tau}$

Working in mA units, $i = 0.4e^{-1/1} = 0.4 \times 0.368 =$ **0.147 mA**

(d) Capacitor voltage, $v_C = V(1 - e^{-t/\tau}) = 20(1 - e^{-2/1})$

$= 20(1 - 0.135) = 20 \times 0.865 =$ **17.3 V**

(e) Resistor voltage, $v_R = Ve^{-t/\tau}$

Thus $15 = 20e^{-t/1}$, $\frac{15}{20} = e^{-t}$, i.e. $e^t = \frac{20}{15} = \frac{4}{3}$

Taking natural logarithms of each side of the equation gives

$$t = \ln\frac{4}{3} = \ln 1.3333$$

i.e, time, $t = 0.288$ s

Problem 4. A circuit consists of a resistor connected in series with a 0.5 μF capacitor and has a time constant of 12 ms. Determine (a) the value of the resistor, and (b) the capacitor voltage 7 ms after connecting the circuit to a 10 V supply

(a) The time constant $\tau = CR$, hence $R = \dfrac{\tau}{C}$

i.e. $R = \dfrac{12 \times 10^{-3}}{0.5 \times 10^{-6}} = 24 \times 10^3 =$ **24 kΩ**

(b) The equation for the growth of capacitor voltage is:

$$v_C = V(1 - e^{-t/\tau})$$

Since $\tau = 12$ ms $= 12 \times 10^{-3}$ s, $V = 10$ V and

$t = 7$ ms $= 7 \times 10^{-3}$ s,

then $v_C = 10\left[1 - e^{-\frac{7\times10^{-3}}{12\times10^{-3}}}\right] = 10(1 - e^{-0.583})$

$= 10(1 - 0.558) =$ **4.42 V**

Alternatively, the value of v_C when t is 7 ms may be determined using the growth characteristic as shown in Problem 1.

17.5 Discharging a capacitor

When a capacitor is charged (i.e. with the switch in position A in Figure 17.7), and the switch is then moved to position B, the electrons stored in the capacitor keep the current flowing for a short time. Initially, at the instant of moving from A to B, the current flow is such that the capacitor voltage v_C is balanced by an equal and opposite voltage $v_R = iR$. Since initially $v_C = v_R = V$, then $i = I = V/R$. During the transient

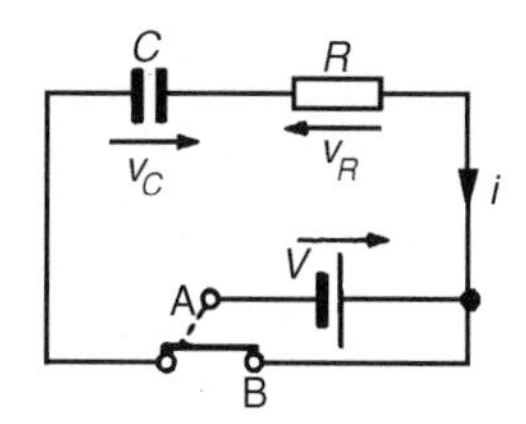

Figure 17.7

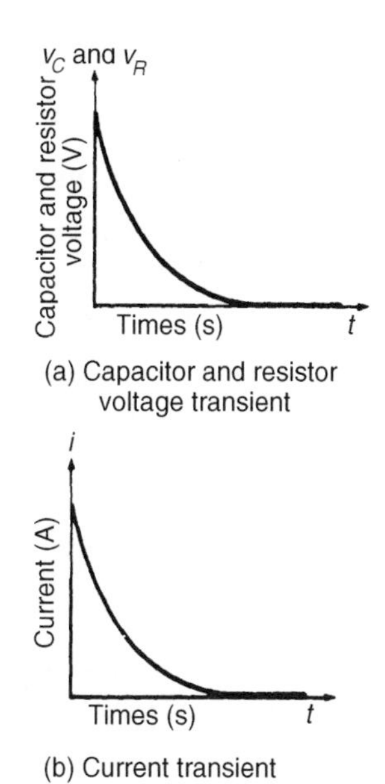

Figure 17.8

decay, by applying Kirchhoff's voltage law to Figure 17.7, $v_C = v_R$. Finally the transients decay exponentially to zero, i.e. $v_C = v_R = 0$. The transient curves representing the voltages and current are as shown in Figure 17.8.

The equations representing the transient curves during the discharge period of a series connected C–R circuit are:

decay of voltage, $v_C = v_R = Ve^{(-t/CR)} = Ve^{(-t/\tau)}$

decay of current, $i = Ie^{(-t/CR)} = Ie^{(-t/\tau)}$

When a capacitor has been disconnected from the supply it may still be charged and it may retain this charge for some considerable time. Thus precautions must be taken to ensure that the capacitor is automatically discharged after the supply is switched off. This is done by connecting a high value resistor across the capacitor terminals.

> Problem 5. A capacitor is charged to 100 V and then discharged through a 50 kΩ resistor. If the time constant of the circuit is 0.8 s, determine: (a) the value of the capacitor, (b) the time for the capacitor voltage to fall to 20 V, (c) the current flowing when the capacitor has been discharging for 0.5 s, and (d) the voltage drop across the resistor when the capacitor has been discharging for one second.

Parts (b), (c) and (d) of this problem may be solved graphically as shown in Problems 1 and 2 or by calculation as shown below.

$V = 100$ V, $\tau = 0.8$ s, $R = 50\ k\Omega = 50 \times 10^3\ \Omega$

(a) Since time constant, $\tau = CR$, $C = \tau/R$

$$\text{i.e. } C = \frac{0.8}{50 \times 10^3} = \mathbf{16\ \mu F}$$

(b) $v_C = Ve^{-t/\tau}$

$20 = 100e^{-t/0.8}$, i.e. $\frac{1}{5} = e^{-t/0.8}$

Thus $e^{t/0.8} = 5$ and taking natural logarithms of each side, gives

$$\frac{t}{0.8} = \ln 5, \text{ i.e., } t = 0.8 \ln 5$$

Hence $t = 1.29$ s

(c) $i = Ie^{-t/\tau}$

The initial current flowing, $I = \dfrac{V}{R} = \dfrac{100}{50 \times 10^3} = 2$ mA

Working in mA units, $i = Ie^{-t/\tau} = 2e^{(-0.5/0.8)} = 2e^{-0.625}$

$= 2 \times 0.535 = \mathbf{1.07\ mA}$

(d) $v_R = v_C = Ve^{-t/\tau}$

$$= 100e^{-1/0.8} = 100e^{-1.25}$$

$$= 100 \times 0.287 = \mathbf{28.7\ V}$$

Problem 6. A 0.1 μF capacitor is charged to 200 V before being connected across a 4 kΩ resistor. Determine (a) the initial discharge current, (b) the time constant of the circuit, and (c) the minimum time required for the voltage across the capacitor to fall to less than 2 V

(a) Initial discharge current, $i = \frac{V}{R} = \frac{200}{4 \times 10^3} = \mathbf{0.05\ A\ or\ 50\ mA}$

(b) Time constant $\tau = CR = 0.1 \times 10^{-6} \times 4 \times 10^3$

$$= \mathbf{0.0004\ s\ or\ 0.4\ ms}$$

(c) The minimum time for the capacitor voltage to fall to less than 2 V, i.e., less than or $\frac{2}{200}$ or 1% of the initial value is given by 5τ.

$5\tau = 5 \times 0.4 = \mathbf{2\ ms}$

In a d.c. circuit, a capacitor blocks the current except during the times that there are changes in the supply voltage.

Further problems on transients in series connected C-R circuits may be found in Section 17.12, problems 1 to 8, page 276.

17.6 Current growth in an L–R circuit

(a) The circuit diagram for a series connected L–R circuit is shown in Figure 17.9. When switch S is closed, then by Kirchhoff's voltage law:

$$V = v_L + v_R \qquad (17.3)$$

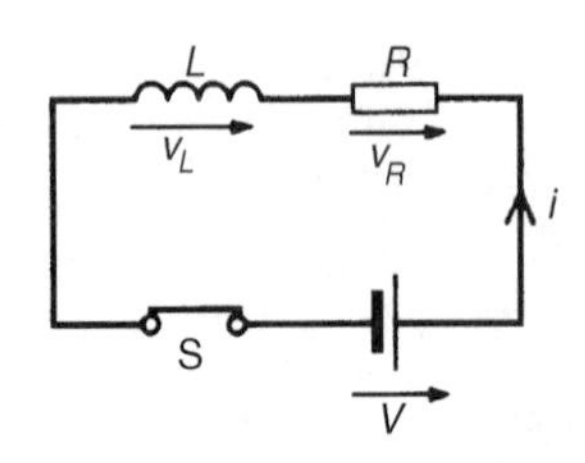

Figure 17.9

(b) The battery voltage V is constant. The voltage across the inductance is the induced voltage, i.e.

$$v_L = L \times \frac{\text{change of current}}{\text{change of time}} = L\frac{di}{dt}$$

The voltage drop across R, v_R is given by iR. Hence, at all times:

$$V = L(di/dt) + iR \qquad (17.4)$$

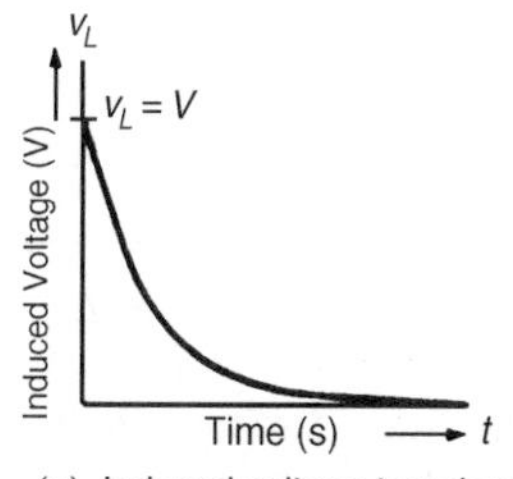

(a) Induced voltage transient

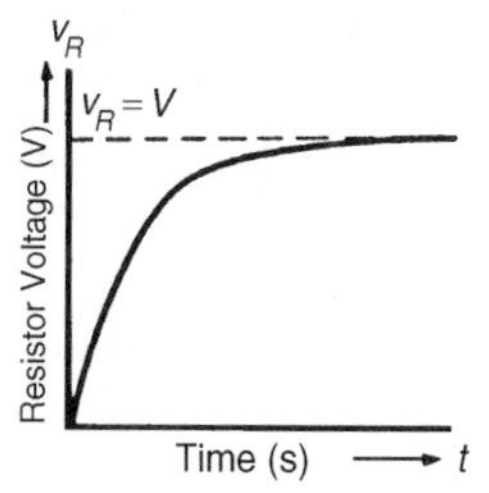

(b) Resistor voltage transient

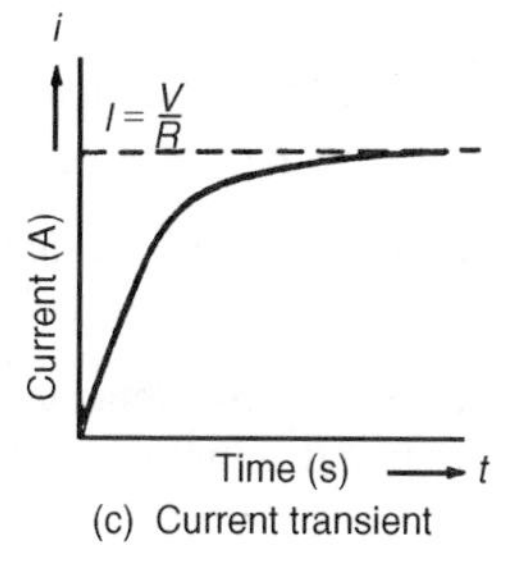

(c) Current transient

Figure 17.10

(c) At the instant of closing the switch, the rate of change of current is such that it induces an e.m.f. in the inductance which is equal and opposite to V, hence $V = v_L + 0$, i.e. $v_L = V$. From equation (17.3), because $v_L = V$, then $v_R = 0$ and $i = 0$

(d) A short time later at time t_1 seconds after closing S, current i_1 is flowing, since there is a rate of change of current initially, resulting in a voltage drop of i_1R across the resistor. Since V (constant) $= v_L + v_R$ the induced e.m.f. is reduced, and equation (17.4) becomes:

$$V = L\frac{di_1}{dt_1} + i_1R$$

(e) A short time later still, say at time t_2 seconds after closing the switch, the current flowing is i_2, and the voltage drop across the resistor increases to i_2R. Since v_R increases, v_L decreases.

(f) Ultimately, a few seconds after closing S, the current flow is entirely limited by R, the rate of change of current is zero and hence v_L is zero. Thus $V = iR$. Under these conditions, steady state current flows, usually signified by I. Thus, $I = V/R$, $v_R = IR$ and $v_L = 0$ at steady state conditions.

(g) Curves showing the changes in v_L, v_R and i with time are shown in Figure 17.10 and indicate that v_L is a maximum value initially (i.e equal to V), decaying exponentially to zero, whereas v_R and i grow exponentially from zero to their steady state values of V and $I = V/R$ respectively.

17.7 Time constant for an L–R circuit

With reference to Section 17.3, the time constant of a series connected L–R circuit is defined in the same way as the time constant for a series connected C–R circuit. Its value is given by:

time constant, $\tau = L/R$ seconds

17.8 Transient curves for an L–R circuit

Transient curves representing the induced voltage/time, resistor voltage/time and current/time characteristics may be drawn graphically, as outlined in Section 17.4. A method of construction is shown in Problem 7. Each of the transient curves shown in Figure 17.10 have mathematical equations, and these are:

decay of induced voltage, $v_L = Ve^{(-Rt/L)} = Ve^{(-t/\tau)}$

growth of resistor voltage, $v_R = V(1 - e^{-Rt/L}) = V(1 - e^{-t/\tau})$

growth of current flow, $i = I(1 - e^{-Rt/L}) = I(1 - e^{-t/\tau})$

The application of these equations is shown in Problem 9.

> Problem 7. A relay has an inductance of 100 mH and a resistance of 20 Ω. It is connected to a 60 V, d.c. supply. Use the 'initial slope and three point' method to draw the current/time characteristic and hence determine the value of current flowing at a time equal to two time constants and the time for the current to grow to 1.5 A

Before the current/time characteristic can be drawn, the time constant and steady-state value of the current have to be calculated.

$$\text{Time constant, } \tau = \frac{L}{R} = \frac{100 \times 10^{-3}}{20} = 5 \text{ ms}$$

$$\text{Final value of current, } I = \frac{V}{R} = \frac{60}{20} = 3 \text{ A}$$

The method used to construct the characteristic is the same as that used in Problem 2.

(a) The scales should span at least five time constants (horizontally), i.e. 25 ms, and 3 A (vertically).

(b) With reference to Figure 17.11, the initial slope is obtained by making AB equal to 1 time constant, (5 ms), and joining OB.

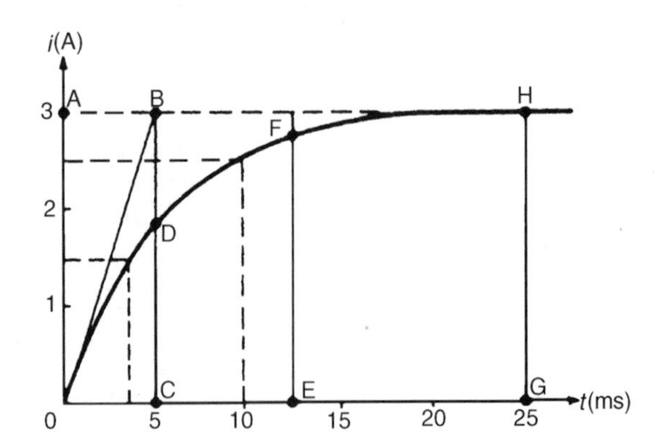

Figure 17.11

(c) At a time of 1 time constant, CD is $0.632 \times I = 0.632 \times 3 = 1.896$ A

At a time of 2.5 time constants, EF is $0.918 \times I = 0.918 \times 3 = 2.754$ A

At a time of 5 time constants, GH is $I = 3$ A

(d) A smooth curve is drawn through points O, D, F and H and this curve is the current/time characteristic.

From the characteristic, when $t = 2\tau$, $i \approx$ **2.6 A**. [This may be checked by calculation using $i = I(1 - e^{-t/\tau})$, where $I = 3$ and $t = 2\tau$, giving $i = 2.59$ A]. Also, when the current is 1.5 A, the corresponding time is about **3.6 ms**. [This may be checked by calculation, using $i = I(1 - e^{-t/\tau})$ where $i = 1.5$, $I = 3$ and $\tau = 5$ ms, giving $t = 3.466$ ms.]

Problem 8. A coil of inductance 0.04 H and resistance 10 Ω is connected to a 120 V, d.c. supply. Determine (a) the final value of current, (b) the time constant of the circuit, (c) the value of current after a time equal to the time constant from the instant the supply voltage is connected, (d) the expected time for the current to rise to within 1% of its final value.

(a) Final steady current, $I = \dfrac{V}{R} = \dfrac{120}{10} = \mathbf{12\ A}$

(b) Time constant of the circuit, $\tau = \dfrac{L}{R} = \dfrac{0.04}{10}$

$= \mathbf{0.004\ s\ or\ 4\ ms}$

(c) In the time τ s the current rises to 63.2% of its final value of 12 A, i.e. in 4 ms the current rises to $0.632 \times 12 =$ **7.58 A**

(d) The expected time for the current to rise to within 1% of its final value is given by 5 τ s, i.e. $5 \times 4 =$ **20 ms**

Problem 9. The winding of an electromagnet has an inductance of 3 H and a resistance of 15 Ω. When it is connected to a 120 V, d.c. supply, calculate:

(a) the steady state value of current flowing in the winding,

(b) the time constant of the circuit,

(c) the value of the induced e.m.f. after 0.1 s,

(d) the time for the current to rise to 85% of its final value, and

(e) the value of the current after 0.3 s

(a) The steady state value of current is $I = V/R$, i.e. $I = 120/15 =$ **8 A**

(b) The time constant of the circuit, $\tau = L/R = 3/15 =$ **0.2 s**

Parts (c), (d) and (e) of this problem may be determined by drawing the transients graphically, as shown in Problem 7 or by calculation as shown below.

(c) The induced e.m.f., v_L is given by $v_L = Ve^{-t/\tau}$. The d.c. voltage V is 120 V, t is 0.1 s and t is 0.2 s, hence

$$v_L = 120e^{-0.1/0.2} = 120e^{-0.5} = 120 \times 0.6065$$

i.e. $v_L =$ **72.78 V**

(d) When the current is 85% of its final value, $i = 0.85I$.

$$\text{Also, } i = I(1 - e^{-t/\tau}), \text{ thus } 0.85I = I(1 - e^{-t/\tau})$$

$$0.85 = 1 - e^{-t/\tau} \text{ and } \tau = 0.2, \text{ hence}$$

$$0.85 = 1 - e^{-t/0.2}$$

$$e^{-t/0.2} = 1 - 0.85 = 0.15$$

$$e^{t/0.2} = \frac{1}{1.15} = 6.\dot{6}$$

Taking natural logarithms of each side of this equation gives:

$$\ln e^{t/0.2} = \ln 6.\dot{6}, \text{ and by the laws of logarithms}$$

$$\frac{t}{0.2} \ln e = \ln 6.\dot{6}. \text{ But } \ln e = 1, \text{ hence}$$

$$t = 0.2 \ln 6.\dot{6} \text{ i.e. } t = \mathbf{0.379\ s}$$

(e) The current at any instant is given by $i = I(1 - e^{-t/\tau})$

When $I = 8$, $t = 0.3$ and $\tau = 0.2$, then

$$i = 8(1 - e^{-0.3/0.2}) = 8(1 - e^{-1.5})$$

$$= 8(1 - 0.2231) = 8 \times 0.7769 \text{ i.e., } i = \mathbf{6.215\ A}$$

17.9 Current decay in an L–R circuit

When a series connected L–R circuit is connected to a d.c. supply as shown with S in position A of Figure 17.12, a current $I = V/R$ flows after a short time, creating a magnetic field ($\Phi \propto I$) associated with the inductor. When S is moved to position B, the current value decreases, causing a decrease in the strength of the magnetic field. Flux linkages occur, generating a voltage v_L, equal to $L(di/dt)$. By Lenz's law, this voltage keeps current i flowing in the circuit, its value being limited by R. Thus $v_L = v_R$. The current decays exponentially to zero and since v_R is proportional to the current flowing, v_R decays exponentially to zero. Since $v_L = v_R$, v_L also decays exponentially to zero. The curves representing these transients are similar to those shown in Figure 17.8.

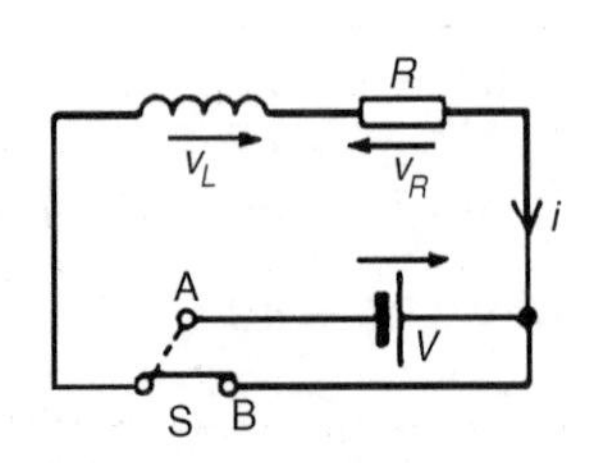

Figure 17.12

The equations representing the decay transient curves are:

decay of voltages, $v_L = v_R = Ve^{(-Rt/L)} = Ve^{(-t/\tau)}$

decay of current, $i = Ie^{(-Rt/L)} = Ie^{(-t/\tau)}$

> Problem 10. The field winding of a 110 V, d.c. motor has a resistance of 15 Ω and a time constant of 2 s. Determine the inductance and use the tangential method to draw the current/time characteristic when the supply is removed and replaced by a shorting link. From the characteristic determine (a) the current flowing in the winding 3 s after being shorted-out and (b) the time for the current to decay to 5 A.

Since the time constant, $\tau = \dfrac{L}{R}$, $\quad L = R\tau$

i.e. inductance $L = 15 \times 2 = \mathbf{30\ H}$

The current/time characteristic is constructed in a similar way to that used in Problem 1.

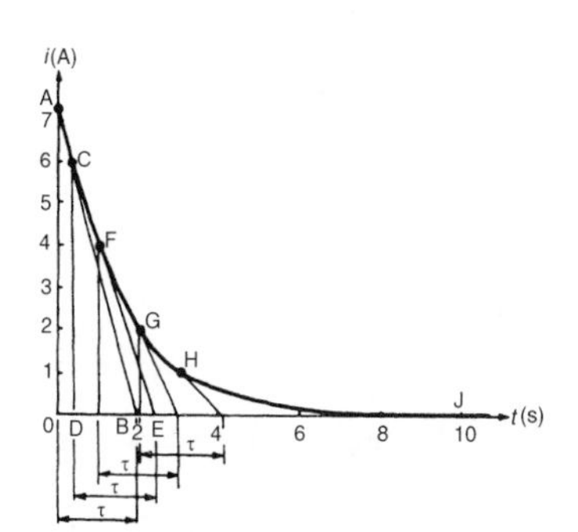

Figure 17.13

(i) The scales should span at least five time constants horizontally, i.e. 10 s, and $I = V/R = 110/15 = 7.\dot{3}$ A vertically.

(ii) With reference to Figure 17.13, the initial slope is obtained by making OB equal to 1 time constant, (2 s), and joining AB.

(iii) At, say, $i = 6$ A, let C be the point on AB corresponding to a current of 6 A. Make DE equal to 1 time constant, (2 s), and join CE.

(iv) Repeat the procedure given in (iii) for current values of, say, 4 A, 2 A and 1 A, giving points F, G and H.

(v) Point J is at five time constants, when the value of current is zero.

(vi) Join points A, C, F, G, H and J with a smooth curve. This curve is the current/time characteristic.

(a) From the current/time characteristic, when $t = 3$ s, $i =$ **1.5 A**. [This may be checked by calculation using $i = Ie^{-t/\tau}$, where $I = 7.\dot{3}$, $t = 3$ and $\tau = 2$, giving $i = 1.64$ A.] The discrepancy between the two results is due to relatively few values, such as C, F, G and H, being taken.

(b) From the characteristic, when $i = 5$ A, $\boldsymbol{t = 0.70}$ s. [This may be checked by calculation using $i = Ie^{-t/\tau}$, where $i = 5$, $I = 7.\dot{3}$, $\tau = 2$, giving $t = 0.766$ s.] Again, the discrepancy between the graphical and calculated values is due to relatively few values such as C, F, G and H being taken.

> Problem 11. A coil having an inductance of 6 H and a resistance of R Ω is connected in series with a resistor of 10 Ω to a 120 V, d.c. supply. The time constant of the circuit is 300 ms. When steady-state conditions have been reached, the supply is replaced instantaneously by a short-circuit. Determine: (a) the resistance of the coil,(b) the current flowing in the circuit one second after the shorting link has been placed in the circuit, and (c) the time taken for the current to fall to 10% of its initial value.

(a) The time constant, $\tau = \dfrac{\text{circuit inductance}}{\text{total circuit resistance}} = \dfrac{L}{R+10}$

Thus $$R = \frac{L}{\tau} - 10 = \frac{6}{0.3} - 10 = \mathbf{10\ \Omega}$$

Parts (b) and (c) may be determined graphically as shown in Problems 7 and 10 or by calculation as shown below.

(b) The steady-state current, $I = \dfrac{V}{R} = \dfrac{120}{10+10} = 6$ A

The transient current after 1 second, $i = Ie^{-t/\tau} = 6e^{-1/0.3}$

Thus $i = 6e^{-3.\dot{3}} = 6 \times 0.03567 = \mathbf{0.214\ A}$

(c) 10% of the initial value of the current is $(10/100) \times 6$, i.e. 0.6 A

Using the equation $i = Ie^{-t/\tau}$ gives

$$0.6 = 6e^{-t/0.3}$$

i.e. $$\frac{0.6}{6} = e^{-t/0.3} \text{ or } e^{t/0.3} = \frac{6}{0.6} = 10$$

Taking natural logarithms of each side of this equation gives:

$$\frac{t}{0.3} = \ln 10$$

$$t = 0.3 \ln 10 = \mathbf{0.691\ s}$$

Problem 12. An inductor has a negligible resistance and an inductance of 200 mH and is connected in series with a 1 kΩ resistor to a 24 V, d.c. supply. Determine the time constant of the circuit and the steady-state value of the current flowing in the circuit. Find (a) the current flowing in the circuit at a time equal to one time constant, (b) the voltage drop across the inductor at a time equal to two time constants and (c) the voltage drop across the resistor after a time equal to three time constants.

The time constant, $\tau = \dfrac{L}{R} = \dfrac{0.2}{1000} = \mathbf{0.2\ ms}$

The steady-state current $I = \dfrac{V}{R} = \dfrac{24}{1000} = \mathbf{24\ mA}$

(a) The transient current, $i = I(1 - e^{-t/\tau})$ and $t = 1\tau$

Working in mA units gives, $i = 24(1 - e^{-(1\tau/\tau)}) = 24(1 - e^{-1})$

$= 24(1 - 0.368) = \mathbf{15.17\ mA}$

(b) The voltage drop across the inductor, $v_L = Ve^{-t/\tau}$

When $t = 2\tau$, $v_L = 24e^{-2\tau/\tau} = 24e^{-2}$

$= \mathbf{3.248\ V}$

(c) The voltage drop across the resistor, $v_R = V(1 - e^{-t/\tau})$

When $t = 3\tau$, $v_R = 24(1 - e^{-3\tau/\tau}) = 24(1 - e^{-3})$

$= \mathbf{22.81\ V}$

Further problems on transients in series L–R circuits may be found in Section 17.12, problems 9 to 12, page 277.

17.10 Switching inductive circuits

Energy stored in the magnetic field of an inductor exists because a current provides the magnetic field. When the d.c. supply is switched off the current falls rapidly, the magnetic field collapses causing a large induced e.m.f. which will either cause an arc across the switch contacts or will break down the insulation between adjacent turns of the coil. The high induced e.m.f. acts in a direction which tends to keep the current flowing, i.e. in the same direction as the applied voltage. The energy from the magnetic field will thus be aided by the supply voltage in maintaining an arc, which could cause severe damage to the switch. To reduce the induced e.m.f. when the supply switch is opened, a discharge resistor R_D is connected in parallel with the inductor as shown in Figure 17.14. The magnetic field energy is dissipated as heat in R_D and R and arcing at the switch contacts is avoided.

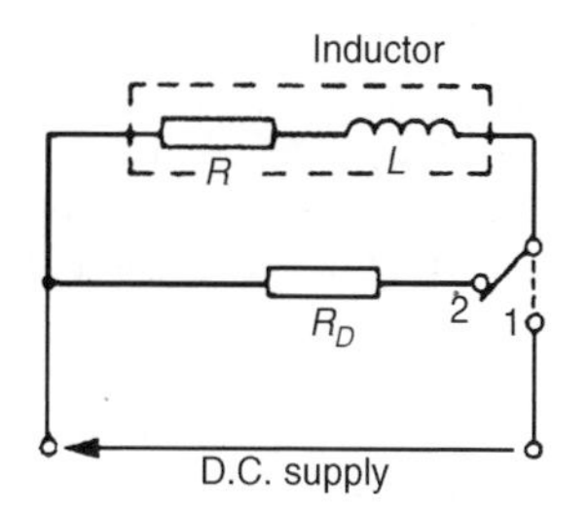

Figure 17.14

17.11 The effects of time constant on a rectangular waveform

Integrator circuit

By varying the value of either C or R in a series connected C–R circuit, the time constant ($\tau = CR$), of a circuit can be varied. If a rectangular waveform varying from $+E$ to $-E$ is applied to a C–R circuit as shown in Figure 17.15, output waveforms of the capacitor voltage have various shapes, depending on the value of R. When R is small, $t = CR$ is small and an output waveform such as that shown in Figure 17.16(a) is obtained. As the value of R is increased, the waveform changes to that shown in Figure 17.16(b). When R is large, the waveform is as shown in Figure 17.16(c), the circuit then being described as an **integrator circuit**.

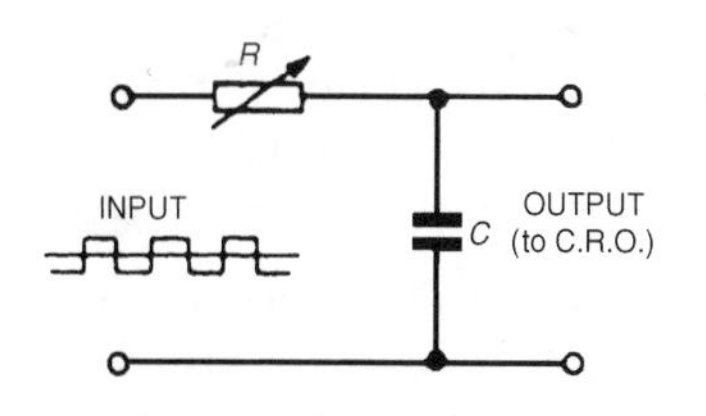

Figure 17.15

Differentiator circuit

If a rectangular waveform varying from $+E$ to $-E$ is applied to a series connected C–R circuit and the waveform of the voltage drop across the resistor is observed, as shown in Figure 17.17, the output waveform alters as R is varied due to the time constant, ($\tau = CR$), altering. When R is small, the waveform is as shown in Figure 17.18(a), the voltage being generated across R by the capacitor discharging fairly quickly. Since the

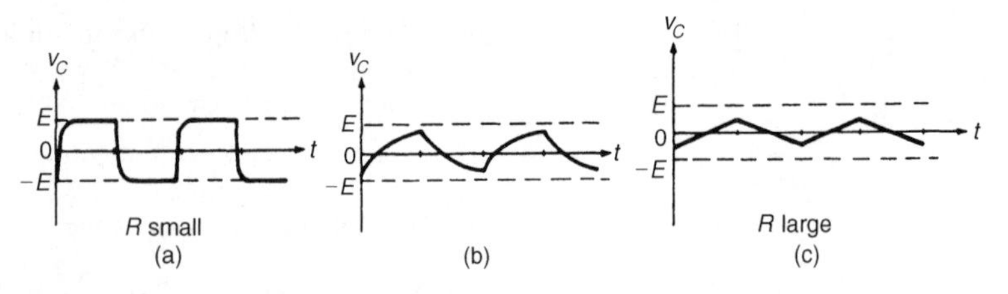

Figure 17.16

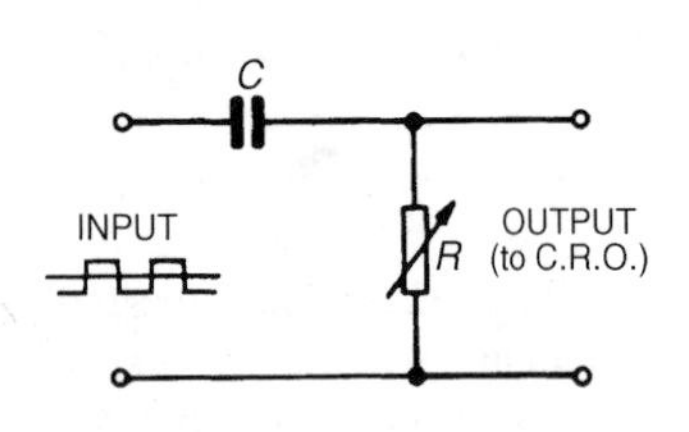

Figure 17.17

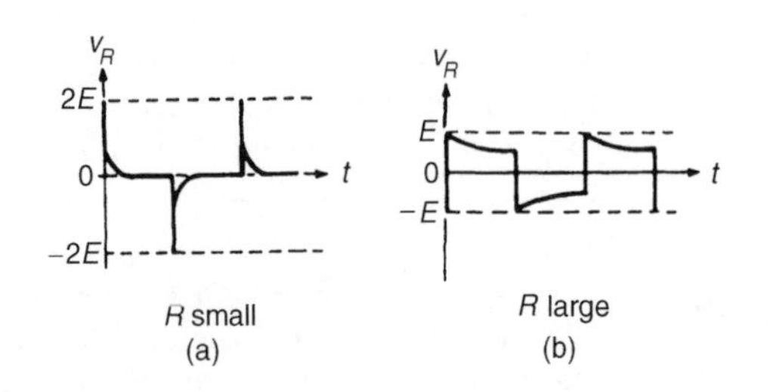

Figure 17.18

change in capacitor voltage is from $+E$ to $-E$, the change in discharge current is $2E/R$, resulting in a change in voltage across the resistor of $2E$. This circuit is called a **differentiator circuit**. When R is large, the waveform is as shown in Figure 17.18(b).

17.12 Further problems on d.c. transients

Transients in series connected C–R circuits

1 An uncharged capacitor of 0.2 μF is connected to a 100 V, d.c. supply through a resistor of 100 kΩ. Determine, either graphically or by calculation the capacitor voltage 10 ms after the voltage has been applied. [39.35 V]

2 A circuit consists of an uncharged capacitor connected in series with a 50 kΩ resistor and has a time constant of 15 ms. Determine either graphically or by calculation (a) the capacitance of the capacitor and (b) the voltage drop across the resistor 5 ms after connecting the circuit to a 20 V, d.c. supply. [(a) 0.3 μF, (b) 14.33 V]

3 A 10 μF capacitor is charged to 120 V and then discharged through a 1.5 MΩ resistor. Determine either graphically or by calculation the capacitor voltage 2 s after discharging has commenced. Also find how long it takes for the voltage to fall to 25 V. [105.0 V, 23.53 s]

4 A capacitor is connected in series with a voltmeter of resistance 750 kΩ and a battery. When the voltmeter reading is steady the battery is replaced with a shorting link. If it takes 17 s for the voltmeter reading to fall to two-thirds of its original value, determine the capacitance of the capacitor. [55.9 μF]

5 When a 3 μF charged capacitor is connected to a resistor, the voltage falls by 70% in 3.9 s. Determine the value of the resistor.
[1.08 MΩ]

6 A 50 μF uncharged capacitor is connected in series with a 1 kΩ resistor and the circuit is switched to a 100 V, d.c. supply. Determine:

(a) the initial current flowing in the circuit,

(b) the time constant,

(c) the value of current when t is 50 ms and

(d) the voltage across the resistor 60 ms after closing the switch.
[(a) 0.1 A (b) 50 ms (c) 36.8 mA (d) 30.1 V]

7 An uncharged 5 μF capacitor is connected in series with a 30 kΩ resistor across a 110 V, d.c. supply. Determine the time constant of the circuit and the initial charging current. Use a graphical method to draw the current/time characteristic of the circuit and hence determine the current flowing 120 ms after connecting to the supply.
[150 ms, 3.67 mA, 1.65 mA]

8 An uncharged 80 μF capacitor is connected in series with a 1 kΩ resistor and is switched across a 110 V supply. Determine the time constant of the circuit and the initial value of current flowing. Derive graphically the current/time characteristic for the transient condition and hence determine the value of current flowing after (a) 40 ms and (b) 80 ms. [80 ms, 0.11 A (a) 66.7 mA (b) 40.5 mA]

Transients in series connected *L–R* circuits

9 A coil has an inductance of 1.2 H and a resistance of 40 Ω and is connected to a 200 V, d.c. supply. Draw the current/time characteristic and hence determine the approximate value of the current flowing 60 ms after connecting the coil to the supply. [4.3 A]

10 A 25 V d.c. supply is connected to a coil of inductance 1 H and resistance 5 Ω. Use a graphical method to draw the exponential growth curve of current and hence determine the approximate value of the current flowing 100 ms after being connected to the supply. [2 A]

11 An inductor has a resistance of 20 Ω and an inductance of 4 H. It is connected to a 50 V d.c. supply. By drawing the appropriate characteristic find (a) the approximate value of current flowing after 0.1 s and (b) the time for the current to grow to 1.5 A.
[(a) 1 A (b) 0.18 s]

12 The field winding of a 200 V d.c. machine has a resistance of 20 Ω and an inductance of 500 mH. Calculate:

(a) the time constant of the field winding,

(b) the value of current flow one time constant after being connected to the supply, and

(c) the current flowing 50 ms after the supply has been switched on. [(a) 25 ms (b) 6.32 A (c) 8.65 A]

18 Operational amplifiers

At the end of this chapter you should be able to:

- recognise the main properties of an operational amplifier
- understand op amp parameters input bias current and offset current and voltage
- define and calculate common-mode rejection ratio
- appreciate slew rate
- explain the principle of operation, draw the circuit diagram symbol and calculate gain for the following operational amplifiers:
  - inverter
  - non-inverter
  - voltage follower (or buffer)
  - summing
  - voltage comparator
  - integrator
  - differentiator
- understand digital to analogue conversion
- understand analogue to digital conversion

18.1 Introduction to operational amplifiers

Operational Amplifiers (usually called **'op amps'**) were originally made from discrete components, being designed to solve mathematical equations electronically, by performing operations such as addition and division in analogue computers. Now produced in integrated-circuit (IC) form, op amps have many uses, with one of the most important being as a high-gain d.c. and a.c. voltage amplifier.

The **main properties** of an op amp include:

(i) a very high open-loop voltage gain A_o of around 10^5 for d.c. and low frequency a.c., which decreases with frequency increase
(ii) a very high input impedance, typically $10^6\ \Omega$ to $10^{12}\ \Omega$, such that current drawn from the device, or the circuit supplying it, is very small and the input voltage is passed on to the op amp with little loss
(iii) a very low output impedance, around $100\ \Omega$, such that its output voltage is transferred efficiently to any load greater than a few kiloohms

The **circuit diagram symbol** for an op amp is shown in Figure 18.1. It has one output, V_o, and two inputs; the **inverting input,** V_1, is marked−, and the non-**inverting input,** V_2, is marked +.

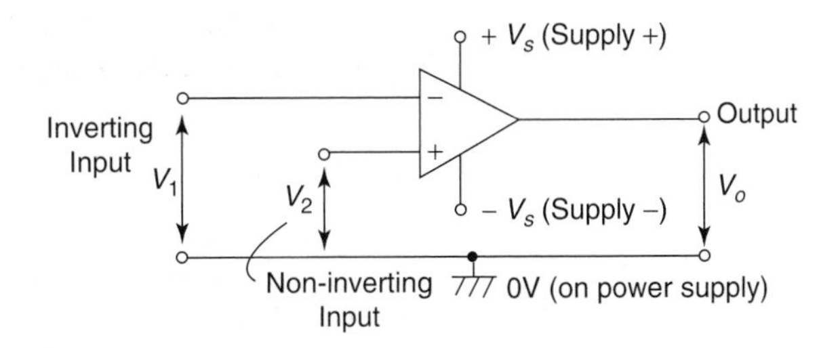

Figure 18.1

The operation of an op amp is most convenient from a dual balanced d.c. power supply $\pm V_S$ (i.e. $+V_S$, 0, $-V_S$); the centre point of the supply, i.e. 0 V, is common to the input and output circuits and is taken as their voltage reference level. The power supply connections are not usually shown in a circuit diagram.

An op amp is basically a **differential** voltage amplifier, i.e. it amplifies the difference between input voltages V_1 and V_2. Three situations are possible:

(i) if $V_2 > V_1$, V_o is positive
(ii) if $V_2 < V_1$, V_o is negative
(iii) if $V_2 = V_1$, V_o is zero

In general, $\boxed{\boldsymbol{V_o = A_o(V_2 - V_1)}}$ or $\boxed{\boldsymbol{A = \frac{V_o}{V_2 - V_1}}}$ (1)

where A_o is the open-loop voltage gain

Problem 1. A differential amplifier has an open-loop voltage gain of 120. The input signals are 2.45 V and 2.35 V. Calculate the output voltage of the amplifier.

From equation (1), **output voltage,**

$$\boldsymbol{V_o} = A_o(V_2 - V_1) = 120(2.45 - 2.35)$$
$$= (120)(0.1) = \mathbf{12\ V}$$

Transfer characteristic

A typical **voltage characteristic** showing how the output V_o varies with the input $(V_2 - V_1)$ is shown in Figure 18.2.

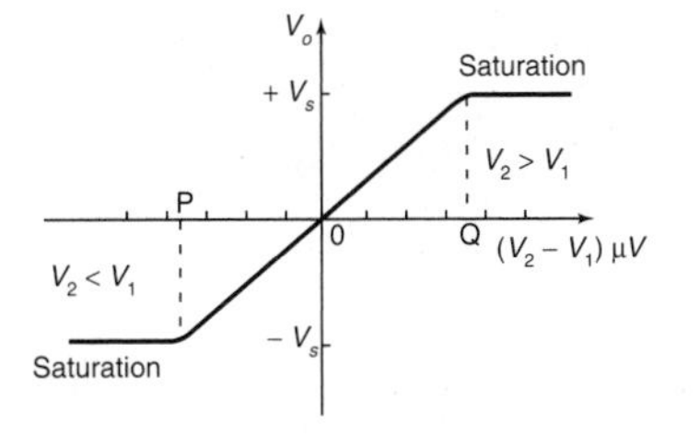

Figure 18.2

It is seen from Figure 18.2 that only within the very small input range P0Q is the output directly proportional to the input; it is in this range that the op amp behaves linearly and there is minimum distortion of the amplifier output. Inputs outside the linear range cause saturation and the output is then close to the maximum value, i.e. $+V_S$ or $-V_S$. The limited linear behaviour is due to the very high open-loop gain A_o, and the higher it is the greater is the limitation.

Negative feedback

Operational amplifiers nearly always use **negative feedback**, obtained by feeding back some, or all, of the output to the inverting ($-$) input (as shown in Figure 18.5 later). The feedback produces an output voltage that opposes the one from which it is taken. This reduces the new output of the amplifier and the resulting closed-loop gain A is then less than the open-loop gain A_o. However, as a result, a wider range of voltages can be applied to the input for amplification. As long as $A_o \gg A$, negative feedback gives:

(i) a constant and predictable voltage gain A, (ii) reduced distortion of the output, and (iii) better frequency response.

The advantages of using negative feedback outweigh the accompanying loss of gain which is easily increased by using two or more op amp stages.

Bandwidth

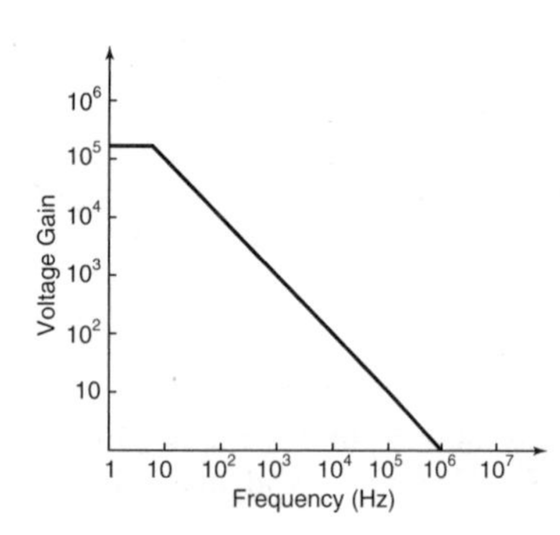

Figure 18.3

The open-loop voltage gain of an op amp is not constant at all frequencies; because of capacitive effects it falls at high frequencies. Figure 18.3 shows the gain/bandwidth characteristic of a 741 op amp. At frequencies below 10 Hz the gain is constant, but at higher frequencies the gain falls at a constant rate of 6 dB/octave (equivalent to a rate of 20 dB per decade) to 0 dB.

The gain-bandwidth product for any amplifier is the linear voltage gain multiplied by the bandwidth at that gain. The value of frequency at which the open-loop gain has fallen to unity is called the transition frequency f_T.

$$f_T = \text{closed-loop voltage gain} \times \text{bandwidth} \qquad (2)$$

In Figure 18.3, $f_T = 10^6$ Hz or 1 MHz; a gain of 20 dB (i.e. $20\log_{10} 10$) gives a 100 kHz bandwidth, whilst a gain of 80 dB (i.e. $20\log_{10} 10^4$) restricts the bandwidth to 100 Hz.

18.2 Some op amp parameters

Input bias current

The input bias current, I_B, is the average of the currents into the two input terminals with the output at zero volts, which is typically around 80 nA (i.e. 80×10^{-9} A) for a 741 op amp. The input bias current causes a volt drop across the equivalent source impedance seen by the op amp input.

Input offset current

The input offset current, I_{os}, of an op amp is the difference between the two input currents with the output at zero volts. In a 741 op amp, I_{os} is typically 20 nA.

Input offset voltage

In the ideal op amp, with both inputs at zero there should be zero output. Due to imbalances within the amplifier this is not always the case and a small output voltage results. The effect can be nullified by applying a small offset voltage, V_{os}, to the amplifier. In a 741 op amp, V_{os} is typically 1 mV.

Common-mode rejection ratio

The output voltage of an op amp is proportional to the difference between the voltages applied to its two input terminals. Ideally, when the two voltages are equal, the output voltages should be zero. A signal applied to both input terminals is called a common-mode signal and it is usually an unwanted noise voltage. The ability of an op amp to suppress common-mode signals is expressed in terms of its common-mode rejection ratio (CMRR), which is defined by:

$$\mathbf{CMRR = 20\log_{10}\left(\frac{differential\ voltage\ gain}{common\ mode\ gain}\right)\ dB} \qquad (3)$$

In a 741 op amp, the CMRR is typically 90 dB.

The common-mode gain, A_{com}, is defined as:

$$\boxed{A_{com} = \frac{V_o}{V_{com}}} \qquad (4)$$

where V_{com} is the common input signal

Problem 2. Determine the common-mode gain of an op amp that has a differential voltage gain of 150×10^3 and a CMRR of 90 dB.

From equation (3),

$$\text{CMRR} = 20\log_{10}\left(\frac{\text{differential voltage gain}}{\text{common mode gain}}\right) \text{dB}$$

Hence $$90 = 20\log_{10}\left(\frac{150 \times 10^3}{\text{common mode gain}}\right)$$

from which $$4.5 = \log_{10}\left(\frac{150 \times 10^3}{\text{common mode gain}}\right)$$

and $$10^{4.5} = \left(\frac{150 \times 10^3}{\text{common mode gain}}\right)$$

Hence, **common-mode gain** $= \dfrac{150 \times 10^3}{10^{4.5}} = \mathbf{4.74}$

Problem 3. A differential amplifier has an open-loop voltage gain of 120 and a common input signal of 3.0 V to both terminals. An output signal of 24 mV results. Calculate the common-mode gain and the CMRR.

From equation (4), the common-mode gain,

$$\boldsymbol{A_{com}} = \frac{V_o}{V_{com}} = \frac{24 \times 10^{-3}}{3.0} = 8 \times 10^{-3} = \mathbf{0.008}$$

From equation (3), the

$$\mathbf{CMRR} = 20 \log_{10}\left(\frac{\text{differential voltage gain}}{\text{common mode gain}}\right) \text{dB}$$

$$= 20 \log_{10}\left(\frac{120}{0.008}\right) = 20 \log_{10} 15000 = \mathbf{83.52\,dB}$$

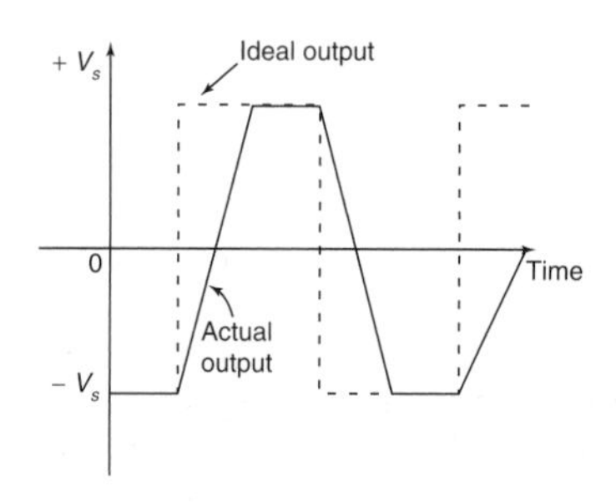

Figure 18.4

Slew rate

The slew rate of an op amp is the maximum rate of change of output voltage following a step input voltage. Figure 18.4 shows the effects of slewing; it causes the output voltage to change at a slower rate that the input, such that the output waveform is a distortion of the input waveform. 0.5 V/μs is a typical value for the slew rate.

18.3 Op amp inverting amplifier

The basic circuit for an inverting amplifier is shown in Figure 18.5 where the input voltage V_i (a.c. or d.c.) to be amplified is applied via resistor R_i to the inverting (−) terminal; the output voltage V_o is therefore in anti-phase with the input. The non-inverting (+) terminal is held at 0 V. Negative feedback is provided by the feedback resistor, R_f, feeding back a certain fraction of the output voltage to the inverting terminal.

Amplifier gain

In an **ideal op amp** two assumptions are made, these being that:

(i) each input draws zero current from the signal source, i.e. their input impedance's are infinite, and

(ii) the inputs are both at the same potential if the op amp is not saturated, i.e. $V_A = V_B$ in Figure 18.5.

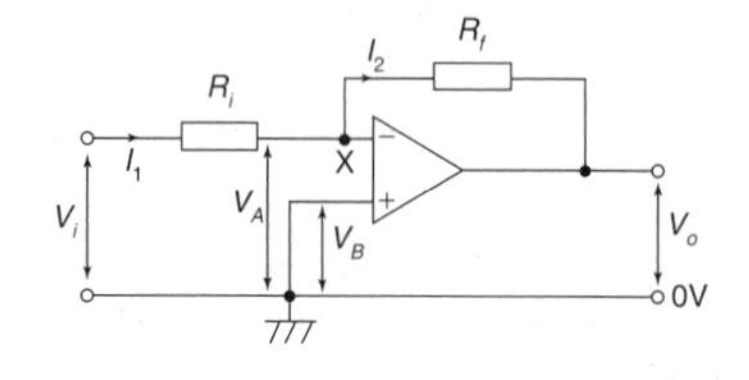

Figure 18.5

In Figure 18.5, $V_B = 0$, hence $V_A = 0$ and point X is called a **virtual earth**.

Thus, $I_1 = \dfrac{V_i - 0}{R_i}$ and $I_2 = \dfrac{0 - V_o}{R_f}$

However, $I_1 = I_2$ from assumption (i) above.

Hence $\frac{V_i}{R_i} = -\frac{V_o}{R_f}$,

the negative sign showing that V_o is negative when V_i is positive, and vice versa.

The **closed-loop gain A** is given by:

$$\boldsymbol{A = \frac{V_o}{V_i} = \frac{-R_f}{R_i}} \qquad (5)$$

This shows that the gain of the amplifier depends only on the two resistors, which can be made with precise values, and not on the characteristics of the op amp, which may vary from sample to sample.
For example, if $R_i = 10\ \text{k}\Omega$ and $R_f = 100\ \text{k}\Omega$, then the closed-loop gain,

$$\mathbf{A} = \frac{-R_f}{R_i} = \frac{-100 \times 10^3}{10 \times 10^3} = \mathbf{-10}$$

Thus an input of 100 mV will cause an output change of 1 V.

Input impedance

Since point X is a virtual earth (i.e. at 0 V), R_i may be considered to be connected between the inverting (−) input terminal and 0 V. The input impedance of the circuit is therefore R_i in parallel with the much greater input impedance of the op amp, i.e. effectively R_i. The circuit input impedance can thus be controlled by simply changing the value of R_i.

Problem 4. In the inverting amplifier of Figure 18.5, $R_i = 1\ \text{k}\Omega$ and $R_f = 2\ \text{k}\Omega$. Determine the output voltage when the input voltage is: (a) +0.4 V (b) −1.2 V

From equation (5), $V_o = \left(\frac{-R_f}{R_i}\right) V_i$

(a) When $V_i = +0.4$ V, $\boldsymbol{V_o} = \left(\frac{-2000}{1000}\right)(+0.4) = \mathbf{-0.8\ V}$

(b) When $V_i = -1.2\ V$, $\mathbf{V_o} = \left(\frac{-2000}{1000}\right)(-1.2) = \mathbf{+2.4\ V}$

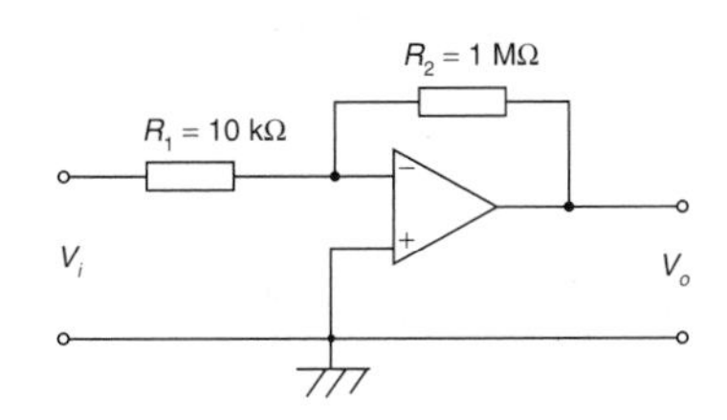

Figure 18.6

Problem 5. The op amp shown in Figure 18.6 has an input bias current of 100 nA at 20°C. Calculate (a) the voltage gain, and (b) the output offset voltage due to the input bias current. (c) How can the effect of input bias current be minimised?

Comparing Figure 18.6 with Figure 18.5, gives $R_i = 10\ \text{k}\Omega$ and $R_f = 1\ \text{M}\Omega$

(a) From equation (5), **voltage gain,**

$$A = \frac{-R_f}{R_i} = \frac{-1 \times 10^6}{10 \times 10^3} = \mathbf{-100}$$

(b) The input bias current, I_B, causes a volt drop across the equivalent source impedance seen by the op amp input, in this case, R_i and R_f in parallel. Hence, the offset voltage, V_{os}, at the input due to the 100 nA input bias current, I_B, is given by:

$$V_{os} = I_B\left(\frac{R_iR_f}{R_i + R_f}\right) = (100 \times 10^{-9})\left(\frac{10 \times 10^3 \times 1 \times 10^6}{(10 \times 10^3) + (1 \times 10^6)}\right)$$

$$= (10^{-7})(9.9 \times 10^3) = 9.9 \times 10^{-4} = \mathbf{0.99\ mV}$$

(c) The effect of input bias current can be minimised by ensuring that both inputs 'see' the same driving resistance. This means that **a resistance of value of 9.9 kΩ** (from part (b)) **should be placed between the non-inverting (+) terminal and earth** in Figure 18.6.

Problem 6. Design an inverting amplifier to have a voltage gain of 40 dB, a closed-loop bandwidth of 5 kHz and an input resistance of 10 kΩ

The voltage gain of an op amp, in decibels, is given by:

gain in decibels $= 20 \log_{10}$ (voltage gain) from chapter 10

Hence $40 = 20 \log_{10} A$

from which, $2 = \log_{10} A$

and $\mathbf{A = 10^2 = 100}$

With reference to Figure 18.5, and from equation (5),

$$A = \left|\frac{R_f}{R_i}\right|$$

i.e. $$100 = \frac{R_f}{10 \times 10^3}$$

Hence $\mathbf{R_f} = 100 \times 10 \times 10^3 = \mathbf{1\ M\Omega}$

From equation (2), Section 18.1,

frequency $=$ gain $\times$ bandwidth $= 100 \times 5 \times 10^3$

$=$ **0.5 MHz** or **500 kHz**

Further problems on the introduction to operational amplifiers may be found in Section 18.12, problems 1 to 6, page 294.

18.4 Op amp non-inverting amplifier

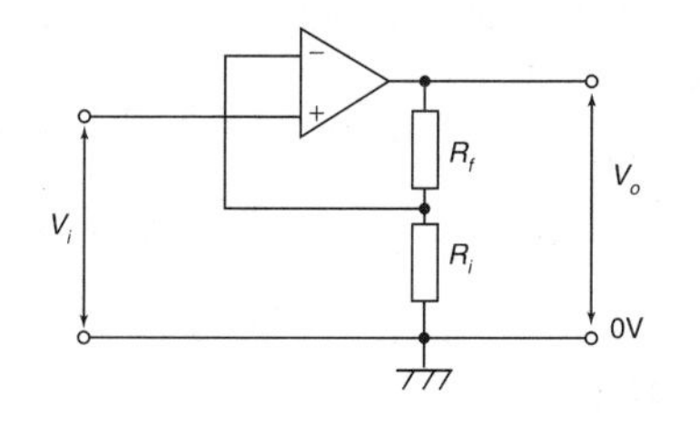

Figure 18.7

The basic circuit for a non-inverting amplifier is shown in Figure 18.7 where the input voltage V_i (a.c. or d.c.) is applied to the non-inverting (+) terminal of the op amp. This produces an output V_o that is in phase with the input. Negative feedback is obtained by feeding back to the inverting (−) terminal, the fraction of V_o developed across R_i in the voltage divider formed by R_f and R_i across V_o

Amplifier gain

In Figure 18.7, let the feedback factor,

$$\beta = \frac{R_i}{R_i + R_f}$$

It may be shown that for an amplifier with open-loop gain A_o, the closed-loop voltage gain A is given by:

$$A = \frac{A_o}{1 + \beta A_o}$$

For a typical op amp, $A_o = 10^5$, thus βA_o is large compared with 1, and the above expression approximates to:

$$\boxed{\boldsymbol{A = \frac{A_o}{\beta A_o} = \frac{1}{\beta}}} \qquad (6)$$

Hence $$\boxed{\boldsymbol{A = \frac{V_o}{V_i} = \frac{R_i + R_f}{R_i} = 1 + \frac{R_f}{R_i}}} \qquad (7)$$

For example, if $R_i = 10\ \text{k}\Omega$ and $R_f = 100\ \text{k}\Omega$,

then $A = 1 + \dfrac{100 \times 10^3}{10 \times 10^3} = 1 + 10 = \mathbf{11}$

Again, the gain depends only on the values of R_i and R_f and is independent of the open-loop gain A_o

Input impedance

Since there is no virtual earth at the non-inverting (+) terminal, the input impedance is much higher (typically 50 **MΩ**) than that of the inverting amplifier. Also, it is unaffected if the gain is altered by changing R_f and/or R_i. This non-inverting amplifier circuit gives good matching when the input is supplied by a high impedance source.

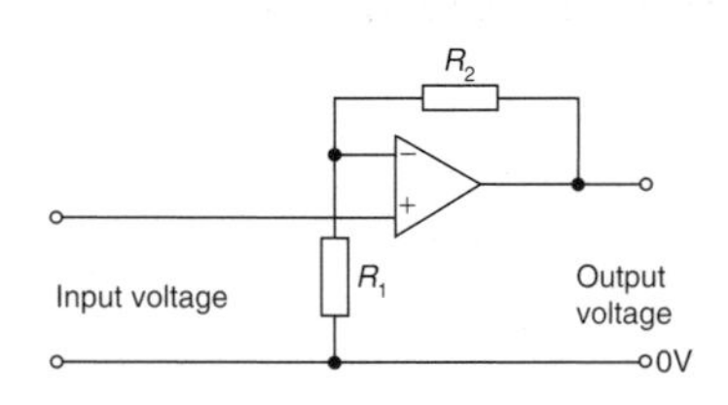

Figure 18.8

Problem 7. For the op amp shown in Figure 18.8, $R_1 = 4.7$ kΩ and $R_2 = 10$ kΩ. If the input voltage is −0.4 V, determine (a) the voltage gain (b) the output voltage

The op amp shown in Figure 18.8 is a non-inverting amplifier, similar to Figure 18.7.

(a) From equation (7), **voltage gain,**

$$A = 1 + \frac{R_f}{R_i} = 1 + \frac{R_2}{R_1} = 1 + \frac{10 \times 10^3}{4.7 \times 10^3}$$
$$= 1 + 2.13 = \mathbf{3.13}$$

(b) Also from equation (7), **output voltage,**

$$V_o = \left(1 + \frac{R}{R_1}\right) V_i$$
$$= (3.13)(-0.4) = \mathbf{-1.25\ V}$$

18.5 Op amp voltage-follower

The **voltage-follower** is a special case of the non-inverting amplifier in which 100% negative feedback is obtained by connecting the output directly to the inverting (−) terminal, as shown in Figure 18.9. Thus R_f in Figure 18.7 is zero and R_i is infinite.

From equation (6), $A = 1/\beta$ (when A_o is very large). Since all of the output is fed back, $\beta = 1$ and $A \approx 1$. Thus the voltage gain is nearly 1 and $V_o = V_i$ to within a few millivolts.

The circuit of Figure 18.9 is called a voltage-follower since, as with its transistor emitter-follower equivalent, V_o follows V_i. It has an extremely high input impedance and a low output impedance. Its main use is as a **buffer amplifier,** giving current amplification, to match a high impedance source to a low impedance load. For example, it is used as the input stage of an analogue voltmeter where the highest possible input impedance is required so as not to disturb the circuit under test; the output voltage is measured by a relatively low impedance moving-coil meter.

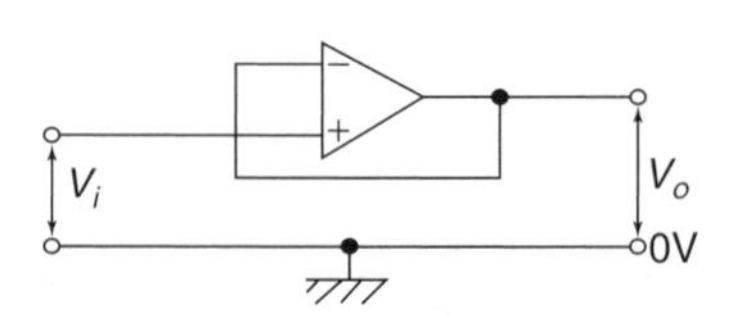

Figure 18.9

18.6 Op amp summing amplifier

Because of the existence of the virtual earth point, an op amp can be used to add a number of voltages (d.c. or a.c.) when connected as a multi-input inverting amplifier. This, in turn, is a consequence of the high value of the open-loop voltage gain A_o. Such circuits may be used as 'mixers' in audio systems to combine the outputs of microphones, electric guitars, pick-ups, etc. They are also used to perform the mathematical process of addition in analogue computing.

The circuit of an op amp summing amplifier having three input voltages V_1, V_2 and V_3 applied via input resistors R_1, R_2 and R_3 is shown in Figure 18.10. If it is assumed that the inverting (−) terminal of the op

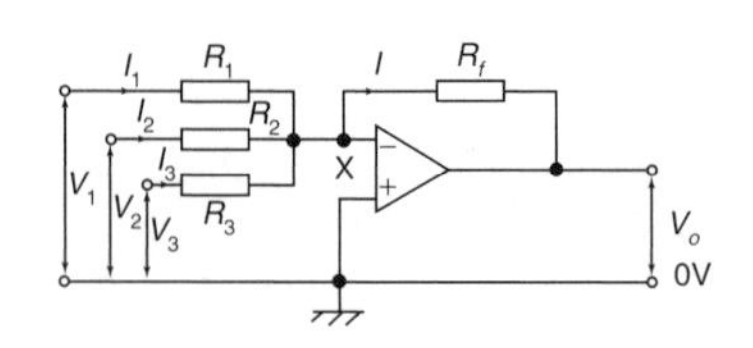

Figure 18.10

amp draws no input current, all of it passing through R_f, then:

$$I = I_1 + I_2 + I_3$$

Since X is a virtual earth (i.e. at 0 V), it follows that:

$$\frac{-V_o}{R_f} = \frac{V_1}{R_1} + \frac{V_2}{R_2} + \frac{V_3}{R_3}$$

Hence

$$\boldsymbol{V_o = -\left(\frac{R_f}{R_1}V_1 + \frac{R_f}{R_2}V_2 + \frac{R_f}{R_3}V_3\right) = -R_f\left(\frac{V_1}{R_1} + \frac{V_2}{R_2} + \frac{V_3}{R_3}\right)} \quad (8)$$

The three input voltages are thus added and amplified if R_f is greater than each of the input resistors; 'weighted' summation is said to have occurred.

Alternatively, the input voltages are added and attenuated if R_f is less than each input resistor.

For example, if $\frac{R_f}{R_1} = 4$, $\frac{R_f}{R_2} = 3$ and $\frac{R_f}{R_3} = 1$ and $V_1 = V_2 = V_3 = +1$ V, then

$$V_o - \left(\frac{R_f}{R_1}V_1 + \frac{R_f}{R_2}V_2 + \frac{R_f}{R_3}V_3\right) = -(4 + 3 + 1) = \mathbf{-8\,V}$$

If $R_1 = R_2 = R_3 = R_i$, the input voltages are amplified or attenuated equally, and

$$V_o = -\frac{R_f}{R_i}(V_1 + V_2 + V_3)$$

If, also, $R_i = R_f$ then $V_o = -(V_1 + V_2 + V_3)$

The virtual earth is also called the **summing point** of the amplifier. It isolates the inputs from one another so that each behaves as if none of the others existed and none feeds any of the other inputs even though all the resistors are connected at the inverting (−) input.

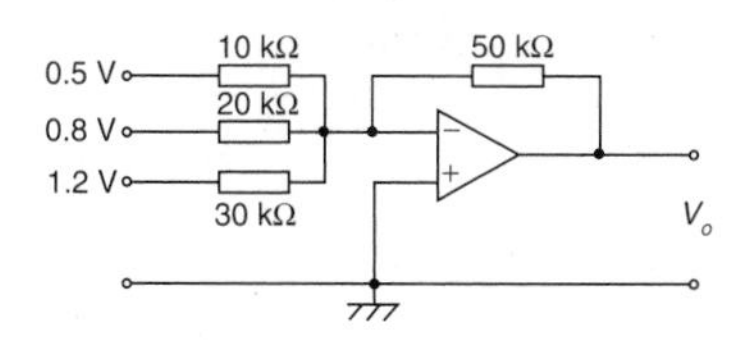

Figure 18.11

Problem 8. For the summing op amp shown in Figure 18.11, determine the output voltage, V_o

From equation (8),

$$V_o = -R_f\left(\frac{V_1}{R_1} + \frac{V_2}{R_2} + \frac{V_3}{R_3}\right)$$

$$= -(50 \times 10^3)\left(\frac{0.5}{10 \times 10^3} + \frac{0.8}{20 \times 10^3} + \frac{1.2}{30 \times 10^3}\right)$$

$$= -(50 \times 10^3)(5 \times 10^{-5} + 4 \times 10^{-5} + 4 \times 10^{-5})$$

$$= -(50 \times 10^3)(13 \times 10^{-5}) = \mathbf{-6.5\,V}$$

18.7 Op amp voltage comparator

If both inputs of the op amp shown in Figure 18.12 are used simultaneously, then from equation (1), page 279, the output voltage is given by:

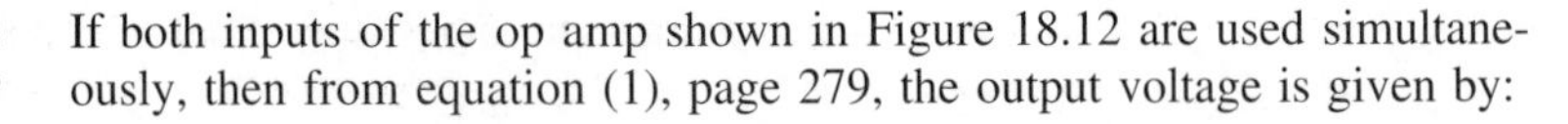

$$V_o = A_o(V_2 - V_1)$$

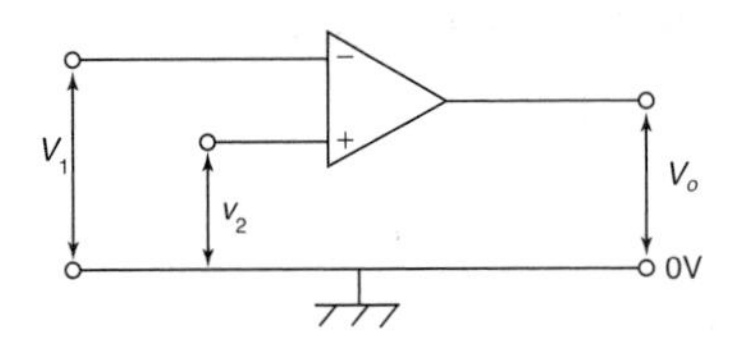

Figure 18.12

When $V_2 > V_1$ then V_o is positive, its maximum value being the positive supply voltage $+V_s$, which it has when $(V_2 - V_1) \geq V_s/A_o$. The op amp is then saturated. For example, if $V_s = +9\,\text{V}$ and $A_o = 10^5$, then saturation occurs when $(V_2 - V_1) \geq 9/10^5$ i.e. when V_2 exceeds V_1 by $90\,\mu\text{V}$ and $V_o \approx 9\,\text{V}$.

When $V_1 > V_2$, then V_o is negative and saturation occurs if V_1 exceeds V_2 by V_s/A_o i.e. around $90\,\mu\text{V}$ in the above example; in this case, $V_o \approx -V_s = -9\,\text{V}$.

A small change in $(V_2 - V_1)$ therefore causes V_o to switch between near $+V_s$ and near to $-V_s$ and enables the op amp to indicate when V_2 is greater or less than V_1, i.e. to act as a **differential amplifier** and compare two voltages. It does this in an electronic digital voltmeter.

Problem 9. Devise a light-operated alarm circuit using an op amp, a LDR, a LED and a $\pm 15\,\text{V}$ supply

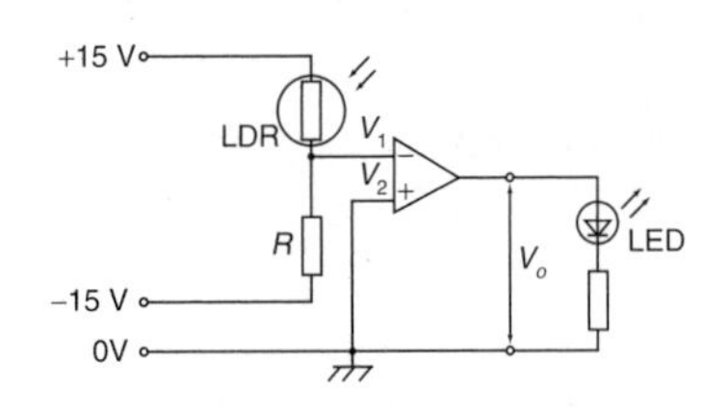

Figure 18.13

A typical light-operated alarm circuit is shown in Figure 18.13.

Resistor R and the light dependent resistor (LDR) form a voltage divider across the $+15/0/-15\,\text{V}$ supply. The op amp compares the voltage V_1 at the voltage divider junction, i.e. at the inverting ($-$) input, with that at the non-inverting ($+$) input, i.e. with V_2, which is 0 V. In the dark the resistance of the LDR is much greater than that of R, so more of the 30 V across the voltage divider is dropped across the LDR, causing V_1 to fall below 0 V. Now $V_2 > V_1$ and the output voltage V_o switches from near $-15\,\text{V}$ to near $+15\,\text{V}$ and the light emitting diode (LED) lights.

18.8 Op amp integrator

The circuit for the op amp integrator shown in Figure 18.14 is the same as for the op amp inverting amplifier shown in Figure 18.5, but feedback occurs via a capacitor C, rather than via a resistor.

The output voltage is given by:

$$\mathbf{V_o = -\frac{1}{CR}\int V_i\,dt} \qquad (9)$$

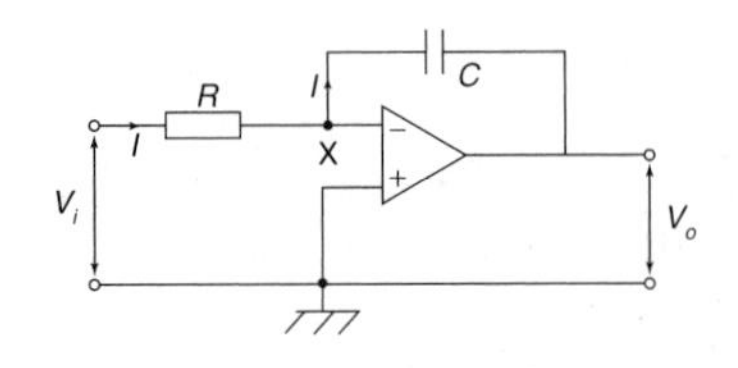

Figure 18.14

Since the inverting ($-$) input is used in Figure 18.15, V_o is negative if V_i is positive, and vice versa, hence the negative sign in equation (9).

Since X is a virtual earth in Figure 18.14, i.e. at 0 V, the voltage across R is V_i and that across C is V_o. Assuming again that none of the input current I enters the op amp inverting ($-$) input, then all of current I flows through C and charges it up. If V_i is constant, I will be a constant value

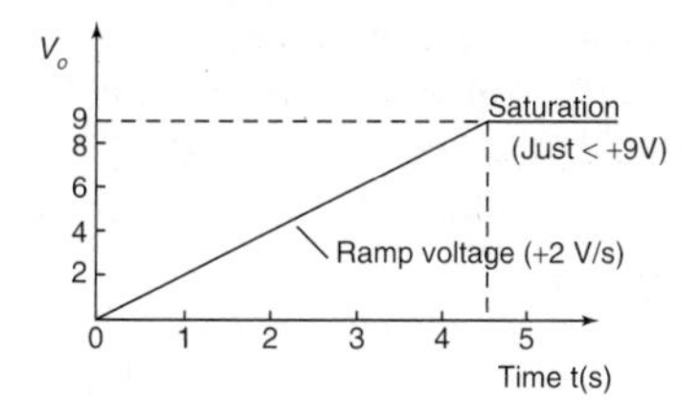

Figure 18.15

given by $I = V_i/R$. Capacitor C therefore charges at a constant rate and the potential of the output side of C ($= V_o$, since its input side is zero) charges so that the feedback path absorbs I. If Q is the charge on C at time t and the p.d. across it (i.e. the output voltage) changes from 0 to V_o in that time then:

$$Q = -V_oC = It$$

(from Chapter 6)

i.e. $$-V_oC = \frac{V_i}{R}t$$

i.e. $$V_o = -\frac{1}{CR}V_it$$

This result is the same as would be obtained from $V_o = -\frac{1}{CR}\int V_i\,dt$ if V_i is a constant value.
For example, if the input voltage $V_i = -2$ V and, say, CR = 1 s, then

$$V_o = -(-2)t = 2t$$

A graph of V_o/t will be a ramp function as shown in Figure 18.15 ($V_o = 2t$ is of the straight line form $y = mx + c$; in this case $y = V_o$ and $x = t$, gradient, $m = 2$ and vertical axis intercept $c = 0$). V_o rises steadily by +2 V/s in Figure 18.15, and if the power supply is, say, ±9 V, then V_o reaches +9 V after 4.5 s when the op amp saturates.

Problem 10. A steady voltage of −0.75 V is applied to an op amp integrator having component values of $R = 200\,\text{k}\Omega$ and $C = 2.5\,\mu\text{F}$. Assuming that the initial capacitor charge is zero, determine the value of the output voltage 100 ms after application of the input.

From equation (9), output voltage,

$$V_o = -\frac{1}{CR}\int V_i\,dt = -\frac{1}{(2.5\times10^{-6})(200\times10^{3})}\int(-0.75)\,dt$$

$$= -\frac{1}{0.5}\int(-0.75)\,dt = -2[-0.75t] = +1.5t$$

When time $t = 100$ ms, **output voltage,**

$$V_o = (1.5)(100\times10^{-3}) = \mathbf{0.15\,V}$$

18.9 Op amp differential amplifier

The circuit for an op amp differential amplifier is shown in Figure 18.16 where voltages V_1 and V_2 are applied to its two input terminals and the difference between these voltages is amplified.

(i) Let V_1 volts be applied to terminal 1 and 0 V be applied to terminal 2. The difference in the potentials at the inverting (−) and non-inverting

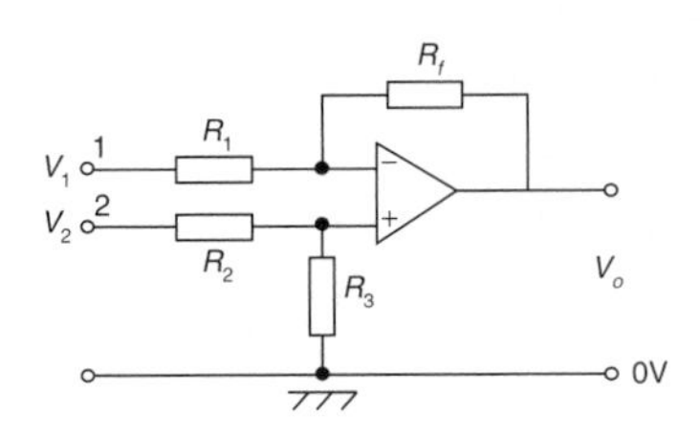

Figure 18.16

(+) op amp inputs is practically zero and hence the inverting terminal must be at zero potential. Then $I_1 = V_1/R_1$. Since the op amp input resistance is high, this current flows through the feedback resistor R_f. The volt drop across R_f, which is the output voltage $V_o = (V_1/R_1)R_f$; hence, the closed loop voltage gain A is given by:

$$\mathbf{A = \frac{V_o}{V_1} = -\frac{R_f}{R_1}} \qquad (10)$$

(ii) By similar reasoning, if V_2 is applied to terminal 2 and 0 V to terminal 1, then the voltage appearing at the non-inverting terminal will be $(R_3/(R_2+R_3))V_2$ volts. This voltage will also appear at the inverting (−) terminal and thus the voltage across R_1 is equal to $-(R_3/(R_2+R_3))V_2$ volts.

Now the output voltage,

$$V_o = \left(\frac{R_3}{R_2+R_3}\right)V_2 + \left[-\left(\frac{R_3}{R_2+R_3}\right)V_2\right]\left(-\frac{R_f}{R_1}\right)$$

and the voltage gain,

$$A = \frac{V_o}{V_2} = \left(\frac{R_3}{R_2+R_3}\right) + \left[-\left(\frac{R_3}{R_2+R_3}\right)\right]\left(-\frac{R_f}{R_1}\right)$$

i.e.

$$\boxed{\mathbf{A = \frac{V_o}{V_2} = \left(\frac{R_3}{R_2+R_3}\right)\left(1+\frac{R_f}{R_1}\right)}} \qquad (11)$$

(iii) Finally, if the voltages applied to terminals 1 and 2 are V_1 and V_2 respectively, then the difference between the two voltages will be amplified.
If $\mathbf{V_1 > V_2}$, then:

$$\mathbf{V_o = (V_1 - V_2)\left(-\frac{R_f}{R_1}\right)} \qquad (12)$$

If $\mathbf{V_2 > V_1}$, then:

$$\mathbf{V_o = (V_2 - V_1)\left(\frac{R_3}{R_2+R_3}\right)\left(1+\frac{R_f}{R_1}\right)} \qquad (13)$$

Problem 11. In the differential amplifier shown in Figure 18.16, $R_1 = 10\,\text{k}\Omega$, $R_2 = 10\,\text{k}\Omega$, $R_3 = 100\,\text{k}\Omega$ and $R_f = 100\,\text{k}\Omega$. Determine the output voltage V_o if:
(a) $V_1 = 5$ mV and $V_2 = 0$
(b) $V_1 = 0$ and $V_2 = 5\,\text{mV}$
(c) $V_1 = 50$ mV and $V_2 = 25\,\text{mV}$
(d) $V_1 = 25$ mV and $V_2 = 50\,\text{mV}$

(a) From equation (10),

$$\mathbf{V_o} = -\frac{R_f}{R_1}V_1 = -\left(\frac{100 \times 10^3}{10 \times 10^3}\right)(5)\ \text{mV} = \mathbf{-50\,mV}$$

(b) From equation (11),

$$V_o = \left(\frac{R_3}{R_2 + R_3}\right)\left(1 + \frac{R_f}{R_1}\right)V_2$$

$$= \left(\frac{100}{110}\right)\left(1 + \frac{100}{10}\right)(5)\ \text{mV} = \mathbf{+50\,mV}$$

(c) $V_1 > V_2$ hence from equation (12),

$$\mathbf{V_o} = (V_1 - V_2)\left(-\frac{R_f}{R_1}\right)$$

$$= (50 - 25)\left(-\frac{100}{10}\right)\ \text{mV} = \mathbf{-250\,mV}$$

(d) $V_2 > V_1$ hence from equation (13),

$$\mathbf{V_o} = (V_2 - V_1)\left(\frac{R_3}{R_2 + R_3}\right)\left(1 + \frac{R_f}{R_1}\right)$$

$$= (50 - 25)\ \text{mV}\left(\frac{100}{100 + 10}\right)\left(1 + \frac{100}{10}\right)\ \text{mV}$$

$$= (25)\left(\frac{100}{110}\right)(11) = \mathbf{+250\ mV}$$

Further problems on operational amplifier calculations may be found in Section 18.12, problems 7 to 11, page 295.

18.10 Digital to analogue (D/A) conversion

There are a number of situations when digital signals have to be converted to analogue ones. For example, a digital computer often needs to produce a graphical display on the screen; this involves using a D/A converter to change the two-level digital output voltage from the computer, into a continuously varying analogue voltage for the input to the cathode ray tube, so that it can deflect the electron beam to produce screen graphics.

A binary weighted resistor D/A converter is shown in Figure 18.17 for a four-bit input. The values of the resistors, R, $2R$, $4R$, $8R$ increase according to the binary scale — hence the name of the converter. The circuit uses an op amp as a **summing amplifier** (see section 18.6) with a feedback resistor R_f. Digitally controlled electronic switches are shown as S_1 to S_4. Each switch connects the resistor in series with it to a fixed reference voltage V_{ref} when the input bit controlling it is a 1 and to ground (0 V) when it is a 0. The input voltages V_1 to V_4 applied to the op amp by the four-bit input via the resistors therefore have one of two values, i.e. either V_{ref} or 0 V.

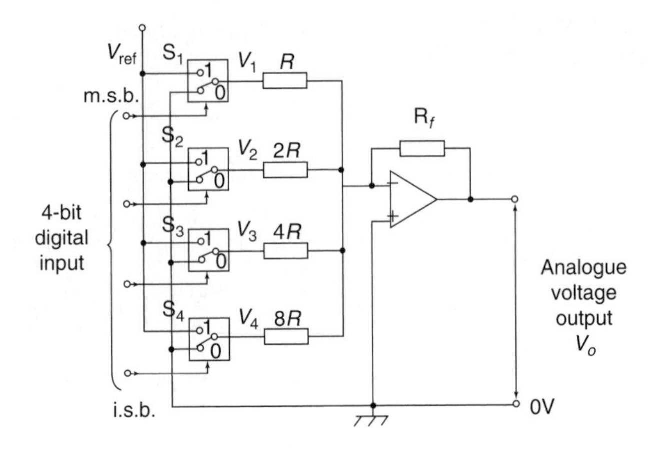

Figure 18.17

From equation (8), page 287, the analogue output voltage V_o is given by:

$$V_o = -\left(\frac{R_f}{R}V_1 + \frac{R_f}{2R}V_2 + \frac{R_f}{4R}V_3 + \frac{R_f}{8R}V_4\right)$$

Let $R_f = R = 1\,\mathrm{k\Omega}$, then:

$$V_o = -\left(V_1 + \frac{1}{2}V_2 + \frac{1}{4}V_3 + \frac{1}{8}V_4\right)$$

With a four-bit input of 0001 (i.e. decimal 1), S_4 connects $8R$ to V_{ref}, i.e. $V_4 = V_{ref}$, and S_1, S_2 and S_3 connect R, $2R$ and $4R$ to 0 V, making $V_1 = V_2 = V_3 = 0$. Let $V_{ref} = -8\,\mathrm{V}$, then output voltage,

$$\mathbf{V_o} = -\left(0 + 0 + 0 + \frac{1}{8}(-8)\right) = \mathbf{+1\,V}$$

With a four-bit input of 0101 (i.e. decimal 5), S_2 and S_4 connects 2R and 4R to V_{ref}, i.e. $V_2 = V_4 = V_{ref}$, and S_1 and S_3 connect R and 4R to 0 V, making $V_1 = V_3 = 0$. Again, if $V_{ref} = -8\,\mathrm{V}$, then output voltage,

$$\mathbf{V_o} = -\left(0 + \frac{1}{2}(-8) + 0 + \frac{1}{8}(-8)\right) = \mathbf{+5\,V}$$

If the input is 0111 (i.e. decimal 7), the output voltage will be 7 V, and so on. From these examples, it is seen that the analogue output voltage, V_o, is directly proportional to the digital input.

V_o has a 'stepped' waveform, the waveform shape depending on the binary input. A typical waveform is shown in Figure 18.18.

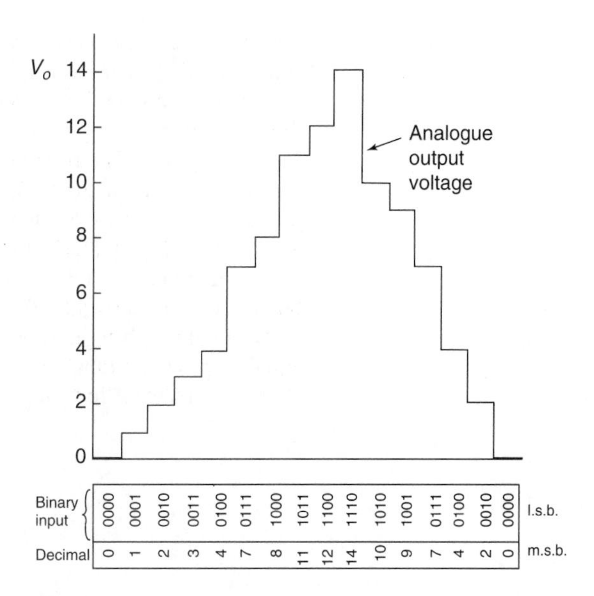

Figure 18.18

18.11 Analogue to digital (A/D) conversion

In a digital voltmeter, its input is in analogue form and the reading is displayed digitally. This is an example where an analogue to digital converter is needed.

A block diagram for a four-bit counter type A/D conversion circuit is shown in Figure 18.19. An op amp is again used, in this case as a **voltage comparator** (see Section 18.7). The analogue input voltage V_2, shown in Figure 18.20(a) as a steady d.c. voltage, is applied to the non-inverting (+) input, whilst a sawtooth voltage V_1 supplies the inverting (−) input.

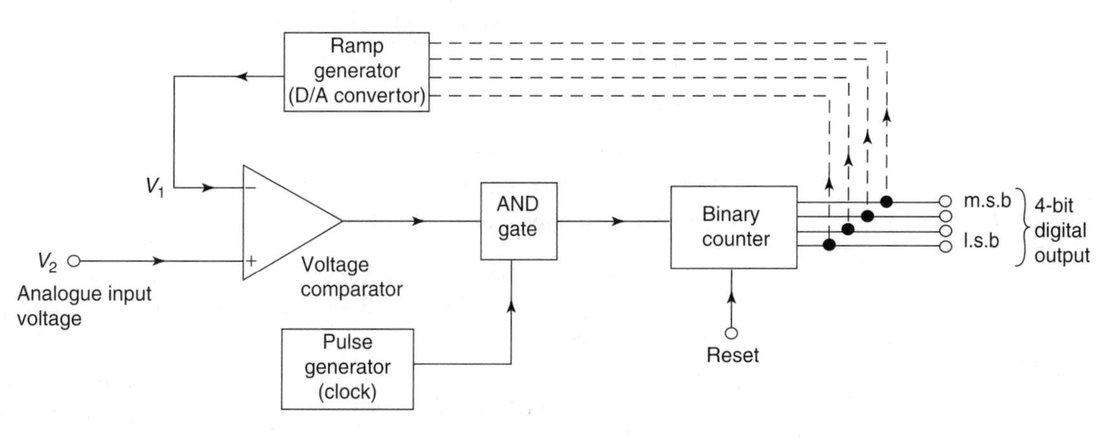

Figure 18.19

The output from the comparator is applied to one input of an AND gate and is a 1 (i.e. 'high') until V_1 equals or exceeds V_2, when it then goes to 0 (i.e. 'low') as shown in Figure 18.20(b). The other input of the AND gate is fed by a steady train of pulses from a pulse generator, as shown in Figure 18.20(c). When both inputs to the AND gate are 'high', the gate 'opens' and gives a 'high' output, i.e. a pulse, as shown in Figure 18.20(d). The time taken by V_1 to reach V_2 is proportional to the analogue voltage if the ramp is linear. The output pulses from the AND gate are recorded by a binary counter and, as shown in Figure 18.20(e), are the digital equivalent of the analogue input voltage V_2. In practise, the ramp generator is a D/A converter which takes its digital input from the binary counter, shown by the broken lines in Figure 18.19. As the counter advances through its normal binary sequence, a staircase waveform with equal steps (i.e. a ramp) is built up at the output of the D/A converter (as shown by the first few steps in Figure 18.18.

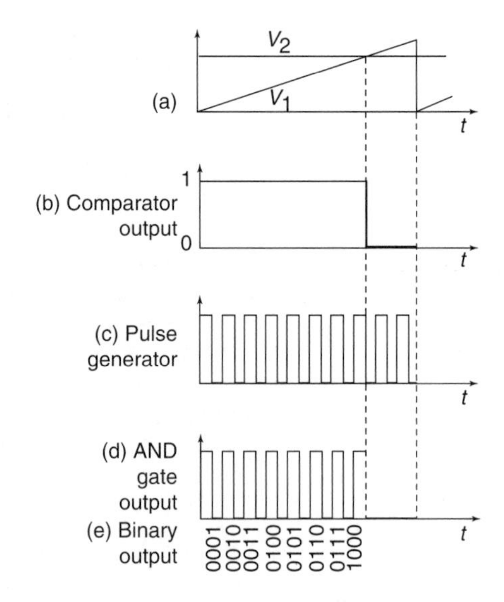

Figure 18.20

18.12 Further problems on operational amplifiers

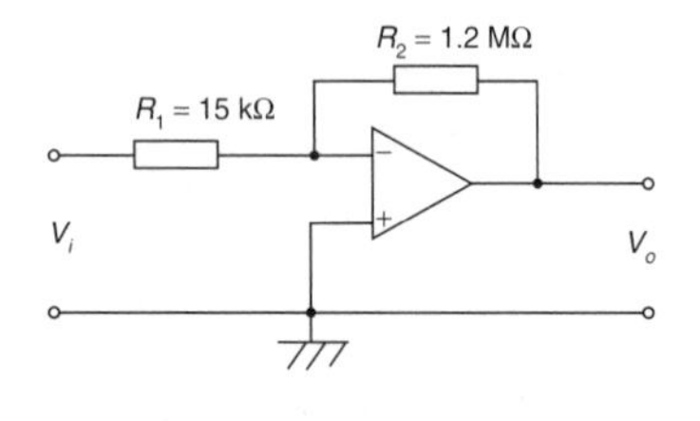

Figure 18.21

Introduction to operational amplifiers

1. A differential amplifier has an open-loop voltage gain of 150 when the input signals are 3.55 V and 3.40 V. Determine the output voltage of the amplifier. [22.5 V]

2. Calculate the differential voltage gain of an op amp that has a common-mode gain of 6.0 and a CMRR of 80 dB. [6×10^4]

3. A differential amplifier has an open-loop voltage gain of 150 and a common input signal of 4.0 V to both terminals. An output signal of 15 mV results. Determine the common-mode gain and the CMRR. [3.75×10^{-3}, 92.04 dB]

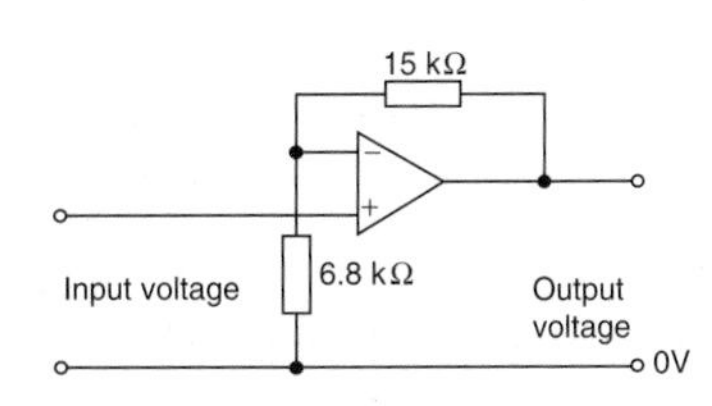

Figure 18.22

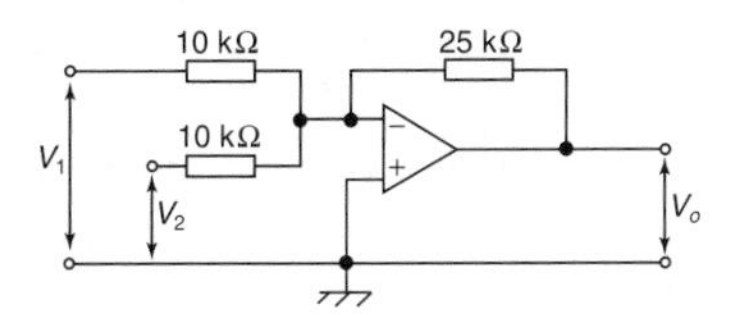

Figure 18.23

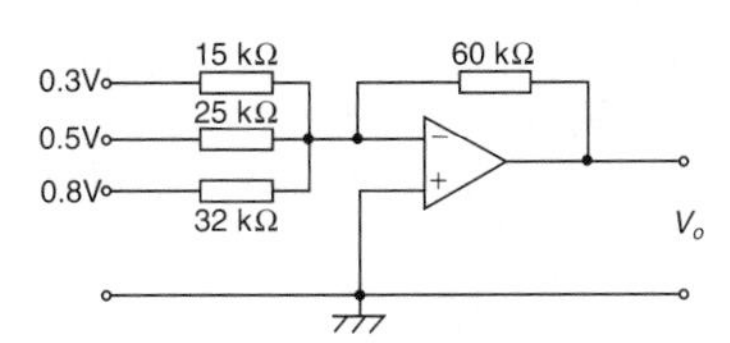

Figure 18.24

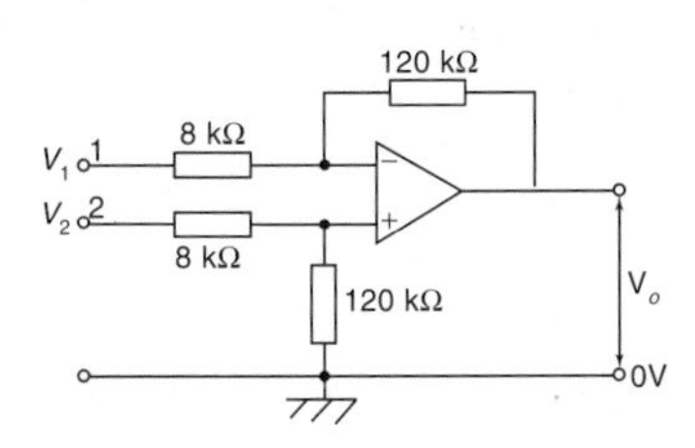

Figure 18.25

4. In the inverting amplifier of Figure 18.5 (on page 282), $R_i = 1.5\,\text{k}\Omega$ and $R_f = 2.5\,\text{k}\Omega$. Determine the output voltage when the input voltage is: (a) $+0.6$ V (b) -0.9 V [(a) -1.0 V (b) $+1.5$ V]

5. The op amp shown in Figure 18.21 has an input bias current of 90 nA at 20°C. Calculate (a) the voltage gain, and (b) the output offset voltage due to the input bias current. [(a) 80 (b) 1.33 mV]

6. Determine (a) the value of the feedback resistor, and (b) the frequency for an inverting amplifier to have a voltage gain of 45 dB, a closed-loop bandwidth of 10 kHz and an input resistance of 20 kΩ. [(a) 3.56 MΩ (b) 1.78 MHz]

Further operational amplifier calculations

7. If the input voltage for the op amp shown in Figure 18.22, is -0.5 V, determine (a) the voltage gain (b) the output voltage [(a) 3.21 (b) -1.60 V]

8. In the circuit of Figure 18.23, determine the value of the output voltage, V_o, when (a) $V_1 = +1$ V and $V_2 = +3$ V (b) $V_1 = +1$ V and $V_2 = -3$ V [(a) -10 V (b) $+5$ V]

9. For the summing op amp shown in Figure 18.24, determine the output voltage, V_o. [-3.9 V]

10. A steady voltage of -1.25 V is applied to an op amp integrator having component values of $R = 125\,\text{k}\Omega$ and $C = 4.0\,\mu\text{F}$. Calculate the value of the output voltage 120 ms after applying the input, assuming that the initial capacitor charge is zero. [0.3 V]

11. In the differential amplifier shown in Figure 18.25, determine the output voltage, V_o, if: (a) $V_1 = 4$ mV and $V_2 = 0$ (b) $V_1 = 0$ and $V_2 = 6$ mV (c) $V_1 = 40$ mV and $V_2 = 30$ mV (d) $V_1 = 25$ mV and $V_2 = 40$ mV

[(a) -60 mV (b) $+90$ mV (c) -150 mV (d) $+225$ mV]

Assignment 5

This assignment covers the material contained in chapters 15 to 18.

The marks for each question are shown in brackets at the end of each question.

1 The power taken by a series inductive circuit when connected to a 100 V, 100 Hz supply is 250 W and the current is 5 A. Calculate (a) the resistance, (b) the impedance, (c) the reactance, (d) the power factor, and (e) the phase angle between voltage and current. (9)

2 A coil of resistance 20 Ω and inductance 200 mH is connected in parallel with a 4 μF capacitor across a 50 V, variable frequency supply. Calculate (a) the resonant frequency, (b) the dynamic resistance, (c) the current at resonance, and (d) the Q-factor at resonance. (10)

3 A series circuit comprises a coil of resistance 30 Ω and inductance 50 mH, and a 2500 pF capacitor. Determine the Q-factor of the circuit at resonance. (4)

4 The winding of an electromagnet has an inductance of 110 mH and a resistance of 5.5 Ω. When it is connected to a 110 V, d.c. supply, calculate (a) the steady state value of current flowing in the winding, (b) the time constant of the circuit, (c) the value of the induced e.m.f. after 0.1 s, (d) the time for the current to rise to 75% of it's final value, and (e) the value of the current after 0.02 s. (11)

5 A single-phase motor takes 30 A at a power factor of 0.65 lagging from a 300 V, 50 Hz supply. Calculate (a) the current taken by a capacitor connected in parallel with the motor to correct the power factor to unity, and (b) the value of the supply current after power factor correction. (7)

6 For the summing operational amplifier shown in Figure A5.1, determine the value of the output voltage, V_o. (3)

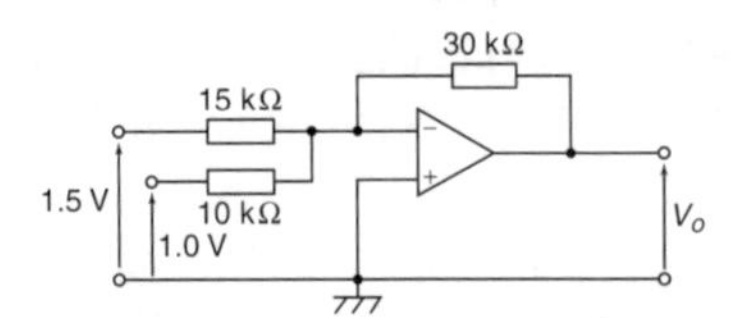

Figure A5.1

7 In the differential amplifier shown in Figure A5.2, determine the output voltage, V_o when: (a) $V_1 = 4$ mV and $V_2 = 0$ (b) $V_1 = 0$ and $V_2 = 5$ mV (c) $V_1 = 20$ mV and $V_2 = 10$ mV. (6)

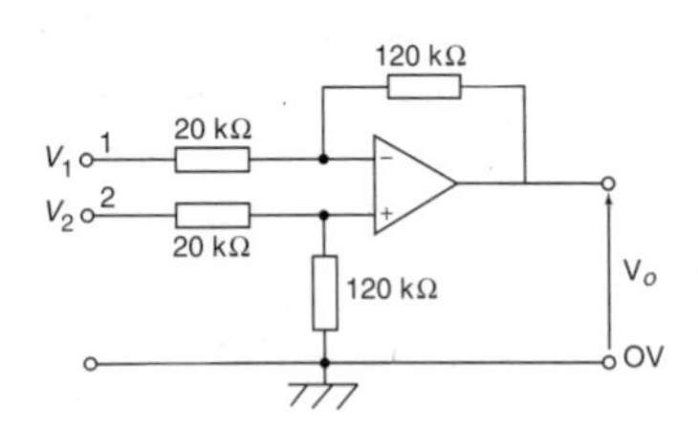

Figure A5.2

19 Three phase systems

At the end of this chapter you should be able to:

- describe a single-phase supply
- describe a three-phase supply
- understand a star connection, and recognize that $I_L = I_p$ and $V_L = \sqrt{3}V_p$
- draw a complete phasor diagram for a balanced, star connected load
- understand a delta connection, and recognize that $V_L = V_p$ and $I_L = \sqrt{3}I_p$
- draw a phasor diagram for a balanced, delta connected load
- calculate power in three-phase systems using $P = \sqrt{3}V_L I_L \cos\phi$
- appreciate how power is measured in a three-phase system, by the one, two and three-wattmeter methods
- compare star and delta connections
- appreciate the advantages of three-phase systems

19.1 Introduction

Generation, transmission and distribution of electricity via the National Grid system is accomplished by three-phase alternating currents.

The voltage induced by a single coil when rotated in a uniform magnetic field is shown in Figure 19.1 and is known as a **single-phase voltage**. Most consumers are fed by means of a single-phase a.c. supply. Two wires are used, one called the live conductor (usually coloured red) and the other is called the neutral conductor (usually coloured black). The neutral is usually connected via protective gear to earth, the earth wire being coloured green. The standard voltage for a single-phase a.c. supply is 240 V. The majority of single-phase supplies are obtained by connection to a three-phase supply (see Figure 19.5, page 299).

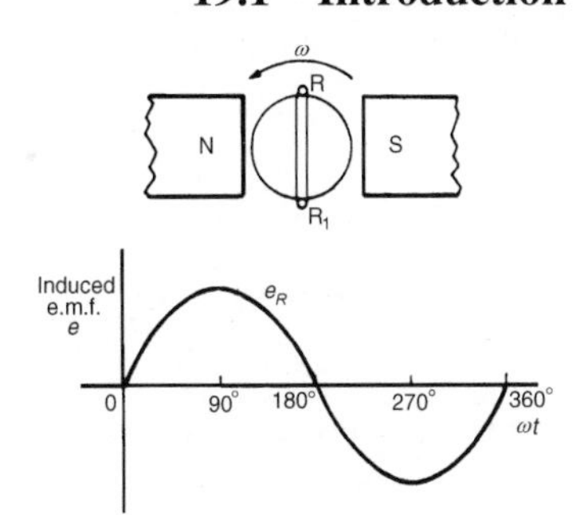

Figure 19.1

19.2 Three-phase supply

A **three-phase supply** is generated when three coils are placed 120° apart and the whole rotated in a uniform magnetic field as shown in Figure 19.2(a). The result is three independent supplies of equal voltages which are each displaced by 120° from each other as shown in Figure 19.2(b).

(i) The convention adopted to identify each of the phase voltages is: R-red, Y-yellow, and B-blue, as shown in Figure 19.2.

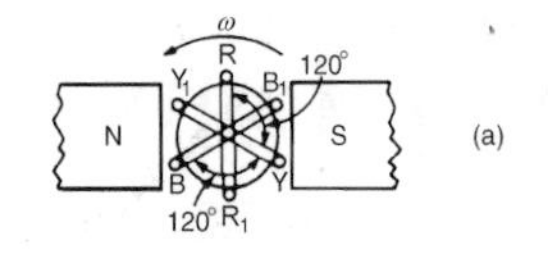
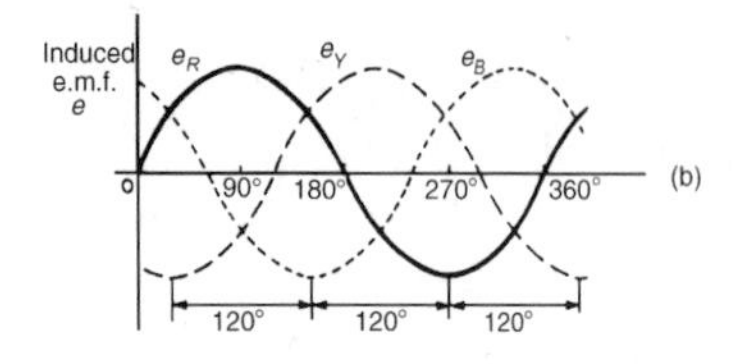

Figure 19.2

(ii) The **phase-sequence** is given by the sequence in which the conductors pass the point initially taken by the red conductor. The national standard phase sequence is R, Y, B.

A three-phase a.c. supply is carried by three conductors, called **'lines'** which are coloured red, yellow and blue. The currents in these conductors are known as line currents (I_L) and the p.d.'s between them are known as line voltages (V_L). A fourth conductor, called the **neutral** (coloured black, and connected through protective devices to earth) is often used with a three-phase supply.

If the three-phase windings shown in Figure 19.2 are kept independent then six wires are needed to connect a supply source (such as a generator) to a load (such as motor). To reduce the number of wires it is usual to interconnect the three phases. There are two ways in which this can be done, these being:
(a) a **star connection**, and (b) a **delta**, or **mesh, connection**. Sources of three-phase supplies, i.e. alternators, are usually connected in star, whereas three-phase transformer windings, motors and other loads may be connected either in star or delta.

19.3 Star connection

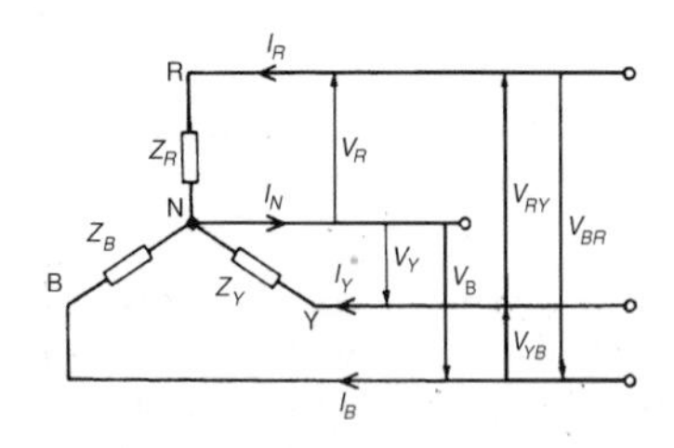

Figure 19.3

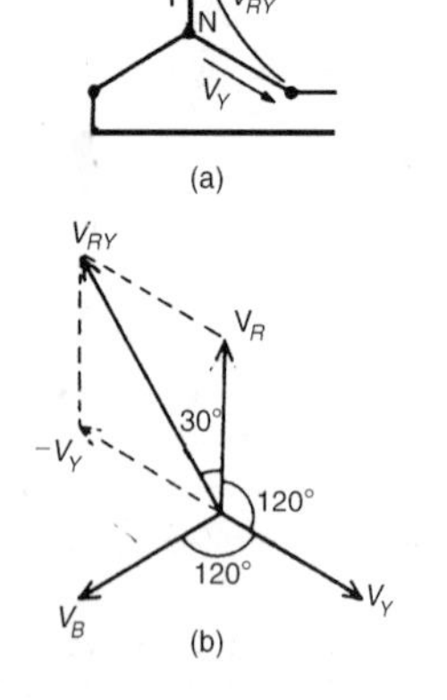

Figure 19.4

(i) A **star-connected load** is shown in Figure 19.3 where the three line conductors are each connected to a load and the outlets from the loads are joined together at N to form what is termed the **neutral point** or the **star point**.

(ii) The voltages, V_R, V_Y and V_B are called **phase voltages** or line to neutral voltages. Phase voltages are generally denoted by V_p

(iii) The voltages, V_{RY}, V_{YB} and V_{BR} are called **line voltages**

(iv) From Figure 19.3 it can be seen that the phase currents (generally denoted by I_p) are equal to their respective line currents I_R, I_Y and I_B, i.e. for a star connection:

$$\boldsymbol{I_L = I_p}$$

(v) For a balanced system: $I_R = I_Y = I_B$, $V_R = V_Y = V_B$

$$V_{RY} = V_{YB} = V_{BR}, \quad Z_R = Z_Y = Z_B$$

and the current in the neutral conductor, $I_N = 0$
When a star connected system is balanced, then the neutral conductor is unnecessary and is often omitted.

(vi) The line voltage, V_{RY}, shown in Figure 19.4(a) is given by $\boldsymbol{V_{RY} = V_R - V_Y}$ (V_Y is negative since it is in the opposite direction to V_{RY}). In the phasor diagram of Figure 19.4(b), phasor V_Y is reversed (shown by the broken line) and then added phasorially to V_R (i.e. $\boldsymbol{V_{RY} = V_R + (-V_Y)}$). By trigonometry, or by measurement, $V_{RY} = \sqrt{3}V_R$, i.e. for a balanced star connection:

$$\boldsymbol{V_L = \sqrt{3}\, V_p}$$

(See problem 3 following for a complete phasor diagram of a star-connected system.)

(vii) The star connection of the three phases of a supply, together with a neutral conductor, allows the use of two voltages — the phase voltage and the line voltage. A 4-wire system is also used when the load is not balanced. The standard electricity supply to consumers in Great Britain is 415/240 V, 50 Hz, 3-phase, 4-wire alternating current, and a diagram of connections is shown in Figure 19.5.

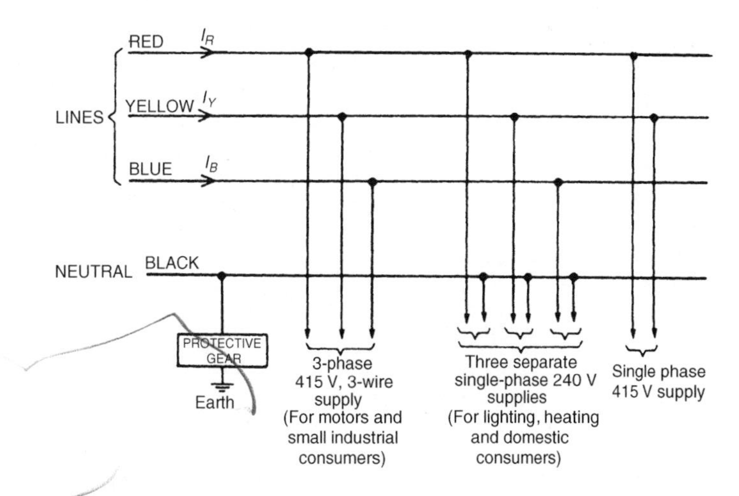

Figure 19.5

Problem 1. Three loads, each of resistance 30 Ω, are connected in star to a 415 V, 3-phase supply. Determine (a) the system phase voltage, (b) the phase current and (c) the line current.

A '415 V, 3-phase supply' means that 415 V is the line voltage, V_L

(a) For a star connection, $V_L = \sqrt{3}V_p$

Hence phase voltage, $V_p = \dfrac{V_L}{\sqrt{3}} = \dfrac{415}{\sqrt{3}} = \mathbf{239.6\ V\ or\ 240\ V}$

correct to 3 significant figures

(b) Phase current, $I_p = \dfrac{V_p}{R_p} = \dfrac{240}{30} = \mathbf{8\ A}$

(c) For a star connection, $I_p = I_L$

Hence the line current, $I_L = \mathbf{8\ A}$

Problem 2. A star-connected load consists of three identical coils each of resistance 30 Ω and inductance 127.3 mH. If the line current is 5.08 A, calculate the line voltage if the supply frequency is 50 Hz

Inductive reactance $X_L = 2\pi f L = 2\pi(50)(127.3 \times 10^{-3}) = 40\ \Omega$

Impedance of each phase $Z_p = \sqrt{(R^2 + X_L^2)} = \sqrt{(30^2 + 40^2)} = 50\ \Omega$

For a star connection $I_L = I_p = \dfrac{V_p}{Z_p}$

Hence phase voltage $V_p = I_p Z_p = (5.08)(50) = 254$ V

Line voltage $V_L = \sqrt{3}V_p = \sqrt{3}(254) =$ **440 V**

> Problem 3. A balanced, three-wire, star-connected, 3-phase load has a phase voltage of 240 V, a line current of 5 A and a lagging power factor of 0.966. Draw the complete phasor diagram.

The phasor diagram is shown in Figure 19.6.

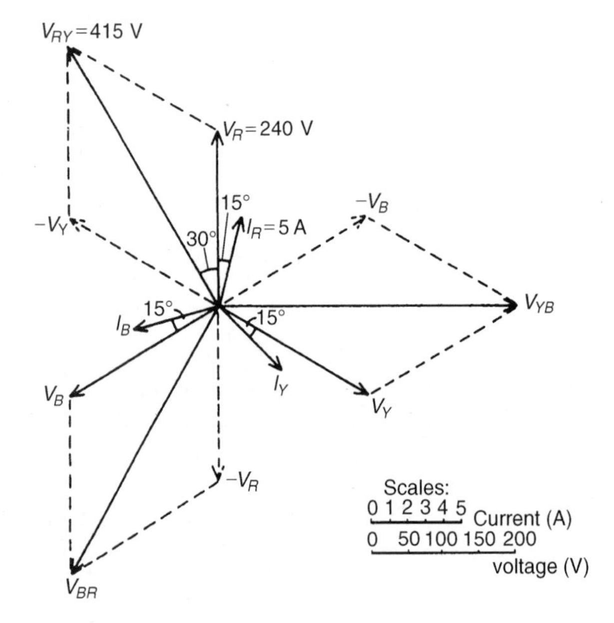

Figure 19.6

Procedure to construct the phasor diagram:

(i) Draw $V_R = V_Y = V_B = 240$ V and spaced 120° apart. (Note that V_R is shown vertically upwards — this however is immaterial for it may be drawn in any direction.)

(ii) Power factor $= \cos\phi = 0.966$ lagging. Hence the load phase angle is given by arccos 0.966, i.e. 15° lagging. Hence $I_R = I_Y = I_B = 5$ A, lagging V_R, V_Y and V_B respectively by 15°

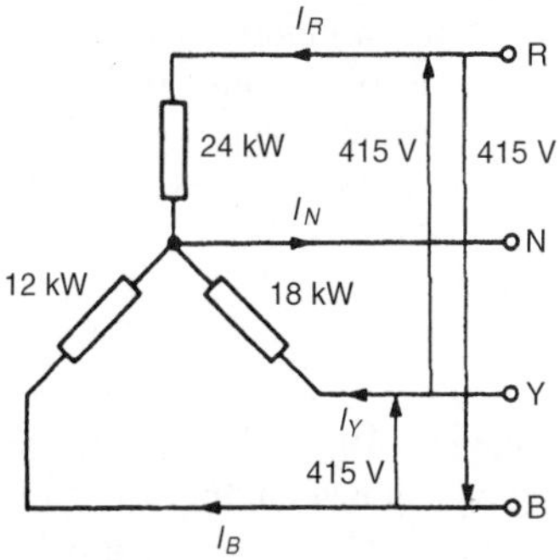

(iii) $V_{RY} = V_R - V_Y$ (phasorially). Hence V_Y is reversed and added phasorially to V_R. By measurement, $V_{RY} = 415$ V (i.e. $\sqrt{3}(240)$) and leads V_R by 30°. Similarly, $V_{YB} = V_Y - V_B$ and $V_{BR} = V_B - V_R$

Figure 19.7

> Problem 4. A 415 V, 3-phase, 4 wire, star-connected system supplies three resistive loads as shown in Figure 19.7. Determine (a) the current in each line and (b) the current in the neutral conductor.

(a) For a star-connected system $V_L = \sqrt{3}V_p$

Hence $$V_p = \frac{V_L}{\sqrt{3}} = \frac{415}{\sqrt{3}} = 240 \text{ V}$$

Since current $I = \dfrac{\text{Power } P}{\text{Voltage } V}$ for a resistive load

then $$I_R = \frac{P_R}{V_R} = \frac{24\,000}{240} = \mathbf{100\ A}$$

$$I_Y = \frac{P_Y}{V_Y} = \frac{18\,000}{240} = \mathbf{75\ A}$$

and $$I_B = \frac{P_B}{V_B} = \frac{12\,000}{240} = \mathbf{50\ A}$$

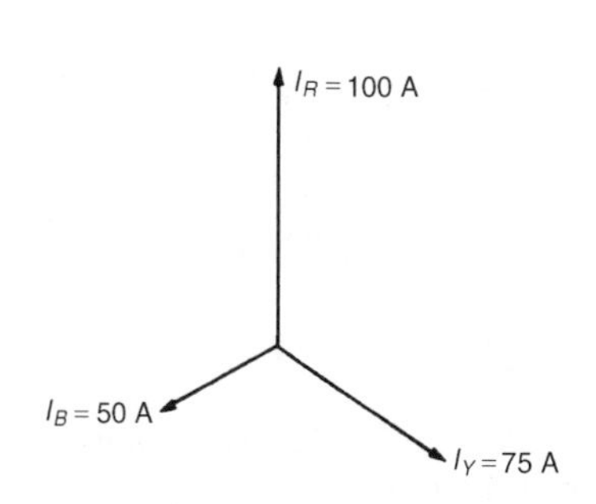

Figure 19.8

(b) The three line currents are shown in the phasor diagram of Figure 19.8. Since each load is resistive the currents are in phase with the phase voltages and are hence mutually displaced by 120°. The current in the neutral conductor is given by:

$$I_N = I_R + I_Y + I_B \text{ phasorially.}$$

Figure 19.9 shows the three line currents added phasorially. oa represents I_R in magnitude and direction. From the nose of oa, ab is drawn representing I_Y in magnitude and direction. From the nose of ab, bc is drawn representing I_B in magnitude and direction. oc represents the resultant, I_N.

By measurement, $\boldsymbol{I_N = 43}$ **A**

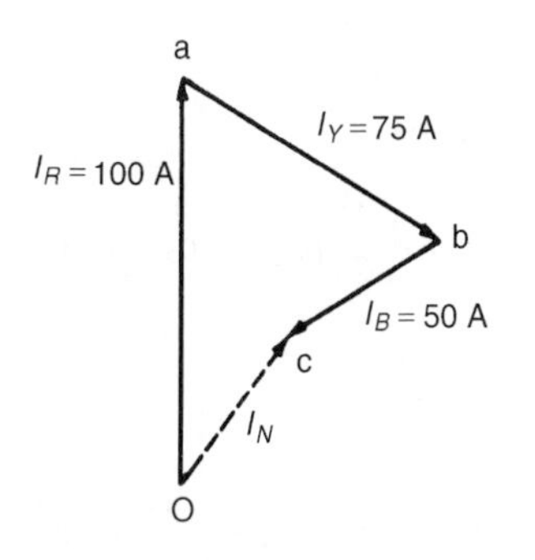

Figure 19.9

Alternatively, by calculation, considering I_R at 90°, I_B at 210° and I_Y at 330°:

Total horizontal component $= 100 \cos 90° + 75 \cos 330° + 50 \cos 210°$

$= 21.65$

Total vertical component $= 100 \sin 90° + 75 \sin 330° + 50 \sin 210°$

$= 37.50$

Hence magnitude of $I_N = \sqrt{(21.65^2 + 37.50^2)} = \mathbf{43.3\ A}$

19.4 Delta connection

(i) A **delta (or mesh) connected load** is shown in Figure 19.10 where the end of one load is connected to the start of the next load.

(ii) From Figure 19.10, it can be seen that the line voltages V_{RY}, V_{YB} and V_{BR} are the respective phase voltages, i.e. for a delta connection:

$$\boldsymbol{V_L = V_p}$$

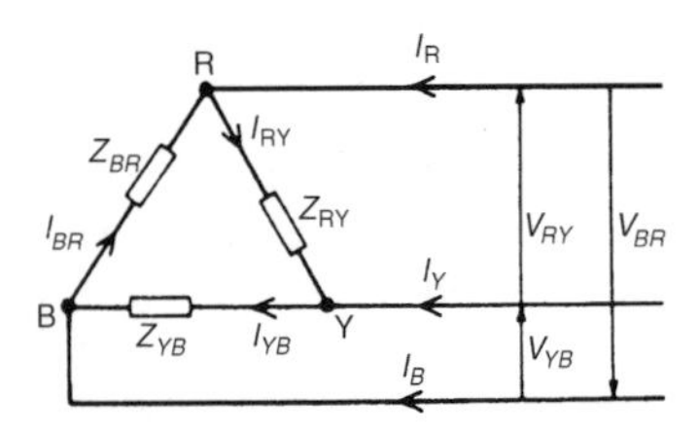

Figure 19.10

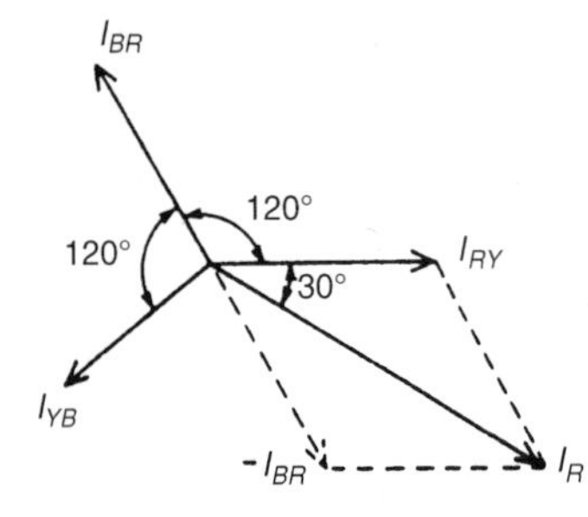

Figure 19.11

(iii) Using Kirchhoff's current law in Figure 19.10, $I_R = I_{RY} - I_{BR} = I_{RY} + (-I_{BR})$. From the phasor diagram shown in Figure 19.11, by trigonometry or by measurement, $I_R = \sqrt{3}I_{RY}$, i.e. for a delta connection:

$$\boldsymbol{I_L = \sqrt{3}I_p}$$

> Problem 5. Three identical coils each of resistance 30 Ω and inductance 127.3 mH are connected in delta to a 440 V, 50 Hz, 3-phase supply. Determine (a) the phase current, and (b) the line current.

Phase impedance, $Z_p = 50\ \Omega$ (from problem 2) and for a delta connection, $V_p = V_L$

(a) Phase current, $I_p = \dfrac{V_p}{Z_p} = \dfrac{V_L}{Z_p} = \dfrac{440}{50} = \mathbf{8.8\ A}$

(b) For a delta connection, $I_L = \sqrt{3}I_p = \sqrt{3}(8.8) = \mathbf{15.24\ A}$

Thus when the load is connected in delta, three times the line current is taken from the supply than is taken if connected in star.

> Problem 6. Three identical capacitors are connected in delta to a 415 V, 50 Hz, 3-phase supply. If the line current is 15 A, determine the capacitance of each of the capacitors.

For a delta connection $I_L = \sqrt{3}I_p$

Hence phase current $I_p = \dfrac{I_L}{\sqrt{3}} = \dfrac{15}{\sqrt{3}} = 8.66\ A$

Capacitive reactance per phase, $X_C = \dfrac{V_p}{I_p} = \dfrac{V_L}{I_p}$ (since for a delta connection $V_L = V_p$)

Hence $X_C = \dfrac{415}{8.66} = 47.92\ \Omega$

$X_C = \dfrac{1}{2\pi f C}$, from which capacitance, $C = \dfrac{1}{2\pi f X_C} = \dfrac{1}{2\pi(50)(47.92)}\,\text{F}$

$= \mathbf{66.43\ \mu F}$

Problem 7. Three coils each having resistance 3 Ω and inductive reactance 4 Ω are connected (i) in star and (ii) in delta to a 415 V, 3-phase supply. Calculate for each connection (a) the line and phase voltages and (b) the phase and line currents.

(i) **For a star connection:** $I_L = I_p$ and $V_L = \sqrt{3}V_p$

(a) A 415 V, 3-phase supply means that the

line voltage, $V_L =$ **415 V**

Phase voltage, $V_p = \dfrac{V_L}{\sqrt{3}} = \dfrac{415}{\sqrt{3}} =$ **240 V**

(b) Impedance per phase, $Z_p = \sqrt{(R^2 + X_L{}^2)} = \sqrt{(3^2 + 4^2)}$

$= 5\ \Omega$

Phase current, $I_p = \dfrac{V_p}{Z_p} = \dfrac{240}{5} =$ **48 A**

Line current, $I_L = I_p =$ **48 A**

(ii) **For a delta connection:** $V_L = V_p$ and $I_L = \sqrt{3}I_p$

(a) Line voltage, $V_L = 415$ V

Phase voltage, $V_p = V_L =$ **415 V**

(b) Phase current, $I_p = \dfrac{V_p}{Z_p} = \dfrac{415}{5} =$ **83 A**

Line current, $I_L = \sqrt{3}I_p = \sqrt{3}(83) =$ **144 A**

Further problems on star and delta connections may be found in Section 19.9, problems 1 to 7, page 312.

19.5 Power in three-phase systems

The power dissipated in a three-phase load is given by the sum of the power dissipated in each phase. If a load is balanced then the total power P is given by: $P = 3 \times$ power consumed by one phase.

The power consumed in one phase $= I_p{}^2R_p$ or $V_pI_p\cos\phi$ (where ϕ is the phase angle between V_p and I_p)

For a star connection, $V_p = \dfrac{V_L}{\sqrt{3}}$ and $I_p = I_L$ hence

$$P = 3\left(\frac{V_L}{\sqrt{3}}\right) I_L\cos\phi = \sqrt{3}V_LI_L\cos\phi$$

For a delta connection, $V_p = V_L$ and $I_p = \dfrac{I_L}{\sqrt{3}}$ hence

$$P = 3V_L\left(\frac{I_L}{\sqrt{3}}\right)\cos\phi = \sqrt{3}V_LI_L\cos\phi$$

Hence for either a star or a delta balanced connection the total power P is given by:

$$\boldsymbol{P = \sqrt{3}V_LI_L\cos\phi\ \text{watts}} \quad \text{or} \quad \boldsymbol{P = 3I_p{}^2R_p\ \text{watts.}}$$

Total volt-amperes, $\boldsymbol{S = \sqrt{3}V_LI_L}$ **volt-amperes**

Problem 8. Three 12 Ω resistors are connected in star to a 415 V, 3-phase supply. Determine the total power dissipated by the resistors.

Power dissipated, $P = \sqrt{3}V_LI_L\cos\phi$ or $P = 3I_p{}^2R_p$

Line voltage, $V_L = 415$ V and phase voltage $V_p = \dfrac{415}{\sqrt{3}} = 240$ V

(since the resistors are star-connected)

Phase current, $I_p = \dfrac{V_p}{Z_p} = \dfrac{V_p}{R_p} = \dfrac{240}{12} = 20$ A

For a star connection $I_L = I_p = 20$ A

For a purely resistive load, the power factor $= \cos\phi = 1$

Hence power $P = \sqrt{3}V_LI_L\cos\phi = \sqrt{3}(415)(20)(1) =$ **14.4 kW**

or power $P = 3I_p{}^2R_p = 3(20)^2(12) =$ **14.4 kW**

Problem 9. The input power to a 3-phase a.c. motor is measured as 5 kW. If the voltage and current to the motor are 400 V and 8.6 A respectively, determine the power factor of the system.

Power, $P = 5000$ W; Line voltage $V_L = 400$ V; Line current, $I_L = 8.6$ A

Power, $P = \sqrt{3}V_LI_L\cos\phi$

Hence power factor $= \cos\phi = \dfrac{P}{\sqrt{3}V_LI_L} = \dfrac{5000}{\sqrt{3}(400)(8.6)} =$ **0.839**

Problem 10. Three identical coils, each of resistance 10 Ω and inductance 42 mH are connected (a) in star and (b) in delta to a 415 V, 50 Hz, 3-phase supply. Determine the total power dissipated in each case.

(a) **Star connection**

Inductive reactance $X_L = 2\pi f L = 2\pi(50)(42 \times 10^{-3})$

$= 13.19\ \Omega$

Phase impedance $Z_p = \sqrt{(R^2 + X_L^2)} = \sqrt{(10^2 + 13.19^2)}$

$= 16.55\ \Omega$

Line voltage $V_L = 415$ V and

phase voltage, $V_p = \dfrac{V_L}{\sqrt{3}} = \dfrac{415}{\sqrt{3}} = 240$ V

Phase current, $I_p = \dfrac{V_p}{Z_p} = \dfrac{240}{16.55} = 14.50$ A

Line current, $I_L = I_p = 14.50$ A

Power factor $= \cos\phi = \dfrac{R_p}{Z_p} = \dfrac{10}{16.55} = 0.6042$ lagging

Power dissipated, $P = \sqrt{3}V_L I_L \cos\phi = \sqrt{3}(415)(14.50)(0.6042)$

$= \mathbf{6.3\ kW}$

(Alternatively, $P = 3{I_p}^2 R_p = 3(14.50)^2(10) =$ **6.3 kW**)

(b) **Delta connection**

$V_L = V_p = 415$ V, $Z_p = 16.55\ \Omega$,

$\cos\phi = 0.6042$ lagging (from above).

Phase current, $I_p = \dfrac{V_p}{Z_p} = \dfrac{415}{16.55} = 25.08$ A

Line current, $I_L = \sqrt{3}I_p = \sqrt{3}(25.08) = 43.44$ A

Power dissipated, $P = \sqrt{3}V_L I_L \cos\phi = \sqrt{3}(415)(43.44)(0.6042)$

$= \mathbf{18.87\ kW}$

(Alternatively, $P = 3{I_p}^2 R_p = 3(25.08)^2(10) =$ **18.87 kW**)

Hence loads connected in delta dissipate three times the power than when connected in star, and also take a line current three times greater.

Problem 11. A 415 V, 3-phase a.c. motor has a power output of 12.75 kW and operates at a power factor of 0.77 lagging and with an efficiency of 85%. If the motor is delta-connected, determine (a) the power input, (b) the line current and (c) the phase current.

(a) Efficiency $= \dfrac{\text{power output}}{\text{power input}}$, hence $\dfrac{85}{100} = \dfrac{12\,750}{\text{power input}}$

from which, **power input** $= \dfrac{12\,750 \times 100}{85} =$ **15 000 W or 15 kW**

(b) Power, $P = \sqrt{3}V_L I_L \cos\phi$, hence

$$\textbf{line current}, I_L = \frac{P}{\sqrt{3}V_L \cos\phi} = \frac{15\,000}{\sqrt{3}(415)(0.77)} = \mathbf{27.10\ A}$$

(c) For a delta connection, $I_L = \sqrt{3}I_p$,

$$\text{hence } \textbf{phase current}, I_p = \frac{I_L}{\sqrt{3}} = \frac{27.10}{\sqrt{3}} = \mathbf{15.65\ A}$$

19.6 Measurement of power in three-phase systems

Power in three-phase loads may be measured by the following methods:

(i) **One-wattmeter method for a balanced load**
Wattmeter connections for both star and delta are shown in Figure 19.12.

Total power = 3 × wattmeter reading

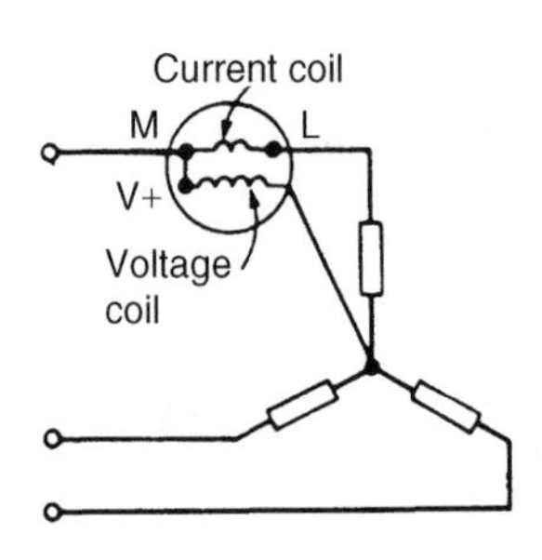

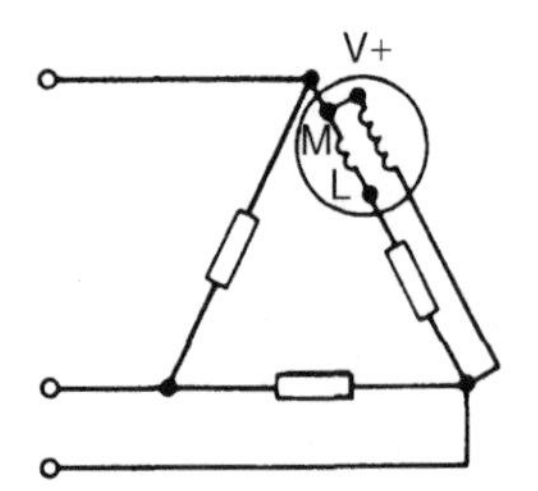

Figure 19.12

(ii) **Two-wattmeter method for balanced or unbalanced loads**
A connection diagram for this method is shown in Figure 19.13 for a star-connected load. Similar connections are made for a delta-connected load.

Total power = sum of wattmeter readings = $P_1 + P_2$

The power factor may be determined from:

$$\tan\phi = \sqrt{3}\left(\frac{P_1 - P_2}{P_1 + P_2}\right)$$ (see Problems 12 and 15 to 18)

It is possible, depending on the load power factor, for one wattmeter to have to be 'reversed' to obtain a reading. In this case it is taken as a negative reading (see Problem 17).

(iii) **Three-wattmeter method for a three-phase, 4-wire system for balanced and unbalanced loads.** (see Figure 19.14).

Total power = $P_1 + P_2 + P_3$

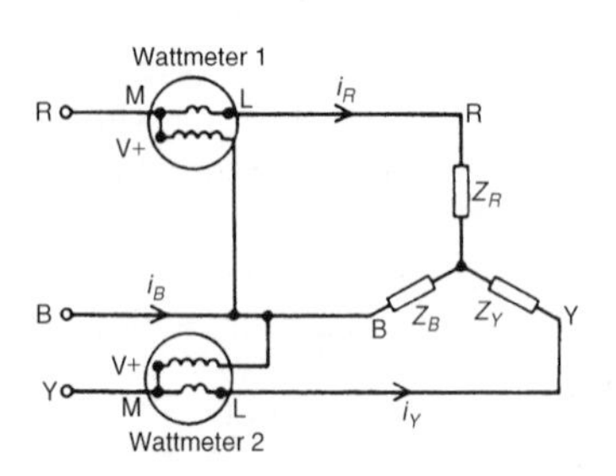

Figure 19.13

Problem 12. (a) Show that the total power in a 3-phase, 3-wire system using the two-wattmeter method of measurement is given by the sum of the wattmeter readings. Draw a connection diagram. (b) Draw a phasor diagram for the two-wattmeter method for a balanced load. (c) Use the phasor diagram of part (b) to derive a formula from which the power factor of a 3-phase system may be determined using only the wattmeter readings.

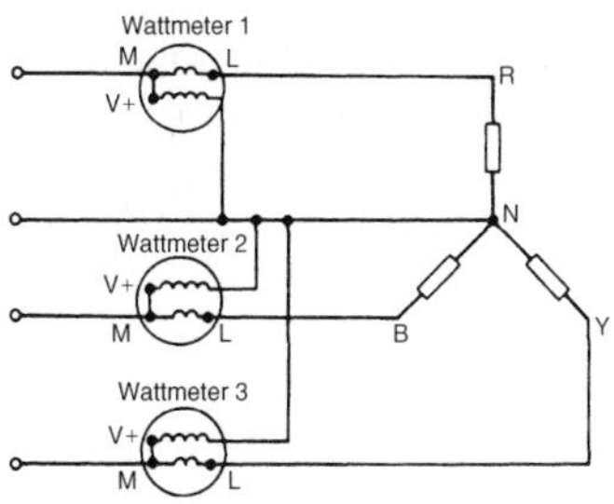

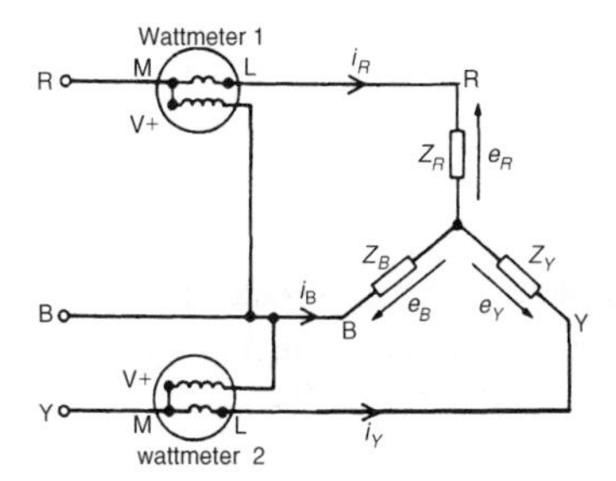

Figure 19.14

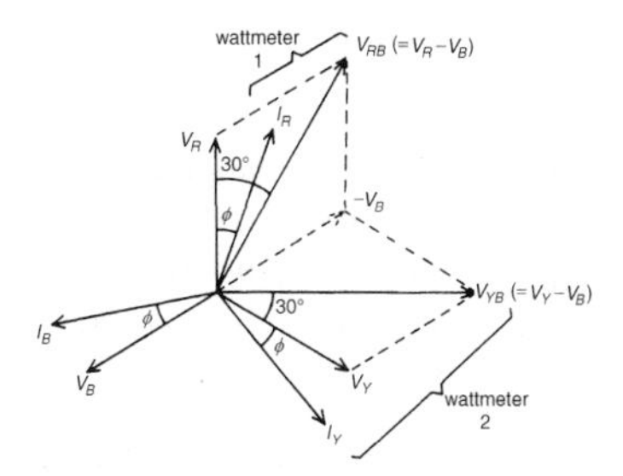

Figure 19.15

Figure 19.16

(a) A connection diagram for the two-wattmeter method of a power measurement is shown in Figure 19.15 for a star-connected load. Total instantaneous power, $p = e_R i_R + e_Y i_Y + e_B i_B$ and in any 3 phase system $i_R + i_Y + i_B = 0$. Hence $i_B = -i_R - i_Y$

Thus, $p = e_R i_R + e_Y i_Y + e_B(-i_R - i_Y)$

$$= (e_R - e_B)i_R + (e_Y - e_B)i_Y$$

However, $(e_R - e_B)$ is the p.d. across wattmeter 1 in Figure 19.15 and $(e_Y - e_B)$ is the p.d. across wattmeter 2.

Hence total instantaneous power,

$p = (\text{wattmeter 1 reading}) + (\text{wattmeter 2 reading}) = p_1 + p_2$

The moving systems of the wattmeters are unable to follow the variations which take place at normal frequencies and they indicate the mean power taken over a cycle. Hence the total power, $P = P_1 + P_2$ for balanced or unbalanced loads.

(b) The phasor diagram for the two-wattmeter method for a balanced load having a lagging current is shown in Figure 19.16, where $V_{RB} = V_R - V_B$ and $V_{YB} = V_Y - V_B$ (phasorially).

(c) Wattmeter 1 reads $V_{RB} I_R \cos(30° - \phi) = P_1$

Wattmeter 2 reads $V_{YB} I_Y \cos(30° + \phi) = P_2$

$$\frac{P_1}{P_2} = \frac{V_{RB} I_R \cos(30° - \phi)}{V_{YB} I_Y \cos(30° + \phi)} = \frac{\cos(30° - \phi)}{\cos(30° + \phi)}$$

since $I_R = I_Y$ and $V_{RB} = V_{YB}$ for a balanced load.

Hence $\dfrac{P_1}{P_2} = \dfrac{\cos 30° \cos\phi + \sin 30° \sin\phi}{\cos 30° \cos\phi - \sin 30° \sin\phi}$

(from compound angle formulae, see *'Higher Engineering Mathematics'*)

Dividing throughout by $\cos 30° \cos\phi$ gives:

$$\frac{P_1}{P_2} = \frac{1 + \tan 30° \tan\phi}{1 - \tan 30° \tan\phi} = \frac{1 + \frac{1}{\sqrt{3}}\tan\phi}{1 - \frac{1}{\sqrt{3}}\tan\phi}, \text{ (since } \frac{\sin\phi}{\cos\phi} = \tan\phi)$$

Cross-multiplying gives: $P_1 - \dfrac{P_1}{\sqrt{3}}\tan\phi = P_2 + \dfrac{P_2}{\sqrt{3}}\tan\phi$

Hence $$P_1 - P_2 = (P_1 + P_2)\frac{\tan\phi}{\sqrt{3}}$$

from which $\mathbf{\tan\phi = \sqrt{3}\left(\frac{P_1 - P_2}{P_1 + P_2}\right)}$

ϕ, $\cos\phi$ and thus power factor can be determined from this formula.

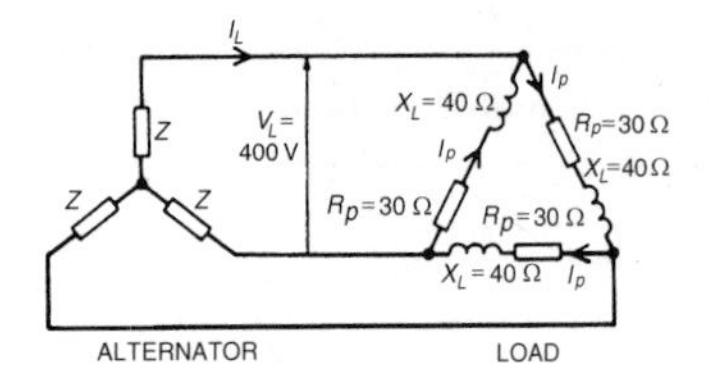

Figure 19.17

Problem 13. A 400 V, 3-phase star connected alternator supplies a delta-connected load, each phase of which has a resistance of 30 Ω and inductive reactance 40 Ω. Calculate (a) the current supplied by the alternator and (b) the output power and the kVA of the alternator, neglecting losses in the line between the alternator and load.

A circuit diagram of the alternator and load is shown in Figure 19.17.

(a) Considering the load: Phase current, $I_p = \frac{V_p}{Z_p}$

$V_p = V_L$ for a delta connection. Hence $V_p = 400$ V

Phase impedance, $Z_p = \sqrt{(R_p{}^2 + X_L{}^2)} = \sqrt{(30^2 + 40^2)} = 50\ \Omega$

Hence $I_p = \frac{V_p}{Z_p} = \frac{400}{50} = 8$ A

For a delta-connection, line current, $I_L = \sqrt{3}I_p = \sqrt{3}(8) = 13.86$ A

Hence 13.86 A is the current supplied by the alternator.

(b) Alternator output power is equal to the power dissipated by the load.

i.e. $P = \sqrt{3}V_L I_L \cos\phi$, where $\cos\phi = \frac{R_p}{Z_p} = \frac{30}{50} = 0.6$

Hence $P = \sqrt{3}(400)(13.86)(0.6) =$ **5.76 kW**

Alternator output kVA, $S = \sqrt{3}V_L I_L = \sqrt{3}(400)(13.86)$

$=$ **9.60 kVA**

Problem 14. Each phase of a delta-connected load comprises a resistance of 30 Ω and an 80 μF capacitor in series. The load is connected to a 400 V, 50 Hz, 3-phase supply. Calculate (a) the phase current, (b) the line current, (c) the total power dissipated and (d) the kVA rating of the load. Draw the complete phasor diagram for the load.

(a) Capacitive reactance, $X_C = \frac{1}{2\pi f C} = \frac{1}{2\pi(50)(80 \times 10^{-6})}$

$= 39.79\ \Omega$

Phase impedance, $Z_p = \sqrt{(R_p{}^2 + X_C{}^2)} = \sqrt{(30^2 + 39.79^2)}$

$= 49.83\ \Omega$

Power factor $= \cos\phi = \frac{R_p}{Z_p} = \frac{30}{49.83} = 0.602$

Hence $\phi = \arccos 0.602 = 52°59'$ leading.

Phase current, $I_p = \frac{V_p}{Z_p}$ and $V_p = V_L$ for a delta connection.

Hence $\quad I_p = \dfrac{400}{49.83} = \mathbf{8.027\ A}$

(b) Line current $I_L = \sqrt{3}I_p$ for a delta-connection

Hence $\quad I_L = \sqrt{3}(8.027) = \mathbf{13.90\ A}$

(c) Total power dissipated, $P = \sqrt{3}V_L I_L \cos\phi$

$$= \sqrt{3}(400)(13.90)(0.602) = \mathbf{5.797\ kW}$$

(d) Total kVA, $S = \sqrt{3}V_L I_L = \sqrt{3}(400)(13.90) = \mathbf{9.630\ kVA}$

The phasor diagram for the load is shown in Figure 19.18

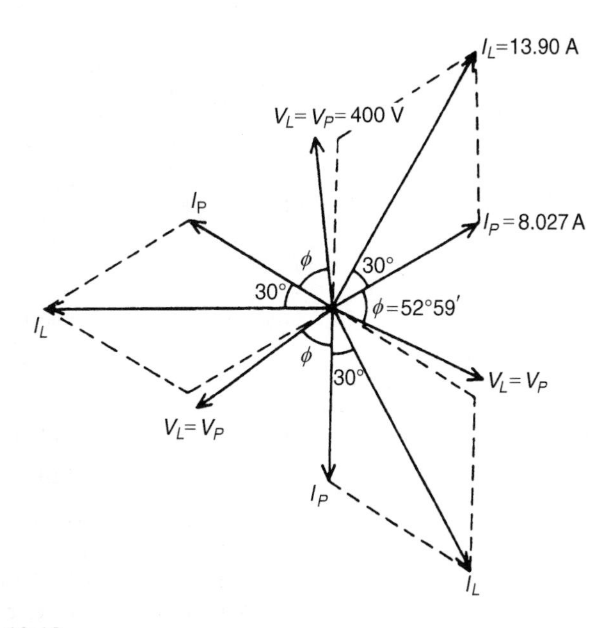

Figure 19.18

Problem 15. Two wattmeters are connected to measure the input power to a balanced 3-phase load by the two-wattmeter method. If the instrument readings are 8 kW and 4 kW, determine (a) the total power input and (b) the load power factor.

(a) Total input power, $P = P_1 + P_2 = 8 + 4 = \mathbf{12\ kW}$

(b) $\tan\phi = \sqrt{3}\left(\dfrac{P_1 - P_2}{P_1 + P_2}\right) = \sqrt{3}\left(\dfrac{8-4}{8+4}\right)$

$$= \sqrt{3}\left(\frac{4}{12}\right) = \sqrt{3}\left(\frac{1}{3}\right) = \frac{1}{\sqrt{3}}$$

Hence $\phi = \arctan \dfrac{1}{\sqrt{3}} = 30°$

Power factor $= \cos\phi = \cos 30° = \mathbf{0.866}$

Problem 16. Two wattmeters connected to a 3-phase motor indicate the total power input to be 12 kW. The power factor is 0.6. Determine the readings of each wattmeter.

If the two wattmeters indicate P_1 and P_2 respectively

$$\text{then } P_1 + P_2 = 12 \text{ kW} \quad (1)$$

$\tan\phi = \sqrt{3}\left(\dfrac{P_1 - P_2}{P_1 + P_2}\right)$ and power factor $= 0.6 = \cos\phi$

Angle $\phi = \arccos 0.6 = 53.13°$ and $\tan 53.13° = 1.3333$

Hence $1.3333 = \dfrac{\sqrt{3}(P_1 - P_2)}{12}$, from which, $P_1 - P_2 = \dfrac{12(1.3333)}{\sqrt{3}}$

i.e., $P_1 - P_2 = 9.237$ kW (2)

Adding equations (1) and (2) gives: $2P_1 = 21.237$,

i.e. $P_1 = \dfrac{21.237}{2} = 10.62$ kW

Hence **wattmeter 1 reads 10.62 kW**

From equation (1), **wattmeter 2 reads (12 − 10.62) = 1.38 kW**

Problem 17. Two wattmeters indicate 10 kW and 3 kW respectively when connected to measure the input power to a 3-phase balanced load, the reverse switch being operated on the meter indicating the 3 kW reading. Determine (a) the input power and (b) the load power factor.

Since the reversing switch on the wattmeter had to be operated the 3 kW reading is taken as −3 kW.

(a) Total input power, $P = P_1 + P_2 = 10 + (-3) = \mathbf{7\ kW}$

(b) $\tan\phi = \sqrt{3}\left(\dfrac{P_1 - P_2}{P_1 + P_2}\right) = \sqrt{3}\left(\dfrac{10 - (-3)}{10 + (-3)}\right)$

$= \sqrt{3}\left(\dfrac{13}{7}\right) = 3.2167$

Angle $\phi = \arctan 3.2167 = 72.73°$

Power factor $= \cos\phi = \cos 72.73° = \mathbf{0.297}$

Problem 18. Three similar coils, each having a resistance of 8 Ω and an inductive reactance of 8 Ω are connected (a) in star and (b) in delta, across a 415 V, 3-phase supply. Calculate for each connection the readings on each of two wattmeters connected to measure the power by the two-wattmeter method.

(a) **Star connection:** $V_L = \sqrt{3}V_p$ and $I_L = I_p$

$$\text{Phase voltage, } V_p = \frac{V_L}{\sqrt{3}} = \frac{415}{\sqrt{3}} \text{ and}$$

$$\text{phase impedance, } Z_p = \sqrt{(R_p{}^2 + X_L{}^2)}$$

$$= \sqrt{(8^2 + 8^2)} = 11.31\ \Omega$$

$$\text{Hence phase current, } I_p = \frac{V_p}{Z_p} = \frac{415/\sqrt{3}}{11.31} = 21.18\ \text{A}$$

Total power, $P = 3I_p{}^2R_p = 3(21.18)^2(8) = 10\,766$ W

If wattmeter readings are P_1 and P_2 then $P_1 + P_2 = 10\,766$ (1)

Since $R_p = 8\ \Omega$ and $X_L = 8\ \Omega$, then phase angle $\phi = 45°$ (from impedance triangle)

$$\tan\phi = \sqrt{3}\left(\frac{P_1 - P_2}{P_1 + P_2}\right), \text{ hence } \tan 45° = \frac{\sqrt{3}(P_1 - P_2)}{10\,766}$$

$$\text{from which } P_1 - P_2 = \frac{10\,766(1)}{\sqrt{3}} = 6\,216\ \text{W} \qquad (2)$$

Adding equations (1) and (2) gives: $2P_1 = 10\,766 + 6\,216$

$$= 16\,982\ \text{W}$$

Hence $P_1 = 8\,491$ W

From equation (1), $P_2 = 10\,766 - 8\,491 = 2\,275$ W

When the coils are star-connected the wattmeter readings are thus 8.491 kW and 2.275 kW.

(b) **Delta connection:** $V_L = V_p$ and $I_L = \sqrt{3}I_p$

$$\text{Phase current, } I_p = \frac{V_p}{Z_p} = \frac{415}{11.31} = 36.69\ \text{A}$$

Total power, $P = 3I_p{}^2R_p = 3(36.69)^2(8) = 32\,310$ W

Hence $P_1 + P_2 = 32\,310$ W (3)

$$\tan\phi = \sqrt{3}\left(\frac{P_1 - P_2}{P_1 + P_2}\right) \text{ thus } 1 = \frac{\sqrt{3}(P_1 - P_2)}{32\,310}$$

$$\text{from which, } P_1 - P_2 = \frac{32\,310}{\sqrt{3}} = 18\,650\ \text{W} \qquad (4)$$

Adding equations (3) and (4) gives:

$$2P_1 = 50\,960, \text{ from which } P_1 = 25\,480 \text{ W}$$

From equation (3), $P_2 = 32\,310 - 25\,480 = 6\,830$ W

When the coils are delta-connected the wattmeter readings are thus 25.48 kW and 6.83 kW.

Further problems on power in 3-phase circuits may be found in Section 19.9, problems 8 to 19, page 313.

19.7 Comparison of star and delta connections

(i) Loads connected in delta dissipate three times more power than when connected in star to the same supply.

(ii) For the same power, the phase currents must be the same for both delta and star connections (since power $= 3I_p{}^2R_p$), hence the line current in the delta-connected system is greater than the line current in the corresponding star-connected system. To achieve the same phase current in a star-connected system as in a delta-connected system, the line voltage in the star system is $\sqrt{3}$ times the line voltage in the delta system.

Thus for a given power transfer, a delta system is associated with larger line currents (and thus larger conductor cross-sectional area) and a star system is associated with a larger line voltage (and thus greater insulation).

19.8 Advantages of three-phase systems

Advantages of three-phase systems over single-phase supplies include:

(i) For a given amount of power transmitted through a system, the three-phase system requires conductors with a smaller cross-sectional area. This means a saving of copper (or aluminium) and thus the original installation costs are less.

(ii) Two voltages are available (see Section 19.3(vii)).

(iii) Three-phase motors are very robust, relatively cheap, generally smaller, have self-starting properties, provide a steadier output and require little maintenance compared with single-phase motors.

19.9 Further problems on three-phase systems

Star and delta connections

1 Three loads, each of resistance 50 Ω are connected in star to a 400 V, 3-phase supply. Determine (a) the phase voltage, (b) the phase current and (c) the line current.

[(a) 231 V (b) 4.62 A (c) 4.62 A]

2 If the loads in question 1 are connected in delta to the same supply determine (a) the phase voltage, (b) the phase current and (c) the line current. [(a) 400 V (b) 8 A (c) 13.86 A]

3 A star-connected load consists of three identical coils, each of inductance 159.2 mH and resistance 50 Ω. If the supply frequency is 50 Hz and the line current is 3 A determine (a) the phase voltage and (b) the line voltage. [(a) 212 V (b) 367 V]

4 Three identical capacitors are connected (a) in star, (b) in delta to a 400 V, 50 Hz, 3-phase supply. If the line current is 12 A determine in each case the capacitance of each of the capacitors. [(a) 165.4 μF (b) 55.13 μF]

5 Three coils each having resistance 6 Ω and inductance L H are connected (a) in star and (b) in delta, to a 415 V, 50 Hz, 3-phase supply. If the line current is 30 A, find for each connection the value of L. [(a) 16.78 mH (b) 73.84 mH]

6 A 400 V, 3-phase, 4 wire, star-connected system supplies three resistive loads of 15 kW, 20 kW and 25 kW in the red, yellow and blue phases respectively. Determine the current flowing in each of the four conductors.
[$I_R = 64.95$ A, $I_Y = 86.60$ A
$I_B = 108.25$ A, $I_N = 37.50$ A]

7 A 3-phase, star-connected alternator delivers a line current of 65 A to a balanced delta-connected load at a line voltage of 380 V. Calculate (a) the phase voltage of the alternator, (b) the alternator phase current and (c) the load phase current. [(a) 219.4 V (b) 65 A (c) 37.53 A]

Power in 3-phase circuits

8 Determine the total power dissipated by three 20 Ω resistors when connected (a) in star and (b) in delta to a 440 V, 3-phase supply. [(a) 9.68 kW (b) 29.04 kW]

9 Determine the power dissipated in the circuit of problem 3. [1.35 kW]

10 A balanced delta-connected load has a line voltage of 400 V, a line current of 8 A and a lagging power factor of 0.94. Draw a complete phasor diagram of the load. What is the total power dissipated by the load? [5.21 kW]

11 Three inductive loads, each of resistance 4 Ω and reactance 9 Ω are connected in delta. When connected to a 3-phase supply the loads consume 1.2 kW. Calculate (a) the power factor of the load, (b) the phase current, (c) the line current and (d) the supply voltage. [(a) 0.406 (b) 10 A (c) 17.32 A (d) 98.49 V]

12 The input voltage, current and power to a motor is measured as 415 V, 16.4 A and 6 kW respectively. Determine the power factor of the system. [0.509]

13 A 440 V, 3-phase a.c. motor has a power output of 11.25 kW and operates at a power factor of 0.8 lagging and with an efficiency of 84%. If the motor is delta connected determine (a) the power input, (b) the line current and (c) the phase current.
[(a) 13.39 kW (b) 21.96 A (c) 12.68 A]

14 Two wattmeters are connected to measure the input power to a balanced 3-phase load. If the wattmeter readings are 9.3 kW and 5.4 kW determine (a) the total output power, and (b) the load power factor. [(a) 14.7 kW (b) 0.909]

15 8 kW is found by the two-wattmeter method to be the power input to a 3-phase motor. Determine the reading of each wattmeter if the power factor of the system is 0.85. [5.431 kW, 2.569 kW]

16 Three similar coils, each having a resistance of 4.0 Ω and an inductive reactance of 3.46 Ω are connected (a) in star and (b) in delta across a 400 V, 3-phase supply. Calculate for each connection the readings on each of two wattmeters connected to measure the power by the two-wattmeter method.
[(a) 17.15 kW, 5.73 kW (b) 51.46 kW, 17.18 kW]

17 A 3-phase, star-connected alternator supplies a delta connected load, each phase of which has a resistance of 15 Ω and inductive reactance 20 Ω. If the line voltage is 400 V, calculate (a) the current supplied by the alternator and (b) the output power and kVA rating of the alternator, neglecting any losses in the line between the alternator and the load. [(a) 27.71 A (b) 11.52 kW, 19.2 kVA]

18 Each phase of a delta-connected load comprises a resistance of 40 Ω and a 40 μF capacitor in series. Determine, when connected to a 415 V, 50 Hz, 3-phase supply (a) the phase current, (b) the line current, (c) the total power dissipated, and (d) the kVA rating of the load. [(a) 4.66 A (b) 8.07 A (c) 2.605 kW (d) 5.80 kVA]

19 Three 24 μF capacitors are connected in star across a 400 V, 50 Hz, 3-phase supply. What value of capacitance must be connected in delta in order to take the same line current?
[8 μF]

20 Transformers

At the end of this chapter you should be able to:

- understand the principle of operation of a transformer
- understand the term 'rating' of a transformer
- use $\frac{V_1}{V_2} = \frac{N_1}{N_2} = \frac{I_2}{I_1}$ in calculations on transformers
- construct a transformer no-load phasor diagram and calculate magnetizing and core loss components of the no-load current
- state the e.m.f. equation for a transformer $E = 4.44 f \Phi_m N$ and use it in calculations
- construct a transformer on-load phasor diagram for an inductive circuit assuming the volt drop in the windings is negligible
- describe transformer construction
- derive the equivalent resistance, reactance and impedance referred to the primary of a transformer
- understand voltage regulation
- describe losses in transformers and calculate efficiency
- appreciate the concept of resistance matching and how it may be achieved
- perform calculations using $R_1 = \left(\frac{N_1}{N_2}\right)^2 R_L$
- describe an auto transformer, its advantages/disadvantages and uses
- describe an isolating transformer, stating uses
- describe a three-phase transformer
- describe current and voltage transformers

20.1 Introduction

A transformer is a device which uses the phenomenon of mutual induction (see Chapter 9) to change the values of alternating voltages and currents. In fact, one of the main advantages of a.c. transmission and distribution is the ease with which an alternating voltage can be increased or decreased by transformers.

Losses in transformers are generally low and thus efficiency is high. Being static they have a long life and are very stable.

Transformers range in size from the miniature units used in electronic applications to the large power transformers used in power stations. The principle of operation is the same for each.

A transformer is represented in Figure 20.1(a) as consisting of two electrical circuits linked by a common ferromagnetic core. One coil is termed the **primary winding** which is connected to the supply of electricity, and the other the **secondary winding**, which may be connected to a load. A circuit diagram symbol for a transformer is shown in Figure 20.1(b).

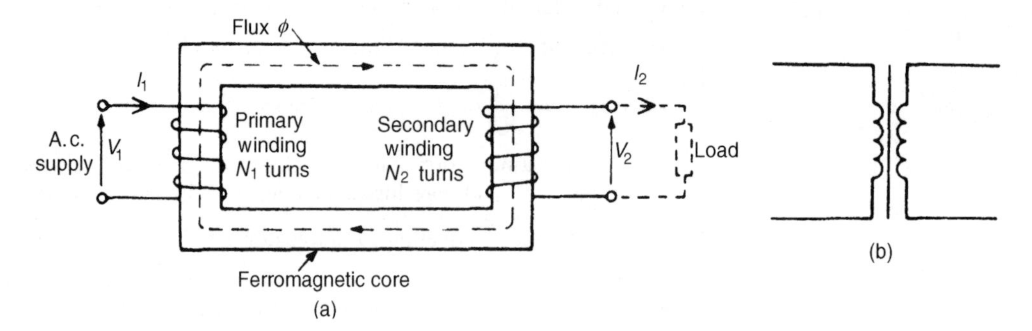

Figure 20.1

20.2 Transformer principle of operation

When the secondary is an open-circuit and an alternating voltage V_1 is applied to the primary winding, a small current — called the no-load current I_0 — flows, which sets up a magnetic flux in the core. This alternating flux links with both primary and secondary coils and induces in them e.m.f.'s of E_1 and E_2 respectively by mutual induction.

The induced e.m.f. E in a coil of N turns is given by

$$E = -N\frac{d\Phi}{dt} \text{ volts,}$$

where $d\Phi/dt$ is the rate of change of flux. In an ideal transformer, the rate of change of flux is the same for both primary and secondary and thus $E_1/N_1 = E_2/N_2$, i.e. **the induced e.m.f. per turn is constant**.

Assuming no losses, $E_1 = V_1$ and $E_2 = V_2$ Hence

$$\frac{V_1}{N_1} = \frac{V_2}{N_2} \text{ or } \frac{V_1}{V_2} = \frac{N_1}{N_2} \qquad (20.1)$$

V_1/V_2 is called the voltage ratio and N_1/N_2 the turns ratio, or the **'transformation ratio'** of the transformer. If N_2 is less than N_1 then V_2 is less than V_1 and the device is termed a **step-down transformer**. If N_2 is greater then N_1 then V_2 is greater than V_1 and the device is termed a **step-up transformer**.

When a load is connected across the secondary winding, a current I_2 flows. In an ideal transformer losses are neglected and a transformer is considered to be 100% efficient.

Hence input power = output power, or $V_1 I_1 = V_2 I_2$, i.e., in an ideal transformer, the **primary and secondary volt-amperes are equal**.

Thus $$\frac{V_1}{V_2} = \frac{I_2}{I_1} \tag{20.2}$$

Combining equations (20.1) and (20.2) gives:

$$\boxed{\frac{V_1}{V_2} = \frac{N_1}{N_2} = \frac{I_2}{I_1}} \tag{20.3}$$

The **rating** of a transformer is stated in terms of the volt-amperes that it can transform without overheating. With reference to Figure 20.1(a), the transformer rating is either $V_1 I_1$ or $V_2 I_2$, where I_2 is the full-load secondary current.

Problem 1. A transformer has 500 primary turns and 3000 secondary turns. If the primary voltage is 240 V, determine the secondary voltage, assuming an ideal transformer.

For an ideal transformer, voltage ratio = turns ratio i.e.,

$$\frac{V_1}{V_2} = \frac{N_1}{N_2}, \text{ hence } \frac{240}{V_2} = \frac{500}{3000}$$

Thus secondary voltage $V_2 = \dfrac{(3000)(240)}{(500)} = \mathbf{1440\ V\ or\ 1.44\ kV}$

Problem 2. An ideal transformer with a turns ratio of 2:7 is fed from a 240 V supply. Determine its output voltage.

A turns ratio of 2:7 means that the transformer has 2 turns on the primary for every 7 turns on the secondary (i.e. a step-up transformer). Thus,

$$\frac{N_1}{N_2} = \frac{2}{7}$$

For an ideal transformer, $\dfrac{N_1}{N_2} = \dfrac{V_1}{V_2}$; hence $\dfrac{2}{7} = \dfrac{240}{V_2}$

Thus the secondary voltage $V_2 = \dfrac{(240)(7)}{(2)} = \mathbf{840\ V}$

Problem 3. An ideal transformer has a turns ratio of 8:1 and the primary current is 3 A when it is supplied at 240 V. Calculate the secondary voltage and current.

A turns ratio of 8:1 means $\frac{N_1}{N_2} = \frac{8}{1}$, i.e. a step-down transformer.

$$\frac{N_1}{N_2} = \frac{V_1}{V_2}, \text{ or secondary voltage } V_2 = V_1\left(\frac{N_2}{N_1}\right) = 240\left(\frac{1}{8}\right) = \mathbf{30\ volts}$$

Also, $\frac{N_1}{N_2} = \frac{I_2}{I_1}$; hence secondary current $I_2 = I_1\left(\frac{N_1}{N_2}\right)$

$$= 3\left(\frac{8}{1}\right) = \mathbf{24\ A}$$

Problem 4. An ideal transformer, connected to a 240 V mains, supplies a 12 V, 150 W lamp. Calculate the transformer turns ratio and the current taken from the supply.

$V_1 = 240$ V, $V_2 = 12$ V, $I_2 = \frac{P}{V_2} = \frac{150}{12} = 12.5$ A

Turns ratio $= \frac{N_1}{N_2} = \frac{V_1}{V_2} = \frac{240}{12} = \mathbf{20}$

$\frac{V_1}{V_2} = \frac{I_2}{I_1}$, from which, $I_1 = I_2\left(\frac{V_2}{V_1}\right) = 12.5\left(\frac{12}{240}\right)$

Hence current taken from the supply, $I_1 = \frac{12.5}{20} = \mathbf{0.625\ A}$

Problem 5. A 5 kVA single-phase transformer has a turns ratio of 10:1 and is fed from a 2.5 kV supply. Neglecting losses, determine (a) the full-load secondary current, (b) the minimum load resistance which can be connected across the secondary winding to give full load kVA, (c) the primary current at full load kVA.

(a) $\frac{N_1}{N_2} = \frac{10}{1}$ and $V_1 = 2.5$ kV $= 2500$ V

Since $\frac{N_1}{N_2} = \frac{V_1}{V_2}$, secondary voltage $V_2 = V_1\left(\frac{N_2}{N_1}\right)$

$$= 2500\left(\frac{1}{10}\right) = 250 \text{ V}$$

The transformer rating in volt-amperes $= V_2 I_2$ (at full load), i.e., $5000 = 250 I_2$

Hence full load secondary current $I_2 = \dfrac{5000}{250} = \mathbf{20\ A}$

(b) Minimum value of load resistance, $R_L = \dfrac{V_2}{I_2} = \dfrac{250}{20} = \mathbf{12.5\ \Omega}$

(c) $\dfrac{N_1}{N_2} = \dfrac{I_2}{I_1}$, from which primary current $I_1 = I_2\left(\dfrac{N_2}{N_1}\right)$

$$= 20\left(\frac{1}{10}\right) = \mathbf{2\ A}$$

Further problems on the transformer principle of operation may be found in Section 20.16, problems 1 to 9, page 344.

20.3 Transformer no-load phasor diagram

(i) The core flux is common to both primary and secondary windings in a transformer and is thus taken as the reference phasor in a phasor diagram. On no-load the primary winding takes a small no-load current I_0 and since, with losses neglected, the primary winding is a pure inductor, this current lags the applied voltage V_1 by 90°. In the phasor diagram assuming no losses, shown in Figure 20.2(a), current I_0 produces the flux and is drawn in phase with the flux. The primary induced e.m.f. E_1 is in phase opposition to V_1 (by Lenz's law) and is shown 180° out of phase with V_1 and equal in magnitude. The secondary induced e.m.f. is shown for a 2:1 turns ratio transformer.

(ii) A no-load phasor diagram for a practical transformer is shown in Figure 20.2(b). If current flows then losses will occur. When losses are considered then the no-load current I_0 is the phasor sum of two components—(i) $\boldsymbol{I_M}$, **the magnetizing component**, in phase with the flux, and (ii) $\boldsymbol{I_C}$, **the core loss component** (supplying the hysteresis and eddy current losses). From Figure 20.2(b):

No-load current, $\boldsymbol{I_0 = \sqrt{(I_M^2 + I_C^2)}}$, where $\boldsymbol{I_M = I_0 \sin\phi_0}$ and

$$\boldsymbol{I_C = I_0 \cos\phi_0}$$

Power factor on no-load $= \cos\phi_0 = \dfrac{I_C}{I_0}$

The total core losses (i.e. iron losses) $= V_1 I_0 \cos\phi_0$

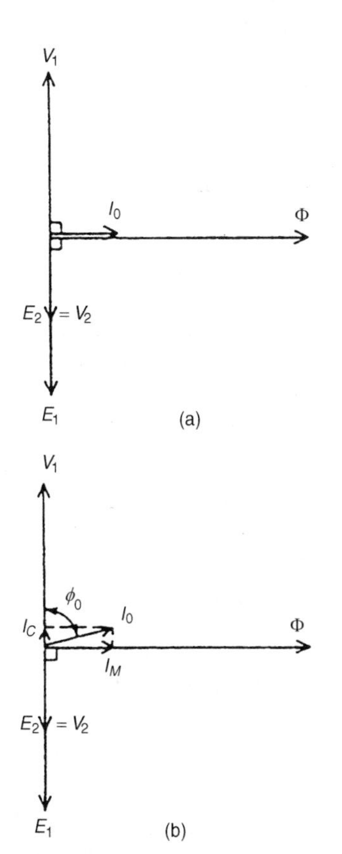

Figure 20.2

Problem 6. A 2400 V/400 V single-phase transformer takes a no-load current of 0.5 A and the core loss is 400 W. Determine the values of the magnetizing and core loss components of the no-load current. Draw to scale the no-load phasor diagram for the transformer.

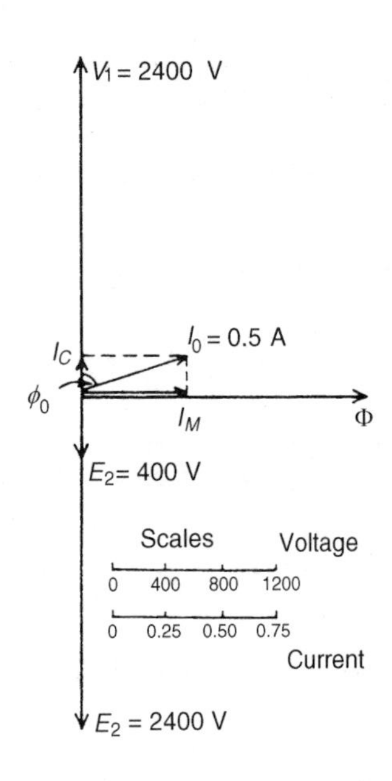

Figure 20.3

$V_1 = 2400$ V, $V_2 = 400$ V, $I_0 = 0.5$ A

Core loss (i.e. iron loss) $= 400 = V_1 I_0 \cos\phi_0$

i.e. $400 = (2400)(0.5)\cos\phi_0$

$$\text{Hence } \cos\phi_0 = \frac{400}{(2400)(0.5)} = 0.3333$$

$$\phi_0 = \arccos 0.3333 = 70.53^\circ$$

The no-load phasor diagram is shown in Figure 20.3.

Magnetizing component, $I_M = I_0 \sin\phi_0 = 0.5 \sin 70.53^\circ =$ **0.471 A**

Core loss component, $I_C = I_0 \cos\phi_0 = 0.5 \cos 70.53^\circ =$ **0.167 A**

Problem 7. A transformer takes a current of 0.8 A when its primary is connected to a 240 volt, 50 Hz supply, the secondary being on open circuit. If the power absorbed is 72 watts, determine (a) the iron loss current, (b) the power factor on no-load, and (c) the magnetizing current.

$I_0 = 0.8$ A, $V = 240$ V

(a) Power absorbed = total core loss $= 72 = V_1 I_0 \cos\phi_0$

Hence $72 = 240\, I_0 \cos\phi_0$

and iron loss current, $I_c = I_0 \cos\phi_0 = \dfrac{72}{240} =$ **0.30 A**

(b) Power factor at no load, $\cos\phi_0 = \dfrac{I_c}{I_0} = \dfrac{0.30}{0.80} =$ **0.375**

(c) From the right-angled triangle in Figure 20.2(b) and using Pythagoras' theorem, $I_0^2 = I_c^2 + I_M^2$

from which, magnetizing current, $I_M = \sqrt{(I_0^2 - I_c^2)}$

$= \sqrt{(0.80^2 - 0.30^2)}$

$=$ **0.74 A**

Further problems on the no-load phasor diagram may be found in Section 20.16, problems 10 to 12, page 344.

20.4 E.m.f. equation of a transformer

The magnetic flux Φ set up in the core of a transformer when an alternating voltage is applied to its primary winding is also alternating and is sinusoidal.

Let Φ_m be the maximum value of the flux and f be the frequency of the supply. The time for 1 cycle of the alternating flux is the periodic time T, where $T = 1/f$ seconds

The flux rises sinusoidally from zero to its maximum value in $\frac{1}{4}$ cycle, and the time for $\frac{1}{4}$ cycle is $1/4f$ seconds.

Hence the average rate of change of flux $= \dfrac{\Phi_m}{(1/4f)} = 4f\Phi_m$ Wb/s, and since 1 Wb/s = 1 volt, the average e.m.f. induced in each turn $= 4f\Phi_m$ volts.

As the flux Φ varies sinusoidally, then a sinusoidal e.m.f. will be induced in each turn of both primary and secondary windings.

For a sine wave, form factor $= \dfrac{\text{rms value}}{\text{average value}} = 1.11$ (see Chapter 14)

$$\text{Hence rms value} = \text{form factor} \times \text{average value}$$
$$= 1.11 \times \text{average value}$$

$$\text{Thus rms e.m.f. induced in each turn} = 1.11 \times 4f\Phi_m \text{ volts}$$
$$= 4.44f\Phi_m \text{ volts}$$

Therefore, rms value of e.m.f. induced in primary,

$$\mathbf{E_1 = 4.44f\Phi_m N_1 \text{ volts}} \qquad (20.4)$$

and rms value of e.m.f. induced in secondary,

$$\mathbf{E_2 = 4.44f\Phi_m N_2 \text{ volts}} \qquad (20.5)$$

Dividing equation (20.4) by equation (20.5) gives:

$\dfrac{E_1}{E_2} = \dfrac{N_1}{N_2}$, as previously obtained in Section 20.2.

Problem 8. A 100 kVA, 4000 V/200 V, 50 Hz single-phase transformer has 100 secondary turns. Determine (a) the primary and secondary current, (b) the number of primary turns, and (c) the maximum value of the flux.

$V_1 = 4000$ V, $V_2 = 200$ V, $f = 50$ Hz, $N_2 = 100$ turns

(a) Transformer rating $= V_1I_1 = V_2I_2 = 100\,000$ VA

Hence primary current, $I_1 = \dfrac{100\,000}{V_1} = \dfrac{100\,000}{4000} = \mathbf{25\ A}$

and secondary current, $I_2 = \dfrac{100\,000}{V_2} = \dfrac{100\,000}{200} = \mathbf{500\ A}$

(b) From equation (20.3), $\dfrac{V_1}{V_2} = \dfrac{N_1}{N_2}$

from which, primary turns, $N_1 = \left(\dfrac{V_1}{V_2}\right)(N_2) = \left(\dfrac{4000}{200}\right)(100)$

i.e., $\mathbf{N_1 = 2000}$ **turns**

(c) From equation (20.5), $E_2 = 4.44 f \Phi_m N_2$

from which, maximum flux $\Phi_m = \dfrac{E_2}{4.44 f N_2} = \dfrac{200}{4.44(50)(100)}$

(assuming $E_2 = V_2$)

$= \mathbf{9.01 \times 10^{-3}}$ **Wb** or **9.01 mWb**

[Alternatively, equation (20.4) could have been used,

where $E_1 = 4.44 f \Phi_m N_1$

from which, $\Phi_m = \dfrac{E_1}{4.44 f N_1} = \dfrac{4000}{4.44(50)(2000)}$ (assuming $E_1 = V_1$)

$= 9.01$ mWb, as above]

Problem 9. A single-phase, 50 Hz transformer has 25 primary turns and 300 secondary turns. The cross-sectional area of the core is 300 cm^2. When the primary winding is connected to a 250 V supply, determine (a) the maximum value of the flux density in the core, and (b) the voltage induced in the secondary winding.

(a) From equation (20.4), e.m.f. $E_1 = 4.44 f \Phi_m N_1$ volts i.e.,

$250 = 4.44(50)\Phi_m(25)$

from which, maximum flux density, $\Phi_m = \dfrac{250}{(4.44)(50)(25)}$ Wb

$= 0.04505$ Wb

However, $\Phi_m = B_m \times A$, where B_m = maximum flux density in the core and A = cross-sectional area of the core (see Chapter 7)

Hence $B_m \times 300 \times 10^{-4} = 0.04505$

from which, **maximum flux density**, $\mathbf{B_m} = \dfrac{0.04505}{300 \times 10^{-4}} = \mathbf{1.50\ T}$

(b) $\dfrac{V_1}{V_2} = \dfrac{N_1}{N_2}$, from which, $V_2 = V_1\left(\dfrac{N_2}{N_1}\right)$

i.e., voltage induced in the secondary winding,

$$V_2 = (250)\left(\frac{300}{25}\right) = \mathbf{3000\ V\ or\ 3\ kV}$$

Problem 10. A single-phase 500 V/100 V, 50 Hz transformer has a maximum core flux density of 1.5 T and an effective core cross-sectional area of 50 cm^2. Determine the number of primary and secondary turns.

The e.m.f. equation for a transformer is $E = 4.44f\Phi_m N$

and maximum flux, $\Phi_m = B \times A = (1.5)(50 \times 10^{-4}) = 75 \times 10^{-4}$ Wb

Since $E_1 = 4.44f\Phi_m N_1$

$$\text{then primary turns, } N_1 = \frac{E_1}{4.44f\Phi_m} = \frac{500}{4.44(50)(75 \times 10^{-4})}$$

$$= \mathbf{300\ turns}$$

Since $E_2 = 4.4f\Phi_m N_2$

$$\text{then secondary turns, } N_2 = \frac{E_2}{4.4f\Phi_m} = \frac{100}{4.44(50)(75 \times 10^{-4})}$$

$$= \mathbf{60\ turns}$$

Problem 11. A 4500 V/225 V, 50 Hz single-phase transformer is to have an approximate e.m.f. per turn of 15 V and operate with a maximum flux of 1.4 T. Calculate (a) the number of primary and secondary turns and (b) the cross-sectional area of the core.

(a) $$\text{E.m.f. per turn} = \frac{E_1}{N_1} = \frac{E_2}{N_2} = 15$$

$$\text{Hence primary turns, } N_1 = \frac{E_1}{15} = \frac{4500}{15} = \mathbf{300}$$

$$\text{and secondary turns, } N_2 = \frac{E_2}{15} = \frac{225}{15} = \mathbf{15}$$

(b) E.m.f. $E_1 = 4.44f\Phi_m N_1$

$$\text{from which, } \Phi_m = \frac{E_1}{4.44fN_1} = \frac{4500}{4.44(50)(300)} = 0.0676\ \text{Wb}$$

Now flux $\Phi_m = B_m \times A$, where A is the cross-sectional area of the core, hence

$$\text{area } A = \frac{\Phi_m}{B_m} = \frac{0.0676}{1.4} = \mathbf{0.0483\ m^2} \text{ or } \mathbf{483\ cm^2}$$

Further problems on the e.m.f. equation may be found in Section 20.16, problems 13 to 16, page 345.

20.5 Transformer on-load phasor diagram

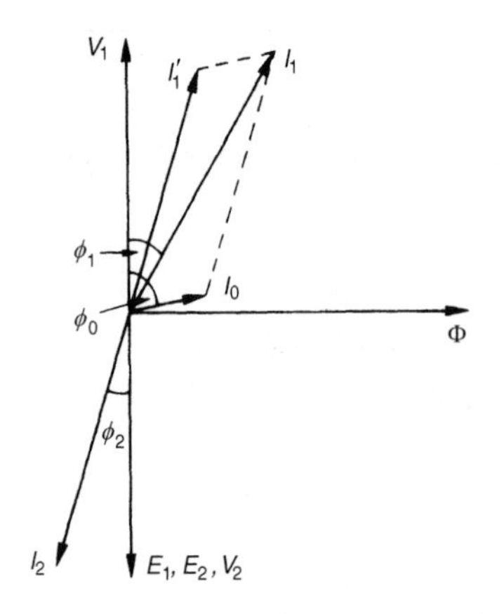

Figure 20.4

If the voltage drop in the windings of a transformer are assumed negligible, then the terminal voltage V_2 is the same as the induced e.m.f. E_2 in the secondary. Similarly, $V_1 = E_1$. Assuming an equal number of turns on primary and secondary windings, then $E_1 = E_2$, and let the load have a lagging phase angle ϕ_2.

In the phasor diagram of Figure 20.4, current I_2 lags V_2 by angle ϕ_2. When a load is connected across the secondary winding a current I_2 flows in the secondary winding. The resulting secondary e.m.f. acts so as to tend to reduce the core flux. However this does not happen since reduction of the core flux reduces E_1, hence a reflected increase in primary current I_1' occurs which provides a restoring mmf. Hence at all loads, primary and secondary mmf's are equal, but in opposition, and the core flux remains constant. I_1' is sometimes called the 'balancing' current and is equal, but in the opposite direction, to current I_2 as shown in Figure 20.4. I_0, shown at a phase angle ϕ_0 to V_1, is the no-load current of the transformer (see Section 20.3).

The phasor sum of I_1' and I_0 gives the supply current I_1 and the phase angle between V_1 and I_1 is shown as ϕ_1.

Problem 12. A single-phase transformer has 2000 turns on the primary and 800 turns on the secondary. Its no-load current is 5 A at a power factor of 0.20 lagging. Assuming the volt drop in the windings is negligible, determine the primary current and power factor when the secondary current is 100 A at a power factor of 0.85 lagging.

Let I_1' be the component of the primary current which provides the restoring mmf. Then

$$I_1'N_1 = I_2N_2$$

$$\text{i.e., } I_1'(2000) = (100)(800)$$

$$\text{from which, } I_1' = \frac{(100)(800)}{2000} = 40 \text{ A}$$

If the power factor of the secondary is 0.85

$$\text{then } \cos\phi_2 = 0.85, \text{ from which, } \phi_2 = \arccos 0.85 = 31.8°$$

If the power factor on no-load is 0.20,

$$\text{then } \cos\phi_0 = 0.2 \text{ and } \phi_0 = \arccos 0.2 = 78.5°$$

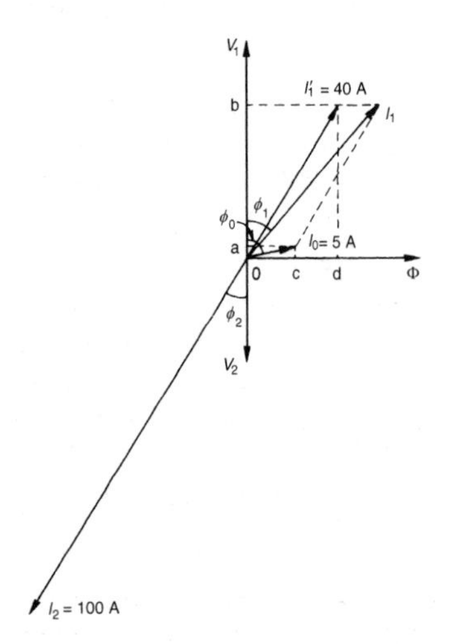

Figure 20.5

In the phasor diagram shown in Figure 20.5, $I_2 = 100$ A is shown at an angle of $\phi_2 = 31.8°$ to V_2 and $I_1' = 40$ A is shown in anti-phase to I_2.

The no-load current $I_0 = 5$ A is shown at an angle of $\phi_0 = 78.5°$ to V_1. Current I_1 is the phasor sum of I_1' and I_0 and by drawing to scale, $I_1 = 44$ A and angle $\phi_1 = 37°$

By calculation,
$$\begin{aligned} I_1 \cos\phi_1 &= oa + ob \\ &= I_0 \cos\phi_0 + I_1' \cos\phi_2 \\ &= (5)(0.2) + (40)(0.85) \\ &= 35.0 \text{ A} \end{aligned}$$

$$\begin{aligned} \text{and } I_1 \sin\phi_1 &= oc + od \\ &= I_0 \sin\phi_0 + I_1' \sin\phi_2 \\ &= (5)\sin 78.5° + (40)\sin 31.8° \\ &= 25.98 \text{ A} \end{aligned}$$

Hence the magnitude of $I_1 = \sqrt{(35.0^2 + 25.98^2)} = \mathbf{43.59\ A}$

and $\tan\phi_1 = \left(\dfrac{25.98}{35.0}\right)$, from which, $\phi_1 = \arctan\left(\dfrac{25.98}{35.0}\right) = \mathbf{36.59°}$

Hence the power factor of the primary $= \cos\phi_1 = \cos 36.59° = \mathbf{0.80}$

A further problem on the transformer on-load may be found in Section 20.16, problem 17, page 345.

20.6 Transformer construction

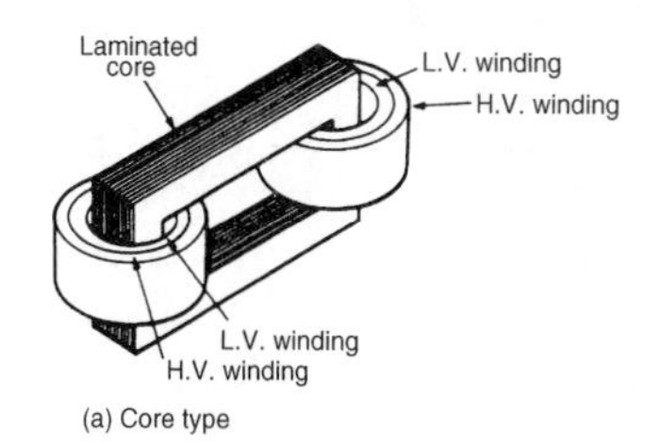

(a) Core type

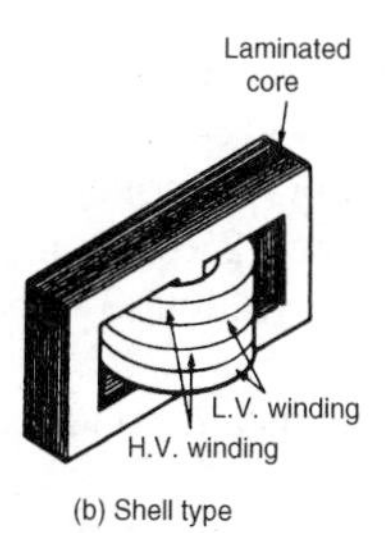

(b) Shell type

Figure 20.6

(i) There are broadly two types of single-phase double-wound transformer constructions — the **core type** and the **shell type**, as shown in Figure 20.6. The low and high voltage windings are wound as shown to reduce leakage flux.

(ii) For **power transformers**, rated possibly at several MVA and operating at a frequency of 50 Hz in Great Britain, the core material used is usually laminated silicon steel or stalloy, the laminations reducing eddy currents and the silicon steel keeping hysteresis loss to a minimum.

Large power transformers are used in the main distribution system and in industrial supply circuits. Small power transformers have many applications, examples including welding and rectifier supplies, domestic bell circuits, imported washing machines, and so on.

(iii) For **audio frequency (a.f.) transformers**, rated from a few mVA to no more than 20 VA, and operating at frequencies up to about 15 kHz, the small core is also made of laminated silicon steel. A typical application of a.f. transformers is in an audio amplifier system.

(iv) **Radio frequency (r.f.) transformers**, operating in the MHz frequency region have either an air core, a ferrite core or a dust core. Ferrite is a ceramic material having magnetic properties similar to silicon steel, but having a high resistivity. Dust cores consist of fine particles of carbonyl iron or permalloy (i.e. nickel and iron), each particle of which is insulated from its neighbour. Applications of r.f. transformers are found in radio and television receivers.

(v) Transformer **windings** are usually of enamel-insulated copper or aluminium.

(vi) **Cooling** is achieved by air in small transformers and oil in large transformers.

20.7 Equivalent circuit of a transformer

Figure 20.7 shows an equivalent circuit of a transformer. R_1 and R_2 represent the resistances of the primary and secondary windings and X_1 and X_2 represent the reactances of the primary and secondary windings, due to leakage flux.

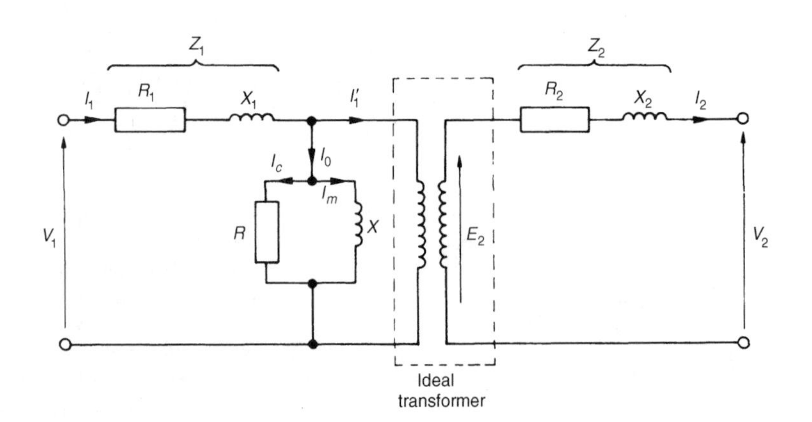

Figure 20.7

The core losses due to hysteresis and eddy currents are allowed for by resistance R which takes a current I_c, the core loss component of the primary current. Reactance X takes the magnetizing component I_M.

In a simplified equivalent circuit shown in Figure 20.8, R and X are omitted since the no-load current I_0 is normally only about 3–5% of the full load primary current.

It is often convenient to assume that all of the resistance and reactance as being on one side of the transformer.

Resistance R_2 in Figure 20.8 can be replaced by inserting an additional resistance R_2' in the primary circuit such that the power absorbed

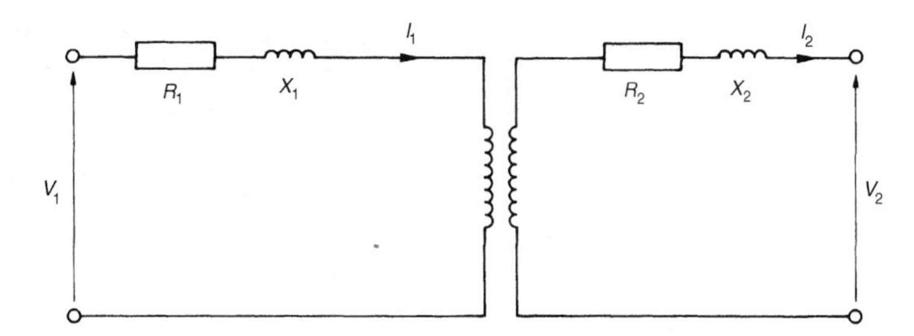

Figure 20.8

in R_2' when carrying the primary current is equal to that in R_2 due to the secondary current, i.e., $I_1^2 R_2' = I_2^2 R_2$

$$\text{from which, } R_2' = R_2 \left(\frac{I_2}{I_1}\right)^2 = R_2 \left(\frac{V_1}{V_2}\right)^2$$

Then the total equivalent resistance in the primary circuit R_e is equal to the primary and secondary resistances of the actual transformer. Hence

$$R_e = R_1 + R_2', \text{ i.e., } \boxed{\boldsymbol{R_e = R_1 + R_2 \left(\frac{V_1}{V_2}\right)^2}} \quad (20.6)$$

By similar reasoning, the equivalent reactance in the primary circuit is given by

$$X_e = X_1 + X_2', \text{ i.e., } \boxed{\boldsymbol{X_e = X_1 + X_2 \left(\frac{V_1}{V_2}\right)^2}} \quad (20.7)$$

The equivalent impedance Z_e of the primary and secondary windings referred to the primary is given by

$$\boxed{\boldsymbol{Z_e = \sqrt{(R_e^2 + X_e^2)}}} \quad (20.8)$$

If ϕ_e is the phase angle between I_1 and the volt drop $I_1 Z_e$ then

$$\boxed{\boldsymbol{\cos \phi_e = \frac{R_e}{Z_e}}} \quad (20.9)$$

The simplified equivalent circuit of a transformer is shown in Figure 20.9.

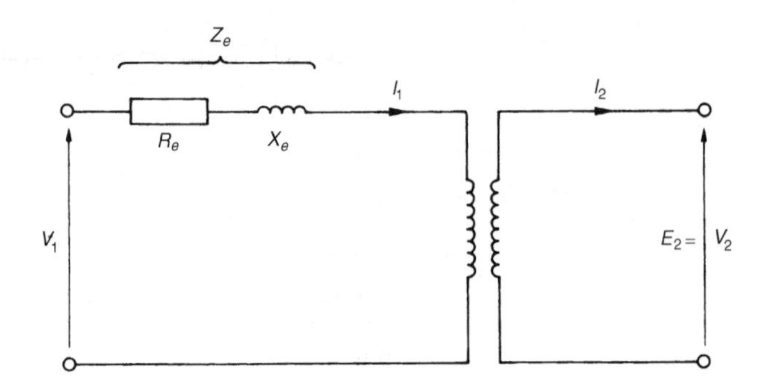

Figure 20.9

> Problem 13. A transformer has 600 primary turns and 150 secondary turns. The primary and secondary resistances are 0.25 Ω and 0.01 Ω respectively and the corresponding leakage reactances are 1.0 Ω and 0.04 Ω respectively. Determine (a) the equivalent resistance referred to the primary winding, (b) the equivalent reactance referred to the primary winding, (c) the equivalent impedance referred to the primary winding, and (d) the phase angle of the impedance.

(a) From equation (20.6), equivalent resistance $R_e = R_1 + R_2 \left(\frac{V_1}{V_2}\right)^2$

i.e., $\boldsymbol{R_e} = 0.25 + 0.01 \left(\frac{600}{150}\right)^2$ since $\frac{V_1}{V_2} = \frac{N_1}{N_2}$

$= \mathbf{0.41\ \Omega}$

(b) From equation (20.7), equivalent reactance, $X_e = X_1 + X_2 \left(\frac{V_1}{V_2}\right)^2$

i.e., $X_e = 1.0 + 0.04 \left(\frac{600}{150}\right)^2 = \mathbf{1.64\ \Omega}$

(c) From equation (20.8), equivalent impedance, $Z_e = \sqrt{(R_e^2 + X_e^2)}$

$= \sqrt{(0.41^2 + 1.64^2)}$

$= \mathbf{1.69\ \Omega}$

(d) From equation (20.9), $\cos\phi_e = \frac{R_e}{Z_e} = \frac{0.41}{1.69}$

Hence $\phi_e = \arccos\left(\frac{0.41}{1.69}\right) = \mathbf{75.96°}$

A further problem on the equivalent circuit of a transformer may be found in Section 20.16, problem 18, page 346.

20.8 Regulation of a transformer

When the secondary of a transformer is loaded, the secondary terminal voltage, V_2, falls. As the power factor decreases, this voltage drop increases. This is called the **regulation of the transformer** and it is usually expressed as a percentage of the secondary no-load voltage, E_2. For full-load conditions:

$$\textbf{Regulation} = \left(\frac{E_2 - V_2}{E_2}\right) \times 100\% \qquad (20.10)$$

The fall in voltage, $(E_2 - V_2)$, is caused by the resistance and reactance of the windings.

Typical values of voltage regulation are about 3% in small transformers and about 1% in large transformers.

Problem 14. A 5 kVA, 200 V/400 V, single-phase transformer has a secondary terminal voltage of 387.6 volts when loaded. Determine the regulation of the transformer.

From equation (20.10):

$$\text{regulation} = \frac{\text{(No-load secondary voltage} - \text{terminal voltage on load)}}{\text{no-load secondary voltage}} \times 100\%$$

$$= \left[\frac{400 - 387.6}{400}\right] \times 100\%$$

$$= \left(\frac{12.4}{400}\right) \times 100\% = \mathbf{3.1\%}$$

Problem 15. The open circuit voltage of a transformer is 240 V. A tap changing device is set to operate when the percentage regulation drops below 2.5%. Determine the load voltage at which the mechanism operates.

$$\text{Regulation} = \frac{\text{(no load voltage} - \text{terminal load voltage)}}{\text{no load voltage}} \times 100\%$$

$$\text{Hence } 2.5 = \left[\frac{240 - V_2}{240}\right] 100\%$$

Therefore $\frac{(2.5)(240)}{100} = 240 - V_2$

i.e, $6 = 240 - V_2$

from which, **load voltage,** $V_2 = 240 - 6 =$ **234 volts**

Further problems on regulation may be found in Section 20.16, problems 19 and 20, page 346.

20.9 Transformer losses and efficiency

There are broadly two sources of **losses in transformers** on load, these being copper losses and iron losses.

(a) **Copper losses** are variable and result in a heating of the conductors, due to the fact that they possess resistance. If R_1 and R_2 are the primary and secondary winding resistances then the total copper loss is $I_1^2R_1 + I_2^2R_2$

(b) **Iron losses** are constant for a given value of frequency and flux density and are of two types—hysteresis loss and eddy current loss.

  (i) **Hysteresis loss** is the heating of the core as a result of the internal molecular structure reversals which occur as the magnetic flux alternates. The loss is proportional to the area of the hysteresis loop and thus low loss nickel iron alloys are used for the core since their hysteresis loops have small areas.(See Chapters 7 and 38)

  (ii) **Eddy current loss** is the heating of the core due to e.m.f.'s being induced not only in the transformer windings but also in the core. These induced e.m.f.'s set up circulating currents, called eddy currents. Owing to the low resistance of the core, eddy currents can be quite considerable and can cause a large power loss and excessive heating of the core. Eddy current losses can be reduced by increasing the resistivity of the core material or, more usually, by laminating the core (i.e., splitting it into layers or leaves) when very thin layers of insulating material can be inserted between each pair of laminations. This increases the resistance of the eddy current path, and reduces the value of the eddy current.

Transformer efficiency, $\eta = \frac{\text{output power}}{\text{input power}} = \frac{\text{input power} - \text{losses}}{\text{input power}}$

$$\boldsymbol{\eta = 1 - \frac{\text{losses}}{\text{input power}}} \qquad (20.11)$$

and is usually expressed as a percentage. It is not uncommon for power transformers to have efficiencies of between 95% and 98%.

Output power $= V_2 I_2 \cos\phi_2$,

total losses $=$ copper loss $+$ iron losses,

and input power $=$ output power $+$ losses

Problem 16. A 200 kVA rated transformer has a full-load copper loss of 1.5 kW and an iron loss of 1 kW. Determine the transformer efficiency at full load and 0.85 power factor.

$$\text{Efficiency } \eta = \frac{\text{output power}}{\text{input power}} = \frac{\text{input power — losses}}{\text{input power}}$$

$$= 1 - \frac{\text{losses}}{\text{input power}}$$

Full-load output power $= VI\cos\phi = (200)(0.85) = 170$ kW

Total losses $= 1.5 + 1.0 = 2.5$ kW

Input power $=$ output power $+$ losses $= 170 + 2.5 = 172.5$ kW

Hence efficiency $= \left(1 - \dfrac{2.5}{172.5}\right) = 1 - 0.01449 = 0.9855$ or **98.55%**

Problem 17. Determine the efficiency of the transformer in Problem 16 at half full-load and 0.85 power factor.

Half full-load power output $= \frac{1}{2}(200)(0.85) = 85$ kW

Copper loss (or I^2R loss) is proportional to current squared.

Hence the copper loss at half full-load is $\left(\frac{1}{2}\right)^2(1500) = 375$ W

Iron loss $= 1000$ W (constant)

Total losses $= 375 + 1000 = 1375$ W or 1.375 kW

Input power at half full-load $=$ output power at half full-load $+$ losses

$= 85 + 1.375 = 86.375$ kW

$$\text{Hence efficiency} = \left(1 - \frac{\text{losses}}{\text{input power}}\right) = \left(1 - \frac{1.375}{86.375}\right)$$

$= 1 - 0.01592 = 0.9841$ or **98.41%**

Problem 18. A 400 kVA transformer has a primary winding resistance of 0.5 Ω and a secondary winding resistance of 0.001 Ω. The iron loss is 2.5 kW and the primary and secondary voltages are 5 kV and 320 V respectively. If the power factor of the load is 0.85, determine the efficiency of the transformer (a) on full load, and (b) on half load.

(a) Rating = 400 kVA = $V_1 I_1 = V_2 I_2$

Hence primary current, $I_1 = \dfrac{400 \times 10^3}{V_1} = \dfrac{400 \times 10^3}{5000} = 80$ A

and secondary current, $I_2 = \dfrac{400 \times 10^3}{V_2} = \dfrac{400 \times 10^3}{320} = 1250$ A

Total copper loss $= I_1^2 R_1 + I_2^2 R_2$,

(where $R_1 = 0.5\ \Omega$ and $R_2 = 0.001\ \Omega$)

$= (80)^2(0.5) + (1250)^2(0.001)$

$= 3200 + 1562.5 = 4762.5$ watts

On full load, total loss = copper loss + iron loss

$= 4762.5 + 2500$

$= 7262.5\ \text{W} = 7.2625\ \text{kW}$

Total output power on full load $= V_2 I_2 \cos\phi_2$

$= (400 \times 10^3)(0.85)$

$= 340$ kW

Input power = output power + losses = 340 kW + 7.2625 kW

= 347.2625 kW

Efficiency, $\eta = \left[1 - \dfrac{\text{losses}}{\text{input power}}\right] \times 100\%$

$= \left[1 - \dfrac{7.2625}{347.2625}\right] \times 100\% = \mathbf{97.91\%}$

(b) Since the copper loss varies as the square of the current, then total copper loss on half load $= 4762.5 \times \left(\frac{1}{2}\right)^2 = 1190.625$ W

Hence total loss on half load $= 1190.625 + 2500$

$= 3690.625$ W or 3.691 kW

Output power on half full load $= \frac{1}{2}(340) = 170$ kW

Input power on half full load = output power + losses

$= 170\text{ kW} + 3.691\text{ kW}$

$= 173.691\text{ kW}$

Hence efficiency at half full load,

$$\eta = \left[1 - \frac{\text{losses}}{\text{input power}}\right] \times 100\%$$

$$= \left[1 - \frac{3.691}{173.691}\right] \times 100\% = \mathbf{97.87\%}$$

Maximum efficiency

It may be shown that the efficiency of a transformer is a maximum when the variable copper loss (i.e., $I_1^2R_1 + I_2^2R_2$) is equal to the constant iron losses.

Problem 19. A 500 kVA transformer has a full load copper loss of 4 kW and an iron loss of 2.5 kW. Determine (a) the output kVA at which the efficiency of the transformer is a maximum, and (b) the maximum efficiency, assuming the power factor of the load is 0.75.

(a) Let x be the fraction of full load kVA at which the efficiency is a maximum.

The corresponding total copper loss $= (4\text{ kW})(x^2)$

At maximum efficiency, copper loss = iron loss Hence

$$4x^2 = 2.5$$

$$\text{from which } x^2 = \frac{2.5}{4} \text{ and } x = \sqrt{\left(\frac{2.5}{4}\right)} = 0.791$$

Hence **the output kVA at maximum efficiency** $= 0.791 \times 500$

$= \mathbf{395.5\ kVA}$

(b) Total loss at maximum efficiency $= 2 \times 2.5 = 5$ kW

Output power $= 395.5\text{ kVA} \times \text{p.f.} = 395.5 \times 0.75 = 296.625$ kW

Input power = output power + losses

$= 296.625 + 5 = 301.625$ kW

$$\textbf{Maximum efficiency, } \eta = \left[1 - \frac{\text{losses}}{\text{input power}}\right] \times 100\%$$

$$= \left[1 - \frac{5}{301.625}\right] \times 100\% = \mathbf{98.34\%}$$

Further problems on losses and efficiency may be found in Section 20.16, problems 21 to 26, page 346.

20.10 Resistance matching

Varying a load resistance to be equal, or almost equal, to the source internal resistance is called **matching**. Examples where resistance matching is important include coupling an aerial to a transmitter or receiver, or in coupling a loudspeaker to an amplifier, where coupling transformers may be used to give maximum power transfer.

With d.c. generators or secondary cells, the internal resistance is usually very small. In such cases, if an attempt is made to make the load resistance as small as the source internal resistance, overloading of the source results.

A method of achieving maximum power transfer between a source and a load (see Section 13.9, page 187), is to adjust the value of the load resistance to 'match' the source internal resistance. A transformer may be used as a **resistance matching device** by connecting it between the load and the source.

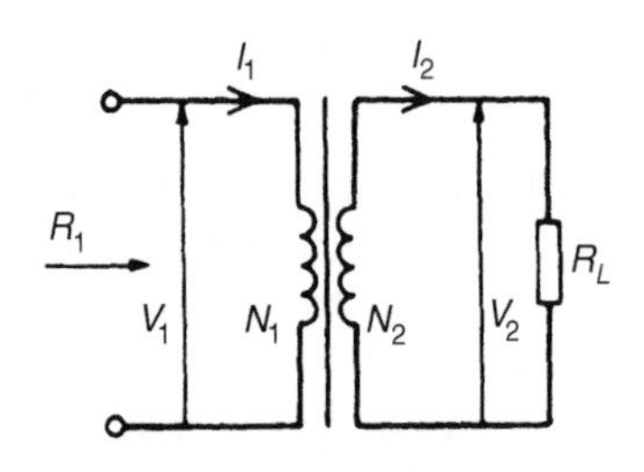

Figure 20.10

The reason why a transformer can be used for this is shown below. With reference to Figure 20.10:

$$R_L = \frac{V_2}{I_2} \text{ and } R_1 = \frac{V_1}{I_1}$$

For an ideal transformer, $V_1 = \left(\frac{N_1}{N_2}\right) V_2$ and $I_1 = \left(\frac{N_2}{N_1}\right) I_2$

Thus the equivalent input resistance R_1 of the transformer is given by:

$$R_1 = \frac{V_1}{I_1} = \frac{\left(\frac{N_1}{N_2}\right) V_2}{\left(\frac{N_2}{N_1}\right) I_2} = \left(\frac{N_1}{N_2}\right)^2 \left(\frac{V_2}{I_2}\right) = \left(\frac{N_1}{N_2}\right)^2 R_L$$

i.e., $$\boxed{\boldsymbol{R_1 = \left(\frac{N_1}{N_2}\right)^2 R_L}}$$

Hence by varying the value of the turns ratio, the equivalent input resistance of a transformer can be 'matched' to the internal resistance of a load to achieve maximum power transfer.

Problem 20. A transformer having a turns ratio of 4:1 supplies a load of resistance 100 Ω. Determine the equivalent input resistance of the transformer.

From above, the equivalent input resistance,

$$R_1 = \left(\frac{N_1}{N_2}\right)^2 R_L = \left(\frac{4}{1}\right)^2 (100) = \mathbf{1600\ \Omega}$$

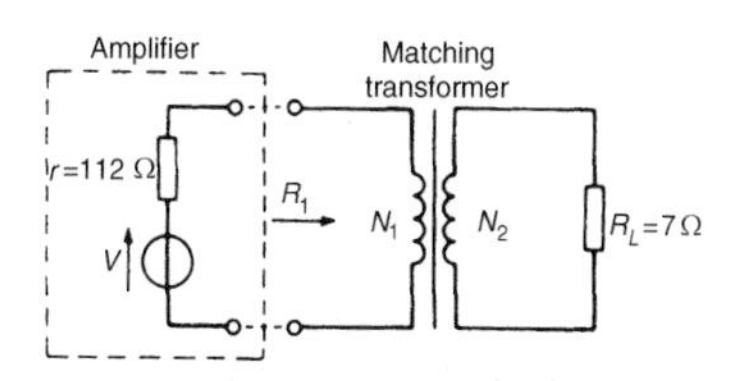

Figure 20.11

Problem 21. The output stage of an amplifier has an output resistance of 112 Ω. Calculate the optimum turns ratio of a transformer which would match a load resistance of 7 Ω to the output resistance of the amplifier.

The circuit is shown in Figure 20.11.

The equivalent input resistance, R_1 of the transformer needs to be 112 Ω for maximum power transfer.

$$R_1 = \left(\frac{N_1}{N_2}\right)^2 R_L$$

$$\text{Hence } \left(\frac{N_1}{N_2}\right)^2 = \frac{R_1}{R_L} = \frac{112}{7} = 16$$

$$\text{i.e., } \frac{N_1}{N_2} = \sqrt{(16)} = 4$$

Hence the optimum turns ratio is 4:1

Problem 22. Determine the optimum value of load resistance for maximum power transfer if the load is connected to an amplifier of output resistance 150 Ω through a transformer with a turns ratio of 5:1.

The equivalent input resistance R_1 of the transformer needs to be 150 Ω for maximum power transfer.

$$R_1 = \left(\frac{N_1}{N_2}\right)^2 R_L, \text{ from which, } R_L = R_1\left(\frac{N_2}{N_1}\right)^2 = 150\left(\frac{1}{5}\right)^2 = \mathbf{6\ \Omega}$$

Problem 23. A single-phase, 220 V/1760 V ideal transformer is supplied from a 220 V source through a cable of resistance 2 Ω. If the load across the secondary winding is 1.28 kΩ determine (a) the primary current flowing and (b) the power dissipated in the load resistor.

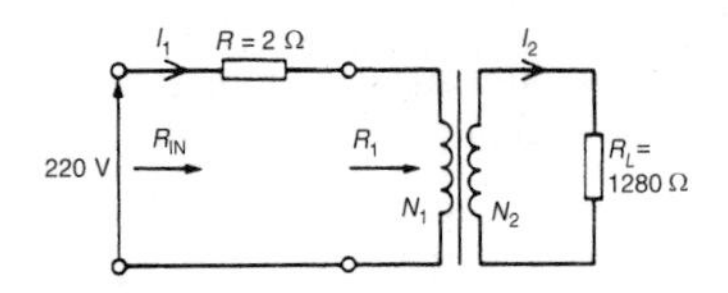

Figure 20.12

The circuit diagram is shown in Figure 20.12.

(a) Turns ratio $\dfrac{N_1}{N_2} = \dfrac{V_1}{V_2} = \dfrac{220}{1760} = \dfrac{1}{8}$

Equivalent input resistance of the transformer,

$$R_1 = \left(\frac{N_1}{N_2}\right)^2 R_L$$

$$= \left(\frac{1}{8}\right)^2 (1.28 \times 10^3) = 20\ \Omega$$

Total input resistance, $R_{IN} = R + R_1 = 2 + 20 = 22\ \Omega$

$$\text{Primary current, } I_1 = \frac{V_1}{R_{IN}} = \frac{220}{22} = \mathbf{10\ A}$$

(b) For an ideal transformer $\dfrac{V_1}{V_2} = \dfrac{I_2}{I_1}$, from which $I_2 = I_1\left(\dfrac{V_1}{V_2}\right)$

$$= 10\left(\frac{220}{1760}\right) = 1.25\ \text{A}$$

Power dissipated in load resistor R_L, $P = I_2^2 R_L$

$$= (1.25)^2(1.28 \times 10^3)$$

$$= \mathbf{2000\ watts\ or\ 2\ kW}$$

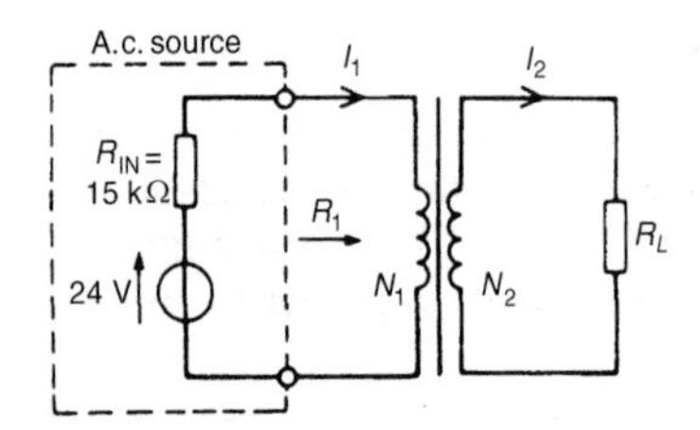

Figure 20.13

Problem 24. An a.c. source of 24 V and internal resistance 15 kΩ is matched to a load by a 25:1 ideal transformer. Determine (a) the value of the load resistance and (b) the power dissipated in the load.

The circuit diagram is shown in Figure 20.13.

(a) For maximum power transfer R_1 needs to be equal to 15 kΩ

$R_1 = \left(\dfrac{N_1}{N_2}\right)^2 R_L$, from which load resistance,

$$R_L = R_1\left(\frac{N_2}{N_1}\right)^2$$

$$= (15\,000)\left(\frac{1}{25}\right)^2 = \mathbf{24\ \Omega}$$

(b) The total input resistance when the source is connected to the matching transformer is $R_{IN} + R_1$, i.e., 15 kΩ + 15 kΩ = 30 kΩ

$$\text{Primary current, } I_1 = \frac{V}{30\,000} = \frac{24}{30\,000} = 0.8 \text{ mA}$$

$$\frac{N_1}{N_2} = \frac{I_2}{I_1}, \text{ from which } I_2 = I_1\left(\frac{N_1}{N_2}\right) = (0.8 \times 10^{-3})\left(\frac{25}{1}\right)$$

$$= 20 \times 10^{-3} \text{ A}$$

$$\text{Power dissipated in the load } R_L,\ P = I_2^2 R_L = (20 \times 10^{-3})^2(24)$$

$$= 9600 \times 10^{-6} \text{ W} = \mathbf{9.6\ mW}$$

Further problems on resistance matching may be found in Section 20.16, problems 27 to 31, page 347.

20.11 Auto transformers

An auto transformer is a transformer which has part of its winding common to the primary and secondary circuits. Figure 20.14(a) shows the circuit for a double-wound transformer and Figure 20.14(b) that for an auto transformer. The latter shows that the secondary is actually part of the primary, the current in the secondary being $(I_2 - I_1)$. Since the current is less in this section, the cross-sectional area of the winding can be reduced, which reduces the amount of material necessary.

Figure 20.15 shows the circuit diagram symbol for an auto transformer.

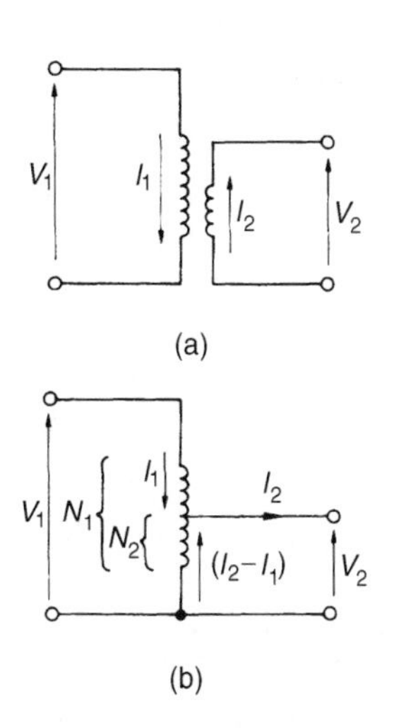

Figure 20.14

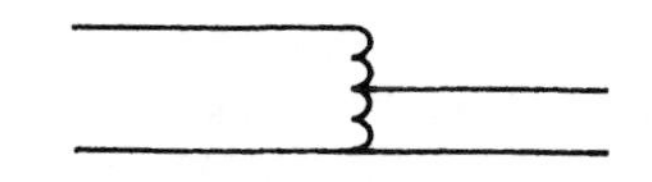

Figure 20.15

> Problem 25. A single-phase auto transformer has a voltage ratio 320 V:250 V and supplies a load of 20 kVA at 250 V. Assuming an ideal transformer, determine the current in each section of the winding.

Rating = 20 kVA = $V_1 I_1 = V_2 I_2$

$$\text{Hence primary current, } I_1 = \frac{20 \times 10^3}{V_1} = \frac{20 \times 10^3}{320} = \mathbf{62.5\ A}$$

$$\text{and secondary current, } I_2 = \frac{20 \times 10^3}{V_2} = \frac{20 \times 10^3}{250} = \mathbf{80\ A}$$

Hence current in common part of the winding = 80 − 62.5 = **17.5 A**

The current flowing in each section of the transformer is shown in Figure 20.16.

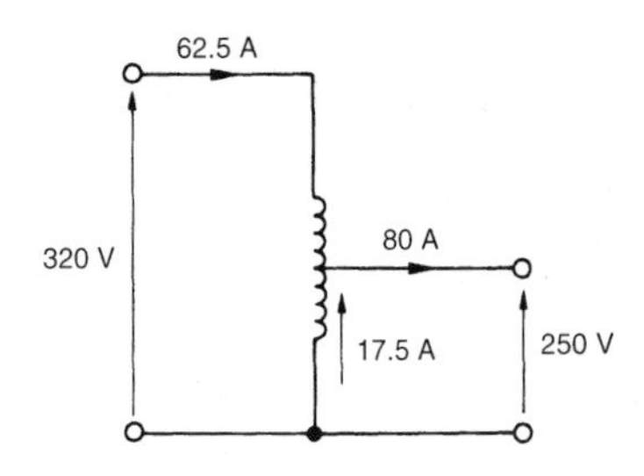

Figure 20.16

Saving of copper in an auto transformer

For the same output and voltage ratio, the auto transformer requires less copper than an ordinary double-wound transformer. This is explained below.

The volume, and hence weight, of copper required in a winding is proportional to the number of turns and to the cross-sectional area of the wire. In turn this is proportional to the current to be carried, i.e., volume of copper is proportional to NI.

Volume of copper in an auto transformer $\propto (N_1 - N_2)I_1 + N_2(I_2 - I_1)$

see Figure 20.14(b)

$$\propto N_1I_1 - N_2I_1 + N_2I_2 - N_2I_1$$

$$\propto N_1I_1 + N_2I_2 - 2N_2I_1$$

$$\propto 2N_1I_1 - 2N_2I_1$$

(since $N_2I_2 = N_1I_1$)

Volume of copper in a double-wound transformer $\propto N_1I_1 + N_2I_2$

$$\propto 2N_1I_1$$

(again, since $N_2I_2 = N_1I_1$)

$$\text{Hence } \frac{\text{volume of copper in an auto transformer}}{\text{volume of copper in a double-wound transformer}}$$

$$= \frac{2N_1I_1 - 2N_2I_1}{2N_1I_1}$$

$$= \frac{2N_1I_1}{2N_1I_1} - \frac{2N_2I_1}{2N_1I_1}$$

$$= 1 - \frac{N_2}{N_1}$$

If $\frac{N_2}{N_1} = x$ then

(volume of copper in an auto transformer)

= (1 − x) (volume of copper in a double-wound transformer) (20.12)

If, say, $x = \frac{4}{5}$ then

(volume of copper in auto transformer)

$= \left(1 - \frac{4}{5}\right)$ (volume of copper in a double-wound transformer)

$= \frac{1}{5}$(volume in double-wound transformer)

i.e., a saving of 80%

Similarly, if $x = \frac{1}{4}$, the saving is 25%, and so on.

The closer N_2 is to N_1, the greater the saving in copper.

Problem 26. Determine the saving in the volume of copper used in an auto transformer compared with a double-wound transformer for (a) a 200 V:150 V transformer, and (b) a 500 V:100 V transformer.

(a) For a 200 V:150 V transformer, $x = \frac{V_2}{V_1} = \frac{150}{200} = 0.75$

Hence from equation (20.12), (volume of copper in auto transformer)

$= (1 - 0.75)$ (volume of copper in double-wound transformer)

$= (0.25)$ (volume of copper in double-wound transformer)

$= 25\%$ of copper in a double-wound transformer

Hence the saving is 75%

(b) For a 500 V:100 V transformer, $x = \frac{V_2}{V_1} = \frac{100}{500} = 0.2$

Hence (volume of copper in auto transformer)

$= (1 - 0.2)$ (volume of copper in double-wound transformer)

$= (0.8)$ (volume in double-wound transformer)

$= 80\%$ of copper in a double-wound transformer

Hence the saving is 20%

Further problems on the auto-transformer may be found in Section 20.16, problems 32 and 33, page 347.

Advantages of auto transformers

The advantages of auto transformers over double-wound transformers include:

1 a saving in cost since less copper is needed (see above)
2 less volume, hence less weight
3 a higher efficiency, resulting from lower I^2R losses
4 a continuously variable output voltage is achievable if a sliding contact is used
5 a smaller percentage voltage regulation.

Disadvantages of auto transformers

The primary and secondary windings are not electrically separate, hence if an open-circuit occurs in the secondary winding the full primary voltage appears across the secondary.

Uses of auto transformers

Auto transformers are used for reducing the voltage when starting induction motors (see Chapter 22) and for interconnecting systems that are operating at approximately the same voltage.

20.12 Isolating transformers

Transformers not only enable current or voltage to be transformed to some different magnitude but provide a means of isolating electrically one part of a circuit from another when there is no electrical connection between primary and secondary windings. An **isolating transformer** is a 1:1 ratio transformer with several important applications, including bathroom shaver-sockets, portable electric tools, model railways, and so on.

20.13 Three-phase transformers

Three-phase double-wound transformers are mainly used in power transmission and are usually of the core type. They basically consist of three pairs of single-phase windings mounted on one core, as shown in Figure 20.17, which gives a considerable saving in the amount of iron used. The primary and secondary windings in Figure 20.17 are wound on top of each other in the form of concentric cylinders, similar to that shown in Figure 20.6(a). The windings may be with the primary delta-connected and the secondary star-connected, or star-delta, star-star or delta-delta, depending on its use.

A delta-connection is shown in Figure 20.18(a) and a star-connection in Figure 20.18(b).

Problem 27. A three-phase transformer has 500 primary turns and 50 secondary turns. If the supply voltage is 2.4 kV find the secondary line voltage on no-load when the windings are connected (a) star-delta, (b) delta-star.

(a) For a star-connection, $V_L = \sqrt{3}V_p$ (see Chapter 19)

Primary phase voltage, $V_{p1} = \dfrac{V_{L1}}{\sqrt{3}} = \dfrac{2400}{\sqrt{3}} = 1385.64$ volts

For a delta-connection, $V_L = V_p$

$\dfrac{N_1}{N_2} = \dfrac{V_1}{V_2}$, from which,

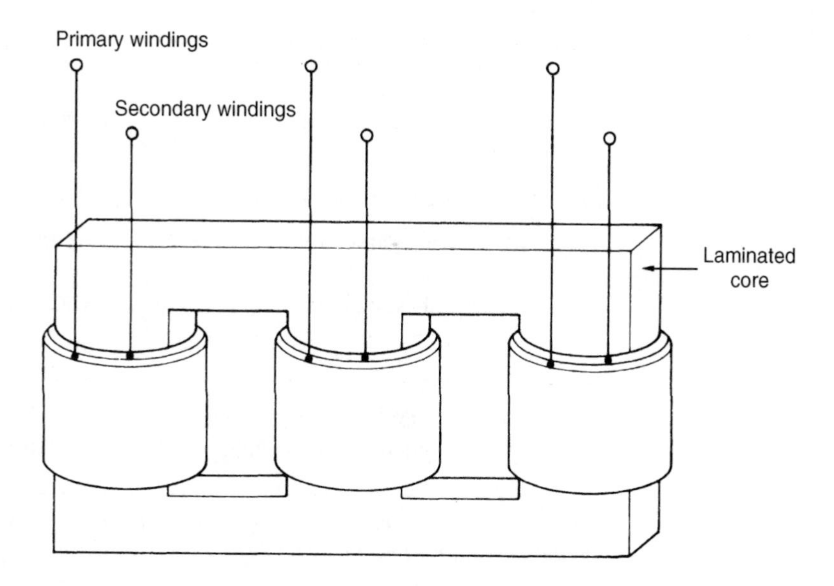

Figure 20.17

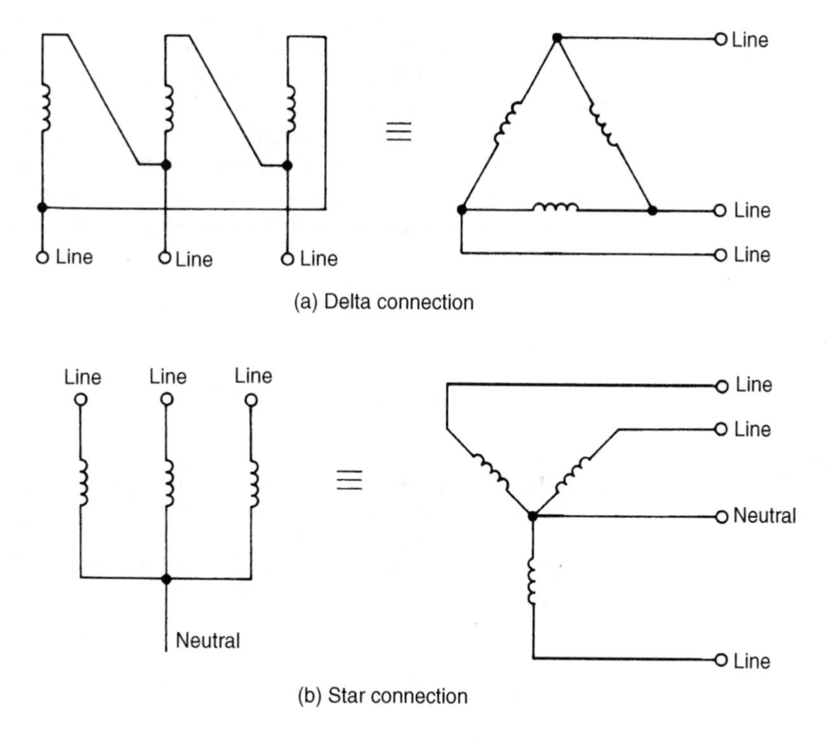

Figure 20.18

$$\text{secondary phase voltage, } V_{p2} = V_{p1}\left(\frac{N_2}{N_1}\right)$$

$$= (1385.64)\left(\frac{50}{500}\right) = \mathbf{138.6\ volts}$$

(b) For a delta-connection, $V_L = V_p$,

hence primary phase voltage $V_{p1} = 2.4$ kV $= 2400$ volts

$$\text{Secondary phase voltage, } V_{p2} = V_{p1}\left(\frac{N_2}{N_1}\right) = (2400)\left(\frac{50}{500}\right)$$

$$= 240 \text{ volts}$$

For a star-connection, $V_L = \sqrt{3}V_p$,
hence the secondary line voltage $= \sqrt{3}(240) =$ **416 volts**

A further problem on the three-phase transformer may be found in Section 20.16, problem 34, page 347.

20.14 Current transformers

For measuring currents in excess of about 100 A a current transformer is normally used. With a d.c. moving-coil ammeter the current required to give full scale deflection is very small — typically a few milliamperes. When larger currents are to be measured a shunt resistor is added to the circuit (see Chapter 10). However, even with shunt resistors added it is not possible to measure very large currents. When a.c. is being measured a shunt cannot be used since the proportion of the current which flows in the meter will depend on its impedance, which varies with frequency.

$$\text{In a double-wound transformer: } \frac{I_1}{I_2} = \frac{N_2}{N_1}$$

$$\text{from which, } \mathbf{secondary\ current\ } I_2 = I_1\left(\frac{N_1}{N_2}\right)$$

In current transformers the primary usually consists of one or two turns whilst the secondary can have several hundred turns. A typical arrangement is shown in Figure 20.19.

If, for example, the primary has 2 turns and the secondary 200 turns, then if the primary current is 500 A,

$$\text{secondary current, } I_2 = I_1\left(\frac{N_1}{N_2}\right) = (500)\left(\frac{2}{200}\right) = 5 \text{ A}$$

Current transformers isolate the ammeter from the main circuit and allow the use of a standard range of ammeters giving full-scale deflections of 1 A, 2 A or 5 A.

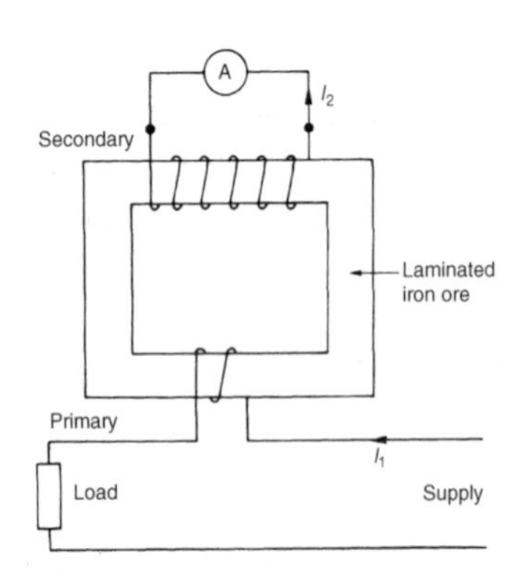

Figure 20.19

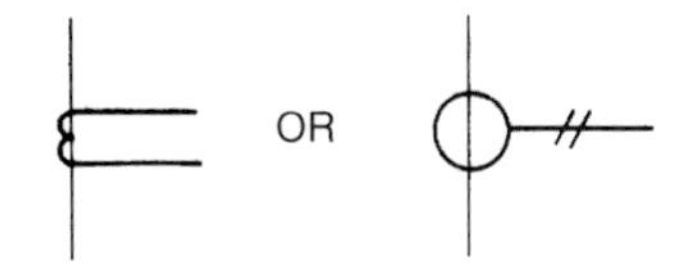

Figure 20.20

For very large currents the transformer core can be mounted around the conductor or bus-bar. Thus the primary then has just one turn. It is very important to short-circuit the secondary winding before removing the ammeter. This is because if current is flowing in the primary, dangerously high voltages could be induced in the secondary should it be open-circuited.

Current transformer circuit diagram symbols are shown in Figure 20.20.

Problem 28. A current transformer has a single turn on the primary winding and a secondary winding of 60 turns. The secondary winding is connected to an ammeter with a resistance of 0.15 Ω. The resistance of the secondary winding is 0.25 Ω. If the current in the primary winding is 300 A, determine (a) the reading on the ammeter, (b) the potential difference across the ammeter and (c) the total load (in VA) on the secondary.

(a) Reading on the ammeter, $I_2 = I_1\left(\frac{N_1}{N_2}\right) = 300\left(\frac{1}{60}\right) = \mathbf{5\ A}$

(b) P.d. across the ammeter $= I_2 R_A$, where R_A is the ammeter resistance

$= (5)(0.15) = \mathbf{0.75\ volts}$

(c) Total resistance of secondary circuit $= 0.15 + 0.25 = 0.40\ \Omega$

Induced e.m.f. in secondary $= (5)(0.40) = 2.0$ V

Total load on secondary $= (2.0)(5) = \mathbf{10\ VA}$

A further problem on the current transformer may be found in *Section 20.16, problem 35, page 348.*

20.15 Voltage transformers

For measuring voltages in excess of about 500 V it is often safer to use a voltage transformer. These are normal double-wound transformers with a large number of turns on the primary, which is connected to a high voltage supply, and a small number of turns on the secondary. A typical arrangement is shown in Figure 20.21.

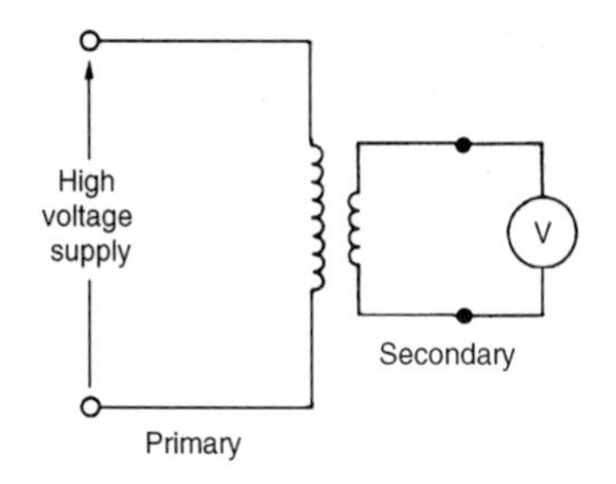

Figure 20.21

Since $\frac{V_1}{V_2} = \frac{N_1}{N_2}$

the **secondary voltage,** $\mathbf{V_2 = V_1\left(\frac{N_2}{N_1}\right)}$

Thus if the arrangement in Figure 20.21 has 4000 primary turns and 20 secondary turns then for a voltage of 22 kV on the primary, the voltage on the secondary,

$$V_2 = V_1 \left(\frac{N_2}{N_1}\right) = 22\,000 \left(\frac{20}{4000}\right) = \mathbf{110\ volts}$$

20.16 Further problems on transformers

Principle of operation

1 A transformer has 600 primary turns connected to a 1.5 kV supply. Determine the number of secondary turns for a 240 V output voltage, assuming no losses. [96]

2 An ideal transformer with a turns ratio of 2:9 is fed from a 220 V supply. Determine its output voltage. [990 V]

3 A transformer has 800 primary turns and 2000 secondary turns. If the primary voltage is 160 V, determine the secondary voltage assuming an ideal transformer. [400 V]

4 An ideal transformer has a turns ratio of 12:1 and is supplied at 192 V. Calculate the secondary voltage. [16 V]

5 An ideal transformer has a turns ratio of 12:1 and is supplied at 180 V when the primary current is 4 A. Calculate the secondary voltage and current. [15 V, 48 A]

6 A step-down transformer having a turns ratio of 20:1 has a primary voltage of 4 kV and a load of 10 kW. Neglecting losses, calculate the value of the secondary current. [50 A]

7 A transformer has a primary to secondary turns ratio of 1:15. Calculate the primary voltage necessary to supply a 240 V load. If the load current is 3 A determine the primary current. Neglect any losses. [16 V, 45 A]

8 A 10 kVA, single-phase transformer has a turns ratio of 12:1 and is supplied from a 2.4 kV supply. Neglecting losses, determine (a) the full load secondary current, (b) the minimum value of load resistance which can be connected across the secondary winding without the kVA rating being exceeded, and (c) the primary current. [(a) 50 A (b) 4 Ω (c) 4.17 A]

9 A 20 Ω resistance is connected across the secondary winding of a single-phase power transformer whose secondary voltage is 150 V. Calculate the primary voltage and the turns ratio if the supply current is 5 A, neglecting losses. [225 V, 3:2]

No-load phasor diagram

10 (a) Draw the phasor diagram for an ideal transformer on no-load.

(b) A 500 V/100 V, single-phase transformer takes a full load primary current of 4 A. Neglecting losses, determine (a) the full load secondary current, and (b) the rating of the transformer.
[(a) 20 A (b) 2 kVA]

11 A 3300 V/440 V, single-phase transformer takes a no-load current of 0.8 A and the iron loss is 500 W. Draw the no-load phasor diagram and determine the values of the magnetizing and core loss components of the no-load current. [0.786 A, 0.152 A]

12 A transformer takes a current of 1 A when its primary is connected to a 300 V, 50 Hz supply, the secondary being on open-circuit. If the power absorbed is 120 watts, calculate (a) the iron loss current,(b) the power factor on no-load, and (c) the magnetizing current. [(a) 0.4 A (b) 0.4 (c) 0.92 A]

E.m.f equation

13 A 60 kVA, 1600 V/100 V, 50 Hz, single-phase transformer has 50 secondary windings. Calculate (a) the primary and secondary current, (b) the number of primary turns, and (c) the maximum value of the flux. [(a) 37.5 A, 600 A (b) 800 (c) 9.0 mWb]

14 A single-phase, 50 Hz transformer has 40 primary turns and 520 secondary turns. The cross-sectional area of the core is 270 cm^2. When the primary winding is connected to a 300 volt supply, determine (a) the maximum value of flux density in the core, and (b) the voltage induced in the secondary winding.
[(a) 1.25 T (b) 3.90 kV]

15 A single-phase 800 V/100 V, 50 Hz transformer has a maximum core flux density of 1.294 T and an effective cross-sectional area of 60 cm^2. Calculate the number of turns on the primary and secondary windings. [464, 8]

16 A 3.3 kV/110 V, 50 Hz, single-phase transformer is to have an approximate e.m.f. per turn of 22 V and operate with a maximum flux of 1.25 T. Calculate (a) the number of primary and secondary turns, and (b) the cross-sectional area of the core.
[(a) 150, 5 (b) 792.8 cm^2]

Transformer on-load

17 A single-phase transformer has 2400 turns on the primary and 600 turns on the secondary. Its no-load current is 4 A at a power factor of 0.25 lagging. Assuming the volt drop in the windings is negligible, calculate the primary current and power factor when the secondary current is 80 A at a power factor of 0.8 lagging.
[23.26 A, 0.73]

Equivalent circuit of a transformer

18 A transformer has 1200 primary turns and 200 secondary turns. The primary and secondary resistances are 0.2 Ω and 0.02 Ω respectively and the corresponding leakage reactances are 1.2 Ω and 0.05 Ω respectively. Calculate (a) the equivalent resistance, reactance and impedance referred to the primary winding, and (b) the phase angle of the impedance. [(a) 0.92 Ω, 3.0 Ω, 3.14 Ω (b) 72.95°]

Regulation

19 A 6 kVA, 100 V/500 V, single-phase transformer has a secondary terminal voltage of 487.5 volts when loaded. Determine the regulation of the transformer. [2.5%]

20 A transformer has an open circuit voltage of 110 volts. A tap-changing device operates when the regulation falls below 3%. Calculate the load voltage at which the tap-changer operates. [106.7 volts]

Losses and efficiency

21 A single-phase transformer has a voltage ratio of 6:1 and the h.v. winding is supplied at 540 V. The secondary winding provides a full load current of 30 A at a power factor of 0.8 lagging. Neglecting losses, find (a) the rating of the transformer, (b) the power supplied to the load, (c) the primary current. [(a) 2.7 kVA, (b) 2.16 kW, (c) 5 A]

22 A single-phase transformer is rated at 40 kVA. The transformer has full-load copper losses of 800 W and iron losses of 500 W. Determine the transformer efficiency at full load and 0.8 power factor. [96.10%]

23 Determine the efficiency of the transformer in problem 22 at half full-load and 0.8 power factor. [95.81%]

24 A 100 kVA, 2000 V/400 V, 50 Hz, single-phase transformer has an iron loss of 600 W and a full-load copper loss of 1600 W. Calculate its efficiency for a load of 60 kW at 0.8 power factor. [97.56%]

25 (a) What are eddy currents? State how their effect is reduced in transformers.

(b) Determine the efficiency of a 15 kVA transformer for the following conditions:

(i) full-load, unity power factor

(ii) 0.8 full-load, unity power factor

(iii) half full-load, 0.8 power factor.

Assume that iron losses are 200 W and the full-load copper loss is 300 W. [(a) 96.77% (ii) 96.84% (iii) 95.62%]

26 A 250 kVA transformer has a full load copper loss of 3 kW and an iron loss of 2 kW. Calculate (a) the output kVA at which the efficiency of the transformer is a maximum, and (b) the maximum efficiency, assuming the power factor of the load is 0.80. [(a) 204.1 kVA (b) 97.61%]

Resistance matching

27 A transformer having a turns ratio of 8:1 supplies a load of resistance 50 Ω. Determine the equivalent input resistance of the transformer. [3.2 kΩ]

28 What ratio of transformer is required to make a load of resistance 30 Ω appear to have a resistance of 270 Ω? [3:1]

29 A single-phase, 240 V/2880 V ideal transformer is supplied from a 240 V source through a cable of resistance 3 Ω. If the load across the secondary winding is 720 Ω determine (a) the primary current flowing and (b) the power dissipated in the load resistance. [(a) 30 A (b) 4.5 kW]

30 A load of resistance 768 Ω is to be matched to an amplifier which has an effective output resistance of 12 Ω. Determine the turns ratio of the coupling transformer. [1:8]

31 An a.c. source of 20 V and internal resistance 20 kΩ is matched to a load by a 16:1 single-phase transformer. Determine (a) the value of the load resistance and (b) the power dissipated in the load. [(a) 78.13 Ω (b) 5 mW]

Auto-transformer

32 A single-phase auto transformer has a voltage ratio of 480 V:300 V and supplies a load of 30 kVA at 300 V. Assuming an ideal transformer, calculate the current in each section of the winding. [$I_1 = 62.5\ A$, $I_2 = 100\ A$, $(I_2 - I_1) = 37.5\ A$]

33 Calculate the saving in the volume of copper used in an auto transformer compared with a double-wound transformer for (a) a 300 V:240 V transformer, and (b) a 400 V:100 V transformer. [(a) 80% (b) 25%]

Three-phase transformer

34 A three-phase transformer has 600 primary turns and 150 secondary turns. If the supply voltage is 1.5 kV determine the secondary line voltage on no-load when the windings are connected (a) delta-star, (b) star-delta. [(a) 649.5 V (b) 216.5 V]

Current transformer

35 A current transformer has two turns on the primary winding and a secondary winding of 260 turns. The secondary winding is connected to an ammeter with a resistance of 0.2 Ω. The resistance of the secondary winding is 0.3 Ω. If the current in the primary winding is 650 A, determine (a) the reading on the ammeter, (b) the potential difference across the ammeter, and (c) the total load in VA on the secondary. [(a) 5 A (b) 1 V (c) 7.5 VA]

21 D.c. machines

At the end of this chapter you should be able to:

- distinguish between the function of a motor and a generator
- describe the action of a commutator
- describe the construction of a d.c. machine
- distinguish between wave and lap windings
- understand shunt, series and compound windings of d.c. machines
- understand armature reaction
- calculate generated e.m.f. in an armature winding using $E = \frac{2p\Phi nZ}{c}$
- describe types of d.c. generator and their characteristics
- calculate generated e.m.f. for a generator using $E = V + I_a R_a$
- state typical applications of d.c. generators
- list d.c. machine losses and calculate efficiency
- calculate back e.m.f. for a d.c. motor using $E = V - I_a R_a$
- calculate the torque of a d.c. motor using $T = \frac{EI_a}{2\pi n}$ and $T = \frac{p\Phi Z I_a}{\pi c}$
- describe types of d.c. motor and their characteristics
- state typical applications of d.c. motors
- describe a d.c. motor starter
- describe methods of speed control of d.c. motors
- list types of enclosure for d.c. motors

21.1 Introduction

When the input to an electrical machine is electrical energy, (seen as applying a voltage to the electrical terminals of the machine), and the output is mechanical energy, (seen as a rotating shaft), the machine is called an electric **motor**. Thus an electric motor converts electrical energy into mechanical energy.

The principle of operation of a motor is explained in Section 8.4, page 96.

When the input to an electrical machine is mechanical energy, (seen as, say, a diesel motor, coupled to the machine by a shaft), and the output is electrical energy, (seen as a voltage appearing at the electrical terminals of the machine), the machine is called a **generator**. Thus, a generator converts mechanical energy to electrical energy.

The principle of operation of a generator is explained in Section 9.2, page 101.

21.2 The action of a commutator

In an electric motor, conductors rotate in a uniform magnetic field. A single-loop conductor mounted between permanent magnets is shown in Figure 21.1. A voltage is applied at points A and B in Figure 21.1(a).

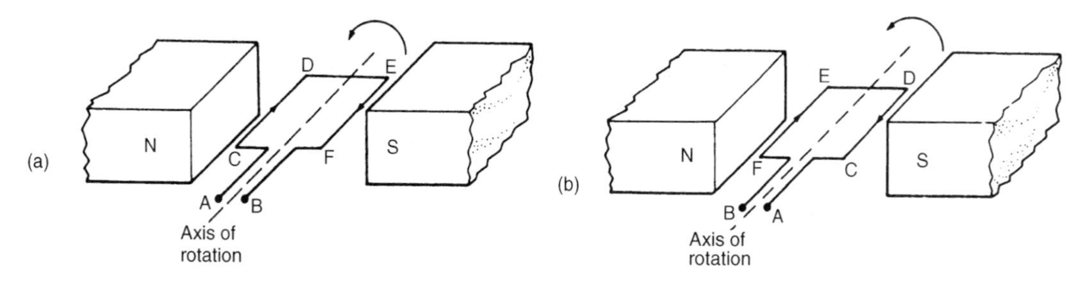

Figure 21.1

A force, F, acts on the loop due to the interaction of the magnetic field of the permanent magnets and the magnetic field created by the current flowing in the loop. This force is proportional to the flux density, B, the current flowing, I, and the effective length of the conductor, l, i.e. $F = BIl$. The force is made up of two parts, one acting vertically downwards due to the current flowing from C to D and the other acting vertically upwards due to the current flowing from E to F (from Fleming's left hand rule). If the loop is free to rotate, then when it has rotated through 180°, the conductors are as shown in Figure 21.1(b). For rotation to continue in the same direction, it is necessary for the current flow to be as shown in Figure 21.1(b), i.e. from D to C and from F to E. This apparent reversal in the direction of current flow is achieved by a process called **commutation**. With reference to Figure 21.2(a), when a direct voltage is applied at A and B, then as the single-loop conductor rotates, current flow will always be away from the commutator for the part of the conductor adjacent to the N-pole and towards the commutator for the part of the conductor adjacent to the S-pole. Thus the forces act to give continuous rotation in an anti-clockwise direction. The arrangement shown in Figure 21.2(a) is called a 'two-segment' commutator and the voltage is applied to the rotating segments by stationary **brushes**, (usually carbon blocks), which slide on the commutator material, (usually copper), when rotation takes place.

In practice, there are many conductors on the rotating part of a d.c. machine and these are attached to many commutator segments. A schematic diagram of a multi-segment commutator is shown in Figure 21.2(b).

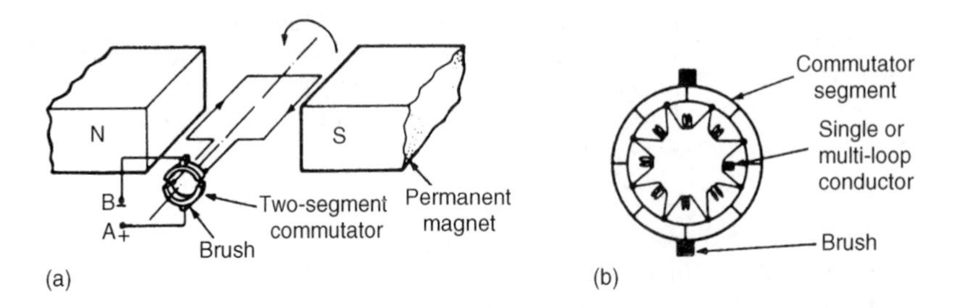

Figure 21.2

Poor commutation results in sparking at the trailing edge of the brushes. This can be improved by using **interpoles** (situated between each pair of main poles), high resistance brushes, or using brushes spanning several commutator segments.

21.3 D.c. machine construction

The basic parts of any d.c. machine are shown in Figure 21.3, and comprise:

(a) a stationary part called the **stator** having,

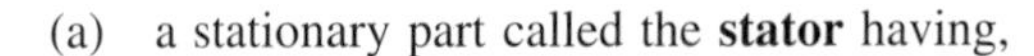

(i) a steel ring called the **yoke**, to which are attached

(ii) the magnetic **poles**, around which are the

(iii) **field windings**, i.e. many turns of a conductor wound round the pole core; current passing through this conductor creates an electromagnet, (rather than the permanent magnets shown in Figures 21.1 and 21.2),

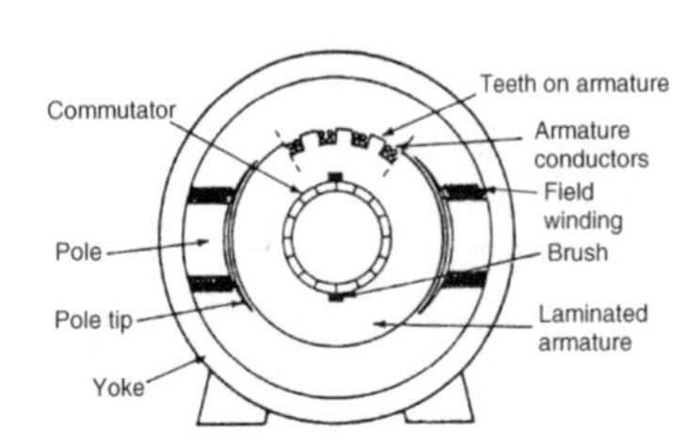

Figure 21.3

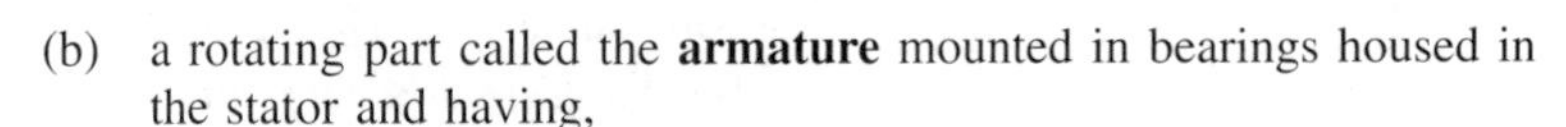

(b) a rotating part called the **armature** mounted in bearings housed in the stator and having,

(iv) a laminated cylinder of iron or steel called the **core**, on which teeth are cut to house the

(v) **armature winding**, i.e. a single or multi-loop conductor system and

(vi) the **commutator**, (see Section 21.2).

Armature windings can be divided into two groups, depending on how the wires are joined to the commutator. These are called **wave windings** and **lap windings**.

(a) In **wave windings** there are two paths in parallel irrespective of the number of poles, each path supplying half the total current output. Wave wound generators produce high voltage, low current outputs.

(b) In **lap windings** there are as many paths in parallel as the machine has poles. The total current output divides equally between them. Lap wound generators produce high current, low voltage output.

21.4 Shunt, series and compound windings

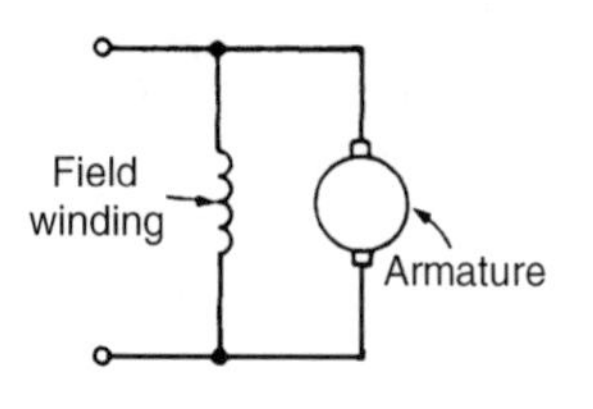

(a) Shunt-wound machine

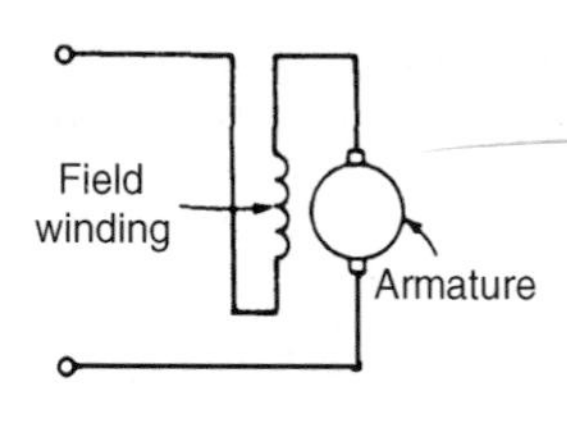

(b) Series-wound machine

Figure 21.4

When the field winding of a d.c. machine is connected in parallel with the armature, as shown in Figure 21.4(a), the machine is said to be **shunt** wound. If the field winding is connected in series with the armature, as shown in Figure 21.4(b), then the machine is said to be **series** wound. A **compound** wound machine has a combination of series and shunt windings.

Depending on whether the electrical machine is series wound, shunt wound or compound wound, it behaves differently when a load is applied. The behaviour of a d.c. machine under various conditions is shown by means of graphs, called characteristic curves or just **characteristics**. The characteristics shown in the following sections are theoretical, since they neglect the effects of armature reaction.

Armature reaction is the effect that the magnetic field produced by the armature current has on the magnetic field produced by the field system. In a generator, armature reaction results in a reduced output voltage, and in a motor, armature reaction results in increased speed.

A way of overcoming the effect of armature reaction is to fit compensating windings, located in slots in the pole face.

21.5 E.m.f. generated in an armature winding

Let Z = number of armature conductors,

Φ = useful flux per pole, in webers

p = number of **pairs** of poles

and n = armature speed in rev/s

The e.m.f. generated by the armature is equal to the e.m.f. generated by one of the parallel paths. Each conductor passes 2p poles per revolution and thus cuts 2pΦ webers of magnetic flux per revolution. Hence flux cut by one conductor per second $= 2p\Phi n$ Wb and so the average e.m.f. E generated per conductor is given by:

$$E = 2p\Phi n \text{ volts} \quad \text{(since 1 volt = 1 Weber per second)}$$

Let c = number of parallel paths through the winding between positive and negative brushes

$c = 2$ for a wave winding

$c = 2p$ for a lap winding

The number of conductors in series in each path $= \dfrac{Z}{c}$

The total e.m.f. between brushes

= (average e.m.f./conductor)(number of conductors in series per path)

$$= 2p\Phi n \frac{Z}{c}$$

i.e., **generated e.m.f.,** $\boldsymbol{E = \frac{2p\Phi nZ}{c}}$ **volts** (21.1)

Since Z, p and c are constant for a given machine, then $E \propto \Phi n$. However $2\pi n$ is the angular velocity ω in radians per second, hence the generated e.m.f. is proportional to Φ and ω, i.e.,

generated e.m.f., $\boldsymbol{E \propto \Phi\omega}$ (21.2)

Problem 1. An 8-pole, wave-connected armature has 600 conductors and is driven at 625 rev/min. If the flux per pole is 20 mWb, determine the generated e.m.f.

$Z = 600$, $c = 2$ (for a wave winding), $p = 4$ pairs

$n = \frac{625}{60}$ rev/s, $\Phi = 20 \times 10^{-3}$ Wb

$$\text{Generated e.m.f., } E = \frac{2p\Phi nZ}{c} = \frac{2(4)(20 \times 10^{-3})\left(\frac{625}{60}\right)(600)}{2} = \mathbf{500\ volts}$$

Problem 2. A 4-pole generator has a lap-wound armature with 50 slots with 16 conductors per slot. The useful flux per pole is 30 mWb. Determine the speed at which the machine must be driven to generate an e.m.f. of 240 V.

$E = 240$ V, $c = 2p$ (for a lap winding), $Z = 50 \times 16 = 800$,

$\Phi = 30 \times 10^{-3}$ Wb.

$$\text{Generated e.m.f. } E = \frac{2p\Phi nZ}{c} = \frac{2p\Phi nZ}{2p} = \Phi nZ$$

$$\text{Rearranging gives, speed, } n = \frac{E}{\Phi Z} = \frac{240}{(30 \times 10^{-3})(800)}$$

$$= \mathbf{10\ rev/s} \text{ or } \mathbf{600\ rev/min}$$

Problem 3. An 8-pole, lap-wound armature has 1200 conductors and a flux per pole of 0.03 Wb. Determine the e.m.f. generated when running at 500 rev/min.

Generated e.m.f., $E = \frac{2p\Phi nZ}{c} = \frac{2p\Phi nZ}{2p}$, for a lap-wound machine, i.e.,

$$E = \Phi nZ = (0.03)\left(\frac{500}{60}\right)(1200) = \mathbf{300\ volts}$$

Problem 4. Determine the generated e.m.f. in problem 3 if the armature is wave-wound.

$$\text{Generated e.m.f. } E = \frac{2p\Phi nZ}{c} = \frac{2p\Phi nZ}{2} \quad \text{(since } c=2 \text{ for wave-wound)}$$

$$= p\Phi nZ = (4)(\Phi nZ)$$

$$= (4)(300) \qquad \text{from problem 3,}$$

$$= \mathbf{1200\ volts}$$

Problem 5. A d.c. shunt-wound generator running at constant speed generates a voltage of 150 V at a certain value of field current. Determine the change in the generated voltage when the field current is reduced by 20%, assuming the flux is proportional to the field current.

The generated e.m.f. E of a generator is proportional to $\Phi\omega$, i.e. is proportional to Φn, where Φ is the flux and n is the speed of rotation.

It follows that $E = k\Phi n$, where k is a constant.

At speed n_1 and flux Φ_1, $E_1 = k\Phi_1 n_1$.

At speed n_2 and flux Φ_2, $E_2 = k\Phi_2 n_2$.

Thus, by division:

$$\frac{E_1}{E_2} = \frac{k\Phi_1 n_1}{k\Phi_2 n_2} = \frac{\Phi_1 n_1}{\Phi_2 n_2}$$

The initial conditions are $E_1 = 150$ V, $\Phi = \Phi_1$ and $n = n_1$. When the flux is reduced by 20%, the new value of flux is 80/100 or 0.8 of the initial value, i.e. $\Phi_2 = 0.8\Phi_1$. Since the generator is running at constant speed, $n_2 = n_1$.

$$\text{Thus} \quad \frac{E_1}{E_2} = \frac{\Phi_1 n_1}{\Phi_2 n_2} = \frac{\Phi_1 n_1}{0.8\Phi_1 n_1} = \frac{1}{0.8}$$

that is, $E_2 = 150 \times 0.8 = 120$ V

Thus, a reduction of 20% in the value of the flux **reduces the generated voltage to 120 V** at constant speed.

> Problem 6. A d.c. generator running at 30 rev/s generates an e.m.f. of 200 V. Determine the percentage increase in the flux per pole required to generate 250 V at 20 rev/s.

From equation (21.2), generated e.m.f., $E \propto \Phi\omega$ and since $\omega = 2\pi n$, $E \propto \Phi n$.

Let $E_1 = 200$ V, $n_1 = 30$ rev/s and flux per pole at this speed be Φ_1

Let $E_2 = 250$ V, $n_1 = 20$ rev/s and flux per pole at this speed be Φ_2

Since $E \propto \Phi n$ then $\dfrac{E_1}{E_2} = \dfrac{\Phi_1 n_1}{\Phi_2 n_2}$

Hence $\dfrac{200}{250} = \dfrac{\Phi_1(30)}{\Phi_2(20)}$

from which, $\Phi_2 = \dfrac{\Phi_1(30)(250)}{(20)(200)} = 1.875\ \Phi_1$

Hence the increase in flux per pole needs to be **87.5%**

Further problems on generated e.m.f. may be found in Section 21.17, problems 1 to 5, page 381.

21.6 D.c. generators

D.c. generators are classified according to the method of their field excitation. These groupings are:

(i) **Separately-excited generators**, where the field winding is connected to a source of supply other than the armature of its own machine.

(ii) **Self-excited generators**, where the field winding receives its supply from the armature of its own machine, and which are sub-divided into (a) shunt, (b) series, and (c) compound wound generators.

21.7 Types of d.c. generator and their characteristics

(a) Separately-excited generator

A typical separately-excited generator circuit is shown in Figure 21.5.

When a load is connected across the armature terminals, a load current I_a will flow. The terminal voltage V will fall from its open-circuit e.m.f. E due to a volt drop caused by current flowing through the armature resistance, shown as R_a, i.e.,

terminal voltage, $\boldsymbol{V = E - I_a R_a}$

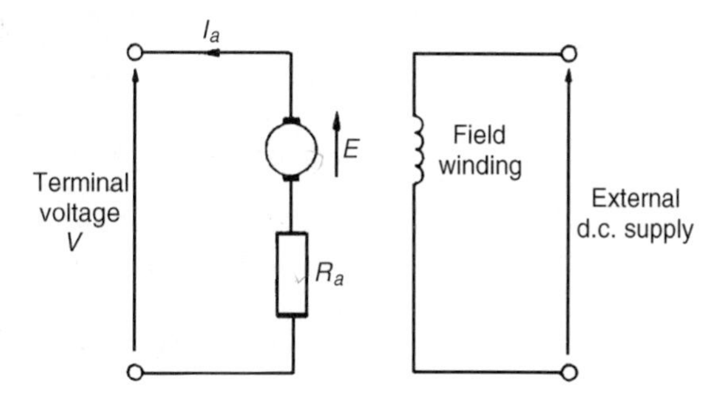

Figure 21.5

or **generated e.m.f., $E = V + I_a R_a$** (21.3)

Problem 7. Determine the terminal voltage of a generator which develops an e.m.f. of 200 V and has an armature current of 30 A on load. Assume the armature resistance is 0.30 Ω

With reference to Figure 21.5, terminal voltage,

$$V = E - I_a R_a = 200 - (30)(0.30) = 200 - 9 = \mathbf{191\ volts}$$

Problem 8. A generator is connected to a 60 Ω load and a current of 8 A flows. If the armature resistance is 1 Ω determine (a) the terminal voltage, and (b) the generated e.m.f.

(a) Terminal voltage, $V = I_a R_L = (8)(60) = \mathbf{480\ volts}$

(b) Generated e.m.f., $E = V + I_a R_a$ from equation (21.3)

$$= 480 + (8)(1) = 480 + 8 = \mathbf{488\ volts}$$

Problem 9. A separately-excited generator develops a no-load e.m.f. of 150 V at an armature speed of 20 rev/s and a flux per pole of 0.10 Wb. Determine the generated e.m.f. when (a) the speed increases to 25 rev/s and the pole flux remains unchanged, (b) the speed remains at 20 rev/s and the pole flux is decreased to 0.08 Wb, and (c) the speed increases to 24 rev/s and the pole flux is decreased to 0.07 Wb.

(a) From Section 21.5, generated e.m.f. $E \propto \Phi n$

from which, $\dfrac{E_1}{E_2} = \dfrac{\Phi_1 n_1}{\Phi_2 n_2}$

Hence $$\frac{150}{E_2} = \frac{(0.10)(20)}{(0.10)(25)}$$

from which, $$E_2 = \frac{(150)(0.10)(25)}{(0.10)(20)} = \textbf{187.5 volts}$$

(b) $$\frac{150}{E_3} = \frac{(0.10)(20)}{(0.08)(20)}$$

from which, e.m.f., $$E_3 = \frac{(150)(0.08)(20)}{(0.10)(20)} = \textbf{120 volts}$$

(c) $$\frac{150}{E_4} = \frac{(0.10)(20)}{(0.07)(24)}$$

from which, e.m.f. $$E_4 = \frac{(150)(0.07)(24)}{(0.10)(20)} = \textbf{126 volts}$$

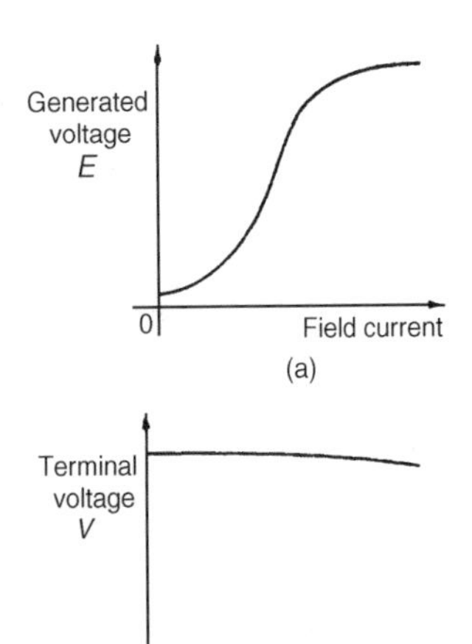

Figure 21.6

Characteristics

The two principal generator characteristics are the generated voltage/field current characteristics, called the **open-circuit characteristic** and the terminal voltage/load current characteristic, called the **load characteristic**. A typical separately-excited generator **open-circuit characteristic** is shown in Figure 21.6(a) and a typical **load characteristic** is shown in Figure 21.6(b).

A separately-excited generator is used only in special cases, such as when a wide variation in terminal p.d. is required, or when exact control of the field current is necessary. Its disadvantage lies in requiring a separate source of direct current.

(b) Shunt-wound generator

In a shunt wound generator the field winding is connected in parallel with the armature as shown in Figure 21.7. The field winding has a relatively high resistance and therefore the current carried is only a fraction of the armature current.

For the circuit shown in Figure 21.7,

terminal voltage $V = E - I_a R_a$

or generated e.m.f., $E = V + I_a R_a$

$I_a = I_f + I$, from Kirchhoff's current law,

where I_a = armature current

I_f = field current $\left(= \dfrac{V}{R_f}\right)$

and I = load current

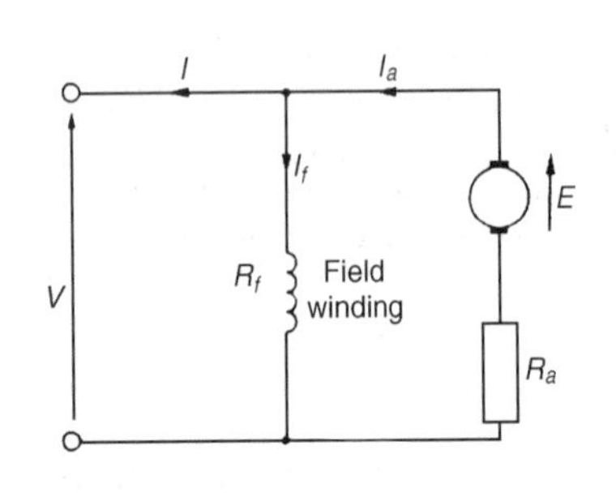

Figure 21.7

Problem 10. A shunt generator supplies a 20 kW load at 200 V through cables of resistance, $R = 100\ \text{m}\Omega$. If the field winding resistance, $R_f = 50\ \Omega$ and the armature resistance, $R_a = 40\ \text{m}\Omega$, determine (a) the terminal voltage, and (b) the e.m.f. generated in the armature.

(a) The circuit is as shown in Figure 21.8.

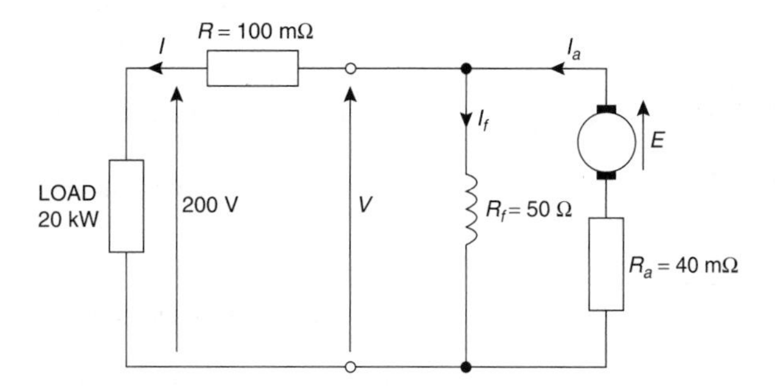

Figure 21.8

$$\text{Load current, } I = \frac{20\,000 \text{ watts}}{200 \text{ volts}} = 100 \text{ A}$$

$$\text{Volt drop in the cables to the load} = IR = (100)(100 \times 10^{-3}) = 10 \text{ V}$$

Hence terminal voltage, $V = 200 + 10 =$ **210 volts**

(b) Armature current $I_a = I_f + I$

$$\text{Field current, } I_f = \frac{V}{R_f} = \frac{210}{50} = 4.2 \text{ A}$$

Hence $I_a = I_f + I = 4.2 + 100 = 104.2$ A

$$\text{Generated e.m.f. } E = V + I_a R_a$$
$$= 210 + (104.2)(40 \times 10^{-3})$$
$$= 210 + 4.168$$
$$= \mathbf{214.17 \text{ volts}}$$

Characteristics

The generated e.m.f., E, is proportional to $\Phi\omega$, (see Section 21.5), hence at constant speed, since $\omega = 2\pi n$, $E \propto \Phi$. Also the flux Φ is proportional

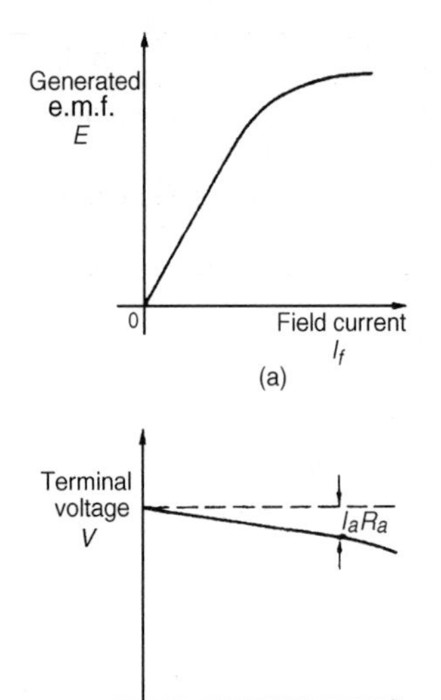

Figure 21.9

to field current I_f until magnetic saturation of the iron circuit of the generator occurs. Hence the open circuit characteristic is as shown in Figure 21.9(a).

As the load current on a generator having constant field current and running at constant speed increases, the value of armature current increases, hence the armature volt drop, $I_a R_a$ increases. The generated voltage E is larger than the terminal voltage V and the voltage equation for the armature circuit is $V = E - I_a R_a$. Since E is constant, V decreases with increasing load. The load characteristic is as shown in Figure 21.9(b). In practice, the fall in voltage is about 10% between no-load and full-load for many d.c. shunt-wound generators.

The shunt-wound generator is the type most used in practice, but the load current must be limited to a value that is well below the maximum value. This then avoids excessive variation of the terminal voltage. Typical applications are with battery charging and motor car generators.

(c) Series-wound generator

In the series-wound generator the field winding is connected in series with the armature as shown in Figure 21.10.

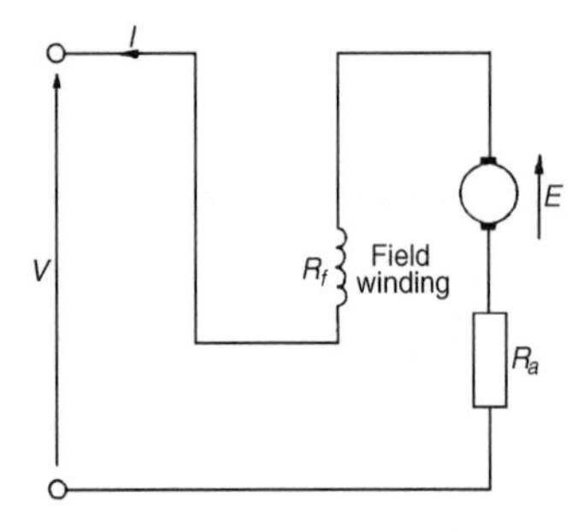

Figure 21.10

Characteristic

The load characteristic is the terminal voltage/current characteristic. The generated e.m.f. E, is proportional to $\Phi\omega$ and at constant speed ω (= $2\pi n$) is a constant. Thus E is proportional to Φ. For values of current below magnetic saturation of the yoke, poles, air gaps and armature core, the flux Φ is proportional to the current, hence $E \propto I$. For values of current above those required for magnetic saturation, the generated e.m.f. is approximately constant. The values of field resistance and armature resistance in a series wound machine are small, hence the terminal voltage V is very nearly equal to E. A typical load characteristic for a series generator is shown in Figure 21.11.

In a series-wound generator, the field winding is in series with the armature and it is not possible to have a value of field current when the terminals are open circuited, thus it is not possible to obtain an open-circuit characteristic.

Series-wound generators are rarely used in practise, but can be used as a 'booster' on d.c. transmission lines.

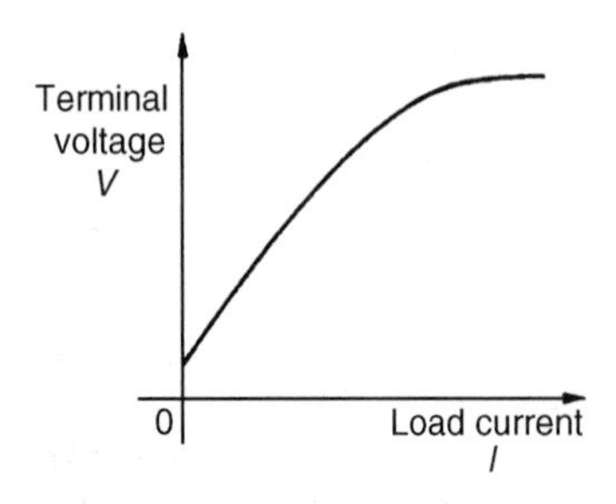

Figure 21.11

(d) Compound-wound generator

In the compound-wound generator two methods of connection are used, both having a mixture of shunt and series windings, designed to combine the advantages of each. Figure 21.12(a) shows what is termed a **long-shunt** compound generator, and Figure 21.12(b) shows a **short-shunt**

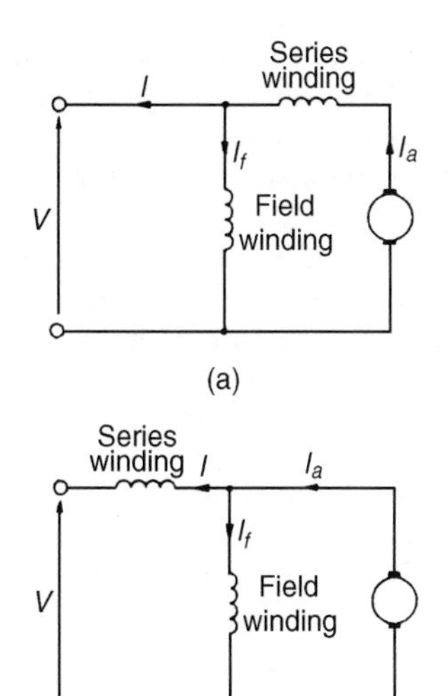

Figure 21.12

compound generator. The latter is the most generally used form of d.c. generator.

Problem 11. A short-shunt compound generator supplies 80 A at 200 V. If the field resistance, $R_f = 40\ \Omega$, the series resistance, $R_{Se} = 0.02\ \Omega$ and the armature resistance, $R_a = 0.04\ \Omega$, determine the e.m.f. generated.

The circuit is shown in Figure 21.13.

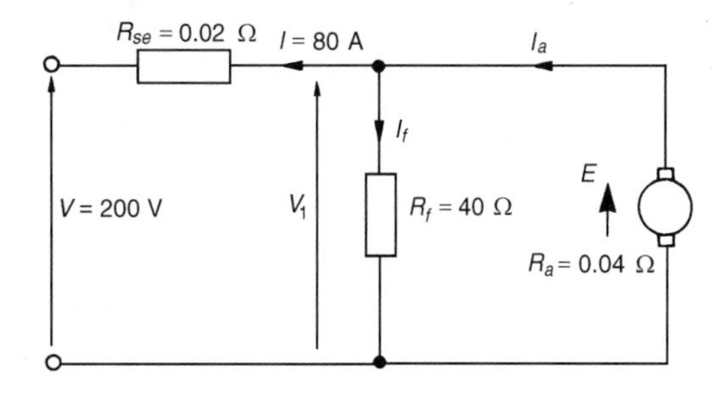

Figure 21.13

Volt drop in series winding $= IR_{Se} = (80)(0.02) = 1.6$ V

P.d. across the field winding = p.d. across armature

$$= V_1 = 200 + 1.6 = 201.6 \text{ V}$$

$$\text{Field current } I_f = \frac{V_1}{R_f} = \frac{201.6}{40} = 5.04 \text{ A}$$

Armature current, $I_a = I + I_f = 80 + 5.04 = 85.04$ A

$$\begin{aligned}\text{Generated e.m.f., } E &= V_1 + I_a R_a \\ &= 201.6 + (85.04)(0.04) \\ &= 201.6 + 3.4016 \\ &= \mathbf{205\ volts}\end{aligned}$$

Characteristics

In cumulative-compound machines the magnetic flux produced by the series and shunt fields are additive. Included in this group are **over-compounded, level-compounded** and **under-compounded machines**—the degree of compounding obtained depending on the number of turns of wire on the series winding.

A large number of series winding turns results in an over-compounded characteristic, as shown in Figure 21.14, in which the full-load terminal voltage exceeds the no-load voltage. A level-compound machine gives a full-load terminal voltage which is equal to the no-load voltage, as shown in Figure 21.14.

An under-compounded machine gives a full-load terminal voltage which is less than the no-load voltage, as shown in Figure 21.14. However even this latter characteristic is a little better than that for a shunt generator alone.

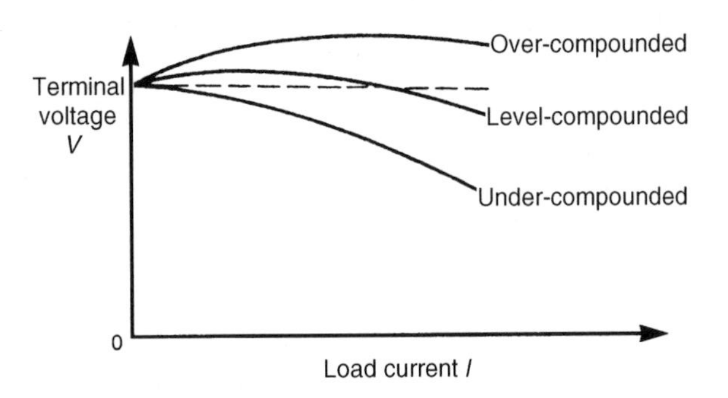

Figure 21.14

Compound-wound generators are used in electric arc welding, with lighting sets and with marine equipment.

Further problems on the d.c. generator may be found in Section 21.17, problems 6 to 11, page 382.

21.8 D.c. machine losses

As stated in Section 21.1, a generator is a machine for converting mechanical energy into electrical energy and a motor is a machine for converting electrical energy into mechanical energy. When such conversions take place, certain losses occur which are dissipated in the form of heat.

The principal **losses of machines** are:

(i) **Copper loss**, due to I^2R heat losses in the armature and field windings.

(ii) **Iron (or core) loss**, due to hysteresis and eddy-current losses in the armature. This loss can be reduced by constructing the armature of silicon steel laminations having a high resistivity and low hysteresis loss. At constant speed, the iron loss is assumed constant.

(iii) **Friction and windage losses**, due to bearing and brush contact friction and losses due to air resistance against moving parts (called windage). At constant speed, these losses are assumed to be constant.

(iv) **Brush contact loss** between the brushes and commutator. This loss is approximately proportional to the load current.

The total losses of a machine can be quite significant and operating efficiencies of between 80% and 90% are common.

21.9 Efficiency of a d.c. generator

The efficiency of an electrical machine is the ratio of the output power to the input power and is usually expressed as a percentage. The Greek letter, 'η' (eta) is used to signify efficiency and since the units are power/power, then efficiency has no units. Thus

$$\textbf{efficiency}, \eta = \left(\frac{\textbf{output power}}{\textbf{input power}}\right) \times \mathbf{100\%}$$

If the total resistance of the armature circuit (including brush contact resistance) is R_a, then **the total loss in the armature circuit is $I_a^2 R_a$**

If the terminal voltage is V and the current in the shunt circuit is I_f, then **the loss in the shunt circuit is $I_f V$**

If the sum of the iron, friction and windage losses is C then **the total losses is given by**:

$$I_a^2 R_a + I_f V + C \quad (I_a^2 R_a + I_f V \text{ is, in fact, the 'copper loss'})$$

If the output current is I, then **the output power is VI**

Total input power $= VI + I_a^2 R_a + I_f V + C$. Hence

$$\textbf{efficiency}, \eta = \frac{\textbf{output}}{\textbf{input}} = \left(\frac{VI}{VI + I_a^2 R_a + I_f V + C}\right) \times \mathbf{100\%} \quad (21.4)$$

The **efficiency of a generator is a maximum** when the load is such that:

$$I_a^2 R_a = VI_f + C$$

i.e., when the variable loss = the constant loss

Problem 12. A 10 kW shunt generator having an armature circuit resistance of 0.75 Ω and a field resistance of 125 Ω, generates a terminal voltage of 250 V at full load. Determine the efficiency of the generator at full load, assuming the iron, friction and windage losses amount to 600 W.

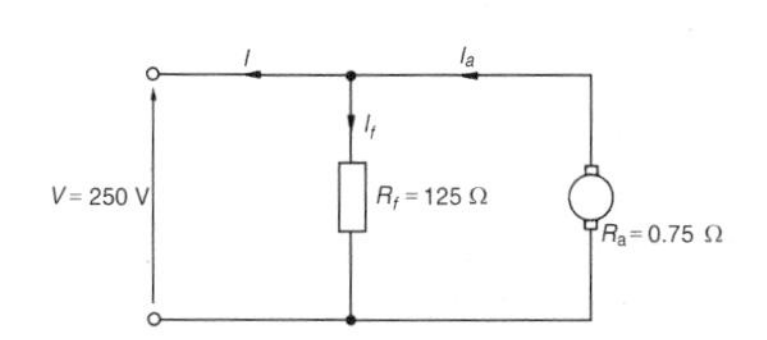

Figure 21.15

The circuit is shown in Figure 21.15.

Output power = 10 000 $W = VI$

from which, load current $I = \frac{10\,000}{V} = \frac{10\,000}{250} = 40$ A

Field current, $I_f = \frac{V}{R_f} = \frac{250}{125} = 2$ A

Armature current, $I_a = I_f + I = 2 + 40 = 42$ A

$$\text{Efficiency, } \eta = \left(\frac{VI}{VI + I_a^2 R_a + I_f V + C}\right) \times 100\%$$

$$= \left(\frac{10\,000}{10\,000 + (42)^2(0.75) + (2)(250) + 600}\right) \times 100\%$$

$$= \frac{10\,000}{12\,423} \times 100\% = \mathbf{80.50\%}$$

A further problem on the efficiency of a d.c. generator may be found in Section 21.17, problem 12, page 382.

21.10 D.c. motors

The construction of a d.c. motor is the same as a d.c. generator. The only difference is that in a generator the generated e.m.f. is greater than the terminal voltage, whereas in a motor the generated e.m.f. is less than the terminal voltage.

D.c. motors are often used in power stations to drive emergency stand-by pump systems which come into operation to protect essential equipment and plant should the normal a.c. supplies or pumps fail.

Back e.m.f.

When a d.c. motor rotates, an e.m.f. is induced in the armature conductors. By Lenz's law this induced e.m.f. E opposes the supply voltage V and is called a **back e.m.f.**, and the supply voltage, V is given by:

$$\boxed{V = E + I_a R_a} \quad \text{or} \quad \boxed{E = V - I_a R_a} \qquad (21.5)$$

Problem 13. A d.c. motor operates from a 240 V supply. The armature resistance is 0.2 Ω. Determine the back e.m.f. when the armature current is 50 A.

For a motor, $V = E + I_a R_a$

hence back e.m.f., $E = V - I_a R_a$

$= 240 - (50)(0.2) = 240 - 10 =$ **230 volts**

Problem 14. The armature of a d.c. machine has a resistance of 0.25 Ω and is connected to a 300 V supply. Calculate the e.m.f. generated when it is running: (a) as a generator giving 100 A, and (b) as a motor taking 80 A.

(a) As a generator, generated e.m.f.,

$$E = V + I_a R_a, \text{ from equation (21.3)},$$
$$= 300 + (100)(0.25)$$
$$= 300 + 25 = \mathbf{325\ volts}$$

(b) As a motor, generated e.m.f. (or back e.m.f.),

$$E = V - I_a R_a, \text{ from equation (21.5)},$$
$$= 300 - (80)(0.25) = \mathbf{280\ volts}$$

Further problems on back e.m.f. may be found in Section 21.17, problems 13 to 15, page 383.

21.11 Torque of a d.c. machine

From equation (21.5), for a d.c. motor, the supply voltage V is given by

$$V = E + I_a R_a$$

Multiplying each term by current I_a gives:

$$VI_a = EI_a + I_a^2 R_a$$

The term $\boldsymbol{VI_a}$ is the **total electrical power supplied to the armature**, the term $\boldsymbol{I_a^2 R_a}$ is the **loss due to armature resistance**, and the term $\boldsymbol{EI_a}$ is the **mechanical power developed by the armature**

If T is the torque, in newton metres, then the mechanical power developed is given by $T\omega$ watts (see Chapter 34, '*Science for Engineering*').

Hence $T\omega = 2\pi nT = EI_a$ from which,

$$\textbf{torque } \boldsymbol{T = \frac{EI_a}{2\pi n}} \textbf{ newton metres} \qquad (21.6)$$

From Section 21.5, equation (21.1), the e.m.f. E generated is given by

$$E = \frac{2p\Phi nZ}{c}$$

Hence $2\pi nT = EI_a = \left(\frac{2p\Phi nZ}{c}\right) I_a$

$$\text{and torque } T = \frac{\left(\dfrac{2p\Phi nZ}{c}\right) I_a}{2\pi n}$$

i.e., $$\boxed{\boldsymbol{T = \frac{p\Phi Z I_a}{\pi c}} \textbf{ newton metres}} \quad (21.7)$$

For a given machine, Z, c and p are fixed values

Hence torque, $$\boxed{\boldsymbol{T \propto \Phi I_a}} \quad (21.8)$$

Problem 15. An 8-pole d.c. motor has a wave-wound armature with 900 conductors. The useful flux per pole is 25 mWb. Determine the torque exerted when a current of 30 A flows in each armature conductor.

$p = 4$, $c = 2$ for a wave winding, $\Phi = 25 \times 10^{-3}$ Wb, $Z = 900$, $I_a = 30$ A

$$\text{From equation (21.7), torque } T = \frac{p\Phi Z I_a}{\pi c}$$

$$= \frac{(4)(25 \times 10^{-3})(900)(30)}{\pi(2)}$$

$$= \mathbf{429.7\ Nm}$$

Problem 16. Determine the torque developed by a 350 V d.c. motor having an armature resistance of 0.5 Ω and running at 15 rev/s. The armature current is 60 A.

$V = 350$ V, $R_a = 0.5\ \Omega$, $n = 15$ rev/s, $I_a = 60$ A

Back e.m.f. $E = V - I_a R_a = 350 - (60)(0.5) = 320$ V

$$\text{From equation (21.6), torque } T = \frac{E I_a}{2\pi n} = \frac{(320)(60)}{2\pi(15)} = \mathbf{203.7\ Nm}$$

Problem 17. A six-pole lap-wound motor is connected to a 250 V d.c. supply. The armature has 500 conductors and a resistance of 1 Ω. The flux per pole is 20 mWb. Calculate (a) the speed and (b) the torque developed when the armature current is 40 A

$V = 250\ V$, $Z = 500$, $R_a = 1\ \Omega$, $\Phi = 20 \times 10^{-3}$ Wb, $I_a = 40$ A, $c = 2p$ for a lap winding

(a) Back e.m.f. $E = V - I_a R_a = 250 - (40)(1) = 210$ V

$$\text{E.m.f. } E = \frac{2p\Phi nZ}{c}$$

$$\text{i.e. } 210 = \frac{2p(20 \times 10^{-3})n(500)}{2p}$$

$$\text{Hence speed } n = \frac{210}{(20 \times 10^{-3})(500)} = \mathbf{21\ rev/s}$$

$$\text{or } (21 \times 60) = \mathbf{1260\ rev/min}$$

(b) $\text{Torque } T = \frac{EI_a}{2\pi n} = \frac{(210)(40)}{2\pi(21)} = \mathbf{63.66\ Nm}$

Problem 18. The shaft torque of a diesel motor driving a 100 V d.c. shunt-wound generator is 25 Nm. The armature current of the generator is 16 A at this value of torque. If the shunt field regulator is adjusted so that the flux is reduced by 15%, the torque increases to 35 Nm. Determine the armature current at this new value of torque.

From equation (21.8), the shaft torque T of a generator is proportional to ΦI_a, where Φ is the flux and I_a is the armature current. Thus, $T = k\Phi I_a$, where k is a constant.

The torque at flux Φ_1 and armature current I_{a1} is $T_1 = k\Phi_1 I_{a1}$.

Similarly, $T_2 = k\Phi_2 I_{a2}$

By division $\frac{T_1}{T_2} = \frac{k\Phi_1 I_{a1}}{k\Phi_2 I_{a2}} = \frac{\Phi_1 I_{a1}}{\Phi_2 I_{a2}}$

Hence $\frac{25}{35} = \frac{\Phi_1 \times 16}{0.85\Phi_1 \times I_{a2}}$

i.e. $I_{a2} = \frac{16 \times 35}{0.85 \times 25} = 26.35\ \text{A}$

That is, **the armature current at the new value of torque is 26.35 A**

Problem 19. A 100 V d.c. generator supplies a current of 15 A when running at 1500 rev/min. If the torque on the shaft driving the generator is 12 Nm, determine (a) the efficiency of the generator and (b) the power loss in the generator.

(a) From Section 21.9, the efficiency of a

$$\text{generator} = \frac{\text{output power}}{\text{input power}} \times 100\%$$

The output power is the electrical output, i.e. VI watts. The input power to a generator is the mechanical power in the shaft driving the generator, i.e. $T\omega$ or $T(2\pi n)$ watts, where T is the torque in Nm and n is speed of rotation in rev/s. Hence, for a generator

$$\text{efficiency,} \quad \eta = \frac{VI}{T(2\pi n)} \times 100\%$$

$$\text{i.e.} \quad \eta = \frac{(100)(15)(100)}{(12)(2\pi)\left(\frac{1500}{60}\right)}$$

i.e. **efficiency = 79.6%**

(b) The input power = output power + losses

Hence, $T(2\pi n) = VI + \text{losses}$

$$\text{i.e.} \quad \text{losses} = T(2\pi n) - VI$$

$$= \left[(12)(2\pi)\left(\frac{1500}{60}\right)\right] - [(100)(15)]$$

i.e. **power loss** = 1885 − 1500 = **385 W**

Further problems on losses, efficiency, and torque may be found in Section 21.17, problems 16 to 21, page 383.

21.12 Types of d.c. motor and their characteristics

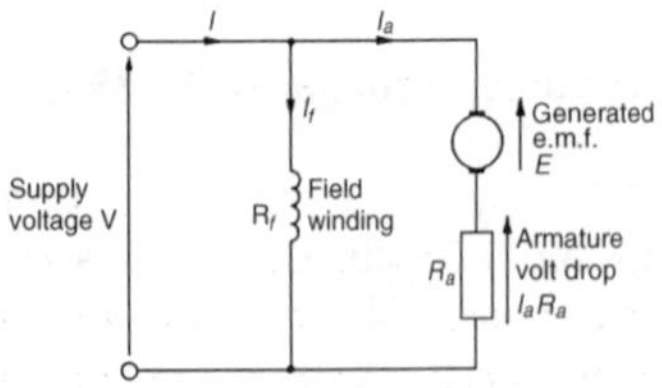

Figure 21.16

(a) **Shunt-wound motor**

In the shunt wound motor the field winding is in parallel with the armature across the supply as shown in Figure 21.16.

For the circuit shown in Figure 21.16,

Supply voltage, $V = E + I_a R_a$

or generated e.m.f., $E = V - I_a R_a$

Supply current, $I = I_a + I_f$, from Kirchhoff's current law.

Problem 20. A 240 V shunt motor takes a total current of 30 A. If the field winding resistance $R_f = 150\ \Omega$ and the armature resistance $R_a = 0.4\ \Omega$ determine (a) the current in the armature, and (b) the back e.m.f.

(a) Field current $I_f = \dfrac{V}{R_f} = \dfrac{240}{150} = 1.6$ A

Supply current $I = I_a + I_f$

Hence armature current, $I_a = I - I_f = 30 - 1.6 = \mathbf{28.4\ A}$

(b) Back e.m.f. $E = V - I_a R_a$

$= 240 - (28.4)(0.4) = \mathbf{228.64\ volts}$

Characteristics

The two principal characteristics are the torque/armature current and speed/armature current relationships. From these, the torque/speed relationship can be derived.

(i) The theoretical torque/armature current characteristic can be derived from the expression $T \propto \Phi I_a$, (see Section 21.11). For a shunt-wound motor, the field winding is connected in parallel with the armature circuit and thus the applied voltage gives a constant field current, i.e. a shunt-wound motor is a constant flux machine. Since Φ is constant, it follows that $T \propto I_a$, and the characteristic is as shown in Figure 21.17.

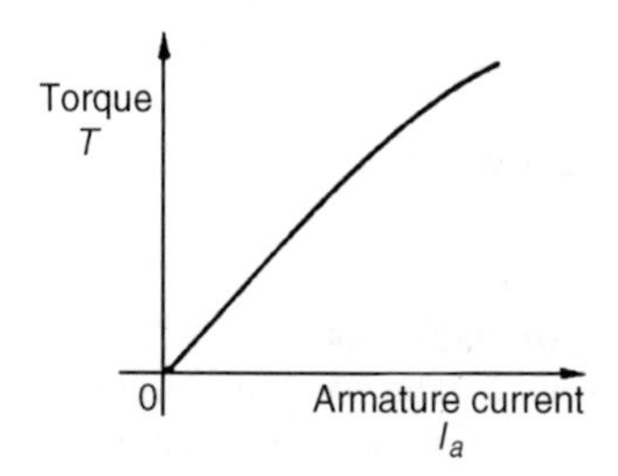

Figure 21.17

(ii) The armature circuit of a d.c. motor has resistance due to the armature winding and brushes, R_a ohms, and when armature current I_a is flowing through it, there is a voltage drop of $I_a R_a$ volts. In Figure 21.16 the armature resistance is shown as a separate resistor in the armature circuit to help understanding. Also, even though the machine is a motor, because conductors are rotating in a magnetic field, a voltage, $E \propto \Phi\omega$, is generated by the armature conductors. From equation (21.5) $V = E + I_a R_a$ or $E = V - I_a R_a$

However, from Section 21.5, $E \propto \Phi n$, hence $n \propto E/\Phi$, i.e.

$$\text{speed of rotation, } n \propto \frac{E}{\Phi} \propto \frac{V - I_a R_a}{\Phi} \qquad (21.9)$$

For a shunt motor, V, Φ and R_a are constants, hence as armature current I_a increases, $I_a R_a$ increases and $V - I_a R_a$ decreases, and the speed is proportional to a quantity which is decreasing and is as shown in Figure 21.18. As the load on the shaft of the motor increases, I_a increases and the speed drops slightly. In practice, the speed falls by about 10% between no-load and full-load on many d.c. shunt-wound motors. Due to this relatively small drop in speed, the d.c. shunt-wound motor is taken as basically being a constant-speed machine and may be used for driving lathes, lines of shafts, fans, conveyor belts, pumps, compressors, drilling machines and so on.

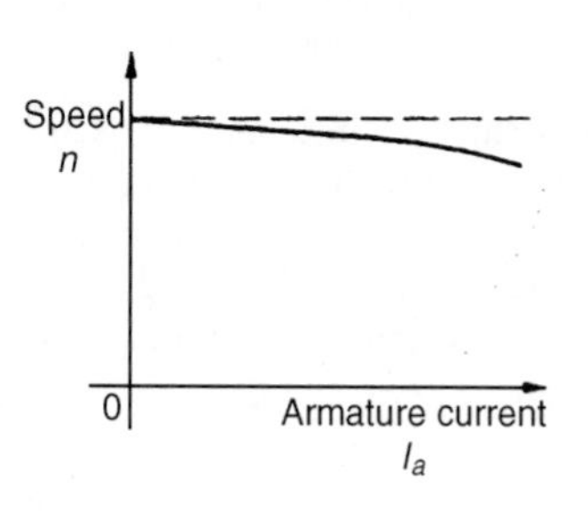

Figure 21.18

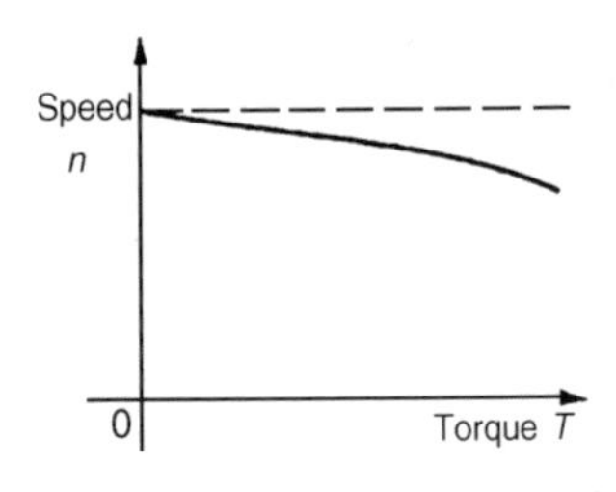

Figure 21.19

(iii) Since torque is proportional to armature current, (see (i) above), the theoretical speed/torque characteristic is as shown in Figure 21.19.

Problem 21. A 200 V, d.c. shunt-wound motor has an armature resistance of 0.4 Ω and at a certain load has an armature current of 30 A and runs at 1350 rev/min. If the load on the shaft of the motor is increased so that the armature current increases to 45 A, determine the speed of the motor, assuming the flux remains constant.

The relationship $E \propto \Phi n$ applies to both generators and motors. For a motor,

$$E = V - I_a R_a, \quad \text{(see equation 21.5)}$$

Hence $E_1 = 200 - 30 \times 0.4 = 188$ V,

and $E_2 = 200 - 45 \times 0.4 = 182$ V.

The relationship, $\dfrac{E_1}{E_2} = \dfrac{\Phi_1 n_1}{\Phi_2 n_2}$

applies to both generators and motors. Since the flux is constant, $\Phi_1 = \Phi_2$

Hence $\dfrac{188}{182} = \dfrac{\Phi_1 \times \left(\dfrac{1350}{60}\right)}{\Phi_1 \times n_2}$, i.e., $n_2 = \dfrac{22.5 \times 182}{188} = 21.78$ rev/s

Thus the speed of the motor when the armature current is 45 A is 21.78×60 rev/min, i.e. **1307 rev/min**

Problem 22. A 220 V, d.c. shunt-wound motor runs at 800 rev/min and the armature current is 30 A. The armature circuit resistance is 0.4 Ω. Determine (a) the maximum value of armature current if the flux is suddenly reduced by 10% and (b) the steady state value of the armature current at the new value of flux, assuming the shaft torque of the motor remains constant.

(a) For a d.c. shunt-wound motor, $E = V - I_a R_a$. Hence initial generated e.m.f., $E_1 = 220 - 30 \times 0.4 = 208$ V. The generated e.m.f. is also such that $E \propto \Phi n$, so at the instant the flux is reduced, the speed has not had time to change, and $E = 208 \times 90/100 = 187.2$ V.

Hence, the voltage drop due to the armature resistance is $220 - 187.2$, i.e., 32.8 V. The **instantaneous value of the current** is 32.8/0.4, i.e., **82 A**. This increase in current is about three times the initial value and causes an increase in torque, ($T \propto \Phi I_a$). The motor accelerates because of the larger torque value until steady state conditions are reached.

(b) $T \propto \Phi I_a$ and since the torque is constant,

$\Phi_1 I_{a1} = \Phi_2 I_{a2}$. The flux Φ is reduced by 10%, hence

$\Phi_2 = 0.9\Phi_1$.

Thus, $\Phi_1 \times 30 = 0.9\Phi_1 \times I_{a2}$

i.e. the steady state value of armature current, $I_{a2} = \dfrac{30}{0.9} = \mathbf{33\frac{1}{3}\ A}$

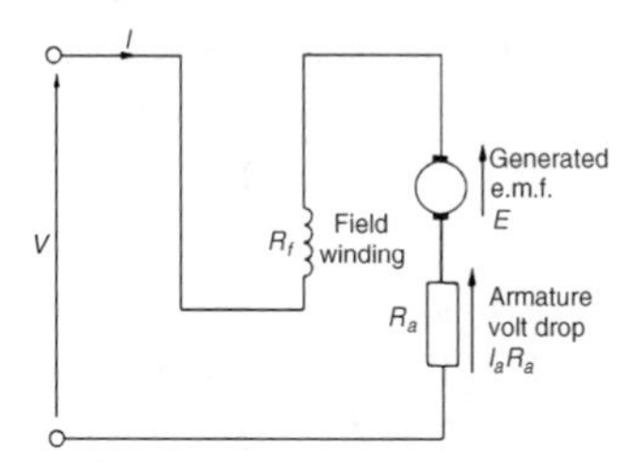

Figure 21.20

(b) Series-wound motor

In the series-wound motor the field winding is in series with the armature across the supply as shown in Figure 21.20.
For the series motor shown in Figure 21.20,

$$\text{Supply voltage V} = E + I(R_a + R_f)$$

$$\text{or generated e.m.f. E} = V - I(R_a + R_f)$$

Characteristics

In a series motor, the armature current flows in the field winding and is equal to the supply current, I.

(i) **The torque/current characteristic**
It is shown in Section 21.11 that torque $T \propto \Phi I_a$. Since the armature and field currents are the same current, I, in a series machine, then $T \propto \Phi I$ over a limited range, before magnetic saturation of the magnetic circuit of the motor is reached, (i.e., the linear portion of the B–H curve for the yoke, poles, air gap, brushes and armature in series). Thus $\Phi \propto I$ and $T \propto I^2$. After magnetic saturation, Φ almost becomes a constant and $T \propto I$. Thus the theoretical torque/current characteristic is as shown in Figure 21.21.

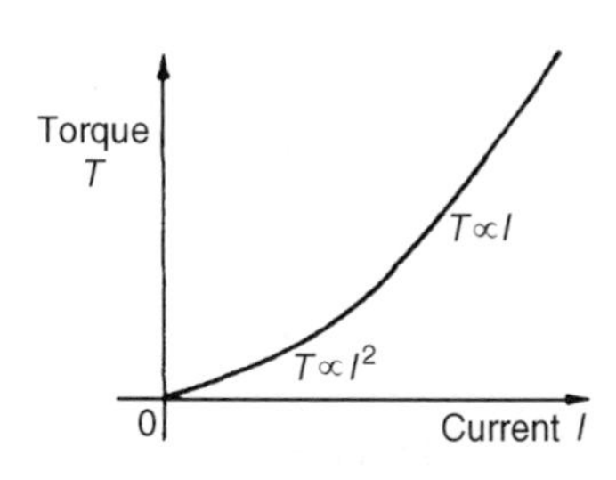

Figure 21.21

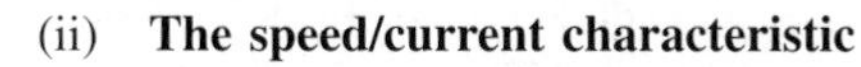

(ii) **The speed/current characteristic**
It is shown in equation (21.9) that $n \propto (V - I_aR_a)/\Phi$. In a series motor, $I_a = I$ and below the magnetic saturation level, $\Phi \propto I$. Thus $n \propto (V - IR)/I$ where R is the combined resistance of the series field and armature circuit. Since IR is small compared with V, then an approximate relationship for the speed is $n \propto V/I \propto 1/I$ since V is constant. Hence the theoretical speed/current characteristic is as shown in Figure 21.22. The high speed at small values of current indicate that this type of motor must not be run on very light loads and invariably, such motors are permanently coupled to their loads.

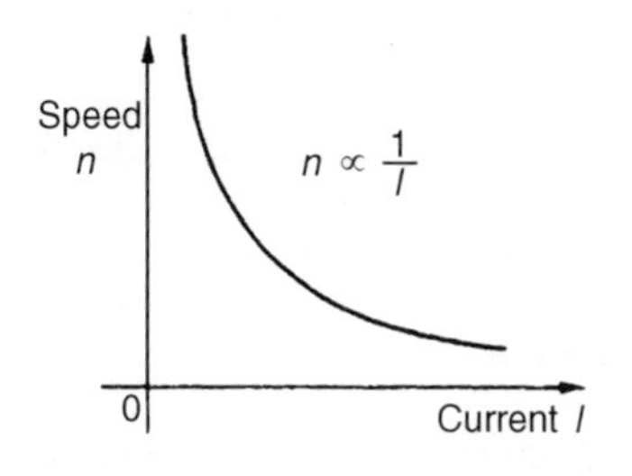

Figure 21.22

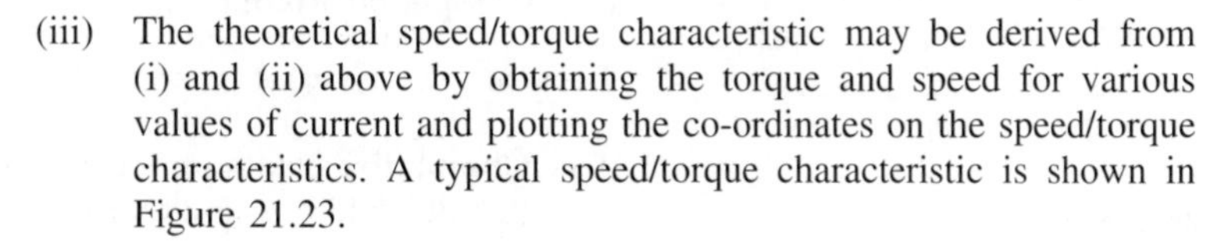

(iii) The theoretical speed/torque characteristic may be derived from (i) and (ii) above by obtaining the torque and speed for various values of current and plotting the co-ordinates on the speed/torque characteristics. A typical speed/torque characteristic is shown in Figure 21.23.

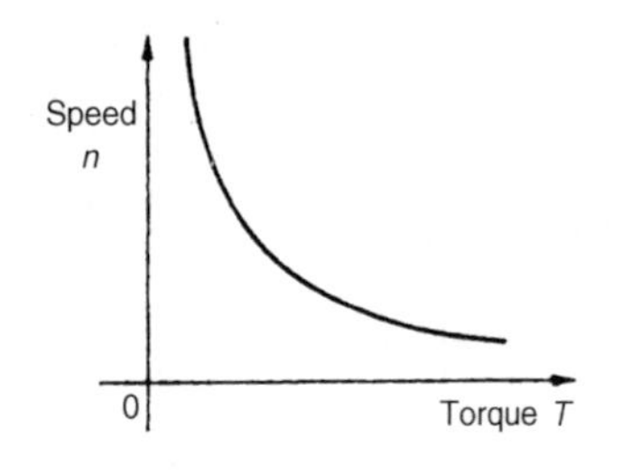

Figure 21.23

A d.c. series motor takes a large current on starting and the characteristic shown in Figure 21.21 shows that the series-wound motor has a large torque when the current is large. Hence these motors are used for traction (such as trains, milk delivery vehicles, etc.), driving fans and for cranes and hoists, where a large initial torque is required.

Problem 23. A series motor has an armature resistance of 0.2 Ω and a series field resistance of 0.3 Ω. It is connected to a 240 V supply and at a particular load runs at 24 rev/s when drawing 15 A from the supply.

(a) Determine the generated e.m.f. at this load.
(b) Calculate the speed of the motor when the load is changed such that the current is increased to 30 A. Assume that this causes a doubling of the flux.

(a) With reference to Figure 21.20, generated e.m.f., E, at initial load, is given by

$$E_1 = V - I_a(R_a + R_f)$$
$$= 240 - (15)(0.2 + 0.3) = 240 - 7.5 = \mathbf{232.5\ volts}$$

(b) When the current is increased to 30 A, the generated e.m.f. is given by:

$$E_2 = V - I_a(R_a + R_f)$$
$$= 240 - (30)(0.2 + 0.3) = 240 - 15 = 225 \text{ volts}$$

Now e.m.f. $E \propto \Phi n$

thus $\dfrac{E_1}{E_2} = \dfrac{\Phi_1 n_1}{\Phi_2 n_2}$

i.e., $\dfrac{232.5}{225} = \dfrac{\Phi_1(24)}{(2\Phi_1)(n_2)}$ since $\Phi_2 = 2\Phi_1$

Hence **speed of motor,** $n_2 = \dfrac{(24)(225)}{(232.5)(2)} = \mathbf{11.6\ rev/s}$

As the current has been increased from 15 A to 30 A, the speed has decreased from 24 rev/s to 11.6 rev/s. Its speed/current characteristic is similar to Figure 21.22.

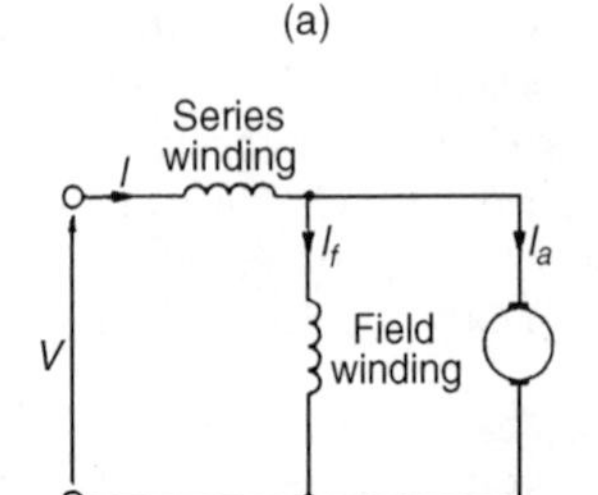

Figure 21.24

(c) Compound-wound motor

There are two types of compound wound motor:

(i) **Cumulative compound,** in which the series winding is so connected that the field due to it assists that due to the shunt winding.

(ii) **Differential compound,** in which the series winding is so connected that the field due to it opposes that due to the shunt winding.

Figure 21.24(a) shows a **long-shunt** compound motor and Figure 21.24(b) a **short-shunt** compound motor.

Characteristics

A compound-wound motor has both a series and a shunt field winding, (i.e. one winding in series and one in parallel with the armature), and is usually wound to have a characteristic similar in shape to a series wound motor (see Figures 21.21–21.23). A limited amount of shunt winding is present to restrict the no-load speed to a safe value. However, by varying the number of turns on the series and shunt windings and the directions of the magnetic fields produced by these windings (assisting or opposing), families of characteristics may be obtained to suit almost all applications. Generally, compound-wound motors are used for heavy duties, particularly in applications where sudden heavy load may occur such as for driving plunger pumps, presses, geared lifts, conveyors, hoists and so on.

Typical compound motor torque and speed characteristics are shown in Figure 21.25.

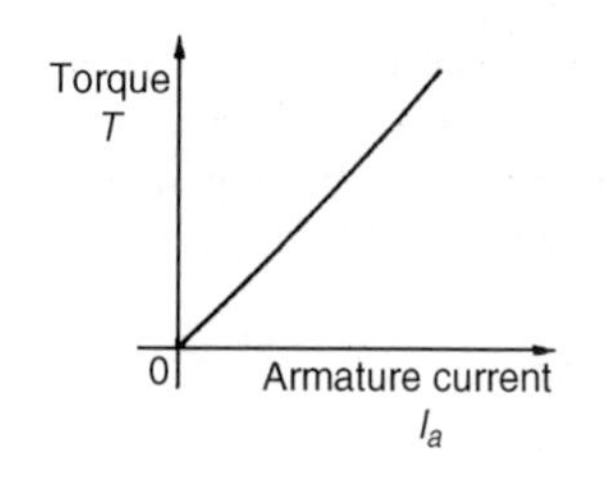

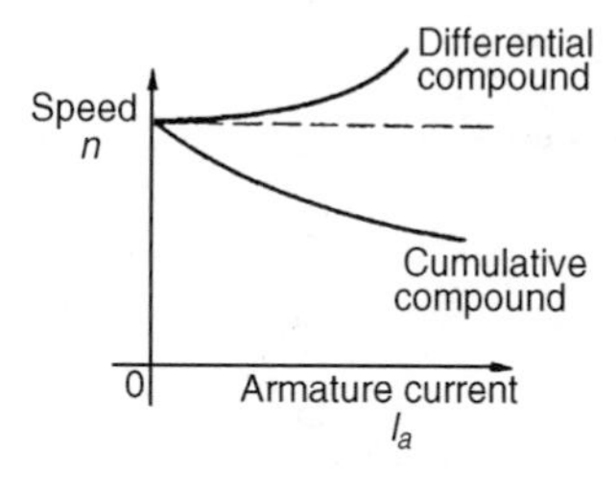

Figure 21.25

21.13 The efficiency of a d.c. motor

It was stated in Section 21.9, that the efficiency of a d.c. machine is given by:

$$\text{efficiency, } \eta = \frac{\text{output power}}{\text{input power}} \times 100\%$$

Also, the total losses $= I_a^2 R_a + I_f V + C$ (for a shunt motor) where C is the sum of the iron, friction and windage losses.

For a motor, the input power $= VI$

$$\text{and the output power} = VI - \text{losses}$$
$$= VI - I_a^2 R_a - I_f V - C$$

Hence $$\boldsymbol{\text{efficiency } \eta = \left(\frac{VI - I_a^2 R_a - I_f V - C}{VI}\right) \times 100\%} \qquad (21.10)$$

The **efficiency of a motor is a maximum** when the load is such that:

$$\boldsymbol{I_a^2 R_a = I_f V + C}$$

Problem 24. A 320 V shunt motor takes a total current of 80 A and runs at 1000 rev/min. If the iron, friction and windage losses amount to 1.5 kW, the shunt field resistance is 40 Ω and the armature resistance is 0.2 Ω, determine the overall efficiency of the motor.

The circuit is shown in Figure 21.26.

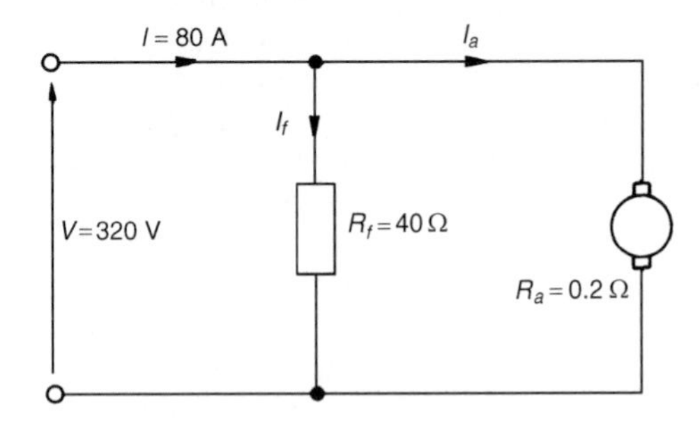

Figure 21.26

Field current, $I_f = \dfrac{V}{R_f} = \dfrac{320}{40} = 8\text{ A}$

Armature current $I_a = I - I_f = 80 - 8 = 72\text{ A}$

C = iron, friction and windage losses = 1500 W

$$\text{Efficiency, } \eta = \left(\frac{VI - I_a^2 R_a - I_f V - C}{VI}\right) \times 100\%$$

$$= \left(\frac{(320)(80) - (72)^2(0.2) - (8)(320) - 1500}{(320)(80)}\right) \times 100\%$$

$$= \left(\frac{25\,600 - 1036.8 - 2560 - 1500}{25\,600}\right) \times 100\%$$

$$= \left(\frac{20\,503.2}{25\,600}\right) \times 100\% = \mathbf{80.1\%}$$

Problem 25. A 250 V series motor draws a current of 40 A. The armature resistance is 0.15 Ω and the field resistance is 0.05 Ω. Determine the maximum efficiency of the motor.

The circuit is as shown in Figure 21.27.

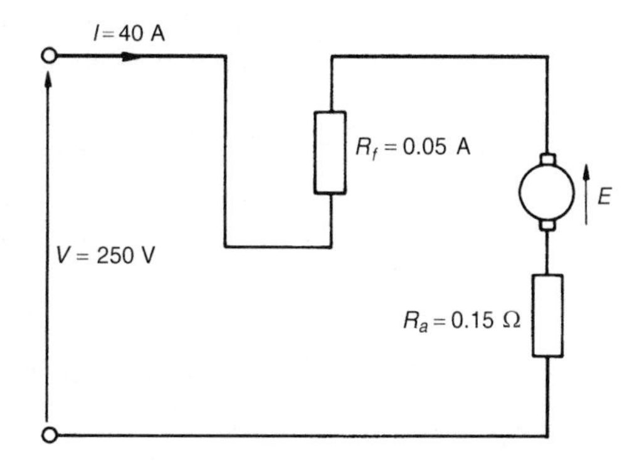

Figure 21.27

From equation (21.10), efficiency,

$$\eta = \left(\frac{VI - I_a^2 R_a - I_f V - C}{VI}\right) \times 100\%$$

However for a series motor, $I_f = 0$ and the $I_a^2 R_a$ loss needs to be $I^2(R_a + R_f)$

$$\text{Hence efficiency, } \eta = \left(\frac{VI - I^2(R_a + R_\text{f}) - C}{VI}\right) \times 100\%$$

For maximum efficiency $I^2(R_a + R_f) = C$

$$\text{Hence efficiency, } \eta = \left(\frac{VI - 2I^2(R_a + R_f)}{VI}\right) \times 100\%$$

$$= \left(\frac{(250)(40) - 2(40)^2(0.15 + 0.05)}{(250)(40)}\right) \times 100\%$$

$$= \left(\frac{10\,000 - 640}{10\,000}\right) \times 100\%$$

$$= \left(\frac{9360}{10\,000}\right) \times 100\% = \mathbf{93.6\%}$$

Problem 26. A 200 V d.c. motor develops a shaft torque of 15 Nm at 1200 rev/min. If the efficiency is 80%, determine the current supplied to the motor.

$$\text{The efficiency of a motor} = \frac{\text{output power}}{\text{input power}} \times 100\%$$

The output power of a motor is the power available to do work at its shaft and is given by $T\omega$ or $T(2\pi n)$ watts, where T is the torque in Nm and n is the speed of rotation in rev/s. The input power is the electrical power in watts supplied to the motor, i.e. VI watts.

Thus for a motor, efficiency, $\eta = \dfrac{T(2\pi n)}{VI} \times 100\%$

i.e., $$80 = \left[\frac{(15)(2\pi)(1200/60)}{(200)(I)}\right](100)$$

Thus the current supplied, $I = \dfrac{(15)(2\pi)(20)(100)}{(200)(80)} = \mathbf{11.8\ A}$

Problem 27. A d.c. series motor drives a load at 30 rev/s and takes a current of 10 A when the supply voltage is 400 V. If the total resistance of the motor is 2 Ω and the iron, friction and windage losses amount to 300 W, determine the efficiency of the motor.

$$\text{Efficiency, } \eta = \left(\frac{VI - I^2R - C}{VI}\right) \times 100\%$$

$$= \left(\frac{(400)(10) - (10)^2(2) - 300}{(400)(10)}\right) \times 100\%$$

$$= \left(\frac{4000 - 200 - 300}{4000}\right) \times 100\%$$

$$= \left(\frac{3500}{4000}\right) \times 100\% = \mathbf{87.5\%}$$

Further problems on d.c. motors may be found in Section 21.17, problems 22 to 30, page 384.

21.14 D.c. motor starter

If a d.c. motor whose armature is stationary is switched directly to its supply voltage, it is likely that the fuses protecting the motor will burn out. This is because the armature resistance is small, frequently being less than one ohm. Thus, additional resistance must be added to the armature circuit at the instant of closing the switch to start the motor.

As the speed of the motor increases, the armature conductors are cutting flux and a generated voltage, acting in opposition to the applied voltage, is produced, which limits the flow of armature current. Thus the value of the additional armature resistance can then be reduced.

When at normal running speed, the generated e.m.f. is such that no additional resistance is required in the armature circuit. To achieve this varying resistance in the armature circuit on starting, a d.c. motor starter is used, as shown in Figure 21.28.

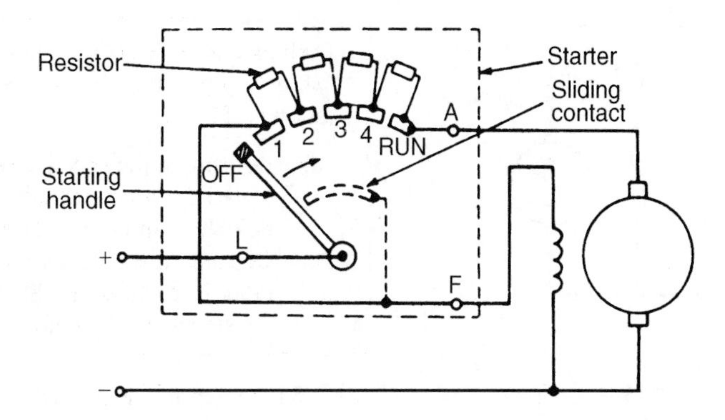

Figure 21.28

The starting handle is moved **slowly** in a clockwise direction to start the motor. For a shunt-wound motor, the field winding is connected to stud 1 or to L via a sliding contact on the starting handle, to give maximum field current, hence maximum flux, hence maximum torque on starting, since $T \propto \Phi I_a$.

A similar arrangement without the field connection is used for series motors.

21.15 Speed control of d.c. motors

Shunt-wound motors

The speed of a shunt-wound d.c. motor, n, is proportional to $(V - I_a R_a)/\Phi$ (see equation (21.9)). The speed is varied either by varying the value of flux, Φ, or by varying the value of R_a. The former is achieved by using a variable resistor in series with the field winding, as shown in Figure 21.29(a) and such a resistor is called the **shunt field regulator**. As the value of resistance of the shunt field regulator is increased, the value of the field current, I_f, is decreased.

This results in a decrease in the value of flux, Φ, and hence an increase in the speed, since $n \propto 1/\Phi$. Thus only speeds **above** that given without a shunt field regulator can be obtained by this method. Speeds **below** those given by $(V - I_a R_a)/\Phi$ are obtained by increasing the resistance in the armature circuit, as shown in Figure 21.29(b), where

$$n \propto \frac{V - I_a(R_a + R)}{\Phi}$$

Since resistor R is in series with the armature, it carries the full armature current and results in a large power loss in large motors where a considerable speed reduction is required for long periods.

These methods of speed control are demonstrated in the following worked problem.

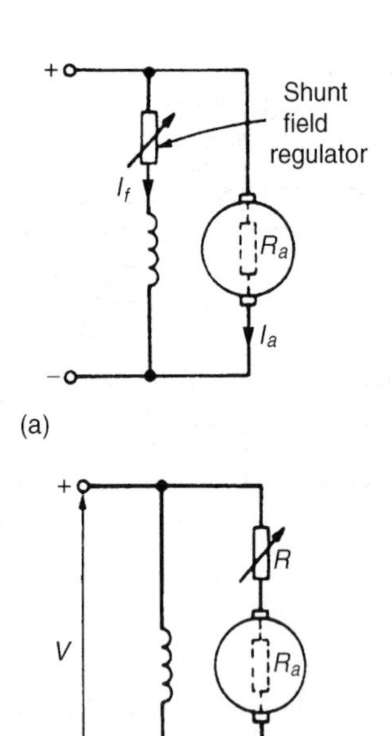

Figure 21.29

> Problem 28. A 500 V shunt motor runs at its normal speed of 10 rev/s when the armature current is 120 A. The armature resistance is 0.2 Ω.
>
> (a) Determine the speed when the current is 60 A and a resistance of 0.5 Ω is connected in series with the armature, the shunt field remaining constant.
>
> (b) Determine the speed when the current is 60 A and the shunt field is reduced to 80% of its normal value by increasing resistance in the field circuit.

(a) With reference to Figure 21.29(b),

back e.m.f. at 120 A, $E_1 = V - I_a R_a = 500 - (120)(0.2)$

$= 500 - 24 = 476$ volts

When $I_a = 60\ A$, $E_2 = 500 - (60)(0.2 + 0.5)$

$= 500 - (60)(0.7)$

$= 500 - 42 = 458$ volts

Now $\dfrac{E_1}{E_2} = \dfrac{\Phi_1 n_1}{\Phi_2 n_2}$

i.e., $\dfrac{476}{458} = \dfrac{\Phi_1(10)}{\Phi_1(n_2)}$ since $\Phi_2 = \Phi_1$

from which, speed $n_2 = \dfrac{(10)(458)}{(476)} =$ **9.62 rev/s**

(b) Back e.m.f. when $I_a = 60$ A, $E_2 = 500 - (60)(0.2)$

$= 500 - 12 = 488$ volts

Now $\dfrac{E_1}{E_2} = \dfrac{\Phi_1 n_1}{\Phi_2 n_2}$

i.e., $\dfrac{476}{488} = \dfrac{(\Phi_1)(10)}{(0.8\Phi_1)(n_3)}$, since $\Phi_2 = 0.8\Phi_1$

from which, speed $n_3 = \dfrac{(10)(488)}{(0.8)(476)} =$ **12.82 rev/s**

Series-wound motors

The speed control of series-wound motors is achieved using either (a) field resistance, or (b) armature resistance techniques.

(a) The speed of a d.c. series-wound motor is given by:

$$n = k\left(\frac{V - IR}{\Phi}\right)$$

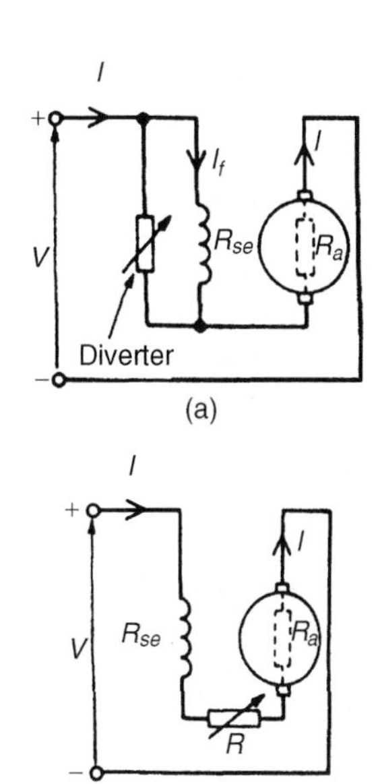

Figure 21.30

where k is a constant, V is the terminal voltage, R is the combined resistance of the armature and series field and Φ is the flux.

Thus, a reduction in flux results in an increase in speed. This is achieved by putting a variable resistance in parallel with the field-winding and reducing the field current, and hence flux, for a given value of supply current. A circuit diagram of this arrangement is shown in Figure 21.30(a). A variable resistor connected in parallel with the series-wound field to control speed is called a **diverter**. Speeds above those given with no diverter are obtained by this method. Problem 29 below demonstrates this method.

(b) Speeds below normal are obtained by connecting a variable resistor in series with the field winding and armature circuit, as shown in Figure 21.30(b). This effectively increases the value of R in the equation

$$n = k\left(\frac{V - IR}{\Phi}\right)$$

and thus reduces the speed. Since the additional resistor carries the full supply current, a large power loss is associated with large motors in which a considerable speed reduction is required for long periods. This method is demonstrated in problem 30.

Problem 29. On full-load a 300 V series motor takes 90 A and runs at 15 rev/s. The armature resistance is 0.1 Ω and the series winding resistance is 50 mΩ. Determine the speed when developing full load torque but with a 0.2 Ω diverter in parallel with the field winding. (Assume that the flux is proportional to the field current.)

$$\begin{aligned}\text{At 300 V, e.m.f. } E_1 &= V - IR\\ &= V - I(R_a + R_{se})\\ &= 300 - (90)(0.1 + 0.05)\\ &= 300 - (90)(0.15)\\ &= 300 - 13.5 = 286.5 \text{ volts}\end{aligned}$$

With the 0.2 Ω diverter in parallel with R_{se} (see Figure 21.30(a)),

$$\text{the equivalent resistance, } R = \frac{(0.2)(0.05)}{(0.2) + (0.05)} = \frac{(0.2)(0.05)}{(0.25)} = 0.04\ \Omega$$

$$\text{By current division, current } I_1 \text{ (in Figure 21.30(a))} = \left(\frac{0.2}{0.2 + 0.05}\right) I = 0.8I$$

Torque, $T \propto I_a\Phi$ and for full load torque, $I_{a1}\Phi_1 = I_{a2}\Phi_2$

Since flux is proportional to field current $\Phi_1 \propto I_{a1}$ and $\Phi_2 \propto 0.8I_{a2}$ then $(90)(90) = (I_{a2})(0.8I_{a2})$

from which, $I_{a2}^2 = \dfrac{(90)^2}{0.8}$ and $I_{a2} = \dfrac{90}{\sqrt{(0.8)}} = 100.62$ A

Hence e.m.f. $E_2 = V - I_{a2}(R_a + R)$

$$= 300 - (100.62)(0.1 + 0.04)$$
$$= 300 - (100.62)(0.14)$$
$$= 300 - 14.087 = 285.9 \text{ volts}$$

Now e.m.f., $E \propto \Phi n$ from which, $\dfrac{E_1}{E_2} = \dfrac{\Phi_1 n_1}{\Phi_2 n_2} = \dfrac{I_{a1}n_1}{0.8I_{a2}n_2}$

Hence $\dfrac{(286.5)}{285.9} = \dfrac{(90)(15)}{(0.8)(100.62)n_2}$

and new speed, $n_2 = \dfrac{(285.9)(90)(15)}{(286.5)(0.8)(100.62)} = \mathbf{16.74\ rev/s}$

Thus the speed of the motor has increased from 15 rev/s (i.e., 900 rev/min) to 16.74 rev/s (i.e., 1004 rev/min) by inserting a 0.2 Ω diverter resistance in parallel with the series winding.

Problem 30. A series motor runs at 800 rev/min when the voltage is 400 V and the current is 25 A. The armature resistance is 0.4 Ω and the series field resistance is 0.2 Ω. Determine the resistance to be connected in series to reduce the speed to 600 rev/min with the same current.

With reference to Figure 21.30(b), at 800 rev/min,

$$\text{e.m.f., } E_1 = V - I(R_a + R_{se}) = 400 - (25)(0.4 + 0.2)$$
$$= 400 - (25)(0.6)$$
$$= 400 - 15 = 385 \text{ volts}$$

At 600 rev/min, since the current is unchanged, the flux is unchanged.

Thus $E \propto \Phi n$, or $E \propto n$, and $\dfrac{E_1}{E_2} = \dfrac{n_1}{n_2}$

Hence $\dfrac{385}{E_2} = \dfrac{800}{600}$

from which, $E_2 = \dfrac{(385)(600)}{(800)} = 288.75$ volts

and $E_2 = V - I(R_a + R_{se} + R)$

Hence $288.75 = 400 - 25(0.4 + 0.2 + R)$

Rearranging gives: $0.6 + R = \dfrac{400 - 288.75}{25} = 4.45$

from which, extra series resistance, $R = 4.45 - 0.6$

i.e., $\mathbf{R = 3.85\ \Omega}$

Thus the addition of a series resistance of 3.85 Ω has reduced the speed from 800 rev/min to 600 rev/min

Further problems on the speed control of d.c. motors may be found in Section 21.17, problems 31 to 33, page 384.

21.16 Motor cooling

Motors are often classified according to the type of enclosure used, the type depending on the conditions under which the motor is used and the degree of ventilation required.

The most common type of protection is the **screen-protected type,** where ventilation is achieved by fitting a fan internally, with the openings at the end of the motor fitted with wire mesh.

A **drip-proof type** is similar to the screen-protected type but has a cover over the screen to prevent drips of water entering the machine.

A **flame-proof type** is usually cooled by the conduction of heat through the motor casing.

With a **pipe-ventilated type,** air is piped into the motor from a dust-free area, and an internally fitted fan ensures the circulation of this cool air.

21.17 Further problems on d.c. machines

Generated e.m.f.

1 A 4-pole, wave-connected armature of a d.c. machine has 750 conductors and is driven at 720 rev/min. If the useful flux per pole is 15 mWb, determine the generated e.m.f. [270 volts]

2 A 6-pole generator has a lap-wound armature with 40 slots with 20 conductors per slot. The flux per pole is 25 mWb. Calculate the speed at which the machine must be driven to generate an e.m.f. of 300 V. [15 rev/s or 900 rev/min]

3 A 4-pole armature of a d.c. machine has 1000 conductors and a flux per pole of 20 mWb. Determine the e.m.f. generated when running at 600 rev/min when the armature is (a) wave-wound, (b) lap-wound. [(a) 400 volts (b) 200 volts]

4 A d.c. generator running at 25 rev/s generates an e.m.f. of 150 V. Determine the percentage increase in the flux per pole required to generate 180 V at 20 rev/s. [50%]

5 Determine the terminal voltage of a generator which develops an e.m.f. of 240 V and has an armature current of 50 A on load. Assume the armature resistance is 40 mΩ. [238 volts]

D.c. generator

6 A generator is connected to a 50 Ω load and a current of 10 A flows. If the armature resistance is 0.5 Ω, determine (a) the terminal voltage, and (b) the generated e.m.f. [(a) 500 volts (b) 505 volts]

7 A separately excited generator develops a no-load e.m.f. of 180 V at an armature speed of 15 rev/s and a flux per pole of 0.20 Wb. Calculate the generated e.m.f. when

(a) the speed increases to 20 rev/s and the flux per pole remaining unchanged,

(b) the speed remains at 15 rev/s and the pole flux is decreased to 0.125 Wb, and

(c) the speed increases to 25 rev/s and the pole flux is decreased to 0.18 Wb. [(a) 240 volts (b) 112.5 volts (c) 270 volts]

8 A shunt generator supplies a 50 kW load at 400 V through cables of resistance 0.2 Ω. If the field winding resistance is 50 Ω and the armature resistance is 0.05 Ω, determine (a) the terminal voltage, (b) the e.m.f. generated in the armature. [(a) 425 volts (b) 431.68 volts]

9 A short-shunt compound generator supplies 50 A at 300 V. If the field resistance is 30 Ω, the series resistance 0.03 Ω and the armature resistance 0.05 Ω, determine the e.m.f. generated. [304.5 volts]

10 A d.c. generator has a generated e.m.f. of 210 V when running at 700 rev/min and the flux per pole is 120 mWb. Determine the generated e.m.f. (a) at 1050 rev/min, assuming the flux remains constant, (b) if the flux is reduced by one-sixth at constant speed, and (c) at a speed of 1155 rev/min and a flux of 132 mWb. [(a) 315 V (b) 175 V (c) 381.2 V]

11 A 250 V d.c. shunt-wound generator has an armature resistance of 0.1 Ω. Determine the generated e.m.f. when the generator is supplying 50 kW, neglecting the field current of the generator. [270 V]

Efficiency of d.c. generator

12 A 15 kW shunt generator having an armature circuit resistance of 0.4 Ω and a field resistance of 100 Ω, generates a terminal voltage of 240 V at full load. Determine the efficiency of the generator at

full load, assuming the iron, friction and windage losses amount to 1 kW. [82.14%]

Back e.m.f.

13 A d.c. motor operates from a 350 V supply. If the armature resistance is 0.4 Ω determine the back e.m.f. when the armature current is 60 A. [326 volts]

14 The armature of a d.c. machine has a resistance of 0.5 Ω and is connected to a 200 V supply. Calculate the e.m.f. generated when it is running (a) as a motor taking 50 A and (b) as a generator giving 70 A. [(a) 175 volts (b) 235 volts]

15 Determine the generated e.m.f. of a d.c. machine if the armature resistance is 0.1 Ω and it (a) is running as a motor connected to a 230 V supply, the armature current being 60 A, and (b) is running as a generator with a terminal voltage of 230 V, the armature current being 80 A. [(a) 224 V (b) 238 V]

Losses, efficiency and torque

16 The shaft torque required to drive a d.c. generator is 18.7 Nm when it is running at 1250 rev/min. If its efficiency is 87% under these conditions and the armature current is 17.3 A, determine the voltage at the terminals of the generator. [123.1 V]

17 A 220 V, d.c. generator supplies a load of 37.5 A and runs at 1550 rev/min. Determine the shaft torque of the diesel motor driving the generator, if the generator efficiency is 78%. [65.2 Nm]

18 A 4-pole d.c. motor has a wave-wound armature with 800 conductors. The useful flux per pole is 20 mWb. Calculate the torque exerted when a current of 40 A flows in each armature conductor. [203.7 Nm]

19 Calculate the torque developed by a 240 V d.c. motor whose armature current is 50 A, armature resistance is 0.6 Ω and is running at 10 rev/s. [167.1 Nm]

20 An 8-pole lap-wound d.c. motor has a 200 V supply. The armature has 800 conductors and a resistance of 0.8 Ω. If the useful flux per pole is 40 mWb and the armature current is 30 A, calculate (a) the speed and (b) the torque developed. [(a) 5.5 rev/s or 330 rev/min (b) 152.8 Nm]

21 A 150 V d.c. generator supplies a current of 25 A when running at 1200 rev/min. If the torque on the shaft driving the generator is 35.8 Nm, determine (a) the efficiency of the generator, and (b) the power loss in the generator. [(a) 83.4% (b) 748.8 W]

D.c. motors

22 A 240 V shunt motor takes a total current of 80 A. If the field winding resistance is 120 Ω and the armature resistance is 0.4 Ω, determine (a) the current in the armature, and (b) the back e.m.f. [(a) 78 A (b) 208.8 V]

23 A d.c. motor has a speed of 900 rev/min when connected to a 460 V supply. Find the approximate value of the speed of the motor when connected to a 200 V supply, assuming the flux decreases by 30% and neglecting the armature volt drop. [559 rev/min]

24 A series motor having a series field resistance of 0.25 Ω and an armature resistance of 0.15 Ω, is connected to a 220 V supply and at a particular load runs at 20 rev/s when drawing 20 A from the supply. Calculate the e.m.f. generated at this load. Determine also the speed of the motor when the load is changed such that the current increases to 25 A. Assume the flux increases by 25%. [212 V, 15.85 rev/s]

25 A 500 V shunt motor takes a total current of 100 A and runs at 1200 rev/min. If the shunt field resistance is 50 Ω, the armature resistance is 0.25 Ω and the iron, friction and windage losses amount to 2 kW, determine the overall efficiency of the motor. [81.95%]

26 A 250 V, series-wound motor is running at 500 rev/min and its shaft torque is 130 Nm. If its efficiency at this load is 88%, find the current taken from the supply. [30.94 A]

27 In a test on a d.c. motor, the following data was obtained.
Supply voltage: 500 V. Current taken from the supply: 42.4 A
Speed: 850 rev/min. Shaft torque: 187 Nm
Determine the efficiency of the motor correct to the nearest 0.5%. [78.5%]

28 A 300 V series motor draws a current of 50 A. The field resistance is 40 mΩ and the armature resistance is 0.2 Ω. Determine the maximum efficiency of the motor. [92%]

29 A series motor drives a load at 1500 rev/min and takes a current of 20 A when the supply voltage is 250 V. If the total resistance of the motor is 1.5 Ω and the iron, friction and windage losses amount to 400 W, determine the efficiency of the motor. [80%]

30 A series-wound motor is connected to a d.c. supply and develops full-load torque when the current is 30 A and speed is 1000 rev/min. If the flux per pole is proportional to the current flowing, find the current and speed at half full-load torque, when connected to the same supply. [21.2 A, 1415 rev/min]

Speed control

31 A 350 V shunt motor runs at its normal speed of 12 rev/s when the armature current is 90 A. The resistance of the armature is 0.3 Ω.

(a) Find the speed when the current is 45 A and a resistance of 0.4 Ω is connected in series with the armature, the shunt field remaining constant. (b) Find the speed when the current is 45 A and the shunt field is reduced to 75% of its normal value by increasing resistance in the field circuit. [(a) 11.83 rev/s (b) 16.67 rev/s]

32 A series motor runs at 900 rev/min when the voltage is 420 V and the current is 40 A. The armature resistance is 0.3 Ω and the series field resistance is 0.2 Ω. Calculate the resistance to be connected in series to reduce the speed to 720 rev/min with the same current. [2 Ω]

33 A 320 V series motor takes 80 A and runs at 1080 rev/min at full load. The armature resistance is 0.2 Ω and the series winding resistance is 0.05 Ω. Assuming the flux is proportional to the field current, calculate the speed when developing full-load torque, but with a 0.15 Ω diverter in parallel with the field winding. [1239 rev/min]

22 Three-phase induction motors

At the end of this chapter you should be able to:

- appreciate the merits of three-phase induction motors
- understand how a rotating magnetic field is produced
- state the synchronous speed, $n_s = (f/p)$ and use in calculations
- describe the principle of operation of a three-phase induction motor
- distinguish between squirrel-cage and wound-rotor types of motor
- understand how a torque is produced causing rotor movement
- understand and calculate slip
- derive expressions for rotor e.m.f., frequency, resistance, reactance, impedance, current and copper loss, and use them in calculations
- state the losses in an induction motor and calculate efficiency
- derive the torque equation for an induction motor, state the condition for maximum torque, and use in calculations
- describe torque-speed and torque-slip characteristics for an induction motor
- state and describe methods of starting induction motors
- state advantages of cage rotor and wound rotor types of induction motor
- describe the double cage induction motor
- state typical applications of three-phase induction motors

22.1 Introduction

In d.c. motors, introduced in Chapter 21, conductors on a rotating armature pass through a stationary magnetic field. In a **three-phase induction motor**, the magnetic field rotates and this has the advantage that no external electrical connections to the rotor need be made. Its name is derived from the fact that the current in the rotor is **induced** by the magnetic field instead of being supplied through electrical connections to the supply. The result is a motor which: (i) is cheap and robust, (ii) is explosion proof, due to the absence of a commutator or slip-rings and brushes with their associated sparking, (iii) requires little or no skilled maintenance, and (iv) has self-starting properties when switched to a supply with no additional expenditure on auxiliary equipment. The principal disadvantage of a three-phase induction motor is that its speed cannot be readily adjusted.

22.2 Production of a rotating magnetic field

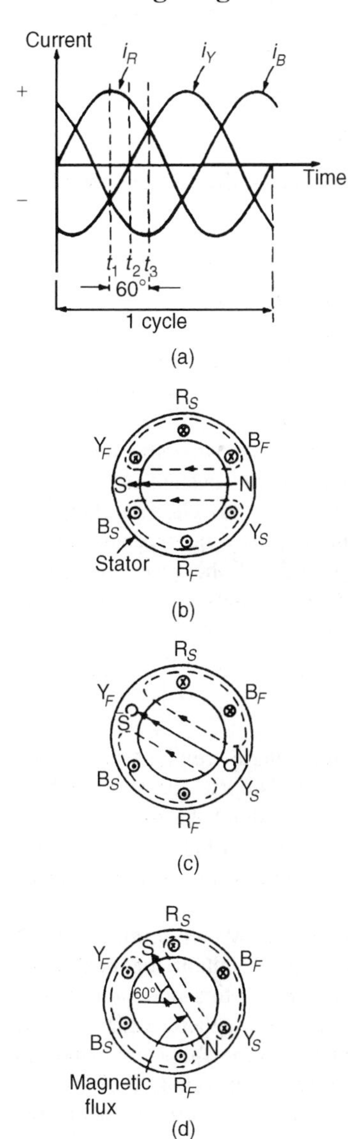

Figure 22.1

When a three-phase supply is connected to symmetrical three-phase windings, the currents flowing in the windings produce a magnetic field. This magnetic field is constant in magnitude and rotates at constant speed as shown below, and is called the **synchronous speed**.

With reference to Figure 22.1, the windings are represented by three single-loop conductors, one for each phase, marked R_SR_F, Y_SY_F and B_SB_F, the S and F signifying start and finish. In practice, each phase winding comprises many turns and is distributed around the stator; the single-loop approach is for clarity only.

When the stator windings are connected to a three-phase supply, the current flowing in each winding varies with time and is as shown in Figure 22.1(a). If the value of current in a winding is positive, the assumption is made that it flows from start to finish of the winding, i.e., if it is the red phase, current flows from R_S to R_F, i.e. away from the viewer in R_S and towards the viewer in R_F. When the value of current is negative, the assumption is made that it flows from finish to start, i.e. towards the viewer in an 'S' winding and away from the viewer in an 'F' winding. At time, say t_1, shown in Figure 22.1(a), the current flowing in the red phase is a maximum positive value. At the same time, t_1, the currents flowing in the yellow and blue phases are both 0.5 times the maximum value and are negative.

The current distribution in the stator windings is therefore as shown in Figure 22.1(b), in which current flows away from the viewer, (shown as $\otimes$) in R_S since it is positive, but towards the viewer (shown as $\odot$) in Y_S and B_S, since these are negative. The resulting magnetic field is as shown, due to the 'solenoid' action and application of the corkscrew rule.

A short time later at time t_2, the current flowing in the red phase has fallen to about 0.87 times its maximum value and is positive, the current in the yellow phase is zero and the current in the blue phase is about 0.87 times its maximum value and is negative. Hence the currents and resultant magnetic field are as shown in Figure 22.1(c). At time t_3, the currents in the red and yellow phases are 0.5 of their maximum values and the current in the blue phase is a maximum negative value. The currents and resultant magnetic field are as shown in Figure 22.1(d).

Similar diagrams to Figure 22.1(b), (c) and (d) can be produced for all time values and these would show that the magnetic field travels through one revolution for each cycle of the supply voltage applied to the stator windings. By considering the flux values rather than the current values, it is shown below that the rotating magnetic field has a constant value of flux. The three coils shown in Figure 22.2(a), are connected in star to a three-phase supply. Let the positive directions of the fluxes produced by currents flowing in the coils, be ϕ_A, ϕ_B and ϕ_C respectively. The directions of ϕ_A, ϕ_B and ϕ_C do not alter, but their magnitudes are proportional to

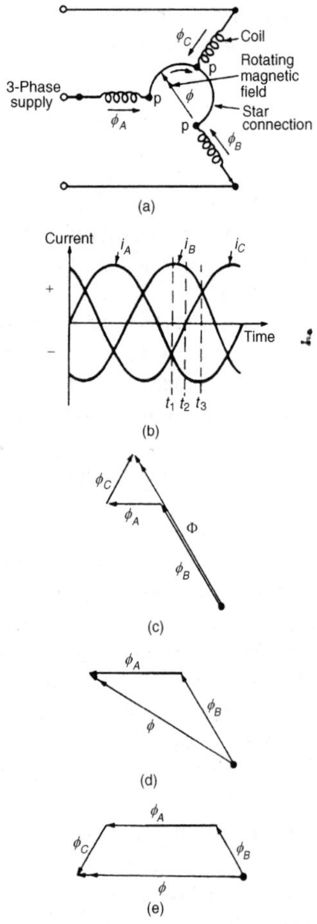

Figure 22.2

the currents flowing in the coils at any particular time. At time t_1, shown in Figure 22.2(b), the currents flowing in the coils are:

i_B, a maximum positive value, i.e., the flux is towards point P;

i_A and i_C, half the maximum value and negative, i.e., the flux is away from point P.

These currents give rise to the magnetic fluxes ϕ_A, ϕ_B and ϕ_C, whose magnitudes and directions are as shown in Figure 22.2(c). The resultant flux is the phasor sum of ϕ_A, ϕ_B and ϕ_C, shown as Φ in Figure 22.2(c). At time t_2, the currents flowing are:

i_B, $0.866 \times$ maximum positive value, i_C, zero, and

i_A, $0.866 \times$ maximum negative value.

The magnetic fluxes and the resultant magnetic flux are as shown in Figure 22.2(d).

At time t_3, i_B is $0.5 \times$ maximum value and is positive

i_A is a maximum negative value, and

i_C is $0.5 \times$ maximum value and is positive.

The magnetic fluxes and the resultant magnetic flux are as shown in Figure 22.2(e).

Inspection of Figures 22.2(c), (d) and (e) shows that the magnitude of the resultant magnetic flux, Φ, in each case is constant and is $1\frac{1}{2} \times$ the maximum value of ϕ_A, ϕ_B or ϕ_C, but that its direction is changing. The process of determining the resultant flux may be repeated for all values of time and shows that the magnitude of the resultant flux is constant for all values of time and also that it rotates at constant speed, making one revolution for each cycle of the supply voltage.

22.3 Synchronous speed

The rotating magnetic field produced by three phase windings could have been produced by rotating a permanent magnet's north and south pole at synchronous speed, (shown as N and S at the ends of the flux phasors in Figures 22.1(b), (c) and (d)). For this reason, it is called a 2-pole system and an induction motor using three phase windings only is called a 2-pole induction motor.

If six windings displaced from one another by 60° are used, as shown in Figure 22.3(a), by drawing the current and resultant magnetic field diagrams at various time values, it may be shown that one cycle of the supply current to the stator windings causes the magnetic field to move through half a revolution. The current distribution in the stator windings are shown in Figure 22.3(a), for the time t shown in Figure 22.3(b).

It can be seen that for six windings on the stator, the magnetic flux produced is the same as that produced by rotating two permanent magnet north poles and two permanent magnet south poles at synchronous speed. This is called a 4-pole system and an induction motor using six phase

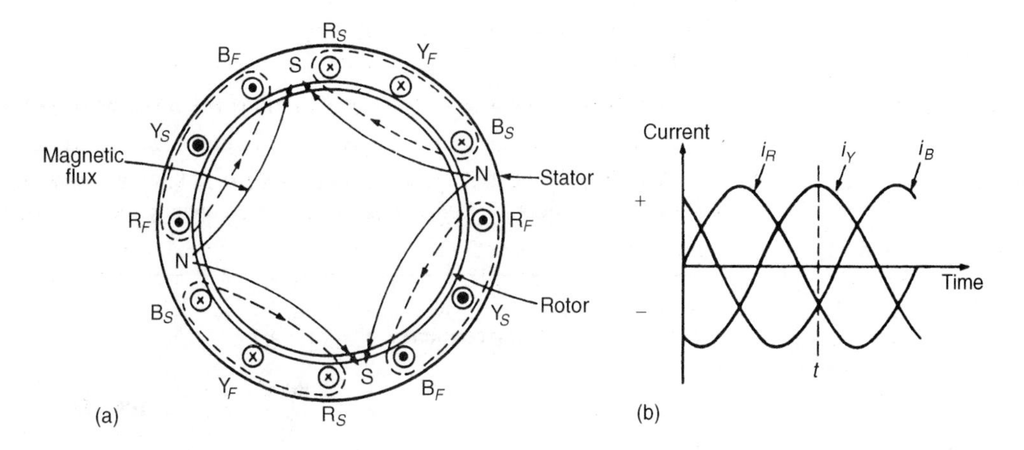

Figure 22.3

windings is called a 4-pole induction motor. By increasing the number of phase windings the number of poles can be increased to any even number.

In general, if f is the frequency of the currents in the stator windings and the stator is wound to be equivalent to p **pairs** of poles, the speed of revolution of the rotating magnetic field, i.e., the synchronous speed, n_s is given by:

$$\boldsymbol{n_s = \frac{f}{p} \text{ rev/s}}$$

Problem 1. A three-phase two-pole induction motor is connected to a 50 Hz supply. Determine the synchronous speed of the motor in rev/min.

From above, $n_s = f/p$ rev/s, where n_s is the synchronous speed, f is the frequency in hertz of the supply to the stator and p is the number of **pairs** of poles. Since the motor is connected to a 50 hertz supply, $f = 50$. The motor has a two-pole system, hence p, the number of pairs of poles is one.

Thus, synchronous speed, $n_s = \dfrac{50}{1} = 50 \text{ rev/s} = 50 \times 60 \text{ rev/min}$

$= \mathbf{3000\ rev/min}$

Problem 2. A stator winding supplied from a three-phase 60 Hz system is required to produce a magnetic flux rotating at 900 rev/min. Determine the number of poles.

Synchronous speed, $n_s = 900 \text{ rev/min} = \dfrac{900}{60} \text{ rev/s} = 15 \text{ rev/s}$

Since $n_s = \frac{f}{p}$ then $p = \frac{f}{n_s} = \frac{60}{15} = 4$

Hence the number of pole pairs is 4 and thus **the number of poles is 8.**

Problem 3. A three-phase 2-pole motor is to have a synchronous speed of 6000 rev/min. Calculate the frequency of the supply voltage.

Since $n_s = \frac{f}{p}$ then **frequency,** $\boldsymbol{f} = (n_s)(p)$

$$= \left(\frac{6000}{60}\right)\left(\frac{2}{2}\right) = \mathbf{100\ Hz}$$

Further problems on synchronous speed may be found in Section 22.18, problems 1 to 3, page 406.

22.4 Construction of a three-phase induction motor

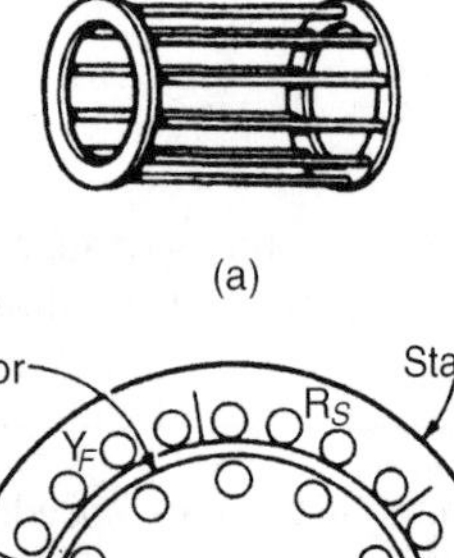

(a)

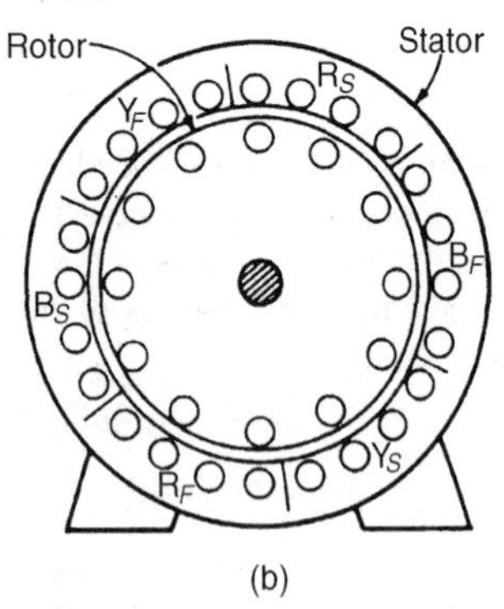

(b)

Figure 22.4

The stator of a three-phase induction motor is the stationary part corresponding to the yoke of a d.c. machine. It is wound to give a 2-pole, 4-pole, 6-pole, ... rotating magnetic field, depending on the rotor speed required. The rotor, corresponding to the armature of a d.c. machine, is built up of laminated iron, to reduce eddy currents.

In the type most widely used, known as a **squirrel-cage rotor**, copper or aluminium bars are placed in slots cut in the laminated iron, the ends of the bars being welded or brazed into a heavy conducting ring, (see Figure 22.4(a)). A cross-sectional view of a three-phase induction motor is shown in Figure 22.4(b).

The conductors are placed in slots in the laminated iron rotor core. If the slots are skewed, better starting and quieter running is achieved. This type of rotor has no external connections which means that slip rings and brushes are not needed. The squirrel-cage motor is cheap, reliable and efficient.

Another type of rotor is the **wound rotor.** With this type there are phase windings in slots, similar to those in the stator. The windings may be connected in star or delta and the connections made to three slip rings. The slip rings are used to add external resistance to the rotor circuit, particularly for starting (see Section 22.13), but for normal running the slip rings are short circuited.

The principle of operation is the same for both the squirrel cage and the wound rotor machines.

22.5 Principle of operation of a three-phase induction motor

When a three-phase supply is connected to the stator windings, a rotating magnetic field is produced. As the magnetic flux cuts a bar on the rotor, an e.m.f. is induced in it and since it is joined, via the end conducting rings, to

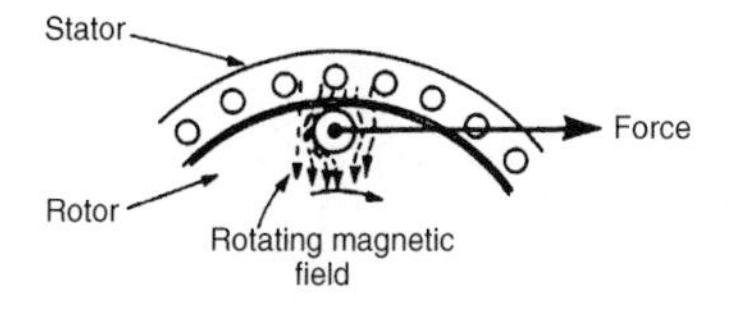

Figure 22.5

another bar one pole pitch away, a current flows in the bars. The magnetic field associated with this current flowing in the bars interacts with the rotating magnetic field and a force is produced, tending to turn the rotor in the same direction as the rotating magnetic field, (see Figure 22.5). Similar forces are applied to all the conductors on the rotor, so that a torque is produced causing the rotor to rotate.

22.6 Slip

The force exerted by the rotor bars causes the rotor to turn in the direction of the rotating magnetic field. As the rotor speed increases, the rate at which the rotating magnetic field cuts the rotor bars is less and the frequency of the induced e.m.f.'s in the rotor bars is less. If the rotor runs at the same speed as the rotating magnetic field, no e.m.f.'s are induced in the rotor, hence there is no force on them and no torque on the rotor. Thus the rotor slows down. For this reason the rotor can never run at synchronous speed.

When there is no load on the rotor, the resistive forces due to windage and bearing friction are small and the rotor runs very nearly at synchronous speed. As the rotor is loaded, the speed falls and this causes an increase in the frequency of the induced e.m.f.'s in the rotor bars and hence the rotor current, force and torque increase. The difference between the rotor speed, n_r, and the synchronous speed, n_s, is called the **slip speed**, i.e.

$$\textbf{slip speed} = n_s - n_r \textbf{ rev/s}$$

The ratio $(n_s - n_r)/n_s$ is called the **fractional slip** or just the **slip**, s, and is usually expressed as a percentage. Thus

$$\textbf{slip, } s = \left(\frac{n_s - n_r}{n_s}\right) \times 100\%$$

Typical values of slip between no load and full load are about 4 to 5% for small motors and 1.5 to 2% for large motors.

Problem 4. The stator of a 3-phase, 4-pole induction motor is connected to a 50 Hz supply. The rotor runs at 1455 rev/min at full load. Determine (a) the synchronous speed and (b) the slip at full load.

(a) The number of pairs of poles, $p = 4/2 = 2$

The supply frequency $f = 50$ Hz

The **synchronous speed,** $n_s = \dfrac{f}{p} = \dfrac{50}{2} = \mathbf{25\ rev/s}$

(b) The rotor speed, $n_r = \dfrac{1455}{60} = 24.25$ rev/s

$$\textbf{The slip}, s = \left(\frac{n_s - n_r}{n_s}\right) \times 100\%$$

$$= \left(\frac{25 - 24.25}{25}\right) \times 100\% = \mathbf{3\%}$$

Problem 5. A 3-phase, 60 Hz induction motor has 2 poles. If the slip is 2% at a certain load, determine (a) the synchronous speed, (b) the speed of the rotor and (c) the frequency of the induced e.m.f.'s in the rotor.

(a) $f = 60$ Hz, $p = \frac{2}{2} = 1$

Hence **synchronous speed,** $n_s = \frac{f}{p} = \frac{60}{1} = \mathbf{60\ rev/s}$

or $60 \times 60 = \mathbf{3600\ rev/min}$

(b) Since slip, $s = \left(\frac{n_s - n_r}{n_s}\right) \times 100\%$

$$2 = \left(\frac{60 - n_r}{60}\right) \times 100$$

Hence $\frac{2 \times 60}{100} = 60 - n_r$

i.e. $n_r = 60 - \frac{2 \times 60}{100} = 58.8$ rev/s

i.e. the rotor runs at $58.8 \times 60 =$ **3528 rev/min**

(c) Since the synchronous speed is 60 rev/s and that of the rotor is 58.8 rev/s, the rotating magnetic field cuts the rotor bars at (60 − 58.8), i.e. 1.2 rev/s.
Thus the frequency of the e.m.f.'s induced in the rotor bars is 1.2 Hz

Problem 6. A three-phase induction motor is supplied from a 50 Hz supply and runs at 1200 rev/min when the slip is 4%. Determine the synchronous speed.

Slip, $s = \left(\frac{n_s - n_r}{n_s}\right) \times 100\%$

Rotor speed, $n_r = \frac{1200}{60} = 20$ rev/s, and $s = 4$

Hence $4 = \left(\frac{n_s - 20}{n_s}\right) \times 100\%$

or $$0.04 = \frac{n_s - 20}{n_s}$$

from which, $n_s(0.04) = n_s - 20$

and $$20 = n_s - 0.04\, n_s = n_s(1 - 0.04)$$

Hence **synchronous speed,** $\boldsymbol{n_s} = \left(\frac{20}{1 - 0.04}\right) = 20.8\dot{3}$ rev/s

$= (20.8\dot{3} \times 60)$ rev/min $=$ **1250 rev/min**

Further problems on slip may be found in Section 22.18, problems 4 to 7, page 406.

22.7 Rotor e.m.f. and frequency

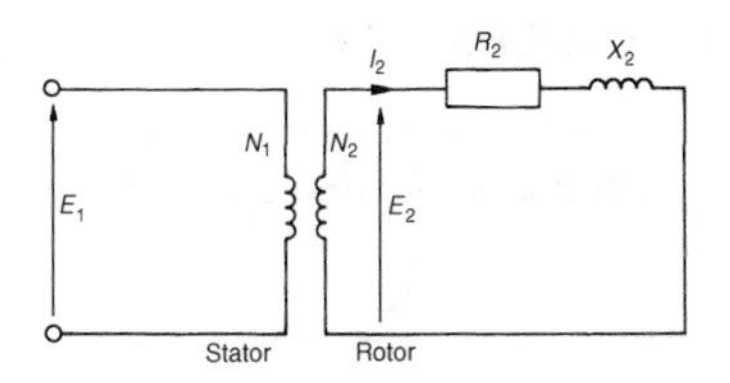

Figure 22.6

Rotor e.m.f.

When an induction motor is stationary, the stator and rotor windings form the equivalent of a transformer as shown in Figure 22.6.

The rotor e.m.f. at standstill is given by $E_2 = \left(\frac{N_2}{N_1}\right) E_1$ (22.1)

where E_1 is the supply voltage per phase to the stator.

When an induction motor is running, the induced e.m.f. in the rotor is less since the relative movement between conductors and the rotating field is less. The induced e.m.f. is proportional to this movement, hence it must be proportional to the slip, s.

Hence when running, rotor e.m.f. per phase $= E_r = sE_2$

$$= s\left(\frac{N_2}{N_1}\right) E_1 \qquad (22.2)$$

Rotor frequency

The rotor e.m.f. is induced by an alternating flux and the rate at which the flux passes the conductors is the slip speed. Thus the frequency of the rotor e.m.f. is given by:

$$f_r = (n_s - n_r)p = \frac{(n_s - n_r)}{n_s}(n_s p)$$

However $\left(\frac{n_s - n_r}{n_s}\right)$ is the slip s and $n_s p$ is the supply frequency f, hence

$$\boxed{f_r = sf} \qquad (22.3)$$

Problem 7. The frequency of the supply to the stator of an 8-pole induction motor is 50 Hz and the rotor frequency is 3 Hz. Determine (a) the slip, and (b) the rotor speed.

(a) From equation (22.3), $f_r = s\,f$

Hence $3 = (s)(50)$

from which, **slip,** $s = \frac{3}{50} = \mathbf{0.06}$ **or 6%**

(b) Synchronous speed, $n_s = \frac{f}{p} = \frac{50}{4} = 12.5$ rev/s

or $(12.5 \times 60) = 750$ rev/min

Slip, $s = \left(\frac{n_s - n_r}{n_s}\right)$, hence $0.06 = \left(\frac{12.5 - n_r}{12.5}\right)$

$(0.06)(12.5) = 12.5 - n_r$

and **rotor speed,** $\boldsymbol{n_r} = 12.5 - (0.06)(12.5)$

$= \mathbf{11.75}$ **rev/s** or **705 rev/min**

Further problems on rotor frequency may be found in Section 22.18, problems 8 and 9, page 407.

22.8 Rotor impedance and current

Rotor resistance

The rotor resistance R_2 is unaffected by frequency or slip, and hence remains constant.

Rotor reactance

Rotor reactance varies with the frequency of the rotor current.

At standstill, reactance per phase, $X_2 = 2\pi f L$

When running, reactance per phase, $X_r = 2\pi f_r L$

$= 2\pi(sf)L$ from equation (22.3)

$= s(2\pi f L)$

i.e. $\boxed{X_r = s\,X_2}$ (22.4)

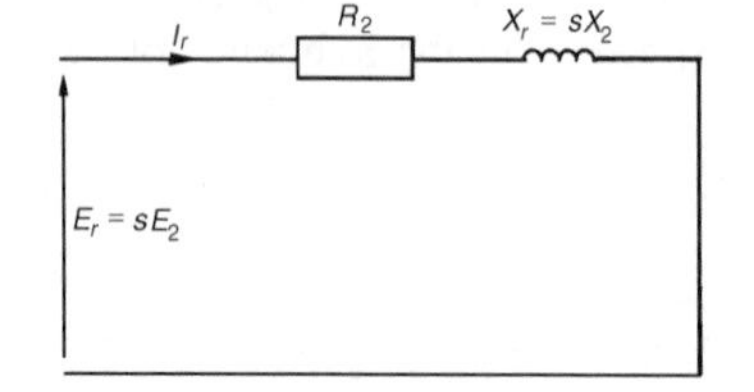

Figure 22.7

Figure 22.7 represents the rotor circuit when running.

Rotor impedance

Rotor impedance per phase, $Z_r = \sqrt{[R_2^{\,2} + (sX_2)^2]}$ (22.5)

At standstill, slip $s = 1$, then $Z_2 = \sqrt{R_2^2 + X_2^2]}$ (22.6)

Rotor current

From Figures 22.6 and 22.7,

at standstill, starting current, $$I_2 = \frac{E_2}{Z_2} = \frac{\left(\frac{N_2}{N_1}\right) E_1}{\sqrt{[R_2^2 + X_2^2]}} \quad (22.7)$$

and when running, current, $$I_r = \frac{E_r}{Z_r} = \frac{s\left(\frac{N_2}{N_1}\right) E_1}{\sqrt{[R_2^2 + (sX_2)^2]}} \quad (22.8)$$

22.9 Rotor copper loss

Power $P = 2\pi nT$, where T is the torque in newton metres, hence torque $T = P/(2\pi n)$

If P_2 is the power input to the rotor from the rotating field, and P_m is the mechanical power output (including friction losses)

then $$T = \frac{P_2}{2\pi n_s} = \frac{P_m}{2\pi n_r}$$

from which, $$\frac{P_2}{n_s} = \frac{P_m}{n_r} \text{ or } \frac{P_m}{P_2} = \frac{n_r}{n_s}$$

Hence $$1 - \frac{P_m}{P_2} = 1 - \frac{n_r}{n_s}$$

$$\frac{P_2 - P_m}{P_2} = \frac{n_s - n_r}{n_s} = s$$

$P_2 - P_m$ is the electrical or copper loss in the rotor, i.e. $P_2 - P_m = I_r^2 R_2$

Hence $$\textbf{slip, } s = \frac{\textbf{rotor copper loss}}{\textbf{rotor input}} = \frac{I_r^2 R_2}{P_2} \quad (22.9)$$

or power input to the rotor, $$P_2 = \frac{I_r^2 R_2}{s} \quad (22.10)$$

22.10 Induction motor losses and efficiency

Figure 22.8 summarizes losses in induction motors.

Motor efficiency, $$\eta = \frac{\text{output power}}{\text{input power}} = \frac{P_m}{P_1} \times 100\%$$

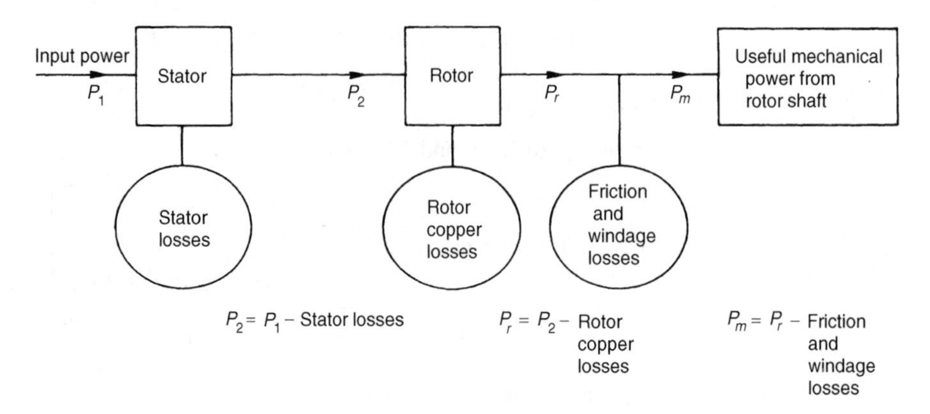

Figure 22.8

Problem 8. The power supplied to a three-phase induction motor is 32 kW and the stator losses are 1200 W. If the slip is 5%, determine (a) the rotor copper loss, (b) the total mechanical power developed by the rotor, (c) the output power of the motor if friction and windage losses are 750 W, and (d) the efficiency of the motor, neglecting rotor iron loss.

(a) Input power to rotor = stator input power – stator losses

$$= 32 \text{ kW} - 1.2 \text{ kW} = 30.8 \text{ kW}$$

From equation (22.9), $\text{slip} = \dfrac{\text{rotor copper loss}}{\text{rotor input}}$

i.e., $\dfrac{5}{100} = \dfrac{\text{rotor copper loss}}{30.8}$

from which, **rotor copper loss** = (0.05)(30.8) = **1.54 kW**

(b) Total mechanical power developed by the rotor

= rotor input power – rotor losses

= 30.8 – 1.54 = **29.26 kW**

(c) Output power of motor

= power developed by the rotor – friction and windage losses

= 29.26 – 0.75 = **28.51 kW**

(d) Efficiency of induction motor, $\eta = \left(\dfrac{\text{output power}}{\text{input power}}\right) \times 100\%$

$$= \left(\frac{28.51}{32}\right) \times 100\% = \mathbf{89.10\%}$$

Problem 9. The speed of the induction motor of Problem 8 is reduced to 35% of its synchronous speed by using external rotor resistance. If the torque and stator losses are unchanged, determine (a) the rotor copper loss, and (b) the efficiency of the motor.

(a) Slip, $s = \left(\frac{n_s - n_r}{n_s}\right) \times 100\% = \left(\frac{n_s - 0.35n_s}{n_s}\right) \times 100\%$

$= (0.65)(100) = 65\%$

Input power to rotor $= 30.8$ kW (from Problem 8)

Since $s = \frac{\text{rotor copper loss}}{\text{rotor input}}$

then **rotor copper loss** $= (s)(\text{rotor input})$

$= \left(\frac{65}{100}\right)(30.8) = \mathbf{20.02\ kW}$

(b) Power developed by rotor

$=$ input power to rotor $-$ rotor copper loss

$= 30.8 - 20.02 = 10.78$ kW

Output power of motor

$=$ power developed by rotor $-$ friction and windage losses

$= 10.78 - 0.75 = 10.03$ kW

Efficiency, $\eta = \frac{\text{output power}}{\text{input power}} 100\% = \left(\frac{10.03}{32}\right) \times 100\%$

$= \mathbf{31.34\%}$

Further problems on losses and efficiency may be found in Section 22.18, problems 10 and 11, page 407.

22.11 Torque equation for an induction motor

Torque $T = \frac{P_2}{2\pi n_s} = \left(\frac{1}{2\pi n_s}\right)\left(\frac{I_r{}^2 R_2}{s}\right)$ (from equation (22.10)

From equation (22.8), $I_r = \frac{s(N_2/N_1)\,E_1}{\sqrt{[R_2{}^2 + (s\,X_2)^2]}}$

Hence torque per phase, $T = \left(\frac{1}{2\pi n_s}\right)\left[\frac{s^2(N_2/N_1)^2 E_1{}^2}{R_2{}^2 + (sX_2)^2}\right]\left(\frac{R_2}{s}\right)$

i.e. $T = \left(\frac{1}{2\pi n_s}\right)\left[\frac{s(N_2/N_1)^2 E_1{}^2 R_2}{R_2{}^2 + (sX_2)^2}\right]$

If there are m phases then

$$\text{torque, } T = \left(\frac{m}{2\pi n_s}\right)\left[\frac{s(N_2/N_1)^2 {E_1}^2 R_2}{{R_2}^2 + (sX_2)^2}\right]$$

i.e., $$\boldsymbol{T = \left[\frac{m(N_2/N_1)^2}{2\pi n_s}\right]\left[\frac{s{E_1}^2 R_2}{{R_2}^2 + (sX_2)^2}\right]} \quad (22.11)$$

$$= k\left(\frac{s{E_1}^2 R_2}{{R_2}^2 + (sX_2)^2}\right),$$ where k is a constant for a particular machine,

i.e., $$\textbf{torque } \boldsymbol{T \propto \frac{s{E_1}^2 R_2}{{R_2}^2 + (sX_2)^2}} \quad (22.12)$$

Under normal conditions, the supply voltage is usually constant, hence equation (22.12) becomes:

$$T \propto \frac{sR_2}{{R_2}^2 + (sX_2)^2} \propto \frac{R_2}{\dfrac{{R_2}^2}{s} + s{X_2}^2}$$

The torque will be a maximum when the denominator is a minimum and this occurs when ${R_2}^2/s = s{X_2}^2$

i.e., when $s = \dfrac{R_2}{X_2}$ or $R_2 = sX_2 = X_r$ from equation (22.4)

Thus **maximum torque** occurs when rotor resistance and rotor reactance are equal, i.e., $\boldsymbol{R_2 = X_r}$

Problems 10 to 13 following illustrate some of the characteristics of three-phase induction motors.

Problem 10. A 415 V, three-phase, 50 Hz, 4 pole, star-connected induction motor runs at 24 rev/s on full load. The rotor resistance and reactance per phase are 0.35 Ω and 3.5 Ω respectively, and the effective rotor-stator turns ratio is 0.85:1. Calculate (a) the synchronous speed, (b) the slip, (c) the full load torque, (d) the power output if mechanical losses amount to 770 W, (e) the maximum torque, (f) the speed at which maximum torque occurs, and (g) the starting torque.

(a) Synchronous speed, $n_s = \dfrac{f}{p} = \dfrac{50}{2} =$ **25 rev/s** or (25×60)

= **1500 rev/min**

(b) Slip, $s = \left(\frac{n_s - n_r}{n_s}\right) = \frac{25 - 24}{25} = \mathbf{0.04}$ or **4%**

(c) Phase voltage, $E_1 = \frac{415}{\sqrt{3}} = 239.6$ volts

Full load torque, $T = \left[\frac{m\ (N_2/N_1)^2}{2\pi n_s}\right] \left[\frac{s\ E_1{}^2\ R_2}{R_2{}^2 + (s\ X_2)^2}\right]$

from equation (22.11)

$$= \left[\frac{3\ (0.85)^2}{2\pi\ (25)}\right] \left[\frac{0.04\ (239.6)^2\ 0.35}{(0.35)^2 + (0.04 \times 3.5)^2}\right]$$

$$= (0.01380)\left(\frac{803.71}{0.1421}\right) = \mathbf{78.05\ Nm}$$

(d) Output power, including friction losses, $P_m = 2\pi n_r T$

$$= 2\pi(24)(78.05)$$

$$= 11\,770 \text{ watts}$$

Hence **power output** $= P_m -$ mechanical losses

$$= 11\,770 - 770 = 11\,000 \text{ W} = \mathbf{11\ kW}$$

(e) Maximum torque occurs when $R_2 = X_r = 0.35\ \Omega$

Slip, $s = \frac{R_2}{X_2} = \frac{0.35}{3.5} = 0.1$

Hence **maximum torque,** $\boldsymbol{T_m} = (0.01380)\left[\frac{s\ E_1{}^2\ R_2}{R_2{}^2 + (s\ X_2)^2}\right]$

from part (c)

$$= (0.01380)\left[\frac{0.1\ (239.6)^2\ 0.35}{0.35^2 + 0.35^2}\right]$$

$$= (0.01380)\left[\frac{2009.29}{0.245}\right]$$

$$= \mathbf{113.18\ Nm}$$

(f) For maximum torque, slip $s = 0.1$

Slip, $s = \left(\frac{n_s - n_r}{n_s}\right)$ i.e., $0.1 = \left(\frac{25 - n_r}{25}\right)$

Hence $(0.1)(25) = 25 - n_r$ and $n_r = 25 - (0.1)(25)$

Thus speed at which maximum torque occurs,

$$n_r = 25 - 2.5$$

$$= \mathbf{22.5\ rev/s} \text{ or } \mathbf{1350\ rev/min}$$

(g) At the start, i.e., at standstill, slip $s = 1$

$$\text{Hence starting torque} = \left[\frac{m\,(N_2/N_1)^2}{2\pi n_s}\right]\left[\frac{E_1{}^2\,R_2}{R_2{}^2 + X_2{}^2}\right]$$

from equation (22.11) with $s = 1$

$$= (0.01380)\left[\frac{(239.6)^2\;0.35}{0.35^2 + 3.5^2}\right]$$

$$= (0.01380)\left(\frac{20092.86}{12.3725}\right)$$

i.e., **starting torque = 22.41 Nm**

(Note that the full load torque (from part (c)) is 78.05 Nm but the starting torque is only 22.41 Nm)

Problem 11. Determine for the induction motor in problem 10 at full load, (a) the rotor current, (b) the rotor copper loss, and (c) the starting current.

(a) From equation (22.8), **rotor current,**

$$\boldsymbol{I_r} = \frac{s\left(\dfrac{N_2}{N_1}\right)E_1}{\sqrt{[R_2{}^2 + (s\,X_2)^2]}}$$

$$= \frac{(0.04)\,(0.85)\,(239.6)}{\sqrt{[0.35^2 + (0.04 \times 3.5)^2]}}$$

$$= \frac{8.1464}{0.37696} = \mathbf{21.61\ A}$$

(b) Rotor copper loss per phase $= I_r{}^2\,R_2 = (21.61)^2(0.35) = 163.45$ W

Total copper loss (for 3 phases) $= 3 \times 163.45 =$ **490.35 W**

(c) From equation (22.7), starting current,

$$I_2 = \frac{\left(\dfrac{N_2}{N_1}\right)E_1}{\sqrt{[R_2{}^2 + X_2{}^2]}} = \frac{(0.85)\,(239.6)}{\sqrt{[0.35^2 + 3.5^2]}} = \mathbf{57.90\ A}$$

(Note that the starting current of 57.90 A is considerably higher than the full load current of 21.61 A)

Problem 12. For the induction motor in problems 10 and 11, if the stator losses are 650 W, determine (a) the power input at full load, (b) the efficiency of the motor at full load and (c) the current taken from the supply at full load, if the motor runs at a power factor of 0.87 lagging.

(a) Output power $P_m = 11.770$ kW from part (d), Problem 10
Rotor copper loss $= 490.35$ W $= 0.49035$ kW from part (b), Problem 11

Stator input power, $P_1 = P_m +$ rotor copper loss + rotor stator loss

$= 11.770 + 0.49035 + 0.650 = \mathbf{12.910\ kW}$

(b) Net power output $= 11$ kW from part (d), Problem 10

$$\text{Hence efficiency, } \eta = \frac{\text{output}}{\text{input}} \times 100\% = \left(\frac{11}{12.910}\right) \times 100\%$$

$$= \mathbf{85.21\%}$$

(c) Power input, $P_1 = \sqrt{3}\ V_L I_L \cos\phi$ (see Chapter 19) and $\cos\phi =$ p.f. $= 0.87$

$$\text{hence, } \textbf{supply current, } I_L = \frac{P_1}{\sqrt{3}\ V_L\ \cos\ \phi}$$

$$= \frac{12.910 \times 1000}{\sqrt{3}\ (415)\ 0.87} = \mathbf{20.64\ A}$$

Problem 13. For the induction motor of Problems 10 to 12, determine the resistance of the rotor winding required for maximum starting torque.

From equation (22.4), rotor reactance $X_r = s\ X_2$
At the moment of starting, slip, $s = 1$

Maximum torque occurs when rotor reactance equals rotor resistance hence for **maximum torque,** $\mathbf{R_2} = X_r = s\ X_2 = X_2 = \mathbf{3.5\ \Omega}$

Thus if the induction motor was a wound rotor type with slip rings then an external star-connected resistance of $(3.5 - 0.35)\ \Omega = 3.15\ \Omega$ per phase could be added to the rotor resistance to give maximum torque at starting (see Section 22.13).

Further problems on the torque equation may be found in Section 22.18, problems 12 to 15, page 407.

22.12 Induction motor torque–speed characteristics

From Problem 10, parts (c) and (g), it is seen that the normal starting torque may be less than the full load torque. Also, from Problem 10, parts (e) and (f), it is seen that the speed at which maximum torque occurs is determined by the value of the rotor resistance. At synchronous speed, slip $s = 0$ and torque is zero. From these observations, the torque-speed and torque-slip characteristics of an induction motor are as shown in Figure 22.9.

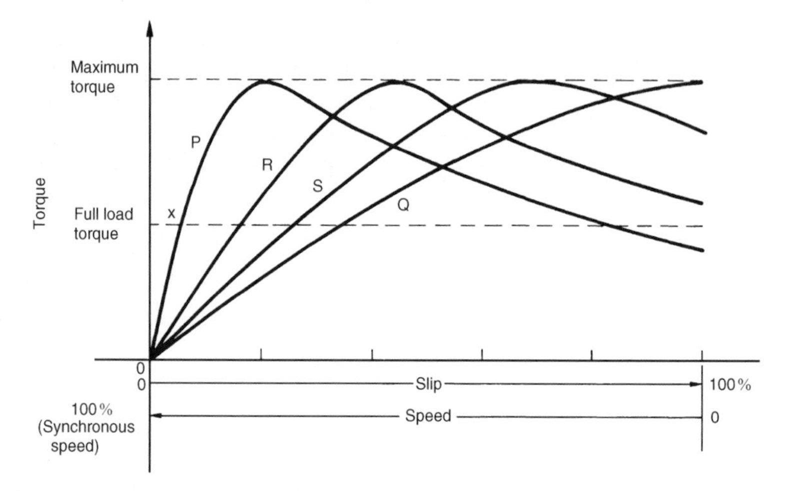

Figure 22.9

The rotor resistance of an induction motor is usually small compared with its reactance (for example, $R_2 = 0.35\ \Omega$ and $X_2 = 3.5\ \Omega$ in the above Problems), so that maximum torque occurs at a high speed, typically about 80% of synchronous speed.

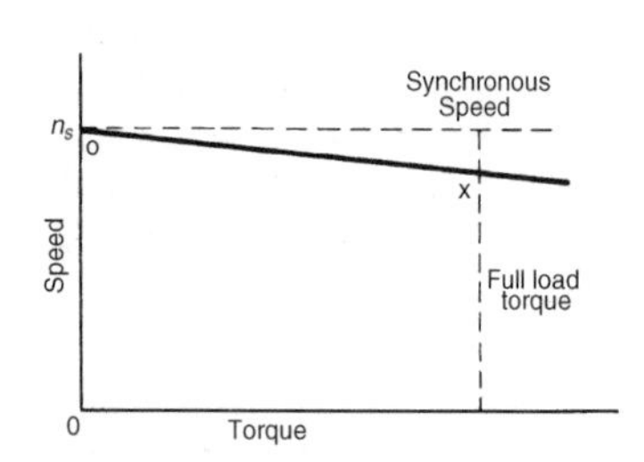

Figure 22.10

Curve P in Figure 22.9 is a typical characteristic for an induction motor. The curve P cuts the full-load torque line at point X, showing that at full load the slip is about 4–5%. The normal operating conditions are between 0 and X, thus it can be seen that for normal operation the speed variation with load is quite small — the induction motor is an almost constant-speed machine. Redrawing the speed-torque characteristic between 0 and X gives the characteristic shown in Figure 22.10, which is similar to a d.c. shunt motor as shown in chapter 21.

If maximum torque is required at starting then a high resistance rotor is necessary, which gives characteristic Q in Figure 22.9. However, as can be seen, the motor has a full load slip of over 30%, which results in a drop in efficiency. Also such a motor has a large speed variation with variations of load. Curves R and S of Figure 22.9 are characteristics for values of rotor resistances between those of P and Q. Better starting torque than for curve P is obtained, but with lower efficiency and with speed variations under operating conditions.

A **squirrel-cage induction motor** would normally follow characteristic P. This type of machine is highly efficient and about constant-speed under normal running conditions. However it has a poor starting torque and must be started off-load or very lightly loaded (see Section 22.13 below). Also, on starting, the current can be four or five times the normal full load current, due to the motor acting like a transformer with secondary short circuited. In problem 11, for example, the current at starting was nearly three times the full load current.

A **wound-rotor induction motor** would follow characteristic P when the slip-rings are short-circuited, which is the normal running condition. However, the slip-rings allow for the addition of resistance to the rotor circuit externally and, as a result, for starting, the motor can have a characteristic similar to curve Q in Figure 22.9 and the high starting current experienced by the cage induction motor can be overcome.

In general, for three-phase induction motors, the power factor is usually between about 0.8 and 0.9 lagging, and the full load efficiency is usually about 80–90%.

From equation (22.12), it is seen that torque is proportional to the square of the supply voltage. Any voltage variations therefore would seriously affect the induction motor performance.

22.13 Starting methods for induction motors

Squirrel-cage rotor

(i) **Direct-on-line starting**
With this method, starting current is high and may cause interference with supplies to other consumers.

(ii) **Auto transformer starting**
With this method, an auto transformer is used to reduce the stator voltage, E_1, and thus the starting current (see equation (22.7)). However, the starting torque is seriously reduced (see equation (22.12)), so the voltage is reduced only sufficiently to give the required reduction of the starting current. A typical arrangement is shown in Figure 22.11. A double-throw switch connects the auto transformer in circuit for starting, and when the motor is up to speed the switch is moved to the run position which connects the supply directly to the motor.

(iii) **Star-delta starting**
With this method, for starting, the connections to the stator phase winding are star-connected, so that the voltage across each phase

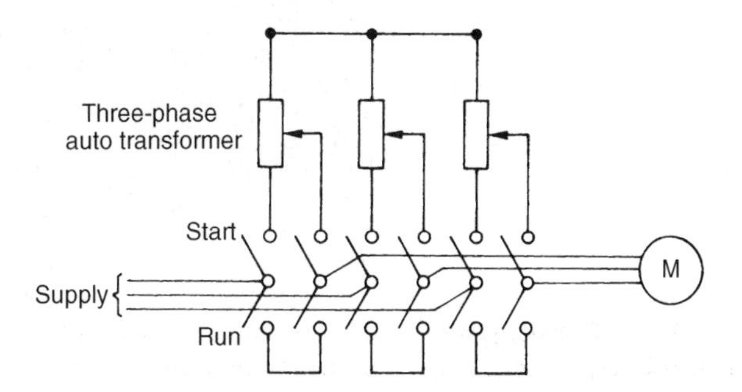

Figure 22.11

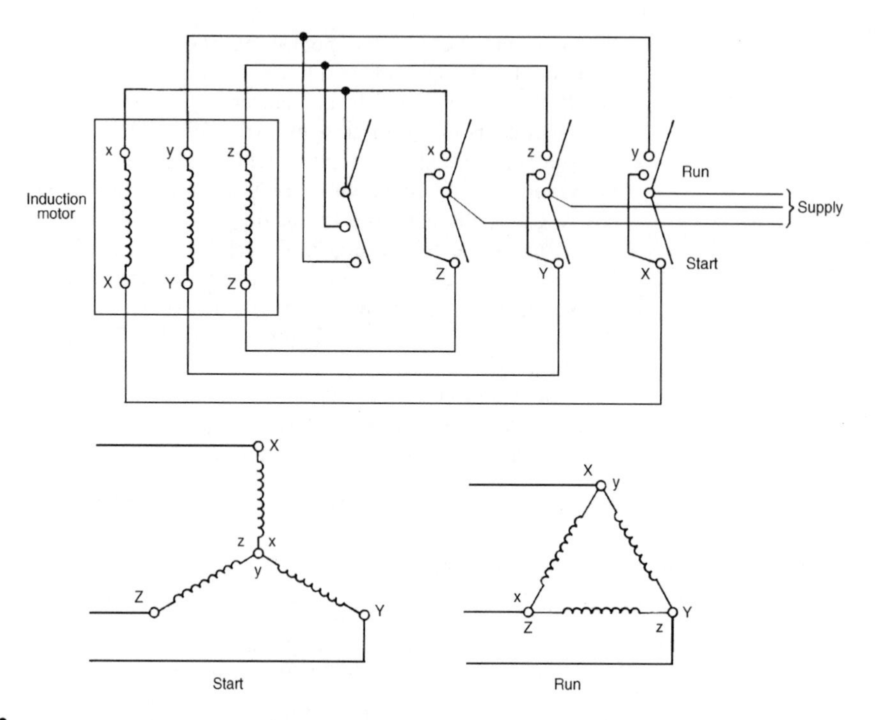

Figure 22.12

winding is $1/\sqrt{3}$ (i.e. 0.577) of the line voltage. For running, the windings are switched to delta-connection. A typical arrangement is shown in Figure 22.12. This method of starting is less expensive than by auto transformer.

Wound rotor

When starting on load is necessary, a wound rotor induction motor must be used. This is because maximum torque at starting can be obtained by adding external resistance to the rotor circuit via slip rings, (see problem 13). A face-plate type starter is used, and as the resistance is gradually reduced, the machine characteristics at each stage will be similar to Q, S, R and P of Figure 22.13. At each resistance step, the motor operation will transfer from one characteristic to the next so that the overall starting characteristic will be as shown by the bold line in Figure 22.13. For very large induction motors, very gradual and smooth starting is achieved by a liquid type resistance.

22.14 Advantages of squirrel-cage induction motors

The advantages of squirrel-cage motors compared with the wound rotor type are that they:

(i) are cheaper and more robust

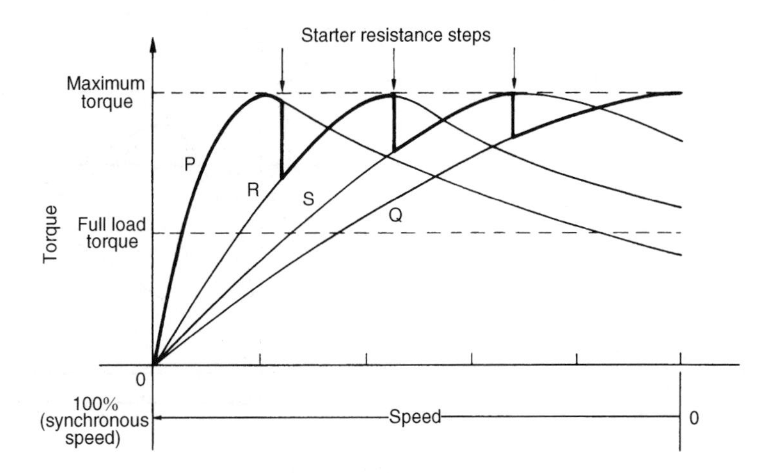

Figure 22.13

(ii) have slightly higher efficiency and power factor
(iii) are explosion-proof, since the risk of sparking is eliminated by the absence of slip rings and brushes.

22.15 Advantages of wound rotor induction motor

The advantages of the wound rotor motor compared with the cage type are that they:

(i) have a much higher starting torque
(ii) have a much lower starting current
(iii) have a means of varying speed by use of external rotor resistance.

22.16 Double cage induction motor

The advantages of squirrel-cage and wound rotor induction motors are combined in the double cage induction motor. This type of induction motor is specially constructed with the rotor having two cages, one inside the other. The outer cage has high resistance conductors so that maximum torque is achieved at or near starting. The inner cage has normal low resistance copper conductors but high reactance since it is embedded deep in the iron core. The torque-speed characteristic of the inner cage is that of a normal induction motor, as shown in Figure 22.14. At starting, the outer cage produces the torque, but when running the inner cage produces the torque. The combined characteristic of inner and outer cages is shown in Figure 22.14. The double cage induction motor is highly efficient when running.

22.17 Uses of three-phase induction motors

Three-phase induction motors are widely used in industry and constitute almost all industrial drives where a nearly constant speed is required, from small workshops to the largest industrial enterprises.

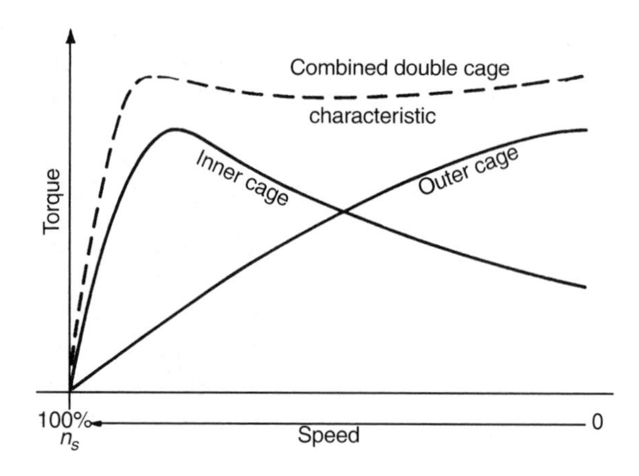

Figure 22.14

Typical applications are with machine tools, pumps and mill motors. The squirrel cage rotor type is the most widely used of all a.c. motors.

22.18 Further problems on three-phase induction motors

Synchronous speed

1 The synchronous speed of a 3-phase, 4-pole induction motor is 60 rev/s. Determine the frequency of the supply to the stator windings. [120 Hz]

2 The synchronous speed of a 3-phase induction motor is 25 rev/s and the frequency of the supply to the stator is 50 Hz. Calculate the equivalent number of pairs of poles of the motor. [2]

3 A 6-pole, 3-phase induction motor is connected to a 300 Hz supply. Determine the speed of rotation of the magnetic field produced by the stator. [100 rev/s]

Slip

4 A 6-pole, 3-phase induction motor runs at 970 rev/min at a certain load. If the stator is connected to a 50 Hz supply, find the percentage slip at this load. [3%]

5 A 3-phase, 50 Hz induction motor has 8 poles. If the full load slip is 2.5%, determine (a) the synchronous speed, (b) the rotor speed, and (c) the frequency of the rotor e.m.f.'s.
[(a) 750 rev/min (b) 731 rev/min (c) 1.25 Hz]

6 A three-phase induction motor is supplied from a 60 Hz supply and runs at 1710 rev/min when the slip is 5%. Determine the synchronous speed. [1800 rev/min]

7 A 4-pole, 3-phase, 50 Hz induction motor runs at 1440 rev/min at full load. Calculate (a) the synchronous speed, (b) the slip and (c) the frequency of the rotor induced e.m.f.'s.
[(a) 1500 rev/min (b) 4% (c) 2 Hz]

Rotor frequency

8 A 12-pole, 3-phase, 50 Hz induction motor runs at 475 rev/min. Determine (a) the slip speed, (b) the percentage slip and (c) the frequency of rotor currents. [(a) 25 rev/min (b) 5% (c) 2.5 Hz]

9 The frequency of the supply to the stator of a 6-pole induction motor is 50 Hz and the rotor frequency is 2 Hz. Determine (a) the slip, and (b) the rotor speed in rev/min. [(a) 0.04 or 4% (b) 960 rev/min]

Losses and efficiency

10 The power supplied to a three-phase induction motor is 50 kW and the stator losses are 2 kW. If the slip is 4%, determine (a) the rotor copper loss, (b) the total mechanical power developed by the rotor, (c) the output power of the motor if friction and windage losses are 1 kW, and (d) the efficiency of the motor, neglecting rotor iron losses. [(a) 1.92 kW (b) 46.08 kW (c) 45.08 kW (d) 90.16%]

11 By using external rotor resistance, the speed of the induction motor in problem 15 is reduced to 40% of its synchronous speed. If the torque and stator losses are unchanged, calculate (a) the rotor copper loss, and (b) the efficiency of the motor. [(a) 28.80 kW (b) 36.40%]

Torque equation

12 A 400 V, three-phase, 50 Hz, 2-pole, star-connected induction motor runs at 48.5 rev/s on full load. The rotor resistance and reactance per phase are 0.4 Ω and 4.0 Ω respectively, and the effective rotor-stator turns ratio is 0.8:1. Calculate (a) the synchronous speed, (b) the slip, (c) the full load torque, (d) the power output if mechanical losses amount to 500 W, (e) the maximum torque, (f) the speed at which maximum torque occurs, and (g) the starting torque.
[(a) 50 rev/s or 3000 rev/min (b) 0.03 or 3%
(c) 22.43 Nm (d) 6.34 kW (e) 40.74 Nm
(f) 45 rev/s or 2700 rev/min (g) 8.07 Nm]

13 For the induction motor in problem 12, calculate at full load (a) the rotor current, (b) the rotor copper loss, and (c) the starting current.
[(a) 10.62 A (b) 135.3 W (c) 45.96 A]

14 If the stator losses for the induction motor in problem 12 are 525 W, calculate at full load (a) the power input, (b) the efficiency of the motor and (c) the current taken from the supply if the motor runs at a power factor of 0.84. [(a) 7.49 kW (b) 84.65% (c) 12.87 A]

15 For the induction motor in problem 12, determine the resistance of the rotor winding required for maximum starting torque. [4.0 Ω]

Assignment 7

This assignment covers the material contained in chapters 21 and 22.

The marks for each question are shown in brackets at the end of each question.

1 A 6-pole armature has 1000 conductors and a flux per pole of 40 mWb. Determine the e.m.f. generated when running at 600 rev/min when (a) lap wound (b) wave wound. (6)

2 The armature of a d.c. machine has a resistance of 0.3 Ω and is connected to a 200 V supply. Calculate the e.m.f. generated when it is running (a) as a generator giving 80 A (b) as a motor taking 80 A (4)

3 A 15 kW shunt generator having an armature circuit resistance of 1 Ω and a field resistance of 160 Ω generates a terminal voltage of 240 V at full-load. Determine the efficiency of the generator at full-load assuming the iron, friction and windage losses amount to 544 W. (6)

4 A 4-pole d.c. motor has a wave-wound armature with 1000 conductors. The useful flux per pole is 40 mWb. Calculate the torque exerted when a current of 25 A flows in each armature conductor. (4)

5 A 400 V shunt motor runs at it's normal speed of 20 rev/s when the armature current is 100 A. The armature resistance is 0.25 Ω. Calculate the speed, in rev/min when the current is 50 A and a resistance of 0.40 Ω is connected in series with the armature, the shunt field remaining constant. (7)

6 The stator of a three-phase, 6-pole induction motor is connected to a 60 Hz supply. The rotor runs at 1155 rev/min at full load. Determine (a) the synchronous speed, and (b) the slip at full load. (6)

7 The power supplied to a three-phase induction motor is 40 kW and the stator losses are 2 kW. If the slip is 4% determine (a) the rotor copper loss, (b) the total mechanical power developed by the rotor, (c) the output power of the motor if frictional and windage losses are 1.48 kW, and (d) the efficiency of the motor, neglecting rotor iron loss. (9)

8 A 400 V, three-phase, 100 Hz, 8-pole induction motor runs at 24.25 rev/s on full load. The rotor resistance and reactance per phase are 0.2 Ω and 2 Ω respectively and the effective rotor-stator turns ratio is 0.80:1. Calculate (a) the synchronous speed, (b) the percentage slip, and (c) the full load torque. (8)

Main formulae for Part 2

A.c. theory:

$T = \dfrac{1}{f}$ or $f = \dfrac{1}{T}$ $\quad I = \sqrt{\left(\dfrac{i_1^2 + i_2^2 + i_3^2 + \cdots + i_n}{n}\right)}$

For a sine wave: $I_{AV} = \dfrac{2}{\pi} I_m$ or $0.637 I_m$ $\quad I = \dfrac{1}{\sqrt{2}} I_m$ or $0.707 I_m$

Form factor $= \dfrac{\text{rms}}{\text{average}}$ $\quad$ Peak factor $= \dfrac{\text{maximum}}{\text{rms}}$

General sinusoidal voltage: $v = V_m \sin(\omega t \pm \phi)$

Single-phase circuits:

$X_L = 2\pi f L \quad X_C = \dfrac{1}{2\pi f C} \quad Z = \dfrac{V}{I} = \sqrt{(R^2 + X^2)}$

Series resonance: $f_r = \dfrac{1}{2\pi\sqrt{LC}}$

$$Q = \frac{V_L}{V} \text{ or } \frac{V_C}{V} = \frac{2\pi f_r L}{R} = \frac{1}{2\pi f_r C R} = \frac{1}{R}\sqrt{\frac{L}{C}}$$

$$Q = \frac{f_r}{f_2 - f_1} \text{ or } (f_2 - f_1) = \frac{f_r}{Q}$$

Parallel resonance (LR-C circuit): $f_r = \dfrac{1}{2\pi}\sqrt{\dfrac{1}{LC} - \dfrac{R^2}{L^2}}$

$$I_r = \frac{VRC}{L} \quad R_D = \frac{L}{CR} \quad Q = \frac{2\pi f_r L}{R} = \frac{I_C}{I_r}$$

$$P = VI\cos\phi \text{ or } I^2R \quad S = VI$$

$$Q = VI\sin\phi \quad \text{power factor} = \cos\phi = \frac{R}{Z}$$

D.c. transients:

C–R circuit $\tau = CR$

Charging: $v_C = V\left(1 - e^{-(t/CR)}\right) \quad v_r = Ve^{-(t/CR)} \quad i = Ie^{-(t/CR)}$

Discharging: $v_C = v_R = Ve^{-(t/CR)} \quad i = Ie^{-(t/CR)}$

L-R circuit $\tau = \dfrac{L}{R}$

Current growth: $v_L = Ve^{-(Rt/L)}$

$$v_R = V\left(1 - e^{-(Rt/L)}\right)$$

$$i = I\left(1 - e^{-(Rt/L)}\right)$$

Current decay: $v_L = v_R = Ve^{-(Rt/L)} \quad i = Ie^{-(Rt/L)}$

Operational amplifiers:

$$\text{CMRR} = 20\log_{10}\left(\frac{\text{differential voltage gain}}{\text{common mode gain}}\right) dB$$

Inverter: $A = \frac{V_o}{V_i} = \frac{-R_f}{R_i}$

Non-inverter: $A = \frac{V_o}{V_i} = 1 + \frac{R_f}{R_i}$

Summing: $V_o = -R_f\left(\frac{V_1}{R_1} + \frac{V_2}{R_2} + \frac{V_3}{R_3}\right)$

Integrator: $V_o = -\frac{1}{\text{CR}}\int V_i\,dt$

Differential: If $V_1 > V_2$: $V_o = (V_1 - V_2)\left(-\frac{R_f}{R_1}\right)$

If $V_2 > V_1$: $V_o = (V_2 - V_1)\left(\frac{R_3}{R_2 + R_3}\right)\left(1 + \frac{R_f}{R_1}\right)$

Three-phase systems: Star $I_L = I_p$ $V_L = \sqrt{3}V_p$ Delta $V_L = V_p$ $I_L = \sqrt{3}I_p$

$P = \sqrt{3}V_L I_L \cos\phi$ or $P = 3I_p^2 R_p$

Two-wattmeter method $P = P_1 + P_2$ $\tan\phi = \sqrt{3}\frac{(P_1 - P_2)}{(P_1 + P_2)}$

Transformers: $\frac{V_1}{V_2} = \frac{N_1}{N_2} = \frac{I_2}{I_1}$ $I_0 = \sqrt{(I_M^2 + I_C^2)}$ $I_M = I_0 \sin\phi_0$ $I_C = I_0\cos\phi_0$

$E = 4.44 f \Phi_m N$ Regulation $= \left(\frac{E_2 - E_1}{E_2}\right) \times 100\%$

Equivalent circuit: $R_e = R_1 + R_2\left(\frac{V_1}{V_2}\right)^2$ $X_e = X_1 + X_2\left(\frac{V_1}{V_2}\right)^2$

$Z_e = \sqrt{(R_e^2 + X_e^2)}$

Efficiency, $\eta = 1 - \frac{\text{losses}}{\text{input power}}$ Output power $= V_2 I_2 \cos\phi_2$

Total loss = copper loss + iron loss

Input power = output power + losses

Resistance matching: $R_1 = \left(\frac{N_1}{N_2}\right)^2 R_L$

D.c. machines: Generated e.m.f. $E = \frac{2p\Phi n Z}{c} \propto \Phi\omega$

($c = 2$ for wave winding, $c = 2p$ for lap winding)

Generator: $E = V + I_a R_a$

Efficiency, $\eta = \left(\frac{VI}{VI + I_a^2 R_a + I_f V + C}\right) \times 100\%$

Motor: $E = V - I_a R_a$

$$\text{Efficiency, } \eta = \left(\frac{VI - I_a^2 R_a - I_f V - C}{VI}\right) \times 100\%$$

$$\text{Torque} = \frac{EI_a}{2\pi n} = \frac{p\Phi Z I_a}{\pi c} \propto I_a \Phi$$

Three-phase induction motors:

$$n_S = \frac{f}{p} \qquad s = \left(\frac{n_S - n_r}{n_S}\right) \times 100 \qquad f_r = sf \qquad X_r = sX_2$$

$$I_r = \frac{E_r}{Z_r} = \frac{s\left(\dfrac{N_2}{N_1}\right)E_1}{\sqrt{[R_2^2 + (sX_2)^2]}} \qquad s = \frac{I_r^2 R_2}{P_2}$$

$$\text{Efficiency, } \eta = \frac{P_m}{P_1}$$

$$= \frac{\text{input} - \text{stator loss} - \text{rotor copper loss} - \text{friction \& windage loss}}{\text{input power}}$$

$$\text{Torque, } T = \left(\frac{m(N_2/N_1)^2}{2\pi n_S}\right)\left(\frac{sE_1^2 R_2}{R_2^2 + (sX_2)^2}\right) \propto \frac{sE_1^2 R_2}{R_2^2 + (sX_2)^2}$$

Part 3 Advanced Circuit Theory and Technology

23 Revision of complex numbers

At the end of this chapter you should be able to:

- define a complex number
- understand the Argand diagram
- perform calculations on addition, subtraction, multiplication, and division in Cartesian and polar forms
- use De Moivres theorem for powers and roots of complex numbers

23.1 Introduction

A **complex number** is of the form $(a + jb)$ where a is a **real number** and jb is an **imaginary number**. Hence $(1 + j2)$ and $(5 - j7)$ are examples of complex numbers.

By definition, $\boxed{j = \sqrt{-1}}$ and $\boxed{j^2 = -1}$

Complex numbers are widely used in the analysis of series, parallel and series-parallel electrical networks supplied by alternating voltages (see Chapters 24 to 26), in deriving balance equations with a.c. bridges (see Chapter 27), in analysing a.c. circuits using Kirchhoff's laws (Chapter 30), mesh and nodal analysis (Chapter 31), the superposition theorem (Chapter 32), with Thévenin's and Norton's theorems (Chapter 33) and with delta-star and star-delta transforms (Chapter 34) and in many other aspects of higher electrical engineering. The advantage of the use of complex numbers is that the manipulative processes become simply algebraic processes.

A complex number can be represented pictorially on an **Argand diagram**. In Figure 23.1, the line OA represents the complex number $(2 + j3)$, OB represents $(3 - j)$, OC represents $(-2 + j2)$ and OD represents $(-4 - j3)$.

A complex number of the form $a + jb$ is called a **Cartesian or rectangular complex number**. The significance of the j operator is shown in Figure 23.2. In Figure 23.2(a) the number 4 (i.e., $4 + j0$) is shown drawn as a phasor horizontally to the right of the origin on the real axis. (Such a phasor could represent, for example, an alternating current, $i = 4 \sin \omega t$ amperes, when time t is zero.)

The number $j4$ (i.e., $0 + j4$) is shown in Figure 23.2(b) drawn vertically upwards from the origin on the imaginary axis. Hence multiplying

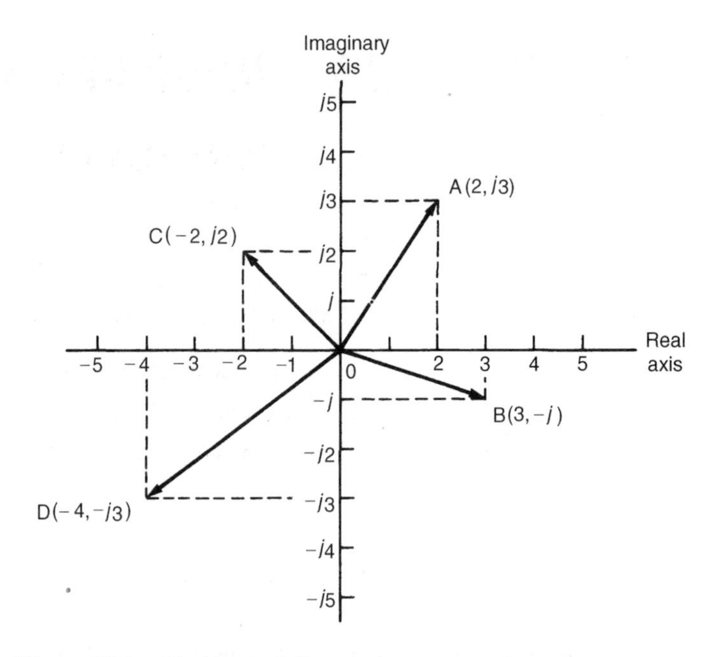

Figure 23.1 *The Argand diagram*

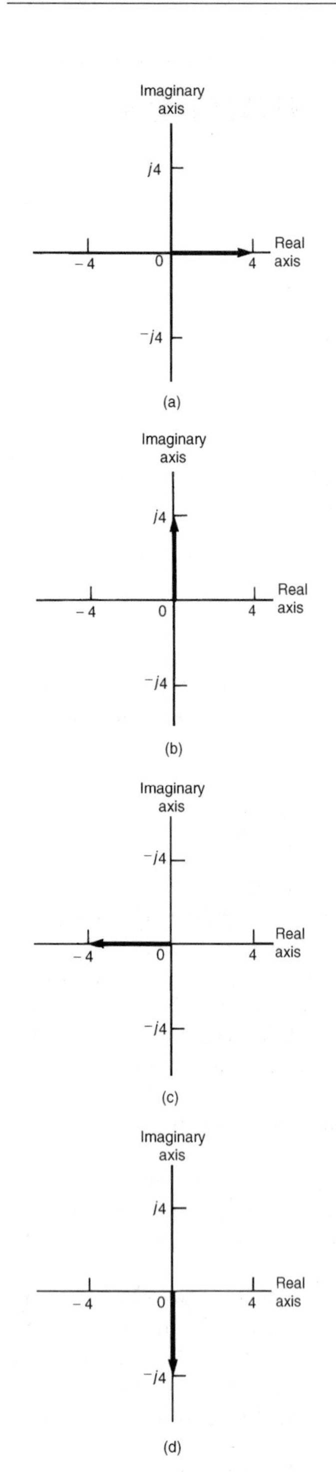

Figure 23.2

the number 4 by the operator j results in an anticlockwise phase-shift of 90° without altering its magnitude.

Multiplying $j4$ by j gives $j^2 4$, i.e. −4, and is shown in Figure 23.2(c) as a phasor four units long on the horizontal real axis to the left of the origin — an anticlockwise phase-shift of 90° compared with the position shown in Figure 23.2(b). Thus multiplying by j^2 reverses the original direction of a phasor.

Multiplying $j^2 4$ by j gives $j^3 4$, i.e. $-j4$, and is shown in Figure 23.2(d) as a phasor four units long on the vertical, imaginary axis downward from the origin — an anticlockwise phase-shift of 90° compared with the position shown in Figure 23.2(c).

Multiplying $j^3 4$ by j gives $j^4 4$, i.e. 4, which is the original position of the phasor shown in Figure 23.2(a).

Summarizing, application of the operator j to any number rotates it 90° anticlockwise on the Argand diagram, multiplying a number by j^2 rotates it 180° anticlockwise, multiplying a number by j^3 rotates it 270° anticlockwise and multiplication by j^4 rotates it 360° anticlockwise, i.e., back to its original position. In each case the phasor is unchanged in its magnitude.

By similar reasoning, if a phasor is operated on by $-j$ then a phase shift of −90° (i.e., clockwise direction) occurs, again without change of magnitude.

In electrical circuits, 90° phase shifts occur between voltage and current with pure capacitors and inductors; this is the key as to why j notation is used so much in the analysis of electrical networks. This is explained in Chapter 24 following.

23.2 Operations involving Cartesian complex numbers

(a) Addition and subtraction

$$(a + jb) + (c + jd) = (a + c) + j(b + d)$$

and $$(a + jb) - (c + jd) = (a - c) + j(b - d)$$

Thus, $(3 + j2) + (2 - j4) = 3 + j2 + 2 - j4 = \mathbf{5 - j2}$

and $(3 + j2) - (2 - j4) = 3 + j2 - 2 + j4 = \mathbf{1 + j6}$

(b) Multiplication

$$\begin{aligned}(a + jb)(c + jd) &= ac + a(jd) + (jb)c + (jb)(jd)\\ &= ac + jad + jbc + j^2bd\end{aligned}$$

But $j^2 = -1$, thus

$$(a + jb)(c + jd) = (ac - bd) + j(ad + bc)$$

For example, $$\begin{aligned}(3 + j2)(2 - j4) &= 6 - j12 + j4 - j^28\\ &= (6 - (-1)8) + j(-12 + 4)\\ &= 14 + j(-8) = \mathbf{14 - j8}\end{aligned}$$

(c) Complex conjugate

The **complex conjugate** of $(a + jb)$ is $(a - jb)$. For example, the conjugate of $(3 - j2)$ is $(3 + j2)$.

The product of a complex number and its complex conjugate is always a real number, and this is an important property used when dividing complex numbers. Thus

$$\begin{aligned}(a + jb)(a - jb) &= a^2 - jab + jab - j^2b^2\\ &= a^2 - (-b^2) = a^2 + b^2 \text{ (i.e. a real number)}\end{aligned}$$

For example, $(1 + j2)(1 - j2) = 1^2 + 2^2 = \mathbf{5}$

and $(3 - j4)(3 + j4) = 3^2 + 4^2 = \mathbf{25}$

(d) Division

The expression of one complex number divided by another, in the form $a + jb$, is accomplished by multiplying the numerator and denominator by the complex conjugate of the denominator. This has the effect of making the denominator a real number. Hence, for example,

$$\begin{aligned}\frac{2 + j4}{3 - j4} = \frac{2 + j4}{3 - j4} \times \frac{3 + j4}{3 + j4} &= \frac{6 + j8 + j12 + j^216}{3^2 + 4^2}\\ &= \frac{6 + j8 + j12 - 16}{25}\end{aligned}$$

$$= \frac{-10 + j20}{25}$$

$$= \frac{\mathbf{-10}}{\mathbf{25}} + j\frac{\mathbf{20}}{\mathbf{25}} \text{ or } -\mathbf{0.4} + j\mathbf{0.8}$$

The elimination of the imaginary part of the denominator by multiplying both the numerator and denominator by the conjugate of the denominator is often termed **'rationalizing'**.

Problem 1. In an electrical circuit the total impedance Z_T is given by

$$Z_T = \frac{Z_1 Z_2}{Z_1 + Z_2} + Z_3$$

Determine Z_T in $(a + jb)$ form, correct to two decimal places, when $Z_1 = 5 - j3$, $Z_2 = 4 + j7$ and $Z_3 = 3.9 - j6.7$

$$Z_1 Z_2 = (5 - j3)(4 + j7) = 20 + j35 - j12 - j^2 21$$

$$= 20 + j35 - j12 + 21 = 41 + j23$$

$$Z_1 + Z_2 = (5 - j3) + (4 + j7) = 9 + j4$$

Hence $\dfrac{Z_1 Z_2}{Z_1 + Z_2} = \dfrac{41 + j23}{9 + j4} = \dfrac{(41 + j23)(9 - j4)}{(9 + j4)(9 - j4)}$

$$= \frac{369 - j164 + j207 - j^2 92}{9^2 + 4^2}$$

$$= \frac{369 - j164 + j207 + 92}{97}$$

$$= \frac{461 + j43}{97} = 4.753 + j0.443$$

Thus $\dfrac{Z_1 Z_2}{Z_1 + Z_2} + Z_3 = (4.753 + j0.443) + (3.9 - j6.7)$

$= \mathbf{8.65 - j6.26}$, correct to two decimal places.

Problem 2. Given $Z_1 = 3 + j4$ and $Z_2 = 2 - j5$ determine in cartesian form correct to three decimal places:

(a) $\dfrac{1}{Z_1}$ (b) $\dfrac{1}{Z_2}$ (c) $\dfrac{1}{Z_1} + \dfrac{1}{Z_2}$ (d) $\dfrac{1}{(1/Z_1) + (1/Z_2)}$

(a) $$\frac{1}{Z_1}=\frac{1}{3+j4}=\frac{3-j4}{(3+j4)(3-j4)}=\frac{3-j4}{3^2+4^2}$$

$$=\frac{3-j4}{25}=\frac{3}{25}-j\frac{4}{25}=\mathbf{0.120-j0.160}$$

(b) $$\frac{1}{Z_2}=\frac{1}{2-j5}=\frac{2+j5}{(2-j5)(2+j5)}=\frac{2+j5}{2^2+5^2}=\frac{2+j5}{29}$$

$$=\frac{2}{29}+j\frac{5}{29}=\mathbf{0.069+j0.172}$$

(c) $$\frac{1}{Z_1}+\frac{1}{Z_2}=(0.120-j0.160)+(0.069+j0.172)$$

$$=\mathbf{0.189+j0.012}$$

(d) $$\frac{1}{(1/Z_1)+(1/Z_2)}=\frac{1}{0.189+j0.012}$$

$$=\frac{0.189-j0.012}{(0.189+j0.012)(0.189-j0.012)}$$

$$=\frac{0.189-j0.012}{0.189^2+0.012^2}=\frac{0.189-j0.012}{0.03587}$$

$$=\frac{0.189}{0.03587}-\frac{j0.012}{0.03587}=\mathbf{5.269-j0.335}$$

Further problems on operations involving Cartesian complex numbers may be found in Section 23.7, problems 1 to 11, page 424.

23.3 Complex equations

If two complex numbers are equal, then their real parts are equal and their imaginary parts are equal. Hence, if $a+jb=c+jd$ then $a=c$ and $b=d$. This is a useful property, since equations having two unknown quantities can be solved from one equation. Complex equations are used when deriving balance equations with a.c. bridges (see Chapter 27).

Problem 3. Solve the following complex equations:

(a) $3(a+jb)=9-j2$

(b) $(2+j)(-2+j)=x+jy$

(c) $(a-j2b)+(b-j3a)=5+j2$

(a) $3(a + jb) = 9 - j2$. Thus $3a + j3b = 9 - j2$

Equating real parts gives: $3a = 9$, i.e. $\boldsymbol{a = 3}$

Equating imaginary parts gives: $3b = -2$, i.e., $\boldsymbol{b = -2/3}$

(b) $(2 + j)(-2 + j) = x + jy$

$$\text{Thus}\quad -4 + j2 - j2 + j^2 = x + jy$$

$$-5 + j0 = x + jy$$

Equating real and imaginary parts gives: $\boldsymbol{x = -5, y = 0}$

(c) $(a - j2b) + (b - j3a) = 5 + j2$

Thus $(a + b) + j(-2b - 3a) = 5 + j2$

$$\text{Hence}\quad a + b = 5 \qquad (1)$$

$$\text{and}\quad 2b - 3a = 2 \qquad (2)$$

We have two simultaneous equations to solve. Multiplying equation (1) by 2 gives:

$$2a + 2b = 10 \qquad (3)$$

Adding equations (2) and (3) gives $-a = 12$, i.e. $\boldsymbol{a = -12}$

From equation (1), $\boldsymbol{b = 17}$

Problem 4. An equation derived from an a.c. bridge network is given by

$$R_1R_3 = (R_2 + j\omega L_2)\left[\frac{1}{(1/R_4) + j\omega C}\right]$$

R_1, R_3, R_4 and C_4 are known values. Determine expressions for R_2 and L_2 in terms of the known components.

Multiplying both sides of the equation by $(1/R_4 + j\omega C_4)$ gives

$$(R_1R_3)(1/R_4 + j\omega C_4) = R_2 + j\omega L_2$$

$$\text{i.e.}\quad R_1R_3/R_4 + jR_1R_3\omega C_4 = R_2 + j\omega L_2$$

Equating the real parts gives: $\boldsymbol{R_2 = R_1R_3/R_4}$

Equating the imaginary parts gives:

$$\omega L_2 = R_1R_3\omega C_4, \text{ from which, } \boldsymbol{L_2 = R_1R_3C_4}$$

Further problems on complex equations may be found in Section 23.7, problems 12 to 16, page 425.

23.4 The polar form of a complex number

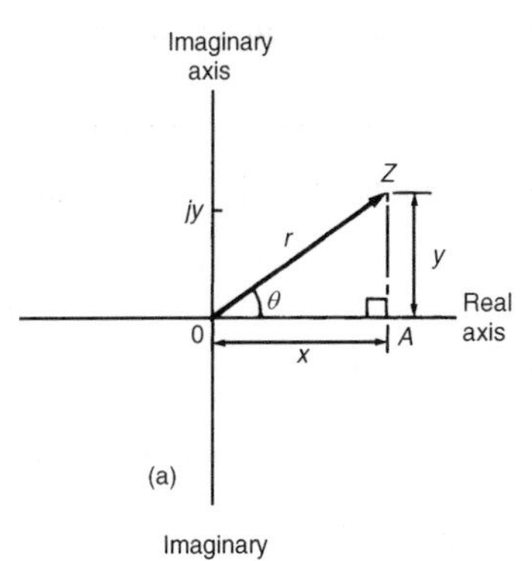

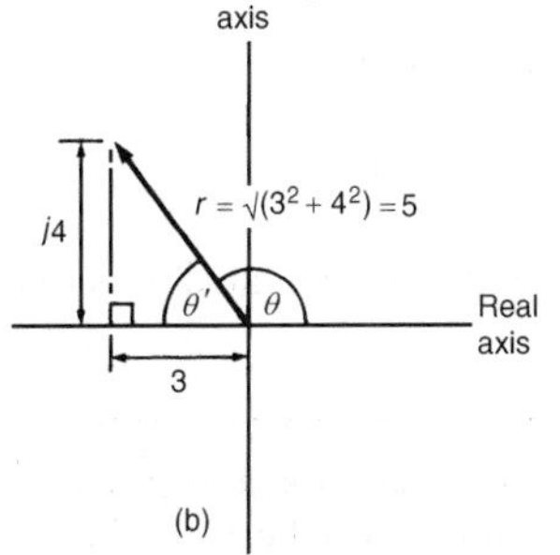

Figure 23.3

In Figure 23.3(a), $Z = x + jy = r\cos\theta + jr\sin\theta$ from trigonometry,

$$= r\,(\cos\theta + j\sin\theta)$$

This latter form is usually abbreviated to $\mathbf{Z = r\angle\theta}$, and is called the **polar form** of a complex number.

r is called the **modulus** (or magnitude of Z) and is written as mod Z or $|Z|$. r is determined from Pythagoras's theorem on triangle OAZ, i.e.

$$|\mathbf{Z}| = \mathbf{r} = \sqrt{(x^2 + y^2)}$$

The modulus is represented on the Argand diagram by the distance OZ. θ is called the **argument** (or amplitude) of Z and is written as arg Z. θ is also deduced from triangle OAZ: **arg Z** $= \boldsymbol{\theta} = \mathbf{arctan}\, y/x$

For example, the cartesian complex number $(3 + j4)$ is equal to $r\angle\theta$ in polar form, where $\mathbf{r} = \sqrt{(3^2 + 4^2)} = \mathbf{5}$ and $\boldsymbol{\theta} = \arctan\frac{4}{3} = \mathbf{53.13°}$

Hence $\mathbf{(3 + j4) = 5\angle 53.13°}$

Similarly, $(-3 + j4)$ is shown in Figure 23.3(b),

where $\quad r = \sqrt{(3^2 + 4^2)} = 5, \quad \theta' = \arctan\frac{4}{3} = 53.13°$

and $\quad \theta = 180° - 53.13° = 126.87°$

Hence $\quad \mathbf{(-3 + j4) = 5\angle 126.87°}$

23.5 Multiplication and division using complex numbers in polar form

(a) Multiplication

$$(r_1\angle\theta_1)(r_2\angle\theta_2) = r_1 r_2\angle(\theta_1 + \theta_2)$$

Thus $3\angle 25° \times 2\angle 32° = 6\angle 57°$, $\quad 4\angle 11° \times 5\angle -18° = 20\angle -7°$, $2\angle(\pi/3) \times 7\angle(\pi/6) = 14\angle(\pi/2)$, and so on.

(b) Division

$$\frac{r_1\angle\theta_1}{r_2\angle\theta_2} = \frac{r_1}{r_2}\angle(\theta_1 + \theta_2)$$

Thus $\quad \dfrac{8\angle 58°}{2\angle 11°} = 4\angle 47°, \quad \dfrac{9\angle 136°}{3\angle -60°} = 3\angle(136° - -60°)$

$$= 3\angle 196° \text{ or } 3\angle -164°,$$

and $\quad \dfrac{10\angle(\pi/2)}{5\angle(-\pi/4)} = 2\angle(3\pi/4)$, and so on.

Conversion from cartesian or rectangular form to polar form, and *vice versa*, may be achieved by using the $R \to P$ and $P \to R$ conversion facility which is available on most calculators with scientific notation. This allows, of course, a great saving of time.

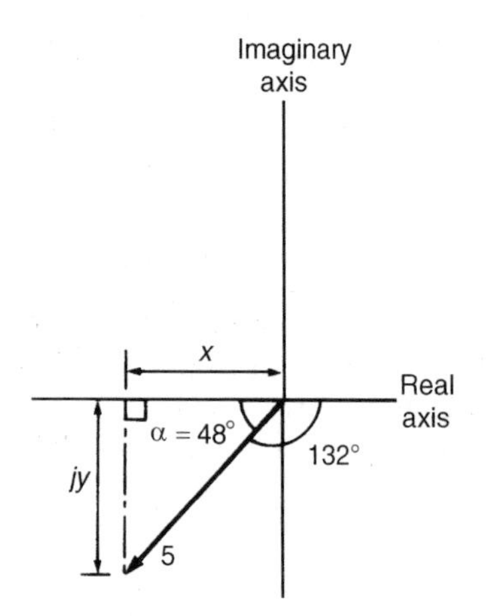

Figure 23.4

Problem 5. Convert $5\angle -132^\circ$ into $a + jb$ form correct to four significant figures.

Figure 23.4 indicates that the polar complex number $5\angle -132^\circ$ lies in the third quadrant of the Argand diagram.

Using trigonometrical ratios,

$$x = 5\cos 48^\circ = 3.346 \text{ and } y = 5\sin 48^\circ = 3.716$$

Hence $\mathbf{5\angle -132^\circ = -3.346 - j3.716}$

Alternatively, $5\angle -132^\circ = 5(\cos -132^\circ + j\sin -132^\circ)$

$$= 5\cos(-132^\circ) + j5\sin(-132^\circ)$$

$$= \mathbf{-3.346 - j3.716}, \text{ as above.}$$

With this latter method the real and imaginary parts are obtained directly, using a calculator.

Problem 6. Two impedances in an electrical network are given by $Z_1 = 4.7\angle 35^\circ$ and $Z_2 = 7.3\angle -48^\circ$. Determine in polar form the total impedance Z_T given that $Z_T = Z_1 Z_2/(Z_1 + Z_2)$

$$Z_1 = 4.7\angle 35^\circ = 4.7\cos 35^\circ + j4.7\sin 35^\circ = 3.85 + j2.70$$

$$Z_2 = 7.3\angle -48^\circ = 7.3\cos(-48^\circ) + j7.3\sin(-48^\circ)$$

$$= 4.88 - j5.42$$

$$Z_1 + Z_2 = (3.85 + j2.70) + (4.88 - j5.42) = 8.73 - j2.72$$

$$= \sqrt{(8.73^2 + 2.72^2)}\angle \arctan\left(\frac{-2.72}{8.73}\right)$$

$$= 9.14\angle -17.31^\circ$$

Hence $Z_T = Z_1 Z_2/(Z_1 + Z_2) = \dfrac{4.7\angle 35^\circ \times 7.3\angle -48^\circ}{9.14\angle -17.31^\circ}$

$$= \frac{4.7 \times 7.3}{9.14}\angle [35^\circ - 48^\circ - (-17.31^\circ)]$$

$$= \mathbf{3.75\angle 4.31^\circ \text{ or } 3.75\angle 4^\circ 19'}$$

Further problems on the polar form of complex numbers may be found in Section 23.7, problems 17 to 31, page 426.

23.6 De Moivre's theorem — powers and roots of complex numbers

De Moivre's theorem, states:

$$[r\angle\theta]^n = r^n\angle n\theta$$

This result is true for all positive, negative or fractional values of n. De Moivre's theorem is thus useful in determining powers and roots of complex numbers. For example,

$$[2\angle 15^\circ]^6 = 2^6\angle(6\times 15^\circ) = \mathbf{64\angle 90^\circ = 0 + j64}$$

A square root of a complex number is determined as follows:

$$\sqrt{[r\angle\theta]} = [r\angle\theta]^{1/2} = r^{1/2}\angle\tfrac{1}{2}\theta$$

However, it is important to realize that a real number has two square roots, equal in size but opposite in sign. On an Argand diagram the roots are 180° apart (see problem 8 following).

Problem 7. Determine $(-2+j3)^5$ in polar and in cartesian form.

$Z = -2 + j3$ is situated in the second quadrant of the Argand diagram.

Thus $r = \sqrt{[(2)^2 + (3)^2]} = \sqrt{13}$ and $\alpha = \arctan 3/2 = 56.31^\circ$

Hence the argument $\theta = 180^\circ - 56.31^\circ = 123.69^\circ$

Thus $-2 + j3$ in polar form is $\sqrt{13}\angle 123.69^\circ$

$$\begin{aligned}(-2+j3)^5 &= [\sqrt{13}\angle 123.69^\circ]^5\\ &= (\sqrt{13})^5\angle(5\times 123.69^\circ) \text{ from De Moivre's theorem}\\ &= 13^{5/2}\angle 618.45^\circ = 13^{5/2}\angle 258.45^\circ (\text{since } 618.45^\circ\\ &\qquad\qquad \equiv 618.45^\circ - 360^\circ)\\ &= 13^{5/2}\angle -101.55^\circ = \mathbf{609.3\angle -101^\circ 33'}\end{aligned}$$

In cartesian form, $609.3\angle -101.55^\circ = 609.3\cos(-101.55^\circ) + j609.3\sin(-101.55^\circ)$

$$= \mathbf{-122 - j597}$$

Problem 8. Determine the two square roots of the complex number $(12 + j5)$ in cartesian and polar form, correct to three significant figures. Show the roots on an Argand diagram.

In polar form $12 + j5 = \sqrt{(12^2 + 5^2)}\angle \arctan(5/12)$, since $12 + j5$ is in the first quadrant of the Argand diagram, i.e. $12 + j5 = 13\angle 22.62°$

Since we are finding the square roots of $13\angle 22.62°$ there will be two solutions. To obtain the second solution it is helpful to express $13\angle 22.62°$ also as $13\angle(360° + 22.62°)$, i.e. $13\angle 382.62°$ (we have merely rotated one revolution to obtain this result). The reason for doing this is that when we divide the angles by 2 we still obtain angles less than 360°, as shown below.

$$\text{Hence } \sqrt{(12 + j5)} = \sqrt{[13\angle 22.62°]} \text{ or } \sqrt{[13\angle 382.62°]}$$

$$= [13\angle 22.62°]^{1/2} \text{ or } [13\angle 382.62°]^{1/2}$$

$$= 13^{1/2}\angle\left(\tfrac{1}{2} \times 22.62°\right) \text{ or } 13^{1/2}\angle\left(\tfrac{1}{2} \times 382.62°\right)$$

from De Moivre's theorem,

$$= \sqrt{13}\angle 11.31° \text{ or } \sqrt{13}\angle 191.31°$$

$$= 3.61\angle 11.31° \text{ or } 3.61\angle -168.69°$$

i.e., $\mathbf{3.61\angle 11°19'}$ **or** $\mathbf{3.61\angle -168°41'}$

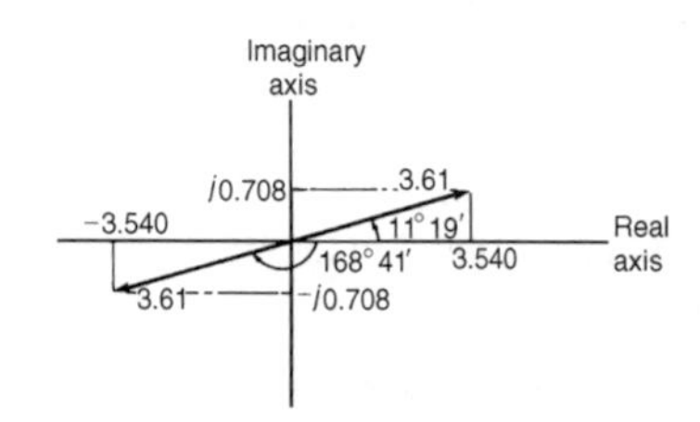

Figure 23.5

These two solutions of $\sqrt{(12 + j5)}$ are shown in the Argand diagram of Figure 23.5. $3.61\angle 11°19'$ is in the first quadrant of the Argand diagram.

Thus $3.61\angle 11°19' = 3.61(\cos 11°19' + j \sin 11°19') = 3.540 + j0.708$

$3.61\angle -168°41'$ is in the third quadrant of the Argand diagram.

$$\text{Thus } 3.61\angle -168°41' = 3.61[\cos(-168°41') + j \sin(-168°41')]$$

$$= -3.540 - j0.708$$

Thus in cartesian form the two roots are $\pm\mathbf{(3.540 + j0.708)}$

From the Argand diagram the roots are seen to be 180° apart, i.e. they lie on a straight line. This is always true when finding square roots of complex numbers.

Further problems on powers and roots of complex numbers may be found in Section 23.7 following, problems 32 to 39, page 428.

23.7 Further problems on complex numbers

Operations on Cartesian complex numbers

In problems 1 to 5, evaluate in $a + jb$ form assuming that $Z_1 = 2 + j3$, $Z_2 = 3 - j4$, $Z_3 = -1 + j2$ and $Z_4 = -2 - j5$

1 (a) $Z_1 - Z_2$ (b) $Z_2 + Z_3 - Z_4$ [(a) $-1 + j7$ (b) $4 + j3$]

2 (a) $Z_1 Z_2$ (b) $Z_3 Z_4$ [(a) $18 + j$ (b) $12 + j$]

3 (a) $Z_1Z_3Z_4$ (b) $Z_2Z_3 + Z_4$ [(a) $21 + j38$ (b) $3 + j5$]

4 (a) $\frac{Z_1}{Z_2}$ (b) $\frac{Z_1 + Z_2}{Z_3 + Z_4}$ $\left[\text{(a)} -\frac{6}{25} + j\frac{17}{25} \text{ (b)} -\frac{2}{3} + j\right]$

5 (a) $\frac{Z_1Z_2}{Z_1 + Z_2}$ (b) $Z_1 + \frac{Z_2}{Z_3} + Z_4$ $\left[\text{(a)} \frac{89}{26} + j\frac{23}{26} \text{ (b)} -\frac{11}{5} - j\frac{12}{5}\right]$

6 Evaluate $\left[\frac{(1+j)^2 - (1-j)^2}{j}\right]$ [4]

7 If $Z_1 = 4 - j3$ and $Z_2 = 2 + j$ evaluate x and y given

$$x + jy = \frac{1}{Z_1 - Z_2} + \frac{1}{Z_1Z_2}$$ [$x = 0.188$, $y = 0.216$]

8 Evaluate (a) $(1 + j)^4$ (b) $\frac{2 - j}{2 + j}$ (c) $\frac{1}{2 + j3}$

$\left[\text{(a)} -4 \text{ (b)} \frac{3}{5} - j\frac{4}{5} \text{ (c)} \frac{2}{13} - j\frac{3}{13}\right]$

9 If $Z = \frac{1 + j3}{1 - j2}$ evaluate Z^2 in $a + jb$ form. [$0 - j2$]

10 In an electrical circuit the equivalent impedance Z is given by

$$Z = Z_1 + \frac{Z_2Z_3}{Z_2 + Z_3}$$

Determine Z is rectangular form, correct to two decimal places, when $Z_1 = 5.91 + j3.15$, $Z_2 = 5 + j12$ and $Z_3 = 8 - j15$

[$Z = 21.62 + j8.39$]

11 Given $Z_1 = 5 - j9$ and $Z_2 = 7 + j2$, determine in $(a + jb)$ form, correct to four decimal places

(a) $\frac{1}{Z_1}$ (b) $\frac{1}{Z_2}$ (c) $\frac{1}{Z_1} + \frac{1}{Z_2}$ (d) $\frac{1}{(1/Z_1) + (1/Z_2)}$

[(a) $0.0472 + j0.0849$ (b) $0.1321 - j0.0377$
(c) $0.1793 + j0.0472$ (d) $5.2158 - j1.3731$]

Complex equations

In problems 12 to 15 solve the given complex equations:

12 $4(a + jb) = 7 - j3$ $\left[a = \frac{7}{4}, b = -\frac{3}{4}\right]$

13 $(3 + j4)(2 - j3) = x + jy$ [$x = 18$, $y = -1$]

14 $(a - j3b) + (b - j2a) = 4 + j6$ $[a = 18, b = -14]$

15 $5 + j2 = \sqrt{(e + jf)}$ $[e = 21, f = 20]$

16 An equation derived from an a.c. bridge circuit is given by

$$(R_3)\left[\frac{-j}{\omega C_1}\right] = \left[R_x - \frac{j}{\omega C_x}\right]\left[\frac{R_4(-j/(\omega C_4))}{R_4 - (j/(\omega C_4))}\right]$$

Components R_3, R_4, C_1 and C_4 have known values. Determine expressions for R_x and C_x in terms of the known components.

$$\left[R_x = \frac{R_3 C_4}{C_1}, C_x = \frac{C_1 R_4}{R_3}\right]$$

Polar form of complex numbers

In problems 17 and 18 determine the modulus and the argument of each of the complex numbers given.

17 (a) $3 + j4$ (b) $2 - j5$ [(a) 5, $53°8'$ (b) 5.385, $-68°12'$]

18 (a) $-4 + j$ (b) $-5 - j3$ [(a) 4.123, $165°58'$ (b) 5.831, $-149°2'$]

In problems 19 and 20 express the given cartesian complex numbers in polar form, leaving answers in surd form.

19 (a) $6 + j5$ (b) $3 - j2$ (c) -3

[(a) $\sqrt{61}\angle 39°48'$ (b) $\sqrt{13}\angle -33°41'$
(c) $3\angle 180°$ or $3\angle \pi$]

20 (a) $-5 + j$ (b) $-4 - j3$ (c) $-j2$

[(a) $\sqrt{26}\angle 168°41'$ (b) $5\angle -143°8'$
(c) $2\angle -90°$ or $2\angle -\pi/2$]

In problems 21 to 23 convert the given polar complex numbers into $(a + jb)$ form, giving answers correct to four significant figures.

21 (a) $6\angle 30°$ (b) $4\angle 60°$ (c) $3\angle 45°$

[(a) $5.196 + j3.000$ (b) $2.000 + j3.464$ (c) $2.121 + j2.121$]

22 (a) $2\angle \pi/2$ (b) $3\angle \pi$ (c) $5\angle (5\pi/6)$

[(a) $0 + j2.000$ (b) $-3.000 + j0$ (c) $-4.330 + j2.500$]

23 (a) $8\angle 150°$ (b) $4.2\angle -120°$ (c) $3.6\angle -25°$

[(a) $-6.928 + j4.000$ (b) $-2.100 - j3.637$
(c) $3.263 - j1.521$]

In problems 24 to 26, evaluate in polar form.

24 (a) $2\angle 40° \times 5\angle 20°$ (b) $2.6\angle 72° \times 4.3\angle 45°$

[(a) $10\angle 60°$ (b) $11.18\angle 117°$]

25 (a) $5.8\angle 35^\circ \div 2\angle -10^\circ$ (b) $4\angle 30^\circ \times 3\angle 70^\circ \div 2\angle -15^\circ$

[(a) $2.9\angle 45^\circ$ (b) $6\angle 115^\circ$]

26 (a) $\dfrac{4.1\angle 20^\circ \times 3.2\angle -62^\circ}{1.2\angle 150^\circ}$ (b) $6\angle 25^\circ + 3\angle -36^\circ - 4\angle 72^\circ$

[(a) $10.93\angle 168^\circ$ (b) $7.289\angle -24^\circ 35'$]

27 Solve the complex equations, giving answers correct to four significant figures.

(a) $\dfrac{12\angle(\pi/2) \times 3\angle(3\pi/4)}{2\angle -(\pi/3)} = x + jy$

(b) $15\angle \pi/3 + 12\angle \pi/2 - 6\angle -\pi/3 = r\angle\theta$

[(a) $x = 4.659$, $y = -17.392$ (b) $r = 30.52$, $\theta = 81^\circ 31'$]

28 The total impedance Z_T of an electrical circuit is given by

$$Z_T = \frac{Z_1 \times Z_2}{Z_1 + Z_2} + Z_3$$

Determine Z_T in polar form correct to three significant figures when $Z_1 = 3.2\angle -56^\circ$, $Z_2 = 7.4\angle 25^\circ$ and $Z_3 = 6.3\angle 62^\circ$ [$6.61\angle 37.24^\circ$]

29 A star-connected impedance Z_1 is given by

$$Z_1 = \frac{Z_A Z_B}{Z_A + Z_B + Z_C}$$

Evaluate Z_1, in both cartesian and polar form, given $Z_A = (20 + j0)\Omega$, $Z_B = (0 - j20)\Omega$ and $Z_C = (10 + j10)\Omega$

[$(4 - j12)\Omega$ or $12.65\angle -71.57^\circ\ \Omega$]

30 The current I flowing in an impedance is given by

$$I = \frac{(8\angle 60^\circ)(10\angle 0^\circ)}{(8\angle 60^\circ + 5\angle 30^\circ)}\ \text{A}$$

Determine the value of current in polar form, correct to two decimal places. [$6.36\angle 11.46^\circ$ A]

31 A delta-connected impedance Z_A is given by

$$Z_A = \frac{Z_1 Z_2 + Z_2 Z_3 + Z_3 Z_1}{Z_2}$$

Determine Z_A, in both cartesian and polar form, given $Z_1 = (10 + j0)\Omega$, $Z_2 = (0 - j10)\Omega$ and $Z_3 = (10 + j10)\Omega$

[$(10 + j20)\Omega$, $22.36\angle 63.43^\circ\ \Omega$]

Powers and roots of complex numbers

In problems 32 to 35, evaluate in cartesian and in polar form.

32 (a) $(2 + j3)^2$ (b) $(4 - j5)^2$

[(a) $-5 + j12$; $13\angle 112°37'$ (b) $-9 - j40$; $41\angle -102°41'$]

33 (a) $(-3 + j2)^5$ (b) $(-2 - j)^3$

[(a) $597 + j122$; $609.3\angle 11°33'$
(b) $-2 - j11$; $11.18\angle -100°17'$]

34 (a) $(4\angle 32°)^4$ (b) $(2\angle 125°)^5$

[(a) $-157.6 + j201.7$; $256\angle 128°$
(b) $-2.789 - j31.88$; $32\angle -95°$]

35 (a) $(3\angle -\pi/3)^3$ (b) $(1.5\angle -160°)^4$

[(a) $-27 + j0$; $27\angle -\pi$ (b) $0.8792 + j4.986$; $5.063\angle 80°$]

In problems 36 to 38, determine the two square roots of the given complex numbers in cartesian form and show the results on an Argand diagram.

36 (a) $2 + j$ (b) $3 - j2$

[(a) $\pm(1.455 + j0.344)$ (b) $\pm(1.818 - j0.550)$]

37 (a) $-3 + j4$ (b) $-1 - j3$ [(a) $\pm(1 + j2)$ (b) $\pm(1.040 - j1.442)$]

38 (a) $5\angle 36°$ (b) $14\angle 3\pi/2$

[(a) $\pm(2.127 + j0.691)$ (b) $\pm(-2.646 + j2.646)$]

39 Convert $2 - j$ into polar form and hence evaluate $(2 - j)^7$ in polar form. [$\sqrt{5}\angle -26°34'$; $279.5\angle 174°3'$]

24 Application of complex numbers to series a.c. circuits

At the end of this chapter you should be able to:

- appreciate the use of complex numbers in a.c. circuits
- perform calculations on series a.c. circuits using complex numbers

24.1 Introduction

Simple a.c. circuits may be analysed by using phasor diagrams. However, when circuits become more complicated analysis is considerably simplified by using complex numbers. It is essential that the basic operations used with complex numbers, as outlined in Chapter 23, are thoroughly understood before proceeding with a.c. circuit analysis. The theory introduced in Chapter 15 is relevant; in this chapter similar circuits will be analysed using j notation and Argand diagrams.

24.2 Series a.c. circuits

(a) Pure resistance

In an a.c. circuit containing resistance R only (see Figure 24.1(a)), the current I_R is **in phase** with the applied voltage V_R as shown in the phasor diagram of Figure 24.1(b). The phasor diagram may be superimposed on the Argand diagram as shown in Figure 24.1(c). The impedance **Z** of the circuit is given by

$$Z = \frac{V_R\angle 0^\circ}{I_R\angle 0^\circ} = \boldsymbol{R}$$

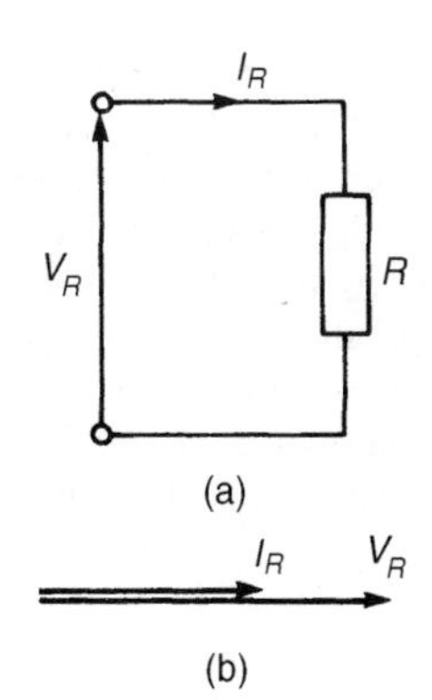

Figure 24.1 *(a) Circuit diagram (b) Phasor diagram (c) Argand diagram*

(b) Pure inductance

In an a.c. circuit containing pure inductance L only (see Figure 24.2(a)), the current I_L **lags** the applied voltage V_L by 90° as shown in the phasor diagram of Figure 24.2(b). The phasor diagram may be superimposed on the Argand diagram as shown in Figure 24.2(c). The impedance Z of the circuit is given by

$$Z = \frac{V_L\angle 90^\circ}{I_L\angle 0^\circ} = \frac{V_L}{I_L}\angle 90^\circ = \boldsymbol{X_L\angle 90^\circ} \text{ or } \boldsymbol{jX_L}$$

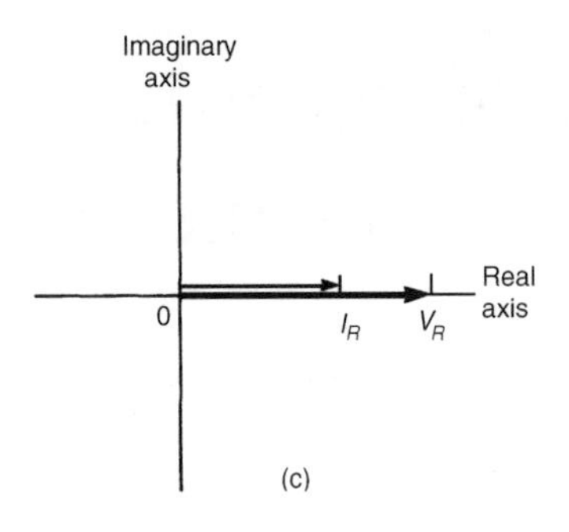

Figure 24.1 *Continued*

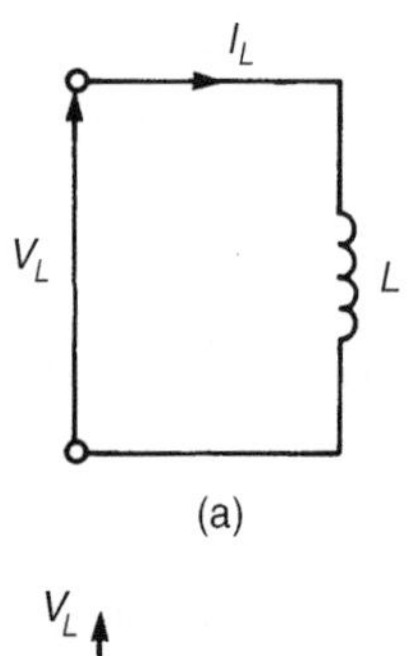

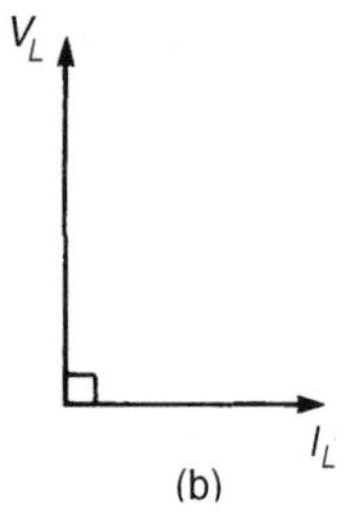

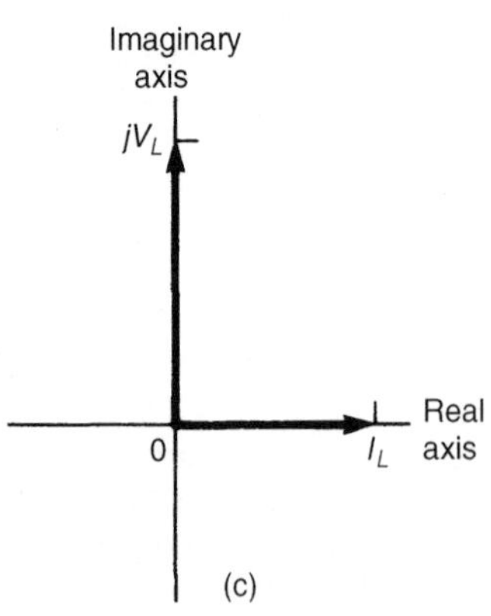

Figure 24.2 *(a) Circuit diagram (b) Phasor diagram (c) Argand diagram*

where X_L is the **inductive reactance** given by

$$\boldsymbol{X_L = \omega L = 2\pi f L \text{ ohms},}$$

where f is the frequency in hertz and L is the inductance in henrys.

(c) Pure capacitance

In an a.c. circuit containing pure capacitance only (see Figure 24.3(a)), the current I_C **leads** the applied voltage V_C by 90° as shown in the phasor diagram of Figure 24.3(b). The phasor diagram may be superimposed on the Argand diagram as shown in Figure 24.3(c). The impedance Z of the circuit is given by

$$Z = \frac{V_C\angle{-90°}}{I_C\angle 0°} = \frac{V_C}{I_C}\angle{-90°} = \boldsymbol{X_C\angle{-90°}} \text{ or } \boldsymbol{-jX_C}$$

where X_C is the **capacitive reactance** given by

$$\boldsymbol{X_C = \frac{1}{\omega C} = \frac{1}{2\pi f C} \text{ ohms}}$$

where C is the capacitance in farads.

$$\left[\text{Note: } -jX_C = \frac{-j}{\omega C} = \frac{-j(j)}{\omega C(j)} = \frac{-j^2}{j\omega C} = \frac{-(-1)}{j\omega C} = \frac{1}{j\omega C}\right]$$

(d) *R–L* series circuit

In an a.c. circuit containing resistance R and inductance L in series (see Figure 24.4(a)), the applied voltage V is the phasor sum of V_R and V_L as shown in the phasor diagram of Figure 24.4(b). The current I lags the applied voltage V by an angle lying between 0° and 90° — the actual value depending on the values of V_R and V_L, which depend on the values of R and L. The circuit phase angle, i.e., the angle between the current and the applied voltage, is shown as angle ϕ in the phasor diagram. In any series circuit the current is common to all components and is thus taken as the reference phasor in Figure 24.4(b). The phasor diagram may be superimposed on the Argand diagram as shown in Figure 24.4(c), where it may be seen that in complex form the supply voltage V is given by:

$$\boldsymbol{V = V_R + jV_L}$$

Figure 24.5(a) shows the voltage triangle that is derived from the phasor diagram of Figure 24.4(b) (i.e. triangle Oab). If each side of the voltage triangle is divided by current I then the impedance triangle of

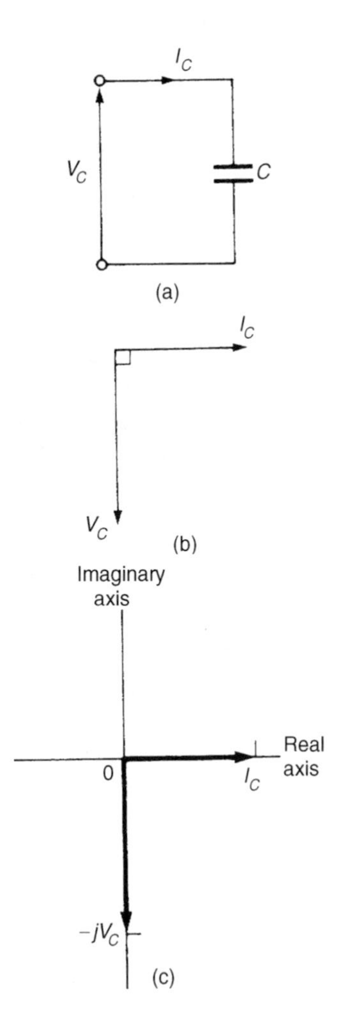

Figure 24.3 *(a) Circuit diagram (b) Phasor diagram (c) Argand diagram*

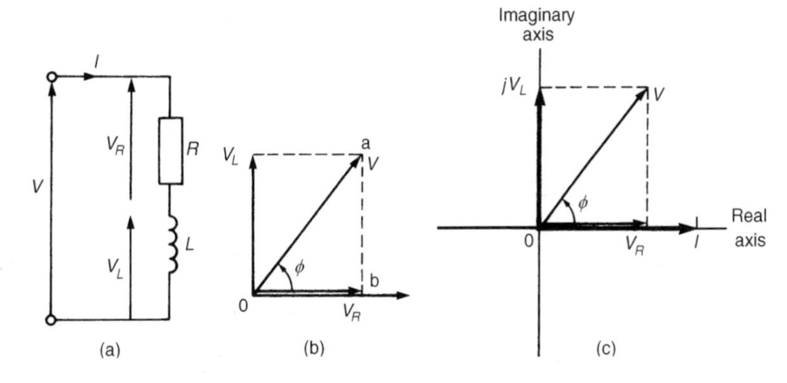

Figure 24.4 *(a) Circuit diagram (b) Phasor diagram (c) Argand diagram*

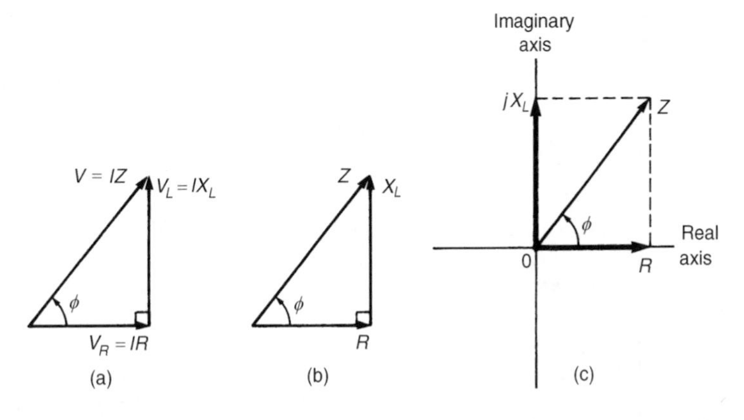

Figure 24.5 *(a) Voltage triangle (b) Impedance triangle (c) Argand diagram*

Figure 24.5(b) is derived. The impedance triangle may be superimposed on the Argand diagram, as shown in Figure 24.5(c), where it may be seen that in complex form the impedance Z is given by:

$$\boldsymbol{Z = R + jX_L}$$

Thus, for example, an impedance expressed as $(3 + j4)\Omega$ means that the resistance is 3 Ω and the inductive reactance is 4 Ω

In polar form, $Z = |Z|\angle\phi$ where, from the impedance triangle, the modulus of impedance $|Z| = \sqrt{(R^2 + X_L^2)}$ and the circuit phase angle $\phi = \arctan(X_L/R)$ lagging

(e) *R–C* series circuit

In an a.c. circuit containing resistance R and capacitance C in series (see Figure 24.6(a)), the applied voltage V is the phasor sum of V_R and

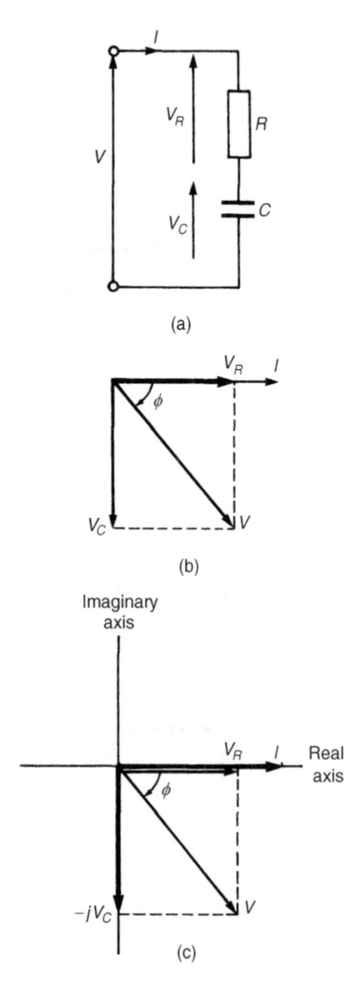

Figure 24.6 *(a) Circuit diagram (b) Phasor diagram (c) Argand diagram*

V_C as shown in the phasor diagram of Figure 24.6(b). The current I leads the applied voltage V by an angle lying between 0° and 90° — the actual value depending on the values of V_R and V_C, which depend on the values of R and C. The circuit phase angle is shown as angle ϕ in the phasor diagram. The phasor diagram may be superimposed on the Argand diagram as shown in Figure 24.6(c), where it may be seen that in complex form the supply voltage V is given by:

$$\boldsymbol{V = V_R - jV_C}$$

Figure 24.7(a) shows the voltage triangle that is derived from the phasor diagram of Figure 24.6(b). If each side of the voltage triangle is divided by current I, the impedance triangle is derived as shown in Figure 24.7(b). The impedance triangle may be superimposed on the Argand diagram as shown in Figure 24.7(c), where it may be seen that in complex form the impedance Z is given by

$$\boldsymbol{Z = R - jX_C}$$

Thus, for example, an impedance expressed as $(9 - j14)\Omega$ means that the resistance is 9 Ω and the capacitive reactance $\boldsymbol{X_C}$ is 14 Ω

In polar form, $Z = |Z|\angle\phi$ where, from the impedance triangle, $|Z| = \sqrt{(R^2 + X_C^2)}$ and $\phi = \arctan(X_C/R)$ leading

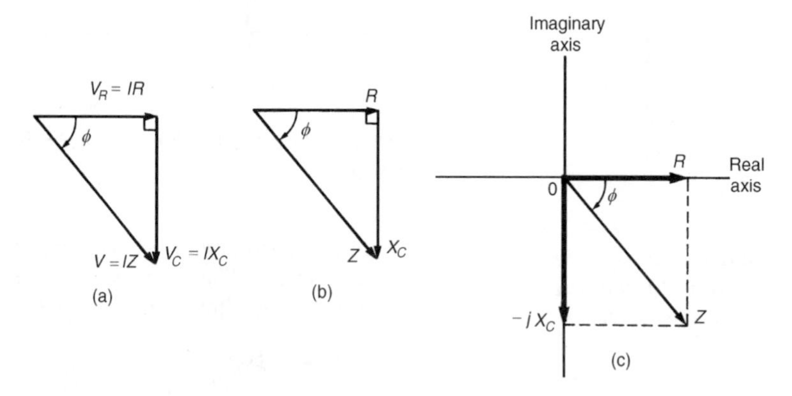

Figure 24.7 *(a) Voltage triangle (b) Impedance triangle (c) Argand diagram*

(f) *R–L–C* series circuit

In an a.c. circuit containing resistance R, inductance L and capacitance C in series (see Figure 24.8(a)), the applied voltage V is the phasor sum of V_R, V_L and V_C as shown in the phasor diagram of Figure 24.8(b) (where the condition $V_L > V_C$ is shown). The phasor diagram may be superimposed on the Argand diagram as shown in Figure 24.8(c), where

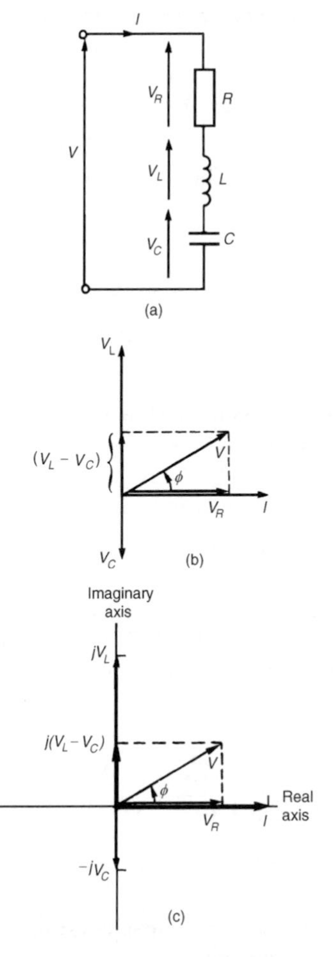

Figure 24.8 *(a) Circuit diagram (b) Phasor diagram (c) Argand diagram*

it may be seen that in complex form the supply voltage V is given by:

$$\mathbf{V = V_R + j(V_L - V_C)}$$

From the voltage triangle the impedance triangle is derived and superimposing this on the Argand diagram gives, in complex form,

$$\text{impedance } \mathbf{Z = R + j(X_L - X_C)} \textbf{ or } \mathbf{Z = |Z|\angle\phi}$$

where, $|Z| = \sqrt{[R^2 + (X_L - X_C)^2]}$ and $\phi = \arctan(X_L - X_C)/R$

When $V_L = V_C$, $X_L = X_C$ and the applied voltage V and the current I are in phase. This effect is called **series resonance** and is discussed separately in Chapter 28.

(g) General series circuit

In an a.c. circuit containing several impedances connected in series, say, $Z_1, Z_2, Z_3, \ldots, Z_n$, then the total equivalent impedance Z_T is given by

$$\mathbf{Z_T = Z_1 + Z_2 + Z_3 + \ldots + Z_n}$$

Problem 1. Determine the values of the resistance and the series-connected inductance or capacitance for each of the following impedances: (a) $(12 + j5)\Omega$ (b) $-j40\ \Omega$ (c) $30\angle 60°\ \Omega$ (d) $2.20 \times 10^6\angle{-30°}\ \Omega$. Assume for each a frequency of 50 Hz.

(a) From Section 24.2(d), for an R–L series circuit, impedance $Z = R + jX_L$.

Thus $Z = (12 + j5)\Omega$ represents a resistance of 12 Ω and an inductive reactance of 5 Ω in series.

Since inductive reactance $X_L = 2\pi fL$,

$$\text{inductance } L = \frac{X_L}{2\pi f} = \frac{5}{2\pi(50)} = 0.0159 \text{ H}$$

i.e., the inductance is 15.9 mH.

Thus an impedance $(12 + j5)\Omega$ represents a resistance of 12 Ω in series with an inductance of 15.9 mH.

(b) From Section 24.2(c), for a purely capacitive circuit, impedance $Z = -jX_c$.

Thus $Z = -j40\ \Omega$ represents zero resistance and a capacitive reactance of 40 Ω.

Since capacitive reactance $X_C = 1/(2\pi fC)$,

$$\text{capacitance } C = \frac{1}{2\pi fX_C} = \frac{1}{2\pi(50)(40)} \text{ F} = \frac{10^6}{2\pi(50)(40)}\ \mu\text{F}$$
$$= 79.6\ \mu\text{F}$$

Thus an impedance $-j40\ \Omega$ represents a pure capacitor of capacitance 79.6 μF

(c) $30\angle 60° = 30(\cos 60° + j\sin 60°) = 15 + j25.98$

Thus $Z = 30\angle 60°\ \Omega = (15 + j25.98)\Omega$ represents a resistance of 15 Ω and an inductive reactance of 25.98 Ω in series (from Section 24.2(d)).

Since $X_L = 2\pi f L$,

$$\text{inductance } L = \frac{X_L}{2\pi f} = \frac{25.98}{2\pi(50)} = 0.0827 \text{ H or } 82.7 \text{ mH}$$

Thus an impedance $30\angle 60°\ \Omega$ represents a resistance of 15 Ω in series with an inductance of 82.7 mH

(d) $$2.20 \times 10^6\angle -30° = 2.20 \times 10^6[\cos(-30°) + j\sin(-30°)]$$
$$= 1.905 \times 10^6 - j1.10 \times 10^6$$

Thus $Z = 2.20 \times 10^6\angle -30°\ \Omega = (1.905 \times 10^6 - j1.10 \times 10^6)\Omega$ represents a resistance of $1.905 \times 10^6\ \Omega$ (i.e. 1.905 MΩ) and a capacitive reactance of $1.10 \times 10^6\ \Omega$ in series (from Section 24.2(e)).

Since capacitive reactance $X_C = 1/(2\pi f C)$,

$$\text{capacitance } C = \frac{1}{2\pi f X_C} = \frac{1}{2\pi(50)(1.10 \times 10^6)} \text{ F}$$
$$= 2.894 \times 10^{-9} \text{ F or } 2.894 \text{ nF}$$

Thus an impedance $2.2 \times 10^6\angle -30°\ \Omega$ represents a resistance of 1.905 MΩ in series with a 2.894 nF capacitor.

Problem 2. Determine, in polar and rectangular forms, the current flowing in an inductor of negligible resistance and inductance 159.2 mH when it is connected to a 250 V, 50 Hz supply.

Inductive reactance $X_L = 2\pi f L = 2\pi(50)(159.2 \times 10^{-3}) = 50\ \Omega$

Thus circuit impedance $Z = (0 + j50)\Omega = 50\angle 90°\ \Omega$

Supply voltage, $V = 250\angle 0°$ V (or $(250 + j0)$V)

(Note that since the voltage is given as 250 V, this is assumed to mean $250\angle 0°$ V or $(250 + j0)$V)

$$\text{Hence current } I = \frac{V}{Z} = \frac{250\angle 0°}{50\angle 90°} = \frac{250}{50}\angle(0° - 90°) = \mathbf{5\angle -90°\ A}$$

$$\text{Alternatively,} \quad I = \frac{V}{Z} = \frac{(250 + j0)}{(0 + j50)} = \frac{250(-j50)}{j50(-j50)}$$

$$= \frac{-j(50)(250)}{50^2} = \mathbf{-j5\ A}$$

which is the same as $5\angle{-90^\circ}$ A

> Problem 3. A 3 μF capacitor is connected to a supply of frequency 1 kHz and a current of $2.83\angle 90^\circ$ A flows. Determine the value of the supply p.d.

$$\text{Capacitive reactance } X_C = \frac{1}{2\pi f C} = \frac{1}{2\pi(1000)(3 \times 10^{-6})} = 53.05\ \Omega$$

Hence circuit impedance $Z = (0 - j53.05)\Omega = 53.05\angle{-90^\circ}\ \Omega$

$$\text{Current } I = 2.83\angle 90^\circ \text{ A (or } (0 + j2.83)\text{A)}$$

$$\text{Supply p.d., } V = IZ = (2.83\angle 90^\circ)(53.05\angle{-90^\circ})$$

$$\text{i.e.} \quad \mathbf{p.d. = 150\angle 0^\circ\ V}$$

$$\text{Alternatively, } V = IZ = (0 + j2.83)(0 - j53.05)$$

$$= -j^2(2.83)(53.05) = \mathbf{150\ V}$$

> Problem 4. The impedance of an electrical circuit is $(30 - j50)$ ohms. Determine (a) the resistance, (b) the capacitance, (c) the modulus of the impedance, and (d) the current flowing and its phase angle, when the circuit is connected to a 240 V, 50 Hz supply.

(a) Since impedance $Z = (30 - j50)\Omega$, **the resistance is 30 ohms** and the capacitive reactance is 50 Ω

(b) Since $X_C = 1/(2\pi f C)$, **capacitance**,

$$\boldsymbol{C} = \frac{1}{2\pi f X_C} = \frac{1}{2\pi(50)(50)} = \mathbf{63.66\ \mu F}$$

(c) The modulus of impedance, $|Z| = \sqrt{(R^2 + X_C^2)} = \sqrt{(30^2 + 50^2)}$

$$= \mathbf{58.31\ \Omega}$$

(d) Impedance $Z = (30 - j50)\Omega = 58.31\angle \arctan \dfrac{X_C}{R}$

$$= 58.31\angle{-59.04^\circ}\ \Omega$$

$$\text{Hence current } I = \frac{V}{Z} = \frac{240\angle 0^\circ}{58.31\angle{-59.04^\circ}} = \mathbf{4.12\angle 59.04^\circ\ A}$$

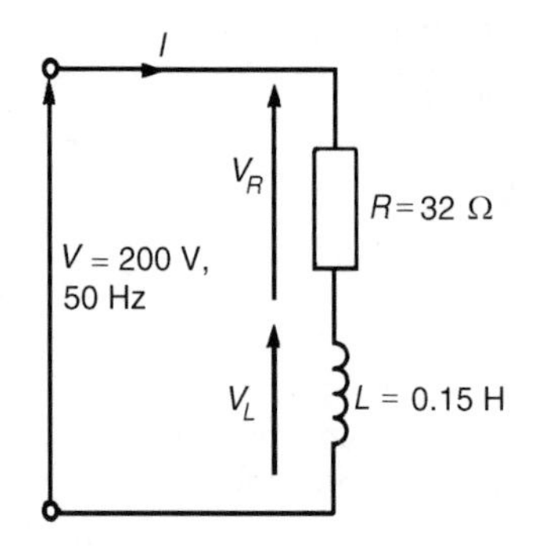

Figure 24.9

Problem 5. A 200 V, 50 Hz supply is connected across a coil of negligible resistance and inductance 0.15 H connected in series with a 32 Ω resistor. Determine (a) the impedance of the circuit, (b) the current and circuit phase angle, (c) the p.d. across the 32 Ω resistor, and (d) the p.d. across the coil.

(a) Inductive reactance $X_L = 2\pi f L = 2\pi(50)(0.15) = 47.1\ \Omega$

Impedance $Z = R + jX_L =$ $\mathbf{(32 + j47.1)\Omega}$ or $\mathbf{57.0\angle 55.81^\circ\ \Omega}$

The circuit diagram is shown in Figure 24.9

(b) Current $I = \dfrac{V}{Z} = \dfrac{200\angle 0^\circ}{57.0\angle 55.81^\circ} = \mathbf{3.51\angle -55.81^\circ\ A}$

i.e., **the current is 3.51 A lagging the voltage by 55.81°**

(c) P.d. across the 32 resistor, $V_R = IR = (3.51\angle -55.81^\circ)(32\angle 0^\circ)$

i.e., $\mathbf{V_R = 112.3\angle -55.81^\circ\ V}$

(d) P.d. across the coil, $V_L = IX_L = (3.51\angle -55.81^\circ)(47.1\angle 90^\circ)$

i.e. $\mathbf{V_L = 165.3\angle 34.19^\circ\ V}$

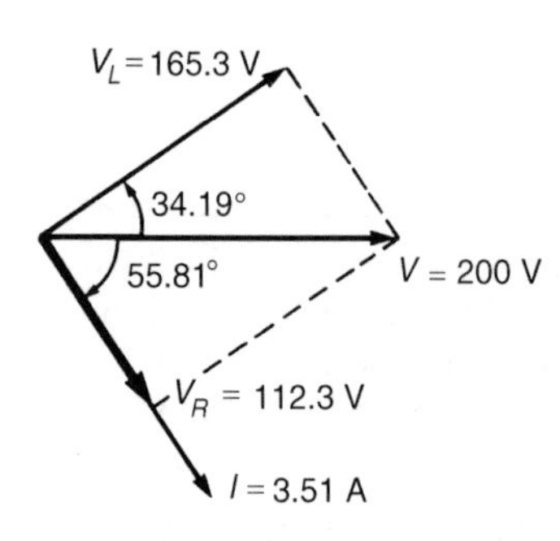

Figure 24.10

The phasor sum of V_R and V_L is the supply voltage V as shown in the phasor diagram of Figure 24.10.

$$V_R = 112.3\angle -55.81^\circ = (63.11 - j92.89)\ V$$

$$V_L = 165.3\angle 34.19^\circ\ V = (136.73 + j92.89)\ V$$

Hence $V = V_R + V_L = (63.11 - j92.89) + (136.73 + j92.89)$

$= (200 + j0)$ V or $200\angle 0^\circ$ V, correct to three significant figures.

Problem 6. Determine the value of impedance if a current of $(7 + j16)$A flows in a circuit when the supply voltage is $(120 + j200)$V. If the frequency of the supply is 5 MHz, determine the value of the components forming the series circuit.

Impedance $Z = \dfrac{V}{I} = \dfrac{(120 + j200)}{(7 + j16)} = \dfrac{233.24\angle 59.04^\circ}{17.464\angle 66.37^\circ}$

$= 13.36\angle -7.33\ \Omega$ or $(13.25 - j1.705)\Omega$

The series circuit thus consists of a **13.25 Ω resistor** and a capacitor of capacitive reactance **1.705 Ω**

Since $X_C = \dfrac{1}{2\pi f C}$, capacitance $C = \dfrac{1}{2\pi f X_C} = \dfrac{1}{2\pi(5 \times 10^6)(1.705)}$

$$= 1.867 \times 10^{-8}\ \text{F} = \mathbf{18.67\ nF}$$

Problem 7. For the circuit shown in Figure 24.11, determine the value of impedance Z_2.

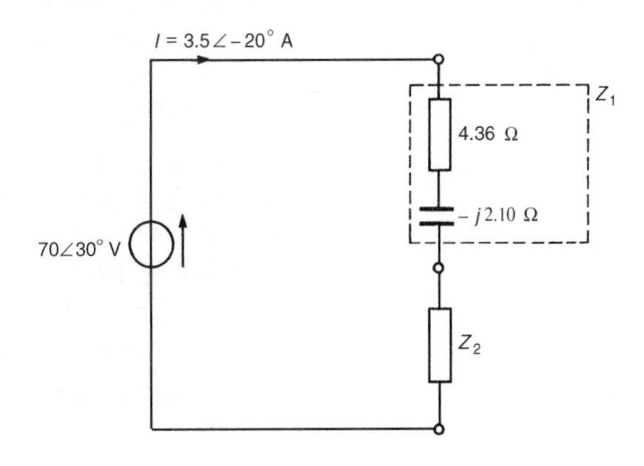

Figure 24.11

Total circuit impedance $Z = \dfrac{V}{I} = \dfrac{70\angle 30^\circ}{3.5\angle -20^\circ}$

$$= 20\angle 50^\circ\ \Omega \text{ or } (12.86 + j15.32)\Omega$$

Total impedance $Z = Z_1 + Z_2$ (see Section 24.2(g)).

Hence $\quad (12.86 + j15.32) = (4.36 - j2.10) + Z_2$

from which, impedance $Z_2 = (12.86 + j15.32) - (4.36 - j2.10)$

$$= \mathbf{(8.50 + j17.42)\Omega \text{ or } 19.38\angle 63.99^\circ\ \Omega}$$

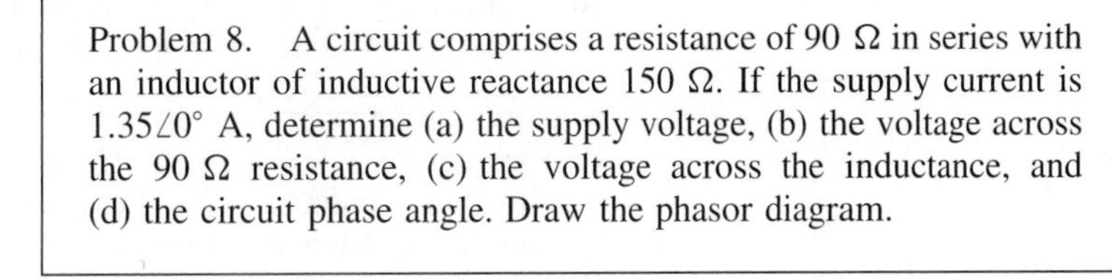

Problem 8. A circuit comprises a resistance of 90 Ω in series with an inductor of inductive reactance 150 Ω. If the supply current is $1.35\angle 0^\circ$ A, determine (a) the supply voltage, (b) the voltage across the 90 Ω resistance, (c) the voltage across the inductance, and (d) the circuit phase angle. Draw the phasor diagram.

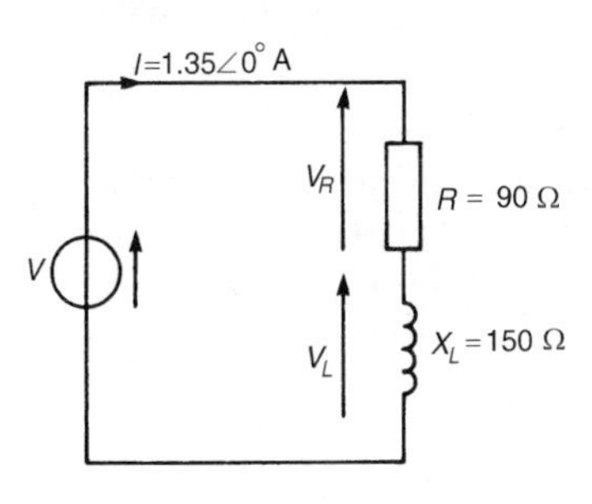

Figure 24.12

The circuit diagram is shown in Figure 24.12

(a) Circuit impedance $Z = R + jX_L = (90 + j150)\Omega$ or $174.93\angle 59.04^\circ\ \Omega$

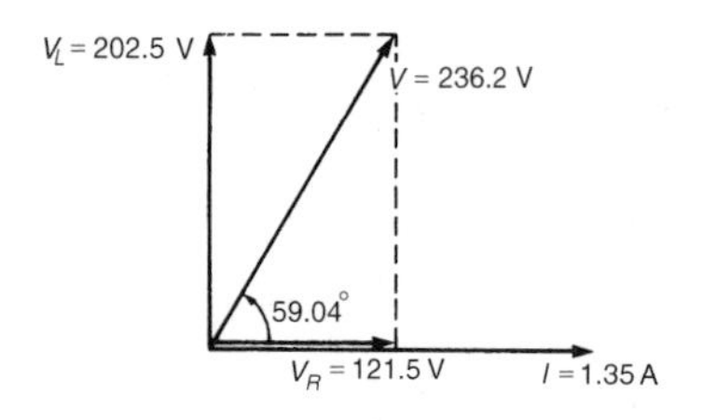

Figure 24.13

Supply voltage, $V = IZ = (1.35\angle 0°)(174.93\angle 59.04°)$

$$= \mathbf{236.2\angle 59.04°\ V\ or\ (121.5 + j202.5)V}$$

(b) Voltage across 90 Ω resistor, V_R = **121.5 V** (since $V = V_R + jV_L$)

(c) Voltage across inductance, V_L = **202.5 V** leading V_R by 90°.

(d) Circuit phase angle is the angle between the supply current and voltage, i.e., **59.04°** lagging(i.e., current lags voltage). The phasor diagram is shown in Figure 24.13.

> Problem 9. A coil of resistance 25 Ω and inductance 20 mH has an alternating voltage given by $v = 282.8 \sin(628.4t + (\pi/3))$ volts applied across it. Determine (a) the rms value of voltage (in polar form), (b) the circuit impedance, (c) the rms current flowing, and (d) the circuit phase angle.

(a) Voltage $v = 282.8 \sin(628.4t + (\pi/3))$ volts means $V_m = 282.8$ V, hence rms voltage

$$V = 0.707 \times 282.8 \left[\text{or } \frac{1}{\sqrt{2}} \times 282.8\right],$$

i.e., $V = 200$ V

In complex form the rms voltage may be expressed as **200∠π/3 V or 200∠60° V**

(b) $\omega = 2\pi f = 628.4$ rad/s, hence frequency

$f = 628.4/(2\pi) = 100$ Hz

Inductive reactance $X_L = 2\pi f L = 2\pi(100)(20 \times 10^{-3}) = 12.57\ \Omega$

Hence circuit impedance $Z = R + jX_L$ = **(25 + j12.57)Ω or 27.98∠26.69° Ω**

(c) Rms current, $I = \dfrac{V}{Z} = \dfrac{200\angle 60°}{27.98\angle 26.69°} = \mathbf{7.148\angle 33.31°\ A}$

(d) Circuit phase angle is the angle between current I and voltage V, i.e., 60° − 33.31° = **26.69° lagging**.

> Problem 10. A 240 V, 50 Hz voltage is applied across a series circuit comprising a coil of resistance 12 Ω and inductance 0.10 H, and 120 μF capacitor. Determine the current flowing in the circuit.

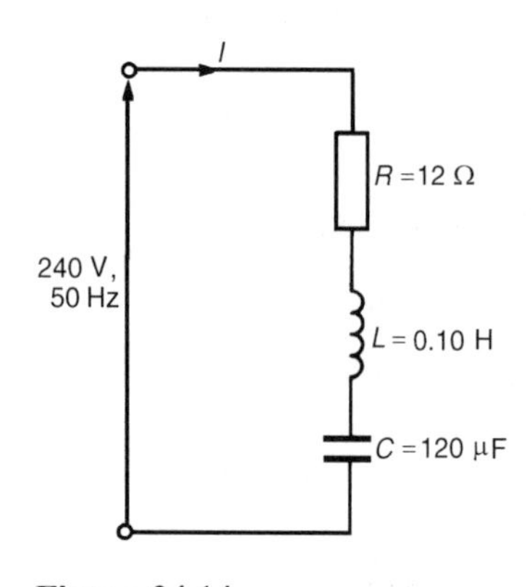

Figure 24.14

The circuit diagram is shown in Figure 24.14.

Inductive reactance, $X_L = 2\pi f L = 2\pi(50)(0.10) = 31.4\ \Omega$

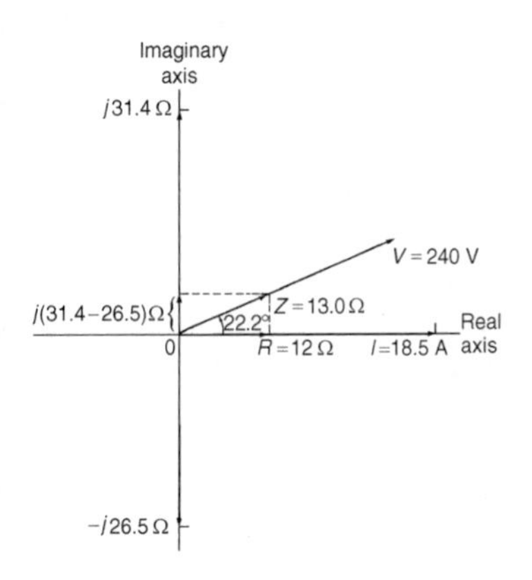

Figure 24.15

Capacitive reactance, $X_C = \dfrac{1}{2\pi fC} = \dfrac{1}{2\pi(50)(120 \times 10^{-6})} = 26.5\ \Omega$

Impedance $Z = R + j(X_L - X_C)$ (see Section 24.2(f))

i.e. $Z = 12 + j(31.4 - 26.5) = (12 + j4.9)\Omega$ or $13.0\angle 22.2^\circ\ \Omega$

Current flowing, $I = \dfrac{V}{Z} = \dfrac{240\angle 0^\circ}{13.0\angle 22.2^\circ} = \mathbf{18.5\angle{-22.2^\circ}\ A,}$

i.e., the current flowing is 18.5 A, lagging the voltage by 22.2°.

The phasor diagram is shown on the Argand diagram in Figure 24.15

Problem 11. A coil of resistance R ohms and inductance L henrys is connected in series with a 50 μF capacitor. If the supply voltage is 225 V at 50 Hz and the current flowing in the circuit is $1.5\angle{-30^\circ}$ A, determine the values of R and L. Determine also the voltage across the coil and the voltage across the capacitor.

Circuit impedance $Z = \dfrac{V}{Z} = \dfrac{225\angle 0^\circ}{1.5\angle{-30^\circ}}$

$$= 150\angle 30^\circ\ \Omega \text{ or } (129.9 + j75.0)\Omega$$

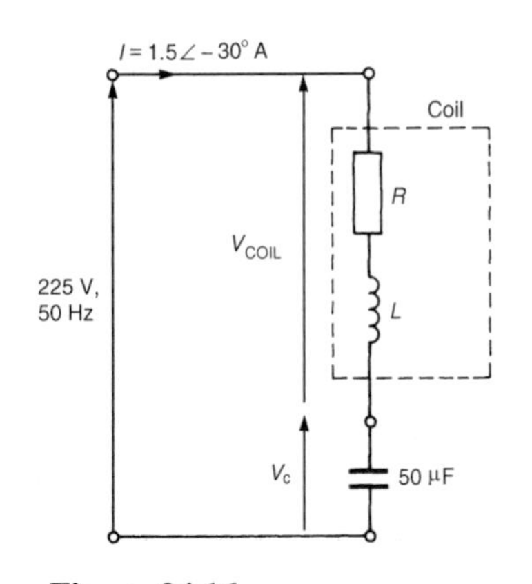

Figure 24.16

Capacitive reactance $X_C = \dfrac{1}{2\pi fC} = \dfrac{1}{2\pi(50)(50 \times 10^{-6})} = 63.66\ \Omega$

Circuit impedance $\quad Z = R + j(X_L - X_C)$

i.e. $\quad 129.9 + j75.0 = R + j(X_L - 63.66)$

Equating the real parts gives: **resistance $R = 129.9\ \Omega$.**

Equating the imaginary parts gives: $75.0 = X_L - 63.66$,

from which, $X_L = 75.0 + 63.66 = 138.66\ \Omega$

Since $X_L = 2\pi fL$, **inductance $L = \dfrac{X_L}{2\pi f} = \dfrac{138.66}{2\pi(50)} = 0.441$ H**

The circuit diagram is shown in Figure 24.16.

Voltage across coil, $V_{COIL} = IZ_{COIL}$

$Z_{COIL} = R + jX_L = (129.9 + j138.66)\Omega$ or $190\angle 46.87^\circ\ \Omega$

Hence $\mathbf{V_{COIL}} = (1.5\angle{-30^\circ})(190\angle 46.87^\circ)$

$$= \mathbf{285\angle 16.87^\circ\ V} \text{ or } \mathbf{(272.74 + j82.71)V}$$

Voltage across capacitor, $V_C = IX_C = (1.5\angle{-30^\circ})(63.66\angle{-90^\circ})$

$$= \mathbf{95.49\angle{-120^\circ}\ V} \text{ or}$$

$$\mathbf{(-47.75 - j82.70)V}$$

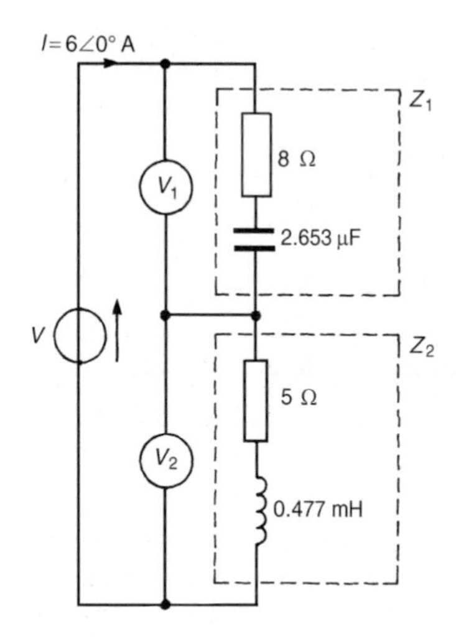

Figure 24.17

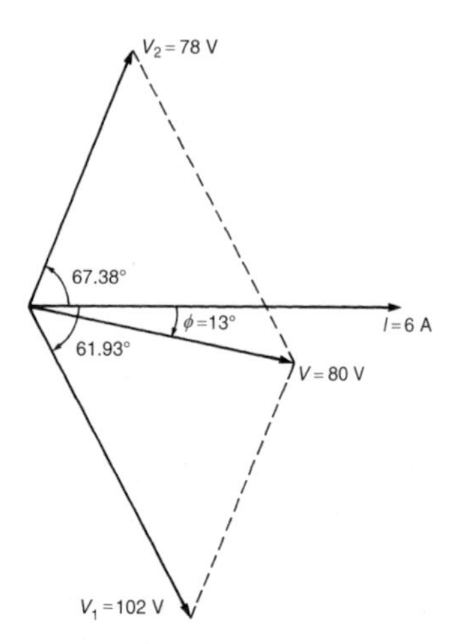

Figure 24.18

[Check: Supply voltage $V = V_{COIL} + V_C$

$$= (272.74 + j82.71) + (-47.75 - j82.70)$$

$$= (225 + j0)\text{V or } 225\angle 0^\circ \text{ V}]$$

Problem 12. For the circuit shown in Figure 24.17, determine the values of voltages V_1 and V_2 if the supply frequency is 4 kHz. Determine also the value of the supply voltage V and the circuit phase angle. Draw the phasor diagram.

For impedance Z_1, $X_C = \dfrac{1}{2\pi fC} = \dfrac{1}{2\pi(4000)(2.653 \times 10^{-6})} = 15\ \Omega$

Hence $Z_1 = (8 - j15)\Omega$ or $17\angle -61.93^\circ\ \Omega$

and **voltage** $\boldsymbol{V_1} = IZ_1 = (6\angle 0^\circ)(17\angle -61.93^\circ)$

$$= \mathbf{102\angle -61.93^\circ\ V\ or\ (48 - j90)V}$$

For impedance Z_2, $X_L = 2\pi fL = 2\pi(4000)(0.477 \times 10^{-3}) = 12\ \Omega$

Hence $Z_2 = (5 + j12)\Omega$ or $13\angle 67.38^\circ\ \Omega$

and **voltage** $\boldsymbol{V_2} = IZ_2 = (6\angle 0^\circ)(13\angle 67.38^\circ)$

$$= \mathbf{78\angle 67.38^\circ\ V} \text{ or } \mathbf{(30 + j72)V}$$

Supply voltage, $V = V_1 + V_2 = (48 - j90) + (30 + j72)$

$$= \mathbf{(78 - j18)V} \text{ or } \mathbf{80\angle -13^\circ\ V}$$

Circuit phase angle, $\boldsymbol{\phi = 13^\circ}$ **leading**. The phasor diagram is shown in Figure 24.18.

Further problems on the application of complex numbers to series a.c. circuits may be found in Section 24.3 following, problems 1 to 20.

24.3 Further problems on series a.c. circuits

1 Determine the resistance R and series inductance L (or capacitance C) for each of the following impedances, assuming the frequency to be 50 Hz. (a) $(4 + j7)\Omega$ (b) $(3 - j20)\Omega$ (c) $j10\ \Omega$ (d) $-j3\ \text{k}\Omega$ (e) $15\angle(\pi/3)\Omega$ (f) $6\angle -45^\circ\ \text{M}\Omega$

[(a) $R = 4\ \Omega, L = 22.3$ mH (b) $R = 3\ \Omega, C = 159.3\ \mu$F
(c) $R = 0, L = 31.8$ mH (d) $R = 0, C = 1.061\ \mu$F
(e) $R = 7.5\ \Omega, L = 41.3$ mH
(f) $R = 4.243\ \text{M}\Omega, C = 0.750$ nF]

2 A 0.4 μF capacitor is connected to a 250 V, 2 kHz supply. Determine the current flowing. [$1.257\angle 90^\circ$ A or $j1.247$ A]

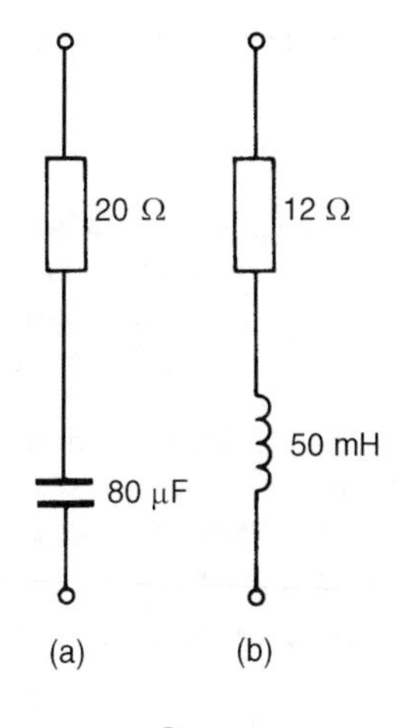

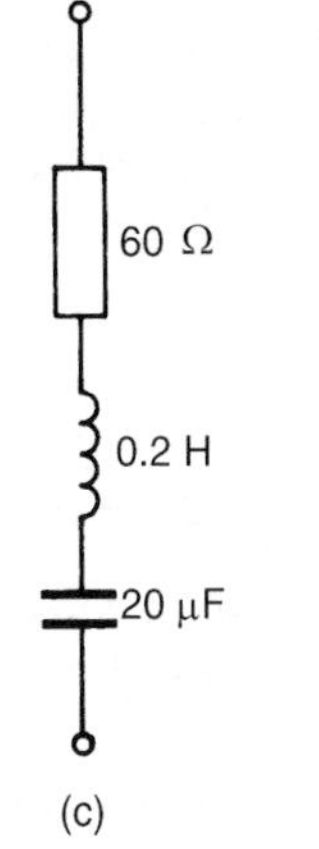

Figure 24.19

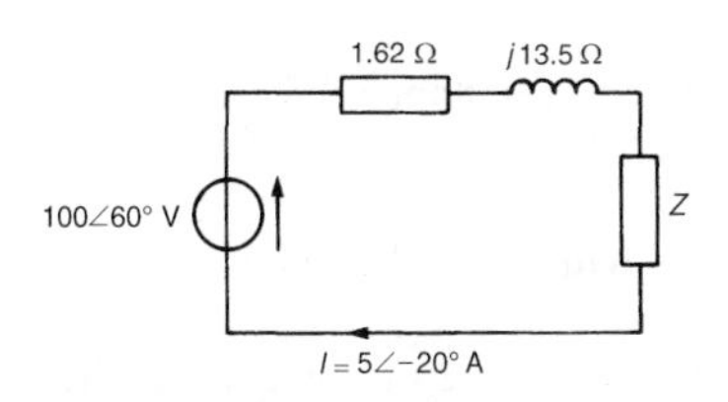

Figure 24.20

3 Two voltages in a circuit are represented by $(15 + j10)$V and $(12 - j4)$V. Determine the magnitude of the resultant voltage when these voltages are added. [27.66 V]

4 A current of $2.5\angle -90^\circ$ A flows in a coil of inductance 314.2 mH and negligible resistance when connected across a 50 Hz supply. Determine the value of the supply p.d. [$246.8\angle 0^\circ$ V]

5 A voltage $(75 + j90)$V is applied across an impedance and a current of $(5 + j12)$A flows. Determine (a) the value of the circuit impedance, and (b) the values of the components comprising the circuit if the frequency is 1 kHz.

[(a) $Z = (8.61 - j2.66)\Omega$ or $9.01\angle -17.19^\circ$ Ω
(b) $R = 8.61$ Ω, $C = 59.83$ μF]

6 Determine, in polar form, the complex impedances for the circuits shown in Figure 24.19 if the frequency in each case is 50 Hz.

[(a) $44.53\angle -63.31^\circ$ Ω (b) $19.77\angle 52.62^\circ$ Ω
(c) $113.5\angle -58.08^\circ$ Ω]

7 For the circuit shown in Figure 24.20 determine the impedance Z in polar and rectangular forms.

[$Z = (1.85 + j6.20)\Omega$ or $6.47\angle 73.39^\circ$ Ω]

8 A 30 μF capacitor is connected in series with a resistance R at a frequency of 200 Hz. The resulting current leads the voltage by 30°. Determine the magnitude of R. [45.95 Ω]

9 A coil has a resistance of 40 Ω and an inductive reactance of 75 Ω. The current in the coil is $1.70\angle 0^\circ$ A. Determine the value of (a) the supply voltage, (b) the p.d. across the 40 Ω resistance, (c) the p.d. across the inductive part of the coil, and (d) the circuit phase angle. Draw the phasor diagram.

[(a) $(68 + j127.5)$ V or $144.5\angle 61.93^\circ$ V (b) $68\angle 0^\circ$ V
(c) $127.5\angle 90^\circ$ V (d) 61.93° lagging]

10 An alternating voltage of 100 V, 50 Hz is applied across an impedance of $(20 - j30)\Omega$. Calculate (a) the resistance, (b) the capacitance, (c) the current, and (d) the phase angle between current and voltage

[(a) 20 Ω (b) 106.1 μF (c) 2.774 A (d) 56.31° leading]

11 A capacitor C is connected in series with a coil of resistance R and inductance 30 mH. The current flowing in the circuit is $2.5\angle -40^\circ$ A when the supply p.d. is 200 V at 400 Hz. Determine the value of (a) resistance R, (b) capacitance C, (c) the p.d. across C, and (d) the p.d., across the coil. Draw the phasor diagram.

[(a) 61.28 Ω (b) 16.59 μF
(c) $59.95\angle -130^\circ$ V (d) $242.9\angle 10.90^\circ$ V]

12 A series circuit consists of a 10 Ω resistor, a coil of inductance 0.09 H and negligible resistance, and a 150 μF capacitor, and is

connected to a 100 V, 50 Hz supply. Calculate the current flowing and its phase relative to the supply voltage.

[8.17 A lagging V by 35.20°]

13 A 150 mV, 5 kHz source supplies an ac. circuit consisting of a coil of resistance 25 Ω and inductance 5 mH connected in series with a capacitance of 177 nF. Determine the current flowing and its phase angle relative to the source voltage. [$4.44\angle 42.31°$ mA]

14 Two impedances, $Z_1 = 5\angle 30°$ Ω and $Z_2 = 10\angle 45°$ Ω draw a current of 3.36 A when connected in series to a certain a.c. supply. Determine (a) the supply voltage, (b) the phase angle between the voltage and current, (c) the p.d. across Z_1, and (d) the p.d. across Z_2.

[(a) 50 V (b) 40.01° lagging (c) $16.8\angle 30°$ V (d) $33.6\angle 45°$ V]

15 A 4500 pF capacitor is connected in series with a 50 Ω resistor across an alternating voltage $v = 212.1 \sin(\pi 10^6 t + \pi/4)$ volts. Calculate (a) the rms value of the voltage, (b) the circuit impedance, (c) the rms current flowing, (d) the circuit phase angle, (e) the voltage across the resistor, and (f) the voltage across the capacitor.

[(a) $150\angle 45°$ V (b) $86.63\angle -54.75°$ Ω
(c) $1.73\angle 99.75°$ A (d) 54.75° leading
(e) $86.50\angle 99.75°$ V (f) $122.38\angle 9.75°$ V]

16 If the p.d. across a coil is $(30 + j20)$V at 60 Hz and the coil consists of a 50 mH inductance and 10 Ω resistance, determine the value of current flowing (in polar and cartesian forms).

[$1.69\angle -28.36°$ A; $(1.49 - j0.80)$A]

17 Three impedances are connected in series across a 120 V, 10 kHz supply. The impedances are:

(i) Z_1, a coil of inductance 200 μH and resistance 8 Ω
(ii) Z_2, a resistance of 12 Ω
(iii) Z_3, a 0.50 μF capacitor in series with a 15 Ω resistor.

Determine (a) the circuit impedance, (b) the circuit current, (c) the circuit phase angle, and (d) the p.d. across each impedance.

[(a) $39.95\angle -28.82°$ Ω (b) $3.00\angle 28.82°$ A (c) 28.82° leading
(d) $V_1 = 44.70\angle 86.35°$ V, $V_2 = 36.00\angle 28.82°$ V,
$V_3 = 105.56\angle -35.95°$ V]

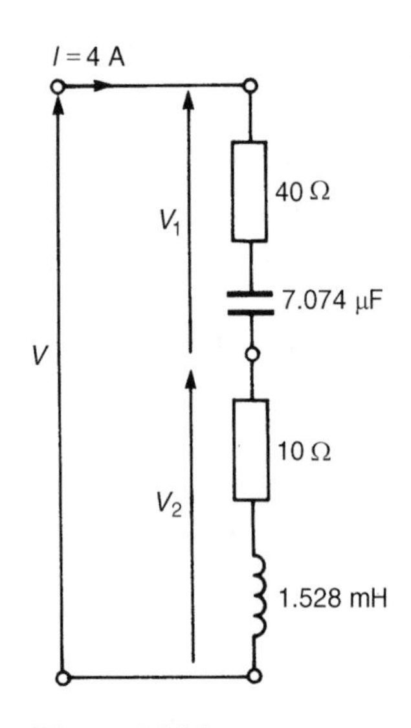

Figure 24.21

18 Determine the value of voltages V_1 and V_2 in the circuit shown in Figure 24.21, if the frequency of the supply is 2.5 kHz. Find also

the value of the supply voltage V and the circuit phase angle. Draw the phasor diagram.

[$V_1 = 164\angle{-12.68°}$ V or $(160 - j36)$V
$V_2 = 104\angle 67.38°$ V or $(40 + j96)$V
$V_3 = 208.8\angle 16.70°$ V or $(200 + j60)$V
Phase angle = 16.70° lagging]

19 A circuit comprises a coil of inductance 40 mH and resistance 20 Ω in series with a variable capacitor. The supply voltage is 120 V at 50 Hz. Determine the value of capacitance needed to cause a current of 2.0 A to flow in the circuit. [46.04 F]

20 For the circuit shown in Figure 24.22, determine (i) the circuit current I flowing, and (ii) the p.d. across each impedance.

[(i) $3.71\angle{-17.35°}$ A (ii) $V_1 = 55.65\angle 12.65°$ V,
$V_2 = 37.10\angle{-77.35°}$ V, $V_3 = 44.52\angle 32.65°$ V]

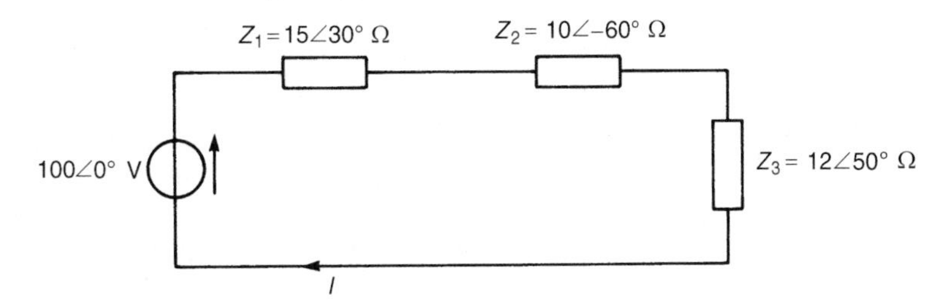

Figure 24.22

25 Application of complex numbers to parallel a.c. networks

At the end of this chapter you should be able to:

- determine admittance, conductance and susceptance in a.c. circuits
- perform calculations on parallel a.c. circuits using complex numbers

25.1 Introduction

As with series circuits, parallel networks may be analysed by using phasor diagrams. However, with parallel networks containing more than two branches this can become very complicated. It is with parallel a.c. network analysis in particular that the full benefit of using complex numbers may be appreciated. The theory for parallel a.c. networks introduced in Chapter 16 is relevant; more advanced networks will be analysed in this chapter using j notation. Before analysing such networks admittance, conductance and susceptance are defined.

25.2 Admittance, conductance and susceptance

Admittance is defined as the current I flowing in an a.c. circuit divided by the supply voltage V (i.e. it is the reciprocal of impedance Z). The symbol for admittance is Y. Thus

$$\boxed{Y = \frac{I}{V} = \frac{1}{Z}}$$

The unit of admittance is the **Siemen, S**.

An impedance may be resolved into a real part R and an imaginary part X, giving $Z = R \pm jX$. Similarly, an admittance may be resolved into two parts — the real part being called the **conductance G**, and the imaginary part being called the **susceptance B** — and expressed in complex form. Thus admittance

$$\boxed{Y = G \pm jB}$$

When an a.c. circuit contains:

(a) **pure resistance**, then

$$Z = R \text{ and } \boldsymbol{Y} = \frac{1}{Z} = \frac{1}{R} = \boldsymbol{G}$$

(b) **pure inductance**, then

$$Z = jX_L \text{ and } \boldsymbol{Y} = \frac{1}{Z} = \frac{1}{jX_L} = \frac{-j}{(jX_L)(-j)} = \frac{-j}{X_L} = \boldsymbol{-jB_L}$$

thus a negative sign is associated with inductive susceptance, B_L

(c) **pure capacitance**, then

$$Z = -jX_C \text{ and } \boldsymbol{Y} = \frac{1}{Z} = \frac{1}{-jX_C} = \frac{j}{(-jX_C)(j)} = \frac{j}{X_C} = \boldsymbol{+jB_C}$$

thus a positive sign is associated with capacitive susceptance, B_C

(d) **resistance and inductance in series**, then

$$Z = R + jX_L \text{ and } Y = \frac{1}{Z} = \frac{1}{R + jX_L} = \frac{(R - jX_L)}{R^2 + X_L^2}$$

i.e. $$Y = \frac{R}{R^2 + X_L^2} - j\frac{X_L}{R^2 + {X_L}^2} \text{ or } \boldsymbol{Y} = \frac{\boldsymbol{R}}{|\boldsymbol{Z}|^2} \boldsymbol{-j}\frac{\boldsymbol{X_L}}{|\boldsymbol{Z}|^2}$$

Thus conductance, $G = R/|Z|^2$ and inductive susceptance, $B_L = -X_L/|Z|^2$.

(Note that in an inductive circuit, the imaginary term of the impedance, X_L, is positive, whereas the imaginary term of the admittance, B_L, is negative.)

(e) **resistance and capacitance in series**, then

$$Z = R - jX_C \text{ and } Y = \frac{1}{Z} = \frac{1}{R - jX_C} = \frac{R + jX_C}{R^2 + X_C^2}$$

i.e. $$Y = \frac{R}{R^2 + X_C^2} + j\frac{X_C}{R^2 + X_C^2} \text{ or } \boldsymbol{Y} = \frac{\boldsymbol{R}}{|\boldsymbol{Z}|^2} \boldsymbol{+j}\frac{\boldsymbol{X_C}}{|\boldsymbol{Z}|^2}$$

Thus conductance, $G = R/|Z|^2$ and capacitive susceptance, $B_C = X_C/|Z|^2$.

(Note that in a capacitive circuit, the imaginary term of the impedance, X_C, is negative, whereas the imaginary term of the admittance, B_C, is positive.)

(f) **resistance and inductance in parallel**, then

$$\frac{1}{Z} = \frac{1}{R} + \frac{1}{jX_L} = \frac{jX_L + R}{(R)(jX_L)}$$

$$\text{from which, } \boldsymbol{Z} = \frac{\boldsymbol{(R)(jX_L)}}{\boldsymbol{R + jX_L}} \left(\text{i.e. } \frac{\text{product}}{\text{sum}}\right)$$

$$\text{and} \quad Y = \frac{1}{Z} = \frac{R + jX_L}{jRX_L} + \frac{R}{jRX_L} + \frac{jX_L}{jRX_L}$$

$$\text{i.e.,} \quad Y = \frac{1}{jX_L} + \frac{1}{R} = \frac{(-j)}{(jX_L)(-j)} + \frac{1}{R}$$

$$\text{or} \quad \boldsymbol{Y = \frac{1}{R} - \frac{j}{X_L}}$$

Thus conductance, $\boldsymbol{G = 1/R}$ and inductive susceptance, $\boldsymbol{B_L = -1/X_L}$

(g) **resistance and capacitance in parallel**, then

$$Z = \frac{(R)(-jX_C)}{R - jX_C} \left(\text{i.e. } \frac{\text{product}}{\text{sum}}\right)$$

$$\text{and} \quad Y = \frac{1}{Z} = \frac{R - jX_C}{-jRX_C} = \frac{R}{-jRX_C} - \frac{jX_C}{-jRX_C}$$

$$\text{i.e.} \quad Y = \frac{1}{-jX_C} + \frac{1}{R} = \frac{(j)}{(-jX_C)(j)} + \frac{1}{R}$$

$$\text{or} \quad \boldsymbol{Y = \frac{1}{R} + \frac{j}{X_C}} \qquad (25.1)$$

Thus conductance, $\boldsymbol{G = 1/R}$ and capacitive susceptance, $\boldsymbol{B_C = 1/X_C}$.

The conclusions that may be drawn from Sections (d) to (g) above are:

(i) that a **series** circuit is more easily represented by an **impedance**,

(ii) that a **parallel** circuit is often more easily represented by an **admittance** especially when more than two parallel impedances are involved.

> Problem 1. Determine the admittance, conductance and susceptance of the following impedances: (a) $-j5\ \Omega$ (b) $(25 + j40)\Omega$ (c) $(3 - j2)\Omega$ (d) $50\angle 40°\ \Omega$

(a) If impedance $Z = -j5\ \Omega$, then

$$\text{admittance } Y = \frac{1}{Z} = \frac{1}{-j5} = \frac{j}{(-j5)(j)} = \frac{j}{5}$$

$$= \boldsymbol{j0.2\ S} \text{ or } \boldsymbol{0.2\angle 90°\ S}$$

Since there is no real part, **conductance, $G = 0$**, and **capacitive susceptance, $B_C = 0.2$ S**

(b) If impedance $Z = (25 + j40)\Omega$ then

$$\text{admittance } Y = \frac{1}{Z} = \frac{1}{(25 + j40)} = \frac{25 - j40}{25^2 + 40^2} = \frac{25}{2225} - \frac{j40}{2225}$$

$$= \mathbf{(0.0112 - j0.0180)S}$$

Thus **conductance, $G = 0.0112$ S** and **inductive susceptance, $B_L = 0.0180$ S**

(c) If impedance $Z = (3 - j2)\Omega$, then

$$\text{admittance } Y = \frac{1}{Z} = \frac{1}{(3 - j2)} = \frac{3 + j2}{3^2 + 2^2} = \left(\frac{3}{13} + j\frac{2}{13}\right) \text{S}$$

or $\mathbf{(0.231 + j0.154)S}$

Thus **conductance, $G = 0.231$ S** and **capacitive susceptance, $B_C = 0.154$ S**

(d) If impedance $Z = 50\angle 40°\ \Omega$, then

$$\text{admittance } Y = \frac{1}{Z} = \frac{1}{50\angle 40°} = \frac{1\angle 0°}{50\angle 40°} = \frac{1}{50}\angle -40°$$

$$= \mathbf{0.02\angle -40°\ S} \text{ or } \mathbf{(0.0153 - j0.0129)S}$$

Thus **conductance, $G = 0.0153$ S** and **inductive susceptance, $B_L = 0.0129$ S**

> Problem 2. Determine expressions for the impedance of the following admittances:
>
> (a) $0.004\angle 30°$ S (b) $(0.001 - j0.002)$S (c) $(0.05 + j0.08)$S

(a) Since admittance $Y = 1/Z$, impedance $Z = 1/Y$.

$$\text{Hence impedance } Z = \frac{1}{0.004\angle 30°} = \frac{1\angle 0°}{0.004\angle 30°}$$

$$= \mathbf{250\angle -30°\ \Omega} \text{ or } \mathbf{(216.5 - j125)\Omega}$$

(b) $$\text{Impedance } Z = \frac{1}{(0.001 - j0.002)} = \frac{0.001 + j0.002}{(0.001)^2 + (0.002)^2}$$

$$= \frac{0.001 + j0.002}{0.000\,005}$$

$$= \mathbf{(200 + j400)\Omega} \text{ or } \mathbf{447.2\angle 63.43°\ \Omega}$$

(c) Admittance $Y = (0.05 + j0.08)\ \text{S} = 0.094\angle 57.99°\ \text{S}$

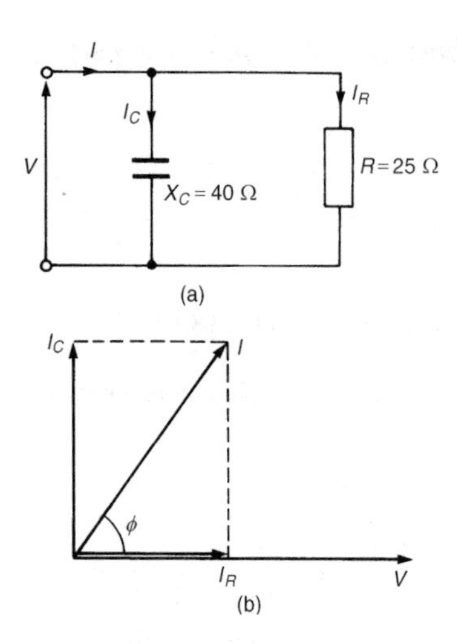

Figure 25.1 *(a) Circuit diagram, (b) Phasor diagram*

$$\text{Hence impedance } Z = \frac{1}{0.0094\angle 57.99^\circ}$$

$$= \mathbf{10.64\angle{-57.99^\circ}\ \Omega} \text{ or } \mathbf{(5.64 - j9.02)\Omega}$$

> Problem 3. The admittance of a circuit is $(0.040 + j0.025)$S. Determine the values of the resistance and the capacitive reactance of the circuit if they are connected (a) in parallel, (b) in series. Draw the phasor diagram for each of the circuits.

(a) *Parallel connection*

Admittance $Y = (0.040 + j0.025)$ S, therefore conductance, $G = 0.040$ S and capacitive susceptance, $B_C = 0.025$ S. From equation (25.1) when a circuit consists of resistance R and capacitive reactance in parallel, then $Y = (1/R) + (j/X_C)$.

$$\text{Hence resistance } R = \frac{1}{G} = \frac{1}{0.040} = \mathbf{25\ \Omega}$$

$$\text{and capacitive reactance } X_C = \frac{1}{B_C} = \frac{1}{0.025} = \mathbf{40\ \Omega}$$

The circuit and phasor diagrams are shown in Figure 25.1.

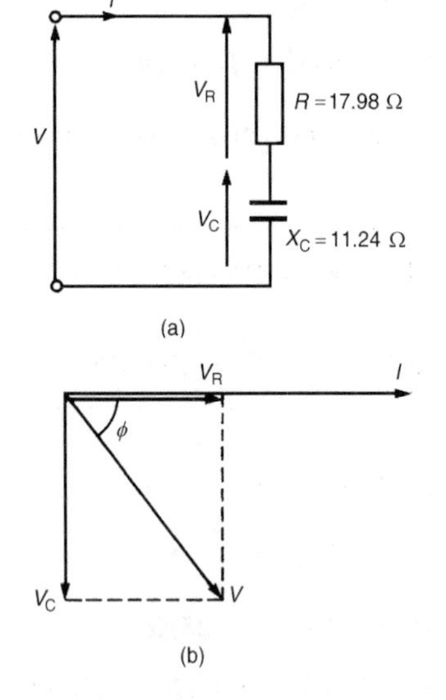

Figure 25.2 *(a) Circuit diagram, (b) Phasor diagram*

(b) *Series connection*

Admittance $Y = (0.040 + j0.025)$ S, therefore

$$\text{impedance } Z = \frac{1}{Y} = \frac{1}{0.040 + j0.025} = \frac{0.040 - j0.025}{(0.040)^2 + (0.025)^2}$$

$$= (17.98 - j11.24)\Omega$$

Thus the **resistance, R= 17.98 Ω** and **capacitive reactance,**

X_C= 11.24 Ω.

The circuit and phasor diagrams are shown in Figure 25.2.

The circuits shown in Figures 25.1(a) and 25.2(a) are equivalent in that they take the same supply current I for a given supply voltage V; the phase angle ϕ between the current and voltage is the same in each of the phasor diagrams shown in Figures 25.1(b) and 25.2(b).

Further problems on admittance, conductance and susceptance may be found in Section 25.4, problems 1 to 6, page 454.

25.3 Parallel a.c. networks

Figure 25.3 shows a circuit diagram containing three impedances, Z_1, Z_2 and Z_3 connected in parallel. The potential difference across each impedance is the same, i.e. the supply voltage V. Current $I_1 = V/Z_1$, $I_2 = V/Z_2$ and $I_3 = V/Z_3$. If Z_T is the total equivalent impedance of the

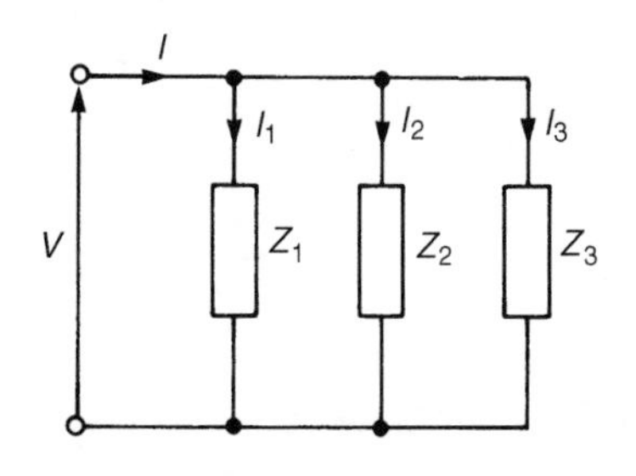

Figure 25.3

circuit then $I = V/Z_T$. The supply current, $I = I_1 + I_2 + I_3$ (phasorially).

Thus $\dfrac{V}{Z_T} = \dfrac{V}{Z_1} + \dfrac{V}{Z_2} + \dfrac{V}{Z_3}$ and $\boxed{\dfrac{1}{Z_T} = \dfrac{1}{Z_1} + \dfrac{1}{Z_2} + \dfrac{1}{Z_3}}$

or total admittance, $Y_T = Y_1 + Y_2 + Y_3$

In general, for n impedances connected in parallel,

$\boxed{\mathbf{Y_T = Y_1 + Y_2 + Y_3 + \ldots + Y_n}}$ (phasorially)

It is in parallel circuit analysis that the use of admittance has its greatest advantage.

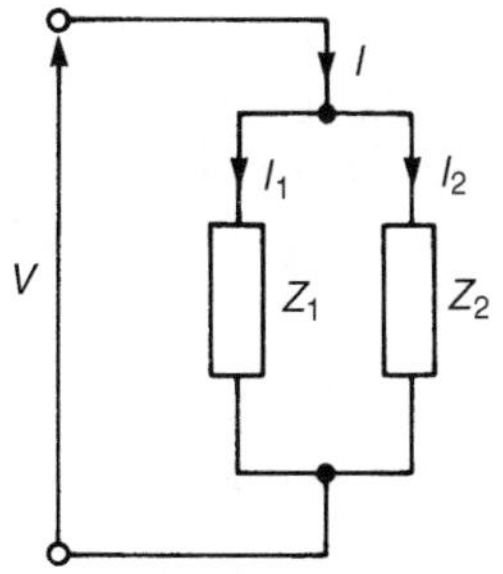

Figure 25.4

Current division in a.c. circuits

For the special case of two impedances, Z_1 and Z_2, connected in parallel (see Figure 25.4),

$$\frac{1}{Z_T} = \frac{1}{Z_1} + \frac{1}{Z_2} = \frac{Z_2 + Z_1}{Z_1 Z_2}$$

The total impedance, $\mathbf{Z_T = Z_1 Z_2/(Z_1 + Z_2)}$ (i.e. product/sum).

From Figure 25.4,

supply voltage, $V = IZ_T = I\left(\dfrac{Z_1 Z_2}{Z_1 + Z_2}\right)$

Also, $V = I_1 Z_1 (\text{and } V = I_2 Z_2)$

Thus, $I_1 Z_1 = I\left(\dfrac{Z_1 Z_2}{Z_1 + Z_2}\right)$

i.e., $\boxed{\textbf{current } \mathbf{I_1 = I\left(\dfrac{Z_2}{Z_1 + Z_2}\right)}}$

Similarly, $\boxed{\textbf{current } \mathbf{I_2 = I\left(\dfrac{Z_1}{Z_1 + Z_2}\right)}}$

Note that all of the above circuit symbols infer complex quantities either in cartesian or polar form.

The following problems show how complex numbers are used to analyse parallel a.c. networks.

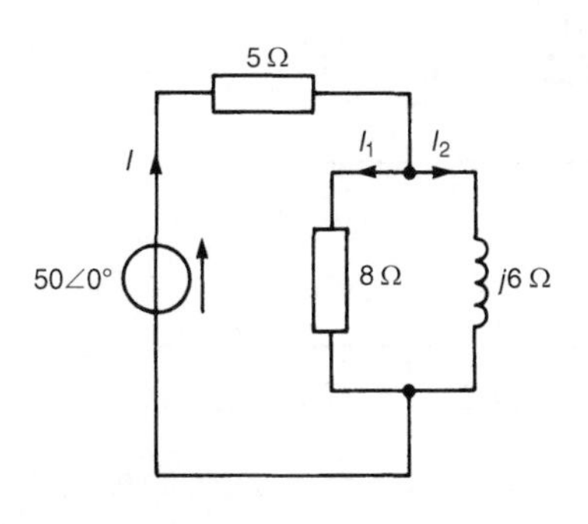

Figure 25.5

Problem 4. Determine the values of currents I, I_1 and I_2 shown in the network of Figure 25.5.

Total circuit impedance,

$$Z_T = 5 + \frac{(8)(j6)}{8 + j6} = 5 + \frac{(j48)(8 - j6)}{8^2 + 6^2}$$

$$= 5 + \frac{j384 + 288}{100}$$

$$= (7.88 + j3.84)\Omega \text{ or } 8.77\angle 25.98^\circ\ \Omega$$

$$\text{Current } I = \frac{V}{Z_T} = \frac{50\angle 0^\circ}{8.77\angle 25.98^\circ} = \mathbf{5.70\angle{-25.98}^\circ\ A}$$

$$\text{Current } I_1 = I\left(\frac{j6}{8 + j6}\right) = (5.70\angle{-25.98}^\circ)\left(\frac{6\angle 90^\circ}{10\angle 36.87^\circ}\right)$$

$$= \mathbf{3.42\angle 27.15^\circ\ A}$$

$$\text{Current } I_2 = I\left(\frac{8}{8 + j6}\right) = (5.70\angle{-25.98}^\circ)\left(\frac{8\angle 0^\circ}{10\angle 36.87^\circ}\right)$$

$$= \mathbf{4.56\angle{-62.85}^\circ\ A}$$

[Note: $I = I_1 + I_2 = 3.42\angle 27.15^\circ + 4.56\angle{-62.85}^\circ$

$$= (3.043 + j1.561) + (2.081 - j4.058)$$

$$= (5.124 - j2.497)\text{ A} = 5.70\angle{-25.98}^\circ\text{ A}]$$

Problem 5. For the parallel network shown in Figure 25.6, determine the value of supply current I and its phase relative to the 40 V supply.

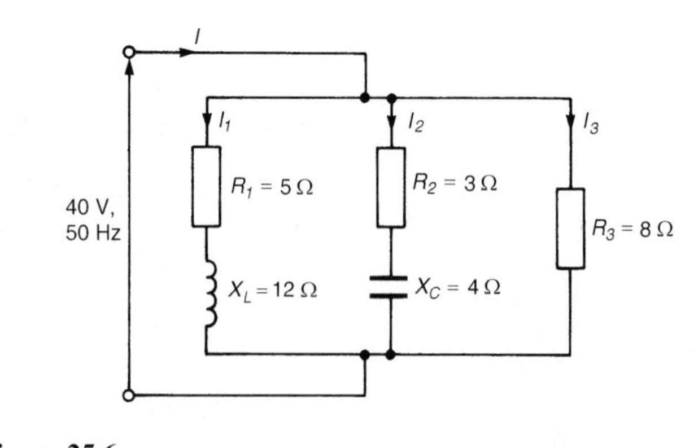

Figure 25.6

Impedance $Z_1 = (5 + j12)\Omega$, $Z_2 = (3 - j4)\Omega$ and $Z_3 = 8\ \Omega$

Supply current $I = \dfrac{V}{Z_T} = VY_T$ where Z_T = total circuit impedance, and Y_T = total circuit admittance.

$$Y_T = Y_1 + Y_2 + Y_3$$
$$= \frac{1}{Z_1} + \frac{1}{Z_2} + \frac{1}{Z_3} = \frac{1}{(5 + j12)} + \frac{1}{(3 - j4)} + \frac{1}{8}$$
$$= \frac{5 - j12}{5^2 + 12^2} + \frac{3 + j4}{3^2 + 4^2} + \frac{1}{8}$$
$$= (0.0296 - j0.0710) + (0.1200 + j0.1600) + (0.1250)$$

i.e. $Y_T = (0.2746 + j0.0890)\text{S or } 0.2887\angle 17.96^\circ \text{ S}$

Current $I = VY_T = (40\angle 0^\circ)(0.2887\angle 17.96^\circ) = 11.55\angle 17.96^\circ$ A

Hence the current I is 11.55 A and is leading the 40 V supply by 17.96°

Alternatively, current $I = I_1 + I_2 + I_3$

$$\text{Current } I_1 = \frac{40\angle 0^\circ}{5 + j12} = \frac{40\angle 0^\circ}{13\angle 67.38^\circ}$$
$$= 3.077\angle -67.38^\circ \text{ A or } (1.183 - j2.840) \text{ A}$$

$$\text{Current } I_2 = \frac{40\angle 0^\circ}{3 - j4} = \frac{40\angle 0^\circ}{5\angle -53.13^\circ} = 8\angle 53.13^\circ \text{ A or } (4.80 + j6.40) \text{ A}$$

$$\text{Current } I_3 = \frac{40\angle 0^\circ}{8\angle 0^\circ} = 5\angle 0^\circ \text{ A or } (5 + j0) \text{ A}$$

Thus current $I = I_1 + I_2 + I_3$

$= (1.183 - j2.840) + (4.80 + j6.40) + (5 + j0)$

$= 10.983 + j3.560 = \mathbf{11.55\angle 17.96^\circ}$ **A**, as previously obtained.

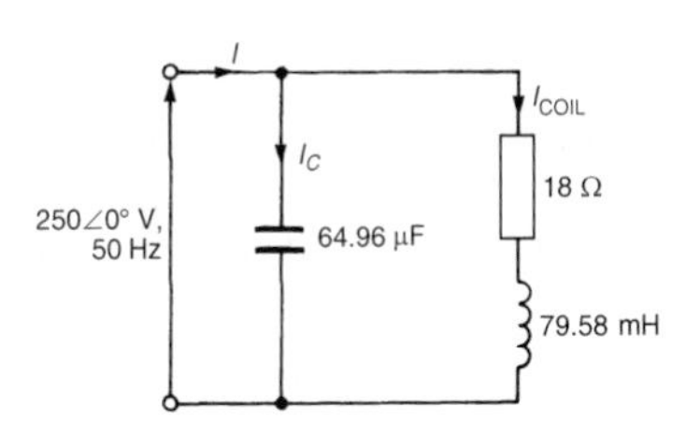

Figure 25.7

Problem 6. An a.c. network consists of a coil, of inductance 79.58 mH and resistance 18 Ω, in parallel with a capacitor of capacitance 64.96 μF. If the supply voltage is 250∠0° V at 50 Hz, determine (a) the total equivalent circuit impedance, (b) the supply current, (c) the circuit phase angle, (d) the current in the coil, and (e) the current in the capacitor.

The circuit diagram is shown in Figure 25.7.

Inductive reactance, $X_L = 2\pi f L = 2\pi(50)(79.58 \times 10^{-3}) = 25\ \Omega$.

Hence the impedance of the coil,

$$Z_{COIL} = (R + jX_L) = (18 + j25)\Omega \text{ or } 30.81\angle 54.25^\circ\ \Omega$$

$$\text{Capacitive reactance, } X_C = \frac{1}{2\pi f C} = \frac{1}{2\pi(50)(64.96 \times 10^{-6})} = 49\ \Omega$$

In complex form, the impedance presented by the capacitor, Z_C is $-jX_C$, i.e., $-j49\ \Omega$ or $49\angle -90^\circ\ \Omega$

(a) Total equivalent circuit impedance,

$$\boldsymbol{Z_T} = \frac{Z_{COIL}X_C}{Z_{COIL}+Z_C}\left(\text{i.e. } \frac{\text{product}}{\text{sum}}\right)$$

$$= \frac{(30.81\angle 54.25^\circ)(49\angle -90^\circ)}{(18+j25)+(-j49)}$$

$$= \frac{(30.81\angle 54.25^\circ)(49\angle -90^\circ)}{18-j24}$$

$$= \frac{(30.81\angle 54.25^\circ)(49\angle -90^\circ)}{30\angle -53.13^\circ}$$

$$= 50.32\angle(54.25^\circ - 90^\circ - (-53.13^\circ))$$

$$= \mathbf{50.32\angle 17.38^\circ\ \Omega} \text{ or } \mathbf{(48.02+j15.03)\ \Omega}$$

(b) Supply current $\boldsymbol{I} = \dfrac{V}{Z_T} = \dfrac{250\angle 0^\circ}{50.32\angle 17.38^\circ}$

$$= \mathbf{4.97\angle -17.38^\circ\ A}$$

(c) Circuit phase angle = **17.38° lagging**, i.e., the current I lags the voltage V by 17.38°

(d) Current in the coil, $\boldsymbol{I_{COIL}} = \dfrac{V}{Z_{COIL}} = \dfrac{250\angle 0^\circ}{30.81\angle 54.25^\circ}$

$$= \mathbf{8.11\angle -54.25^\circ\ A}$$

(e) Current in the capacitor, $\boldsymbol{I_C} = \dfrac{V}{Z_C} = \dfrac{250\angle 0^\circ}{49\angle -90^\circ}$

$$= \mathbf{5.10\angle 90^\circ\ A}$$

Problem 7. (a) For the network diagram of Figure 25.8, determine the value of impedance Z_1 (b) If the supply frequency is 5 kHz, determine the value of the components comprising impedance Z_1

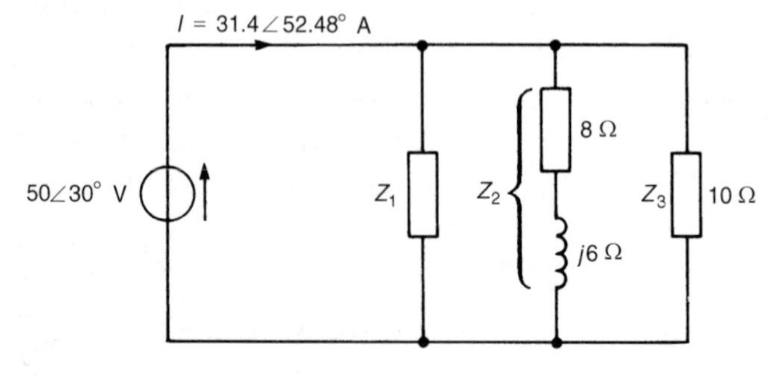

Figure 25.8

(a) Total circuit admittance,

$$Y_T = \frac{I}{V} = \frac{31.4\angle 52.48^\circ}{50\angle 30^\circ}$$

$$= \mathbf{0.628\angle 25.48^\circ\ S} \text{ or } \mathbf{(0.58 + j0.24)S}$$

$$Y_T = Y_1 + Y_2 + Y_3$$

$$\text{Thus } (0.58 + j0.24) = Y_1 + \frac{1}{(8 + j6)} + \frac{1}{10}$$

$$= Y_1 + \frac{8 - j6}{8^2 + 6^2} + 0.1$$

$$\text{i.e., } 0.58 + j0.24 = Y_1 + 0.08 - j0.06 + 0.1$$

$$\text{Hence } Y_1 = (0.58 - 0.08 - 0.1) + j(0.24 + j0.06)$$

$$= (0.4 + j0.3)\text{S or } 0.5\angle 36.87^\circ\ \text{S}$$

$$\text{Thus impedance, } Z_1 = \frac{1}{Y_1} = \frac{1}{0.5\angle 36.87^\circ}$$

$$= \mathbf{2\angle -36.87^\circ\ \Omega} \text{ or } \mathbf{(1.6 - j1.2)\Omega}$$

(b) Since $Z_1 = (1.6 - j1.2)\Omega$, **resistance = 1.6 Ω** and capacitive reactance, $X_C = 1.2\ \Omega$.

$$\text{Since } X_C = \frac{1}{2\pi f C}, \text{ capacitance } C = \frac{1}{2\pi f X_C} = \frac{1}{2\pi(5000)(1.2)}\ \text{F}$$

i.e., **capacitance = 26.53 μF**

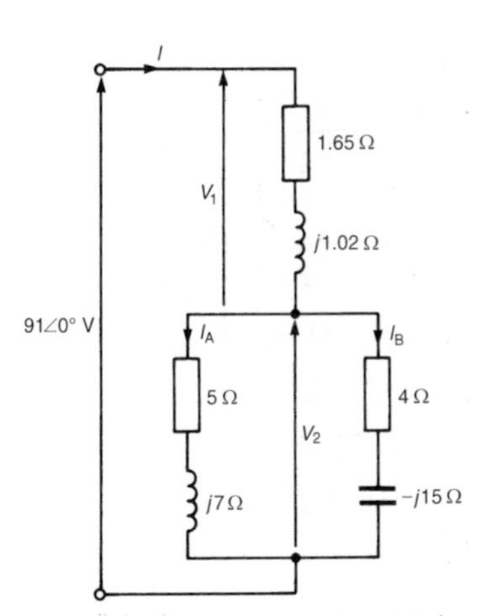

Figure 25.9

Problem 8. For the series-parallel arrangement shown in Figure 25.9, determine (a) the equivalent series circuit impedance, (b) the supply current I, (c) the circuit phase angle, (d) the values of voltages V_1 and V_2, and (e) the values of currents I_A and I_B

(a) The impedance, Z, of the two branches connected in parallel is given by:

$$Z = \frac{(5 + j7)(4 - j15)}{(5 + j7) + (4 - j15)} = \frac{20 - j75 + j28 - j^2 105}{9 - j8}$$

$$= \frac{125 - j47}{9 - j8} = \frac{133.54\angle -20.61^\circ}{12.04\angle -41.63^\circ}$$

$$= 11.09\angle 21.02^\circ\ \Omega \text{ or } (10.35 + j3.98)\Omega$$

Equivalent series circuit impedance,

$$Z_T = (1.65 + j1.02) + (10.35 + j3.98)$$

$$= \mathbf{(12 + j5)\Omega} \text{ or } \mathbf{13\angle 22.62^\circ\ \Omega}$$

(b) Supply current, $I = \dfrac{V}{Z} = \dfrac{91\angle 0°}{13\angle 22.62°} = \mathbf{7\angle -22.62°\ A}$

(c) Circuit phase angle = **22.62° lagging**

(d) Voltage $V_1 = IZ_1$, where $Z_1 = (1.65 + j1.02)\Omega$ or $1.94\angle 31.72°\ \Omega$.

Hence $\boldsymbol{V_1} = (7\angle -22.62°)(1.94\angle 31.72°) = \mathbf{13.58\angle 9.10°\ V}$

Voltage $V_2 = IZ$, where Z is the equivalent impedance of the two branches connected in parallel.

Hence $\boldsymbol{V_2} = (7\angle -22.62°)(11.09\angle 21.02°) = \mathbf{77.63\angle -1.60°\ V}$

(e) Current $I_A = V_2/Z_A$, where $Z_A = (5 + j7)\Omega$ or $8.60\angle 54.46°\ \Omega$.

Thus $\quad \boldsymbol{I_A} = \dfrac{77.63\angle -1.60°}{8.60\angle 54.46°} = \mathbf{9.03\angle - 56.06°\ A}$

Current $\boldsymbol{I_B} = V_2/Z_B$,

where $Z_B = (4 - j15)\Omega$ or $15.524\angle -75.07°\ \Omega$

Thus $\quad I_B = \dfrac{77.63\angle -1.60°}{15.524\angle -75.07°} = \mathbf{5.00\angle 73.47°\ A}$

[Alternatively, by current division,

$$\boldsymbol{I_A} = I\left(\frac{Z_B}{Z_A + Z_B}\right) = 7\angle -22.62°\left(\frac{15.524\angle -75.07°}{(5 + j7) + (4 - j15)}\right)$$
$$= 7\angle -22.62°\left(\frac{15.524\angle -75.07°}{9 - j8}\right)$$
$$= 7\angle -22.62°\left(\frac{15.524\angle -75.07°}{12.04\angle -41.63°}\right)$$
$$= \mathbf{9.03\angle -56.06°A}$$

$$\boldsymbol{I_B} = I\left(\frac{Z_A}{Z_A + Z_B}\right) = 7\angle -22.62°\left(\frac{8.60\angle 54.46°}{12.04\angle -41.63°}\right)$$
$$= \mathbf{5.00\angle 73.47°A}]$$

Further problems on parallel a.c. networks may be found in Section 25.4 following, problems 7 to 21, page 455.

25.4 Further problems on parallel a.c. networks

Admittance, conductance and susceptance

1 Determine the admittance (in polar form), conductance and susceptance of the following impedances: (a) $j10\ \Omega$ (b) $-j40\ \Omega$ (c) $32\angle -30°\ \Omega$ (d) $(5 + j9)\Omega$ (e) $(16 - j10)\Omega$

[(a) $0.1\angle -90°$ S, 0, 0.1 S
(b) $0.025\angle 90°$ S, 0, 0.025 S
(c) $0.03125\angle 30°$ S, 0.0271 S, 0.0156 S
(d) $0.0971\angle -60.95°$S, 0.0472 S, 0.0849 S
(e) $0.0530\angle 32.01°$ S, 0.0449 S, 0.0281 S]

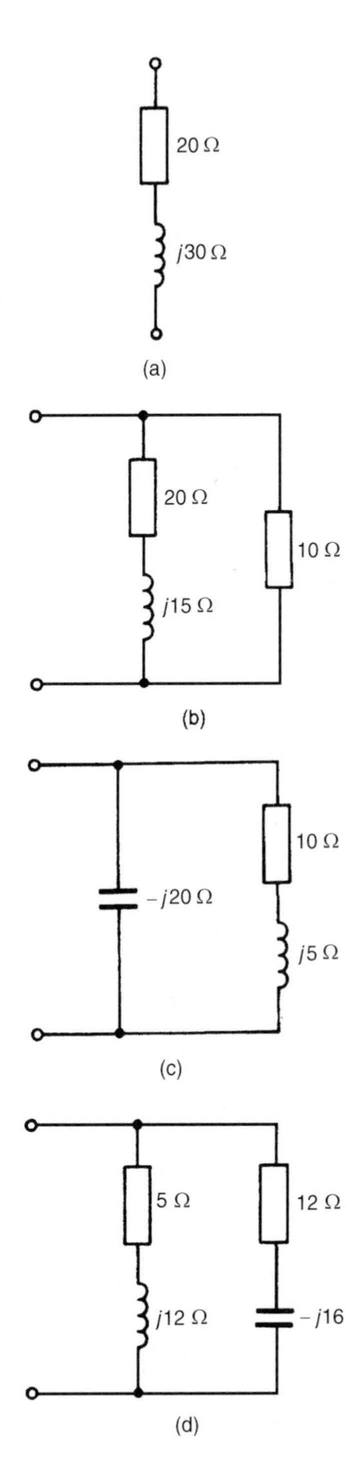

Figure 25.10

2 Derive expressions, in polar form, for the impedances of the following admittances: (a) $0.05\angle 40^\circ$ S (b) $0.0016\angle -25^\circ$ S (c) $(0.1 + j0.4)$S (d) $(0.025 - j0.040)$S

[(a) $20\angle -40^\circ\ \Omega$ (b) $625\angle 25^\circ\ \Omega$ (c) $2.425\angle -75.96^\circ\ \Omega$ (d) $21.20\angle 57.99^\circ\ \Omega$]

3 The admittance of a series circuit is $(0.010 - j0.004)$S. Determine the values of the circuit components if the frequency is 50 Hz.

[$R = 86.21\ \Omega$, $L = 109.8$ mH]

4 The admittance of a network is $(0.05 - j0.08)$S. Determine the values of resistance and reactance in the circuit if they are connected (a) in series, (b) in parallel.

[(a) $R = 5.62\ \Omega$, $X_L = 8.99\ \Omega$ (b) $R = 20\ \Omega$, $X_L = 12.5\ \Omega$]

5 The admittance of a two-branch parallel network is $(0.02 + j0.05)$S. Determine the circuit components if the frequency is 1 kHz.

[$R = 50\ \Omega$, $C = 7.958\ \mu$F]

6 Determine the total admittance, in rectangular and polar forms, of each of the networks shown in Figure 25.10.

[(a) $(0.0154 - j0.0231)$S or $0.0278\angle -56.31^\circ$ S
(b) $(0.132 - j0.024)$S or $0.134\angle -10.30^\circ$ S
(c) $(0.08 + j0.01)$S or $0.0806\angle 7.125^\circ$ S
(d) $(0.0596 - j0.0310)$S or $0.0672\angle -27.48^\circ$ S]

Parallel a.c. networks

7 Determine the equivalent circuit impedances of the parallel networks shown in Figure 25.11.

[(a) $(4 - j8)\Omega$ or $8.94\angle -63.43^\circ\ \Omega$
(b) $(7.56 + j1.95)\Omega$ or $7.81\angle 14.46^\circ\ \Omega$
(c) $(14.04 - j0.74)\Omega$ or $14.06\angle -3.02^\circ\ \Omega$]

8 Determine the value and phase of currents I_1 and I_2 in the network shown in Figure 25.12.

[$I_1 = 8.94\angle -10.30^\circ$ A, $I_2 = 17.89\angle 79.70^\circ$ A]

9 For the series-parallel network shown in Figure 25.13, determine (a) the total network impedance across AB, and (b) the supply current flowing if a supply of alternating voltage $30\angle 20^\circ$ V is connected across AB. [(a) $10\angle 36.87^\circ\ \Omega$ (b) $3\angle -16.87^\circ$ A]

10 For the parallel network shown in Figure 25.14, determine (a) the equivalent circuit impedance, (b) the supply current I, (c) the circuit phase angle, and (d) currents I_1 and I_2

[(a) $10.33\angle -6.31^\circ\ \Omega$ (b) $4.84\angle 6.31^\circ$ A (c) 6.31° leading
(d) $I_1 = 0.953\angle -73.38^\circ$ A, $I_2 = 4.765\angle 17.66^\circ$ A]

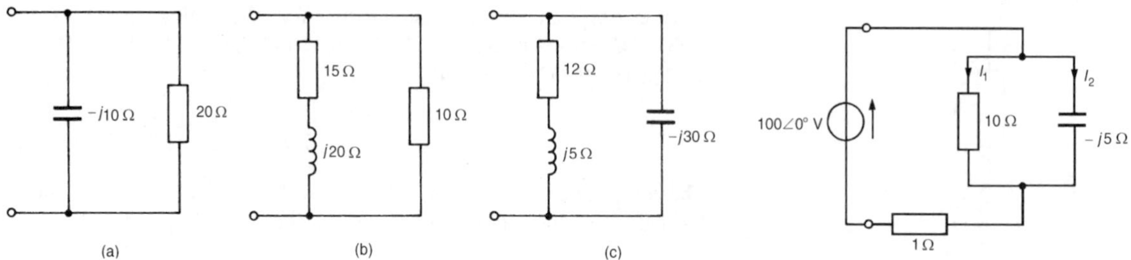

Figure 25.11

Figure 25.12

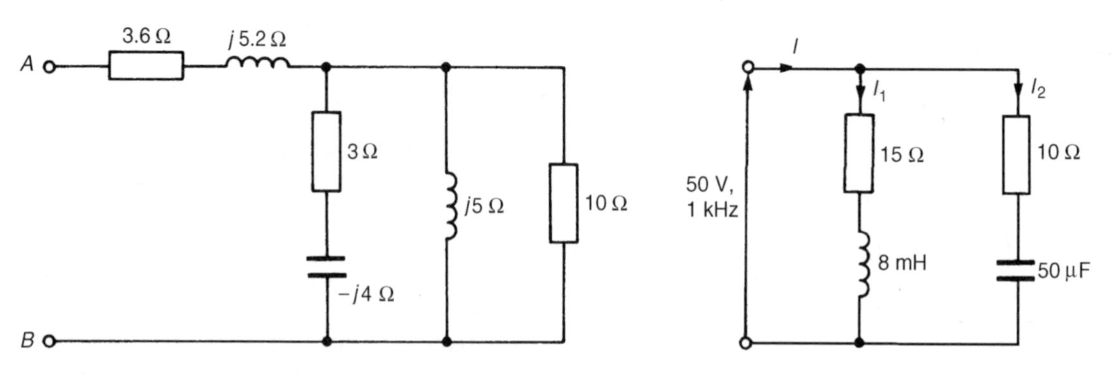

Figure 25.13

Figure 25.14

11 For the network shown in Figure 25.15, determine (a) current I_1, (b) current I_2, (c) current I, (d) the equivalent input impedance, and (e) the supply phase angle.
[(a) $15.08\angle 90°$ A (b) $3.39\angle -45.15°$ A (c) $12.90\angle 79.33°$ A (d) $9.30\angle -79.33°$ Ω (e) 79.33° leading]

12 Determine, for the network shown in Figure 25.16, (a) the total network admittance, (b) the total network impedance, (c) the supply current I, (d) the network phase angle, and (e) currents I_1, I_2, I_3 and I_4
[(a) $0.0733\angle 43.39°$ S (b) $13.64\angle -43.39°$ Ω (c) $1.833\angle 43.39°$ A (d) 43.39° leading (e) $I_1 = 0.455\angle -43.30°$ A, $I_2 = 1.863\angle 57.50°$ A, $I_3 = 1\angle 0°$ A, $I_4 = 1.570\angle 90°$ A]

13 Four impedances of $(10 - j20)\Omega$, $(30 + j0)\Omega$, $(2 - j15)\Omega$ and $(25 + j12)\Omega$ are connected in parallel across a 250 V ac. supply. Find the supply current and its phase angle. [$32.62\angle 43.55°$ A]

14 In the network shown in Figure 25.17, the voltmeter indicates 24 V. Determine the reading on the ammeter. [7.53 A]

15 Three impedances are connected in parallel to a 100 V, 50 Hz supply. The first impedance is $(10 + j12.5)\Omega$ and the second impedance is

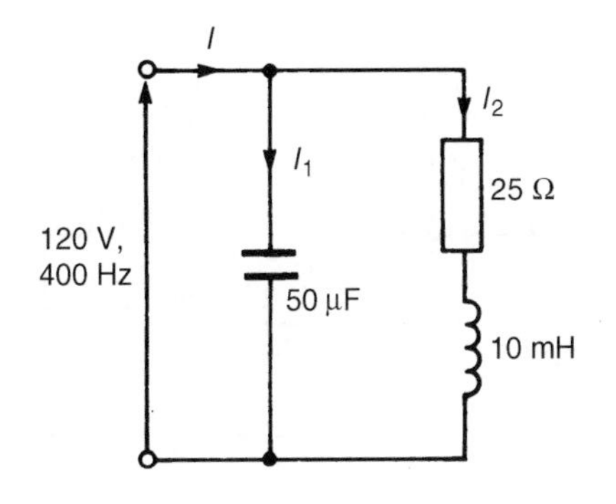

Figure 25.15

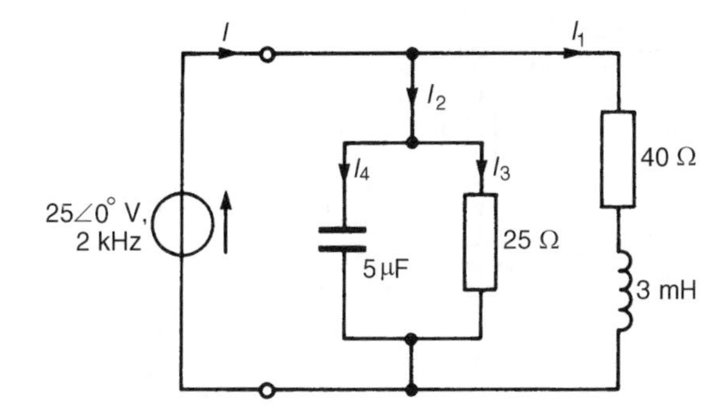

Figure 25.16

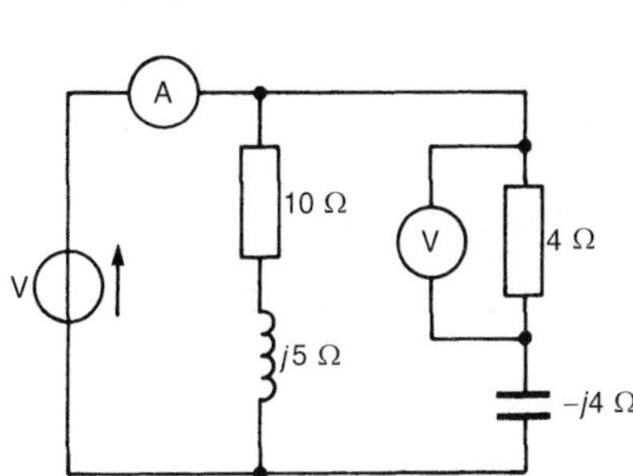
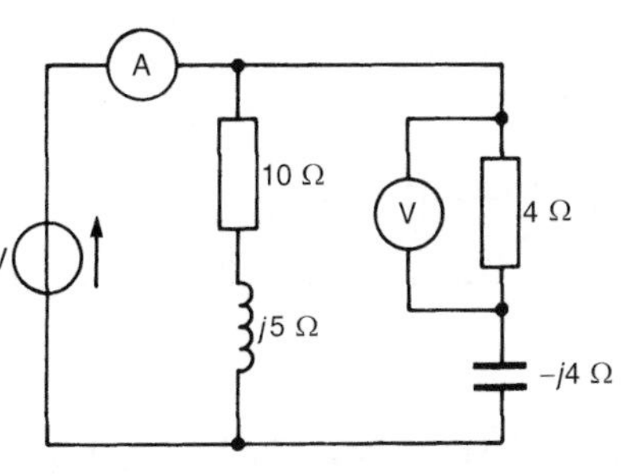

Figure 25.17

Figure 25.18

$(20 + j8)\Omega$. Determine the third impedance if the total current is $20\angle{-25^\circ}$ A [$(9.74 + j1.82)\Omega$ or $9.91\angle 10.56^\circ\ \Omega$]

16 For each of the network diagrams shown in Figure 25.18, determine the supply current I and their phase relative to the applied voltages.

[(a) $1.632\angle{-17.10^\circ}$ A (b) $5.411\angle{-8.46^\circ}$ A]

17 Determine the value of current flowing in the $(12 + j9)\Omega$ impedance in the network shown in Figure 25.19. [$7.66\angle 33.63^\circ$ A]

18 In the series-parallel network shown in Figure 25.20 the p.d. between points A and B is $50\angle{-68.13^\circ}$ V. Determine (a) the supply current I, (b) the equivalent input impedance, (c) the supply voltage V, (d) the supply phase angle, (e) the p.d. across points B and C, and (f) the value of currents I_1 and I_2

[(a) $11.99\angle{-31.81^\circ}$ A (b) $8.54\angle 20.56^\circ\ \Omega$
(c) $102.4\angle{-11.25^\circ}$ V (d) 20.56° lagging (e) $86.0\angle 17.91^\circ$ V
(f) $I_1 = 7.37\angle{-13.05^\circ}$ A $I_2 = 5.54\angle{-57.16^\circ}$ A]

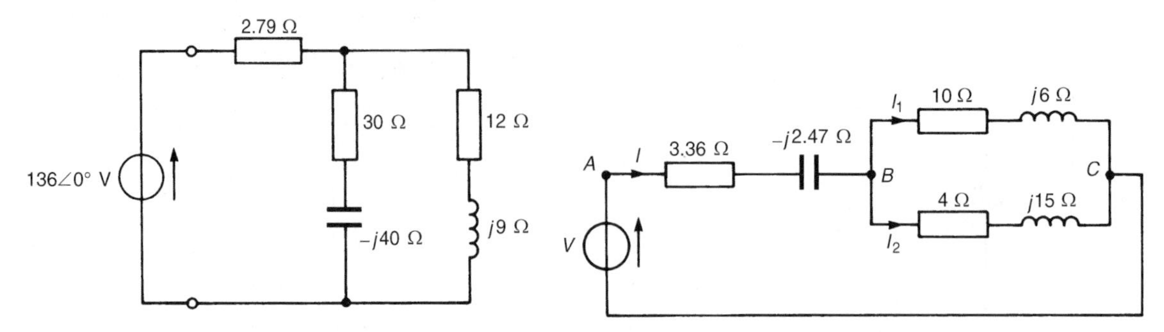

Figure 25.19 **Figure 25.20**

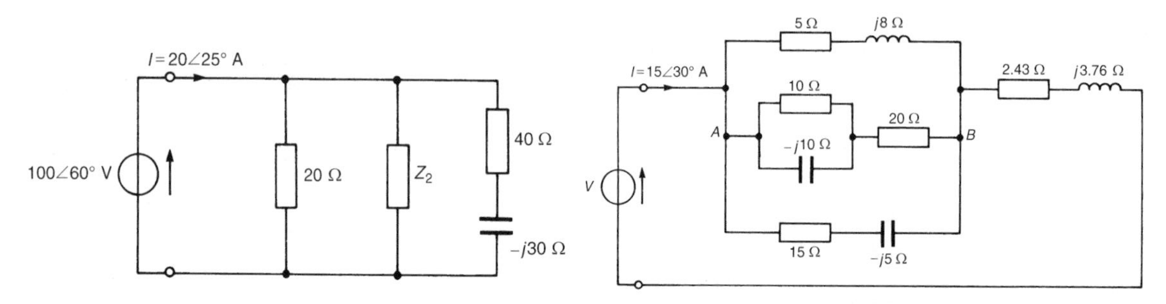

Figure 25.21 **Figure 25.22**

19 For the network shown in Figure 25.21, determine (a) the value of impedance Z_2, (b) the current flowing in Z_2, and (c) the components comprising Z_2 if the supply frequency is 2 kHz

[(a) 6.25∠52.34° Ω (b) 16.0∠7.66° A
(c) $R = 3.819$ Ω, $L = 0.394$ mH]

20 Coils of impedance $(5 + j8)$Ω and $(12 + j16)$Ω are connected in parallel. In series with this combination is an impedance of $(15 - j40)$Ω. If the alternating supply pd. is 150∠0° V, determine (a) the equivalent network impedance, (b) the supply current, (c) the supply phase angle, (d) the current in the $(5 + j8)$Ω impedance, and (e) the current in the $(12 + j16)$Ω impedance.

[(a) 39.31∠−61.84° Ω (b) 3.816∠61.84° A
(c) 61.84° leading (d) 2.595∠60.28° A
(e) 1.224∠65.15° A]

21 For circuit shown in Figure 25.22, determine (a) the input impedance, (b) the source voltage V, (c) the p.d. between points A and B, and (d) the current in the 10 Ω resistor.

[(a) 10.0∠36.87° Ω (b) 150∠66.87° V
(c) 90∠51.92° V (d) 2.50∠18.23° A]

26 Power in a.c. circuits

At the end of this chapter you should be able to:

- determine active, apparent and reactive power in a.c. series/parallel networks
- appreciate the need for power factor improvement
- perform calculations involving power factor improvement

26.1 Introduction

Alternating currents and voltages change their polarity during each cycle. It is not surprising therefore to find that power also pulsates with time. The product of voltage v and current i at any instant of time is called instantaneous power p, and is given by:

$$\boldsymbol{p = vi}$$

26.2 Determination of power in a.c. circuits

(a) Purely resistive a.c. circuits

Let a voltage $v = V_m \sin \omega t$ be applied to a circuit comprising resistance only. The resulting current is $i = I_m \sin \omega t$, and the corresponding instantaneous power, p, is given by:

$$p = vi = (V_m \sin \omega t)(I_m \sin \omega t)$$

i.e., $p = V_m I_m \sin^2 \omega t$

From trigonometrical double angle formulae, $\cos 2A = 1 - 2\sin^2 A$, from which,

$$\sin^2 A = \tfrac{1}{2}(1 - \cos 2A)$$

Thus $\sin^2 \omega t = \tfrac{1}{2}(1 - \cos 2\omega t)$

Then power $p = V_m I_m \left[\tfrac{1}{2}(1 - \cos 2\omega t)\right]$, i.e., $\boldsymbol{p = \tfrac{1}{2} V_m I_m (1 - \cos 2\omega t)}$.

The waveforms of v, i and p are shown in Figure 26.1. The waveform of power repeats itself after π/ω seconds and hence the power has a frequency twice that of voltage and current. The power is always positive, having a maximum value of $V_m I_m$. The average or mean value of the power is $\tfrac{1}{2} V_m I_m$.

The rms value of voltage $V = 0.707\, V_m$, i.e. $V = V_m/\sqrt{2}$, from which, $V_m = \sqrt{2}\, V$. Similarly, the rms value of current, $I = I_m/\sqrt{2}$, from which, $I_m = \sqrt{2}\, I$. Hence the average power, P, developed in a purely resistive

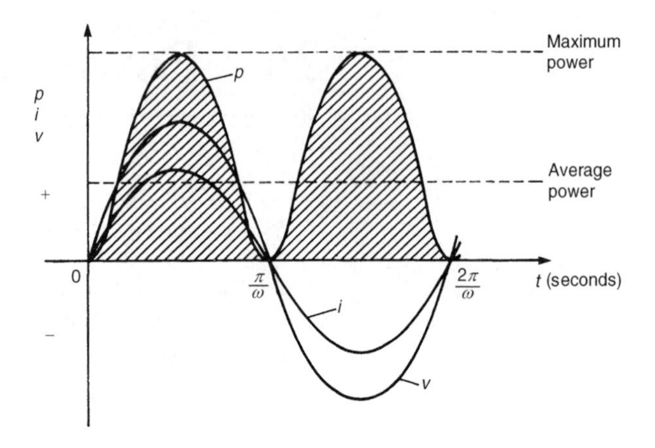

Figure 26.1 *The waveforms of v, i and p*

a.c. circuit is given by

$$P = \tfrac{1}{2}V_m I_m = \tfrac{1}{2}(\sqrt{2}\,V)(\sqrt{2}\,I) = VI \text{ watts}$$

Also, power $P = I^2R$ or V^2/R as for a d.c. circuit, since $V = IR$.

Summarizing, the average power P in a purely resistive a.c. circuit is given by

$$\boldsymbol{P = VI = I^2R = \frac{V^2}{R} \text{ watts}}$$

where V and I are rms values.

(b) Purely inductive a.c. circuits

Let a voltage $v = V_m \sin \omega t$ be applied to a circuit containing pure inductance (theoretical case). The resulting current is $i = I_m \sin(\omega t - (\pi/2))$ since current lags voltage by 90° in a purely inductive circuit, and the corresponding instantaneous power, p, is given by:

$$p = vi = (V_m \sin \omega t) I_m \sin(\omega t - (\pi/2))$$

i.e., $\quad p = V_m I_m \sin \omega t \sin(\omega t - (\pi/2))$

However, $\sin(\omega t - (\pi/2)) = -\cos \omega t$

Thus $p = -V_m I_m \sin \omega t \cos \omega t$

Rearranging gives: $p = -\frac{1}{2}V_m I_m (2 \sin \omega t \cos \omega t)$. However, from the double-angle formulae, $2 \sin \omega t \cos \omega t = \sin 2\omega t$.

Thus **power,** $\boldsymbol{p = -\frac{1}{2}V_m I_m \sin 2\omega t}$

The waveforms of v, i and p are shown in Figure 26.2. The frequency of power is twice that of voltage and current. For the power curve shown in Figure 26.2, the area above the horizontal axis is equal to the area below, thus over a complete cycle the average power P is zero. It is noted that when v and i are both positive, power p is positive and energy is delivered from the source to the inductance; when v and i have opposite signs, power p is negative and energy is returned from the inductance to the source.

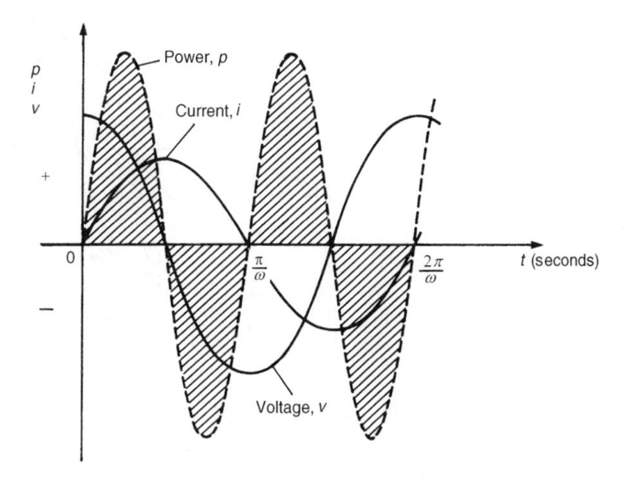

Figure 26.2 *Power in a purely inductive a.c. circuit*

In general, when the current through an inductance is increasing, energy is transferred from the circuit to the magnetic field, but this energy is returned when the current is decreasing.

Summarizing, the average power *P* in a purely inductive a.c. circuit is zero.

(c) Purely capacitive a.c. circuits

Let a voltage $v = V_m \sin \omega t$ be applied to a circuit containing pure capacitance. The resulting current is $i = I_m \sin(\omega t + (\pi/2))$, since current leads voltage by 90° in a purely capacitive circuit, and the corresponding instantaneous power, p, is given by:

$$p = vi = (V_m \sin \omega t) I_m \sin(\omega t + (\pi/2))$$

i.e., $p = V_m I_m \sin \omega t \sin(\omega t + (\pi/2))$

However, $\sin(\omega t + (\pi/2)) = \cos \omega t$.

Thus $P = V_m I_m \sin \omega t \cos \omega t$

Rearranging gives $p = \frac{1}{2} V_m I_m (2 \sin \omega t \cos \omega t)$.

Thus **power, $p = \frac{1}{2} V_m I_m \sin 2\omega t$**

The waveforms of v, i and p are shown in Figure 26.3. Over a complete cycle the average power P is zero. When the voltage across a capacitor is increasing, energy is transferred from the circuit to the electric field, but this energy is returned when the voltage is decreasing.

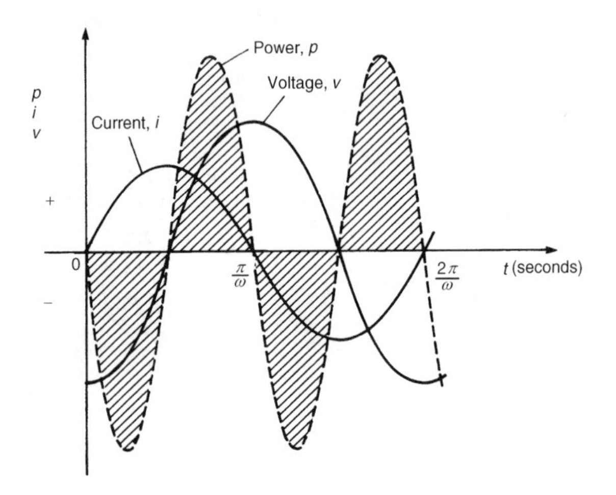

Figure 26.3 *Power in a purely capacitive a.c. circuit*

Summarizing, the average power P in a purely capacitive a.c. circuit is zero.

(d) R–L or R–C a.c. circuits

Let a voltage $v = V_m \sin \omega t$ be applied to a circuit containing resistance and inductance or resistance and capacitance. Let the resulting current be $i = I_m \sin(\omega t + \phi)$, where phase angle ϕ will be positive for an R–C circuit and negative for an R–L circuit. The corresponding instantaneous power, p, is given by:

$$p = vi = (V_m \sin \omega t)(I_m \sin(\omega t + \phi))$$

i.e., $$p = V_m I_m \sin \omega t \sin(\omega t + \phi)$$

Products of sine functions may be changed into differences of cosine functions by using: $\sin A \sin B = -\frac{1}{2}[\cos(A + B) - \cos(A - B)]$

Substituting $\omega t = A$ and $(\omega t + \phi) = B$ gives:

power, $$p = V_m I_m \left\{-\tfrac{1}{2}[\cos(\omega t + \omega t + \phi) - \cos(\omega t - (\omega t + \phi))]\right\}$$

i.e., $$p = \tfrac{1}{2} V_m I_m[\cos(-\phi) - \cos(2\omega t + \phi)]$$

However, $\cos(-\phi) = \cos \phi$.

Thus $\boldsymbol{p = \tfrac{1}{2} V_m I_m[\cos \phi - \cos(2\omega t + \phi)]}$

The instantaneous power p thus consists of

(i) a sinusoidal term, $-\frac{1}{2}V_m I_m \cos(2\omega t + \phi)$, which has a mean value over a cycle of zero, and

(ii) a constant term, $\frac{1}{2}V_m I_m \cos\phi$ (since ϕ is constant for a particular circuit).

Thus the average value of power, $P = \frac{1}{2}V_m I_m \cos\phi$.

Since $V_m = \sqrt{2}\,V$ and $I_m = \sqrt{2}\,I$,

average power, $\quad P = \frac{1}{2}(\sqrt{2}\,V)(\sqrt{2}\,I)\cos\phi$

i.e., $\boxed{\mathbf{P = VI\cos\phi \text{ watts}}}$

The waveforms of v, i and p, are shown in Figure 26.4 for an R–L circuit. The waveform of power is seen to pulsate at twice the supply frequency. The areas of the power curve (shown shaded) above the horizontal time axis represent power supplied to the load; the small areas below the axis represent power being returned to the supply from the inductance as the magnetic field collapses.

A similar shape of power curve is obtained for an R–C circuit, the small areas below the horizontal axis representing power being returned to the supply from the charged capacitor. The difference between the areas above and below the horizontal axis represents the heat loss due to the circuit resistance. Since power is dissipated only in a pure resistance, the alternative equations for power, $P = I_R^2 R$, may be used, where I_R is the rms current flowing through the resistance.

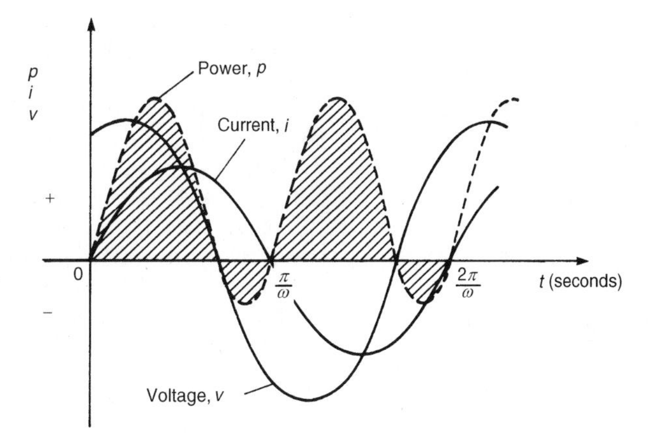

Figure 26.4 *Power in a.c. circuit containing resistance and inductive reactance*

Summarizing, the average power P in a circuit containing resistance and inductance and/or capacitance, whether in series or in parallel, is given by $P = VI \cos\phi$ or $P = I_R^2 R$ (V, I and I_R being rms values).

26.3 Power triangle and power factor

A phasor diagram in which the current I lags the applied voltage V by angle ϕ (i.e., an inductive circuit) is shown in Figure 26.5(a). The horizontal component of V is $V\cos\phi$, and the vertical component of V is $V\sin\phi$. If each of the voltage phasors of triangle Oab is multiplied by I, Figure 26.5(b) is produced and is known as the **'power triangle'**. Each side of the triangle represents a particular type of power:

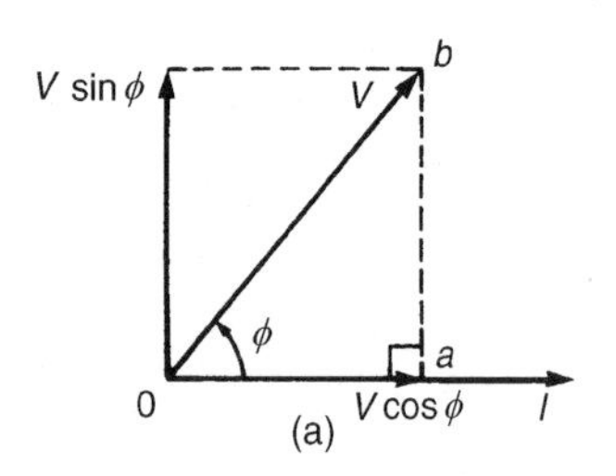

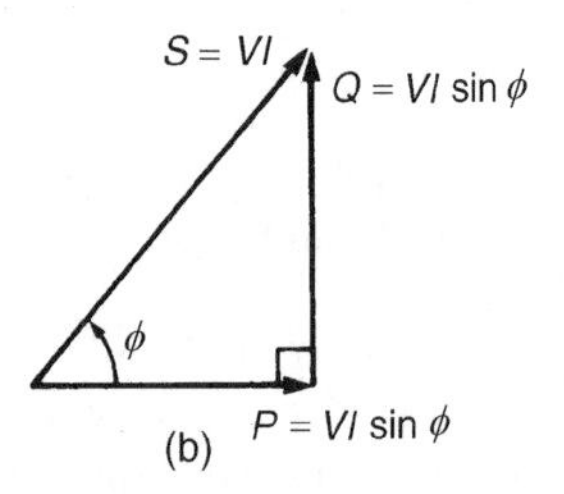

Figure 26.5 (a) Phasor diagram, (b) Power triangle for inductive circuit

True or active power $P = VI\cos\phi$ watts (W)
Apparent power $S = VI$ voltamperes (VA)
Reactive power $Q = VI\sin\phi$ vars (var)

The power triangle is **not** a phasor diagram since quantities P, Q and S are mean values and not rms values of sinusoidally varying quantities.

Superimposing the power triangle on an Argand diagram produces a relationship between P, S and Q in complex form, i.e.,

$$S = P + jQ$$

Apparent power, S, is an important quantity since a.c. apparatus, such as generators, transformers and cables, is usually rated in voltamperes rather than in watts. The allowable output of such apparatus is usually limited not by mechanical stress but by temperature rise, and hence by the losses in the device. The losses are determined by the voltage and current and are almost independent of the power factor. Thus the amount of electrical equipment installed to supply a certain load is essentially determined by the voltamperes of the load rather than by the power alone. The **rating** of a machine is defined as the maximum apparent power that it is designed to carry continuously without overheating.

The **reactive power**, Q, contributes nothing to the net energy transfer and yet it causes just as much loading of the equipment as if it did so. Reactive power is a term much used in power generation, distribution and utilization of electrical energy.

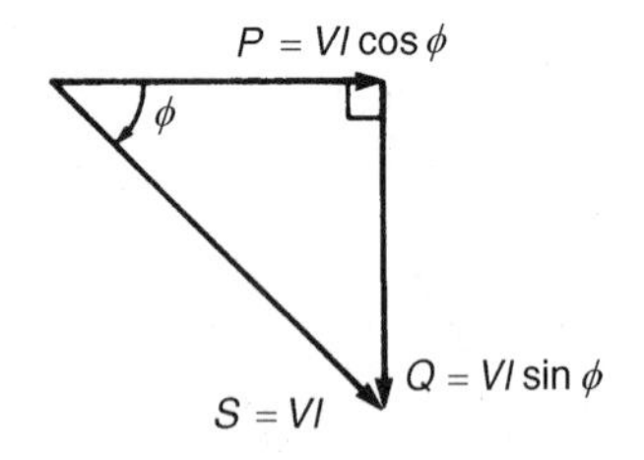

Figure 26.6 Power triangle for capacitive circuit

Inductive reactive power, by convention, is defined as positive reactive power; capacitive reactive power, by convention, is defined as negative reactive power. The above relationships derived from the phasor diagram of an inductive circuit may be shown to be true for a capacitive circuit, the power triangle being as shown in Figure 26.6.

Power factor is defined as:

$$\textbf{power factor} = \frac{\textbf{active power } P}{\textbf{apparent power } S}$$

For sinusoidal voltages and currents,

$$\text{power factor} = \frac{P}{S} = \frac{VI\cos\phi}{VI}$$

$$= \mathbf{\cos\phi = \frac{R}{Z}} \text{ (from the impedance triangle)}$$

A circuit in which current lags voltage (i.e., an inductive circuit) is said to have a lagging power factor, and indicates a lagging reactive power Q.

A circuit in which current leads voltage (i.e., a capacitive circuit) is said to have a leading power factor, and indicates a leading reactive power Q.

26.4 Use of complex numbers for determination of power

Let a circuit be supplied by an alternating voltage $V\angle\alpha$, where

$$V\angle\alpha = V(\cos\alpha + j\sin\alpha) = V\cos\alpha + jV\sin\alpha = a + jb \qquad (26.1)$$

Let the current flowing in the circuit be $I\angle\beta$, where

$$I\angle\beta = I(\cos\beta + j\sin\beta) = I\cos\beta + j\,I\sin\beta = c + jd \qquad (26.2)$$

From Sections 26.2 and 26.3, power $P = VI\cos\phi$, where ϕ is the angle between the voltage V and current I. If the voltage is $V\angle\alpha^\circ$ and the current is $I\angle\beta^\circ$, then the angle between voltage and current is $(\alpha - \beta)^\circ$

Thus power, $P = VI\cos(\alpha - \beta)$

From compound angle formulae, $\cos(\alpha - \beta) = \cos\alpha\cos\beta + \sin\alpha\sin\beta$.

Hence power, $P = VI[\cos\alpha\cos\beta + \sin\alpha\sin\beta]$

Rearranging gives $P = (V\cos\alpha)(I\cos\beta) + (V\sin\alpha)(I\sin\beta)$, i.e.,

$P = (a)(c) + (b)(d)$ from equations (26.1) and (26.2)

Summarizing, if $V = (a + jb)$ and $I = (c + jd)$, then

$$\boxed{\textbf{power, } P = ac + bd} \qquad (26.3)$$

Thus power may be calculated from the sum of the products of the real components and imaginary components of voltage and current.

Reactive power, $Q = VI\sin(\alpha - \beta)$

From compound angle formulae, $\sin(\alpha - \beta) = \sin\alpha\cos\beta - \cos\alpha\sin\beta$.

Thus $Q = VI[\sin\alpha\cos\beta - \cos\alpha\sin\beta]$

Rearranging gives $Q = (V\sin\alpha)(I\cos\beta) - (V\cos\alpha)(I\sin\beta)$ i.e.,

$Q = (b)(c) - (a)(d)$ from equations (26.1) and (26.2).

Summarizing, if $V = (a + jb)$ and $I = (c + jd)$, then

reactive power, $\boxed{Q = bc - ad}$ (26.4)

Expressions (26.3) and (26.4) provide an alternative method of determining true power P and reactive power Q when the voltage and current are complex quantities. From Section 26.3, apparent power $S = P + jQ$. However, merely multiplying V by I in complex form will not give this result, i.e. (from above)

$$S = VI = (a + jb)(c + jd) = (ac - bd) + j(bc + ad)$$

Here the real part is not the expression for power as given in equation (26.3) and the imaginary part is not the expression of reactive power given in equation (26.4)

The correct expression may be derived by multiplying the voltage V by the conjugate of the current, i.e. $(c - jd)$, denoted by I^*. Thus

apparent power $S = VI^* = (a + jb)(c - jd)$
$= (ac + bd) + j(bc - ad)$

i.e., $\boxed{S = P + jQ}$, from equations (26.3) and (26.4).

Thus the active and reactive powers may be determined if, and only if, the voltage V is multiplied by the conjugate of current I. As stated in Section 26.3, a positive value of Q indicates an inductive circuit, i.e., a circuit having a lagging power factor, whereas a negative value of Q indicates a capacitive circuit, i.e., a circuit having a leading power factor.

Problem 1. A coil of resistance 5 Ω and inductive reactance 12 Ω is connected across a supply voltage of $52\angle 30°$ volts. Determine the active power in the circuit.

The circuit diagram is shown in Figure 26.7.

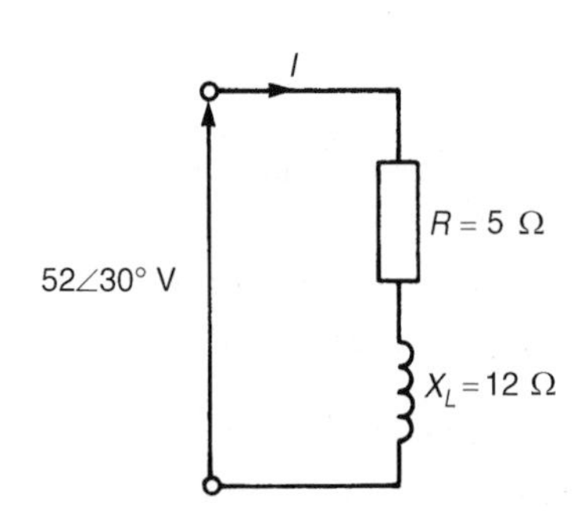

Figure 26.7

Impedance $Z = (5 + j12)\Omega$ or $13\angle 67.38°\ \Omega$

Voltage $V = 52\angle 30°$ V or $(45.03 + j26.0)$V

$$\text{Current } I = \frac{V}{Z} = \frac{52\angle 30°}{13\angle 67.38°}$$
$$= 4\angle -37.38° \text{ A or } (3.18 - j2.43)\text{A}$$

There are three methods of calculating power.

Method 1. Active power, $P = VI\cos\phi$, where ϕ is the angle between voltage V and current I. Hence

$$P = (52)(4)\cos[30^\circ - (-37.38^\circ)] = (52)(4)\cos 67.38^\circ = \mathbf{80\ W}$$

Method 2. Active power, $P = I_R^2 R = (4)^2(5) = \mathbf{80\ W}$

Method 3. Since $V = (45.03 + j26.0)$V and

$I = (3.18 - j2.43)$A, then active power,

$P = (45.03)(3.18) + (26.0)(-2.43)$ from equation (26.3), i.e.,

$P = 143.2 - 63.2 = \mathbf{80\ W}$

Problem 2. A current of $(15 + j8)$A flows in a circuit whose supply voltage is $(120 + j200)$V. Determine (a) the active power, and (b) the reactive power.

(a) *Method 1.* Active power $P = (120)(15) + (200)(8)$, from equation (26.3), i.e.,

$P = 1800 + 1600 = \mathbf{3400\ W}$ or $\mathbf{3.4\ kW}$

Method 2. Current $I = (15 + j8)\text{A} = 17\angle 28.07^\circ$ A and

Voltage $V = (120 + j200)\text{V} = 233.24\angle 59.04^\circ\ V$

Angle between voltage and current $= 59.04^\circ - 28.07^\circ = 30.97^\circ$

Hence power, $P = VI\cos\phi = (233.24)(17)\cos 30.97^\circ = \mathbf{3.4\ kW}$

(b) *Method 1.* Reactive power, $Q = (200)(15) - (120)(8)$ from equation (26.4), i.e.,

$Q = 3000 - 960 = \mathbf{2040\ var}$ or $\mathbf{2.04\ kvar}$

Method 2. Reactive power, $Q = VI\sin\phi$

$= (233.24)(17)\sin 30.97^\circ$

$= \mathbf{2.04\ kvar}$

Alternatively, parts (a) and (b) could have been obtained directly, using

$$\text{Apparent power, } S = VI^* = (120 + j200)(15 - j8)$$
$$= (1800 + 1600) + j(3000 - 960)$$
$$= 3400 + j2040 = P + jQ$$

from which, **power $P = 3400$ W** and **reactive power, $Q = 2040$ var**

> Problem 3. A series circuit possesses resistance R and capacitance C. The circuit dissipates a power of 1.732 kW and has a power factor of 0.866 leading. If the applied voltage is given by $v = 141.4\sin(10^4t + (\pi/9))$ volts, determine (a) the current flowing and its phase, (b) the value of resistance R, and (c) the value of capacitance C.

(a) Since $v = 141.4\sin(10^4t + (\pi/9))$ volts, then 141.4 V represents the maximum value, from which the rms voltage, $V = 141.4/\sqrt{2} = 100$ V, and the phase angle of the voltage $= +\pi/9$ rad or 20° leading. Hence as a phasor the voltage V is written as $\mathbf{100\angle 20°}$ **V**.

Power factor $= 0.866 = \cos\phi$, from which $\phi = \arccos 0.866 = 30°$. Hence the angle between voltage and current is 30°.

Power $P = VI\cos\phi$. Hence $1732 = (100)I\cos 30°$ from which,

$$\text{current, } |I| = \frac{1732}{(100)(0.866)} = \mathbf{20\ A}$$

Since the power factor is leading, the current phasor leads the voltage—in this case by 30°. Since the voltage has a phase angle of 20°,

$$\textbf{current, } \boldsymbol{I} = 20\angle(20° + 30°)\ \text{A} = \mathbf{20\angle 50°\ A}$$

(b) Impedance $Z = \dfrac{V}{I} = \dfrac{100\angle 20°}{20\angle 50°} = 5\angle -30°\ \Omega$ or $(4.33 - j2.5)\Omega$

Hence the **resistance,** $\boldsymbol{R} = \mathbf{4.33\ \Omega}$ and the capacitive reactance, $X_C = 2.5\ \Omega$.

Alternatively, the resistance may be determined from active power, $P = I^2R$. Hence $1732 = (20)^2R$, from which,

$$\textbf{resistance } \boldsymbol{R} = \frac{1732}{(20)^2} = \mathbf{4.33\ \Omega}$$

(c) Since $v = 141.4\sin(10^4t + (\pi/9))$ volts, angular velocity

$\omega = 10^4$ rad/s. Capacitive reactance, $X_C = 2.5\ \Omega$, thus

$$2.5 = \frac{1}{2\pi fC} = \frac{1}{\omega C}$$

from which, **capacitance,** $\boldsymbol{C} = \dfrac{1}{2.5\omega} = \dfrac{1}{(2.5)(10^4)}\ \text{F} = \mathbf{40\ \mu F}$

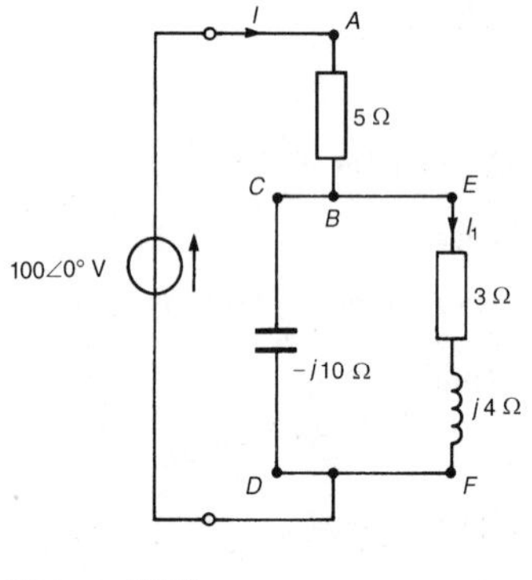

Figure 26.8

> Problem 4. For the circuit shown in Figure 26.8, determine the active power developed between points (a) A and B, (b) C and D, (c) E and F.

$$\text{Circuit impedance, } Z = 5 + \frac{(3 + j4)(-j10)}{(3 + j4 - j10)} = 5 + \frac{(40 - j30)}{(3 - j6)}$$

$$= 5 + \frac{50\angle -36.87^\circ}{6.71\angle -63.43^\circ} = 5 + 7.45\angle 26.56^\circ$$

$$= 5 + 6.66 + j3.33 = (11.66 + j3.33)\Omega \text{ or } 12.13\angle 15.94^\circ\ \Omega$$

$$\text{Current } I = \frac{V}{Z} = \frac{100\angle 0^\circ}{12.13\angle 15.94^\circ} = 8.24\angle -15.94^\circ \text{ A}$$

(a) Active power developed between points A and $B = I^2R = (8.24)^2(5) =$ **339.5 W**

(b) Active power developed between points C and D **is zero,** since no power is developed in a pure capacitor.

(c) Current, $I_1 = I\left(\frac{Z_{CD}}{Z_{CD} + Z_{EF}}\right) = 8.24\angle -15.94^\circ \left(\frac{-j10}{3 - j6}\right)$

$$= 8.24\angle -15.94^\circ \left(\frac{10\angle -90^\circ}{6.71\angle -63.43^\circ}\right)$$

$$= 12.28\angle -42.51^\circ \text{ A}$$

Hence the active power developed between points E and F $= I_1^2R = (12.28)^2(3) = 452.4$ W

[Check: Total active power developed $= 339.5 + 452.4 = 791.9$ W or 792 W, correct to three significant figures.

Total active power, $P = I^2R_T = (8.24)^2(11.66) = 792$ W (since 11.66 Ω is the total circuit equivalent resistance)

or $P = VI\cos\phi = (100)(8.24)\cos 15.94^\circ = 792$ W]

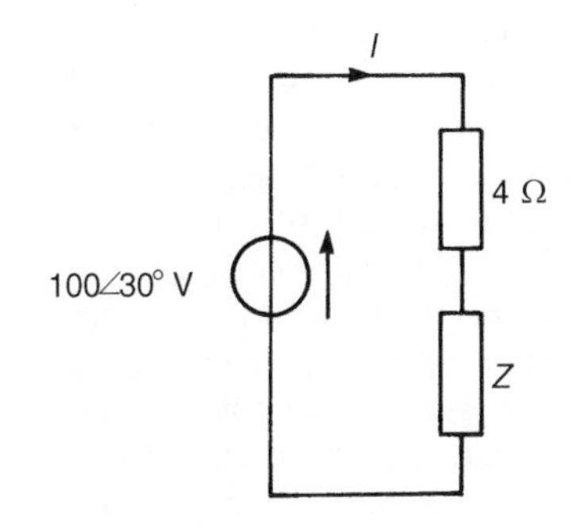

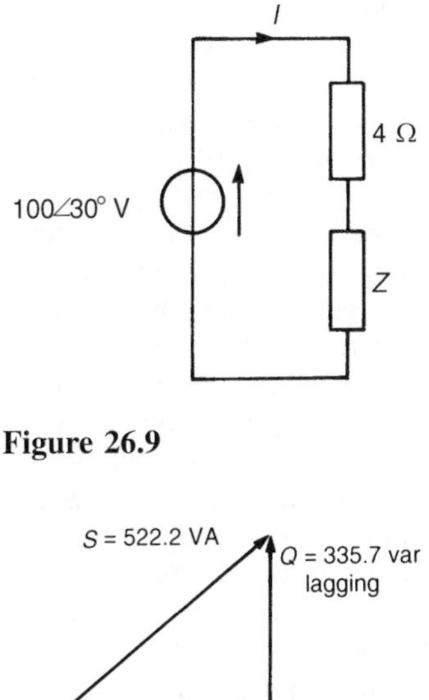

Figure 26.9

> Problem 5. The circuit shown in Figure 26.9 dissipates an active power of 400 W and has a power factor of 0.766 lagging. Determine (a) the apparent power, (b) the reactive power, (c) the value and phase of current I, and (d) the value of impedance Z.

Since power factor = 0.766 lagging, the circuit phase angle $\phi = \arccos 0.766$, i.e., $\phi = 40^\circ$ lagging which means that the current I lags voltage V by 40°.

Figure 26.10

(a) Since power, $P = VI\cos\phi$, the magnitude of apparent power,

$$S = VI = \frac{P}{\cos\phi} = \frac{400}{0.766} = \mathbf{522.2\ VA}$$

(b) Reactive power $Q = VI\sin\phi = (522.2)(\sin 40^\circ) =$ **335.7 var lagging**. (The reactive power is lagging since the circuit is inductive,

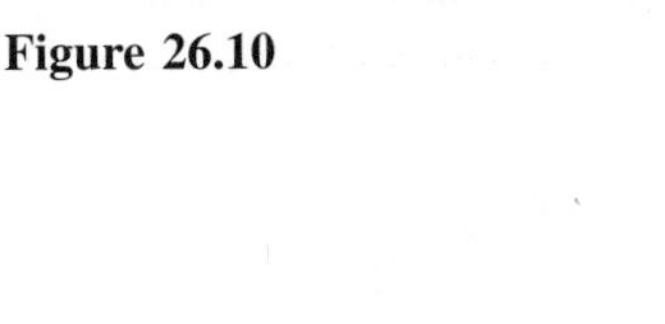

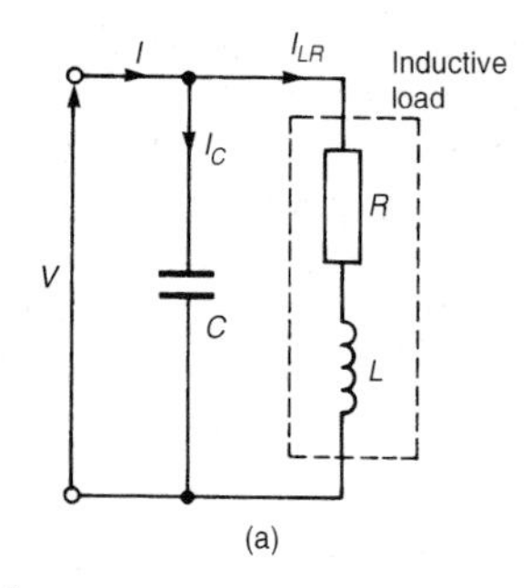

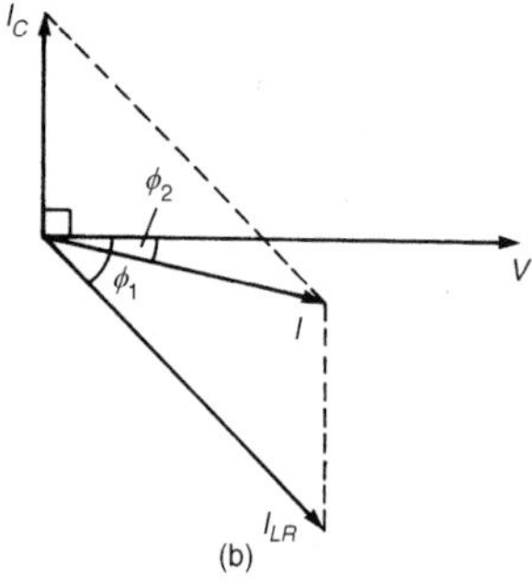

Figure 26.11 *(a) Circuit diagram (b) Phasor diagram*

which is indicated by the lagging power factor.) The power triangle is shown in Figure 26.10.

(c) Since $VI = 522.2$ VA,

$$\text{magnitude of current } |I| = \frac{522.2}{V} = \frac{522.2}{100} = \mathbf{5.222\ A}$$

Since the voltage is at a phase angle of 30° (see Figure 26.9) and current lags voltage by 40°, the phase angle of current is $30° - 40° = -10°$. Hence **current $I = 5.222\angle -10°$ A**

(d) Total circuit impedance $Z_T = \dfrac{V}{I} = \dfrac{100\angle 30°}{5.222\angle -10°}$

$$= 19.15\angle 40°\ \Omega \text{ or } (14.67 + j12.31)\Omega$$

Hence **impedance** $Z = Z_T - 4 = (14.67 + j12.31) - 4$

$$= \mathbf{(10.67 + j12.31)\Omega} \text{ or } \mathbf{16.29\angle 49.08°\ \Omega}$$

Further problems on power in a.c. circuits may be found in Section 26.6, problems 1 to 12, page 472.

26.5 Power factor improvement

For a particular active power supplied, a high power factor reduces the current flowing in a supply system and therefore reduces the cost of cables, transformers, switchgear and generators, as mentioned in Section 16.7, page 252. Supply authorities use tariffs which encourage consumers to operate at a reasonably high power factor. One method of improving the power factor of an inductive load is to connect a bank of capacitors in parallel with the load. Capacitors are rated in reactive voltamperes and the effect of the capacitors is to reduce the reactive power of the system without changing the active power. Most residential and industrial loads on a power system are inductive, i.e. they operate at a lagging power factor.

A simplified circuit diagram is shown in Figure 26.11(a) where a capacitor C is connected across an inductive load. Before the capacitor is connected the circuit current is I_{LR} and is shown lagging voltage V by angle ϕ_1 in the phasor diagram of Figure 26.11(b). When the capacitor C is connected it takes a current I_C which is shown in the phasor diagram leading voltage V by 90°. The supply current I in Figure 26.11(a) is now the phasor sum of currents I_{LR} and I_C as shown in Figure 26.11(b). The circuit phase angle, i.e., the angle between V and I, has been reduced from ϕ_1 to ϕ_2 and the power factor has been improved from $\cos\phi_1$ to $\cos\phi_2$.

Figure 26.12(a) shows the power triangle for an inductive circuit with a lagging power factor of $\cos\phi_1$. In Figure 26.12(b), the angle ϕ_1 has been reduced to ϕ_2, i.e., the power factor has been improved from $\cos\phi_1$ to $\cos\phi_2$ by introducing leading reactive voltamperes (shown as length ab) which is achieved by connecting capacitance in parallel with

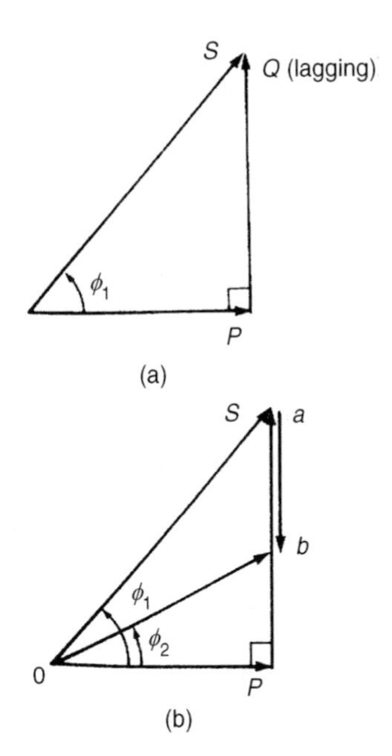

Figure 26.12 *Effect of connecting capacitance in parallel with the inductive load*

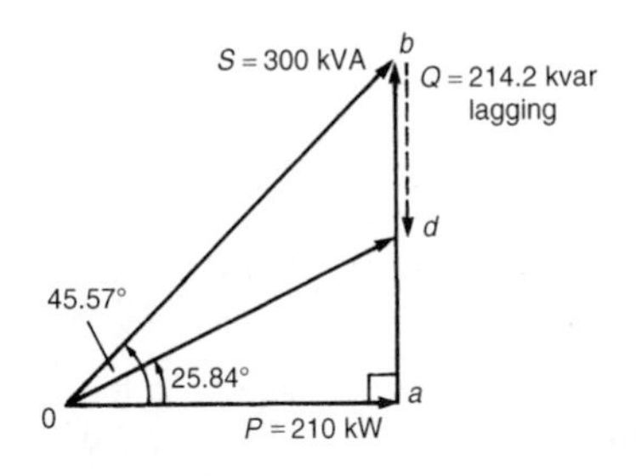

Figure 26.13

the inductive load. The power factor has been improved by reducing the reactive voltamperes; the active power P has remained unaffected.

Power factor correction results in the apparent power S decreasing (from 0a to 0b in Figure 26.12(b)) and thus the current decreasing, so that the power distribution system is used more efficiently.

Another method of power factor improvement, besides the use of static capacitors, is by using synchronous motors; such machines can be made to operate at leading power factors.

> Problem 6. A 300 kVA transformer is at full load with an overall power factor of 0.70 lagging. The power factor is improved by adding capacitors in parallel with the transformer until the overall power factor becomes 0.90 lagging. Determine the rating (in kilovars) of the capacitors required.

At full load, active power, $P = VI\cos\phi = (300)(0.70) = 210$ kW.

Circuit phase angle $\phi = \arccos 0.70 = 45.57°$

Reactive power, $Q = VI\sin\phi = (300)(\sin 45.57°) = 214.2$ kvar lagging.

The power triangle is shown as triangle 0ab in Figure 26.13. When the power factor is 0.90, the circuit phase angle $\phi = \arccos 0.90 = 25.84°$. The capacitor rating needed to improve the power factor to 0.90 is given by length bd in Figure 26.13.

Tan $25.84° = ad/210$, from which, $ad = 210\tan 25.84° = 101.7$ kvar. Hence the capacitor rating, i.e., $bd = ab - ad = 214.2 - 101.7 =$ **112.5 kvar leading**.

> Problem 7. A circuit has an impedance $Z = (3 + j4)\Omega$ and a source p.d. of $50\angle 30°$ V at a frequency of 1.5 kHz. Determine (a) the supply current, (b) the active, apparent and reactive power, (c) the rating of a capacitor to be connected in parallel with impedance Z to improve the power factor of the circuit to 0.966 lagging, and (d) the value of capacitance needed to improve the power factor to 0.966 lagging.

(a) Supply current, $I = \dfrac{V}{Z} = \dfrac{50\angle 30°}{(3 + j4)} = \dfrac{50\angle 30°}{5\angle 53.13°} = \mathbf{10\angle -23.13°\ A}$

(b) Apparent power,
$$S = VI^* = (50\angle 30°)(10\angle 23.13°)$$
$$= 500\angle 53.13°\ \text{VA}$$
$$= (300 + j400)\text{VA} = P + jQ$$

Hence **active power, $P = 300$ W**

apparent power, $S = 500$ VA and

reactive power, $Q = 400$ var lagging.

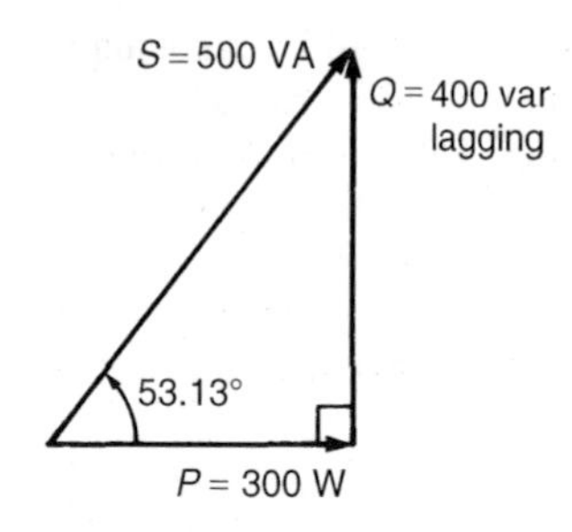

Figure 26.14

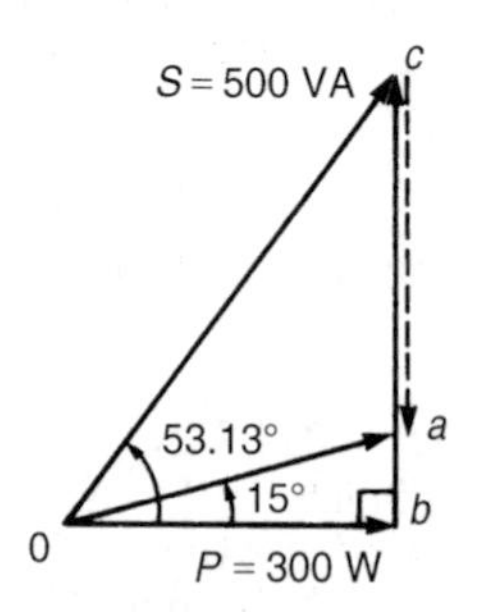

Figure 26.15

The power triangle is shown in Figure 26.14.

(c) A power factor of 0.966 means that $\cos\phi = 0.966$.
Hence angle $\phi = \arccos 0.966 = 15°$
To improve the power factor from $\cos 53.13°$, i.e. 0.60, to 0.966, the power triangle will need to change from Ocb (see Figure 26.15) to 0ab, the length ca representing the rating of a capacitor connected in parallel with the circuit. From Figure 26.15, $\tan 15° = ab/300$, from which, $ab = 300 \tan 15° = 80.38$ var.

Hence the **rating of the capacitor**, ca $= cb - ab$

$= 400 - 80.38$

$= \mathbf{319.6}$ **var leading**.

(d) Current in capacitor, $I_C = \dfrac{Q}{V} = \dfrac{319.6}{50} = 6.39$ A

Capacitive reactance, $X_C = \dfrac{V}{I_C} = \dfrac{50}{6.39} = 7.82\ \Omega$

Thus $7.82 = 1/(2\pi f C)$, from which,

$$\textbf{required capacitance } C = \frac{1}{2\pi(1500)(7.82)}\ \text{F} \equiv \mathbf{13.57\ \mu F}$$

Further problems on power factor improvement may be found in Section 26.6 following, problems 13 to 16, page 473.

26.6 Further problems on power in a.c. circuits

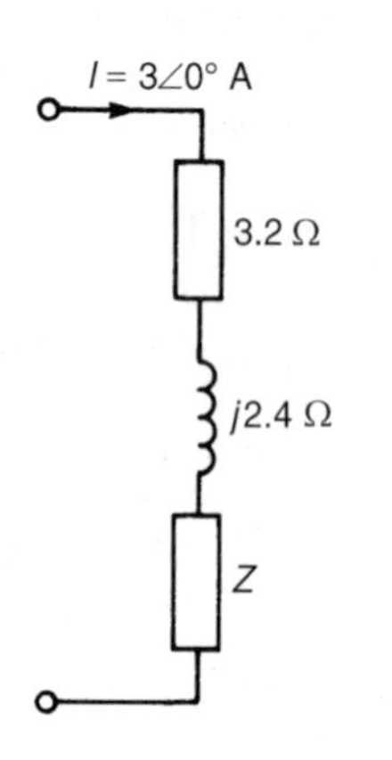

Figure 26.16

Power in a.c. circuits

1 When the voltage applied to a circuit is given by $(2 + j5)$V, the current flowing is given by $(8 + j4)$A. Determine the power dissipated in the circuit. [36 W]

2 A current of $(12 + j5)$A flows in a circuit when the supply voltage is $(150 + j220)$V. Determine (a) the active power, (b) the reactive power, and (c) the apparent power. Draw the power triangle. [(a) 2.90 kW (b) 1.89 kvar lagging (c) 3.46 kVA]

3 A capacitor of capacitive reactance 40 Ω and a resistance of 30 Ω are connected in series to a supply voltage of $200\angle 60°$ V. Determine the active power in the circuit. [480 W]

4 The circuit shown in Figure 26.16 takes 81 VA at a power factor of 0.8 lagging. Determine the value of impedance Z. [$(4 + j3)\Omega$ or $5\angle 36.87°\ \Omega$]

5 A series circuit possesses inductance L and resistance R. The circuit dissipates a power of 2.898 kW and has a power factor of 0.966 lagging. If the applied voltage is given by $v = 169.7\sin(100t - (\pi/4))$ volts, determine (a) the current flowing

and its phase, (b) the value of resistance R, and (c) the value of inductance L.

[(a) $25\angle-60°$ A (b) 4.64 Ω (c) 12.4 mH]

6 The p.d. across and the current in a certain circuit are represented by $(190 + j40)$V and $(9 - j4)$A respectively. Determine the active power and the reactive power, stating whether the latter is leading or lagging. [1550 W; 1120 var lagging]

7 Two impedances, $Z_1 = 6\angle40°$ Ω and $Z_2 = 10\angle30°$ Ω are connected in series and have a total reactive power of 1650 var lagging. Determine (a) the average power, (b) the apparent power, and (c) the power factor. [(a) 2469 W (b) 2970 VA (c) 0.83 lagging]

8 A current $i = 7.5\sin(\omega t - (\pi/4))$ A flows in a circuit which has an applied voltage $v = 180\sin(\omega t + (\pi/12))$V. Determine (a) the circuit impedance, (b) the active power, (c) the reactive power, and (d) the apparent power. Draw the power triangle.

[(a) $24\angle60°$ Ω (b) 337.5 W
(c) 584.6 var lagging (d) 675 VA]

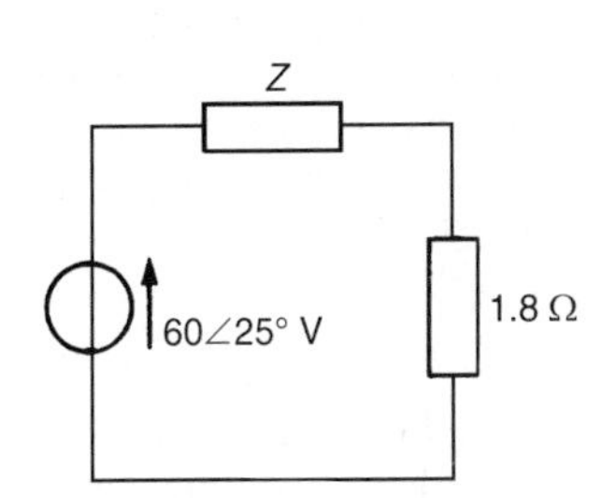

Figure 26.17

9 The circuit shown in Figure 26.17 has a power of 480 W and a power factor of 0.8 leading. Determine (a) the apparent power, (b) the reactive power, and (c) the value of impedance Z.

[(a) 600 VA (b) 360 var leading
(c) $(3 - j3.6)$Ω or $4.69\angle-50.19°$ Ω]

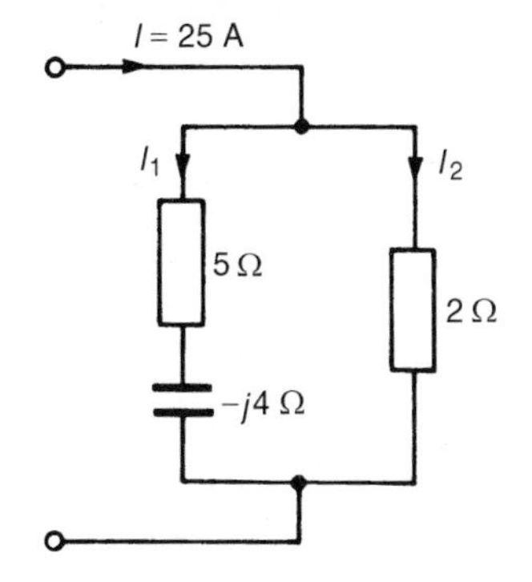

Figure 26.18

10 For the network shown in Figure 26.18, determine (a) the values of currents I_1 and I_2, (b) the total active power, (c) the reactive power, and (d) the apparent power.

[(a) $I_1 = 6.20\angle29.74°$ A, $I_2 = 19.86\angle-8.92°$ A (b) 981 W
(c) 153.9 var leading (d) 992.8 VA]

11 A circuit consists of an impedance $5\angle-45°$ Ω in parallel with a resistance of 10 Ω. The supply current is 4 A. Determine for the circuit (a) the active power, (b) the reactive power, and (c) the power factor. [(a) 49.34 W (b) 28.90 var leading (c) 0.863 leading]

12 For the network shown in Figure 26.19, determine the active power developed between points (a) A and B, (b) C and D, (c) E and F

[(a) 254.1 W (b) 0 (c) 65.92 W]

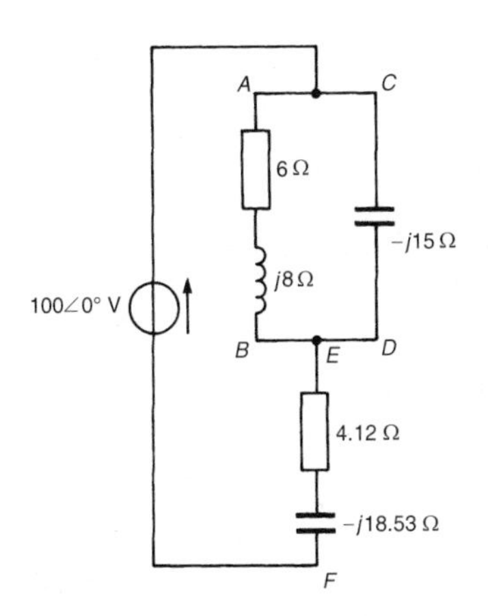

Figure 26.19

Power factor improvement

13 A 600 kVA transformer is at full load with an overall power factor of 0.64 lagging. The power factor is improved by adding capacitors in parallel with the transformer until the overall power factor becomes 0.95 lagging. Determine the rating (in kvars) of the capacitors needed. [334.8 kvar leading]

14 A source p.d. of $130\angle40°$ V at 2 kHz is applied to a circuit having an impedance of $(5 + j12)$Ω. Determine (a) the supply current, (b) the active, apparent and reactive powers, (c) the rating of the capacitor to be connected in parallel with the impedance to improve the

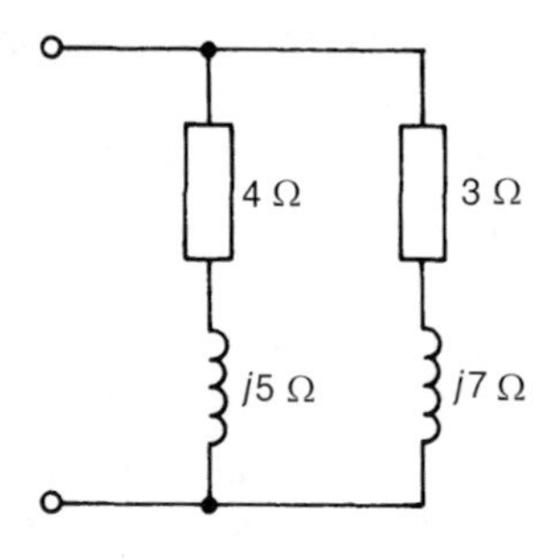

Figure 26.20

power factor of the circuit to 0.940 lagging, and (d) the value of the capacitance of the capacitor required.

[(a) $10\angle -27.38°$ A (b) 500 W, 1300 VA, 1200 var lagging (c) 1018.5 var leading (d) 4.797 μF]

15 The network shown in Figure 26.20 has a total active power of 2253 W. Determine (a) the total impedance, (b) the supply current, (c) the apparent power, (d) the reactive power, (e) the circuit power factor, (f) the capacitance of the capacitor to be connected in parallel with the network to improve the power factor to 0.90 lagging, if the supply frequency is 50 Hz.

[(a) $3.51\angle 58.40°$ Ω (b) 35.0 A (c) 4300 VA (d) 3662 var lagging (e) 0.524 lagging (f) 542.3 μF]

16 The power factor of a certain load is improved to 0.92 lagging with the addition of a 30 kvar bank of capacitors. If the resulting supply apparent power is 200 kVA, determine (a) the active power, (b) the reactive power before power factor correction, and (c) the power factor before correction.

[(a) 184 kW (b) 108.4 kvar lagging (c) 0.862 lagging]

Assignment 8

This assignment covers the material contained in chapters 23 to 26.

The marks for each question are shown in brackets at the end of each question.

1 The total impedance Z_T of an electrical circuit is given by:

$$Z_T = Z_1 + \frac{Z_2 \times Z_3}{Z_2 + Z_3}$$

Determine Z_T in polar form, correct to 3 significant figures, when

$Z_1 = 5.5\angle{-21^\circ}\ \Omega$, $Z_2 = 2.6\angle 30^\circ\ \Omega$ and $Z_3 = 4.8\angle 71^\circ\ \Omega$ (10)

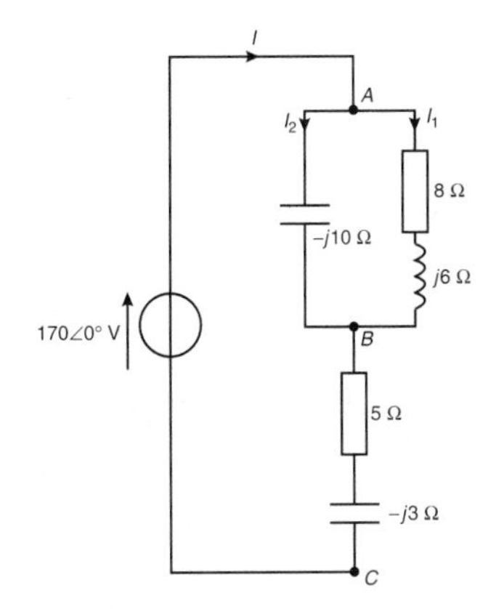

Figure A8.1

2 For the network shown in Figure A8.1, determine

(a) the equivalent impedance of the parallel branches
(b) the total circuit equivalent impedance
(c) current I
(d) the circuit phase angle
(e) currents I_1 and I_2
(f) the p.d. across points A and B
(g) the p.d. across points B and C
(h) the active power developed in the inductive branch
(i) the active power developed across the $-j10\ \Omega$ capacitor
(j) the active power developed between points B and C
(k) the total active power developed in the network
(l) the total apparent power developed in the network
(m) the total reactive power developed in the network (30)

3 A 400 kVA transformer is at full load with an overall power factor of 0.72 lagging. The power factor is improved by adding capacitors in parallel with the transformer until the overall power factor becomes 0.92 lagging. Determine the rating (in kilovars) of the capacitors required (10)

27 A.c. bridges

At the end of this chapter you should be able to:

- derive the balance equations of any a.c. bridge circuit
- state types of a.c. bridge circuit
- calculate unknown components when using an a.c. bridge circuit

27.1 Introduction

A.C. bridges are electrical networks, based upon an extension of the Wheatstone bridge principle, used for the determination of an unknown impedance by comparison with known impedances and for the determination of frequency. In general, they contain four impedance arms, an a.c. power supply and a balance detector which is sensitive to alternating currents. It is more difficult to achieve balance in an a.c. bridge than in a d.c. bridge because both the magnitude and the phase angle of impedances are related to the balance condition. Balance equations are derived by using complex numbers. A.C. bridges provide precise methods of measurement of inductance and capacitance, as well as resistance.

27.2 Balance conditions for an a.c. bridge

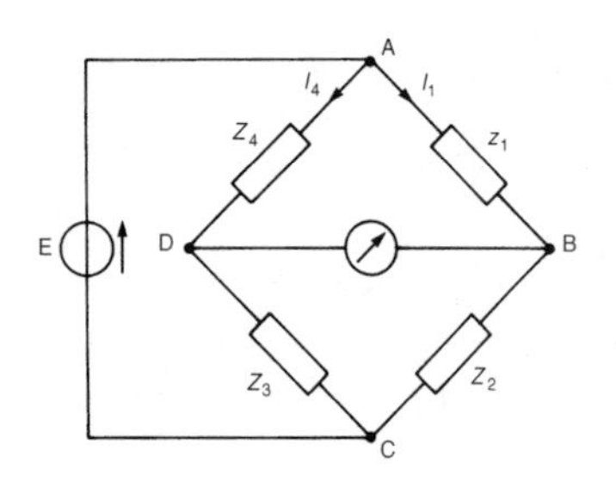

Figure 27.1 *Four-arm bridge*

The majority of well known a.c. bridges are classified as four-arm bridges and consist of an arrangement of four impedances (in complex form, $Z = R \pm jX$) as shown in Figure 27.1. As with the d.c. Wheatstone bridge circuit, an a.c. bridge is said to be 'balanced' when the current through the detector is zero (i.e., when no current flows between B and D of Figure 27.1). If the current through the detector is zero, then the current I_1 flowing in impedance Z_1 must also flow in impedance Z_2. Also, at balance, the current I_4 flowing in impedance Z_4, must also flow through Z_3.

At balance:

(i) the volt drop between A and B is equal to the volt drop between A and D,

i.e., $V_{AB} = V_{AD}$

i.e., $I_1Z_1 = I_4Z_4$ (both in magnitude and in phase) (27.1)

(ii) the volt drop between B and C is equal to the volt drop between D and C,

i.e., $V_{BC} = V_{DC}$

i.e., $I_1Z_2 = I_4Z_3$ (both in magnitude and in phase) (27.2)

Dividing equation (27.1) by equation (27.2) gives

$$\frac{I_1Z_1}{I_1Z_2} = \frac{I_4Z_4}{I_4Z_3}$$

from which $\dfrac{Z_1}{Z_2} = \dfrac{Z_4}{Z_3}$

or $\boxed{\mathbf{Z_1Z_3 = Z_2Z_4}}$ (27.3)

Equation (27.3) shows that at balance the products of the impedances of opposite arms of the bridge are equal.

If in polar form, $Z_1 = |Z_1|\angle\alpha_1$, $Z_2 = |Z_2|\angle\alpha_2$, $Z_3 = |Z_3|\angle\alpha_3$, and $Z_4 = |Z_4|\angle\alpha_4$, then from equation (27.3), $(|Z_1|\angle\alpha_1)(|Z_3|\angle\alpha_3) = (|Z_2|\angle\alpha_2)(|Z_4|\angle\alpha_4)$, which shows that there are **two** conditions to be satisfied simultaneously for balance in an a.c. bridge, i.e.,

$$\mathbf{|Z_1|\,|Z_3| = |Z_2|\,|Z_4| \quad \text{and} \quad \alpha_1 + \alpha_3 = \alpha_2 + \alpha_4}$$

When deriving balance equations of a.c. bridges, where at least two of the impedances are in complex form, it is important to appreciate that for a complex equation $a + jb = c + jd$ the real parts are equal, i.e. $a = c$, and the imaginary parts are equal, i.e., $b = d$.

Usually one arm of an a.c. bridge circuit contains the unknown impedance while the other arms contain known fixed or variable components. Normally only two components of the bridge are variable. When balancing a bridge circuit, the current in the detector is gradually reduced to zero by successive adjustments of the two variable components. At balance, the unknown impedance can be expressed in terms of the fixed and variable components.

Procedure for determining the balance equations of any a.c. bridge circuit

(i) Determine for the bridge circuit the impedance in each arm in complex form and write down the balance equation as in equation (27.3). Equations are usually easier to manipulate if L and C are initially expressed as X_L and X_C, rather than ωL or $1/(\omega C)$.

(ii) Isolate the unknown terms on the left-hand side of the equation in the form $a + jb$.

(iii) Manipulate the terms on the right-hand side of the equation into the form $c + jd$.

(iv) Equate the real parts of the equation, i.e., $a = c$, and equate the imaginary parts of the equation, i.e., $b = d$.

(v) Substitute ωL for X_L and $1/(\omega C)$ for X_c where appropriate and express the final equations in their simplest form.

Types of detector used with a.c. bridges vary with the type of bridge and with the frequency at which it is operated. Common detectors used include:

(i) a C.R.O., which is suitable for use with a very wide range of frequencies;

(ii) earphones (or telephone headsets), which are suitable for frequencies up to about 10 kHz and are used often at about 1 kHz, in which region the human ear is very sensitive;

(iii) various electronic detectors, which use tuned circuits to detect current at the correct frequency; and

(iv) vibration galvanometers, which are usually used for mains-operated bridges. This type of detector consists basically of a narrow moving coil which is suspended on a fine phosphor bronze wire between the poles of a magnet. When a current of the correct frequency flows through the coil, it is set into vibration. This is because the mechanical resonant frequency of the suspension is purposely made equal to the electrical frequency of the coil current. A mirror attached to the coil reflects a spot of light on to a scale, and when the coil is vibrating the spot appears as an extended beam of light. When the band reduces to a spot the bridge is balanced. Vibration galvanometers are available in the frequency range 10 Hz to 300 Hz.

27.3 Types of a.c. bridge circuit

A large number of bridge circuits have been developed, each of which has some particular advantage under certain conditions. Some of the most important a.c. bridges include the Maxwell, Hay, Owen and Maxwell-Wien bridges for measuring inductance, the De Sauty and Schering bridges for measuring capacitance, and the Wien bridge for measuring frequency. Obviously a large number of combinations of components in bridges is possible.

In many bridges it is found that two of the balancing impedances will be of the same nature, and often consist of standard non-inductive resistors.

For a bridge to balance quickly the requirement is either:

(i) the adjacent arms are both pure components (i.e. either both resistors, or both pure capacitors, or one of each) — this type of bridge being called a **ratio-arm bridge** (see, for example, paras (a), (c), (e) and (g) below); or

(ii) a pair of opposite arms are pure components — this type of bridge being called a **product-arm bridge** (see, for example, paras (b), (d) and (f) below).

A ratio-arm bridge can only be used to measure reactive quantities of the same type. When using a product-arm bridge the reactive component of the balancing impedance must be of opposite sign to the unknown reactive component.

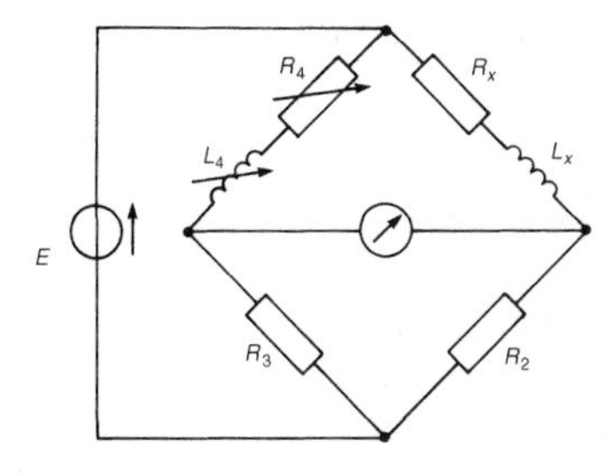

Figure 27.2 *Simple Maxwell bridge*

A commercial or universal bridge is available and can be used to measure resistance, inductance or capacitance.

(a) The simple Maxwell bridge

This bridge is used to measure the resistance and inductance of a coil having a high Q-factor (where Q-factor $= \omega L/R$, see Chapters 15 and 28).

A coil having unknown resistance R_x and inductance L_x is shown in the circuit diagram of a simple Maxwell bridge in Figure 27.2. R_4 and L_4 represent a standard coil having known variable values. At balance, expressions for R_x and L_x may be derived in terms of known components R_2, R_3, R_4 and L_4.

The procedure for determining the balance equations given in Section 27.2 may be followed.

(i) From Figure 27.2, $Z_x = R_x + jX_{L_x}$, $Z_2 = R_2$, $Z_3 = R_3$ and $Z_4 = R_4 + jX_{L_4}$.

At balance, $(Z_x)(Z_3) = (Z_2)(Z_4)$, from equation (27.3),

i.e., $(R_x + jX_{L_x})(R_3) = (R_2)(R_4 + jX_{L_4})$

(ii) Isolating the unknown impedance on the left-hand side of the equation gives

$$(R_x + jX_{L_x}) = \frac{R_2}{R_3}(R_4 + jX_{L_4})$$

(iii) Manipulating the right-hand side of the equation into $(a + jb)$ form gives

$$(R_x + jX_{L_x}) = \frac{R_2 R_4}{R_3} + j\frac{R_2 X_{L_4}}{R_3}$$

(iv) Equating the real parts gives $R_x = \dfrac{R_2 R_4}{R_3}$

Equating the imaginary parts gives $X_{L_x} = \dfrac{R_2 X_{L_4}}{R_3}$

(v) Since $X_L = \omega L$, then

$$\omega L_x = \frac{R_2(\omega L_4)}{R_3} \text{ from which } L_x = \frac{R_2 L_4}{R_3}$$

Thus at balance the unknown components in the simple Maxwell bridge are given by

$$\boldsymbol{R_x = \frac{R_2 R_4}{R_3} \quad \text{and} \quad L_x = \frac{R_2 L_4}{R_3}}$$

These are known as the **'balance equations'** for the bridge.

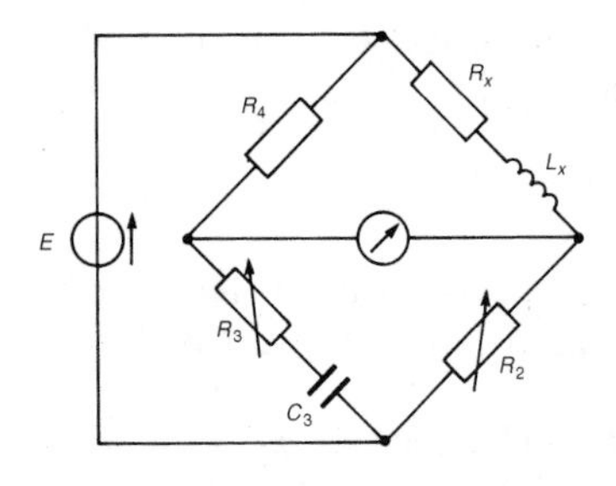

Figure 27.3 *Hay bridge*

(b) The Hay bridge

This bridge is used to measure the resistance and inductance of a coil having a very high Q-factor. A coil having unknown resistance R_x and inductance L_x is shown in the circuit diagram of a Hay bridge in Figure 27.3.

Following the procedure of Section 27.2 gives:

(i) From Figure 27.3, $Z_x = R_x + jX_{L_x}$, $Z_2 = R_2$, $Z_3 = R_3 - jX_{C_3}$, and $Z_4 = R_4$.

At balance $(Z_x)(Z_3) = (Z_2)(Z_4)$, from equation (27.3),

i.e., $(R_x + jX_{L_x})(R_3 - jX_{C_3}) = (R_2)(R_4)$

(ii) $$(R_x + jX_{L_x}) = \frac{R_2R_4}{R_3 - jX_{C_3}}$$

(iii) Rationalizing the right-hand side gives

$$(R_x + jX_{L_x}) = \frac{R_2R_4(R_3 + jX_{C_3})}{(R_3 - jX_{C_3})(R_3 + jX_{C_3})} = \frac{R_2R_4(R_3 + jX_{C_3})}{R_3^2 + X_{C_3}^2}$$

$$\text{i.e. } (R_x + jX_{L_x}) = \frac{R_2R_3R_4}{R_3^2 + X_{C_3}^2} + j\frac{R_2R_4X_{C_3}}{R_3^2 + X_{C_3}^2}$$

(iv) Equating the real parts gives $R_x = \dfrac{R_2R_3R_4}{R_3^2 + X_{C_3}^2}$

Equating the imaginary parts gives $X_{L_x} = \dfrac{R_2R_4X_{C_3}}{R_3^2 + X_{C_3}^2}$

(v) Since $X_{C_3} = \dfrac{1}{\omega C_3}$,

$$R_x = \frac{R_2R_3R_4}{R_3^2 + (1/(\omega^2C_3^2))} = \frac{R_2R_3R_4}{(\omega^2C_3^2R_3^2 + 1)/(\omega^2C_3^2)}$$

$$\text{i.e.} \quad R_x = \frac{\omega^2C_3^2R_2R_3R_4}{1 + \omega^2C_3^2R_3^2}$$

Since $X_{L_x} = \omega L_x$,

$$\omega L_x = \frac{R_2R_4(1/(\omega C_3))}{(\omega^2C_3^2R_3^2 + 1)/(\omega^2C_3^2)} = \frac{\omega^2C_3^2R_2R_4}{\omega C_3(1 + \omega^2C_3^2R_3^2)}$$

$$\text{i.e.} \quad L_x = \frac{C_3R_2R_4}{(1 + \omega^2C_3^2R_3^2)} \text{ by cancelling.}$$

Thus at balance the unknown components in the Hay bridge are given by

$$\boldsymbol{R_x = \frac{\omega^2C_3^2R_2R_3R_4}{(1 + \omega^2C_3^2R_3^2)} \text{ and } L_x = \frac{C_3R_2R_4}{(1 + \omega^2C_3^2R_3^2)}}$$

Since $\omega(=2\pi f)$ appears in the balance equations, the bridge is **frequency-dependent.**

(c) The Owen bridge

This bridge is used to measure the resistance and inductance of coils possessing a large value of inductance. A coil having unknown resistance R_x and inductance L_x is shown in the circuit diagram of an Owen bridge in Figure 27.4, from which $Z_x = R_x + jX_{L_x}$, $Z_2 = R_2 - jX_{C_2}$, $Z_3 = -jX_{C_3}$ and $Z_4 = R_4$.

At balance $(Z_x)(Z_3) = (Z_2)(Z_4)$, from equation (27.3), i.e.,

$$(R_x + jX_{L_x})(-jX_{C_3}) = (R_2 - jX_{C_2})(R_4).$$

Rearranging gives $R_x + jX_{L_x} = \dfrac{(R_2 - jX_{C_2})R_4}{-jX_{C_3}}$

By rationalizing and equating real and imaginary parts it may be shown that at balance the unknown components in the Owen bridge are given by

$$\boldsymbol{R_x = \frac{R_4 C_3}{C_2}} \text{ and } \boldsymbol{L_x = R_2 R_4 C_3}$$

(d) The Maxwell-Wien bridge

This bridge is used to measure the resistance and inductance of a coil having a low Q-factor. A coil having unknown resistance R_x and inductance L_x is shown in the circuit diagram of a Maxwell-Wien bridge in Figure 27.5, from which $Z_x = R_x + jX_{L_x}$,$Z_2 = R_2$ and $Z_4 = R_4$.

Arm 3 consists of two parallel-connected components. The equivalent impedance Z_3, is given either

(i) by $\dfrac{\text{product}}{\text{sum}}$, i.e., $Z_3 = \dfrac{(R_3)(-jX_{C_3})}{(R_3 - jX_{C_3})}$, or

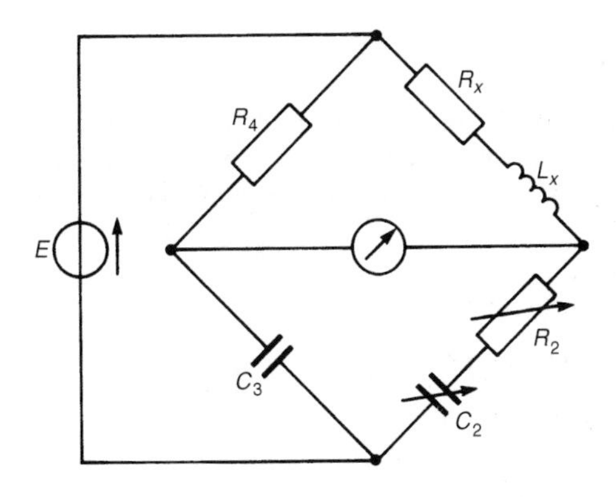

Figure 27.4 *Owen bridge*

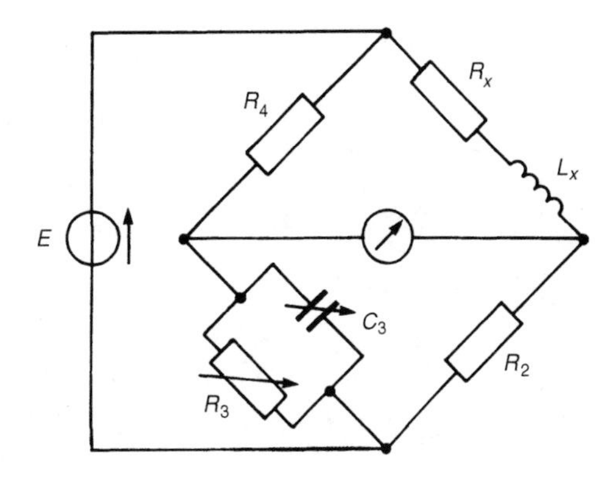

Figure 27.5 *Maxwell-Wien bridge*

(ii) by using the reciprocal impedance expression,

$$\frac{1}{Z_3} = \frac{1}{R_3} + \frac{1}{-jX_{C_3}}$$

from which $Z_3 = \dfrac{1}{(1/R_3) + (1/(-jX_{C_3}))} = \dfrac{1}{(1/R_3) + (j/X_{C_3})}$

or $\quad Z_3 = \dfrac{1}{\dfrac{1}{R_3} + j\omega C_3}$, since $X_{C_3} = \dfrac{1}{\omega C_3}$

Whenever an arm of an a.c. bridge consists of two branches in parallel, either method of obtaining the equivalent impedance may be used.

For the Maxwell-Wien bridge of Figure 27.5, at balance

$$(Z_x)(Z_3) = (Z_2)(Z_4), \text{ from equation (27.3)}$$

i.e., $\quad (R_x + jX_{L_x})\dfrac{(R_3)(-jX_{C_3})}{(R_3 - jX_{C_3})} = R_2R_4$

using method (i) for Z_3. Hence

$$(R_x + jX_{L_x}) = R_2R_4\frac{(R_3 - jX_{C_3})}{(R_3)(-jX_{C_3})}$$

By rationalizing and equating real and imaginary parts it may be shown that at balance the unknown components in the Maxwell-Wien bridge are given by

$$\boldsymbol{R_x = \frac{R_2R_4}{R_3}} \text{ and } \boldsymbol{L_x = C_3R_2R_4}$$

(e) The de Sauty bridge

This bridge provides a very simple method of measuring a capacitance by comparison with another known capacitance. In the de Sauty bridge shown in Figure 27.6, C_x is an unknown capacitance and C_4 is a standard capacitor.

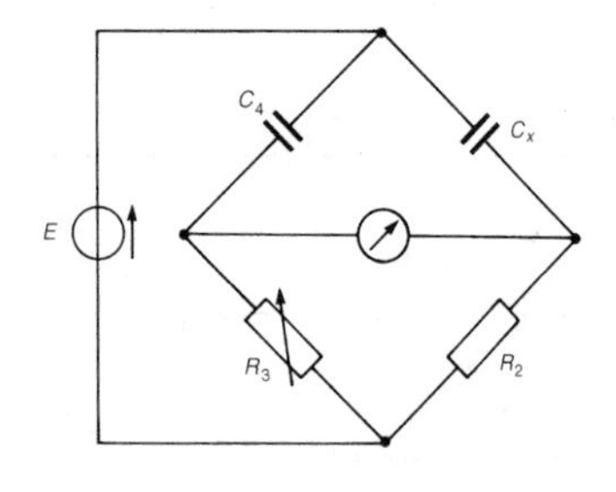

Figure 27.6 *De Sauty bridge*

At balance $\quad (Z_x)(Z_3) = (Z_2)(Z_4)$

i.e. $\quad (-jX_{C_x})(R_3) = (R_2)(-jX_{C_4})$

Hence $\quad (X_{C_x})(R_3) = (R_2)(X_{C_4})$

$$\left(\frac{1}{\omega C_x}\right)(R_3) = (R_2)\left(\frac{1}{\omega C_4}\right)$$

from which $\frac{R_3}{C_x} = \frac{R_2}{C_4}$ or $\boxed{\mathbf{C_x = \frac{R_3 C_4}{R_2}}}$

This simple bridge is usually inadequate in most practical cases. The power factor of the capacitor under test is significant because of internal dielectric losses — these losses being the dissipation within a dielectric material when an alternating voltage is applied to a capacitor.

(f) The Schering bridge

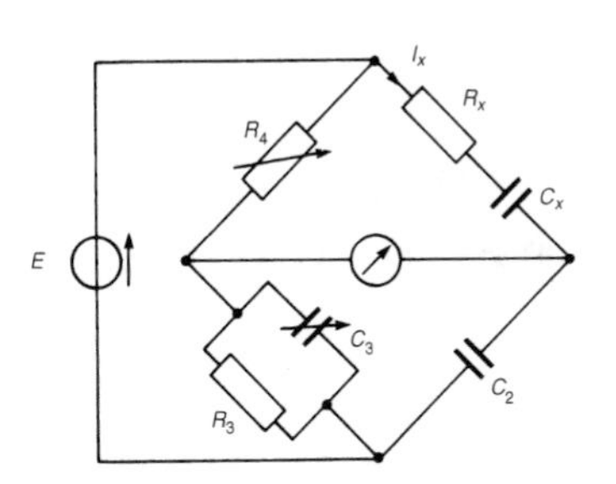

Figure 27.7 *Schering bridge*

This bridge is used to measure the capacitance and equivalent series resistance of a capacitor. From the measured values the power factor of insulating materials and dielectric losses may be determined. In the circuit diagram of a Schering bridge shown in Figure 27.7, C_x is the unknown capacitance and R_x its equivalent series resistance.

From Figure 27.7, $Z_x = R_x - jX_{C_x}$, $Z_2 = -jX_{C_2}$

$$Z_3 = \frac{(R_3)(-jX_{C_3})}{(R_3 - jX_{C_3})} \text{ and } Z_4 = R_4$$

At balance, $(Z_x)(Z_3) = (Z_2)(Z_4)$ from equation (27.3), i.e.,

$$(R_x - jX_{C_x})\frac{(R_3)(-jX_{C_3})}{R_3 - jX_{C_3}} = (-jX_{C_2})(R_4)$$

from which $$(R_x - jX_{C_x}) = \frac{(-jX_{C_2}R_4)(R_3 - jX_{C_3})}{-jX_{C_3}R_3}$$

$$= \frac{X_{C_2}R_4}{X_{C_3}R_3}(R_3 - jX_{C_3})$$

Equating the real parts gives

$$R_x = \frac{X_{C_2}R_4}{X_{C_3}} = \frac{(1/\omega C_2)R_4}{(1/\omega C_3)} = \frac{C_3 R_4}{C_2}$$

Equating the imaginary parts gives

$$-X_{C_x} = \frac{-X_{C_2}R_4}{R_3}$$

i.e. $$\frac{1}{\omega C_x} = \frac{(1/\omega C_2)R_4}{R_3} = \frac{R_4}{\omega C_2 R_3}$$

from which $$C_x = \frac{C_2 R_3}{R_4}$$

Thus at balance the unknown components in the Schering bridge are given by

$$\boxed{\mathbf{R_x = \frac{C_3 R_4}{C_2} \text{ and } C_x = \frac{C_2 R_3}{R_4}}}$$

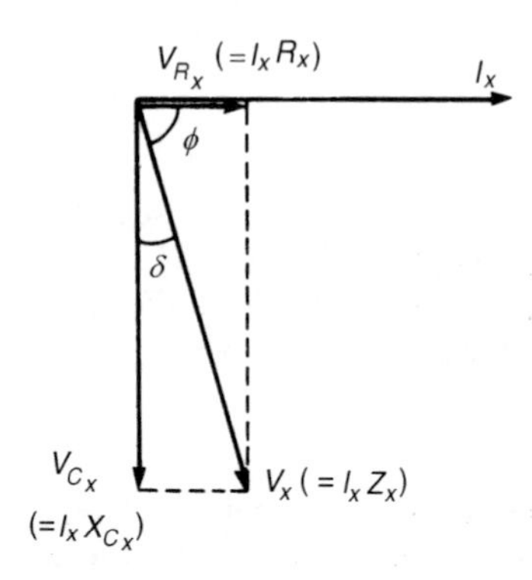

Figure 27.8 *Phasor diagram for the unknown arm in the Schering bridge*

The loss in a dielectric may be represented by either (a) a resistance in parallel with a capacitor, or (b) a lossless capacitor in series with a resistor.

If the dielectric is represented by an R-C circuit, as shown by R_x and C_x in Figure 27.7, the phasor diagram for the unknown arm is as shown in Figure 27.8. Angle ϕ is given by

$$\phi = \arctan \frac{V_{C_x}}{V_{R_x}} = \arctan \frac{I_x X_{C_x}}{I_x R_x}$$

i.e., $$\phi = \arctan\left(\frac{1}{\omega C_x R_x}\right)$$

The power factor of the unknown arm is given by $\cos\phi$.

The angle δ $(= 90° - \phi)$ is called the **loss angle** and is given by

$$\delta = \arctan \frac{V_{R_x}}{V_{C_x}} = \mathbf{arctan\ \omega C_x R_x} \text{ and}$$

$$\delta = \arctan\left[\omega\left(\frac{C_2R_3}{R_4}\right)\left(\frac{C_3R_4}{C_2}\right)\right]$$

$$= \mathbf{arctan\ (\omega R_3 C_3)}$$

(See also Chapter 39, page 716)

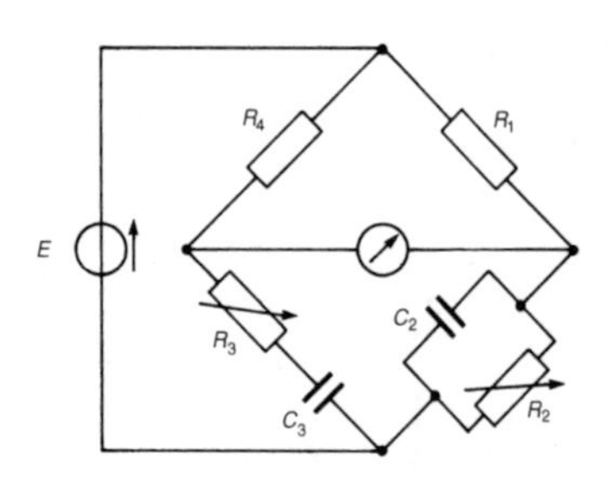

Figure 27.9 *Wien bridge*

(g) The Wien bridge

This bridge is used to measure frequency in terms of known components (or, alternatively, to measure capacitance if the frequency is known). It may also be used as a frequency-stabilizing network.

A typical circuit diagram of a Wien bridge is shown in Figure 27.9, from which

$$Z_1 = R_1,\ Z_2 = \frac{1}{(1/R_2) + j\omega C_2} \text{ (see (ii), para (d), page 482),}$$

$$Z_3 = R_3 - jX_{C_3} \text{ and } Z_4 = R_4.$$

At balance, $(Z_1)(Z_3) = (Z_2)(Z_4)$ from equation (27.3), i.e.,

$$(R_1)(R_3 - jX_{C_3}) = \left(\frac{1}{(1/R_2) + j\omega C_2}\right)(R_4)$$

Rearranging gives

$$\left(R_3 - \frac{j}{\omega C_3}\right)\left(\frac{1}{R_2} + j\omega C_2\right) = \frac{R_4}{R_1}$$

$$\frac{R_3}{R_2} + \frac{C_2}{C_3} - j\left(\frac{1}{\omega C_3 R_2}\right) + j\omega C_2 R_3 = \frac{R_4}{R_1}$$

Equating real parts gives

$$\boxed{\frac{\mathbf{R_3}}{\mathbf{R_2}} + \frac{\mathbf{C_2}}{\mathbf{C_3}} = \frac{\mathbf{R_4}}{\mathbf{R_1}}} \qquad (27.4)$$

Equating imaginary parts gives

$$-\frac{1}{\omega C_3 R_2} + \omega C_2 R_3 = 0$$

i.e. $$\omega C_2 R_3 = \frac{1}{\omega C_3 R_2}$$

from which $\omega^2 = \dfrac{1}{C_2 C_3 R_2 R_3}$

Since $\omega = 2\pi f$, $$\boxed{\textbf{frequency, } f = \frac{\mathbf{1}}{\mathbf{2\pi\sqrt{(C_2C_3R_2R_3)}}}} \qquad (27.5)$$

Note that if $C_2 = C_3 = C$ and $R_2 = R_3 = R$,

$$\text{frequency, } f = \frac{1}{2\pi\sqrt{(C^2R^2)}} = \frac{1}{2\pi CR}$$

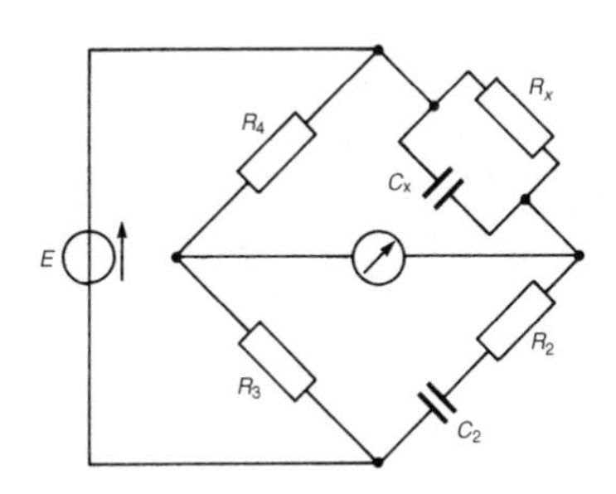

Figure 27.10

Problem 1. The a.c. bridge shown in Figure 27.10 is used to measure the capacitance C_x and resistance R_x. (a) Derive the balance equations of the bridge. (b) Given $R_3 = R_4$, $C_2 = 0.2\ \mu F$, $R_2 = 2.5\ k\Omega$ and the frequency of the supply is 1 kHz, determine the values of R_x and C_x at balance.

(a) Since C_x and R_x are the unknown values and are connected in parallel, it is easier to use the reciprocal impedance form for this branch $\left(\text{rather than } \dfrac{\text{product}}{\text{sum}}\right)$,

i.e. $$\frac{1}{Z_x} = \frac{1}{R_x} + \frac{1}{-jX_{C_x}} = \frac{1}{R_x} + \frac{j}{X_{C_x}}$$

from which $$Z_x = \frac{1}{(1/R_x) + j\omega C_x}$$

From Figure 27.10, $Z_2 = R_2 - jX_{C_2}$, $Z_3 = R_3$and $Z_4 = R_4$.

At balance, $(Z_x)(Z_3) = (Z_2)(Z_4)$

$$\left(\frac{1}{(1/R_x) + j\omega C_x}\right)(R_3) = (R_2 - j\omega X_{C_2})(R_4)$$

hence $\dfrac{R_3}{R_4(R_2 - jX_{C_2})} = \dfrac{1}{R_x} + j\omega C_x$

Rationalizing gives $\dfrac{R_3(R_2 + jX_{C_2})}{R_4(R_2^2 + X_{C_2}^2)} = \dfrac{1}{R_x} + j\omega C_x$

Hence $\dfrac{1}{R_x} + j\omega C_x = \dfrac{R_3 R_2}{R_4(R_2^2 + (1/\omega^2 C_2^2)} + \dfrac{jR_3(1/\omega C_2)}{R_3(R_2^2 + (1/\omega^2 C_2^2))}$

Equating the real parts gives

$$\frac{1}{R_x} = \frac{R_3 R_2}{R_4(R_2^2 + (1/\omega^2 C_2^2)}$$

i.e. $R_x = \dfrac{R_4}{R_2 R_3}\left(\dfrac{R_2^2\omega^2 C_2^2 + 1}{\omega^2 C_2^2}\right)$

and $\boldsymbol{R_x = \dfrac{R_4(1 + \omega^2 C_2^2 R_2^2)}{R_2 R_3 \omega^2 C_2^2}}$

Equating the imaginary parts gives

$$\omega C_x = \frac{R_3(1/\omega C_2)}{R_4(R_2^2 + (1/\omega^2 C_2^2))}$$

$$= \frac{R_3}{\omega C_2 R_4((R_2^2\omega^2 C_2^2 + 1)/\omega^2 C_2^2)}$$

i.e. $\omega C_x = \dfrac{R_3 \omega^2 C_2^2}{\omega C_2 R_4(1 + \omega^2 C_2^2 R_2^2)}$

and $\boldsymbol{C_x = \dfrac{R_3 C_2}{R_4(1 + \omega^2 C_2^2 R_2^2)}}$

(b) Substituting the given values gives

$R_x = \dfrac{(1 + \omega^2 C_2^2 R_2^2)}{R_2 \omega^2 C_2^2}$ since $R_3 = R_4$

i.e. $R_x = \dfrac{1 + (2\pi 1000)^2(0.2 \times 10^{-6})^2(2.5 \times 10^3)^2}{(2.5 \times 10^3)(2\pi 1000)^2(0.2 \times 10^{-6})^2}$

$= \dfrac{1 + 9.8696}{3.9478 \times 10^{-3}} \equiv 2.75\ \text{k}\Omega$

$C_x = \dfrac{C_2}{(1 + \omega^2 C_2^2 R_2^2)}$ since $R_3 = R_4$

$= \dfrac{(0.2 \times 10^{-6})}{1 + 9.8696}\ \mu\text{F} = 0.01840\ \mu\text{F}$ or 18.40 nF

Hence at balance $R_x = 2.75\ \text{k}\Omega$ and $C_x = 18.40$ nF

Problem 2. For the Wien bridge shown in Figure 27.9, $R_2 = R_3 = 30\ \text{k}\Omega$, $R_4 = 1\ \text{k}\Omega$ and $C_2 = C_3 = 1$ nF. Determine, when the bridge is balanced, (a) the value of resistance R_1, and (b) the frequency of the bridge.

(a) From equation (27.4)

$$\frac{R_3}{R_2} + \frac{C_2}{C_3} = \frac{R_4}{R_1}$$

i.e., $1 + 1 = 1000/R_1$, since $R_2 = R_3$ and $C_2 = C_3$, from which

$$\textbf{resistance } \boldsymbol{R_1} = \frac{1000}{2} = \mathbf{500\ \Omega}$$

(b) From equation (27.5),

$$\text{frequency, } f = \frac{1}{2\pi\sqrt{(C_2C_3R_2R_3)}} = \frac{1}{2\pi\sqrt{[(10^{-9})^2(30\times 10^3)^2]}}$$

$$= \frac{1}{2\pi(10^{-9})(30\times 10^3)} \equiv \mathbf{5.305\ kHz}$$

Problem 3. A Schering bridge network is as shown in Figure 27.7, page 480. Given $C_2 = 0.2\ \mu\text{F}$, $R_4 = 200\ \Omega$, $R_3 = 600\ \Omega$, $C_3 = 4000$ pF and the supply frequency is 1.5 kHz, determine, when the bridge is balanced, (a) the value of resistance R_x, (b) the value of capacitance C_x, (c) the phase angle of the unknown arm, (d) the power factor of the unknown arm and (e) its loss angle.

From para (f), the equations for R_x and C_x at balance are given by

$$R_x = \frac{R_4C_3}{C_2} \text{ and } C_x = \frac{C_2R_3}{R_4}$$

(a) Resistance, $R_x = \dfrac{R_4C_3}{C_2} = \dfrac{(200)(4000\times 10^{-12})}{0.2\times 10^{-6}} = \mathbf{4\ \Omega}$

(b) Capacitance, $C_x = \dfrac{C_2R_3}{R_4} = \dfrac{(0.2\times 10^{-6})(600)}{(200)} F = \mathbf{0.6\ \mu F}$

(c) The phasor diagram for R_x and C_x in series is shown in Figure 27.11.

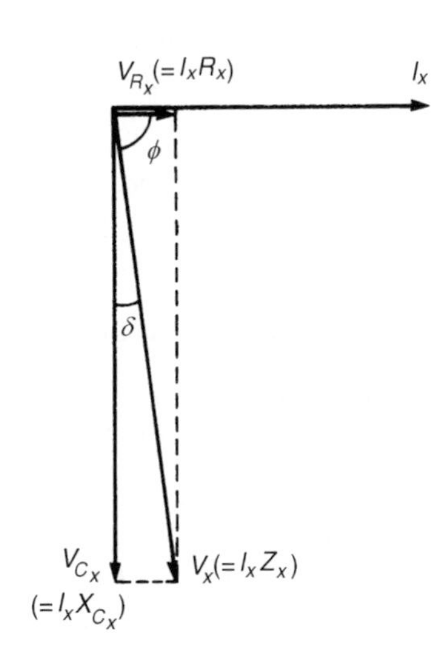

Figure 27.11

Phase angle, $\phi = \arctan \dfrac{V_{C_x}}{V_{R_x}} = \arctan \dfrac{I_xX_{C_x}}{I_xR_x} = \arctan \dfrac{1}{\omega C_xR_x}$

i.e. $\quad \Phi = \arctan\left(\dfrac{1}{(2\pi 1500)(0.6\times 10^{-6})(4)}\right)$

$= \arctan 44.21 = \mathbf{88.7^\circ}$ **lead**

(d) Power factor of capacitor = $\cos\phi = \cos 88.7° = \mathbf{0.0227}$

(e) Loss angle, shown as δ in Figure 27.11, is given by
$\delta = 90° - 88.7° = \mathbf{1.3°}$

Alternatively, loss angle $\delta = \arctan \omega C_x R_x$ (see para (f), page 483)

$$= \arctan\left(\frac{1}{44.21}\right) \text{ from (c) above,}$$

i.e., $\delta = \mathbf{1.3°}$

Further problems on a.c. bridges may be found in Section 27.4 following, problems 1 to 13.

27.4 Further problems on a.c. bridges

1 A Maxwell-Wien bridge circuit ABCD has the following arm impedances: AB, 250 Ω resistance; BC, 2 μF capacitor in parallel with a 10 kΩ resistor; CD, 400 Ω resistor; DA, unknown inductor having inductance L in series with resistance R. Determine the values of L and R if the bridge is balanced. [$L = 0.20$ H, $R = 10$ Ω]

2 In a four-arm de Sauty a.c. bridge, arm 1 contains a 2 kΩ non-inductive resistor, arm 3 contains a loss-free 2.4 μF capacitor, and arm 4 contains a 5 kΩ non-inductive resistor. When the bridge is balanced, determine the value of the capacitor contained in arm 2. [6 μF]

3 A four-arm bridge ABCD consists of: AB — fixed resistor R_1; BC — variable resistor R_2 in series with a variable capacitor C_2; CD — fixed resistor R_3; DA — coil of unknown resistance R and inductance L. Determine the values of R and L if, at balance, $R_1 = 1$ kΩ, $R_2 = 2.5$ kΩ, $C_2 = 4000$ pF, $R_3 = 1$ kΩ and the supply frequency is 1.6 kHz. [$R = 4.00$ Ω, $L = 3.96$ mH]

4 The bridge shown in Figure 27.12 is used to measure capacitance C_x and resistance R_x. Derive the balance equations of the bridge and determine the values of C_x and R_x when $R_1 = R_4$, $C_2 = 0.1$ μF, $R_2 = 2$ kΩ and the supply frequency is 1 kHz. [$C_x = 38.77$ nF, $R_x = 3.27$ kΩ]

5 In a Schering bridge network ABCD, the arms are made up as follows: AB — a standard capacitor C_1; BC — a capacitor C_2 in parallel with a resistor R_2; CD — a resistor R_3; DA — the capacitor under test, represented by a capacitor C_x in series with a resistor R_x. The detector is connected between B and D and the a.c. supply is connected between A and C. Derive the equations for R_x and C_x when the bridge is balanced. Evaluate R_x and C_x if, at balance, $C_1 = 1$ nF, $R_2 = 100$ Ω, $R_3 = 1$ kΩand $C_2 = 10$ nF. [$R_x = 10$ kΩ, $C_x = 100$ pF]

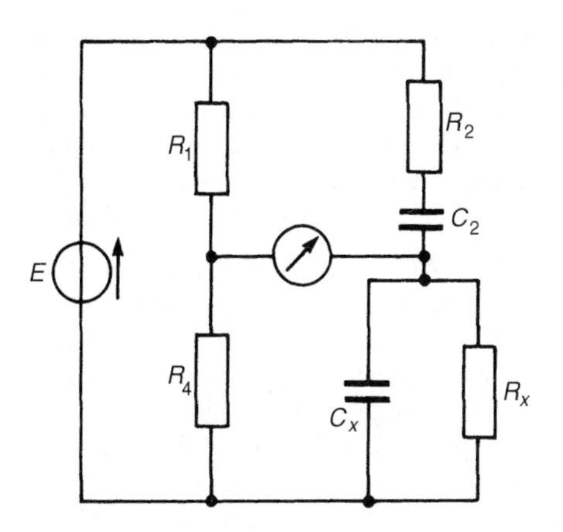

Figure 27.12

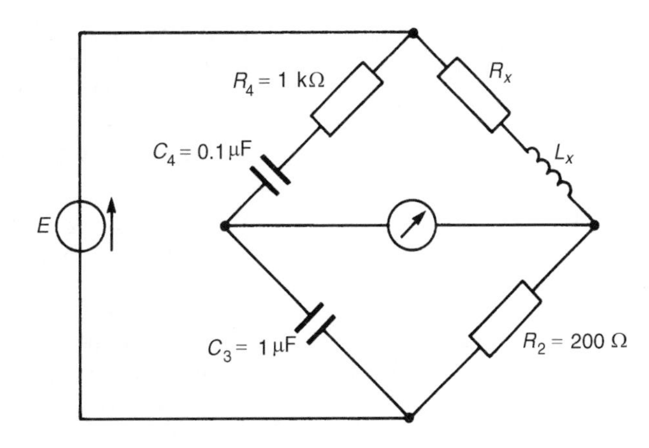

Figure 27.13

6 The a.c. bridge shown in Figure 27.13 is balanced when the values of the components are shown. Determine at balance, the values of R_x and L_x. [$R = 2$ kΩ, $L_x = 0.2$ H]

7 An a.c. bridge has, in arm AB, a pure capacitor of 0.4 μF; in arm BC, a pure resistor of 500 Ω; in arm CD, a coil of 50 Ω resistance and 0.1 H inductance; in arm DA, an unknown impedance comprising resistance R_x and capacitance C_x in series. If the frequency of the bridge at balance is 800 Hz, determine the values of R_x and C_x. [$R_x = 500$ Ω, $C_x = 4$ μF]

8 When the Wien bridge shown in Figure 27.9 is balanced, the components have the following values: $R_2 = R_3 = 20$ kΩ, $R_4 =$ 500 Ω, $C_2 = C_3 = 800$ pF. Determine for the balance condition (a) the value of resistance R_1 and (b) the frequency of the bridge supply. [(a) 250 Ω (b) 9.95 kHz]

9 The conditions at balance of a Schering bridge ABCD used to measure the capacitance and loss angle of a paper capacitor are as follows: AB — a pure capacitance of 0.2 μF; BC — a pure capacitance of 3000 pF in parallel with a 400 Ω resistance; CD — a pure resistance of 200 Ω; DA — the capacitance under test which may be considered as a capacitance C_x in series with a resistance R_x. If the supply frequency is 1 kHz determine (a) the value of R_x, (b) the value of C_x, (c) the power factor of the capacitor, and (d) its loss angle. [(a) 3 Ω (b) 0.4 μF (c) 0.0075 (d) 0.432°]

10 At balance, an a.c. bridge PQRS used to measure the inductance and resistance of an inductor has the following values: PQ — a non-inductive 400 Ω resistor; QR — the inductor with unknown inductance L_x in series with resistance R_x; RS — a 3 μF capacitor in series with a non-inductive 250 Ω resistor; SP — a 15 μF capacitor. A detector is connected between Q and S and the a.c. supply is connected between P and R. Derive the balance equations for R_x and L_x and determine their values. [$R_x = 2$ kΩ, $L_x = 1.5$ H]

11 A 1 kHz a.c. bridge ABCD has the following components in its four arms: AB — a pure capacitor of 0.2 μF; BC — a pure resistance of 500 Ω; CD — an unknown impedance; DA — a 400 Ω resistor in parallel with a 0.1 μF capacitor. If the bridge is balanced, determine the series components comprising the impedance in arm CD.

[$R = 59.41$ Ω, $L = 37.6$ mH]

12 An a.c. bridge ABCD has in arm AB a standard lossless capacitor of 200 pF; arm BC, an unknown impedance, represented by a lossless capacitor C_x in series with a resistor R_x; arm CD, a pure 5 kΩ resistor; arm DA, a 6 Ω resistor in parallel with a variable capacitor set at 250 pF. The frequency of the bridge supply is 1500 Hz. Determine for the condition when the bridge is balanced (a) the values of R_x and C_x, and (b) the loss angle.

[(a) $R_x = 6.25$ kΩ, $C_x = 240$ pF; (b) 0.81°]

13 An a.c. bridge ABCD has the following components: AB — *a* 1 kΩ resistance in parallel with a 0.2 μF capacitor; BC — a 1.2 kΩ resistance; CD — a 750 Ω resistance; DA — a 0.8 μF capacitor in series with an unknown resistance. Determine (a) the frequency for which the bridge is in balance, and (b) the value of the unknown resistance in arm DA to produce balance.

[(a) 649.7 Hz (b) 375 Ω]

28 Series resonance and Q-factor

At the end of this chapter you should be able to:

- state the conditions for resonance in an a.c. series circuit
- calculate the resonant frequency in an a.c. series circuit, $f_r = \dfrac{1}{2\pi\sqrt{(LC)}}$
- define Q-factor as $\dfrac{X}{R}$ and as $\dfrac{V_L}{V}$ or $\dfrac{V_C}{V}$
- determine the maximum value of V_C and V_{COIL} and the frequency at which this occurs
- determine the overall Q-factor for two components in series
- define bandwidth and selectivity
- calculate Q-factor and bandwidth in an a.c. series circuit
- determine the current and impedance when the frequency deviates from the resonant frequency

28.1 Introduction

When the voltage V applied to an electrical network containing resistance, inductance and capacitance is in phase with the resulting current I, the circuit is said to be **resonant**. The phenomenon of **resonance** is of great value in all branches of radio, television and communications engineering, since it enables small portions of the communications frequency spectrum to be selected for amplification independently of the remainder.

At resonance, the equivalent network impedance Z is purely resistive since the supply voltage and current are in phase. The power factor of a resonant network is unity,(i.e., power factor $= \cos\phi = \cos 0 = 1$).

In electrical work there are two types of resonance — one associated with series circuits,(which was introduced in Chapter 15), when the input impedance is a minimum, (which is discussed further in this chapter), and the other associated with simple parallel networks, when the input impedance is a maximum (which is discussed in Chapter 29).

28.2 Series resonance

Figure 28.1 shows a circuit comprising a coil of inductance L and resistance R connected in series with a capacitor C. The R–L–C series circuit has a total impedance Z given by $Z = R + j(X_L - X_C)$ ohms, or $Z =$

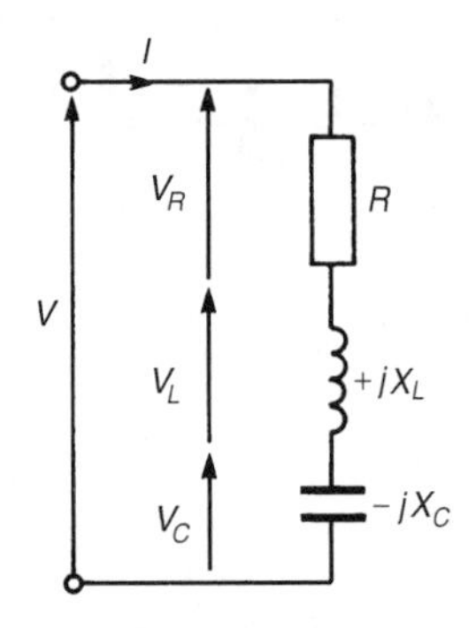

Figure 28.1 *R – L – C series circuit*

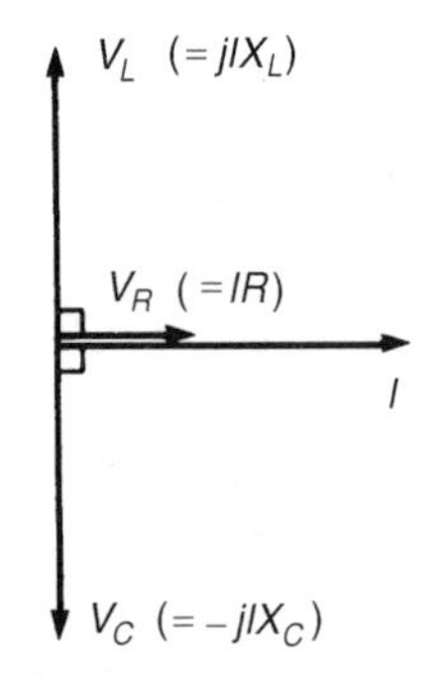

Figure 28.2 *Phasor diagram* $|V_L| = |V_C|$

$R + j(\omega L - 1/\omega C)$ ohms where $\omega = 2\pi f$. The circuit is at resonance when $(X_L - X_C) = 0$, i.e., when $X_L = X_C$ or $\omega L = 1/(\omega C)$. The phasor diagram for this condition is shown in Figure 28.2, where $|V_L| = |V_C|$.

Since at resonance $\omega_r L = \dfrac{1}{\omega_r C}$, $\omega_r^2 = \dfrac{1}{LC}$ and $\omega = \dfrac{1}{\sqrt{(LC)}}$

Thus resonant frequency, $\boldsymbol{f_r = \dfrac{1}{2\pi\sqrt{(LC)}}}$ **hertz,** since $\omega_r = 2\pi f_r$

Figure 28.3 shows how inductive reactance X_L and capacitive reactance X_C vary with the frequency. At the resonant frequency f_r, $|X_L| = |X_C|$. Since impedance $Z = R + j(X_L - X_C)$ and, at resonance, $(X_L - X_C) = 0$, then **impedance *Z* = *R* at resonance**. This is the **minimum** value possible for the impedance as shown in the graph of the modulus of impedance, $|Z|$, against frequency in Figure 28.4.

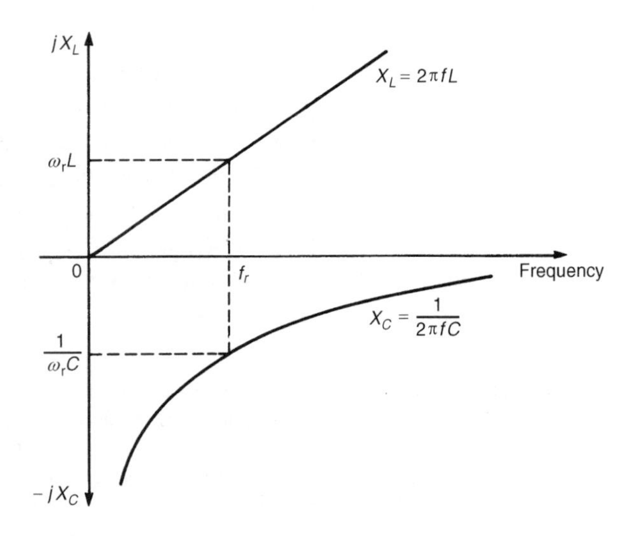

Figure 28.3 *Variation of X_L and X_C with frequency*

At frequencies less than f_r, $X_L < X_C$ and the circuit is capacitive; at frequencies greater than f_r, $X_L > X_C$ and the circuit is inductive.

Current $I = V/Z$. Since impedance Z is a minimum value at resonance, the **current *I* has a maximum value**. At resonance, current $I = V/R$. A graph of current against frequency is shown in Figure 28.4.

Problem 1. A coil having a resistance of 10 Ω and an inductance of 75 mH is connected in series with a 40 μF capacitor across a 200 V a.c. supply. Determine at what frequency resonance occurs, and (b) the current flowing at resonance.

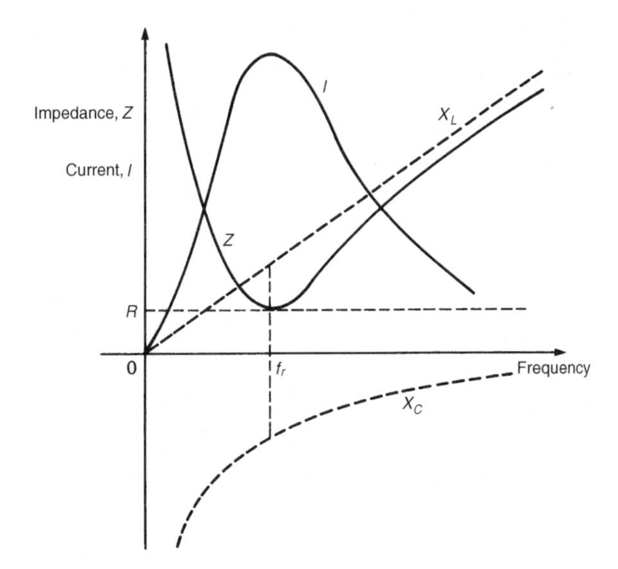

Figure 28.4 *|Z| and I plotted against frequency*

(a) Resonant frequency,

$$f_r = \frac{1}{2\pi\sqrt{(LC)}} = \frac{1}{2\pi\sqrt{[(75 \times 10^{-3})(40 \times 10^{-6})]}}$$

i.e., $f_r = \mathbf{91.9\ Hz}$

(b) Current at resonance, $\boldsymbol{I} = \frac{V}{R} = \frac{200}{10} = \mathbf{20\ A}$

> Problem 2. An R–L–C series circuit is comprised of a coil of inductance 10 mH and resistance 8 Ω and a variable capacitor C. The supply frequency is 1 kHz. Determine the value of capacitor C for series resonance.

At resonance, $\omega_r L = 1/(\omega_r C)$, from which, capacitance, $C = 1/(\omega_r^2 L)$

Hence **capacitance** $\boldsymbol{C} = \frac{1}{(2\pi 1000)^2(10 \times 10^{-3})} = \mathbf{2.53\ \mu F}$

> Problem 3. A coil having inductance L is connected in series with a variable capacitor C. The circuit possesses stray capacitance C_S which is assumed to be constant and effectively in parallel with the variable capacitor C. When the capacitor is set to 1000 pF the resonant frequency of the circuit is 92.5 kHz, and when the capacitor is set to 500 pF the resonant frequency is 127.8 kHz. Determine the values of (a) the stray capacitance C_S, and (b) the coil inductance L.

For a series R–L–C circuit the resonant frequency f_r is given by:

$$f_r = \frac{1}{2\pi\sqrt{(LC)}}$$

The total capacitance of C in parallel with C_S is given by $(C + C_S)$
At 92.5 kHz, $C = 1000$ pF. Hence

$$92.5 \times 10^3 = \frac{1}{2\pi\sqrt{[L(1000 + C_S)10^{-12}]}} \qquad (1)$$

At 127.8 kHz, $C = 500$ pF. Hence

$$127.8 \times 10^3 = \frac{1}{2\pi\sqrt{[L(500 + C_S)10^{-12}]}} \qquad (2)$$

(a) Dividing equation (2) by equation (1) gives:

$$\frac{127.8 \times 10^3}{92.5 \times 10^3} = \frac{\dfrac{1}{2\pi\sqrt{[L(500 + C_S)10^{-12}]}}}{\dfrac{1}{2\pi\sqrt{[L(1000 + C_S)10^{-12}]}}}$$

i.e.,
$$\frac{127.8}{92.5} = \frac{\sqrt{[L(1000 + C_S)10^{-12}]}}{\sqrt{[L(500 + C_S)10^{-12}]}} = \sqrt{\left(\frac{1000 + C_S}{500 + C_S}\right)}$$

where C_S is in picofarads, from which,

$$\left(\frac{127.8}{92.5}\right)^2 = \frac{1000 + C_S}{500 + C_S}$$

i.e.,
$$1.909 = \frac{1000 + C_S}{500 + C_S}$$

Hence
$$1.909(500 + C_S) = 1000 + C_S$$

$$954.5 + 1.909C_S = 1000 + C_S$$

$$1.909C_S - C_S = 1000 - 954.5$$

$$0.909C_S = 45.5$$

Thus **stray capacitance C_S** $= 45.5/0.909 =$ **50 pF**

(b) Substituting $C_S = 50$ pF in equation (1) gives:

$$92.5 \times 10^3 = \frac{1}{2\pi\sqrt{[L(1050 \times 10^{-12})]}}$$

$$\text{Hence} \quad (92.5 \times 10^3 \times 2\pi)^2 = \frac{1}{L(1050 \times 10^{-12})}$$

$$\text{from which, } \textbf{inductance } \boldsymbol{L} = \frac{1}{(1050 \times 10^{-12})(92.5 \times 10^3 \times 2\pi)^2}\ \text{H}$$

$$= \mathbf{2.82\ mH}$$

Further problems on series resonance may be found in Section 28.8, problems 1 to 5, page 512

28.3 Q-factor

Q-factor is a figure of merit for a resonant device such as an L–C–R circuit.

Such a circuit resonates by cyclic interchange of stored energy, accompanied by energy dissipation due to the resistance.

$$\text{By definition, at resonance } Q = 2\pi \left(\frac{\text{maximum energy stored}}{\text{energy loss per cycle}}\right)$$

Since the energy loss per cycle is equal to (the average power dissipated) × (periodic time),

$$Q = 2\pi \left(\frac{\text{maximum energy stored}}{\text{average power dissipated} \times \text{periodic time}}\right)$$

$$= 2\pi \left(\frac{\text{maximum energy stored}}{\text{average power dissipated} \times (1/f_r)}\right)$$

since the periodic time $T = 1/f_r$.

$$\text{Thus} \quad Q = 2\pi f_r \left(\frac{\text{maximum energy stored}}{\text{average power dissipated}}\right)$$

$$\text{i.e.,} \quad Q = \omega_r \left(\frac{\text{maximum energy stored}}{\text{average power dissipated}}\right)$$

where ω_r is the angular frequency at resonance.

In an L–C–R circuit both of the reactive elements store energy during a quarter cycle of the alternating supply input and return it to the circuit source during the following quarter cycle. An inductor stores energy in its magnetic field, then transfers it to the electric field of the capacitor and then back to the magnetic field, and so on. Thus the inductive and capacitive elements transfer energy from one to the other successively with the source of supply ideally providing no additional energy at all. Practical reactors both store and dissipate energy.

Q-factor is an abbreviation for **quality factor** and refers to the 'goodness' of a reactive component.

For an **inductor**, $Q = \omega_r \left(\dfrac{\text{maximum energy stored}}{\text{average power dissipated}}\right)$

$$= \omega_r \left(\frac{\frac{1}{2}LI_m^2}{I^2R}\right) = \frac{\omega_r\left(\frac{1}{2}LI_m^2\right)}{(I_m/\sqrt{2})^2R} = \frac{\omega_r L}{R} \qquad (28.1)$$

For a **capacitor** $Q = \dfrac{\omega_r\left(\frac{1}{2}CV_m^2\right)}{(I_m/\sqrt{2})^2R} = \dfrac{\omega_r \frac{1}{2}C(I_mX_C)^2}{(I_m/\sqrt{2})^2R}$

$$= \frac{\omega_r \frac{1}{2}CI_m^2(1/\omega_r C)^2}{(I_m/\sqrt{2})^2R}$$

i.e., $$Q = \frac{1}{\omega_r CR} \qquad (28.2)$$

From expressions (28.1) and (28.2) it can be deduced that

$$Q = \frac{X_L}{R} = \frac{X_C}{R} = \frac{\text{reactance}}{\text{resistance}}$$

In fact, Q-factor can also be defined as

$$\text{Q-factor} = \frac{\text{reactance power}}{\text{resistance}} = \frac{Q}{P}$$

where Q is the reactive power which is also the peak rate of energy storage, and P is the average energy dissipation rate. Hence

$$\text{Q-factor} = \frac{Q}{P} = \frac{I^2X_L(\text{or } I^2X_C)}{I^2R} = \frac{X_L}{R}\left(\text{or } \frac{X_C}{R}\right)$$

i.e., $$\boldsymbol{Q = \frac{\textbf{reactance}}{\textbf{resistance}}}$$

In an R–L–C series circuit the amount of energy stored at resonance is constant.

When the capacitor voltage is a maximum, the inductor current is zero, and vice versa, i.e., $\frac{1}{2}LI_m^2 = \frac{1}{2}CV_m^2$

Thus the Q-factor at resonance, Q_r is given by

$$\boldsymbol{Q_r = \frac{\omega_r L}{R} = \frac{1}{\omega_r CR}} \qquad (28.3)$$

However, at resonance $\omega_r = 1/\sqrt{(LC)}$

Hence $Q_r = \dfrac{\omega_r L}{R} = \dfrac{1}{\sqrt{(LC)}}\left(\dfrac{L}{R}\right)$ i.e, $\boldsymbol{Q_r = \dfrac{1}{R}\sqrt{\left(\dfrac{L}{C}\right)}}$

It should be noted that when Q-factor is referred to, it is nearly always assumed to mean 'the Q-factor at resonance'.

With reference to Figures 28.1 and 28.2, at resonance, $V_L = V_C$

$$V_L = IX_L = I\omega_r L = \frac{V}{R}\omega_r L = \left(\frac{\omega_r L}{R}\right) V = Q_r V$$

and $$V_C = IX_C = \frac{I}{\omega_r C} = \frac{V/R}{\omega_r C} = \left(\frac{1}{\omega_r CR}\right) V = Q_r V$$

Hence, at resonance, $V_L = V_C = Q_r V$

or $$\boxed{\boldsymbol{Q_r = \frac{V_L \text{ (or } V_C)}{V}}}$$

The voltages V_L and V_C at resonance may be much greater than that of the supply voltage V. For this reason Q is often called the **circuit magnification factor**. It represents a measure of the number of times V_L or V_C is greater than the supply voltage.

The Q-factor at resonance can have a value of several hundreds. Resonance is usually of interest only in circuits of Q-factor greater than about 10; circuits having Q considerably below this value are effectively merely operating at unity power factor.

Problem 4. A series circuit comprises a 10 Ω resistance, a 5 μF capacitor and a variable inductance L. The supply voltage is $20\angle 0°$ volts at a frequency of 318.3 Hz. The inductance is adjusted until the p.d. across the 10 Ω resistance is a maximum. Determine for this condition (a) the value of inductance L, (b) the p.d. across each component and (c) the Q-factor.

(a) The maximum voltage across the resistance occurs at resonance when the current is a maximum. At resonance, $\omega_r L = 1/(\omega_r C)$, from which

$$\textbf{inductance } \boldsymbol{L} = \frac{1}{\omega_r^2 C} = \frac{1}{(2\pi 318.3)^2(5 \times 10^{-6})}$$

$$= \mathbf{0.050\ H} \text{ or } \mathbf{50\ mH}$$

(b) Current at resonance $I_r = \dfrac{V}{R} = \dfrac{20\angle 0°}{10\angle 0°} = 2.0\angle 0°$ A

p.d. across resistance, $V_R = I_r R = (2.0\angle 0°)(10) = \mathbf{20\angle 0°\ V}$

p.d. across inductance, $V_L = IX_L$

$$X_L = 2\pi(318.3)(0.050) = 100\ \Omega$$

Hence $$V_L = (2.0\angle 0°(100\angle 90°) = \mathbf{200\angle 90°\ V}$$

$$\text{p.d. across capacitor, } \boldsymbol{V_C} = IX_C = (2.0\angle 0^\circ)(100\angle -90^\circ)$$
$$= \mathbf{200\angle -90^\circ\ V}$$

(c) Q-factor at resonance, $\boldsymbol{Q_r} = \dfrac{V_L(\text{or } V_C)}{V} = \dfrac{200}{20} = \mathbf{10}$

$$\left[\text{Alternatively, } Q_r = \frac{\omega_r L}{R} = \frac{100}{10} = \mathbf{10}\right.$$

$$\text{or} \qquad Q_r = \frac{1}{\omega_r CR} = \frac{1}{2\pi(318.3)(5\times 10^{-6})(10)} = \mathbf{10}$$

$$\left.\text{or} \qquad Q_r = \frac{1}{R}\sqrt{\left(\frac{L}{C}\right)} = \frac{1}{10}\sqrt{\left(\frac{0.050}{5\times 10^{-6}}\right)} = \mathbf{10}\right]$$

28.4 Voltage magnification

For a circuit with a high value of Q (say, exceeding 100), the maximum volt-drop across the coil, V_{COIL}, and the maximum volt-drop across the capacitor, V_C, coincide with the maximum circuit current at the resonant frequency f_r, as shown in 28.5(a). However, if a circuit of low Q (say, less than 10) is used, it may be shown experimentally that the maximum value of V_C occurs at a frequency less than f_r while the maximum value of V_{COIL} occurs at a frequency higher than f_r, as shown in Figure 28.5(b). The maximum current, however, still occurs at the resonant frequency with low Q. This is analysed below.

Since $Q_r = \dfrac{V_C}{V}$ then $V_C = VQ_r$

However $V_C = IX_C = I\left(\dfrac{-j}{\omega C}\right) = I\left(\dfrac{1}{j\omega C}\right)$ and since $I = \dfrac{V}{Z}$,

$$V_C = \frac{V}{Z}\left(\frac{1}{j\omega C}\right) = \frac{V}{(j\omega C)Z}$$

$$Z = R + j\left(\omega L - \frac{1}{\omega C}\right)$$

$$\text{thus} \qquad V_C = \frac{V}{(j\omega C)\left[R + j\left(\omega L - \dfrac{1}{\omega C}\right)\right]}$$

$$= \frac{V}{j\omega CR + j^2\omega^2 CL - j^2\dfrac{\omega C}{\omega C}}$$

$$= \frac{V}{j\omega CR - \omega^2 LC + 1} = \frac{V}{(1-\omega^2 LC) + j\omega CR}$$

$$= \frac{V[(1-\omega^2 LC) - j\omega CR]}{[(1-\omega^2 LC) + j\omega CR][(1-\omega^2 LC) - j\omega CR]}$$

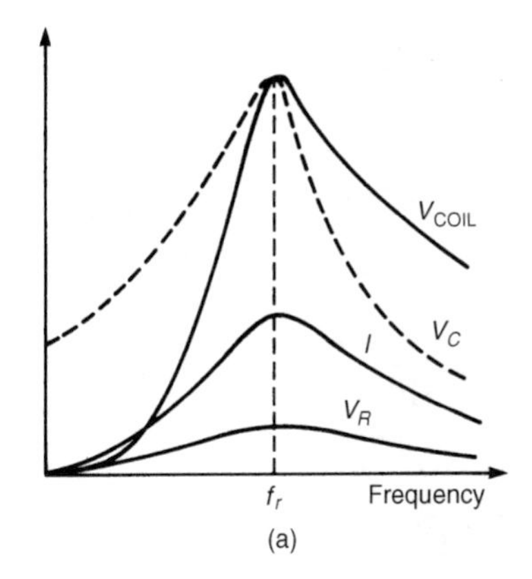

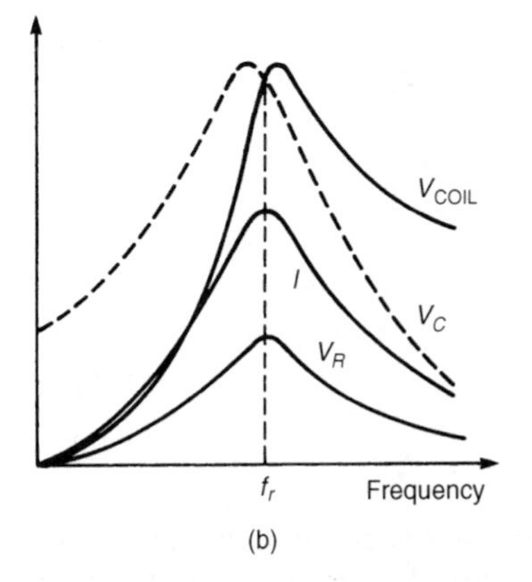

Figure 28.5 *(a) High Q-factor (b) Low Q-factor*

$$= \frac{V[(1-\omega^2LC) - j\omega CR]}{[(1-\omega^2LC)^2 + (\omega CR)^2}$$

The magnitude of V_C, $|V_C| = \dfrac{V\sqrt{[(1-\omega^2LC)^2 + (\omega CR)^2]}}{[(1-\omega^2LC)^2 + (\omega CR)^2]}$

from the Argand diagram

$$= \frac{V}{\sqrt{[(1-\omega^2LC)^2 + (\omega CR)^2]}} \qquad (28.4)$$

To find the maximum value of V_C, equation (28.4) is differentiated with respect to ω, equated to zero and then solved—this being the normal procedure for maximum/minimum problems. Thus, using the quotient and function of a function rules:

$$\frac{dV_C}{d\omega} = \frac{\sqrt{[(1-\omega^2LC)^2 + (\omega CR)^2]}[0] - [V]\frac{1}{2}[(1-\omega^2LC)^2 + (\omega CR)^2]^{-1/2}2(1-\omega^2LC)(-2\omega LC) + 2\omega C^2R^2}{\left\{\sqrt{[(1-\omega^2LC)^2 + (\omega CR)^2]}\right\}^2}$$

$$= \frac{0 - \dfrac{V}{2}[(1-\omega^2LC)^2 + (\omega CR)^2]^{-1/2}2(1-\omega^2LC) \times(-2\omega LC) + 2\omega C^2R^2}{(1-\omega^2LC)^2 + (\omega CR)^2}$$

$$= \frac{-\dfrac{V}{2}[2(1-\omega^2LC)(-2\omega LC) + 2\omega C^2R^2]}{[(1-\omega^2LC)^2 + (\omega CR)^2]^{3/2}} = 0$$

for a maximum value

Hence $\quad -\dfrac{V}{2}[2(1-\omega^2LC)(-2\omega LC) + 2\omega C^2R^2] = 0$

and $\quad -V[(1-\omega^2LC)(-2\omega LC) + \omega C^2R^2] = 0$

and $\quad (1-\omega^2LC)(-2\omega LC) + \omega C^2R^2 = 0$

from which, $\omega C^2R^2 = (1-\omega^2LC)(2\omega LC)$

i.e., $\quad C^2R^2 = 2LC(1-\omega^2LC)$

$$\frac{C^2R^2}{LC} = 2 - 2\omega^2LC \text{ and } 2\omega^2LC = 2 - \frac{CR^2}{L}$$

Hence $\quad \omega^2 = \dfrac{2}{2LC} - \dfrac{\dfrac{CR^2}{L}}{2LC} = \dfrac{1}{LC} - \dfrac{1}{2}\left(\dfrac{R}{L}\right)^2$

The resonant frequency, $\omega_r = \dfrac{1}{\sqrt{(LC)}}$ from which, $\omega_r^2 = \dfrac{1}{LC}$

Thus $$\omega^2 = \omega_r^2 - \frac{1}{2}\left(\frac{R}{L}\right)^2 \qquad (28.5)$$

$$Q = \frac{\omega_r L}{R} \text{ from which } \frac{R}{L} = \frac{\omega_r}{Q} \text{ and } \left(\frac{R}{L}\right)^2 = \frac{\omega_r^2}{Q^2}$$

Hence, from equation (28.5) $$\omega^2 = \omega_r^2 - \frac{1}{2}\frac{\omega_r^2}{Q^2}$$

i.e., $$\omega^2 = \omega_r^2\left(1 - \frac{1}{2Q^2}\right) \qquad (28.6)$$

or $$\omega = \omega_r\sqrt{\left(1 - \frac{1}{2Q^2}\right)}$$

or $$\boxed{\boldsymbol{f = f_r\sqrt{\left(1 - \frac{1}{2Q^2}\right)}}} \qquad (28.7)$$

Hence the maximum p.d. across the capacitor does not occur at the resonant frequency, but at a frequency slightly less than f_r as shown in Figure 28.5(b). If Q is large, then $f \approx f_r$ as shown in Figure 28.5(a).

From equation (28.4), $|V_C| = \dfrac{V}{\sqrt{[(1 - \omega^2 LC)^2 + (\omega CR)^2]}}$

and substituting $\omega^2 = \omega_r^2\left(1 - \dfrac{1}{2Q^2}\right)$ from equation (28.6) gives:

maximum value of V_c,

$$V_{C_m} = \frac{V}{\sqrt{\left[\left(1 - \omega_r^2\left(1 - \frac{1}{2Q^2}\right)LC\right)^2 + \omega_r^2\left(1 - \frac{1}{2Q^2}\right)C^2R^2\right]}}$$

$\omega_r^2 = \dfrac{1}{LC}$ hence

$$V_{C_m} = \frac{V}{\sqrt{\left[\left(1 - \frac{1}{LC}\left(1 - \frac{1}{2Q^2}\right)LC\right)^2 + \frac{1}{LC}\left(1 - \frac{1}{2Q^2}\right)C^2R^2\right]}}$$

$$= \frac{V}{\sqrt{\left[\left(1-\left(1-\frac{1}{2Q^2}\right)\right)^2+\frac{CR^2}{L}\left(1-\frac{1}{2Q^2}\right)\right]}}$$

$$= \frac{V}{\sqrt{\left[\frac{1}{4Q^4}+\frac{CR^2}{L}-\frac{CR^2}{L}\left(\frac{1}{2Q^2}\right)\right]}} \qquad (28.8)$$

$$Q = \frac{\omega_r L}{R} = \frac{1}{\omega_r CR} \text{ hence } Q^2 = \left(\frac{\omega_r L}{R}\right)\left(\frac{1}{\omega_r CR}\right) = \frac{L}{CR^2}$$

from which, $\frac{CR^2}{L} = \frac{1}{Q^2}$

Substituting in equation (28.8),

$$V_{C_m} = \frac{V}{\sqrt{\left(\frac{1}{4Q^4}+\frac{1}{Q^2}-\frac{1}{2Q^4}\right)}} = \frac{V}{\sqrt{\left(\frac{1}{Q^2}\left[\frac{1}{4Q^2}+1-\frac{1}{2Q^2}\right]\right)}}$$

$$= \frac{V}{\frac{1}{Q}\sqrt{\left[1-\frac{1}{4Q^2}\right]}}$$

i.e.,

$$\boxed{V_{C_m} = \frac{QV}{\sqrt{\left[1-\left(\frac{1}{2Q}\right)^2\right]}}} \qquad (28.9)$$

From equation (28.9), when Q is large, $V_{C_m} \approx QV$

If a similar exercise is undertaken for the voltage across the inductor it is found that the maximum value is given by:

$$\boxed{V_{L_m} = \frac{QV}{\sqrt{\left[1-\left(\frac{1}{2Q}\right)^2\right]}}},$$

i.e., the same equation as for V_{C_m}, and frequency,

$$\boxed{f = \frac{f_r}{\sqrt{\left[\left(1-\frac{1}{2Q^2}\right)\right]}}},$$

showing that the maximum p.d. across the coil does not occur at the resonant frequency but at a value slightly greater than f_r, as shown in Figure 28.5(b).

Problem 5. A series L–R–C circuit has a sinusoidal input voltage of maximum value 12 V. If inductance, $L = 20$ mH, resistance, $R = 80\ \Omega$, and capacitance, $C = 400$ nF, determine (a) the resonant frequency, (b) the value of the p.d. across the capacitor at the resonant frequency, (c) the frequency at which the p.d. across the capacitor is a maximum, and (d) the value of the maximum voltage across the capacitor.

(a) The resonant frequency,

$$f_r = \frac{1}{2\pi\sqrt{(LC)}} = \frac{1}{2\pi\sqrt{[(20\times10^{-3})(400\times10^{-9})]}}$$

$$= \mathbf{1779.4\ Hz}$$

(b) $V_C = QV$ and $Q = \dfrac{\omega_r L}{R}\left(\text{or } \dfrac{1}{\omega_r CR} \text{ or } \dfrac{1}{R}\sqrt{\dfrac{L}{C}}\right)$

Hence $\quad Q = \dfrac{(2\pi 1779.4)(20\times10^{-3})}{80} = 2.80$

Thus $\quad V_C = QV = (2.80)(12) = \mathbf{33.60\ V}$

(c) From equation (28.7), the frequency f at which V_C is a maximum value,

$$f = f_r\sqrt{\left(1 - \frac{1}{2Q^2}\right)} = (1779.4)\sqrt{\left(1 - \frac{1}{2(2.80)^2}\right)}$$

$$= \mathbf{1721.7\ Hz}$$

(d) From equation (28.9), the maximum value of the p.d. across the capacitor is given by:

$$V_{C_m} = \frac{QV}{\sqrt{\left[1 - \left(\frac{1}{2Q}\right)^2\right]}} = \frac{(2.80)(12)}{\sqrt{\left[1 - \left(\frac{1}{2(2.80)}\right)^2\right]}} = \mathbf{34.15\ V}$$

28.5 Q-factors in series

If the losses of a capacitor are not considered as negligible, the overall Q-factor of the circuit will depend on the Q-factor of the individual components. Let the Q-factor of the inductor be Q_L and that of the capacitor be Q_C

The overall Q-factor, $Q_T = \frac{1}{R_T}\sqrt{\frac{L}{C}}$ from Section (28.3),

where $R_T = R_L + R_C$

Since $Q_L = \frac{\omega_r L}{R_L}$ then $R_L = \frac{\omega_r L}{Q_L}$ and since

$$Q_C = \frac{1}{\omega_r C R_C} \text{ then } R_C = \frac{1}{Q_C \omega_r C}$$

$$\text{Hence } Q_T = \frac{1}{R_L + R_C}\sqrt{\frac{L}{C}} = \frac{1}{\left(\frac{\omega_r L}{Q_L} + \frac{1}{Q_C \omega_r C}\right)}\sqrt{\frac{L}{C}}$$

$$= \frac{1}{\left[\frac{\left(\frac{1}{\sqrt{(LC)}}\right)L}{Q_L} + \frac{1}{Q_C\left(\frac{1}{\sqrt{(LC)}}\right)C}\right]}\sqrt{\frac{L}{C}}$$

since $\omega_r = \frac{1}{\sqrt{(LC)}}$

$$= \frac{1}{\frac{L}{Q_L L^{1/2} C^{1/2}} + \frac{L^{1/2} C^{1/2}}{Q_C C}}\sqrt{\frac{L}{C}} = \frac{1}{\frac{1}{Q_L}\frac{L^{1/2}}{C^{1/2}} + \frac{1}{Q_C}\frac{L^{1/2}}{C^{1/2}}}\sqrt{\frac{L}{C}}$$

$$= \frac{1}{\frac{1}{Q_L}\sqrt{\frac{L}{C}} + \frac{1}{Q_C}\sqrt{\frac{L}{C}}}\sqrt{\frac{L}{C}} = \frac{1}{\sqrt{\frac{L}{C}}\left(\frac{1}{Q_L} + \frac{1}{Q_C}\right)}\sqrt{\frac{L}{C}}$$

$$= \frac{1}{\frac{1}{Q_L} + \frac{1}{Q_C}} = \frac{1}{\frac{Q_C + Q_L}{Q_L Q_C}}$$

i.e., the overall Q-factor,

$$\boldsymbol{Q_T = \frac{Q_L Q_C}{Q_L + Q_C}}$$

Problem 6. An inductor of Q-factor 60 is connected in series with a capacitor having a Q-factor of 390. Determine the overall Q-factor of the circuit.

From above, overall Q-factor,

$$Q_T = \frac{Q_L Q_C}{Q_L + Q_C} = \frac{(60)(390)}{60 + 390} = \frac{23400}{450} = \mathbf{52}$$

28.6 Bandwidth

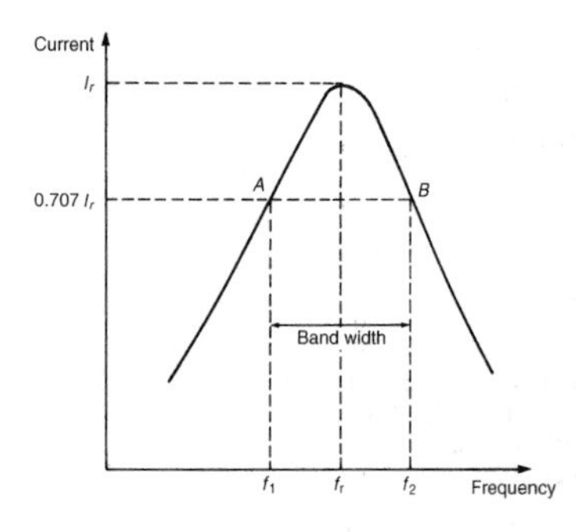

Figure 28.6 *Bandwidth and half-power points f_1 and f_2*

Figure 28.6 shows how current I varies with frequency f in an R–L–C series circuit. At the resonant frequency f_r, current is a maximum value, shown as I_r Also shown are the points A and B where the current is 0.707 of the maximum value at frequencies f_1 and f_2. The power delivered to the circuit is I^2R. At $I = 0.707I_r$, the power is $(0.707I_r)^2R = 0.5\ I_r^2R$, i.e., half the power that occurs at frequency f_r. The points corresponding to f_1 and f_2 are called the **half-power points**. The distance between these points, i.e., $(f_2 - f_1)$, is called the **bandwidth.**

When the ratio of two powers P_1 and P_2 is expressed in decibel units, the number of decibels X is given by:

$$X = 10\ \lg\left(\frac{P_2}{P_1}\right)\ \text{dB (see Section 10.14, page 126)}$$

Let the power at the half-power points be $(0.707I_r)^2\ R = (I_r^2R)/2$ and let the peak power be I_r^2R. then the ratio of the power in decibels is given by:

$$10\ \lg\left[\frac{I_r^2R/2}{I_r^2R}\right] = 10\ \lg\frac{1}{2} = \mathbf{-3\ dB}$$

It is for this reason that the half-power points are often referred to as **'the −3 dB points'**.

At the half-power frequencies, $I = 0.707\ I_r$, thus impedance

$$Z = \frac{V}{I} = \frac{V}{0.707I_r} = 1.414\left(\frac{V}{I_r}\right) = \sqrt{2}Z_r = \sqrt{2}R$$

(since at resonance $Z_r = R$)

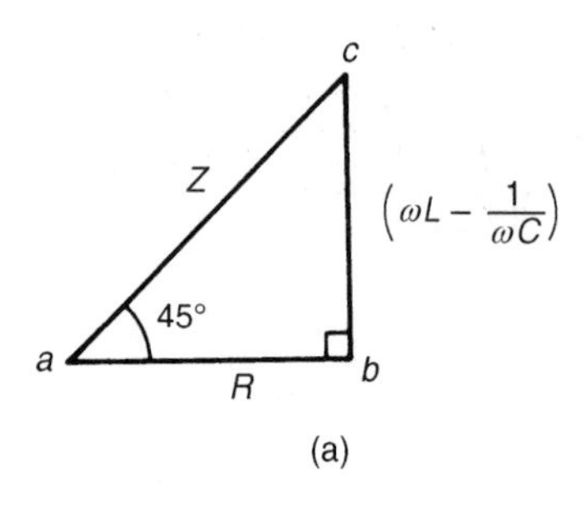

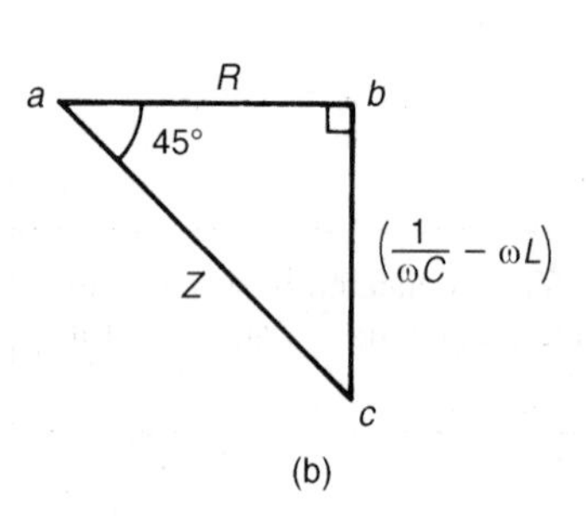

Figure 28.7 *(a) Inductive impedance triangle (b) Capacitive impedance triangle*

Since $Z = \sqrt{2}R$, an isosceles triangle is formed by the impedance triangles, as shown in Figure 28.7, where $ab = bc$. From the impedance triangles it can be seen that the equivalent circuit reactance is equal to the circuit resistance at the half-power points.

At f_1, the lower half-power frequency $|X_C| > |X_L|$ (see Figure 28.4)

Thus $\dfrac{1}{2\pi f_1 C} - 2\pi f_1 L = R$

from which, $1 - 4\pi^2 f_1^2 LC = 2\pi f_1 CR$

i.e., $(4\pi^2 LC)f_1^2 + (2\pi CR)f_1 - 1 = 0$

This is a quadratic equation in f_1. Using the quadratic formula gives:

$$f_1 = \frac{-(2\pi CR) \pm \sqrt{[(2\pi CR)^2 - 4(4\pi^2 LC)(-1)]}}{2(4\pi^2 LC)}$$

$$= \frac{-(2\pi CR) \pm \sqrt{[4\pi^2 C^2 R^2 + 16\pi^2 LC]}}{8\pi^2 LC}$$

$$= \frac{-(2\pi CR) \pm \sqrt{[4\pi^2 C^2 (R^2 + (4L/C))]}}{8\pi^2 LC}$$

$$= \frac{-(2\pi CR) \pm 2\pi C\sqrt{[R^2 + (4L/C)]}}{8\pi^2 LC}$$

Hence $f_1 = \dfrac{-R \pm \sqrt{[R^2 + (4L/C)]}}{4\pi L} = \boldsymbol{\dfrac{-R + \sqrt{[R^2 + (4L/C)]}}{4\pi L}}$

(since $\sqrt{[R^2 + (4L/C)]} > R$ and f_1 cannot be negative).

At f_2, the upper half-power frequency $|X_L| > |X_C|$ (see Figure 28.4)

Thus $2\pi f_2 L - \dfrac{1}{2\pi f_2 C} = R$

from which, $4\pi^2 f_2^2 LC - 1 = R(2\pi f_2 C)$

i.e., $(4\pi^2 LC) f_2^2 - (2\pi CR) f_2 - 1 = 0$

This is a quadratic equation in f_2 and may be solved using the quadratic formula as for f_1, giving:

$$\boldsymbol{f_2 = \frac{R + \sqrt{[R^2 + (4L/C)]}}{4\pi L}}$$

Bandwidth $= (f_2 - f_1)$

$$= \left\{\frac{R + \sqrt{[R^2 + (4L/C)]}}{4\pi L}\right\} - \left\{\frac{-R + \sqrt{[R^2 + (4L/C)]}}{4\pi L}\right\}$$

$$= \frac{2R}{4\pi L} = \frac{R}{2\pi L} = \frac{1}{2\pi L/R}$$

$$= \frac{f_r}{2\pi f_r L/R} = \frac{f_r}{Q_r}$$

from equation (28.3). Hence for a series R–L–C circuit

$$\boxed{\boldsymbol{Q_r = \frac{f_r}{f_2 - f_1}}} \qquad (28.10)$$

Problem 7. A filter in the form of a series L–R–C circuit is designed to operate at a resonant frequency of 10 kHz. Included within the filter is a 10 mH inductance and 5 Ω resistance. Determine the bandwidth of the filter.

Q-factor at resonance is given by

$$Q_r = \frac{\omega_r L}{R} = \frac{(2\pi\ 10\,000)(10 \times 10^{-3})}{5} = 125.66$$

Since $Q_r = f_r/(f_2 - f_1)$,

$$\textbf{Bandwidth, } (f_2 - f_1) = \frac{f_r}{Q_r} = \frac{10\,000}{125.66} = \textbf{79.6 Hz}$$

An alternative equation involving f_r

At the lower half-power frequency f_1: $\dfrac{1}{\omega_1 C} - \omega_1 L = R$

At the higher half-power frequency f_2: $\omega_2 L - \dfrac{1}{\omega_2 C} = R$

Equating gives: $\dfrac{1}{\omega_1 C} - \omega_1 L = \omega_2 L - \dfrac{1}{\omega_2 C}$

Multiplying throughout by C gives: $\dfrac{1}{\omega_1} - \omega_1 LC = \omega_2 LC - \dfrac{1}{\omega_2}$

However, for series resonance, $\omega_r^2 = 1/(LC)$

Hence $\dfrac{1}{\omega_1} - \dfrac{\omega_1}{\omega_r^2} = \dfrac{\omega_2}{\omega_r^2} - \dfrac{1}{\omega_2}$

i.e., $\dfrac{1}{\omega_1} + \dfrac{1}{\omega_2} = \dfrac{\omega_2}{\omega_r^2} + \dfrac{\omega_1}{\omega_r^2} = \dfrac{\omega_1 + \omega_2}{\omega_r^2}$

Therefore $\dfrac{\omega_2 + \omega_1}{\omega_1 \omega_2} = \dfrac{\omega_1 + \omega_2}{\omega_r^2}$,

from which, $\omega_r^2 = \omega_1 \omega_2$ or $\omega_r = \sqrt{(\omega_1 \omega_2)}$

Hence $2\pi f_r = \sqrt{[(2\pi f_1)(2\pi f_2)]}$ and $\boxed{f_r = \sqrt{(f_1 f_2)}}$ (28.11)

Selectivity is the ability of a circuit to respond more readily to signals of a particular frequency to which it is tuned than to signals of other frequencies. The response becomes progressively weaker as the frequency departs from the resonant frequency. Discrimination against other signals becomes more pronounced as circuit losses are reduced, i.e., as the Q-factor is increased. Thus $Q_r = f_r/(f_2 - f_1)$ is a measure of the circuit selectivity in terms of the points on each side of resonance where the circuit current has fallen to 0.707 of its maximum value reached at resonance. The higher the Q-factor, the narrower the bandwidth and the more selective is the circuit. Circuits having high Q-factors (say, in the order 300) are therefore useful in communications engineering. A high Q-factor in a series power circuit has disadvantages in that it can lead to dangerously high voltages across the insulation and may result in electrical breakdown.

For example, suppose that the working voltage of a capacitor is stated as 1 kV and is used in a circuit having a supply voltage of 240 V. The maximum value of the supply will be $\sqrt{2}(240)$, i.e., 340 V. The working voltage of the capacitor would appear to be ample. However, if the Q-factor is, say, 10, the voltage across the capacitor will reach 2.4 kV.

Since the capacitor is rated only at 1 kV, dielectric breakdown is more than likely to occur.

Low Q-factors, say, in the order of 5 to 25, may be found in power transformers using laminated iron cores.

A capacitor-start induction motor, as used in domestic appliances such as washing machines and vacuum-cleaners, having a Q-factor as low as 1.5 at starting would result in a voltage across the capacitor 1.5 times that of the supply voltage; hence the cable joining the capacitor to the motor would require extra insulation.

Problem 8. An R–L–C series circuit has a resonant frequency of 1.2 kHz and a Q-factor at resonance of 30. If the impedance of the circuit at resonance is 50 Ω determine the values of (a) the inductance, and (b) the capacitance. Find also (c) the bandwidth, (d) the lower and upper half-power frequencies and (e) the value of the circuit impedance at the half-power frequencies.

(a) At resonance the circuit impedance, $Z = R$, i.e., $R = 50\ \Omega$.

Q-factor at resonance, $Q_r = \omega_r L/R$

Hence **inductance**, $$L = \frac{Q_r R}{\omega_r} = \frac{(30)(50)}{(2\pi 1200)} = \mathbf{0.199\ H} \text{ or } \mathbf{199\ mH}$$

(b) At resonance $\omega_r L = 1/(\omega_r C)$

Hence **capacitance**, $$C = \frac{1}{\omega_r^2 L} = \frac{1}{(2\pi 1200)^2 (0.199)}$$

$$= \mathbf{0.088\ \mu F} \text{ or } \mathbf{88\ nF}$$

(c) Q-factor at resonance is also given by $Q_r = f_r/(f_2 - f_1)$, from which,

$$\textbf{bandwidth,}(f_2 - f_1) = \frac{f_r}{Q_r} = \frac{1200}{30} = \mathbf{40\ Hz}$$

(d) From equation (28.11), resonant frequency, $f_r = \sqrt{(f_1 f_2)}$, i.e., $1200 = \sqrt{(f_1 f_2)}$

from which, $$f_1 f_2 = (1200)^2 = 1.44 \times 10^6 \qquad (28.12)$$

From part(c), $$f_2 - f_1 = 40 \qquad (28.13)$$

From equation (28.12), $f_1 = (1.44 \times 10^6)/f_2$

Substituting in equation (28.13) gives:

$$f_2 - \frac{1.44 \times 10^6}{f_2} = 40$$

Multiplying throughout by f_2 gives:

$$f_2^2 - 1.44 \times 10^6 = 40f_2$$

i.e., $$f_2^2 - 40f_2 - 1.44 \times 10^6 = 0$$

This is a quadratic equation in f_2. Using the quadratic formula gives:

$$f_2 = \frac{40 \pm \sqrt{[(40)^2 - 4(1.44 \times 10^6)]}}{2} = \frac{40 \pm 2400}{2}$$

$$= \frac{40 + 2400}{2} \text{ (since } f_2 \text{ cannot be negative)}$$

Hence **the upper half-power frequency, $f_2 = 1220$ Hz**

From equation (28.12), **the lower half-power frequency**,

$$f_1 = f_2 - 40 = 1220 - 40 = \mathbf{1180\ Hz}$$

Note that the upper and lower half-power frequency values are symmetrically placed about the resonance frequency. This is usually the case when the Q-factor has a high value (say, >10).

(e) At the half-power frequencies, current $I = 0.707\ I_r$

Hence impedance,

$$Z = \frac{V}{I} = \frac{V}{0.707\ I_r} = 1.414\left(\frac{V}{I_r}\right) = \sqrt{2}Zr = \sqrt{2}R$$

Thus **impedance at the half-power frequencies,**
$Z = \sqrt{2}R = \sqrt{2}(50) = \mathbf{70.71\ \Omega}$

Problem 9. A series R–L–C circuit is connected to a 0.2 V supply and the current is at its maximum value of 4 mA when the supply frequency is adjusted to 3 kHz. The Q-factor of the circuit under these conditions is 100. Determine the value of (a) the circuit resistance, (b) the circuit inductance, (c) the circuit capacitance, and (d) the voltage across the capacitor

Since the current is at its maximum, the circuit is at resonance and the resonant frequency is 3 kHz.

(a) At resonance, impedance, $Z = R = \frac{V}{I} = \frac{0.2}{4 \times 10^{-3}} = 50\ \Omega$

Hence **the circuit resistance in 50 Ω**

(b) Q-factor at resonance is given by $Q_r = \omega_r L/R$, from which,

$$\textbf{inductance, } L = \frac{Q_r R}{\omega_r} = \frac{(100)(50)}{2\pi 3000} = \mathbf{0.265\ H} \text{ or } \mathbf{265\ mH}$$

(c) Q-factor at resonance is also given by $Q_r = 1/(\omega_r CR)$, from which,

$$\textbf{capacitance, } C = \frac{1}{\omega_r R Q_r} = \frac{1}{(2\pi 3000)(50)(100)}$$
$$= \mathbf{0.0106\ \mu F} \text{ or } \mathbf{10.6\ nF}$$

(d) Q-factor at resonance in a series circuit represents the voltage magnification, i.e., $Q_r = V_C/V$, from which, $V_C = Q_r V = (100)(0.2) = 20$ V.

Hence **the voltage across the capacitor is 20 V**

$$\text{(Alternatively, } V_C = IX_C = \frac{I}{\omega_r C} = \frac{4 \times 10^{-3}}{(2\pi 3000)(0.0106 \times 10^{-6})}$$
$$= \mathbf{20\ V})$$

Problem 10. A coil of inductance 351.8 mH and resistance 8.84 Ω is connected in series with a 20 μF capacitor. Determine (a) the resonant frequency, (b) the Q-factor at resonance, (c) the bandwidth, and (d) the lower and upper −3dB frequencies.

(a) Resonant frequency,

$$f_r = \frac{1}{2\pi\sqrt{(LC)}} = \frac{1}{2\pi\sqrt{[(0.3518)(20 \times 10^{-6})]}}$$
$$= \mathbf{60.0\ Hz}$$

(b) Q-factor at resonance,

$$Q_r = \frac{1}{R}\sqrt{\frac{L}{C}} = \frac{1}{8.84}\sqrt{\left(\frac{0.3518}{20 \times 10^{-6}}\right)} = \mathbf{15}$$

$$\left[\text{Alternatively, } Q_r = \frac{\omega_r L}{R} = \frac{2\pi(60.0)(0.3518)}{8.84} = \mathbf{15}\right.$$

$$\left.\text{or} \quad Q_r = \frac{1}{\omega_r CR} = \frac{1}{(2\pi 60.0)(20 \times 10^{-6})(8.84)} = \mathbf{15}\right]$$

(c) Bandwidth, $(f_2 - f_1) = \frac{f_r}{Q_r} = \frac{60.0}{15} = \mathbf{4\ Hz}$

(d) With a Q-factor of 15 it may be assumed that the lower and upper −3 dB frequencies, f_1 and f_2 are symmetrically placed about the resonant frequency of 60.0 Hz. Hence **the lower −3 dB frequency, $f_1 = 58$ Hz, and the upper −3 dB frequency, $f_2 = 62$ Hz.**

[This may be checked by using $(f_2 - f_1) = 4$ and $f_r = \sqrt{(f_1 f_2)}$]

28.7 Small deviations from the resonant frequency

Let ω_1 be a frequency below the resonant frequency ω_r in an L–R–C series circuit, and ω_2 be a frequency above ω_r by the same amount as ω_1 is below, i.e., $\omega_r - \omega_1 = \omega_2 - \omega_r$

Let the fractional deviation from the resonant frequency be δ where

$$\delta = \frac{\omega_r - \omega_1}{\omega_r} = \frac{\omega_2 - \omega_r}{\omega_r}$$

Hence $\omega_r\delta = \omega_r - \omega_1$ and $\omega_r\delta = \omega_2 - \omega_r$

from which, $\omega_1 = \omega_r - \omega_r\delta$ and $\omega_2 = \omega_r + \omega_r\delta$

i.e., $\omega_1 = \omega_r(1 - \delta)$ (28.14)

and $\omega_2 = \omega_r(1 + \delta)$ (28.15)

At resonance, $I_r = \dfrac{V}{R}$, and at other frequencies, $I = \dfrac{V}{Z}$ where Z is the circuit impedance.

Hence $\dfrac{I}{I_r} = \dfrac{V/Z}{V/R} = \dfrac{R}{Z} = \dfrac{R}{R + j\left(\omega L - \dfrac{1}{\omega C}\right)}$

From equation (28.15), **at frequency ω_2**,

$$\frac{I}{I_r} = \frac{R}{R + j\left[\omega_r(1+\delta)L - \dfrac{1}{\omega_r(1+\delta)C}\right]}$$

$$= \frac{R/R}{\dfrac{R}{R} + j\left[\dfrac{\omega_r L}{R}(1+\delta) - \dfrac{1}{\omega_r RC(1+\delta)}\right]}$$

At resonance, $\dfrac{1}{\omega_r C} = \omega_r L$ hence

$$\frac{I}{I_r} = \frac{1}{1 + j\left[\dfrac{\omega_r L}{R}(1+\delta) - \dfrac{\omega_r L}{R(1+\delta)}\right]}$$

$$= \frac{1}{1 + j\dfrac{\omega_r L}{R}\left[(1+\delta) - \dfrac{1}{(1+\delta)}\right]}$$

Since $\dfrac{\omega_r L}{R} = Q$ then

$$\frac{I}{I_r} = \frac{1}{1 + jQ\left[\dfrac{(1+\delta)^2 - 1}{(1+\delta)}\right]} = \frac{1}{1 + jQ\left[\dfrac{1 + 2\delta + \delta^2 - 1}{(1+\delta)}\right]}$$

$$= \frac{1}{1 + jQ\left[\dfrac{2\delta + \delta^2}{1+\delta}\right]} = \frac{1}{1 + j\delta Q\left[\dfrac{2+\delta}{1+\delta}\right]}$$

If the deviation from the resonant frequency δ is very small such that $\delta \ll 1$

then $$\frac{I}{I_r} \approx \frac{1}{1 + j\delta Q\left[\frac{2}{1}\right]} = \frac{\mathbf{1}}{\mathbf{1} + \boldsymbol{j2\delta Q}} \qquad (28.16)$$

and $$\frac{I}{I_r} = \frac{V/Z}{V/Z_r} = \frac{Z_r}{Z} = \frac{1}{1 + j2\delta Q}$$

from which, $$\boxed{\frac{\boldsymbol{Z}}{\boldsymbol{Z_r}} = \mathbf{1} + \boldsymbol{j2\delta Q}} \qquad (28.17)$$

It may be shown that **at frequency** $\boldsymbol{\omega_1}$, $\dfrac{I}{I_r} = \dfrac{\mathbf{1}}{\mathbf{1} - \boldsymbol{j2\delta Q}}$ and

$$\boxed{\frac{\boldsymbol{Z}}{\boldsymbol{Z_r}} = \mathbf{1} - \boldsymbol{j2\delta Q}}$$

Problem 11. In an L–R–C series network, the inductance, $L = 8$ mH, the capacitance, $C = 0.3$ μF, and the resistance, $R = 15\ \Omega$. Determine the current flowing in the circuit when the input voltage is $7.5\angle 0°$ V and the frequency is (a) the resonant frequency, (b) a frequency 3% above the resonant frequency. Find also (c) the impedance of the circuit when the frequency is 3% above the resonant frequency.

(a) At resonance, $Z_r = R = 15\ \Omega$

Current at resonance, $\boldsymbol{I_r} = \dfrac{V}{Z_r} = \dfrac{7.5\angle 0°}{15\angle 0°} = \mathbf{0.5\angle 0°\ A}$

(b) If the frequency is 3% above the resonant frequency, then $\delta = 0.03$

From equation (28.16), $\dfrac{I}{I_r} = \dfrac{1}{1 + j2\delta Q}$

$$Q = \frac{1}{R}\sqrt{\frac{L}{C}} = \frac{1}{15}\sqrt{\left(\frac{8 \times 10^{-3}}{0.3 \times 10^{-6}}\right)} = 10.89$$

Hence $$\frac{I}{0.5\angle 0°} = \frac{1}{1 + j2(0.03)(10.89)} = \frac{1}{1 + j0.6534}$$

$$= \frac{1}{1.1945\angle 33.16°}$$

and $$\boldsymbol{I} = \frac{0.5\angle 0°}{1.1945\angle 33.16°} = \mathbf{0.4186\angle -33.16°\ A}$$

(c) From equation (28.17), $\dfrac{Z}{Z_r} = 1 + j2\delta Q$

$$\text{hence } \mathbf{Z} = Z_r(1 + j2\delta Q) = R(1 + j2\delta Q)$$
$$= 15(1 + j2(0.03)(10.89))$$
$$= 15(1 + j0.6534)$$
$$= 15(1.1945\angle 33.16^\circ)$$
$$= \mathbf{17.92\angle 33.16^\circ\ \Omega}$$

$$\text{Alternatively, } Z = \frac{V}{I} = \frac{7.5\angle 0^\circ}{0.4186\angle -33.16^\circ} = 17.92\angle 33.16^\circ\ \Omega$$

Further problems on Q-factor and bandwidth may be found in Section 28.8 following, problems 6 to 16, page 513

28.8 Further problems on series resonance and Q-factor

Series resonance

1 A coil having an inductance of 50 mH and resistance 8.0 Ω is connected in series with a 25 μF capacitor across a 100 V a.c. supply. Determine (a) the resonant frequency of the circuit, and (b) the current flowing at resonance. [(a) 142.4 Hz (b) 12.5 A]

2 The current at resonance in a series R–L–C circuit is 0.12 mA. The circuit has an inductance of 0.05 H and the supply voltage is 24 mV at a frequency of 40 kHz. Determine (a) the circuit resistance, and (b) the circuit capacitance. [(a) 200 Ω (b) 316.6 pF]

3 A coil of inductance 2.0 mH and resistance 4.0 Ω is connected in series with a 0.3 μF capacitor. The circuit is connected to a 5.0 V, variable frequency supply. Calculate (a) the frequency at which resonance occurs, (b) the voltage across the capacitance at resonance, and (c) the voltage across the coil at resonance.

[(a) 6.50 kHz (b) 102.1 V (c) 102.2 V]

4 A series R–L–C circuit having an inductance of 0.40 H has an instantaneous voltage, $v = 60\sin(4000t - (\pi/6))$ volts and an instantaneous current, $i = 2.0\sin 4000t$ amperes. Determine (a) the values of the circuit resistance and capacitance, and (b) the frequency at which the circuit will be resonant.

[(a) 26 Ω; 154.8 nF (b) 639.6 Hz]

5 A variable capacitor C is connected in series with a coil having inductance L. The circuit possesses stray capacitance C_S which is assumed to be constant and effectively in parallel with the variable capacitor C. When the capacitor is set to 2.0 nF the resonant frequency of the circuit is 86.85 kHz, and when the capacitor is set to 1.0 nF the resonant frequency is 120 kHz. Determine the values of (a) the stray circuit capacitance C_S, and (b) the coil inductance L.

[(a) 100 pF (b) 1.60 mH]

Q-factor and bandwidth

6 A series R–L–C circuit comprises a 5 μF capacitor, a 4 Ω resistor and a variable inductance L. The supply voltage is $10\angle 0^\circ$ V at a frequency of 159.1 Hz. The inductance is adjusted until the p.d. across the 4 Ω resistance is a maximum. Determine for this condition (a) the value of inductance, (b) the p.d. across each component, and (c) the Q-factor of the circuit.
[(a) 200 mH (b) $V_R = 10\angle 0^\circ$ V; $V_L = 500\angle 90^\circ$ V; $V_C = 500\angle -90^\circ$ V (c) 50]

7 A coil of resistance 10.05 Ω and inductance 400 mH is connected in series with a 0.396 μF capacitor. Determine (a) the resonant frequency, (b) the resonant Q-factor, (c) the bandwidth, and (d) the lower and upper half-power frequencies.
[(a) 400 Hz (b) 100 (c) 4 Hz (d) 398 Hz and 402 Hz]

8 An R–L–C series circuit has a resonant frequency of 2 kHz and a Q-factor at resonance of 40. If the impedance of the circuit at resonance is 30 Ω determine the values of (a) the inductance and (b) the capacitance. Find also (c) the bandwidth, (d) the lower and upper −3 dB frequencies, and (e) the impedance at the −3 dB frequencies.
[(a) 95.5 mH (b) 66.3 nF (c) 50 Hz (d) 1975 Hz and 2025 Hz (e) 42.43 Ω]

9 A filter in the form of a series L–C–R circuit is designed to operate at a resonant frequency of 20 kHz and incorporates a 20 mH inductor and 30 Ω resistance. Determine the bandwidth of the filter.
[238.7 Hz]

10 A series L–R–C circuit has a supply input of 5 volts. Given that inductance, $L = 5$ mH, resistance, $R = 75$ Ω and capacitance, $C = 0.2$ μF, determine (a) the resonant frequency, (b) the value of voltage across the capacitor at the resonant frequency, (c) the frequency at which the p.d. across the capacitance is a maximum, and (d) the value of the maximum voltage across the capacitor.
[(a) 5033 Hz (b) 10.54 V (c) 4741 Hz (d) 10.85 V]

11 A capacitor having a Q-factor of 250 is connected in series with a coil which has a Q-factor of 80. Calculate the overall Q-factor of the circuit. [60.61]

12 An R–L–C series circuit has a maximum current of 2 mA flowing in it when the frequency of the 0.1 V supply is 4 kHz. The Q-factor of the circuit under these conditions is 90. Determine (a) the voltage across the capacitor, and (b) the values of the circuit resistance, inductance and capacitance. [(a) 9 V (b) 50 Ω; 0.179 H; 8.84 nF]

13 Calculate the inductance of a coil which must be connected in series with a 4000 pF capacitor to give a resonant frequency of 200 kHz. If the coil has a resistance of 12 Ω, determine the circuit Q-factor. [158.3 μH; 16.58]

14 A circuit consists of a coil of inductance 200 μH and resistance 8.0 Ω in series with a lossless 500 pF capacitor. Determine (a) the resonant Q-factor, and (b) the bandwidth of the circuit.

[(a) 79.06 (b) 6366 Hz]

15 A coil of inductance 200 μH and resistance 50.27 Ω and a variable capacitor are connected in series to a 5 mV supply of frequency 2 MHz. Determine (a) the value of capacitance to tune the circuit to resonance, (b) the supply current at resonance, (c) the p.d. across the capacitor at resonance, (d) the bandwidth, and (e) the half-power frequencies.

[(a) 31.66 pF (b) 99.46 μA (c) 250 mV (d) 40 kHz (e) 2.02 MHz; 1.98 MHz]

16 A supply voltage of 3 V is applied to a series R–L–C circuit whose resistance is 12 Ω, inductance is 7.5 mH and capacitance is 0.5 μF. Determine (a) the current flowing at resonance, (b) the current flowing at a frequency 2.5% below the resonant frequency, and (c) the impedance of the circuit when the frequency is 1% lower than the resonant frequency.

[(a) 0.25 A (b) $0.223\angle 27.04°$A (c) $13.47\angle -27.04°$ Ω]

29 Parallel resonance and Q-factor

At the end of this chapter you should be able to:

- state the condition for resonance in an a.c. parallel network
- calculate the resonant frequency in a.c. parallel networks
- calculate dynamic resistance $R_D = \dfrac{L}{CR}$ in an a.c. parallel network
- calculate Q-factor and bandwidth in an a.c. parallel network
- determine the overall Q-factor for capacitors connected in parallel
- determine the impedance when the frequency deviates from the resonant frequency

29.1 Introduction

A parallel network containing resistance R, pure inductance L and pure capacitance C connected in parallel is shown in Figure 29.1. Since the inductance and capacitance are considered as pure components, this circuit is something of an 'ideal' circuit. However, it may be used to highlight some important points regarding resonance which are applicable to any parallel circuit.

From Figure 29.1,

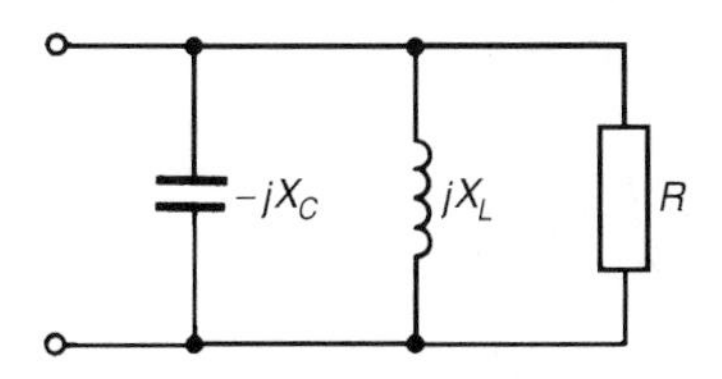

Figure 29.1 *Parallel R–L–C circuit*

the admittance of the resistive branch, $G = \dfrac{1}{R}$

the admittance of the inductive branch, $B_L = \dfrac{1}{jX_L} = \dfrac{-j}{\omega L}$

the admittance of the capacitive branch, $B_C = \dfrac{1}{-jX_C} = \dfrac{j}{1/\omega C} = j\omega C$

Total circuit admittance, $Y = G + j(B_C - B_L)$,

i.e., $$Y = \frac{1}{R} + j\left(\omega C - \frac{1}{\omega L}\right)$$

The circuit is at resonance when the imaginary part is zero, i.e., when $\omega C - (1/\omega L) = 0$. Hence at resonance $\omega_r C = 1/(\omega_r L)$ and $\omega_r^2 = 1/(LC)$,

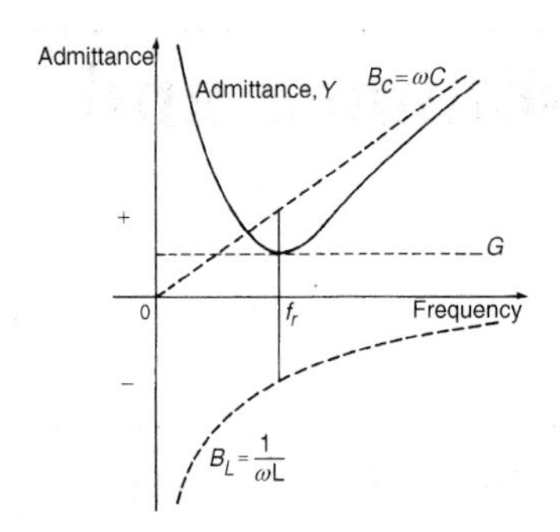

Figure 29.2 *$|Y|$ plotted against frequency*

from which $\omega_r = 1/\sqrt{(LC)}$ and the resonant frequency

$$\boxed{f_r = \frac{1}{2\pi\sqrt{(LC)}} \text{ hertz}}$$

the same expression as for a series R–L–C circuit.

Figure 29.2 shows typical graphs of B_C, B_L, G and Y against frequency f for the circuit shown in Figure 29.1. At resonance, $B_C = B_L$ and admittance $Y = G = 1/R$. This represents the condition of **minimum admittance** for the circuit and thus **maximum impedance.**

Since current $I = V/Z = VY$, the **current** is at a **minimum** value at resonance in a parallel network.

From the ideal circuit of Figure 29.1 we have therefore established the following facts which apply to any parallel circuit. At resonance:

(i) admittance Y is a minimum
(ii) impedance Z is a maximum
(iii) current I is a minimum
(iv) an expression for the resonant frequency f_r may be obtained by making the 'imaginary' part of the complex expression for admittance equal to zero.

29.2 The LR–C parallel network

A more practical network, containing a coil of inductance L and resistance R in parallel with a pure capacitance C, is shown in Figure 29.3.

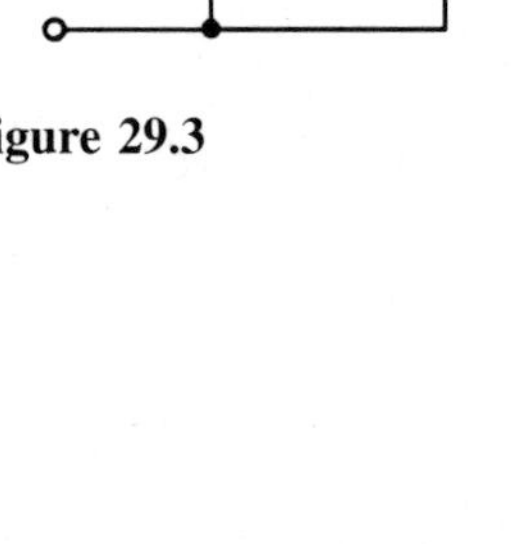

Figure 29.3

$$\text{Admittance of coil, } Y_{\text{COIL}} = \frac{1}{R + jX_L} = \frac{R - jX_L}{R^2 + X_L^2}$$

$$= \frac{R}{R^2 + \omega^2 L^2} - \frac{j\omega L}{R^2 + \omega^2 L^2}$$

$$\text{Admittance of capacitor, } Y_C = \frac{1}{-jX_C} = \frac{j}{X_c} = j\omega C$$

$$\text{Total circuit admittance, } Y = Y_{\text{COIL}} + Y_C$$

$$= \frac{R}{R^2 + \omega^2 L^2} - \frac{j\omega L}{R^2 + \omega^2 L^2} + j\omega C \qquad (29.1)$$

At resonance, the total circuit admittance Y is real ($Y = R/(R^2 + \omega^2 L^2)$), i.e., the imaginary part is zero. Hence, at resonance:

$$\frac{-\omega_r L}{R^2 + \omega_r^2 L^2} + \omega_r C = 0$$

Therefore $\dfrac{\omega_r L}{R^2 + \omega_r^2 L^2} = \omega_r C$ and $\dfrac{L}{C} = R^2 + \omega_r^2 L^2$

Thus $\omega_r^2 L^2 = \dfrac{L}{C} - R^2$

and $$\omega_r^2 = \frac{L}{CL^2} - \frac{R^2}{L^2} = \frac{1}{LC} - \frac{R^2}{L^2} \qquad (29.2)$$

Hence $$\omega_r = \sqrt{\left(\frac{1}{LC} - \frac{R^2}{L^2}\right)}$$

and $$\textbf{resonant frequency, } f_r = \frac{1}{2\pi}\sqrt{\left(\frac{1}{LC} - \frac{R^2}{L^2}\right)} \qquad (29.3)$$

Note that when $R^2/L^2 \ll 1/(LC)$ then $f_r = 1/(2\pi\sqrt{(LC)}$, as for the series R–L–C circuit. Equation (29.3) is the same as obtained in Chapter 16, page 248; however, the above method may be applied to any parallel network as demonstrated in Section 29.4 below.

29.3 Dynamic resistance

Since the current at resonance is in phase with the voltage, the impedance of the network acts as a resistance. This resistance is known as the **dynamic resistance, R_D**. Impedance at resonance, $R_D = V/I_r$, where I_r is the current at resonance.

$$I_r = VY_r = V\left(\frac{R}{R^2 + \omega_r^2 L^2}\right)$$

from equation (29.1) with the j terms equal to zero.

$$\text{Hence } R_D = \frac{V}{I_r} = \frac{V}{VR/(R^2 + \omega_r^2 L^2)} = \frac{R^2 + \omega_r^2 L^2}{R}$$

$$= \frac{R^2 + L^2(1/LC) - (R^2/L^2)}{R} \text{ from equation (29.2)}$$

$$= \frac{R^2 + (L/C) - R^2}{R} = \frac{L/C}{R} = \frac{L}{CR}$$

Hence $$\textbf{dynamic resistance, } R_D = \frac{L}{CR} \qquad (29.4)$$

29.4 The *LR–CR* parallel network

A more general network comprising a coil of inductance L and resistance R_L in parallel with a capacitance C and resistance R_C in series is shown in Figure 29.4.

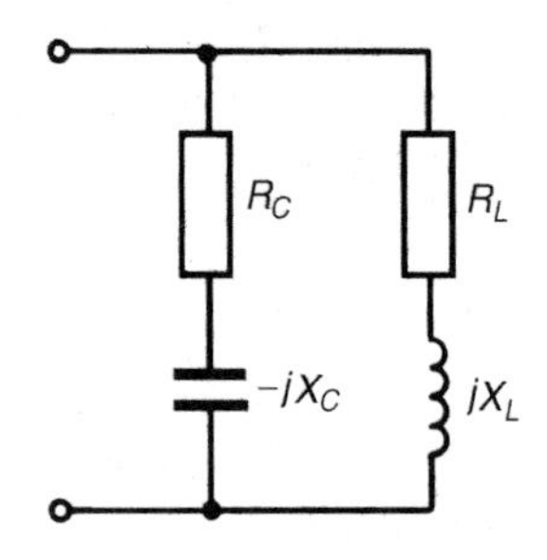

Figure 29.4

Admittance of inductive branch,

$$Y_L = \frac{1}{R_L + jX_L} = \frac{R_L - jX_L}{R_L^2 + X_L^2} = \frac{R_L}{R_L^2 + X_L^2} - \frac{jX_L}{R_L^2 + X_L^2}$$

Admittance of capacitive branch,

$$Y_C = \frac{1}{R_C - jX_C} = \frac{R_C + jX_C}{R_C^2 + X_C^2} = \frac{R_C}{R_C^2 + X_C^2} + \frac{jX_C}{R_C^2 + X_C^2}$$

Total network admittance,

$$Y = Y_L + Y_C = \frac{R_L}{R_L^2 + jX_L^2} - \frac{jX_L}{R_L^2 + X_L^2} + \frac{R_C}{R_C^2 + X_C^2} + \frac{jX_C}{R_C^2 + X_C^2}$$

At resonance the admittance is a minimum, i.e., when the imaginary part of Y is zero. Hence, at resonance,

$$\frac{-X_L}{R_L^2 + X_L^2} + \frac{X_C}{R_C^2 + X_C^2} = 0$$

i.e.,
$$\frac{\omega_r L}{R_L^2 + \omega^2 L^2} = \frac{1/(\omega_r C)}{R_C^2 + 1/\omega_r^2 C^2)} \tag{29.5}$$

Rearranging gives: $\omega_r L\left(R_C^2 + \dfrac{1}{\omega_r^2 C^2}\right) = \dfrac{1}{\omega_r C}(R_L^2 + \omega_r^2 L^2)$

$$\omega_r L R_C^2 + \frac{L}{\omega_r C^2} = \frac{R_L^2}{\omega_r C} + \frac{\omega_r L^2}{C}$$

Multiplying throughout by $\omega_r C^2$ gives:

$$\omega_r^2 C^2 L R_C^2 + L = R_L^2 C + \omega_r^2 L^2 C$$

$$\omega_r^2 (C^2 L R_C^2 - L^2 C) = R_L^2 C - L$$

$$\omega_r^2 CL(CR_C^2 - L) = R_L^2 C - L$$

Hence $\omega_r^2 = \dfrac{(CR_L^2 - L)}{LC(CR_C^2 - L)}$

i.e., $\omega_r = \dfrac{1}{\sqrt{(LC)}}\sqrt{\left(\dfrac{R_L^2 - (L/C)}{R_C^2 - (L/C)}\right)}$

Hence **resonant frequency,** $\boldsymbol{f_r = \dfrac{1}{2\pi\sqrt{(LC)}}\sqrt{\left(\dfrac{R_L^2 - (L/C)}{R_C^2 - (L/C)}\right)}}$ (29.6)

It is clear from equation (29.5) that parallel resonance may be achieved in such a circuit in several ways — by varying either the frequency f, the inductance L, the capacitance C, the resistance R_L or the resistance R_C.

29.5 Q-factor in a parallel network

The Q-factor in the series R–L–C circuit is a measure of the voltage magnification. In a parallel circuit, currents higher than the supply current can circulate within the parallel branches of a parallel resonant network, the current leaving the capacitor and establishing the magnetic field of the inductance, this then collapsing and recharging the capacitor, and so on. The Q-factor of a parallel resonant circuit is the ratio of the current circulating in the parallel branches of the circuit to the supply current, i.e. in a parallel circuit, Q-factor is a measure of the **current magnification.**

Circulating currents may be several hundreds of times greater than the supply current at resonance. For the parallel network of Figure 29.5, the Q-factor at resonance is given by:

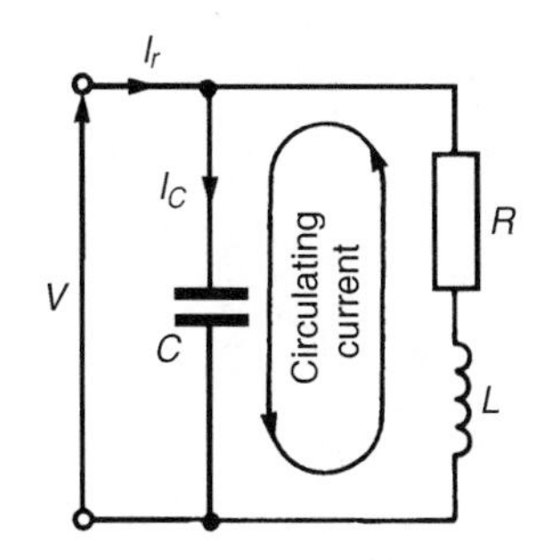

Figure 29.5

$$\boldsymbol{Q_r = \frac{\text{circulating current}}{\text{current at resonance}} = \frac{\text{capacitor current}}{\text{current at resonance}} = \frac{I_C}{I_r}}$$

Current in capacitor, $I_C = V/X_C = V\omega_r C$

Current at resonance, $I_r = \dfrac{V}{R_D} = \dfrac{V}{L/CR} = \dfrac{VCR}{L}$

Hence $\boldsymbol{Q_r} = \dfrac{I_C}{I_r} = \dfrac{V\omega_r C}{VCR/L}$ **i.e.,** $\boxed{\boldsymbol{Q_r = \dfrac{\omega_r L}{R}}}$

the same expression as for series resonance.

The difference between the resonant frequency of a series circuit and that of a parallel circuit can be quite small. The resonant frequency of a coil in parallel with a capacitor is shown in Equation (29.3); however, around the closed loop comprising the coil and capacitor the energy would naturally resonate at a frequency given by that for a series R–L–C circuit, as shown in Chapter 28. This latter frequency is termed the **natural frequency**, $\boldsymbol{f_n}$, and the frequency of resonance seen at the terminals of Figure 29.5 is often called the **forced resonant frequency**, $\boldsymbol{f_r}$. (For a series circuit, the forced and natural frequencies coincide.)

From the coil-capacitor loop of Figure 29.5, $f_n = \dfrac{1}{2\pi\sqrt{(LC)}}$

and the forced resonant frequency, $f_r = \dfrac{1}{2\pi}\sqrt{\left(\dfrac{1}{LC} - \dfrac{R^2}{L^2}\right)}$

$$\text{Thus } \frac{f_r}{f_n} = \frac{\frac{1}{2\pi}\sqrt{\left(\frac{1}{LC} - \frac{R^2}{L^2}\right)}}{\frac{1}{2\pi\sqrt{(LC)}}} = \frac{\sqrt{\left(\frac{1}{LC} - \frac{R^2}{L^2}\right)}}{\frac{1}{\sqrt{(LC)}}}$$

$$= \sqrt{\left(\frac{1}{LC} - \frac{R^2}{L^2}\right)}\sqrt{(LC)} = \sqrt{\left(\frac{LC}{LC} - \frac{LCR^2}{L^2}\right)} = \sqrt{\left(1 - \frac{R^2C}{L}\right)}$$

From Chapter 28, $Q = \frac{1}{R}\sqrt{\left(\frac{L}{C}\right)}$ from which

$$Q^2 = \frac{1}{R^2}\left(\frac{L}{C}\right) \text{ or } \frac{R^2C}{L} = \frac{1}{Q^2}$$

$$\text{Hence } \frac{f_r}{f_n} = \sqrt{\left(1 - \frac{R^2C}{L}\right)} = \sqrt{\left(1 - \frac{1}{Q^2}\right)}$$

i.e.,
$$\boldsymbol{f_r = f_n\sqrt{\left(1 - \frac{1}{Q^2}\right)}}$$

Thus it is seen that even with small values of Q the difference between f_r and f_n tends to be very small. A high value of Q makes the parallel resonant frequency tend to the same value as that of the series resonant frequency.

The expressions already obtained in Chapter 28 for bandwidth and resonant frequency, also apply to parallel circuits,

i.e.,
$$\boldsymbol{Q_r = f_r/(f_2 - f_1)} \tag{29.7}$$

and
$$\boldsymbol{f_r = \sqrt{(f_1 f_2)}} \tag{29.8}$$

The overall Q-factor Q_T of two parallel components having different Q-factors is given by:

$$\boldsymbol{Q_T = \frac{Q_L Q_C}{Q_L + QC}} \tag{29.9}$$

as for the series circuit.

By similar reasoning to that of the series R–L–C circuit it may be shown that at the half-power frequencies the admittance is $\sqrt{2}$ times its minimum value at resonance and, since $Z = 1/Y$, the value of impedance at the half-power frequencies is $1/\sqrt{2}$ or 0.707 times its maximum value at resonance.

By similar analysis to that given in Chapter 28, it may be shown that for a parallel network:

$$\frac{Y}{Y_r} = \frac{R_D}{Z} = 1 + j2\delta Q \qquad (29.10)$$

where Y is the circuit admittance, Y_r is the admittance at resonance, Z is the network impedance and R_D is the dynamic resistance (i.e., the impedance at resonance) and δ is the fractional deviation from the resonant frequency.

Problem 1. A coil of inductance 5 mH and resistance 10 Ω is connected in parallel with a 250 nF capacitor across a 50 V variable-frequency supply. Determine (a) the resonant frequency, (b) the dynamic resistance, (c) the current at resonance, and (d) the circuit Q-factor at resonance.

(a) Resonance frequency

$$f_r = \frac{1}{2\pi}\sqrt{\left(\frac{1}{LC} - \frac{R^2}{L^2}\right)} \quad \text{from equation (29.3),}$$

$$= \frac{1}{2\pi}\sqrt{\left(\frac{1}{5\times10^{-3}\times250\times10^{-9}} - \frac{10^2}{(5\times10^{-3})^2}\right)}$$

$$= \frac{1}{2\pi}\sqrt{(800\times10^6 - 4\times10^6)} = \frac{1}{2\pi}\sqrt{(796\times10^6)} = \mathbf{4490\ Hz}$$

(b) From equation (29.4), dynamic resistance,

$$R_D = \frac{L}{CR} = \frac{5\times10^{-3}}{(250\times10^{-9})(10)} = \mathbf{2000\ \Omega}$$

(c) Current at resonance, $I_r = \dfrac{V}{R_D} = \dfrac{50}{2000} = \mathbf{25\ mA}$

(d) Q-factor at resonance, $Q_r = \dfrac{\omega_r L}{R} = \dfrac{(2\pi4490)(5\times10^{-3})}{10} = \mathbf{14.1}$

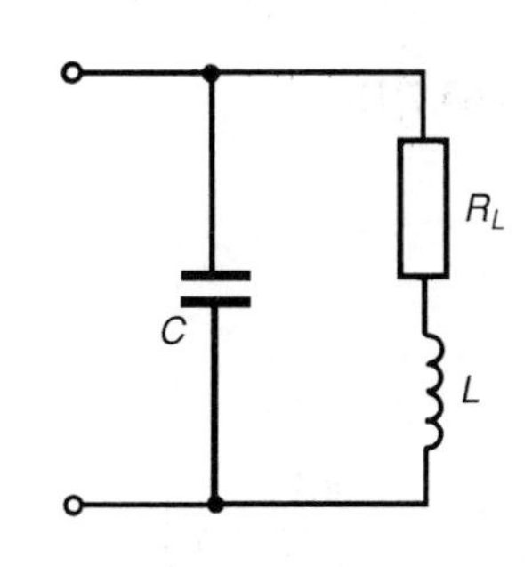

Figure 29.6

Problem 2. In the parallel network of Figure 29.6, inductance, $L = 100$ mH and capacitance, $C = 40$ μF. Determine the resonant frequency for the network if (a) $R_L = 0$ and (b) $R_L = 30$ Ω

Total circuit admittance,

$$Y = \frac{1}{R_L + jX_L} + \frac{1}{-jX_C} = \frac{R_L - jX_L}{R_L^2 + X_L^2} + \frac{j}{X_C}$$

$$= \frac{R_L}{R_L^2 + X_L^2} - \frac{jX_L}{R_L^2 + X_L^2} + \frac{j}{X_C}$$

The network is at resonance when the admittance is at a minimum value, i.e., when the imaginary part is zero. Hence, at resonance,

$$\frac{-X_L}{R_L^2 + X_L^2} + \frac{1}{X_C} = 0 \quad \text{or} \quad \omega_r C = \frac{\omega_r L}{R_L^2 + \omega_r^2 L^2} \tag{29.11}$$

(a) When $R_L = 0$, $\omega_r C = \dfrac{\omega_r L}{\omega_r^2 L^2}$

from which, $\omega_r^2 = \dfrac{1}{LC}$ and $\omega_r = \dfrac{1}{\sqrt{(LC)}}$

Hence resonant frequency,

$$f_r = \frac{1}{2\pi\sqrt{(LC)}} = \frac{1}{2\pi\sqrt{(100 \times 10^{-3} \times 40 \times 10^{-6})}} = \mathbf{79.6\ Hz}$$

(b) When $R_L = 30\Omega$, $\omega_r C = \dfrac{\omega_r L}{30^2 + \omega_r^2 L^2}$ from equation (29.11) above

from which, $$30^2 + \omega_r^2 L^2 = \frac{L}{C}$$

i.e., $$\omega_r^2 (100 \times 10^{-3})^2 = \frac{100 \times 10^{-3}}{40 \times 10^{-6}} - 900$$

i.e., $$\omega_r^2 (0.01) = 2500 - 900 = 1600$$

Thus, $\omega_r^2 = 1600/0.01 = 160\,000$ and $\omega_r = \sqrt{160\,000} = 400$ rad/s

Hence resonant frequency, $f_r = \dfrac{400}{2\pi} = \mathbf{63.7\ Hz}$

[Alternatively, from equation (29.3),

$$f_r = \frac{1}{2\pi}\sqrt{\left(\frac{1}{LC} - \frac{R^2}{L^2}\right)}$$

$$= \frac{1}{2\pi}\sqrt{\left(\frac{1}{(100 \times 10^{-3})(40 \times 10^{-6})} - \frac{30^2}{(100 \times 10^{-3})^2}\right)}$$

$$= \frac{1}{2\pi}\sqrt{(250\,000 - 90\,000)} = \frac{1}{2\pi}\sqrt{160\,000} = \frac{1}{2\pi}(400) = \mathbf{63.7\ Hz}]$$

Hence, as the resistance of a coil increases, the resonant frequency decreases in the circuit of Figure 29.6.

> Problem 3. A coil of inductance 120 mH and resistance 150 Ω is connected in parallel with a variable capacitor across a 20 V, 4 kHz supply. Determine for the condition when the supply current is a minimum, (a) the capacitance of the capacitor, (b) the dynamic resistance, (c) the supply current, (d) the Q-factor, (e) the bandwidth, (f) the upper and lower −3 dB frequencies, and (g) the value of the circuit impedance at the −3 dB frequencies.

(a) The supply current is a minimum when the parallel network is at resonance.

Resonant frequency, $\boldsymbol{f_r} = \dfrac{1}{2\pi}\sqrt{\left(\dfrac{1}{LC} - \dfrac{R^2}{L^2}\right)}$ from equation (29.3),

from which, $(2\pi f_r)^2 = \dfrac{1}{LC} - \dfrac{R^2}{L^2}$

Hence $\dfrac{1}{LC} = (2\pi f_r)^2 + \dfrac{R^2}{L^2}$ and

$$\text{capacitance } \boldsymbol{C} = \frac{1}{L[(2\pi f_r)^2 + (R^2/L^2)]}$$

$$= \frac{1}{120 \times 10^{-3}[(2\pi 4000)^2 + (150^2/(120 \times 10^{-3})^2)]}$$

$$= \frac{1}{0.12(631.65 \times 10^6 + 1.5625 \times 10^6)}$$

$$= \mathbf{0.01316\ \mu F} \text{ or } \mathbf{13.16\ nF}$$

(b) Dynamic resistance, $\mathbf{R_D} = \dfrac{L}{CR} = \dfrac{120 \times 10^{-3}}{(13.16 \times 10^{-9})(150)}$

$= \mathbf{60.79\ k\Omega}$

(c) Supply current at resonance,

$$\boldsymbol{I_r} = \frac{V}{R_D} = \frac{20}{60.79 \times 10^{-3}} = \mathbf{0.329\ mA} \text{ or } \mathbf{329\ \mu A}$$

(d) Q-factor at resonance, $Q_r = \dfrac{\omega_r L}{R} = \dfrac{(2\pi 4000)(120 \times 10^{-3})}{150} = \mathbf{20.11}$

[Note that the expressions $Q_r = \dfrac{1}{\omega_r CR}$ or $Q_r = \dfrac{1}{R}\sqrt{\left(\dfrac{L}{C}\right)}$

used for the R–L–C series circuit may also be used in parallel circuits when the resistance of the coil is much smaller than the inductive reactance of the coil.

In this case $R = 150\ \Omega$ and $X_L = 2\pi(4000)(120 \times 10^{-3}) = 3016\ \Omega$.

Hence, alternatively,

$$Q_r = \frac{1}{\omega_r CR} = \frac{1}{(2\pi 4000)(13.16 \times 10^{-9})(150)} = \mathbf{20.16}$$

$$\text{or } Q_r = \frac{1}{R}\sqrt{\left(\frac{L}{C}\right)} = \frac{1}{150}\sqrt{\left(\frac{120 \times 10^{-3}}{13.16 \times 10^{-9}}\right)} = \mathbf{20.13}]$$

(e) If the lower and upper −3 dB frequencies are f_1 and f_2 respectively then the bandwidth is $(f_2 - f_1)$. Q-factor at resonance is given by $Q_r = f_r/(f_2 - f_1)$, from which, bandwidth,

$$(f_2 - f_1) = \frac{f_r}{Q_r} = \frac{4000}{20.11} = \mathbf{199\ Hz}$$

(f) Resonant frequency, $f_r = \sqrt{(f_1 f_2)}$, from which

$$f_1 f_2 = f_r^2 = (4000)^2 = 16 \times 10^6 \qquad (29.12)$$

Also, from part (e), $f_2 - f_1 = 199$ (29.13)

From equation (29.12), $f_1 = \dfrac{16 \times 10^6}{f_2}$

Substituting in equation (29.13) gives: $f_2 - \dfrac{16 \times 10^6}{f_2} = 199$

i.e., $f_2^2 - 16 \times 10^6 = 199 f_2$ from which,

$$f_2^2 - 199 f_2 - 16 \times 10^6 = 0.$$

Solving this quadratic equation gives:

$$f_2 = \frac{199 \pm \sqrt{[(199)^2 - 4(-16 \times 10^6)]}}{2} = \frac{199 \pm 8002.5}{2}$$

i.e., **the upper 3 dB frequency, f_2 = 4100 Hz** (neglecting the negative answer).

From equation (29.12),

$$\textbf{the lower −3 dB frequency, } f_1 = \frac{10 \times 10^6}{f_2} = \frac{16 \times 10^6}{4100}$$

$$= \mathbf{3900\ Hz}$$

(Note that f_1 and f_2 are equally displaced about the resonant frequency, f_r, as they always will be when Q is greater than about 10 — just as for a series circuit)

(g) The value of the circuit impedance, Z, at the -3 dB frequencies is given by

$$Z = \frac{1}{\sqrt{2}} Z_r$$

where Z_r is the impedance at resonance.

The impedance at resonance $Z_r = R_D$, the dynamic resistance.

Hence **impedance at the −3 dB frequencies** $= \frac{1}{\sqrt{2}}(60.79 \times 10^3)$

$= \mathbf{42.99\ k\Omega}$

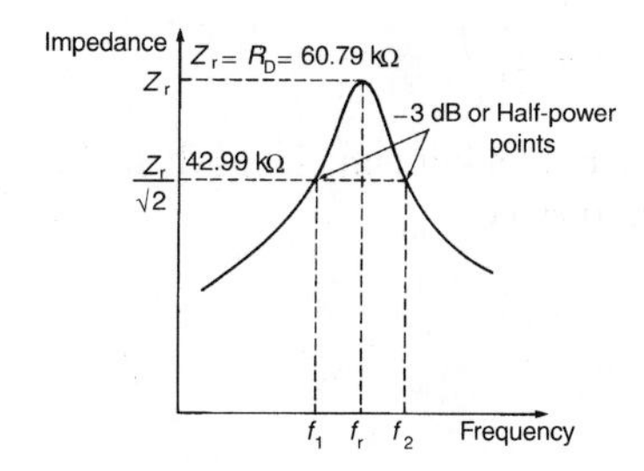

Figure 29.7

Figure 29.7 shows impedance plotted against frequency for the circuit in the region of the resonant frequency.

Problem 4. A two-branch parallel network is shown in Figure 29.8. Determine the resonant frequency of the network.

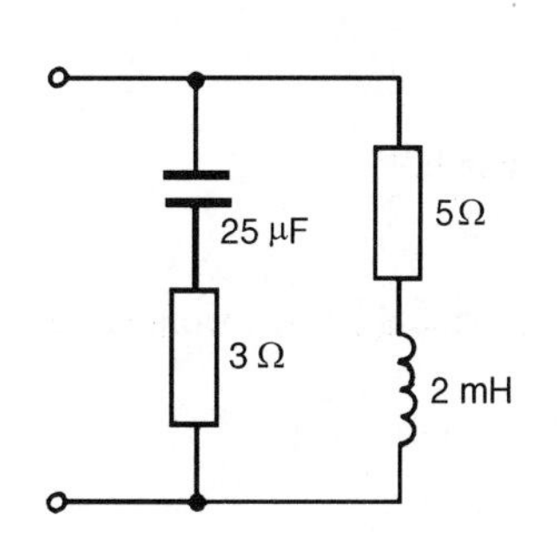

Figure 29.8

From equation (29.6),

$$\text{resonant frequency, } f_r = \frac{1}{2\pi\sqrt{(LC)}}\sqrt{\left(\frac{R_L^2 - (L/C)}{R_C^2 - (L/C)}\right)}$$

where $R_L = 5\ \Omega$, $R_C = 3\ \Omega$, $L = 2$ mH and $C = 25\ \mu$F. Thus

$$f_r = \frac{1}{2\pi\sqrt{[(2 \times 10^{-3})(25 \times 10^{-6})]}}\sqrt{\left(\frac{5^2 - ((2 \times 10^{-3})/(25 \times 10^{-6}))}{3^2 - ((2 \times 10^{-3})/(25 \times 10^{-6}))}\right)}$$

$$= \frac{1}{2\pi\sqrt{(5 \times 10^{-8})}}\sqrt{\left(\frac{25 - 80}{9 - 80}\right)}$$

$$= \frac{10^4}{2\pi\sqrt{5}}\sqrt{\left(\frac{-55}{-71}\right)} = \mathbf{626.5\ Hz}$$

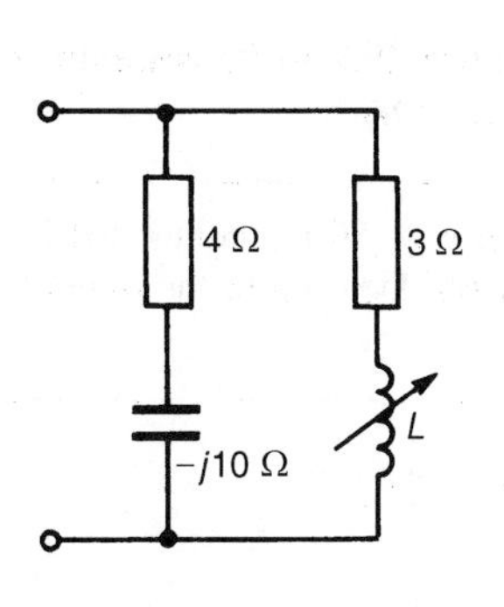

Figure 29.9

Problem 5. Determine for the parallel network shown in Figure 29.9 the values of inductance L for which the network is resonant at a frequency of 1 kHz.

The total network admittance, Y, is given by

$$Y = \frac{1}{3+jX_L} + \frac{1}{4-j10} = \frac{3-jX_L}{3^2+X_L^2} + \frac{4+j10}{4^2+10^2}$$

$$= \frac{3}{3^2+X_L^2} - \frac{jX_L}{3^2+X_L^2} + \frac{4}{116} + \frac{j10}{116}$$

$$= \left(\frac{3}{3^2+X_L^2} + \frac{4}{116}\right) + j\left(\frac{10}{116} - \frac{X_L}{3^2+X_L^2}\right)$$

Resonance occurs when the admittance is a minimum, i.e., when the imaginary part of Y is zero. Hence, at resonance,

$$\frac{10}{116} - \frac{X_L}{3^2+X_L^2} = 0 \text{ i.e., } \frac{10}{116} = \frac{X_L}{3^2+X_L^2}$$

Therefore $10(9+X_L^2) = 116\,X_L$ i.e., $10\,X_L^2 - 116\,X_L + 90 = 0$

from which, $X_L^2 - 11.6\,X_L + 9 = 0$

Solving the quadratic equation gives:

$$X_L = \frac{11.6 \pm \sqrt{[(-11.6)^2 - 4(9)]}}{2} = \frac{11.6 \pm 9.93}{2}$$

i.e., $X_L = 10.765\ \Omega$ or $0.835\ \Omega$. Hence $10.765 = 2\pi f_r L_1$, from which,

$$\text{inductance } L_1 = \frac{10.765}{2\pi(1000)} = 1.71 \text{ mH}$$

and $0.835 = 2\pi f_r L_2$ from which,

$$\text{inductance, } L_2 = \frac{0.835}{2\pi(1000)} = 0.13 \text{ mH}$$

Thus the conditions for the circuit of Figure 29.9 to be resonant are that inductance L is either 1.71 mH or 0.13 mH

Problem 6. A capacitor having a Q-factor of 300 is connected in parallel with a coil having a Q-factor of 60. Determine the overall Q-factor of the parallel combination.

From equation (29.9), the overall Q-factor is given by:

$$Q_T = \frac{Q_L Q_C}{Q_L + Q_C} = \frac{(60)(300)}{60+300} = \frac{18000}{360} = \mathbf{50}$$

Problem 7. In an $LR-C$ network, the capacitance is 10.61 nF, the bandwidth is 500 Hz and the resonant frequency is 150 kHz. Determine for the circuit (a) the Q-factor, (b) the dynamic resistance, and (c) the magnitude of the impedance when the supply frequency is 0.4% greater than the tuned frequency.

(a) From equation (29.7), $\boldsymbol{Q} = \dfrac{f_r}{f_2 - f_1} = \dfrac{150 \times 10^3}{500} = \mathbf{300}$

(b) From equation (29.4), dynamic resistance, $R_D = \dfrac{L}{CR}$

Also, in an $LR-C$ network, $Q = \dfrac{\omega_r L}{R}$ from which, $R = \dfrac{\omega_r L}{Q}$

Hence, $\boldsymbol{R_D} = \dfrac{L}{CR} = \dfrac{L}{C\left(\dfrac{\omega_r L}{Q}\right)} = \dfrac{LQ}{C\omega_r L} = \dfrac{Q}{\omega_r C}$

$$= \frac{300}{(2\pi 150 \times 10^3)(10.61 \times 10^{-9})} = \mathbf{30\ k\Omega}$$

(c) From equation (29.10), $\dfrac{R_D}{Z} = 1 + j2\delta Q$ from which, $Z = \dfrac{R_D}{1 + j2\delta Q}$

$\delta = 0.4\% = 0.004$ hence $Z = \dfrac{30 \times 10^3}{1 + j2(0.004)(300)}$

$$= \frac{30 \times 10^3}{1 + j2.4} = \frac{30 \times 10^3}{2.6\angle 67.38^\circ}$$

$$= 11.54\angle -67.38^\circ\ k\Omega$$

Hence **the magnitude of the impedance** when the frequency is 0.4% greater than the tuned frequency is **11.54 kΩ**.

Further problems on parallel resonance may be found in the Section 29.6 following, problems 1 to 14.

29.6 Further problems on parallel resonance and Q-factor

1 A coil of resistance 20 Ω and inductance 100 mH is connected in parallel with a 50 μF capacitor across a 30 V variable-frequency supply. Determine (a) the resonant frequency of the circuit, (b) the dynamic resistance, (c) the current at resonance, and (d) the circuit Q-factor at resonance. [(a) 63.66 Hz (b) 100 Ω (c) 0.30 A (d) 2]

2 A 25 V, 2.5 kHz supply is connected to a network comprising a variable capacitor in parallel with a coil of resistance 250 Ω and inductance 80 mH. Determine for the condition when the supply

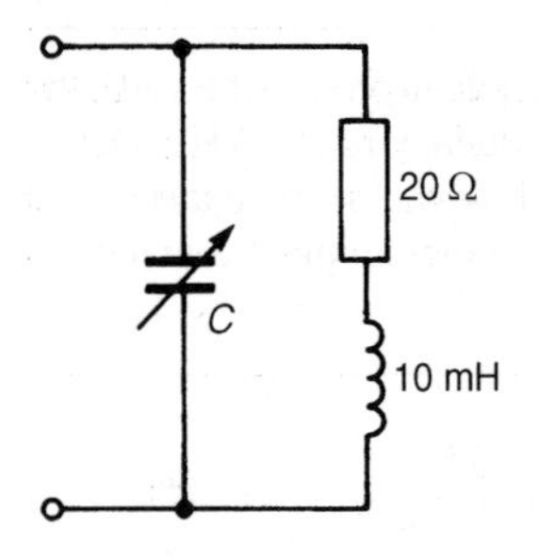

Figure 29.10

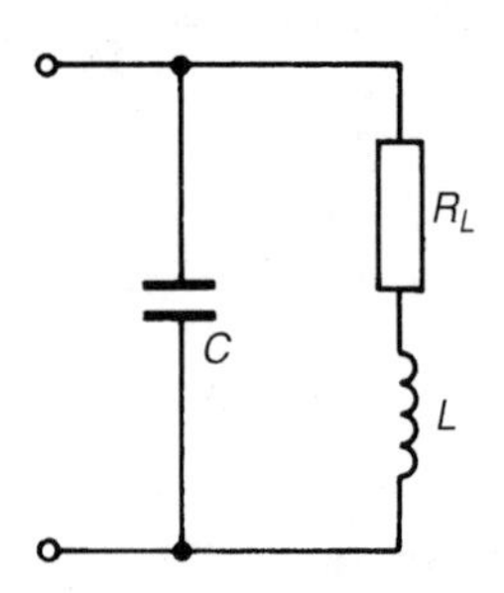

Figure 29.11

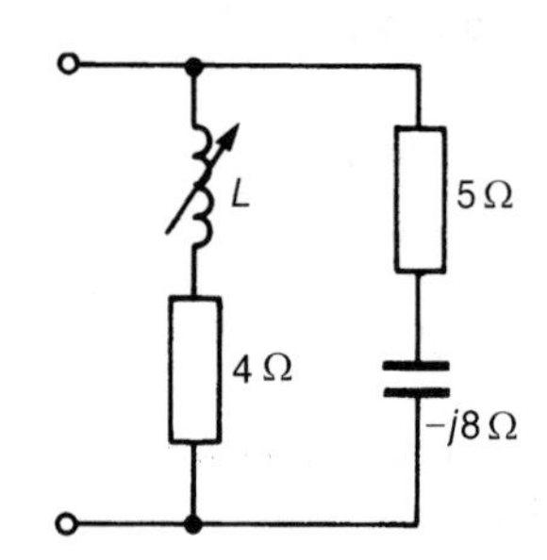

Figure 29.12

current is a minimum (a) the capacitance of the capacitor, (b) the dynamic resistance, (c) the supply current, (d) the Q-factor, (e) the bandwidth, (f) the upper and lower half-power frequencies and (g) the value of the circuit impedance at the −3 dB frequencies.

[(a) 48.73 nF (b) 6.57 kΩ (c) 3.81 mA (d) 5.03 (e) 497.3 Hz (f) 2761 Hz; 2264 Hz (g) 4.64 kΩ]

3 A 0.1 μF capacitor and a pure inductance of 0.02 H are connected in parallel across a 12 V variable-frequency supply. Determine (a) the resonant frequency of the circuit, and (b) the current circulating in the capacitance and inductance at resonance.

[(a) 3.56 kHz (b) 26.84 mA]

4 A coil of resistance 300 Ω and inductance 100 mH and a 4000 pF capacitor are connected (i) in series and (ii) in parallel. Find for each connection (a) the resonant frequency, (b) the Q-factor, and (c) the impedance at resonance.

[(i) (a) 7958 Hz (b) 16.67 (c) 300 Ω]
[(ii) (a) 7943 Hz (b) 16.64 (c) 83.33 kΩ]

5 A network comprises a coil of resistance 100 Ω and inductance 0.8 H and a capacitor having capacitance 30 μF. Determine the resonant frequency of the network when the capacitor is connected (a) in series with, and (b) in parallel with the coil.

[(a) 32.5 Hz (b) 25.7 Hz]

6 Determine the value of capacitor *C* shown in Figure 29.10 for which the resonant frequency of the network is 1 kHz. [2.30 μF]

7 In the parallel network shown in Figure 29.11, inductance *L* is 40 mH and capacitance *C* is 5 μF. Determine the resonant frequency of the circuit if (a) $R_L = 0$ and (b) $R_L = 40\ \Omega$.

[(a) 355.9 Hz (b) 318.3 Hz]

8 A capacitor of reactance 5 Ω is connected in series with a 10 Ω resistor. The whole circuit is then connected in parallel with a coil of inductive reactance 20 Ω and a variable resistor. Determine the value of this resistance for which the parallel network is resonant.

[10 Ω]

9 Determine, for the parallel network shown in Figure 29.12, the values of inductance *L* for which the circuit is resonant at a frequency of 600 Hz. [2.50 mH or 0.45 mH]

10 Find the resonant frequency of the two-branch parallel network shown in Figure 29.13. [667 Hz]

11 Determine the value of the variable resistance *R* in Figure 29.14 for which the parallel network is resonant. [11.87 Ω]

12 For the parallel network shown in Figure 29.15, determine the resonant frequency. Find also the value of resistance to be connected in series with the 10 μF capacitor to change the resonant frequency to 1 kHz. [928 Hz; 5.27 Ω]

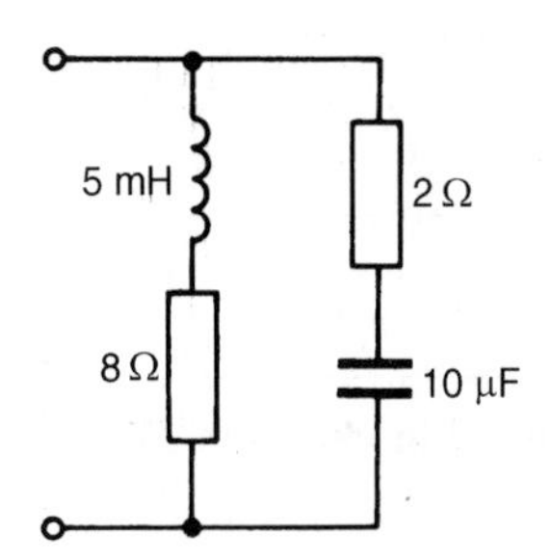

Figure 29.13

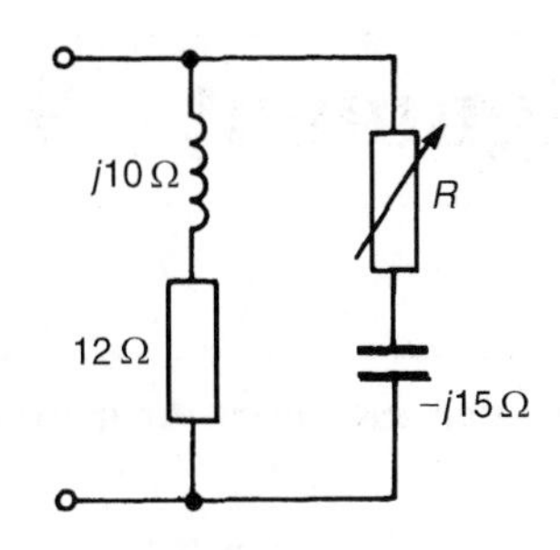

Figure 29.14

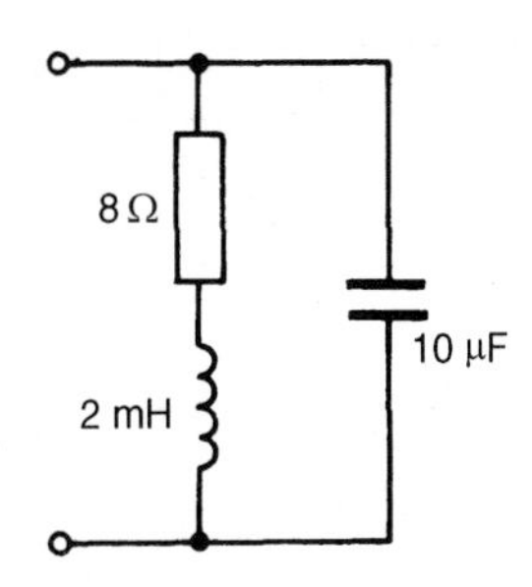

Figure 29.15

13 Determine the overall Q-factor of a parallel arrangement consisting of a capacitor having a Q-factor of 410 and an inductor having a Q-factor of 90. [73.8]

14 The value of capacitance in an $LR-C$ parallel network is 49.74 nF. If the resonant frequency of the circuit is 200 kHz and the bandwidth is 800 Hz, determine for the network (a) the Q-factor, (b) the dynamic resistance, and (c) the magnitude of the impedance when the supply frequency is 0.5% smaller than the tuned frequency.
[(a) 250 (b) 4 kΩ (c) 1.486 kΩ]

Assignment 9

This assignment covers the material contained in chapters 27 to 29.

The marks for each part of the question are shown in brackets at the end of each question.

1 In a Schering bridge network PQRS, the arms are made up as follows: PQ — a standard capacitor C_1, QR — a capacitor C_2 in parallel with a resistor R_2, RS — a resistor R_3, SP — the capacitor under test, represented by a capacitor C_x in series with a resistor R_x. The detector is connected between Q and S and the a.c. supply is connected between P and R.

(a) Sketch the bridge and derive the equations for R_x and C_x when the bridge is balanced.

(b) Evaluate R_x and C_x if, at balance $C_1 = 5$ nF, $R_2 = 300\ \Omega$, $C_2 = 30$ nF and $R_3 = 1.5$ kΩ. (16)

2 A coil of inductance 25 mH and resistance 5 Ω is connected in series with a variable capacitor C. If the supply frequency is 1 kHz and the current flowing is 2 A, determine, for series resonance, (a) the value of capacitance C, (b) the supply p.d., and (c) the p.d. across the capacitor. (8)

3 An L–R–C series circuit has a peak current of 5 mA flowing in it when the frequency of the 200 mV supply is 5 kHz. The Q-factor of the circuit under these conditions is 75. Determine (a) the voltage across the capacitor, and (b) the values of the circuit resistance, inductance and capacitance. (8)

4 A coil of resistance 15 Ω and inductance 150 mH is connected in parallel with a 4 μF capacitor across a 50 V variable-frequency supply. Determine (a) the resonant frequency of the circuit, (b) the dynamic resistance (c) the current at resonance, and (d) the circuit Q-factor at resonance. (10)

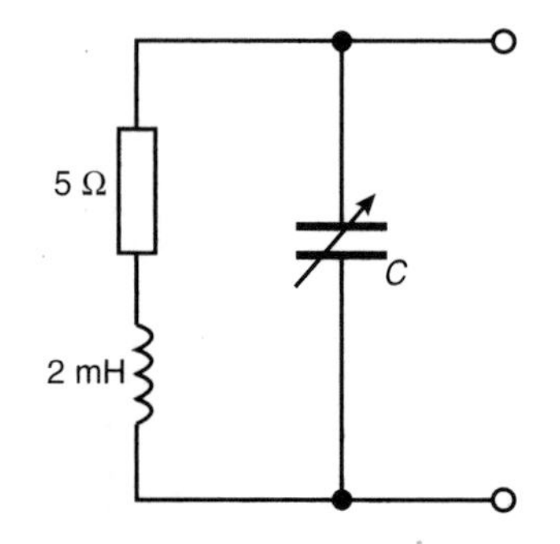

Figure A9.1

5 For the parallel network shown in Figure A9.1, determine the value of C for which the resonant frequency is 2 kHz. (8)

30 Introduction to network analysis

At the end of this chapter you should be able to:

- appreciate available methods of analysing networks
- solve simultaneous equations in two and three unknowns using determinants
- analyse a.c. networks using Kirchhoff's laws

30.1 Introduction

Voltage sources in series-parallel networks cause currents to flow in each branch of the circuit and corresponding volt-drops occur across the circuit components. A.c. circuit (or network) analysis involves the determination of the currents in the branches and/or the voltages across components.

The laws which determine the currents and voltage drops in a.c. networks are:

(a) **current, $I = V/Z$**, where Z is the complex impedance and V the voltage across the impedance;

(b) **the laws for impedances in series and parallel,** i.e., total impedance,

$$Z_T = Z_1 + Z_2 + Z_3 + \ldots + Z_n \text{ for } n \text{ impedances connected in series,}$$

and $$\frac{1}{Z_T} = \frac{1}{Z_1} + \frac{1}{Z_2} + \frac{1}{Z_3} + \ldots + \frac{1}{Z_n} \text{ for } n \text{ impedances connected in parallel; and}$$

(c) **Kirchhoff's laws,** which may be stated as:

(i) *'At any point in an electrical circuit the phasor sum of the currents flowing towards that junction is equal to the phasor sum of the currents flowing away from the junction.'*

(ii) *'In any closed loop in a network, the phasor sum of the voltage drops (i.e., the products of current and impedance) taken around the loop is equal to the phasor sum of the e.m.f.'s acting in that loop.'*

In any circuit the currents and voltages at any point may be determined by applying Kirchhoff's laws (as demonstrated in this chapter), or by extensions of Kirchhoff's laws, called mesh-current analysis and nodal analysis (see Chapter 31).

However, for more complicated circuits, a number of circuit theorems have been developed as alternatives to the use of Kirchhoff's laws to solve problems involving both d.c. and a.c. electrical networks. These include:

(a) the superposition theorem (see Chapter 32)
(b) Thévenin's theorem (see Chapter 33)
(c) Norton's theorem (see Chapter 33),
(d) the maximum power transfer theorems (see Chapter 35).

In addition to these theorems, and often used as a preliminary to using circuit theorems, star-delta (or $T-\pi$) and delta-star (or $\pi-T$) transformations provide a method for simplifying certain circuits (see Chapter 34).

In a.c. circuit analysis involving Kirchhoff's laws or circuit theorems, the use of complex numbers is essential.

The above laws and theorems apply to linear circuits, i.e., circuits containing impedances whose values are independent of the direction and magnitude of the current flowing in them.

30.2 Solution of simultaneous equations using determinants

When Kirchhoff's laws are applied to electrical circuits, simultaneous equations result which require solution. If two loops are involved, two simultaneous equations containing two unknowns need to be solved; if three loops are involved, three simultaneous equations containing three unknowns need to be solved and so on. The elimination and substitution methods of solving simultaneous equations may be used to solve such equations. However a more convenient method is to use **determinants.**

Two unknowns

When solving linear simultaneous equations in two unknowns using determinants:

(i) the equations are initially written in the form:

$$a_1x + b_1y + c_1 = 0$$

$$a_2x + b_2y + c_2 = 0$$

(ii) the solution is given by:

$$\frac{x}{D_x} = \frac{-y}{D_y} = \frac{1}{D}$$

where $$D_x = \begin{vmatrix} b_1 & c_1 \\ b_2 & c_2 \end{vmatrix}$$

i.e., the determinant of the coefficients left when the x-column is 'covered up',

$$D_y = \begin{vmatrix} a_1 & c_1 \\ a_2 & c_2 \end{vmatrix}$$

i.e., the determinant of the coefficients left when the y-column is 'covered up',

and $$D = \begin{vmatrix} a_1 & b_1 \\ a_2 & b_2 \end{vmatrix}$$

i.e., the determinant of the coefficients left when the constants-column is 'covered up'.

A '2 × 2' determinant $\begin{vmatrix} a & d \\ b & c \end{vmatrix}$ is evaluated as $ad - bc$

Three unknowns

When solving linear simultaneous equations in three unknowns using determinants:

(i) the equations are initially written in the form:

$$a_1x + b_1y + c_1z + d_1 = 0$$

$$a_2x + b_2y + c_2z + d_2 = 0$$

$$a_3x + b_3y + c_3z + d_3 = 0$$

(ii) the solution is given by:

$$\frac{x}{D_x} = \frac{-y}{D_y} = \frac{z}{D_z} = \frac{-1}{D}$$

where $$D_x = \begin{vmatrix} b_1 & c_1 & d_1 \\ b_2 & c_2 & d_2 \\ b_3 & c_3 & d_3 \end{vmatrix}$$

i.e., the determinant of the coefficients left when the x-column is 'covered up',

$$D_y = \begin{vmatrix} a_1 & c_1 & d_1 \\ a_2 & c_2 & d_2 \\ a_3 & c_3 & d_3 \end{vmatrix}$$

i.e., the determinant of the coefficients left when the y-column is 'covered up',

$$D_z = \begin{vmatrix} a_1 & b_1 & d_1 \\ a_2 & b_2 & d_2 \\ a_3 & b_3 & d_3 \end{vmatrix}$$

i.e., the determinant of the coefficients left when the z-column is 'covered up', and $D = \begin{vmatrix} a_1 & b_1 & c_1 \\ a_2 & b_2 & c_2 \\ a_3 & b_3 & c_3 \end{vmatrix}$

i.e., the determinant of the coefficients left when the constants-column is 'covered up'.

To evaluate a 3 × 3 determinant:

(a) The **minor** of an element of a 3 by 3 matrix is the value of the 2 by 2 determinant obtained by covering up the row and column containing that element.

Thus for the matrix $\begin{pmatrix} 1 & 2 & 3 \\ 4 & 5 & 6 \\ 7 & 8 & 9 \end{pmatrix}$ the minor of element 4 is the determinant $\begin{vmatrix} 2 & 3 \\ 8 & 9 \end{vmatrix}$, i.e., $(2 \times 9) - (3 \times 8) = 18 - 24 = -6$. Similarly, the minor of element 3 is $\begin{vmatrix} 4 & 5 \\ 7 & 8 \end{vmatrix}$, i.e., $(4 \times 8) - (5 \times 7) = 32 - 35 = -3$

(b) The sign of the minor depends on its position within the matrix, the sign pattern being $\begin{pmatrix} + & - & + \\ - & + & - \\ + & - & + \end{pmatrix}$. Thus the signed minor of element 4 in the above matrix is $-\begin{vmatrix} 2 & 3 \\ 8 & 9 \end{vmatrix} = -(-6) = 6$

The signed-minor of an element is called the **cofactor** of the element.

Thus the cofactor of element 2 is $-\begin{vmatrix} 4 & 6 \\ 7 & 9 \end{vmatrix} = -(36 - 42) = 6$

(c) **The value of a 3 by 3 determinant is the sum of the products of the elements and their cofactors of any row or any column of the corresponding 3 by 3 matrix.**

Thus a 3 by 3 determinant $\begin{vmatrix} a & b & c \\ d & e & f \\ g & h & j \end{vmatrix}$ is evaluated as

$$a\begin{vmatrix} e & f \\ h & j \end{vmatrix} - b\begin{vmatrix} d & f \\ g & j \end{vmatrix} + c\begin{vmatrix} d & e \\ g & h \end{vmatrix} \quad \text{using the top row,}$$

or $$-b\begin{vmatrix} d & f \\ g & j \end{vmatrix} + e\begin{vmatrix} a & c \\ g & j \end{vmatrix} - h\begin{vmatrix} a & c \\ d & f \end{vmatrix} \quad \text{using the second column.}$$

There are thus six ways of evaluating a 3 by 3 determinant.

Determinants are used to solve simultaneous equations in some of the following problems and in Chapter 31.

30.3 Network analysis using Kirchhoff's laws

Kirchhoff's laws may be applied to both d.c. and a.c. circuits. The laws are introduced in Chapter 13 for d.c. circuits. To demonstrate the method of analysis, consider the d.c. network shown in Figure 30.1. If the current flowing in each branch is required, the following three-step procedure may be used:

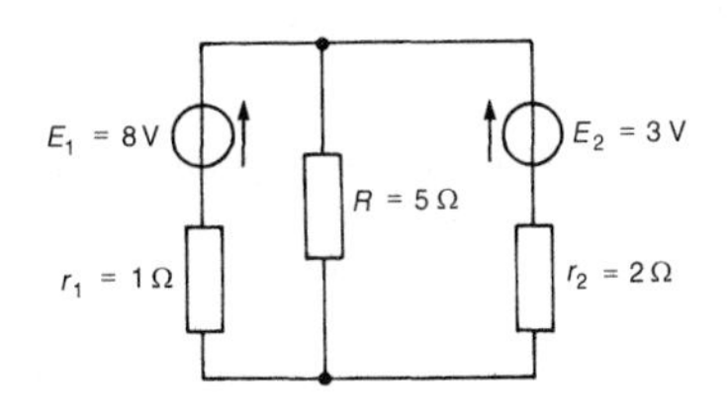

Figure 30.1

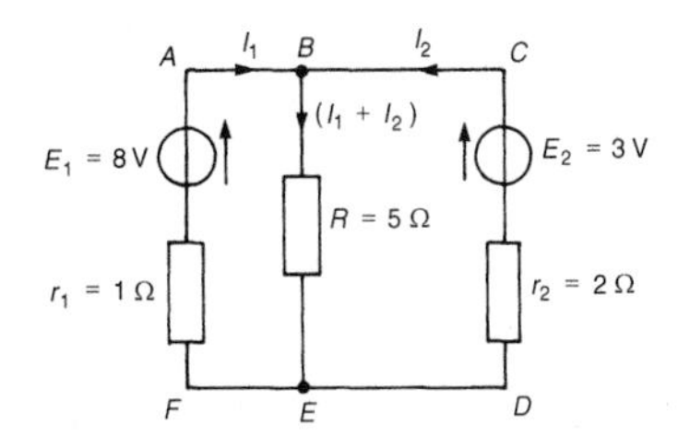

Figure 30.2

(i) Label branch currents and their directions on the circuit diagram. The directions chosen are arbitrary but, as a starting-point, a useful guide is to assume that current flows from the positive terminals of the voltage sources. This is shown in Figure 30.2 where the three branch currents are expressed in terms of I_1 and I_2 only, since the current through resistance R, by Kirchhoff's current law, is $(I_1 + I_2)$

(ii) Divide the circuit into loops — two in this ease (see Figure 30.2) and then apply Kirchhoff's voltage law to each loop in turn. From loop ABEF, and moving in a clockwise direction (the choice of loop direction is arbitrary), $E_1 = I_1 r + (I_1 + I_2)R$ (note that the two voltage drops are positive since the loop direction is the same as the current directions involved in the volt drops). Hence

$$8 = I_1 + 5(I_1 + I_2)$$

or $\quad 6I_1 + 5I_2 = 8 \qquad (1)$

From loop BCDE in Figure 30.2, and moving in an anticlockwise direction, (note that the direction does not have to be the same as that used for the first loop), $E_2 = I_2 r_2 + (I_1 + I_2)R$,

i.e., $\quad 3 = 2I_2 + 5(I_1 + I_2)$

or $\quad 5I_1 + 7I_2 = 3 \qquad (2)$

(iii) Solve simultaneous equations (1) and (2) for I_1 and I_2

Multiplying equation (1) by 7 gives: $\quad 42I_1 + 35I_2 = 56 \qquad (3)$

Multiplying equation (2) by 5 gives: $\quad 25I_1 + 35I_2 = 15 \qquad (4)$

Equation (3) − equation (4) gives: $\quad 17I_1 = 41$

from which, current $\boldsymbol{I_1} = 41/17 = 2.412\text{ A} = \mathbf{2.41\ A}$, correct to two decimal places.

From equation (1): $6(2.412) + 5I_2 = 8$, from which,

$$\text{current } \boldsymbol{I_2} = \frac{8 - 6(2.412)}{5} = -1.294\ A$$

$$= \mathbf{-1.29\ A}, \text{ correct to two decimal places.}$$

The minus sign indicates that current I_2 flows in the opposite direction to that shown in Figure 30.2.

The current flowing through resistance R is

$$(I_1 + I_2) = 2.412 + (-1.294)$$
$$= 1.118\ A = \mathbf{1.12\ A}, \text{ correct to two decimal places.}$$

[A third loop may be selected in Figure 30.2, (just as a check), moving clockwise around the outside of the network. Then $E_1 - E_2 = I_1 r_1 - I_2 r_2$, i.e. $8 - 3 = I_1 - 2I_2$. Thus $5 = 2.412 - 2(-1.294) = 5$]

An alternative method of solving equations (1) and (2) is shown below using determinants. Since

$$6I_1 + 5I_2 - 8 = 0 \quad (1)$$
$$5I_1 + 7I_2 - 3 = 0 \quad (2)$$

then
$$\frac{I_1}{\begin{vmatrix} 5 & -8 \\ 7 & -3 \end{vmatrix}} = \frac{-I_2}{\begin{vmatrix} 6 & -8 \\ 5 & -3 \end{vmatrix}} = \frac{1}{\begin{vmatrix} 6 & 5 \\ 5 & 7 \end{vmatrix}}$$

i.e.,
$$\frac{I_1}{-15 + 56} = \frac{-I_2}{-18 + 40} = \frac{1}{42 - 25}$$
$$\frac{I_1}{41} = \frac{-I_2}{22} = \frac{1}{17}$$

from which, $\boldsymbol{I_1} = 41/17 = \mathbf{2.41\ A}$ and $\boldsymbol{I_2} = -22/17 = \mathbf{-1.29\ A}$, as obtained previously.

The above procedure is shown for a simple d.c. circuit having two unknown values of current. The procedure however applies equally well to a.c. networks and/or to circuits where three unknown currents are involved. This is illustrated in the following problems.

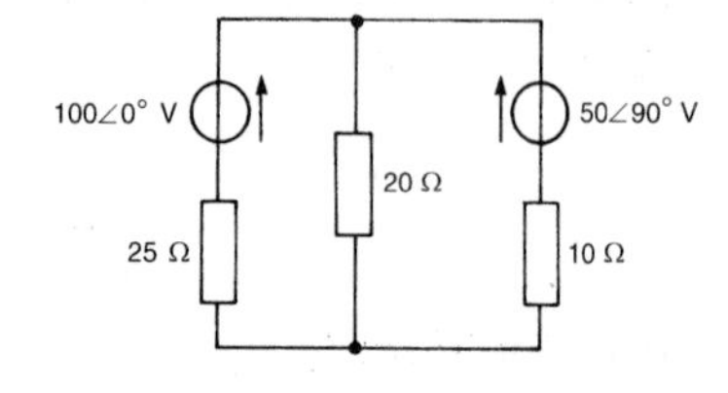

Figure 30.3

Figure 30.4

Problem 1. Use Kirchhoff's laws to find the current flowing in each branch of the network shown in Figure 30.3.

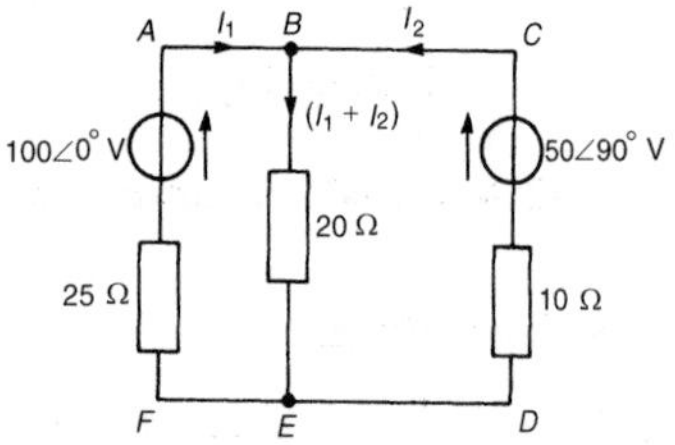

(i) The branch currents and their directions are labelled as shown in Figure 30.4

(ii) Two loops are chosen. From loop ABEF, and moving clockwise,

$$25I_1 + 20(I_1 + I_2) = 100\angle 0^\circ$$

i.e.,
$$45I_1 + 20I_2 = 100 \quad (1)$$

From loop BCDE, and moving anticlockwise,

$$10I_2 + 20(I_1 + I_2) = 50\angle 90^\circ$$

i.e.,
$$20I_1 + 30I_2 = j50 \quad (2)$$

3 × equation (1) gives: $135I_1 + 60I_2 = 300$ (3)

2 × equation (2) gives: $40I_1 + 60I_2 = j100$ (4)

Equation (3) — equation (4) gives: $95I_1 = 300 - j100$,

from which, current $I_1 = \dfrac{300 - j100}{95}$ $=$ **3.329∠−18.43° A** or **(3.158 − *j*1.052)A**

Substituting in equation (1) gives:

$45(3.158 - j1.052) + 20I_2 = 100$, from which,

$$I_2 = \frac{100 - 45(3.158 - j1.052)}{20}$$

= **(−2.106 + *j*2.367)A** or **3.168∠131.66° A**

Thus

$$I_1 + I_2 = (3.158 - j1.052) + (-2.106 + j2.367)$$

= **(1.052 + *j*1.315) A** or **1.684∠51.34° A**

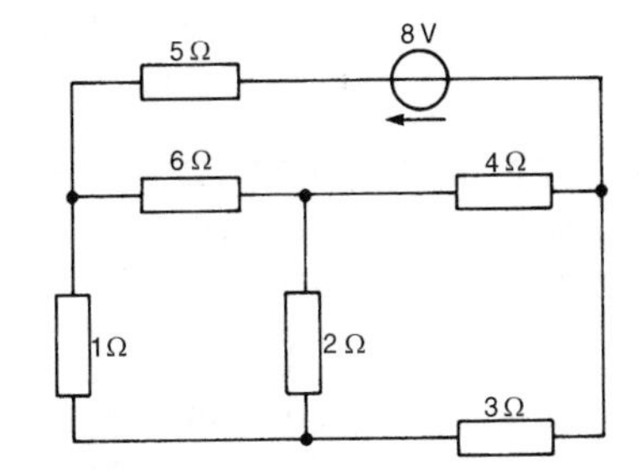

Figure 30.5

Problem 2. Determine the current flowing in the 2 Ω resistor of the circuit shown in Figure 30.5 using Kirchhoff's laws. Find also the power dissipated in the 3 Ω resistance.

(i) Currents and their directions are assigned as shown in Figure 30.6.

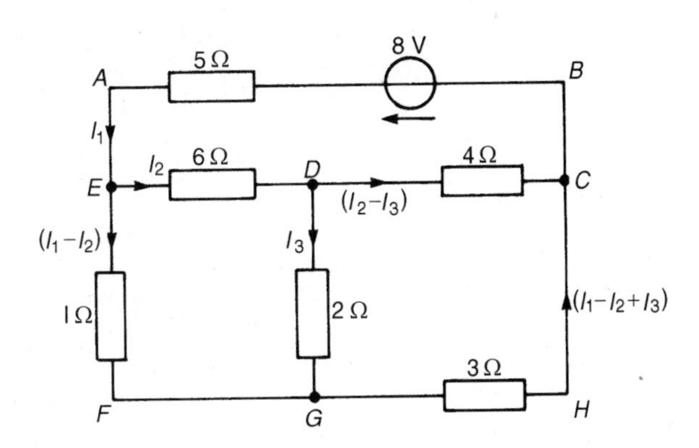

Figure 30.6

(ii) Three loops are chosen since three unknown currents are required. The choice of loop directions is arbitrary. From loop ABCDE, and moving anticlockwise,

$$5I_1 + 6I_2 + 4(I_2 - I_3) = 8$$

i.e., $$5I_1 + 10I_2 - 4I_3 = 8 \qquad (1)$$

From loop EDGF, and moving clockwise,

$$6I_2 + 2I_3 - 1(I_1 - I_2) = 0$$

i.e., $$-I_1 + 7I_2 + 2I_3 = 0 \qquad (2)$$

From loop DCHG, and moving anticlockwise,

$$2I_3 + 3(I_1 - I_2 + I_3) - 4(I_2 - I_3) = 0$$

i.e., $$3I_1 - 7I_2 + 9I_3 = 0 \qquad (3)$$

(iii) Thus $5I_1 + 10I_2 - 4I_3 - 8 = 0$

$$-I_1 + 7I_2 + 2I_3 + 0 = 0$$

$$3I_1 - 7I_2 + 9I_3 + 0 = 0$$

Hence, using determinants,

$$\frac{I_1}{\begin{vmatrix} 10 & -4 & -8 \\ 7 & 2 & 0 \\ -7 & 9 & 0 \end{vmatrix}} = \frac{-I_2}{\begin{vmatrix} 5 & -4 & -8 \\ -1 & 2 & 0 \\ 3 & 9 & 0 \end{vmatrix}} = \frac{I_3}{\begin{vmatrix} 5 & 10 & -8 \\ -1 & 7 & 0 \\ 3 & -7 & 0 \end{vmatrix}} = \frac{-1}{\begin{vmatrix} 5 & 10 & -4 \\ -1 & 7 & 2 \\ 3 & -7 & 9 \end{vmatrix}}$$

Thus

$$\frac{I_1}{-8\begin{vmatrix} 7 & 2 \\ -7 & 9 \end{vmatrix}} = \frac{-I_2}{-8\begin{vmatrix} -1 & 2 \\ 3 & 9 \end{vmatrix}} = \frac{I_3}{-8\begin{vmatrix} -1 & 7 \\ 3 & -7 \end{vmatrix}}$$

$$= \frac{-1}{5\begin{vmatrix} 7 & 2 \\ -7 & 9 \end{vmatrix} - 10\begin{vmatrix} -1 & 2 \\ 3 & 9 \end{vmatrix} - 4\begin{vmatrix} -1 & 7 \\ 3 & -7 \end{vmatrix}}$$

$$\frac{I_1}{-8(63 + 14)} = \frac{-I_2}{-8(-9 - 6)} = \frac{I_3}{-8(7 - 21)}$$

$$= \frac{-1}{5(63 + 14) - 10(-9 - 6) - 4(7 - 21)}$$

$$\frac{I_1}{-616} = \frac{-I_2}{120} = \frac{I_3}{112} = \frac{-1}{591}$$

Hence $$I_1 = \frac{616}{591} = 1.042 \text{ A},$$

$$I_2 = \frac{120}{591} = 0.203 \text{ A and}$$

$$I_3 = \frac{-112}{591} = -0.190 \text{ A}$$

Thus **the current flowing in the 2 Ω resistance is 0.190 A** in the opposite direction to that shown in Figure 30.6.

$$\text{Current in the } 3\Omega \text{ resistance} = I_1 - I_2 + I_3$$
$$= 1.042 - 0.203 + (-0.190) = 0.649 \text{ A.}$$

Hence **power dissipated in the 3 Ω resistance**, $I^2(3) = (0.649)^2(3) =$ **1.26W**

Problem 3. For the a.c. network shown in Figure 30.7, determine the current flowing in each branch using Kirchhoff's laws.

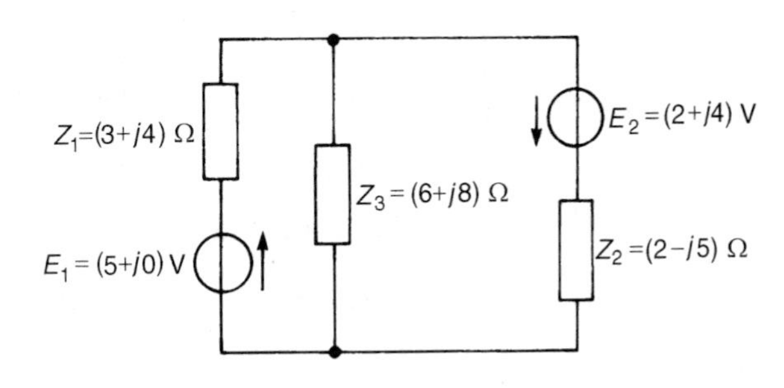

Figure 30.7

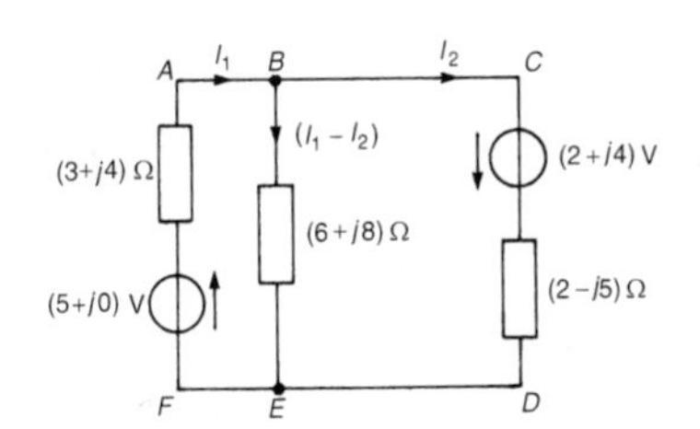

Figure 30.8

(i) Currents I_1 and I_2 with their directions are shown in Figure 30.8.

(ii) Two loops are chosen with their directions both clockwise.

From loop ABEF, $(5 + j0) = I_1(3 + j4) + (I_1 - I_2)(6 + j8)$

i.e., $$5 = (9 + j12)\, I_1 - (6 + j8)I_2 \qquad (1)$$

From loop BCDE, $(2 + j4) = I_2(2 - j5) - (I_1 - I_2)(6 + j8)$

i.e., $$(2 + j4) = -(6 + j8)\, I_1 + (8 + j3)I_2 \qquad (2)$$

(iii) Multiplying equation (1) by $(8 + j3)$ gives:

$$5(8 + j3) = (8 + j3)(9 + j12)I_1 - (8 + j3)(6 + j8)I_2 \qquad (3)$$

Multiplying equation (2) by $(6 + j8)$ gives:

$$(6 + j8)(2 + j4) = -(6 + j8)(6 + j8)I_1 + (6 + j8)(8 + j3)I_2 \qquad (4)$$

Adding equations (3) and (4) gives:

$$5(8 + j3) + (6 + j8)(2 + j4) = [(8 + j3)(9 + j12) - (6 + j8)(6 + j8)]I_1$$

i.e., $$(20 + j55) = (64 + j27)I_1$$

$$\text{from which, } \boldsymbol{I_1} = \frac{20 + j55}{64 + j27} = \frac{58.52\angle 70.02^\circ}{69.46\angle 22.87^\circ} = \mathbf{0.842\angle 47.15^\circ\ A}$$

$$= (0.573 + j0.617)\text{ A}$$

$$= \mathbf{(0.57 + j0.62)\ A,}\text{ correct to two decimal places.}$$

$$\text{From equation (1), } 5 = (9 + j12)(0.573 + j0.617) - (6 + j8)I_2$$

$$5 = (-2.247 + j12.429) - (6 + j8)I_2$$

$$\text{from which, } \boldsymbol{I_2} = \frac{-2.247 + j12.429 - 5}{6 + j8}$$

$$= \frac{14.39\angle 120.25^\circ}{10\angle 53.13^\circ}$$

$$= \mathbf{1.439\angle 67.12^\circ A} = (0.559 + j1.326)\text{ A}$$

$$= \mathbf{(0.56 + j1.33)A,}\text{ correct to two decimal places.}$$

The current in the $(6 + j8)\Omega$ impedance,

$$\boldsymbol{I_1} - \boldsymbol{I_2} = (0.573 + j0.617) - (0.559 + j1.326)$$

$$= \mathbf{(0.014 - j0.709)A}\text{ or }\mathbf{0.709\angle -88.87^\circ\ A}$$

An alternative method of solving equations (1) and (2) is shown below, using determinants.

$$(9 + j12)I_1 - (6 + j8)I_2 - 5 = 0 \quad (1)$$

$$-(6 + j8)I_1 + (8 + j3)I_2 - (2 + j4) = 0 \quad (2)$$

$$\text{Thus } \frac{I_1}{\begin{vmatrix} -(6 + j8) & -5 \\ (8 + j3) & -(2 + j4) \end{vmatrix}} = \frac{-I_2}{\begin{vmatrix} (9 + j12) & -5 \\ -(6 + j8) & -(2 + j4) \end{vmatrix}}$$

$$= \frac{1}{\begin{vmatrix} (9 + j12) & -(6 + j8) \\ -(6 + j8) & (8 + j3) \end{vmatrix}}$$

$$\frac{I_1}{(-20 + j40) + (40 + j15)} = \frac{-I_2}{(30 - j60) - (30 + j40)}$$

$$= \frac{1}{(36 + j123) - (-28 + j96)}$$

$$\frac{I_1}{20 + j55} = \frac{-I_2}{-j100} = \frac{1}{64 + j27}$$

$$\text{Hence } \boldsymbol{I_1} = \frac{20 + j55}{64 + j27} = \frac{58.52\angle 70.02^\circ}{69.46\angle 22.87^\circ}$$

$$= \mathbf{0.842\angle 47.15^\circ\ A}$$

and $\quad I_2 = \dfrac{100\angle 90^\circ}{69.46\angle 22.87^\circ} = \mathbf{1.440\angle 67.13^\circ\ A}$

The current flowing in the $(6 + j8)\ \Omega$ impedance is given by:

$$I_1 - I_2 = 0.842\angle 47.15^\circ - 1.440\angle 67.13^\circ A$$
$$= \mathbf{(0.013 - j0.709)\ A} \text{ or } \mathbf{0.709\angle -88.95^\circ\ A}$$

Problem 4. For the network shown in Figure 30.9, use Kirchhoff's laws to determine the magnitude of the current in the $(4 + j3)\ \Omega$ impedance.

(i) Currents I_1, I_2 and I_3 with their directions are shown in Figure 30.10. The current in the $(4 + j3)\ \Omega$ impedance is specified by one symbol only (i.e., I_3), which means that the three equations formed need to be solved for only one unknown current.

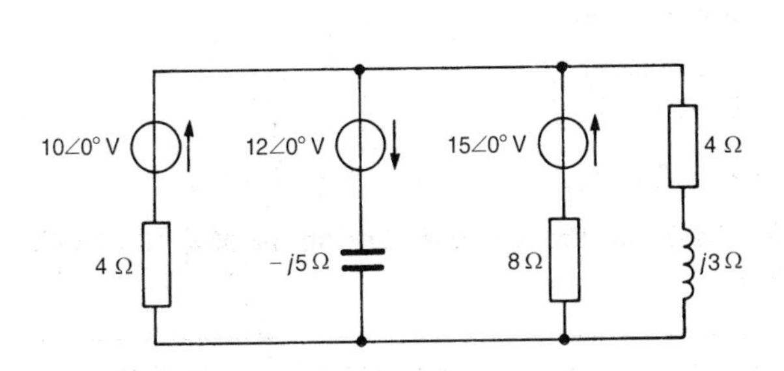

Figure 30.9

Figure 30.10

(ii) Three loops are chosen. From loop ABGH, and moving clockwise,

$$4I_1 - j5I_2 = 10 + 12 \quad (1)$$

From loop BCFG, and moving anticlockwise,

$$-j5I_2 - 8(I_1 - I_2 - I_3) = 15 + 12 \quad (2)$$

From loop CDEF, and moving clockwise,

$$-8(I_1 - I_2 - I_3) + (4 + j3)(I_3) = 15 \quad (3)$$

Hence

$$4I_1 - j5I_2 + 0I_3 - 22 = 0$$
$$-8I_1 + (8 - j5)I_2 + 8I_3 - 27 = 0$$
$$-8I_1 + 8I_2 + (12 + j3)I_3 - 15 = 0$$

Solving for I_3 using determinants gives:

$$\frac{I_3}{\begin{vmatrix} 4 & -j5 & -22 \\ -8 & (8-j5) & -27 \\ -8 & 8 & -15 \end{vmatrix}} = \frac{-1}{\begin{vmatrix} 4 & -j5 & 0 \\ -8 & (8-j5) & 8 \\ -8 & 8 & (12+j3) \end{vmatrix}}$$

Thus
$$\frac{I_3}{4\begin{vmatrix} (8-j5) & -27 \\ 8 & -15 \end{vmatrix} + j5\begin{vmatrix} -8 & -27 \\ -8 & -15 \end{vmatrix} - 22\begin{vmatrix} -8 & (8-j5) \\ -8 & 8 \end{vmatrix}}$$
$$= \frac{-1}{4\begin{vmatrix} (8-j5) & 8 \\ 8 & (12+j3) \end{vmatrix} + j5\begin{vmatrix} -8 & 8 \\ -8 & (12+j3) \end{vmatrix}}$$

Hence

$$\frac{I_3}{384+j700} = \frac{-1}{308-j304} \text{ from which,}$$

$$I_3 = \frac{-(384+j700)}{(308-j304)}$$
$$= \frac{798.41\angle -118.75}{432.76\angle -44.63^\circ}$$
$$= 1.85\angle -74.12^\circ \text{ A}$$

Hence the magnitude of the current flowing in the $(4+j3)\Omega$ impedance is 1.85 A

Further problems on network analysis using Kirchhoff's laws may be found in Section 30.4 following, problems 1 to 10.

30.4 Further problems on Kirchhoff's laws

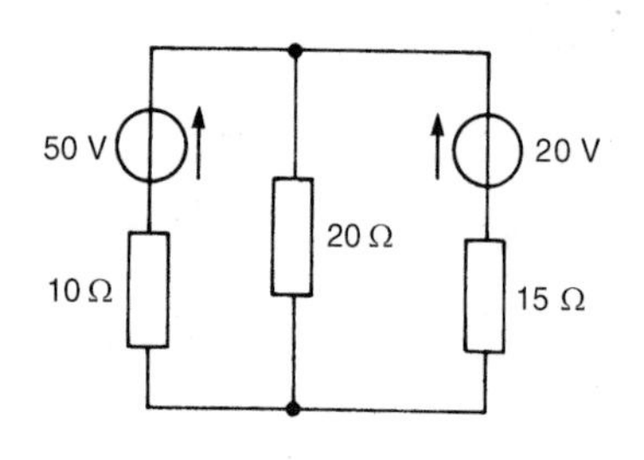

Figure 30.11

1 For the network shown in Figure 30.11, determine the current flowing in each branch.

[50 V source discharges at 2.08 A, 20 V source charges at 0.62 A, current through 20 Ω resistor is 1.46 A]

2. Determine the value of currents I_A, I_B and I_C for the network shown in Figure 30.12. [$I_A = 5.38$ A, $I_B = 4.81$ A, $I_C = 0.58$ A]

3. For the bridge shown in Figure 30.13, determine the current flowing in (a) the 5 Ω resistance, (b) the 22 Ω resistance, and (c) the 2 Ω resistance. [(a) 4 A (b) 1 A (c) 7 A]

4. For the circuit shown in Figure 30.14, determine (a) the current flowing in the 10 V source, (b) the p.d. across the 6 Ω resistance, and (c) the active power dissipated in the 4 Ω resistance.

[(a) 1.59 A (b) 3.71 V (c) 3.79 W]

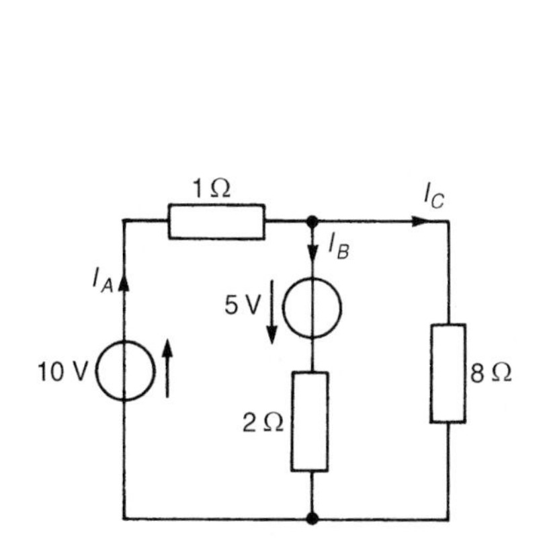

Figure 30.12

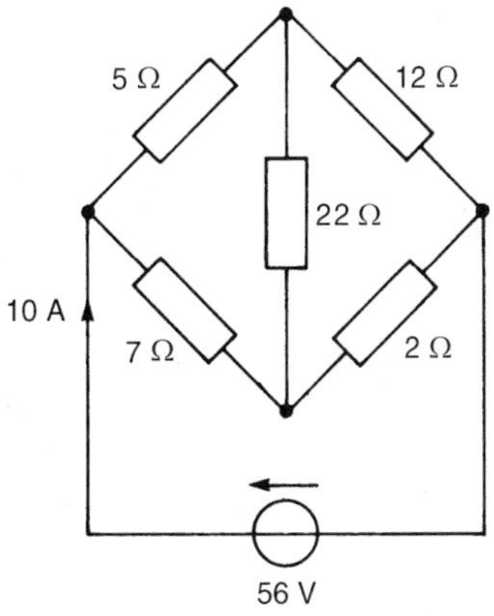

Figure 30.13

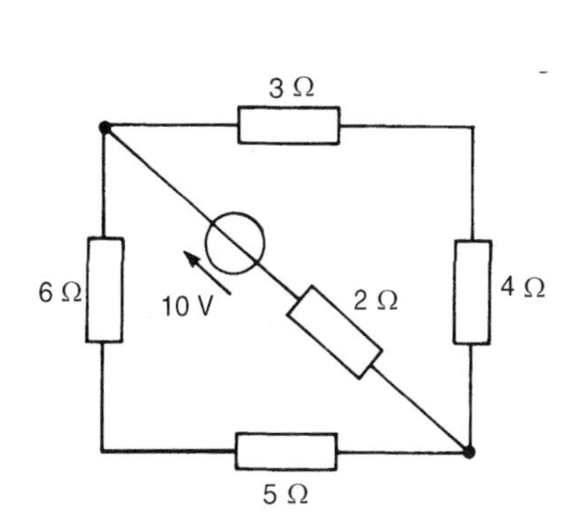

Figure 30.14

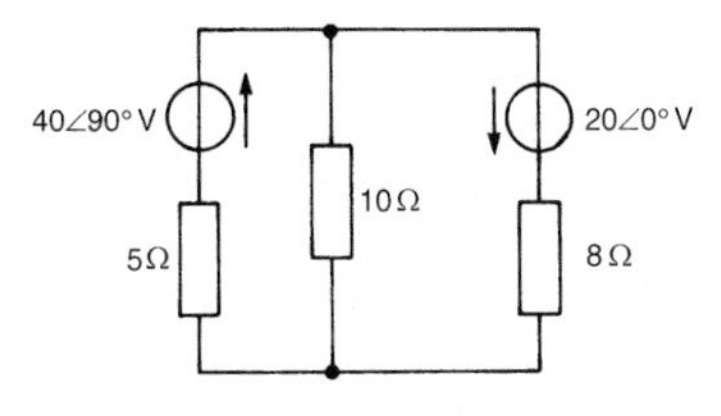

Figure 30.15

5. Use Kirchhoff's laws to determine the current flowing in each branch of the network shown in Figure 30.15.

   [40∠90° V source discharges at 4.40∠74.48° A
   20∠0° V source discharges at 2.94∠53.13° A
   current in 10 Ω resistance is 1.97∠107.35° A
   (downward)]

6. For the network shown in Figure 30.16, use Kirchhoff's laws to determine the current flowing in the capacitive branch. [1.58 A]

7. Use Kirchhoff's laws to determine, for the network shown in Figure 30.17, the current flowing in (a) the 20 Ω resistance, and (b) the 4 Ω resistance. Determine also (c) the p.d. across the 8 Ω resistance, and (d) the active power dissipated in the 10 Ω resistance.

   [(a) 0.14 A (b) 10.1 A (c) 2.27 V (d) 1.81 W]

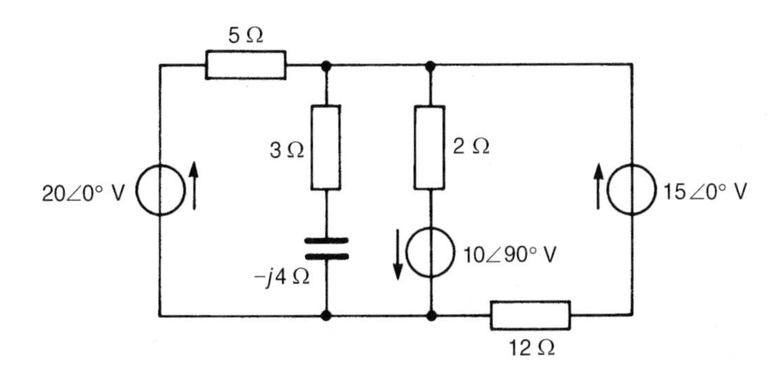

Figure 30.16

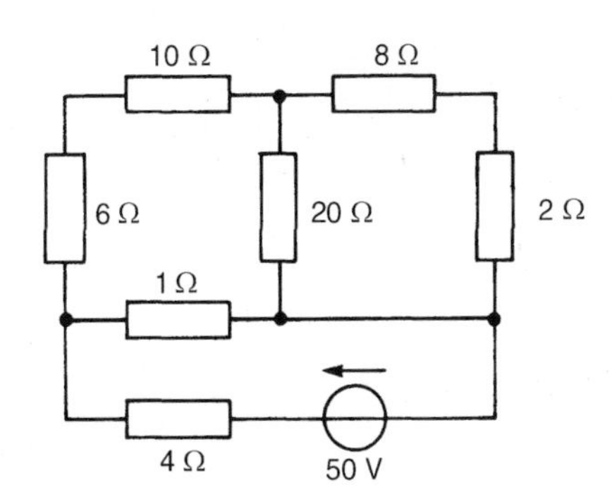

Figure 30.17

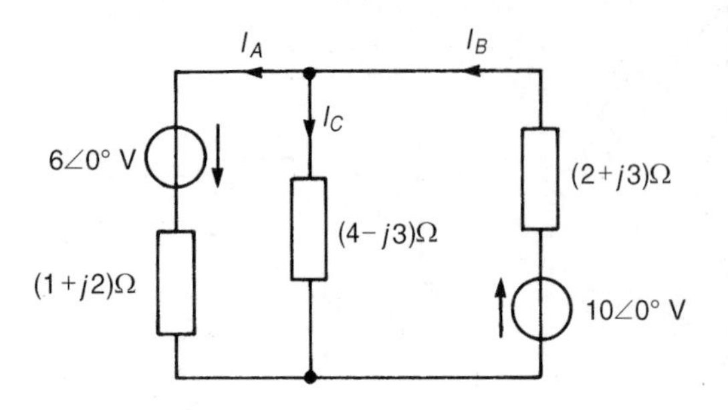

Figure 30.18

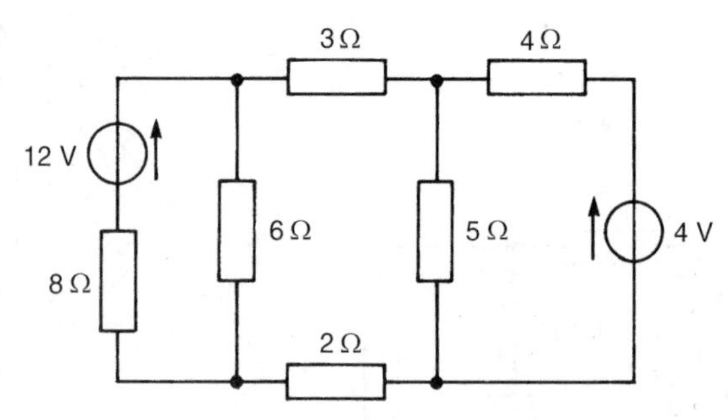

Figure 30.19

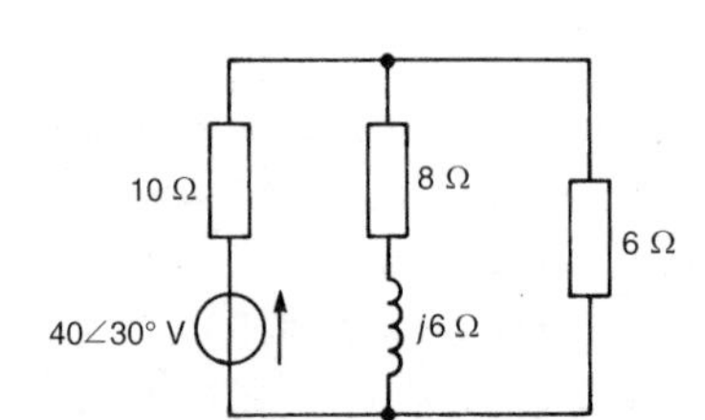

Figure 30.20

8. Determine the value of currents I_A, I_B and I_C shown in the network of Figure 30.18, using Kirchhoff's laws.
   [$I_A = 2.80\angle -59.59°$ A, $I_B = 2.71\angle -58.78°$ A, $I_C = 0.097\angle 97.13°$ A]

9. Use Kirchhoff's laws to determine the currents flowing in (a) the 3 Ω resistance, (b) the 6 Ω resistance and (c) the 4 V source of the network shown in Figure 30.19. Determine also the active power dissipated in the 5 Ω resistance.
   [(a) 0.27 A (b) 0.70 A (c) 0.29 A discharging (d) 1.60 W]

10. Determine the magnitude of the p.d. across the $(8 + j6)$ Ω impedance shown in Figure 30.20 by using Kirchhoff's laws. [11.37 V]

31 Mesh-current and nodal analysis

At the end of this chapter you should be able to:

- solve d.c. and a.c. networks using mesh-current analysis
- solve d.c. and a.c. networks using nodal analysis

31.1 Mesh-current analysis

Mesh-current analysis is merely an extension of the use of Kirchhoff's laws, explained in Chapter 30. Figure 31.1 shows a network whose circulating currents I_1, I_2 and I_3 have been assigned to closed loops in the circuit rather than to branches. Currents I_1, I_2 and I_3 are called **mesh-currents** or **loop-currents**.

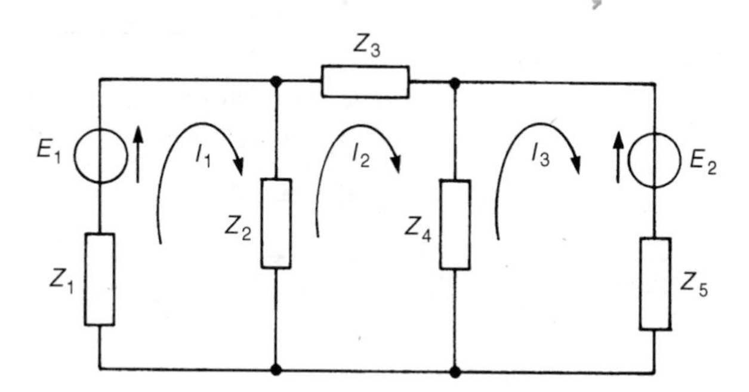

Figure 31.1

In mesh-current analysis the loop-currents are all arranged to circulate in the same direction (in Figure 31.1, shown as clockwise direction). Kirchhoff's second law is applied to each of the loops in turn, which in the circuit of Figure 31.1 produces three equations in three unknowns which may be solved for I_1, I_2 and I_3. The three equations produced from Figure 31.1 are:

$$I_1(Z_1 + Z_2) - I_2Z_2 = E_1$$

$$I_2(Z_2 + Z_3 + Z_4) - I_1Z_2 - I_3Z_4 = 0$$

$$I_3(Z_4 + Z_5) - I_2Z_4 = -E_2$$

The branch currents are determined by taking the phasor sum of the mesh currents common to that branch. For example, the current flowing in impedance Z_2 of Figure 31.1 is given by $(I_1 - I_2)$ phasorially. The method of mesh-current analysis, called **Maxwell's theorem**, is demonstrated in the following problems.

Problem 1. Use mesh-current analysis to determine the current flowing in (a) the 5 Ω resistance, and (b) the 1 Ω resistance of the d.c. circuit shown in Figure 31.2.

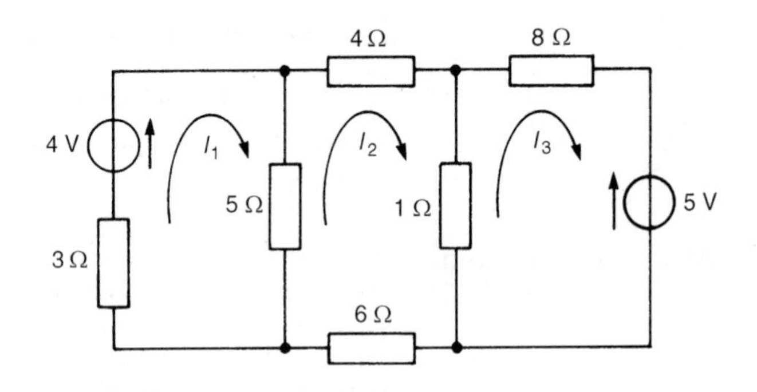

Figure 31.2

The mesh currents I_1, I_2 and I_3 are shown in Figure 31.2. Using Kirchhoff's voltage law:

For loop 1, $(3+5)I_1 - 5I_2 = 4$ (1)

For loop 2, $(4+1+6+5)I_2 - (5)I_1 - (1)I_3 = 0$ (2)

For loop 3, $(1+8)I_3 - (1)I_2 = -5$ (3)

Thus

$$8I_1 - 5I_2 \qquad - 4 = 0 \qquad (1')$$

$$-5I_1 + 16I_2 - I_3 \qquad = 0 \qquad (2')$$

$$-I_2 \quad + 9I_3 + 5 = 0 \qquad (3')$$

Using determinants,

$$\frac{I_1}{\begin{vmatrix} -5 & 0 & -4 \\ 16 & -1 & 0 \\ -1 & 9 & 5 \end{vmatrix}} = \frac{-I_2}{\begin{vmatrix} 8 & 0 & -4 \\ -5 & -1 & 0 \\ 0 & 9 & 5 \end{vmatrix}} = \frac{I_3}{\begin{vmatrix} 8 & -5 & -4 \\ -5 & 16 & 0 \\ 0 & -1 & 5 \end{vmatrix}}$$

$$= \frac{-1}{\begin{vmatrix} 8 & -5 & 0 \\ -5 & 16 & -1 \\ 0 & -1 & 9 \end{vmatrix}}$$

$$\frac{I_1}{-5\begin{vmatrix}-1 & 0\\ 9 & 5\end{vmatrix} - 4\begin{vmatrix}16 & -1\\ -1 & 9\end{vmatrix}} = \frac{-I_2}{8\begin{vmatrix}-1 & 0\\ 9 & 5\end{vmatrix} - 4\begin{vmatrix}-5 & -1\\ 0 & 9\end{vmatrix}}$$

$$= \frac{I_3}{-4\begin{vmatrix}-5 & 16\\ 0 & -1\end{vmatrix} + 5\begin{vmatrix}8 & -5\\ -5 & 16\end{vmatrix}}$$

$$= \frac{-1}{8\begin{vmatrix}16 & -1\\ -1 & 9\end{vmatrix} + 5\begin{vmatrix}-5 & -1\\ 0 & 9\end{vmatrix}}$$

$$\frac{I_1}{-5(-5)-4(143)} = \frac{-I_2}{8(-5)-4(-45)} = \frac{I_3}{-4(5)+5(103)}$$

$$= \frac{-1}{8(143)+5(-45)}$$

$$\frac{I_1}{-547} = \frac{-I_2}{140} = \frac{I_3}{495} = \frac{-1}{919}$$

Hence $I_1 = \dfrac{547}{919} = 0.595$ A, $\quad I_2 = \dfrac{140}{919} = 0.152$ A, and

$$I_3 = \frac{-495}{919} = -0.539 \text{ A}$$

Thus **current in the 5 Ω resistance** $= I_1 - I_2 = 0.595 - 0.152$

$=$ **0.44 A,**

and **current in the 1 Ω resistance** $= I_2 - I_3 = 0.152 - (-0.539)$

$=$ **0.69 A**

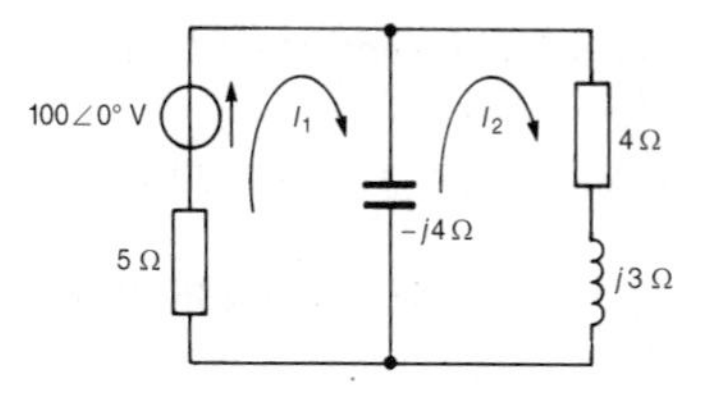

Figure 31.3

Problem 2. For the a.c. network shown in Figure 31.3 determine, using mesh-current analysis, (a) the mesh currents I_1 and I_2 (b) the current flowing in the capacitor, and (c) the active power delivered by the $100\angle 0°$ V voltage source.

(a) For the first loop $(5 - j4)I_1 - (-j4I_2) = 100\angle 0°$ (1)

For the second loop $(4 + j3 - j4)I_2 - (-j4I_1) = 0$ (2)

Rewriting equations (1) and (2) gives:

$$(5 - j4)I_1 + j4I_2 - 100 = 0 \quad (1')$$

$$j4I_1 + (4 - j)I_2 + 0 = 0 \quad (2')$$

Thus, using determinants,

$$\frac{I_1}{\begin{vmatrix} j4 & -100 \\ (4-j) & 0 \end{vmatrix}} = \frac{-I_2}{\begin{vmatrix} (5-j4) & -100 \\ j4 & 0 \end{vmatrix}} = \frac{1}{\begin{vmatrix} (5-j4) & j4 \\ j4 & (4-j) \end{vmatrix}}$$

$$\frac{I_1}{(400-j100)} = \frac{-I_2}{j400} = \frac{1}{(32-j21)}$$

Hence $\quad \boldsymbol{I_1} = \dfrac{(400-j100)}{(32-j21)} = \dfrac{412.31\angle-14.04°}{38.28\angle-33.27°}$

$= 10.77\angle19.23°$ A $=$ **10.8∠−19.2° A**, correct to one decimal place

$$\boldsymbol{I_2} = \frac{400\angle-90°}{38.28\angle-33.27°} = 10.45\angle-56.73° \text{ A}$$

$=$ **10.5∠−56.7° A**, correct to one decimal place

(b) Current flowing in capacitor $= I_1 - I_2$

$= 10.77\angle19.23° - 10.45\angle-56.73°$

$= 4.44 + j12.28 = 13.1\angle70.12°$ A,

i.e., **the current in the capacitor is 13.1 A**

(c) Source power $P = \text{VI}\cos\phi = (100)(10.77)\cos 19.23°$

$=$ **1016.9 W** $= 1020$ W, correct to three significant figures.

(Check: power in 5 Ω resistor $= I_1^2(5) = (10.77)^2(5) = 579.97$ W

and power in 4 Ω resistor $= I_2^2(4) = (10.45)^2(4) = 436.81$ W

Thus total power dissipated $= 579.97 + 436.81$

$= 1016.8$ W $= 1020$ W, correct to three significant figures.)

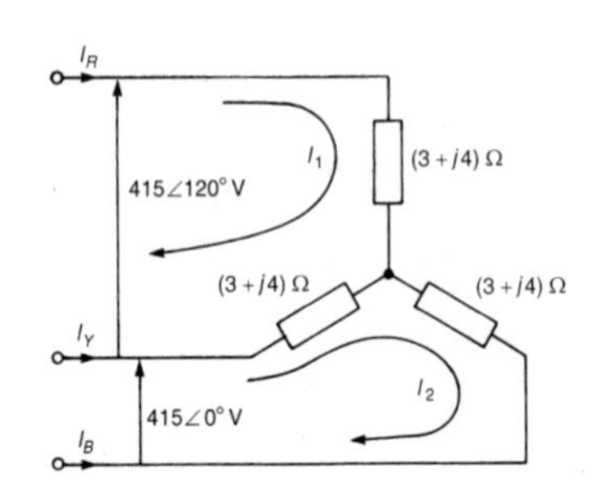

Figure 31.4

Problem 3. A balanced star-connected 3-phase load is shown in Figure 31.4. Determine the value of the line currents I_R, I_Y and I_B using mesh-current analysis.

Two mesh currents I_1 and I_2 are chosen as shown in Figure 31.4.

From loop 1, $I_1(3+j4)+I_1(3+j4)-I_2(3+j4)=415\angle 120^\circ$

$$\text{i.e., } (6+j8)I_1-(3+j4)I_2-415\angle 120^\circ=0 \qquad (1)$$

From loop 2, $I_2(3+j4)-I_1(3+j4)+I_2(3+j4)=415\angle 0^\circ$

$$\text{i.e., } -(3+j4)I_1+(6+j8)I_2-415\angle 0^\circ=0 \qquad (2)$$

Solving equations (1) and (2) using determinants gives:

$$\frac{I_1}{\begin{vmatrix}-(3+j4) & -415\angle 120^\circ\\ (6+j8) & -415\angle 0^\circ\end{vmatrix}}=\frac{-I_2}{\begin{vmatrix}(6+j8) & -415\angle 120^\circ\\ -(3+j4) & -415\angle 0^\circ\end{vmatrix}}$$

$$=\frac{1}{\begin{vmatrix}(6+j8) & -(3+j4)\\ -(3+j4) & (6+j8)\end{vmatrix}}$$

$$\frac{I_1}{2075\angle 53.13^\circ+4150\angle 173.13^\circ}=\frac{-I_2}{-4150\angle 53.13^\circ-2075\angle 173.13^\circ}$$

$$=\frac{1}{100\angle 106.26^\circ-25\angle 106.26^\circ}$$

$$\frac{I_1}{3594\angle 143.13^\circ}=\frac{I_2}{3594\angle 83.13^\circ}=\frac{1}{75\angle 106.26^\circ}$$

Hence $I_1=\dfrac{3594\angle 143.13^\circ}{75\angle 106.26^\circ}=47.9\angle 36.87^\circ$ A

and $I_2=\dfrac{3594\angle 83.13^\circ}{75\angle 106.26^\circ}=47.9\angle -23.13^\circ$ A

Thus line current $\boldsymbol{I_R}=I_1=\mathbf{47.9\angle 36.87^\circ}$ **A**

$$\boldsymbol{I_B}=-I_2=-(47.9\angle -23.23^\circ\text{ A})$$

$$=\mathbf{47.9\angle 156.87^\circ\ A}$$

and $$\boldsymbol{I_Y}=I_2-I_1=47.9\angle -23.13^\circ-47.9\angle 36.87^\circ$$

$$=\mathbf{47.9\angle -83.13^\circ\ A}$$

Further problems on mesh-current analysis may be found in Section 31.3, problems 1 to 9, page 559.

31.2 Nodal analysis

A **node** of a network is defined as a point where two or more branches are joined. If three or more branches join at a node, then that node is called a **principal node** or **junction**. In Figure 31.5, points 1, 2, 3, 4 and 5 are nodes, and points 1, 2 and 3 are principal nodes.

A node voltage is the voltage of a particular node with respect to a node called the reference node. If in Figure 31.5, for example, node 3 is chosen as the reference node then V_{13} is assumed to mean the voltage at node 1 with respect to node 3 (as distinct from V_{31}). Similarly, V_{23} would be assumed to mean the voltage at node 2 with respect to node 3, and so on. However, since the node voltage is always determined with respect to a particular chosen reference node, the notation V_1 for V_{13} and V_2 for V_{23} would always be used in this instance.

The object of nodal analysis is to determine the values of voltages at all the principal nodes with respect to the reference node, e.g., to find voltages V_1 and V_2 in Figure 31.5. When such voltages are determined, the currents flowing in each branch can be found.

Kirchhoff's current law is applied to nodes 1 and 2 in turn in Figure 31.5 and two equations in unknowns V_1 and V_2 are obtained which may be simultaneously solved using determinants.

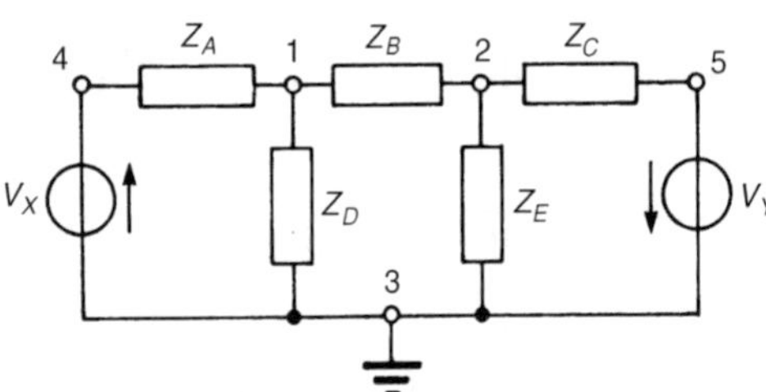

Figure 31.5

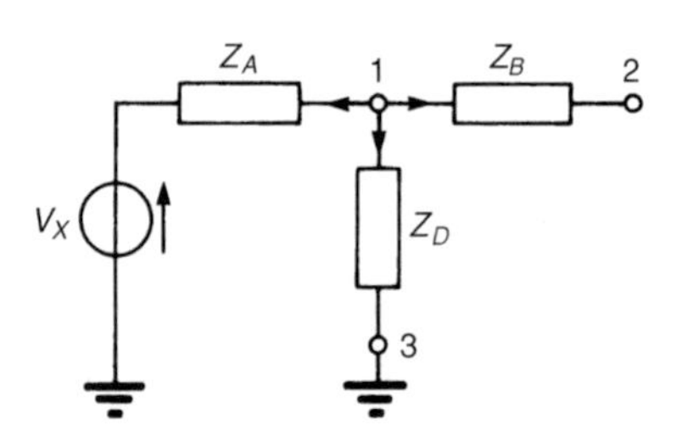

Figure 31.6

The branches leading to node 1 are shown separately in Figure 31.6. Let us assume that all branch currents are leaving the node as shown. Since the sum of currents at a junction is zero,

$$\frac{V_1 - V_x}{Z_A} + \frac{V_1}{Z_D} + \frac{V_1 - V_2}{Z_B} = 0 \qquad (1)$$

Similarly, for node 2, assuming all branch currents are leaving the node as shown in Figure 31.7,

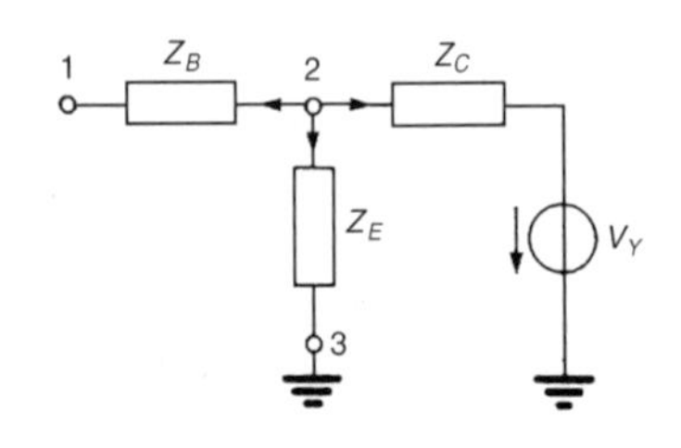

Figure 31.7

$$\frac{V_2 - V_1}{Z_B} + \frac{V_2}{Z_E} + \frac{V_2 + V_Y}{Z_C} = 0 \qquad (2)$$

In equations (1) and (2), the currents are all assumed to be leaving the node. In fact, any selection in the direction of the branch currents may be made — the resulting equations will be identical. (For example, if for node 1 the current flowing in Z_B is considered as flowing towards node 1 instead of away, then the equation for node 1 becomes

$$\frac{V_1 - V_x}{Z_A} + \frac{V_1}{Z_D} = \frac{V_2 - V_1}{Z_B}$$

which if rearranged is seen to be exactly the same as equation (1).)

Rearranging equations (1) and (2) gives:

$$\left(\frac{1}{Z_A} + \frac{1}{Z_B} + \frac{1}{Z_D}\right) V_1 - \left(\frac{1}{Z_B}\right) V_2 - \left(\frac{1}{Z_A}\right) V_x = 0 \quad (3)$$

$$-\left(\frac{1}{Z_B}\right) V_1 + \left(\frac{1}{Z_B} + \frac{1}{Z_C} + \frac{1}{Z_E}\right) V_2 + \left(\frac{1}{Z_C}\right) V_Y = 0 \quad (4)$$

Equations (3) and (4) may be rewritten in terms of admittances (where admittance $Y = l/Z$):

$$(Y_A + Y_B + Y_D)V_1 - Y_B V_2 - Y_A V_x = 0 \quad (5)$$

$$-Y_B V_1 + (Y_B + Y_C + Y_E)V_2 + Y_C V_Y = 0 \quad (6)$$

Equations (5) and (6) may be solved for V_1 and V_2 by using determinants. Thus

$$\frac{V_1}{\begin{vmatrix} -Y_B & -Y_A \\ (Y_B + Y_C + Y_E) & Y_C \end{vmatrix}} = \frac{-V_2}{\begin{vmatrix} (Y_A + Y_B + Y_D) & -Y_A \\ -Y_B & Y_C \end{vmatrix}}$$

$$= \frac{1}{\begin{vmatrix} (Y_A + Y_B + Y_D) & -Y_B \\ -Y_B & (Y_B + Y_C + Y_E) \end{vmatrix}}$$

Current equations, and hence voltage equations, may be written at each principal node of a network with the exception of a reference node. The number of equations necessary to produce a solution for a circuit is, in fact, always one less than the number of principal nodes.

Whether mesh-current analysis or nodal analysis is used to determine currents in circuits depends on the number of loops and nodes the circuit contains, Basically, the method that requires the least number of equations is used. The method of nodal analysis is demonstrated in the following problems.

Problem 4. For the network shown in Figure 31.8, determine the voltage V_{AB}, by using nodal analysis.

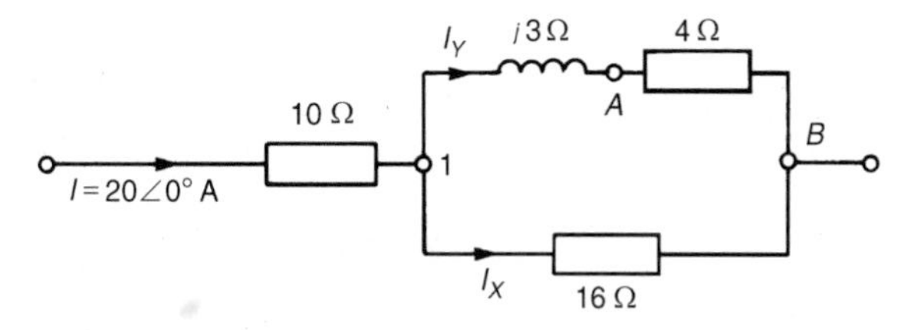

Figure 31.8

Figure 31.8 contains two principal nodes (at 1 and B) and thus only one nodal equation is required. B is taken as the reference node and the equation for node 1 is obtained as follows. Applying Kirchhoff's current law to node 1 gives:

$$I_X + I_Y = I$$

i.e., $$\frac{V_1}{16} + \frac{V_1}{(4+j3)} = 20\angle 0^\circ$$

Thus $$V_1\left(\frac{1}{16} + \frac{1}{4+j3}\right) = 20$$

$$V_1\left(0.0625 + \frac{4-j3}{4^2+3^2}\right) = 20$$

$$V_1(0.0625 + 0.16 - j0.12) = 20$$

$$V_1(0.2225 - j0.12) = 20$$

from which, $$V_1 = \frac{20}{(0.2225 - j0.12)} = \frac{20}{0.2528\angle -28.34^\circ}$$

i.e., voltage $V_1 = 79.1\angle 28.34^\circ$ V

The current through the $(4+j3)\Omega$ branch, $I_y = V_1/(4+j3)$

Hence the voltage drop between points A and B, i.e., across the 4 Ω resistance, is given by:

$$V_{AB} = (I_y)(4) = \frac{V_1(4)}{(4+j3)} = \frac{79.1\angle 28.34^\circ}{5\angle 36.87^\circ}(4) = \mathbf{63.3\angle -8.53^\circ\ V}$$

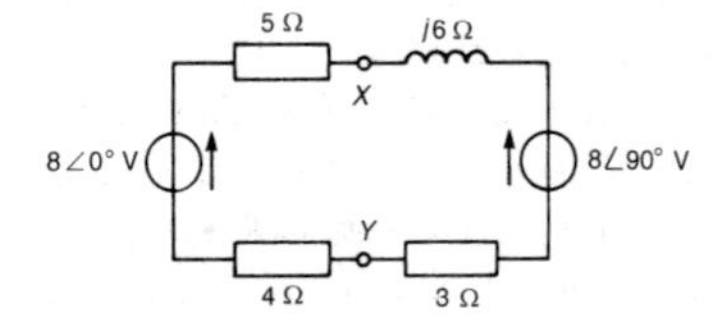

Figure 31.9

Problem 5. Determine the value of voltage V_{XY} shown in the circuit of Figure 31.9.

The circuit contains no principal nodes. However, if point Y is chosen as the reference node then an equation may be written for node X assuming that current leaves point X by both branches.

Thus $$\frac{V_X - 8\angle 0^\circ}{(5+4)} + \frac{V_x - 8\angle 90^\circ}{(3+j6)} = 0$$

from which, $$V_X\left(\frac{1}{9} + \frac{1}{3+j6}\right) = \frac{8}{9} + \frac{j8}{3+j6}$$

$$V_X\left(\frac{1}{9} + \frac{3-j6}{3^2+6^2}\right) = \frac{8}{9} + \frac{j8(3-j6)}{3^2+6^2}$$

$$V_X(0.1778 - j0.1333) = 0.8889 + \frac{48 + j24}{45}$$

$$V_X(0.2222\angle -36.86^\circ) = 1.9556 + j0.5333$$

$$= 2.027\angle 15.25^\circ$$

Since point Y is the reference node,

$$\text{voltage } V_X = V_{XY} = \frac{2.027\angle 15.25^\circ}{0.2222\angle -36.86^\circ} = \mathbf{9.12\angle 52.11^\circ\ V}$$

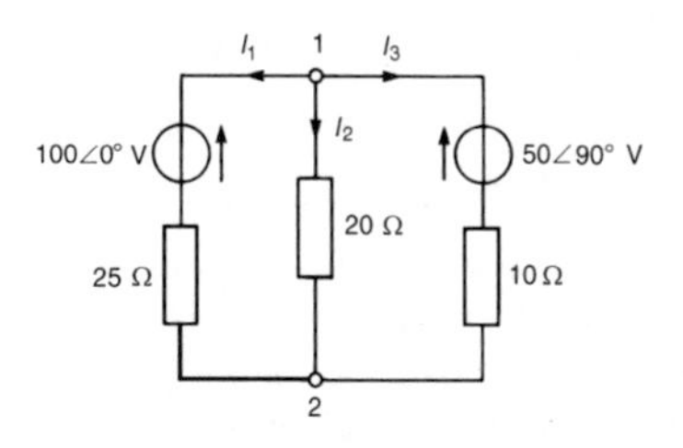

Figure 31.10

Problem 6. Use nodal analysis to determine the current flowing in each branch of the network shown in Figure 31.10.

This is the same problem as problem 1 of Chapter 30, page 536, which was solved using Kirchhoff's laws. A comparison of methods can be made.

There are only two principal nodes in Figure 31.10 so only one nodal equation is required. Node 2 is taken as the reference node.

The equation at node 1 is $I_1 + I_2 + I_3 = 0$

i.e., $$\frac{V_1 - 100\angle 0^\circ}{25} + \frac{V_1}{20} + \frac{V_1 - 50\angle 90^\circ}{10} = 0$$

i.e., $$\left(\frac{1}{25} + \frac{1}{20} + \frac{1}{10}\right) V_1 - \frac{100\angle 0^\circ}{25} - \frac{50\angle 90^\circ}{10} = 0$$

$$0.19\ V_1 = 4 + j5$$

Thus the voltage at node 1, $V_1 = \dfrac{4 + j5}{0.19} = 33.70\angle 51.34^\circ$ V

or $(21.05 + j26.32)$V

Hence the current in the 25 Ω resistance,

$$I_1 = \frac{V_1 - 100\angle 0^\circ}{25} = \frac{21.05 + j26.32 - 100}{25}$$

$$= \frac{-78.95 + j26.32}{25}$$

$= \mathbf{3.33\angle 161.56^\circ\ A}$ flowing away from node 1

(or $3.33\angle(161.56^\circ - 180^\circ)$A $= \mathbf{3.33\angle -18.44^\circ\ A}$ flowing toward node 1)

The current in the 20 Ω resistance,

$$I_2 = \frac{V_1}{20} = \frac{33.70\angle 51.34^\circ}{20} = \mathbf{1.69\angle 51.34^\circ\ A}$$

flowing from node 1 to node 2

The current in the 10 Ω resistor,

$$I_3 = \frac{V_1 - 50\angle 90^\circ}{10} = \frac{21.05 + j26.32 - j50}{10} = \frac{21.05 - j23.68}{10}$$

$$= \mathbf{3.17\angle{-48.36^\circ}\ A} \text{ away from node 1}$$

(or $3.17\angle(-48.36^\circ - 180^\circ) = 3.17\angle{-228.36^\circ}$ A $=$ $\mathbf{3.17\angle 131.64^\circ\ A}$ toward node 1)

Problem 7. In the network of Figure 31.11 use nodal analysis to determine (a) the voltage at nodes 1 and 2, (b) the current in the $j4$ Ω inductance, (c) the current in the 5 Ω resistance, and (d) the magnitude of the active power dissipated in the 2.5 Ω resistance.

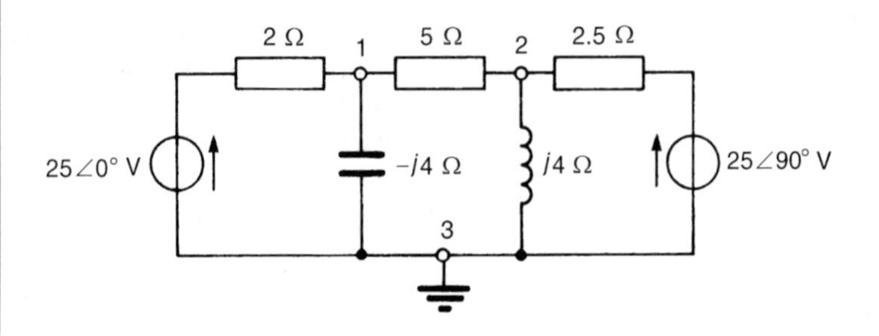

Figure 31.11

(a) At node 1, $\frac{V_1 - 25\angle 0^\circ}{2} + \frac{V_1}{-j4} + \frac{V_1 - V_2}{5} = 0$

Rearranging gives:

$$\left(\frac{1}{2} + \frac{1}{-j4} + \frac{1}{5}\right)V_1 - \left(\frac{1}{5}\right)V_2 - \frac{25\angle 0^\circ}{2} = 0$$

i.e., $$(0.7 + j0.25)V_1 - 0.2V_2 - 12.5 = 0 \quad (1)$$

At node 2, $\frac{V_2 - 25\angle 90^\circ}{2.5} + \frac{V_2}{j4} + \frac{V_2 - V_1}{5} = 0$

Rearranging gives:

$$-\left(\frac{1}{5}\right)V_1 + \left(\frac{1}{2.5} + \frac{1}{j4} + \frac{1}{5}\right)V_2 - \frac{25\angle 90^\circ}{2.5} = 0$$

i.e., $$-0.2V_1 + (0.6 - j0.25)V_2 - j10 = 0 \quad (2)$$

Thus two simultaneous equations have been formed with two unknowns, V_1 and V_2. Using determinants, if

$$(0.7 + j0.25)V_1 - 0.2V_2 - 12.5 = 0 \quad (1)$$

and $\quad -0.2V_1 + (0.6 - j0.25)V_2 - j10 = 0 \qquad (2)$

then
$$\frac{V_1}{\begin{vmatrix} -0.2 & -12.5 \\ (0.6 - j0.25) & -j10 \end{vmatrix}} = \frac{-V_2}{\begin{vmatrix} (0.7 + j0.25) & -12.5 \\ -0.2 & -j10 \end{vmatrix}}$$
$$= \frac{1}{\begin{vmatrix} (0.7 + j0.25) & -0.2 \\ -0.2 & (0.6 - j0.25) \end{vmatrix}}$$

i.e.,
$$\frac{V_1}{(j2 + 7.5 - j3.125)} = \frac{-V_2}{(-j7 + 2.5 - 2.5)}$$
$$= \frac{1}{(0.42 - j0.175 + j0.15 + 0.0625 - 0.04)}$$

and
$$\frac{V_1}{7.584\angle -8.53^\circ} = \frac{-V_2}{-7\angle 90^\circ} = \frac{1}{0.443\angle -3.23^\circ}$$

Thus **voltage,** $\boldsymbol{V_1} = \dfrac{7.584\angle -8.53^\circ}{0.443\angle -3.23^\circ} = 17.12\angle -5.30^\circ$ V

$= \mathbf{17.1\angle -5.3^\circ}$ **V**, correct to one decimal place,

and **voltage,** $\boldsymbol{V_2} = \dfrac{7\angle 90^\circ}{0.443\angle -3.23^\circ} = 15.80\angle 93.23^\circ$ V

$= \mathbf{15.8\angle 93.2^\circ}$ **V**, correct to one decimal place.

(b) The current in the $j4\ \Omega$ inductance is given by:

$$\frac{V_2}{j4} = \frac{15.80\angle 93.23^\circ}{4\angle 90^\circ} = \mathbf{3.95\angle 3.23^\circ\ A} \text{ flowing away from node 2}$$

(c) The current in the $5\ \Omega$ resistance is given by:

$$I_5 = \frac{V_1 - V_2}{5} = \frac{17.12\angle -5.30^\circ - 15.80\angle 93.23^\circ}{5}$$

i.e., $\quad I_5 = \dfrac{(17.05 - j1.58) - (-0.89 + j15.77)}{5}$

$$= \frac{17.94 - j17.35}{5} = \frac{24.96\angle -44.04^\circ}{5}$$

$= \mathbf{4.99\angle -44.04^\circ}$ **A** flowing from node 1 to node 2

(d) The active power dissipated in the $2.5\ \Omega$ resistor is given by

$$P_{2.5} = (I_{2.5})^2(2.5) = \left(\frac{V_2 - 25\angle 90^\circ}{2.5}\right)^2 (2.5)$$

$$= \frac{(0.89 + j15.77 - j25)^2}{2.5} = \frac{(9.273\angle -95.51^\circ)^2}{2.5}$$

$$= \frac{85.99\angle -191.02^\circ}{2.5} \text{ by de Moivre's theorem}$$
$$= 34.4\angle 169^\circ \text{ W}$$

Thus the magnitude of the active power dissipated in the 2.5 Ω resistance is 34.4 W

Problem 8. In the network shown in Figure 31.12 determine the voltage V_{XY} using nodal analysis.

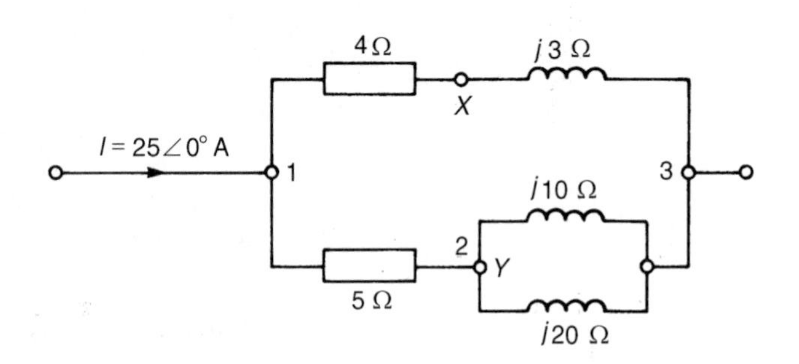

Figure 31.12

Node 3 is taken as the reference node.

At node 1, $$25\angle 0^\circ = \frac{V_1}{4 + j3} + \frac{V_1 - V_2}{5}$$

i.e., $$\left(\frac{4 - j3}{25} + \frac{1}{5}\right) V_1 - \frac{1}{5}V_2 - 25 = 0$$

or $$(0.379\angle -18.43^\circ)V_1 - 0.2V_2 - 25 = 0 \qquad (1)$$

At node 2, $$\frac{V_2}{j10} + \frac{V_2}{j20} + \frac{V_2 - V_1}{5} = 0$$

i.e., $$-0.2V_1 + \left(\frac{1}{j10} + \frac{1}{j20} + \frac{1}{5}\right) V_2 = 0$$

or $$-0.2V_1 + (-j0.1 - j0.05 + 0.2)V_2 = 0$$

i.e., $$-0.2V_1 + (0.25\angle -36.87^\circ)V_2 + 0 = 0 \qquad (2)$$

Simultaneous equations (1) and (2) may be solved for V_1 and V_2 by using determinants. Thus,

$$\frac{V_1}{\begin{vmatrix} -0.2 & -25 \\ 0.25\angle-36.87^\circ & 0 \end{vmatrix}} = \frac{-V_2}{\begin{vmatrix} 0.379\angle-18.43^\circ & -25 \\ -0.2 & 0 \end{vmatrix}}$$

$$= \frac{1}{\begin{vmatrix} 0.379\angle-18.43^\circ & -0.2 \\ -0.2 & 0.25\angle-36.87^\circ \end{vmatrix}}$$

i.e., $$\frac{V_1}{6.25\angle-36.87^\circ} = \frac{-V_2}{-5} = \frac{1}{0.09475\angle-55.30^\circ - 0.04}$$

$$= \frac{1}{0.079\angle-79.85^\circ}$$

Hence voltage, $$\mathbf{V}_1 = \frac{6.25\angle-36.87^\circ}{0.079\angle-79.85^\circ} = \mathbf{79.11\angle42.98^\circ\ V}$$

and voltage, $$\mathbf{V}_2 = \frac{5}{0.079\angle-79.85^\circ} = \mathbf{63.29\angle79.85^\circ\ V}$$

The current flowing in the $(4 + j3)\Omega$ branch is $V_1/(4 + j3)$. Hence the voltage between point X and node 3 is:

$$\frac{V_1}{(4 + j3)}(j3) = \frac{(79.11\angle42.98^\circ)(3\angle90^\circ)}{5\angle36.87^\circ}$$

$$= 47.47\angle96.11^\circ\ \text{V}$$

Thus the voltage

$$V_{XY} = V_X - V_Y = V_X - V_2 = 47.47\angle96.11^\circ - 63.29\angle79.85^\circ$$

$$= -16.21 - j15.10 = \mathbf{22.15\angle-137^\circ\ V}$$

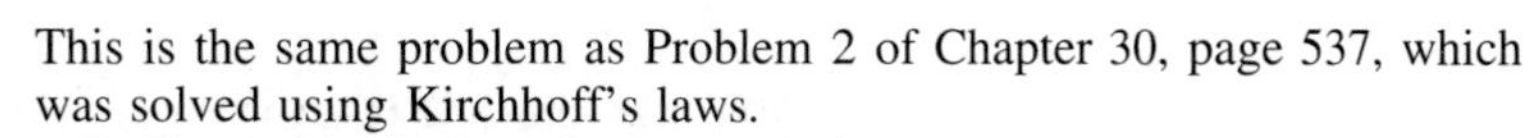

Problem 9. Use nodal analysis to determine the voltages at nodes 2 and 3 in Figure 31.13 and hence determine the current flowing in the 2 Ω resistor and the power dissipated in the 3 Ω resistor.

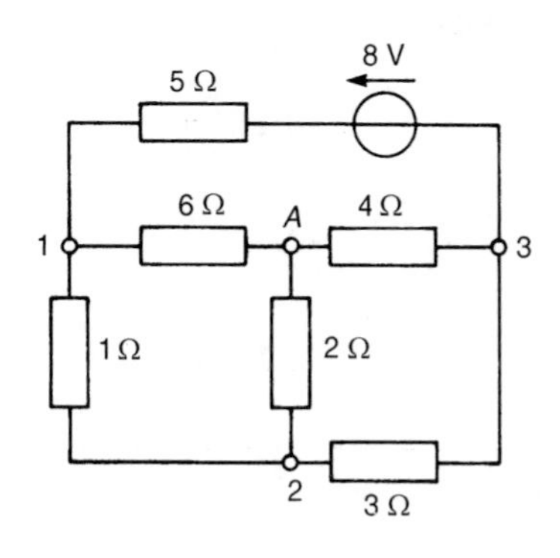

Figure 31.13

This is the same problem as Problem 2 of Chapter 30, page 537, which was solved using Kirchhoff's laws.

In Figure 31.13, the reference node is shown at point A.

At node 1, $$\frac{V_1 - V_2}{1} + \frac{V_1}{6} + \frac{V_1 - 8 - V_3}{5} = 0$$

i.e., $$1.367V_1 - V_2 - 0.2V_3 - 1.6 = 0 \qquad (1)$$

At node 2, $$\frac{V_2}{2} + \frac{V_2 - V_1}{1} + \frac{V_2 - V_3}{3} = 0$$

i.e., $$-V_1 + 1.833V_2 - 0.333V_3 + 0 = 0 \qquad (2)$$

At node 3, $$\frac{V_3}{4} + \frac{V_3 - V_2}{3} + \frac{V_3 + 8 - V_1}{5} = 0$$

i.e., $$-0.2V_1 - 0.333V_2 + 0.783V_3 + 1.6 = 0 \qquad (3)$$

Equations (1) to (3) can be solved for V_1, V_2 and V_3 by using determinants. Hence

$$\frac{V_1}{\begin{vmatrix} -1 & -0.2 & -1.6 \\ 1.833 & -0.333 & 0 \\ -0.333 & 0.783 & 1.6 \end{vmatrix}} = \frac{-V_2}{\begin{vmatrix} 1.367 & -0.2 & -1.6 \\ -1 & -0.333 & 0 \\ -0.2 & 0.783 & 1.6 \end{vmatrix}}$$

$$= \frac{V_3}{\begin{vmatrix} 1.367 & -1 & -1.6 \\ -1 & 1.833 & 0 \\ -0.2 & -0.333 & 1.6 \end{vmatrix}} = \frac{-1}{\begin{vmatrix} 1.367 & -1 & -0.2 \\ -1 & 1.833 & -0.333 \\ -0.2 & -0.333 & 0.783 \end{vmatrix}}$$

Solving for V_2 gives: $$\frac{-V_2}{-1.6(-0.8496) + 1.6(-0.6552)}$$

$$= \frac{-1}{1.367(1.3244) + 1(-0.8496) - 0.2(0.6996)}$$

hence $\frac{-V_2}{0.31104} = \frac{-1}{0.82093}$ from which, **voltage,** $\mathbf{V_2} = \frac{0.31104}{0.82093}$

$$= \mathbf{0.3789\ V}$$

Thus the current in the 2 Ω resistor $= \frac{V_2}{2} = \frac{0.3789}{2} = \mathbf{0.19\ A,}$ flowing from node 2 to node A.

Solving for V_3 gives: $$\frac{V_3}{-1.6(0.6996) + 1.6(1.5057)} = \frac{-1}{0.82093}$$

hence $\frac{V_3}{1.2898} = \frac{-1}{0.82093}$ from which, **voltage,** $\mathbf{V_3} = \frac{-1.2898}{0.82093}$

$$= \mathbf{-1.571\ V}$$

Power in the 3 Ω resistor $= (I_3)^2(3) = \left(\frac{V_2 - V_3}{3}\right)^2(3)$

$$= \frac{(0.3789 - (-1.571))^2}{3} = \mathbf{1.27\ W}$$

Further problems on nodal analysis may be found in Section 31.3 following, problems 10 to 15, page 560.

31.3 Further problems on mesh-current and nodal analysis

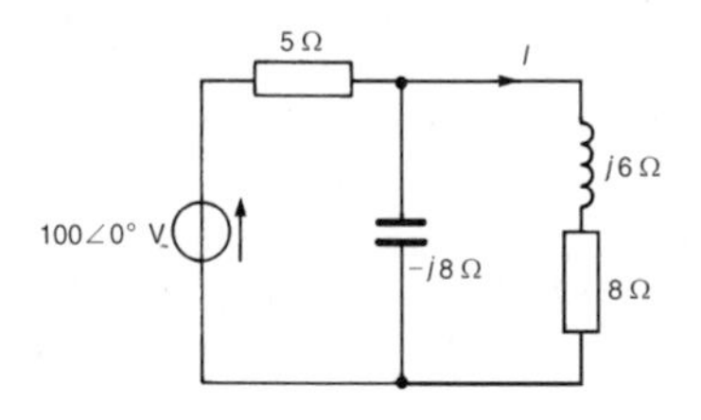

Figure 31.14

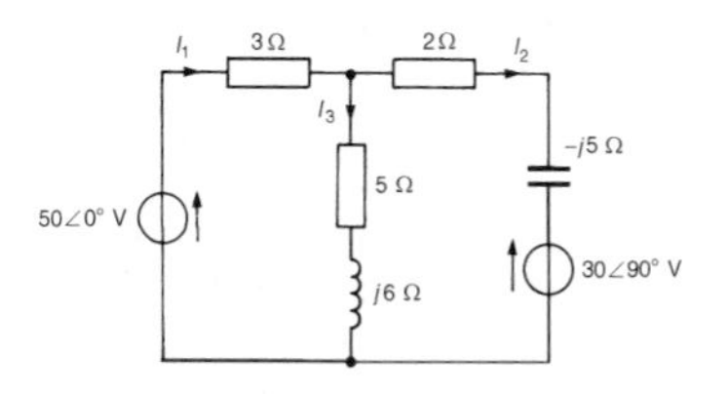

Figure 31.15

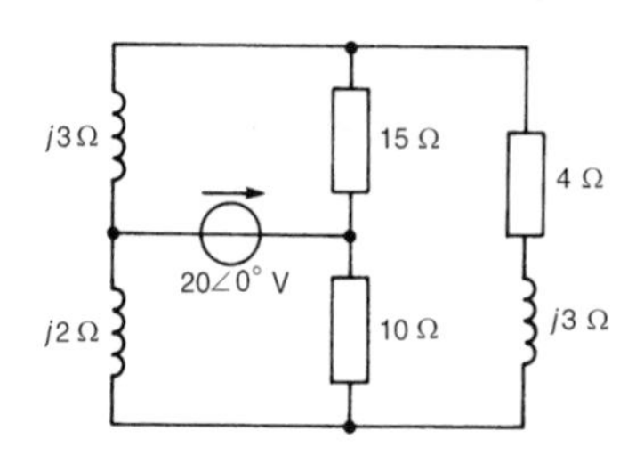

Figure 31.16

Mesh-current analysis

1 Repeat problems 1 to 10, page 542, of Chapter 30 using mesh-current analysis.

2 For the network shown in Figure 31.14, use mesh-current analysis to determine the value of current I and the active power output of the voltage source. [6.96∠−49.94° A; 644 W]

3 Use mesh-current analysis to determine currents I_1, I_2 and I_3 for the network shown in Figure 31.15.
[$I_1 = 8.73\angle{-1.37°}$ A, $I_2 = 7.02\angle 17.25°$ A,
$I_3 = 3.05\angle{-48.67°}$ A]

4 For the network shown in Figure 31.16, determine the current flowing in the $(4 + j3)\Omega$ impedance. [0]

5 For the network shown in Figure 31.17, use mesh-current analysis to determine (a) the current in the capacitor, I_C, (b) the current in the inductance, I_L, (c) the p.d. across the 4 Ω resistance, and (d) the total active circuit power.
[(a) 14.5 A (b) 11.5 A (c) 71.8 V (d) 2499 W]

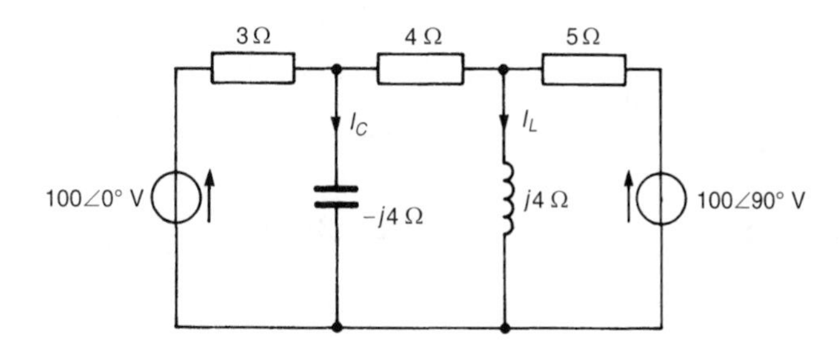

Figure 31.17

6 Determine the value of the currents I_R, I_Y and I_B in the network shown in Figure 31.18 by using mesh-current analysis.
[$I_R = 7.84\angle 71.19°$ A; $I_Y = 9.04\angle{-37.50°}$ A;
$I_B = 9.89\angle{-168.81°}$ A]

7 In the network of Figure 31.19, use mesh-current analysis to determine (a) the current in the capacitor, (b) the current in the 5 Ω resistance, (c) the active power output of the 15∠0° V source, and (d) the magnitude of the p.d. across the $j2$ Ω inductance.
[(a) 1.03 A (b) 1.48 A
(c) 16.28 W (d) 3.47 V]

8 A balanced 3-phase delta-connected load is shown in Figure 31.20. Use mesh-current analysis to determine the values of mesh currents

I_1, I_2 and I_3 shown and hence find the line currents I_R, I_Y and I_B.

[$I_1 = 83\angle 173.13°$ A, $I_2 = 83\angle 53.13°$ A,
$I_3 = 83\angle -66.87°$ A $I_R = 143.8\angle 143.13°$ A,
$I_Y = 143.8\angle 23.13°$ A, $I_B = 143.8\angle -96.87°$ A]

9 Use mesh-circuit analysis to determine the value of currents I_A to I_E in the circuit shown in Figure 31.21.

[$I_A = 2.40\angle 52.52°$ A; $I_B = 1.02\angle 46.19°$ A;
$I_C = 1.39\angle 57.17°$ A; $I_D = 0.67\angle 15.57°$ A;
$I_E = 0.996\angle 83.74°$ A]

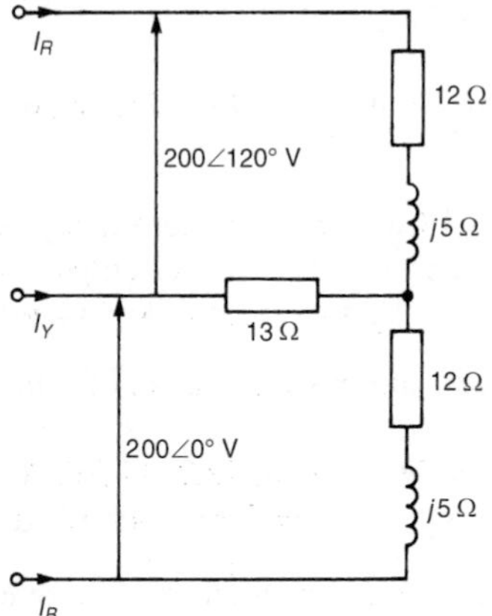
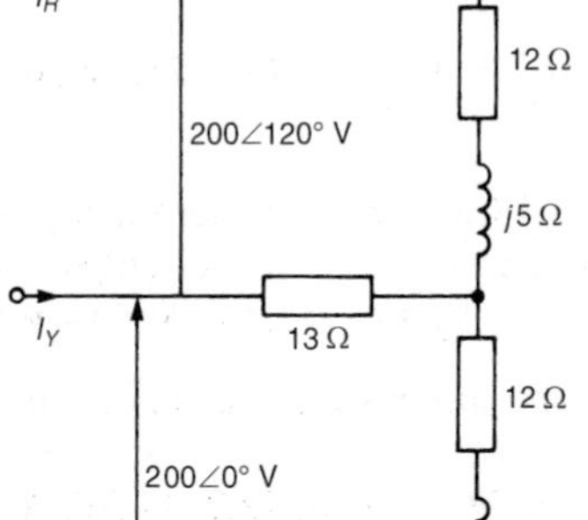

Figure 31.18

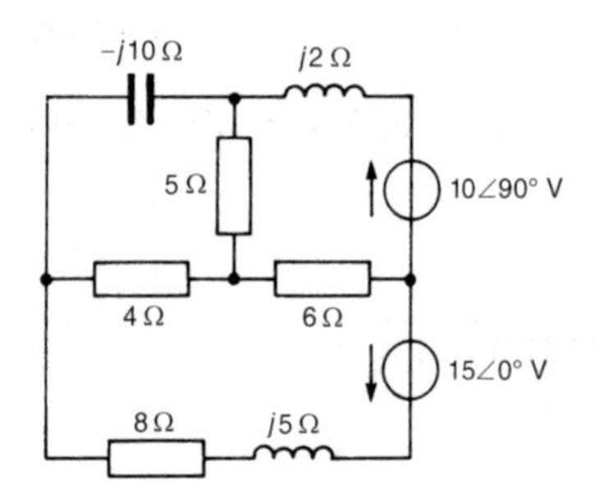

Figure 31.19

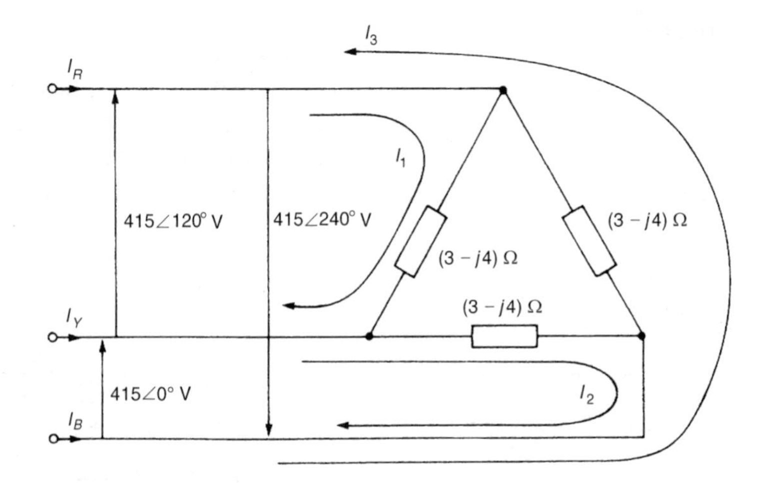

Figure 31.20

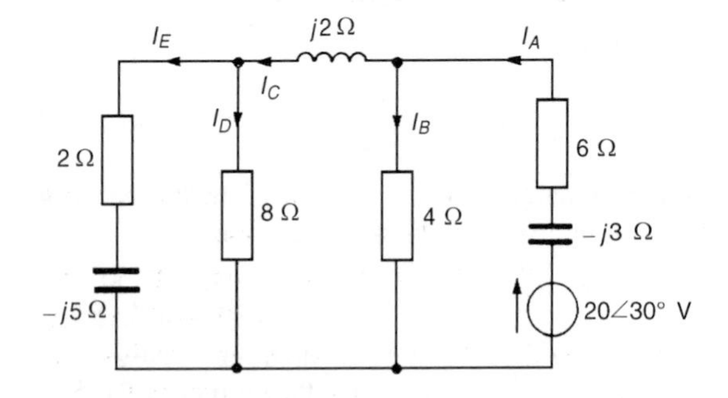

Figure 31.21

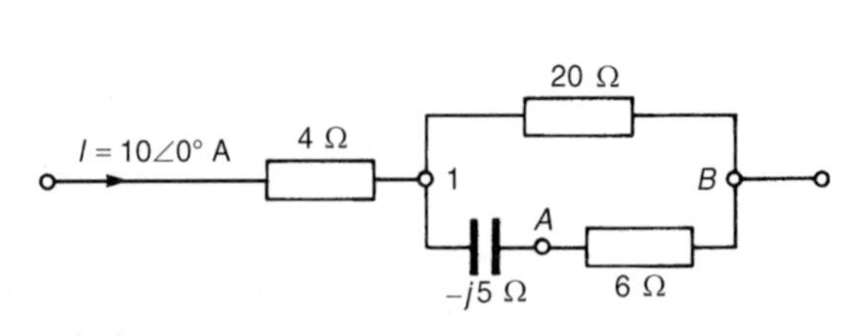

Figure 31.22

Nodal analysis

10 Repeat problems 1, 2, 5, 8 and 10 on page 542 of Chapter 30, and problems 2, 3, 5, and 9 above, using nodal analysis.

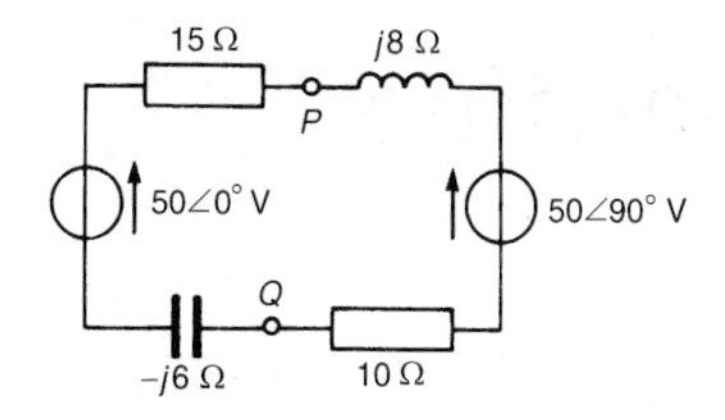

Figure 31.23

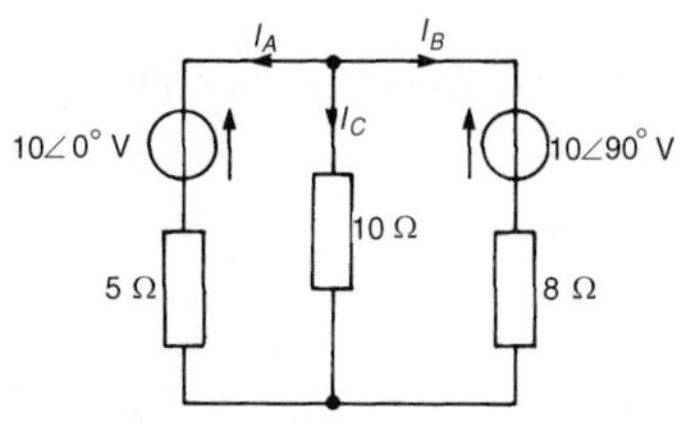

Figure 31.24

11 Determine for the network shown in Figure 31.22 the voltage at node 1 and the voltage V_{AB}
[$V_1 = 59.0\angle -28.92°$ V; $V_{AB} = 45.3\angle 10.89°$ V]

12 Determine the voltage V_{PQ} in the network shown in Figure 31.23.
[$V_{PQ} = 55.87\angle 50.60°$ V]

13 Use nodal analysis to determine the currents I_A, I_B and I_C shown in the network of Figure 31.24.
[$I_A = 1.21\angle 150.96°$ A; $I_B = 1.06\angle -56.32°$ A; $I_C = 0.55\angle 32.01°$ A]

14 For the network shown in Figure 31.25 determine (a) the voltages at nodes 1 and 2, (b) the current in the 40 Ω resistance, (c) the current in the 20 Ω resistance, and (d) the magnitude of the active power dissipated in the 10 Ω resistance
[(a) $V_1 = 88.12\angle 33.86°$ V, $V_2 = 58.72\angle 72.28°$ V
(b) $2.20\angle 33.86°$ A, away from node 1,
(c) $2.80\angle 118.65°$ A, away from node 1, (d) 223 W]

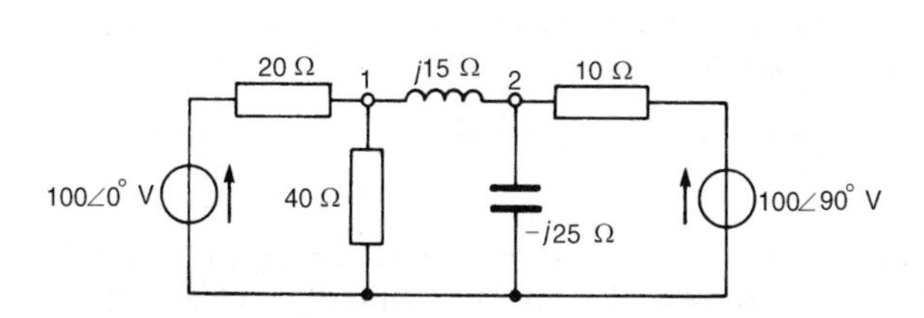

Figure 31.25

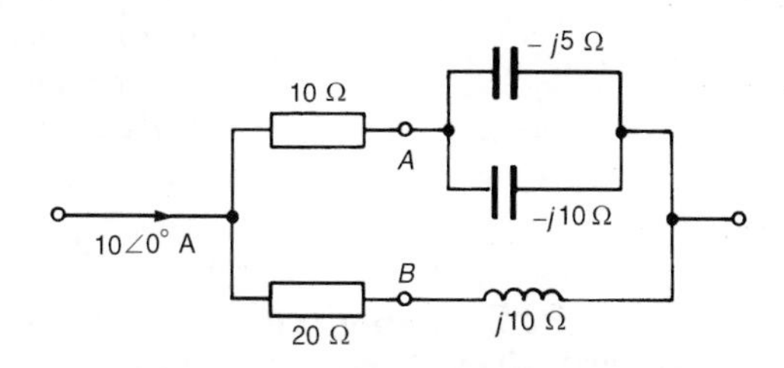

Figure 31.26

15 Determine the voltage V_{AB} in the network of Figure 31.26, using nodal analysis.
[$V_{AB} = 54.23\angle -102.52°$ V]

32 The superposition theorem

At the end of this chapter you should be able to:

- solve d.c. and a.c. networks using the superposition theorem

32.1 Introduction

The superposition theorem states:

'In any network made up of linear impedances and containing more than one source of e.m.f. the resultant current flowing in any branch is the phasor sum of the currents that would flow in that branch if each source were considered separately, all other sources being replaced at that time by their respective internal impedances.'

32.2 Using the superposition theorem

The superposition theorem, which was introduced in Chapter 13 for d.c. circuits, may be applied to both d.c. and a.c. networks. A d.c. network is shown in Figure 32.1 and will serve to demonstrate the principle of application of the superposition theorem.

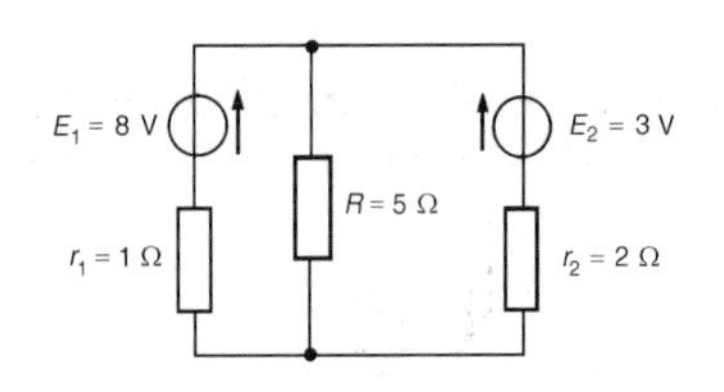

Figure 32.1

To find the current flowing in each branch of the circuit, the following six-step procedure can be adopted:

(i) Redraw the original network with one of the sources, say E_2, removed and replaced by r_2 only, as shown in Figure 32.2.

(ii) Label the current in each branch and its direction as shown in Figure 32.2, and then determine its value. The choice of current direction for I_1 depends on the source polarity which, by convention, is taken as flowing from the positive terminal as shown.

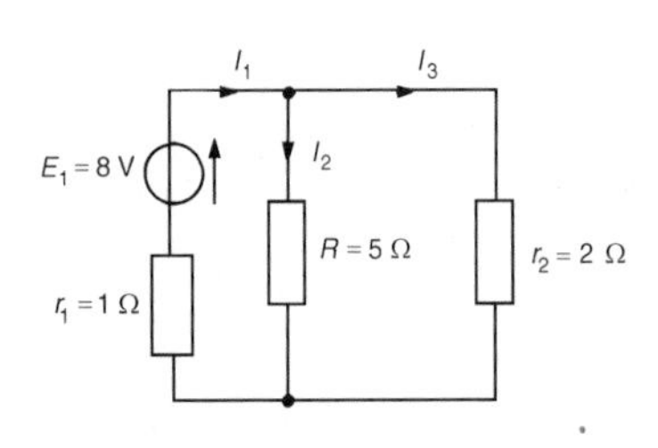

Figure 32.2

R in parallel with r_2 gives an equivalent resistance of

$$(5 \times 2)/(5 + 2) = 10/7 = 1.429\ \Omega$$

as shown in the equivalent network of Figure 32.3. From Figure 28.3,

$$\text{current } I_1 = \frac{E_1}{(r_1 + 1.429)} = \frac{8}{2.429} = 3.294 \text{ A}$$

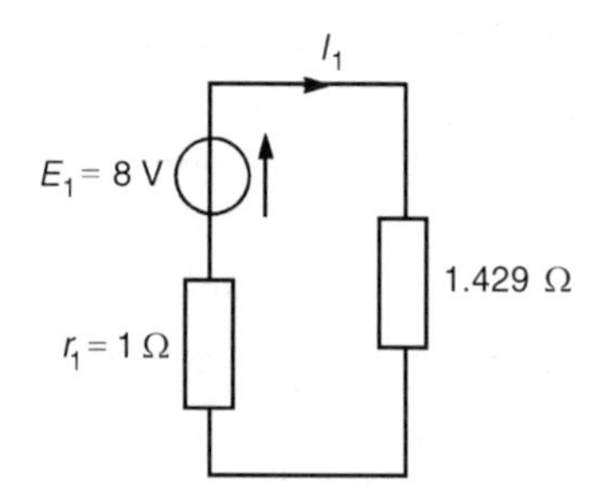

Figure 32.3

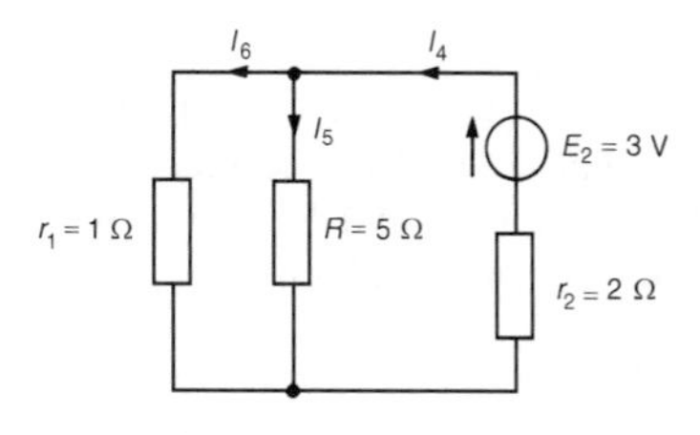

Figure 32.4

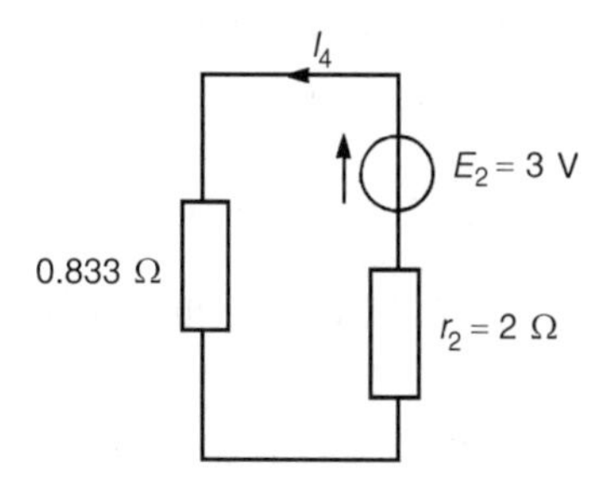

Figure 32.5

From Figure 32.2,

$$\text{current } I_2 = \left(\frac{r_2}{R + r_2}\right)(I_1) = \left(\frac{2}{5+2}\right)(3.294) = 0.941 \text{ A}$$

$$\text{and current } I_3 = \left(\frac{5}{5+2}\right)(3.294) = 2.353 \text{ A}$$

(iii) Redraw the original network with source E_1 removed and replaced by r_1 only, as shown in Figure 32.4.

(iv) Label the currents in each branch and their directions as shown in Figure 32.4, and determine their values.

R and r_1 in parallel gives an equivalent resistance of

$$(5 \times 1)/(5 + 1) = 5/6\ \Omega \text{ or } 0.833\ \Omega,$$

as shown in the equivalent network of Figure 32.5. From Figure 32.5,

$$\text{current } I_4 = \frac{E_2}{r_2 + 0.833} = \frac{3}{2.833} = 1.059 \text{ A}$$

From Figure 32.4,

$$\text{current } I_5 = \left(\frac{1}{1+5}\right)(1.059) = 0.177 \text{ A}$$

$$\text{and current } I_6 = \left(\frac{5}{1+5}\right)(1.059) = 0.8825 \text{ A}$$

(v) Superimpose Figure 32.2 on Figure 32.4, as shown in Figure 32.6.

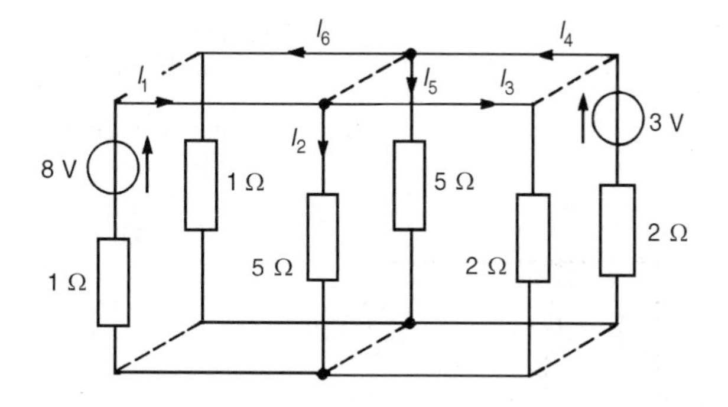

Figure 32.6

(vi) Determine the algebraic sum of the currents flowing in each branch. (Note that in an a.c. circuit it is the phasor sum of the currents that is required.)

From Figure 32.6, the resultant current flowing through the 8 V source is given by

$I_1 - I_6 = 3.294 - 0.8825 =$ **2.41 A** (discharging, i.e., flowing from the positive terminal of the source).

The resultant current flowing in the 3 V source is given by

$I_3 - I_4 = 2.353 - 1.059 =$ **1.29 A** (charging, i.e., flowing into the positive terminal of the source).

The resultant current flowing in the 5 Ω resistance is given by

$I_2 + I_5 = 0.941 + 0.177 =$ **1.12 A**

The values of current are the same as those obtained on page 536 by using Kirchhoff's laws.

The following problems demonstrate further the use of the superposition theorem in analysing a.c. as well as d.c. networks. The theorem is straightforward to apply, but is lengthy. Thévenin's and Norton's theorems (described in Chapter 33) produce results more quickly.

Problem 1. A.c. sources of $100\angle 0°$ V and internal resistance 25 Ω, and $50\angle 90°$ V and internal resistance 10 Ω, are connected in parallel across a 20 Ω load. Determine using the superposition theorem, the current in the 20 Ω load and the current in each voltage source.

(This is the same problem as problem 1 on page 536 and problem 6 on page 553 and a comparison of methods may be made.)

The circuit diagram is shown in Figure 32.7. Following the above procedure:

(i) The network is redrawn with the $50\angle 90°$ V source removed as shown in Figure 32.8

(ii) Currents I_1, I_2 and I_3 are labelled as shown in Figure 32.8.

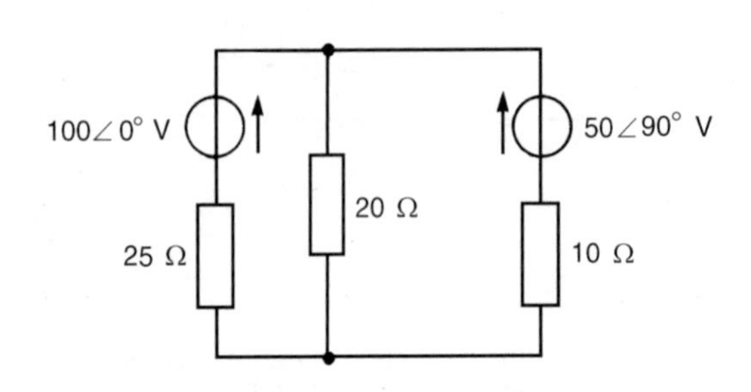

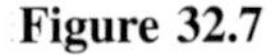

Figure 32.7

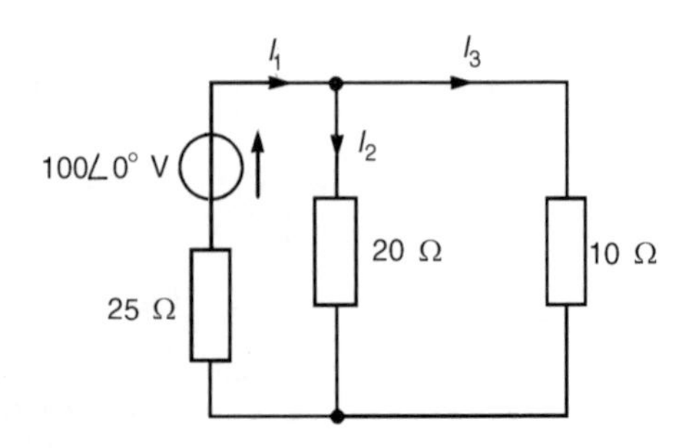

Figure 32.8

$$I_1 = \frac{100\angle 0^\circ}{25 + (10 \times 20)/(10 + 20)} = \frac{100\angle 0^\circ}{25 + 6.667} = 3.158\angle 0^\circ \text{ A}$$

$$I_2 = \left(\frac{10}{10 + 20}\right)(3.158\angle 0^\circ) = 1.053\angle 0^\circ \text{ A}$$

$$I_3 = \left(\frac{20}{10 + 20}\right)(3.158\angle 0^\circ) = 2.105\angle 0^\circ \text{ A}$$

(iii) The network is redrawn with the $100\angle 0^\circ$ V source removed as shown in Figure 32.9

(iv) Currents I_4, I_5 and I_6 are labelled as shown in Figure 32.9.

$$I_4 = \frac{50\angle 90^\circ}{10 + (25 \times 20)/(25 + 20)} = \frac{50\angle 90^\circ}{10 + 11.111}$$

$$= 2.368\angle 90^\circ \text{ A or } j2.368 \text{ A}$$

$$I_5 = \left(\frac{25}{20 + 25}\right)(j2.368) = j1.316 \text{ A}$$

$$I_6 = \left(\frac{20}{20 + 25}\right)(j2.368) = j1.052 \text{ A}$$

(v) Figure 32.10 shows Figure 32.9 superimposed on Figure 32.8, giving the currents shown.

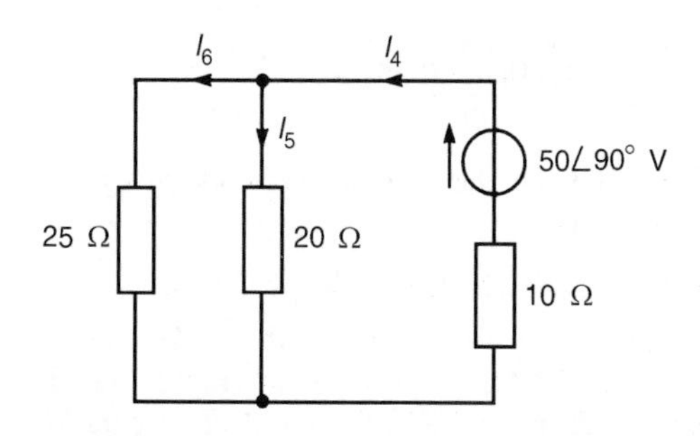

Figure 32.9

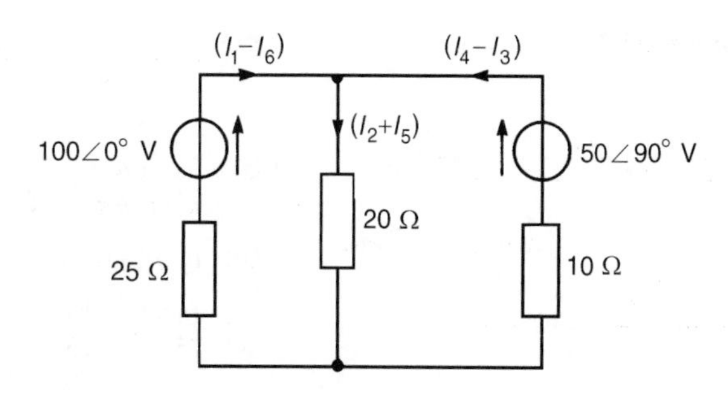

Figure 32.10

(vi) Current in the 20 Ω load, $I_2 + I_5 = (1.053 + j1.316)$ A or **1.69∠51.33° A**

Current in the $100\angle 0^\circ$ V source, $I_1 - I_6 = (3.158 - j1.052)$ A or **3.33∠−18.42° A**

Current in the $50\angle 90^\circ$ V source, $I_4 - I_3 = (j2.368 - 2.105)$ or **3.17∠131.64° A**

> Problem 2. Use the superposition theorem to determine the current in the 4 Ω resistor of the network shown in Figure 32.11.

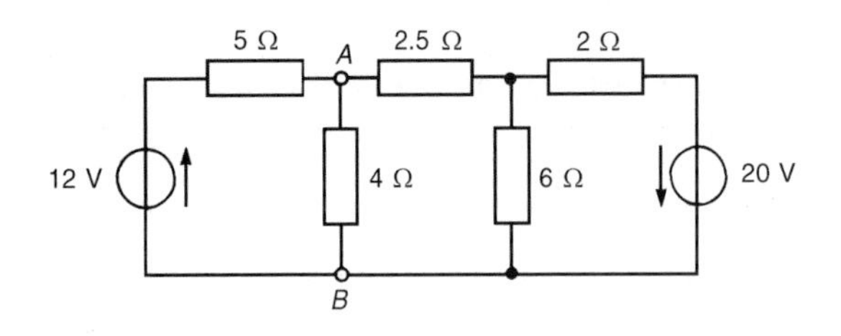

Figure 32.11

(i) Removing the 20 V source gives the network shown in Figure 32.12.

Figure 32.12

(ii) Currents I_1 and I_2 are shown labelled in Figure 32.12. It is unnecessary to determine the currents in all the branches since only the current in the 4 Ω resistance is required.

From Figure 32.12, 6 Ω in parallel with 2 Ω gives $(6 \times 2)/(6 + 2) = 1.5$ Ω, as shown in Figure 32.13. 2.5 Ω in series with 1.5 Ω gives 4 Ω, 4 Ω in parallel with 4 Ω gives 2 Ω, and 2 Ω in series with 5 Ω gives 7 Ω.

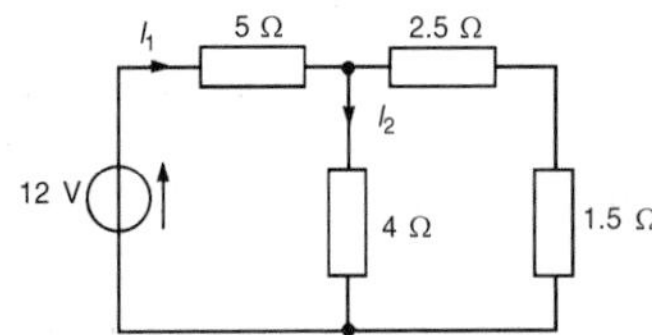
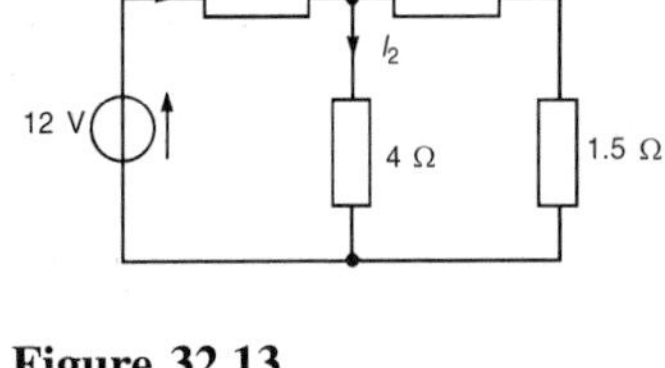

Figure 32.13

$$\text{Thus current } I_1 = \frac{12}{7} = 1.714 \text{ A and}$$

$$\text{current } I_2 = \left(\frac{4}{4+4}\right)(1.714) = 0.857 \text{ A}$$

(iii) Removing the 12 V source from the original network gives the network shown in Figure 32.14.

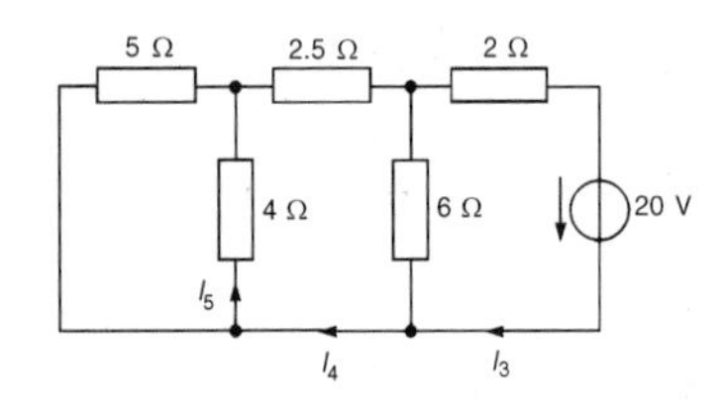

Figure 32.14

(iv) Currents I_3, I_4 and I_5 are shown labelled in Figure 32.14.

From Figure 32.14, 5 Ω in parallel with 4 Ω gives $(5 \times 4)/(5 + 4) = 20/9 = 2.222$ Ω, as shown in Figure 32.15, 2.222 Ω in series with 2.5 Ω gives 4.722 Ω, 4.722 Ω in parallel with 6 Ω gives $(4.722 \times 6)/(4.722 + 6) = 2.642$ Ω, 2.642 Ω in series with 2 Ω gives 4.642 Ω.

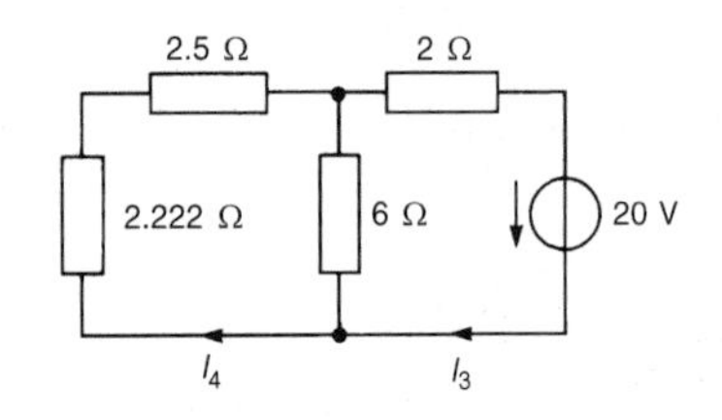

Figure 32.15

$$\text{Hence } I_3 = \frac{20}{4642} = 4.308 \text{ A}$$

$$I_4 = \left(\frac{6}{6 + 4.722}\right)(4.308) = 2.411 \text{ A, from Figure 32.15}$$

$$I_5 = \left(\frac{5}{4+5}\right)(2.411) = 1.339 \text{ A, from Figure 32.14}$$

(v) Superimposing Figure 32.14 on Figure 32.12 shows that the current flowing in the 4 Ω resistor is given by $I_5 - I_2$

(vi) $I_5 - I_2 = 1.339 - 0.857 =$ **0.48 A, flowing from B toward A** (see Figure 32.11)

Problem 3. Use the superposition theorem to obtain the current flowing in the $(4 + j3)\Omega$ impedance of Figure 32.16.

(i) The network is redrawn with V_2 removed, as shown in Figure 32.17.

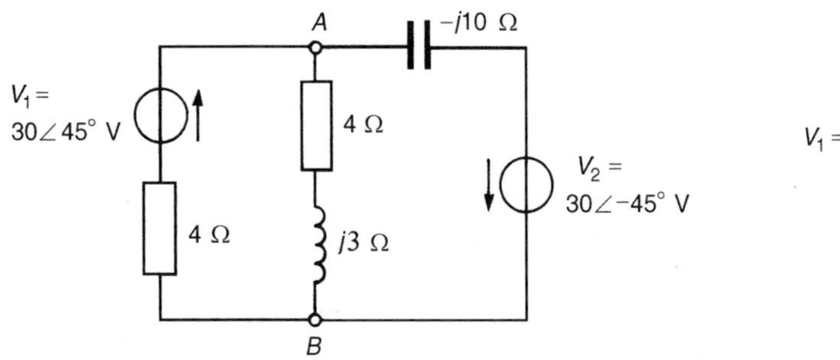

Figure 32.16

Figure 32.17

(ii) Current I_1 and I_2 are shown in Figure 32.17. From Figure 32.17, $(4 + j3)\Omega$ in parallel with $-j10\ \Omega$ gives an equivalent impedance of

$$\frac{(4 + j3)(-j10)}{(4 + j3 - j10)} = \frac{30 - j40}{4 - j7} = \frac{50\angle -53.13^\circ}{8.062\angle -60.26^\circ}$$

$$= 6.202\angle 7.13^\circ \text{ or } (6.154 + j0.770)\Omega$$

Total impedance of Figure 32.17 is

$$6.154 + j0.770 + 4 = (10.154 + j0.770)\Omega \text{ or } 10.183\angle 4.34^\circ\ \Omega$$

$$\text{Hence current } I_1 = \frac{30\angle 45^\circ}{10.183\angle 4.34^\circ} = 2.946\angle 40.66^\circ \text{ A}$$

$$\text{and current } I_2 = \left(\frac{-j10}{4 - j7}\right)(2.946\angle 40.66^\circ)$$

$$= \frac{(10\angle -90^\circ)(2.946\angle 40.66^\circ)}{8.062\angle -60.26^\circ}$$

$$= 3.654\angle 10.92^\circ \text{ A or } (3.588 + j0.692)\text{A}$$

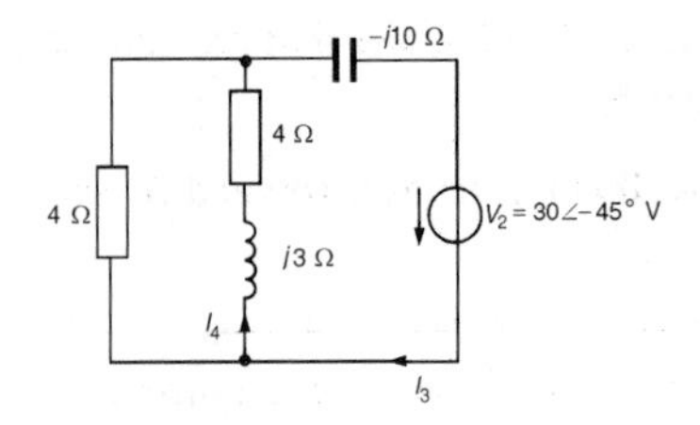

Figure 32.18

(iii) The original network is redrawn with V_1 removed, as shown in Figure 32.18.

(iv) Currents I_3 and I_4 are shown in Figure 32.18. From Figure 32.18, $4\ \Omega$ in parallel with $(4 + j3)\Omega$ gives an equivalent impedance of

$$\frac{4(4+j3)}{4+4+j3} = \frac{16+j12}{8+j3} = \frac{20\angle 36.87°}{8.544\angle 20.56°}$$
$$= 2.341\angle 16.31°\ \Omega \text{ or } (2.247 + j0.657)\Omega$$

Total impedance of Figure 32.18 is

$$2.247 + j0.657 - j10 = (2.247 - j9.343)\Omega \text{ or }$$
$$9.609\angle -76.48°\ \Omega$$

Hence current $I_3 = \dfrac{30\angle -45°}{9.609\angle -76.48°} = 3.122\angle 31.48°$ A

and current $I_4 = \left(\dfrac{4}{8+j3}\right)(3.122\angle 31.48°)$

$$= \frac{(4\angle 0°)(3.122\angle 31.48°)}{8.544\angle 20.56°}$$

$$= 1.462\angle 10.92° \text{ A or } (1.436 + j0.277)\text{A}$$

(v) If the network of Figure 32.18 is superimposed on the network of Figure 32.17, it can be seen that the current in the $(4 + j3)\Omega$ impedance is given by $I_2 - I_4$

(vi) $I_2 - I_4 = (3.588 + j0.692) - (1.436 + j0.277)$

$= \mathbf{(2.152 + j0.415)A}$ or $\mathbf{2.192\angle 10.92°\ A}$, flowing from **A to B** in Figure 32.16.

Problem 4. For the a.c. network shown in Figure 32.19 determine, using the superposition theorem, (a) the current in each branch, (b) the magnitude of the voltage across the $(6 + j8)\Omega$ impedance, and (c) the total active power delivered to the network.

(a) (i) The original network is redrawn with E_2 removed, as shown in Figure 32.20.

(ii) Currents I_1, I_2 and I_3 are labelled as shown in Figure 32.20. From Figure 32.20, $(6 + j8)\Omega$ in parallel with $(2 - j5)\Omega$ gives an equivalent impedance of

$$\frac{(6+j8)(2-j5)}{(6+j8)+(2-j5)} = (5.123 - j3.671)\Omega$$

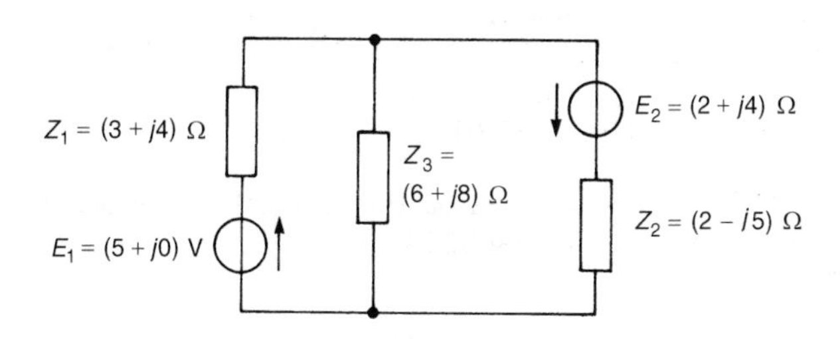

Figure 32.19

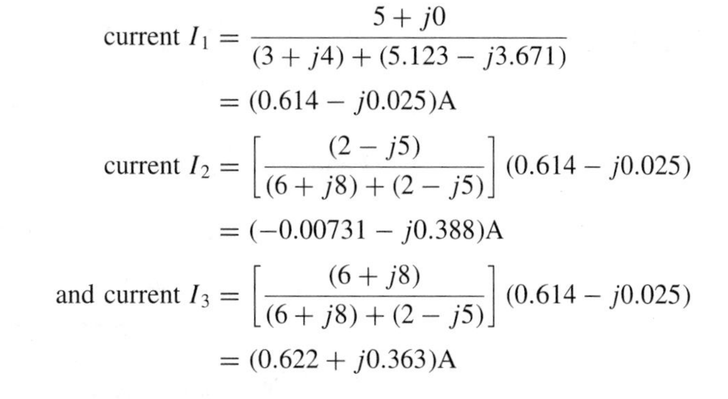

Figure 32.20

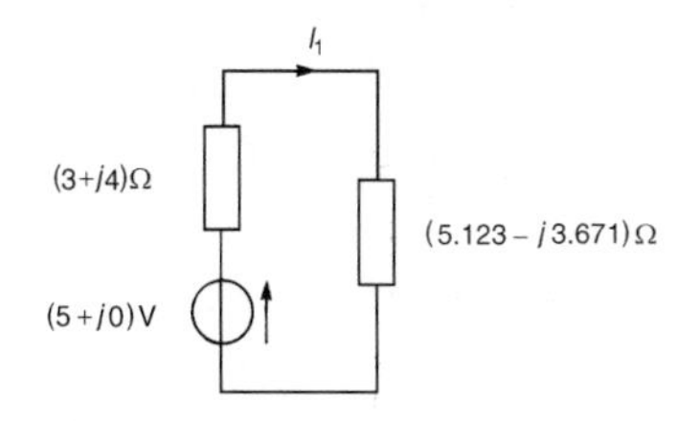

Figure 32.21

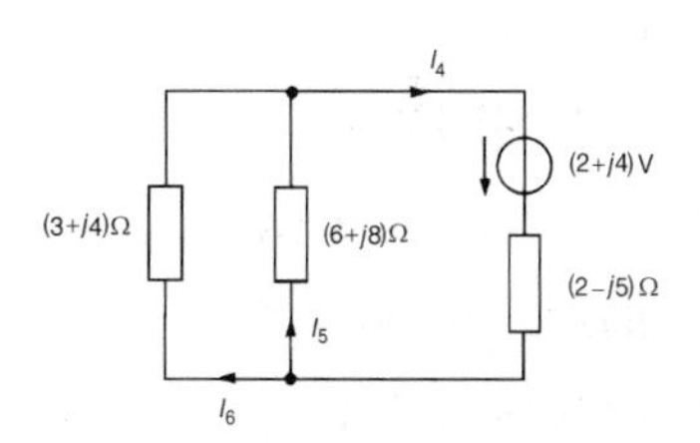

Figure 32.22

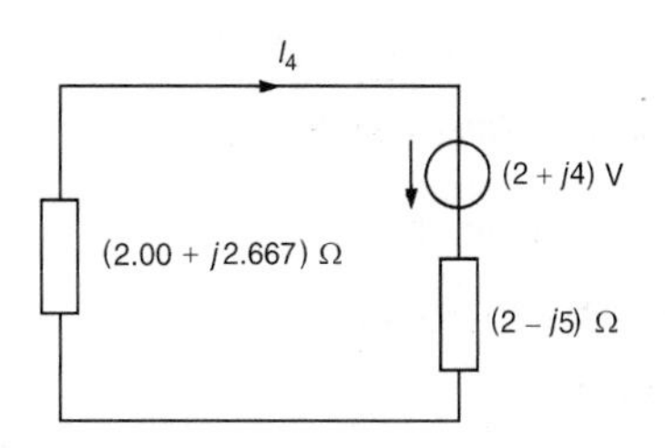

Figure 32.23

From the equivalent network of Figure 32.21,

$$\text{current } I_1 = \frac{5 + j0}{(3 + j4) + (5.123 - j3.671)}$$
$$= (0.614 - j0.025)\text{A}$$
$$\text{current } I_2 = \left[\frac{(2 - j5)}{(6 + j8) + (2 - j5)}\right](0.614 - j0.025)$$
$$= (-0.00731 - j0.388)\text{A}$$
$$\text{and current } I_3 = \left[\frac{(6 + j8)}{(6 + j8) + (2 - j5)}\right](0.614 - j0.025)$$
$$= (0.622 + j0.363)\text{A}$$

(iii) The original network is redrawn with E_1 removed, as shown in Figure 32.22.

(iv) Currents I_4, I_5 and I_6 are shown labelled in Figure 32.22 with I_4 flowing away from the positive terminal of the $(2 + j4)$V source.

From Figure 32.22, $(3 + j4)\Omega$ in parallel with $(6 + j8)\Omega$ gives an equivalent impedance of

$$\frac{(3 + j4)(6 + j8)}{(3 + j4) + (6 + j8)} = (2.00 + j2.667)\Omega$$

From the equivalent network of Figure 32.23,

$$\text{current } I_4 = \frac{(2 + j4)}{(2.00 + j2.667) + (2 - j5)}$$
$$= (-0.062 + j0.964)\text{A}$$

From Figure 32.22,

$$\text{current } I_5 = \left[\frac{(3 + j4)}{(3 + j4) + (6 + j8)}\right](-0.062 + j0.964)$$
$$= (-0.0207 + j0.321)\text{A}$$

$$\text{and current } I_6 = \left[\frac{6+j8}{(3+j4)+(6+j8)}\right](-0.062+j0.964)$$
$$= (-0.041 + j0.643)\text{A}$$

(v) If Figure 32.22 is superimposed on Figure 32.20, the resultant currents are as shown in Figure 32.24.

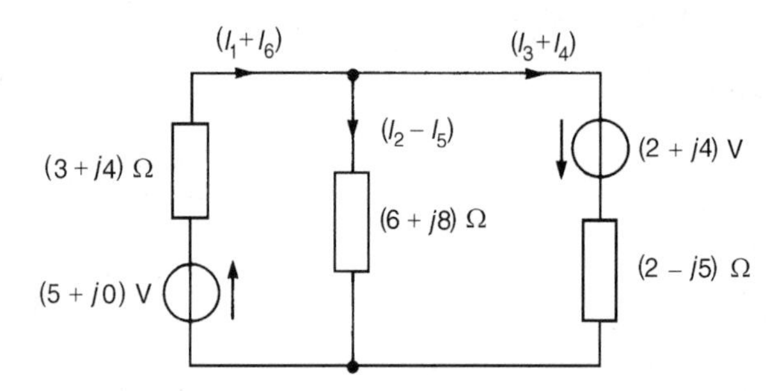

Figure 32.24

(vi) Resultant current flowing from $(5 + j0)$V source is given by

$$I_1 + I_6 = (0.614 - j0.025) + (-0.041 + j0.643)$$
$$= \mathbf{(0.573 + j0.618)A} \text{ or } \mathbf{0.843\angle 47.16^\circ\ A}$$

Resultant current flowing from $(2 + j4)$V source is given by

$$I_3 + I_4 = (0.622 + j0.363) + (-0.062 + j0.964)$$
$$= \mathbf{(0.560 + j1.327)A} \text{ or } \mathbf{1.440\angle 67.12^\circ\ A}$$

Resultant current flowing through the $(6 + j8)\Omega$ impedance is given by

$$I_2 - I_5 = (-0.00731 - j0.388) - (-0.0207 + j0.321)$$
$$= \mathbf{(0.0134 - j0.709)A} \text{ or } \mathbf{0.709\angle -88.92^\circ\ A}$$

(b) Voltage across $(6 + j8)\Omega$ impedance is given by

$$(I_2 - I_5)(6 + j8) = (0.709\angle -88.92^\circ)(10\angle 53.13^\circ)$$
$$= 7.09\angle -35.79^\circ \text{ V}$$

i.e., the magnitude of the voltage across the $(6 + j8)\Omega$ impedance is **7.09 V**

(c) Total active power P delivered to the network is given by

$$P = E_1(I_1 + I_6)\cos\phi_1 + E_2(I_3 + I_4)\cos\phi_2$$

where ϕ_1 is the phase angle between E_1 and $(I_1 + I_6)$ and ϕ_2 is the phase angle between E_2 and $(I_3 + I_4)$, i.e.,

$$P = (5)(0.843)\cos(47.16° - 0°)$$
$$+ (\sqrt{(2^2 + 4^2)})(1.440)\cos\left(67.12° - \arctan\tfrac{4}{2}\right)$$
$$= 2.866 + 6.427 = 9.293 \text{ W}$$
$$= \mathbf{9.3\ W}, \text{ correct to one dec. place.}$$

(This value may be checked since total active power dissipated is given by:

$$P = (I_1 + I_6)^2(3) + (I_2 - I_5)^2(6) + (I_3 + I_4)^2(2)$$
$$= (0.843)^2(3) + (0.709)^2(6) + (1.440)^2(2)$$
$$= 2.132 + 3.016 + 4.147 = 9.295 \text{ W}$$
$$= \mathbf{9.3\ W}, \text{ correct to one dec. place.)}$$

Problem 5. Use the superposition theorem to determine, for the network shown in Figure 32.25, (a) the magnitude of the current flowing in the capacitor, (b) the p.d. across the 5 Ω resistance, (c) the active power dissipated in the 20 Ω resistance and (d) the total active power taken from the supply.

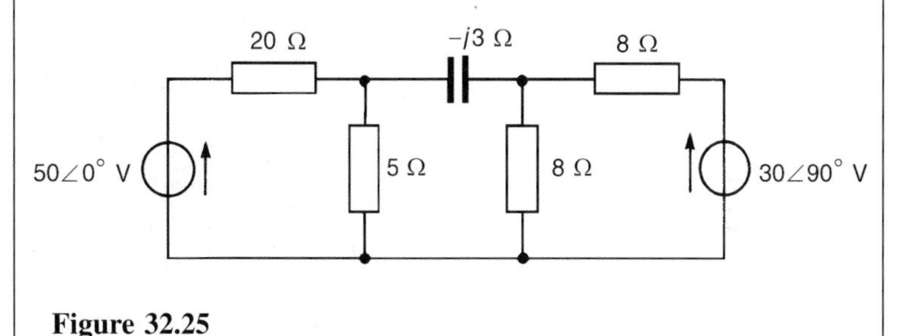

Figure 32.25

(i) The network is redrawn with the 30∠90° V source removed, as shown in Figure 32.26.

(ii) Currents I_1 to I_5 are shown labelled in Figure 32.26. From Figure 32.26, two 8 Ω resistors in parallel give an equivalent resistance of 4 Ω.

$$\text{Hence } I_1 = \frac{50\angle 0°}{20 + (5(4 - j3)/(5 + 4 - j3))} = 2.220\angle 2.12° \text{ A}$$

$$I_2 = \frac{(4 - j3)}{(5 + 4 - j3)} I_1 = 1.170\angle -16.32° \text{ A}$$

$$I_3 = \left(\frac{5}{5 + 4 - j3}\right) I_1 = 1.170\angle 20.55° \text{ A}$$

$$I_4 = \left(\frac{8}{8 + 8}\right) I_3 = 0.585\angle 20.55° \text{ A} = I_5$$

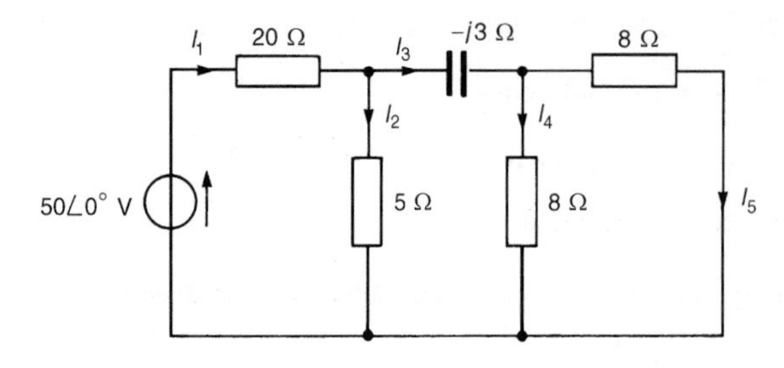

Figure 32.26

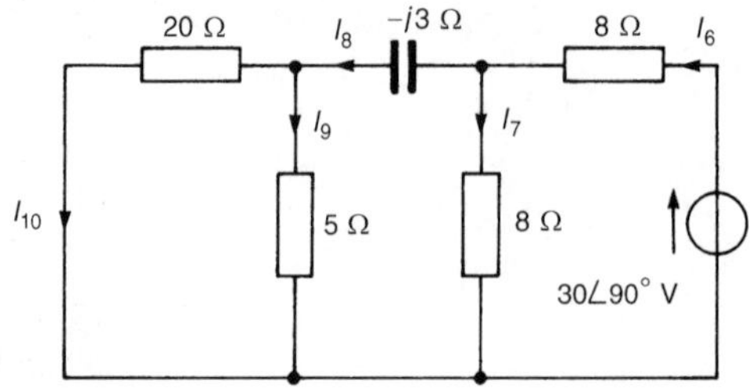

Figure 32.27

(iii) The original network is redrawn with the $50\angle 0°$ V source removed, as shown in Figure 32.27.

(iv) Currents I_6 to I_{10} are shown labelled in Figure 32.27. From Figure 32.27, 20 Ω in parallel with 5 Ω gives an equivalent resistance of $(20 \times 5)/(20+5) = 4\ \Omega$.

$$\text{Hence } I_6 = \frac{30\angle 90°}{8 + (8(4-j3)/(8+4-j3))} = 2.715\angle 96.52°\ \text{A}$$

$$I_7 = \frac{(4-j3)}{(8+4-j3)} I_6 = 1.097\angle 73.69°\ \text{A}$$

$$I_8 = \left(\frac{8}{8+4-j3}\right) I_6 = 1.756\angle 110.56°\ \text{A}$$

$$I_9 = \left(\frac{20}{20+5}\right) I_8 = 1.405\angle 110.56°\ \text{A}$$

$$\text{and} \quad I_{10} = \left(\frac{5}{20+5}\right) I_8 = 0.351\angle 110.56°\ \text{A}$$

(a) The current flowing in the capacitor is given by

$$(I_3 - I_8) = 1.170\angle 20.55° - 1.756\angle 110.56°$$
$$= (1.712 - j1.233)\text{A or } 2.11\angle -35.76°\ \text{A}$$

i.e., **the magnitude of the current in the capacitor is 2.11 A**

(b) The p.d. across the 5 Ω resistance is given by $(I_2 + I_9)$ (5).

$$(I_2 + I_9) = 1.170\angle -16.32° + 1.405\angle 110.56°$$
$$= (0.629 + j0.987)\text{A or } 1.17\angle 57.49°\ \text{A}$$

Hence **the magnitude of the pd. across the 5 Ω resistance** is $(1.17)(5) =$ **5.85 V**

(c) Active power dissipated in the 20 Ω resistance is given by $(I_1 - I_{10})^2(20)$.

$$(I_1 - I_{10}) = 2.220\angle 2.12° - 0.351\angle 110.56°$$
$$= (2.342 - j0.247)\text{A or } 2.355\angle -6.02°\ \text{A}$$

Hence **the active power dissipated in the 20 Ω resistance** is $(2.355)^2(20) =$ **111 W**

(d) Active power developed by the $50\angle 0°$ V source

$$P_1 = V(I_1 - I_{10})\cos\phi_1 = (50)(2.355)\cos(6.02° - 0°)$$
$$= 117.1 \text{ W}$$

Active power developed by $30\angle 90$ V source,

$$P_2 = 30(I_6 - I_5)\cos\phi_2$$
$$(I_6 - I_5) = 2.715\angle 96.52° - 0.585\angle 20.55°$$
$$= (-0.856 + j2.492)\text{A or } 2.635\angle 108.96° \text{ A}$$

Hence $P_2 = (30)(2.635)\cos(108.96° - 90°) = 74.8$ W.

Total power developed, $P = P_1 + P_2 = 117.1 + 74.8 =$ **191.9 W**

(This value may be checked by summing the I^2R powers dissipated in the four resistors.)

Further problems on the superposition theorem may be found in Section 32.3 following, problems 1 to 8.

32.3 Further problems on the superposition theorem

1 Repeat problems 1, 5, 8 and 9 on page 542, of Chapter 30, and problems 3, 5 and 13 on page 559, of Chapter 31, using the superposition theorem.

2 Two batteries each of e.m.f. 15 V are connected in parallel to supply a load of resistance 2.0 Ω. The internal resistances of the batteries are 0.5 Ω and 0.3 Ω. Determine, using the superposition theorem, the current in the load and the current supplied by each battery.
[6.86 A; 2.57 A; 4.29 A]

3 Use the superposition theorem to determine the magnitude of the current flowing in the capacitive branch of the network shown in Figure 32.28. [2.584 A]

4 A.c. sources of $20\angle 90°$ V and internal resistance 10 Ω and $30\angle 0°$ V and internal resistance 12 Ω are connected in parallel across an 8 Ω load. Use the superposition theorem to determine (a) the current in the 8 Ω load, and (b) the current in each voltage source.
[(a) 1.30 A (b) $20\angle 90°$ V source discharges at $1.58\angle 120.98°$ A, $30\angle 0°$ V source discharges at $1.90\angle -16.49°$ A]

5 Use the superposition theorem to determine current I_x flowing in the 5 Ω resistance of the network shown in Figure 32.29.
[$0.529\angle 5.71°$ A]

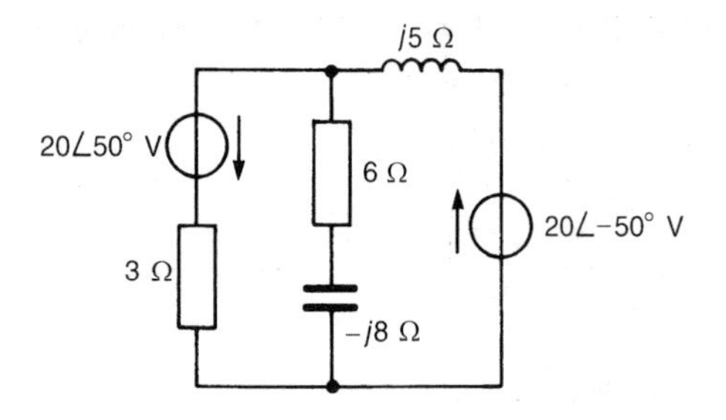

Figure 32.28

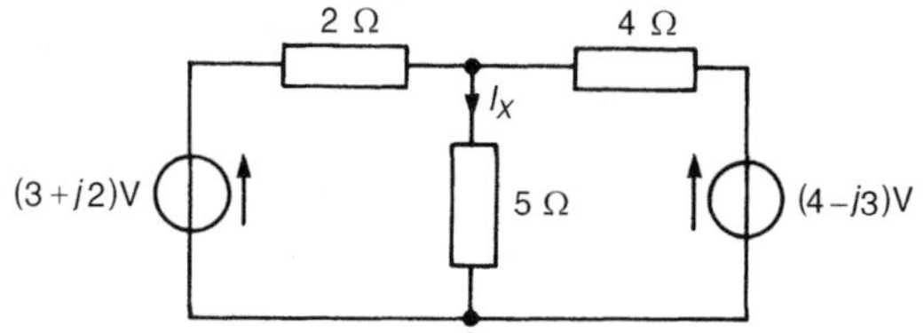

Figure 32.29

6 For the network shown in Figure 32.30, determine, using the superposition theorem, (a) the current flowing in the capacitor, (b) the current flowing in the 2 Ω resistance, (c) the p.d. across the 5 Ω resistance, and (d) the total active circuit power.
[(a) 1.28 A (b) 0.74 A (c) 3.01 V (d) 2.91 W]

7 (a) Use the superposition theorem to determine the current in the 12 Ω resistance of the network shown in Figure 32.31. Determine also the p.d. across the 8 Ω resistance and the power dissipated in the 20 Ω resistance.

(b) If the 37.5 V source in Figure 32.31 is reversed in direction, determine the current in the 12 Ω resistance.
[(a) 0.375 A, 8.0 V, 57.8 W (b) 0.625 A]

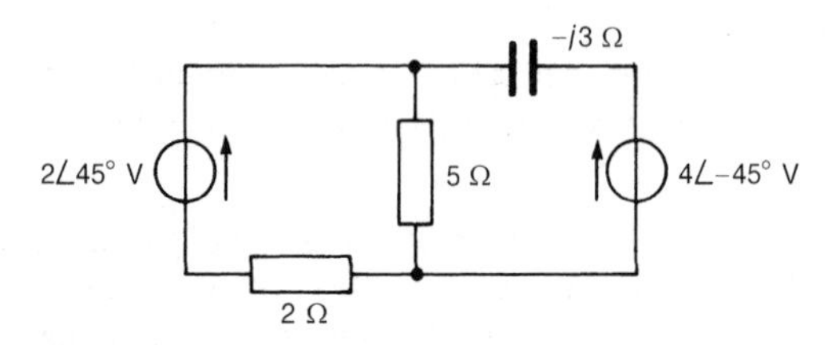

Figure 32.30

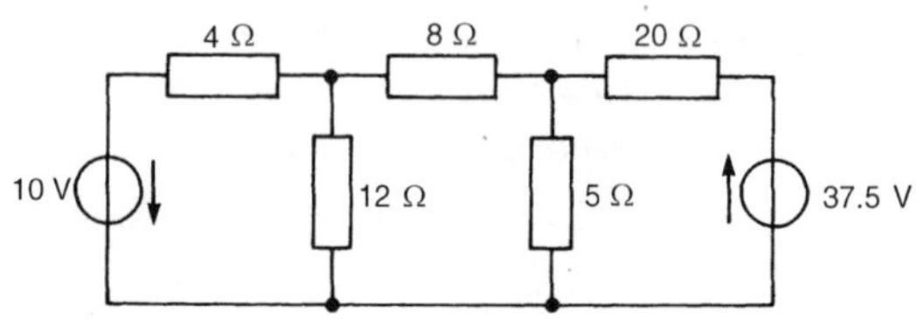

Figure 32.31

8 For the network shown in Figure 32.32, use the superposition theorem to determine (a) the current in the capacitor, (b) the pd. across the 10 Ω resistance, (c) the active power dissipated in the 20 Ω resistance, and (d) the total active circuit power.
[(a) 3.97 A (b) 28.7 V (c) 36.4 W (d) 371.6 W]

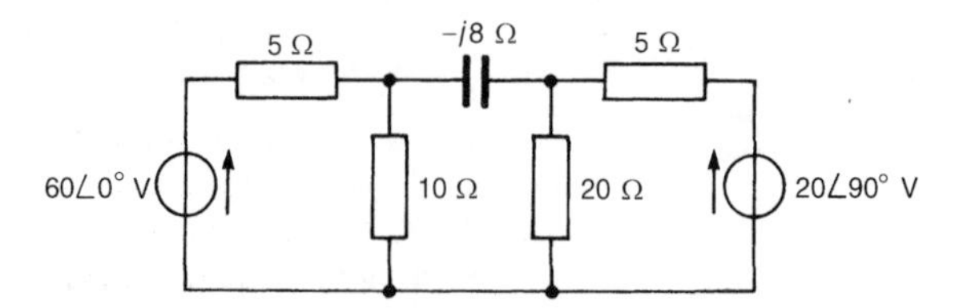

Figure 32.32

33 Thévenin's and Norton's theorems

At the end of this chapter you should be able to:

- understand and use Thévenin's theorem to analyse a.c. and d.c. networks
- understand and use Norton's theorem to analyse a.c. and d.c. networks
- appreciate and use the equivalence of Thévenin and Norton networks

33.1 Introduction

Many of the networks analysed in Chapters 30, 31 and 32 using Kirchhoff's laws, mesh-current and nodal analysis and the superposition theorem can be analysed more quickly and easily by using Thévenin's or Norton's theorems. Each of these theorems involves replacing what may be a complicated network of sources and linear impedances with a simple equivalent circuit. A set procedure may be followed when using each theorem, the procedures themselves requiring a knowledge of basic circuit theory. (It may be worth checking some general d.c. circuit theory in Section 13.4. page 174, before proceeding)

33.2 Thévenin's theorem

Thévenin's theorem states:

'The current which flows in any branch of a network is the same as that which would flow in the branch if it were connected across a source of electrical energy, the e.m.f. of which is equal to the potential difference which would appear across the branch if it were open-circuited, and the internal impedance of which is equal to the impedance which appears across the open-circuited branch terminals when all sources are replaced by their internal impedances.'

The theorem applies to any linear active network ('linear' meaning that the measured values of circuit components are independent of the direction and magnitude of the current flowing in them, and 'active' meaning that it contains a source, or sources, of e.m.f.)

The above statement of Thévenin's theorem simply means that a complicated network with output terminals AB, as shown in Figure 33.1(a), can be replaced by a single voltage source E in series with an impedance z, as shown in Figure 33.1(b). E is the open-circuit

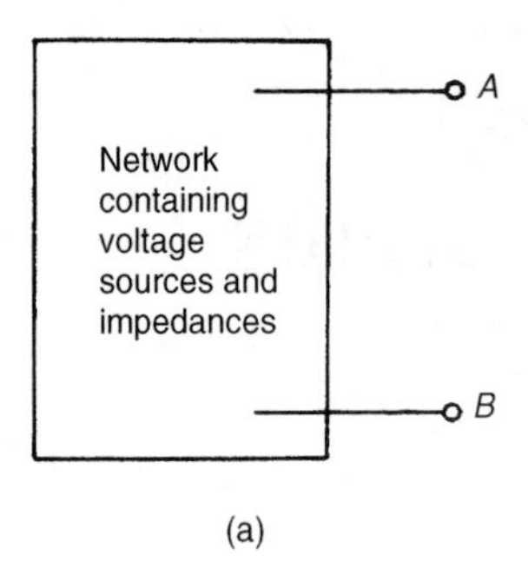

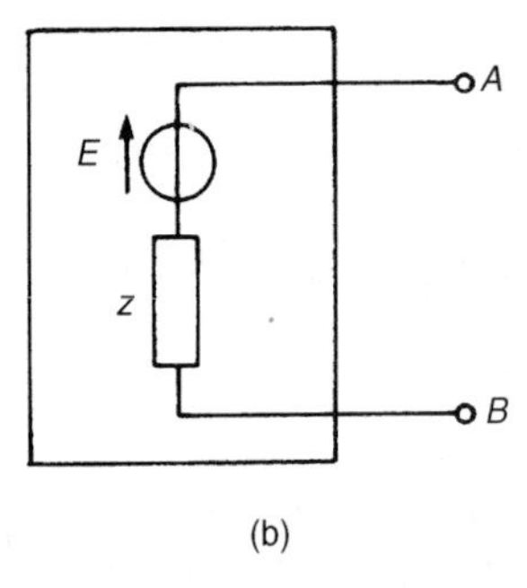

Figure 33.1 *The Thévenin equivalent circuit*

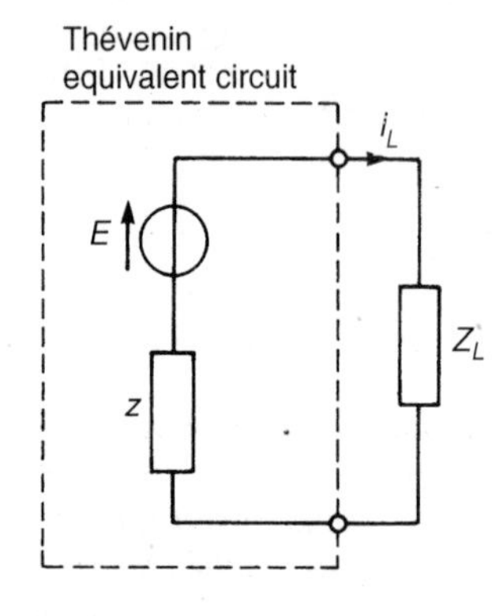

Figure 33.2

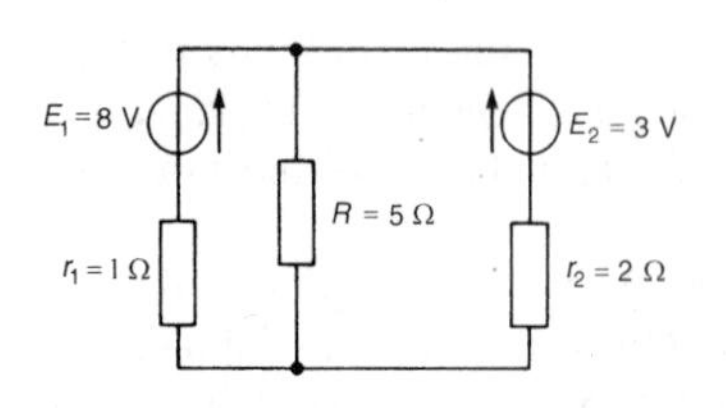

Figure 33.3

voltage measured at terminals AB and z is the equivalent impedance of the network at the terminals AB when all internal sources of e.m.f. are made zero. The polarity of voltage E is chosen so that the current flowing through an impedance connected between A and B will have the same direction as would result if the impedance had been connected between A and B of the original network. Figure 33.1(b) is known as the **Thévenin equivalent circuit**, and was initially introduced in Section 13.4, page 174 for d.c. networks.

The following four-step **procedure** can be adopted when determining, by means of Thévenin's theorem, the current flowing in a branch containing impedance Z_L of an active network:

(i) remove the impedance Z_L from that branch;

(ii) determine the open-circuit voltage E across the break;

(iii) remove each source of e.m.f. and replace it by its internal impedance (if it has zero internal impedance then replace it by a short-circuit), and then determine the internal impedance, z, 'looking in' at the break;

(iv) determine the current from the Thévenin equivalent circuit shown in Figure 33.2, i.e.

$$\textbf{current } \boldsymbol{i_L} = \frac{\boldsymbol{E}}{\boldsymbol{Z_L + z}}.$$

A simple d.c. network (Figure 33.3) serves to demonstrate how the above procedure is applied to determine the current flowing in the 5 Ω resistance by using Thévenin's theorem. This is the same network as used in Chapter 30 when it was solved using Kirchhoff's laws (see page 535), and by means of the superposition theorem in Chapter 32 (see page 562). A comparison of methods may be made.

Using the above procedure:

(i) the 5 Ω resistor is removed, as shown in Figure 33.4(a).

(ii) The open-circuit voltage E across the break is now required. The network of Figure 33.4(a) is redrawn for convenience as shown in Figure 33.4(b), where current,

$$I_1 = \frac{E_1 - E_2}{r_1 + r_2} = \frac{8-3}{1+2} = \frac{5}{3} \quad \text{or} \quad 1\frac{2}{3} \text{ A}$$

Hence the open-circuit voltage E is given by

$$E = E_1 - I_1 r_1 \text{ i.e., } E = 8 - \left(1\tfrac{2}{3}\right)(1) = 6\tfrac{1}{3} \text{ V}$$

$$\text{(Alternatively, } E = E_2 - (-I_1)r_2 = 3 + \left(1\tfrac{2}{3}\right)(2) = 6\tfrac{1}{3} \text{ V.)}$$

(iii) Removing each source of e.m.f. gives the network of Figure 33.5. The impedance, z, 'looking in' at the break AB is given by

$$z = (1 \times 2)/(1 + 2) = \tfrac{2}{3}\ \Omega$$

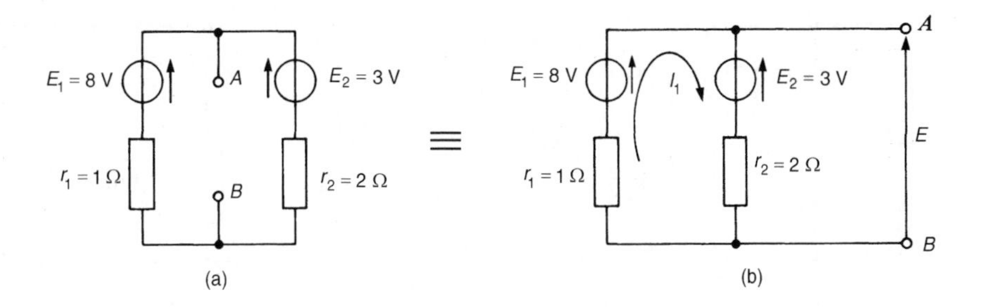

Figure 33.4

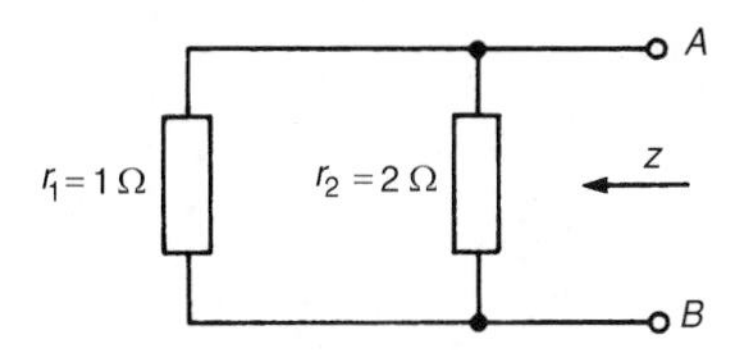

Figure 33.5

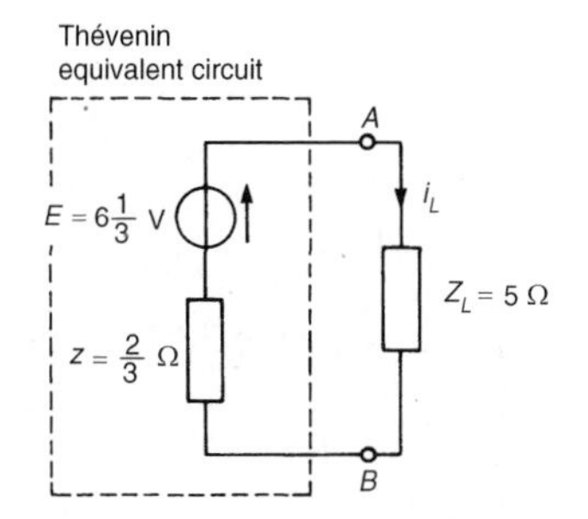

Figure 33.6

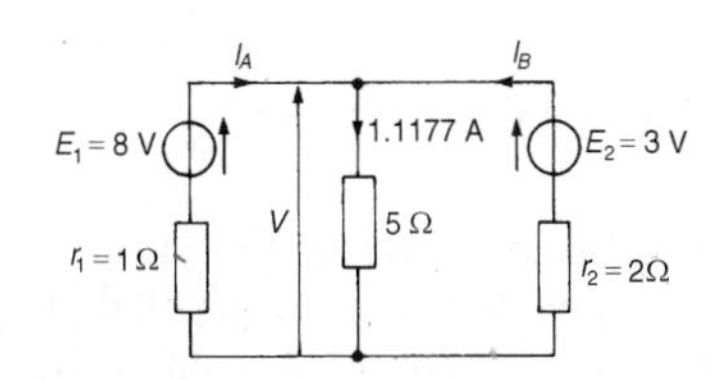

Figure 33.7

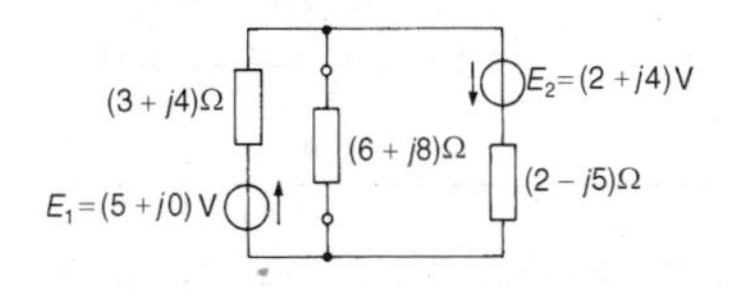

Figure 33.8

(iv) The Thévenin equivalent circuit is shown in Figure 33.6, where current i_L is given by

$$i_L = \frac{E}{Z_L + z} = \frac{6\frac{1}{3}}{5 + \frac{2}{3}} = 1.1177$$

$$= \mathbf{1.12\ A}, \text{ correct to two decimal places}$$

To determine the currents flowing in the other two branches of the circuit of Figure 33.3, basic circuit theory is used. Thus, from Figure 33.7, voltage $V = (1.1177)(5) = 5.5885$ V.

Then $V = E_1 - I_A r_1$, i.e., $5.5885 = 8 - I_A(1)$, from which

current $I_A = 8 - 5.5885 = \mathbf{2.41\ A}$.

Similarly, $V = E_2 - I_B r_2$, i.e., $5.5885 = 3 - I_B(2)$, from which

$$\text{current } I_B = \frac{3 - 5.5885}{2} = \mathbf{-1.29\ A}$$

(i.e., flowing in the direction opposite to that shown in Figure 33.7).

The Thévenin theorem procedure used above may be applied to a.c. as well as d.c. networks, as shown below.

An a.c. network is shown in Figure 33.8 where it is required to find the current flowing in the $(6 + j8)\Omega$ impedance by using Thévenin's theorem. Using the above procedure

(i) The $(6 + j8)\Omega$ impedance is removed, as shown in Figure 33.9(a).

(ii) The open-circuit voltage across the break is now required. The network is redrawn for convenience as shown in Figure 33.9(b), where current

$$I_1 = \frac{(5 + j0) + (2 + j4)}{(3 + j4) + (2 - j5)} = \frac{(7 + j4)}{(5 - j)}$$

$$= 1.581\angle 41.05^\circ \text{ A}$$

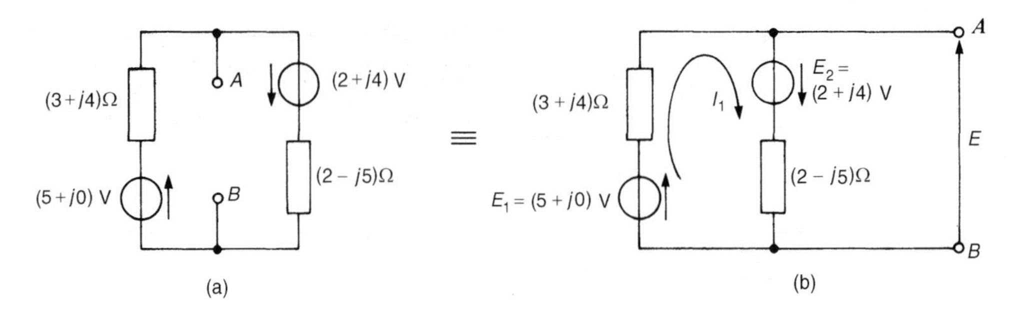

Figure 33.9

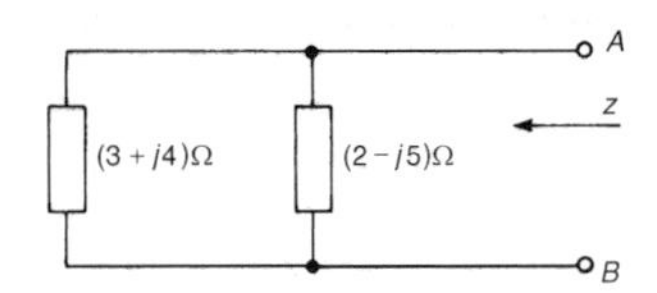

Figure 33.10

Hence open-circuit voltage across AB,

$$E = E_1 - I_1(3 + j4), \text{ i.e.,}$$

$$E = (5 + j0) - (1.581\angle 41.05^\circ)(5\angle 53.13^\circ)$$

from which $E = 9.567\angle -54.73^\circ$ V

(iii) From Figure 33.10, the impedance z 'looking in' at terminals AB is given by

$$z = \frac{(3 + j4)(2 - j5)}{(3 + j4) + (2 - j5)}$$

$$= 5.281\angle -3.76^\circ\ \Omega \quad \text{or} \quad (5.270 - j0.346)\Omega$$

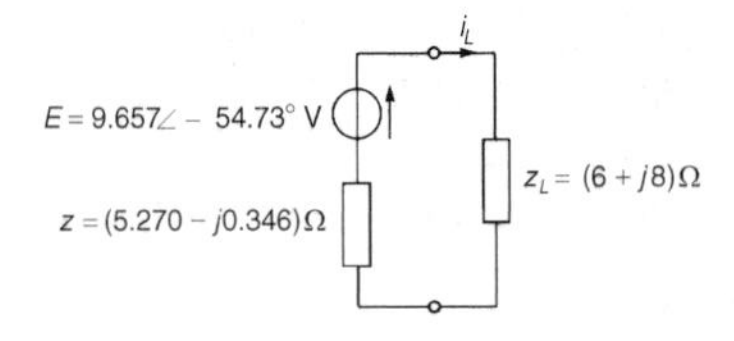

Figure 33.11

(iv) The Thévenin equivalent circuit is shown in Figure 33.11, from which current

$$i_L = \frac{E}{Z_L + z} = \frac{9.657\angle -54.73^\circ}{(6 + j8) + (5.270 - j0.346)}$$

Thus, current in $(6 + j8)\Omega$ impedance,

$$i_L = \frac{9.657\angle -54.73^\circ}{13.623\angle 34.18^\circ} = \mathbf{0.71\angle -88.91^\circ\ A}$$

The network of Figure 33.8 is analysed using Kirchhoff's laws in problem 3, page 539, and by the superposition theorem in problem 4, page 568. The above analysis using Thévenin's theorem is seen to be much quicker.

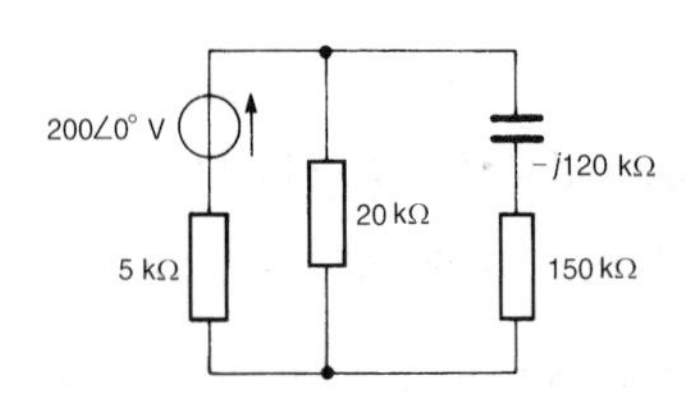

Figure 33.12

> Problem 1. For the circuit shown in Figure 33.12, use Thévenin's theorem to determine (a) the current flowing in the capacitor, and (b) the p.d. across the 150 kΩ resistor.

(a) (i) Initially the $(150 - j120)$kΩ impedance is removed from the circuit as shown in Figure 33.13.

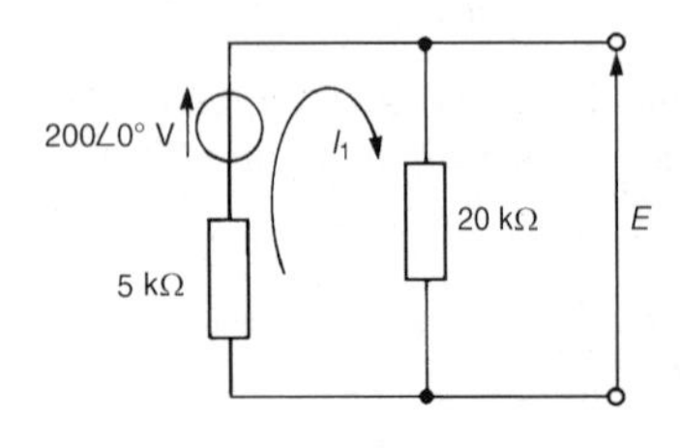

Figure 33.13

Note that, to find the current in the capacitor, only the capacitor need have been initially removed from the circuit. However, removing each of the components from the branch through which the current is required will often result in a simpler solution.

(ii) From Figure 33.13,

$$\text{current } I_1 = \frac{200\angle 0°}{(5000 + 20000)} = 8 \text{ mA}$$

The open-circuit e.m.f. E is equal to the p.d. across the 20 kΩ resistor, i.e.

$$E = (8 \times 10^{-3})(20 \times 10^3) = \mathbf{160\ V}.$$

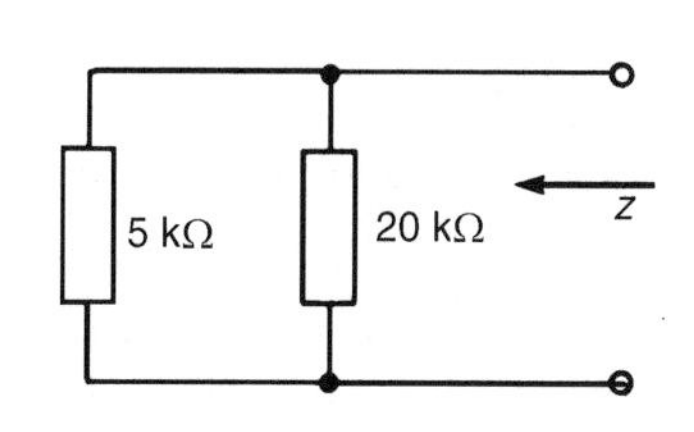

Figure 33.14

(iii) Removing the $200\angle 0°$ V source gives the network shown in Figure 33.14.

The impedance, z, 'looking in' at the open-circuited terminals is given by

$$z = \frac{5 \times 20}{5 + 20} \text{ k}\Omega = \mathbf{4\ k\Omega}$$

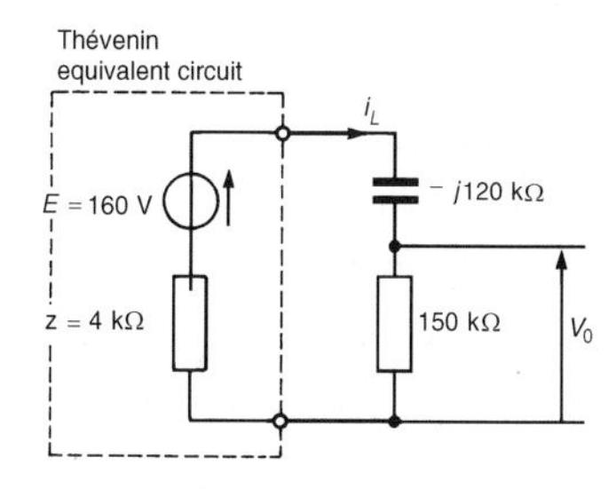

Figure 33.15

(iv) The Thévenin equivalent circuit is shown in Figure 33.15, where current i_L is given by

$$i_L = \frac{E}{Z_L + z} = \frac{160}{(150 - j120) \times 10^3 + 4 \times 10^3}$$

$$= \frac{160}{195.23 \times 10^3\angle -37.93°}$$

$$= 0.82\angle 37.93° \text{ mA}$$

Thus the current flowing in the capacitor is 0.82 mA.

(b) P.d. across the 150 kΩ resistor,

$$V_0 = i_L R = (0.82 \times 10^{-3})(150 \times 10^3) = \mathbf{123\ V}$$

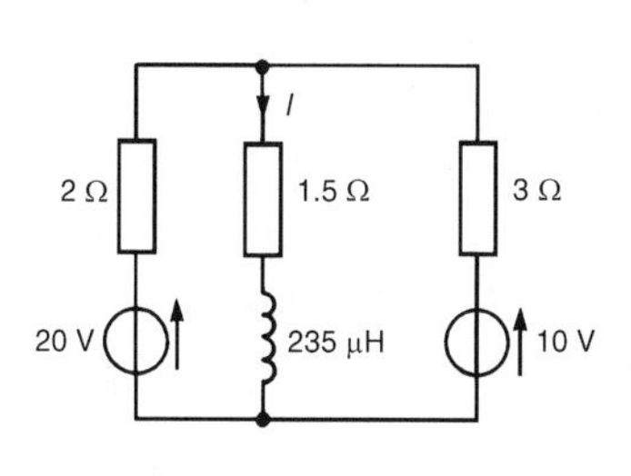

Figure 33.16

Problem 2. Determine, for the network shown in Figure 33.16, the value of current I. Each of the voltage sources has a frequency of 2 kHz.

(i) The impedance through which current I is flowing is initially removed from the network, as shown in Figure 33.17.

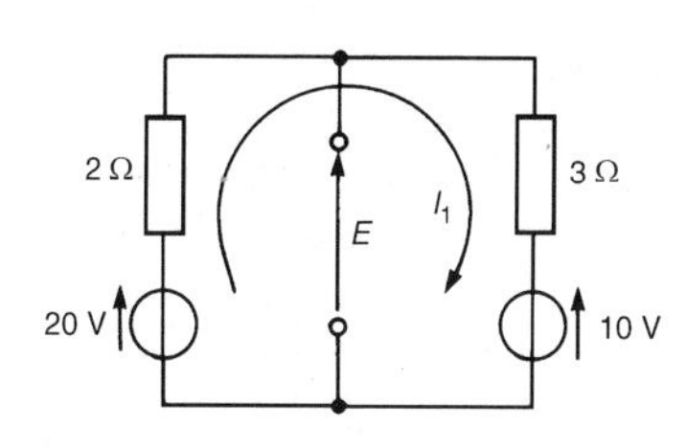

Figure 33.17

(ii) From Figure 33.17,

$$\text{current, } I_1 = \frac{20 - 10}{2 + 3} = 2 \text{ A}$$

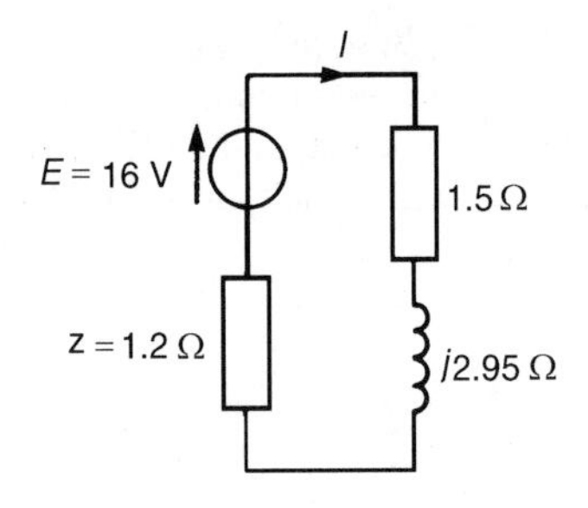

Figure 33.18

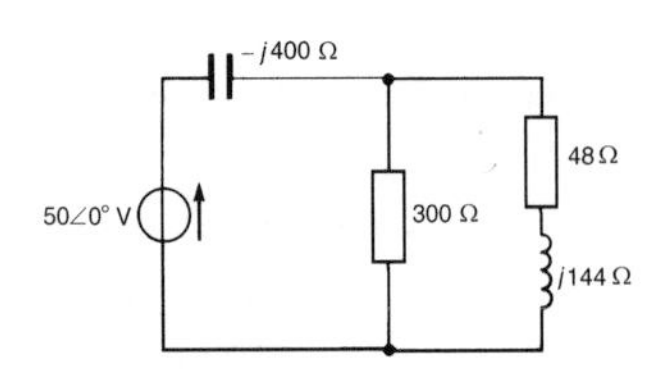

Figure 33.19

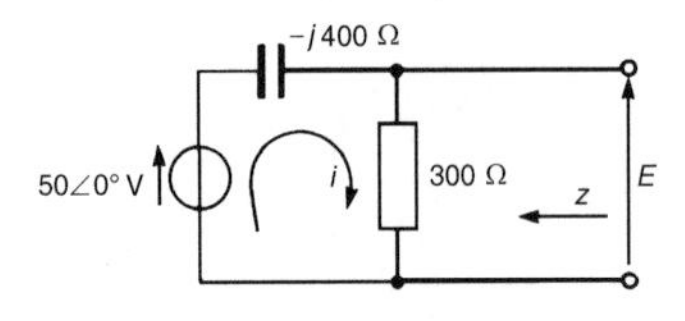

Figure 33.20

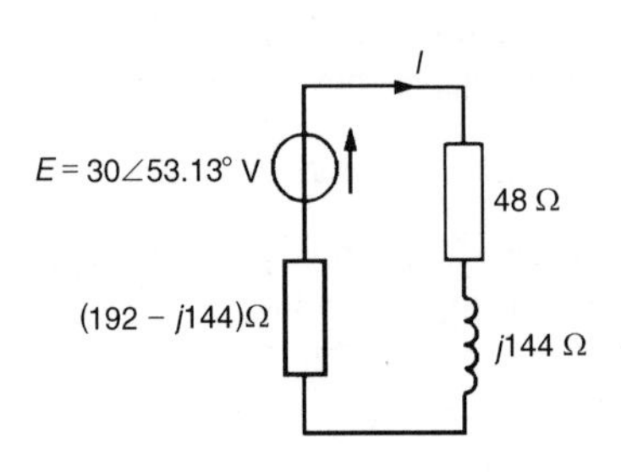

Figure 33.21

Hence the open circuit e.m.f. $\boldsymbol{E} = 20 - I_1(2) = 20 - 2(2) = \mathbf{16\ V}$.

(Alternatively, $E = 10 + I_1(3) = 10 + (2)(3) = 16$ V.)

(iii) When the sources of e.m.f. are removed from the circuit, the impedance, z, 'looking in' at the break is given by

$$z = \frac{2 \times 3}{2 + 3} = \mathbf{1.2\ \Omega}$$

(iv) The Thévenin equivalent circuit is shown in Figure 33.18, where inductive reactance,

$$X_L = 2\pi f L = 2\pi(2000)(235 \times 10^{-6}) = 2.95\ \Omega$$

Hence current

$$I = \frac{16}{(1.2 + 1.5 + j2.95)} = \frac{16}{4.0\angle 47.53^\circ}$$

$$= \mathbf{4.0\angle{-47.53^\circ}\ A} \quad \text{or} \quad \mathbf{(2.70 - j2.95)\ A}$$

Problem 3. Use Thévenin's theorem to determine the power dissipated in the 48 Ω resistor of the network shown in Figure 33.19.

The power dissipated by a current I flowing through a resistor R is given by I^2R, hence initially the current flowing in the 48 Ω resistor is required.

(i) The $(48 + j144)\Omega$ impedance is initially removed from the network as shown in Figure 33.20.

(ii) From Figure 33.20,

$$\text{current, } i = \frac{50\angle 0^\circ}{(300 - j400)} = 0.1\angle 53.13^\circ \text{ A}$$

Hence the open-circuit voltage

$$\boldsymbol{E} = i(300) = (0.1\angle 53.13^\circ)(300) = \mathbf{30\angle 53.13^\circ\ V}$$

(iii) When the $50\angle 0^\circ$ V source shown in Figure 33.20 is removed, the impedance, z, is given by

$$z = \frac{(-j400)(300)}{(300 - j400)} = \frac{(400\angle{-90^\circ})(300)}{500\angle{-53.13^\circ}}$$

$$= 240\angle{-36.87^\circ}\ \Omega \quad \text{or} \quad \mathbf{(192 - j144)\Omega}$$

(iv) The Thévenin equivalent circuit is shown in Figure 33.21 connected to the $(48 + j144)\Omega$ load.

$$\text{Current } I = \frac{30\angle 53.13^\circ}{(192 - j144) + (48 + j144)} = \frac{30\angle 53.13^\circ}{240\angle 0^\circ}$$

$$= 0.125\angle 53.13^\circ \text{ A}$$

Hence the power dissipated in the 48 Ω resistor

$$= I^2R = (0.125)^2(48) = \mathbf{0.75\ W}$$

Problem 4. For the network shown in Figure 33.22, use Thévenin's theorem to determine the current flowing in the 80 Ω resistor.

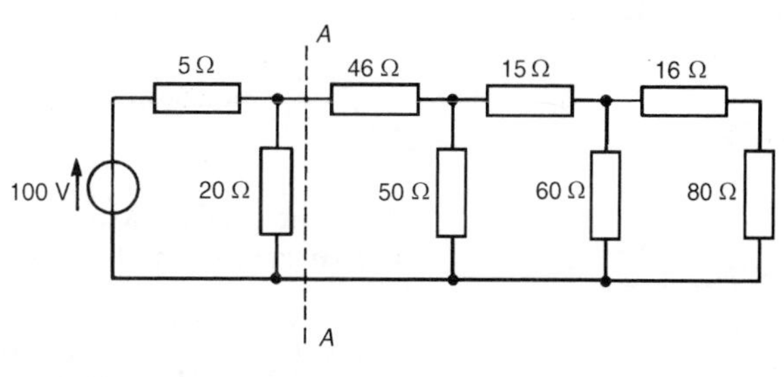

Figure 33.22

One method of analysing a multi-branch network as shown in Figure 33.22 is to use Thévenin's theorem on one part of the network at a time. For example, the part of the circuit to the left of AA may be reduced to a Thévenin equivalent circuit.
From Figure 33.23,

$$E_1 = \left(\frac{20}{20+5}\right)100 = 80\text{ V, by voltage division}$$

and $$z_1 = \frac{20 \times 5}{20+5} = 4\ \Omega$$

Thus the network of Figure 33.22 reduces to that of Figure 33.24. The part of the network shown in Figure 33.24 to the left of BB may be reduced to a Thévenin equivalent circuit, where

$$E_2 = \left(\frac{50}{50+46+4}\right)(80) = 40\text{ V}$$

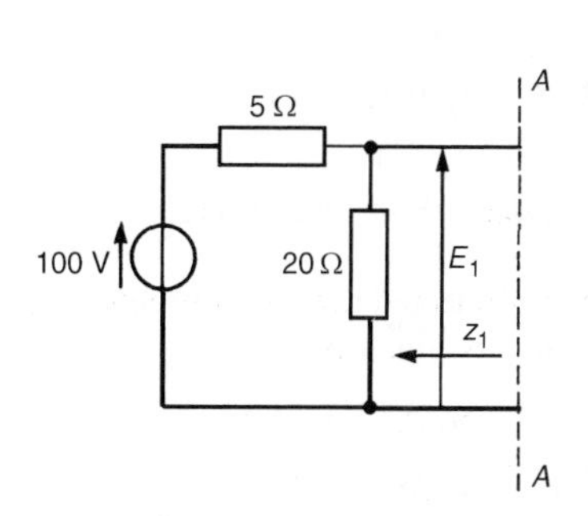

Figure 33.23

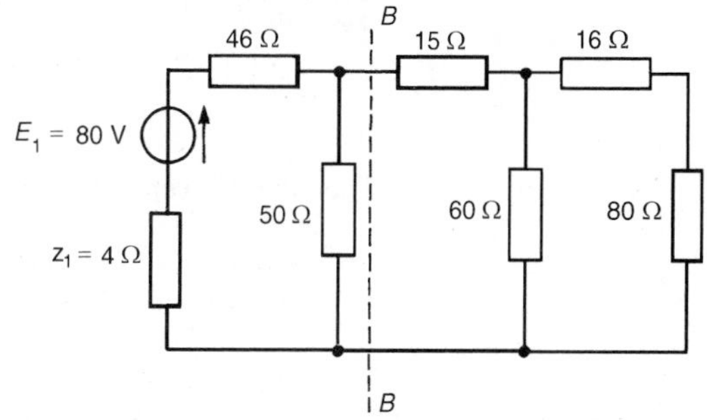

Figure 33.24

and $z_2 = \dfrac{50 \times 50}{50 + 50} = 25\ \Omega$

Thus the original network reduces to that shown in Figure 33.25.

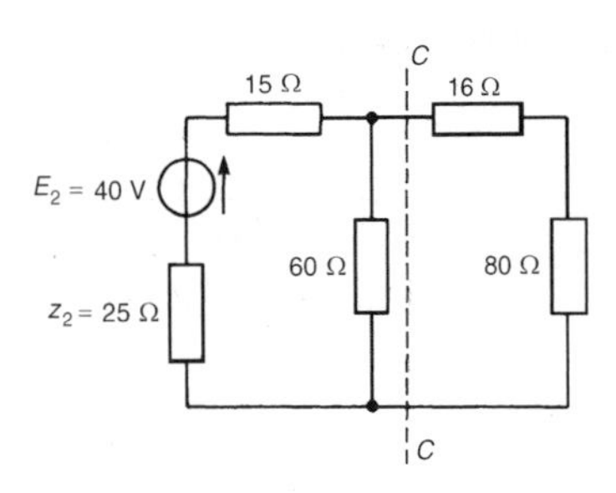

Figure 33.25

The part of the network shown in Figure 33.25 to the left of CC may be reduced to a Thévenin equivalent circuit, where

$$E_3 = \left(\frac{60}{60 + 25 + 15}\right)(40) = 24\ \text{V}$$

and $z_3 = \dfrac{(60)(40)}{(60 + 40)} = 24\ \Omega$

Thus the original network reduces to that of Figure 33.26, from which **the current in the 80 Ω resistor** is given by

$$I = \left(\frac{24}{80 + 16 + 24}\right) = \mathbf{0.20\ A}$$

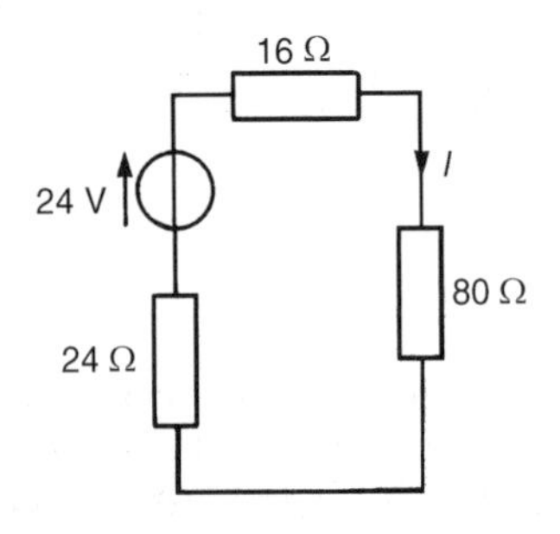

Figure 33.26

> Problem 5. Determine the Thévenin equivalent circuit with respect to terminals AB of the circuit shown in Figure 33.27. Hence determine (a) the magnitude of the current flowing in a $(3.75 + j11)\Omega$ impedance connected across terminals AB, and (b) the magnitude of the p.d. across the $(3.75 + j11)\Omega$ impedance.

Current I_1 shown in Figure 33.27 is given by

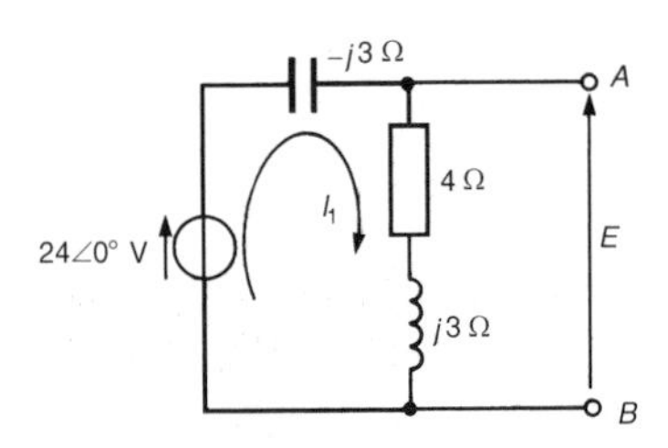

Figure 33.27

$$I_1 = \frac{24\angle 0^\circ}{(4 + j3 - j3)} = \frac{24\angle 0^\circ}{4\angle 0^\circ} = 6\angle 0^\circ\ \text{A}$$

The Thévenin equivalent voltage, i.e., the open-circuit voltage across terminals AB, is given by

$$E = I_1(4 + j3) = (6\angle 0^\circ)(5\angle 36.87^\circ) = \mathbf{30\angle 36.87^\circ\ V}$$

When the $24\angle 0^\circ$ V source is removed, the impedance z 'looking in' at AB is given by

$$z = \frac{(4 + j3)(-j3)}{(4 + j3 - j3)} = \frac{9 - j12}{4} = \mathbf{(2.25 - j3.0)\Omega}$$

Thus the Thévenin equivalent circuit is as shown in Figure 33.28.

(a) When a $(3.75 + j11)\Omega$ impedance is connected across terminals AB, the current I flowing in the impedance is given by

$$I = \frac{30\angle 36.87^\circ}{(3.75 + j11) + (2.25 - j3.0)} = \frac{30\angle 36.87^\circ}{10\angle 53.13^\circ}$$

$$= 3\angle -16.26^\circ\ \text{A}$$

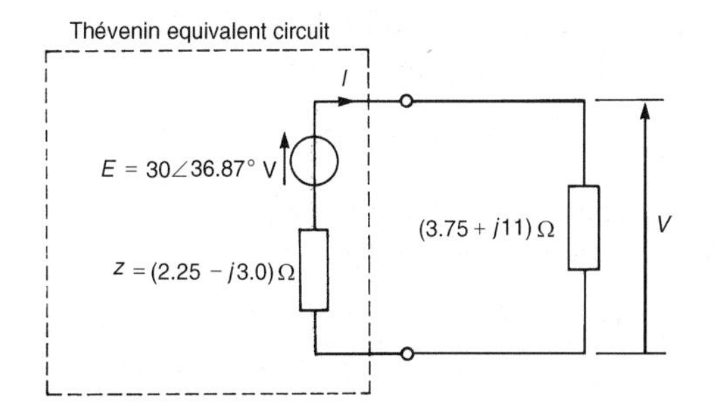

Figure 33.28

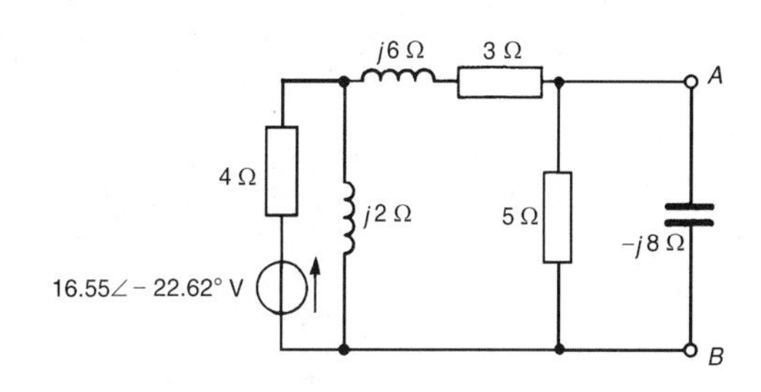

Figure 33.29

Hence the current flowing in the (3.75 + j11)Ω impedance is 3 A.

(b) P.d. across the $(3.75 + j11)\Omega$ impedance is given by

$$V = (3\angle -16.26°)(3.75 + j11) = 3\angle -16.26°)(11.62\angle 71.18°)$$

$$= 34.86\angle 54.92° \text{ V}$$

Hence the magnitude of the p.d. across the impedance is 34.9 V.

Problem 6. Use Thévenin's theorem to determine the current flowing in the capacitor of the network shown in Figure 33.29.

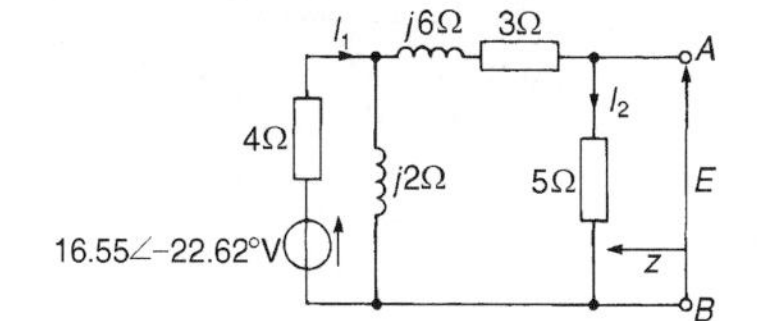

Figure 33.30

(i) The capacitor is removed from branch AB, as shown in Figure 33.30.

(ii) The open-circuit voltage, E, shown in Figure 33.30, is given by $(I_2)(5)$. I_2 may be determined by current division if I_1 is known. (Alternatively, E may be determined by the method used in problem 4.)

Current $I_1 = V/Z$, where Z is the total circuit impedance and $V = 16.55\angle -22.62°$ V.

$$\text{Impedance, } Z = 4 + \frac{(j2)(8 + j6)}{j2 + 8 + j6} = 4 + \frac{-12 + j16}{8 + j8}$$

$$= 4.596\angle 22.38° \ \Omega$$

$$\text{Hence} \quad I_1 = \frac{16.55\angle -22.62°}{4.596\angle 22.38°} = 3.60\angle -45° \text{ A}$$

$$\text{and} \quad I_2 = \left(\frac{j2}{j2 + 3 + j6 + 5}\right) I_1 = \frac{(2\angle 90°)(3.60\angle -45°)}{11.314\angle 45°}$$

$$= 0.636\angle 0° \text{ A}$$

(An alternative method of finding I_2 is to use Kirchhoff's laws or mesh-current or nodal analysis on Figure 33.30.)

Hence $\quad \boldsymbol{E} = (I_2)(5) = (0.636\angle 0^\circ)(5) = \mathbf{3.18\angle 0^\circ\ V}$

(iii) If the $16.55\angle -22.62^\circ$ V source is removed from Figure 33.30, the impedance, z, 'looking in' at AB is given by

$$z = \frac{5[((4\times j2)/(4+j2)) + (3+j6)]}{5+[((4\times j2)/(4+j2)) + 3 + j6]} = \frac{5(3.8+j7.6)}{8.8+j7.6}$$

i.e. $\quad z = 3.654\angle 22.61^\circ\ \Omega \quad$ or $\quad \mathbf{(3.373 + j1.405)\Omega}$

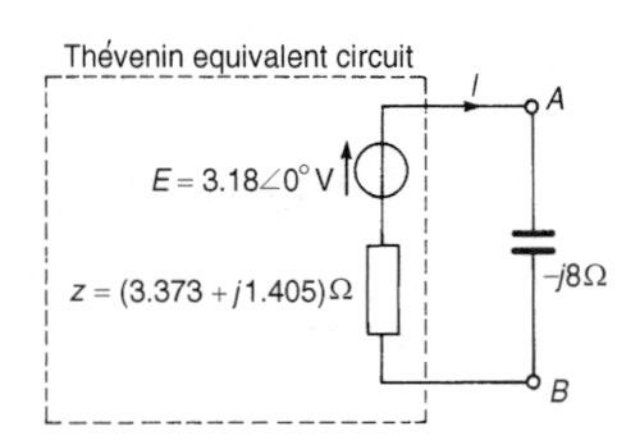

Figure 33.31

(iv) The Thévenin equivalent circuit is shown in Figure 33.31, where the current flowing in the capacitor, I, is given by

$$I = \frac{3.18\angle 0^\circ}{(3.373 + j1.405) - j8} = \frac{3.18\angle 0^\circ}{7.408\angle -62.91^\circ}$$

$= \mathbf{0.43\angle 62.91^\circ}$ **A in the direction from A to B**

Problem 7. For the network shown in Figure 33.32, derive the Thévenin equivalent circuit with respect to terminals PQ, and hence determine the power dissipated by a 2 Ω resistor connected across PQ.

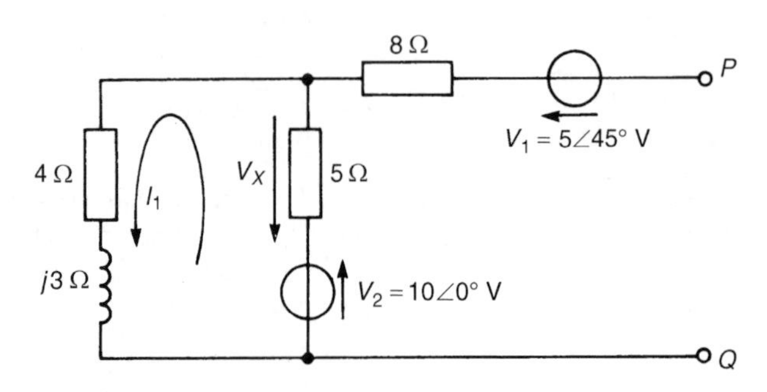

Figure 33.32

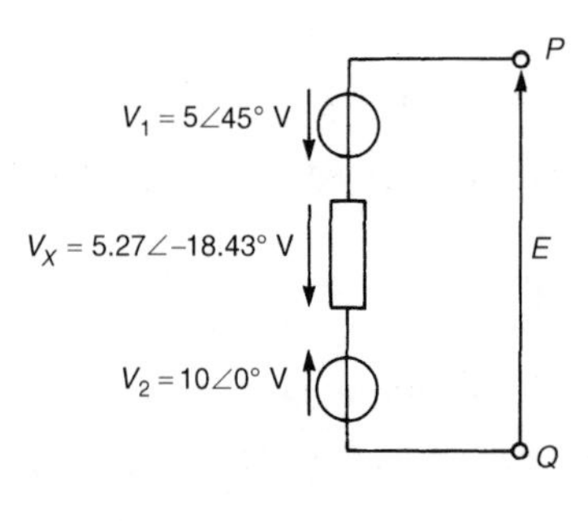

Figure 33.33

Current I_1 shown in Figure 33.32 is given by

$$I_1 = \frac{10\angle 0^\circ}{(5+4+j3)} = 1.054\angle -18.43^\circ\ \text{A}$$

Hence the voltage drop across the 5 Ω resistor is given by $V_X = (I_1)(5) = 5.27\angle -18.43^\circ$ V, and is in the direction shown in Figure 33.32, i.e., the direction opposite to that in which I_1 is flowing.

The open-circuit voltage E across PQ is the phasor sum of V_1, V_x and V_2, as shown in Figure 33.33.

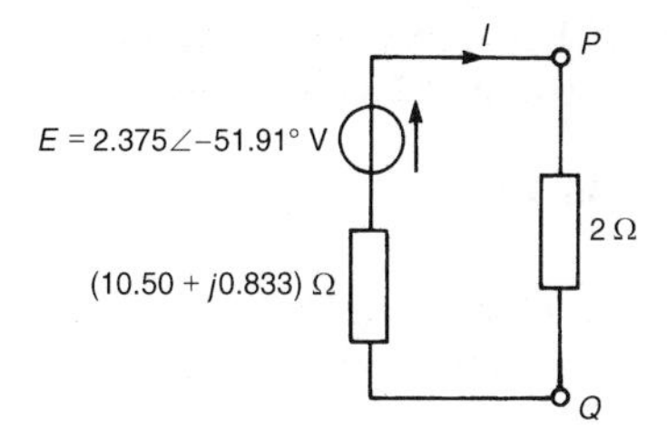

Figure 33.34

Thus $\boldsymbol{E} = 10\angle 0^\circ - 5\angle 45^\circ - 5.27\angle{-18.43^\circ}$

$= (1.465 - j1.869)\text{V}$ or $\mathbf{2.375\angle{-51.91^\circ}\ V}$

The impedance, z, 'looking in' at terminals PQ with the voltage sources removed is given by

$$z = 8 + \frac{5(4+j3)}{(5+4+j3)} = 8 + 2.635\angle 18.44^\circ = \mathbf{(10.50 + j0.833)\Omega}$$

The Thévenin equivalent circuit is shown in Figure 33.34 with the 2 Ω resistance connected across terminals PQ.
The current flowing in the 2 Ω resistance is given by

$$I = \frac{2.375\angle{-51.91^\circ}}{(10.50 + j0.833) + 2} = 0.1896\angle{-55.72^\circ}\ \text{A}$$

The power P dissipated in the 2 Ω resistor is given by

$P = I^2R = (0.1896)^2(2) = \mathbf{0.0719\ W \equiv 72\ mW}$, correct to two significant figures.

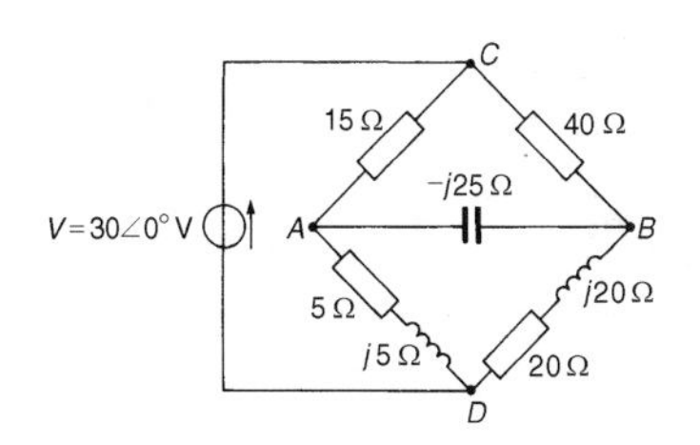

Figure 33.35

Problem 8. For the a.c. bridge network shown in Figure 33.35, determine the current flowing in the capacitor, and its direction, by using Thévenin's theorem. Assume the 30∠0° V source to have negligible internal impedance.

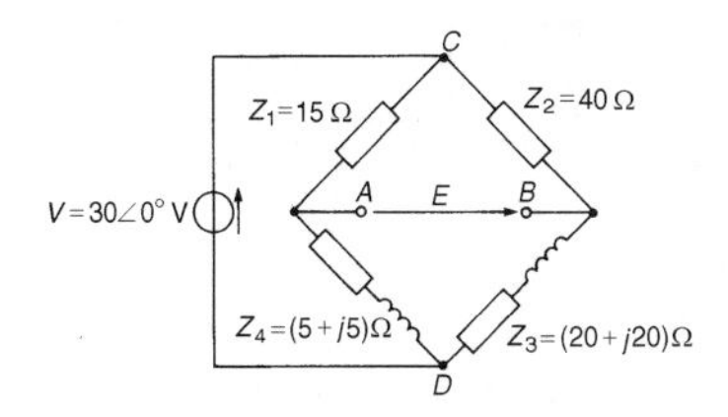

Figure 33.36

(i) The $-j25$ Ω capacitor is initially removed from the network, as shown in Figure 33.36.

(ii) P.d. between A and C,

$$V_{AC} = \left(\frac{Z_1}{Z_1 + Z_4}\right)V = \left(\frac{15}{15+5+j5}\right)(30\angle 0^\circ)$$

$$= 21.83\angle{-14.04^\circ}\ \text{V}$$

P.d. between B and C,

$$V_{BC} = \left(\frac{Z_2}{Z_2 + Z_3}\right)V = \left(\frac{40}{40+20+j20}\right)(30\angle 0^\circ)$$

$$= 18.97\angle{-18.43^\circ}\ \text{V}$$

Assuming that point A is at a higher potential than point B, then the p.d. between A and B is

$21.83\angle - 14.04^\circ - 18.97\angle{-18.43^\circ}$

$= (3.181 + j0.701)\text{V}$ or $3.257\angle 12.43^\circ$ V,

i.e., the open-circuit voltage across AB is given by

$E = 3.257\angle 12.43^\circ$ V.

Point C is at a potential of $30\angle 0°$ V. Between C and A is a volt drop of $21.83\angle -14.04°$ V. Hence the **voltage at point A is**

$$30\angle 0° - 21.83\angle -14.04° = \mathbf{10.29\angle 30.98°\ V}$$

Between points C and B is a voltage drop of $18.97\angle -18.43°$ V. Hence the **voltage at point B** is $30\angle 0° - 18.97\angle -18.43° =$ **$13.42\angle 26.55°$ V**.

Since the magnitude of the voltage at B is higher than at A, current must flow in the direction B to A.

(iii) Replacing the $30\angle 0°$ V source with a short-circuit (i.e., zero internal impedance) gives the network shown in Figure 33.37(a). The network is shown redrawn in Figure 33.37(b) and simplified in Figure 33.37(c). Hence the impedance, z, 'looking in' at terminals AB is given by

$$z = \frac{(15)(5 + j5)}{(15 + 5 + j5)} + \frac{(40)(20 + j20)}{(40 + 20 + j20)}$$

$$= 5.145\angle 30.96° + 17.889\angle 26.57°$$

i.e., $z = (20.41 + j10.65)\Omega$

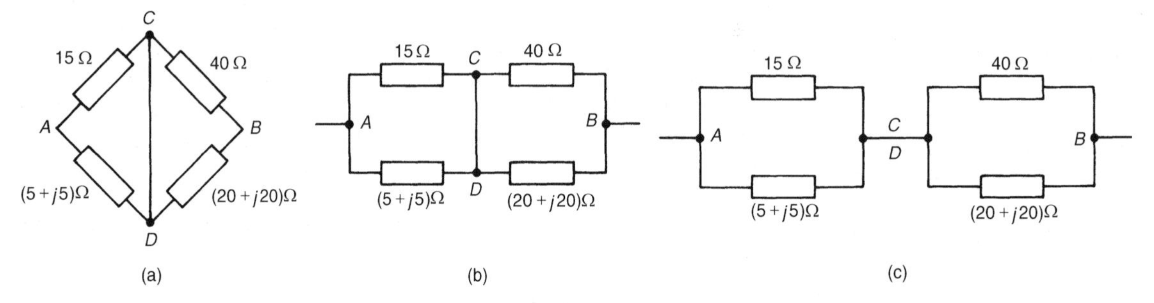

Figure 33.37

(iv) The Thévenin equivalent circuit is shown in Figure 33.38, where current I is given by

$$I = \frac{3.257\angle 12.43°}{(20.41 + j10.65) - j25} = \frac{3.257\angle 12.43°}{24.95\angle -35.11°}$$

$$= 0.131\angle 47.54°\ \text{A}$$

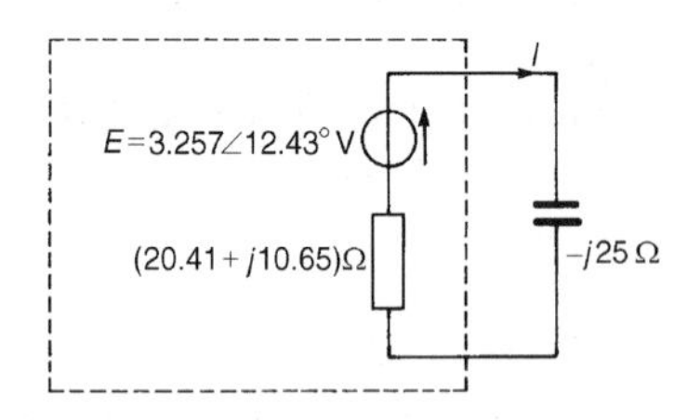

Figure 33.38

Thus a current of 131 mA flows in the capacitor in a direction from B to A.

Further problems on Thévenin's theorem may be found in Section 33.5, problems 1 to 10, page 598.

33.3 Norton's theorem

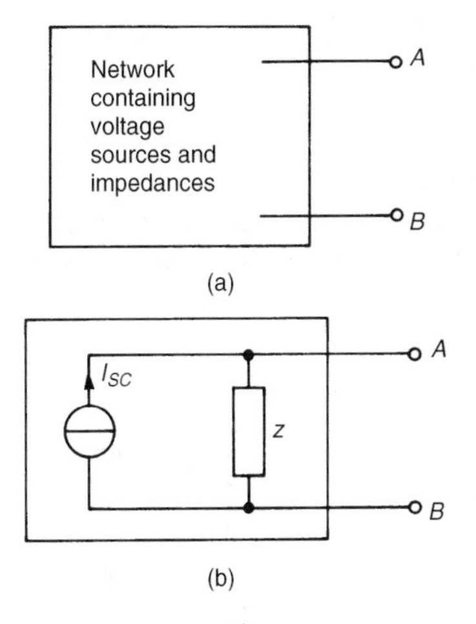

Figure 33.39 *The Norton equivalent circuit*

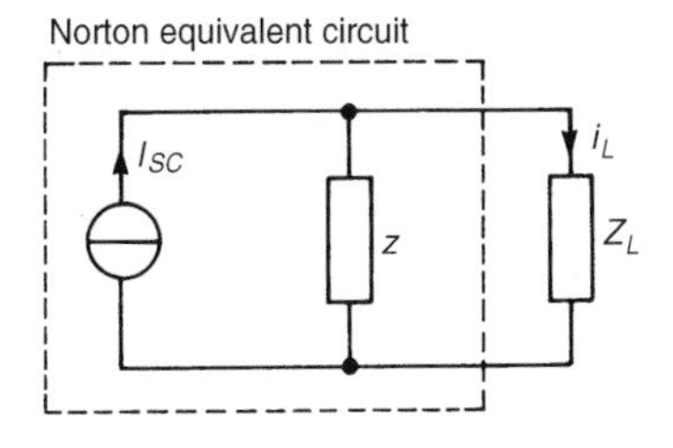

Figure 33.40

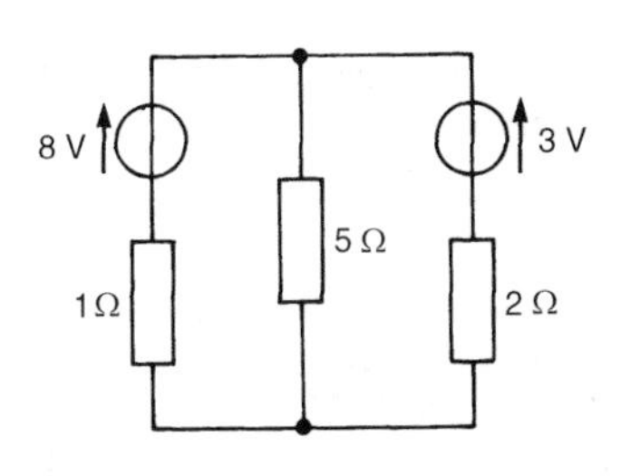

Figure 33.41

A source of electrical energy can be represented by a source of e.m.f. in series with an impedance. In Section 33.2, the Thévenin constant-voltage source consisted of a constant e.m.f. E, which may be alternating or direct, in series with an internal impedance, z. However, this is not the only form of representation. A source of electrical energy can also be represented by a constant-current source, which may be alternating or direct, in parallel with an impedance. It is shown in Section 33.4 that the two forms are in fact equivalent.

Norton's theorem states:

'The current that flows in any branch of a network is the same as that which would flow in the branch if it were connected across a source of electrical energy, the short-circuit current of which is equal to the current that would flow in a short-circuit across the branch, and the internal impedance of which is equal to the impedance which appears across the open-circuited branch terminals.'

The above statement simply means that any linear active network with output terminals AB, as shown in Figure 33.39(a), can be replaced by a current source in parallel with an impedance z as shown in Figure 33.39(b). The equivalent current source I_{SC} (note the symbol in Figure 33.39(b) as per BS 3939:1985) is the current through a short-circuit applied to the terminals of the network. The impedance z is the equivalent impedance of the network at the terminals AB when all internal sources of e.m.f. are made zero. Figure 33.39(b) is known as the **Norton equivalent circuit**, and was initially introduced in Section 13.7, page 181 for d.c. networks.

The following four-step procedure may be adopted when determining the current flowing in an impedance Z_L of a branch AB of an active network, using Norton's theorem:

(i) short-circuit branch AB;

(ii) determine the short-circuit current I_{SC};

(iii) remove each source of e.m.f. and replace it by its internal impedance (or, if a current source exists, replace with an open circuit), then determine the impedance, z, 'looking in' at a break made between A and B;

(iv) determine the value of the current i_L flowing in impedance Z_L from the Norton equivalent network shown in Figure 33.40, i.e.,

$$i_L = \left(\frac{z}{Z_L + z}\right) I_{SC}$$

A simple d.c. network (Figure 33.41) serves to demonstrate how the above procedure is applied to determine the current flowing in the 5 Ω resistance by using Norton's theorem:

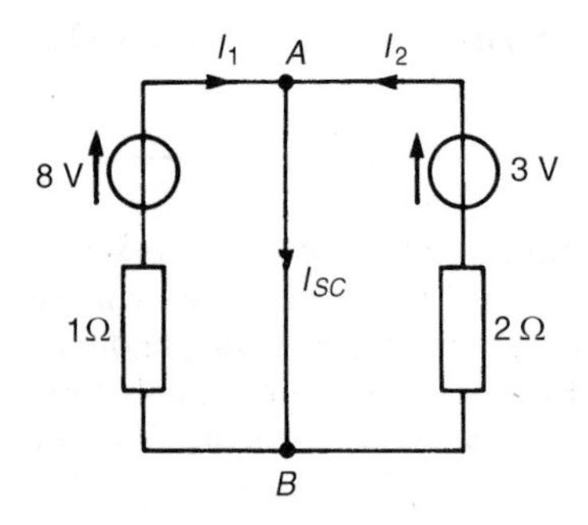

Figure 33.42

(i) The 5 Ω branch is short-circuited, as shown in Figure 33.42.

(ii) From Figure 33.42, $I_{SC} = I_1 + I_2 = \frac{8}{1} + \frac{3}{2} = 9.5$ A

(iii) If each source of e.m.f. is removed the impedance 'looking in' at a break made between A and B is given by $z = (1 \times 2)/(1 + 2) = \frac{2}{3}\ \Omega$.

(iv) From the Norton equivalent network shown in Figure 33.43, the current in the 5 Ω resistance is given by $I_L = \left(\frac{2}{3}\Big/\left(5 + \frac{2}{3}\right)\right) 9.5 =$ **1.12 A**, as obtained previously using Kirchhoff's laws, the superposition theorem and by Thévenin's theorem.

As with Thévenin's theorem, Norton's theorem may be used with a.c. as well as d.c. networks, as shown below.

An a.c. network is shown in Figure 33.44 where it is required to find the current flowing in the $(6 + j8)\Omega$ impedance by using Norton's theorem. Using the above procedure:

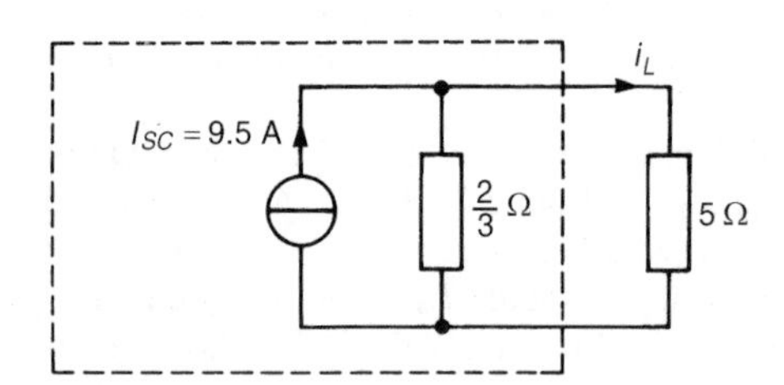

Figure 33.43

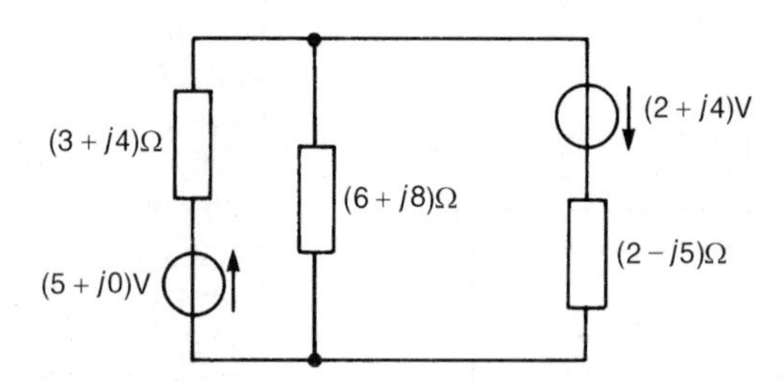

Figure 33.44

(i) Initially the $(6 + j8)\Omega$ impedance is short-circuited, as shown in Figure 33.45.

(ii) From Figure 33.45,

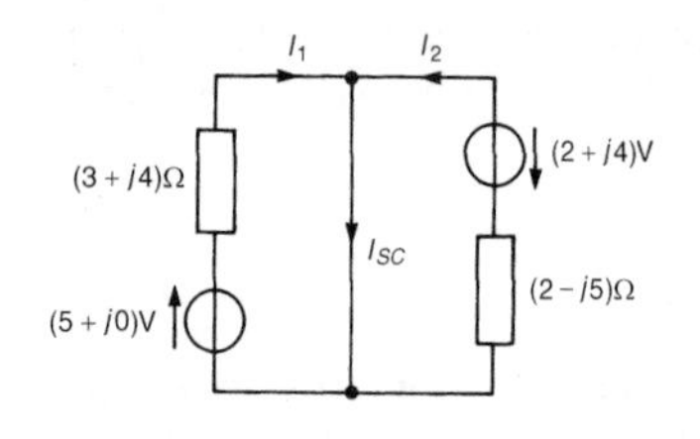

Figure 33.45

$$\boldsymbol{I_{SC}} = I_1 + I_2 = \frac{(5 + j0)}{(3 + j4)} + \frac{(-(2 + j4))}{(2 - j5)}$$

$$= 1\angle -53.13^\circ - \frac{4.472\angle 63.43^\circ}{5.385\angle -68.20^\circ}$$

$$= (1.152 - j1.421)\text{A or } \mathbf{1.829\angle -50.97^\circ\ A}$$

(iii) If each source of e.m.f. is removed, the impedance, z, 'looking in' at a break made between A and B is given by

$$z = \frac{(3 + j4)(2 - j5)}{(3 + j4) + (2 - j5)}$$

$$= \mathbf{5.28\angle -3.76^\circ\ \Omega} \quad \text{or} \quad \mathbf{(5.269 - j0.346)\Omega}$$

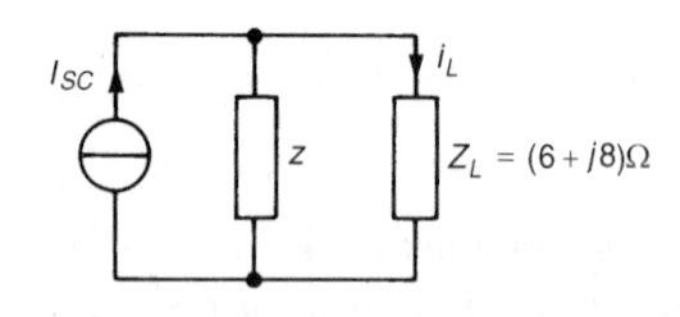

Figure 33.46

(iv) From the Norton equivalent network shown in Figure 33.46, the current is given by

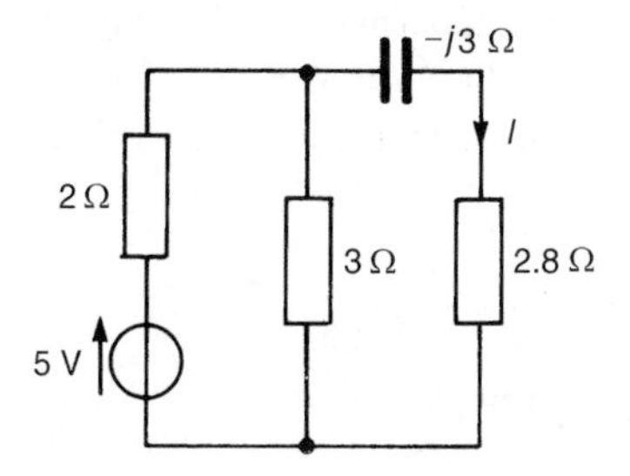

Figure 33.47

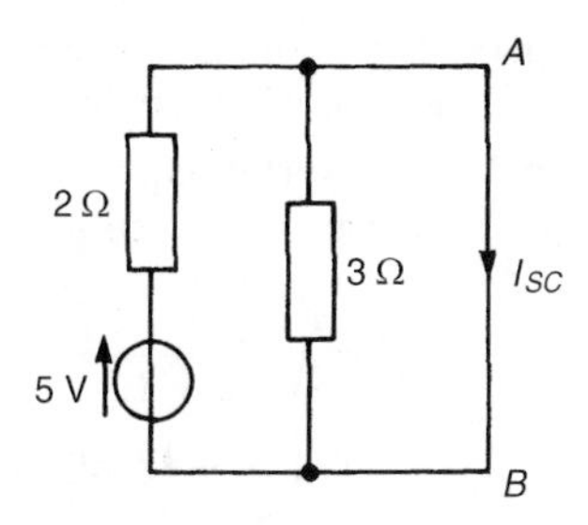

Figure 33.48

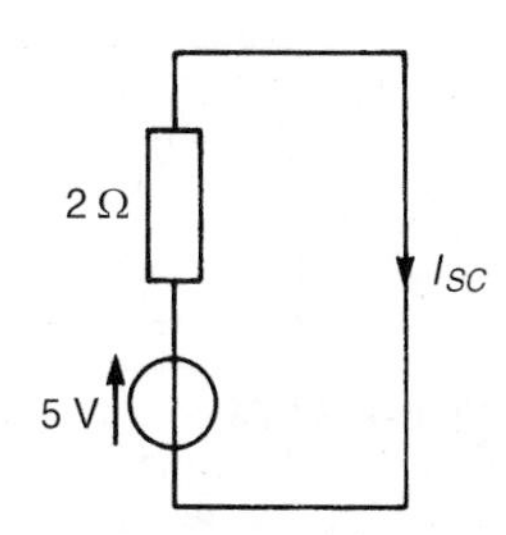

Figure 33.49

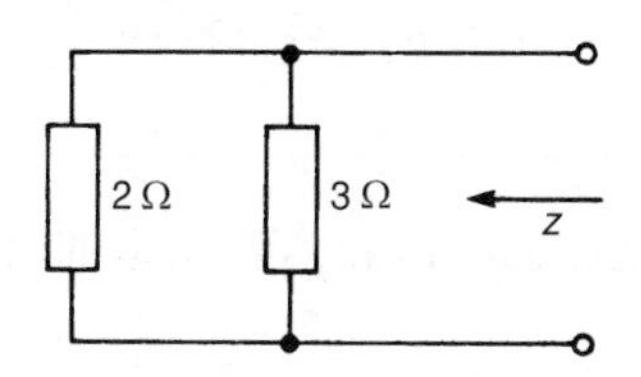

Figure 33.50

$$i_L = \left(\frac{z}{Z_L + z}\right) I_{SC}$$

$$= \left(\frac{5.28\angle -3.76^\circ}{(6 + j8) + (5.269 - j0.346)}\right) 1.829\angle -50.97^\circ$$

i.e., **current in $(6 + j8)\Omega$ impedance, $i_L = 0.71\angle -88.91^\circ$ A**

Problem 9. Use Norton's theorem to determine the value of current I in the circuit shown in Figure 33.47.

(i) The branch containing the 2.8 Ω resistor is short-circuited, as shown in Figure 33.48.

(ii) The 3 Ω resistor in parallel with a short-circuit is the same as 3 Ω in parallel with 0 giving an equivalent impedance of $(3 \times 0)/(3 + 0) = 0$. Hence the network reduces to that shown in Figure 33.49, where $I_{SC} = 5/2 = \mathbf{2.5\ A}$.

(iii) If the 5 V source is removed from the network the input impedance, z, 'looking-in' at a break made in AB of Figure 33.48 gives $z = (2 \times 3)/(2 + 3) = \mathbf{1.2\ \Omega}$ (see Figure 33.50).

(iv) The Norton equivalent network is shown in Figure 33.51, where current I is given by

$$I = \left(\frac{1.2}{1.2 + (2.8 - j3)}\right)(2.5) = \frac{3}{4 - j3} = \mathbf{0.60\angle 36.87^\circ\ A}$$

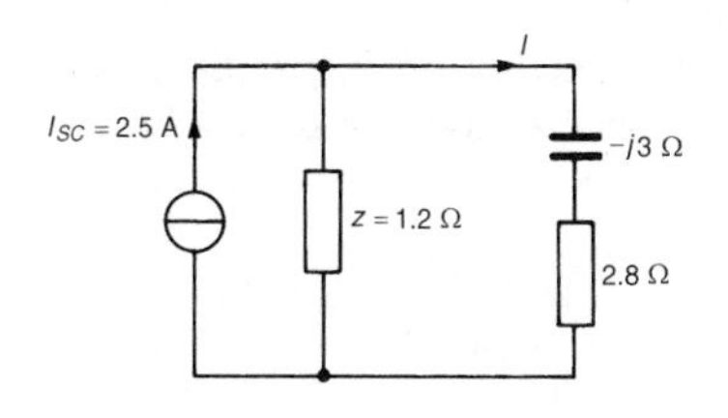

Figure 33.51

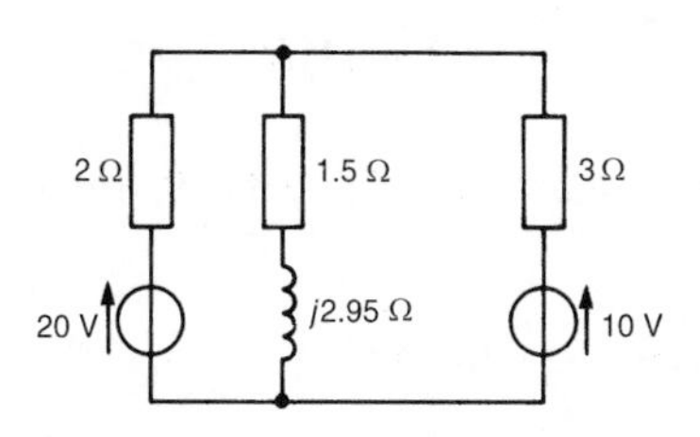

Figure 33.52

Problem 10. For the circuit shown in Figure 33.52 determine the current flowing in the inductive branch by using Norton's theorem.

(i) The inductive branch is initially short-circuited, as shown in Figure 33.53.

(ii) From Figure 33.53,

$$I_{SC} = I_1 + I_2 = \frac{20}{2} + \frac{10}{3} = \mathbf{13.\dot{3}\ A}$$

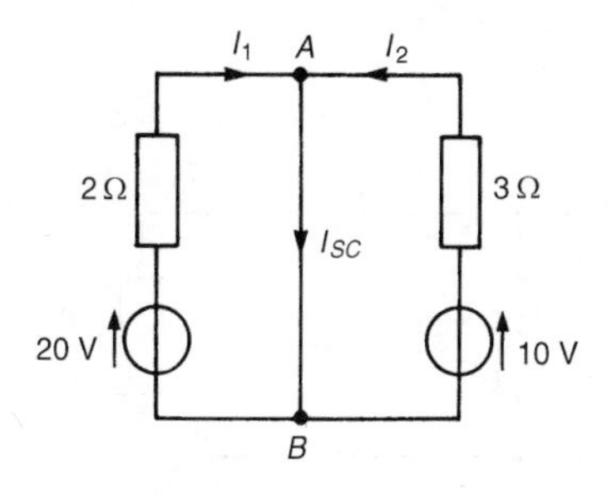

Figure 33.53

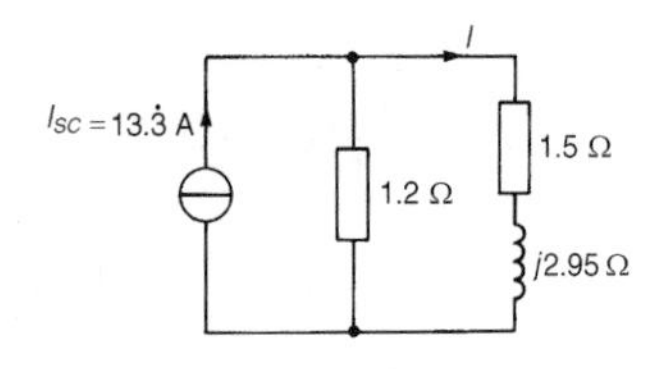

Figure 33.54

(iii) If the voltage sources are removed, the impedance, z, 'looking in' at a break made in AB is given by $z = (2 \times 3)/(2+3) = \mathbf{1.2\ \Omega}$.

(iv) The Norton equivalent network is shown in Figure 33.54, where current I is given by

$$I = \left(\frac{1.2}{1.2 + 1.5 + j2.95}\right)(13.\dot{3}) = \frac{16}{2.7 + j2.95}$$

$$= \mathbf{4.0\angle{-47.53^\circ}\ A} \quad \text{or} \quad \mathbf{(2.7 - j2.95)A}$$

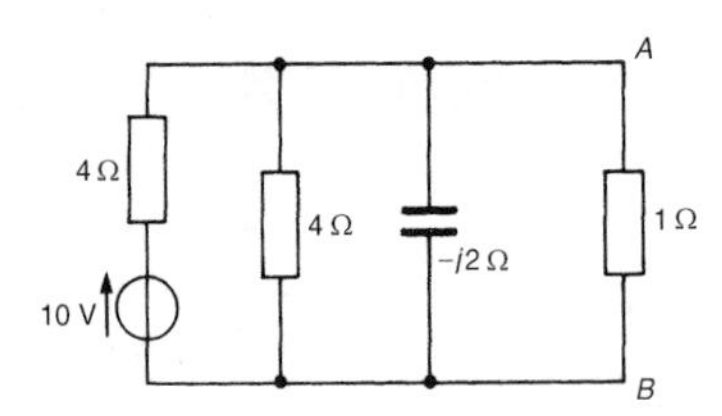

Figure 33.55

Problem 11. Use Norton's theorem to determine the magnitude of the p.d. across the 1 Ω resistance of the network shown in Figure 33.55.

(i) The branch containing the 1 Ω resistance is initially short-circuited, as shown in Figure 33.56.

(ii) 4 Ω in parallel with $-j2$ Ω in parallel with 0 Ω (i.e., the short-circuit) is equivalent to 0, giving the equivalent circuit of Figure 33.57. Hence $\boldsymbol{I_{SC}} = 10/4 = \mathbf{2.5\ A}$.

(iii) The 10 V source is removed from the network of Figure 33.55, as shown in Figure 33.58, and the impedance z, 'looking in' at a break made in AB is given by

$$\frac{1}{z} = \frac{1}{4} + \frac{1}{4} + \frac{1}{-j2} = \frac{-j - j + 2}{-j4} = \frac{2 - j2}{-j4}$$

from which

$$z = \frac{-j4}{2 - j2} = \frac{-j4(2 + j2)}{2^2 + 2^2} = \frac{8 - j8}{8} = \mathbf{(1 - j1)\Omega}$$

(iv) The Norton equivalent network is shown in Figure 33.59, from which current I is given by

$$I = \left(\frac{1 - j1}{(1 - j1) + 1}\right)(2.5) = 1.58\angle{-18.43^\circ}\ \text{A}$$

Hence the magnitude of the p.d. across the 1 Ω resistor is given by

$$IR = (1.58)(1) = \mathbf{1.58\ V}.$$

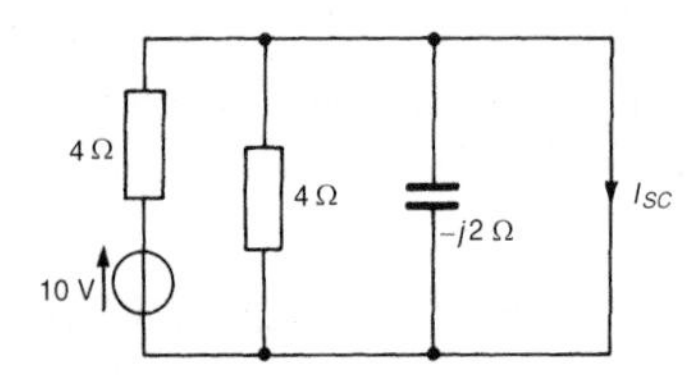

Figure 33.56

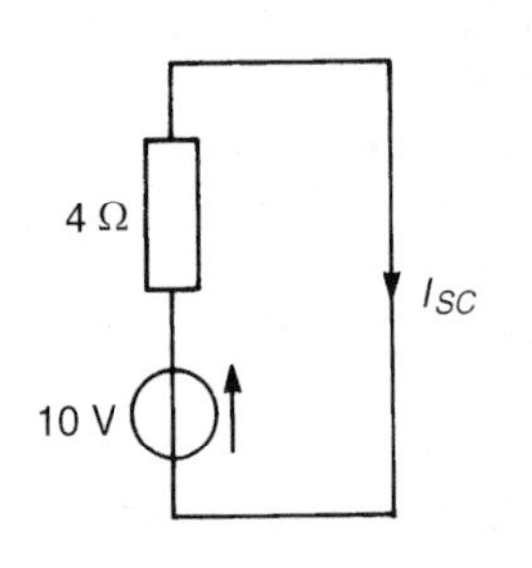

Figure 33.57

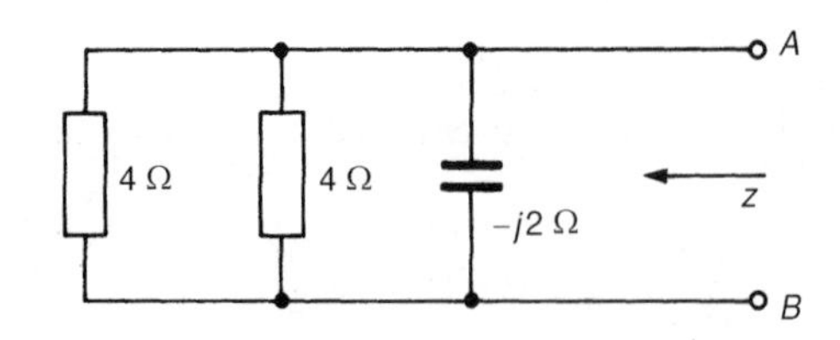

Figure 33.58

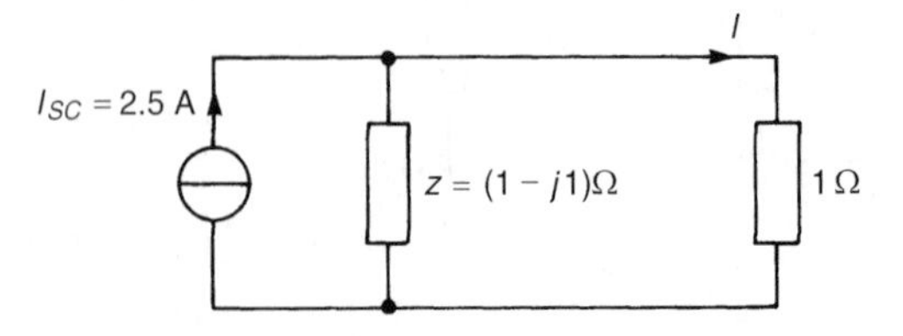

Figure 33.59

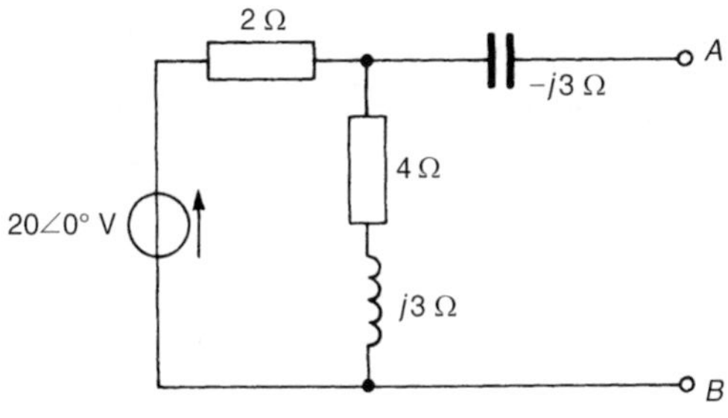

Figure 33.60

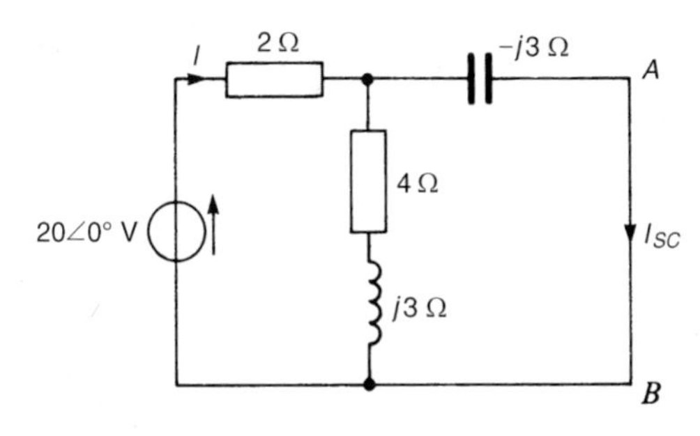

Figure 33.61

Problem 12. For the network shown in Figure 33.60, obtain the Norton equivalent network at terminals AB. Hence determine the power dissipated in a 5 Ω resistor connected between A and B.

(i) Terminals AB are initially short-circuited, as shown in Figure 33.61.

(ii) The circuit impedance Z presented to the $20\angle 0°$ V source is given by

$$Z = 2 + \frac{(4 + j3)(-j3)}{(4 + j3) + (-j3)} = 2 + \frac{9 - j12}{4}$$
$$= (4.25 - j3)\Omega \quad \text{or} \quad 5.202\angle -35.22° \ \Omega$$

Thus current I in Figure 33.61 is given by

$$I = \frac{20\angle 0°}{5.202\angle -35.22°} = 3.845\angle 35.22° \text{ A}$$

Hence

$$I_{SC} = \left(\frac{(4 + j3)}{(4 + j3) - j3}\right)(3.845\angle 35.22°)$$
$$= \mathbf{4.806\angle 72.09° \ A}$$

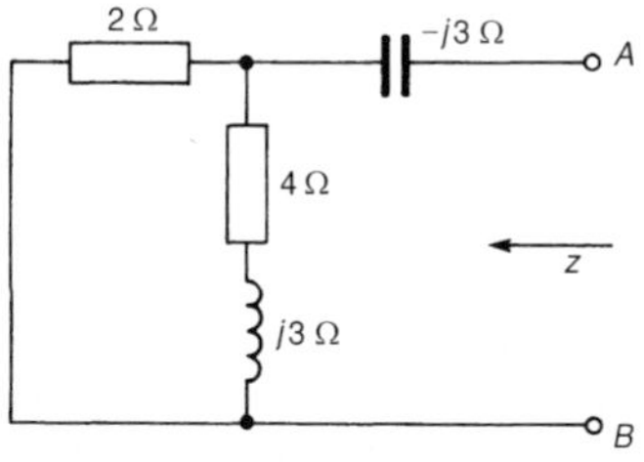

Figure 33.62

(iii) Removing the $20\angle 0°$ V source of Figure 33.60 gives the network of Figure 33.62.

Impedance, z, 'looking in' at terminals AB is given by

$$z = -j3 + \frac{2(4 + j3)}{2 + 4 + j3} = -j3 + 1.491\angle 10.3°$$
$$= \mathbf{(1.467 - j2.733)\Omega} \quad \text{or} \quad \mathbf{3.102\angle -61.77°\Omega}$$

(iv) The Norton equivalent network is shown in Figure 33.63.

$$\text{Current } I_L = \left(\frac{3.102\angle -61.77°}{1.467 - j2.733 + 5}\right)(4.806\angle 72.09°)$$
$$= 2.123\angle 33.23° \text{ A}$$

Norton equivalent circuit

$I_{SC} = 4.806\angle 72.09°$ A

$z = (1.467 - j2.733)\,\Omega$

I_L A B 5 Ω

Figure 33.63

Hence the power dissipated in the 5 Ω resistor is

$$I_L^2 R = (2.123)^2(5) = \mathbf{22.5\ W}$$

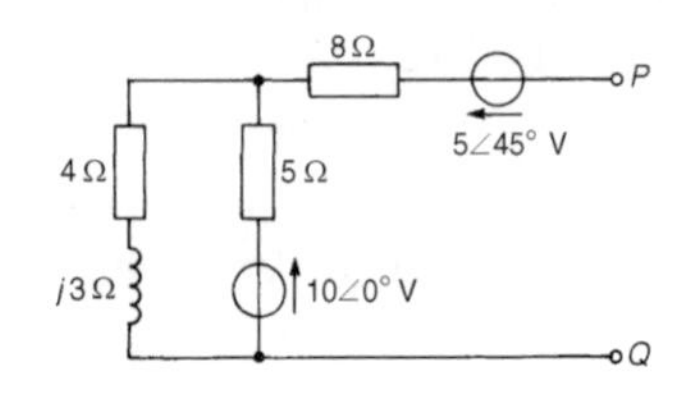

Figure 33.64

> Problem 13. Derive the Norton equivalent network with respect to terminals PQ for the network shown in Figure 33.64 and hence determine the magnitude of the current flowing in a 2 Ω resistor connected across PQ.

This is the same problem as problem 7 on page 584 which was solved by Thévenin's theorem.

A comparison of methods may thus be made.

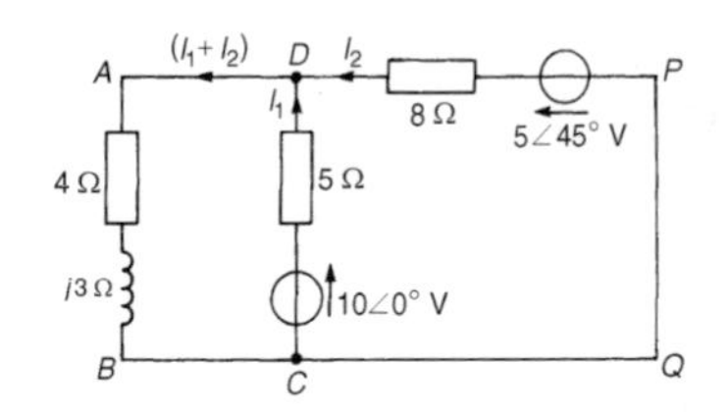

Figure 33.65

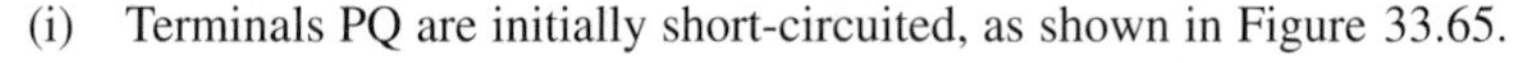

(i) Terminals PQ are initially short-circuited, as shown in Figure 33.65.

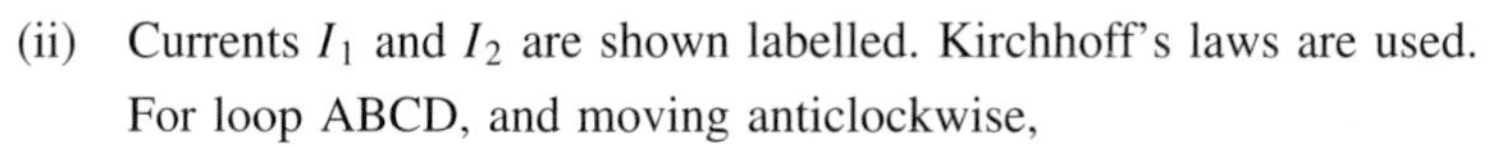

(ii) Currents I_1 and I_2 are shown labelled. Kirchhoff's laws are used. For loop ABCD, and moving anticlockwise,

$$10\angle 0° = 5I_1 + (4 + j3)(I_1 + I_2),$$

i.e., $$(9 + j3)I_1 + (4 + j3)I_2 - 10 = 0 \qquad (1)$$

For loop DPQC, and moving clockwise,

$$10\angle 0° - 5\angle 45° = 5I_1 - 8I_2,$$

i.e., $$5I_1 - 8I_2 + (5\angle 45° - 10) = 0 \qquad (2)$$

Solving Equations (1) and (2) by using determinants gives

$$\frac{I_1}{\begin{vmatrix} (4 + j3) & -10 \\ -8 & (5\angle 45° - 10) \end{vmatrix}} = \frac{-I_2}{\begin{vmatrix} (9 + j3) & -10 \\ 5 & (5\angle 45° - 10) \end{vmatrix}}$$

$$= \frac{I}{\begin{vmatrix} (9 + j3) & (4 + j3) \\ 5 & -8 \end{vmatrix}}$$

from which

$$I_2 = \frac{-\begin{vmatrix} (9+j3) & -10 \\ 5 & (5\angle 45^\circ - 10) \end{vmatrix}}{\begin{vmatrix} (9+j3) & (4+j3) \\ 5 & -8 \end{vmatrix}} = \frac{-[(9+j3)(5\angle 45^\circ - 10) + 50]}{[-72 - j24 - 20 - j15]}$$

$$= \frac{-[22.52\angle 146.50^\circ]}{[99.925\angle -157.03^\circ]} = -0.225\angle 303.53^\circ \text{ or } -0.225\angle -56.47^\circ$$

Hence the short-circuit current $I_{SC} = 0.225\angle -56.47^\circ$ A flowing from P to Q.

(iii) The impedance, z, 'looking in' at a break made between P and Q is given by

$z = (10.50 + j0.833)\Omega$ (see problem 7, page 584).

(iv) The Norton equivalent circuit is shown in Figure 33.66, where current I is given by

$$I = \left(\frac{10.50 + j0.833}{10.50 + j0.833 + 2}\right)(0.225\angle -56.47^\circ)$$

$$= 0.19\angle -55.74^\circ \text{ A}$$

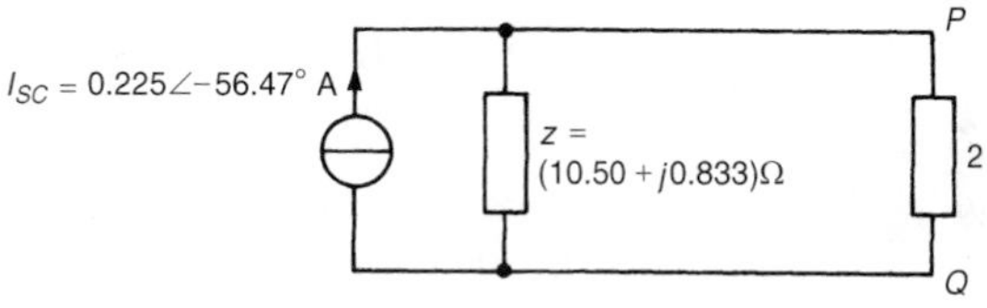

$I_{SC} = 0.225\angle -56.47^\circ$ A

$z = (10.50 + j0.833)\Omega$

2 Ω

P

Q

Figure 33.66

Hence the magnitude of the current flowing in the 2 Ω resistor is 0.19 A.

Further problems on Norton's theorem may be found in Section 33.5, problems 11 to 15, page 600

33.4 Thévenin and Norton equivalent networks

It is seen in Sections 33.2 and 33.3 that when Thévenin's and Norton's theorems are applied to the same circuit, identical results are obtained. Thus the Thévenin and Norton networks shown in Figure 33.67 are equivalent to each other. The impedance 'looking in' at terminals AB is the same in each of the networks; i.e., z.

If terminals AB in Figure 33.67(a) are short-circuited, the short-circuit current is given by E/z.

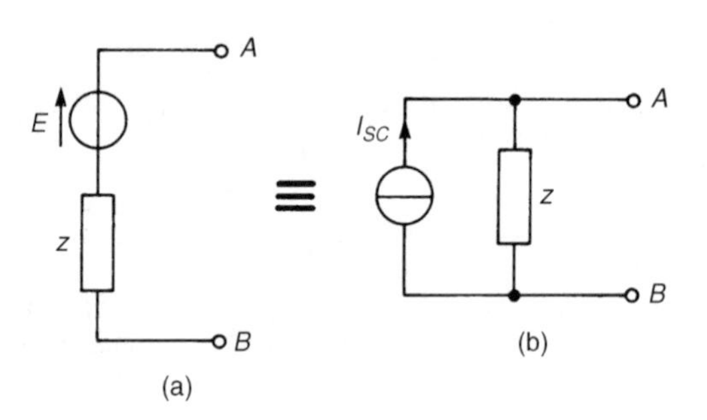

Figure 33.67 *Equivalent Thévenin and Norton circuits*

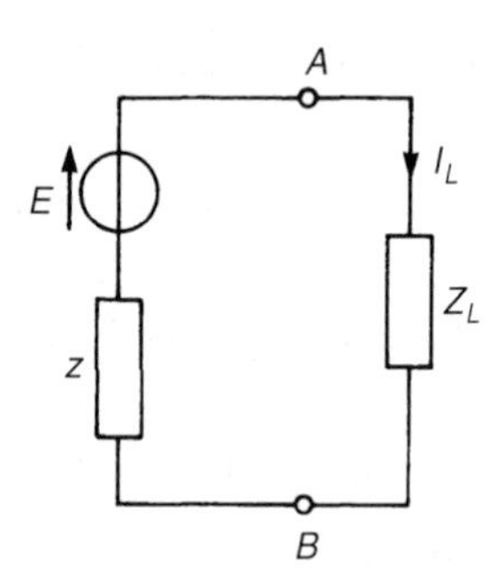

Figure 33.68

If terminals AB in Figure 33.67(b) are short-circuited, the short-circuit current is I_{SC}.

Thus $\quad \boldsymbol{I_{SC} = E/z}$.

Figure 33.68 shows a source of e.m.f. E in series with an impedance z feeding a load impedance Z_L. From Figure 33.68,

$$I_L = \frac{E}{z + Z_L} = \frac{E/z}{(z + Z_L)/z} = \left(\frac{z}{z + Z_L}\right)\frac{E}{z}$$

i.e., $\quad \boldsymbol{I_L = \left(\frac{z}{z + Z_L}\right) I_{SC}}$, from above.

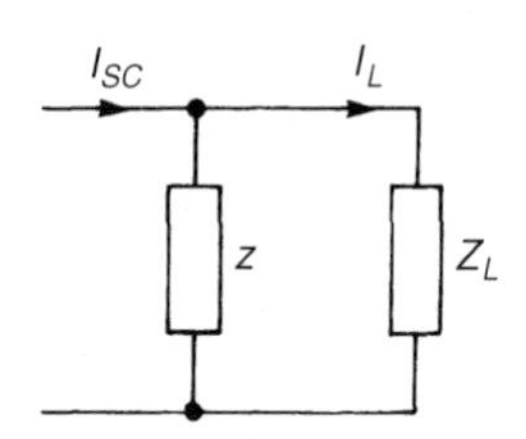

Figure 33.69

From Figure 33.69 it can be seen that, when viewed from the load, the source appears as a source of current I_{SC} which is divided between z and Z_L connected in parallel.

Thus it is shown that the two representations shown in Figure 33.67 are equivalent.

Problem 14. (a) Convert the circuit shown in Figure 33.70(a) to an equivalent Norton network. (b) Convert the network shown in Figure 33.70(b) to an equivalent Thévenin circuit.

(a) If the terminals AB of Figure 33.70(a) are short circuited, the short-circuit current, $I_{SC} = 20/4 = 5$ A. The impedance 'looking in' at terminals AB is 4 Ω. Hence the equivalent Norton network is as shown in Figure 33.71(a).

(b) The open-circuit voltage E across terminals AB in Figure 33.70(b) is given by $E = (I_{SC})(z) = (3)(2) = 6$ V. The impedance 'looking in' at terminals AB is 2 Ω.

Hence the equivalent Thévenin circuit is as shown in Figure 33.71(b).

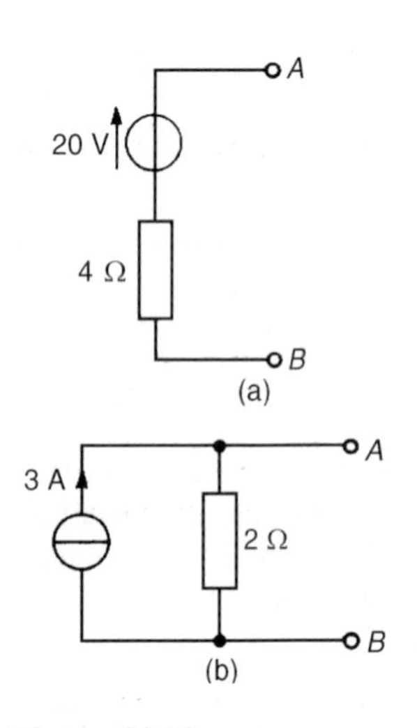

Figure 33.70

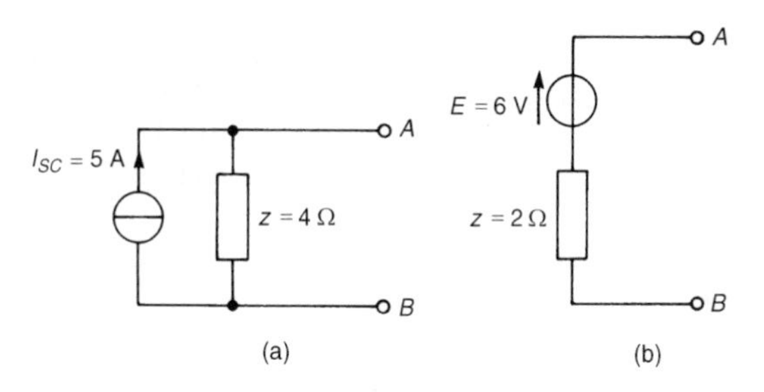

Figure 33.71

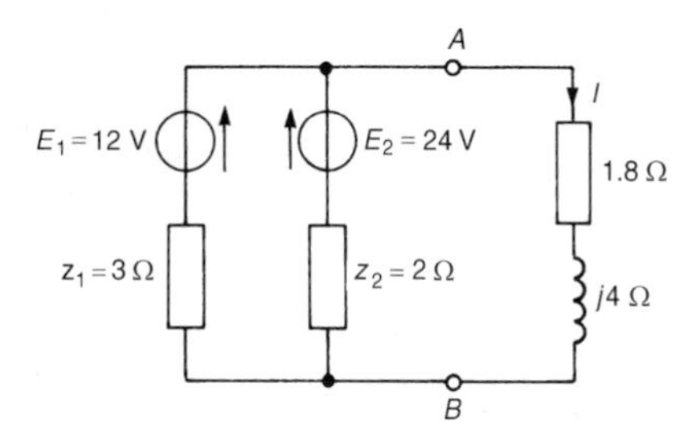

Figure 33.72

Problem 15. (a) Convert the circuit to the left of terminals AB in Figure 33.72 to an equivalent Thévenin circuit by initially converting to a Norton equivalent circuit. (b) Determine the magnitude of the current flowing in the $(1.8 + j4)\Omega$ impedance connected between terminals A and B of Figure 33.72.

(a) For the branch containing the 12 V source, conversion to a Norton equivalent network gives $I_{SC_1} = 12/3 = 4$ A and $z_1 = 3\ \Omega$. For the branch containing the 24 V source, conversion to a Norton equivalent circuit gives $I_{SC_2} = 24/2 = 12$ A and $z_2 = 2\ \Omega$.

Thus Figure 33.73 shows a network equivalent to Figure 33.72. From Figure 33.73, the total short-circuit current is $4 + 12 = 16$ A, and the total impedance is given by $(3 \times 2)/(3 + 2) = 1.2\ \Omega$. Thus Figure 33.73 simplifies to Figure 33.74.

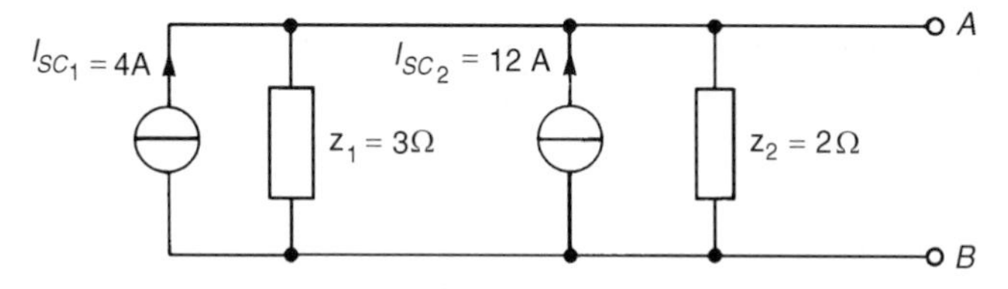

Figure 33.73

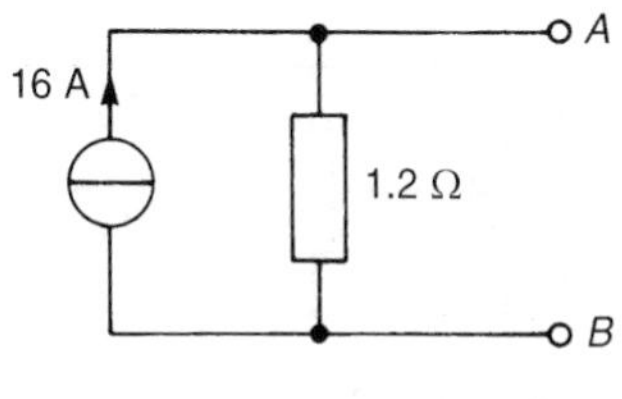

Figure 33.74

The open-circuit voltage across AB of Figure 33.74, $E = (16)(1.2) = 19.2$ V, and the impedance 'looking in' at AB, $z = 1.2\ \Omega$. Hence the Thévenin equivalent circuit is as shown in Figure 33.75.

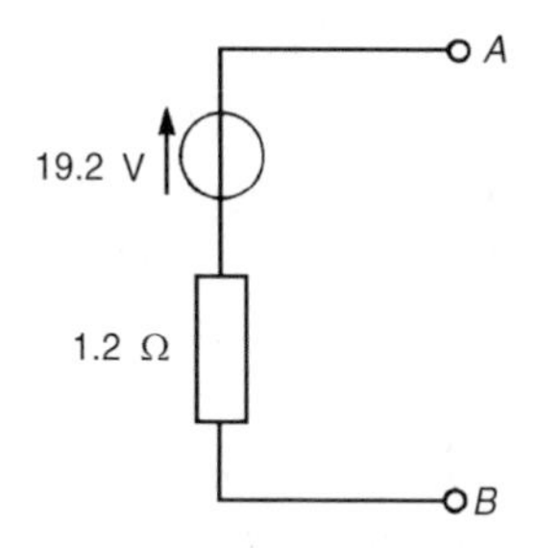

Figure 33.75

(b) When the $(1.8 + j4)\Omega$ impedance is connected to terminals AB of Figure 33.75, the current I flowing is given by

$$I = \frac{19.2}{(1.2 + 1.8 + j4)} = 3.84\angle -53.13^\circ \text{ A}$$

Hence the current flowing in the $(1.8+j4)\Omega$ impedance is 3.84 A.

Problem 16. Determine, by successive conversions between Thévenin's and Norton's equivalent networks, a Thévenin equivalent circuit for terminals AB of Figure 33.76. Hence determine the magnitude of the current flowing in the capacitive branch connected to terminals AB.

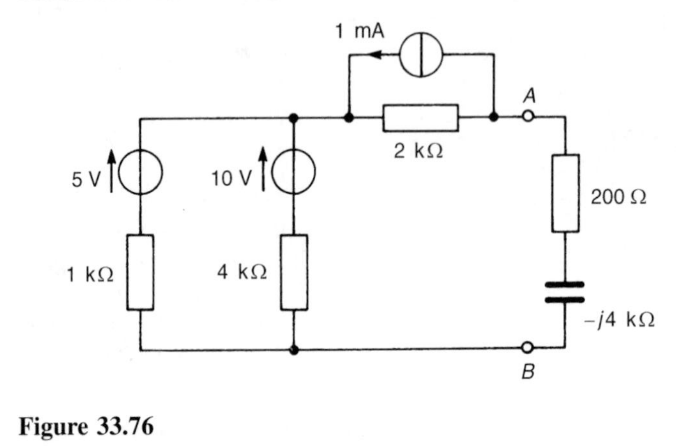

Figure 33.76

For the branch containing the 5 V source, converting to a Norton equivalent network gives $I_{SC} = 5/1000 = 5$ mA and $z = 1$ kΩ. For the branch containing the 10 V source, converting to a Norton equivalent network gives $I_{SC} = 10/4000 = 2.5$ mA and $z = 4$ kΩ. Thus the circuit of Figure 33.76 converts to that of Figure 33.77.

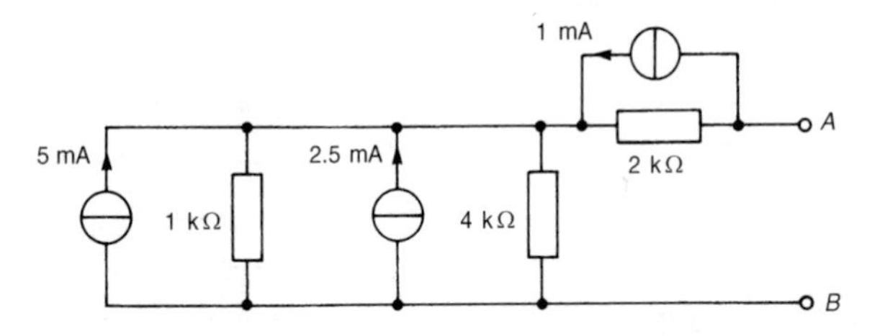

Figure 33.77

Figure 33.78

The above two Norton equivalent networks shown in Figure 33.77 may be combined, since the total short-circuit current is $(5 + 2.5) = 7.5$ mA and the total impedance z is given by $(1 \times 4)/(1 + 4) = 0.8$ kΩ. This results in the network of Figure 33.78.

Both of the Norton equivalent networks shown in Figure 33.78 may be converted to Thévenin equivalent circuits. Open-circuit voltage across CD is

$$(7.5 \times 10^{-3})(0.8 \times 10^{3}) = 6 \text{ V}$$

and the impedance 'looking in' at CD is 0.8 kΩ. Open-circuit voltage across EF is $(1 \times 10^{-3})(2 \times 10^{2}) = 2$ V and the impedance 'looking in' at EF is 2 kΩ. Thus Figure 33.78 converts to Figure 33.79.

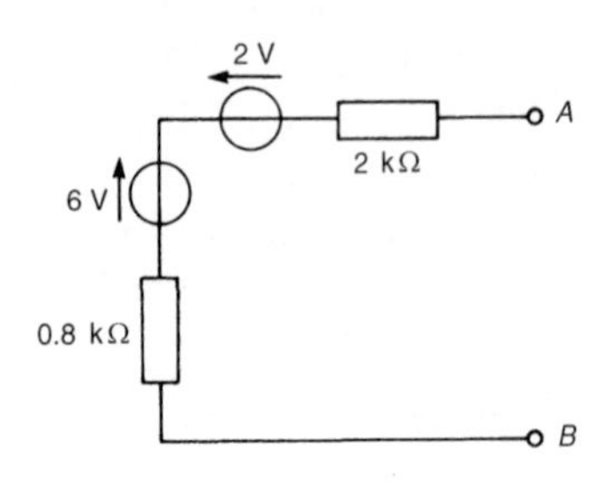

Figure 33.79

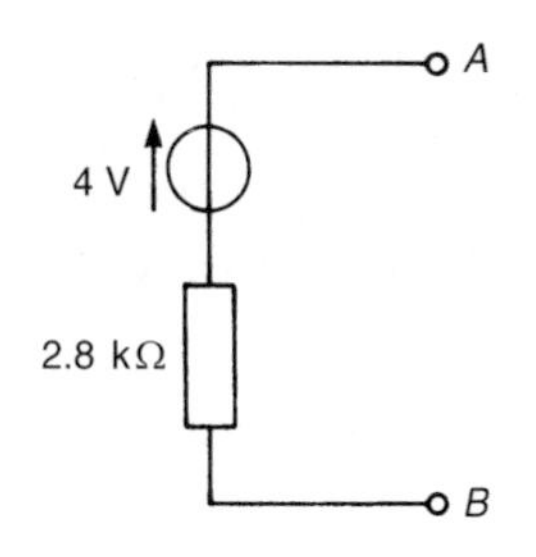

Figure 33.80

Combining the two Thévenin circuits gives e.m.f. $E = 6 - 2 = \mathbf{4\ V}$, and impedance $z = (0.8 + 2) = \mathbf{2.8\ k\Omega}$. Thus the Thévenin equivalent circuit for terminals AB of Figure 33.76 is as shown in Figure 33.80.

If an impedance $(200 - j4000)\Omega$ is connected across terminals AB, then the current I flowing is given by

$$I = \frac{4}{2800 + (200 - j4000)} = \frac{4}{5000\angle -53.13^\circ} = 0.80\angle 53.13^\circ \text{ mA}$$

i.e., **the current in the capacitive branch is 0.80 mA.**

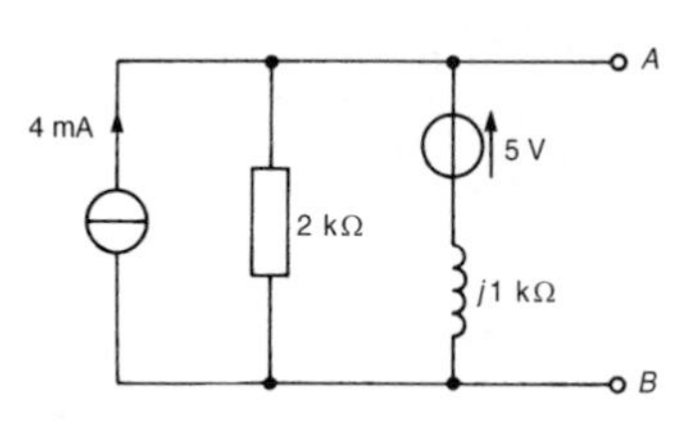

Figure 33.81

Problem 17. (a) Determine an equivalent Thévenin circuit for terminals AB of the network shown in Figure 33.81. (b) Calculate the power dissipated in a $(600 - j800)\Omega$ impedance connected between A and B of Figure 33.81.

(a) Converting the Thévenin circuit to a Norton network gives

$$I_{SC} = \frac{5}{j1000} = -j5 \text{ mA} \quad \text{or} \quad 5\angle -90^\circ \text{ mA and } z = j1 \text{ k}\Omega$$

Thus Figure 33.81 converts to that shown in Figure 33.82. The two Norton equivalent networks may be combined, giving

$$I_{SC} = 4 + 5\angle -90^\circ = (4 - j5)\text{mA or } 6.403\angle -51.34^\circ \text{ mA}$$

$$\text{and} \quad z = \frac{(2)(j1)}{(2 + j1)} = (0.4 + j0.8)\text{k}\Omega \quad \text{or} \quad 0.894\angle 63.43^\circ \text{ k}\Omega$$

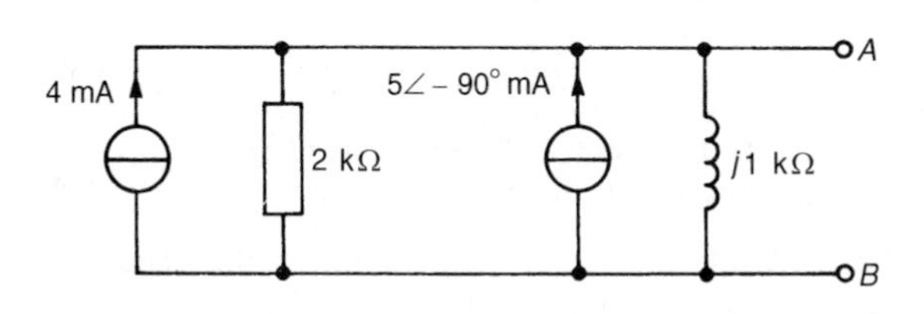

Figure 33.82

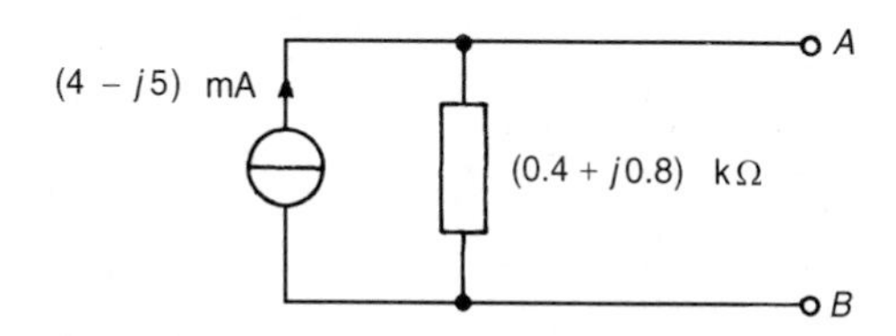

Figure 33.83

This results in the equivalent network shown in Figure 33.83. Converting to an equivalent Thévenin circuit gives open circuit e.m.f. across AB,

$$\begin{aligned} E &= (6.403 \times 10^{-3}\angle -51.34^\circ)(0.894 \times 10^3\angle 63.43^\circ) \\ &= \mathbf{5.724\angle 12.09^\circ\ V} \end{aligned}$$

and

$$\text{impedance } z = 0.894\angle 63.43^\circ \text{ k}\Omega \quad \text{or} \quad \mathbf{(400 + j800)\Omega}$$

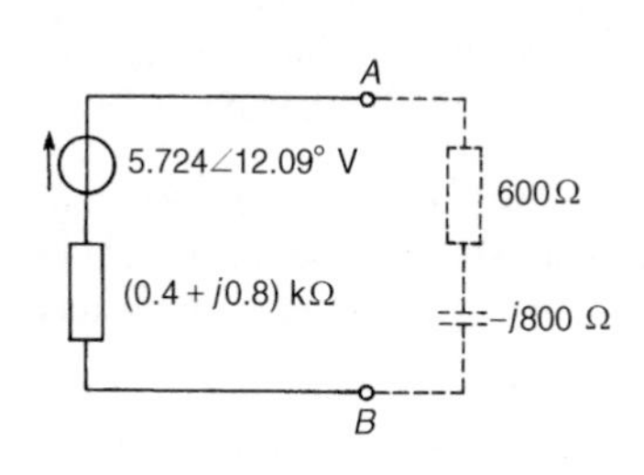

Figure 33.84

Thus the Thévenin equivalent circuit is as shown in Figure 33.84.

(b) When a $(600 - j800)\Omega$ impedance is connected across AB, the current I flowing is given by

$$I = \frac{5.724\angle 12.09^\circ}{(400 + j800) + (600 - j800)} = 5.724\angle 12.09^\circ \text{ mA}$$

Hence the power P dissipated in the $(600 - j800)\Omega$ impedance is given by

$$P = I^2 R = (5.724 \times 10^{-3})^2(600) = \mathbf{19.7\ mW}$$

Further problems on Thévenin and Norton equivalent networks may be found in Section 33.5 following, problems 16 to 21, page 600

33.5 Further problems on Thévenin's and Norton's theorem

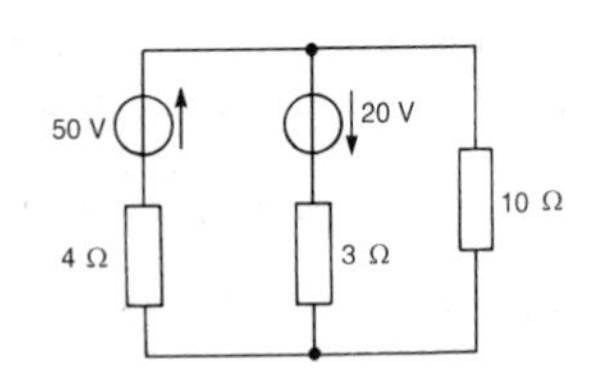

Figure 33.85

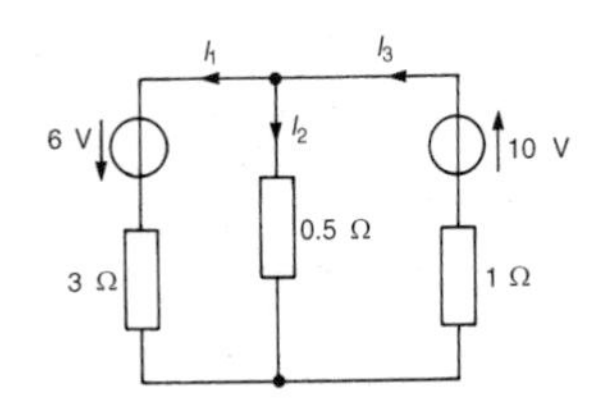

Figure 33.86

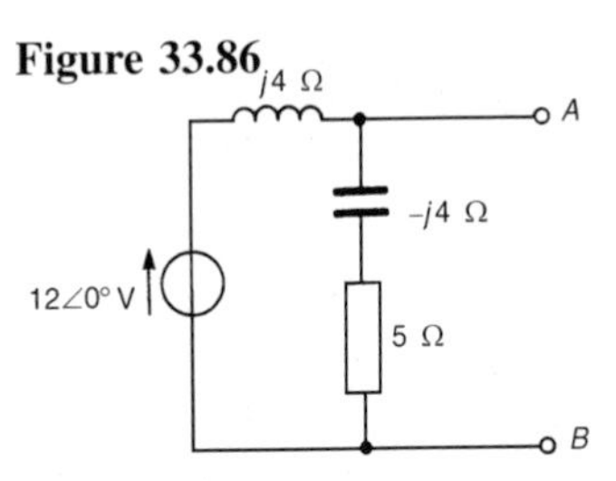

Figure 33.87

Thévenin's theorem

1 Use Thévenin's theorem to determine the current flowing in the 10 Ω resistor of the d.c. network shown in Figure 33.85. [0.85 A]

2 Determine, using Thévenin's theorem, the values of currents I_1, I_2 and I_3 of the network shown in Figure 33.86. [$I_1 = 2.8$ A, $I_2 = 4.8$ A, $I_3 = 7.6$ A]

3 Determine the Thévenin equivalent circuit with respect to terminals AB of the network shown in Figure 33.87. Hence determine the magnitude of the current flowing in a $(4 - j7)\Omega$ impedance connected across terminals AB and the power delivered to this impedance. [$E = 15.37\angle -38.66^\circ$, $z = (3.20 + j4.00)\Omega$; 1.97 A; 15.5 W]

4 For the network shown in Figure 33.88 use Thévenin's theorem to determine the current flowing in the 3 Ω resistance. [1.17 A]

5 Derive for the network shown in Figure 33.89 the Thévenin equivalent circuit at terminals AB, and hence determine the current flowing in a 20 Ω resistance connected between A and B. [$E = 2.5$ V, $r = 5$ Ω; 0.10 A]

6 Determine for the network shown in Figure 33.90 the Thévenin equivalent circuit with respect to terminals AB, and hence determine the current flowing in the $(5 + j6)\Omega$ impedance connected between A and B. [$E = 14.32\angle 6.38^\circ$, $z = (3.99 + j0.55)\Omega$; 1.29 A]

7 For the network shown in Figure 33.91, derive the Thévenin equivalent circuit with respect to terminals AB, and hence determine the magnitude of the current flowing in a $(2 + j13)\Omega$ impedance connected between A and B. [1.157 A]

8 Use Thévenin's theorem to determine the power dissipated in the 4 Ω resistance of the network shown in Figure 33.92. [0.24 W]

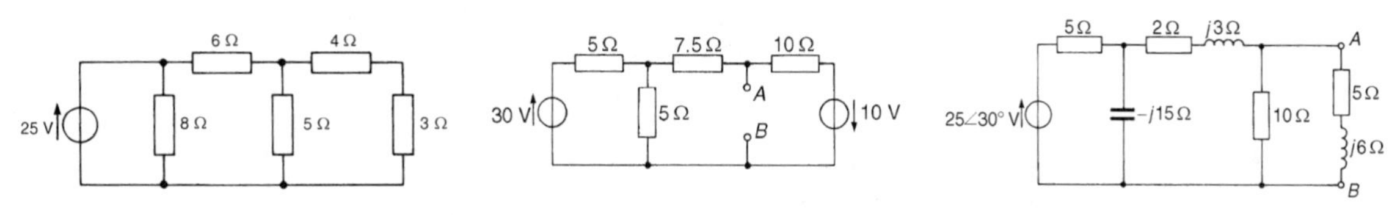

Figure 33.88 **Figure 33.89** **Figure 33.90**

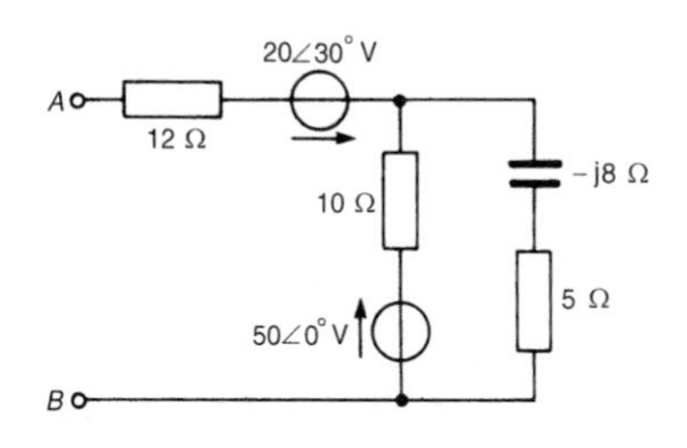

Figure 33.91

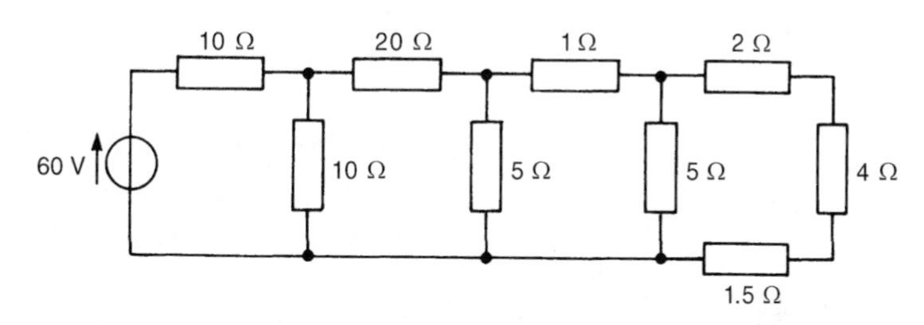

Figure 33.92

9 For the bridge network shown in Figure 33.93 use Thévenin's theorem to determine the current flowing in the $(4 + j3)\Omega$ impedance and its direction. Assume that the $20\angle 0^\circ$ V source has negligible internal impedance. [0.12 A from Q to P]

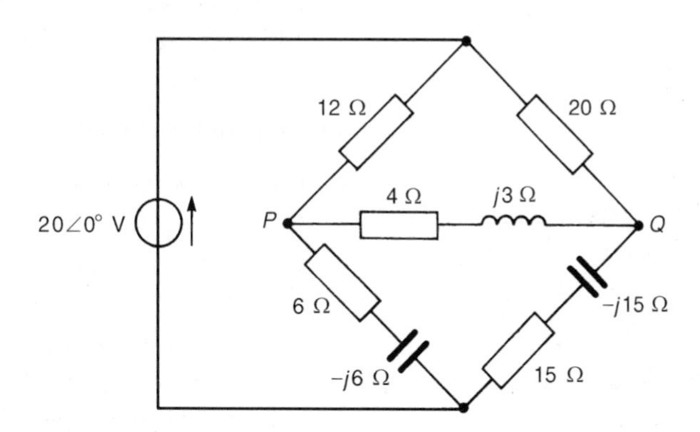

Figure 33.93

10 Repeat problems 1 to 10, page 542 of Chapter 30, 2 and 3 and 11 to 15, page 559 of Chapter 31 and 2 to 8, page 573 of Chapter 32, using Thévenin's theorem and compare the method of solution with that used for Kirchhoff's laws, mesh-current and nodal analysis and the superposition theorem.

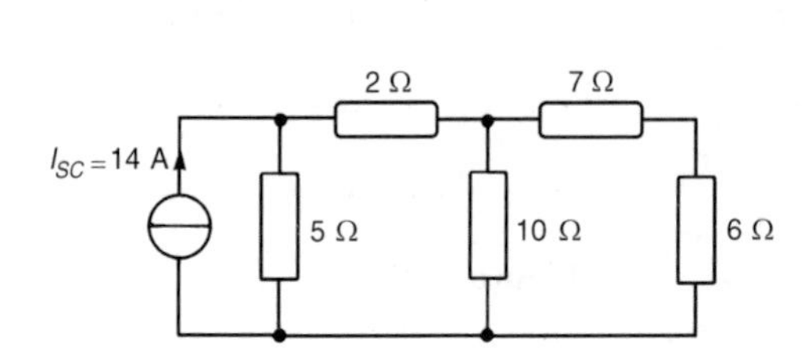

Figure 33.94

Figure 33.95

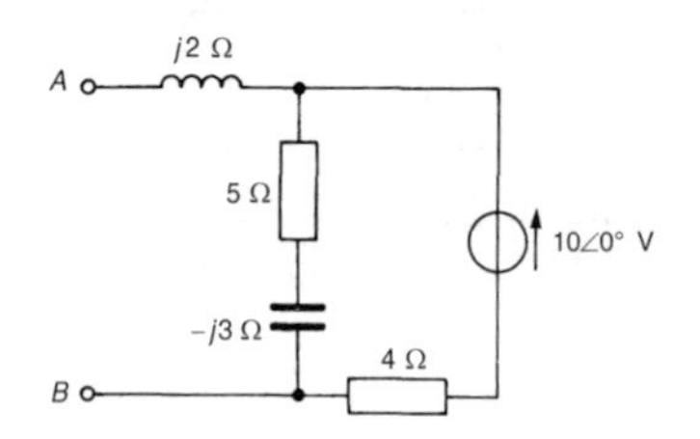

Figure 33.96

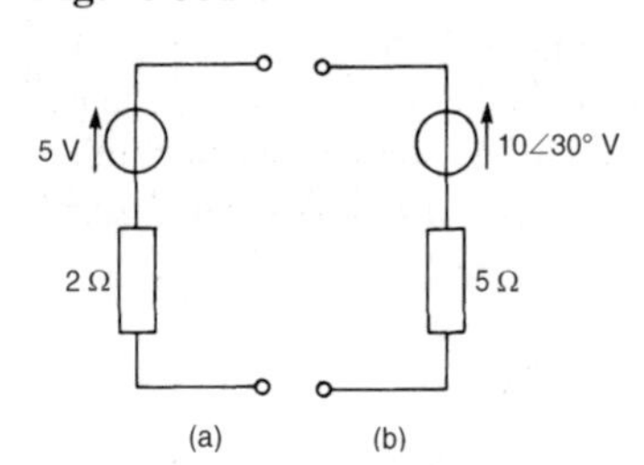

Figure 33.97

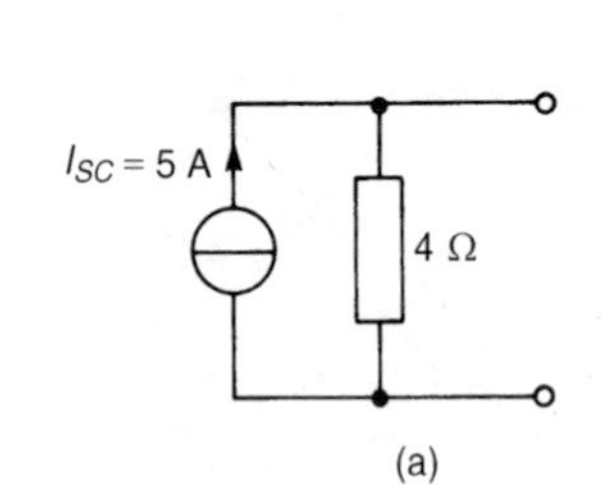

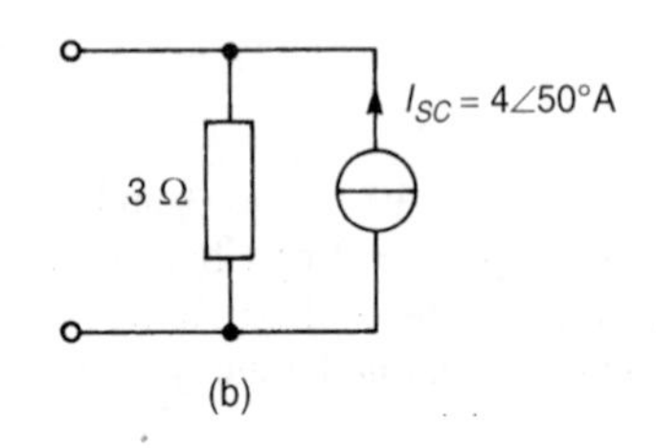

Figure 33.98

Norton's theorem

11 Repeat problems 1 to 4 and 6 to 8 above using Norton's theorem instead of Thévenin's theorem.

12 Determine the current flowing in the 10 Ω resistance of the network shown in Figure 33.94 by using Norton's theorem. [3.13 A]

13 For the network shown in Figure 33.95, use Norton's theorem to determine the current flowing in the 10 Ω resistance. [1.08 A]

14 Determine for the network shown in Figure 33.96 the Norton equivalent network at terminals AB. Hence determine the current flowing in a $(2 + j4)\Omega$ impedance connected between A and B.
[$I_{SC} = 2.185\angle -43.96°$ A, $z = (2.40 + j1.47)\Omega$; 0.88 A]

15 Repeat problems 1 to 10, page 542 of Chapter 30 and 2 and 3 and 12 to 15, page 559 of Chapter 31, using Norton's theorem.

Thévenin and Norton equivalent networks

16 Convert the circuits shown in Figure 33.97 to Norton equivalent networks. [(a) $I_{SC} = 2.5$ A, $z = 2\ \Omega$ (b) $I_{SC} = 2\angle 30°$, $z = 5\ \Omega$]

17 Convert the networks shown in Figure 33.98 to Thévenin equivalent circuits. [(a) $E = 20$ V, $z = 4\ \Omega$; (b) $E = 12\angle 50°$ V, $z = 3\ \Omega$]

18 (a) Convert the network to the left of terminals AB in Figure 33.99 to an equivalent Thévenin circuit by initially converting to a Norton equivalent network.

(b) Determine the current flowing in the $(2.8 - j3)\Omega$ impedance connected between A and B in Figure 33.99.
[(a) $E = 18$ V, $z = 1.2\ \Omega$ (b) 3.6 A]

19 Determine, by successive conversions between Thévenin and Norton equivalent networks, a Thévenin equivalent circuit for terminals AB of Figure 33.100. Hence determine the current flowing in a $(2 + j4)\Omega$ impedance connected between A and B.
$\left[E = 9\frac{1}{3}\text{ V}, z = 1\ \Omega; 1.87\angle -53.13°\text{ A}\right]$

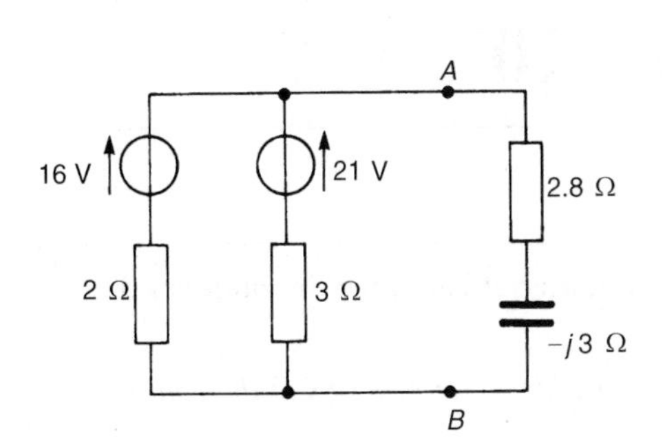

Figure 33.99

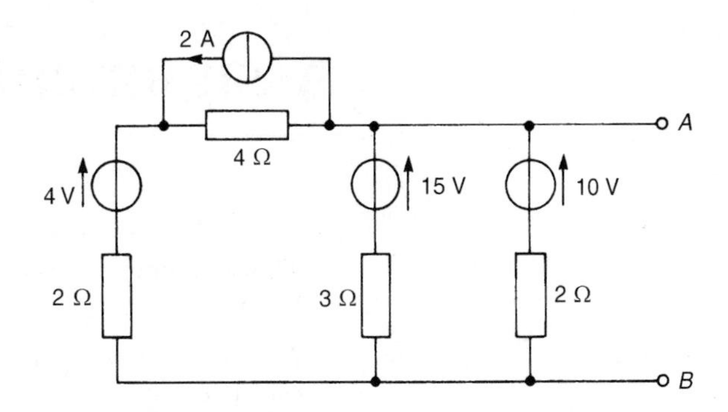

Figure 33.100

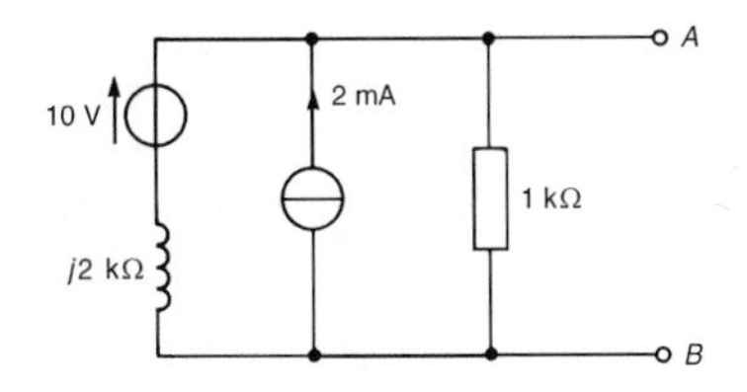

Figure 33.101

20 Derive an equivalent Thévenin circuit for terminals AB of the network shown in Figure 33.101. Hence determine the p.d. across AB when a $(3 + j4)\text{k}\Omega$ impedance is connected between these terminals.

[$E = 4.82\angle{-41.63°}$ V, $z = (0.8 + j0.4)$ kΩ; 4.15 V]

21 For the network shown in Figure 33.102, derive (a) the Thévenin equivalent circuit, and (b) the Norton equivalent network. (c) A 6 Ω resistance is connected between A and B. Determine the current flowing in the 6 Ω resistance by using both the Thévenin and Norton equivalent circuits.

[(a) $E = 6.71\angle{-26.57°}$ V, $z = (4.50 + j3.75)\Omega$
(b) $I_{SC} = 1.15\angle{-66.38°}$, $z = (4.50 + j3.75)\Omega$
(c) 0.60 A

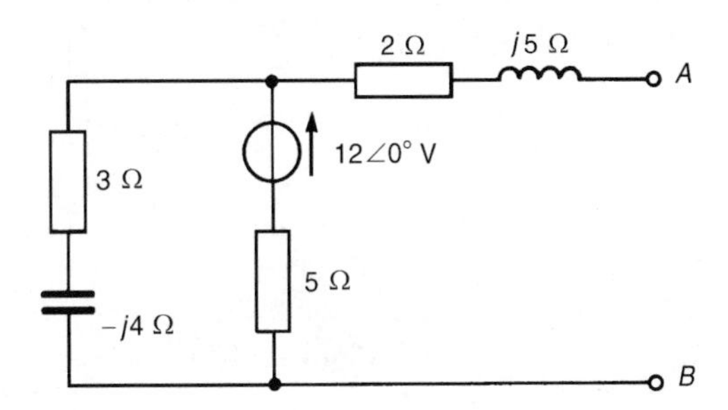

Figure 33.102

Assignment 10

This assignment covers the material contained in chapters 30 to 33.

The marks for each question are shown in brackets at the end of each question.

For the network shown in Figure A10.1, determine the current flowing in each branch using:

(a) Kirchhoff's laws (10)
(b) Mesh-current analysis (12)
(c) Nodal analysis (12)
(d) the Superposition theorem (22)
(e) Thévenin's theorem (14)
(f) Norton's theorem (10)

Demonstrate that each method gives the same value for each of the branch currents.

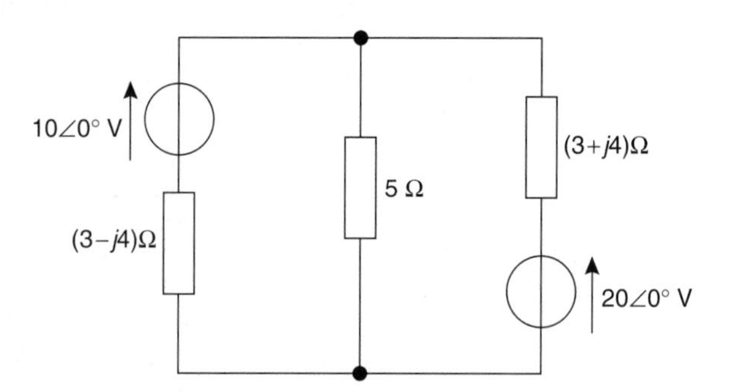

Figure A10.1

34 Delta-star and star-delta transformations

At the end of this chapter you should be able to:

- recognize delta (or π) and star (or T) connections
- apply the delta-star and star-delta transformations in appropriate a.c. and d.c. networks

34.1 Introduction

By using Kirchhoff's laws, mesh-current analysis, nodal analysis or the superposition theorem, currents and voltages in many network can be determined as shown in Chapters 30 to 32. Thevenin's and Norton's theorems, introduced in Chapter 33, provide an alternative method of solving networks and often with considerably reduced numerical calculations. Also, these latter theorems are especially useful when only the current in a particular branch of a complicated network is required. Delta-star and star-delta transformations may be applied in certain types of circuit to simplify them before application of circuit theorems.

34.2 Delta and star connections

The network shown in Figure 34.1(a) consisting of three impedances Z_A, Z_B and Z_C is said to be ***π*-connected**. This network can be redrawn as shown in Figure 34.1(b), where the arrangement is referred to as **delta-connected** or **mesh-connected**.

The network shown in Figure 34.2(a), consisting of three impedances, Z_1, Z_2 and Z_3, is said to be ***T*-connected**. This network can be redrawn as shown in Figure 34.2(b), where the arrangement is referred to as **star-connected**.

34.3 Delta-star transformation

It is possible to replace the delta connection shown in Figure 34.3(a) by an equivalent star connection as shown in Figure 34.3(b) such that the impedance measured between any pair of terminals (1–2, 2–3 or 3–1) is the same in star as in delta. The equivalent star network will consume the same power and operate at the same power factor as the original delta network. A delta-star transformation may alternatively be termed 'π to T transformation'.

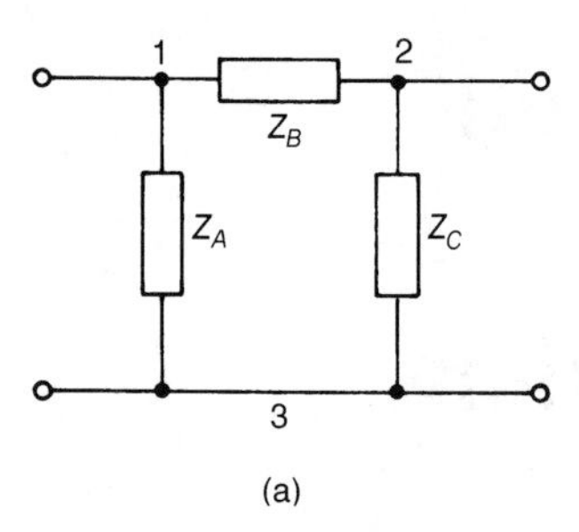

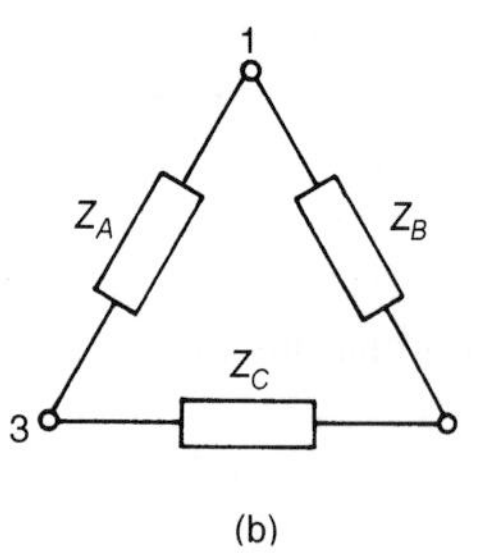

Figure 34.1 *(a) π-connected network, (b) Delta-connected network*

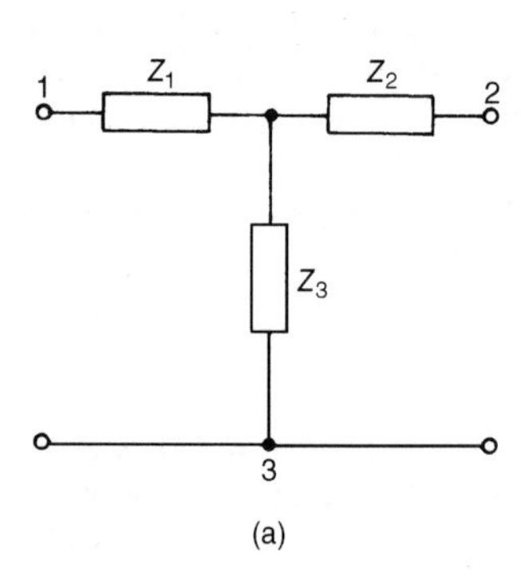

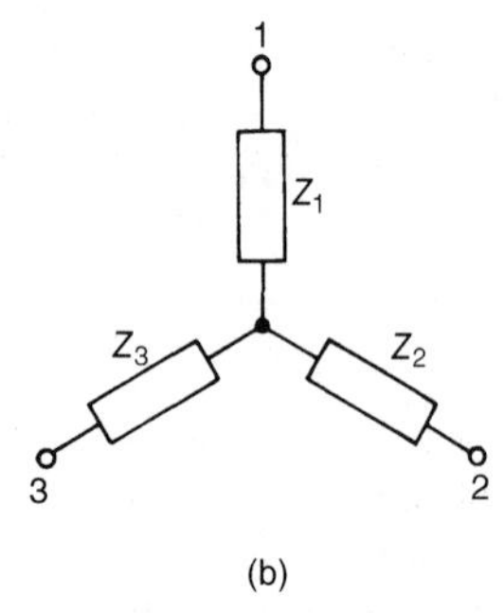

Figure 34.2 *(a) T-connected network, (b) Star-connected network*

Considering terminals 1 and 2 of Figure 34.3(a), the equivalent impedance is given by the impedance Z_B in parallel with the series combination of Z_A and Z_C,

i.e., $$\frac{Z_B(Z_A + Z_C)}{Z_B + Z_A + Z_C}$$

In Figure 34.3(b), the equivalent impedance between terminals 1 and 2 is Z_1 and Z_2 in series, i.e., $Z_1 + Z_2$ Thus,

Delta Star

$$Z_{12} = \frac{Z_B(Z_A + Z_C)}{Z_B + Z_A + Z_C} = Z_1 + Z_2 \qquad (34.1)$$

By similar reasoning, $$Z_{23} = \frac{Z_C(Z_A + Z_B)}{Z_C + Z_A + Z_B} = Z_2 + Z_3 \qquad (34.2)$$

and $$Z_{31} = \frac{Z_A(Z_B + Z_C)}{Z_A + Z_B + Z_C} = Z_3 + Z_1 \qquad (34.3)$$

Hence we have three simultaneous equations to be solved for Z_1, Z_2 and Z_3.

Equation (34.1) – equation (34.2) gives:

$$\frac{Z_A Z_B - Z_A Z_C}{Z_A + Z_B + Z_C} = Z_1 - Z_3 \qquad (34.4)$$

Equation (34.3) + equation (34.4) gives:

$$\frac{2Z_A Z_B}{Z_A + Z_B + Z_C} = 2Z_1$$

from which $$Z_1 = \frac{Z_A Z_B}{Z_A + Z_B + Z_C}$$

Similarly, equation (34.2) – equation (34.3) gives:

$$\frac{Z_B Z_C - Z_A Z_B}{Z_A + Z_B + Z_C} = Z_2 - Z_1 \qquad (34.5)$$

Equation (34.1) + equation (34.5) gives:

$$\frac{2Z_B Z_C}{Z_A + Z_B + Z_C} = 2Z_2$$

from which $$Z_2 = \frac{Z_B Z_C}{Z_A + Z_B + Z_C}$$

Finally, equation (34.3) – equation (34.1) gives:

$$\frac{Z_A Z_C - Z_B Z_C}{Z_A + Z_B + Z_C} = Z_3 - Z_2 \qquad (34.6)$$

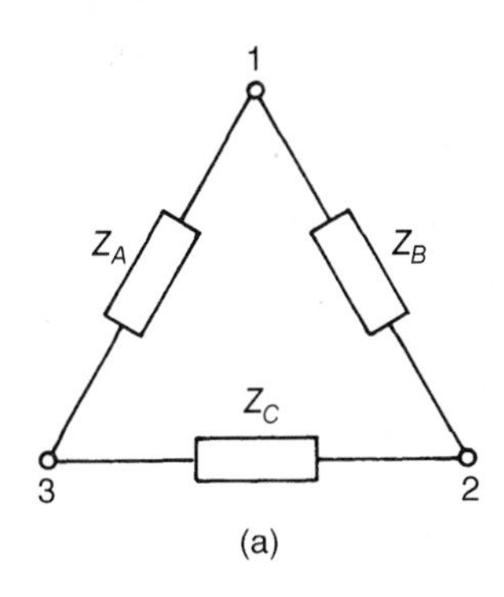

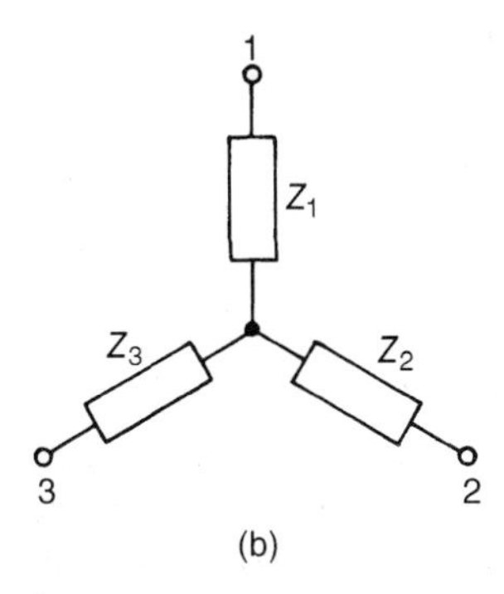

Figure 34.3

Equation (34.2) + equation (34.6) gives:

$$\frac{2Z_A Z_C}{Z_A + Z_B + Z_C} = 2Z_3$$

from which $Z_3 = \dfrac{Z_A Z_C}{Z_A + Z_B + Z_C}$

Summarizing, the star section shown in Figure 34.3(b) is equivalent to the delta section shown in Figure 34.3(a) when

$$\boldsymbol{Z_1 = \frac{Z_A Z_B}{Z_A + Z_B + Z_C}} \qquad (34.7)$$

$$\boldsymbol{Z_2 = \frac{Z_B Z_C}{Z_A + Z_B + Z_C}} \qquad (34.8)$$

and

$$\boldsymbol{Z_3 = \frac{Z_A Z_C}{Z_A + Z_B + Z_C}} \qquad (34.9)$$

It is noted that impedance Z_1 is given by the product of the two impedances in delta joined to terminal 1 (i.e., Z_A and Z_B), divided by the sum of the three impedances; impedance Z_2 is given by the product of the two impedances in delta joined to terminal 2 (i.e., Z_B and Z_C), divided by the sum of the three impedances; and impedance Z_3 is given by the product of the two impedances in delta joined to terminal 3 (i.e., Z_A and Z_C), divided by the sum of the three impedances.

Thus, for example, the star equivalent of the resistive delta network shown in Figure 34.4 is given by

$$Z_1 = \frac{(2)(3)}{2+3+5} = \mathbf{0.6\ \Omega},$$

$$Z_2 = \frac{(3)(5)}{2+3+5} = \mathbf{1.5\ \Omega}$$

and $Z_3 = \dfrac{(2)(5)}{2+3+5} = \mathbf{1.0\ \Omega}$

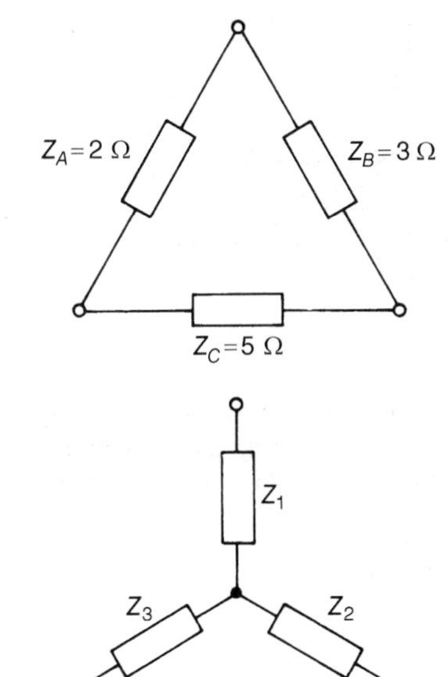

Figure 34.4

Problem 1. Replace the delta-connected network shown in Figure 34.5 by an equivalent star connection.

Let the equivalent star network be as shown in Figure 34.6. Then, from equation (34.7),

$$Z_1 = \frac{Z_A Z_B}{Z_A + Z_B + Z_C}$$

$$= \frac{(20)(10 + j10)}{20 + 10 + j10 - j20}$$

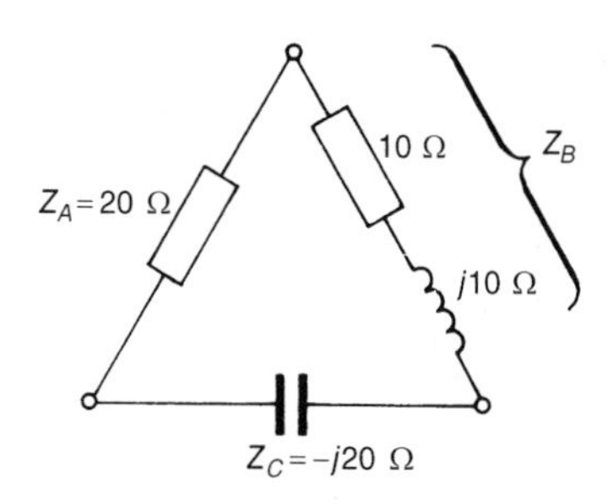

Figure 34.5

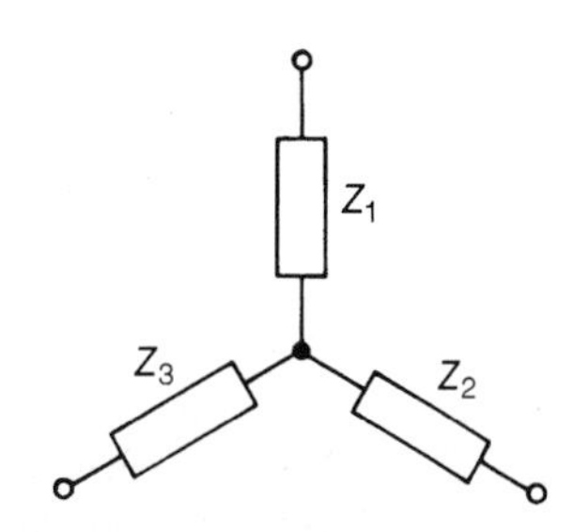

Figure 34.6

$$= \frac{(20)(10 + j10)}{(30 - j10)} = \frac{(20)(1.414\angle 45°)}{31.62\angle -18.43°}$$

$$= \mathbf{8.944\angle 63.43°\ \Omega} \text{ or } \mathbf{(4 + j8)\Omega}$$

From equation (34.8),

$$Z_2 = \frac{Z_B Z_C}{Z_A + Z_B + Z_C} = \frac{(10 + j10)(-j20)}{31.62\angle -18.43°}$$

$$= \frac{(1.414\angle 45°)(20\angle -90°)}{31.62\angle -18.43°}$$

$$= \mathbf{8.944\angle -26.57°\ \Omega} \text{ or } \mathbf{(8 - j4)\Omega}$$

From equation (34.9),

$$Z_3 = \frac{Z_A Z_C}{Z_A + Z_B + Z_C} = \frac{(20)(-j20)}{31.62\angle -18.43°}$$

$$= \frac{(400\angle -90°)}{31.62\angle -18.43°}$$

$$= \mathbf{12.650\angle -71.57°\ \Omega} \text{ or } \mathbf{(4 - j12)\Omega}$$

Problem 2. For the network shown in Figure 34.7, determine (a) the equivalent circuit impedance across terminals AB, (b) supply current I and (c) the power dissipated in the 10 Ω resistor.

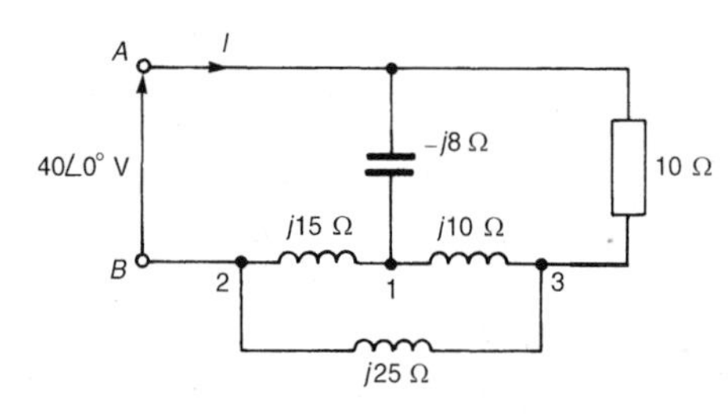

Figure 34.7

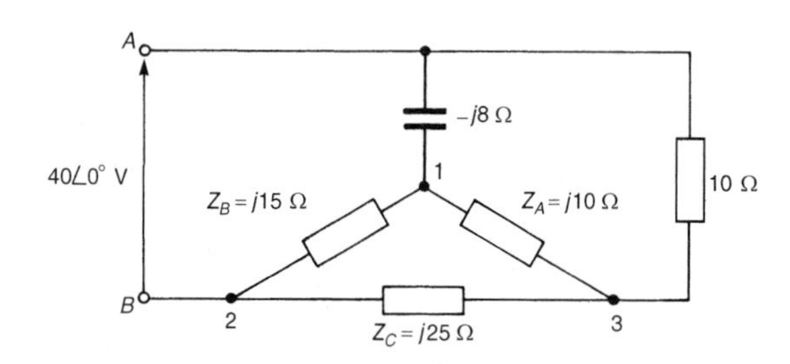

Figure 34.8

(a) The network of Figure 34.7 is redrawn, as in Figure 34.8, showing more clearly the part of the network 1, 2, 3 forming a delta connection. This may he transformed into a star connection as shown in Figure 34.9.

From equation (34.7),

$$Z_1 = \frac{Z_A Z_B}{Z_A + Z_B + Z_C} = \frac{(j10)(j15)}{j10 + j15 + j25}$$

$$= \frac{(j10)(j15)}{(j50)} = \mathbf{j3\ \Omega}$$

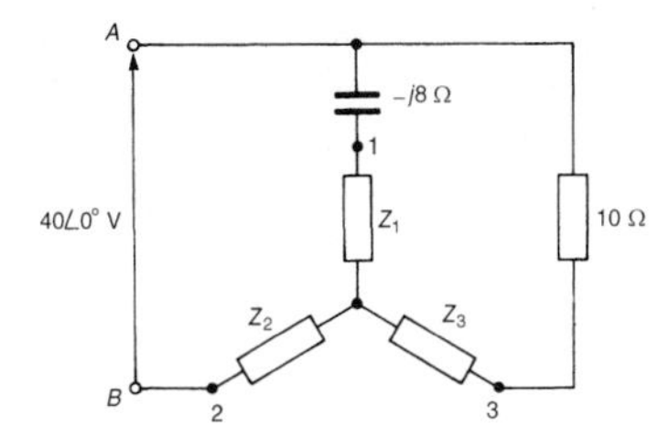

Figure 34.9

From equation (34.8),

$$Z_2 = \frac{Z_B Z_C}{Z_A + Z_B + Z_C} = \frac{(j15)(j25)}{j50} = \boldsymbol{j7.5\ \Omega}$$

From equation (34.9),

$$Z_3 = \frac{Z_A Z_C}{Z_A + Z_B + Z_C} = \frac{(j10)(j25)}{(j50)} = \boldsymbol{j5\ \Omega}$$

The equivalent network is shown in Figure 34.10 and is further simplified in Figure 34.11.

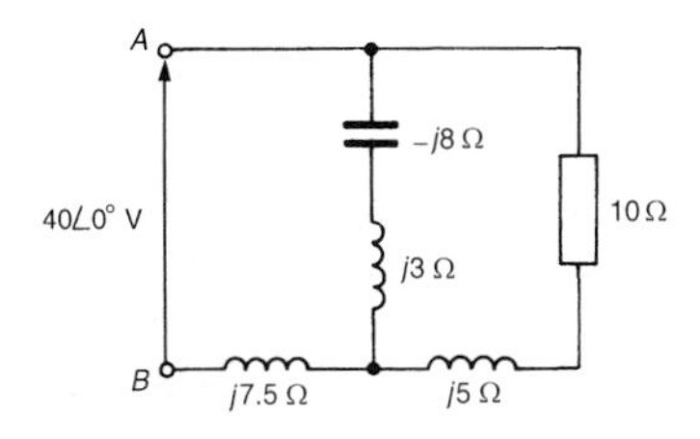

Figure 34.10

$(10 + j5)\Omega$ in parallel with $-j5\ \Omega$ gives an equivalent impedance of

$$\frac{(10 + j5)(-j5)}{(10 + j5 - j5)} = (2.5 - j5)\Omega$$

Hence the total circuit equivalent impedance across terminals AB is given by $\boldsymbol{Z_{AB}} = (2.5 - j5) + j7.5 = \mathbf{(2.5 + j2.5)\Omega}$ or $\mathbf{3.54\angle 45°\ \Omega}$

(b) Supply current $\boldsymbol{I} = \dfrac{V}{Z_{AB}} = \dfrac{40\angle 0°}{3.54\angle 45°} = \mathbf{11.3\angle{-45°}\ A}$

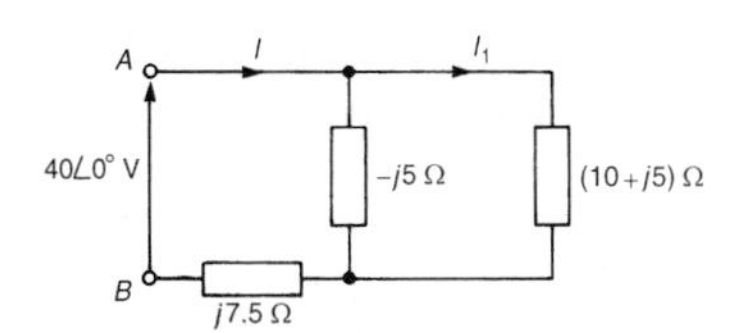

Figure 34.11

(c) Power P dissipated in the $10\ \Omega$ resistance of Figure 34.7 is given by $(I_1)^2(10)$, where I_1 (see Figure 34.11) is given by:

$$I_1 = \left[\frac{-j5}{10 + j5 - j5}\right](11.3\angle{-45°}) = 5.65\angle{-135°}\ \text{A}$$

Hence power $\boldsymbol{P} = (5.65)^2(10) = \mathbf{319\ W}$

Problem 3. Determine, for the bridge network shown in Figure 34.12, (a) the value of the single equivalent resistance that replaces the network between terminals A and B, (b) the current supplied by the 52 V source, and (c) the current flowing in the $8\ \Omega$ resistance.

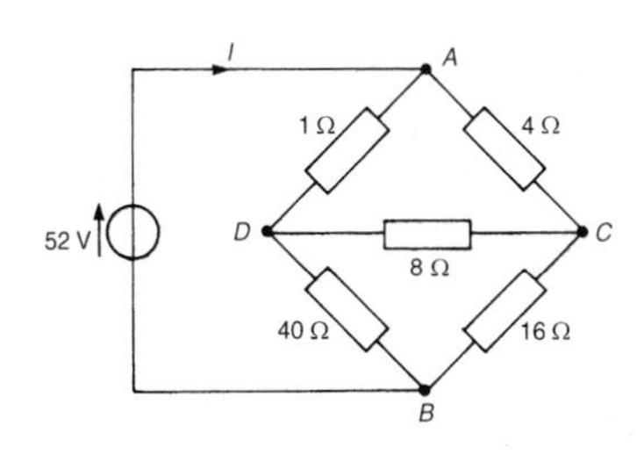

Figure 34.12

(a) In Figure 34.12, no resistances are directly in parallel or directly in series with each other. However, ACD and BCD are both delta connections and either may be converted into an equivalent star connection. The delta network BCD is redrawn in Figure 34.13(a) and is transformed into an equivalent star connection as shown in Figure 34.13(b), where

$$Z_1 = \frac{(8)(16)}{8 + 16 + 40} = 2\ \Omega \qquad \text{(from equation (34.7))}$$

$$Z_2 = \frac{(16)(40)}{8 + 16 + 40} = 10\ \Omega \qquad \text{(from equation (34.8))}$$

$$Z_3 = \frac{(8)(40)}{8 + 16 + 40} = 5\ \Omega \qquad \text{(from equation (34.9))}$$

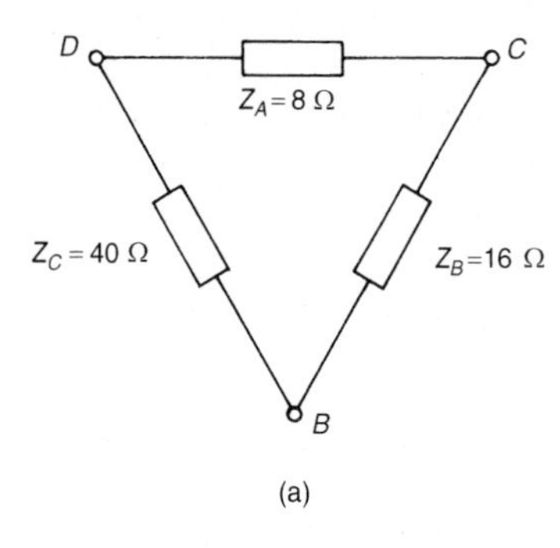

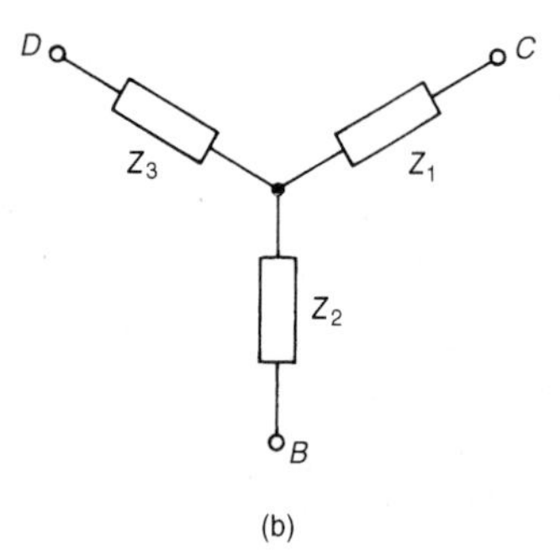

Figure 34.13

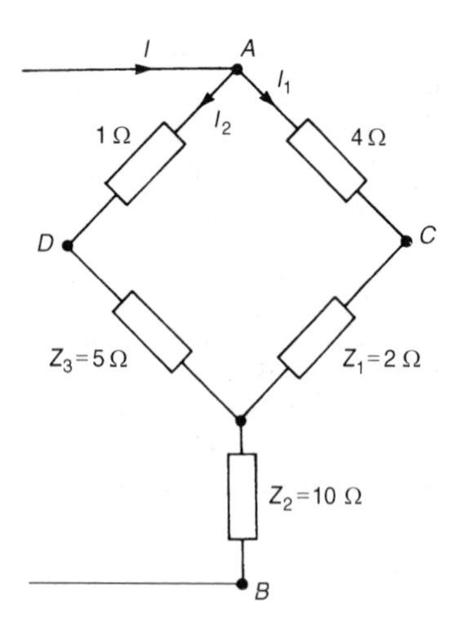

Figure 34.14

The network of Figure 34.12 may thus be redrawn as shown in Figure 34.14. The 4 Ω and 2 Ω resistances are in series with each other, as are the 1 Ω and 5 Ω resistors. Hence the equivalent network is as shown in Figure 34.15. The total equivalent resistance across terminals A and B is given by

$$R_{AB} = \frac{(6)(6)}{(6)+(6)} + 10 = \mathbf{13\ \Omega}$$

(b) Current supplied by the 52 V source, i.e., current I in Figure 34.15, is given by

$$I = \frac{V}{Z_{AB}} = \frac{52}{13} = \mathbf{4\ A}$$

(c) From Figure 34.15, current $I_1 = [6/(6+6)](I) = 2$ A, and current $I_2 = 2$ A also. From Figure 34.14, p.d. across AC, $V_{AC} = (I_1)(4) = 8$ V and p.d. across AD, $V_{AD} = (I_2)(1) = 2$ V. Hence p.d. between C and D (i.e., p.d. across the 8 Ω resistance of Figure 34.12) is given by $(8 - 2) = 6$ V.

Thus **the current in the 8 Ω resistance** is given by $V_{CD}/8 = 6/8 =$ **0.75 A**

Problem 4. Figure 34.16 shows an Anderson bridge used to measure, with high accuracy, inductance L_X and series resistance R_X

(a) Transform the delta ABD into its equivalent star connection and hence determine the balance equations for R_X and L_X

(b) If $R_2 = R_3 = 1$ kΩ, $R_4 = 500$ Ω, $R_5 = 200$ Ω and $C = 2$ μF, determine the values of R_X and L_X at balance.

(a) The delta ABD is redrawn separately in Figure 34.17, together with its equivalent star connection comprising impedances Z_1, Z_2 and Z_3. From equation (34.7),

$$Z_1 = \frac{(R_5)(-jX_C)}{R_5 - jX_C + R_3} = \frac{-jR_5X_C}{(R_3+R_5) - jX_C}$$

From equation (34.8),

$$Z_2 = \frac{(-jX_C)(R_3)}{R_5 - jX_C + R_3} = \frac{-jR_3X_C}{(R_3+R_5) - jX_C}$$

From equation (34.9),

$$Z_3 = \frac{R_5R_3}{(R_3+R_5) - jX_C}$$

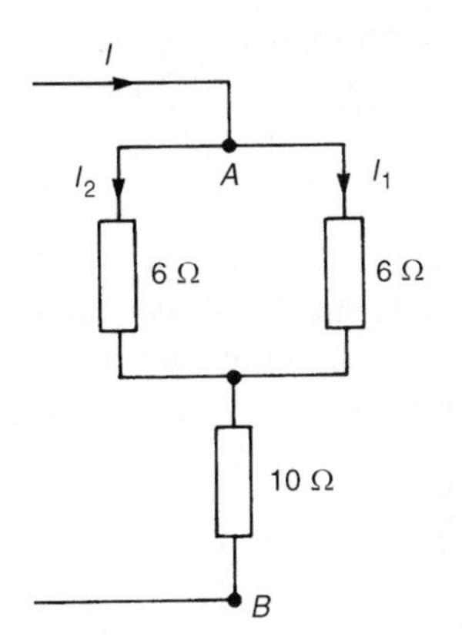

Figure 34.15

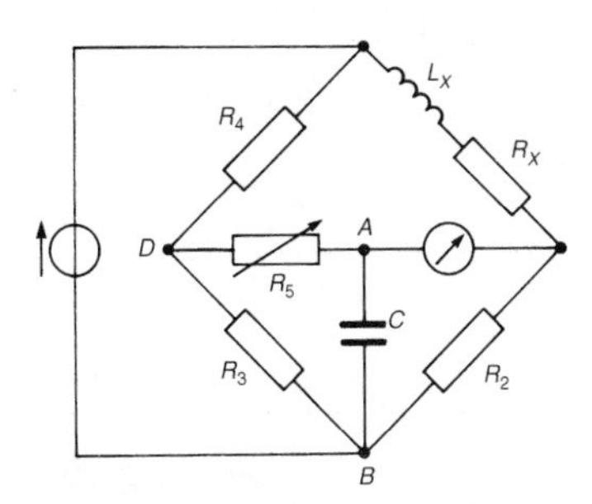

Figure 34.16

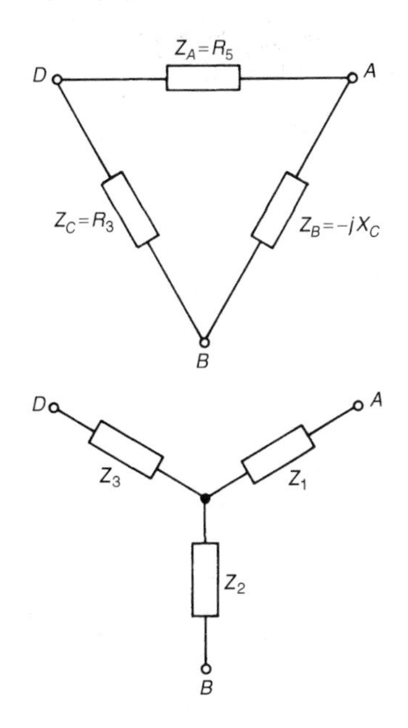

Figure 34.17

The network of Figure 34.16 is redrawn with the star replacing the delta as shown in Figure 34.18, and further simplified in Figure 34.19. (Note that impedance Z_1 does not affect the balance of the bridge since it is in series with the detector.)

At balance,

$$(R_X + jX_{L_X})(Z_2) = (R_2)(R_4 + Z_3) \qquad \text{from Chapter 27,}$$

from which,

$$(R_X + jX_{L_X}) = \frac{R_2}{Z_2}(R_4 + Z_3) = \frac{R_2R_4}{Z_2} + \frac{R_2Z_3}{Z_2}$$

$$= \frac{R_2R_4}{-jR_3X_C/((R_3+R_5)-jX_C)} + \frac{R_2(R_5R_3/((R_3+R_5)-jX_C))}{-jR_3X_C/((R_3+R_5)-jX_C)}$$

$$= \frac{R_2R_4((R_3+R_5)-jX_C)}{-jR_3X_C} + \frac{R_2R_5R_3}{-jR_3X_C}$$

$$= \frac{jR_2R_4((R_3+R_5)-jX_C)}{R_3X_C} + \frac{jR_2R_5}{X_C}$$

i.e., $$(R_X + jX_{L_X}) = \frac{jR_2R_4(R_3+R_5)}{R_3X_C} + \frac{R_2R_4X_C}{R_3X_C} + \frac{jR_2R_5}{X_C}$$

Equating the real parts gives:

$$\boldsymbol{R_X = \frac{R_2R_4}{R_3}}$$

Equating the imaginary parts gives:

$$X_{L_X} = \frac{R_2R_4(R_3+R_5)}{R_3X_C} + \frac{R_2R_5}{X_C}$$

i.e., $$\omega L_X = \frac{R_2R_4R_3}{R_3(1/\omega C)} + \frac{R_2R_4R_5}{R_3(1/\omega C)} + \frac{R_2R_5}{(1/\omega C)}$$

$$= \omega CR_2R_4 + \frac{\omega CR_2R_4R_5}{R_3} + \omega CR_2R_5$$

Hence $$\boldsymbol{L_X = R_2C\left(R_4 + \frac{R_4R_5}{R_3} + R_5\right)}$$

(b) When $R_2 = R_3 = 1$ kΩ, $R_4 = 500$ Ω, $R_5 = 200$ Ω and $C = 2$ μF, then, at balance

$$\boldsymbol{R_X} = \frac{R_2R_4}{R_3} = \frac{(1000)(500)}{(1000)} = \mathbf{500\ \Omega}$$

and $$\boldsymbol{L_X} = R_2C\left(R_4 + \frac{R_4R_5}{R_3} + R_5\right)$$

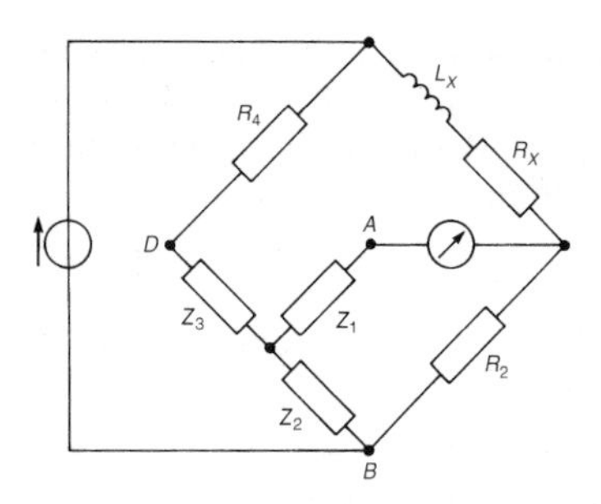

Figure 34.18

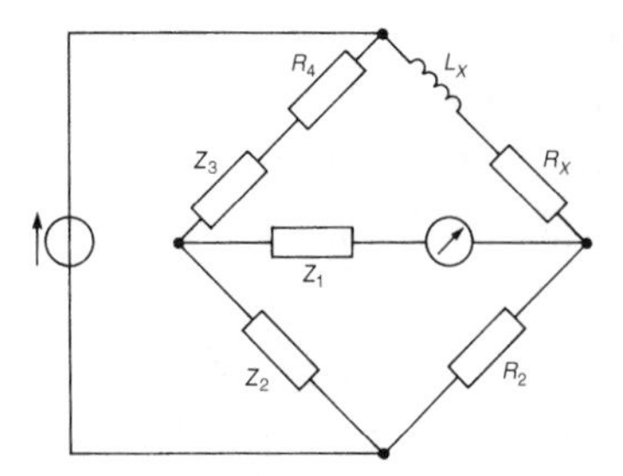

Figure 34.19

$$= (1000)(2 \times 10^{-6})\left[500 + \frac{(500)(200)}{(1000)} + 200\right]$$

$$= \mathbf{1.60\ H}$$

Problem 5. For the network shown in Figure 34.20, determine (a) the current flowing in the $(0 + j10)\Omega$ impedance, and (b) the power dissipated in the $(20 + j0)\Omega$ impedance.

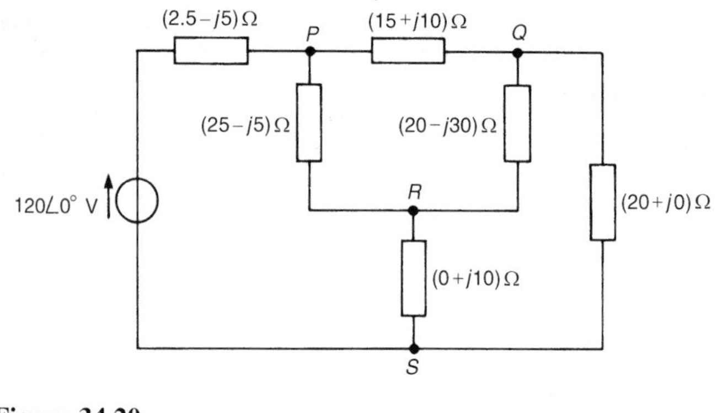

Figure 34.20

(a) The network may initially be simplified by transforming the delta PQR to its equivalent star connection as represented by impedances Z_1, Z_2 and Z_3 in Figure 34.21. From equation (34.7),

$$Z_1 = \frac{(15 + j10)(25 - j5)}{(15 + j10) + (25 - j5) + (20 - j30)}$$

$$= \frac{(15 + j10)(25 - j5)}{(60 - j25)}$$

$$= \frac{(18.03\angle 33.69^\circ)(25.50\angle -11.31^\circ)}{65\angle -22.62^\circ}$$

$$= 7.07\angle 45^\circ\ \Omega \text{ or } (5 + j5)\Omega$$

From equation (34.8),

$$Z_2 = \frac{(15 + j10)(20 - j30)}{(65\angle -22.62^\circ)}$$

$$= \frac{(18.03\angle 33.69^\circ)(36.06\angle -56.31^\circ)}{65\angle -22.62^\circ}$$

$$= 10.0\angle 0^\circ \text{ or } (10 + j0)\Omega$$

From equation (34.9),

$$Z_3 = \frac{(25 - j5)(20 - j30)}{(65\angle -22.62^\circ)}$$

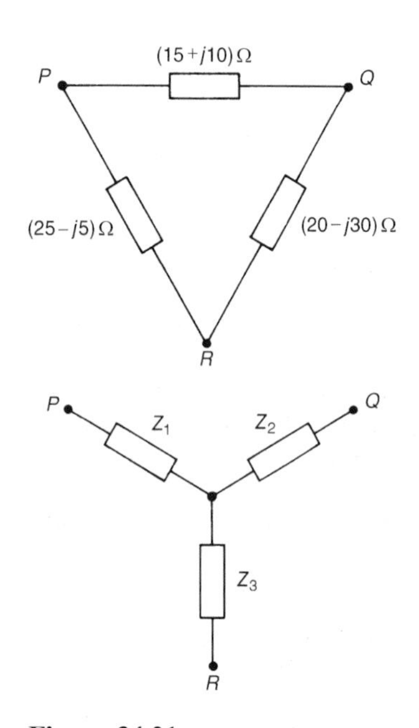

Figure 34.21

$$= \frac{(25.50\angle -11.31^\circ)(36.06\angle -56.31^\circ)}{65\angle -22.62^\circ}$$
$$= 14.15\angle -45^\circ\ \Omega \text{ or } (10 - j10)\Omega$$

The network is shown redrawn in Figure 34.22 and further simplified in Figure 34.23, from which,

$$\text{current } I_1 = \frac{120\angle 0^\circ}{7.5 + ((10)(30)/(10+30))}$$
$$= \frac{120\angle 0^\circ}{15} = 8 \text{ A}$$
$$\text{current } I_2 = \left(\frac{10}{10+30}\right)(8) = 2 \text{ A}$$
$$\text{current } I_3 = \left(\frac{30}{10+30}\right)(8) = 6 \text{ A}$$

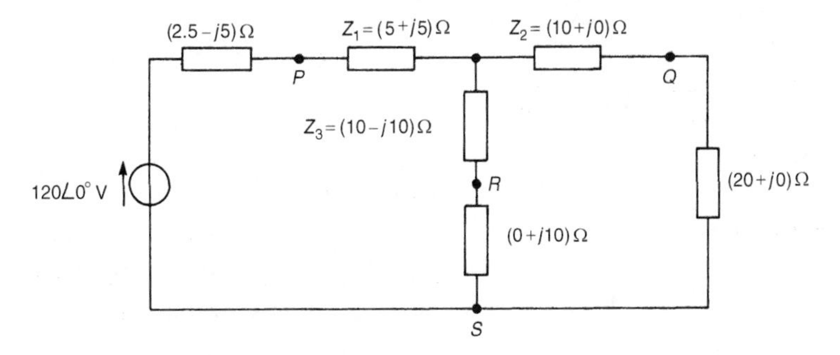

Figure 34.22

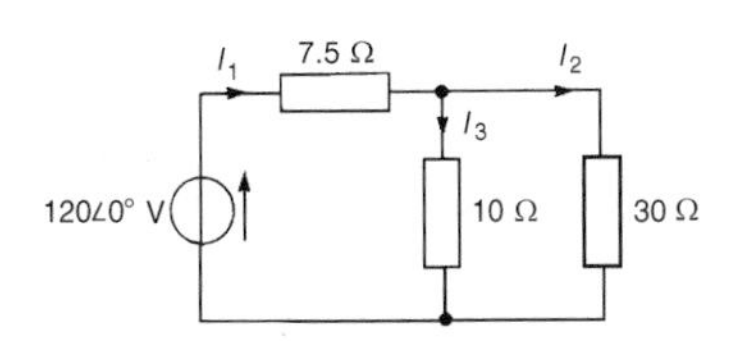

Figure 34.23

The current flowing in the $(0 + j10)\Omega$ impedance of Figure 34.20 is the current I_3 shown in Figure 34.23, i.e., **6 A**

(b) The power P dissipated in the $(20 + j0)\Omega$ impedance of Figure 34.20 is given by $P = I_2^2(20) = (2)^2(20) = $ **80 W**

34.4 Star-delta transformation

It is possible to replace the star section shown in Figure 34.24(a) by an equivalent delta section as shown in Figure 34.24(b). Such a transformation is also known as a 'T to π transformation'.

From equations (34.7), (34.8) and (34.9),

$$Z_1Z_2 + Z_2Z_3 + Z_3Z_1 = \frac{Z_AZ_B^2Z_C + Z_AZ_BZ_C^2 + Z_A^2Z_BZ_C}{(Z_A+Z_B+Z_C)^2}$$
$$= \frac{Z_AZ_BZ_C(Z_B+Z_C+Z_A)}{(Z_A+Z_B+Z_C)^2}$$
$$= \frac{Z_AZ_BZ_C}{(Z_A+Z_B+Z_C)} \qquad (34.10)$$

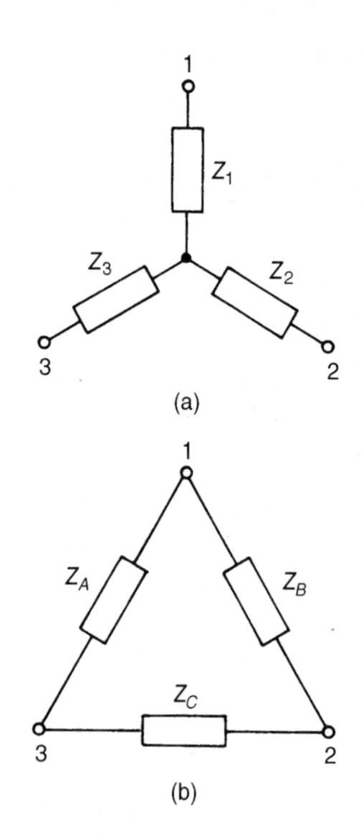

Figure 34.24

i.e., $$Z_1Z_2 + Z_2Z_3 + Z_3Z_1 = Z_A\left(\frac{Z_BZ_C}{Z_A + Z_B + Z_C}\right) = Z_A(Z_2)$$

from equation (34.8)

Hence $$Z_A = \frac{Z_1Z_2 + Z_2Z_3 + Z_3Z_1}{Z_2}$$

From equation (34.10),

$$Z_1Z_2 + Z_2Z_3 + Z_3Z_1 = Z_B\left(\frac{Z_AZ_C}{Z_A + Z_B + Z_C}\right) = Z_B(Z_3)$$

from equation (34.9)

Hence $$Z_B = \frac{Z_1Z_2 + Z_2Z_3 + Z_3Z_1}{Z_3}$$

Also from equation (34.10),

$$Z_1Z_2 + Z_2Z_3 + Z_3Z_1 = Z_C\left(\frac{Z_AZ_B}{Z_A + Z_B + Z_C}\right) = Z_C(Z_1)$$

from equation (34.7)

Hence $$Z_C = \frac{Z_1Z_2 + Z_2Z_3 + Z_3Z_1}{Z_1}$$

Summarizing, the delta section shown in Figure 34.24(b) is equivalent to the star section shown in Figure 34.24(a) when

$$\boxed{Z_A = \frac{Z_1Z_2 + Z_2Z_3 + Z_3Z_1}{Z_2}} \quad (34.11)$$

$$\boxed{Z_B = \frac{Z_1Z_2 + Z_2Z_3 + Z_3Z_1}{Z_3}} \quad (34.12)$$

and $$\boxed{Z_C = \frac{Z_1Z_2 + Z_2Z_3 + Z_3Z_1}{Z_1}} \quad (34.13)$$

It is noted that the numerator in each expression is the sum of the products of the star impedances taken in pairs. The denominator of the expression for Z_A, which is connected between terminals 1 and 3 of Figure 34.24(b), is Z_2, which is connected to terminal 2 of Figure 34.24(a). Similarly, the denominator of the expression for Z_B which is connected between terminals 1 and 2 of Figure 34.24(b), is Z_3, which is connected to terminal 3 of Figure 34.24(a). Also the denominator of the expression for Z_C which is connected between terminals 2 and 3 of Figure 34.24(b), is Z_1, which is connected to terminal 1 of Figure 34.24(a).

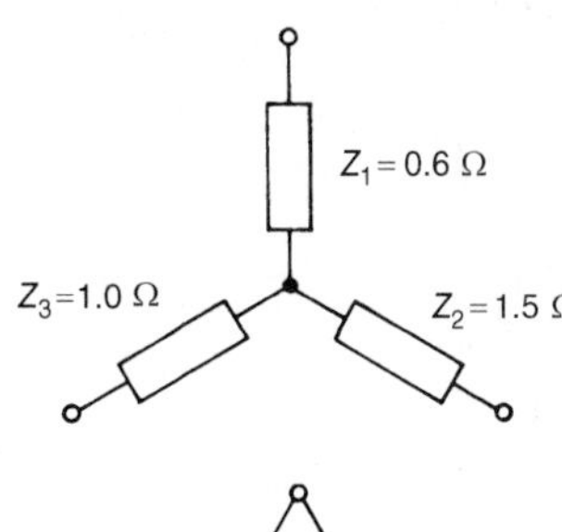

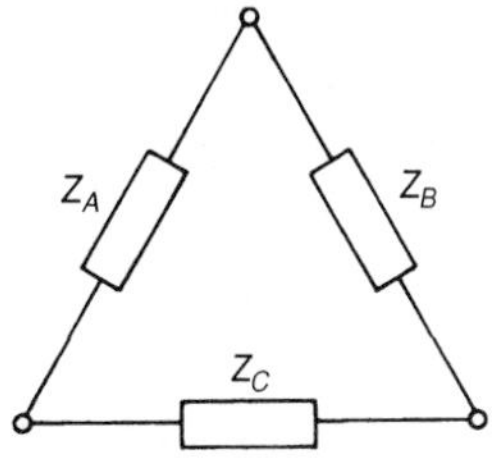

Figure 34.25

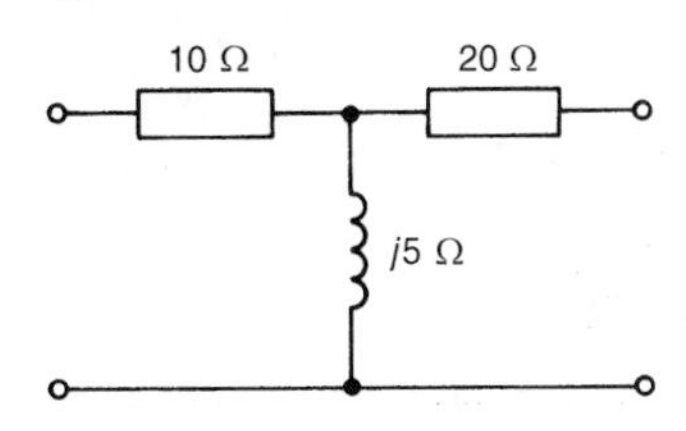

Figure 34.26

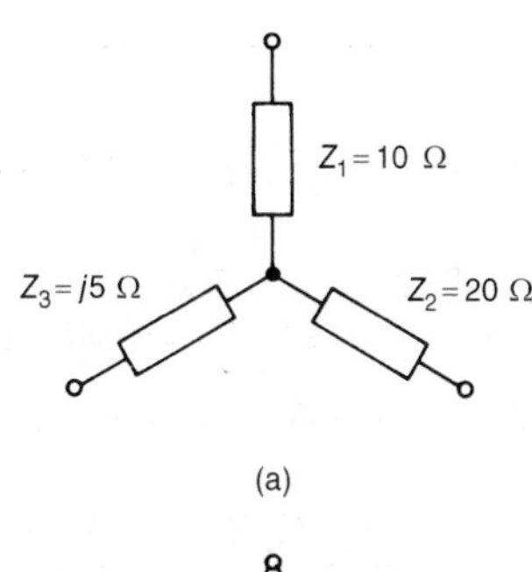

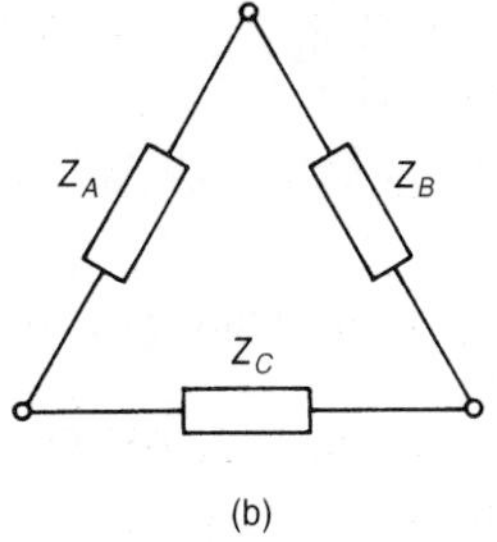

Figure 34.27

Thus, for example, the delta equivalent of the resistive star circuit shown in Figure 34.25 is given by:

$$\boldsymbol{Z_A} = \frac{(0.6)(1.5) + (1.5)(1.0) + (1.0)(0.6)}{1.5} = \frac{3.0}{1.5} = \mathbf{2\ \Omega},$$

$$\boldsymbol{Z_B} = \frac{3.0}{1.0} = \mathbf{3\ \Omega}, \quad \boldsymbol{Z_C} = \frac{3.0}{0.6} = \mathbf{5\ \Omega}$$

Problem 6. Determine the delta-connected equivalent network for the star-connected impedances shown in Figure 34.26

Figure 34.27(a) shows the network of Figure 34.26 redrawn and Figure 34.27(b) shows the equivalent delta connection containing impedances Z_A, Z_B and Z_C. From equation (34.11),

$$Z_A = \frac{Z_1Z_2 + Z_2Z_3 + Z_3Z_1}{Z_2} = \frac{(10)(20) + (20)(j5) + (j5)(10)}{20}$$

$$= \frac{200 + j150}{20} = \mathbf{(10 + j7.5)\Omega}$$

From equation (34.12),

$$Z_B = \frac{(200 + j150)}{Z_3} = \frac{(200) + (j150)}{j5}$$

$$= \frac{-j5(200 + j150)}{25} = \mathbf{(30 - j40)\Omega}$$

From equation (34.13),

$$Z_C = \frac{(200 + j150)}{Z_1} = \frac{(200) + (j150)}{10} = \mathbf{(20 + j15)\Omega}$$

Problem 7. Three impedances, $Z_1 = 100\angle 0°\ \Omega$, $Z_2 = 63.25\angle 18.43°\ \Omega$ and $Z_3 = 100\angle -90°\ \Omega$ are connected in star. Convert the star to an equivalent delta connection.

The star-connected network and the equivalent delta network comprising impedances Z_A, Z_B and Z_C are shown in Figure 34.28. From equation (34.11),

$$Z_A = \frac{Z_1Z_2 + Z_2Z_3 + Z_3Z_1}{Z_2}$$

$$= \frac{(100\angle 0°)(63.25\angle 18.43°) + (63.25\angle 18.43°)(100\angle -90°) + (100\angle -90°)(100\angle 0°)}{63.25\angle 18.43°}$$

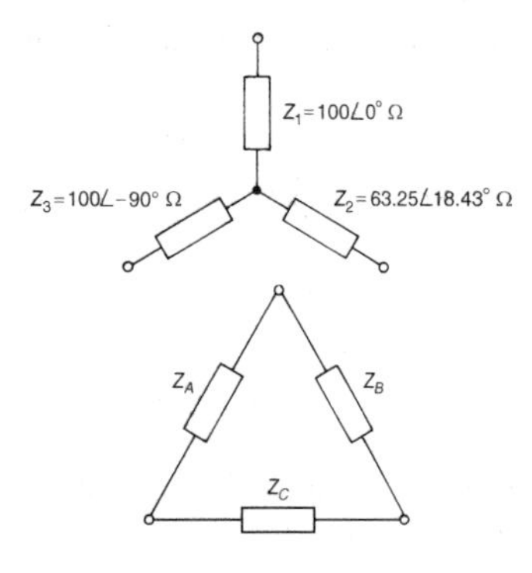

Figure 34.28

$$= \frac{6325\angle 18.43° + 6325\angle -71.57° + 100\,00\angle -90°}{63.25\angle 18.43°}$$

$$= \frac{6000 + j2000 + 2000 - j6000 - j100\,00}{63.25\angle 18.43°}$$

$$= \frac{8000 - j14\,000}{63.25\angle 18.43°} = \frac{16\,124.5\angle -60.26°}{63.25\angle 18.43°}$$

$$= \mathbf{254.93\angle -78.69°\ \Omega \text{ or } (50 - j250)\Omega}$$

From equation (34.12),

$$Z_B = \frac{Z_1Z_2 + Z_2Z_3 + Z_3Z_1}{Z_3}$$

$$= \frac{16\,124.5\angle -60.26}{100\angle -90°}$$

$$= \mathbf{161.25\angle 29.74°\ \Omega \text{ or } (140 + j80)\Omega}$$

From equation (34.13),

$$Z_C = \frac{Z_1Z_2 + Z_2Z_3 + Z_3Z_1}{Z_1} = \frac{16\,124.5\angle -60.26}{100\angle 0°}$$

$$= \mathbf{161.25\angle -60.26°\ \Omega \text{ or } (80 + j140)\Omega}$$

Further problems on delta-star and star-delta transformations may be found in Section 34.5 following, problems 1 to 10.

34.5 Further problems on delta-star and star-delta transformations

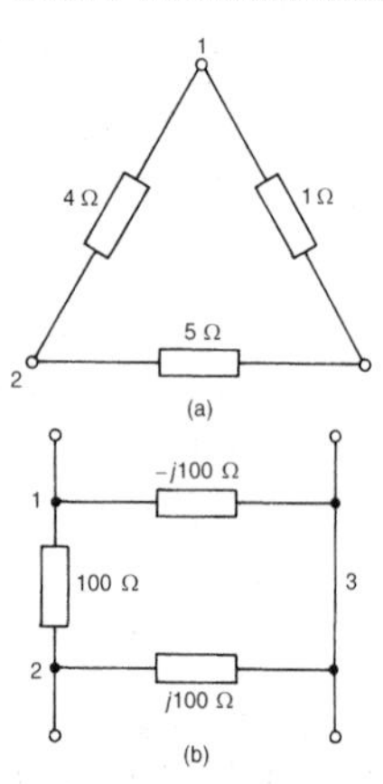

Figure 34.29

1 Transform the delta connected networks shown in Figure 34.29 to their equivalent star-connected networks.
[(a) $Z_1 = 0.4\ \Omega, Z_2 = 2\ \Omega, Z_3 = 0.5\ \Omega$
(b) $Z_1 = -j100\ \Omega, Z_2 = j100\ \Omega, Z_3 = 100\ \Omega$]

2 Determine the delta-connected equivalent networks for the star-connected impedances shown in Figure 34.30
[(a) $Z_{12} = 18\ \Omega, Z_{23} = 9\ \Omega, Z_{31} = 13.5\ \Omega$
(b) $Z_{12} = (10 + j0)\Omega, Z_{23} = (5 + j5)\Omega,$
$Z_{31} = (0 - j10)\Omega$]

3 (a) Transform the π network shown in Figure 34.31(a) to its equivalent star-connected network.
(b) Change the T-connected network shown in Figure 34.31(b) to its equivalent delta-connected network.
[(a) $Z_1 = 5.12\angle 78.35°\ \Omega, Z_2 = 6.82\angle -26.65°\ \Omega,$
$Z_3 = 10.23\angle -11.65°\ \Omega$
(b) $Z_{12} = 35.93\angle 40.50°\ \Omega, Z_{23} = 53.89\angle -19.50°\ \Omega,$
$Z_{31} = 26.95\angle -49.50°\ \Omega$]

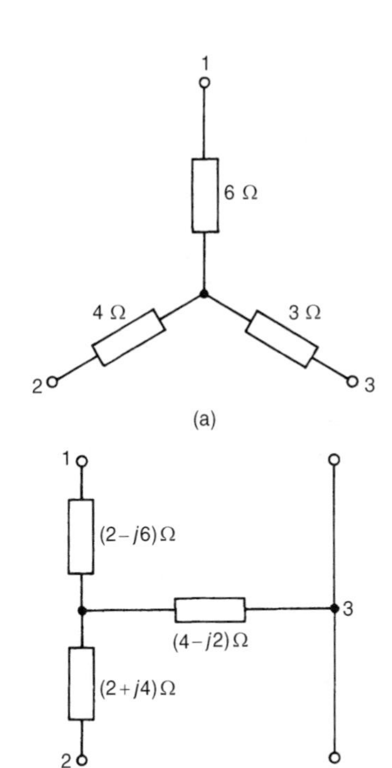

Figure 34.30

4 For the network shown in Figure 34.32 determine (a) current I, and (b) the power dissipated in the 10 Ω resistance.
[(a) $7.32\angle 24.06°$ A (b) 668 W]

5 (a) A delta-connected network contains three $24\angle 60°$ Ω impedances. Determine the impedances of the equivalent star-connected network.

(b) Three impedances, each of $(2 + j3)\Omega$, are connected in star. Determine the impedances of the equivalent delta-connected network.
[(a) Each impedance = $8\angle 60°$ Ω
(b) Each impedance = $(6 + j9)\Omega$]

6 (a) Derive the star-connected network of three impedances equivalent to the network shown in Figure 34.33.

(b) Obtain the delta-connected equivalent network for Figure 34.33.
[(a) 5 Ω, 6 Ω, 3 Ω
(b) 21 Ω, 12.6 Ω, 10.5 Ω]

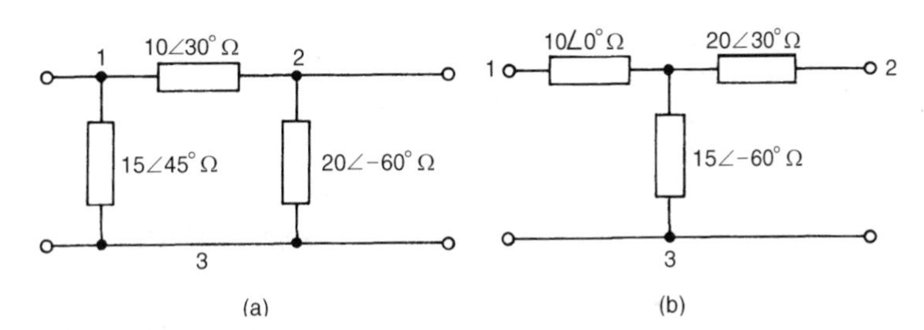

Figure 34.31

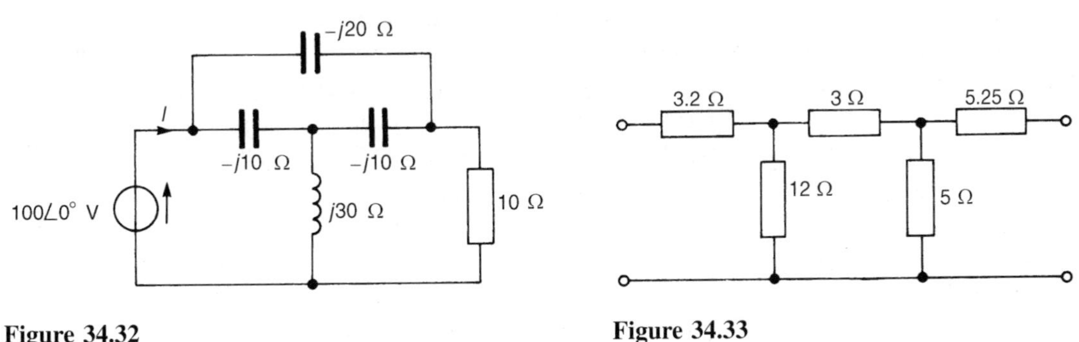

Figure 34.32

Figure 34.33

7 For the a.c. bridge network shown in Figure 34.34, transform the delta-connected network ABC into an equivalent star, and hence determine the current flowing in the capacitor. [131 mA]

8 For the network shown in Figure 34.35 transform the delta-connected network ABC to an equivalent star-connected network, convert the

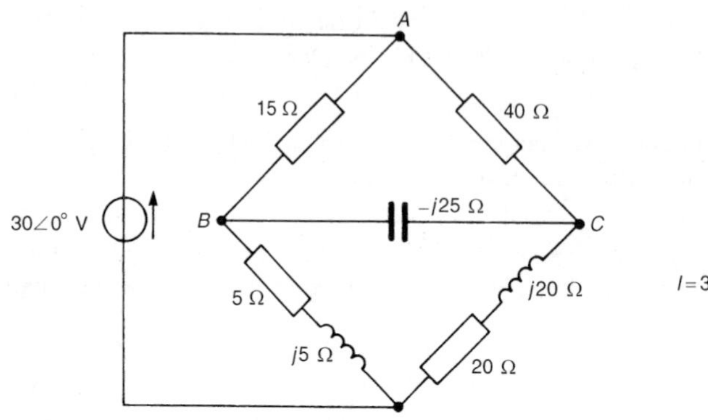

Figure 34.34

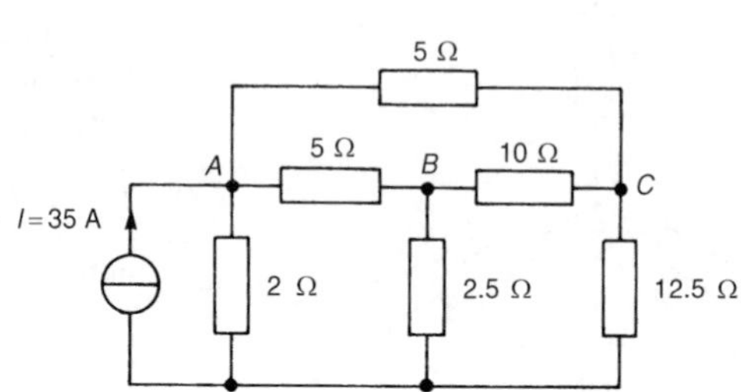

Figure 34.35

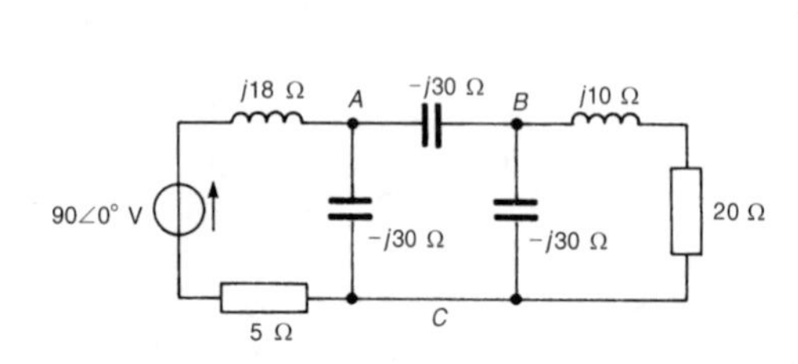

Figure 34.36

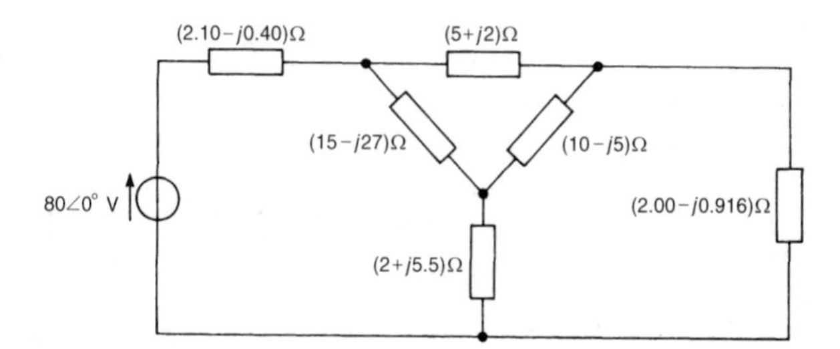

Figure 34.37

35 A, 2 Ω Norton circuit to an equivalent Thévenin circuit and hence determine the p.d. across the 12.5 Ω resistor. [31.25 V]

9 Transform the delta-connected network ABC shown in Figure 34.36 and hence determine the magnitude of the current flowing in the 20 Ω resistance. [4.47 A]

10 For the network shown in Figure 34.37 determine (a) the current supplied by the $80\angle 0°$ V source, and (b) the power dissipated in the $(2.00 - j0.916)\Omega$ impedance. [(a) 9.73 A (b) 98.6 W]

35 Maximum power transfer theorems and impedance matching

At the end of this chapter you should be able to:

- appreciate the conditions for maximum power transfer in a.c. networks
- apply the maximum power transfer theorems to a.c. networks
- appreciate advantages of impedance matching in a.c. networks
- perform calculations involving matching transformers for impedance matching in a.c. networks

35.1 Maximum power transfer theorems

A network that contains linear impedances and one or more voltage or current sources can be reduced to a Thévenin equivalent circuit as shown in Chapter 33. When a load is connected to the terminals of this equivalent circuit, power is transferred from the source to the load.

A Thévenin equivalent circuit is shown in Figure 35.1 with source internal impedance, $z = (r + jx)\Omega$ and complex load $Z = (R + jX)\Omega$.

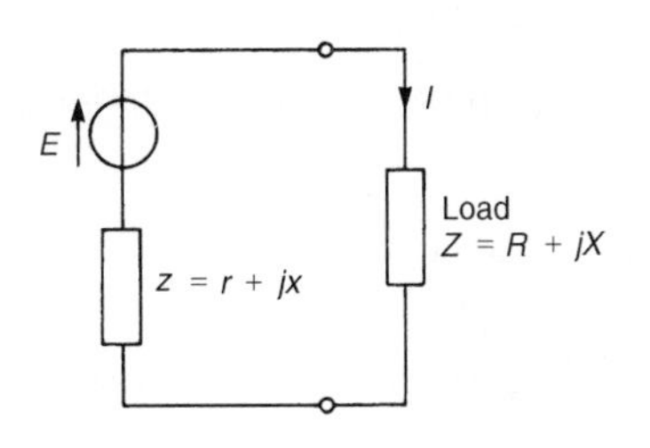

Figure 35.1

The maximum power transferred from the source to the load depends on the following four conditions.

Condition 1. Let the load consist of a pure variable resistance R (i.e. let $X = 0$). Then current I in the load is given by:

$$I = \frac{E}{(r + R) + jx}$$

and the magnitude of current, $|I| = \dfrac{E}{\sqrt{[(r + R)^2 + x^2]}}$

The active power P delivered to load R is given by

$$P = |I|^2 R = \frac{E^2 R}{(r + R)^2 + x^2}$$

To determine the value of R for maximum power transferred to the load, P is differentiated with respect to R and then equated to zero (this being the normal procedure for finding maximum or minimum values using calculus). Using the quotient rule of differentiation,

$$\frac{dP}{dR} = E^2 \left\{ \frac{[(r+R)^2 + x^2](1) - (R)(2)(r+R)}{[(r+R)^2 + x^2]^2} \right\}$$

$= 0$ for a maximum (or minimum) value.

For $\frac{dP}{dR}$ to be zero, the numerator of the fraction must be zero.

Hence $\quad (r+R)^2 + x^2 - 2R(r+R) = 0$

i.e., $\quad r^2 + 2rR + R^2 + x^2 - 2Rr - 2R^2 = 0$

from which, $r^2 + x^2 = R^2 \qquad (35.1)$

or $\quad \boldsymbol{R = \sqrt{(r^2 + x^2)} = |z|}$

Thus, with a variable purely resistive load, the maximum power is delivered to the load if the load resistance R is made equal to the magnitude of the source impedance.

Condition 2. Let both the load and the source impedance be purely resistive (i.e., let $x = X = 0$). From equation (35.1) it may be seen that the maximum power is transferred when $\boldsymbol{R = r}$ (this is, in fact, the d.c. condition explained in Chapter 13, page 187)

Condition 3. Let the load Z have both variable resistance R and variable reactance X. From Figure 35.1,

$$\text{current } I = \frac{E}{(r+R) + j(x+x)} \text{ and } |I| = \frac{E}{\sqrt{[(r+R)^2 + (x+X)^2]}}$$

The active power P delivered to the load is given by $P = |I|^2 R$ (since power can only be dissipated in a resistance) i.e.,

$$P = \frac{E^2 R}{(r+R)^2 + (x+X)^2}$$

If X is adjusted such that $X = -x$ then the value of power is a maximum.

If $X = -x$ then $P = \frac{E^2 R}{(r+R)^2}$

$$\frac{dP}{dR} = E^2 \left\{ \frac{(r+R)^2(1) - (R)(2)(r+R)}{(r+R)^4} \right\} = 0 \text{ for a maximum value}$$

Hence $\quad (r+R)^2 - 2R(r+R) = 0$

i.e., $\quad r^2 + 2rR + R^2 - 2Rr - 2R^2 = 0$

from which, $\quad r^2 - R^2 = 0$ and $R = r$

Thus with the load impedance Z consisting of variable resistance R and variable reactance X, maximum power is delivered to the load when $\boxed{\mathbf{X = -x \text{ and } R = r,}}$ i.e., when $R + jX = r - jx$. Hence maximum power is delivered to the load when the load impedance is the complex conjugate of the source impedance.

Condition 4. Let the load impedance Z have variable resistance R and fixed reactance X. From Figure 35.1, the magnitude of current,

$$|I| = \frac{E}{\sqrt{[(r + R)^2 + (x + X)^2]}}$$

and the power dissipated in the load,

$$P = \frac{E^2 R}{(r + R)^2 + (x + X)^2}$$

$$\frac{dP}{dR} = E^2 \left\{ \frac{[(r + R)^2 + (x + X)^2](1) - (R)(2)(r + R)}{[(r + R)^2 + (x + X)^2]^2} \right\}$$

$$= 0 \quad \text{for a maximum value}$$

Hence $\quad (r + R)^2 + (x + X)^2 - 2R(r + R) = 0$

$$r^2 + 2rR + R^2 + (x + X)^2 - 2Rr - 2R^2 = 0$$

from which, $R^2 = r^2 + (x + X)^2$ and $\boxed{\mathbf{R = \sqrt{[r^2 + (x + X)^2]}}}$

Summary

With reference to Figure 35.1:

1 When the load is purely resistive (i.e., $X = 0$) and adjustable, maximum power transfer is achieved when $\boxed{\mathbf{R = |z| = \sqrt{(r^2 + x^2)}}}$

2 When both the load and the source impedance are purely resistive (i.e., $X = x = 0$), maximum power transfer is achieved when $\boxed{\mathbf{R = r}}$

3 When the load resistance R and reactance X are both independently adjustable, maximum power transfer is achieved when

$$\boxed{\mathbf{X = -x \text{ and } R = r}}$$

4 When the load resistance R is adjustable with reactance X fixed, maximum power transfer is achieved when

$$\boxed{\mathbf{R = \sqrt{[r^2 + (x + X)^2]}}}$$

The maximum power transfer theorems are primarily important where a small source of power is involved — such as, for example, the output from a telephone system (see Section 35.2)

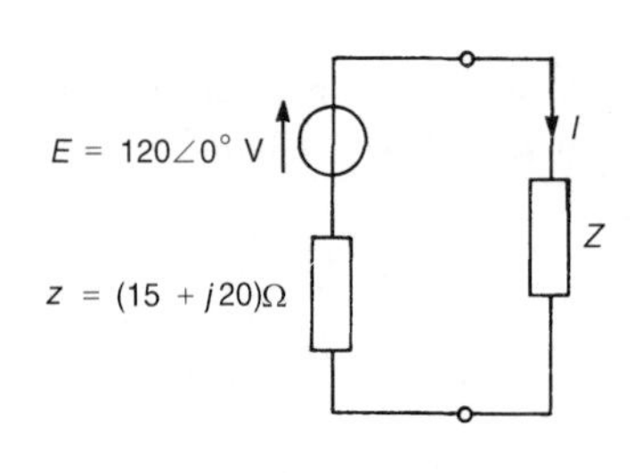

Figure 35.2

> Problem 1. For the circuit shown in Figure 35.2 the load impedance Z is a pure resistance. Determine (a) the value of R for maximum power to be transferred from the source to the load, and (b) the value of the maximum power delivered to R.

(a) From condition 1, maximum power transfer occurs when $R = |z|$, i.e., when

$$R = |15 + j20| = \sqrt{(15^2 + 20^2)} = \mathbf{25\ \Omega}$$

(b) Current I flowing in the load is given by $I = E/Z_T$, where the total circuit impedance $Z_T = z + R = 15 + j20 + 25 = (40 + j20)\Omega$ or $44.72\angle 26.57°\ \Omega$

$$\text{Hence} \quad I = \frac{120\angle 0°}{44.72\angle 26.57°} = 2.683\angle -26.57°\ \text{A}$$

Thus **maximum power delivered**, $\boldsymbol{P} = I^2R = (2.683)^2(25)$

$$= \mathbf{180\ W}$$

> Problem 2. If the load impedance Z in Figure 35.2 of problem 1 consists of variable resistance R and variable reactance X, determine (a) the value of Z that results in maximum power transfer, and (b) the value of the maximum power.

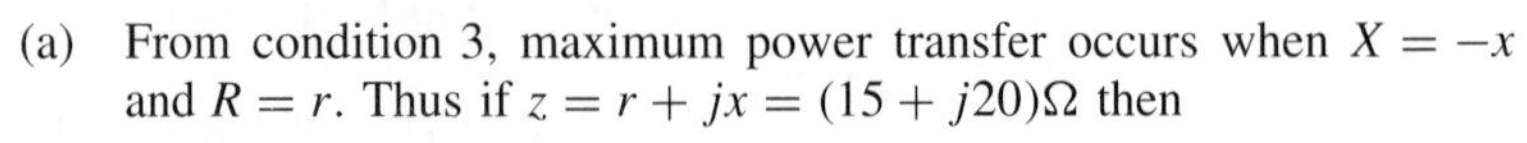

(a) From condition 3, maximum power transfer occurs when $X = -x$ and $R = r$. Thus if $z = r + jx = (15 + j20)\Omega$ then

$$\mathbf{Z = (15 - j20)\Omega} \text{ or } \mathbf{25\angle -53.13°\ \Omega}$$

(b) Total circuit impedance at maximum power transfer condition, $Z_T = z + Z$, i.e.,

$$Z_T = (15 + j20) + (15 - j20) = 30\ \Omega$$

$$\text{Hence current in load, } I = \frac{E}{Z_T} = \frac{120\angle 0°}{30} = 4\angle 0°\ \text{A}$$

and **maximum power** transfer in the load, $P = I^2R = (4)^2(15)$

$$= \mathbf{240\ W}$$

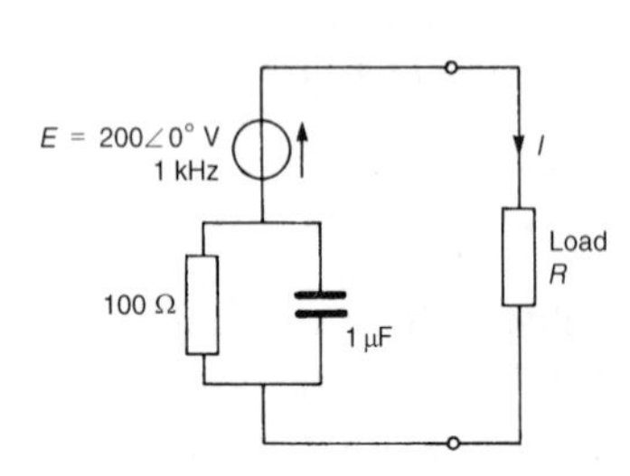

Figure 35.3

> Problem 3. For the network shown in Figure 35.3, determine (a) the value of the load resistance R required for maximum power transfer, and (b) the value of the maximum power transferred.

(a) This problem is an example of condition 1, where maximum power transfer is achieved when $R = |z|$. Source impedance z is composed of a 100 Ω resistance in parallel with a 1 μF capacitor.

$$\text{Capacitive reactance, } X_C = \frac{1}{2\pi f C} = \frac{1}{2\pi(1000)(1 \times 10^{-6})} = 159.15\ \Omega$$

Hence source impedance,

$$z = \frac{(100)(-j159.15)}{(100 - j159.15)} = \frac{15915\angle{-90°}}{187.96\angle - 57.86°}$$
$$= 84.67\angle{-32.14°}\ \Omega \text{ or } (71.69 - j45.04)\Omega$$

Thus the value of **load resistance** for maximum power transfer is **84.67 Ω** (i.e., $|z|$)

(b) With $z = (71.69 - j45.04)\Omega$ and $R = 84.67\ \Omega$ for maximum power transfer, the total circuit impedance,

$$Z_T = 71.69 + 84.67 - j45.04$$
$$= (156.36 - j45.04)\Omega \text{ or } 162.72\angle{-16.07°}\ \Omega$$

$$\text{Current flowing in the load, } I = \frac{V}{Z_T} = \frac{200\angle 0°}{162.72\angle{-16.07°}} = 1.23\angle 16.07°\ \text{A}$$

Thus the **maximum power** transferred, $P = I^2R = (1.23)^2(84.67)$ $= \mathbf{128\ W}$

Problem 4. In the network shown in Figure 35.4 the load consists of a fixed capacitive reactance of 7 Ω and a variable resistance R. Determine (a) the value of R for which the power transferred to the load is a maximum, and (b) the value of the maximum power.

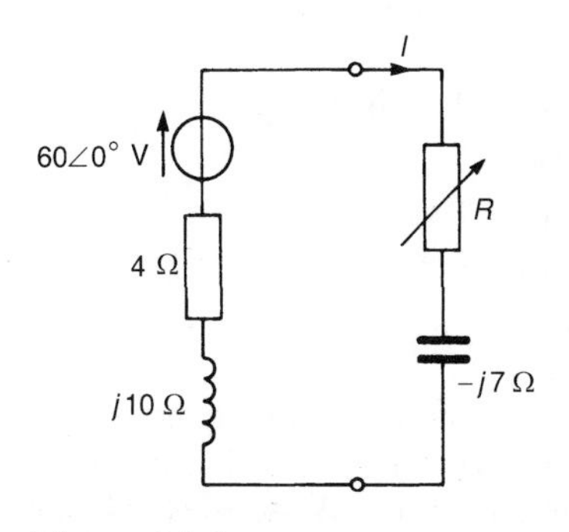

Figure 35.4

(a) From condition (4), maximum power transfer is achieved when

$$\boldsymbol{R} = \sqrt{[r^2 + (x + X)^2]} = \sqrt{[4^2 + (10 - 7)^2]}$$
$$= \sqrt{(4^2 + 3^2)} = \mathbf{5\ \Omega}$$

(b) $$\text{Current } I = \frac{60\angle 0°}{(4 + j10) + (5 - j7)} = \frac{60\angle 0°}{(9 + j3)}$$
$$= \frac{60\angle 0°}{9.487\angle 18.43°} = 6.324\angle{-18.43°}\ \text{A}$$

Thus the **maximum power** transferred, $P = I^2R = (6.324)^2(5)$ $= \mathbf{200\ W}$

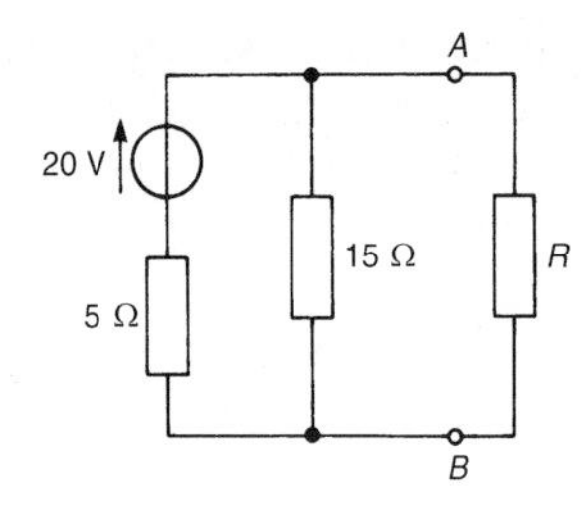

Figure 35.5

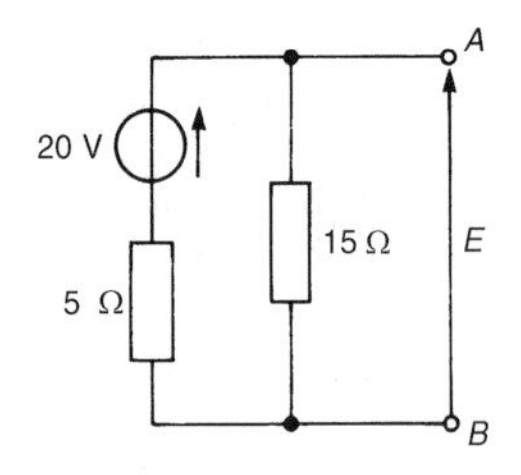

Figure 35.6

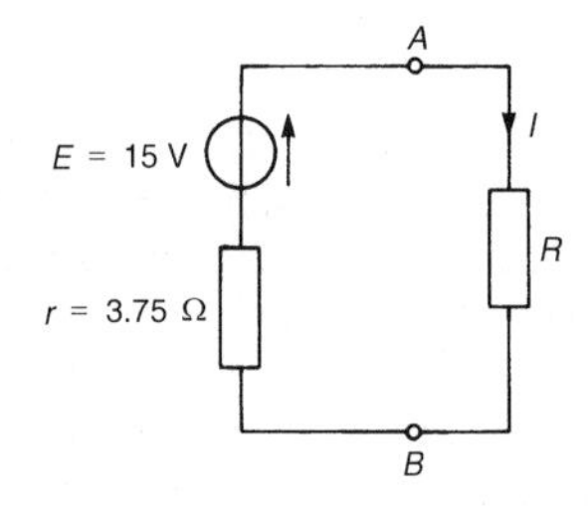

Figure 35.7

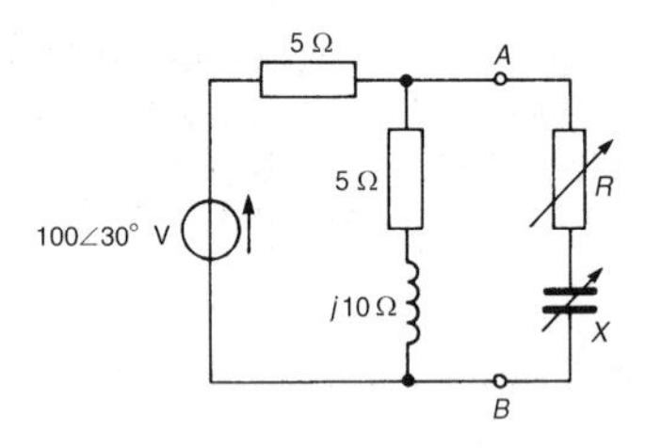

Figure 35.8

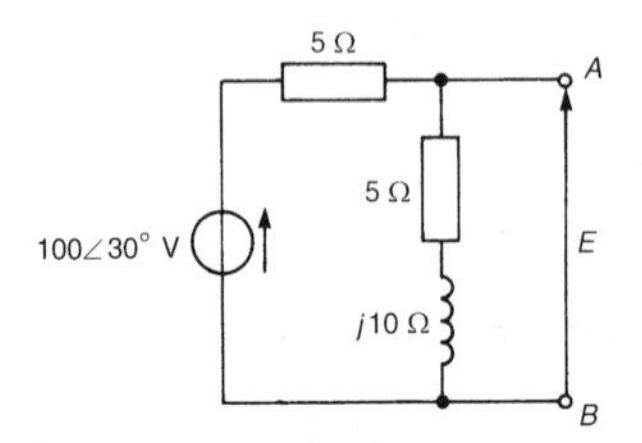

Figure 35.9

Problem 5. Determine the value of the load resistance R shown in Figure 35.5 that gives maximum power dissipation and calculate the value of this power.

Using the procedure of Thévenin's theorem (see page 576):

(i) R is removed from the network as shown in Figure 35.6

(ii) P.d. across AB, $E = (15/(15+5))(20) = 15$ V

(iii) Impedance 'looking-in' at terminals AB with the 20 V source removed is given by $r = (5 \times 15)/(5+15) = 3.75\ \Omega$

(iv) The equivalent Thévenin circuit supplying terminals AB is shown in Figure 35.7. From condition (2), for maximum power transfer, $R = r$, i.e., $\boldsymbol{R = 3.75\ \Omega}$

$$\text{Current } I = \frac{E}{R+r} = \frac{15}{3.75+3.75} = 2 \text{ A}$$

Thus the **maximum power** dissipated in the load,

$$P = I^2R = (2)^2(3.75) = \mathbf{15\ W}$$

Problem 6. Determine, for the network shown in Figure 35.8, (a) the values of R and X that will result in maximum power being transferred across terminals AB, and (b) the value of the maximum power.

(a) Using the procedure for Thévenin's theorem:

(i) Resistance R and reactance X are removed from the network as shown in Figure 35.9

(ii) P.d. across AB,

$$E = \left(\frac{5+j10}{5+j10+5}\right)(100\angle 30^\circ) = \frac{(11.18\angle 63.43^\circ)(100\angle 30^\circ)}{14.14\angle 45^\circ}$$

$$= 79.07\angle 48.43^\circ \text{ V}$$

(iii) With the $100\angle 30^\circ$ V source removed the impedance, z, 'looking in' at terminals AB is given by:

$$z = \frac{(5)(5+j10)}{(5+5+j10)} = \frac{(5)(11.18\angle 63.43^\circ)}{(14.14\angle 45^\circ)}$$

$$= 3.953\angle 18.43^\circ\ \Omega \text{ or } (3.75 + j1.25)\Omega$$

(iv) The equivalent Thévenin circuit is shown in Figure 35.10. From condition 3, maximum power transfer is achieved when $X = -x$ and $R = r$, i.e., in this case when $\boldsymbol{X = -1.25\ \Omega}$ and $\boldsymbol{R = 3.75\ \Omega}$

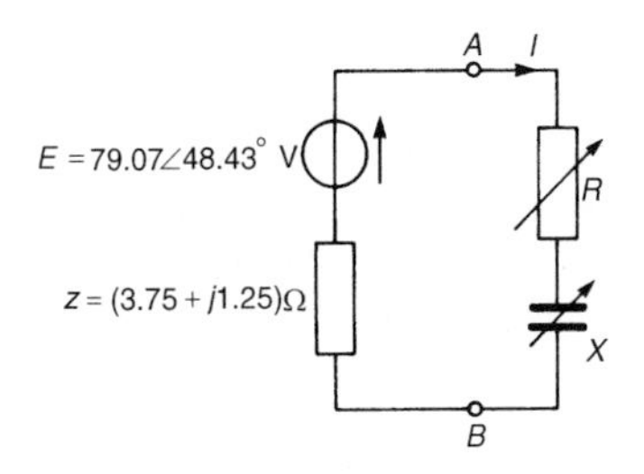

Figure 35.10

(b) Current $I = \dfrac{E}{z+Z} = \dfrac{79.07\angle 48.43^\circ}{(3.75 + j1.25) + (3.75 - j1.25)}$

$$= \frac{79.07\angle 48.43^\circ}{7.5} = 10.543\angle 48.43^\circ \text{ A}$$

Thus the **maximum power** transferred, $P = I^2R = (10.543)^2(3.75)$

$$= \mathbf{417\ W}$$

Further problems on the maximum power transfer theorems may be found in Section 35.3, problems 1 to 10, page 626.

35.2 Impedance matching

It is seen from Section 35.1 that when it is necessary to obtain the maximum possible amount of power from a source, it is advantageous if the circuit components can be adjusted to give equality of impedances. This adjustment is called **'impedance matching'** and is an important consideration in electronic and communications devices which normally involve small amounts of power. Examples where matching is important include coupling an aerial to a transmitter or receiver, or coupling a loudspeaker to an amplifier.

The mains power supply is considered as infinitely large compared with the demand upon it, and under such conditions it is unnecessary to consider the conditions for maximum power transfer. With transmission lines (see Chapter 44), the lines are 'matched', ideally, i.e., terminated in their characteristic impedance.

With d.c. generators, motors or secondary cells, the internal impedance is usually very small and in such cases, if an attempt is made to make the load impedance as small as the source internal impedance, overloading of the source results.

A method of achieving maximum power transfer between a source and a load is to adjust the value of the load impedance to match the source impedance, which can be done using a **'matching-transformer'**.

A transformer is represented in Figure 35.11 supplying a load impedance Z_L.

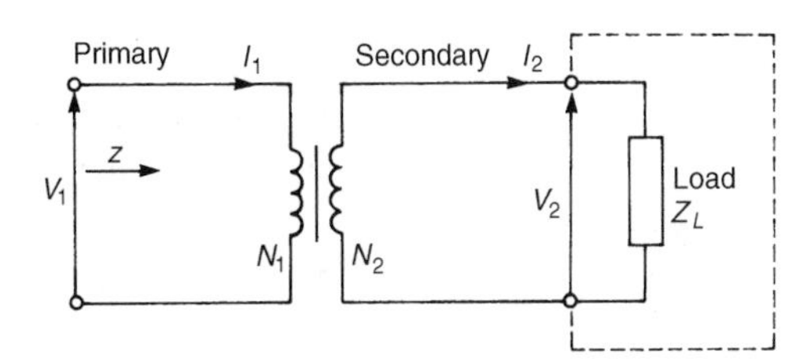

Figure 35.11 *Matching impedance by means of a transformer*

Small transformers used in low power networks are usually regarded as ideal (i.e., losses are negligible), such that

$$\frac{V_1}{V_2} = \frac{N_1}{N_2} = \frac{I_2}{I_1}$$

From Figure 35.11, the primary input impedance $|z|$ is given by

$$|z| = \frac{V_1}{I_1} = \frac{(N_1/N_2)V_2}{(N_2/N_1)I_2} = \left(\frac{N_1}{N_2}\right)^2 \frac{V_2}{I_2}$$

Since the load impedance $|Z_L| = V_2/I_2$,

$$\boldsymbol{|z| = \left(\frac{N_1}{N_2}\right)^2 |Z_L|} \qquad (35.2)$$

If the input impedance of Figure 35.11 is purely resistive (say, r) and the load impedance is purely resistive (say, R_L) then equation (35.2) becomes

$$\boldsymbol{r = \left(\frac{N_1}{N_2}\right)^2 R_L} \qquad (35.3)$$

(This is the case introduced in Section 20.10, page 334).

Thus by varying the value of the transformer turns ratio, the equivalent input impedance of the transformer can be 'matched' to the impedance of a source to achieve maximum power transfer.

Problem 7. Determine the optimum value of load resistance for maximum power transfer if the load is connected to an amplifier of output resistance 448 Ω through a transformer with a turns ratio of 8:1

The equivalent input resistance r of the transformer must be 448 Ω for maximum power transfer. From equation (35.3), $r = (N_1/N_2)^2 R_L$, from which, load resistance $\boldsymbol{R_L} = r(N_2/N_1)^2 = 448(1/8)^2 = \mathbf{7\ \Omega}$

Problem 8. A generator has an output impedance of $(450 + j60)\Omega$. Determine the turns ratio of an ideal transformer necessary to match the generator to a load of $(40 + j19)\Omega$ for maximum transfer of power.

Let the output impedance of the generator be z, where $z = (450 + j60)\Omega$ or $453.98\angle 7.59^\circ\ \Omega$, and the load impedance be Z_L, where $Z_L = (40 + j19)\Omega$ or $44.28\angle 25.41^\circ\ \Omega$. From Figure 35.11 and equation (35.2), $z = (N_1/N_2)^2 Z_L$. Hence

transformer turns ratio $\left(\frac{N_1}{N_2}\right) = \sqrt{\frac{z}{Z_L}} = \sqrt{\frac{453.98}{44.28}} = \sqrt{(10.25)} = \mathbf{3.20}$

Problem 9. A single-phase, 240 V/1920 V ideal transformer is supplied from a 240 V source through a cable of resistance 5 Ω. If the load across the secondary winding is 1.60 kΩ determine (a) the primary current flowing, and (b) the power dissipated in the load resistance.

The network is shown in Figure 35.12.

(a) Turns ratio, $\dfrac{N_1}{N_2} = \dfrac{V_1}{V_2} = \dfrac{240}{1920} = \dfrac{1}{8}$

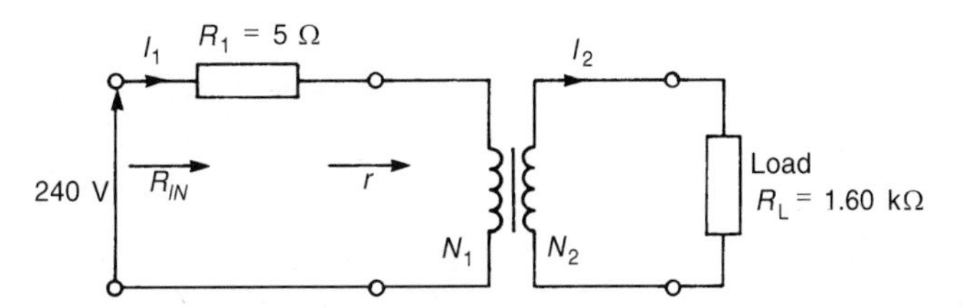

Figure 35.12

Equivalent input resistance of the transformer,

$$r = \left(\frac{N_1}{N_2}\right)^2 R_L = \left(\frac{1}{8}\right)^2 (1600) = 25\ \Omega$$

Total input resistance, $R_{IN} = R_1 + r = 5 + 25 = 30\ \Omega$. Hence the primary current, $\boldsymbol{I_1} = V_1/R_{IN} = 240/30 = \mathbf{8\ A}$

(b) For an ideal transformer, $\dfrac{V_1}{V_2} = \dfrac{I_2}{I_1}$

from which, $I_2 = I_1\left(\dfrac{V_1}{V_2}\right) = (8)\left(\dfrac{240}{1920}\right) = 1\ \text{A}$

Power dissipated in the load resistance, $P = I_2^2 R_L = (1)^2(1600)$

$= \mathbf{1.6\ kW}$

Problem 10. An ac. source of 30∠0° V and internal resistance 20 kΩ is matched to a load by a 20:1 ideal transformer. Determine for maximum power transfer (a) the value of the load resistance, and (b) the power dissipated in the load.

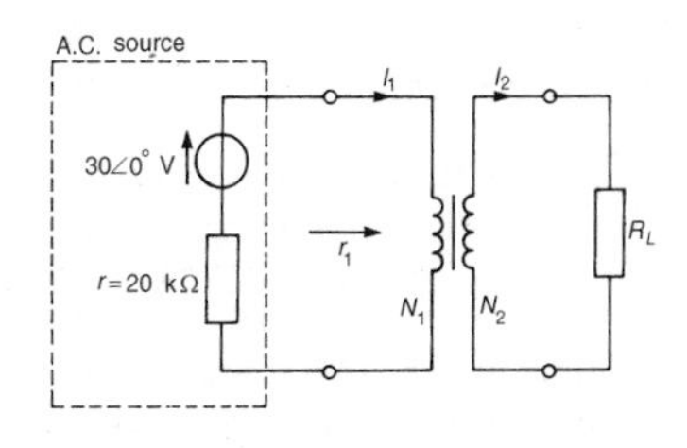

Figure 35.13

The network diagram is shown in Figure 35.13.

(a) For maximum power transfer, r_1 must be equal to 20 kΩ. From equation (35.3), $r_1 = (N_1/N_2)^2 R_L$ from which,

$$\text{load resistance } \boldsymbol{R_L} = r_1\left(\frac{N_2}{N_1}\right)^2 = (20\,000)\left(\frac{1}{20}\right)^2 = \mathbf{50\ \Omega}$$

(b) The total input resistance when the source is connected to the matching transformer is $(r + r_1)$, i.e., $20\ \text{k}\Omega + 20\ \text{k}\Omega = 40\ \text{k}\Omega$. Primary current,

$$I_1 = V/40\,000 = 30/40\,000 = 0.75\ \text{mA}$$

$$\frac{N_1}{N_2} = \frac{I_2}{I_1} \text{ from which, } I_2 = I_1\left(\frac{N_1}{N_2}\right) = (0.75 \times 10^{-3})\left(\frac{20}{1}\right)$$

$$= 15\ \text{mA}$$

Power dissipated in load resistance R_L is given by

$$\boldsymbol{P} = I_2^2 R_L = (15 \times 10^{-3})^2(50) = \mathbf{0.01125\ W} \text{ or } \mathbf{11.25\ mW}$$

Further problems on impedance matching may be found in Section 35.3 following, problems 11 to 15, page 627.

35.3 Further problems on maximum power transfer theorems and impedance matching

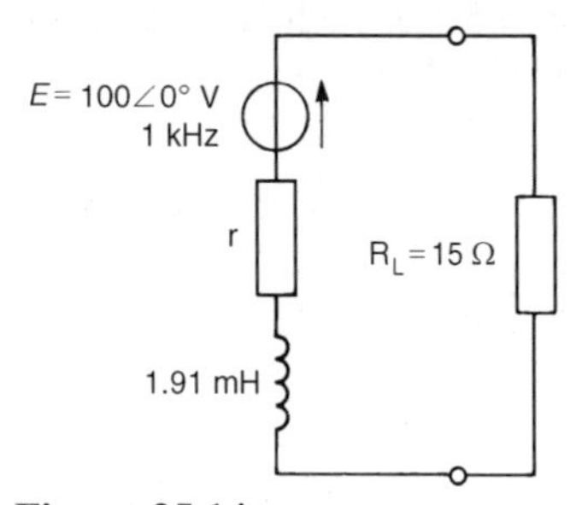

Figure 35.14

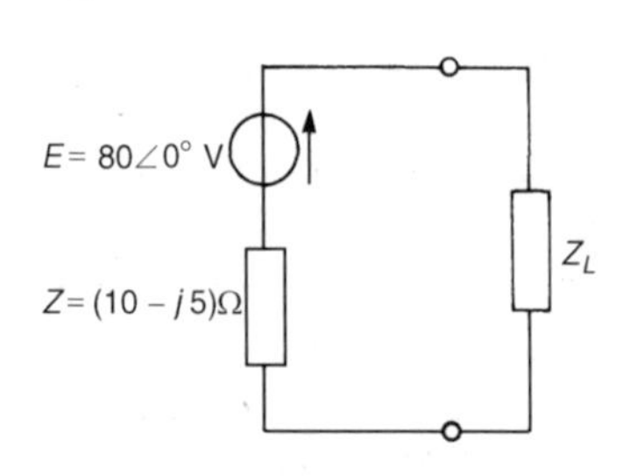

Figure 35.15

Maximum power transfer theorems

1 For the circuit shown in Figure 35.14 determine the value of the source resistance r if the maximum power is to he dissipated in the 15 Ω load. Determine the value of this maximum power. [$r = 9\ \Omega, P = 208.4$ W]

2 In the circuit shown in Figure 35.15 the load impedance Z_L is a pure resistance R. Determine (a) the value of R for maximum power to be transferred from the source to the load, and (b) the value of the maximum power delivered to R. [(a) 11.18 Ω (b) 151.1 W]

3 If the load impedance Z_L in Figure 35.15 of problem 2 consists of a variable resistance R and variable reactance X, determine (a) the value of Z_L which results in maximum power transfer, and (b) the value of the maximum power. [(a) $(10 + j5)\Omega$ (b) 160 W]

4 For the network shown in Figure 35.16 determine (a) the value of the load resistance R_L required for maximum power transfer, and (b) the value of the maximum power. [(a) 26.83 Ω (b) 35.4 W]

5 Find the value of the load resistance R_L shown in Figure 35.17 that gives maximum power dissipation, and calculate the value of this power. [$R_L = 2.1\ \Omega, P = 23.3$ W]

6 For the circuit shown in Figure 35.18 determine (a) the value of load resistance R_L which results in maximum power transfer, and (b) the value of the maximum power. [(a) 16 Ω (b) 48 W]

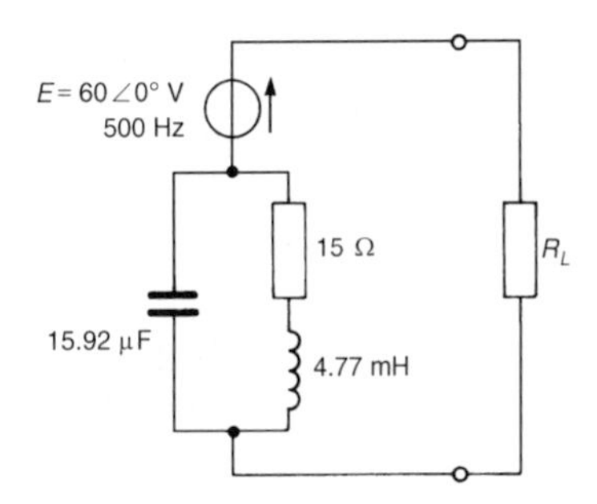

Figure 35.16

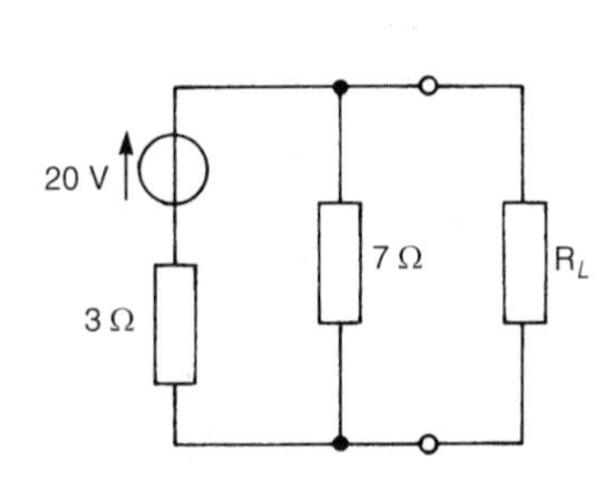

Figure 35.17

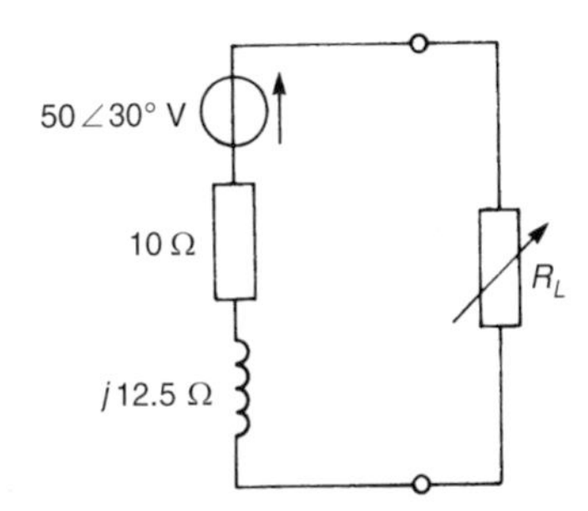

Figure 35.18

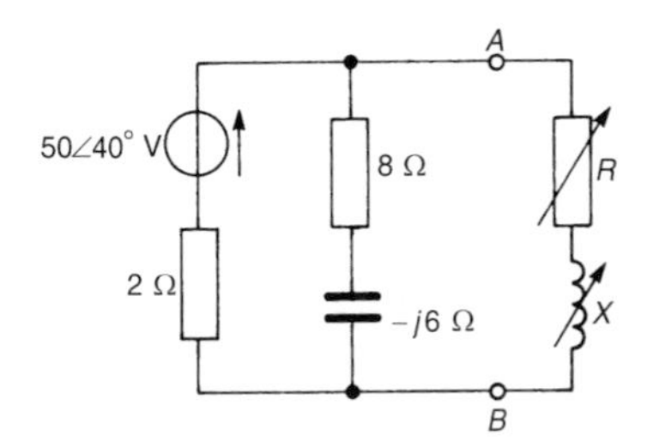

Figure 35.19

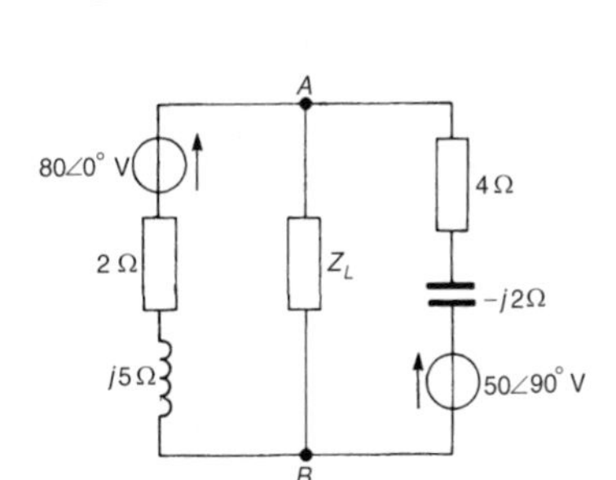

Figure 35.20

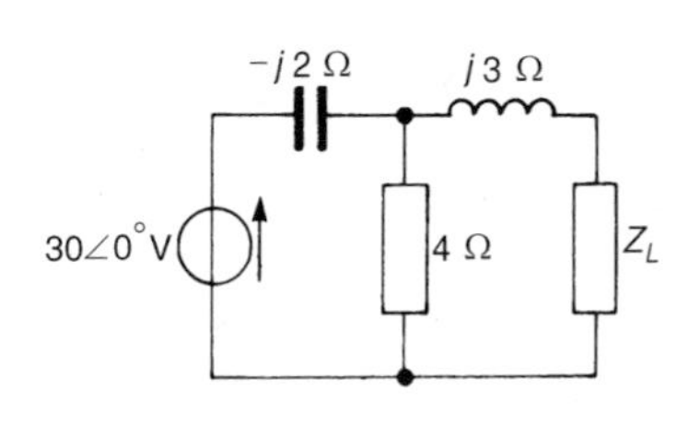

Figure 35.21

7 Determine, for the network shown in Figure 35.19, (a) the values of R and X which result in maximum power being transferred across terminals AB, and (b) the value of the maximum power.
[(a) $R = 1.706\ \Omega$, $X = 0.177\ \Omega$ (b) 269 W]

8 A source of $120\angle 0°$ V and impedance $(5 + j3)\Omega$ supplies a load consisting of a variable resistor R in series with a fixed capacitive reactance of 8 Ω. Determine (a) the value of R to give maximum power transfer, and (b) the value of the maximum power.
[(a) 7.07 Ω (b) 596.5 W]

9 If the load Z_L between terminals A and B of Figure 35.20 is variable in both resistance and reactance determine the value of Z_L such that it will receive maximum power. Calculate the value of the maximum power. [$R = 3.47\ \Omega$, $X = -0.93\ \Omega$, 13.6 W]

10 For the circuit of Figure 35.21, determine the value of load impedance Z_L for maximum load power if (a) Z_L comprises a variable resistance R and variable reactance X, and (b) Z_L is a pure resistance R. Determine the values of load power in each case
[(a) $R = 0.80\ \Omega$, $X = -1.40\ \Omega$, $P = 225$ W
(b) $R = 1.61\ \Omega$, $P = 149.2$ W]

Impedance matching

11 The output stage of an amplifier has an output resistance of 144 Ω. Determine the optimum turns ratio of a transformer that would match a load resistance of 9 Ω to the output resistance of the amplifier for maximum power transfer. [4:1]

12 Find the optimum value of load resistance for maximum power transfer if a load is connected to an amplifier of output resistance 252 Ω through a transformer with a turns ratio of 6:1 [7 Ω]

13 A generator has an output impedance of $(300 + j45)\Omega$. Determine the turns ratio of an ideal transformer necessary to match the generator to a load of $(37 + j19)\Omega$ for maximum power transfer.
[2.70]

14 A single-phase, 240 V/2880 V ideal transformer is supplied from a 240 V source through a cable of resistance 3.5 Ω. If the load across the secondary winding is 1.8 kΩ, determine (a) the primary current flowing, and (b) the power dissipated in the load resistance.
[(a) 15 A (b) 2.81 kW]

15 An a.c. source of $20\angle 0^\circ$ V and internal resistance 10.24 kΩ is matched to a load for maximum power transfer by a 16:1 ideal transformer. Determine (a) the value of the load resistance, and (b) the power dissipated in the load. [(a) 40 Ω (b) 9.77 mW]

Assignment 11

This assignment covers the material in chapters 34 and 35.

The marks for each question are shown in brackets at the end of each question.

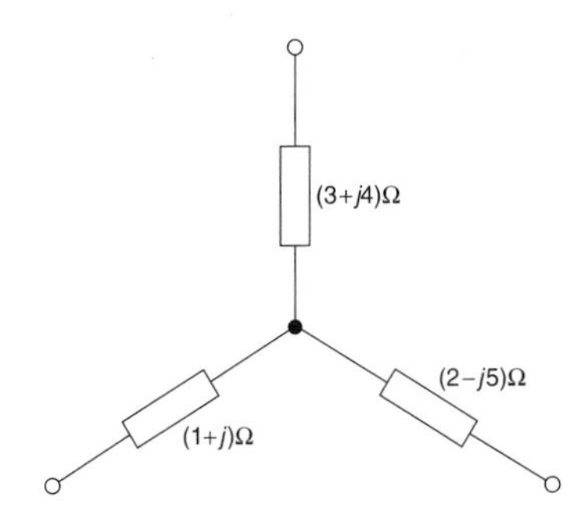

Figure A11.1

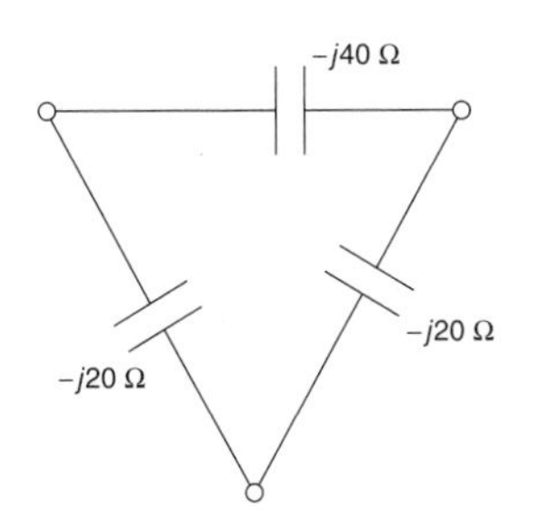

Figure A11.2

1 Determine the delta-connected equivalent network for the star-connected impedances shown in Figure A11.1 (9)

2 Transform the delta-connection in Figure A11.2 to it's equivalent star connection. Hence determine for the network shown in Figure A11.3

(a) the total circuit impedance

(b) the current I

(c) the current in the 20 Ω resistor

(d) the power dissipated in the 20 Ω resistor. (17)

3 If the load impedance Z in Figure A11.4 consists of variable resistance and variable reactance, find (a) the value of Z that results in maximum power transfer, and (b) the value of the maximum power. (6)

4 Determine the value of the load resistance R in Figure A11.5 that gives maximum power dissipation and calculate the value of power. (9)

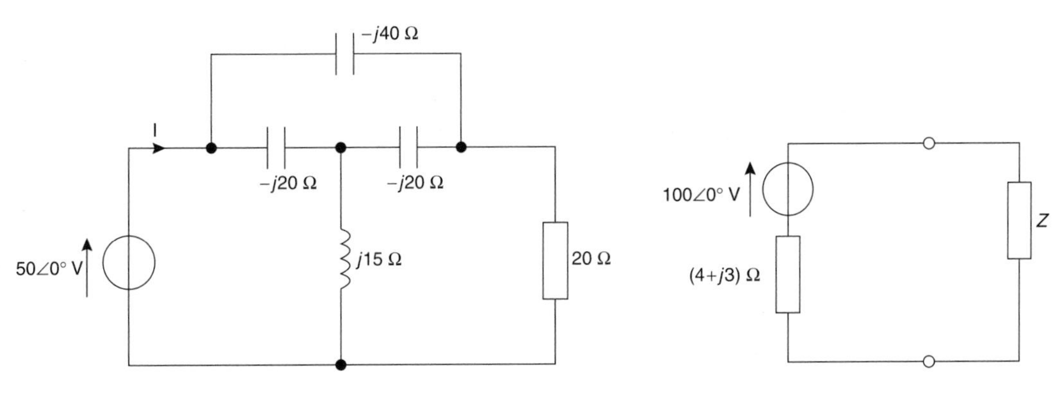

Figure A11.3

Figure A11.4

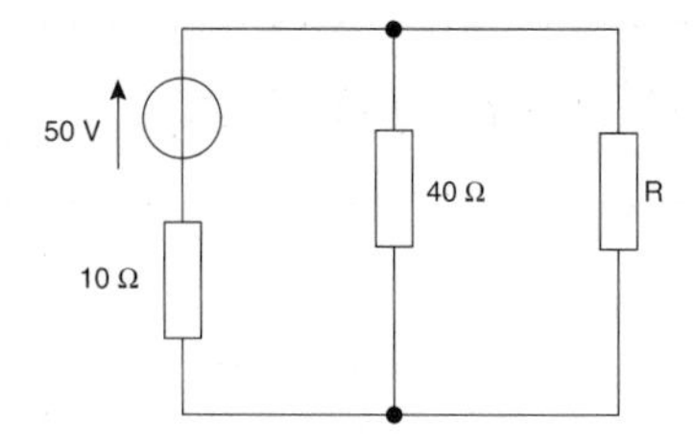

Figure A11.5

5 An a.c. source of $10\angle 0°$ V and internal resistance 5 kΩ is matched to a load for maximum power transfer by a 5:1 ideal transformer. Determine (a) the value of the load resistance, and (b) the power dissipated in the load. (9)

36 Complex Waveforms

At the end of this chapter you should be able to:

- define a complex wave
- recognize periodic functions
- recognize the general equation of a complex waveform
- use harmonic synthesis to build up a complex wave
- recognize characteristics of waveforms containing odd, even or odd and even harmonics, with or without phase change
- calculate rms and mean values, and form factor of a complex wave
- calculate power associated with complex waves
- perform calculations on single phase circuits containing harmonics
- define and perform calculations on harmonic resonance
- list and explain some sources of harmonics

36.1 Introduction

In preceding chapters a.c. supplies have been assumed to be sinusoidal, this being a form of alternating quantity commonly encountered in electrical engineering. However, many supply waveforms are **not** sinusoidal. For example, sawtooth generators produce ramp waveforms, and rectangular waveforms may be produced by multivibrators. A waveform that is not sinusoidal is called a **complex wave.** Such a waveform may be shown to be composed of the sum of a series of sinusoidal waves having various interrelated periodic times.

A function $f(t)$ is said to be **periodic** if $f(t+T) = f(t)$ for all values of t, where T is the interval between two successive repetitions and is called the **period** of the function $f(t)$. A sine wave having a period of $2\pi/\omega$ is a familiar example of a periodic function.

A typical complex periodic-voltage waveform, shown in Figure 36.1, has period T seconds and frequency f hertz. A complex wave such as this can be resolved into the sum of a number of sinusoidal waveforms, and each of the sine waves can have a different frequency, amplitude and phase.

The initial, major sine wave component has a frequency f equal to the frequency of the complex wave and this frequency is called the **fundamental frequency**. The other sine wave components are known as **harmonics,** these having frequencies which are integer multiples of frequency f. Hence the second harmonic has a frequency of $2f$, the third harmonic has a frequency of $3f$, and so on. Thus if the fundamental (i.e.,

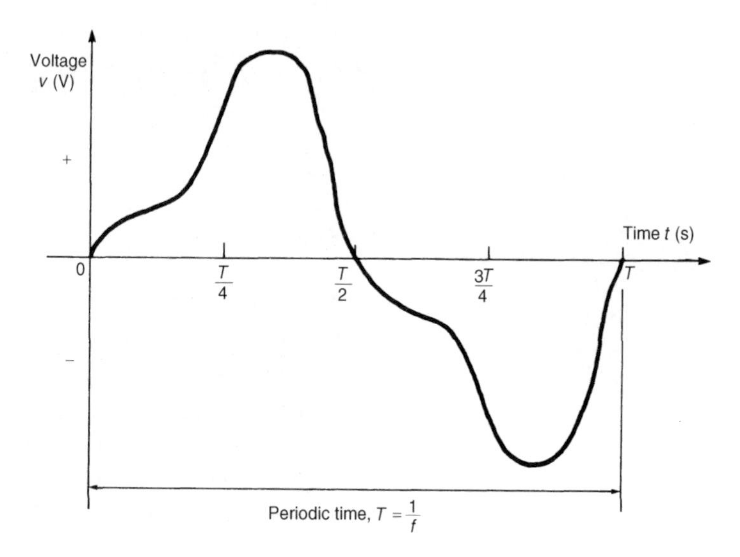

Figure 36.1 *Typical complex periodic voltage waveform*

supply) frequency of a complex wave is 50 Hz, then the third harmonic frequency is 150 Hz, the fourth harmonic frequency is 200 Hz, and so on.

36.2 The general equation for a complex waveform

The instantaneous value of a complex voltage wave v acting in a linear circuit may be represented by the general equation

$$\boldsymbol{v = V_m \sin(\omega t + \Psi_1) + V_{2m} \sin(2\omega t + \Psi_2) + \cdots + V_{nm} \sin(n\omega t + \Psi_n) \text{volts}} \quad (36.1)$$

Here $V_{1m} \sin(\omega t + \Psi_1)$ represents the fundamental component of which V_{1m} is the maximum or peak value, frequency, $f = \omega/2\pi$ and Ψ_1 is the phase angle with respect to time, $t = 0$.

Similarly, $V_{2m} \sin(2\omega t + \Psi_2)$ represents the second harmonic component, and $V_{nm} \sin(n\omega t + \Psi_n)$ represents the nth harmonic component, of which V_{nm} is the peak value, frequency $= n\omega/2\pi (= nf)$ and Ψ_n is the phase angle.

In the same way, the instantaneous value of a complex current i may be represented by the general equation

$$\boldsymbol{i = I_{1m} \sin(\omega t + \theta_1) + I_{2m} \sin(2\omega t + \theta_2) + \cdots + I_{nm} \sin(n\omega t + \theta_n) \text{amperes}} \quad (36.2)$$

Where equations (36.1) and (36.2) refer to the voltage across and the current flowing through a given linear circuit, the phase angle between the fundamental voltage and current is $\phi_1 = (\Psi_1 - \theta_1)$, the phase angle between the second harmonic voltage and current is $\phi_2 = (\Psi_2 - \theta_2)$, and so on.

It often occurs that not all harmonic components are present in a complex waveform. Sometimes only the fundamental and odd harmonics are present, and in others only the fundamental and even harmonics are present.

36.3 Harmonic synthesis

Harmonic analysis is the process of resolving a complex periodic waveform into a series of sinusoidal components of ascending order of frequency. Many of the waveforms met in practice can be represented by mathematical expressions similar to those of equations (36.1) and (36.2), and the magnitude of their harmonic components together with their phase may be calculated using **Fourier series** (see *Higher Engineering Mathematics*). **Numerical methods** are used to analyse waveforms for which simple mathematical expressions cannot be obtained. A numerical method of harmonic analysis is explained in Chapter 37. In a laboratory, waveform analysis may be performed using a **waveform analyser** which produces a direct readout of the component waves present in a complex wave.

By adding the instantaneous values of the fundamental and progressive harmonics of a complex wave for given instants in time, the shape of a complex waveform can be gradually built up. This graphical procedure is known as **harmonic synthesis** (synthesis meaning 'the putting together of parts or elements so as to make up a complex whole').

A number of examples of harmonic synthesis will now be considered.

Example 1

Consider the complex voltage expression given by

$$v_a = \mathbf{100 \sin \omega t + 30 \sin 3\omega t \text{ volts}}$$

The waveform is made up of a fundamental wave of maximum value 100 V and frequency, $f = \omega/2\pi$ hertz and a third harmonic component of maximum value 30 V and frequency $= 3\omega/2\pi(= 3f)$, the fundamental and third harmonics being initially in phase with each other. Since the maximum value of the third harmonic is 30 V and that of the fundamental is 100 V, the resultant waveform v_a is said to contain 30/100, i.e., '30% third harmonic'. In Figure 36.2, the fundamental waveform is shown by the broken line plotted over one cycle, the periodic time being $2\pi/\omega$ seconds. On the same axis is plotted $30 \sin 3\omega t$, shown by the dotted line, having a maximum value of 30 V and for which three cycles are completed in time T seconds. At zero time, $30 \sin 3\omega t$ is in phase with $100 \sin \omega t$.

The fundamental and third harmonic are combined by adding ordinates at intervals to produce the waveform for v_a as shown. For example, at time $T/12$ seconds, the fundamental has a value of 50 V and the third harmonic a value of 30 V. Adding gives a value of 80 V for waveform v_a, at time $T/12$ seconds. Similarly, at time $T/4$ seconds, the fundamental has a value of 100 V and the third harmonic a value of -30 V. After addition, the resultant waveform v_a is 70 V at time $T/4$. The procedure

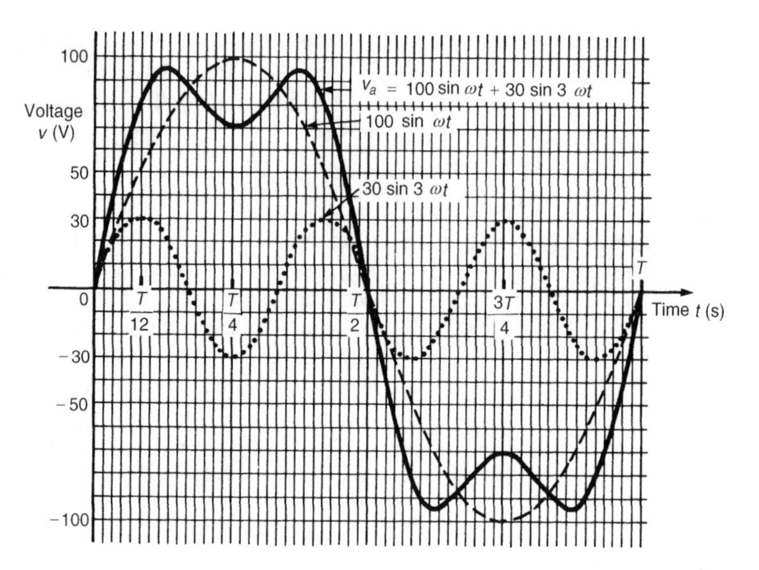

Figure 36.2

is continued between $t = 0$ and $t = T$ to produce the complex waveform for v_a. The negative half-cycle of waveform v_a is seen to be identical in shape to the positive half-cycle.

Example 2

Consider the addition of a fifth harmonic component to the complex waveform of Figure 36.2, giving a resultant waveform expression

$$\boldsymbol{v_b = 100 \sin \omega t + 30 \sin 3\omega t + 20 \sin 5\omega t \text{ volts}}$$

Figure 36.3 shows the effect of adding $(100 \sin \omega t + 30 \sin 3\omega t)$ obtained from Figure 36.2 to $20 \sin 5\omega t$. The shapes of the negative and positive half-cycles are still identical. If further odd harmonics of the appropriate amplitude and phase were added to v_b, a good approximation to **a square wave** would result.

Example 3

Consider the complex voltage expression given by

$$\boldsymbol{v_c = 100 \sin \omega t + 30 \sin \left(3\omega t + \frac{\pi}{2}\right) \text{volts}}$$

This expression is similar to voltage v_a in that the peak value of the fundamental and third harmonic are the same. However the third harmonic has a phase displacement of $\pi/2$ radian leading (i.e., leading $30 \sin 3\omega t$ by $\pi/2$ radian). Note that, since the periodic time of the fundamental is T seconds, the periodic time of the third harmonic is $T/3$ seconds, and a phase displacement of $\pi/2$ radian or $\frac{1}{4}$ cycle of the third harmonic represents a time interval of $(T/3) \div 4$, i.e., $T/12$ seconds.

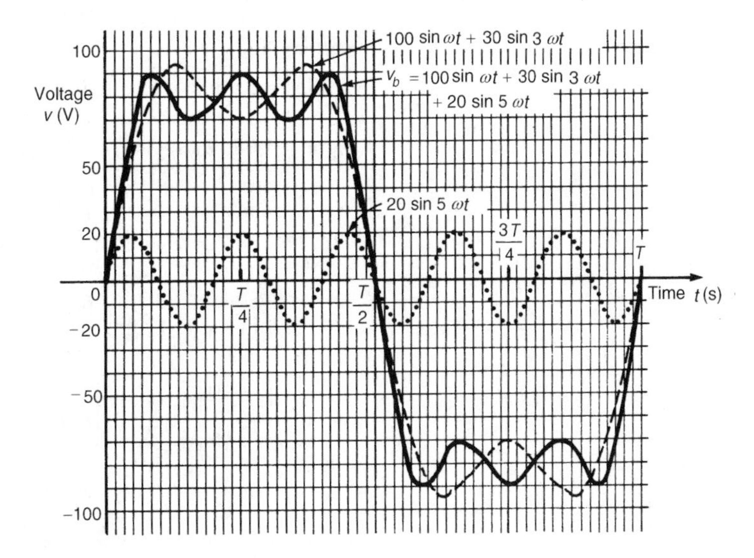

Figure 36.3

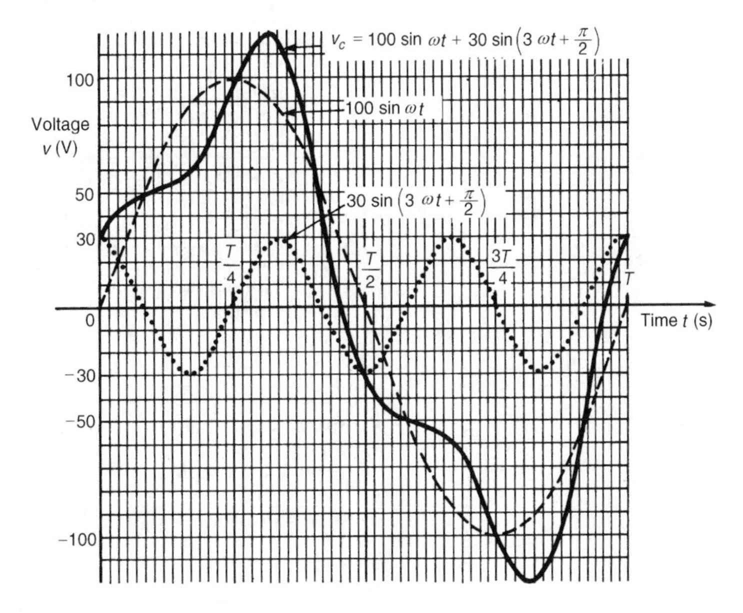

Figure 36.4

Figure 36.4 shows graphs of $100 \sin \omega t$ and $30 \sin(3\omega t + (\pi/2))$ over the time for one cycle of the fundamental. When ordinates of the two graphs are added at intervals, the resultant waveform v_c is as shown. The shape of the waveform v_c is quite different from that of waveform v_a shown in Figure 36.2, even though the percentage third harmonic is the same. If the negative half-cycle in Figure 36.4 is reversed it can be seen that the shape of the positive and negative half-cycles are identical.

Example 4

Consider the complex voltage expression given by

$$v_d = \mathbf{100 \sin \omega t + 30 \sin \left(3\omega t - \frac{\pi}{2}\right) volts}$$

The fundamental, $100 \sin \omega t$, and the third harmonic component, $30 \sin(3\omega t - (\pi/2))$, are plotted in Figure 36.5, the latter lagging $30 \sin 3\omega t$ by $\pi/2$ radian or $T/12$ seconds. Adding ordinates at intervals gives the resultant waveform v_d as shown. The negative half-cycle of v_d is identical in shape to the positive half-cycle.

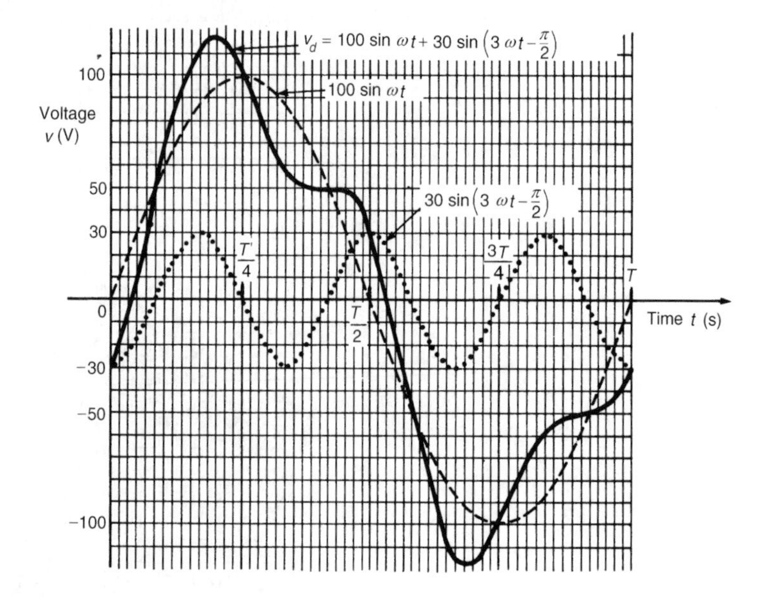

Figure 36.5

Example 5

Consider the complex voltage expression given by

$$v_e = \mathbf{100 \sin \omega t + 30 \sin(3\omega t + \pi) volts}$$

The fundamental, $100 \sin \omega t$, and the third harmonic component, $30 \sin(3\omega t + \pi)$, are plotted as shown in Figure 36.6, the latter leading $30 \sin 3\omega t$ by π radian or $T/6$ seconds. Adding ordinates at intervals gives the resultant waveform v_e as shown. The negative half-cycle of v_e is identical in shape to the positive half-cycle.

Example 6

Consider the complex voltage expression given by

$$v_f = 100 \sin \omega t - 30 \sin \left(3\omega t + \frac{\pi}{2}\right) \text{ volts}$$

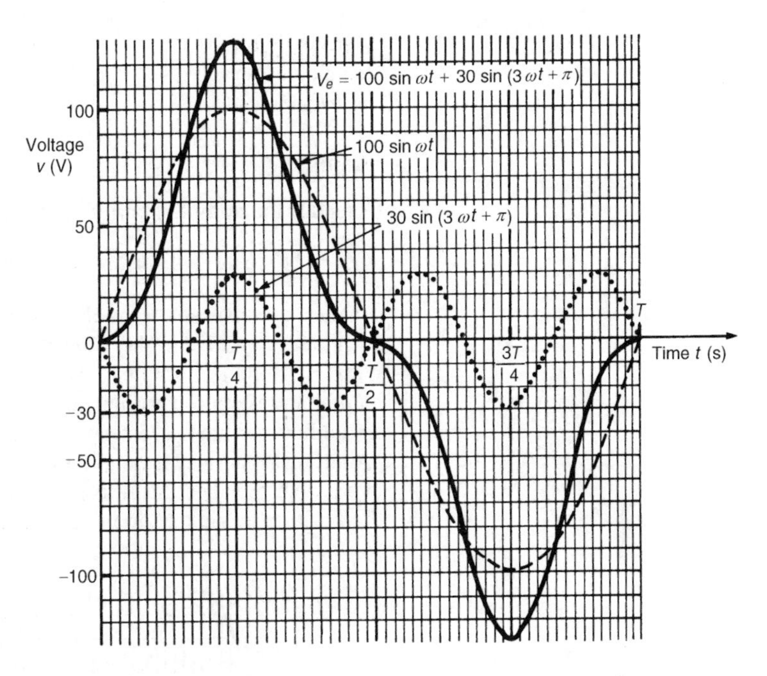

Figure 36.6

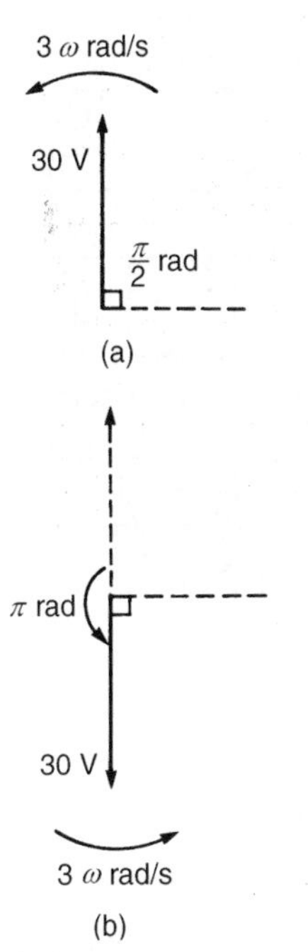

Figure 36.7

The phasor representing $30\sin(3\omega t + (\pi/2))$ is shown in Figure 36.7(a) at time $t = 0$. The phasor representing $-30\sin(3\omega t + (\pi/2))$ is shown in Figure 36.7(b) where it is seen to be in the opposite direction to that shown in Figure 36.7(a).

$-30\sin(3\omega t + (\pi/2))$ is the same as $30\sin(3\omega t - (\pi/2))$. Thus

$$v_f = 100\sin\omega t - 30\sin\left(3\omega t + \frac{\pi}{2}\right) = 100\sin\omega t + 30\sin\left(3\omega t - \frac{\pi}{2}\right)$$

The waveform representing this expression has already been plotted in Figure 36.5.

General conclusions on examples 1 to 6

Whenever odd harmonics are added to a fundamental waveform, whether initially in phase with each other or not, the positive and negative half-cycles of the resultant complex wave are identical in shape (i.e., in Figures 36.2 to 36.6, the values of voltage in the third quadrant — between $T/2$ seconds and $3T/4$ seconds — are identical to the voltage values in the first quadrant — between 0 and $T/4$ seconds, except that they are negative, and the values of voltage in the second and fourth quadrants are identical, except for the sign change). This is a feature of waveforms containing a fundamental and odd harmonics and is true whether harmonics are added or subtracted from the fundamental.

From Figures 36.2 to 36.6, it is seen that a waveform can change its shape considerably as a result of changes in both phase and magnitude of the harmonics.

Example 7

Consider the complex current expression given by

$$i_a = \mathbf{10 \sin \omega t + 4 \sin 2\omega t} \textbf{ amperes}$$

Current i_a consists of a fundamental component, $10 \sin \omega t$, and a second harmonic component, $4 \sin 2\omega t$, the components being initially in phase with each other. Current i_a contains 40% second harmonic. The fundamental and second harmonic are shown plotted separately in Figure 36.8. By adding ordinates at intervals, the complex waveform representing i_a is produced as shown. It is noted that if all the values in the negative half-cycle were reversed then this half-cycle would appear as a mirror image of the positive half-cycle about a vertical line drawn through time, $t = T/2$.

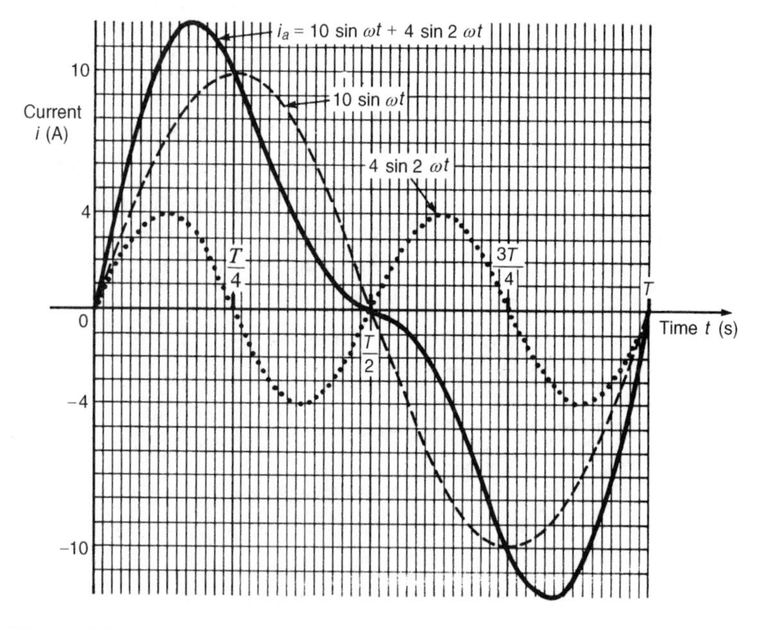

Figure 36.8

Example 8

Consider the complex current expression given by

$$i_b = \mathbf{10 \sin \omega t + 4 \sin 2\omega t + 3 \sin 4\omega t} \textbf{ amperes}$$

The waveforms representing $(10 \sin \omega t + 4 \sin 2\omega t)$ and the fourth harmonic component, $3 \sin 4\omega t$, are each shown separately in Figure 36.9, the former waveform having been produced in Figure 36.8. By adding

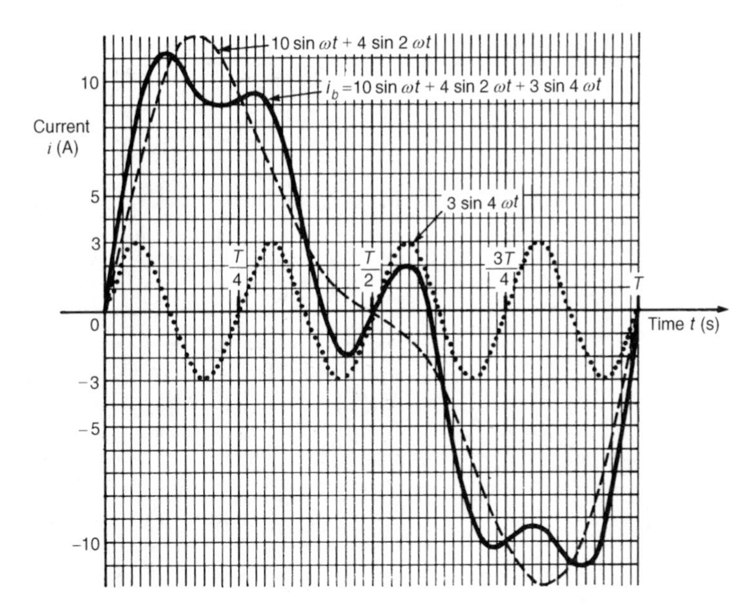

Figure 36.9

ordinates at intervals, the complex waveform for i_b is produced as shown in Figure 36.9. If the half-cycle between times $T/2$ and T is reversed then it is seen to be a mirror image of the half-cycle lying between 0 and $T/2$ about a vertical line drawn through the time, $t = T/2$.

Example 9

Consider the complex current expressions given by

$$i_c = \mathbf{10 \sin \omega t + 4 \sin \left(2\omega t + \frac{\pi}{2}\right) amperes}$$

The fundamental component, $10 \sin \omega t$, and the second harmonic component, having an amplitude of 4 A and a phase displacement of $\pi/2$ radian leading (i.e., leading $4 \sin 2\omega t$ by $\pi/2$ radian or $T/8$ seconds), are shown plotted separately in Figure 36.10. By adding ordinates at intervals, the complex waveform for i_c is produced as shown. The positive and negative half-cycles of the resultant waveform i_c are seen to be quite dissimilar.

Example 10

Consider the complex current expression given by

$$i_d = \mathbf{10 \sin \omega t + 4 \sin(2\omega t + \pi) amperes}$$

The fundamental, $10 \sin \omega t$, and the second harmonic component which leads $4 \sin 2\omega t$ by π rad are shown separately in Figure 36.11. By adding ordinates at intervals, the resultant waveform i_d is produced as shown. If the negative half-cycle is reversed, it is seen to be a mirror image of the positive half-cycle about a line drawn vertically through time, $t = T/2$.

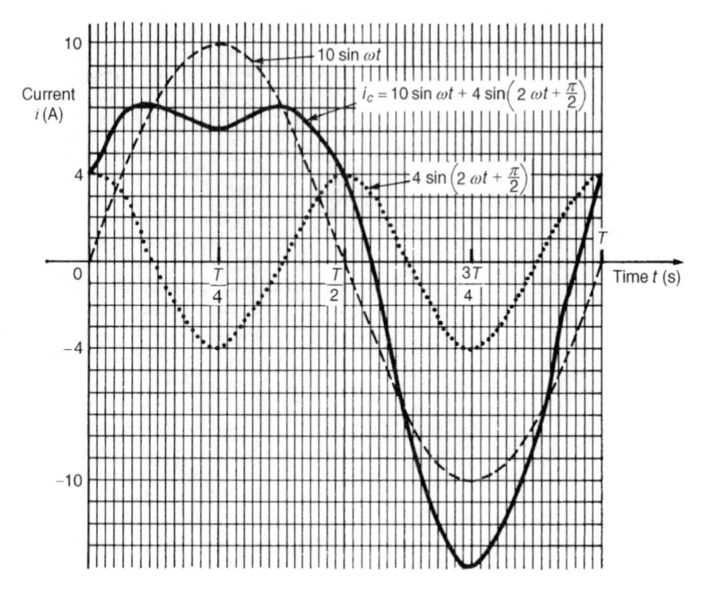

Figure 36.10

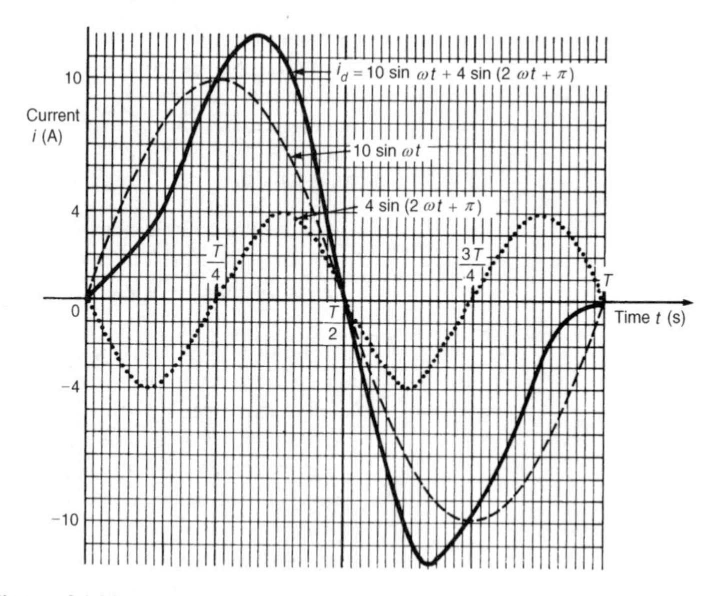

Figure 36.11

General conclusions on examples 7 to 10

Whenever even harmonics are added to a fundamental component:

(a) if the harmonics are initially in phase or if there is a phase-shift of π rad, the negative half-cycle, when reversed, is a mirror image of the positive half-cycle about a vertical line drawn through time, $t = T/2$;

(b) if the harmonics are initially out of phase with each other (i.e., other than π rad), the positive and negative half-cycles are dissimilar.

These are features of waveforms containing the fundamental and even harmonics.

Example 11

Consider the complex voltage expression given by

$$v_g = \mathbf{50 \sin \omega t + 25 \sin 2\omega t + 15 \sin 3\omega t \text{ volts}}$$

The fundamental and the second and third harmonics are each shown separately in Figure 36.12. By adding ordinates at intervals, the resultant waveform v_g is produced as shown. If the negative half-cycle is reversed, it appears as a mirror image of the positive half-cycle about a vertical line drawn through time $= T/2$.

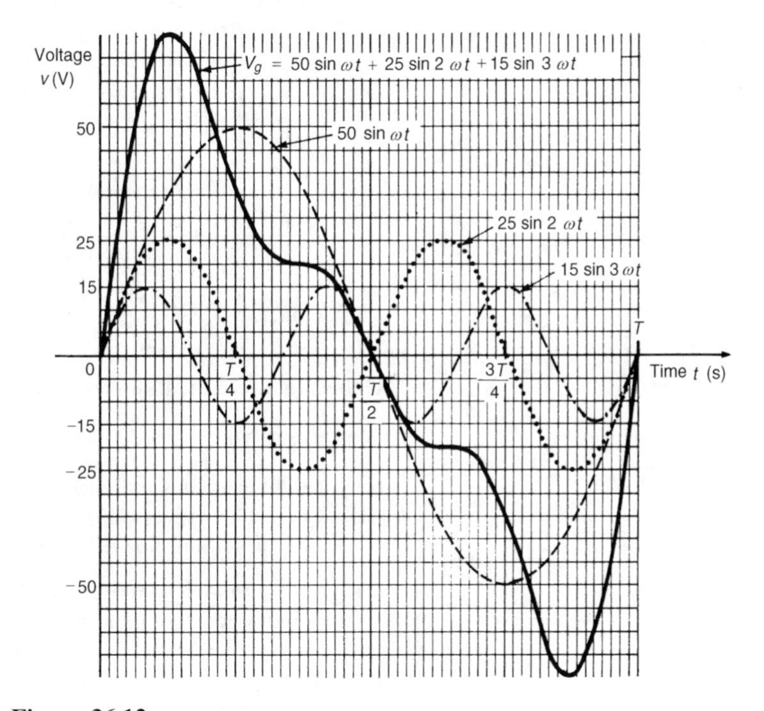

Figure 36.12

Example 12

Consider the complex voltage expression given by

$$v_h = \mathbf{50 \sin \omega t + 25 \sin(2\omega t - \pi) + 15 \sin \left(3\omega t + \frac{\pi}{2}\right) \text{volts}}$$

The fundamental, the second harmonic lagging by π radian and the third harmonic leading by $\pi/2$ radian are initially plotted separately, as shown in Figure 36.13. Adding ordinates at intervals gives the resultant waveform v_h as shown. The positive and negative half-cycles are seen to be quite dissimilar.

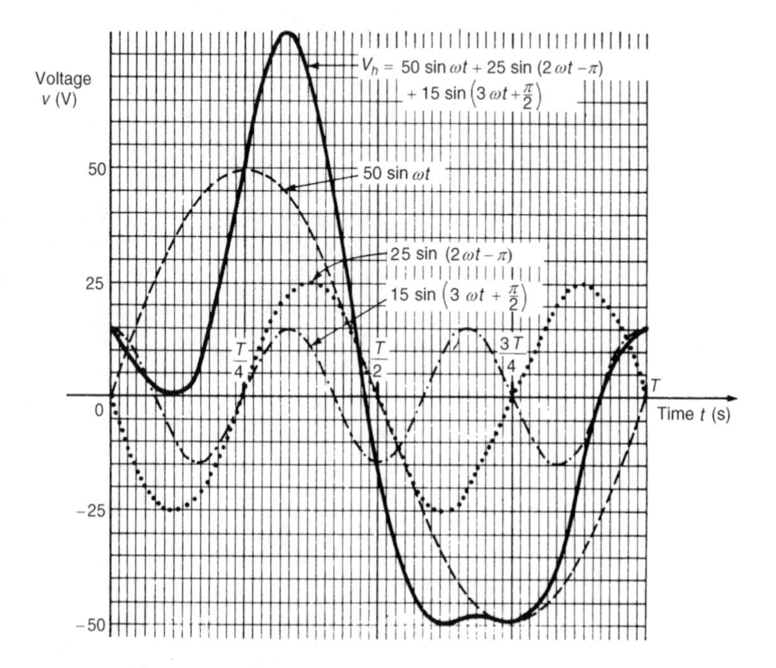

Figure 36.13

General conclusions on examples 11 and 12

Whenever a waveform contains both odd and even harmonics:

(a) if the harmonics are initially in phase with each other, the negative cycle, when reversed, is a mirror image of the positive half-cycle about a vertical line drawn through time, $t = T/2$;

(b) if the harmonics are initially out of phase with each other, the positive and negative half-cycles are dissimilar.

Example 13

Consider the complex current expression given by

$$i = 32 + 50 \sin \omega t + 20 \sin \left(2\omega t - \frac{\pi}{2}\right) \text{mA}$$

The current i comprises three components — a 32 mA d.c. component, a fundamental of amplitude 50 mA and a second harmonic of amplitude 20 mA, lagging by $\pi/2$ radian. The fundamental and second harmonic are shown separately in Figure 36.14. Adding ordinates at intervals gives the complex waveform $50 \sin \omega t + 20 \sin(2\omega t - (\pi/2))$.

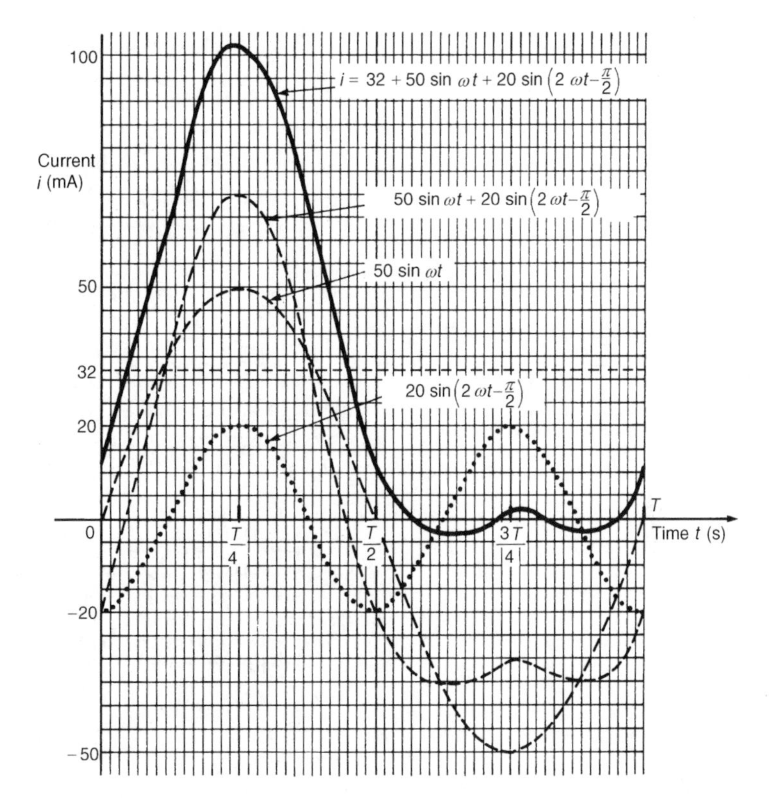

Figure 36.14

This waveform is then added to the 32 mA d.c. component to produce the waveform i as shown. The effect of the d.c. component is seen to be to shift the whole wave 32 mA upward. The waveform approaches that expected from a **half-wave rectifier** (see Section 36.8).

Problem 1. A complex waveform v comprises a fundamental voltage of 240 V rms and frequency 50 Hz, together with a 20% third harmonic which has a phase angle lagging by $3\pi/4$ rad at time $= 0$. (a) Write down an expression to represent voltage v. (b) Use harmonic synthesis to sketch the complex waveform representing voltage v over one cycle of the fundamental component.

(a) A fundamental voltage having an rms value of 240 V has a maximum value, or amplitude of $(\sqrt{2})(240)$, i.e., 339.4 V.

If the fundamental frequency is 50 Hz then angular velocity, $\omega = 2\pi f = 2\pi(50) = 100\pi$ rad/s. Hence the fundamental voltage is represented by $339.4 \sin 100\pi t$ volts. Since the fundamental frequency is 50 Hz, the time for one cycle of the fundamental is given by $T = 1/f = 1/50$ s or 20 ms.

The third harmonic has an amplitude equal to 20% of 339.4 V, i.e., 67.9 V. The frequency of the third harmonic component is $3 \times 50 = 150$ Hz, thus the angular velocity is 2π (150), i.e., 300π rad/s. Hence the third harmonic voltage is represented by $67.9\sin(300\pi t - (3\pi/4))$ volts. Thus

$$\textbf{voltage, } v = \mathbf{339.4\sin 100\pi t + 67.9\sin\left(300\pi t - \frac{3\pi}{4}\right)} \textbf{ volts}$$

(b) One cycle of the fundamental, $339.4\sin 100\pi t$, is shown sketched in Figure 36.15, together with three cycles of the third harmonic component, $67.9\sin(300\pi t - (3\pi/4))$ initially lagging by $3\pi/4$ rad. By adding ordinates at intervals, the complex waveform representing voltage is produced as shown. If the negative half-cycle is reversed, it is seen to be identical to the positive half-cycle, which is a feature of waveforms containing the fundamental and odd harmonics.

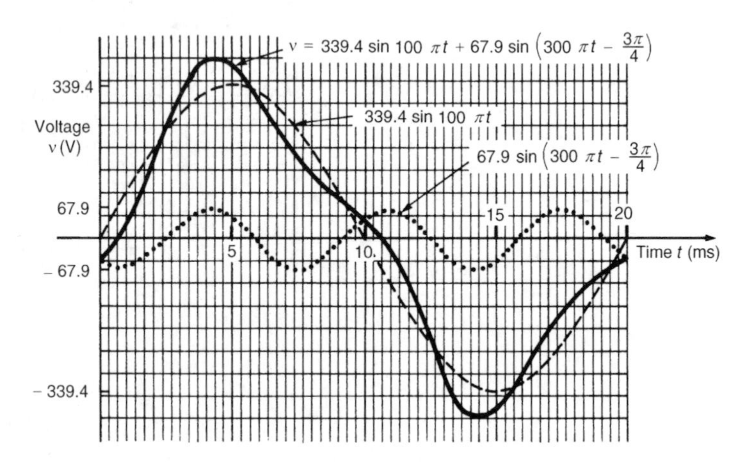

Figure 36.15

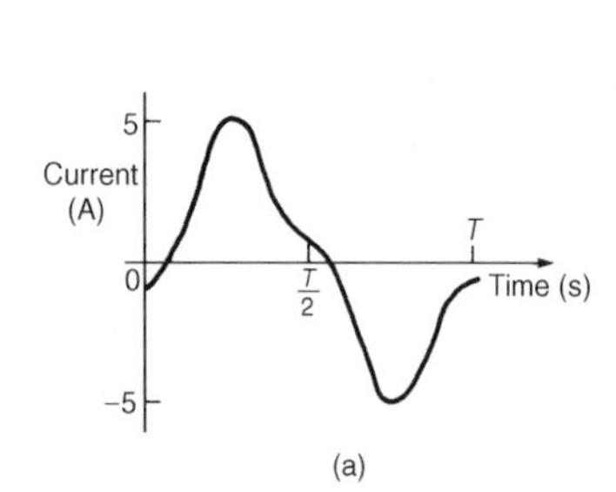

(a)

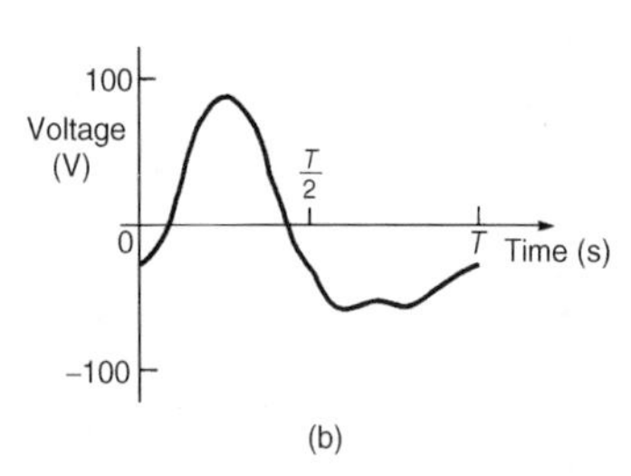

(b)

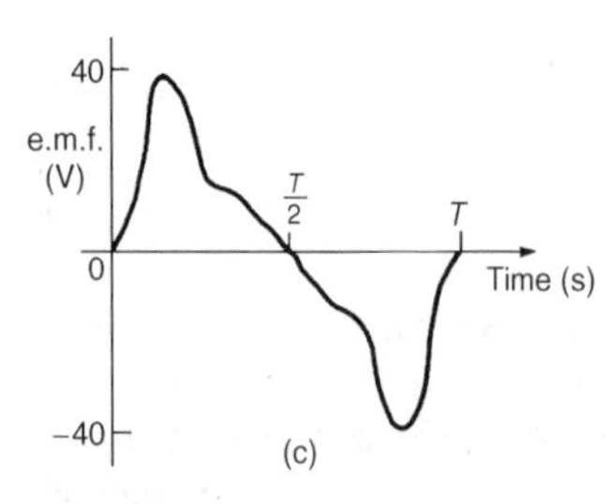

(c)

Figure 36.16

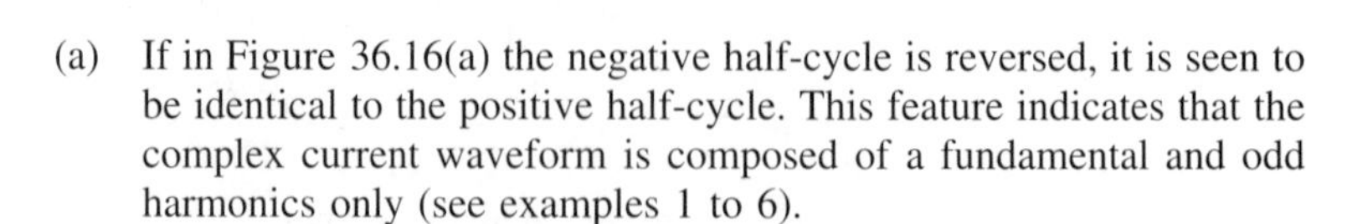

Problem 2. For each of the periodic complex waveforms shown in Figure 36.16, suggest whether odd or even harmonics (or both) are likely to be present.

(a) If in Figure 36.16(a) the negative half-cycle is reversed, it is seen to be identical to the positive half-cycle. This feature indicates that the complex current waveform is composed of a fundamental and odd harmonics only (see examples 1 to 6).

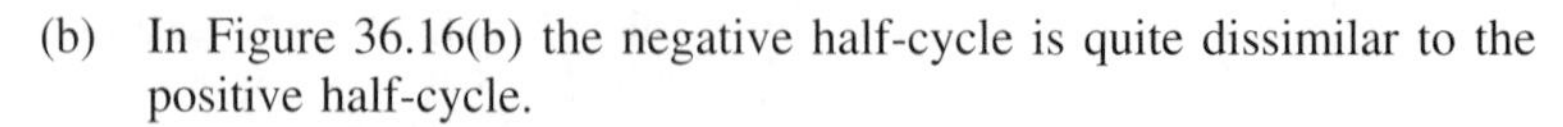

(b) In Figure 36.16(b) the negative half-cycle is quite dissimilar to the positive half-cycle.

This indicates that the complex voltage waveform comprises either

(i) a fundamental and even harmonics, initially out of phase with each other (see example 9), or

(ii) a fundamental and odd and even harmonics, one or more of the harmonics being initially out of phase (see example 12).

(c) If in Figure 36.16(c) the negative half-cycle is reversed, it is seen to be a mirror image of the positive half-cycle about a vertical line drawn through time $T/2$. This feature indicates that the complex e.m.f. waveform comprises either:

(i) a fundamental and even harmonics initially in phase with each other (see examples 7 and 8), or

(ii) a fundamental and odd and even harmonics, each initially in phase with each other (see example 11).

Further problems on harmonic synthesis may be found in Section 36.9, problems 1 to 6, page 671

36.4 Rms value, mean value and the form factor of a complex wave

Rms value

Let the instantaneous value of a complex current, i, be given by

$$i = I_{1m}\sin(\omega t + \theta_1) + I_{2m}\sin(2\omega t + \theta_2) + \cdots + I_{nm}\sin(n\omega t + \theta_n)\text{amperes}$$

The effective or rms value of this current is given by

$$I = \sqrt{(\text{mean value of } i^2)}$$

$$i^2 = [I_{1m}\sin(\omega t + \theta_1) + I_{2m}\sin(2\omega t + \theta_2) + \cdots + I_{nm}\sin(n\omega t + \theta_n)]^2$$

i.e., $$i_2 = I_{1m}^2\sin^2(\omega t + \theta_1) + I_{2m}^2\sin^2(2\omega t + \theta_2) + \cdots + I_{nm}^2\sin^2(n\omega t + \theta_n) + 2I_{1m}I_{2m}\sin(\omega t + \theta_1)\sin(2\omega t + \theta_2) + \cdots \quad (36.3)$$

Without writing down all terms involved when squaring current i, it can be seen that two types of term result, these being:

(i) terms such as $I_{1m}^2\sin^2(\omega t + \theta_1)$, $I_{2m}^2\sin^2(2\omega t + \theta_2)$, and so on, and

(ii) terms such as $2I_{1m}I_{2m}\sin(\omega t + \theta_1)\sin(2\omega t + \theta_2)$, i.e., products of different harmonics.

The mean value of i^2 is the sum of the mean values of each term in equation (36.3).

Taking an example of the first type, say $I_{1m}^2\sin^2(\omega t + \theta_1)$, the mean value over one cycle of the fundamental is determined using integral calculus:

$$\text{Mean value of } I_{1m}^2 \sin^2(\omega t + \theta_1) = \frac{1}{2\pi}\int_0^{2\pi} I_{1m}^2 \sin^2(\omega t + \theta_1)\, d(\omega t)$$

(since the mean value of $y = f(x)$ between $x = a$ and $x = b$ is given by $\frac{1}{b-a}\int_a^b y\,dx$)

$$= \frac{I_{1m}^2}{2\pi}\int_0^{2\pi}\left\{\frac{1 - \cos 2(\omega t + \theta_1)}{2}\right\} d(\omega t),$$

(since $\cos 2x = 1 - 2\sin^2 x$, from which $\sin^2 x = (1 - \cos 2x)/2$),

$$= \frac{I_{1m}^2}{4\pi}\left[\omega t - \frac{\sin 2(\omega t + \theta_1)}{2}\right]_0^{2\pi}$$

$$= \frac{I_{1m}^2}{4\pi}\left[\left(2\pi - \frac{\sin 2(2\pi + \theta_1)}{2}\right) - \left(0 - \frac{\sin 2(0 + \theta_1)}{2}\right)\right]$$

$$= \frac{I_{1m}^2}{4\pi}\left[2\pi - \frac{\sin 2(2\pi + \theta_1)}{2} + \frac{\sin 2\theta_1}{2}\right] = \frac{I_{1m}^2}{4\pi}(2\pi) = \mathbf{\frac{I_{1m}^2}{2}}$$

Hence it follows that the mean value of $I_{nm}^2 \sin^2(n\omega t + \theta_n)$ is given by $I_{nm}^2/2$.

Taking an example of the second type, say,

$$2I_{1m}I_{2m}\sin(\omega t + \theta_1)\sin(2\omega t + \theta_2),$$

the mean value over one cycle of the fundamental is also determined using integration:

Mean value of $2I_{1m}I_{2m}\sin(\omega t + \theta_1)\sin(2\omega t + \theta_2)$

$$= \frac{1}{2\pi}\int_0^{2\pi} 2I_{1m}I_{2m}\sin(\omega t + \theta_1)\sin(2\omega t + \theta_2)\, d(\omega t)$$

$$= \frac{I_{1m}I_{2m}}{\pi}\int_0^{2\pi}\frac{1}{2}\{\cos(\omega t + \theta_2 - \theta_1) - \cos(3\omega t + \theta_2 + \theta_1)\}\, d(\omega t)$$

(since $\sin A \sin B = \frac{1}{2}[\cos(A - B) - \cos(A + B)]$, and taking $A = (2\omega t + \theta_2)$ and $B = (\omega t + \theta_1)$)

$$= \frac{I_{1m}I_{2m}}{2\pi}\left[\sin(\omega t + \theta_2 - \theta_1) - \frac{\sin(3\omega t + \theta_2 + \theta_1)}{3}\right]_0^{2\pi}$$

$$= \frac{I_{1m}I_{2m}}{2\pi}\left[\left(\sin(2\pi + \theta_2 - \theta_1) - \frac{\sin(6\pi + \theta_2 + \theta_1)}{3}\right)\right.$$

$$\left. - \left(\sin(\theta_2 - \theta_1) - \frac{\sin(\theta_2 + \theta_1)}{3}\right)\right] = \frac{I_{1m}I_{2m}}{2\pi}[0] = \mathbf{0} \qquad (36.4)$$

Hence it follows that all such products of different harmonics will have a mean value of zero. Thus

$$\text{mean value of } i^2 = \frac{I_{1m}^2}{2} + \frac{I_{2m}^2}{2} + \cdots + \frac{I_{nm}^2}{2}$$

Hence the rms value of current,

$$I = \sqrt{\left(\frac{I_{1m}^2}{2} + \frac{I_{2m}^2}{2} + \cdots + \frac{I_{nm}^2}{2}\right)}$$

i.e., $$\boldsymbol{I = \sqrt{\left(\frac{I_{1m}^2 + I_{2m}^2 + \cdots + I_{nm}^2}{2}\right)}} \qquad (36.5)$$

For a sine wave, rms value $= (1/\sqrt{2})$ maximum value, i.e., maximum value $= \sqrt{2}$ rms value. Hence, for example, $I_{1m} = \sqrt{2}I_1$, where I_1 is the rms value of the fundamental component, and $(I_{1m})^2 = (\sqrt{2}I_1)^2 = 2I_1^2$. Thus, from equation (36.5), rms current

$$I = \sqrt{\left(\frac{2I_1^2 + 2I_2^2 + \cdots + 2I_n^2}{2}\right)}$$

i.e., $$\boldsymbol{I = \sqrt{(I_1^2 + I_2^2 + \cdots + I_n^2)}} \qquad (36.6)$$

where $I_1, I_2, \ldots, I_n$ are the rms values of the respective harmonics.

By similar reasoning, for a complex voltage waveform represented by

$$v = V_{1m}\sin(\omega t + \Psi_1) + V_{2m}\sin(2\omega t + \Psi_2) + \cdots + V_{nm}\sin(n\omega t + \Psi_n)\text{volts}$$

the rms value of voltage, V, is given by

$$\boldsymbol{V = \sqrt{\left(\frac{V_{1m}^2 + V_{2m}^2 + \cdots + V_{nm}^2}{2}\right)}} \qquad (36.7)$$

or $$\boldsymbol{V = \sqrt{(V_1^2 + V_2^2 + \cdots + V_n^2)}} \qquad (36.8)$$

where $V_1, V_2, \ldots, V_n$ are the rms values of the respective harmonics.

From equations (36.5) to (36.8) it is seen that the rms value of a complex wave is unaffected by the relative phase angles of the harmonic components. For a d.c. current or voltage, the instantaneous value, the mean value and the maximum value are equal. Thus, if a complex waveform should contain a d.c. component I_0, then the rms current I is given by

$$\boldsymbol{I = \sqrt{\left(I_0^2 + \frac{I_{1m}^2 + I_{2m}^2 + \cdots + I_{nm}^2}{2}\right)}} \quad (36.9)$$

or

$$\boldsymbol{I = \sqrt{(I_0^2 + I_1^2 + I_2^2 + \cdots + I_n^2)}} \quad (36.9)$$

Mean value

The mean or average value of a complex quantity whose negative half-cycle is similar to its positive half-cycle is given, for current, by

$$\boldsymbol{I_{av} = \frac{1}{\pi}\int_0^{\pi} i\, d(\omega t)} \quad (36.10)$$

and for voltage by

$$\boldsymbol{v_{av} = \frac{1}{\pi}\int_0^{\pi} v\, d(\omega t)} \quad (36.11)$$

each waveform being taken over half a cycle.

Unlike rms values, mean values **are** affected by the relative phase angles of the harmonic components.

Form factor

The form factor of a complex waveform whose negative half-cycle is similar in shape to its positive half-cycle is defined as:

$$\text{form factor} = \frac{\text{rms value of the waveform}}{\text{mean value}} \quad (36.12)$$

where the mean value is taken over half a cycle.

Changes in the phase displacement of the harmonics may appreciably alter the form factor of a complex waveform.

Problem 3. Determine the rms value of the current waveform represented by

$$i = 100\sin\omega t + 20\sin(3\omega t + \pi/6) + 10\sin(5\omega t + 2\pi/3)\text{mA}$$

From equation (36.5), the rms value of current is given by

$$I = \sqrt{\left(\frac{100^2 + 20^2 + 10^2}{2}\right)} = \sqrt{\left(\frac{10000 + 400 + 100}{2}\right)} = \mathbf{72.46\ mA}$$

Problem 4. A complex voltage is represented by

$$v = (10\sin\omega t + 3\sin 3\omega t + 2\sin 5\omega t)\text{volts}$$

Determine for the voltage, (a) the rms value, (b) the mean value and (c) the form factor.

(a) From equation (36.7), the rms value of voltage is given by

$$V = \sqrt{\left(\frac{10^2 + 3^2 + 2^2}{2}\right)} = \sqrt{\left(\frac{113}{2}\right)} = \mathbf{7.52\ V}$$

(b) From equation (36.11), the mean value of voltage is given by

$$V_{av} = \frac{1}{\pi}\int_0^{\pi}(10\sin\omega t + 3\sin 3\omega t + 2\sin 5\omega t)\,d(\omega t)$$

$$= \frac{1}{\pi}\left[-10\cos\omega t - \frac{3\cos 3\omega t}{3} - \frac{2\cos 5\omega t}{5}\right]_0^{\pi}$$

$$= \frac{1}{\pi}\left[\left(-10\cos\pi - \cos 3\pi - \frac{2}{5}\cos 5\pi\right)\right.$$

$$\left. - \left(-10\cos 0 - \cos 0 - \frac{2}{5}\cos 0\right)\right]$$

$$= \frac{1}{\pi}\left[\left(10 + 1 + \frac{2}{5}\right) - \left(-10 - 1 - \frac{2}{5}\right)\right] = \frac{22.8}{\pi} = \mathbf{7.26\ V}$$

(c) From equation (36.12), form factor is given by

$$\text{form factor} = \frac{\text{rms value of the waveform}}{\text{mean value}} = \frac{7.52}{7.26} = \mathbf{1.036}$$

Problem 5. A complex voltage waveform which has an rms value of 240 V contains 30% third harmonic and 10% fifth harmonic, both of the harmonics being initially in phase with each other. (a) Determine the rms value of the fundamental and each harmonic. (b) Write down an expression to represent the complex voltage waveform if the frequency of the fundamental is 31.83 Hz.

(a) From equation (36.8), rms voltage $V = \sqrt{(V_1^2 + V_3^2 + V_5^2)}$.

Since $V_3 = 0.30\ V_1,\ V_5 = 0.10\ V_1$ and $V = 240$ V, then

$$240 = \sqrt{[V_1^2 + (0.30\ V_1)^2 + (0.10\ V_1)^2]}$$

i.e., $240 = \sqrt{(1.10\ V_1^2)} = 1.049\ V_1$

from which the rms value of the fundamental,

$$V_1 = 240/1.049 = \mathbf{228.8\ V}.$$

Rms value of the third harmonic,

$$V_3 = 0.30\ V_1 = (0.30)(228.8) = \mathbf{68.64\ V}$$

and the rms value of the fifth harmonic,

$$V_5 = 0.10\ V_1 = (0.10)(228.8) = \mathbf{22.88\ V}$$

(b) Maximum value of the fundamental,

$$V_{1m} = \sqrt{2}V_1 = \sqrt{2}(228.8) = 323.6\ \text{V}$$

Maximum value of the third harmonic,

$$V_{3m} = \sqrt{2}V_3 = \sqrt{2}(68.64) = 97.07\ \text{V}$$

Maximum value of the fifth harmonic,

$$V_{5m} = \sqrt{2}V_5 = \sqrt{2}(22.88) = 32.36\ \text{V}$$

Since the fundamental frequency is 31.83 Hz, the fundamental voltage may be written as $323.6 \sin 2\pi(31.83)t$, i.e., $323.6 \sin 200t$ volts

The third harmonic component is $97.07 \sin 600t$ volts and the fifth harmonic component is $32.36 \sin 1000t$ volts. Hence an expression representing the complex voltage waveform is given by

$$\boldsymbol{v = (323.6 \sin 200t + 97.07 \sin 600t + 32.36 \sin 1000t)\ \text{volts}}$$

Further problems on rms values, mean values and form factor of complex waves may be found in Section 36.9, problems 7 to 11, page 672.

36.5 Power associated with complex waves

Let a complex voltage wave be represented by

$$v = V_{1m} \sin \omega t + V_{2m} \sin 2\omega t + V_{3m} \sin 3\omega t + \cdots,$$

and when this is applied to a circuit let the resulting current be represented by

$$i = I_{1m} \sin(\omega t - \phi_1) + I_{2m} \sin(2\omega t - \phi_2) + I_{3m} \sin(3\omega t - \phi_3) + \cdots$$

(Since the phase angles are lagging, the circuit in this case is inductive.)
At any instant in time the power p supplied to the circuit is given by $p = vi$, i.e.,

$$p = (V_{1m} \sin \omega t + V_{2m} \sin 2\omega t + \cdots)(I_{1m} \sin(\omega t - \phi_1) + I_{2m} \sin(2\omega t - \phi_2) + \cdots)$$

$$= V_{1m} I_{1m} \sin \omega t \sin(\omega t - \phi_1) + V_{1m} I_{2m} \sin \omega t \sin(2\omega t - \phi_2) = \cdots \quad (36.13)$$

The average or active power supplied over one cycle is given by the sum of the average values of each individual product term taken over one cycle. It is seen from equation (36.4) that the average value of product terms involving harmonics of different frequencies is always zero. This means therefore that only products of voltage and current harmonics of the same frequency need be considered in equation (36.13).

Taking the first term, for example, the average power P_1 over one cycle of the fundamental is given by

$$P_1 = \frac{1}{2\pi}\int_0^{2\pi} V_{1m} I_{1m} \sin\omega t \sin(\omega t - \phi_1)\, d(\omega t)$$

$$= \frac{V_{1m} I_{1m}}{2\pi}\int_0^{2\pi} \frac{1}{2}\{\cos\phi_1 - \cos(2\omega t - \phi_1)\}\, d(\omega t)$$

since $\sin A \sin B = \frac{1}{2}\{\cos(A-B) - \cos(A+B)\}$,

$$= \frac{V_{1m} I_{1m}}{4\pi}\left[(\omega t)\cos\phi_1 - \frac{\sin(2\omega t - \phi_1)}{2}\right]_0^{2\pi}$$

$$= \frac{V_{1m} I_{1m}}{4\pi}\left[\left(2\pi\cos\phi_1 - \frac{\sin(4\pi - \phi_1)}{2}\right) - \left(0 - \frac{\sin(-\phi_1)}{2}\right)\right]$$

$$= \frac{V_{1m} I_{1m}}{4\pi}[2\pi\cos\phi_1] = \frac{V_{1m} I_{1m}}{2}\cos\phi_1$$

$V_{1m} = \sqrt{2}V_1$ and $I_{1m} = \sqrt{2}I_1$, where V_1 and I_1 are rms values, hence

$$P_1 = \frac{(\sqrt{2}V_1)(\sqrt{2}I_1)}{2}\cos\phi_1$$

i.e., $P_1 = V_1 I_1 \cos\phi_1$ watts

Similarly, the average power supplied over one cycle of the fundamental for the second harmonic is $V_2 I_2 \cos\Phi_2$, and so on. Hence the total power supplied by complex voltages and currents is the sum of the powers supplied by each harmonic component acting on its own. The average power P supplied for one cycle of the fundamental is given by

$$\boldsymbol{P = V_1 I_1 \cos\phi_1 + V_2 I_2 \cos\phi_2 + \cdots + V_n I_n \cos\phi_n} \qquad (36.14)$$

If the voltage waveform contains a d.c. component V_0 which causes a direct current component I_0, then the average power supplied by the d.c. component is $V_0 I_0$ and the total average power P supplied is given by

$$\boldsymbol{P = V_0 I_0 + V_1 I_1 \cos\phi_1 + V_2 I_2 \cos\phi_2 + \cdots + V_n I_n \cos\phi_n} \qquad (36.15)$$

Alternatively, if R is the equivalent series resistance of a circuit then the total power is given by

$$P = I_0^2 R + I_1^2 R + I_2^2 R + I_3^2 R + \cdots$$

i.e., $$\boxed{P = I^2R} \tag{36.16}$$

where I is the rms value of current i.

Power factor

When harmonics are present in a waveform the overall circuit power factor is defined as

$$\text{overall power factor} = \frac{\text{total power supplied}}{\text{total rms voltage} \times \text{total rms current}} = \frac{\text{total power}}{\text{volt amperes}}$$

i.e., $$\boxed{p.f. = \frac{V_1I_1\cos\phi_1 + V_2I_2\cos\phi_2 + \cdots}{VI}} \tag{36.17}$$

Problem 6. Determine the average power in a 20 Ω resistance if the current i flowing through it is of the form

$$i = (12\sin\omega t + 5\sin 3\omega t + 2\sin 5\omega t)\text{amperes}$$

From equation (36.5), rms current,

$$I = \sqrt{\left(\frac{12^2 + 5^2 + 2^2}{2}\right)} = 9.30\ \text{A}$$

From equation (36.16), average power,

$$P = I^2R = (9.30)^2(20) = \mathbf{1730\ W\ or\ 1.73\ kW}$$

Problem 7. A complex voltage v given by

$$v = 60\sin\omega t + 15\sin\left(3\omega t + \frac{\pi}{4}\right) + 10\sin\left(5\omega t - \frac{\pi}{2}\right)\ \text{volts}$$

is applied to a circuit and the resulting current i is given by

$$i = 2\sin\left(\omega t - \frac{\pi}{6}\right) + 0.30\sin\left(3\omega t - \frac{\pi}{12}\right) + 0.1\sin\left(5\omega t - \frac{8\pi}{9}\right)\ \text{amperes}$$

Determine (a) the total active power supplied to the circuit, and (b) the overall power factor.

(a) From equation (36.14), total power supplied,

$$P = V_1 I_1 \cos\phi_1 + V_3 I_3 \cos\phi_3 + V_5 I_5 \cos\phi_5$$

$$= \left(\frac{60}{\sqrt{2}}\right)\left(\frac{2}{\sqrt{2}}\right)\cos\left(0 - \left(-\frac{\pi}{6}\right)\right)$$

$$+ \left(\frac{15}{\sqrt{2}}\right)\left(\frac{0.3}{\sqrt{2}}\right)\cos\left(\frac{\pi}{4} - \left(-\frac{\pi}{12}\right)\right)$$

$$+ \left(\frac{10}{\sqrt{2}}\right)\left(\frac{0.1}{\sqrt{2}}\right)\cos\left(-\frac{\pi}{2} - \left(-\frac{8\pi}{9}\right)\right)$$

$$= 51.96 + 1.125 + 0.171 = \mathbf{53.26\ W}$$

(b) From equation (36.5), rms current,

$$I = \sqrt{\left(\frac{2^2 + 0.3^2 + 0.1^2}{2}\right)} = 1.43\ \text{A}$$

and from equation (36.7), rms voltage,

$$V = \sqrt{\left(\frac{60^2 + 15^2 + 10^2}{2}\right)} = 44.30\ \text{V}$$

From equation (36.17),

$$\text{overall power factor} = \frac{53.26}{(44.30)(1.43)} = \mathbf{0.841}$$

(With a sinusoidal waveform,

$$\text{power factor} = \frac{\text{power}}{\text{volt-amperes}} = \frac{VI\cos\phi}{VI} = \cos\phi$$

Thus power factor depends upon the value of phase angle ϕ, and is lagging for an inductive circuit and leading for a capacitive circuit. However, with a complex waveform, power factor is not given by $\cos\phi$. In the expression for power in equation (36.14), there are n phase-angle terms, $\phi_1, \phi_2, \ldots, \phi_n$, all of which may be different. It is for this reason that it is not possible to state whether the overall power factor is lagging or leading when harmonics are present.)

Further problems on power associated with complex waves may be found in Section 36.9, problems 12 to 15, page 673.

36.6 Harmonics in single-phase circuits

When a complex alternating voltage wave, i.e., one containing harmonics, is applied to a single-phase circuit containing resistance, inductance and/or capacitance (i.e., linear circuit elements), then the resulting current will also be complex and contain harmonics.

Let a complex voltage v be represented by

$$v = V_{1m} \sin \omega t + V_{2m} \sin 2\omega t + V_{3m} \sin 3\omega t + \cdots$$

(a) Pure resistance

The impedance of a pure resistance R is independent of frequency and the current and voltage are in phase for each harmonic. Thus the general expression for current i is given by

$$i = \frac{v}{R} = \frac{V_{1m}}{R} \sin \omega t + \frac{V_{2m}}{R} \sin 2\omega t + \frac{V_{3m}}{R} \sin 3\omega t + \cdots \qquad (36.18)$$

The percentage harmonic content in the current wave is the same as that in the voltage wave. For example, the percentage second harmonic content from equation (36.18) is

$$\frac{V_{2m}/R}{V_{1m}/R} \times 100\%, \text{ i.e., } \frac{V_{2m}}{V_{1m}} \times 100\%$$

the same as for the voltage wave. The current and voltage waveforms will therefore be identical in shape.

(b) Pure inductance

The impedance of a pure inductance L, i.e., inductive reactance $X_L(= 2\pi f L)$, varies with the harmonic frequency when voltage v is applied to it. Also, for every harmonic term, the current will lag the voltage by 90° or $\pi/2$ rad. The current i is given by

$$i = \frac{v}{X_L} = \frac{V_{1m}}{\omega L} \sin\left(\omega t - \frac{\pi}{2}\right) + \frac{V_{2m}}{2\omega L} \sin\left(2\omega t - \frac{\pi}{2}\right) + \frac{V_{3m}}{3\omega L} \sin\left(3\omega t - \frac{\pi}{2}\right) + \cdots \qquad (36.19)$$

since for the nth harmonic the reactance is $n\omega L$.

Equation (36.19) shows that for, say, the nth harmonic, the percentage harmonic content in the current waveform is only $1/n$ of the corresponding harmonic content in the voltage waveform.

If a complex current contains a d.c. component then the direct voltage drop across a pure inductance is zero.

(c) Pure capacitance

The impedance of a pure capacitance C, i.e., capacitive reactance $X_C(= 1/(2\pi f C))$, varies with the harmonic frequency when voltage v is applied to it. Also, for each harmonic term the current will lead the voltage by 90° or $\pi/2$ rad. The current i is given by

$$i = \frac{v}{X_C} = \frac{V_{1m}}{1/\omega C}\sin\left(\omega t + \frac{\pi}{2}\right) + \frac{V_{2m}}{1/2\omega C}\sin\left(2\omega t + \frac{\pi}{2}\right) + \frac{V_{3m}}{1/3\omega C}\sin\left(3\omega t + \frac{\pi}{2}\right) + \cdots,$$

since for the nth harmonic the reactance is $1/(n\omega C)$. Hence current,

$$\boldsymbol{i = V_{1m}(\omega C)\sin\left(\omega t + \frac{\pi}{2}\right) + V_{2m}(2\omega C)\sin\left(2\omega t + \frac{\pi}{2}\right) + V_{3m}(3\omega C)\sin\left(3\omega t + \frac{\pi}{2}\right) + \cdots} \qquad (36.20)$$

Equation (36.20) shows that the percentage harmonic content of the current waveform is n times larger for the nth harmonic than that of the corresponding harmonic voltage.

If a complex current contains a d.c. component then none of this direct current will flow through a pure capacitor, although the alternating components of the supply still operate.

Problem 8. A complex voltage waveform represented by

$$v = 100\sin\omega t + 30\sin\left(3\omega t + \frac{\pi}{3}\right) + 10\sin\left(5\omega t - \frac{\pi}{6}\right) \text{ volts}$$

is applied across (a) a pure 40 Ω resistance, (b) a pure 7.96 mH inductance, and (c) a pure 25 μF capacitor. Determine for each case an expression for the current flowing if the fundamental frequency is 1 kHz.

(a) From equation (36.18),

$$\text{current } i = \frac{v}{R} = \frac{100}{40}\sin\omega t + \frac{30}{40}\sin\left(3\omega t + \frac{\pi}{3}\right) + \frac{10}{40}\sin\left(5\omega t - \frac{\pi}{6}\right)$$

i.e.
$$\boldsymbol{i = 2.5\sin\omega t + 0.75\sin\left(3\omega t + \frac{\pi}{3}\right) + 0.25\sin\left(5\omega t - \frac{\pi}{6}\right) \text{ amperes}}$$

(b) At the fundamental frequency, $\omega L = 2\pi(1000)(7.96 \times 10^{-3}) = 50\ \Omega$. From equation (36.19),

$$\text{current } i = \frac{100}{50}\sin\left(\omega t - \frac{\pi}{2}\right) + \frac{30}{3 \times 50}\sin\left(3\omega t + \frac{\pi}{3} - \frac{\pi}{2}\right) + \frac{10}{5 \times 50}\sin\left(5\omega t - \frac{\pi}{6} - \frac{\pi}{2}\right)$$

i.e. **current** $i = 2\sin\left(\omega t - \frac{\pi}{2}\right) + 0.20\sin\left(3\omega t - \frac{\pi}{6}\right)$

$$+ 0.04\sin\left(5\omega t - \frac{2\pi}{3}\right)\text{ amperes}$$

(c) At the fundamental frequency, $\omega C = 2\pi(1000)(25 \times 10^{-6}) = 0.157$. From equation (36.20),

$$\text{current } i = 100(0.157)\sin\left(\omega t + \frac{\pi}{2}\right) + 30(3 \times 0.157)\sin\left(3\omega t + \frac{\pi}{3} + \frac{\pi}{2}\right) + 10(5 \times 0.157)\sin\left(5\omega t - \frac{\pi}{6} + \frac{\pi}{2}\right)$$

i.e., $i = 15.70\sin\left(\omega t + \frac{\pi}{2}\right) + 14.13\sin\left(3\omega t + \frac{5\pi}{6}\right) + 7.85\sin\left(5\omega t + \frac{\pi}{3}\right)$ **amperes**

Problem 9. A supply voltage v given by

$$v = (240\sin 314t + 40\sin 942t + 30\sin 1570t)\text{volts}$$

is applied to a circuit comprising a resistance of 12 Ω connected in series with a coil of inductance 9.55 mH. Determine (a) an expression to represent the instantaneous value of the current, (b) the rms voltage, (c) the rms current, (d) the power dissipated, and (e) the overall power factor.

(a) The supply voltage comprises a fundamental, $240\sin 314t$, a third harmonic, $40\sin 942t$ (third harmonic since 942 is 3 × 314) and a fifth harmonic, $30\sin 1570t$.

Fundamental

Since the fundamental frequency, $\omega_1 = 314$ rad/s, inductive reactance,

$$X_{L1} = \omega_1 L = (314)(9.55 \times 10^{-3}) = 3.0\ \Omega.$$

Hence impedance at the fundamental frequency,

$$Z_1 = (12 + j3.0)\Omega = 12.37\angle 14.04°\ \Omega$$

Maximum current at fundamental frequency

$$I_{1m} = \frac{V_{1m}}{Z_1} = \frac{240\angle 0°}{12.37\angle 14.04°} = 19.40\angle -14.04°\ \text{A}$$

$14.04° = 14.04 \times (\pi/180)$ rad $= 0.245$ rad, thus

$$I_{1m} = 19.40\angle -0.245\ \text{A}$$

Hence the fundamental current $i_1 = 19.40\sin(314t - 0.245)$A.

(Note that with an expression of the form $R\sin(\omega t \pm \alpha)$, ωt is an angle measured in radians, thus the phase displacement, α, should also be expressed in radians.)

Third harmonic

Since the third harmonic frequency, $\omega_3 = 942$ rad/s, inductive reactance,

$$X_{L3} = 3X_{L1} = 9.0\ \Omega.$$

Hence impedance at the third harmonic frequency,

$$Z_3 = (12 + j9.0)\Omega = 15\angle 36.87^\circ\ \Omega$$

Maximum current at the third harmonic frequency,

$$\begin{aligned} I_{3m} &= \frac{V_{3m}}{Z_3} = \frac{40\angle 0^\circ}{15\angle 36.87^\circ} \\ &= 2.67\angle -36.87^\circ\ \text{A} \\ &= 2.67\angle -0.644\ \text{A} \end{aligned}$$

Hence the third harmonic current, $i_3 = 2.67\sin(942t - 0.644)$A.

Fifth harmonic

Inductive reactance, $X_{L5} = 5X_{L1} = 15\ \Omega$

Impedance $Z_5 = (12 + j15)\Omega = 19.21\angle 51.34^\circ\ \Omega$

$$\text{Current,}\quad I_{5m} = \frac{V_{5m}}{Z_5} = \frac{30\angle 0^\circ}{19.21\angle 51.34^\circ} = 1.56\angle -51.34^\circ\ \text{A} = 1.56\angle -0.896\ \text{A}$$

Hence the fifth harmonic current, $i_5 = 1.56\sin(1570t - 0.896)$A

Thus an expression to represent the instantaneous current, i, is given by $i = i_1 + i_3 + i_5$ i.e.,

$$\mathbf{i = 19.40\sin(314t - 0.245) + 2.67\sin(942t - 0.644) + 1.56\sin(1570t - 0.896)\,amperes}$$

(b) From equation (36.7), rms voltage,

$$V = \sqrt{\left(\frac{240^2 + 40^2 + 30^2}{2}\right)} = \mathbf{173.35\ V}$$

(c) From equation (36.5), rms current,

$$I = \sqrt{\left(\frac{19.40^2 + 2.67^2 + 1.56^2}{2}\right)} = \mathbf{13.89\ A}$$

(d) From equation (36.16), power dissipated,

$$P = I^2R = (13.89)^2(12) = \mathbf{2315\ W} \text{ or } \mathbf{2.315\ kW}$$

(Alternatively, equation (36.14) may be used to determine power.)

(e) From equation (36.17),

$$\text{overall power factor} = \frac{2315}{(173.35)(13.89)} = \mathbf{0.961}$$

Problem 10. An e.m.f. is represented by

$$e = 50 + 200\sin\omega t + 40\sin\left(2\omega t - \frac{\pi}{2}\right) + 5\sin\left(4\omega t + \frac{\pi}{4}\right) \text{ volts,}$$

the fundamental frequency being 50 Hz. The e.m.f. is applied across a circuit comprising a 100 μF capacitor connected in series with a 50 Ω resistor. Obtain an expression for the current flowing and hence determine the rms value of current.

D.c. component

In a d.c. circuit no current will flow through a capacitor. The current waveform will not possess a d.c. component even though the e.m.f. waveform has a 50 V d.c. component. Hence $i_0 = 0$.

Fundamental

Capacitive reactance,

$$X_{C1} = \frac{1}{2\pi fC} = \frac{1}{2\pi(50)(100 \times 10^{-6})} = 31.83\ \Omega$$

Impedance $Z_1 = (50 - j31.83)\Omega = 59.27\angle -32.48^\circ\ \Omega$

$$I_{1m} = \frac{V_{1m}}{Z_1} = \frac{200\angle 0^\circ}{59.27\angle -32.48^\circ} = 3.374\angle 32.48^\circ\ \text{A} = 3.374\angle 0.567\ \text{A}$$

Hence the fundamental current, $i_1 = 3.374\sin(\omega t + 0.567)\text{A}$

Second harmonic

Capacitive reactance,

$$X_{C2} = \frac{1}{2(2\pi fC)} = \frac{31.83}{2} = 15.92\ \Omega$$

Impedance $Z_2 = (50 - j15.92)\Omega = 52.47\angle -17.66^\circ\ \Omega$

$$I_{2m} = \frac{V_{2m}}{Z_2} = \frac{40\angle -\pi/2}{52.47\angle -17.66^\circ} = 0.762\angle\left(-\frac{\pi}{2} - (-17.66^\circ)\right)$$

$$= 0.762\angle -72.34^\circ\ \text{A}$$

Hence the second harmonic current, $i_2 = 0.762 \sin(2\omega t - 72.34^\circ)\text{A}$

$$= 0.762 \sin(2\omega t - 1.263)\text{A}$$

Fourth harmonic

Capacitive reactance, $X_{C4} = \frac{1}{4} X_{C1} = \frac{31.83}{4} = 7.958\ \Omega$

Impedance, $Z_4 = (50 - j7.958)\Omega = 50.63\angle -9.04^\circ\ \Omega$

$$I_{4m} = \frac{V_{4m}}{Z_4} = \frac{5\angle \pi/4}{50.63\angle -9.04^\circ} = 0.099\angle(\pi/4 - (-9.04^\circ))$$

$$= 0.099\angle 54.04^\circ\ \text{A}$$

Hence the fourth harmonic current, $i_4 = 0.099 \sin(4\omega t + 54.04^\circ)\text{A}$

$$= 0.099 \sin(4\omega t + 0.943)\text{A}$$

An expression for current flowing is therefore given by

$$i = i_0 + i_1 + i_2 + i_4$$

i.e., $\boldsymbol{i = 3.374 \sin(\omega t + 0.567) + 0.762 \sin(2\omega t - 1.263)}$

$$\boldsymbol{+ 0.099 \sin(4\omega t + 0.943)\ \text{amperes}}$$

From equation (36.5), rms current,

$$I = \sqrt{\left(\frac{3.374^2 + 0.762^2 + 0.099^2}{2}\right)} = \mathbf{2.45\ A}$$

Problem 11. A complex voltage v is represented by:

$$v = 25 + 100 \sin \omega t + 40 \sin\left(3\omega t + \frac{\pi}{6}\right) + 20 \sin\left(5\omega t + \frac{\pi}{12}\right) \text{ volts}$$

where $\omega = 10^4$ rad/s. The voltage is applied to a series circuit comprising a 5.0 Ω resistance and a 500 μH inductance.

Determine (a) an expression to represent the current flowing in the circuit, (b) the rms value of current, correct to two decimal places, and (c) the power dissipated in the circuit, correct to three significant figures.

(a) **d.c. component**

Inductance has no effect on a steady current. Hence the d.c. component of the current, i_0, is given by

$$i_0 = \frac{v_0}{R} = \frac{25}{5.0} = 5.0\ \text{A}$$

Fundamental

Inductive reactance, $X_{L1} = \omega L = (10^4)(500 \times 10^{-6}) = 5\ \Omega$

Impedance, $Z_1 = (5 + j5)\Omega = 7.071\angle 45^\circ\ \Omega$

$$I_{1m} = \frac{V_{1m}}{Z_1} = \frac{100\angle 0^\circ}{7.071\angle 45^\circ} = 14.14\angle -45^\circ\ \text{A}$$
$$= 14.14\angle -\pi/4\ \text{A or } 14.14\angle -0.785\ \text{A}$$

Hence fundamental current, $i_1 = 14.14\sin(\omega t - 0.785)\text{A}$

Third harmonic

Inductive reactance at third harmonic frequency,

$$X_{L3} = 3X_{L1} = 15\ \Omega$$

Impedance, $Z_3 = (5 + j15)\Omega = 15.81\angle 71.57^\circ\ \Omega$

$$I_{3m} = \frac{V_{3m}}{Z_3} = \frac{40\angle \pi/6}{15.81\angle 71.57^\circ} = 2.53\angle -41.57^\circ\ \text{A}$$
$$= 2.53\angle -0.726\ \text{A}$$

Hence the third harmonic current, $i_3 = 2.53\sin(3\omega t - 41.57^\circ)\text{A}$
$= 2.53\sin(3\omega t - 0.726)\text{A}$

Fifth harmonic

Inductive reactance at fifth harmonic frequency, $X_{L5} = 5X_{L1} = 25\ \Omega$

Impedance, $Z_5 = (5 + j25)\Omega = 25.495\angle 78.69^\circ\ \Omega$

$$I_5 = \frac{V_{5m}}{Z_5} = \frac{20\angle \pi/12}{25.495\angle 78.69^\circ} = 0.784\angle -63.69^\circ\ \text{A}$$
$$= 0.784\angle -1.112\ \text{A}$$

Hence the fifth harmonic current, $i_5 = 0.784\sin(5\omega t - 63.69^\circ)\text{A}$
$= 0.784\sin(5\omega t - 1.112)\text{A}$

Thus current, $i = i_0 + i_1 + i_3 + i_5$, i.e.,

$$\boldsymbol{i = 5 + 14.14\sin(\omega t - 0.785) + 2.43\sin(3\omega t - 0.726) + 0.784\sin(5\omega t - 1.112)\text{A}}$$

(b) From equation (36.9), rms current,

$$I = \sqrt{\left(5.0^2 + \frac{14.14^2 + 2.53^2 + 0.784^2}{2}\right)}$$
$= 11.3348\ \text{A} =$ **11.33 A**, correct to two decimal places.

(c) From equation (36.16), power dissipated,

$$P = I^2R = (11.3348)^2(5.0) = 642.4 \text{ W}$$
$$= \mathbf{642\ W}, \text{ correct to three significant figures}$$

(Alternatively, from equation (36.15),

$$\text{power } P = (25)(5.0) + \left(\frac{100}{\sqrt{2}}\right)\left(\frac{14.14}{\sqrt{2}}\right)\cos 45^\circ + \left(\frac{40}{\sqrt{2}}\right)\left(\frac{2.53}{\sqrt{2}}\right)\cos 71.57^\circ + \left(\frac{20}{\sqrt{2}}\right)\left(\frac{0.784}{\sqrt{2}}\right)\cos 78.69^\circ$$
$$= 125 + 499.92 + 16.00 + 1.54$$
$$= 642.46 \text{ W or } \mathbf{642\ W},$$

correct to three significant figures, as above.)

> Problem 12. The voltage applied to a particular circuit comprising two components connected in series is given by
>
> $$v = (30 + 40\sin 10^3t + 25\sin 2\times 10^3t + 15\sin 4\times 10^3t)\text{volts}$$
>
> and the resulting current is given by
>
> $$i = 0.743\sin(10^3t + 1.190) + 0.781\sin(2\times 10^3t + 0.896) + 0.636\sin(4\times 10^3t + 0.559)\text{A}$$
>
> Determine (a) the average power supplied, (b) the type of components present, and (c) the values of the components.

(a) From equation (36.15), the average power P is given by

$$P = (30)(0) + \left(\frac{40}{\sqrt{2}}\right)\left(\frac{0.743}{\sqrt{2}}\right)\cos 1.190 + \left(\frac{25}{\sqrt{2}}\right)\left(\frac{0.781}{\sqrt{2}}\right)\cos 0.896 + \left(\frac{15}{\sqrt{2}}\right)\left(\frac{0.636}{\sqrt{2}}\right)\cos 0.559$$

i.e., $P = 0 + 5.523 + 6.099 + 4.044 = \mathbf{15.67\ W}$

(b) The expression for the voltage contains a d.c. component of 30 V. However there is no corresponding term in the expression for current. This indicates that one of the components is a **capacitor** (since in a d.c. circuit a capacitor offers an infinite impedance to a direct current). Since power is delivered to the circuit the other component is a **resistor**.

(c) From equation (36.5), rms current,

$$I = \sqrt{\left(\frac{0.743^2 + 0.781^2 + 0.636^2}{2}\right)} = 0.885 \text{ A}$$

Average power $P = I^2R$, from which,

$$\text{resistance } R = \frac{P}{I^2} = \frac{15.67}{(0.885)^2} = \mathbf{20\ \Omega}$$

At the fundamental frequency, $\omega = 10^3$ rad/s

$$\text{impedance } |Z_1| = \frac{V_{1m}}{I_{1m}} = \frac{40}{0.743} = 53.84\ \Omega$$

Impedance $|Z_1| = \sqrt{(R^2 + X_{C1}^2)}$, from which

$$X_{C1} = \sqrt{(Z_1^2 - R^2)} = \sqrt{(53.84^2 - 20^2)} = 50\ \Omega$$

Hence $1/\omega C = 50$, from which

$$\textbf{capacitance } C = \frac{1}{\omega(50)} = \frac{1}{10^3(50)} = \mathbf{20\ \mu F}$$

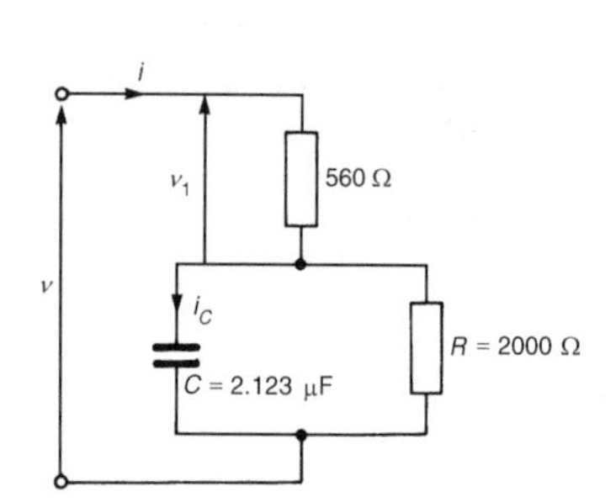

Figure 36.17

Problem 13. In the circuit shown in Figure 36.17 the supply voltage v is given by $v = 300 \sin 314t + 120 \sin(942t + 0.698)$ volts. Determine (a) an expression for the supply current, i, (b) the percentage harmonic content of the supply current, (c) the total power dissipated, (d) an expression for the p.d. shown as v_1, and (e) an expression for current i_c

(a) Capacitive reactance of the 2.123 μF capacitor at the fundamental frequency is given by

$$X_{C1} = \frac{1}{(314)(2.123 \times 10^{-6})} = 1500\ \Omega$$

At the fundamental frequency the total circuit impedance, Z_1, is given by

$$Z_1 = 560 + \frac{(2000)(-j1500)}{(2000 - j1500)} = 560 + \frac{3 \times 10^6 \angle -90^\circ}{2500 \angle -36.87^\circ}$$

$$= 560 + 1200 \angle -53.13^\circ = 560 + 720 - j960$$

$$= (1280 - j960)\Omega = 1600 \angle -36.87^\circ\ \Omega$$

$$= 1600 \angle -0.644\ \Omega$$

Since for the nth harmonic the capacitive reactance is $1/(n\omega C)$, the capacitive reactance of the third harmonic is $\frac{1}{3}X_{C1} = \frac{1}{3}(1500) = 500\ \Omega$. Hence at the third harmonic frequency the total circuit impedance, Z_3, is given by

$$Z_1 = 560 + \frac{(2000)(-j500)}{(2000 - j500)} = 560 + \frac{10^6\angle -90^\circ}{2061.55\angle -14.04^\circ}$$

$$= 560 + 485.07\angle -75.96^\circ = 560 + 117.68 - j470.58$$

$$= (677.68 - j470.58)\Omega = 825\angle -34.78^\circ\ \Omega$$

$$= 825\angle -0.607\ \Omega$$

The fundamental current

$$i_1 = \frac{v_1}{Z_1} = \frac{300\angle 0}{1600\angle -0.644} = 0.188\angle 0.644\ \text{A}$$

The third harmonic current

$$i_3 = \frac{v_3}{Z_3} = \frac{120\angle 0.698}{825\angle -0.607} = 0.145\angle 1.305\ \text{A}$$

Thus, **supply current,** $\boldsymbol{i = 0.188\sin(314t + 0.644) + 0.145\sin(942t + 1.305)\text{A}}$

(b) Percentage harmonic content of the supply current is given by

$$\frac{0.145}{0.188} \times 100\% = \mathbf{77\%}$$

(c) From equation (36.14), total active power

$$P = \left(\frac{300}{\sqrt{2}}\right)\left(\frac{0.188}{\sqrt{2}}\right)\cos 0.644 + \left(\frac{120}{\sqrt{2}}\right)\left(\frac{0.145}{\sqrt{2}}\right)\cos 0.607$$

i.e., $P = 22.55 + 7.15 = \mathbf{29.70\ W}$

(d) Voltage $v_1 = iR = 560[0.188\sin(314t + 0.644) + 0.145\sin(942t + 1.305)]$,

i.e., $\boldsymbol{v_1 = 105.3\sin(314t + 0.644) + 81.2\sin(942t + 1.305)\ \text{volts}}$

(e) Current, $i_c = i_1\left(\frac{R}{R - jX_{C1}}\right) + i_3\left(\frac{R}{R - jX_{C3}}\right)$ by current division

$$= (0.188\angle 0.644)\left(\frac{2000}{2000 - j1500}\right) + (0.145\angle 1.305)\left(\frac{2000}{2000 - j500}\right)$$

$$= (0.188\angle 0.644)\left(\frac{2000}{2500\angle -0.644}\right)$$

$$+ (0.145\angle 1.305)\left(\frac{2000}{2061.55\angle -0.245}\right)$$

$$= 0.150\angle 1.288 + 0.141\angle 1.550$$

Hence $\mathbf{i_c = 0.150\sin(314t + 1.288) + 0.141\sin(942t + 1.550)A}$

Further problems on harmonics in single phase circuits may be found in Section 36.9, problems 16 to 24, page 673.

36.7 Resonance due to harmonics

In industrial circuits at power frequencies the typical values of L and C involved make resonance at the fundamental frequency very unlikely. (An exception to this is with the capacitor-start induction motor where the start-winding can achieve unity power factor during run-up.)

However, if the voltage waveform is not a pure sine wave it is quite possible for the resonant frequency to be near the frequency of one of the harmonics. In this case the magnitude of the particular harmonic in the current waveform is greatly increased and may even exceed that of the fundamental. The effect of this is a great distortion of the resultant current waveform so that dangerous volt drops may occur across the inductance and capacitance in the circuit.

When a circuit resonates at one of the harmonic frequencies of the supply voltage, the effect is called **selective or harmonic resonance.**

For resonance with the fundamental, the condition is $\omega L = 1/(\omega C)$; for resonance at, say, the third harmonic, the condition is $3\,\omega L = 1/(3\omega C)$; for resonance at the nth harmonic, the condition is

$$\boxed{\mathbf{n\omega L = 1/(n\omega C)}}.$$

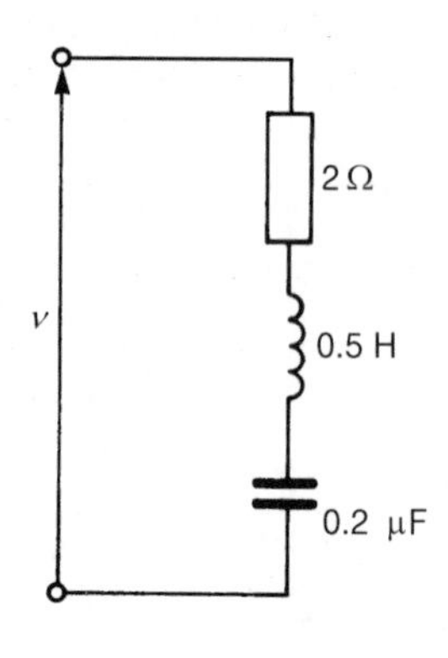

Figure 36.18

Problem 14. A voltage waveform having a fundamental of maximum value 400 V and a third harmonic of maximum value 10 V is applied to the circuit shown in Figure 36.18. Determine (a) the fundamental frequency for resonance with the third harmonic, and (b) the maximum value of the fundamental and third harmonic components of current.

(a) Resonance with the third harmonic means that $3\omega L = 1/(3\omega C)$, i.e.,

$$\omega = \sqrt{\left(\frac{1}{9LC}\right)} = \frac{1}{3\sqrt{(0.5)(0.2\times 10^{-6})}} = 1054 \text{ rad/s}$$

from which, **fundamental frequency**, $f = \frac{\omega}{2\pi} = \frac{1054}{2\pi}$

$= \mathbf{167.7\ Hz}$

(b) At the fundamental frequency,

$$\text{impedance} \quad Z_1 = R + j\left(\omega L - \frac{1}{\omega C}\right)$$

$$= 2 + j\left[(1054)(0.5) - \frac{1}{(1054)(0.2 \times 10^{-6})}\right]$$

$$= (2 - j4217)\Omega$$

i.e., $Z_1 = 4217\angle -89.97°\ \Omega$

Maximum value of current at the fundamental frequency,

$$I_{1m} = \frac{V_{1m}}{Z_1} = \frac{400}{4217} = \mathbf{0.095\ A}$$

At the third harmonic frequency,

$$Z_3 = R + j\left(3\omega L - \frac{1}{3\omega C}\right) = R$$

since resonance occurs at the third harmonic, i.e., $Z_3 = 2\Omega$

Maximum value of current at the third harmonic frequency,

$$I_{3m} = \frac{V_{3m}}{Z_3} = \frac{10}{2} = \mathbf{5\ A}$$

(Note that the magnitude of I_{3m} compared with I_{1m} is 5/0.095, i.e., $\times$ **52.6 greater**.)

Problem 15. A voltage wave has an amplitude of 800 V at the fundamental frequency of 50 Hz and its nth harmonic has an amplitude 1.5% of the fundamental. The voltage is applied to a series circuit containing resistance 5 Ω, inductance 0.369 H and capacitance 0.122 μF. Resonance occurs at the nth harmonic. Determine (a) the value of n, (b) the maximum value of current at the nth harmonic, (c) the p.d. across the capacitor at the nth harmonic and (d) the maximum value of the fundamental current.

(a) For resonance at the nth harmonic, $n\omega L = 1/(n\omega C)$, from which

$$n^2 = \frac{1}{\omega^2 LC} \text{ and } n = \frac{1}{\omega\sqrt{(LC)}}$$

Hence $$n = \frac{1}{2\pi 50\sqrt{(0.369)(0.122 \times 10^{-6})}} = \mathbf{15}$$

Thus resonance occurs at the 15th harmonic.

(b) At resonance, impedance $Z_{15} = R = 5\ \Omega$. Hence the maximum value of current at the 15th harmonic,

$$I_{15m} = \frac{V_{15m}}{R} = \frac{(1.5/100) \times 800}{5} = \mathbf{2.4\ A}$$

(c) At the 15th harmonic, capacitive reactance,

$$X_{C15} = \frac{1}{15\omega C} = \frac{1}{15(2\pi 50)(0.122 \times 10^{-6})} = 1739\ \Omega$$

Hence the p.d. across the capacitor at the 15th harmonic

$$= (I_{15m})(X_{C15}) = (2.4)(1739) = \mathbf{4.174\ kV}$$

(d) At the fundamental frequency, inductive reactance,

$$X_{L1} = \omega L = (2\pi 50)(0.369) = 115.9\ \Omega,$$

and capacitive reactance,

$$X_{Cl} = \frac{1}{\omega C} = \frac{1}{(2\pi 50)(0.122 \times 10^{-6})} = 26091\ \Omega$$

Impedance at the fundamental frequency,

$$|Z| = \sqrt{[R^2 + (X_C - X_L)^2]}$$
$$= 25975\ \Omega$$

Maximum value of current at the fundamental frequency,

$$I_{1m} = \frac{V_{1m}}{Z_1} = \frac{800}{25975} = \mathbf{0.031\ A}\ \text{or}\ \mathbf{31\ mA}$$

Further problems on harmonic resonance may be found in Section 36.9, problems 25 to 29, page 676.

36.8 Sources of harmonics

(i) Harmonics may be produced in the **output waveform of an a.c. generator**. This may be due either to 'tooth-ripple', caused by the effect of the slots that accommodate the windings, or to the nonsinusoidal airgap flux distribution.

Great care is taken to ensure a sinusoidal output from generators in large supply systems; however, non linear loads will cause harmonics to appear in the load current waveform. Thus harmonics are produced in devices that have a non linear response to their inputs. Non linear circuit elements (i.e., those in which the current flowing through them is not proportional to the applied voltage) include rectifiers and any large-signal electronic amplifier in which diodes, transistors, valves or iron-cored inductors are used.

(ii) A **rectifier** is a device for converting an alternating or an oscillating current into a unidirectional or approximate direct current. A rectifier has a low impedance to current flow in one direction and a nearly infinite impedance to current flow in the opposite direction. Thus when an alternating current is applied to a rectifier, current will flow through it during the positive half-cycles only; the current is zero during the negative half-cycles. A typical current waveform is shown in Figure 36.19. This 'half-wave rectification' is produced by using a single diode. The waveform is similar in shape to that shown in Figure 36.14, page 643, where the d.c. component brought the negative half-cycle up to the zero current point. The waveform shown in Figure 36.19 is typical of one containing a fairly large second harmonic.

Figure 36.19 *Typical current waveform containing a fairly large second harmonic*

(iii) **Transistors** and **valves** are non linear devices in that sinusoidal input results in different positive and negative half-cycle amplifications. This means that the output half-cycles have different amplitudes. Since they have a different shape, even harmonic distortion is suggested (see Section 36.3).

(iv) **Ferromagnetic-cored coils** are a source of harmonic generation in a.c. circuits because of the non-linearity of the B/H curve and the hysteresis loop, especially if saturation occurs. Let a sinusoidal voltage $v = V_m \sin \omega t$ be applied to a ferromagnetic-cored coil (having low resistance relative to inductive reactance) of cross-section area A square metres and possessing N turns.

If ϕ is the flux produced in the core then the instantaneous voltage is given by $\boldsymbol{v = N(d\phi/dt)}$.

If B is the flux density of the core, then, since $\Phi = BA$,

$$v = N\frac{d}{dt}(BA) = \boldsymbol{NA\frac{dB}{dt}},$$

since area A is a constant for a particular core.

Separating the variables gives

$$\int dB = \frac{1}{NA}\int v\,dt$$

i.e., $$B = \frac{1}{NA}\int V_m \sin \omega t\,dt = \frac{-V_m}{\omega NA}\cos \omega t$$

Since $-\cos\omega t = \sin(\omega t - 90°)$,

$$\boldsymbol{B} = \frac{\boldsymbol{V_m}}{\boldsymbol{\omega NA}} \mathbf{\sin(\omega t - 90°)} \qquad (36.21)$$

Equation (36.21) shows that if the applied voltage is sinusoidal, the flux density B in the iron core must also be sinusoidal but lagging by 90°.

The condition of low resistance relative to inductive reactance, giving a sinusoidal flux from a sinusoidal supply voltage, is called **free magnetization**.

Consider the application of a sinusoidal voltage to a coil wound on a core with a hysteresis loop as shown in Figure 36.20(a). The horizontal axis of a hysteresis loop is magnetic field strength H, but since $H = Ni/l$ and N and l (the length of the flux path) are constant, the axis may be directly scaled as current i (i.e., $i = Hl/N$). Figure 36.20(b) shows sinusoidal voltage v and flux density B waveforms, B lagging v by 90°.

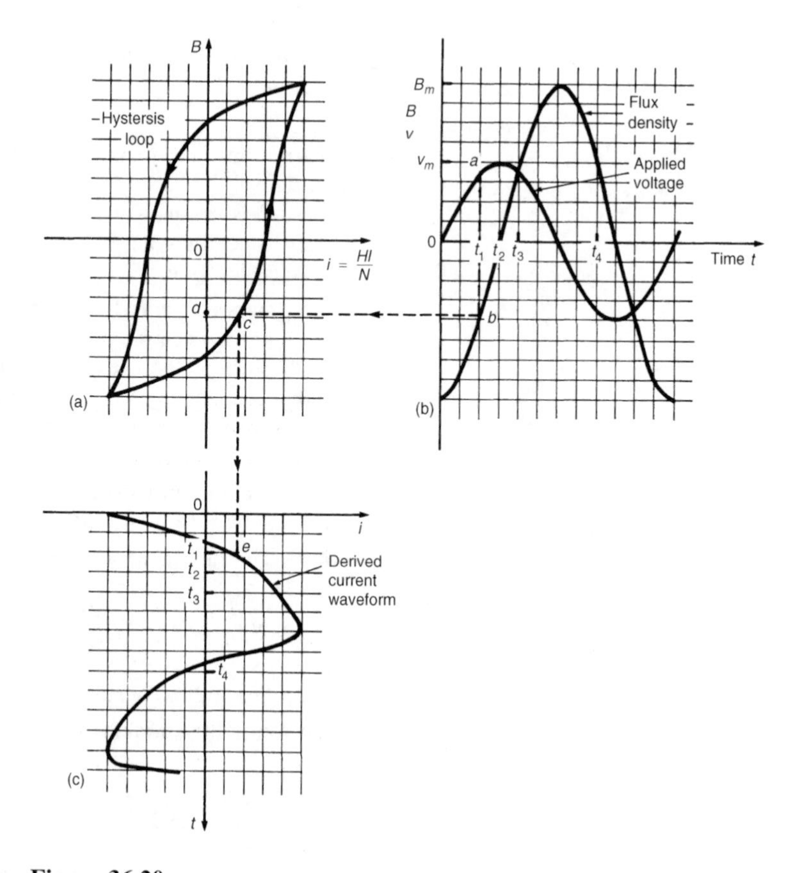

Figure 36.20

The current waveform is shown in Figure 36.20(c) and is derived as follows. At time t_1, point a on the voltage curve corresponds to point b on the flux density curve and the point c on the hysteresis loop. The current at time t_1 is given by the distance dc. Plotting this current on a vertical time-scale gives the derived point e on the current curve. A similar procedure is adopted for times t_2, t_3 and so on over one cycle of the voltage.

(Note that it is important to move around the hysteresis loop in the correct direction.) It is seen from the current curve that it is non-sinusoidal and that the positive and negative half-cycles are identical. This indicates that the waveform contains only odd harmonics (see Section (36.3)).

(v) If, in a circuit containing a ferromagnetic-cored coil, the resistance is high compared with the inductive reactance, then the current flowing from a sinusoidal supply will tend to be sinusoidal. This means that the flux density B of the core cannot be sinusoidal since it is related to the current by the hysteresis loop. This means, in turn, that the induced voltage due to the alternating flux (i.e., $v = NA(dB/dt)$) will not be sinusoidal. This condition is called **forced magnetization**.

The shape of the induced voltage waveform under forced magnetization is obtained as follows. The current waveform is shown on a vertical axis in Figure 36.21(a). The hysteresis loop corresponding to the maximum value of circuit current is drawn as shown in Figure 36.21(b). The flux density curve which is derived from the sinusoidal current waveform is shown in Figure 36.21(c). Point a on the current wave at time t_1 corresponds to point b on the hysteresis loop and to point c on the flux density curve. By taking other points throughout the current cycle the flux density curve is derived as shown.

The relationship between the induced voltage v and the flux density B is given by $v = NA(dB/dt)$. Here dB/dt represents the rate of change of flux density with respect to time, i.e., the gradient of the B/t curve. At point d the gradient of the B/t curve is a maximum in the positive direction. Thus v will be maximum positive as shown by point d' in Figure 36.21(d). At point e the gradient (i.e., dB/dt) is zero, thus v is zero, as shown by point e'. At point f the gradient is maximum in a negative direction, thus v is maximum negative, as shown by point f'. If all such points are taken around the B/t curve, the curve representing induced voltage, shown in Figure 36.21(d), is produced. The resulting voltage waveform is nonsinusoidal. The positive and negative half-cycles are identical in shape, indicating that the waveform contains a fundamental and a prominent third harmonic.

(vi) The amount of power delivered to a load can be controlled using a **thyristor**, which is a semi-conductor device. Examples of applications of controlled rectification include lamp and heater controls and the control of motor speeds. A basic circuit used for single-phase power control is shown in Figure 36.22(a). The trigger module contains circuitry to produce the necessary gate current to turn

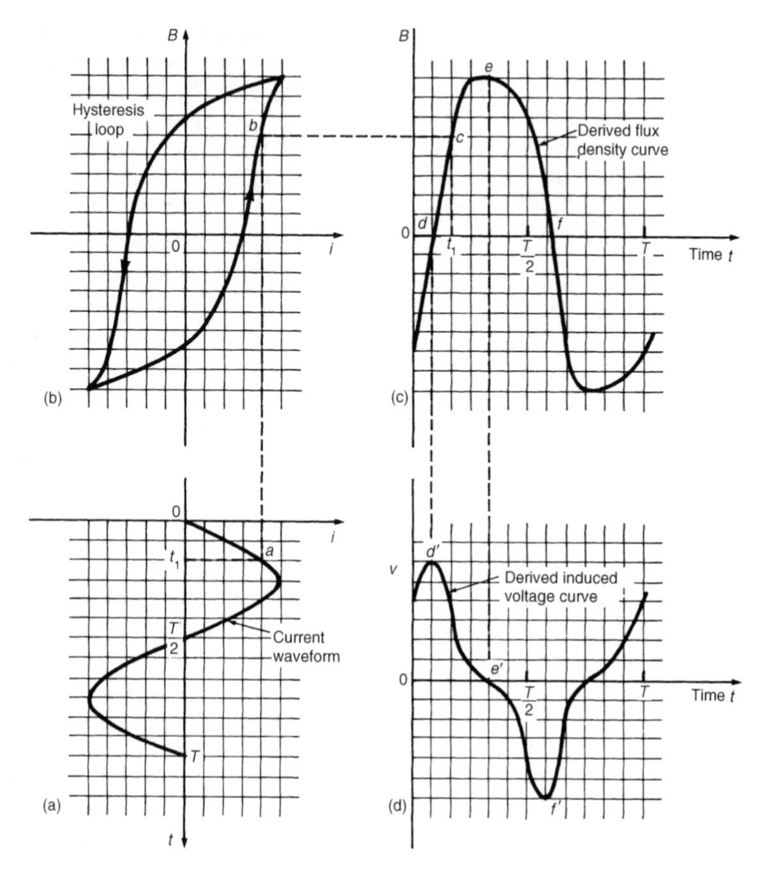

Figure 36.21

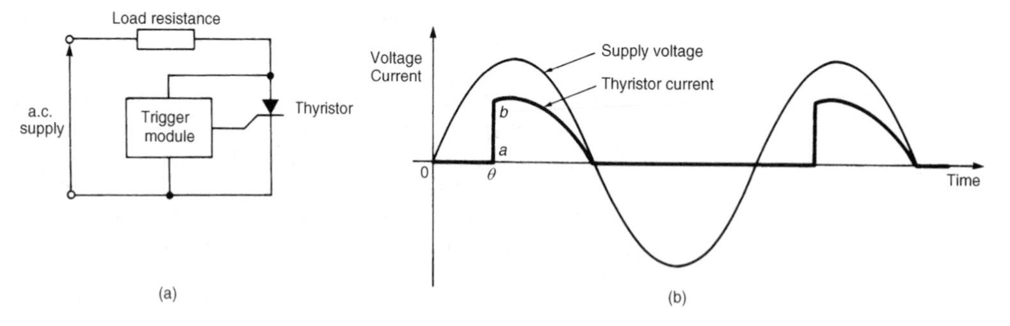

Figure 36.22

the thyristor on. If the pulse is applied at time θ/ω, where θ is the firing or triggering angle, then the current flowing in the load resistor has a waveform as shown in Figure 36.22(b). The sharp rise-time (shown as *ab* in Figure 36.22(b)), however, gives rise to harmonics.

(vii) In **microelectronic systems** rectangular waveforms are common. Again, fast rise-times give rise to harmonics, especially at high frequency. These harmonics can be fed back to the mains if not filtered.

There are thus a large number of sources of harmonics.

36.9 Further problems on complex waveforms

Harmonic synthesis

1 A complex current waveform i comprises a fundamental current of 50 A rms and frequency 100 Hz, together with a 24% third harmonic, both being in phase with each other at zero time. (a) Write down an expression to represent current, i. (b) Sketch the complex waveform of current using harmonic synthesis over one cycle of the fundamental.

[(a) $i = (70.71 \sin 628.3t + 16.97 \sin 1885t)$A]

2 A complex voltage waveform v is comprised of a 212.1 V rms fundamental voltage at a frequency of 50 Hz, a 30% second harmonic component lagging the fundamental voltage at zero time by $\pi/2$ rad, and a 10% fourth harmonic component leading the fundamental at zero time by $\pi/3$ rad. (a) Write down an expression to represent voltage v. (b) Sketch the complex voltage waveform using harmonic synthesis over one cycle of the fundamental waveform.

[(a) $v = 300 \sin 314.2t + 90 \sin(628.3t - (\pi/2)) + 30 \sin(1256.6t + (\pi/3))$volts]

3 A voltage waveform is represented by

$$v = 20 + 50 \sin \omega t + 20 \sin(2\omega t - \pi/2)\text{volts}$$

Draw the complex waveform over one cycle of the fundamental by using harmonic synthesis.

4 Write down an expression representing a current having a fundamental component of amplitude 16 A and frequency 1 kHz, together with its third and fifth harmonics being respectively one-fifth and one-tenth the amplitude of the fundamental, all components being in phase at zero time. Sketch the complex current waveform for one cycle of the fundamental using harmonic synthesis.

[$i = (16 \sin 2\pi 10^3 t + 3.2 \sin 6\pi 10^3 t + 1.6 \sin \pi 10^4 t)$A]

5 For each of the waveforms shown in Figure 36.23, state which harmonics are likely to be present.

[(a) Fundamental and even harmonics, or all harmonics present, initially in phase with each other. (b) Fundamental and odd harmonics only. (c) Fundamental and even harmonics, initially out of phase with each other (or all harmonics present), some being initially out of phase with each other.]

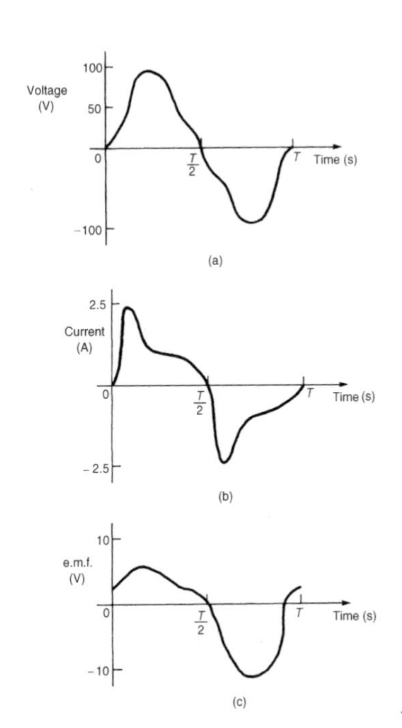

Figure 36.23

6 A voltage waveform is described by

$$v = 200 \sin 377t + 80 \sin(1131t + (\pi/4)) + 20 \sin(1885t - (\pi/3))\text{volts}$$

Determine (a) the fundamental and harmonic frequencies of the waveform, (b) the percentage third harmonic and (c) the percentage fifth harmonic. Sketch the voltage waveform using harmonic synthesis over one cycle of the fundamental.

[(a) 60 Hz, 180 Hz, 300 Hz (b) 40% (c) 10%]

Rms values, mean values and form factor of complex waves

7 Determine the rms value of a complex current wave represented by

$$i = 3.5 \sin \omega t + 0.8 \sin\left(3\omega t - \frac{\pi}{3}\right) + 0.2 \sin\left(5\omega t + \frac{\pi}{2}\right) \text{A}$$

[2.54 A]

8 Derive an expression for the rms value of a complex voltage waveform represented by

$$v = V_0 + V_{1m} \sin(\omega t + \phi_1) + V_{3m} \sin(3\omega t + \phi_3)\text{volts}$$

Calculate the rms value of a voltage waveform given by

$$v = 80 + 240 \sin \omega t + 50 \sin\left(2\omega t + \frac{\pi}{4}\right) + 20 \sin\left(4\omega t - \frac{\pi}{3}\right) \text{volts}$$

[191.4 V]

9 A complex voltage waveform is given by

$$v = 150 \sin 314t + 40 \sin\left(942t - \frac{\pi}{2}\right) + 30 \sin(1570t + \pi)\text{volts}$$

Determine for the voltage (a) the third harmonic frequency, (b) its rms value, (c) its mean value and (d) the form factor.

[(a) 150 Hz (b) 111.8 V (c) 91.7 V (d) 1.22]

10 A complex voltage waveform has an rms value of 220 V and it contains 25% third harmonic and 15% fifth harmonic. (a) Determine the rms value of the fundamental and each harmonic. (b) Write down an expression to represent the complex voltage waveform if the frequency of the fundamental is 60 Hz.

[(a) 211.2 V, 52.8 V, 31.7 V;

[(b) $v = 298.7 \sin 377t + 74.7 \sin 1131t + 44.8 \sin 1885t$V]

11 Define the term 'form factor' when applied to a symmetrical complex waveform. Calculate the form factor of an alternating voltage which is represented by

$$v = (50 \sin 314t + 15 \sin 942t + 6 \sin 1570t)\text{volts}$$

[1.038]

Power associated with complex waves

12 Determine the average power in a 50 Ω resistor if the current i flowing through it is represented by

$$i = (140 \sin \omega t + 40 \sin 3\omega t + 20 \sin 5\omega t)\text{mA}$$

[0.54 W]

13 A voltage waveform represented by

$$v = 100 \sin \omega t + 22 \sin\left(3\omega t - \frac{\pi}{6}\right) + 8 \sin\left(5\omega t - \frac{\pi}{4}\right) \text{ volts}$$

is applied to a circuit and the resulting current i is given by

$$i = 5 \sin\left(\omega t + \frac{\pi}{3}\right) + 1.91 \sin 3\omega t + 0.76 \sin(5\omega t - 0.452)\text{amperes}$$

Calculate (a) the total active power supplied to the circuit, and (b) the overall power factor.

[(a) 146.1 W (b) 0.526]

14 Determine the rms voltage, rms current and average power supplied to a network if the applied voltage is given by

$$v = 100 + 50 \sin\left(400t - \frac{\pi}{3}\right) + 40 \sin\left(1200t - \frac{\pi}{6}\right) \text{ volts}$$

and the resulting current is given by

$$i = 0.928 \sin(400t + 0.424) + 2.14 \sin(1200t + 0.756)\text{amperes}$$

[109.8 V, 1.65 A, 14.60 W]

15 A voltage $v = 40 + 20 \sin 300t + 8 \sin 900t + 3 \sin 1500t$ volts is applied to the terminals of a circuit and the resulting current is given by

$$i = 4 + 1.715 \sin(300t - 0.540) + 0.389 \sin(900t - 1.064)$$
$$+ 0.095 \sin(1500t - 1.249)\text{A}$$

Determine (a) the rms voltage, (b) the rms current and (c) the average power.

[(a) 42.85 V (b) 4.189 A (c) 175.5 W]

Harmonics in single-phase circuits

16 A complex voltage waveform represented by

$$v = 240 \sin \omega t + 60 \sin\left(3\omega t - \frac{\pi}{4}\right) + 30 \sin\left(5\omega t + \frac{\pi}{3}\right) \text{ volts}$$

is applied across (a) a pure 50 Ω resistance, (b) a pure 4.974 μF capacitor, and (c) a pure 15.92 mH inductance. Determine for each case an expression for the current flowing if the fundamental frequency is 400 Hz.

$$\left[\text{(a) } i = 4.8 \sin \omega t + 1.2 \sin\left(3\omega t - \frac{\pi}{4}\right) + 0.6 \sin\left(5\omega t + \frac{\pi}{3}\right) \text{A}\right.$$

$$\text{(b) } i = 3 \sin\left(\omega t + \frac{\pi}{2}\right) + 2.25 \sin\left(3\omega t + \frac{\pi}{4}\right)$$

$$+ 1.875 \sin\left(5\omega t + \frac{5\pi}{6}\right) \text{A}$$

$$\text{(c) } i = 6 \sin\left(\omega t - \frac{\pi}{2}\right) + 0.5 \sin\left(3\omega t - \frac{3\pi}{4}\right)$$

$$\left. + 0.15 \sin\left(5\omega t - \frac{\pi}{6}\right) \text{A}\right]$$

17 A complex current given by

$$i = 5 \sin\left(\omega t + \frac{\pi}{3}\right) + 8 \sin\left(3\omega t + \frac{2\pi}{3}\right) \text{mA}$$

flows through a pure 2000 pF capacitor. If the frequency of the fundamental component is 4 kHz, determine (a) the rms value of current, (b) an expression for the p.d. across the capacitor, and (c) the rms value of voltage.

[(a) 6.671 mA (b) $v = 99.47 \sin(\omega t - (\pi/6))$
$+53.05 \sin(3\omega t + (\pi/6))$V (c) 79.71 V]

18 A complex voltage, v, given by

$$v = 200 \sin \omega t + 42 \sin 3\omega t + 25 \sin 5\omega t \text{ volts}$$

is applied to a circuit comprising a 6 Ω resistance in series with a coil of inductance 5 mH. Determine, for a fundamental frequency of 50 Hz, (a) an expression to represent the instantaneous value of the current flowing, (b) the rms voltage, (c) the rms current, (d) the power dissipated, and (e) the overall power factor.

[(a) $i = 32.25 \sin(314t - 0.256) + 5.50 \sin(942t - 0.666)$
$+ 2.53 \sin(1570t - 0.918)$A
(b) 145.6 V (c) 23.20 A (d) 3.23 kW (e) 0.956]

19 An e.m.f. e is given by

$$e = 40 + 150 \sin \omega t + 30 \sin\left(2\omega t - \frac{\pi}{4}\right)$$

$$+ 10 \sin\left(4\omega t - \frac{\pi}{3}\right) \text{volts}$$

the fundamental frequency being 50 Hz. The e.m.f. is applied across a circuit comprising a 100 Ω resistance in series with a 15 μF capacitor. Determine (i) the rms value of voltage, (ii) an expression for the current flowing and (iii) the rms value of current.

[(i) 115.5 V
(ii) $i = 0.639\sin(\omega t + 1.130) + 0.206\sin(2\omega t + 0.030) + 0.088\sin(4\omega t - 0.559)$A
(iii) 0.479 A]

20 A circuit comprises a 100 Ω resistance in series with a 1 mH inductance. The supply voltage is given by

$$v = 40 + 200\sin\omega t + 50\sin\left(3\omega t + \frac{\pi}{4}\right) + 15\sin\left(5\omega t + \frac{\pi}{6}\right)\text{ volts}$$

where $\omega = 10^5$ rad/s. Determine for the circuit (a) an expression to represent the current flowing, (b) the rms value of current and (c) the power dissipated.

[(a) $i = 0.40 + 1.414\sin(\omega t - (\pi/4)) + 0.158\sin(3\omega t - 0.464) + 0.029\sin(5\omega t - 0.850)$
(b) 1.08 A (c) 117 W]

21 The e.m.f. applied to a circuit comprising two components connected in series is given by

$$v = 50 + 150\sin(2\times 10^3 t) + 40\sin(4\times 10^3 t) + 20\sin(8\times 10^3 t)\text{volts}$$

and the resulting current is given by

$$i = 1.011\sin(2\times 10^3 t + 1.001) + 0.394\sin(4\times 10^3 t + 0.663) + 0.233\sin(8\times 10^3 t + 0.372)\text{A}$$

Determine for the circuit (a) the average power supplied, and (b) the value of the two circuit components.

[(a) 49.3 W (b) $R = 80$ Ω, $C = 4$ μF]

22 A coil having inductance L and resistance R is supplied with a complex voltage given by

$$v = 240\sin\omega t + V_3\sin\left(3\omega t + \frac{\pi}{3}\right) + V_5\sin\left(5\omega t - \frac{\pi}{12}\right)\text{ volts}$$

The resulting current is given by

$$i = 4.064\sin(\omega t - 0.561) + 0.750\sin(3\omega t - 0.036) + 0.182\sin(5\omega t - 1.525)\text{A}$$

The fundamental frequency is 500 Hz. Determine (a) the impedance of the circuit at the fundamental frequency, and hence the values of R and L, (b) the values of V_3 and V_5, (c) the rms voltage, (d) the rms current, (e) the circuit power, and (f) the power factor.

[(a) 59.06 Ω, $R = 50$ Ω, $L = 10$ mH (b) 80 V, 30 V (c) 180.1 V (d) 2.93 A (e) 427.8 W (f) 0.811]

23 An alternating supply voltage represented by

$$v = (240\sin 300t - 40\sin 1500t + 60\sin 2100t)\text{volts}$$

is applied to the terminals of a circuit containing a 40 Ω resistor, a 200 mH inductor and a 25 μF capacitor in series. (a) Derive the expression for the current waveform and (b) calculate the power dissipated by the circuit.

[(a) $i = 2.873\sin(300t + 1.071) - 0.145\sin(1500t - 1.425) + 0.149\sin(2100t - 1.471)$A

(b) 166 W]

24 A voltage v represented by

$$v = 120\sin 314t + 25\sin\left(942t + \frac{\pi}{6}\right)\text{ volts}$$

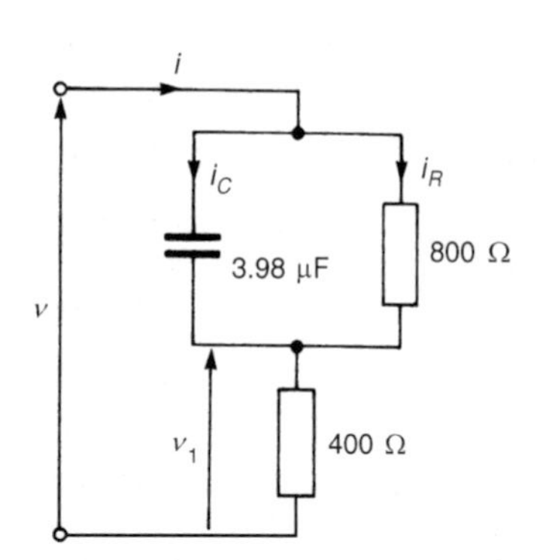

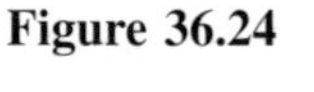
Figure 36.24

is applied to the circuit shown in Figure 36.24. Determine (a) an expression for current i, (b) the percentage harmonic content of the supply current, (c) the total power dissipated, (d) an expression for the p.d. shown as v_1 and (e) expressions for the currents shown as i_R and i_C.

[(a) $i = 0.134\sin(314t + 0.464) + 0.047\sin(942t + 0.988)$A

(b) 35.07% (c) 7.72 W

(d) $v_1 = 53.6\sin(314t + 0.464) + 18.8\sin(942t + 0.988)$V

(e) $i_R = 0.095\sin(314t - 0.321) + 0.015\sin(942t - 0.261)$A

$i_C = 0.095\sin(314t + 1.249) + 0.045\sin(942t + 1.310)$A]

Harmonic resonance

25 A voltage waveform having a fundamental of maximum value 250 V and a third harmonic of maximum value 20 V is applied to a series circuit comprising a 5 Ω resistor, a 400 mH inductance and a 0.5 μF capacitor. Determine (a) the fundamental frequency for resonance with the third harmonic and (b) the maximum values of the fundamental and third harmonic components of the current.

[(a) 118.6 Hz (b) 0.105 A, 4 A]

26 A complex voltage waveform has a maximum value of 500 V at the fundamental frequency of 60 Hz and contains a 17th harmonic having an amplitude of 2% of the fundamental. The voltage is applied

to a series circuit containing resistance 2 Ω, inductance 732 mH and capacitance 36.26 nF. Determine (a) the maximum value of the 17th harmonic current, (b) the maximum value of the 17th harmonic p.d. across the capacitor, and (c) the amplitude of the fundamental current.

[(a) 5 A (b) 23.46 kV (c) 6.29 mA]

27 A complex voltage waveform v is given by the expression

$$v = 150 \sin \omega t + 25 \sin\left(3\omega t - \frac{\pi}{6}\right) + 10 \sin\left(5\omega t + \frac{\pi}{3}\right) \text{ volts}$$

where $\omega = 314$ rad/s. The voltage is applied to a circuit consisting of a coil of resistance 10 Ω and inductance 50 mH in series with a variable capacitor.

(a) Calculate the value of the capacitance which will give resonance with the triple frequency component of the voltage. (b) Write down the corresponding equation for the current waveform. (c) Determine the rms value of current. (d) Find the power dissipated in the circuit.

[(a) 22.54 μF

(b) $i = 1.191 \sin(314t + 1.491) + 2.500 \sin(942t - 0.524)$
$+ 0.195 \sin(1570t - 0.327)$A

(c) 1.963 A (d) 38.56 W]

28 A complex voltage of fundamental frequency 50 Hz is applied to a series circuit comprising resistance 20 Ω, inductance 800 μH and capacitance 74.94 μF. Resonance occurs at the nth harmonic. Determine the value of n. [13]

29 A complex voltage given by $v = 1200 \sin \omega t + 300 \sin 3\omega t + 100 \sin 5\omega t$ volts is applied to a circuit containing a 25 Ω resistor, a 12 μF capacitor and a 37 mH inductance connected in series. The fundamental frequency is 79.62 Hz. Determine (a) the rms value of the voltage, (b) an expression for the current waveform, (c) the rms value of current, (d) the amplitude of the third harmonic voltage across the capacitor, (e) the circuit power, and (f) the overall power factor.

[(a) 877.5 V

(b) $i = 7.991 \sin(\omega t + 1.404) + 12 \sin 3\omega t$
$+ 1.555 \sin(5\omega t - 1.171)$A

(c) 10.25 A (d) 666.4 V (e) 2626 W (f) 0.292]

37 A numerical method of harmonic analysis

At the end of this chapter you should be able to:

- use a tabular method to determine the Fourier series for a complex waveform
- predict the probable harmonic content of a waveform on inspection

37.1 Introduction

Many practical waveforms can be represented by simple mathematical expressions, and, by using Fourier series, the magnitude of their harmonic components determined. For waveforms not in this category, analysis may be achieved by numerical methods. **Harmonic analysis** is the process of resolving a periodic, non-sinusoidal quantity into a series of sinusoidal components of ascending order of frequency.

37.2 Harmonic analysis on data given in tabular or graphical form

A Fourier series is merely a trigonometric series of the form:

$$f(x) = a_0 + a_1 \cos x + a_2 \cos 2x + \cdots + b_1 \sin x + b_2 \sin 2x + \cdots$$

i.e. $$f(x) = a_0 + \sum_{n=1}^{\infty} (a_n \cos nx + b_n \sin nx)$$

The Fourier coefficients a_0, a_n and b_n all require functions to be integrated, i.e.,

$$a_0 = \frac{1}{2\pi}\int_{-\pi}^{\pi} f(x)dx = \frac{1}{2\pi}\int_{0}^{2\pi} f(x)\,dx$$

$= \text{mean value of } f(x) \text{ in the range } -\pi \text{ to } \pi \text{ or } 0 \text{ to } 2\pi$

$$a_n = \frac{1}{\pi}\int_{-\pi}^{\pi} f(x)\cos nx\,dx = \frac{1}{\pi}\int_{0}^{2\pi} f(x)\cos nx\,dx$$

$= \text{twice the mean value of } f(x)\cos nx \text{ in the range } 0 \text{ to } 2\pi$

$$b_n = \frac{1}{\pi}\int_{-\pi}^{\pi} f(x)\sin nx\,dx = \frac{1}{\pi}\int_{0}^{2\pi} f(x)\sin nx\,dx$$

$= \text{twice the mean value of } f(x)\sin nx \text{ in the range } 0 \text{ to } 2\pi$

However, irregular waveforms are not usually defined by mathematical expressions and thus the Fourier coefficients cannot be determined by using calculus. In these cases, approximate methods, such as the **trapezoidal rule**, can be used to evaluate the Fourier coefficients.

Most practical waveforms to be analysed are periodic. Let the period of a waveform be 2π and be divided into p equal parts as shown in Figure 37.1. The width of each interval is thus $2\pi/p$. Let the ordinates be labelled $y_0, y_1, y_2, \ldots, y_p$ (note that $y_0 = y_p$). The trapezoidal rule states:

$$\text{Area} \approx (\text{width of interval})\left[\frac{1}{2}(\text{first} + \text{last ordinate}) + \text{sum of remaining ordinates}\right]$$

$$\approx \frac{2\pi}{p}\left[\frac{1}{2}(y_0 + y_p) + y_1 + y_2 + y_3 + \cdots\right]$$

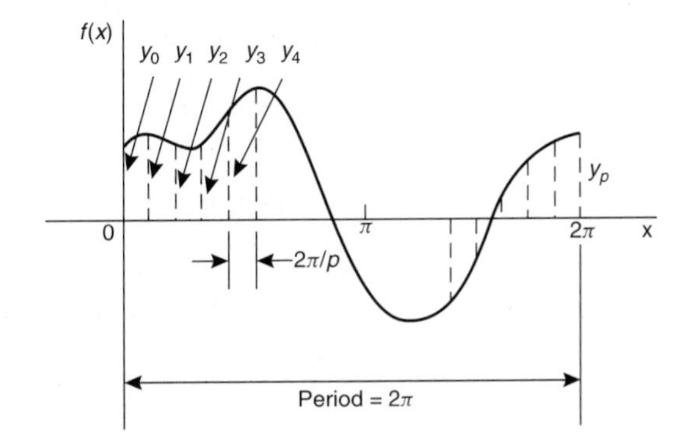

Figure 37.1

Since $y_0 = y_p$, then $\frac{1}{2}(y_0 + y_p) = y_0 = y_p$. Hence area $\approx \frac{2\pi}{p}\sum_{k=1}^{p} y_k$

$$\text{Mean value} = \frac{\text{area}}{\text{length of base}} \approx \frac{1}{2\pi}\left(\frac{2\pi}{p}\right)\sum_{k=1}^{p} y_k \approx \frac{1}{p}\sum_{k=1}^{p} y_k$$

However, a_0 = mean value of $f(x)$ in the range 0 to 2π. Thus

$$\boxed{a_0 \approx \frac{1}{p}\sum_{k=1}^{p} y_k} \qquad (37.1)$$

Similarly, a_n = twice the mean value of $f(x)\cos nx$ in the range 0 to 2π, thus,

$$a_n \approx \frac{2}{p}\sum_{k=1}^{p} y_k \cos nx_k \qquad (37.2)$$

and b_n = twice the mean value of $f(x)\sin nx$ in the range 0 to 2π, thus

$$b_n \approx \frac{2}{p}\sum_{k=1}^{p} y_k \sin nx_k \qquad (37.3)$$

Problem 1. The values of the voltage v volts at different moments in a cycle are given by:

θ *degrees*	30	60	90	120	150	180	210	240	270	300	330	360
v *(volts)*	62	35	−38	−64	−63	−52	−28	24	80	96	90	70

Draw the graph of voltage v against angle θ and analyse the voltage into its first three constituent harmonics, each coefficient correct to 2 decimal places.

The graph of voltage v against angle θ is shown in Figure 37.2. The range 0 to 2π is divided into 12 equal intervals giving an interval width of $2\pi/12$, i.e. $\pi/6$ or 30°. The values of the ordinates y_1, y_2, y_3, ... are 62, 35, −38, ... from the given table of values. If a larger number of intervals are used, results having a greater accuracy are achieved. The data is tabulated in the proforma shown in Table 37.1.

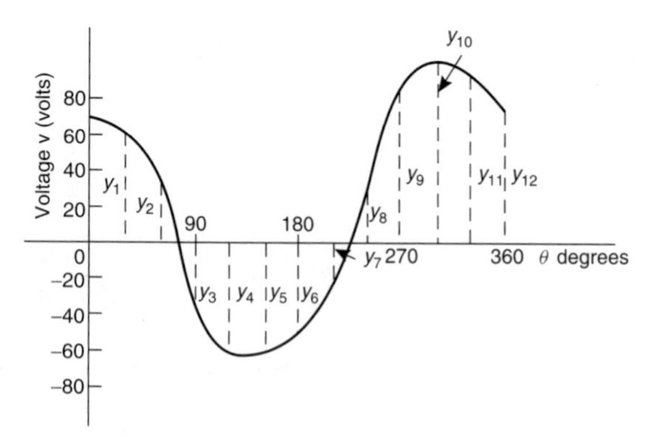

Figure 37.2

From equation (37.1), $a_0 \approx \frac{1}{p}\sum_{k=1}^{p} y_k = \frac{1}{12}(212)$

$= 17.67$ (since $p = 12$)

TABLE 37.1

Ordinates	θ°	v	$\cos\theta$	$v\cos\theta$	$\sin\theta$	$v\sin\theta$	$\cos 2\theta$	$v\cos 2\theta$	$\sin 2\theta$	$v\sin 2\theta$	$\cos 3\theta$	$v\cos 3\theta$	$\sin 3\theta$	$v\sin 3\theta$
y_1	30	62	0.866	53.69	0.5	31	0.5	31	0.866	53.69	0	0	1	62
y_2	60	35	0.5	17.5	0.866	30.31	−0.5	−17.5	0.866	30.31	−1	−35	0	0
y_3	90	−38	0	0	1	−38	−1	38	0	0	0	0	−1	38
y_4	120	−64	−0.5	32	0.866	−55.42	−0.5	32	−0.866	55.42	1	−64	0	0
y_5	150	−63	−0.866	54.56	0.5	−31.5	0.5	−31.5	−0.866	54.56	0	0	1	−63
y_6	180	−52	−1	52	0	0	1	−52	0	0	−1	52	0	0
y_7	210	−28	−0.866	24.25	−0.5	14	0.5	−14	0.866	−24.25	0	0	−1	28
y_8	240	24	−0.5	−12	−0.866	−20.78	−0.5	−12	0.866	−20.78	1	24	0	0
y_9	270	80	0	0	−1	−80	−1	−80	0	0	0	0	1	80
y_{10}	300	96	0.5	48	−0.866	−83.14	−0.5	−48	−0.866	−83.14	−1	−96	0	0
y_{11}	330	90	0.866	77.94	−0.5	−45	0.5	45	−0.866	−77.94	0	0	−1	−90
y_{12}	360	70	1	70	0	0	1	70	0	0	1	70	0	0
$\sum_{k=1}^{12} y_k = 212$			$\sum_{k=1}^{12} y_k \cos\theta_k = 417.94$		$\sum_{k=1}^{12} y_k \sin\theta_k = -278.53$		$\sum_{k=1}^{12} y_k \cos 2\theta_k = -39$		$\sum_{k=1}^{12} y_k \sin 2\theta_k = 29.43$		$\sum_{k=1}^{12} y_k \cos 3\theta_k = -49$		$\sum_{k=1}^{12} y_k \sin 3\theta_k = 55$	

From equation (37.2), $a_n \approx \frac{2}{p}\sum_{k=1}^{p} \cos nx_k$

Hence $a_1 \approx \frac{2}{12}(417.94) = 69.66;$

$a_2 \approx \frac{2}{12}(-39) = -6.50;$

and $a_3 \approx \frac{2}{12}(-49) = -8.17$

From equation (37.3), $b_n \approx \frac{2}{p}\sum_{k=1}^{p} y_k \sin nx_k$

Hence $b_1 \approx \frac{2}{12}(-278.53) = -46.42;$

$b_2 \approx \frac{2}{12}(29.43) = 4.91;$

and $b_3 \approx \frac{2}{12}(55) = 9.17$

Substituting these values into the Fourier series:

$$f(x) = a_0 + \sum_{n=1}^{\infty}(a_n \cos nx + b_n \sin nx)$$

gives: $$\mathbf{v = 17.67 + 69.66\cos\theta - 6.50\cos 2\theta - 8.17\cos 3\theta + \cdots}$$
$$\mathbf{-46.42\sin\theta + 4.91\sin 2\theta + 9.17\sin 3\theta + \cdots} \quad (37.4)$$

Note that in equation (37.4), $(-46.42\sin\theta + 69.66\cos\theta)$ comprises the fundamental, $(4.91\sin 2\theta - 6.50\cos 2\theta)$ comprises the second harmonic and $(9.17\sin 3\theta - 8.17\cos 3\theta)$ comprises the third harmonic.

It is shown in *Higher Engineering Mathematics* that

$$a\sin\omega t + b\cos\omega t \equiv R\sin(\omega t + \alpha)$$

where $a = R\cos\alpha$, $b = R\sin\alpha$, $R = \sqrt{(a^2+b^2)}$ and $\alpha = \arctan\frac{b}{a}$

For the fundamental, $R = \sqrt{[(-46.42)^2 + (69.66)^2]} = 83.71$

If $a = R\cos\alpha$, then $\cos\alpha = \frac{a}{R} = \frac{-46.42}{83.71}$ which is negative,

and if $b = R\sin\alpha$, then $\sin\alpha = \frac{b}{R} = \frac{69.66}{83.71}$ which is positive.

The only quadrant where $\cos\alpha$ is negative *and* $\sin\alpha$ is positive is the second quadrant.

Hence $\alpha = \arctan\frac{b}{a} = \arctan\frac{69.66}{-46.42} = 123.68°$ or 2. 16 rad

Thus $(-46.42\sin\theta + 69.66\cos\theta) = 83.71\sin(\theta + 2.16)$

By a similar method it may be shown that the second harmonic $(4.91 \sin 2\theta - 6.50 \cos 2\theta) \equiv 8.15 \sin(2\theta - 0.92)$ and the third harmonic $(9.17 \sin 3\theta - 8.17 \cos 3\theta) \equiv 12.28 \sin(3\theta - 0.73)$

Hence equation (37.4) may be re-written as:

$$\mathbf{v = 17.67 + 83.71 \sin(\theta + 2.16) + 8.15 \sin(2\theta - 0.92) + 12.28 \sin(3\theta - 0.73)\ volts}$$

which is the form used in Chapter 36 with complex waveforms.

37.3 Complex waveform considerations

It is sometimes possible to predict the harmonic content of a waveform on inspection of particular waveform characteristics.

(i) If a periodic waveform is such that the area above the horizontal axis is equal to the area below then the mean value is zero. Hence $a_0 = 0$ (see Figure 37.3(a)).

(ii) An **even function** is symmetrical about the vertical axis and contains **no sine terms** (see Figure 37.3(b)).

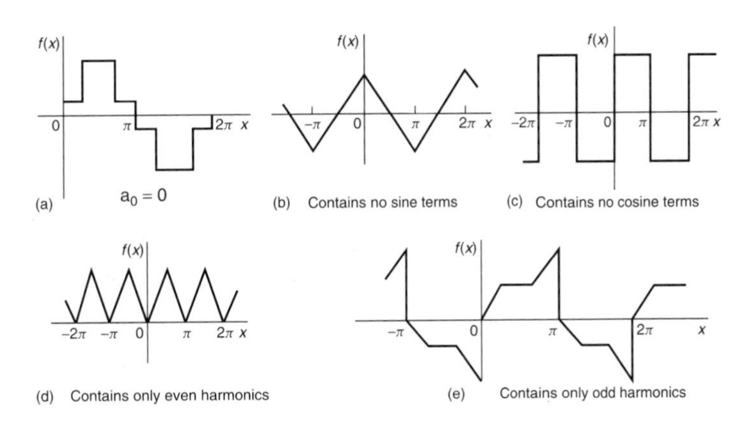

Figure 37.3

(iii) An **odd function** is symmetrical about the origin and contains no **cosine terms** (see Figure 37.3(c)).

(iv) $f(x) = f(x + \pi)$ represents a waveform which repeats after half a cycle and **only even harmonics** are present (see Figure 37.3(d)).

(v) $f(x) = -f(x + \pi)$ represents a waveform for which the positive and negative cycles are identical in shape and **only odd harmonics** are present (see Figure 37.3(e)).

Problem 2. Without calculating Fourier coefficients state which harmonics will be present in the waveforms shown in Figure 37.4.

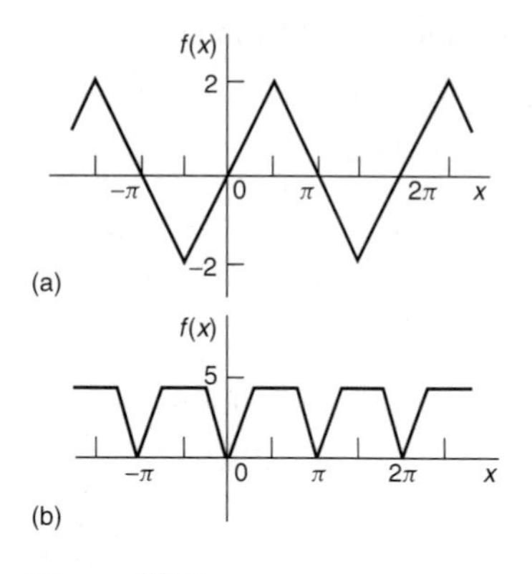

Figure 37.4

(a) The waveform shown in Figure 37.4(a) is symmetrical about the origin and is thus an odd function. An odd function contains no cosine terms. Also, the waveform has the characteristic $f(x) = -f(x + \pi)$, i.e. the positive and negative half cycles are identical in shape. Only odd harmonics can be present in such a waveform. Thus the waveform shown in Figure 37.4(a) contains **only odd sine terms**. Since the area above the x-axis is equal to the area below, $a_0 = 0$.

(b) The waveform shown in Figure 37.4(b) is symmetrical about the $f(x)$ axis and is thus an even function. An even function contains no sine terms. Also, the waveform has the characteristic $f(x) = f(x + \pi)$, i.e., the waveform repeats itself after half a cycle. Only even harmonics can be present in such a waveform. Thus the waveform shown in Figure 37.4(b) contains **only even cosine terms** (together with a constant term, a_0).

Problem 3. An alternating current i amperes is shown in Figure 37.5. Analyse the waveform into its constituent harmonics as far as and including the fifth harmonic, correct to 2 decimal places, by taking 30° intervals.

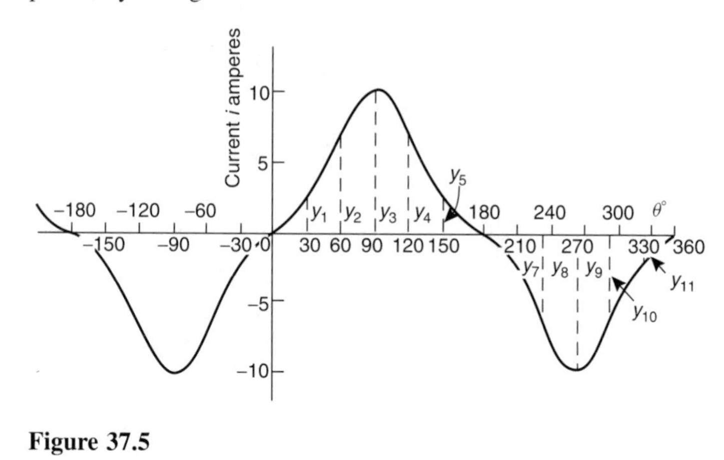

Figure 37.5

With reference to Figure 37.5, the following characteristics are noted:

(i) The mean value is zero since the area above the θ axis is equal to the area below it. Thus the constant term, or d.c. component, $a_0 = 0$.

(ii) Since the waveform is symmetrical about the origin the function i is odd, which means that there are no cosine terms present in the Fourier series.

(iii) The waveform is of the form $f(\theta) = -f(\theta + \pi)$ which means that only odd harmonics are present.

Investigating waveform characteristics has thus saved unnecessary calculations and in this case the Fourier series has only odd sine terms present, i.e.

$$i = b_1 \sin\theta + b_3 \sin 3\theta + b_5 \sin 5\theta + \cdots$$

A proforma, similar to Table 37.1, but without the 'cosine terms' columns and without the 'even sine terms' columns in shown in Table 37.2 up to, and including, the fifth harmonic, from which the Fourier coefficients b_1, b_3 and b_5 can be determined. Twelve coordinates are chosen and labelled $y_1, y_2, y_3, \ldots y_{12}$ as shown in Figure 37.5.

TABLE 37.2

Ordinate	$\theta°$	i	$\sin\theta$	$i\sin\theta$	$\sin 3\theta$	$i\sin 3\theta$	$\sin 5\theta$	$i\sin 5\theta$
y_1	30	2	0.5	1	1	2	0.5	1
y_2	60	7	0.866	6.06	0	0	−0.866	−6.06
y_3	90	10	1	10	−1	−10	1	10
y_4	120	7	0.866	6.06	0	0	−0.866	−6.06
y_5	150	2	0.5	1	1	2	0.5	1
y_6	180	0	0	0	0	0	0	0
y_7	210	−2	−0.5	1	−1	2	−0.5	1
y_8	240	−7	−0.866	6.06	0	0	0.866	−6.06
y_9	270	−10	−1	10	1	−10	−1	10
y_{10}	300	−7	−0.866	6.06	0	0	0.866	−6.06
y_{11}	330	−2	−0.5	1	−1	2	−0.5	1
y_{12}	360	0	0	0	0	0	0	0
			$\sum_{k=1}^{12} i_k \sin\theta_k = 48.24$		$\sum_{k=1}^{12} i_k \sin 3\theta_k = -12$		$\sum_{k=1}^{12} i_k \sin 5\theta_k = -0.24$	

From equation (37.3), Section 37.2, $b_n \approx \frac{2}{p}\sum_{k=1}^{p} i_k \sin n\theta_k$, where $p = 12$.

Hence $\quad b_1 \approx \frac{2}{12}(48.24) = 8.04;$

$$b_3 \approx \frac{2}{12}(-12) = -2.00;$$

and $\quad b_5 \approx \frac{2}{12}(-0.24) = -0.04$

Thus the Fourier series for current i is given by:

$$\mathbf{i = 8.04\sin\theta - 2.00\sin 3\theta - 0.04\sin 5\theta}$$

37.4 Further problems on a numerical method of harmonic analysis

Determine the Fourier series to represent the periodic functions given by the tables of values in Problems 1 to 3, up to and including the third harmonics and each coefficient correct to 2 decimal places. Use 12 ordinates in each case.

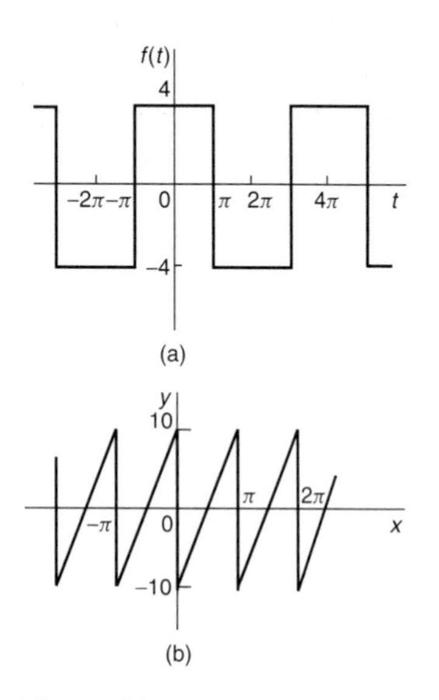

Figure 37.6

1

Angle $\theta°$	30	60	90	120	150	180	210	240	270	300	330	360
Displacement y	40	43	38	30	23	17	11	9	10	13	21	32

$[y = 23.92 + 7.81\cos\theta + 14.61\sin\theta + 0.17\cos 2\theta$
$+ 2.31\sin 2\theta - 0.33\cos 3\theta + 0.50\sin 3\theta]$

2

Angle $\theta°$	0	30	60	90	120	150	180	210	240	270	300	330
Voltage v	−5.0	−1.5	6.0	12.5	16.0	16.5	15.0	12.5	6.5	−4.0	−7.0	−7.5

$[v = 5.00 - 10.78\cos\theta + 6.83\sin\theta - 1.96\cos 2\theta$
$+ 0.80\sin 2\theta + 0.58\cos 3\theta - 1.08\sin 3\theta]$

3

Angle $\theta°$	30	60	90	120	150	180	210	240	270	300	330	360
Current i	0	−1.4	−1.8	−1.9	−1.8	−1.3	0	2.2	3.8	3.9	3.5	2.5

$[i = 0.64 + 1.58\cos\theta - 2.73\sin\theta - 0.23\cos 2\theta$
$- 0.42\sin 2\theta + 0.27\cos 3\theta + 0.05\sin 3\theta]$

4 Without performing calculations, state which harmonics will be present in the waveforms shown in Figure 37.6.

[(a) only odd cosine terms present
(b) only even sine terms present]

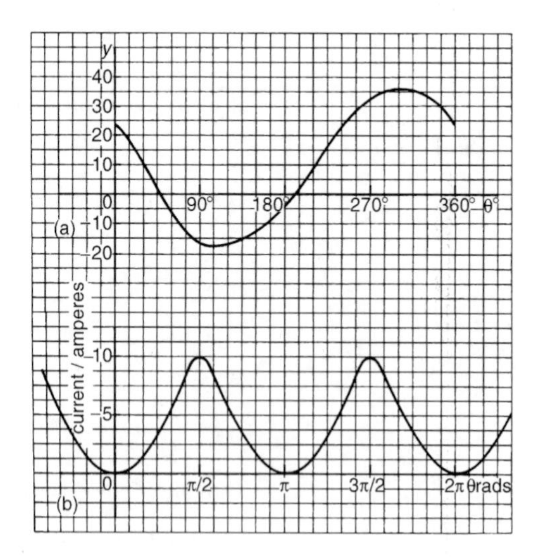

Figure 37.7

5 Analyse the periodic waveform of displacement y against angle θ in Figure 37.7(a) into its constituent harmonics as far as and including the third harmonic, by taking 30° intervals.

$$[y = 9.4 + 13.2\cos\theta - 24.4\sin\theta + 1.75\cos 2\theta - 0.58\sin 2\theta + 1.33\cos 3\theta + 0.67\sin 3\theta]$$

6 For the waveform of current shown in Figure 37.7(b) state why only a d.c. component and even cosine terms will appear in the Fourier series and determine the series, using $\pi/6$ rad intervals, up to and including the sixth harmonic.

$$[I = 3.83 - 4.50\cos 2\theta + 1.17\cos 4\theta - 1.00\cos 6\theta]$$

38 Magnetic materials

At the end of this chapter you should be able to:

- recognize terms associated with magnetic circuits
- appreciate magnetic properties of materials
- categorize materials as ferromagnetic, diamagnetic and paramagnetic
- explain hysteresis and calculate hysteresis loss
- explain and calculate eddy current loss
- explain a method of separation of hysteresis and eddy current loss and determine the separate losses from given data
- distinguish between non-permanent and permanent magnetic materials.

38.1 Revision of terms and units used with magnetic circuits

In Chapter 7, page 74, a number of terms used with magnetic circuits are defined. These are summarized below.

(a) A **magnetic field** is the state of the space in the vicinity of a permanent magnet or an electric current throughout which the magnetic forces produced by the magnet or current are discernible.

(b) **Magnetic flux Φ** is the amount of magnetic field produced by a magnetic source. The unit of magnetic flux is the **weber, Wb.** If the flux linking one turn in a circuit changes by one weber in one second, a voltage of one volt will be induced in that turn.

(c) **Magnetic flux density B** is the amount of flux passing through a defined area that is perpendicular to the direction of the flux.

$$\text{Magnetic flux density} = \frac{\text{magnetic flux}}{\text{area}}$$

i.e., $\boxed{B = \Phi/A,}$ where A is the area in square metres. The unit of magnetic flux density is the **tesla T,** where $1\ T = 1\ \text{Wb/m}^2$.

(d) **Magnetomotive force (mmf)** is the cause of the existence of a magnetic flux in a magnetic circuit. $\boxed{\textbf{mmf, } F_m = NI \textbf{ amperes,}}$

where N is the number of conductors (or turns) and I is the current in amperes. The unit of mmf is sometimes expressed as 'ampere-turns'. However since 'turns' have no dimension, the S.I. unit of mmf is the ampere.

(e) **Magnetic field strength (or magnetizing force),**

$$\boldsymbol{H = NI/l} \textbf{ ampere per metre,}$$

where l is the mean length of the flux path in metres.

Thus $$\textbf{mmf} = \boldsymbol{NI = Hl} \textbf{ amperes.}$$

(f) μ_0 is a constant called the **permeability of free space** (or the magnetic space constant). The value of μ_0 is $4\pi \times 10^{-7}$ H/m.

For air, or any nonmagnetic medium, the ratio $\boldsymbol{B/H = \mu_0}$

(Although all nonmagnetic materials, including air, exhibit slight magnetic properties, these can effectively be neglected.)

(g) μ_r is the **relative permeability** and is defined as

$$\frac{\text{flux density in material}}{\text{flux density in a vacuum}}$$

μ_r varies with the type of magnetic material and, since it is a ratio of flux densities, it has no unit. From its definition, μ_r for a vacuum is 1.

For all media other than free space, $\boldsymbol{B/H = \mu_0\mu_r}$

(h) Absolute permeability $\boldsymbol{\mu = \mu_0\mu_r}$

(i) By plotting measured values of flux density B against magnetic field strength H a **magnetization curve** (or B/H curve) is produced. For nonmagnetic materials this is a straight line having the approximate gradient of μ_0. B/H curves for four materials are shown on page 78.

(j) From (g), $\mu_r = B/(\mu_0 H)$. Thus the relative permeability μ_r of a ferromagnetic material is proportional to the gradient of the B/H curve and varies with the magnetic field strength H.

(k) **Reluctance** $\boldsymbol{S}$ (or R_M) is the 'magnetic resistance' of a magnetic circuit to the presence of magnetic flux.

$$\text{Reluctance } S = \frac{F_m}{\Phi} = \frac{NI}{\Phi} = \frac{Hl}{BA} = \frac{1}{(B/H)A} = \frac{1}{\mu_0\mu_r A}$$

The unit of reluctance is $1/H$ (or H^{-1}) or A/Wb

(l) **Permeance** is the magnetic flux per ampere of total magnetomotive force in the path of a magnetic field. It is the reciprocal of reluctance.

38.2 Magnetic properties of materials

The full theory of magnetism is one of the most complex of subjects. However the phenomenon may be satisfactorily explained by the use of a simple model. Bohr and Rutherford, who discovered atomic structure, suggested that electrons move around the nucleus confined to a plane, like planets around the sun. An even better model is to consider each electron as having a surface, which may be spherical or elliptical or something more complicated.

Magnetic effects in materials are due to the electrons contained in them, the electrons giving rise to magnetism in the following two ways:

(i) by revolving around the nucleus
(ii) by their angular momentum about their own axis, called spin.

In each of these cases the charge of the electron can be thought of as moving round in a closed loop and therefore acting as a current loop.

The main measurable quantity of an atomic model is the **magnetic moment.** When applied to a loop of wire carrying a current,

magnetic moment = current × area of the loop

Electrons associated with atoms possess magnetic moment which gives rise to their magnetic properties.

Diamagnetism is a phenomenon exhibited by materials having a relative permeability less than unity. When electrons move more or less in a spherical orbit around the nucleus, the magnetic moment due to this orbital is zero, all the current due to moving electrons being considered as averaging to zero. If the net magnetic moment of the electron spins were also zero then there would be no tendency for the electron motion to line up in the presence of a magnetic field. However, as a field is being turned on, the flux through the electron orbitals increases. Thus, considering the orbital as a circuit, there will be, by Faraday's laws, an e.m.f. induced in it which will change the current in the circuit. The flux change will accelerate the electrons in its orbit, causing an induced magnetic moment. By Lenz's law the flux due to the induced magnetic moment will be such as to oppose the applied flux. As a result, the net flux through the material becomes less than in a vacuum. Since relative permeability is defined as

$$\frac{\text{flux density in material}}{\text{flux density in vacuum}}$$

with diamagnetic materials the relative permeability is less than one.

Paramagnetism is a phenomenon exhibited by materials where the relative permeability is greater than unity. Paramagnetism occurs in substances where atoms have a permanent magnetic moment. This may be caused by the orbitals not being spherical or by the spin of the electrons. Electron spins tend to pair up and cancel each other. However, there are many atoms with odd numbers of electrons, or in which pairing is incomplete. Such atoms have what is called a permanent dipole moment. When a field is applied to them they tend to line up with the field, like compass needles, and so strengthen the flux in that region. (Diamagnetic materials do not tend to line up with the field in this way.) When this effect is stronger than the diamagnetic effect, the overall effect is to make the relative permeability greater than one. Such materials are called paramagnetic.

Ferromagnetic materials

Ferromagnetism is the phenomenon exhibited by materials having a relative permeability which is considerably greater than 1 and which varies with flux density. Iron, cobalt and nickel are the only elements that are ferromagnetic at ordinary working temperatures, but there are several alloys containing one or more of these metals as constituents, with widely varying ferromagnetic properties.

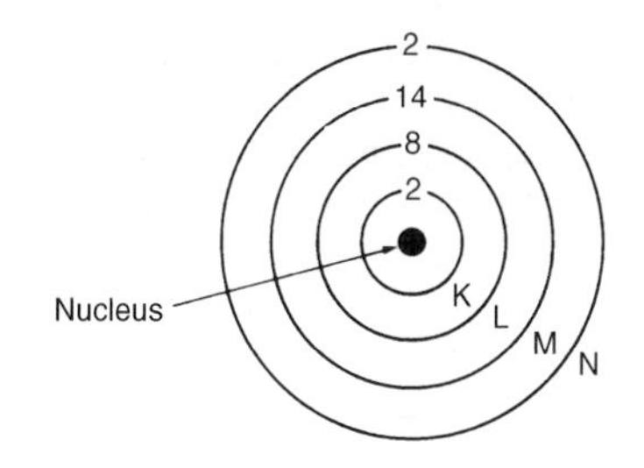

Figure 38.1 *Single iron atom*

Consider the simple model of a single iron atom represented in Figure 38.1. It consists of a small heavy central nucleus surrounded by a total of 26 electrons. Each electron has an orbital motion about the nucleus in a limited region, or shell, such shells being represented by circles K, L, M and N. The numbers in Figure 38.1 represent the number of electrons in each shell.

The outer shell N contains two loosely held electrons, these electrons becoming the carriers of electric current, making iron electrically conductive. There are 14 electrons in the M shell and it is this group that is responsible for magnetism. An electron carries a negative charge and a charge in motion constitutes an electric current with which is associated a magnetic field. Magnetism would therefore result from the orbital motion of each electron in the atom. However, experimental evidence indicates that the resultant magnetic effect due to all the orbital motions in the metal solid is zero; thus the orbital currents may be disregarded.

In addition to the orbital motion, each electron spins on its own axis. A rotating charge is equivalent to a circular current and gives rise to a magnetic field. In any atom, all the axes about which the electrons spin are parallel, but rotation may be in either direction. In the single atom shown in Figure 38.1, in each of the K, L and N shells equal numbers of electrons spin in the clockwise and anticlockwise directions respectively and therefore these shells are magnetically neutral. However, in shell M, nine of the electrons spin in one direction while five spin in the opposite direction. There is therefore a resultant effect due to four electrons.

The atom of cobalt has 15 electrons in the M shell, nine spinning in one direction and six in the other. Thus with cobalt there is a resultant effect due to 3 electrons. A nickel atom has a resultant effect due to 2 electrons. The atoms of the paramagnetic elements, such as manganese, chromium or aluminium, also have a resultant effect for the same reasons as that of iron, cobalt and nickel. However, in the diamagnetic materials there is an exact equality between the clockwise and anticlockwise spins.

The total magnetic field of the resultant effect due to the four electrons in the iron atom is large enough to influence other atoms. Thus the orientation of one atom tends to spread through the material, with atoms acting together in groups instead of behaving independently. These groups of atoms, called **domains** (which tend to remain permanently magnetized), act as units. Thus, when a field is applied to a piece of iron, these domains as a whole tend to line up and large flux densities can be produced. This means that the relative permeability of such materials is much greater than one. As the applied field is increased, more and more domains align and the induced flux increases.

The overall magnetic properties of iron alloys and materials containing iron, such as ferrite (ferrite is a mixture of iron oxide together with other oxides — lodestone is a ferrite), depend upon the structure and composition of the material. However, the presence of iron ensures marked magnetic properties of some kind in them. Ferromagnetic effects decrease with temperature, as do those due to paramagnetism. The loss of ferromagnetism with temperature is more sudden, however; the temperature at which it has all disappeared is called the **Curie temperature**. The ferromagnetic properties reappear on cooling, but any magnetism will have disappeared. Thus a permanent magnet will be demagnetized by heating above the Curie temperature (1040 K for iron) but can be remagnetized after cooling. Above the Curie temperature, ferromagnetics behave as paramagnetics.

38.3 Hysteresis and hysteresis loss

Hysteresis loop

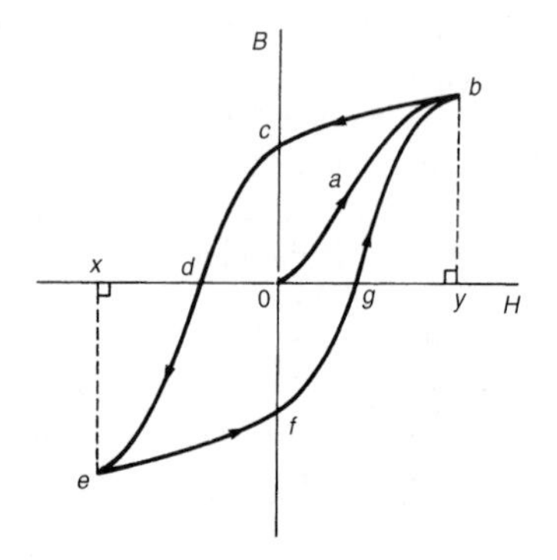

Figure 38.2

Let a ferromagnetic material which is completely demagnetized, i.e., one in which $B = H = 0$ (either by heating the sample above its Curie temperature or by reversing the magnetizing current a large number of times while at the same time gradually reducing the current to zero) be subjected to increasing values of magnetic field strength H and the corresponding flux density B measured. The domains begin to align and the resulting relationship between B and H is shown by the curve Oab in Figure 38.2. At a particular value of H, shown as Oy, most of the domains will be aligned and it becomes difficult to increase the flux density any further. The material is said to be saturated. Thus by is the **saturation flux density.**

If the value of H is now reduced it is found that the flux density follows curve bc, i.e., the domains will tend to stay aligned even when the field is removed. When H is reduced to zero, flux remains in the iron. This **remanent flux density** or **remanence** is shown as Oc in Figure 38.2. When H is increased in the opposite direction, the domains begin to realign in the opposite direction and the flux density decreases until, at a value shown as Od, the flux density has been reduced to zero. The magnetic field strength Od required to remove the residual magnetism, i.e., reduce B to zero, is called the **coercive force**.

Further increase of H in the reverse direction causes the flux density to increase in the reverse direction until saturation is reached, as shown by curve de. If the reversed magnetic field strength Ox is adjusted to the same value of Oy in the initial direction, then the final flux density xe is the same as yb. If H is varied backwards from Ox to Oy, the flux density follows the curve $efgb$, similar to curve $bcde$.

It is seen from Figure 38.2 that the flux density changes lag behind the changes in the magnetic field strength. This effect is called **hysteresis.** The closed figure *bcdefgb* is called the **hysteresis loop** (or the B/H loop).

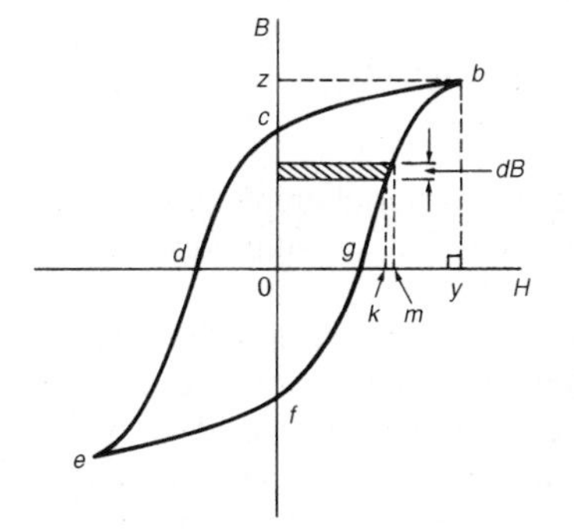

Figure 38.3

Hysteresis loss

A disturbance in the alignment of the domains of a ferromagnetic material causes energy to be expended in taking it through a cycle of magnetization. This energy appears as heat in the specimen and is called the **hysteresis loss**. Let the hysteresis loop shown in Figure 38.3 be that obtained for an iron ring of mean circumference l and cross-sectional area a m^2 and let the number of turns on the magnetizing coil be N.

Let the increase of flux density be dB when the magnetic field strength H is increased by a very small amount km (see Figure 38.3) in time dt second, and let the current corresponding to Ok be i amperes. Thus since $H = NI/l$ then $Ok = Ni/l$, from which,

$$i = \frac{l(Ok)}{N} \tag{38.1}$$

The instantaneous e.m.f. e induced in the winding is given by

$$e = -N\frac{d\Phi}{dt} = -N\frac{d(Ba)}{dt} = -aN\frac{dB}{dt}$$

The applied voltage to neutralize this e.m.f., $v = aN\frac{dB}{dt}$

The instantaneous power supplied to a magnetic field,

$$p = vi = i\left(aN\frac{dB}{dt}\right) \text{ watts}$$

Energy supplied to the magnetic field in time dt seconds

$$= \text{power} \times \text{time} = iaN\frac{dB}{dt}\,dt$$

$$= iaNdB \text{ joules} = \left(\frac{l(Ok)}{N}\right) aN\ dB \quad \text{from equation (38.1)}$$

$$= (Ok)\,dB\,(la) \text{ joules} = (\text{area of shaded strip})\ (\text{volume of ring})$$

i.e., energy supplied in time dt seconds = (area of shaded strip)J/m^3.

Hence the energy supplied to the magnetic field when H is increased from zero to Oy = (area *fgbzf*)J/m^3.

Similarly, the energy returned from the magnetic field when H is reduced from Oy to zero = (area *bzcb*)J/m^3.

Hence net energy absorbed by the magnetic field = (area *fgbcf*)J/m^3

Thus the hysteresis loss for a complete cycle

= **area of loop *efgbcde* J/m^3**

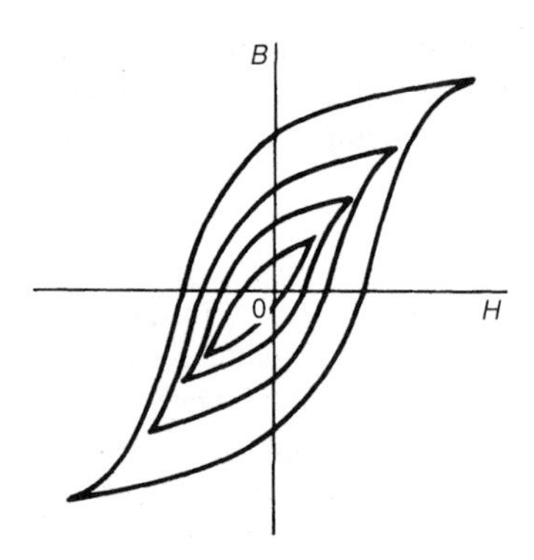

Figure 38.4

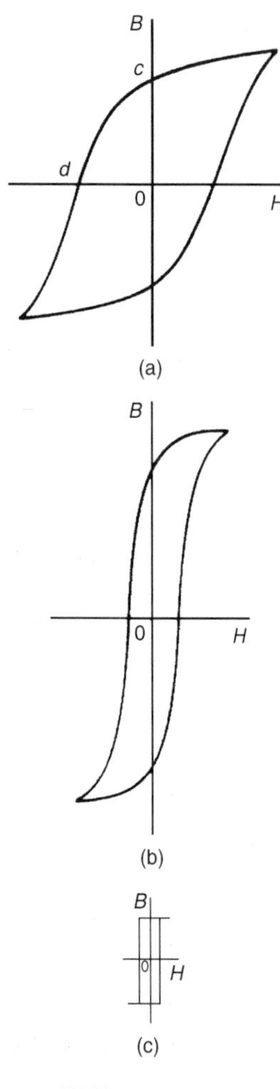

Figure 38.5

If the hysteresis loop is plotted to a scale of 1 cm $= \alpha$ ampere/metre along the horizontal axis and 1 cm $= \beta$ tesla along the vertical axis, and if A represents the area of the loop in square centimetres, then

$$\textbf{hysteresis loss/cycle} = \boldsymbol{A\alpha\beta} \textbf{ joules per metre}^3 \qquad (38.2)$$

If hysteresis loops for a given ferromagnetic material are determined for different maximum values of H, they are found to lie within one another as shown in Figure 38.4.

The maximum sized hysteresis loop for a particular material is obtained at saturation. If, for example, the maximum flux density is reduced to half its value at saturation, the area of the resulting loop is considerably less than half the area of the loop at saturation. From the areas of a number of such hysteresis loops, as shown in Figure 38.4, the hysteresis loss per cycle was found by Steinmetz (an American electrical engineer) to be proportional to $(B_m)^n$, where n is called the **Steinmetz index** and can have a value between about 1.6 and 3.0, depending on the quality of the ferromagnetic material and the range of flux density over which the measurements are made.

From the above it is found that the hysteresis loss is proportional to the volume of the specimen and the number of cycles through which the magnetization is taken. Thus

$$\textbf{hysteresis loss, } \boldsymbol{P_h = k_h v f (B_m)^n} \textbf{ watts} \qquad (38.3)$$

where $v =$ volume in cubic metres, $f =$ frequency in hertz, and k_h is a constant for a given specimen and given range of B.

The magnitude of the hysteresis loss depends on the composition of the specimen and on the heat treatment and mechanical handling to which the specimen has been subjected.

Figure 38.5 shows typical hysteresis loops for (a) hard steel, which has a high remanence Oc and a large coercivity Od, (b) soft steel, which has a large remanence and small coercivity and (c) ferrite, this being a ceramic-like magnetic substance made from oxides of iron, nickel, cobalt, magnesium, aluminium and manganese. The hysteresis of ferrite is very small.

Problem 1. The area of a hysteresis loop obtained from a ferromagnetic specimen is 12.5 cm^2. The scales used were: horizontal axis 1 cm = 500 A/m; vertical axis 1 cm = 0.2 T. Determine (a) the hysteresis loss per m^3 per cycle, and (b) the hysteresis loss per m^3 at a frequency of 50 Hz.

(a) From equation (38.2), hysteresis loss per cycle

$$= A\alpha\beta = (12.5)(500)(0.2) = \mathbf{1250\ J/m^3}$$

(Note that, since $\alpha = 500$ A/m per centimetre and $\beta = 0.2$ T per centimetre, then 1 cm^2 of the loop represents

$$500\ \frac{A}{m} \times 0.2\ T = 100\ \frac{A\ Wb}{m\ m^2} = 100\ \frac{AVs}{m^3} = 100\ \frac{Ws}{m^3} = 100\ J/m^3$$

Hence 12.5 cm^2 represents $12.5 \times 100 =$ **1250 J/m³**)

(b) At 50 Hz frequency, hysteresis loss

$$= (1250\ J/m^3)(50\ 1/s) = \mathbf{62\,500\ W/m^3}$$

> Problem 2. If in problem 1, the maximum flux density is 1.5 T at a frequency of 50 Hz, determine the hysteresis loss per m^3 for a maximum flux density of 1.1 T and frequency of 25 Hz. Assume the Steinmetz index to be 1.6

From equation (38.3), hysteresis loss $P_h = k_h v f (B_m)^n$

The loss at $f = 50$ Hz and $B_m = 1.5$ T is 62 500 W/m^3, from problem 1.

Thus $62\,500 = k_h(1)(50)(1.5)^{1.6}$,

from which, constant $k_h = \dfrac{62\,500}{(50)(1.5)^{1.6}} = 653.4$

When $f = 25$ Hz and $B_m = 1.1$ T,

hysteresis loss, $P_h = k_h v f (B_m)^n$

$$= (653.4)(1)(25)(1.1)^{1.6} = \mathbf{19\,026\ W/m^3}$$

> Problem 3. A ferromagnetic ring has a uniform cross-sectional area of 2000 mm^2 and a mean circumference of 1000 mm. A hysteresis loop obtained for the specimen is plotted to scales of 10 mm = 0.1 T and 10 mm = 400 A/m and is found to have an area of 10^4 mm^2. Determine the hysteresis loss at a frequency of 80 Hz.

From equation (38.2), hysteresis loss per cycle

$$= A\alpha\beta$$

$$= (10^4 \times 10^{-6}\ m^2)\left(\frac{400\ A/m}{10 \times 10^{-3}\ m}\right)\left(\frac{0.1T}{10 \times 10^{-3}\ m}\right)$$

$$= 4000\ J/m^3$$

At a frequency of 80 Hz,

hysteresis loss $= (4000\ J/m)(80\ 1/s) = 320\,000\ W/m^3$

Volume of ring = (cross-sectional area) (mean circumference)

$$= (2000 \times 10^{-6}\ m^2)(1000 \times 10^{-3}\ m) = 2 \times 10^{-3}\ m^3$$

Thus hysteresis loss, $P_h = (320\,000\ W/m^3)(2 \times 10^{-3}\ m^3) =$ **640 W**

Problem 4. The cross-sectional area of a transformer limb is 80 cm^2 and the volume of the transformer core is 5000 cm^3. The maximum value of the core flux is 10 mWb at a frequency of 50 Hz. Taking the Steinmetz constant as 1.7, the hysteresis loss is found to be 100 W. Determine the value of the hysteresis loss when the maximum core flux is 8 mWb and the frequency is 50 Hz.

When the maximum core flux is 10 mWb and the cross-sectional area is 80 cm^2,

$$\text{maximum flux density, } B_{m1} = \frac{\Phi_1}{A} = \frac{10 \times 10^{-3}}{80 \times 10^{-4}} = 1.25 \text{ T}$$

From equation (38.3), hysteresis loss, $P_{h1} = k_h v f (B_{m1})^n$

Hence $100 = k_h(5000 \times 10^{-6})(50)(1.25)^{1.7}$

$$\text{from which, constant } k_h = \frac{100}{(5000 \times 10^{-6})(50)(1.25)^{1.7}} = 273.7$$

When the maximum core flux is 8 mWb,

$$B_{m2} = \frac{8 \times 10^{-3}}{80 \times 10^{-4}} = 1 \text{ T}$$

Hence hysteresis loss, $P_{h2} = k_h v f (B_{m2})^n$

$$= (273.7)(5000 \times 10^{-6})(50)(1)^{1.7} = \mathbf{68.4\ W}$$

Further problems on hysteresis loss may be found in Section 38.8, problems 1 to 6, page 707.

38.4 Eddy current loss

If a coil is wound on a ferromagnetic core (such as in a transformer) and alternating current is passed through the coil, an alternating flux is set up in the core. The alternating flux induces an e.m.f. e in the coil given by $e = N(d\phi/dt)$ However, in addition to the desirable effect of inducing an e.m.f. in the coil, the alternating flux induces undesirable voltages in the iron core. These induced e.m.f.s set up circulating currents in the core, known as **eddy currents.** Since the core possesses resistance, the eddy currents heat the core, and this represents wasted energy.

Eddy currents can be reduced by laminating the core, i.e., splitting it into thin sheets with very thin layers of insulating material inserted between each pair of the laminations (this may be achieved by simply varnishing one side of the lamination or by placing paper between each lamination). The insulation presents a high resistance and this reduces any induced circulating currents.

The eddy current loss may be determined as follows. Let Figure 38.6 represent one strip of the core, having a thickness of t metres, and consider just a rectangular prism of the strip having dimensions t m × 1 m × 1 m as shown. The area of the front face $ABCD$ is $(t \times 1)\text{m}^2$ and, since the flux enters this face at right angles, the eddy currents will flow along paths parallel to the long sides.

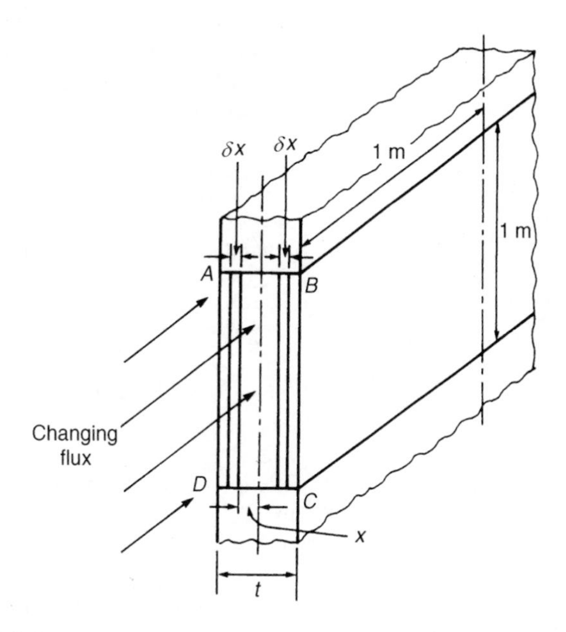

Figure 38.6

Consider two such current paths each of width δx and distance x m from the centre line of the front face. The area of the rectangle enclosed by the two paths, $A = (2x)(1) = 2x\ \text{m}^2$. Hence the maximum flux entering the rectangle,

$$\Phi_m = (B_m)(A) = (B_m)(2x) \text{ weber} \qquad (38.4)$$

Induced e.m.f. e is given by $e = N(d\phi/dt)$. Since the flux varies sinusoidally, $\phi = \Phi_m \sin \omega t$. Thus

$$\text{e.m.f. } e = N\frac{d}{dt}(\Phi_m \sin \omega t) = N\omega\Phi_m \cos \omega t$$

The maximum value of e.m.f. occurs when $\cos \omega t = 1$, i.e., $E_m = N\omega\Phi_m$

$$\text{Rms value of e.m.f., } E = \frac{E_m}{\sqrt{2}} = \frac{N\omega\Phi_m}{\sqrt{2}}$$

Now $\omega = 2\pi f$ hence

$$E = \left(\frac{2\pi}{\sqrt{2}}\right) fN\Phi_m = 4.44 fN\Phi_m$$

$$\text{i.e., } E = 4.44 fN(B_m)(A) \qquad (38.5)$$

From equation (38.4), $\Phi_m = (B_m)(2x)$. Hence induced e.m.f. $E = 4.44 f N(B_m)(2x)$ and, since the number of turns $N = 1$,

$$E = 8.88\ B_m f x \text{ volts} \qquad (38.6)$$

Resistance R is given by $R = \rho l/a$, where ρ is the resistivity of the lamination material. Since the current set up is confined to the two loop sides (thus $l = 2$ m and $a = (\delta x \times 1)\text{m}^2$), the total resistance of the path is given by

$$R = \frac{\rho(2)}{\delta x} = \frac{2\rho}{\delta x} \qquad (38.7)$$

The eddy current loss in the two strips is given by

$$\frac{E^2}{R} = \frac{8.88^2 {B_m}^2 f^2 x^2}{2\rho/\delta x} \quad \text{from equations (38.6) and (38.7)}$$

$$= \frac{8.88^2 {B_m}^2 f^2 x^2 \delta x}{2\rho}$$

The total eddy current loss P_e in the rectangular prism considered is given by

$$P_e = \int_0^{t/2} \left(\frac{8.88^2 {B_m}^2 f^2}{2\rho}\right) x^2 dx = \left(\frac{8.88^2 {B_m}^2 f^2}{2\rho}\right)\left[\frac{x^3}{3}\right]_0^{t/2}$$

$$= \left(\frac{8.88^2 {B_m}^2 f^2}{2\rho}\right)\left(\frac{t^3}{24}\right) \text{ watts}$$

i.e.,

$$\boldsymbol{P_e = k_e (B_m)^2 f^2 t^3} \textbf{ watts} \qquad (38.8)$$

where k_e is a constant.

The volume of the prism is $(t \times 1 \times 1)$ m^3. Hence the eddy current loss per m^3 is given by

$$\boldsymbol{P_e = k_e (B_m)^2 f^2 t^2} \textbf{ watts per m}^3 \qquad (38.9)$$

From equation (38.9) it is seen that eddy current loss is proportional to the square of the thickness of the core strip. It is therefore desirable to make lamination strips as thin as possible. However, at high frequencies where it is not practicable to make very thin laminations, core losses may be reduced by using ferrite cores or dust cores. Ferrite is a ceramic material having magnetic properties similar to silicon steel, and dust cores consist of fine particles of carbonyl iron or permalloy (i.e. nickel and iron), each particle of which is insulated from its neighbour by a binding material. Such materials have a very high value of resistivity.

Problem 5. The eddy current loss in a particular magnetic circuit is 10 W/m^3. If the frequency of operation is reduced from 50 Hz to 30 Hz with the flux density remaining unchanged, determine the new value of eddy current loss per cubic metre.

From equation (38.9), eddy current loss per cubic metre,
$P_e = k_e(B_m)^2 f^2 t^2$ or $P_e = kf^2$, where $k = k_e(B_m)^2 t^2$, since B_m and t are constant.

When the eddy current loss is 10 W/m^3, frequency f is 50 Hz. Hence $10 = k(50)^2$, from which

$$\text{constant } k = \frac{10}{(50)^2}$$

When the frequency is 30 Hz, eddy current loss,

$$P_e = k(30)^2 = \frac{10}{(50)^2}(30)^2 = \mathbf{3.6\ W/m^3}$$

Problem 6. The core of a transformer operating at 50 Hz has an eddy current loss of 100 W/m^3 and the core laminations have a thickness of 0.50 mm. The core is redesigned so as to operate with the same eddy current loss but at a different voltage and at a frequency of 250 Hz. Assuming that at the new voltage the maximum flux density is one-third of its original value and the resistivity of the core remains unaltered, determine the necessary new thickness of the laminations.

From equation (38.9), $P_e = k_e(B_m)^2 f^2 t^2$ watts per m^3.

Hence, at 50 Hz frequency, $100 = k_e(B_m)^2(50)^2(0.50 \times 10^{-3})^2$, from which

$$k_e = \frac{100}{(B_m)^2(50)^2(0.50 \times 10^{-3})^2}$$

At 250 Hz frequency, $100 = k_e\left(\frac{B_m}{3}\right)(250)^2(t)^2$

i.e., $$100 = \left(\frac{100}{(B_m)^2(50)^2(0.50 \times 10^{-3})^2}\right)\left(\frac{B_m}{3}\right)^2(250)^2(t)^2$$

$$= \frac{100(250)^2(t)^2}{(3)^2(50)^2(0.50 \times 10^{-3})^2}$$

from which $$t^2 = \frac{(100)(3)^2(50)^2(0.50 \times 10^{-3})^2}{(100)(250)^2}$$

i.e., **lamination thickness,** $\boldsymbol{t} = \frac{(3)(50)(0.50 \times 10^{-3})}{250} = 0.3 \times 10^{-3}$ m or **0.30 mm**

Problem 7. The core of an inductor has a hysteresis loss of 40 W and an eddy current loss of 20 W when operating at 50 Hz frequency. (a) Determine the values of the losses if the frequency is increased to 60 Hz. (b) What will be the total core loss if the frequency is 50 Hz and the lamination are made one-half of their original thickness? Assume that the flux density remains unchanged in each case

(a) From equation (38.3). hysteresis loss, $P_h = k_h v f(B_m)^n = k_1 f$ (where $k_1 = k_h v(B_m)^n$), since the flux density and volume are constant. Thus when the hysteresis is 40 W and the frequency 50 Hz,

$$40 = k_1(50)$$

$$\text{from which, } k_1 = \frac{40}{50} = 0.8$$

If the frequency is increased to 60 Hz,

hysteresis loss, $\boldsymbol{P_h} = k_1(60) = (0.8)(60) = \mathbf{48\ W}$

From equation (38.8),

$$\text{eddy current loss, } P_e = k_e(B_m)^2 f^2 t^3$$
$$= k_2 f^2 (\text{where } k_2 = k_e(B_m)^2 t^3),$$

since the flux density and lamination thickness are constant.

When the eddy current loss is 20 W the frequency is 50 Hz. Thus $20 = k_2(50)^2$

$$\text{from which } k_2 = \frac{20}{(50)^2} = 0.008$$

If the frequency is increased to 60 Hz,

eddy current loss, $\boldsymbol{P_e} = k_2(60)^2 = (0.008)(60)^2 = \mathbf{28.8\ W}$

(b) The hysteresis loss, $P_h = k_h v f(B_m)^n$, is independent of the thickness of the laminations. Thus, if the thickness of the laminations is halved, the hysteresis loss remains at **40 W**

Eddy current loss $P_e = k_e(B_m)^2 f^2 t^3$, i.e. $P_e = k_3 f^2 t^3$, where $k_3 = k_e(B_m)^2$.

$$\text{Thus} \qquad 20 = k_3(50)^2 t^3$$

$$\text{from which } k_3 = \frac{20}{(50)^2 t^3}$$

When the thickness is $t/2$, $P_e = k_3(50)^2(t/2)^3$

$$= \left(\frac{20}{(50)^2 t^3}\right)(50)^2(t/2)^3 = \mathbf{2.5\ W}$$

Hence the **total core loss** when the thickness of the laminations is halved is given by hysteresis loss + eddy current loss = 40 + 2.5 = **42.5 W**

Problem 8. When a transformer is connected to a 500 V, 50 Hz supply, the hysteresis and eddy current losses are 400 W and 150 W respectively. The applied voltage is increased to 1 kV and the frequency to 100 Hz. Assuming the Steinmetz index to be 1.6, determine the new total core loss.

From equation (38.3), the hysteresis loss, $P_h = k_h v f (B_m)^n$. From equation (38.5), e.m.f., $E = 4.44 f N (B_m)(A)$, from which, $B_m \alpha (E/f)$ since turns N and cross-sectional area, A are constants. Hence $P_h = k_1 f (E/f)^{1.6} = k_1 f^{-0.6} E^{1.6}$

At 500 V and 50 Hz, $400 = k_1(50)^{-0.6}(500)^{1.6}$,

from which, $$k_1 = \frac{400}{(50)^{-0.6}(500)^{1.6}} = 0.20095$$

At 1000 V and 100 Hz,

$$\text{hysteresis loss, } P_h = k_1(100)^{-0.6}(1000)^{1.6} = (0.20095)(100)^{-0.6}(1000)^{1.6} = 800 \text{ W}$$

From equation (38.8)

$$\text{eddy current loss, } P_e = k_e(B_m)^2 f^2 t^3 = k_2(E/f)^2 f^2 = k_2 E^2$$

At 500 V, $150 = k_2(500)^2$, from which

$$k_2 = \frac{150}{(500)^2} = 6 \times 10^{-4}$$

At 1000 V,

$$\text{eddy current loss, } P_e = k_2(1000)^2 = (6 \times 10^{-4})(1000)^2 = 600 \text{ W}$$

Hence the new **total core loss** = 800 + 600 = **1400 W**

Further problems on eddy current loss may be found in Section 38.8, problems 7 to 12, page 708.

38.5 Separation of hysteresis and eddy current losses

From equation (38.3), hysteresis loss, $P_h = k_h v f (B_m)^n$

From equation (38.8), eddy current loss, $P_e = k_e(B_m)^2 f^2 t^3$

The total core loss P_c is given by $P_c = P_h + P_e$

If for a particular inductor or transformer, the core flux density is maintained constant, then $P_h = k_1 f$, where constant $k_1 = k_h v (B_m)^n$, and

$P_e = k_2 f^2$, where constant $k_2 = k_e(B_m)^2 t^3$. Thus the total core loss $P_c = k_1 f + k_2 f^2$ and

$$\boldsymbol{\frac{P_c}{f} = k_1 + k_2 f}$$

which is of the straight line form $y = mx + c$. Thus if P_c/f is plotted vertically against f horizontally, a straight line graph results having a gradient k_2 and a vertical-axis intercept k_1.

If the total core loss P_c is measured over a range of frequencies, then k_1 and k_2 may be determined from the graph of P_c/f against f. Hence the hysteresis loss $P_h(= k_1 f)$ and the eddy current loss $P_e(= k_2 f^2)$ at a given frequency may be determined.

The above method of separation of losses is an approximate one since the Steinmetz index n is not a constant value but tends to increase with increase of frequency. However, a reasonable indication of the relative magnitudes of the hysteresis and eddy current losses in an iron core may be determined.

Problem 9. The total core loss of a ferromagnetic cored transformer winding is measured at different frequencies and the results obtained are:

Total core loss, P_c (watts)	45	105	190	305
Frequency, f (hertz)	30	50	70	90

Determine the separate values of the hysteresis and eddy current losses at frequencies of (a) 50 Hz and (b) 60 Hz.

To obtain a straight line graph, values of P_c/f are plotted against f.

f(Hz)	30	50	70	90
P_c/f	1.5	2.1	2.7	3.4

A graph of P_c/f against f is shown in Figure 38.7. The graph is a straight line of the form $P_c/f = k_1 + k_2 f$

The vertical axis intercept at $f = 0$, $\boldsymbol{k_1 = 0.5}$

The gradient of the graph, $\boldsymbol{k_2} = \dfrac{a}{b} = \dfrac{3.7 - 0.5}{100} = \mathbf{0.032}$

Since $P_c/f = k_1 + k_2 f$, then $P = k_1 f + k_2 f^2$, i.e.,

total core losses = hysteresis loss + eddy current loss.

(a) At a frequency of 50 Hz,

$$\text{hysteresis loss} = k_1 f = (0.5)(50) = \mathbf{25\ W}$$

$$\text{eddy current loss} = k_2 f^2 = (0.032)(50)^2 = \mathbf{80\ W}$$

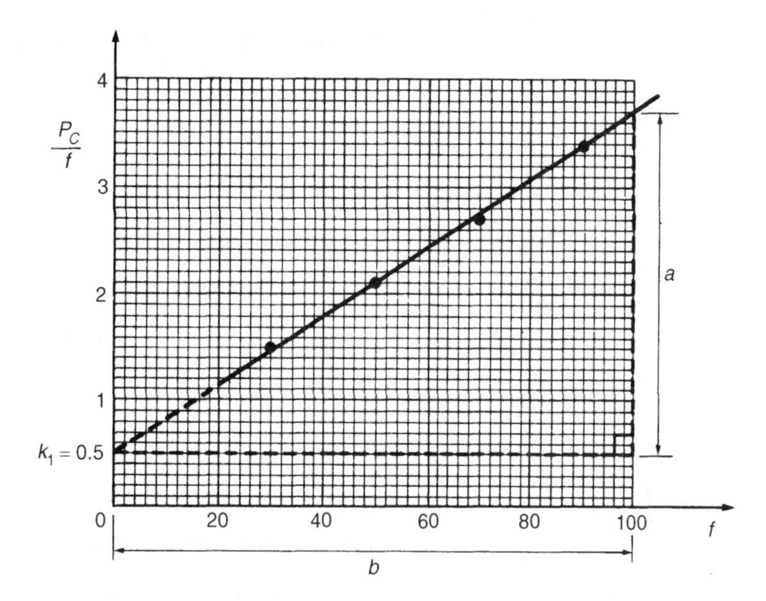

Figure 38.7

(b) At a frequency of 60 Hz,

$$\text{hysteresis loss} = k_1 f = (0.5)(60) = \mathbf{30\ W}$$

$$\text{eddy current loss} = k_2 f^2 = (0.032)(60)^2 = \mathbf{115.2\ W}$$

Problem 10. The core of a synchrogenerator has total losses of 400 W at 50 Hz and 498W at 60 Hz, the flux density being constant for the two tests. (a) Determine the hysteresis and eddy current losses at 50 Hz (b) If the flux density is increased by 25% and the lamination thickness is increased by 40%, determine the hysteresis and eddy current losses at 50 Hz. Assume the Steinmetz index to be 1.7

(a) From equation (38.3),

$$\text{hysteresis loss, } P_h = k_h v f (B_m)^n = k_1 f$$

(if volume v and the maximum flux density are constant)

From equation (38.8),

$$\text{eddy current loss, } P_e = k_e (B_m)^2 f^2 t^3 = k_2 f^2$$

(if the maximum flux density and the lamination thickness are constant)

Hence the total core loss $P_c = P_h + P_e$ i.e., $P_c = k_1 f + k_2 f^2$

$$\text{At 50 Hz frequency, } 400 = k_1(50) + k_2(50)^2 \quad (1)$$

$$\text{At 60 Hz frequency, } 498 = k_1(60) + k_2(60)^2 \quad (2)$$

Solving equations (1) and (2) gives the values of k_1 and k_2.

$$6 \times \text{equation (1) gives: } 2400 = 300k_1 + 15\,000k_2 \quad (3)$$

$$5 \times \text{equation (2) gives: } 2490 = 300k_1 + 18\,000k_2 \quad (4)$$

Equation (4)–equation (3) gives: $90 = 3000k_2$ from which,

$$k_2 = 90/3000 = \mathbf{0.03}$$

Substituting $k_2 = 0.03$ in equation (1) gives $400 = 50k_1 + 75$, from which $\boldsymbol{k_1 = 6.5}$ Thus, at 50 Hz frequency,

$$\text{hysteresis loss } P_h = k_1 f = (6.5)(50) = \mathbf{325\ W}$$

$$\text{eddy current loss } P_e = k_2 f^2 = (0.03)(50)^2 = \mathbf{75\ W}$$

(b) Hysteresis loss, $P_h = k_h v f (B_m)^n$. Since at 50 Hz the flux density is increased by 25%, the new hysteresis loss is $(1.25)^{1.7}$ times greater than 325 W,

i.e., $P_h = (1.25)^{1.7}(325) = \mathbf{474.9\ W}$

Eddy current loss, $P_e = k_e(B_m)^2 f^2 t^3$. Since at 50 Hz the flux density is increased by 25%, and the lamination thickness is increased by 40%, the new eddy current loss is $(1.25)^2(1.4)^3$ times greater than 75 W,

i.e., $P_e = (1.25)^2(1.4)^3(75) = \mathbf{321.6\ W}$

Further problems on the separation of hysteresis and eddy current losses may be found in Section 38.8, problems 13 to 16, page 709.

38.6 Nonpermanent magnetic materials

General

Nonpermanent magnetic materials are those in which magnetism may be induced. With the magnetic circuits of electrical machines, transformers and heavy current apparatus a high value of flux density B is desirable so as to limit the cross-sectional area A ($\Phi = BA$) and therefore the weight and cost involved. At the same time the magnetic field strength H ($= NI/l$) should be as small as possible so as to limit the I^2R losses in the exciting coils. The relative permeability ($\mu_r = B/(\mu_0 H)$) and the saturation flux density should therefore be high. Also, when flux is continually varying, as in transformers, inductors and armature cores, low hysteresis and eddy current losses are essential.

Silicon-iron alloys

In the earliest electrical machines the magnetic circuit material used was iron with low content of carbon and other impurities. However, it was later discovered that the deliberate addition of silicon to the iron brought about a great improvement in magnetic properties. The laminations now used in electrical machines and in transformers at supply frequencies are made of silicon-steel in which the silicon in different grades of the material varies in amounts from about 0.5% to 4.5% by weight. The silicon added to iron increases the resistivity. This in turn increases the resistance ($R = \rho l/A$) and thus helps to reduce eddy current loss. The hysteresis loss is also reduced; however, the silicon reduces the saturation flux density.

A limit to the amount of silicon which may be added in practice is set by the mechanical properties of the material, since the addition of silicon causes a material to become brittle. Also the brittleness of a silicon-iron alloy depends on temperature. About 4.5% silicon is found to be the upper practical limit for silicon-iron sheets. Lohys is a typical example of a silicon-iron alloy and is used for the armatures of d.c. machines and for the rotors and stators of a.c. machines. Stalloy, which has a higher proportion of silicon and lower losses, is used for transformer cores.

Silicon steel sheets are often produced by a hot-rolling process. In these finished materials the constituent crystals are not arranged in any particular manner with respect, for example, to the direction of rolling or the plane of the sheet. If silicon steel is reduced in thickness by rolling in the cold state and the material is then annealed it is possible to obtain a finished sheet in which the crystals are nearly all approximately parallel to one another. The material has strongly directional magnetic properties, the rolling direction being the direction of highest permeability. This direction is also the direction of lowest hysteresis loss. This type of material is particularly suitable for use in transformers, since the axis of the core can be made to correspond with the rolling direction of the sheet and thus full use is made of the high permeability, low loss direction of the sheet.

With silicon-iron alloys a maximum magnetic flux density of about 2 T is possible. With cold-rolled silicon steel, used for large machine construction, a maximum flux density of 2.5 T is possible, whereas the maximum obtainable with the hot-rolling process is about 1.8 T. (In fact, with any material, only under the most abnormal of conditions will the value of flux density exceed 3 T.)

It should be noted that the term 'iron-core' implies that the core is made of iron; it is, in fact, almost certainly made from steel, pure iron being extremely hard to come by. Equally, an iron alloy is generally a steel and so it is preferred to describe a core as being a steel rather than an iron core.

Nickel-iron alloys

Nickel and iron are both ferromagnetic elements and when they are alloyed together in different proportions a series of useful magnetic alloys is obtained. With about 25%–30% nickel content added to iron, the alloy tends to be very hard and almost nonmagnetic at room temperature. However, when the nickel content is increased to, say, 75%–80% (together with small amounts of molybdenum and copper), very high values of initial and maximum permeabilities and very low values of hysteresis loss are obtainable if the alloys are given suitable heat treatment. For example, Permalloy, having a content of 78% nickel, 3% molybdenum and the remainder iron, has an initial permeability of 20 000 and a maximum permeability of 100 000 compared with values of 250 and 5000 respectively for iron. The maximum flux density for Permalloy is about 0.8 T. Mumetal (76% nickel, 5% copper and 2% chromium) has similar characteristics. Such materials are used for the cores of current and a.f. transformers, for magnetic amplifiers and also for magnetic screening. However, nickel-iron alloys are limited in that they have a low saturation value when compared with iron. Thus, in applications where it is necessary to work at a high flux density, nickel-iron alloys are inferior to both iron and silicon-iron. Also nickel-iron alloys tend to be more expensive than silicon-iron alloys.

Eddy current loss is proportional to the thickness of lamination squared, thus such losses can be reduced by using laminations as thin as possible. Nickel-iron alloy strip as thin as 0.004 mm, wound in a spiral, may be used.

Dust cores

In many circuits high permeability may be unnecessary or it may be more important to have a very high resistivity. Where this is so, metal powder or dust cores are widely used up to frequencies of 150 MHz. These consist of particles of nickel-iron-molybdenum for lower frequencies and iron for the higher frequencies. The particles, which are individually covered with an insulating film, are mixed with an insulating, resinous binder and pressed into shape.

Ferrites

Magnetite, or ferrous ferrite, is a compound of ferric oxide and ferrous oxide and possesses magnetic properties similar to those of iron. However, being a semiconductor, it has a very high resistivity. Manufactured ferrites are compounds of ferric oxide and an oxide of some other metal such as manganese, nickel or zinc. Ferrites are free from eddy current losses at all but the highest frequencies (i.e., >100 MHz) but have a much lower initial permeability compared with nickel-iron alloys or silicon-iron alloys. Ferrites have typically a maximum flux density of about 0.4 T. Ferrite cores are used in audio-frequency transformers and inductors.

38.7 Permanent magnetic materials

A permanent magnet is one in which the material used exhibits magnetism without the need for excitation by a current-carrying coil. The silicon-iron and nickel-iron alloys discussed in Section 38.6 are 'soft' magnetic materials having high permeability and hence low hysteresis loss. The opposite characteristics are required in the 'hard' materials used to make permanent magnets. In permanent magnets, high remanent flux density and high coercive force, after magnetization to saturation, are desirable in order to resist demagnetization. The hysteresis loop should embrace the maximum possible area. Possibly the best criterion of the merit of a permanent magnet is its maximum energy product $(BH)_m$, i.e., the maximum value of the product of the flux density B and the magnetic field strength H along the demagnetization curve (shown as cd in Figure 38.2). A rough criterion is the product of coercive force and remanent flux density, i.e. $(Od)(Oc)$ in Figure 38.2. The earliest materials used for permanent magnets were tungsten and chromium steel, followed by a series of cobalt steels, to give both a high remanent flux density and a high value of $(BH)_m$

Alni was the first of the aluminium-nickel-iron alloys to be discovered, and with the addition of cobalt, titanium and niobium, the Alnico series of magnets was developed, the properties of which vary according to composition. These materials are very hard and brittle. Many alloys with other compositions and trade names are commercially available.

A considerable advance was later made when it was found that directional magnetic properties could be induced in alloys of suitable composition if they were heated in a strong magnetic field. This discovery led to the powerful Alcomex and Hycomex series of magnets. By using special casting techniques to give a grain-oriented structure, even better properties are obtained if the field applied during heat treatment is parallel to the columnar crystals in the magnet. The values of coercivity, the remanent flux density and hence $(BH)_m$ are high for these alloys.

The most recent and most powerful permanent magnets discovered are made by powder metallurgy techniques and are based on an intermetallic compound of cobalt and samarium. These are very expensive and are only available in a limited range of small sizes.

38.8 Further problems on magnetic materials

Hysteresis loss

1 The area of a hysteresis loop obtained from a specimen of steel is 2000 mm^2. The scales used are: horizontal axis 1 cm = 400 A/m; vertical axis 1 cm = 0.5 T. Determine (a) the hysteresis loss per m^3 per cycle, (b) the hysteresis loss per m^3 at a frequency of 60 Hz. (c) If the maximum flux density is 1.2 T at a frequency of 60 Hz, determine the hysteresis loss per m^3 for a maximum flux density of 1 T and a frequency of 20 Hz, assuming the Steinmetz index to be 1.7.
[(a) 4 kJ/m^3 (b) 240 kW/m^3 (c) 58.68 kW/m^3]

2 A steel ring has a uniform cross-sectional area of 1500 mm^2 and a mean circumference of 800 mm. A hysteresis loop obtained for the specimen is plotted to scales of 1 cm = 0.05 T and 1 cm = 100 A/m and it is found to have an area of 720 cm^2. Determine the hysteresis loss at a frequency of 50 Hz. [216 W]

3 What is hysteresis? Explain how a hysteresis loop is produced for a ferromagnetic specimen and how its area is representative of the hysteresis loss.

The area of a hysteresis loop plotted for a ferromagnetic material is 80 cm^2, the maximum flux density being 1.2 T. The scales of B and H are such that 1 cm = 0.15 T and 1 cm = 10 A/m. Determine the loss due to hysteresis if 1.25 kg of the material is subjected to an alternating magnetic field of maximum flux density 1.2 T at a frequency of 50 Hz. The density of the material is 7700 kg/m^3 [0.974 W]

4 The cross-sectional area of a transformer limb is 8000 mm^2 and the volume of the transformer core is 4×10^6 mm^3. The maximum value of the core flux is 12 mWb and the frequency is 50 Hz. Assuming the Steinmetz constant is 1.6, the hysteresis loss is found to be 250 W. Determine the hysteresis loss when the maximum core flux is 9 mWb, the frequency remaining unchanged. [157.8 W]

5 The hysteresis loss in a transformer is 200 W when the maximum flux density is 1 T and the frequency is 50 Hz. Determine the hysteresis loss if the maximum flux density is increased to 1.2 T and the frequency reduced to 32 Hz. Assume the hysteresis loss over this range to he proportional to $(B_m)^{1.6}$. [171.4 W]

6 A hysteresis loop is plotted to scales of 1 cm = 0.004 T and 1 cm = 10 A/m and has an area of 200 cm^2. If the ferromagnetic circuit for the loop has a volume of 0.02 m^3 and operates at 60 Hz frequency, determine the hysteresis loss for the ferromagnetic specimen. [9.6 W]

Eddy current loss

7 In a magnetic circuit operating at 60 Hz, the eddy current loss is 25 W/m^3. If the frequency is reduced to 30 Hz with the flux density remaining unchanged, determine the new value of eddy current loss per cubic metre. [6.25 W/m^3]

8 A transformer core operating at 50 Hz has an eddy current loss of 150 W/m^3 and the core laminations are 0.4 mm thick. The core is redesigned so as to operate with the same eddy current loss but at a different voltage and at 200 Hz frequency. Assuming that at the new voltage the flux density is half of its original value and the resistivity of the core remains unchanged, determine the necessary new thickness of the laminations [0.20 mm]

9 An inductor core has an eddy current loss of 25 W and a hysteresis loss of 35 W when operating at 50 Hz frequency. Assuming that the flux density remains unchanged, determine (a) the value of the losses if the frequency is increased to 75 Hz, and (b) the total core loss if the frequency is 50 Hz and the laminations are 2/5 of their original thickness. [(a) $P_h = 52.5$ W, $P_e = 56.25$ W (b) 36.6 W]

10 A transformer is connected to a 400 V, 50 Hz supply. The hysteresis loss is 250 W and the eddy current loss is 120 W. The supply voltage

is increased to 1.2 kV and the frequency to 80 Hz. Determine the new total core loss if the Steinmetz index is assumed to be 1.6

[2173.6 W]

11 The hysteresis and eddy current losses in a magnetic circuit are 5 W and 8 W respectively. If the frequency is reduced from 50 Hz to 30 Hz, the flux density remaining the same, determine the new values of hysteresis and eddy current loss. [3 W; 2.88 W]

12 The core loss in a transformer connected to a 600 V, 50 Hz supply is 1.5 kW of which 60% is hysteresis loss and 40% eddy current loss. Determine the total core loss if the same winding is connected to a 750 V, 60 Hz supply. Assume the Steinmetz constant to be 1.6

[2090 W]

Separation of hysteresis and eddy current losses

13 Tests to determine the total loss of the steel core of a coil at different frequencies gave the following results:

Frequency (Hz)	40	50	70	100
Total core loss (W)	40	57.5	101.5	190

Determine the hysteresis and eddy current losses at (a) 50 Hz and (b) 80 Hz. [(a) 20 W; 37.5 W (b) 32 W; 96 W]

14 Explain why, when steel is subjected to alternating magnetization energy, losses occur due to both hysteresis and eddy currents.

The core loss in a transformer core at normal flux density was measured at frequencies of 40 Hz and 50 Hz, the results being 40 W and 52.5 W respectively. Calculate, at a frequency of 50 Hz, (a) the hysteresis loss and (b) the eddy current loss.

[(a) 40 W (b) 12.5 W]

15 Results of a test used to separate the hysteresis and eddy current losses in the core of a transformer winding gave the following results:

Total core loss (W)	48	96	160	240
Frequency (Hz)	40	60	80	100

If the flux density is held constant throughout the test, determine the values of the hysteresis and eddy current losses at 50 Hz.

[20 W; 50 W]

16 A transformer core has a total core loss of 275 W at 50 Hz and 600 W at 100 Hz, the flux density being constant for the two tests. (a) Determine the hysteresis and eddy current losses at 75 Hz. (b) If the flux density is increased by 40% and the lamination thickness is increased by 20% determines the hysteresis and eddy current losses at 75 Hz. Assume the Steinmetz index to be 1.6

[(a) 375 W; 56.25 W (b) 642.4 W; 190.5 W]

Assignment 12

This assignment covers the material in chapters 36 to 38.

The marks for each question are shown in brackets at the end of each question.

1 A voltage waveform represented by

$$v = 50\sin\omega t + 20\sin\left(3\omega t + \frac{\pi}{3}\right) + 5\sin\left(5\omega t + \frac{\pi}{6}\right) \text{ volts}$$

is applied to a circuit and the resulting current i is given by

$$i = 2.0\sin\left(\omega t - \frac{\pi}{6}\right) + 0.462\sin 3\omega t$$
$$+ 0.0756\sin(5\omega t - 0.71) \text{ amperes.}$$

Calculate (a) the r.m.s. voltage, (b) the mean value of voltage, (c) the form factor for the voltage, (d) the r.m.s. value of current, (e) the mean value of current, (f) the form factor for the current, (g) the total active power supplied to the circuit, and (h) the overall power factor. (24)

2 The value of the current i (in mA) at different moments in a cycle are given by:

θ degrees	0	30	60	90	120	150	180
i mA	50	75	165	190	170	100	−150

θ degrees	210	240	270	300	330	360
i mA	−210	−185	−90	−10	35	50

Draw the graph of current i against θ and analyse the current into it's first three constituent components, each coefficient correct to 2 decimal places. (30)

3 The cross-sectional area of a transformer limb is 8000 mm^2 and the volume of the transformer core is 4×10^6 mm^3. The maximum value of the core flux is 12 mWb at a frequency of 50 Hz. Taking the Steinmetz index as 1.6, the hysteresis loss is found to be 80 W. Determine the value of the hysteresis loss when the maximum core flux is 9 mWb and the frequency is 50 Hz. (6)

4 The core of an inductor has a hysteresis loss of 25 W and an eddy current loss of 15 W when operating at 50 Hz frequency. Determine (a) the values of the losses if the frequency is increased to 70 Hz, and (b) the total core loss if the frequency is 50 Hz and the laminations are made three quarters of their original thickness. Assume that the flux density remains unchanged in each case. (10)

39 Dielectrics and dielectric loss

At the end of this chapter you should be able to:

- understand electric fields, capacitance and permittivity
- assess the dielectric properties of materials
- determine dielectric loss, loss angle, Q-factor and dissipation factor of capacitors

39.1 Electric fields, capacitance and permittivity

Any region in which an electric charge experiences a force is called an electrostatic field. Electric fields, Coulombs law, capacitance and permittivity are discussed in Chapter 6 — refer back to page 55. Summarizing the main formulae:

Electric field strength, $E = \dfrac{V}{d}$ **volts/metre**

Capacitance $C = \dfrac{Q}{V}$ **farads**

Electric flux density, $D = \dfrac{Q}{A}$ **coulombs/metre**^{2}

$$\frac{D}{E} = \varepsilon_0 \varepsilon_r = \varepsilon$$

Relative permittivity $\varepsilon_r = \dfrac{\text{flux density in material}}{\text{flux density in vacuum}}$

The insulating medium separating charged surfaces is called a **dielectric**. Compared with conductors, dielectric materials have very high resistivities (and hence low conductance, since $\rho = 1/\sigma$). They are therefore used to separate conductors at different potentials, such as capacitor plates or electric power lines.

For a parallel-plate capacitor, **capacitance** $C = \dfrac{\varepsilon_0 \varepsilon_r A(n-1)}{d}$

39.2 Polarization

When a dielectric is placed between charged plates, the capacitance of the system increases. The mechanism by which a dielectric increases capacitance is called **polarization**. In an electric field the electrons and atomic nuclei of the dielectric material experience forces in opposite directions. Since the electrons in an insulator cannot flow, each atom becomes a tiny dipole (i.e., an arrangement of two electric charges of opposite polarity) with positive and negative charges slightly separated, i.e., the material becomes polarised.

Within the material this produces no discernible effects. However, on the surfaces of the dielectric, layers of charge appear. Electrons are drawn towards the positive potential, producing a negative charge layer, and away from the negative potential, leaving positive surface charge behind. Therefore the dielectric becomes a volume of neutral insulator with surface charges of opposite polarity on opposite surfaces. The result of this is that the electric field inside the dielectric is less than the electric field causing the polarization, because these two charge layers give rise to a field which opposes the electric field causing it. Since electric field strength, $E = V/d$, the p.d. between the plates, $V = Ed$. Thus, if E decreases when the dielectric is inserted, then V falls too and this drop in p.d. occurs without change of charge on the plates. Thus, since capacitance $C = Q/V$, capacitance increases, this increase being by a factor equal to ε_r above that obtained with a vacuum dielectric.

There are two main ways in which polarization takes place:

(i) The electric field, as explained above, pulls the electrons and nucleii in opposite directions because they have opposite charges, which makes each atom into an electric dipole. The movement is only small and takes place very fast since the electrons are very light. Thus, if the applied electric field is varied periodically, the polarization, and hence the permittivity due to these induced dipoles, is independent of the frequency of the applied field.

(ii) Some atoms have a permanent electric dipole as a result of their structure and, when an electric field is applied, they turn and tend to align along the field. The response of the permanent dipoles is slower than the response of the induced dipoles and that part of the relative permittivity which arises from this type of polarization decreases with increase of frequency.

Most materials contain both induced and permanent dipoles, so the relative permittivity usually tends to decrease with increase of frequency.

39.3 Dielectric strength

The maximum amount of field strength that a dielectric can withstand is called the dielectric strength of the material. When an electric field is established across the faces of a material, molecular alignment and distortion of the electron orbits around the atoms of the dielectric occur. This produces a mechanical stress which in turn generates heat. The production of heat represents a dissipation of power, such a loss being present in all practical dielectrics, especially when used in high-frequency systems where the field polarity is continually and rapidly changing.

A dielectric whose conductivity is not zero between the plates of a capacitor provides a conducting path along which charges can flow and thus discharge the capacitor. The resistance R of the dielectric is given by $R = \rho l/a$, l being the thickness of the dielectric film (which may be as small as 0.001 mm) and a being the area of the capacitor plates. The resistance R of the dielectric may be represented as a leakage resistance across an ideal capacitor (see Section 39.8 on dielectric loss). The required lower limit for acceptable resistance between the plates varies with the use to which the capacitor is put. High-quality capacitors have high shunt-resistance values. A measure of dielectric quality is the time taken for a capacitor to discharge a given amount through the resistance of the dielectric. This is related to the product CR.

$$\text{Capacitance, } C \propto \frac{\text{area}}{\text{thickness}} \quad \text{and} \quad \frac{1}{R} \propto \frac{\text{area}}{\text{thickness}}$$

thus CR is a characteristic of a given dielectric. In practice, circuit design is considerably simplified if the shunt conductance of a capacitor can be ignored (i.e. $R \to \infty$) and the capacitor therefore regarded as an open circuit for direct current.

Since capacitance C of a parallel plate capacitor is given by $C = \varepsilon_0 \varepsilon_r A/d$, reducing the thickness d of a dielectric film increases the capacitance, but decreases the resistance. It also reduces the voltage the capacitor can withstand without breakdown (since $V = Q/C$). Any material will eventually break down, usually destructively, when subjected to a sufficiently large electric field. A spark may occur at breakdown which produces a hole through the film. The metal film forming the metal plates may be welded together at the point of breakdown.

Breakdown depends on electric field strength E (where $E = V/d$), so thinner films will break down with smaller voltages across them. This is the main reason for limiting the voltage that may be applied to a capacitor. All practical capacitors have a safe working voltage stated on them, generally at a particular maximum temperature. Figure 39.1 shows the typical shapes of graphs expected for electric field strength E plotted against thickness and for breakdown voltage plotted against thickness. The shape of the curves depend on a number of factors, and these include:

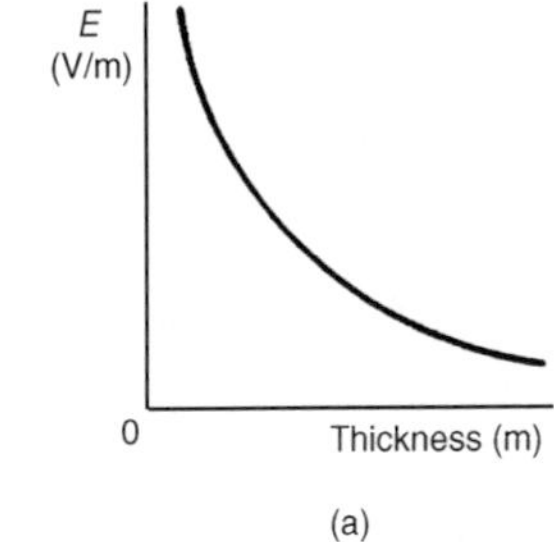

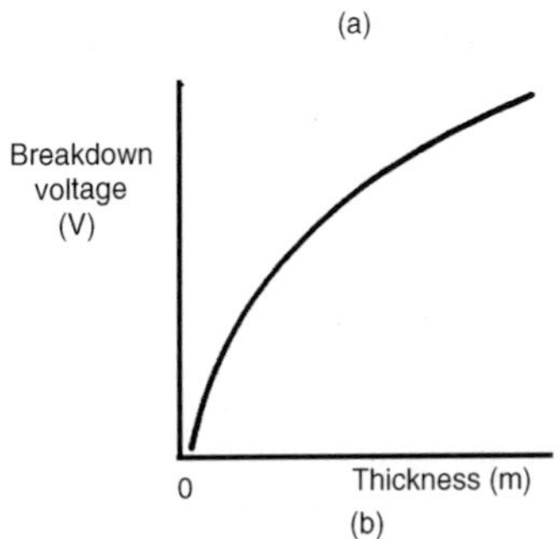

Figure 39.1

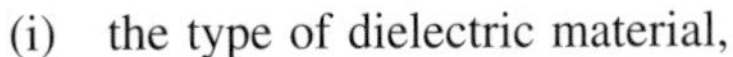

(i) the type of dielectric material,
(ii) the shape and size of the conductors associated with it,
(iii) the atmospheric pressure,
(iv) the humidity/moisture content of the material,
(v) the operating temperature.

Dielectric strength is an important factor in the design of capacitors as well as transformers and high voltage insulators, and in motors and generators. Dielectrics vary in their ability to withstand large fields. Some typical values of dielectric strength, together with resistivity and relative permittivity are shown in Table 39.1. The ceramics have very high relative permittivities and they tend to be 'ferroelectric', i.e., they do not lose their polarities when the electric field is removed. When ferroelectric effects are present, the charge on a capacitor is given by $Q = (CV) +$ (remanent polarization). These dielectrics often possess an appreciable

TABLE 39.1 *Dielectric properties of some common materials*

Material	*Resistivity, ρ* (Ωm)	*Relative permittivity,* ε_r	*Dielectric strength* (V/m)
Air		1.0	3×10^6
Paper	10^{10}	3.7	1.6×10^7
Mica	5×10^{11}	5.4	$10^8 - 10^9$
Titaniumdioxide	10^{12}	100	6×10^6
Polythene	$>10^{11}$	2.3	4×10^7
Polystyrene	$>10^{13}$	2.5	2.5×10^7
Ceramic (type 1)	4×10^{11}	6–500	4.5×10^7
Ceramic (type 2)	$10^6 - 10^{13}$	500–1000	$2 \times 10^6 - 10^7$

negative temperature coefficient of resistance. Despite this, a high permittivity is often very desirable and ceramic dielectrics are widely used.

39.4 Thermal effects

As the temperature of most dielectrics is increased, the insulation resistance falls rapidly. This causes the leakage current to increase, which generates further heat. Eventually a condition known as thermal avalanche or thermal runaway may develop, when the heat is generated faster than it can be dissipated to the surrounding environment. The dielectric will burn and thus fail.

Thermal effects may often seriously influence the choice and application of insulating materials. Some important factors to be considered include:

(i) the melting-point (for example, for waxes used in paper capacitors),
(ii) aging due to heat,
(iii) the maximum temperature that a material will withstand without serious deterioration of essential properties,
(iv) flash-point or ignitability,
(v) resistance to electric arcs,
(vi) the specific heat capacity of the material,
(vii) thermal resistivity,
(viii) the coefficient of expansion,
(ix) the freezing-point of the material.

39.5 Mechanical properties

Mechanical properties determine, to varying degrees, the suitability of a solid material for use as an insulator: tensile strength, transverse strength, shearing strength and compressive strength are often specified. Most solid insulations have a degree of inelasticity and many are quite brittle, thus it is often necessary to consider features such as compressibility, deformation under bending stresses, impact strength and extensibility, tearing strength, machinability and the ability to fold without damage.

39.6 Types of practical capacitor

Practical types of capacitor are characterized by the material used for their dielectric. The main types include: variable air, mica, paper, ceramic, plastic, titanium oxide and electrolytic. Refer back to Chapter 6, Section 11, page 69, for a description of each type.

39.7 Liquid dielectrics and gas insulation

Liquid dielectrics used for insulation purposes are refined mineral oils, silicone fluids and synthetic oils such as chlorinated diphenyl. The principal uses of liquid dielectrics are as a filling and cooling medium for transformers, capacitors and rheostats, as an insulating and arc-quenching medium in switchgear such as circuit breakers, and as an impregnant of absorbent insulations — for example, wood, slate, paper and pressboard, used mainly in transformers, switchgear, capacitors and cables.

Two **gases** used as insulation are nitrogen and sulphur hexafluoride. Nitrogen is used as an insulation medium in some sealed transformers and in power cables, and sulphur hexafluoride is finding increasing use in switchgear both as an insulant and as an arc-extinguishing medium.

39.8 Dielectric loss and loss angle

In capacitors with solid dielectrics, losses can be attributed to two causes:

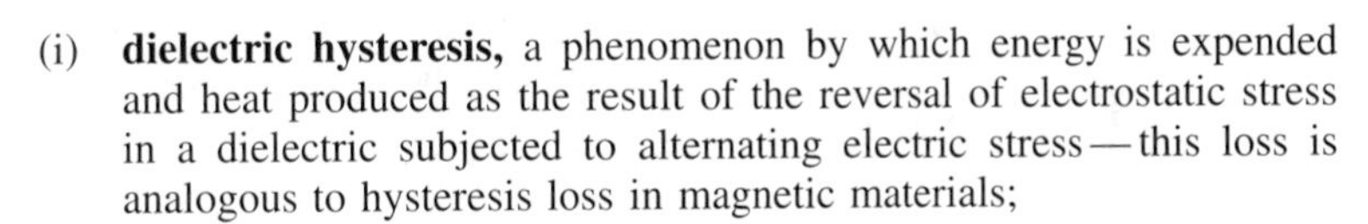

(i) **dielectric hysteresis,** a phenomenon by which energy is expended and heat produced as the result of the reversal of electrostatic stress in a dielectric subjected to alternating electric stress — this loss is analogous to hysteresis loss in magnetic materials;

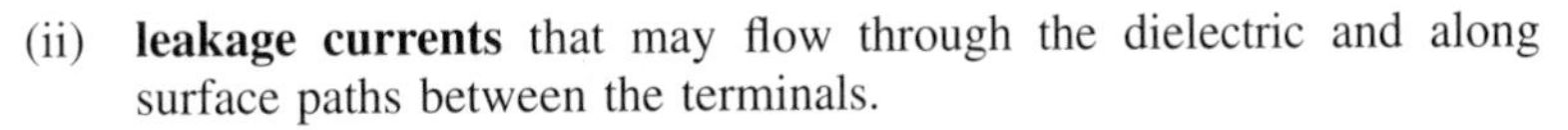

(ii) **leakage currents** that may flow through the dielectric and along surface paths between the terminals.

The total dielectric loss may be represented as the loss in an additional resistance connected between the plates. This may be represented as either a small resistance in series with an ideal capacitor or as a large resistance in parallel with an ideal capacitor.

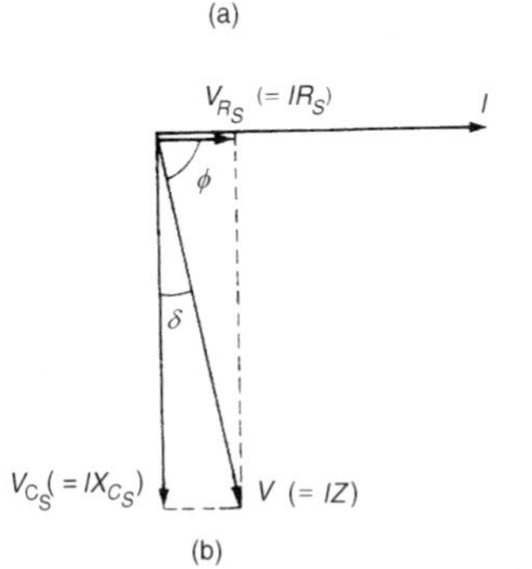

Figure 39.2 *(a) Circuit diagram (b) Phasor diagram*

Series representation

The circuit and phasor diagrams for the series representation are shown in Figure 39.2. The circuit phase angle is shown as angle ϕ. If resistance R_S is zero then current I would lead voltage V by 90°, this being the case of a perfect capacitor. The difference between 90° and the circuit phase angle ϕ is the angle shown as δ. This is known as the **loss angle** of the capacitor, i.e.,

$$\textbf{loss angle, } \delta = (90^\circ - \phi)$$

For the equivalent series circuit,

$$\tan\delta = \frac{V_{R_S}}{V_{C_S}} = \frac{IR_S}{IX_{C_S}}$$

i.e., $$\tan\delta = \frac{R_S}{1/(\omega C_S)} = R_S\omega C_S$$

Since from Chapter 28, $Q = \dfrac{1}{\omega CR}$ then

$$\boldsymbol{\tan\delta = R_S\omega C_S = \frac{1}{Q}} \quad (39.1)$$

Power factor of capacitor,

$$\cos\phi = \frac{V_{R_S}}{V} = \frac{IR_S}{IZ_S} = \frac{R_S}{Z_S} \approx \frac{R_S}{X_{C_S}}$$

since $X_{C_S} \approx Z_S$ when δ is small. Hence **power factor** $= \boldsymbol{\cos\phi \approx R_S\omega C_S}$, i.e.,

$$\boldsymbol{\cos\phi \approx \tan\delta} \quad (39.2)$$

Dissipation factor, $\boldsymbol{D}$ is defined as the reciprocal of Q-factor and is an indication of the quality of the dielectric, i.e.,

$$\boldsymbol{D = \frac{1}{Q} = \tan\delta} \quad (39.3)$$

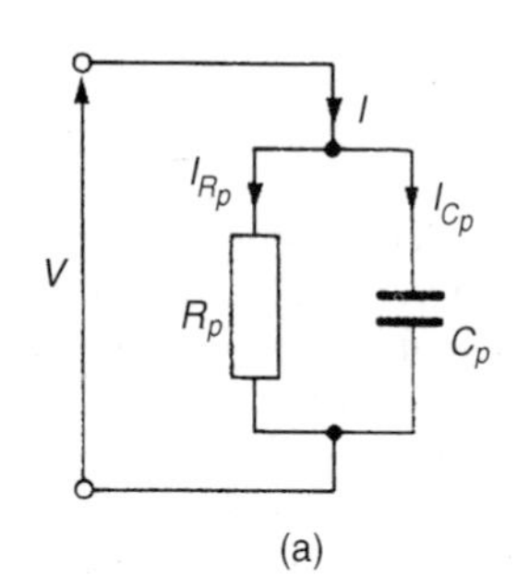

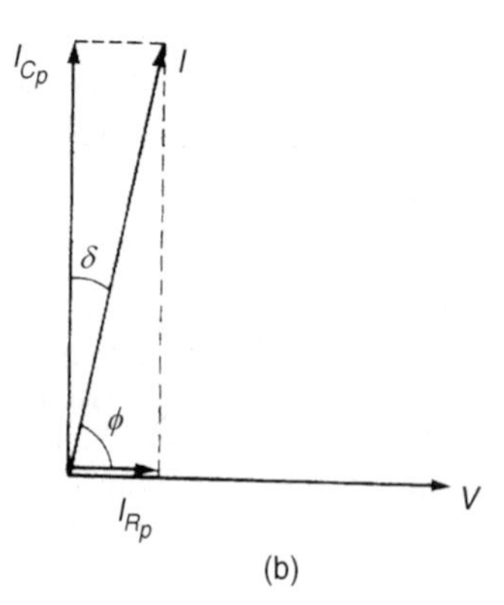

Figure 39.3 *(a) Circuit diagram (b) Phasor diagram*

Parallel representation

The circuit and phasor diagrams for the parallel representation are shown in Figure 39.3. From the phasor diagram,

$$\tan\delta = \frac{I_{R_P}}{I_{C_P}} = \frac{V/R_P}{V/X_{C_P}} = \frac{X_{C_P}}{R_P}$$

i.e., $$\boldsymbol{\tan\delta = \frac{1}{R_P\omega C_P}} \quad (39.4)$$

Power factor of capacitor,

$$\cos\phi = \frac{I_{R_P}}{I} = \frac{V/R_P}{V/Z_P} = \frac{Z_P}{R_P} \approx \frac{X_{C_P}}{R_P}$$

since $X_{C_P} \approx Z_P$, when δ is small. Hence

$$\text{power factor} = \cos\phi \approx \frac{1}{R_P\omega C_P}$$

i.e., $$\boldsymbol{\cos\phi \approx \tan\delta}$$

(For equivalence between the series and the parallel circuit representations,

$$C_S \approx C_P = C \text{ and } R_S\omega C_S \approx \frac{1}{R_P\omega C_P}$$

from which $R_S \approx 1/R_P\omega^2 C^2$)

Power loss in the dielectric $= VI\cos\phi$. From the phasor diagram of Figure 39.3

$$\cos\delta = \frac{I_{C_P}}{I} = \frac{V/X_{C_P}}{I} = \frac{V\omega C}{I} \text{ or } I = \frac{V\omega C}{\cos\delta}$$

Hence power loss $= VI\cos\phi = V\left(\dfrac{V\omega C}{\cos\delta}\right)\cos\phi$

However, $\cos\phi = \sin\delta$ (complementary angles), thus

$$\text{power loss} = V\left(\frac{V\omega C}{\cos\delta}\right)\sin\delta = V^2\omega C\tan\delta$$

(since $\sin\delta/\cos\delta = \tan\delta$)

Hence $$\textbf{dielectric power loss} = \boldsymbol{V^2\omega C\tan\delta} \qquad (39.5)$$

Problem 1. The equivalent series circuit for a particular capacitor consists of a 1.5 Ω resistance in series with a 400 pF capacitor. Determine for the capacitor, at a frequency of 8 MHz, (a) the loss angle, (b) the power factor, (c) the Q-factor, and (d) the dissipation factor.

(a) From equation (39.1), for a series equivalent circuit,

$$\tan\delta = R_S\omega C_S$$
$$= (1.5)(2\pi \times 8 \times 10^6)(400 \times 10^{-12}) = 0.030159$$

Hence **loss angle,** $\boldsymbol{\delta}$ = arctan(0.030159) = **1.727° or 0.030 rad**.

(b) From equation (39.2), **power factor** $= \cos\phi \approx \tan\delta =$ **0.030**

(c) From equation (39.1), $\tan\delta = \dfrac{1}{Q}$ hence $\boldsymbol{Q} = \dfrac{1}{\tan\delta}$

$$= \frac{1}{0.030159} = \mathbf{33.16}$$

(d) From equation (39.3), **dissipation factor**,

$$D = \frac{1}{Q} = 0.030159 \text{ or } \mathbf{0.030}, \text{ correct to 3 decimal places.}$$

> Problem 2. A capacitor has a loss angle of 0.025 rad, and when it is connected across a 5 kV, 50 Hz supply, the power loss is 20 W. Determine the component values of the equivalent parallel circuit.

From equation (39.5),

$$\text{power loss} = V^2\omega C \tan\delta$$

i.e.,
$$20 = (5000)^2(2\pi 50)(C)\tan(0.025)$$

from which **capacitance** $\boldsymbol{C} = \dfrac{20}{(5000)^2(2\pi 50)\tan(0.025)} = \mathbf{0.102\ \mu F}$

(Note tan(0.025) means 'the tangent of 0.025 rad')
From equation (39.4), for a parallel equivalent circuit,

$$\tan\delta = \frac{1}{R_P\omega C_P}$$

from which, parallel resistance,

$$R_P = \frac{1}{\omega C_P \tan\delta} = \frac{1}{(2\pi 50)(0.102\times 10^{-6})\tan 0.025}$$

i.e., $\boldsymbol{R_P = 1.248\ M\Omega}$

> Problem 3. A 2000 pF capacitor has an alternating voltage of 20 V connected across it at a frequency of 10 kHz. If the power dissipated in the dielectric is 500 μW, determine (a) the loss angle, (b) the equivalent series loss resistance, and (c) the equivalent parallel loss resistance.

(a) From equation (39.5), power loss $= V^2\omega C\tan\delta$, i.e.,

$$500\times 10^{-6} = (20)^2(2\pi 10\times 10^3)(2000\times 10^{-12})\tan\delta$$

Hence
$$\tan\delta = \frac{500\times 10^{-6}}{(20)^2(2\pi 10\times 10^3)(2000\times 10^{-12})} = 9.947\times 10^{-3}$$

from which, **loss angle,** $\boldsymbol{\delta = 0.57^\circ}$ **or** $\boldsymbol{9.95\times 10^{-3}}$ **rad.**

(b) From equation (39.1), for an equivalent series circuit, $\tan\delta = R_S\omega C_S$, from which equivalent series resistance,

$$R_S = \frac{\tan\delta}{\omega C_S} = \frac{9.947\times 10^{-3}}{(2\pi 10\times 10^3)(2000\times 10^{-12})}$$

i.e., $\boldsymbol{R_S = 79.16\ \Omega}$

(c) From equation (39.4), for an equivalent parallel circuit,

$$\tan\delta = \frac{1}{R_P \omega C_P}$$

from which equivalent parallel resistance,

$$R_P = \frac{1}{(\tan\delta)\omega C_P}$$

$$= \frac{1}{(9.947 \times 10^{-3})(2\pi 10 \times 10^3)(2000 \times 10^{-12})}$$

i.e., $\mathbf{R_P = 800\ k\Omega}$

Further problems on dielectric loss and loss angle may be found in Section 39.9 following, problems 1 to 5.

39.9 Further problems on dielectric loss and loss angle

1 The equivalent series circuit for a capacitor consists of a 3 Ω resistance in series with a 250 pF capacitor. Determine the loss angle of the capacitor at a frequency of 5 MHz, giving the answer in degrees and in radians. Find also for the capacitor, (a) the power factor, (b) the Q-factor, and (c) the dissipation factor.
[1.35° or 0.024 rad (a) 0.0236 (b) 42.4 (c) 0.0236]

2 A capacitor has a loss angle of 0.008 rad and when it is connected across a 4 kV, 60 Hz supply the power loss is 15 W. Determine the component values of (a) the equivalent parallel circuit, and (b) the equivalent series circuit.
[(a) 0.311 μF, 1.066 MΩ (b) 0.311 μF, 68.24 Ω]

3 A coaxial cable has a capacitance of 4 μF and a dielectric power loss of 12 kW when operated at 50 kV and frequency 50 Hz. Calculate (a) the value of the loss angle, and (b) the equivalent parallel resistance of the cable.
[(a) 0.219° or 3.82×10^{-3} rad (b) 208.3 kΩ]

4 What are the main reasons for power loss in capacitors with solid dielectrics? Explain the term 'loss angle'.
A voltage of 10 V and frequency 20 kHz is connected across a 1 nF capacitor. If the power dissipated in the dielectric is 0.2 mW, determine (a) the loss angle, (b) the equivalent series loss resistance, and (c) the equivalent parallel loss resistance.
[(a) 0.912° or 0.0159 rad (b) 126.7 Ω (c) 0.5 MΩ]

5 The equivalent series circuit for a capacitor consists of a 0.5 Ω resistor in series with a capacitor of reactance 2 kΩ. Determine for the capacitor (a) the loss angle, (b) the power factor, and (c) the equivalent parallel resistance.
[(a) 0.014° or 2.5×10^{-4} rad (b) 2.5×10^{-4} (c) 8 MΩ]

40 Field theory

At the end of this chapter you should be able to:

- understand field plotting by curvilinear squares
- show that the capacitance between concentric cylinders, $C = \dfrac{2\pi\varepsilon_0\varepsilon_r}{\ln(b/a)}$ and calculate C given values of radii a and b
- calculate dielectric stress $E = \dfrac{V}{r\ln(b/a)}$
- appreciate dimensions of the most economical cable
- show that the capacitance of an isolated twin line, $C = \dfrac{\pi\varepsilon_0\varepsilon_r}{\ln(D/a)}$ and calculate C given values of a and D
- calculate energy stored in an electric field
- show that the inductance of a concentric cylinder, $L = \dfrac{\mu_0\mu_r}{2\pi}\left(\dfrac{1}{4} + \ln\dfrac{b}{a}\right)$ and calculate L given values of a and b
- show that the inductance of an isolated twin line, $C = \dfrac{\mu_0\mu_r}{\pi}\left(\dfrac{1}{4} + \ln\dfrac{D}{a}\right)$ and calculate L given values of a and D
- calculate energy stored in an electromagnetic field

40.1 Field plotting by curvilinear squares

Electric fields, magnetic fields and conduction fields (i.e., a region in which an electric current flows) are analogous, i.e., they all exhibit similar characteristics. Thus they may all be analysed by similar processes. In the following the electric field is analysed

Figure 40.1 shows two parallel plates A and B. Let the potential on plate A be $+V$ volts and that on plate B be $-V$ volts. The force acting on a point charge of 1 coulomb placed between the plates is the electric field strength E. It is measured in the direction of the field and its magnitude depends on the p.d. between the plates and the distance between the plates. In Figure 40.1, moving along a line of force from plate B to plate A means moving from $-V$ to $+V$ volts. The p.d. between the plates is therefore 2 V volts and this potential changes linearly when moving from one plate to the other. Hence a potential gradient is followed which changes by equal amounts for each unit of distance moved.

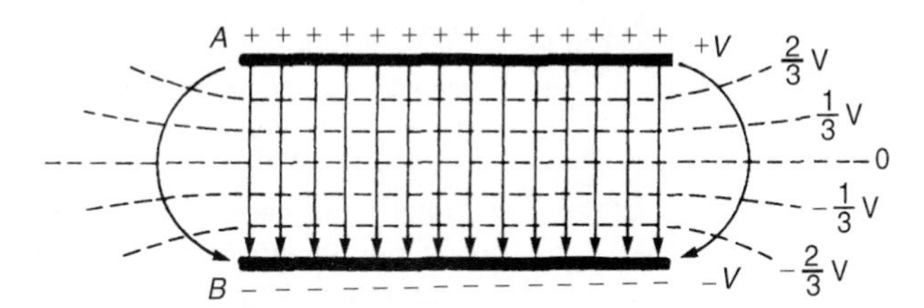

Figure 40.1 *Lines of force intersecting equipotential lines in an electric field*

Lines may be drawn connecting together all points within the field having equal potentials. These lines are called **equipotential lines** and these have been drawn in Figure 40.1 for potentials of $\frac{2}{3}$ V, $\frac{1}{3}$ V, 0, $-\frac{1}{3}$ V and $-\frac{2}{3}$ V. The zero equipotential line represents earth potential and the potentials on plates A and B are respectively above and below earth potential. Equipotential lines form part of an equipotential surface. Such surfaces are parallel to the plates shown in Figure 40.1 and the plates themselves are equipotential surfaces. There can be no current flow between any given points on such a surface since all points on an equipotential surface have the same potential. Thus a line of force (or flux) must intersect an equipotential surface at right angles. A line of force in an electrostatic field is often termed a **streamline**.

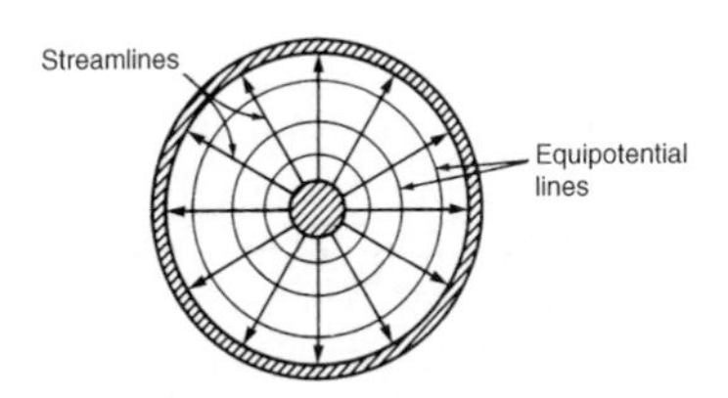

Figure 40.2 *Electric field distribution for a concentric cylinder capacitor*

An electric field distribution for a concentric cylinder capacitor is shown in Figure 40.2. An electric field is set up in the insulating medium between two good conductors. Any volt drop within the conductors can usually be neglected compared with the p.d.'s across the insulation since the conductors have a high conductivity. All points on the conductors are thus at the same potential so that the conductors form the boundary equipotentials for the electrostatic field. Streamlines (or lines of force) which must cut all equipotentials at right angles leave one boundary at right angles, pass across the field, and enter the other boundary at right angles.

In a magnetic field, a streamline is a line so drawn that its direction is everywhere parallel to the direction of the magnetic flux. An equipotential surface in a magnetic field is the surface over which a magnetic pole may be moved without the expenditure of work or energy.

In a conduction field, a streamline is a line drawn with a direction which is everywhere parallel to the direction of the current flow.

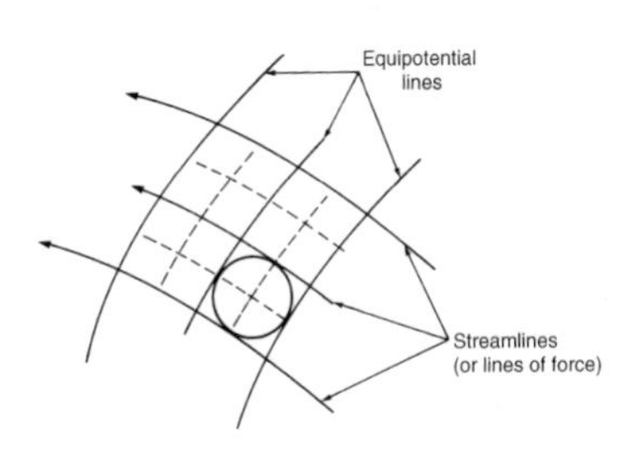

Figure 40.3 *Curvilinear square*

A method of solving certain field problems by a form of graphical estimation is available which may only be applied, however, to plane linear fields; examples include the field existing between parallel plates or between two long parallel conductors. In general, the plane of a field may be divided into a number of squares formed between the line of force (i.e. streamline) and the equipotential. Figure 40.3 shows a typical pattern. In most cases true squares will not exist, since the streamlines and equipotentials are curved. However, since the streamlines and the equipotentials intersect at right angles, square-like figures are formed, and these are usually called **'curvilinear squares'**. The square-like figure shown in Figure 40.3 is a curvilinear square since, on successive subdivision by equal numbers of intermediate streamlines and equipotentials, the smaller figures are seen to approach a true square form.

When subdividing to give a field in detail, and in some cases for the initial equipotentials, **'Moores circle' technique** can be useful in that it tends to eliminate the trial and error process. If, say, two flux lines and an equipotential are given and it is required to draw a neighbouring equipotential, a circle tangential to the three given lines is constructed. The new equipotential is then approximately tangential to the circle, as shown in Figure 40.3.

Consider the electric field established between two parallel metal plates, as shown in Figure 40.4. The streamlines and the equipotential lines are shown sketched and are seen to form curvilinear squares. Consider a true square *abcd* lying between equipotentials AB and CD. Let this square be the end of x metres depth of the field forming a flux tube between adjacent equipotential surfaces *abfe* and *cdhg* as shown in Figure 40.5. Let l be the length of side of the squares. Then the capacitance C_1 of the flux tube is given by

$$C_1 = \frac{\varepsilon_0 \varepsilon_r \text{ (area of plate)}}{\text{plate separation}}$$

$$\text{i.e., } \boldsymbol{C_1} = \frac{\varepsilon_0 \varepsilon_r (lx)}{l} = \boldsymbol{\varepsilon_0 \varepsilon_r x} \qquad (40.1)$$

Thus the capacitance of the flux tube whose end is a true square is independent of the size of the square.

Let the distance between the plates of a capacitor be divided into an exact number of parts, say n (in Figure 40.4, $n = 4$). Using the same scale, the breadth of the plate is divided into a number of parts (which is not always an integer value), say m (in Figure 40.4, $m = 10$, neglecting fringing). Thus between equipotentials AB and CD in Figure 40.4 there are m squares in parallel and so there are m capacitors in parallel. For m capacitors connected in parallel, the equivalent capacitance C_T is given by $C_T = C_1 + C_2 + C_3 + \cdots + C_m$. If the capacitors have the same value,

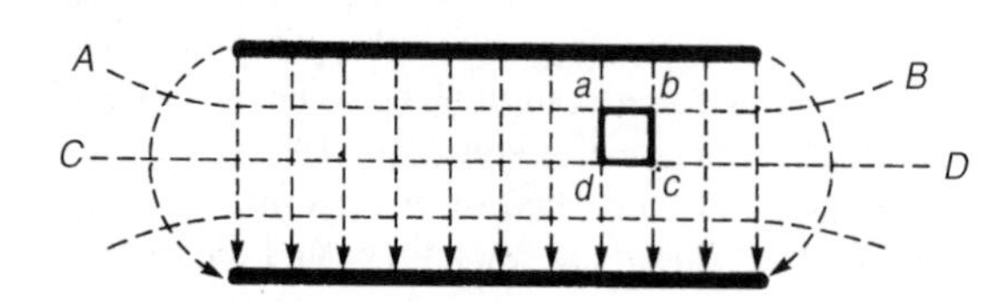

Figure 40.4

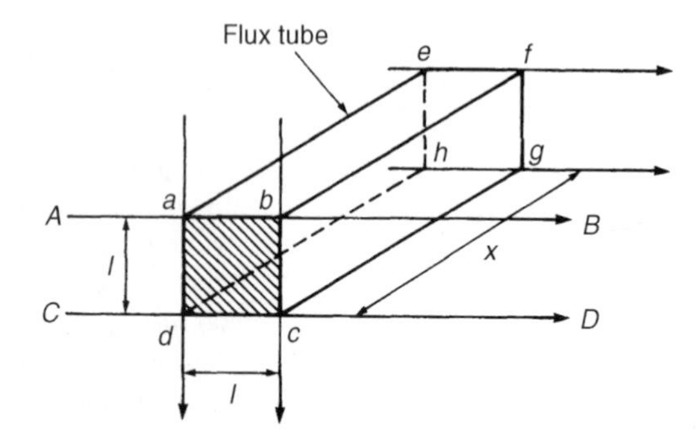

Figure 40.5

i.e., $C_1 = C_2 = C_3 = \cdots = C_m = C_t$, then

$$C_T = mC_t \tag{40.2}$$

Similarly, there are n squares in series in Figure 40.4 and thus n capacitors in series.

For n capacitors connected in series, the equivalent capacitance C_T is given by

$$\frac{1}{C_T} = \frac{1}{C_1} + \frac{1}{C_2} + \cdots + \frac{1}{C_n}$$

If $C_1 = C_2 = \cdots = C_n = C_t$ then $1/C_T = n/C_t$, from which

$$C_T = \frac{C_t}{n} \tag{40.3}$$

Thus if m is the number of parallel squares measured along each equipotential and n is the number of series squares measured along each streamline (or line of force), then the total capacitance C of the field is given, from equations (40.1)–(40.3), by

$$\boxed{\boldsymbol{C = \varepsilon_0 \varepsilon_r x \frac{m}{n} \text{ farads}}} \tag{40.4}$$

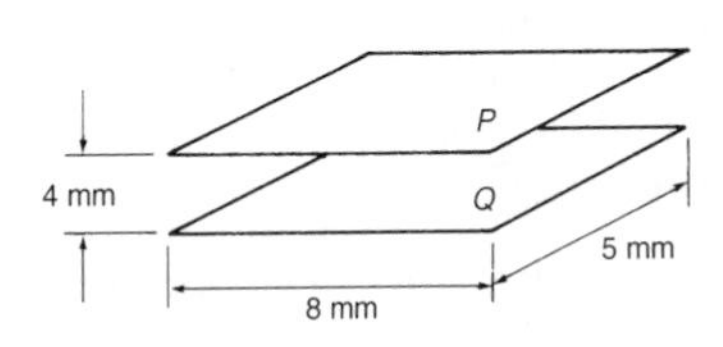

Figure 40.6

For example, let a parallel-plate capacitor have plates 8 mm × 5 mm and spaced 4 mm apart (See Figure 40.6). Let the dielectric have a relative permittivity 3.5. If the distance between the plates is divided into, say, four equipotential lines, then each is 1 mm apart. Hence $n = 4$.

Using the same scale, the number of lines of force from plate P to plate Q must be 8, i.e. $m = 8$. This is, of course, neglecting any fringing. From equation (40.4), capacitance $C = \varepsilon_0 \varepsilon_r x(m/n)$, where $x = 5$ mm or 0.005 m in this case. Hence

$$C = (8.85 \times 10^{-12})(3.5)(0.005)\left(\tfrac{8}{4}\right) = \mathbf{0.31\ pF}$$

(Using the normal equation for capacitance of a parallel-plate capacitor,

$$C = \frac{\varepsilon_0 \varepsilon_r A}{d} = \frac{(8.85 \times 10^{-12})(3.5)(0.008 \times 0.005)}{0.004} = \mathbf{0.31\ pF}$$

The capacitance found by each method gives the same value; this is expected since the field is uniform between the plates, giving a field plot of true squares.)

The effect of fringing may be considered by estimating the capacitance by field plotting. This is described below.

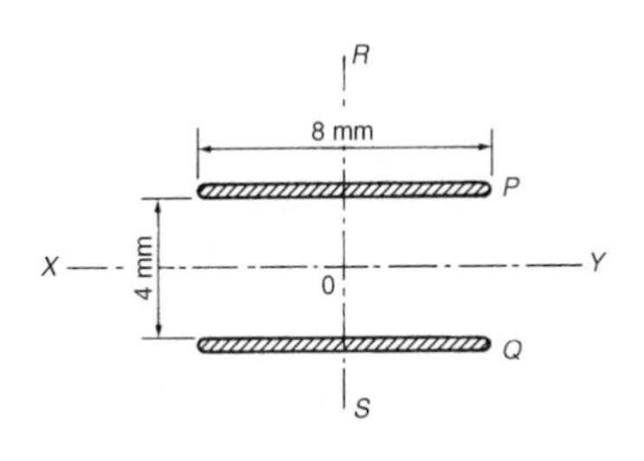

Figure 40.7

In the side view of the plates shown in Figure 40.7, RS is the medial line of force or medial streamline, by symmetry. Also XY is the medial equipotential. The field may thus be divided into four separate symmetrical parts.

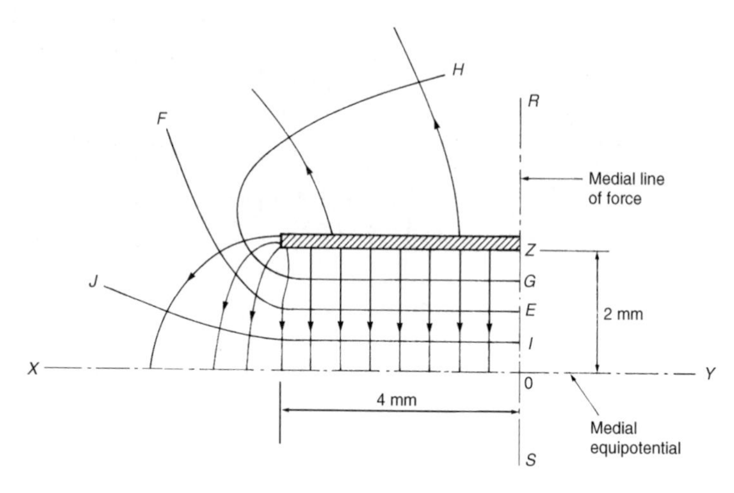

Figure 40.8

Considering just the top left part of the field, the field plot is estimated as follows, with reference to Figure 40.8:

(i) Estimate the position of the equipotential EF which has the mean potential between that of the plate and that of the medial equipotential XO. F is not taken too far since it is difficult to estimate. Point E will lie slightly closer to point Z than point O.

(ii) Estimate the positions of intermediate equipotentials GH and IJ.

(iii) All the equipotential lines plotted are $\frac{2}{4}$, i.e., 0.5 mm apart. Thus a series of streamlines, cutting the equipotential at right angles, are drawn, the streamlines being spaced 0.5 mm apart, with the object of forming, as far as possible, curvilinear squares.

It may be necessary to erase the equipotentials and redraw them to fit the lines of force. The field between the plates is almost uniform, giving a field plot of true squares in this region. At the corner of the plates the squares are smaller, this indicating a great stress in this region.

On the top of the plate the squares become very large, indicating that the main field exists between the plates.
From equation (40.4),

$$\text{total capacitance, } C = \varepsilon_0\varepsilon_r x \frac{m}{n} \text{ farads}$$

The number of parallel squares measured along each equipotential is about 13 in this case and the number of series squares measured along each line of force is 4. Thus, for the plates shown in Figure 40.7, $m = 2 \times 13 = 26$ and $n = 2 \times 4 = 8$. Since x is 5 mm,

$$\text{total capacitance} = \varepsilon_0\varepsilon_r x \frac{m}{n} = (8.85 \times 10^{-12})(3.5)(0.005)\frac{26}{8}$$

$$= \mathbf{0.50\ pF}$$

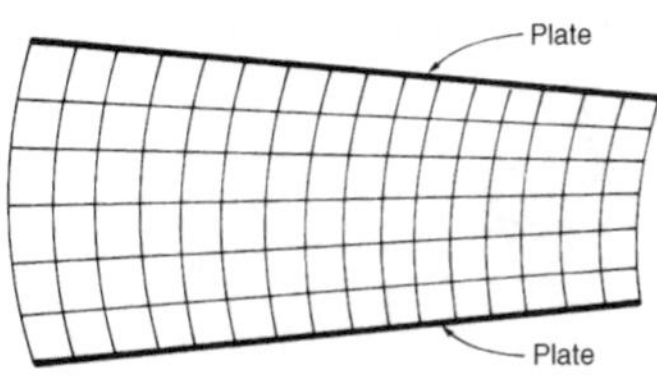

Figure 40.9

Problem 1. A field plot between two metal plates is shown in Figure 40.9. The relative permeability of the dielectric is 2.8. Determine the capacitance per metre length of the system.

From equation (40.4), capacitance $C = \varepsilon_0 \varepsilon_r x(m/n)$. From Figure 40.9, $m = 16$, i.e., the number of parallel squares measured along each equipotential, and $n = 6$, i.e., the number of series squares measured along each line of force. Hence capacitance for a 1 m length,

$$C = (8.85 \times 10^{-12})(2.8)(1)\frac{16}{6} = \mathbf{66.08\ pF}$$

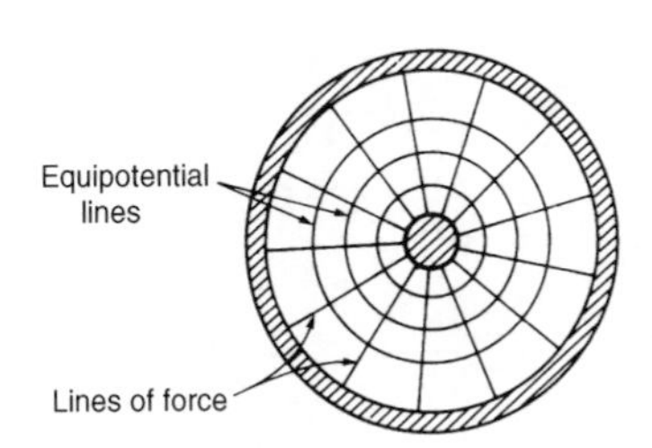

Figure 40.10

Problem 2. A field plot for a cross-section of a concentric cable is shown in Figure 40.10. If the relative permeability of the dielectric is 3.4, determine the capacitance of a 100 m length of the cable.

From equation (40.4), capacitance $C = \varepsilon_0 \varepsilon_r x(m/n)$. In this case, $m = 13$ and $n = 4$. Also $x = 100$ m. Thus

$$\text{capacitance } C = (8.85 \times 10^{-12})(3.4)(100)\frac{13}{4} = \mathbf{9780\ pF} \text{ or } \mathbf{9.78\ nF}$$

Further problems on field plotting by curvilinear squares may be found in Section 40.9, problems 1 to 3, page 753.

40.2 Capacitance between concentric cylinders

A **concentric cable** is one which contains two or more separate conductors, arranged concentrically (i.e., having a common centre), with insulation between them. In a **coaxial cable**, the central conductor, which may be either solid or hollow, is surrounded by an outer tubular conductor, the space in between being occupied by a dielectric. If air is the dielectric then concentric insulating discs are used to prevent the conductors touching each other. The two kinds of cable serve different purposes. The main feature they have in common is a complete absence of external flux and therefore a complete absence of interference with and from other circuits.

The electric field between two concentric cylinders (i.e., a coaxial cable) is shown in the cross-section of Figure 40.11. The conductors form the boundary equipotentials for the field, the boundary equipotentials in Figure 40.11 being concentric cylinders of radii a and b. The streamlines, or lines of force, are radial lines cutting the equipotentials at right angles.

Let Q be the charge per unit length of the inner conductor. Then the total flux across the dielectric per unit length is Q coulombs/metre. This total flux will pass through the elemental cylinder of width δr at radius r (shown in Figure 40.11) and a distance of 1 m into the plane of the paper.

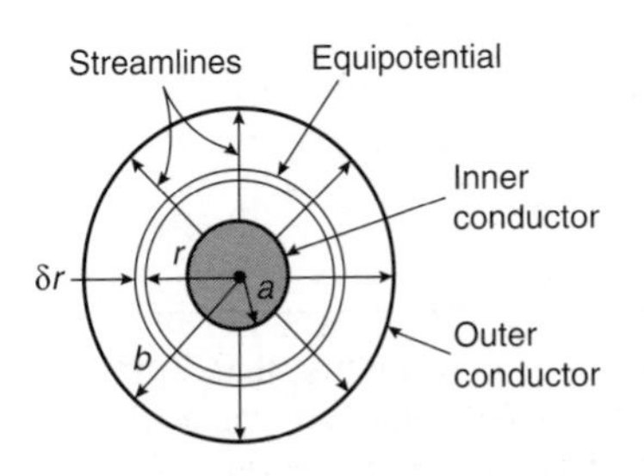

Figure 40.11 *Electric field between two concentric cylinders*

The surface area of a cylinder of length 1 m within the dielectric with radius r is $(2\pi r \times 1)\ \text{m}^2$.

Hence the electric flux density at radius r,

$$D = \frac{Q}{A} = \frac{Q}{2\pi r}$$

The electric field strength or electric stress E, at radius r is given by

$$E = \frac{D}{\varepsilon_0 \varepsilon_r} = \frac{Q}{2\pi r \varepsilon_0 \varepsilon_r} \text{ volts/metre} \tag{40.5}$$

Let the p.d. across the element be δV volts. Since

$$E = \frac{\text{voltage}}{\text{thickness}}$$

voltage $= E \times$ thickness. Therefore

$$\delta V = E\delta r = \frac{Q}{2\pi r \varepsilon_0 \varepsilon_r}\delta r$$

The total p.d. between the boundaries,

$$V = \int_a^b \frac{Q}{2\pi r \varepsilon_0 \varepsilon_r} dr = \frac{Q}{2\pi \varepsilon_0 \varepsilon_r} \int_a^b \frac{1}{r} dr$$

$$= \frac{Q}{2\pi \varepsilon_0 \varepsilon_r}[\ln r]_a^b = \frac{Q}{2\pi \varepsilon_0 \varepsilon_r}[\ln b - \ln a]$$

i.e., $$V = \frac{Q}{2\pi \varepsilon_0 \varepsilon_r} \ln \frac{b}{a} \text{ volts} \tag{40.6}$$

The capacitance per unit length,

$$C = \frac{\text{charge per unit length}}{\text{p.d.}}$$

Hence capacitance,

$$C = \frac{Q}{V} = \frac{Q}{(Q/(2\pi \varepsilon_0 \varepsilon_r))\ln(b/a)}$$

i.e., $$\boldsymbol{C = \frac{2\pi \varepsilon_0 \varepsilon_r}{\ln(b/a)} \textbf{ farads/metre}} \tag{40.7}$$

Problem 3. A coaxial cable has an inner core radius of 0.5 mm and an outer conductor of internal radius 6.0 mm. Determine the capacitance per metre length of the cable if the dielectric has a relative permittivity of 2.7.

From equation (40.7),

$$\text{capacitance } C = \frac{2\pi\varepsilon_0\varepsilon_r}{\ln(b/a)} = \frac{2\pi(8.85 \times 10^{-12})(2.7)}{\ln(6.0/0.5)} = \mathbf{60.4\ pF}$$

Problem 4. A single-core concentric cable has a capacitance of 80 pF per metre length. The relative permittivity of the dielectric is 3.5 and the core diameter is 8.0 mm. Determine the internal diameter of the sheath.

From equation (40.7), capacitance

$$C = \frac{2\pi\varepsilon_0\varepsilon_r}{\ln(b/a)} \text{ F/m}$$

$$\text{from which } \ln\frac{b}{a} = \frac{2\pi\varepsilon_0\varepsilon_r}{C} = \frac{2\pi(8.85 \times 10^{-12})(3.5)}{(80 \times 10^{-12})}$$

$$= 2.433$$

Since the core radius, $a = 8.0/2 = 4.0$ mm, $\ln(b/4.0) = 2.433$ and $b/4.0 = e^{2.433}$.

Thus the internal radius of the sheath, $b = 4.0e^{2.433} = 45.57$ mm. Hence the internal diameter of the sheath $2 \times 45.57 = \mathbf{91.14}$ mm.

Dielectric stress

Rearranging equation (40.6) gives:

$$\frac{Q}{2\pi\varepsilon_0\varepsilon_r} = \frac{V}{\ln(b/a)}$$

However, from equation (40.5),

$$E = \frac{Q}{2\pi r\varepsilon_0\varepsilon_r}$$

Thus dielectric stress,

$$\boldsymbol{E = \frac{V}{r \ln(b/a)} \textbf{ volts/metre}} \qquad (40.8)$$

From equation (40.8), the dielectric stress at any point is seen to be inversely proportional to r, i.e., $E \propto 1/r$.

The dielectric stress E will have a maximum value when r is at its minimum, i.e., when $r = a$. Thus

$$\boldsymbol{E_{max} = \frac{V}{a \ln(b/a)}} \qquad (40.9)$$

It follows that

$$\boldsymbol{E_{min} = \frac{V}{b \ln(b/a)}} \qquad (40.9')$$

Problem 5. A concentric cable has a core diameter of 32 mm and an inner sheath diameter of 80 mm. The core potential is 40 kV and the relative permittivity of the dielectric is 3.5. Determine (a) the capacitance per kilometre length of the cable, (b) the dielectric stress at a radius of 30 mm, and (c) the maximum and minimum values of dielectric stress.

(a) From equation (40.7), capacitance per metre length,

$$C = \frac{2\pi\varepsilon_0\varepsilon_r}{\ln(b/a)}$$

$$= \frac{2\pi(8.85 \times 10^{-12})(3.5)}{\ln(40/16)} = 212.4 \times 10^{-12} \text{ F/m}$$

$$= 212.4 \times 10^{-12} \times 10^3 \text{ F/km}$$

$$= \mathbf{212\ nF/km} \text{ or } \mathbf{0.212\ \mu F/km}$$

(b) From equation (40.8), dielectric stress at radius r,

$$E = \frac{V}{r \ln(b/a)} = \frac{40 \times 10^3}{(30 \times 10^{-3}) \ln(40/16)}$$

$$= \mathbf{1.46 \times 10^6\ V/m} \text{ or } \mathbf{1.46\ MV/m}$$

(c) From equation (40.9), maximum dielectric stress,

$$E_{max} = \frac{V}{a \ln(b/a)} = \frac{40 \times 10^3}{16 \times 10^{-3} \ln(40/16)} = \mathbf{2.73\ MV/m}$$

From equation (40.9′), minimum dielectric stress,

$$E_{min} = \frac{V}{b \ln(b/a)} = \frac{40 \times 10^3}{40 \times 10^{-3} \ln(40/16)} = \mathbf{1.09\ MV/m}$$

Dimensions of most economical cable

It is important to obtain the most economical dimensions when designing a cable. A relationship between a and b may be obtained as follows. If E_{max} and V are both fixed values, then, from equation (40.9),

$$\frac{V}{E_{max}} = a \ln\frac{b}{a}$$

Letting $V/E_{max} = k$, a constant, gives

$$a \ln \frac{b}{a} = k$$

from which $\ln(b/a) = k/a, \quad b/a = e^{k/a}$ and $b = ae^{k/a}$ (40.10)

For the most economical cable, b will be a minimum value. Using the product rule of calculus,

$$\frac{db}{da} = (e^{k/a})(1) + (a)\left(-\frac{k}{a^2}e^{k/a}\right) = 0 \text{ for a minimum value.}$$

(Note, to differentiate $e^{k/a}$ with respect to a, an algebraic substitution may be used, letting $u = 1/a$).

$$e^{k/a} - \frac{k}{a}e^{k/a} = 0$$

Therefore $$e^{k/a}\left(1 - \frac{k}{a}\right) = 0$$

from which $a = k$. Thus

$$\boxed{\boldsymbol{a = \frac{V}{E_{max}}}} \qquad (40.11)$$

From equation (40.10), internal sheath radius, $b = ae^{k/a} = ae^1 = ae$, i.e.,

$$\boxed{\boldsymbol{b = 2.718a}} \qquad (40.12)$$

Problem 6. A single-core concentric cable is to be manufactured for a 60 kV, 50 Hz transmission system. The dielectric used is paper which has a maximum permissible safe dielectric stress of 10 MV/m rms and a relative permittivity of 3.5. Calculate (a) the core and inner sheath radii for the most economical cable, (b) the capacitance per metre length, and (c) the charging current per kilometre run.

(a) From equation (40.11),

$$\text{core radius, } a = \frac{V}{E_m} = \frac{60 \times 10^3 \text{ V}}{10 \times 10^6 \text{ V/m}}$$

$$= 6 \times 10^{-3} \text{ m} = \mathbf{6.0 \text{ mm}}$$

From equation (40.12), internal sheath radius, $b = ae = 6.0e = \mathbf{16.3 \text{ mm}}$

(b) From equation (40.7),

$$\text{capacitance } C = \frac{2\pi\varepsilon_0\varepsilon_r}{\ln(b/a)} \text{ F/m}$$

Since $b = ae$,

$$C = \frac{2\pi\varepsilon_0\varepsilon_r}{\ln e} = 2\pi\varepsilon_0\varepsilon_r = 2\pi(8.85 \times 10^{-12})(3.5)$$

$$= \mathbf{195 \times 10^{-12}\ F/m} \text{ or } \mathbf{195\ pF/m}$$

(c) $$\text{Charging current} = \frac{V}{X_C} = \frac{V}{1/(\omega C)} = \omega CV$$

$$= (2\pi 50)(195 \times 10^{-12})(60 \times 10^3)$$

$$= 3.68 \times 10^{-3} \text{ A/m}$$

Hence the charging current per kilometre = **3.68 A**

Problem 7. A concentric cable has a core diameter of 25 mm and an inside sheath diameter of 80 mm. The relative permittivity of the dielectric is 2.5, the loss angle is 3.5×10^{-3} rad and the working voltage is 132 kV at 50 Hz frequency. Determine for a 1 km length of the cable (a) the capacitance, (b) the charging current and (c) the power loss.

(a) From equation (40.7),

$$\text{capacitance, } C = \frac{2\pi\varepsilon_0\varepsilon_r}{\ln(b/a)} \text{ F/m}$$

$$= \frac{2\pi(8.85 \times 10^{-12})(2.5)}{\ln(40/12.5)} \times 10^3 \text{ F/km}$$

$$= 0.120\ \mu\text{F/km}$$

Thus the capacitance for a 1 km length of the cable is **0.120 μF**

(b) $$\text{Charging current } I = \frac{V}{X_C} = \frac{V}{1/(\omega C)} = \omega CV$$

$$= (2\pi 50)(0.120 \times 10^{-6})(132 \times 10^3)$$

$$= \mathbf{4.98\ A/km}$$

(c) From equation (39.5), Chapter 39,

$$\text{power loss} = V^2\omega C \tan\delta$$

$$= (132 \times 10^3)^2(2\pi 50)(0.120 \times 10^{-6})\tan(3.5 \times 10^{-3})$$

$$= \mathbf{2300\ W}$$

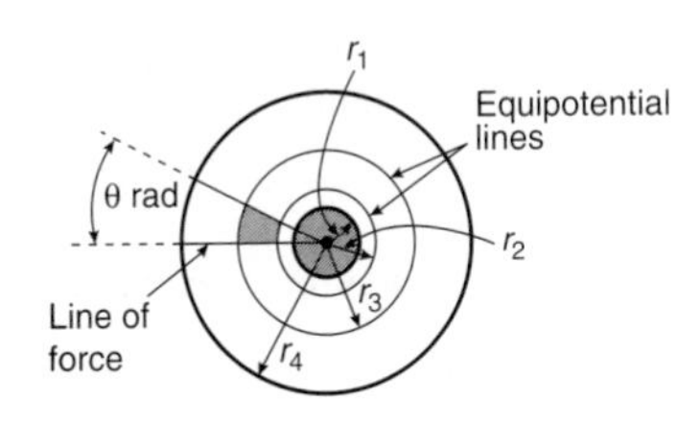

Figure 40.12

Concentric cable field plotting

Figure 40.12 shows a cross-section of a concentric cable having a core radius r_1 and a sheath radius r_4. It was shown in Section 40.1 that the capacitance of a true square is given by $C = \varepsilon_0\varepsilon_r$ farads/metre.

A curvilinear square is shown shaded in Figure 40.12. Such squares can be made to have the same capacitance as a true square by the correct choice of spacing between the lines of force and the equipotential surfaces in the field plot.

From equation (40.7), the capacitance between cylindrical equipotential lines at radii r_a and r_b is given by

$$C = \frac{2\pi\varepsilon_0\varepsilon_r}{\ln(r_b/r_a)} \text{ farads/metre}$$

Thus for a sector of θ radians (see Figure 40.12) the capacitance is given by

$$C = \frac{\theta}{2\pi}\left(\frac{2\pi\varepsilon_0\varepsilon_r}{\ln(r_b/r_a)}\right) = \frac{\theta\varepsilon_0\varepsilon_r}{\ln(r_b/r_a)} \text{ farads/metre}$$

Now if $\theta = \ln(r_b/r_a)$ then $C = \varepsilon_0\varepsilon_r$ F/m, the same as for a true square. If $\theta = \ln(r_b/r_a)$, then $e^\theta = (r_b/r_a)$. Thus if, say, two equipotential surfaces are chosen within the dielectric as shown in Figure 40.12, then $e^\theta = r_2/r_1$, $e^\theta = r_3/r_2$ and $e^\theta = r_4/r_3$. Hence

$$(e^\theta)^3 = \frac{r_2}{r_1} \times \frac{r_3}{r_2} \times \frac{r_4}{r_3}, \quad \text{i.e.,} \quad \boldsymbol{e^{3\theta} = \frac{r_4}{r_1}} \qquad (40.13)$$

It follows that $e^{2\theta} = r_3/r_1$.

Equation (40.13) is used to determine the value of θ and hence the number of sectors. Thus, for a concentric cable having a core radius 8 mm and inner sheath radius 32 mm, if two equipotential surfaces within the dielectric are chosen (and therefore form three capacitors in series in each sector).

$$e^{3\theta} = \frac{r_4}{r_1} = \frac{32}{8} = 4$$

Hence $3\theta = \ln 4$ and $\theta = \frac{1}{3}\ln 4 = 0.462$ rad (or 26.47°). Thus there will be $2\pi/0.462 = 13.6$ sectors in the field plot. (Alternatively, $360°/26.47° = 13.6$ sectors.) From above,

$$e^{2\theta} = r_3/r_1, \text{ i.e., } r_3 = r_1e^{2\theta} = 8e^{2(0.462)} = 20.15 \text{ mm}$$

$$e^\theta = \frac{r_2}{r_1}$$

from which

$$r_2 = r_1e^\theta = 8e^{0.462} = 12.70 \text{ mm}$$

The field plot is shown in Figure 40.13. The number of parallel squares measured along each equipotential is 13.6 and the number of series squares measured along each line of force is 3. Hence in equation (40.4), where $C = \varepsilon_0 \varepsilon_r x(m/n)$, $m = 13.6$ and $n = 3$.

If the dielectric has a relative permittivity of, say, 2.5, then the capacitance per metre length,

$$C = (8.85 \times 10^{-12})(2.5)(1)\frac{13.6}{3} = \mathbf{100\ pF}$$

(From equation (40.7),

$$C = \frac{2\pi\varepsilon_0\varepsilon_r}{\ln(r_4/r_1)}\ \text{F/m} = \frac{2\pi(8.85 \times 10^{-12})(2.5)}{\ln(32/8)} = \mathbf{100\ F/m})$$

Thus field plotting using curvilinear squares provides an alternative method of determining the capacitance between concentric cylinders.

> Problem 8. A concentric cable has a core diameter of 20 mm and a sheath inside diameter of 60 mm. The permittivity of the dielectric is 3.2. Using three equipotential surfaces within the dielectric, determine the capacitance of the cable per metre length by the method of curvilinear squares. Draw the field plot for the cable.

The field plot consists of radial lines of force dividing the cable cross-section into a number of sectors, the lines of force cutting the equipotential surfaces at right angles. Since three equipotential surfaces are required in the dielectric, four capacitors in series are found in each sector of θ radians.

In Figure 40.14, $r_1 = 20/2 = 10$ mm and $r_5 = 60/2 = 30$ mm. It follows from equation (40.13) that $e^{4\theta} = r_5/r_1 = 30/10 = 3$, from which $4\theta = \ln 3$ and $\theta = \frac{1}{4}\ln 3 = 0.2747$ rad.

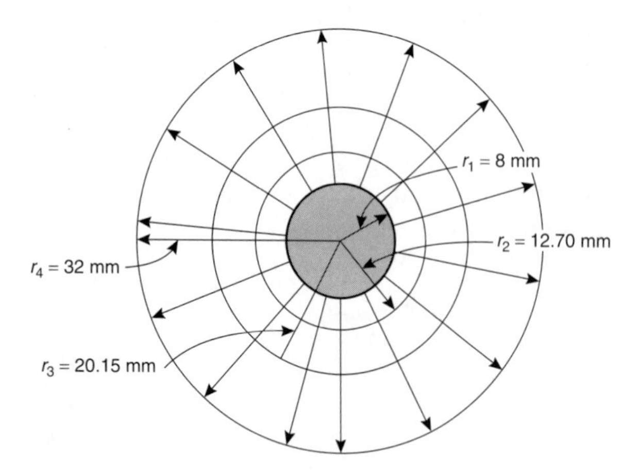

Figure 40.13

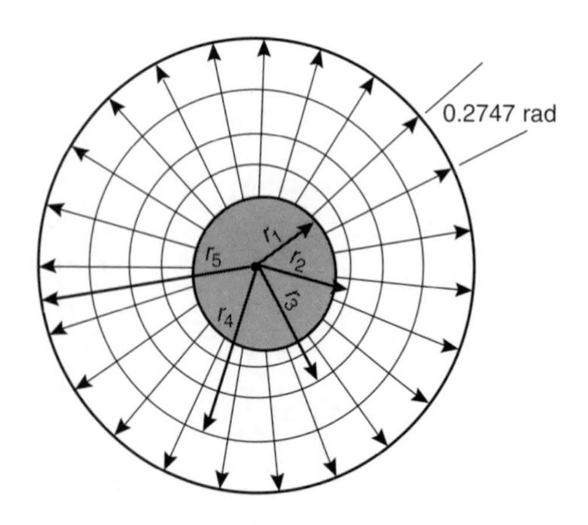

Figure 40.14

Thus the number of sectors in the plot shown in Figure 40.14 is $2\pi/0.2747 = \mathbf{22.9}$.

The three equipotential lines are shown in Figure 40.14 at radii of r_2, r_3 and r_4.

From equation (40.13),

$$e^{3\theta} = \frac{r_4}{r_1}, \text{ from which } r_4 = r_1 e^{3\theta} = 10e^{3(0.2747)} = 22.80 \text{ mm}$$

$$e^{2\theta} = \frac{r_3}{r_1}, \text{ from which } r_3 = r_1 e^{2\theta} = 10e^{2(0.2747)} = 17.32 \text{ mm}$$

$$e^{\theta} = \frac{r_2}{r_1}, \text{ from which } r_2 = r_1 e^{\theta} = 10e^{0.2747} = 13.16 \text{ mm}$$

Thus the field plot for the cable is as shown in Figure 40.14.

From equation (40.4), capacitance $C = \varepsilon_0 \varepsilon_r x(m/n)$. The number of parallel squares along each equipotential, $m = 22.9$ and the number of series squares measured along each line of force, $n = 4$. Thus

$$\text{capacitance } C = (8.85 \times 10^{-12})(3.2)(1)\frac{22.9}{4} = \mathbf{162\ pF}$$

(Checking, from equation (40.7),

$$\text{capacitance } C = \frac{2\pi\varepsilon_0\varepsilon_r}{\ln(r_5/r_1)} = \frac{2\pi(8.85 \times 10^{-12})(3.2)}{\ln(30/10)} = \mathbf{162\ pF})$$

Further problems on the capacitance between concentric cylinders may be found in Section 40.9, problems 4 to 10, page 753.

40.3 Capacitance of an isolated twin line

The field distribution with two oppositely charged, long conductors, A and B, each of radius a is shown in Figure 40.15. The distance D between the centres of the two conductors is such that D is much greater than a. Figure 40.16 shows the field of each conductor separately.

Initially, let conductor A carry a charge of $+Q$ coulombs per metre while conductor B is uncharged. Consider a cylindrical element of radius r about conductor A having a depth of 1 m and a thickness δr as shown in Figure 40.16.

The electric flux density D at the element (i.e. at radius r) is given by

$$D = \frac{\text{charge}}{\text{area}} = \frac{Q}{(2\pi \times 1)} \text{ coulomb/metre}^2$$

The electric field strength at the element,

$$E = \frac{D}{\varepsilon_0\varepsilon_r} = \frac{Q/2\pi r}{\varepsilon_0\varepsilon_r} = \frac{Q}{2\pi r\varepsilon_0\varepsilon_r} \text{ volts/metre}$$

Since $E = V/d$, potential difference, $V = Ed$. Thus

$$\text{p.d. at the element} = E\delta r = \frac{Q\delta r}{2\pi r\varepsilon_0\varepsilon_r} \text{ volts}$$

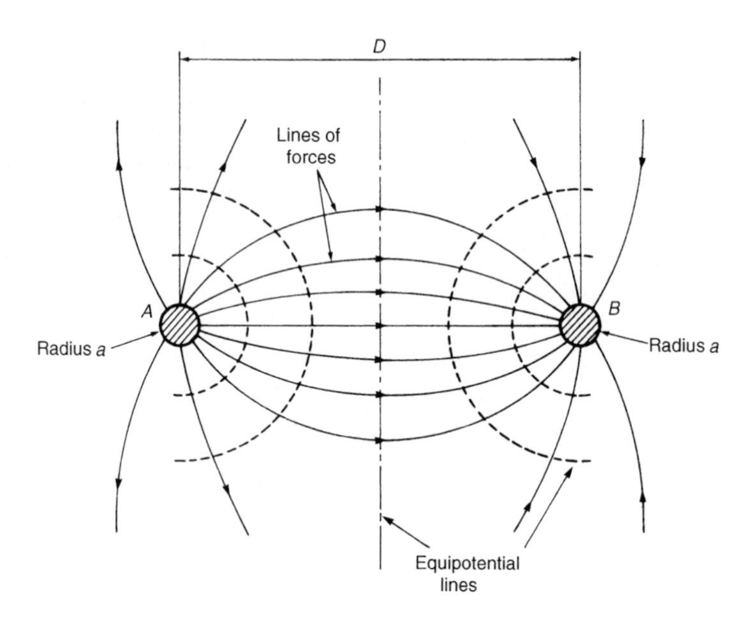

Figure 40.15

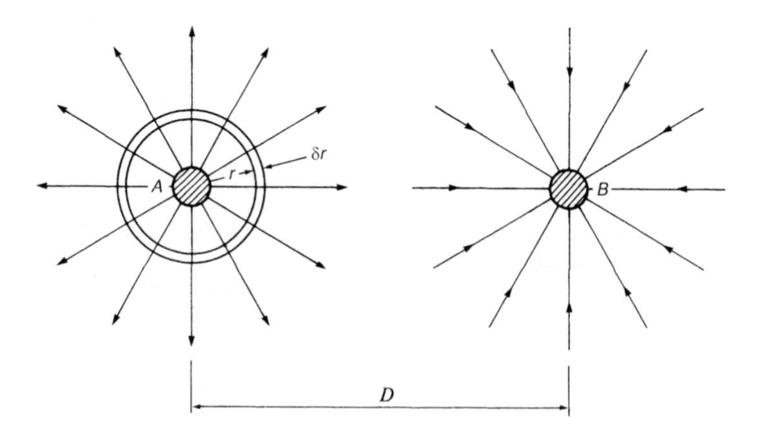

Figure 40.16

The potential may be considered as zero at a large distance from the conductor. Let this be at radius R. Then the potential of conductor A above zero, V_{A_1}, is given by

$$V_{A_1} = \int_a^R \frac{Q\,dr}{2\pi r\varepsilon_0\varepsilon_r} = \frac{Q}{2\pi\varepsilon_0\varepsilon_r}\int_a^R \frac{1}{r}dr = \frac{Q}{2\pi\varepsilon_0\varepsilon_r}[\ln r]_a^R$$

$$= \frac{Q}{2\pi\varepsilon_0\varepsilon_r}[\ln R - \ln a]$$

i.e., $$V_{A_1} = \frac{Q}{2\pi\varepsilon_0\varepsilon_r} \ln \frac{R}{a}$$

Since conductor B lies in the field of conductor A, by reasoning similar to that above, the potential at conductor B above zero, V_{B_1}, is given by

$$V_{B_1} = \int_D^R \frac{Q\,dr}{2\pi r\varepsilon_0\varepsilon_r} = \frac{Q}{2\pi\varepsilon_0\varepsilon_r}[\ln r]_D^R = \frac{Q}{2\pi\varepsilon_0\varepsilon_r} \ln \frac{R}{D}$$

Repeating the above procedure, this time assuming that conductor B carries a charge of $-Q$ coulombs per metre, while conductor A is uncharged, gives

potential of conductor B below zero, $V_{B_2} = \dfrac{-Q}{2\pi\varepsilon_0\varepsilon_r} \ln \dfrac{R}{a}$

and the potential of conductor A below zero, due to the charge on conductor B,

$$V_{A_2} = \frac{-Q}{2\pi\varepsilon_0\varepsilon_r} \ln \frac{R}{D}$$

When both conductors carry equal and opposite charges, the total potential of A above zero is given by

$$\begin{aligned} V_{A_1} + V_{A_2} &= \left(\frac{Q}{2\pi\varepsilon_0\varepsilon_r} \ln \frac{R}{a}\right) + \left(\frac{-Q}{2\pi\varepsilon_0\varepsilon_r} \ln \frac{R}{D}\right) \\ &= \frac{Q}{2\pi\varepsilon_0\varepsilon_r}\left(\ln \frac{R}{a} - \ln \frac{R}{D}\right) \\ &= \frac{Q}{2\pi\varepsilon_0\varepsilon_r}\left(\ln \frac{R/a}{R/D}\right) = \frac{Q}{2\pi\varepsilon_0\varepsilon_r} \ln \frac{D}{a} \end{aligned}$$

and the total potential of B below zero is given by

$$\begin{aligned} V_{B_1} + V_{B_2} &= \frac{Q}{2\pi\varepsilon_0\varepsilon_r}\left(\ln \frac{R}{D} - \ln \frac{R}{a}\right) \\ &= \frac{Q}{2\pi\varepsilon_0\varepsilon_r} \ln \frac{a}{D} = \frac{-Q}{2\pi\varepsilon_0\varepsilon_r} \ln \frac{D}{a} \end{aligned}$$

Hence the p.d. between A and B is

$$2\left(\frac{Q}{2\pi\varepsilon_0\varepsilon_r} \ln \frac{D}{a}\right) \text{ volts/metre}$$

The capacitance between A and B per metre length,

$$C = \frac{\text{charge per metre}}{\text{p.d.}} = \frac{Q}{2(Q/(2\pi\varepsilon_0\varepsilon_r))\ln(D/a)}$$

i.e., $\mathbf{C = \dfrac{1}{2}\dfrac{2\pi\varepsilon_0\varepsilon_r}{\ln(D/a)}}$ **farads/metre**

or $$\boxed{C = \frac{\pi\varepsilon_0\varepsilon_r}{\ln(D/a)} \textbf{ farads/metre}} \qquad (40.14)$$

> Problem 9. Two parallel wires, each of diameter 5 mm, are uniformly spaced in air at a distance of 50 mm between centres. Determine the capacitance of the line if the total length is 200 m.

From equation (40.14). capacitance per metre length,

$$C = \frac{\pi\varepsilon_0\varepsilon_r}{\ln(D/a)} = \frac{\pi(8.85\times10^{-12})(1)}{\ln(50/(5/2))} \text{ since } \varepsilon_r = 1 \text{ for air,}$$

$$= \frac{\pi(8.85\times10^{-12})}{\ln 20} = 9.28\times10^{-12}\text{ F}$$

Hence the capacitance of a 200 m length is $(9.28\times10^{-12}\times200)$ F $=$ **1860 pF** or **1.86 nF**

> Problem 10. A single-phase circuit is composed of two parallel conductors, each of radius 4 mm, spaced 1.2 m apart in air. The p.d. between the conductors at a frequency of 50 Hz is 15 kV. Determine, for a 1 km length of line, (a) the capacitance of the conductors, (b) the value of charge carried by each conductor, and (c) the charging current.

(a) From equation (40.14),

$$\text{capacitance } C = \frac{\pi\varepsilon_0\varepsilon_r}{\ln(D/a)} = \frac{\pi(8.85\times10^{-12})(1)}{\ln(1.2/4\times10^{-3})}$$

$$= \frac{\pi(8.85\times10^{-12})}{\ln 300}$$

$$= 4.875\text{ pF/m}$$

Hence the capacitance per kilometre length is $(4.875\times10^{-12})(10^3)$ F $=$ **4.875 nF**

(b) Charge $Q = CV = (4.875\times10^{-9})(15\times10^3) =$ **73.1 μC**

(c) Charging current $= \dfrac{V}{X_C} = \dfrac{V}{(1/\omega C)} = \omega CV$

$$= (2\pi50)(4.875\times10^{-9})(15\times10^3)$$

$=$ **0.023 A** or **23 mA**

Problem 11. The charging current for an 800 m run of isolated twin line is not to exceed 15 mA. The voltage between the lines is 10 kV at 50 Hz. If the line is air-insulated, determine (a) the maximum value required for the capacitance per metre length, and (b) the maximum diameter of each conductor if their distance between centres is 1.25 m.

(a) Charging current $I = \frac{V}{X_C} = \frac{V}{(1/\omega C)} = \omega CV$

from which,

$$\text{capacitance } C = \frac{I}{\omega V} = \frac{15 \times 10^{-3}}{(2\pi 50)(10 \times 10^3)} \text{ farads per 800 metre run}$$

$$= 4.775 \text{ nF}$$

Hence the required maximum value of capacitance

$$= \frac{4.775 \times 10^{-9}}{800} \text{ F/m} = \mathbf{5.97\ pF/m}$$

(b) From equation (40.14)

$$C = \frac{\pi \varepsilon_0 \varepsilon_r}{\ln(D/a)},$$

$$\text{thus} \quad 5.97 \times 10^{-12} = \frac{\pi(8.85 \times 10^{-12})(1)}{\ln(1.25/a)}$$

$$\text{from which,} \quad \ln\left(\frac{1.25}{a}\right) = \frac{\pi 8.85}{5.97} = 4.657$$

$$\text{Hence} \quad \frac{1.25}{a} = e^{4.657} = 105.3$$

$$\text{and} \quad \text{radius } a = \frac{1.25}{105.3} \text{ m} = 0.01187 \text{ m or } 11.87 \text{ mm}$$

Thus **the maximum diameter of each conductor is** 2×11.87, i.e., **23.7 mm**

Further problems on capacitance of an isolated twin line may be found in Section 40.9, problems 11 to 15, page 754.

40.4 Energy stored in an electric field

Consider the p.d. across a parallel-plate capacitor of capacitance C farads being increased by dv volts in dt seconds. If the corresponding increase in charge is dq coulombs, then $dq = C\,dv$. If the charging current at that instant is i amperes, then $dq = i\,dt$. Thus $i\,dt = C\,dv$, i.e.,

$$i = C\frac{dv}{dt}$$

(i.e., instantaneous current = capacitance × rate of change of p.d.)

The instantaneous value of power to the capacitor,

$$p = vi \text{ watts} = v\left(C\frac{dv}{dt}\right) \text{ watts}$$

The energy supplied to the capacitor during time dt

$$= \text{power} \times \text{time} = \left(vC\frac{dv}{dt}\right)(dt)$$

$$= Cv\,dv \text{ joules}$$

Thus the total energy supplied to the capacitor when the p.d. is increased from 0 to V volts is given by

$$W_f = \int_0^V Cv\,dv = C\left[\frac{v^2}{2}\right]_0^V$$

i.e., **energy stored in the electric field,**

$$\boxed{\mathbf{W_f = \frac{1}{2}\,CV^2 \text{ joules}}} \qquad (40.15)$$

Consider a capacitor with dielectric of relative permittivity ε_r, thickness d metres and area A square metres. Capacitance $C = Q/V$, hence energy stored $= \frac{1}{2}(Q/V)V^2 = \frac{1}{2}QV$ joules.

The electric flux density, $D = Q/A$, from which $Q = DA$.

Hence the energy stored $= \frac{1}{2}(DA)V$ joules.

The electric field strength, $E = V/d$, from which $V = Ed$.

Hence the energy stored $= \frac{1}{2}(DA)(Ed)$ joules. However Ad is the volume of the field.

Hence **energy stored per unit volume,**

$$\boxed{\boldsymbol{\omega_f = \frac{1}{2}\,DE \text{ joules/cubic metre}}} \qquad (40.16)$$

Since $D/E = \varepsilon_0\varepsilon_r$, then $D = \varepsilon_0\varepsilon_r E$. Hence, from equation (40.16), the energy stored per unit volume,

$$\omega_f = \frac{1}{2}(\varepsilon_0\varepsilon_r E)E$$

i.e.,
$$\boxed{\boldsymbol{\omega_f = \frac{1}{2}\varepsilon_0\varepsilon_r E^2 \text{ joules/cubic metre}}} \qquad (40.17)$$

Also, since $D/E = \varepsilon_0\varepsilon_r$, then $E = D/(\varepsilon_0\varepsilon_r)$. Hence from equation (40.16), the energy stored per unit volume,

$$\omega_f = \frac{1}{2}D\left(\frac{D}{\varepsilon_0\varepsilon_r}\right)$$

i.e., $$\boldsymbol{\omega_f = \frac{D^2}{2\varepsilon_0\varepsilon_r}} \textbf{ joules/cubic metre} \qquad (40.18)$$

Summarizing,

energy stored in a capacitor $= \frac{1}{2}\boldsymbol{CV^2}$ **joules**

and energy stored per unit volume of dielectric

$$= \tfrac{1}{2}\boldsymbol{DE} = \tfrac{1}{2}\boldsymbol{\varepsilon_0\varepsilon_r E^2}$$
$$= \boldsymbol{\frac{D^2}{2\varepsilon_0\varepsilon_r}} \textbf{ joules/cubic metre}$$

Problem 12. Determine the energy stored in a 10 nF capacitor when charged to 1 kV, and the average power developed if this energy is dissipated in 10 μs.

From equation (40.15),

$$\text{energy stored, } W_f = \tfrac{1}{2}CV^2 = \tfrac{1}{2}(10 \times 10^{-9})(10^3)^2$$
$$= \mathbf{5\ mJ}$$

$$\text{average power developed} = \frac{\text{energy dissipated, } W}{\text{time, } t}$$
$$= \frac{5 \times 10^{-3}\text{ J}}{10 \times 10^{-6}\text{ s}} = \mathbf{500\ W}$$

Problem 13. A capacitor is charged with 5 mC. If the energy stored is 625 mJ, determine (a) the voltage across the plates and (b) the capacitance of the capacitor.

(a) From equation (40.15),

$$\text{energy stored, } W_f = \frac{1}{2}CV^2 = \frac{1}{2}\left(\frac{Q}{V}\right)V^2 = \frac{1}{2}QV$$

from which voltage across the plates,

$$V = \frac{2 \times \text{energy stored}}{Q} = \frac{2 \times 0.625}{5 \times 10^{-3}} = \mathbf{250\ V}$$

(b) Capacitance $C = \dfrac{Q}{V} = \dfrac{5 \times 10^{-3}}{250}$ F $= 20 \times 10^{-6}$ F $= \mathbf{20\ \mu F}$

Problem 14. A ceramic capacitor is to be constructed to have a capacitance of 0.01 μF and to have a steady working potential of 2.5 kV maximum. Allowing a safe value of field stress of 10 MV/m, determine (a) the required thickness of the ceramic dielectric, (b) the area of plate required if the relative permittivity of the ceramic is 10, and (c) the maximum energy stored by the capacitor.

(a) Field stress $E = V/d$, from which thickness of ceramic dielectric,

$$d = \frac{V}{E} = \frac{2.5 \times 10^3}{10 \times 10^6} = 2.5 \times 10^{-4}\ \text{m} = \mathbf{0.25\ mm}$$

(b) Capacitance $C = \varepsilon_0 \varepsilon_r A/d$ for a two-plate parallel capacitor. Hence cross-sectional area of plate,

$$A = \frac{Cd}{\varepsilon_0 \varepsilon_r} = \frac{(0.01 \times 10^{-6})(0.25 \times 10^{-3})}{(8.85 \times 10^{-12})(10)}$$

$$= \mathbf{0.0282\ m^2} \text{ or } \mathbf{282\ cm^2}$$

(c) Maximum energy stored,

$$W_f = \tfrac{1}{2}CV^2 = \tfrac{1}{2}(0.01 \times 10^{-6})(2.5 \times 10^3)^2$$

$$= \mathbf{0.0313\ J} \text{ or } \mathbf{31.3\ mJ}$$

Problem 15. A 400 pF capacitor is charged to a p.d. of 100 V. The dielectric has a cross-sectional area of 200 cm^2 and a relative permittivity of 2.3. Calculate the energy stored per cubic metre of the dielectric.

From equation (40.18), energy stored per unit volume of dielectric,

$$\omega_f = \frac{D^2}{2\varepsilon_0 \varepsilon_r}$$

Electric flux density

$$D = \frac{Q}{A} = \frac{CV}{A} = \frac{(400 \times 10^{-12})(100)}{200 \times 10^{-4}} = 2 \times 10^{-6}\ \text{C/m}^2$$

Hence energy stored,

$$\omega_f = \frac{D^2}{2\varepsilon_0 \varepsilon_r} = \frac{(2 \times 10^{-6})^2}{2(8.85 \times 10^{-12})(2.3)}$$

$$= \mathbf{0.0983\ J/m^3} \text{ or } \mathbf{98.3\ mJ/m^3}$$

Further problems on energy stored in electric fields may be found in Section 40.9, problems 16 to 23, page 755.

40.5 Induced e.m.f. and inductance

A current flowing in a coil of wire is accompanied by a magnetic flux linking with the coil. If the current changes, the flux linkage (i.e., the product of flux and the number of turns) changes and an e.m.f. is induced in the coil. The magnitude of the induced e.m.f. e in a coil of N turns is given by

$$\boldsymbol{e = N\frac{d\phi}{dt}} \textbf{ volts}$$

where $d\phi/dt$ is the rate of change of flux.

Inductance is the name given to the property of a circuit whereby there is an e.m.f. induced into the circuit by the change of flux linkages produced by a current change. The unit of inductance is the **henry, H**. A circuit has an inductance of 1 H when an e.m.f. of 1 V is induced in it by a current changing uniformly at the rate of 1 A/s.
The magnitude of the e.m.f. induced in a coil of inductance L henry is given by

$$\boldsymbol{e = L\frac{di}{dt}} \textbf{ volts}$$

where di/dt is the rate of change of current.

If a current changing uniformly from zero to I amperes produces a uniform flux change from zero to ϕ webers in t seconds then (from above) average induced e.m.f., $E_{av} = N\phi/t = LI/t$, from which

$$\text{inductance of coil, } \boldsymbol{L = \frac{N\phi}{I}} \textbf{ henry}$$

Flux linkage means the product of flux, in webers, and the number of turns with which the flux is linked. Hence flux linkage $= N\phi$. Thus since $L = N\phi/I$, **inductance = flux linkages per ampere**.

40.6 Inductance of a concentric cylinder (or coaxial cable)

Skin effect

When a direct current flows in a uniform conductor the current will tend to distribute itself uniformly over the cross-section of the conductor. However, with alternating current, particularly if the frequency is high, the current carried by the conductor is not uniformly distributed over the available cross-section, but tends to be concentrated at the conductor surface. This is called **skin effect**. When current is flowing through a conductor, the magnetic flux that results is in the form of concentric circles. Some of this flux exists within the conductor and links with the current more strongly near the centre. The result is that the inductance of the central part of the conductor is greater than the inductance of the conductor near the surface. This is because of the greater number of flux linkages existing in the central region. At high frequencies the reactance $(X_L = 2\pi f L)$ of the extra inductance is sufficiently large to seriously affect the flow of current, most of which flows along the surface of the conductor where the impedance is low rather than near the centre where the impedance is high.

Inductance due to internal linkages at low frequency

When a conductor is used at high frequency the depth of penetration of the current is small compared with the conductor cross-section. Thus the internal linkages may be considered as negligible and the circuit inductance is that due to the fields in the surrounding space. However, at very low frequency the current distribution is considered uniform over the conductor cross-section and the inductance due to flux linkages has its maximum value.

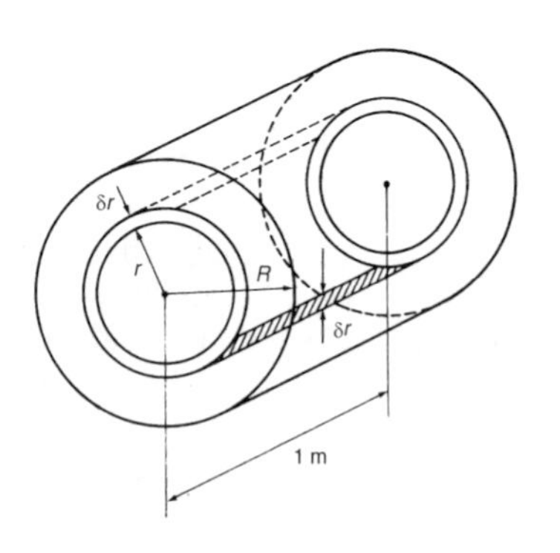

Figure 40.17

Consider a conductor of radius R, as shown in Figure 40.17, carrying a current I amperes uniformly distributed over the cross-section. At all points on the conductor cross-section

$$\text{current density, } J = \frac{\text{current}}{\text{area}} = \left(\frac{I}{\pi R^2}\right) \text{ amperes/metre}^2$$

Consider a thin elemental ring at radius r and width δr contained within the conductor, as shown in Figure 40.17. The current enclosed by the ring,

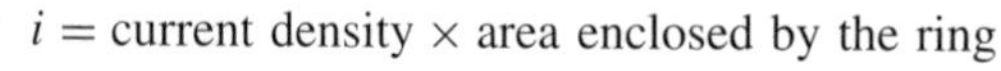

$$i = \text{current density} \times \text{area enclosed by the ring}$$

$$= \left(\frac{I}{\pi R^2}\right)(\pi r^2)$$

i.e. $$i = \frac{Ir^2}{R^2} \text{ amperes}$$

Magnetic field strength, $H = Ni/l$ amperes/metre.

At radius r, the mean length of the flux path, $l = 2\pi r$ (i.e., the circumference of the elemental ring) and $N = 1$ turn.

Hence at radius r,

$$H_r = \frac{Ni}{l} = \frac{(1)(Ir^2/R^2)}{2\pi r} = \frac{Ir}{2\pi R^2} \text{ ampere/metre}$$

and the flux density, $B_r = \mu_0\mu_r H_r = \mu_0\mu_r\left(\dfrac{Ir}{2\pi R^2}\right)$ tesla

Flux $\phi = BA$ webers. For a 1 m length of the conductor, the cross-sectional area A of the element is $(\delta r \times 1)$ m^2 (see Figure 40.17). Thus the flux within the element of thickness δr,

$$\phi = \left(\frac{\mu_0\mu_r Ir}{2\pi R^2}\right)(\delta r) \text{ webers}$$

The flux in the element links the portion $\pi r^2/\pi R^2$, i.e., r^2/R^2 of the total conductor. Hence linkages due to the flux within radius r

$$= \left(\frac{\mu_0\mu_r I r}{2\pi R^2}\delta r\right)\frac{r^2}{R^2} = \frac{\mu_0\mu_r I r^3}{2\pi R^4}\delta r \text{ weber turns}$$

Total linkages per metre due to the flux in the conductor

$$= \int_0^R \frac{\mu_0\mu_r I r^3}{2\pi R^4} dr = \frac{\mu_0\mu_r I}{2\pi R^4}\int_0^R r^3 dr$$

$$= \frac{\mu_0\mu_r I}{2\pi R^4}\left[\frac{r^4}{4}\right]_0^R = \frac{\mu_0\mu_r I}{2\pi R^4}\left[\frac{R^4}{4}\right]$$

$$= \frac{1}{4}\left(\frac{\mu_0\mu_r I}{2\pi}\right) \text{ weber turns}$$

Inductance per metre due to the internal flux = internal flux linkages per ampere

$$= \mathbf{\frac{1}{4}\left(\frac{\mu_0\mu_r}{2\pi}\right) \text{ or } \frac{\mu}{8\pi} \text{ henry/metre}}$$

It is seen that the inductance is independent of the conductor radius R.

Inductance of a pair of concentric cylinders

The cross-section of a concentric (or coaxial) cable is shown in Figure 40.18. Let a current of I amperes flow in one direction in the core and a current of I amperes flow in the opposite direction in the outer sheath conductor.

Consider an element of width δr at radius r, and let the radii of the inner and outer conductor be a and b respectively as shown. The magnetic field strength at radius r,

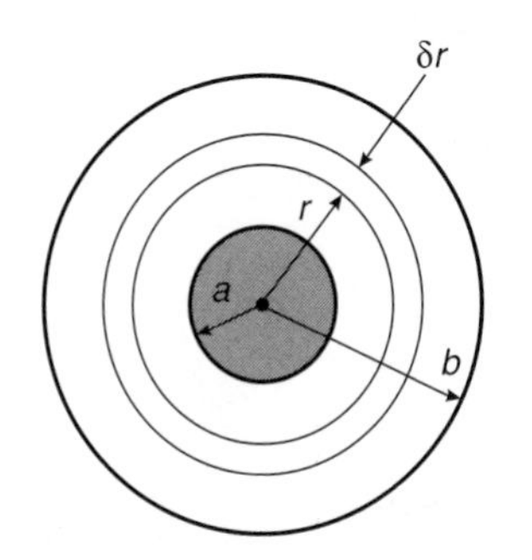

Figure 40.18 *Cross-section of a concentric cable*

$$H_r = \frac{Ni}{l} = \frac{(1)(I)}{2\pi r} = \frac{I}{2\pi r}$$

The flux density at radius r, $B_r = \mu_0\mu_r H_r = \dfrac{\mu_0\mu_r I}{2\pi r}$

For a 1 m length of the cable, the flux ϕ within the element of width δr is given by

$$\Phi = B_r A = \left(\frac{\mu_0\mu_r I}{2\pi r}\right)(\delta r \times 1) = \frac{\mu_0\mu_r I}{2\pi r}dr \text{ webers}$$

This flux links the loop of the cable formed by the core and the outer sheath. Thus the flux linkage per metre length of the cable is $(\mu_0\mu_r I/(2\pi r))\delta r$ weber turns, and

$$\text{total flux linkages per metre} = \int_a^b \frac{\mu_0 \mu_r I}{2\pi r} dr = \frac{\mu_0 \mu_r I}{2\pi} \int_a^b \frac{1}{r} dr$$

$$= \frac{\mu_0 \mu_r I}{2\pi} [\ln r]_a^b = \frac{\mu_0 \mu_r I}{2\pi} [\ln b - \ln a]$$

$$= \frac{\mu_0 \mu_r I}{2\pi} \ln \frac{b}{a} \text{ weber turns}$$

Thus inductance per metre length = flux linkages per ampere

$$= \boldsymbol{\frac{\mu_0 \mu_r}{2\pi} \ln \frac{b}{a}} \textbf{ henry/metre} \qquad (40.19)$$

At low frequencies the inductance due to the internal linkages is added to this result.

Hence the total inductance per metre at low frequency is given by

$$\boldsymbol{L = \frac{1}{4}\left(\frac{\mu_0 \mu_r}{2\pi}\right) + \frac{\mu_0 \mu_r}{2\pi} \ln \frac{b}{a}} \textbf{ henry/metre} \qquad (40.20)$$

or

$$\boxed{\boldsymbol{L = \frac{\mu}{2\pi}\left(\frac{1}{4} + \ln \frac{b}{a}\right)} \textbf{ henry/metre}} \qquad (40.21)$$

Problem 16. A coaxial cable has an inner core of radius 1.0 mm and an outer sheath of internal radius 4.0 mm. Determine the inductance of the cable per metre length.

Assume that the relative permeability is unity.

From equation (40.21),

$$\text{inductance } L = \frac{\mu}{2\pi}\left(\frac{1}{4} + \ln \frac{b}{a}\right) \text{ H/m}$$

$$= \frac{\mu_0 \mu_r}{2\pi}\left(\frac{1}{4} + \ln \frac{4.0}{1.0}\right) = \frac{(4\pi \times 10^{-7})(1)}{2\pi}(0.25 + \ln 4)$$

$$= \mathbf{3.27 \times 10^{-7}\ H/m} \text{ or } \mathbf{0.327\ \mu H/m}$$

Problem 17. A concentric cable has a core diameter of 10 mm. The inductance of the cable is 4×10^{-7} H/m. Ignoring inductance due to internal linkages, determine the diameter of the sheath. Assume that the relative permeability is 1.

From equation (40.19),

$$\text{inductance per metre length} = \frac{\mu_0 \mu_r}{2\pi} \ln \frac{b}{a}$$

where b = sheath radius and a = core radius. Hence

$$4 \times 10^{-7} = \frac{(4\pi \times 10^{-7})(1)}{2\pi} \ln\left(\frac{b}{5}\right)$$

from which $2 = \ln\left(\frac{b}{5}\right)$ and $e^2 = \frac{b}{5}$

Hence radius $b = 5e^2 = 36.95$ mm

Thus the diameter of the sheath is $2 \times 36.95 =$ **73.9 mm**

Problem 18. A coaxial cable 7.5 km long has a core 10 mm diameter and a sheath 25 mm diameter, the sheath having negligible thickness. Determine for the cable (a) the inductance, assuming nonmagnetic materials, and (b) the capacitance, assuming a dielectric of relative permittivity 3.

(a) From equation (40.21),

$$\text{inductance per metre length} = \frac{\mu}{2\pi}\left(\frac{1}{4} + \ln\frac{b}{a}\right)$$

$$= \frac{\mu_0\mu_r}{2\pi}\left[\frac{1}{4} + \ln\left(\frac{12.5}{5}\right)\right]$$

$$= \frac{(4\pi \times 10^{-7})(1)}{2\pi}(0.25 + \ln 2.5)$$

$$= 2.33 \times 10^{-7} \text{ H/m}$$

Since the cable is 7500 m long,

the inductance $= 7500 \times 2.33 \times 10^{-7} =$ **1.75 mH**

(b) From equation (40.7),

$$\text{capacitance, } C = \frac{2\pi\varepsilon_0\varepsilon_r}{\ln(b/a)} = \frac{2\pi(8.85 \times 10^{-12})(3)}{\ln(12.5/5)}$$

$$= 182.06 \text{ pF/m}$$

Since the cable is 7500 m long,

the capacitance $= 7500 \times 182.06 \times 10^{-12} =$ **1.365 μF**

Further problems on the inductance of concentric cables may be found in Section 40.9, problems 24 to 27, page 756.

40.7 Inductance of an isolated twin line

Consider two isolated, long, parallel, straight conductors A and B, each of radius a metres, spaced D metres apart. Let the current in each be I amperes but flowing in opposite directions. Distance D is assumed to be much greater than radius a. The magnetic field associated with the conductors is as shown in Figure 40.19. There is a force of repulsion between conductors A and B.

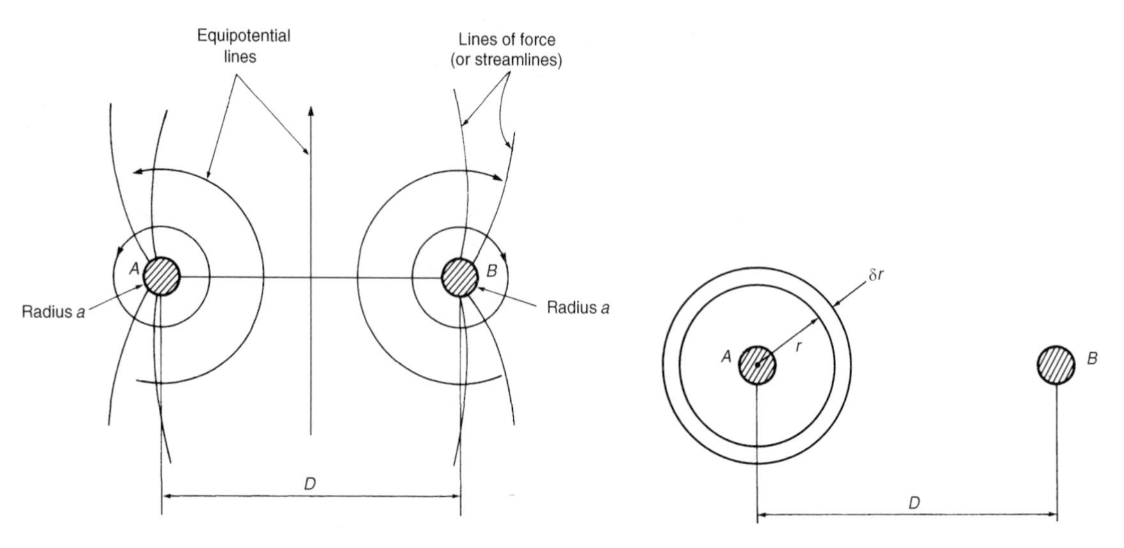

Figure 40.19

Figure 40.20

It is easier to analyse the field by initially considering each conductor alone (as in Section 40.3). At any radius r from conductor A (see Figure 40.20),

$$\text{magnetic field strength, } H_r = \frac{Ni}{l} = \frac{I}{2\pi r} \text{ ampere/metre}$$

$$\text{and flux density, } B_r = \mu_0\mu_r H_r = \frac{\mu_0\mu_r I}{2\pi r} \text{ tesla}$$

The total flux in 1 m of the conductor,

$$\Phi = B_r A = \left(\frac{\mu_0\mu_r I}{2\pi r}\right)(\delta r \times 1) = \frac{\mu_0\mu_r I}{2\pi r}\delta r \text{ webers}$$

Since this flux links conductor A once, the linkages with conductor A due to this flux $= \dfrac{\mu_0\mu_r I}{2\pi r}\delta r$ weber turns.

There is, in fact, no limit to the distance from conductor A at which a magnetic field may be experienced. However, let R be a very large radius at which the magnetic field strength may be regarded as zero. Then the total linkages with conductor A due to current in conductor A is given by

$$\int_a^R \frac{\mu_0\mu_r I}{2\pi r}dr = \frac{\mu_0\mu_r I}{2\pi}\int_a^R \frac{dr}{r} = \frac{\mu_0\mu_r I}{2\pi}[\ln r]_a^R$$

$$= \frac{\mu_0\mu_r I}{2\pi}[\ln R - \ln a] = \frac{\mu_0\mu_r I}{2\pi}\ln\left(\frac{R}{a}\right)$$

Similarly, the total linkages with conductor B due to the current in A

$$= \int_D^R \frac{\mu_0\mu_r I}{2\pi r}dr = \frac{\mu_0\mu_r I}{2\pi}\ln\frac{R}{D}$$

Now consider conductor B alone, carrying a current of $-I$ amperes. By similar reasoning to above, total linkages with

$$\text{conductor B due to the current in B} = \frac{-\mu_0\mu_r I}{2\pi}\ln\left(\frac{R}{a}\right)$$

and total linkages with conductor A due to the current in B

$$= \frac{-\mu_0\mu_r I}{2\pi}\ln\frac{R}{D}$$

Hence total linkages with conductor A

$$= \left(\frac{\mu_0\mu_r I}{2\pi}\ln\frac{R}{a}\right) + \left(\frac{-\mu_0\mu_r I}{2\pi}\ln\frac{R}{D}\right)$$

$$= \frac{\mu_0\mu_r I}{2\pi}\left[\ln\frac{R}{a} - \ln\frac{R}{D}\right]$$

$$= \frac{\mu_0\mu_r I}{2\pi}\left[\ln\frac{R/a}{R/D}\right]$$

$$= \frac{\mu_0\mu_r I}{2\pi}\ln\frac{D}{a}\ \text{weber-turns/metre}$$

Similarly, total linkages with conductor B

$$= -\frac{\mu_0\mu_r I}{2\pi}\ln\frac{D}{a}\ \text{weber-turns metre}$$

For a 1 m length of the two conductors,

$$\text{total inductance} = \text{flux linkages per ampere}$$

$$= 2\left(\frac{\mu_0\mu_r}{2\pi}\ln\frac{D}{a}\right)\ \text{henry/metre}$$

i.e., **total inductance** $= \boldsymbol{\frac{\mu_0\mu_r}{\pi}\ln\frac{D}{a}}$ **henry/metre** (40.22)

Equation (40.22) does not take into consideration the internal linkages of each line.

From Section (40.6), inductance per metre due to internal linkages

$$= \frac{1}{4}\left(\frac{\mu_0\mu_r}{2\pi}\right) \text{ henry/metre}$$

Thus inductance per metre due to internal linkages of two conductors

$$= 2\left(\frac{1}{4}\left(\frac{\mu_0\mu_r}{2\pi}\right)\right) = \frac{\mu_0\mu_r}{4\pi} \text{ henry/metre}$$

Therefore, at low frequency, total inductance per metre of the two conductors

$$= \frac{\mu_0\mu_r}{4\pi} + \frac{\mu_0\mu_r}{\pi}\ln\frac{D}{a}$$

i.e., $$\boldsymbol{L = \frac{\mu_0\mu_r}{\pi}\left(\frac{1}{4} + \ln\frac{D}{a}\right) \textbf{ henry/metre}} \qquad (40.23)$$

(This is often referred to as the **'loop inductance'**).
In most practical lines the relative permeability, $\mu_r = 1$.

Problem 19. A single-phase power line comprises two conductors each with a radius 8.0 mm and spaced 1.2 m apart in air. Determine the inductance of the line per metre length ignoring internal linkages. Assume the relative permeability, $\mu_r = 1$.

From equation (40.22), inductance

$$= \frac{\mu_0\mu_r}{\pi}\ln\frac{D}{a}$$

$$= \frac{(4\pi \times 10^{-7})(1)}{\pi}\ln\left(\frac{1.2}{8.0 \times 10^{-3}}\right) = 4 \times 10^{-7}\ln 150$$

$$= \mathbf{20.0 \times 10^{-7}\ H/m} \text{ or } \mathbf{2.0\ \mu H/m}$$

Problem 20. Determine (a) the loop inductance, and (b) the capacitance of a 1 km length of single-phase twin line having conductors of diameter 10 mm and spaced 800 mm apart in air.

(a) From equation (40.23), total inductance per loop metre

$$= \frac{\mu_0\mu_r}{\pi}\left(\frac{1}{4} + \ln\frac{D}{a}\right)$$

$$= \frac{(4\pi \times 10^{-7})(1)}{\pi}\left(\frac{1}{4} + \ln\frac{800}{10/2}\right)$$

$$= (4 \times 10^{-7})(0.25 + \ln 160)$$

$$= 21.3 \times 10^{-7} \text{ H/m}$$

Hence loop inductance of a 1 km length of line

$$= 21.3 \times 10^{-7} \text{ H/m} \times 10^3 \text{ m}$$

$$= \mathbf{21.3 \times 10^{-4} \text{ H}} \text{ or } \mathbf{2.13 \text{ mH}}$$

(b) From equation (40.14), capacitance per metre length

$$= \frac{\pi \varepsilon_0 \varepsilon_r}{\ln(D/a)}$$

$$= \frac{\pi(8.85 \times 10^{-12})(1)}{\ln(800/5)}$$

$$= 5.478 \times 10^{-12} \text{ F/m}$$

Hence capacitance of a 1 km length of line

$$= 5.478 \times 10^{-12} \text{ F/m} \times 10^3 \text{ m}$$

$$= \mathbf{5.478 \text{ nF}}$$

Problem 21. The total loop inductance of an isolated twin power line is 2.185 μH/m. The diameter of each conductor is 12 mm. Determine the distance between their centres.

From equation (40.23),

$$\text{total loop inductance} = \frac{\mu_0 \mu_r}{\pi}\left(\frac{1}{4} + \ln \frac{D}{a}\right)$$

Hence

$$2.185 \times 10^{-6} = \frac{(4\pi \times 10^{-7})(1)}{\pi}\left(\frac{1}{4} + \ln \frac{D}{6}\right)$$

where D is the distance between centres in millimetres.

$$\frac{2.185 \times 10^{-6}}{4 \times 10^{-7}} = \left(0.25 + \ln \frac{D}{6}\right)$$

$$\ln \frac{D}{6} = 5.4625 - 0.25 = 5.2125$$

$$\frac{D}{6} = e^{5.2125}$$

from which, distance $D = 6e^{5.2125}$ = **1100 mm** or **1.10 m**

Further problems on the inductance of an isolated twin line may be found in Section 40.9, problems 28 to 32, page 756.

40.8 Energy stored in an electromagnetic field

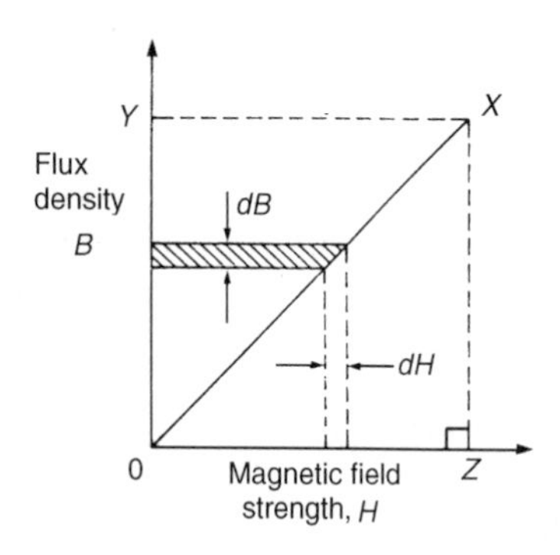

Figure 40.21

Magnetic energy in a nonmagnetic medium

For a nonmagnetic medium the relative permeability, $\mu_r = 1$ and $B = \mu_0 H$.

Thus the magnetic field strength H is proportional to the flux density B and a graph of B against H is a straight line, as shown in Figure 40.21.

It was shown in Section 38.3 that, when the flux density is increased by an amount dB due to an increase dH in the magnetic field strength, then

energy supplied to the magnetic circuit = area of shaded strip (in joules per cubic metre)

Thus, for a maximum flux density OY in Figure 40.21,

$$\text{total energy stored in the magnetic field} = \text{area of triangle } OYX = \tfrac{1}{2} \times \text{base} \times \text{height} = \tfrac{1}{2}(OZ)(OY)$$

If $OY = B$ teslas and $OZ = H$ ampere/metre, then the total energy stored in a non-magnetic medium,

$$\boldsymbol{\omega_f = \tfrac{1}{2} HB \textbf{ joules/metre}^3} \qquad (40.24)$$

Since $B = \mu_0$H for a non-magnetic medium, the energy stored,

$$\omega_f = \tfrac{1}{2} H(\mu_0 H)$$

i.e., $$\boldsymbol{\omega_f = \tfrac{1}{2}\mu_0 H^2 \textbf{ joules/metre}^3} \qquad (40.25)$$

Alternatively, $H = B/\mu_0$, thus the energy stored,

$$\omega_f = \frac{1}{2} HB = \frac{1}{2}\left(\frac{B}{\mu_0}\right) B$$

i.e., $$\boldsymbol{\omega_f = \frac{B^2}{2\mu_0} \textbf{ joules/metre}^3} \qquad (40.26)$$

Magnetic energy stored in an inductor

Establishing a magnetic field requires energy to be expended. However, once the field is established, the only energy expended is that supplied to maintain the flow of current in opposition to the circuit resistance, i.e., the I^2R loss, which is dissipated as heat.

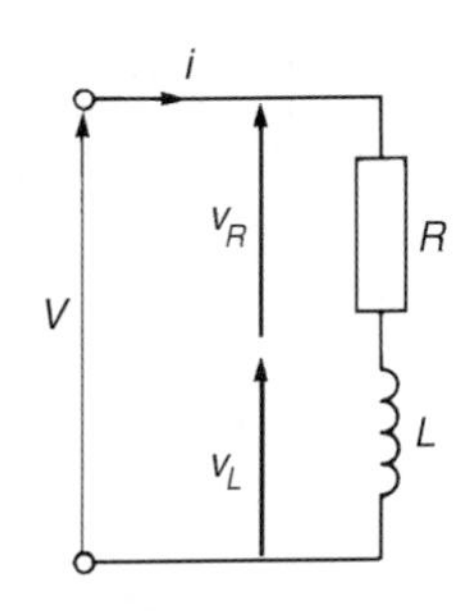

Figure 40.22

For an inductive circuit containing resistance R and inductance L (see Figure 40.22) the applied voltage V at any instant is given by $V = v_R + v_L$

i.e., $V = iR + L\dfrac{di}{dt}$

Multiplying throughout by current i gives the power equation:

$$Vi = i^2R + Li\frac{di}{dt}$$

Multiplying throughout by time dt seconds gives the energy equation:

$$Vi\,dt = i^2Rdt + Li\,di$$

$Vi\,dt$ is the energy supplied by the source in time dt, $i^2R\,dt$ is the energy dissipated in the resistance and $Lidi$ is the energy supplied in establishing the magnetic field or the energy absorbed by the magnetic field in time dt seconds.

Hence the total energy stored in the field when the current increases from 0 to I amperes is given by

$$\text{energy stored, } W_f = \int_0^I Lidi = L\left[\frac{i^2}{2}\right]_0^I$$

i.e., **total energy stored,** $\boxed{\mathbf{W_f = \tfrac{1}{2}LI^2 \text{ joules}}}$ (40.27)

From Section 40.5, inductance $L = N\phi/I$, hence

$$\text{total energy stored} = \frac{1}{2}\left(\frac{N\phi}{I}\right)I^2 = \frac{1}{2}N\phi I \text{ joules}$$

Also $H = NI/l$, from which, $N = Hl/I$, and $\phi = BA$. Thus the total energy stored,

$$W_f = \frac{1}{2}N\phi I = \frac{1}{2}\left(\frac{Hl}{I}\right)(BA)I$$

$$= \frac{1}{2}HBlA \text{ joules}$$

or $\omega_f = \dfrac{1}{2}HB$ joules/metre^3

since lA is the volume of the magnetic field. This latter expression has already been derived in equation (40.24).

Summarizing, the energy stored in a nonmagnetic medium,

$$\boldsymbol{\omega_f = \frac{1}{2}BH = \frac{1}{2}\mu_0H^2 = \frac{B^2}{2\mu_0}} \textbf{ joules/metre}^3$$

and the energy stored in an inductor,

$$\mathbf{W_f = \tfrac{1}{2}LI^2 \text{ joules}}$$

Problem 22. Calculate the value of the energy stored when a current of 50 mA is flowing in a coil of inductance 200 mH. What value of current would double the energy stored?

From equation (40.27), energy stored in inductor,

$$W_f = \frac{1}{2}LI^2 = \frac{1}{2}(200 \times 10^{-3})(50 \times 10^{-3})^2$$

$$= \mathbf{2.5 \times 10^{-4}\ J} \text{ or } \mathbf{0.25\ mJ} \text{ or } \mathbf{250\ \mu J}$$

If the energy stored is doubled, then $(2)(2.5 \times 10^{-4}) = \frac{1}{2}(200 \times 10^{-3})I^2$ from which

$$\text{current } I = \sqrt{\left(\frac{(4)(2.5 \times 10^{-4})}{(200 \times 10^{-3})}\right)} = \mathbf{70.71\ mA}$$

Problem 23. The airgap of a moving coil instrument is 2.0 mm long and has a cross-sectional area of 500 mm^2. If the flux density is 50 mT, determine the total energy stored in the magnetic field of the airgap.

From equation (40.26), energy stored,

$$\omega_f = \frac{B^2}{2\mu_0} = \frac{(50 \times 10^{-3})^2}{2(4\pi \times 10^{-7})} = 9.95 \times 10^2 \text{ J/m}^3$$

Volume of airgap $= Al = (500 \times 2.0)$ mm^3 $= 500 \times 2.0 \times 10^{-9}$ m^3. Hence the energy stored in the airgap,

$$W_f = 9.95 \times 10^2 \text{ J/m}^3 \times 500 \times 2.0 \times 10^{-9} \text{ m}^3$$

$$= \mathbf{9.95 \times 10^{-4}\ J \equiv 0.995\ mJ \equiv 995\ \mu J}$$

Problem 24. Determine the strength of a uniform electric field if it is to have the same energy as that established by a magnetic field of flux density 0.8 T. Assume that the relative permeability of the magnetic field and the relative permittivity of the electric field are both unity.

From equation (40.26), energy stored in magnetic field,

$$\omega_f = \frac{B^2}{2\mu_0} = \frac{(0.8)^2}{2(4\pi \times 10^{-7})} = 2.546 \times 10^5 \text{ J/m}^3$$

From equation (40.17), energy stored in electric field,

$$\omega_f = \frac{1}{2}\varepsilon_0\varepsilon_r E^2$$

Hence, if the current stored in the magnetic and electric fields is to be the same, then $\frac{1}{2}\varepsilon_0\varepsilon_r E^2 = 2.546 \times 10^5$, i.e.,

$$\frac{1}{2}(8.85 \times 10^{-12})(1)E^2 = 2.546 \times 10^5$$

from which electric field strength,

$$E = \sqrt{\left(\frac{(2)(2.546 \times 10^5)}{(8.85 \times 10^{-12})}\right)} = \sqrt{(5.75 \times 10^{16})}$$

$$= \mathbf{2.40 \times 10^8\ V/m} \text{ or } \mathbf{240\ MV/m}$$

Further problems on energy stored in an electromagnetic field may be found in Section 40.9 following, problems 33 to 37, page 756.

40.9 Further problems on field theory

Field plotting by curvilinear squares

1 (a) Explain the meaning of the terms (i) streamline (ii) equipotential, with reference to an electric field.

(b) A field plot between two metal plates is shown in Figure 40.23. If the relative permittivity of the dielectric is 2.4, determine the capacitance of a 50 cm length of the system. [23.4 pF]

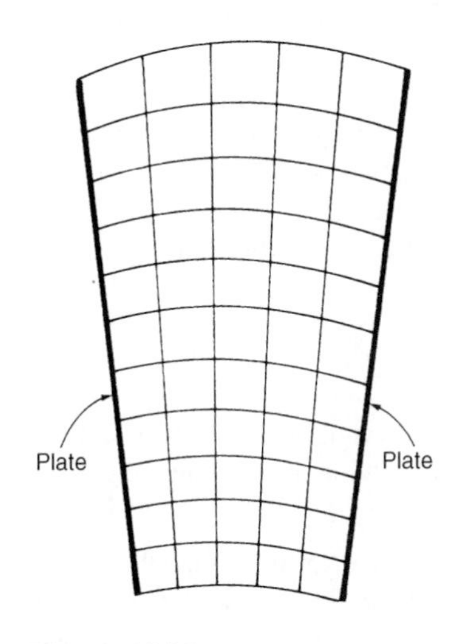

Figure 40.23

2 A field plot for a concentric cable is shown in Figure 40.24. The relative permittivity of the dielectric is 5. Determine the capacitance of a 10 m length of the cable. [1.66 nF]

3 The plates of a capacitor are 10 mm long and 6 mm wide and are separated by a dielectric 3 mm thick and of relative permittivity 2.5. Determine the capacitance of the capacitor (a) when neglecting any fringing at the edges, (b) by producing a field plot taking fringing into consideration.

[(a) 0.44 pF (b) 0.60 pF – 0.70 pF, depending on the accuracy of the plot]

Capacitance between concentric cylinders

4 A coaxial cable has an inner conductor of radius 0.4 mm and an outer conductor of internal radius 4 mm. Determine the capacitance per metre length of the cable if the dielectric has a relative permittivity of 2. [48.30 pF]

5 A concentric cable has a core diameter of 40 mm and an inner sheath diameter of 100 mm. The relative permittivity of the dielectric is 2.5 and the core potential is 50 kV. Determine (a) the capacitance per kilometre length of the cable and (b) the dielectric stress at radii of 30 mm and 40 mm.

[(a) 0.1517 μF (b) 1.819 MV/m, 1.364 MV/m]

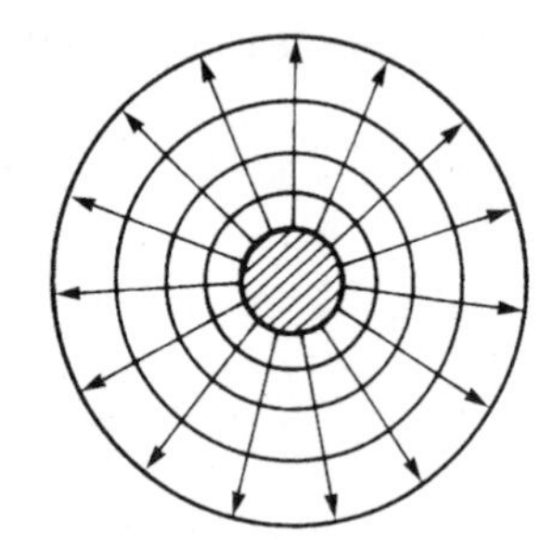

Figure 40.24

6 A coaxial cable has a capacitance of 100 pF per metre length. The relative permittivity of the dielectric is 3.2 and the core diameter is 1.0 mm. Determine the required inside diameter of the sheath.
[5.93 mm]

7 A single-core concentric cable is to be manufactured for a 100 kV, 50 Hz transmission system. The dielectric used is paper which has a maximum safe dielectric stress of 10 MV/m and a relative permittivity of 3.2. Calculate (a) the core and inner sheath radii for the most economical cable, (b) the capacitance per metre length and (c) the charging current per kilometre run.
[(a) 10 mm; 27.2 mm (b) 177.9 pF (c) 5.59 A]

8 A concentric cable has a core diameter of 30 mm and an inside sheath diameter of 75 mm. The relative permittivity is 2.6, the loss angle is 2.5×10^{-3} rad and the working voltage is 100 kV at 50 Hz frequency. Determine for a 1 km length of cable (a) the capacitance, (b) the charging current, and (c) the power loss.
[(a) 0.1578 μF (b) 4.957 A (c) 1239 W]

9 A concentric cable operates at 200 kV and 50 Hz. The maximum electric field strength within the cable is not to exceed 5 MV/m. Determine (a) the radius of the core and the inner radius of the sheath for ideal operation, and (b) the stress on the dielectric at the surface of the core and at the inner surface of the sheath.
[(a) 40 mm, 108.7 mm (b) 5 MV/m, 1.84 MV/m]

10 A concentric cable has a core radius of 20 mm and a sheath inner radius of 40 mm. The permittivity of the dielectric is 2.5. Using two equipotential surfaces within the dielectric, determine the capacitance of the cable per metre length by the method of curvilinear squares. Draw the field plot for the cable. [200.6 pF]

Capacitance of an isolated twin line

11 Two parallel wires, each of diameter 5.0 mm, are uniformly spaced in air at a distance of 40 mm between centres. Determine the capacitance of a 500 m run of the line. [5.014 nF]

12 A single-phase circuit is comprised of two parallel conductors each of radius 5.0 mm and spaced 1.5 m apart in air. The p.d. between the conductors is 20 kV at 50 Hz. Determine (a) the capacitance per metre length of the conductors, and (b) the charging current per kilometre run. [(a) 4.875 pF (b) 30.63 mA]

13 The capacitance of a 300 m length of an isolated twin line is 1522 pF. The line comprises two air conductors which are spaced 1200 mm between centres. Determine the diameter of each conductor.
[10 mm]

14 An isolated twin line is comprised of two air-insulated conductors, each of radius 8.0 mm, which are spaced 1.60 m apart. The voltage between the lines is 7 kV at a frequency of 50 Hz. Determine for a

1 km length (a) the line capacitance, (b) the value of charge carried by each wire, and (c) the charging current.
[(a) 5.248 nF (b) 36.74 μC (c) 11.54 mA]

15 The charging current for a 1 km run of isolated twin line is not to exceed 30 mA. The p.d. between the lines is 20 kV at 50 Hz. If the line is air insulated and the conductors are spaced 1 m apart, determine (a) the maximum value required for the capacitance per metre length, and (b) the maximum diameter of each conductor.
[(a) 4.775 pF (b) 5.92 mm]

Energy stored in an electric field

16 Determine the energy stored in a 5000 pF capacitor when charged to 800 V and the average power developed if this energy is dissipated in 20 μs. [1.6 mJ; 80 W]

17 A 0.25 μF capacitor is required to store 2 J of energy. Determine the p.d. to which the capacitor must be charged. [4 kV]

18 A capacitor is charged with 6 mC. If the energy stored is 1.5 J determine (a) the voltage across the plates, and (b) the capacitance of the capacitor. [(a) 500 V (b) 12 μF]

19 After a capacitor is connected across a 250 V d.c. supply the charge is 5 μC. Determine (a) the capacitance, and (b) the energy stored.
[(a) 20 nF (b) 0.625 mJ]

20 A capacitor consisting of two metal plates each of area 100 cm^2 and spaced 0.1 mm apart in air is connected across a 200 V supply. Determine (a) the electric flux density, (b) the potential gradient and (c) the energy stored in the capacitor.
[(a) 17.7 $\mu C/m^2$ (b) 2 MV/m (c) 17.7 μJ]

21 A mica capacitor is to be constructed to have a capacitance of 0.05 μF and to have a steady working potential of 2 kV maximum. Allowing a safe value of field stress of 20 MV/m, determine (a) the required thickness of the mica dielectric, (b) the area of plate required if the relative permittivity of the mica is 5, (c) the maximum energy stored by the capacitor, and (d) the average power developed if this energy is dissipated in 25 μs.
[(a) 0.1 mm (b) 0.113 m^2 (c) 0.1 J (d) 4 kW]

22 A 500 pF capacitor is charged to a p.d. of 100 V. The dielectric has a cross-sectional area of 200 cm^2 and a relative permittivity of 2.4. Determine the energy stored per cubic metre in the dielectric.
[0.147 J/m^3]

23 Two parallel plates each having dimensions 30 mm by 50 mm are spaced 8 mm apart in air. If a voltage of 40 kV is applied across the plates determine the energy stored in the electric field. [1.328 mJ]

Inductance of a concentric cable

24 A coaxial cable has an inner core of radius 0.8 mm and an outer sheath of internal radius 4.8 mm. Determine the inductance of 25 m of the cable. Assume that the relative permeability of the material used is 1. [10.2 μH]

25 A concentric cable has a core 12 mm diameter and a sheath 40 mm diameter, the sheath having negligible thickness. Determine the inductance and the capacitance of the cable per metre assuming nonmagnetic materials and a dielectric of relative permittivity 3.2. [0.291 μH/m, 147.8 pF/m]

26 A concentric cable has an inner sheath radius of 4.0 cm. The inductance of the cable is 0.5 μH/m. Ignoring inductance due to internal linkages, determine the radius of the core. Assume that the relative permeability of the material is unity. [3.28 mm]

27 The inductance of a concentric cable of core radius 8 mm and inner sheath radius of 35 mm is measured as 2.0 mH. Determine (a) the length of the cable, and (b) the capacitance of the cable. Assume that nonmagnetic materials are used and the relative permittivity of the dielectric is 2.5. [(a) 5.794 km (b) 0.546 μF]

Inductance of an isolated twin line

28 A single-phase power line comprises two conductors each with a radius of 15 mm and spaced 1.8 m apart in air. Determine the inductance per metre length, ignoring internal linkages and assuming the relative permeability, $\mu_r = 1$. [1.915 μH/m]

29 Determine (a) the loop inductance, and (b) the capacitance of a 500 m length of single-phase twin line having conductors of diameter 8 mm and spaced 60 mm apart in air. [(a) 0.592 mH (b) 5.133 nF]

30 An isolated twin power line has conductors 7.5 mm radius. Determine the distance between centres if the total loop inductance of 1 km of the line is 1.95 mH. [765 mm]

31 An isolated twin line has conductors of diameter $d \times 10^{-3}$ metres and spaced D millimetres apart in air. Derive an expression for the total loop inductance L of the line per metre length.

$$\left[L = \frac{\mu_0}{\pi}\left(\frac{1}{4} + \ln\frac{2D}{d}\right)\right]$$

32 A single-phase power line comprises two conductors spaced 2 m apart in air. The loop inductance of 2 km of the line is measured as 3.65 mH. Determine the diameter of the conductors. [53.6 mm]

Energy stored in an electromagnetic field

33 Determine the value of the energy stored when a current of 120 mA flows in a coil of 500 mH. What value of current is required to double the energy stored? [3.6 mJ, 169.7 mA]

34 A moving-coil instrument has two airgaps each 2.5 mm long and having a cross-sectional area of 8.0 cm^2. Determine the total energy stored in the magnetic field of the airgap if the flux density is 100 mT.
[15.92 mJ]

35 Determine the flux density of a uniform magnetic field if it is to have the same energy as that established by a uniform electric field of strength 45 MV/m. Assume the relative permeability of the magnetic field and the relative permittivity of the electric field are both unity.
[0.15 T]

36 A long single core concentric cable has inner and outer conductors of diameters D_1 and D_2 respectively. The conductors each carry a current of I amperes but in opposite directions. If the relative permeability of the material is unity and the inductance due to internal linkages is negligible, show that the magnetic energy stored in a 4 m length of the cable is given by

$$\frac{\mu_0 I^2}{\pi} \ln\left(\frac{D_2}{D_1}\right) \text{ joules}$$

37 1 mJ of energy is stored in a uniform magnetic field having dimensions 20 mm by 10 mm by 1.0 mm. Determine for the field (a) the flux density, and (b) the magnetic field strength.
[(a) 0.112 T (b) 89200 A/m]

41 Attenuators

At the end of this chapter you should be able to:

- understand the function of an attenuator
- understand characteristic impedance and calculate for given values
- appreciate and calculate logarithmic ratios
- design symmetrical T and symmetrical π attenuators given required attenuation and characteristic impedance
- appreciate and calculate insertion loss
- determine iterative and image impedances for asymmetrical T and π networks
- appreciate and design the L-section attenuator
- calculate attenuation for two-port networks in cascade

41.1 Introduction

An **attenuator** is a device for introducing a specified loss between a signal source and a matched load without upsetting the impedance relationship necessary for matching. The loss introduced is constant irrespective of frequency; since reactive elements (L or C) vary with frequency, it follows that ideal attenuators are networks containing pure resistances. A fixed attenuator section is usually known as a 'pad'.

Attenuation is a reduction in the magnitude of a voltage or current due to its transmission over a line or through an attenuator. Any degree of attenuation may be achieved with an attenuator by suitable choice of resistance values but the input and output impedances of the pad must be such that the impedance conditions existing in the circuit into which it is connected are not disturbed. Thus an attenuator must provide the correct input and output impedances as well as providing the required attenuation.

Attenuation sections are made up of resistances connected as T or π arrangements (as introduced in Chapter 34).

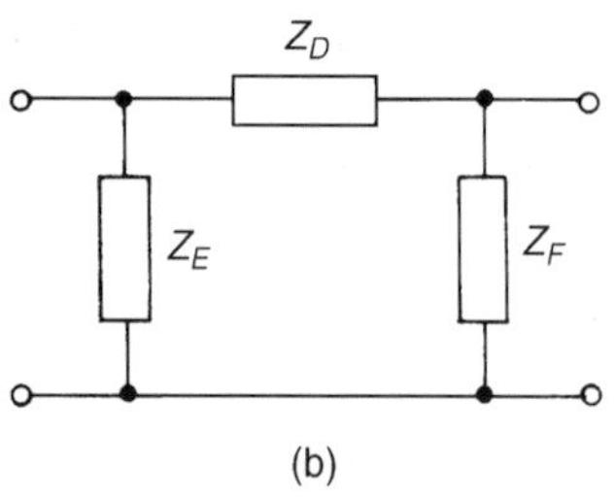

Z_A Z_B Z_C

(a)

Z_D Z_E Z_F

(b)

Figure 41.1 *(a) T-network, (b) π-network*

Two-port networks

Networks in which electrical energy is fed in at one pair of terminals and taken out at a second pair of terminals are called two-port networks. Thus an attenuator is a two-port network, as are transmission lines, transformers and electronic amplifiers. The network between the input port and the output port is a transmission network for which a known relationship exists between the input and output currents and voltages. If

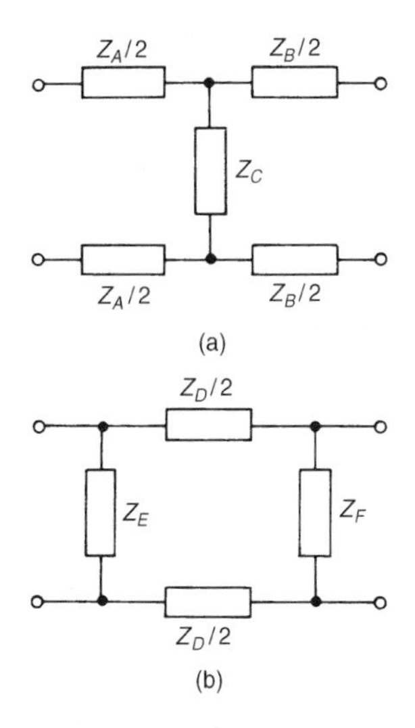

Figure 41.2 *(a) Balanced T-network, (b) Balanced π-network*

a network contains only passive circuit elements, such as in an attenuator, the network is said to be **passive**; if a network contains a source of e.m.f., such as in an electronic amplifier, the network is said to be **active**.

Figure 41.1(a) shows a T-network, which is termed **symmetrical** if $Z_A = Z_B$ and Figure 41.1(b) shows a π-network which is symmetrical if $Z_E = Z_F$. If $Z_A \neq Z_B$ in Figure 41.1(a) and $Z_E \neq Z_F$ in Figure 41.1(b), the sections are termed **asymmetrical**. Both networks shown have one common terminal, which may be earthed, and are therefore said to be **unbalanced.** The **balanced** form of the T-network is shown in Figure 41.2(a) and the balanced form of the π-network is shown in Figure 41.2(b).

Symmetrical T- and π-attenuators are discussed in Section 41.4 and asymmetrical attenuators are discussed in Sections 41.6 and 41.7. Before this it is important to understand the concept of characteristic impedance, which is explained generally in Section 41.2 (characteristic impedances will be used again in Chapter 44), and logarithmic units, discussed in Section 41.3. Another important aspect of attenuators, that of insertion loss, is discussed in Section 41.5. To obtain greater attenuation, sections may be connected in cascade, and this is discussed in Section 41.8.

41.2 Characteristic impedance

The input impedance of a network is the ratio of voltage to current (in complex form) at the input terminals. With a two-port network the input impedance often varies according to the load impedance across the output terminals. For any passive two-port network it is found that a particular value of load impedance can always be found which will produce an input impedance having the same value as the load impedance. This is called the **iterative impedance** for an asymmetrical network and its value depends on which pair of terminals is taken to be the input and which the output (there are thus two values of iterative impedance, one for each direction). For a symmetrical network there is only one value for the iterative impedance and this is called the **characteristic impedance** of the symmetrical two-port network. Let the characteristic impedance be denoted by Z_0. Figure 41.3 shows a **symmetrical *T*-network** terminated in an impedance Z_0.

Let the impedance 'looking-in' at the input port also be Z_0. Then $V_1/I_1 = Z_0 = V_2/I_2$ in Figure 41.3. From circuit theory,

$$Z_0 = \frac{V_1}{I_1} = Z_A + \frac{Z_B(Z_A + Z_0)}{Z_B + Z_A + Z_0}, \text{ since } (Z_A + Z_0) \text{ is in parallel with } Z_B,$$

$$= \frac{Z_A^2 + Z_AZ_B + Z_AZ_0 + Z_AZ_B + Z_BZ_0}{Z_A + Z_B + Z_0}$$

i.e. $$Z_0 = \frac{Z_A^2 + 2Z_AZ_B + Z_AZ_0 + Z_BZ_0}{Z_A + Z_B + Z_0}$$

Thus $$Z_0(Z_A + Z_B + Z_0) = Z_A^2 + 2Z_AZ_B + Z_AZ_0 + Z_BZ_0$$

$$Z_0Z_A + Z_0Z_B + Z_0^2 = Z_A^2 + 2Z_AZ_B + Z_AZ_0 + Z_BZ_0$$

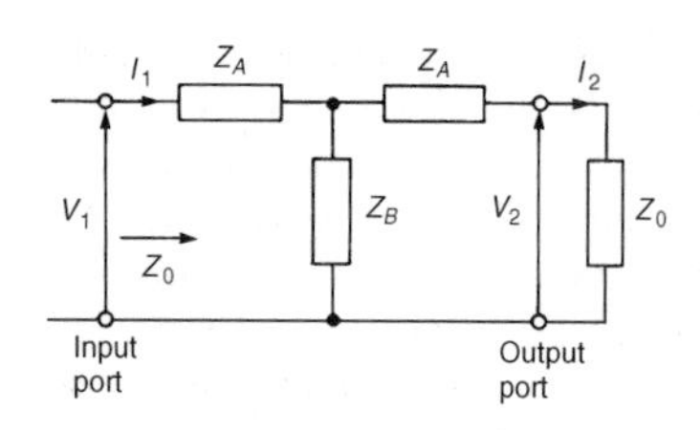

Figure 41.3

i.e., $Z_0^2 = Z_A^2 + 2Z_AZ_B$, from which

$$\text{characteristic impedance,} \quad \boxed{\mathbf{Z_0 = \sqrt{(Z_A^2 + 2Z_AZ_B)}}} \qquad (41.1)$$

If the output terminals of Figure 41.3 are open-circuited, then the open-circuit impedance, $Z_{OC} = Z_A + Z_B$. If the output terminals of Figure 41.3 are short-circuited, then the short-circuit impedance,

$$Z_{SC} = Z_A + \frac{Z_AZ_B}{Z_A + Z_B} = \frac{Z_A^2 + 2Z_AZ_B}{Z_A + Z_B}$$

$$\text{Thus} \quad Z_{OC}Z_{SC} = (Z_A + Z_B)\left(\frac{Z_A^2 + 2Z_AZ_B}{Z_A + z_B}\right) = Z_A^2 + 2Z_AZ_B$$

Comparing this with equation (41.1) gives

$$\boxed{\mathbf{Z_0 = \sqrt{(Z_{OC}Z_{SC})}},} \qquad (41.2)$$

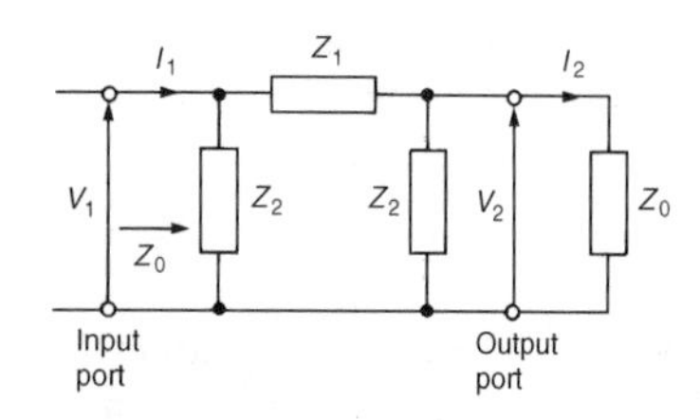

Figure 41.4

Figure 41.4 shows a symmetrical π-network terminated in an impedance Z_0.
If the impedance 'looking in' at the input port is also Z_0, then

$$\frac{V_1}{I_1} = Z_0 = (Z_2) \text{ in parallel with } [Z_1 \text{ in series with } (Z_0 \text{ and } Z_2) \text{ in parallel}]$$

$$= (Z_2) \text{ in parallel with } \left[Z_1 + \frac{Z_0Z_2}{Z_0 + Z_2}\right]$$

$$= (Z_2) \text{ in parallel with } \left[\frac{Z_1Z_0 + Z_1Z_2 + Z_0Z_2}{Z_0 + Z_2}\right]$$

$$\text{i.e.,} \quad Z_0 = \frac{(Z_2)((Z_1Z_0 + Z_1Z_2 + Z_0Z_2)/(Z_0 + Z_2))}{Z_2 + ((Z_1Z_0 + Z_1Z_2 + Z_0Z_2)/(Z_0 + Z_2))}$$

$$= \frac{(Z_1Z_2Z_0 + Z_1Z_2^2 + Z_0Z_2^2)/(Z_0 + Z_2)}{(Z_2Z_0 + Z_2^2 + Z_1Z_0 + Z_1Z_2 + Z_0Z_2)/(Z_0 + Z_2)}$$

$$\text{i.e.} \quad Z_0 = \frac{Z_1Z_2Z_0 + Z_1Z_2^2 + Z_0Z_2^2}{Z_2^2 + 2Z_2Z_0 + Z_1Z_0 + Z_1Z_2}$$

$$\text{Thus} \quad Z_0(Z_2^2 + 2Z_2Z_0 + Z_1Z_0 + Z_1Z_2) = Z_1Z_2Z_0 + Z_1Z_2^2 + Z_0Z_2^2$$

$$2Z_2Z_0^2 + Z_1Z_0^2 = Z_1Z_2^2$$

from which

$$\boxed{\textbf{characteristic impedance, } \mathbf{Z_0 = \sqrt{\left(\frac{Z_1Z_2^2}{Z_1 + 2Z_2}\right)}}} \qquad (41.3)$$

If the output terminals of Figure 41.4 are open-circuited, then the open-circuit impedance,

$$Z_{OC} = \frac{Z_2(Z_1 + Z_2)}{Z_2 + Z_1 + Z_2} = \frac{Z_2(Z_1 + Z_2)}{Z_1 + 2Z_2}$$

If the output terminals of Figure 41.4 are short-circuited, then the short-circuit impedance,

$$Z_{SC} = \frac{Z_2 Z_1}{Z_1 + Z_2}$$

Thus

$$Z_{OC} Z_{SC} = \frac{Z_2(Z_1 + Z_2)}{(Z_1 + 2Z_2)} \left(\frac{Z_2 Z_1}{Z_1 + Z_2} \right) = \frac{Z_1 Z_2^2}{Z_1 + 2Z_2}$$

Comparing this expression with equation (41.3) gives

$$\mathbf{Z_0 = \sqrt{(Z_{OC} Z_{SC})},} \qquad (41.2')$$

which is the same as equation (41.2).

Thus the characteristic impedance Z_0 is given by $Z_0 = \sqrt{(Z_{OC} Z_{SC})}$ whether the network is a symmetrical T or a symmetrical π.

Equations (41.1) to (41.3) are used later in this chapter.

41.3 Logarithmic ratios

The ratio of two powers P_1 and P_2 may be expressed in logarithmic form as shown in Chapter 10.

Let P_1 be the input power to a system and P_2 the output power.

If logarithms to base 10 are used, then the ratio is said to be in **bels**, i.e., power ratio in bels $= \lg(P_2/P_1)$. The bel is a large unit and the **decibel (dB)** is more often used, where 10 decibels = 1 bel, i.e.,

$$\textbf{power ratio in decibels} = \mathbf{10 \lg \frac{P_2}{P_1}} \qquad (41.4)$$

For example:

P_2/P_1	Power ratio (dB)
1	$10 \lg 1 = 0$
100	$10 \lg 100 = +20$ (power gain)
$\frac{1}{10}$	$10 \lg \frac{1}{10} = -10$ (power loss or attenuation)

If **logarithms to base *e*** (i.e., natural or Napierian logarithms) are used, then the ratio of two powers is said to be in **nepers (Np)**, i.e.,

$$\textbf{power ratio in nepers} = \frac{1}{2}\ln\frac{P_2}{P_1} \qquad (41.5)$$

Thus when the power ratio $P_2/P_1 = 5$, the power ratio in nepers $= \frac{1}{2}\ln 5 = 0.805$ Np, and when the power ratio $P_2/P_1 = 0.1$, the power ratio in nepers $= \frac{1}{2}\ln 0.1 = -1.15$ Np.

The attenuation of filter sections and along a transmission line are of an exponential form and it is in such applications that the unit of the neper is used (see Chapters 42 and 44).

If the powers P_1 and P_2 refer to power developed in two equal resistors, R, then $P_1 = V_1^2/R$ and $P_2 = V_2^2/R$. Thus the ratio (from equation (41.4)) can be expressed, by the laws of logarithms, as

$$\textbf{ratio in decibels} = 10\lg\frac{P_2}{P_1} = 10\lg\left(\frac{V_2^2/R}{V_1^2/R}\right) = 10\lg\frac{V_2^2}{V_1^2}$$

$$= 10\lg\left(\frac{V_2}{V_1}\right)^2$$

i.e.

$$\textbf{ratio in decibels} = \mathbf{20\lg\frac{V_2}{V_1}} \qquad (41.6)$$

Although this is really a power ratio, it is called the **logarithmic voltage ratio**.

Alternatively, (from equation (41.5)),

$$\text{ratio in nepers} = \frac{1}{2}\ln\frac{P_2}{P_1} = \frac{1}{2}\ln\left(\frac{V_2^2/R}{V_1^2/R}\right) = \frac{1}{2}\ln\left(\frac{V_2}{V_1}\right)^2$$

i.e.,

$$\textbf{ratio in nepers} = \mathbf{\ln\frac{V_2}{V_1}} \qquad (41.7)$$

Similarly, if currents I_1 and I_2 in two equal resistors R give powers P_1 and P_2 then (from equation (41.4))

$$\text{ratio in decibels} = 10\lg\frac{P_2}{P_1} = 10\lg\left(\frac{I_2^2R}{I_1^2R}\right) = 10\lg\left(\frac{I_2}{I_1}\right)^2$$

i.e.,

$$\textbf{ratio in decibels} = \mathbf{20\lg\frac{I_2}{I_1}} \qquad (41.8)$$

Alternatively (from equation (41.5)),

$$\text{ratio in nepers} = \frac{1}{2}\ln\frac{P_2}{P_1} = \frac{1}{2}\ln\left(\frac{I_2^2R}{I_1^2R}\right)^2 = \frac{1}{2}\ln\left(\frac{I_2}{I_1}\right)^2$$

i.e., **ratio in nepers** $= \ln \dfrac{I_2}{I_1}$ (41.9)

In equations (41.4) to (41.9) the output-to-input ratio has been used. However, the input-to-output ratio may also be used. For example, in equation (41.6), the output-to-input voltage ratio is expressed as $20\lg(V_2/V_1)$ dB. Alternatively, the input-to-output voltage ratio may be expressed as $20\lg(V_1/V_2)$ dB, the only difference in the values obtained being a difference in sign.

If $20\lg(V_2/V_1) = 10$ dB, say, then $20\lg(V_1/V_2) = -10$ dB. Thus if an attenuator has a voltage input V_1 of 50 mV and a voltage output V_2 of 5 mV, the voltage ratio V_2/V_1 is 5/50 or 1/10. Alternatively, this may be expressed as **'an attenuation of 10'**, i.e., $V_1/V_2 = 10$.

Problem 1. The ratio of output power to input power in a system is

(a) 2 (b) 25 (c) 1000 and (d) $\frac{1}{100}$

Determine the power ratio in each case (i) in decibels and (ii) in nepers.

(i) From equation (41.4), power ratio in decibels $= 10\lg(P_2/P_1)$.

(a) When $P_2/P_1 = 2$, power ratio $= 10\lg 2 =$ **3 dB**

(b) When $P_2/P_1 = 25$, power ratio $= 10\lg 25 =$ **14 dB**

(c) When $P_2/P_1 = 1000$, power ratio $= 10\lg 1000 =$ **30 dB**

(d) When $P_2/P_1 = \frac{1}{100}$, power ratio $= 10\lg \frac{1}{100} =$ **−20 dB**

(ii) From equation (41.5), power ratio in nepers $= \frac{1}{2}\ln(P_2/P_1)$.

(a) When $P_2/P_1 = 2$, power ratio $= \frac{1}{2}\ln 2 =$ **0.347 Np**

(b) When $P_2/P_1 = 25$, power ratio $= \frac{1}{2}\ln 25 =$ **1.609 Np**

(c) When $P_2/P_1 = 1000$, power ratio $= \frac{1}{2}\ln 1000 =$ **3.454 Np**

(d) When $P_2/P_1 = \frac{1}{100}$, power ratio $= \frac{1}{2}\ln \frac{1}{100} =$ **−2.303 Np**

The power ratios in (a), (b) and (c) represent power gains, since the ratios are positive values; the power ratio in (d) represents a power loss or attenuation, since the ratio is a negative value.

Problem 2. 5% of the power supplied to a cable appears at the output terminals. Determine the attenuation in decibels.

If $P_1 =$ input power and $P_2 =$ output power, then

$$\frac{P_2}{P_1} = \frac{5}{100} = 0.05$$

From equation (41.4), power ratio in decibels

$$= 10\lg(P_2/P_1) = 10\lg 0.05 = -13 \text{ dB}.$$

Hence the attenuation (i.e., power loss) is 13 dB.

> Problem 3. An amplifier has a gain of 15 dB. If the input power is 12 mW, determine the output power.

From equation (41.4), decibel power ratio $= 10\lg(P_2/P_1)$. Hence $15 = 10\lg(P_2/12)$, where P_2 is the output power in milliwatts.

$$1.5 = \lg\left(\frac{P_2}{12}\right)$$

$$\frac{P_2}{12} = 10^{1.5}$$

from the definition of a logarithm. Thus the output power,

$$P_2 = 12(10)^{1.5} = \mathbf{379.5\ mW}$$

> Problem 4. The current output of an attenuator is 50 mA. If the current ratio of the attenuator is −1.32 Np, determine (a) the current input and (b) the current ratio expressed in decibels. Assume that the input and load resistances of the attenuator are equal.

(a) From equation (41.9), current ratio in nepers $= \ln(I_2/I_1)$. Hence $-1.32 = \ln(50/I_1)$, where I_1 is the input current in mA.

$$e^{-1.32} = \frac{50}{I_1}$$

from which, **current input**, $I_1 = \dfrac{50}{e^{-1.32}} = 50e^{1.32} = \mathbf{187.2\ mA}$

(b) From equation (41.8),

$$\text{current ratio in decibels} = 20\lg\frac{I_2}{I_1} = 20\lg\left(\frac{50}{187.2}\right)$$
$$= \mathbf{-11.47\ dB}$$

Further problems on logarithmic ratios may be found in Section 41.9, problems 1 to 5, page 785.

41.4 Symmetrical T-and π-attenuators

(a) Symmetrical T-attenuator

As mentioned in Section 41.1, the ideal attenuator is made up of pure resistances. A symmetrical T-pad attenuator is shown in Figure 41.5 with a termination R_0 connected as shown. From equation (41.1),

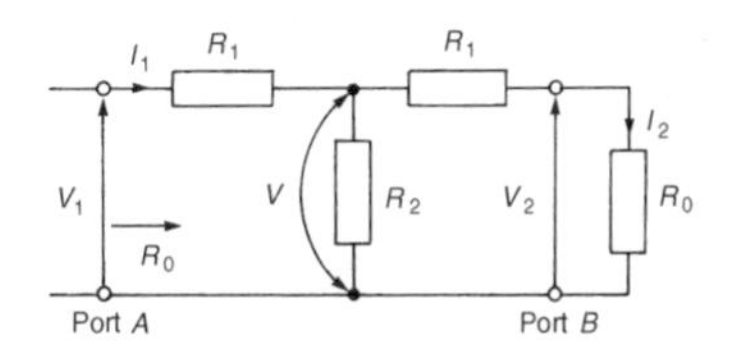

Figure 41.5 *Symmetrical T-pad attenuator*

$$\boxed{\boldsymbol{R_0 = \sqrt{(R_1^2 + 2R_1R_2)}}} \qquad (41.10)$$

and from equation (41.2) $$\boxed{\boldsymbol{R_0 = \sqrt{(R_{OC}R_{SC})}}} \qquad (41.11)$$

With resistance R_0 as the termination, the input resistance of the pad will also be equal to R_0. If the terminating resistance R_0 is transferred to port A then the input resistance looking into port B will again be R_0.

The pad is therefore symmetrical in impedance in both directions of connection and may thus be inserted into a network whose impedance is also R_0. The value of R_0 is the characteristic impedance of the section.

As stated in Section 41.3, attenuation may be expressed as a voltage ratio V_1/V_2 (see Figure 41.5) or quoted in decibels as $20\lg(V_1/V_2)$ or, alternatively, as a power ratio as $10\lg(P_1/P_2)$. If a T-section is symmetrical, i.e., the terminals of the section are matched to equal impedances, then

$$10\lg\frac{P_1}{P_2} = 20\lg\frac{V_1}{V_2} = 20\lg\frac{I_1}{I_2}$$

since $R_{IN} = R_{LOAD} = R_0$, i.e.,

$$10\lg\frac{P_1}{P_2} = 10\lg\left(\frac{V_1}{V_2}\right)^2 = 10\lg\left(\frac{I_1}{I_2}\right)^2$$

from which $$\frac{P_1}{P_2} = \left(\frac{V_1}{V_2}\right)^2 = \left(\frac{I_1}{I_2}\right)^2$$

or $$\sqrt{\left(\frac{P_1}{P_2}\right)} = \left(\frac{V_1}{V_2}\right) = \left(\frac{I_1}{I_2}\right)$$

Let $N = V_1/V_2$ or I_1/I_2 or $\sqrt{(P_1/P_2)}$, where N is the attenuation. In Section 41.5, page 772, it is shown that, for a matched network, i.e., one terminated in its characteristic impedance, N is in fact the insertion loss ratio. (Note that in an asymmetrical network, only the expression $N = \sqrt{(P_1/P_2)}$ may be used—see Section 41.7 on the L-section attenuator)

From Figure 41.5,

$$\text{current } I_1 = \frac{V_1}{R_0}$$

$$\text{Voltage } V = V_1 - I_1R_1 = V_1 - \left(\frac{V_1}{R_0}\right)R_1$$

$$\text{i.e., } V = V_1\left(1 - \frac{R_1}{R_0}\right)$$

$$\text{Voltage } V_2 = \left(\frac{R_0}{R_1 + R_0}\right)V \text{ by voltage division}$$

$$\text{i.e.,}\quad V_2 = \left(\frac{R_0}{R_1 + R_0}\right) V_1 \left(1 - \frac{R_1}{R_0}\right)$$

$$= V_1 \left(\frac{R_0}{R_1 + R_0}\right)\left(\frac{R_0 - R_1}{R_0}\right)$$

Hence
$$\frac{V_2}{V_1} = \frac{R_0 - R_1}{R_0 + R_1} \text{ or } \frac{V_1}{V_2} = N = \frac{R_0 + R_1}{R_0 - R_1} \qquad (41.12)$$

From equation (41.12) and also equation (41.10), it is possible to derive expressions for R_1 and R_2 in terms of N and R_0, thus enabling an attenuator to be designed to give a specified attenuation and to be matched symmetrically into the network. From equation (41.12),

$$N(R_0 - R_1) = R_0 + R_1$$

$$NR_0 - NR_1 = R_0 + R_1$$

$$NR_0 - R_0 = R_1 + NR_1$$

$$R_0(N - 1) = R_1(1 + N)$$

from which
$$\boldsymbol{R_1 = R_0 \frac{(N - 1)}{(N + 1)}} \qquad (41.13)$$

From equation (41.10), $R_0 = \sqrt{(R_1^2 + 2R_1R_2)}$ i.e., $R_0^2 = R_1^2 + 2R_1R_2$,

from which, $R_2 = \dfrac{R_0^2 - R_1^2}{2R_1}$

Substituting for R_1 from equation (41.13) gives

$$R_2 = \frac{R_0^2 - [R_0(N - 1)/(N + 1)]^2}{2[R_0(N - 1)/(N + 1)]}$$

$$= \frac{[R_0^2(N + 1)^2 - R_0^2(N - 1)^2]/(N + 1)^2}{2R_0(N - 1)/(N + 1)}$$

$$\text{i.e.,}\quad R_2 = \frac{R_0^2[(N + 1)^2 - (N - 1)^2]}{2R_0(N - 1)(N + 1)}$$

$$= \frac{R_0[(N^2 + 2N + 1) - (N^2 - 2N + 1)]}{2(N^2 - 1)}$$

$$= \frac{R_0(4N)}{2(N^2 - 1)}$$

Hence
$$\boldsymbol{R_2 = R_0 \left(\frac{2N}{N^2 - 1}\right)} \qquad (41.14)$$

Thus if the characteristic impedance R_0 and the attenuation N $(= V_1/V_2)$ are known for a symmetrical T-network then values of R_1 and R_2 may be

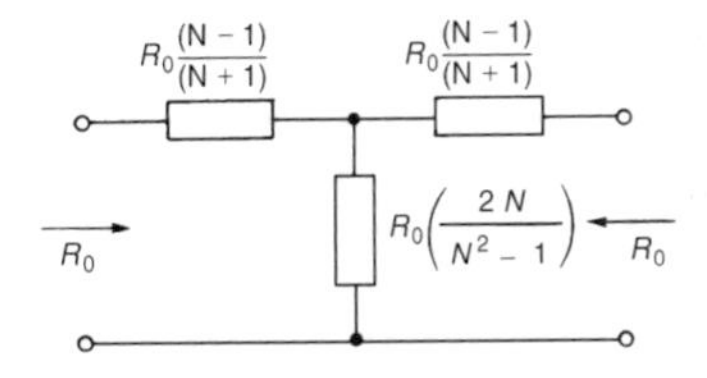

Figure 41.6

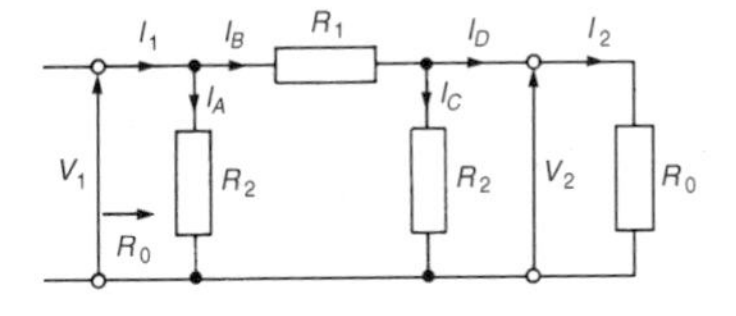

Figure 41.7 *Symmetrical π-attenuator*

calculated. Figure 41.6 shows a T-pad attenuator having input and output impedances of R_0 with resistances R_1 and R_2 expressed in terms of R_0 and N.

(b) Symmetrical π-attenuator

A symmetrical π-attenuator is shown in Figure 41.7 terminated in R_0. From equation (41.3),

$$\text{characteristic impedance} \quad \boldsymbol{R_0 = \sqrt{\left(\frac{R_1 R_2^2}{R_1 + 2R_2}\right)}} \qquad (41.15)$$

$$\text{and from equation (41.2}'\text{),} \quad \boldsymbol{R_0 = \sqrt{(R_{OC} R_{SC})}} \qquad (41.16)$$

Given the attenuation factor $N = \frac{V_1}{V_2}\left(= \frac{I_1}{I_2}\right)$

and the characteristic impedance R_0, it is possible to derive expressions for R_1 and R_2, in a similar way to the T-pad attenuator, to enable a π-attenuator to be effectively designed.

Since $N = V_1/V_2$ then $V_2 = V_1/N$. From Figure 41.7,

current $I_1 = I_A + I_B$ and current $I_B = I_C + I_D$. Thus

$$\text{current } I_1 = \frac{V_1}{R_0} = I_A + I_C + I_D$$

$$= \frac{V_1}{R_2} + \frac{V_2}{R_2} + \frac{V_2}{R_0} = \frac{V_1}{R_2} + \frac{V_1}{NR_2} + \frac{V_1}{NR_0}$$

since $V_2 = V_1/N$, i.e.,

$$\frac{V_1}{R_0} = V_1\left(\frac{1}{R_2} + \frac{1}{NR_2} + \frac{1}{NR_0}\right)$$

$$\text{Hence} \quad \frac{1}{R_0} = \frac{1}{R_2} + \frac{1}{NR_2} + \frac{1}{NR_0}$$

$$\frac{1}{R_0} - \frac{1}{NR_0} = \frac{1}{R_2} + \frac{1}{NR_2}$$

$$\frac{1}{R_0}\left(1 - \frac{1}{N}\right) = \frac{1}{R_2}\left(1 + \frac{1}{N}\right)$$

$$\frac{1}{R_0}\left(\frac{N-1}{N}\right) = \frac{1}{R_2}\left(\frac{N+1}{N}\right)$$

$$\text{Thus} \quad \boldsymbol{R_2 = R_0 \frac{(N+1)}{(N-1)}} \qquad (41.17)$$

From Figure 41.7, current $I_1 = I_A + I_B$, and since the p.d. across R_1 is $(V_1 - V_2)$,

$$\frac{V_1}{R_0} = \frac{V_1}{R_2} + \frac{V_1 - V_2}{R_1}$$

$$\frac{V_1}{R_0} = \frac{V_1}{R_2} + \frac{V_1}{R_1} - \frac{V_2}{R_1}$$

$$\frac{V_1}{R_0} = \frac{V_1}{R_2} + \frac{V_1}{R_1} - \frac{V_1}{NR_1} \quad \text{since } V_2 = V_1/N$$

$$\frac{1}{R_0} = \frac{1}{R_2} + \frac{1}{R_1} - \frac{1}{NR_1}$$

$$\frac{1}{R_0} - \frac{1}{R_2} = \frac{1}{R_1}\left(1 - \frac{1}{N}\right)$$

$$\frac{1}{R_0} - \frac{(N-1)}{R_0(N+1)} = \frac{1}{R_1}\left(\frac{N-1}{N}\right) \quad \text{from equation (41.17),}$$

$$\frac{1}{R_0}\left(1 - \frac{N-1}{N+1}\right) = \frac{1}{R_1}\left(\frac{N-1}{N}\right)$$

$$\frac{1}{R_0}\left(\frac{(N+1)-(N-1)}{(N+1)}\right) = \frac{1}{R_1}\left(\frac{N-1}{N}\right)$$

$$\frac{1}{R_0}\left(\frac{2}{N+1}\right) = \frac{1}{R_1}\left(\frac{N-1}{N}\right)$$

$$R_1 = R_0\left(\frac{N-1}{N}\right)\left(\frac{N+1}{2}\right)$$

Hence $$\boxed{\mathbf{R_1 = R_0\left(\frac{N^2-1}{2N}\right)}} \qquad (41.18)$$

Figure 41.8 shows a π-attenuator having input and output impedances of R_0 with resistances R_1 and R_2 expressed in terms of R_0 and N.

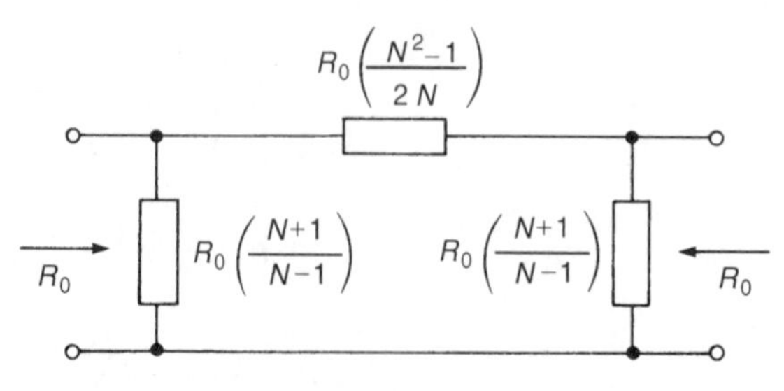

Figure 41.8

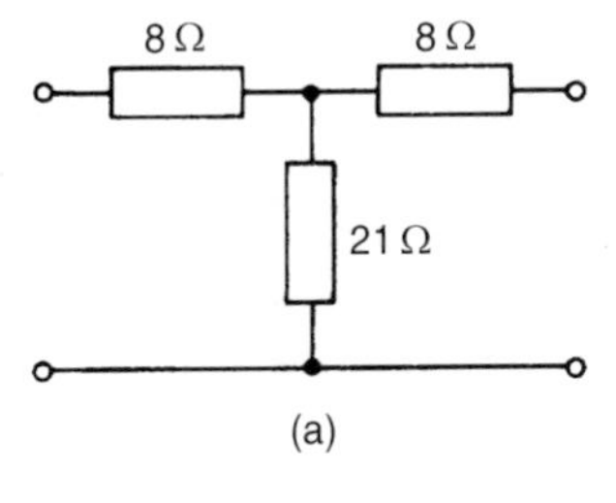

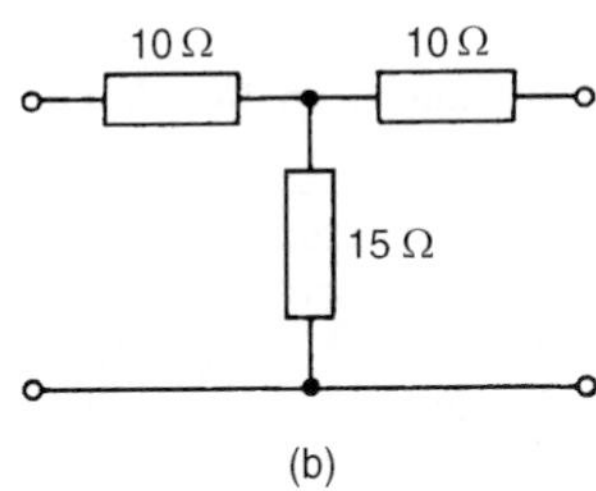

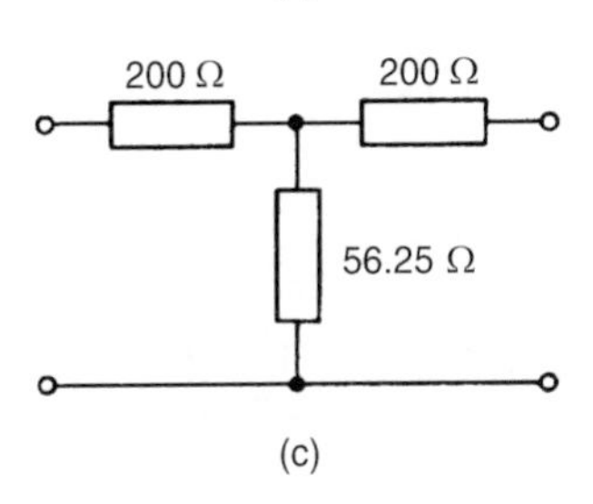

Figure 41.9

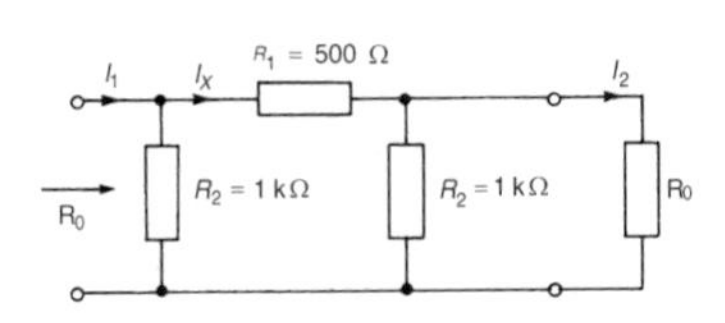

Figure 41.10

There is no difference in the functions of the T- and π-attenuator pads and either may be used in a particular situation.

> Problem 5. Determine the characteristic impedance of each of the attenuator sections shown in Figure 41.9.

From equation (41.10), for a T-section attenuator the characteristic impedance,

$$R_0 = \sqrt{(R_1^2 + 2R_1R_2)}.$$

(a) $R_0 = \sqrt{(8^2 + (2)(8)(21))} = \sqrt{400} = \mathbf{20\ \Omega}$

(b) $R_0 = \sqrt{(10^2 + (2)(10)(15))} = \sqrt{400} = \mathbf{20\ \Omega}$

(c) $R_0 = \sqrt{(200^2 + (2)(200)(56.25))} = \sqrt{62500} = \mathbf{250\ \Omega}$

It is seen that the characteristic impedance of parts (a) and (b) is the same. In fact, there are numerous combinations of resistances R_1 and R_2 which would give the same value for the characteristic impedance.

> Problem 6. A symmetrical π-attenuator pad has a series arm of 500 Ω resistance and each shunt arm of 1 kΩ resistance. Determine (a) the characteristic impedance, and (b) the attenuation (in dB) produced by the pad.

The π-attenuator section is shown in Figure 41.10 terminated in its characteristic impedance, R_0.

(a) From equation (41.15), for a symmetrical π-attenuator section,

$$\text{characteristic impedance, } R_0 = \sqrt{\left(\frac{R_1R_2^2}{R_1 + 2R_2}\right)}$$

$$\text{Hence} \quad R_0 = \sqrt{\left[\frac{(500)(1000)^2}{500 + 2(1000)}\right]} = 447\ \Omega$$

(b) Attenuation $= 20\lg(I_1/I_2)$ dB. From Figure 41.10,

$$\text{current } I_X = \left(\frac{R_2}{R_2 + R_1 + (R_2R_0/(R_2 + R_0))}\right)(I_1),$$

by current division

$$\text{i.e.,} \quad I_X = \left(\frac{1000}{1000 + 500 + ((1000)(447)/(1000 + 447))}\right)I_1$$

$$= 0.553I_1$$

$$\text{and current } I_2 = \left(\frac{R_2}{R_2 + R_0}\right)I_X = \left(\frac{1000}{1000 + 447}\right)I_X = 0.691I_X$$

Hence $I_2 = 0.691(0.553I_1) = 0.382I_1$ and $I_1/I_2 = 1/0.382$

$= 2.617$. Thus

attenuation $= 20 \lg 2.617 =$ **8.36 dB**

(Alternatively, since $I_1/I_2 = N$, then the formula

$$R_2 = R_0\left(\frac{N+1}{N-1}\right)$$

may be transposed for N, from which **attenuation = 20 lg** $\boldsymbol{N}$.)

> Problem 7. For each of the attenuator networks shown in Figure 41.11, determine (a) the input resistance when the output port is open-circuited, (b) the input resistance when the output port is short-circuited, and (c) the characteristic impedance.

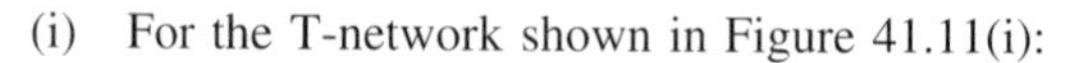

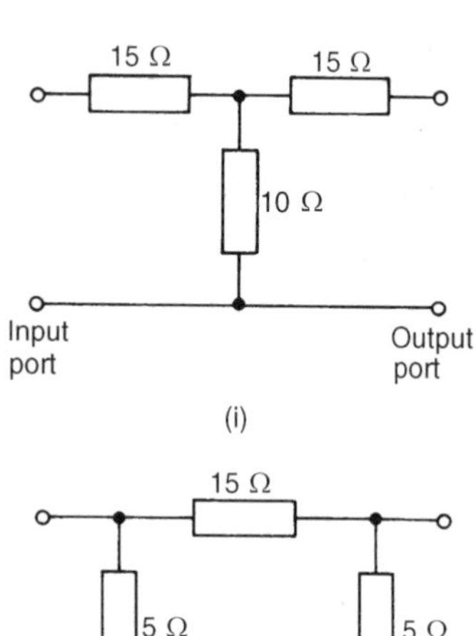

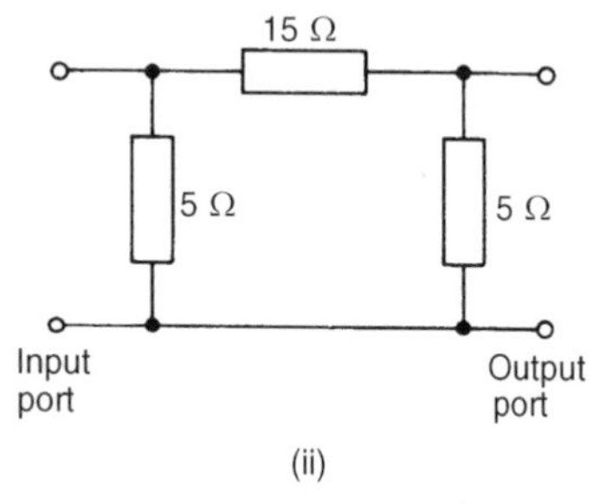

Figure 41.11

(i) For the T-network shown in Figure 41.11(i):

(a) $R_{OC} = 15 + 10 = \mathbf{25\ \Omega}$

(b) $R_{SC} = 15 + \dfrac{10 \times 15}{10 + 15} = 15 + 6 = \mathbf{21\ \Omega}$

(c) From equation (41.11), $R_0 = \sqrt{(R_{OC}R_{SC})} = \sqrt{[(25)(21)]}$
$= \mathbf{22.9\ \Omega}$

(Alternatively, from equation (41.10),

$R_0 = \sqrt{(R_1^2 + 2R_1R_2)} = \sqrt{(15^2 + (2)(15)(10))} = \mathbf{22.9\ \Omega}$)

(ii) For the π-network shown in Figure 41.11(ii):

(a) $R_{OC} = \dfrac{5 \times (15 + 5)}{5 + (15 + 5)} = \dfrac{100}{25} = \mathbf{4\ \Omega}$

(b) $R_{SC} = \dfrac{5 \times 15}{5 + 15} = \dfrac{75}{20} = \mathbf{3.75\ \Omega}$

(c) From equation (41.16),

$R_0 = \sqrt{(R_{OC}R_{SC})}$ as for a T-network

$= \sqrt{[(4)(3.75)]} = \sqrt{15} = \mathbf{3.87\ \Omega}$

(Alternatively, from equation (41.15),

$$R_0 = \sqrt{\left(\frac{R_1R_2^2}{R_1 + 2R_2}\right)} = \sqrt{\left(\frac{15(5)^2}{15 + 2(5)}\right)} = \mathbf{3.87\ \Omega})$$

> Problem 8. Design a T-section symmetrical attenuator pad to provide a voltage attenuation of 20 dB and having a characteristic impedance of 600 Ω.

Voltage attenuation in decibels $= 20\lg(V_1/V_2)$.
Attenuation, $N = V_1/V_2$, hence $20 = 20\lg N$, from which $N = 10$.
Characteristic impedance, $R_0 = 600\ \Omega$
From equation (41.13),

$$\text{resistance } R_1 = \frac{R_0(N-1)}{(N+1)} = \frac{600(10-1)}{(10+1)} = \mathbf{491\ \Omega}$$

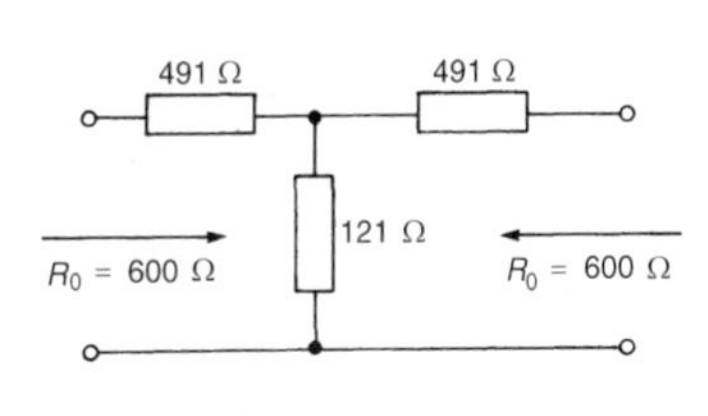

Figure 41.12

From equation (41.14),

$$\text{resistance } R_2 = R_0\left(\frac{2N}{N^2-1}\right) = 600\left(\frac{(2)(10)}{10^2-1}\right) = \mathbf{121\ \Omega}$$

Thus the T-section attenuator shown in Figure 41.12 has a voltage attenuation of 20 dB and a characteristic impedance of 600 Ω.
(Check: From equation (41.10)),

$$R_0 = \sqrt{(R_1^2 + 2R_1R_2)} = \sqrt{[491^2 + 2(491)(121)]} = 600\ \Omega)$$

Problem 9. Design a π-section symmetrical attenuator pad to provide a voltage attenuation of 20 dB and having a characteristic impedance of 600 Ω.

From problem 8, $N = 10$ and $R_0 = 600\ \Omega$
From equation (41.18),

$$\text{resistance } R_1 = R_0\left(\frac{N^2-1}{2N}\right) = 600\left(\frac{10^2-1}{(2)(10)}\right)$$
$$= \mathbf{2970\ \Omega} \text{ or } \mathbf{2.97\ k\Omega}$$

From equation (41.17),

$$R_2 = R_0\left(\frac{N+1}{N-1}\right) = 600\left(\frac{10+1}{10-1}\right) = \mathbf{733\ \Omega}$$

Thus the π-section attenuator shown in Figure 41.13 has a voltage attenuation of 20 dB and a characteristic impedance of 600 Ω.

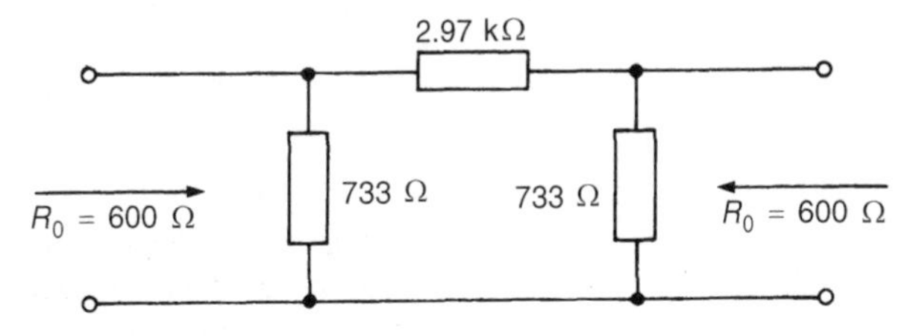

Figure 41.13

(Check: From equation (41.15),

$$R_0 = \sqrt{\left(\frac{R_1 R_2^2}{R_1 + 2R_2}\right)} = \sqrt{\left(\frac{(2970)(733)^2}{2970 + (2)(733)}\right)} = 600\ \Omega)$$

Further problems on symmetrical T- and π-attenuators may be found in Section 41.9, problems 6 to 15, page 785.

41.5 Insertion loss

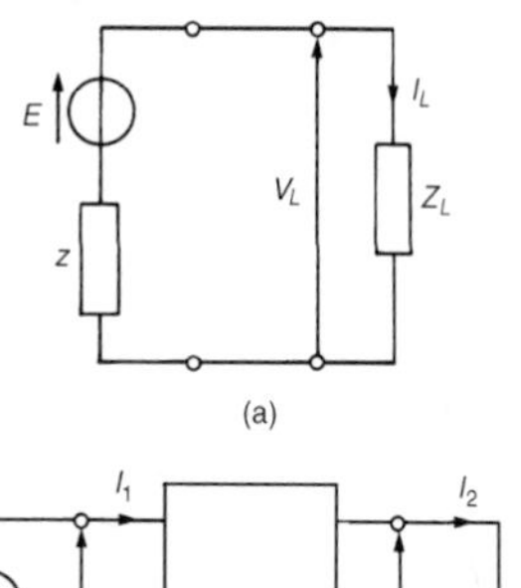

(a)

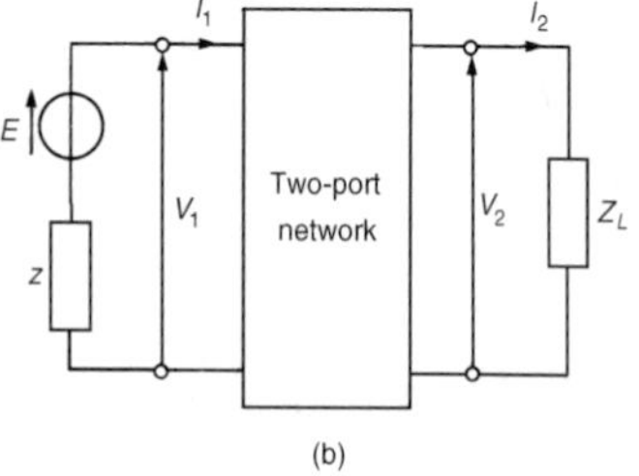

(b)

Figure 41.14

Figure 41.14(a) shows a generator E connected directly to a load Z_L. Let the current flowing be I_L and the p.d. across the load V_L. z is the internal impedance of the source.

Figure 41.14(b) shows a two-port network connected between the generator E and load Z_L.

The current through the load, shown as I_2, and the p.d. across the load, shown as V_2, will generally be less than current I_L and voltage V_L of Figure 41.14(a), as a result of the insertion of the two-port network between generator and load.

The **insertion loss ratio, A_L**, is defined as

$$A_L = \frac{\text{voltage across load when connected directly to the generator}}{\text{voltage across load when the two-port network is connected}}$$

i.e., $$\boldsymbol{A_L = V_L/V_2 = I_L/I_2} \qquad (41.19)$$

since $V_L = I_L Z_L$ and $V_2 = I_2 Z_L$. Since both V_L and V_2 refer to p.d.'s across the same impedance Z_L, the insertion loss ratio may also be expressed (from Section 41.3) as

$$\textbf{insertion loss ratio} = 20\lg\left(\frac{V_L}{V_2}\right)\ \textbf{dB} \text{ or } 20\lg\left(\frac{I_L}{I_2}\right)\ \textbf{dB} \qquad (41.20)$$

When the two-port network is terminated in its characteristic impedance Z_0 the network is said to be **matched**. In such circumstances the input impedance is also Z_0, thus the insertion loss is simply the ratio of input to output voltage (i.e., V_1/V_2). Thus, **for a network terminated in its characteristic impedance**,

$$\textbf{insertion loss} = 20\ \lg\left(\frac{V_1}{V_2}\right)\textbf{dB} \text{ or } 20\ \lg\left(\frac{I_1}{I_2}\right)\textbf{dB} \qquad (41.21)$$

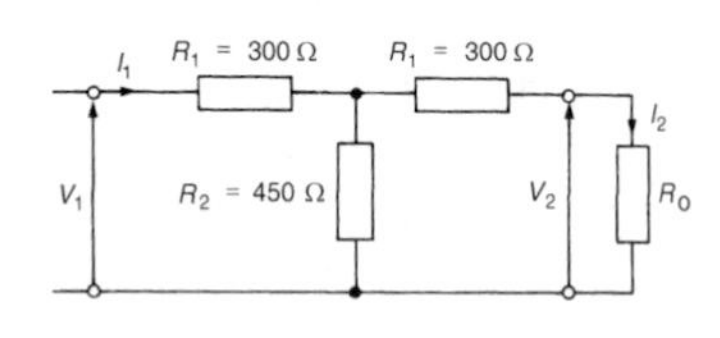

Figure 41.15

Problem 10. The attenuator shown in Figure 41.15 feeds a matched load. Determine (a) the characteristic impedance R_0, and (b) the insertion loss in decibels.

(a) From equation (41.10), the characteristic impedance of a symmetric T-pad attenuator is given by

$$R_0 = \sqrt{(R_1^2 + 2R_1R_2)} = \sqrt{[300^2 + 2(300)(450)]} = \mathbf{600\ \Omega}.$$

(b) Since the T-network is terminated in its characteristic impedance, then from equation (41.21),

$$\text{insertion loss} = 20\lg(V_1/V_2)\ \text{dB or}\ 20\lg(I_1/I_2)\ \text{dB}.$$

By current division in Figure 41.15,

$$I_2 = \left(\frac{R_2}{R_2 + R_1 + R_0}\right)(I_1)$$

Hence

$$\textbf{insertion loss} = 20\lg\frac{I_1}{I_2} = 20\lg\left(\frac{I_1}{(R_2/(R_2+R_1+R_0))I_1}\right)$$

$$= 20\lg\left(\frac{R_2 + R_1 + R_0}{R_2}\right)$$

$$= 20\lg\left(\frac{450 + 300 + 600}{450}\right)$$

$$= 20\lg 3 = \mathbf{9.54\ dB}$$

Problem 11. A 0–3 kΩ rheostat is connected across the output of a signal generator of internal resistance 500 Ω. If a load of 2 kΩ is connected across the rheostat, determine the insertion loss at a tapping of (a) 2 kΩ, (b) 1 kΩ.

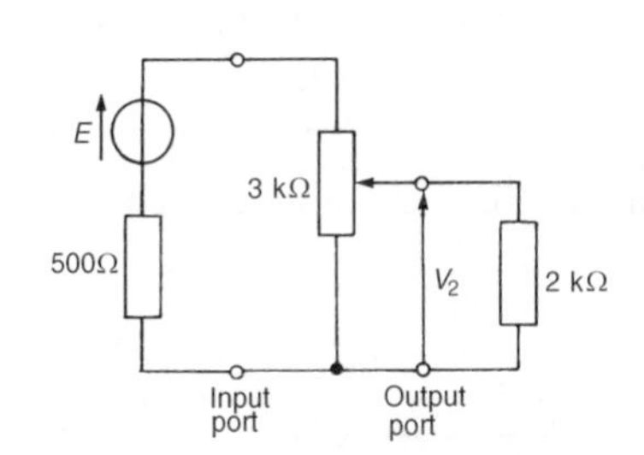

Figure 41.16

The circuit diagram is shown in Figure 41.16. Without the rheostat in the circuit the voltage across the 2 kΩ load, V_L (see Figure 41.17), is given by

$$V_L = \left(\frac{2000}{2000 + 500}\right)E = 0.8\ \text{E}$$

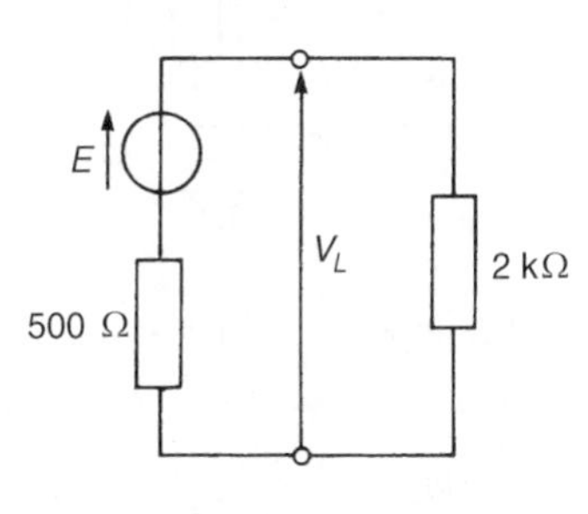

Figure 41.17

(a) With the 2 kΩ tapping, the network of Figure 41.16 may be redrawn as shown in Figure 41.18, which in turn is simplified as shown in Figure 41.19. From Figure 41.19,

$$\text{voltage } V_2 = \left(\frac{1000}{1000 + 1000 + 500}\right)E = 0.4\ \text{E}$$

Hence, from equation (41.19), insertion loss ratio,

$$A_L = \frac{V_L}{V_2} = \frac{0.8E}{0.4E} = \mathbf{2}$$

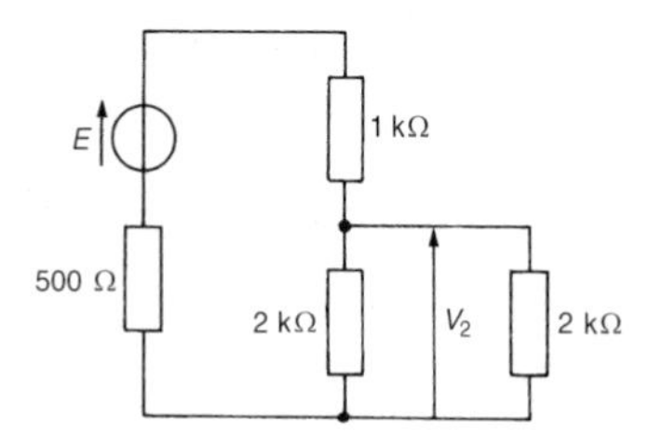

Figure 41.18

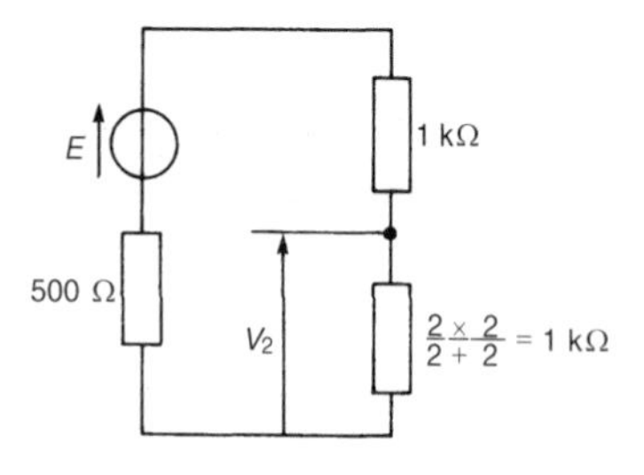

Figure 41.19

or, from equation (41.20),

$$\text{insertion loss} = 20 \lg(V_L/V_2) = 20 \lg 2 = \mathbf{6.02\ dB}$$

(b) With the 1 kΩ tapping, voltage V_2 is given by

$$V_2 = \left(\frac{(1000 \times 2000)/(1000 + 2000)}{((1000 \times 2000)/(1000 + 2000)) + 2000 + 500} \right) E$$

$$= \left(\frac{666.7}{666.7 + 2000 + 500} \right) E = 0.211\ \text{E}$$

Hence, from equation (41.19),

$$\text{insertion loss ratio } A_L = \frac{V_L}{V_2} = \frac{0.8E}{0.211E} = \mathbf{3.79}$$

or, from equation (41.20),

$$\text{insertion loss in decibels} = 20 \lg \left(\frac{V_L}{V_2} \right) = 20 \lg 3.79$$

$$= \mathbf{11.57\ dB}$$

(Note that the insertion loss is not doubled by halving the tapping.)

Problem 12. A symmetrical π-attenuator pad has a series arm of resistance 1000 Ω and shunt arms each of 500 Ω. Determine (a) its characteristic impedance, and (b) the insertion loss (in decibels) when feeding a matched load.

The π-attenuator pad is shown in Figure 41.20, terminated in its characteristic impedance, R_0.

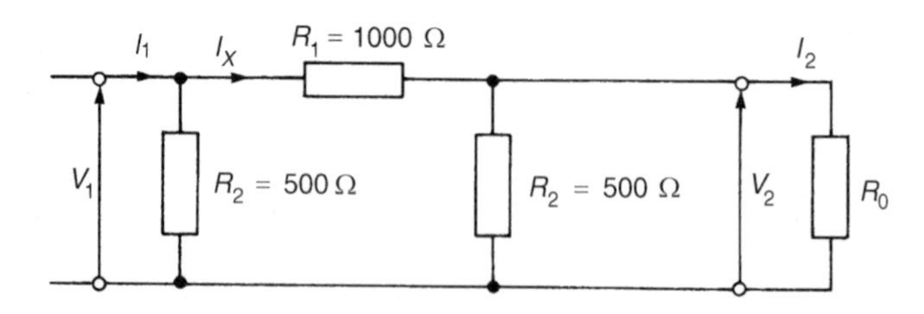

Figure 41.20

(a) From equation (41.15), the characteristic impedance of a symmetrical attenuator is given by

$$R_0 = \sqrt{\left(\frac{R_1 R_2^2}{R_1 + 2R_2} \right)} = \sqrt{\left(\frac{(1000)(500)^2}{1000 + 2(500)} \right)} = \mathbf{354\ \Omega}$$

(b) Since the attenuator network is feeding a matched load, from equation (41.21),

$$\text{insertion loss} = 20\lg\left(\frac{V_1}{V_2}\right)\ \text{dB} = 20\ \lg\left(\frac{I_1}{I_2}\right)\ \text{dB}$$

From Figure 41.20, by current division,

$$\text{current } I_X = \left\{\frac{R_2}{R_2 + R_1 + (R_2R_0/(R_2 + R_0))}\right\}(I_1)$$

$$\text{and current } I_2 = \left(\frac{R_2}{R_2 + R_0}\right) I_X$$

$$= \left(\frac{R_2}{R_2 + R_0}\right)\left(\frac{R_2}{R_2 + R_1 + (R_2R_0/(R_2 + R_0))}\right) I_1$$

i.e.,

$$I_2 = \left(\frac{500}{500 + 354}\right)\left(\frac{500}{500 + 1000 + ((500)(354)/(500 + 354))}\right) I_1$$

$$= (0.5855)(0.2929)I_1 = 0.1715I_1$$

Hence $I_1/I_2 = 1/0.1715 = 5.83$

Thus the insertion loss in decibels $= 20\lg(I_1/I_2)$

$= 20\lg 5.83 =$ **15.3 dB**

Further problems on insertion loss may be found in Section 41.9, problems 16 to 18, page 786.

41.6 Asymmetrical T- and π-sections

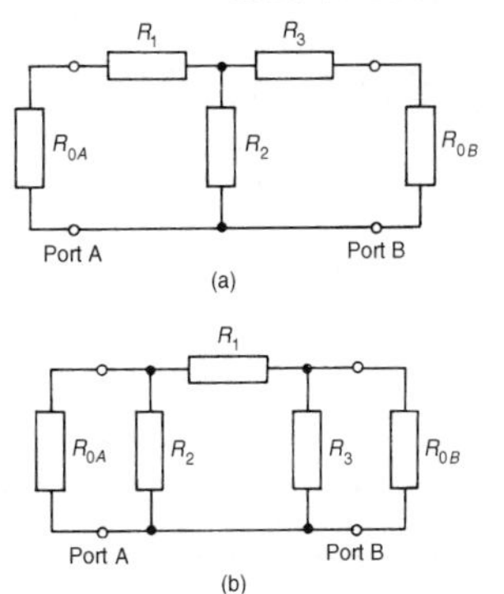

Figure 41.21 *(a) Asymmetrical T-pad section, (b) Asymmetrical π-section*

Figure 41.21(a) shows an asymmetrical T-pad section where resistance $R_1 \neq R_3$. Figure 41.21(b) shows an asymmetrical π-section where $R_2 \neq R_3$.

When viewed from port A, in each of the sections, the output impedance is R_{OB}; when viewed from port B, the input impedance is R_{OA}. Since the sections are asymmetrical R_{OA} does not have the same value as R_{OB}.

Iterative impedance is the term used for the impedance measured at one port of a two-port network when the other port is terminated with an impedance of the same value. For example, the impedance looking into port 1 of Figure 41.22(a) is, say, 500 Ω when port 2 is terminated in 500 Ω and the impedance looking into port 2 of Figure 41.22(b) is, say, 600 Ω when port 1 is terminated in 600 Ω. (In symmetric T- and π-sections the two iterative impedances are equal, this value being the characteristic impedance of the section.)

An **image impedance** is defined as the impedance which, when connected to the terminals of a network, equals the impedance presented to it at the opposite terminals. For example, the impedance looking into

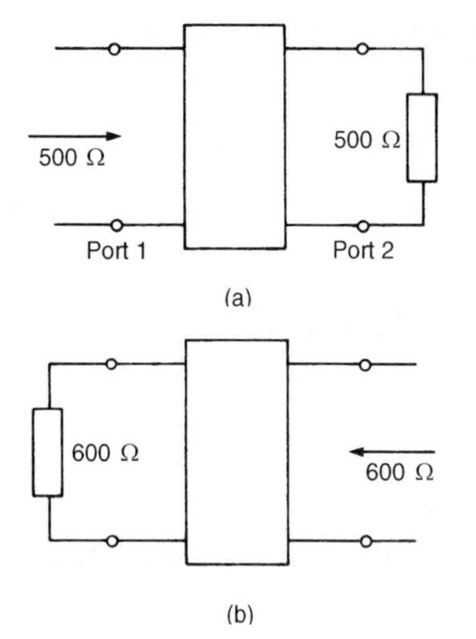

Figure 41.22

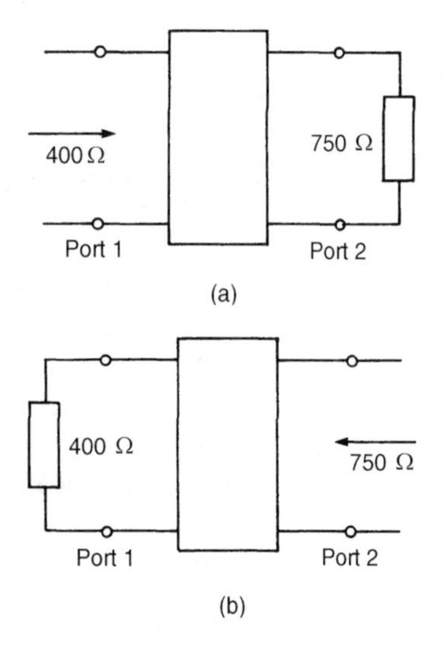

Figure 41.23

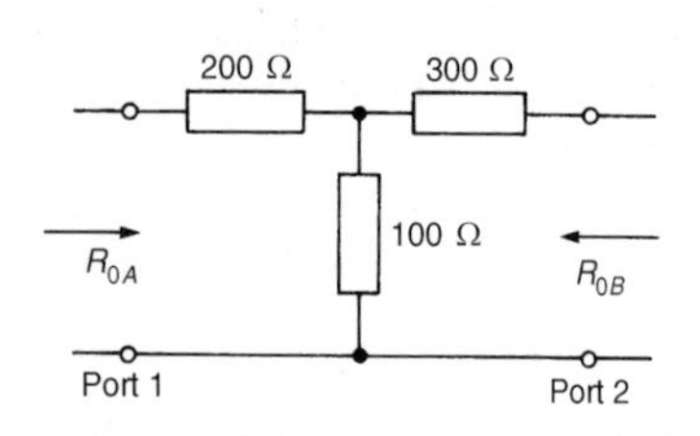

Figure 41.24

port 1 of Figure 41.23(a) is, say, 400 Ω when port 2 is terminated in, say 750 Ω, and the impedance seen looking into port 2 (Figure 41.23(b)) is 750 Ω when port 1 is terminated in 400 Ω. An asymmetrical network is correctly terminated when it is terminated in its image impedance. (If the image impedances are equal, the value is the characteristic impedance.)

The following worked problems show how the iterative and image impedances are determined for asymmetrical T- and π-sections.

Problem 13. An asymmetrical T-section attenuator is shown in Figure 41.24. Determine for the section (a) the image impedances, and (b) the iterative impedances.

(a) The image impedance R_{OA} seen at port 1 in Figure 41.24 is given by equation (41.11): $R_{OA} = \sqrt{(R_{OC})(R_{SC})}$, where R_{OC} and R_{SC} refer to port 2 being respectively open-circuited and short-circuited.

$$R_{OC} = 200 + 100 = 300\ \Omega$$

and $$R_{SC} = 200 + \frac{(100)(300)}{100 + 300} = 275\ \Omega$$

Hence $$R_{OA} = \sqrt{[(300)(275)]} = \mathbf{287.2\ \Omega}$$

Similarly, $R_{OB} = \sqrt{(R_{OC})(R_{SC})}$, where R_{OC} and R_{SC} refer to port 1 being respectively open-circuited and short-circuited.

$$R_{OC} = 300 + 100 = 400\ \Omega$$

and $$R_{SC} = 300 + \frac{(200)(100)}{200 + 100} = 366.7\ \Omega$$

Hence $$R_{OB} = \sqrt{[(400)(366.7)]} = \mathbf{383\ \Omega}.$$

Thus the image impedances are 287.2 Ω and 383 Ω and are shown in the circuit of Figure 41.25.

(Checking:

$$R_{OA} = 200 + \frac{(100)(300 + 383)}{100 + 300 + 383} = 287.2\ \Omega$$

and $$R_{OB} = 300 + \frac{(100)(200 + 287.2)}{100 + 200 + 287.2} = 383\ \Omega)$$

(b) The iterative impedance at port 1 in Figure 41.26, is shown as R_1. Hence

$$R_1 = 200 + \frac{(100)(300 + R_1)}{100 + 300 + R_1} = 200 + \frac{30\,000 + 100R_1}{400 + R_1}$$

from which $$400R_1 + R_1^2 = 80\,000 + 200R_1 + 30\,000 + 100R_1$$

and $$R_1^2 + 100R_1 - 110\,000 = 0$$

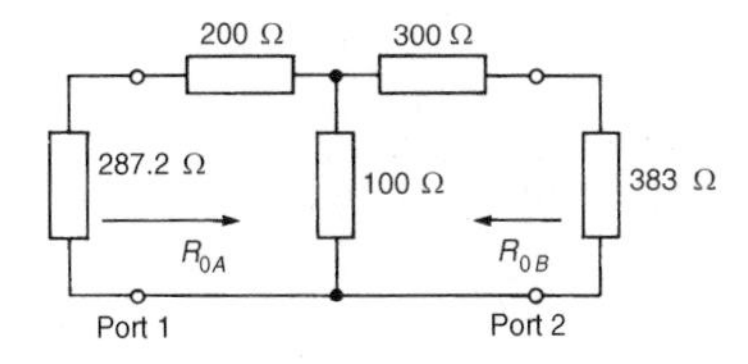

Figure 41.25

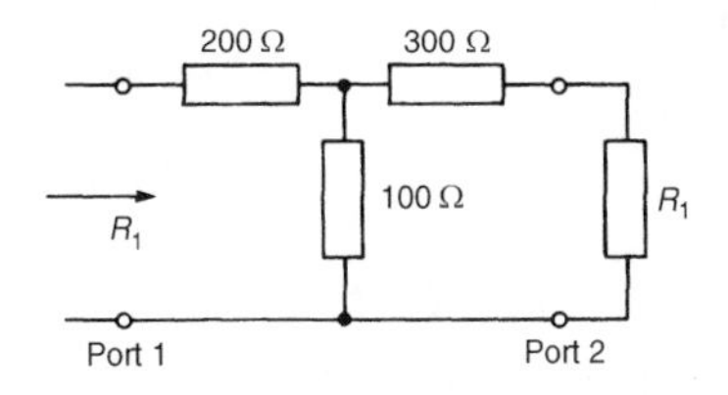

Figure 41.26

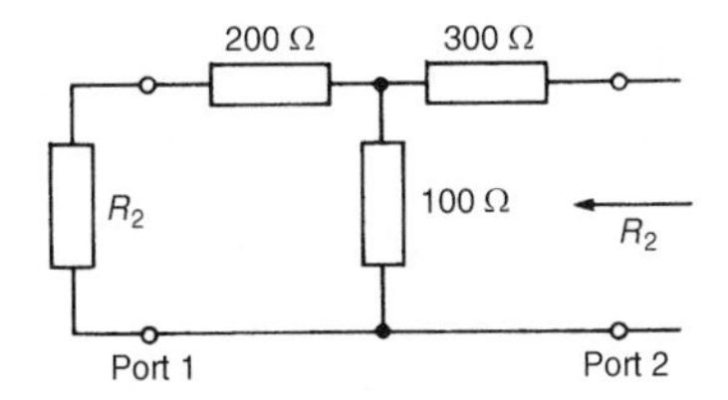

Figure 41.27

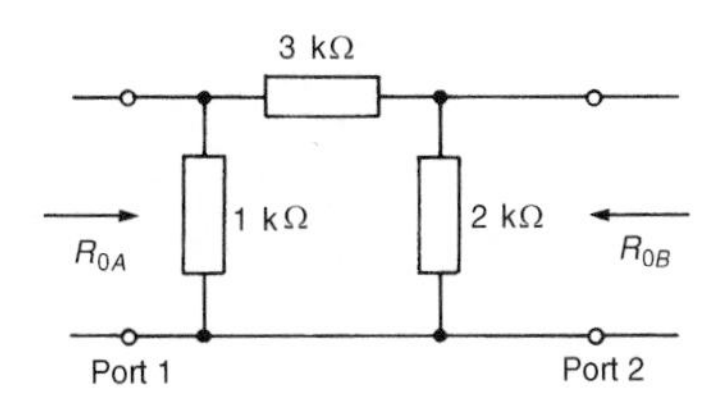

Figure 41.28

Solving by the quadratic formula gives

$$R_1 = \frac{-100 \pm \sqrt{[100^2 - (4)(1)(-110\,000)]}}{2}$$

$$= \frac{-100 \pm 670.8}{2} = \mathbf{285.4\ \Omega}$$

(neglecting the negative value).

The iterative impedance at port 2 in Figure 41.27 is shown as R_2. Hence

$$R_2 = 300 + \frac{(100)(200 + R_2)}{100 + 200 + R_2} = 300 + \frac{20\,000 + 100R_2}{300 + R_2}$$

from which $300R_2 + R_2^2 = 90\,000 + 300R_2 + 20\,000 + 100R_2$

and $R_2^2 - 100R_2 - 110\,000 = 0$

Thus $$R_2 = \frac{100 \pm \sqrt{[(-100)^2 - (4)(1)(-110\,000)]}}{2}$$

$$= \frac{100 \pm 670.8}{2} = \mathbf{385.4\ \Omega}$$

Thus the iterative impedances of the section shown in Figure 41.24 are 285.4 Ω and 385.4 Ω.

Problem 14. An asymmetrical π-section attenuator is shown in Figure 41.28. Determine for the section (a) the image impedances, and (b) the iterative impedances.

(a) The image resistance R_{OA} seen at port 1 is given by

$$R_{OA} = \sqrt{(R_{OC})(R_{SC})},$$

where the impedance at port 1 with port 2 open-circuited,

$$R_{OC} = \frac{(1000)(5000)}{1000 + 5000} = 833\ \Omega$$

and the impedance at port 1, with port 2 short-circuited,

$$R_{SC} = \frac{(1000)(3000)}{1000 + 3000} = 750\ \Omega$$

Hence $R_{OA} = \sqrt{[(833)(750)} = \mathbf{790\ \Omega}$.

Similarly, $R_{OB} = \sqrt{(R_{OC})(R_{SC})}$, where the impedance at port 2 with port 1 open-circuited,

$$R_{OC} = \frac{(2000)(4000)}{2000 + 4000} = 1333\ \Omega$$

and the impedance at port 2 with port 1 short-circuited,

$$R_{SC} = \frac{(2000)(3000)}{2000 + 3000} = 1200\ \Omega$$

Hence $R_{OB} = \sqrt{[(1333)(1200)]} = \mathbf{1265\ \Omega}$

Thus the image impedances are 790 Ω and 1265 Ω.

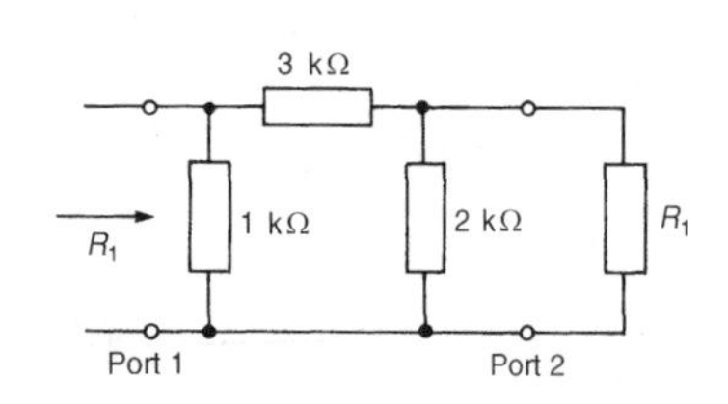

Figure 41.29

(b) The iterative impedance at port 1 in Figure 41.29 is shown as R_1. From circuit theory,

$$R_1 = \frac{1000[3000 + (2000R_1/(2000 + R_1))]}{1000 + 3000 + (2000R_1/(2000 + R_1))}$$

i.e., $$R_1 = \frac{3 \times 10^6 + (2 \times 10^6 R_1/(2000 + R_1))}{4000 + (2000R_1/(2000 + R_1))}$$

$$4000R_1 + \frac{2000R_1^2}{2000 + R_1} = 3 \times 10^6 + \frac{2 \times 10^6 R_1}{2000 + R_1}$$

$$8 \times 10^6 R_1 + 4000R_1^2 + 2000R_1^2 = 6 \times 10^9 + 3 \times 10^6 R_1 + 2 \times 10^6 R_1$$

$$6000R_1^2 + 3 \times 10^6 R_1 - 6 \times 10^9 = 0$$

$$2R_1^2 + 1000R_1 - 2 \times 10^6 = 0$$

Using the quadratic formula gives

$$R_1 = \frac{-1000 \pm \sqrt{[(1000)^2 - (4)(2)(-2 \times 10^6)]}}{4}$$

$$= \frac{-1000 \pm 4123}{4} = \mathbf{781\ \Omega}$$

(neglecting the negative value).

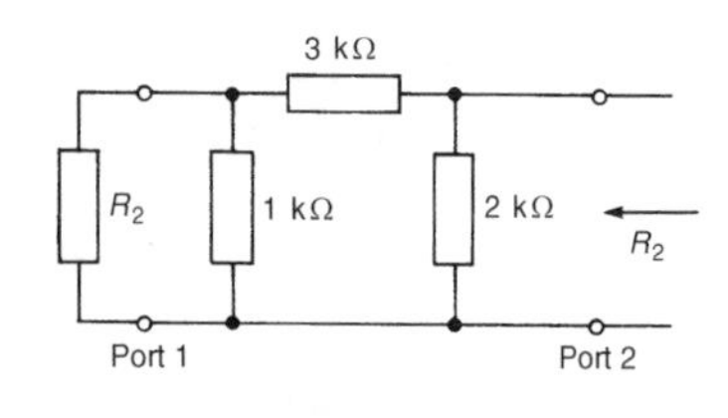

Figure 41.30

The iterative impedance at port 2 in Figure 41.30 is shown as R_2.

$$R_2 = \frac{2000[3000 + (1000R_2/(1000 + R_2))]}{2000 + 3000 + (1000R_2/(1000 + R_2))}$$

$$= \frac{6 \times 10^6 + (2 \times 10^6 R_2/(1000 + R_2))}{5000 + (1000R_2/(1000 + R_2))}$$

Hence

$$5000R_2 + \frac{1000R_2^2}{1000 + R_2} = 6 \times 10^6 + \frac{2 \times 10^6 R_2}{1000 + R_2}$$

$$5 \times 10^6 R_2 + 5000R_2^2 + 1000R_2^2 = 6 \times 10^9 + 6 \times 10^6 R_2 + 2 \times 10^6 R_2$$

$$6000R_2^2 - 3 \times 10^6 R_2 - 6 \times 10^9 = 0$$

$$2R_2^2 - 1000R_2 - 2 \times 10^6 = 0$$

from which

$$R_2 = \frac{1000 \pm \sqrt{[(-1000)^2 - (4)(2)(-2 \times 10^6)]}}{4}$$

$$= \frac{1000 \pm 4123}{4} = \mathbf{1281\ \Omega}$$

Thus the iterative impedances of the section shown in Figure 41.28 are 781 Ω and 1281 Ω.

Further problems on asymmetrical T—and π-sections may be found in Section 41.9, problems 19 to 21, page 787.

41.7 The L-section attenuator

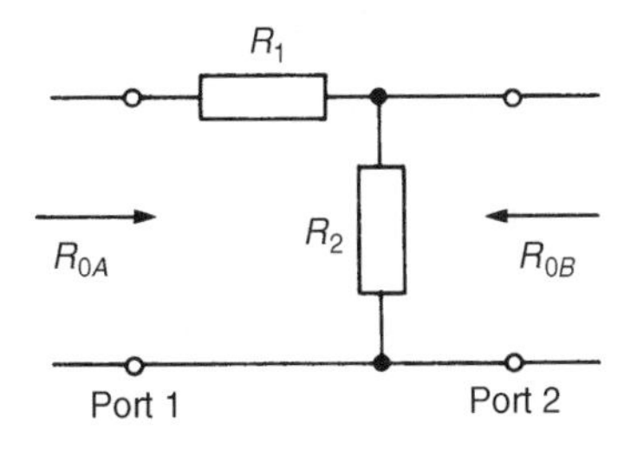

Figure 41.31 *L-section attenuator pad*

A typical L-section attenuator pad is shown in Figure 41.31. Such a pad is used for matching purposes only, the design being such that the attenuation introduced is a minimum. In order to derive values for R_1 and R_2, consider the resistances seen from either end of the section.

Looking in at port 1,

$$R_{OA} = R_1 + \frac{R_2 R_{OB}}{R_2 + R_{OB}}$$

from which

$$R_{OA}R_2 + R_{OA}R_{OB} = R_1R_2 + R_1R_{OB} + R_2R_{OB} \qquad (41.22)$$

Looking in at port 2,

$$R_{OB} = \frac{R_2(R_1 + R_{OA})}{R_1 + R_{OA} + R_2}$$

from which

$$R_{OB}R_1 + R_{OA}R_{OB} + R_{OB}R_2 = R_1R_2 + R_2R_{OA} \qquad (41.23)$$

Adding equations (41.22) and (41.23) gives

$$R_{OA}R_2 + 2R_{OA}R_{OB} + R_{OB}R_1 + R_{OB}R_2 = 2R_1R_2 + R_1R_{OB} + R_2R_{OB} + R_2R_{OA}$$

i.e., $2R_{OA}R_{OB} = 2R_1R_2$

and $$R_1 = \frac{R_{OA}R_{OB}}{R_2} \qquad (41.24)$$

Substituting this expression for R_1 into equation (41.22) gives

$$R_{OA}R_2 + R_{OA}R_{OB} = \left(\frac{R_{OA}R_{OB}}{R_2}\right)R_2 + \left(\frac{R_{OA}R_{OB}}{R_2}\right)R_{OB} + R_2R_{OB}$$

i.e., $$R_{OA}R_2 + R_{OA}R_{OB} = R_{OA}R_{OB} + \frac{R_{OA}R_{OB}^2}{R_2} + R_2R_{OB}$$

from which $$R_2(R_{OA} - R_{OB}) = \frac{R_{OA}R_{OB}^2}{R_2}$$

$$R_2^2(R_{OA} - R_{OB}) = R_{OA}R_{OB}^2$$

and resistance, $$\boldsymbol{R_2 = \sqrt{\left(\frac{R_{OA}R_{OB}^2}{R_{OA} - R_{OB}}\right)}} \qquad (41.25)$$

Thus, from equation (41.24),

$$R_1 = \frac{R_{OA}R_{OB}}{\sqrt{(R_{OA}R_{OB}^2/(R_{OA} - R_{OB}))}} = \frac{R_{OA}R_{OB}}{R_{OB}\sqrt{(R_{OA}/(R_{OA} - R_{OB}))}}$$

$$= \frac{R_{OA}}{\sqrt{R_{OA}}}\sqrt{(R_{OA} - R_{OB})}$$

Hence resistance, $$\boldsymbol{R_1 = \sqrt{[R_{OA}(R_{OA} - R_{OB})]}} \qquad (41.26)$$

Figure 41.32 shows an L-section attenuator pad with its resistances expressed in terms of the input and output resistances, R_{OA} and R_{OB}.

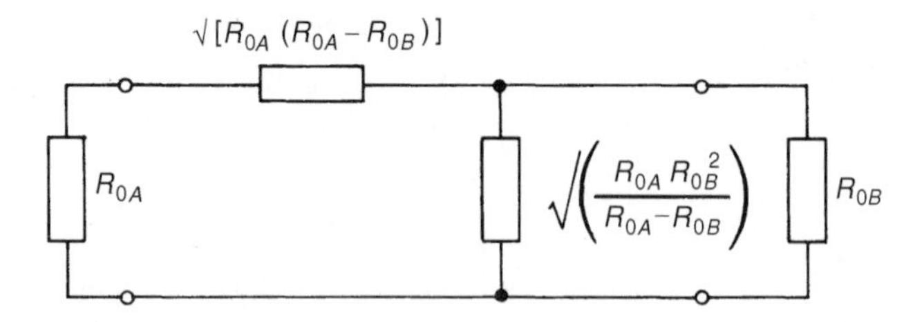

Figure 41.32

Problem 15. A generator having an internal resistance of 500 Ω is connected to a 100 Ω load via an impedance-matching resistance pad as shown in Figure 41.33. Determine (a) the values of resistance R_1 and R_2, (b) the attenuation of the pad in decibels, and (c) its insertion loss.

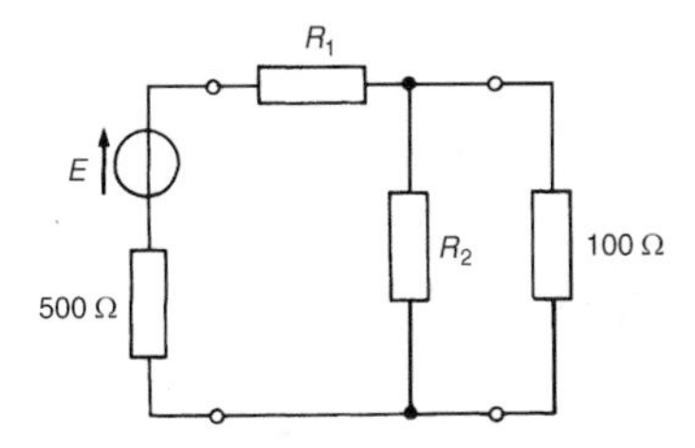

Figure 41.33

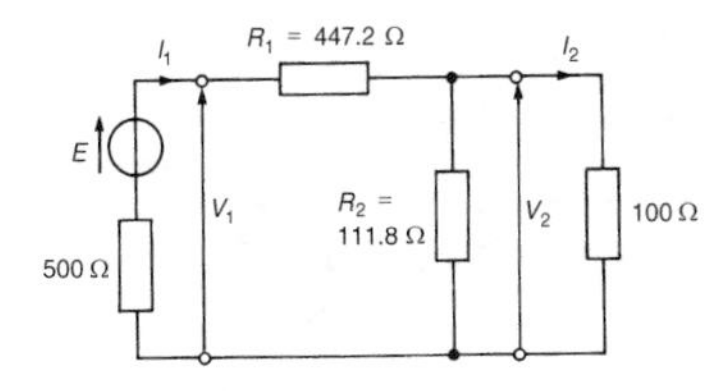

Figure 41.34

(a) From equation (41.26), $R_1 = \sqrt{[500(500 - 100)]} = \mathbf{447.2\ \Omega}$

From equation (41.25), $R_2 = \sqrt{\left(\dfrac{(500)(100)^2}{500 - 100}\right)} = \mathbf{111.8\ \Omega}$

(b) From section 41.3, the attenuation is given by $10\lg(P_1/P_2)$ dB. Note that, for an asymmetrical section such as that shown in Figure 41.33, the expression 20 $\lg(V_1/V_2)$ or 20 $\lg(I_1/I_2)$ may **not** be used for attenuation since the terminals of the pad are not matched to equal impedances. In Figure 41.34,

$$\text{current } I_1 = \frac{E}{500 + 447.2 + (111.8 \times 100/(111.8 + 100))}$$
$$= \frac{E}{1000}$$

and current

$$I_2 = \left(\frac{111.8}{111.8 + 100}\right) I_1 = \left(\frac{111.8}{211.8}\right)\left(\frac{E}{1000}\right) = \frac{E}{1894.5}$$

Thus input power,

$$P_1 = I_1^2(500) = \left(\frac{E}{1000}\right)^2 (500)$$

and output power,

$$P_2 = I_2^2(100) = \left(\frac{E}{1894.5}\right)^2 (100)$$

Hence

$$\text{attenuation} = 10\lg\frac{P_1}{P_2} = 10\lg\left\{\frac{[E/(1000)]^2(500)}{[E/(1894.5]^2(100)}\right\}$$
$$= 10\lg\left\{\left(\frac{1894.5}{1000}\right)^2 (5)\right\} \text{ dB}$$

i.e., **attenuation = 12.54 dB**

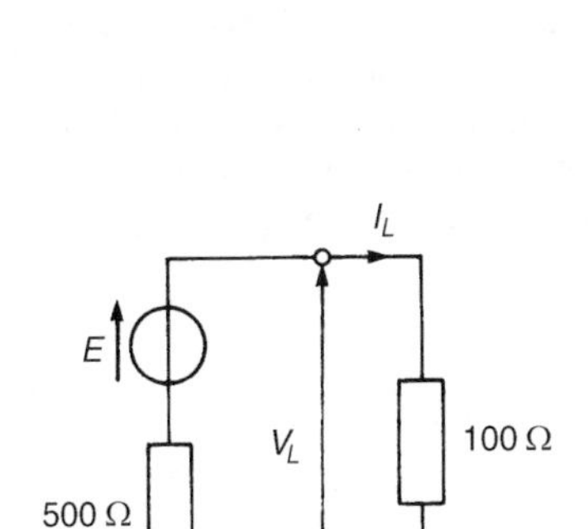

Figure 41.35

(c) Insertion loss A_L is defined as

$$\frac{\text{voltage across load when connected directly to the generator}}{\text{voltage across load when the two-port network is connected}}$$

Figure 41.35 shows the generator connected directly to the load.

$$\text{Load current, } I_L = \frac{E}{500 + 100} = \frac{E}{600}$$

$$\text{and voltage, } V_L = I_L(100) = \frac{E}{600}(100) = \frac{E}{6}$$

From Figure 41.34 voltage, $V_1 = E - I_1(500) = E - (E/1000)500$ from part (b)

i.e., $V_1 = 0.5$ E

voltage, $V_2 = V_1 - I_1R_1 = 0.5\text{ E} - \left(\frac{E}{1000}\right)(447.2) = 0.0528\text{ E}$

$$\textbf{insertion loss, } A_L = \frac{V_L}{V_2} = \frac{E/6}{0.0528E} = \mathbf{3.157}$$

$$\textbf{In decibels, the insertion loss} = 20\lg\frac{V_L}{V_2}$$

$$= 20\lg 3.157 = \mathbf{9.99\ dB}$$

Further problems on L-section attenuators may be found in Section 41.9, problems 22 and 23, page 787.

41.8 Two-port networks in cascade

Often two-port networks are connected in cascade, i.e., the output from the first network becomes the input to the second network, and so on, as shown in Figure 41.36. Thus an attenuator may consist of several cascaded sections so as to achieve a particular desired overall performance.

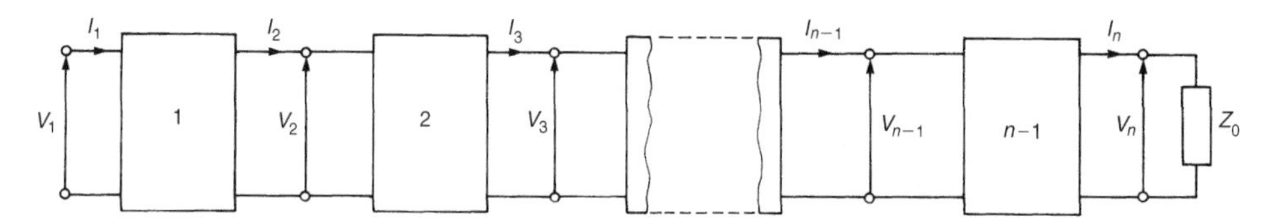

Figure 41.36 *Two-port networks connected in cascade*

If the cascade is arranged so that the impedance measured at one port and the impedance with which the other port is terminated have the same value, then each section (assuming they are symmetrical) will have the same characteristic impedance Z_0 and the last network will be terminated in Z_0. Thus each network will have a matched termination and hence the attenuation in decibels of section 1 in Figure 41.36 is given by $a_1 = 20\lg(V_1/V_2)$. Similarly, the attenuation of section 2 is given by $a_2 = 20\lg(V_2/V_3)$, and so on.

The overall attenuation is given by

$$a = 20\frac{V_1}{V_n}$$

$$= 20\lg\left(\frac{V_1}{V_2} \times \frac{V_2}{V_3} \times \frac{V_3}{V_4} \times \cdots \times \frac{V_{n-1}}{V_n}\right)$$

$$= 20\lg\frac{V_1}{V_2} + 20\lg\frac{V_2}{V_3} + \cdots + 20\lg\frac{V_{n-1}}{V_n}$$

by the laws of logarithms, i.e.,

$$\textbf{overall attenuation, } a = a_1 + a_2 + \cdots + a_{n-1} \tag{41.27}$$

Thus the overall attenuation is the sum of the attenuations (in decibels) of the matched sections.

Problem 16. Five identical attenuator sections are connected in cascade. The overall attenuation is 70 dB and the voltage input to the first section is 20 mV. Determine (a) the attenuation of each individual attenuation section, (b) the voltage output of the final stage, and (c) the voltage output of the third stage.

(a) From equation (41.27), the overall attenuation is equal to the sum of the attenuations of the individual sections and, since in this case each section is identical, **the attenuation of each section** = 70/5 = **14 dB**.

(b) If V_1 = the input voltage to the first stage and V_0 = the output of the final stage, then the overall attenuation $= 20\lg(V_1/V_0)$, i.e.,

$$70 = 20\lg\left(\frac{20}{V_0}\right) \text{ where } V_0 \text{ is in millivolts}$$

$$3.5 = \lg\left(\frac{20}{V_0}\right)$$

$$10^{3.5} = \frac{20}{V_0}$$

from which

$$\textbf{output voltage of final stage, } V_0 = \frac{20}{10^{3.5}} = 6.32 \times 10^{-3}\,\text{mV}$$

$$= \mathbf{6.32\ \mu V}$$

(c) The overall attenuation of three identical stages is $3 \times 14 = 42$ dB. Hence $42 = 20\lg(V_1/V_3)$, where V_3 is the voltage output of the third stage. Thus

$$\frac{42}{20} = \lg\left(\frac{20}{V_3}\right),\ 10^{42/20} = \frac{20}{V_3}$$

from which **the voltage output of the third stage,** $V_3 = 20/10^{2.1}$ = **0.159 mV**

Problem 17. A d.c. generator has an internal resistance of 450 Ω and supplies a 450 Ω load.

(a) Design a T-network attenuator pad having a characteristic impedance of 450 Ω which, when connected between the generator and the load, will reduce the load current to $\frac{1}{8}$ of its initial value.

(b) If two such networks as designed in (a) were connected in series between the generator and the load, determine the fraction of the initial current that would now flow in the load.

(c) Determine the attenuation in decibels given by four such sections as designed in (a).

The T-network attenuator is shown in Figure 41.37 connected between the generator and the load. Since it is matching equal impedances, the network is symmetrical.

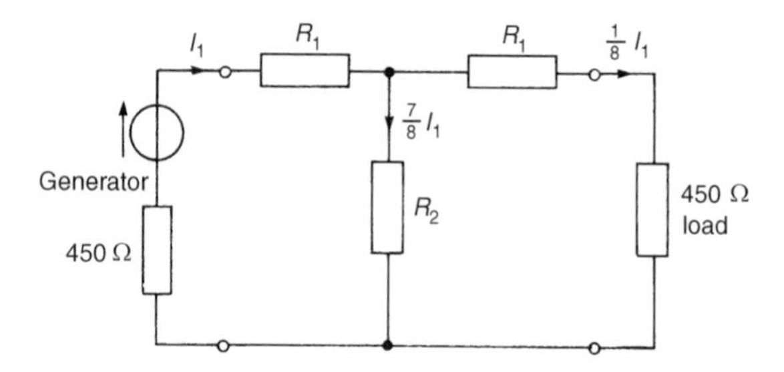

Figure 41.37

(a) Since the load current is to be reduced to $\frac{1}{8}$ of its initial value, the attenuation $N = 8$. From equation (41.13),

$$\text{resistance, } R_1 = \frac{R_0(N-1)}{(N+1)} = 450\frac{(8-1)}{(8+1)} = \mathbf{350\ \Omega}$$

and from equation (41.14),

$$\text{resistance, } R_2 = R_0\left(\frac{2N}{N^2-1}\right) = 450\left(\frac{2\times 8}{8^2-1}\right) = \mathbf{114\ \Omega}$$

(b) When two such networks are connected in series, as shown in Figure 41.38, current I_1 flows into the first stage and $\frac{1}{8}I_1$ flows out of the first stage into the second.

Again, $\frac{1}{8}$ of this current flows out of the second stage, i.e.,

$\frac{1}{8} \times \frac{1}{8}I_1$, i.e., $\frac{1}{64}$ of I_1 flows into the load.

Thus $\frac{1}{64}$ of the original current flows in the load.

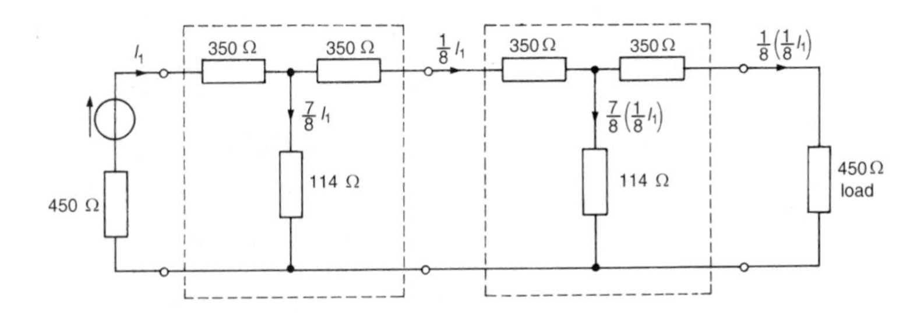

Figure 41.38

(c) The attenuation of a single stage is 8. Expressed in decibels, the attenuation is $20\lg(I_1/I_2) = 20\lg 8 = 18.06$ dB. From equation (41.27), the overall attenuation of four identical stages is given by $18.06 + 18.06 + 18.06 + 18.06$, i.e., **72.24 dB**.

Further problems on cascading two-port networks may be found in Section 41.9 following, problems 24 to 26, page 787.

41.9 Further problems on attenuators

Logarithmic ratios

1 The ratio of two powers is (a) 3, (b) 10, (c) 30, (d) 10000. Determine the decibel power ratio for each.
[(a) 4.77 dB (b) 10 dB (c) 14.8 dB (d) 40 dB]

2 The ratio of two powers is (a) $\frac{1}{10}$, (b) $\frac{1}{2}$, (c) $\frac{1}{40}$, (d) $\frac{1}{1000}$. Determine the decibel power ratio for each.
[(a) −10 dB (b) −3 dB (c) −16 dB (d) −30 dB]

3 An amplifier has (a) a gain of 25 dB, (b) an attenuation of 25 dB. If the input power is 12 mW, determine the output power in each case.
[(a) 3795 mW (b) 37.9 μW]

4 7.5% of the power supplied to a cable appears at the output terminals. Determine the attenuation in decibels. [11.25 dB]

5 The current input of a system is 250 mA. If the current ratio of the system is (i) 15 dB, (ii) −8 dB, determine (a) the current output and (b) the current ratio expressed in nepers.
[(i) (a) 1.406 A (b) 1.727 Np
(ii) (a) 99.53 mA (b) − 0.921 Np]

Symmetrical T — and π-attenuators

6 Determine the characteristic impedances of the T-network attenuator sections shown in Figure 41.39.
[(a) 26.46 Ω (b) 244.9 Ω (c) 1.342 kΩ]

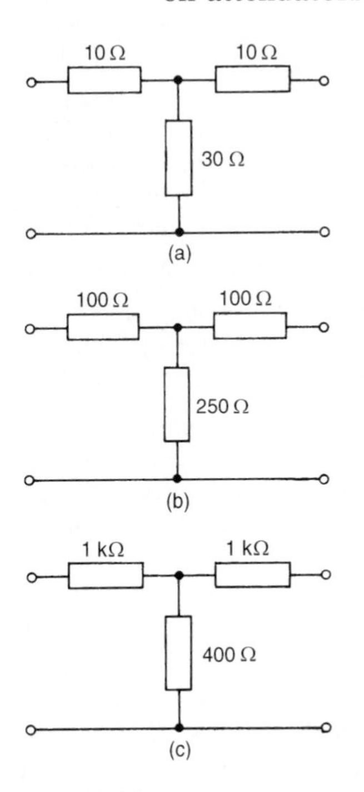

Figure 41.39

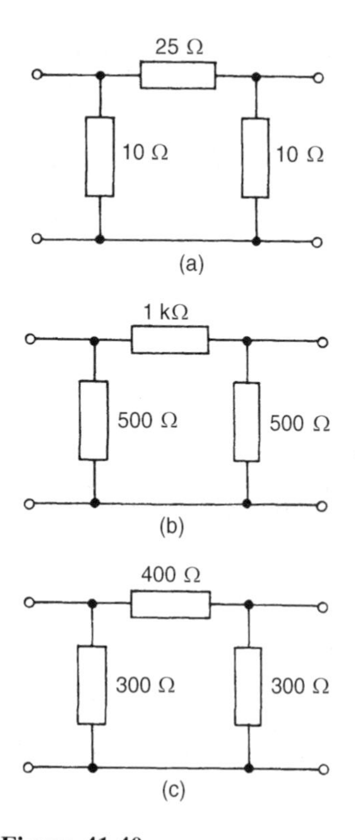

Figure 41.40

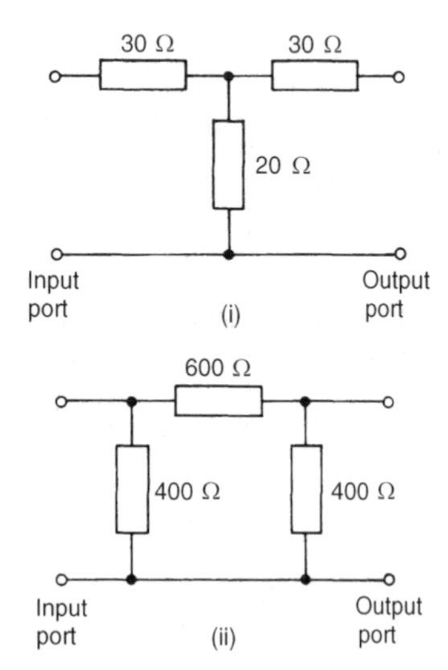

Figure 41.41

7 Determine the characteristic impedances of the π-network attenuator pads shown in Figure 41.40. [(a) 7.45 Ω (b) 353.6 Ω (c) 189.7 Ω]

8 A T-section attenuator is to provide 18 dB voltage attenuation per section and is to match a 1.5 kΩ line. Determine the resistance values necessary per section. [$R_1 = 1165$ Ω, $R_2 = 384$ Ω]

9 A π-section attenuator has a series resistance of 500 Ω and shunt resistances of 2 kΩ. Determine (a) the characteristic impedance, and (b) the attenuation produced by the network. [(a) 667 Ω (b) 6 dB]

10 For each of the attenuator pads shown in Figure 41.41 determine (a) the input resistance when the output port is open-circuited, (b) the input resistance when the output port is short-circuited, and (c) the characteristic impedance.

[(i) (a) 50 Ω (b) 42 Ω (c) 45.83 Ω
(ii) (a) 285.7 Ω (b) 240 Ω (c) 261.9 Ω]

11 A television signal received from an aerial through a length of coaxial cable of characteristic impedance 100 Ω has to be attenuated by 15 dB before entering the receiver. If the input impedance of the receiver is also 100 Ω, design a suitable T-attenuator network to give the necessary reduction. [$R_1 = 69.8$ Ω, $R_2 = 36.7$ Ω]

12 Design (a) a T-section symmetrical attenuator pad, and (b) a π-section symmetrical attenuator pad, to provide a voltage attenuation of 15 dB and having a characteristic impedance of 500 Ω.

[(a) $R_1 = 349$ Ω, $R_2 = 184$ Ω
(b) $R_1 = 1.36$ kΩ, $R_2 = 716$ Ω]

13 Determine the values of the shunt and series resistances for T-pad attenuators of characteristic impedance 400 Ω to provide the following voltage attenuations: (a) 12 dB (b) 25 dB (c) 36 dB

[(a) $R_1 = 239.4$ Ω, $R_2 = 214.5$ Ω
(b) $R_1 = 357.4$ Ω, $R_2 = 45.13$ Ω
(c) $R_1 = 387.5$ Ω, $R_2 = 12.68$ Ω]

14 Design a π-section symmetrical attenuator network to provide a voltage attenuation of 24 dB and having a characteristic impedance of 600 Ω. [$R_1 = 4.736$ kΩ, $R_2 = 680.8$ Ω]

15 A d.c. generator has an internal resistance of 600 Ω and supplies a 600 Ω load. Design a symmetrical (a) T-network and (b) π-network attenuator pad, having a characteristic impedance of 600 Ω which when connected between the generator and load will reduce the load current to $\frac{1}{4}$ its initial value.

[(a) $R_1 = 360$ Ω, $R_2 = 320$ Ω
(b) $R_1 = 1125$ Ω, $R_2 = 1000$ Ω]

Insertion loss

16 The attenuator section shown in Figure 41.42 feeds a matched load. Determine (a) the characteristic impedance R_0 and (b) the insertion loss. [(a) 282.8 Ω (b) 15.31 dB]

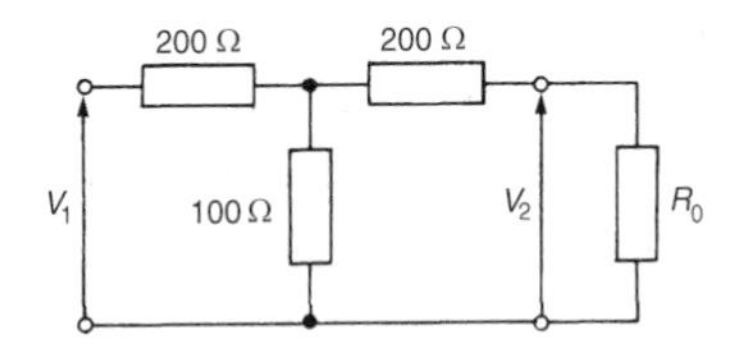

Figure 41.42

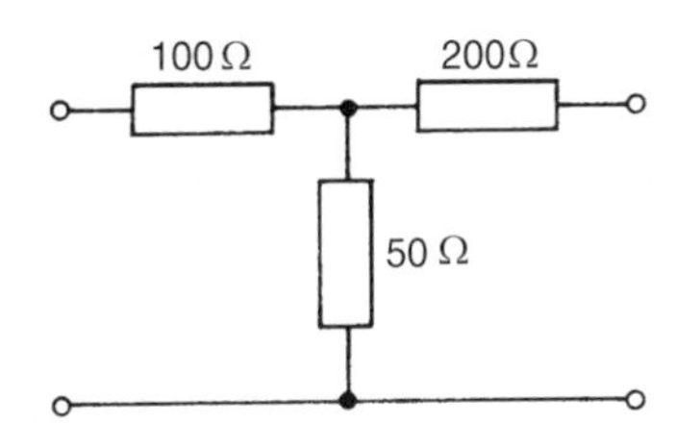

Figure 41.43

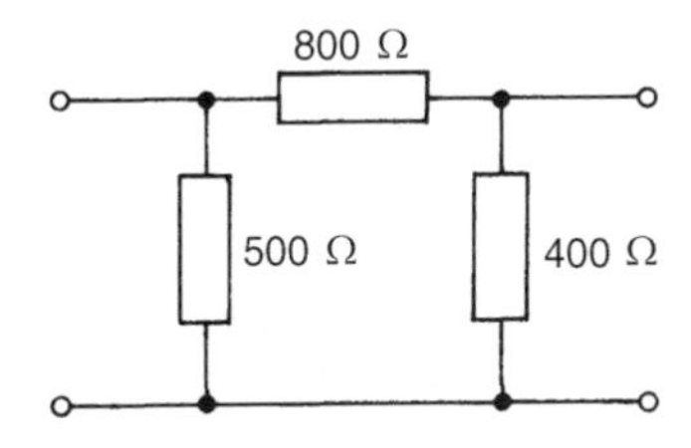

Figure 41.44

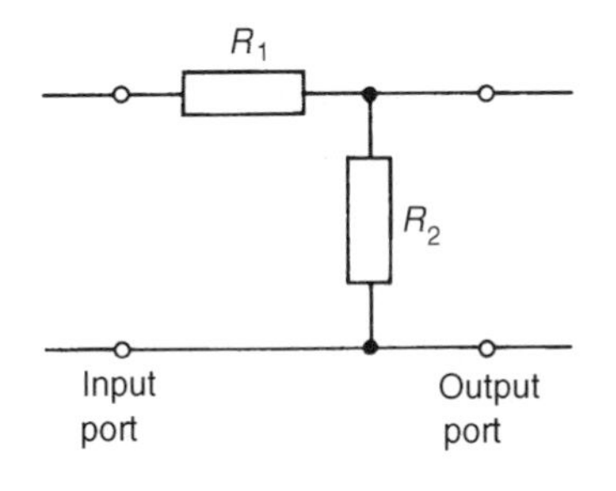

Figure 41.45

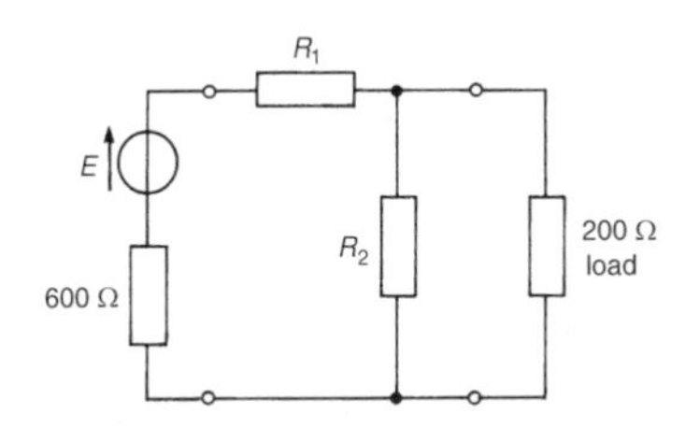

Figure 41.46

17 A 0–10 kΩ variable resistor is connected across the output of a generator of internal resistance 500 Ω. If a load of 1500 Ω is connected across the variable resistor, determine the insertion loss in decibels at a tapping of (a) 7.5 kΩ, (b) 2.5 kΩ

[(a) 8.13 dB (b) 17.09 dB]

18 A symmetrical π attenuator pad has a series arm resistance of 800 Ω and shunt arms each of 250 Ω. Determine (a) the characteristic impedance of the section, and (b) the insertion loss when feeding a matched load. [(a) 196.1 Ω (b) 18.36 dB]

Asymmetric T — and π-attenuators

19 An asymmetric section is shown in Figure 41.43. Determine for the section (a) the image impedances, and (b) the iterative impedances.

[(a) 144.9 Ω, 241.5 Ω (b) 143.6 Ω, 243.6 Ω]

20 An asymmetric π-section is shown in Figure 41.44. Determine for the section (a) the image impedances, and (b) the iterative impedances.

[(a) 329.5 Ω, 285.6 Ω (b) 331.2 Ω, 284.2 Ω]

21 Distinguish between image and iterative impedances of a network. An asymmetric T-attenuator section has series arms of resistance 200 Ω and 400 Ω respectively, and a shunt arm of resistance 300 Ω. Determine the image and iterative impedances of the section.

[(a) 430.9 Ω, 603.3 Ω; 419.6 Ω, 619.6 Ω]

L-section attenuators

22 Figure 41.45 shows an L-section attenuator. The resistance across the input terminals is 250 Ω and the resistance across the output terminals is 100 Ω. Determine the values R_1 and R_2.

[$R_1 = 193.6$ Ω, $R_2 = 129.1$ Ω]

23 A generator having an internal resistance of 600 Ω is connected to a 200 Ω load via an impedance-matching resistive pad as shown in Figure 41.46. Determine (a) the values of resistances R_1 and R_2, (b) the attenuation of the matching pad, and (c) its insertion loss.

[(a) $R_1 = 489.9$ Ω, $R_2 = 249.9$ Ω
(b) 9.96 dB (c) 8.71 dB]

Cascading two-port networks

24 The input to an attenuator is 24 V and the output is 4 V. Determine the attenuation in decibels. If five such identical attenuators are cascaded, determine the overall attenuation.

[(a) 15.56 dB, 77.80 dB]

25 Four identical attenuator sections are connected in cascade. The overall attenuation is 60 dB. The input to the first section is 50 mV.

Determine (a) the attenuation of each stage, (b) the output of the final stage, and (c) the output of the second stage.
[(a) 15 dB (b) 50 μV (c) 1.58 mV]

26 A d.c. generator has an internal resistance of 300 Ω and supplies a 300 Ω load.

(a) Design a symmetrical T network attenuator pad having a characteristic impedance of 300 Ω which, when connected between the generator and the load, will reduce the load current to $\frac{1}{3}$ its initial value.

(b) If two such networks as in (a) were connected in series between the generator and the load, what fraction of the initial current would the load take?

(c) Determine the fraction of the initial current that the load would take if six such networks were cascaded between the generator and the load.

(d) Determine the attenuation in decibels provided by five such identical stages as in (a).

[(a) $R_1 = 150\ \Omega$, $R_2 = 225\ \Omega$
(b) $\frac{1}{9}$ (c) $\frac{1}{729}$ (d) 44.71 dB]

Assignment 13

This assignment covers the material contained in chapters 39 to 41.

The marks for each question are shown in brackets at the end of each question.

1 The equivalent series circuit for a particular capacitor consists of a 2 Ω resistor in series with a 250 pF capacitor. Determine, at a frequency of 10 MHz (a) the loss angle of the capacitor, and (b) the power factor of the capacitor. (3)

2 A 50 V, 20 kHz supply is connected across a 500 pF capacitor and the power dissipated in the dielectric is 200 μW. Determine (a) the loss angle, (b) the equivalent series loss resistance, and (c) the equivalent parallel loss resistance. (9)

3 A coaxial cable, which has a core of diameter 12 mm and a sheath diameter of 30 mm, is 10 km long. Calculate for the cable (a) the inductance, assuming non-magnetic materials, and (b) the capacitance, assuming a dielectric of relative permittivity 5. (8)

4 A 50 km length single-phase twin line has conductors of diameter 20 mm and spaced 1.25 m apart in air. Determine for the line (a) the loop inductance, and (b) the capacitance. (8)

5 Find the strength of a uniform electric field if it is to have the same energy as that established by a magnetic field of flux density 1.15 T. (Assume that the relative permeability of the magnetic field and the relative permittivity of the electric field are both unity) (5)

6 8% of the power supplied to a cable appears at the output terminals. Determine the attenuation in decibels. (3)

7 Design (a) a T-section attenuator, and (b) a π-attenuator to provide a voltage attenuation of 25 dB and having a characteristic impedance of 620 Ω. (14)

42 Filter networks

At the end of this chapter you should be able to:

- appreciate the purpose of a filter network
- understand basic types of filter sections, i.e., low-pass, high-pass, band-pass and band-stop filters
- understand characteristic impedance and attenuation of filter sections
- understand low and high pass ladder networks
- design a low and high pass filter section
- calculate propagation coefficient and time delay in filter sections
- understand and design '*m*-derived' filter sections
- understand and design practical composite filters

42.1 Introduction

A **filter** is a network designed to pass signals having frequencies within certain bands (called **passbands**) with little attenuation, but greatly attenuates signals within other bands (called **attenuation bands** or **stopbands**).

As explained in the previous chapter, an attenuator network pad is composed of resistances only, the attenuation resulting being constant and independant of frequency. However, a filter is frequency sensitive and is thus composed of reactive elements. Since certain frequencies are to be passed with minimal loss, ideally the inductors and capacitors need to be pure components since the presence of resistance results in some attenuation at all frequencies.

Between the pass band of a filter, where ideally the attenuation is zero, and the attenuation band, where ideally the attenuation is infinite, is the **cut-off frequency**, this being the frequency at which the attenuation changes from zero to some finite value.

A filter network containing no source of power is termed **passive**, and one containing one or more power sources is known as an **active** filter network.

The filters considered in this chapter are symmetrical unbalanced T and π sections, the reactances used being considered as ideal.

Filters are used for a variety of purposes in nearly every type of electronic communications and control equipment. The bandwidths of filters used in communications systems vary from a fraction of a hertz to many megahertz, depending on the application.

42.2 Basic types of filter sections

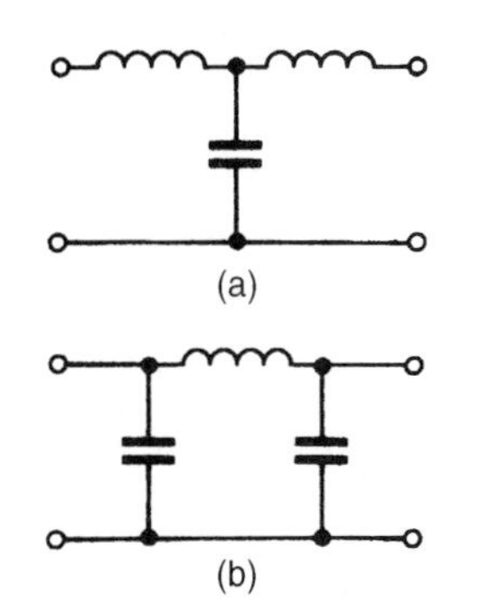

Figure 42.1

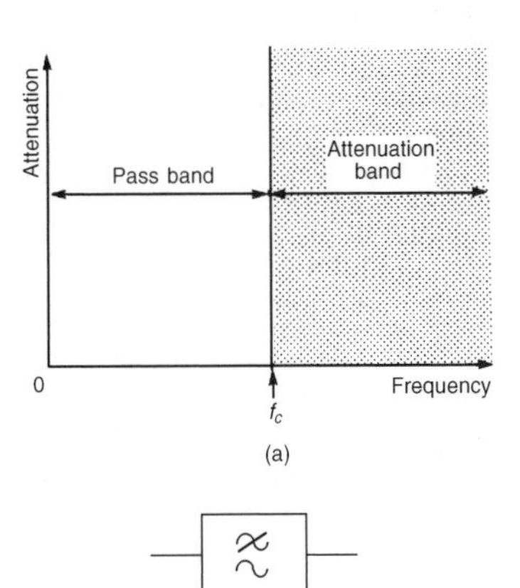

Figure 42.2

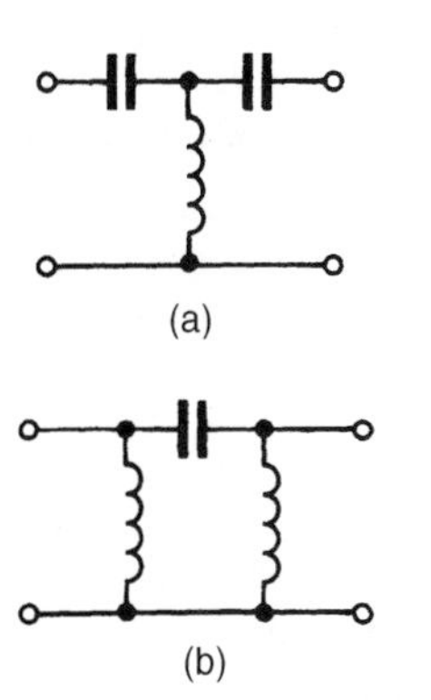

Figure 42.3

(a) Low-pass filters

Figure 42.1 shows simple unbalanced T and π section filters using series inductors and shunt capacitors. If either section is connected into a network and a continuously increasing frequency is applied, each would have a frequency-attenuation characteristic as shown in Figure 42.2(a). This is an ideal characteristic and assumes pure reactive elements. All frequencies are seen to be passed from zero up to a certain value without attenuation, this value being shown as f_c, the cut-off frequency; all values of frequency above f_c are attenuated. It is for this reason that the networks shown in Figures 42.1(a) and (b) are known as **low-pass filters**. The electrical circuit diagram symbol for a low-pass filter is shown in Figure 42.2(b).

Summarizing, a low-pass filter is one designed to pass signals at frequencies below a specified cut-off frequency.

When rectifiers are used to produce the d.c. supplies of electronic systems, a large ripple introduces undesirable noise and may even mask the effect of the signal voltage. Low-pass filters are added to smooth the output voltage waveform, this being one of the most common applications of filters in electrical circuits.

Filters are employed to isolate various sections of a complete system and thus to prevent undesired interactions. For example, the insertion of low-pass decoupling filters between each of several amplifier stages and a common power supply reduces interaction due to the common power supply impedance.

(b) High-pass filters

Figure 42.3 shows simple unbalanced T and π section filters using series capacitors and shunt inductors. If either section is connected into a network and a continuously increasing frequency is applied, each would have a frequency-attenuation characteristic as shown in Figure 42.4(a).

Once again this is an ideal characteristic assuming pure reactive elements. All frequencies below the cut-off frequency f_c are seen to be attenuated and all frequencies above f_c are passed without loss. It is for this reason that the networks shown in Figures 42.3(a) and (b) are known as **high-pass filters**. The electrical circuit-diagram symbol for a high-pass filter is shown in Figure 42.4(b).

Summarizing, a high-pass filter is one designed to pass signals at frequencies above a specified cut-off frequency.

The characteristics shown in Figures 42.2(a) and 42.4(a) are ideal in that they have assumed that there is no attenuation at all in the pass-bands and infinite attenuation in the attenuation bands. Both of these conditions are impossible to achieve in practice. Due to resistance, mainly in the inductive elements the attenuation in the pass-band will not be zero, and in a practical filter section the attenuation in the attenuation band will have a finite value. Practical characteristics for low-pass and high-pass filters are discussed in Sections 42.5 and 42.6. In addition to the resistive loss there is often an added loss due to mismatching. Ideally when a filter is inserted into a network it is matched to the impedance of that network.

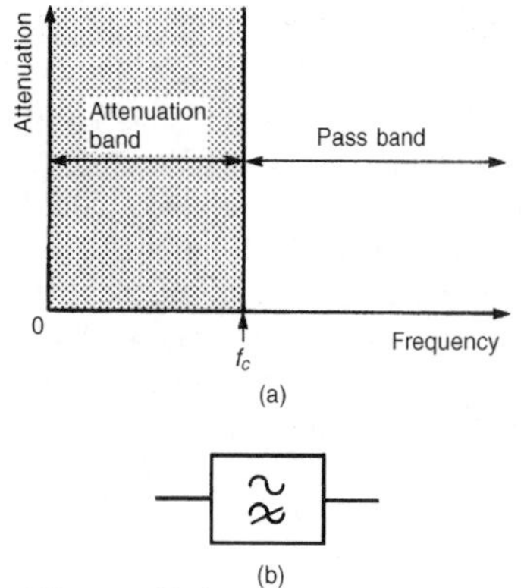

Figure 42.4

However the characteristic impedance of a filter section will vary with frequency and the termination of the section may be an impedance that does not vary with frequency in the same way. To minimize losses due to resistance and mismatching, filters are used under image impedance conditions as far as possible (see Chapter 41).

(c) Band-pass filters

A band-pass filter is one designed to pass signals with frequencies between two specified cut-off frequencies. The characteristic of an ideal band-pass filter is shown in Figure 42.5.

Such a filter may be formed by cascading a high-pass and a low-pass filter. f_{C_H} is the cut-off frequency of the high-pass filter and f_{C_L} is the cut-off frequency of the low-pass filter. As can be seen, $f_{C_L} > f_{C_H}$ for a band-pass filter, the pass-band being given by the difference between these values. The electrical circuit diagram symbol for a band-pass filter is shown in Figure 42.6.

Crystal and ceramic devices are used extensively as band-pass filters. They are common in the intermediate-frequency amplifiers of vhf radios where a precisely-defined bandwidth must be maintained for good performance.

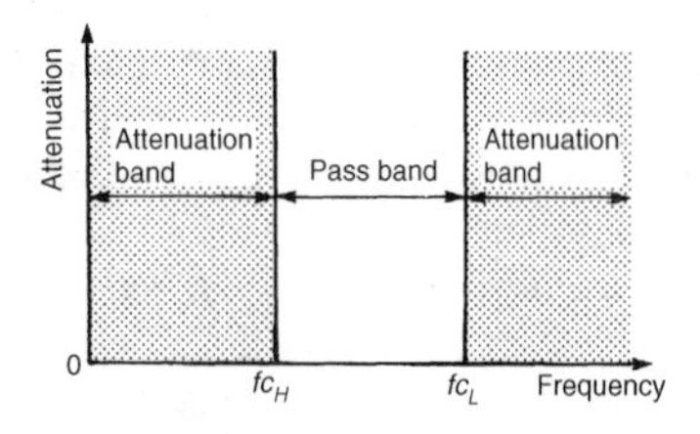

Figure 42.5

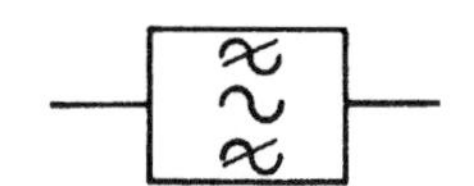

Figure 42.6

(d) Band-stop filters

A band-stop filter is one designed to pass signals with all frequencies except those between two specified cut-off frequencies. The characteristic of an ideal band-stop filter is shown in Figure 42.7. Such a filter may be formed by connecting a high-pass and a low-pass filter in parallel. As can be seen, for a band-stop filter $f_{C_H} > f_{C_L}$, the stop-band being given by the difference between these values. The electrical circuit diagram symbol for a band-stop filter is shown in Figure 42.8.

Sometimes, as in the case of interference from 50 Hz power lines in an audio system, the exact frequency of a spurious noise signal is known. Usually such interference is from an odd harmonic of 50 Hz, for example, 250 Hz. A sharply tuned band-stop filter, designed to attenuate the 250 Hz noise signal, is used to minimize the effect of the output. A high-pass filter with cut-off frequency greater than 250 Hz would also remove the interference, but some of the lower frequency components of the audio signal would be lost as well.

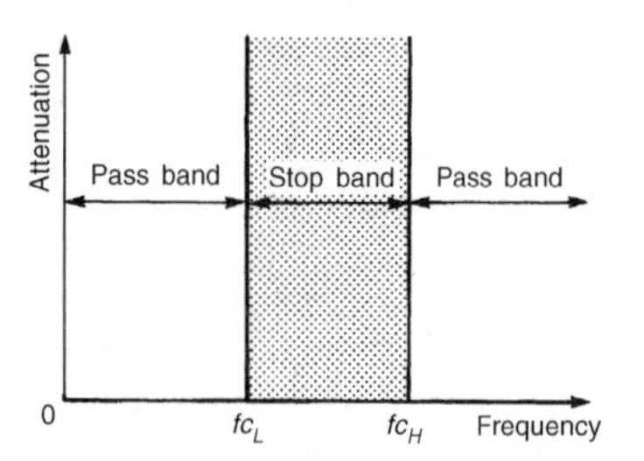

Figure 42.7

42.3 The characteristic impedance and the attenuation of filter sections

Nature of the input impedance

Let a symmetrical filter section be terminated in an impedance Z_O. If the input impedance also has a value of Z_O, then Z_O is the characteristic impedance of the section.

Figure 42.9 shows a T section composed of reactive elements X_A and X_B. If the reactances are of opposite kind, then the input impedance of the section, shown as Z_O, when the output port is open or short-circuited

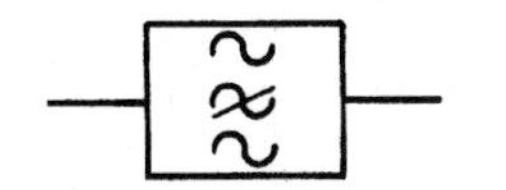

Figure 42.8

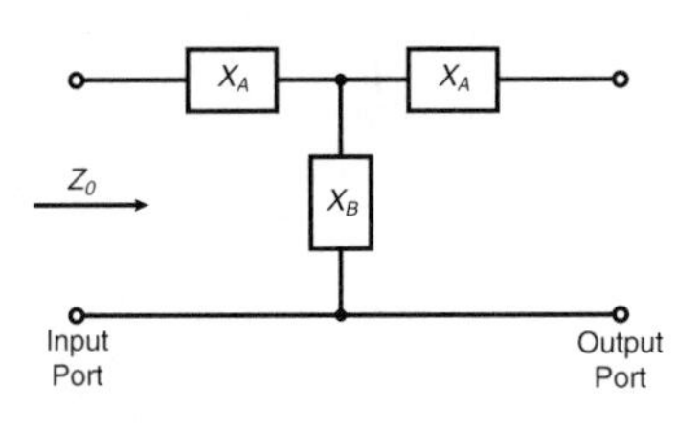

Figure 42.9

can be either inductive or capacitive depending on the frequency of the input signal.

For example, if X_A is inductive, say jX_L, and X_B is capacitive, say, $-jX_C$, then from Figure 42.9,

$$Z_{OC} = jX_L - jX_C = j(X_L - X_C)$$

and
$$Z_{SC} = jX_L + \frac{(jX_L)(-jX_C)}{(jX_L) + (-jX_C)} = jX_L + \frac{(X_L X_C)}{j(X_L - X_C)}$$
$$= jX_L - j\left(\frac{X_L X_C}{X_L - X_C}\right) = j\left(X_L - \frac{X_L X_C}{X_L - X_C}\right)$$

Since $X_L = 2\pi f L$ and $X_C = (1/2\pi f C)$ then Z_{OC} and Z_{SC} can be inductive, (i.e., positive reactance) or capacitive (i.e., negative reactance) depending on the value of frequency, f.

Let the magnitude of the reactance on open-circuit be X_{OC} and the magnitude of the reactance on short-circuit be X_{SC}. Since the filter elements are all purely reactive they may be expressed as jX_{OC} or jX_{SC}, where X_{OC} and X_{SC} are real, being positive or negative in sign. Four combinations of Z_{OC} and Z_{SC} are possible, these being:

(i) $Z_{OC} = +jX_{OC}$ and $Z_{SC} = -jX_{SC}$

(ii) $Z_{OC} = -jX_{OC}$ and $Z_{SC} = +jX_{SC}$

(iii) $Z_{OC} = +jX_{OC}$ and $Z_{SC} = +jX_{SC}$

and (iv) $Z_{OC} = -jX_{OC}$ and $Z_{SC} = -jX_{SC}$

From general circuit theory, input impedance Z_O is given by:

$$Z_O = \sqrt{(Z_{OC} Z_{SC})}$$

Taking either of combinations (i) and (ii) above gives:

$$Z_O = \sqrt{(-j^2 X_{OC} X_{SC})} = \sqrt{(X_{OC} X_{SC})},$$

which is real, thus the input impedance will be **purely resistive**.

Taking either of combinations (iii) and (iv) above gives:

$$Z_O = \sqrt{(j^2 X_{OC} X_{SC})} = +j\sqrt{(X_{OC} X_{SC})},$$

which is imaginary, thus the input impedance will be **purely reactive**.

Thus since the magnitude and nature of Z_{OC} and Z_{SC} depend upon frequency then so also will the magnitude and nature of the input impedance Z_O depend upon frequency.

Characteristic impedance

Figure 42.10 shows a low-pass T section terminated in its characteristic impedance, Z_O.

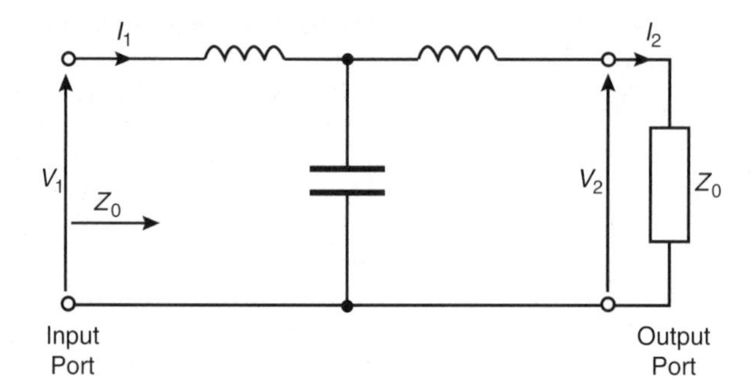

Figure 42.10

From equation (41.2), page 760, the characteristic impedance is given by $Z_O = \sqrt{(Z_{OC}Z_{SC})}$.

The following statements may be demonstrated to be true for any filter:

(a) *The attenuation is zero throughout the frequency range for which the characteristic impedance is purely resistive.*

(b) *The attenuation is finite throughout the frequency range for which the characteristic impedance is purely reactive.*

To demonstrate statement (a) above:

Let the filter shown in Figure 42.10 be operating over a range of frequencies such that Z_O is purely resistive.

From Figure 42.10, $Z_O = \dfrac{V_1}{I_1} = \dfrac{V_2}{I_2}$

Power dissipated in the output termination, $P_2 = V_2 I_2 \cos\phi_2 = V_2 I_2$ (since $\phi_2 = 0$ with a purely resistive load).

Power delivered at the input terminals,

$$P_1 = V_1 I_1 \cos\phi_1 = V_1 I_1 (\text{since } \phi_1 = 0)$$

No power is absorbed by the filter elements since they are purely reactive.

Hence $\quad P_2 = P_1, V_2 = V_1$ and $I_2 = I_1$.

Thus if the filter is terminated in Z_O and operating in a frequency range such that Z_O is purely resistive, then all the power delivered to the input is passed to the output and there is therefore no attenuation.

To demonstrate statement (b) above:

Let the filter be operating over a range of frequencies such that Z_O is purely reactive.

Then, from Figure 42.10, $\dfrac{V_1}{I_1} = jZ_O = \dfrac{V_2}{I_2}$.

Thus voltage and current are at 90° to each other which means that the circuit can neither accept nor deliver any active power from the source to the load ($P = VI\cos\phi = VI\cos 90° = VI(0) = 0$). There is therefore infinite attenuation, theoretically. (In practise, the attenuation is finite, for the condition $(V_1/I_1) = (V_2/I_2)$ can hold for $V_2 < V_1$ and $I_2 < I_1$, since the voltage and current are 90° out of phase.)

Statements (a) and (b) above are important because they can be applied to determine the cut-off frequency point of any filter section simply from a knowledge of the nature of Z_O. **In the pass band, Z_O is real, and in the attenuation band, Z_O is imaginary**. The cut-off frequency is therefore at the point on the frequency scale at which Z_O changes from a real quantity to an imaginary one, or vice versa (see Sections 42.5 and 42.6).

42.4 Ladder networks

Low-pass networks

Figure 42.11 shows a low-pass network arranged as a ladder or repetitive network. Such a network may be considered as a number of T or π sections in cascade. In Figure 42.12(a), a T section may be taken from the ladder by removing ABED, producing the low-pass filter section shown in Figure 42.13(a). The ladder has been cut in the centre of each of its inductive elements hence giving $L/2$ as the series arm elements in Figure 42.13(a).

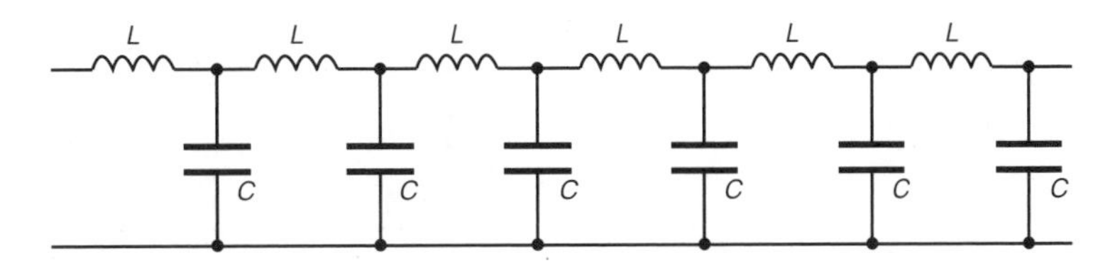

Figure 42.11

Similarly, a π section may be taken from the ladder shown in Figure 42.12(a) by removing FGJH, producing the low-pass filter section

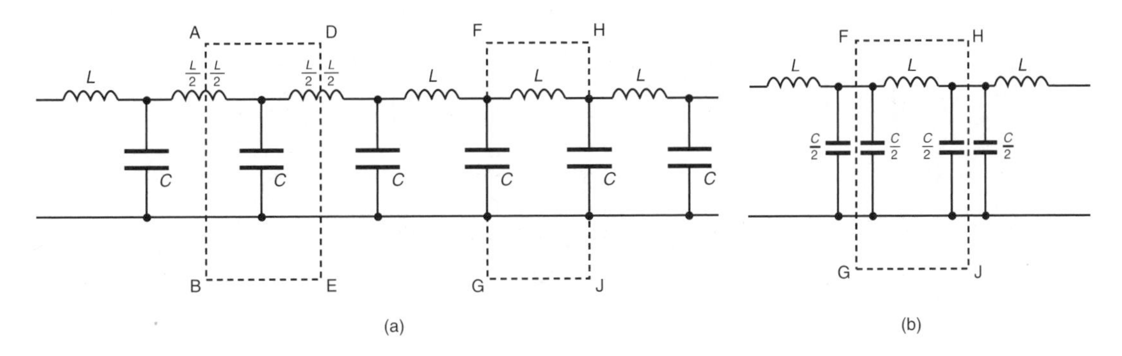

Figure 42.12

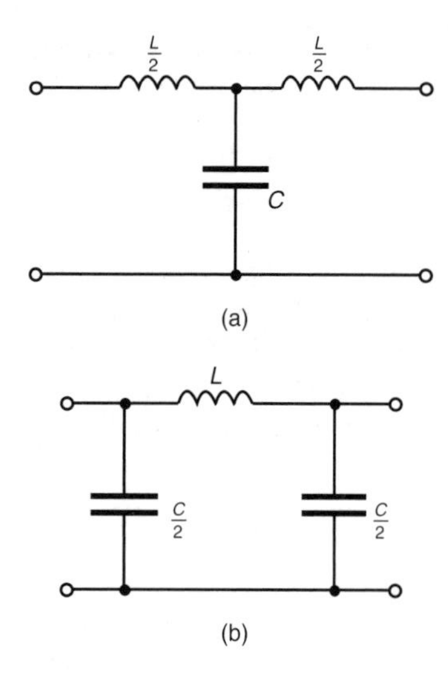

Figure 42.13

shown in Figure 42.13(b). The shunt element C in Figure 42.12(a) may be regarded as two capacitors in parallel, each of value $C/2$ as shown in the part of the ladder redrawn in Figure 42.12(b). (Note that for parallel capacitors, the total capacitance C_T is given by

$$C_T = C_1 + C_2 + \cdots. \text{ In this case } \frac{C}{2} + \frac{C}{2} = C).$$

The ladder network of Figure 42.11 can thus either be considered to be a number of the T networks shown in Figure 42.13(a) connected in cascade, or a number of the π networks shown in Figure 42.13(b) connected in cascade.

It is shown in Section 44.3, page 871, that an infinite transmission line may be reduced to a repetitive low-pass filter network.

High-pass networks

Figure 42.14 shows a high-pass network arranged as a ladder. As above, the repetitive network may be considered as a number of T or π sections in cascade.

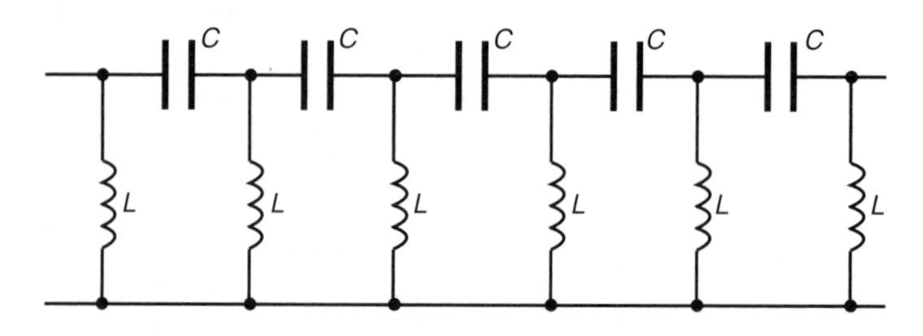

Figure 42.14

In Figure 42.15, a T section may be taken from the ladder by removing ABED, producing the high-pass filter section shown in Figure 42.16(a).

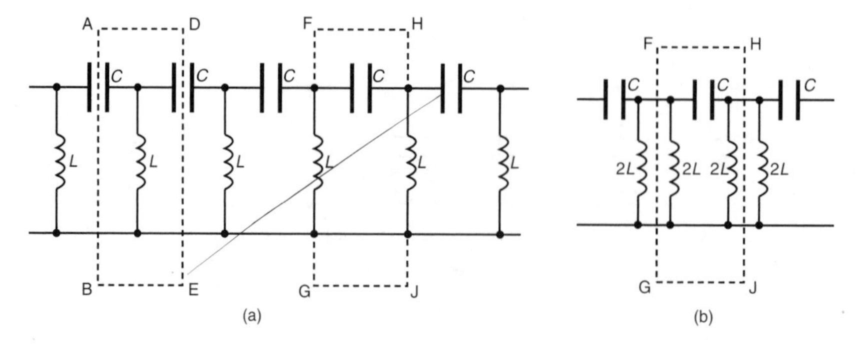

Figure 42.15

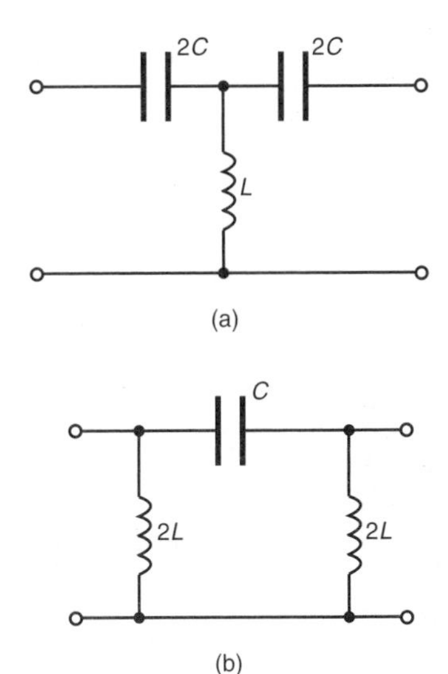

Figure 42.16

Note that the series arm elements are each $2C$. This is because two capacitors each of value $2C$ connected in series gives a total equivalent value of C, (i.e., for series capacitors, the total capacitance C_T is given by

$$\frac{1}{C_T} = \frac{1}{C_1} + \frac{1}{C_2} + \cdots)$$

Similarly, a π section may be taken from the ladder shown in Figure 42.15 by removing FGJH, producing the high-pass filter section shown in Figure 42.16(b). The shunt element L in Figure 42.15(a) may be regarded as two inductors in parallel, each of value $2L$ as shown in the part of the ladder redrawn in Figure 42.15(b). (Note that for parallel inductance, the total inductance L_T is given by

$$\frac{1}{L_T} = \frac{1}{L_1} + \frac{1}{L_2} + \cdots. \text{ In this case, } \frac{1}{2L} + \frac{1}{2L} = \frac{1}{L}.)$$

The ladder network of Figure 42.14 can thus be considered to be either a number of T networks shown in Figure 42.16(a) connected in cascade, or a number of the π networks shown in Figure 42.16(b) connected in cascade.

42.5 Low-pass filter sections

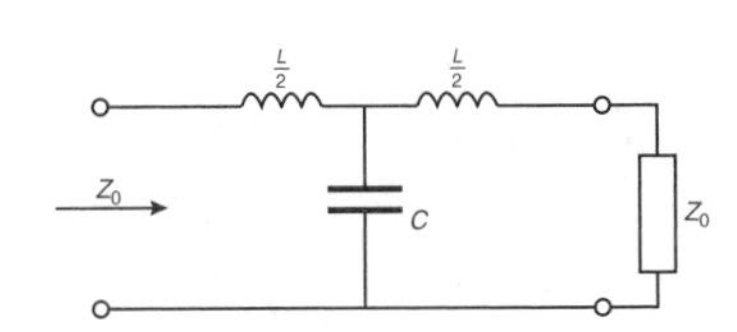

Figure 42.17

(a) The cut-off frequency

From equation (41.1), the characteristic impedance Z_0 for a symmetrical T network is given by: $Z_0 = \sqrt{(Z_A^2 + 2Z_AZ_B)}$. Applying this to the low-pass T section shown in Figure 42.17,

$$Z_A = \frac{j\omega L}{2} \text{ and } Z_B = \frac{1}{j\omega C}$$

$$\text{Thus}\quad Z_0 = \sqrt{\left[\frac{j^2\omega^2L^2}{4} + 2\left(\frac{j\omega L}{2}\right)\left(\frac{1}{j\omega C}\right)\right]}$$

$$= \sqrt{\left(\frac{-\omega^2L^2}{4} + \frac{L}{C}\right)}$$

$$\text{i.e.,}\quad Z_0 = \sqrt{\left(\frac{L}{C} - \frac{\omega^2L^2}{4}\right)} \qquad (42.1)$$

Z_0 will be real if $\dfrac{L}{C} > \dfrac{\omega^2L^2}{4}$

Thus attenuation will commence when $\dfrac{L}{C} = \dfrac{\omega^2L^2}{4}$

$$\text{i.e.,}\quad \text{when } \omega_c^2 = \frac{4}{LC} \qquad (42.2)$$

where $\omega_c = 2\pi f_c$ and f_c is the cut-off frequency.

Thus $\quad (2\pi f_c)^2 = \dfrac{4}{LC}$

$$2\pi f_c = \sqrt{\left(\frac{4}{LC}\right)} = \frac{2}{\sqrt{(LC)}}$$

and $$f_c = \frac{2}{2\pi\sqrt{(LC)}} = \frac{1}{\pi\sqrt{(LC)}}$$

i.e., **the cut-off frequency,** $\boxed{f_c = \dfrac{1}{\pi\sqrt{(LC)}}}$ (42.3)

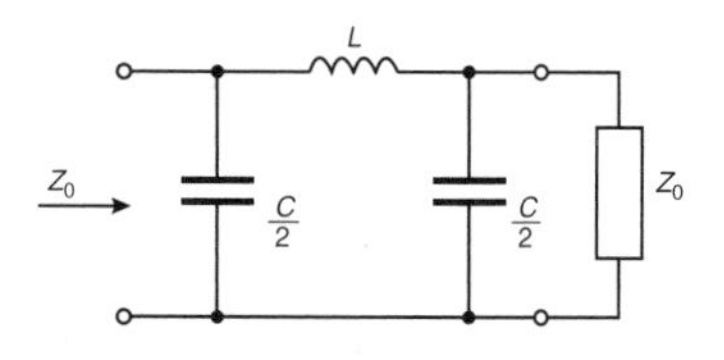

Figure 42.18

The same equation for the cut-off frequency is obtained for the low-pass π network shown in Figure 42.18 as follows:
From equation (41.3), for a symmetrical π network,

$$Z_0 = \sqrt{\left(\frac{Z_1 Z_2^2}{Z_1 + 2Z_2}\right)}$$

Applying this to Figure 42.18 $Z_1 = j\omega L$ and $Z_2 = \dfrac{1}{j\omega\dfrac{C}{2}} = \dfrac{2}{j\omega C}$

Thus $$Z_0 = \sqrt{\left\{\frac{(j\omega L)\left(\dfrac{2}{j\omega C}\right)^2}{j\omega L + 2\left(\dfrac{2}{j\omega C}\right)}\right\}} = \sqrt{\left\{\frac{(j\omega L)\left(\dfrac{4}{-\omega^2 C^2}\right)}{j\omega L - j\left(\dfrac{4}{\omega C}\right)}\right\}}$$

$$= \sqrt{\left\{\frac{-j\dfrac{4L}{\omega C^2}}{j\left(\omega L - \dfrac{4}{\omega C}\right)}\right\}} = \sqrt{\left\{\frac{\dfrac{4L}{\omega C^2}}{\dfrac{4}{\omega C} - \omega L}\right\}}$$

$$= \sqrt{\left\{\frac{4L}{\omega C^2\left(\dfrac{4}{\omega C} - \omega L\right)}\right\}} = \sqrt{\left(\frac{4L}{4C - \omega^2 L C^2}\right)}$$

i.e., $$Z_0 = \sqrt{\left(\frac{1}{\dfrac{C}{L} - \dfrac{\omega^2 C^2}{4}}\right)} \qquad (42.4)$$

Z_0 will be real if $\dfrac{C}{L} > \dfrac{\omega^2 C^2}{4}$

Thus attenuation will commence when $\dfrac{C}{L} = \dfrac{\omega^2 C^2}{4}$

i.e., when $\omega_c^2 = \dfrac{4}{LC}$

from which, **cut-off frequency,** $\boxed{f_c = \dfrac{1}{\pi\sqrt{(LC)}}}$ as in equation (42.3)).

(b) Nominal impedance

When the frequency is very low, ω is small and the term $(\omega^2 L^2/4)$ in equation (42.1) (or the term $(\omega^2 C^2/4)$ in equation (42.4)) may be neglected. The characteristic impedance then becomes equal to $\sqrt{(L/C)}$, which is purely resistive. This value of the characteristic impedance is known as the **design impedance** or the **nominal impedance** of the section and is often given the symbol R_0,

i.e., $$\boxed{R_0 = \sqrt{\frac{L}{C}}} \qquad (42.5)$$

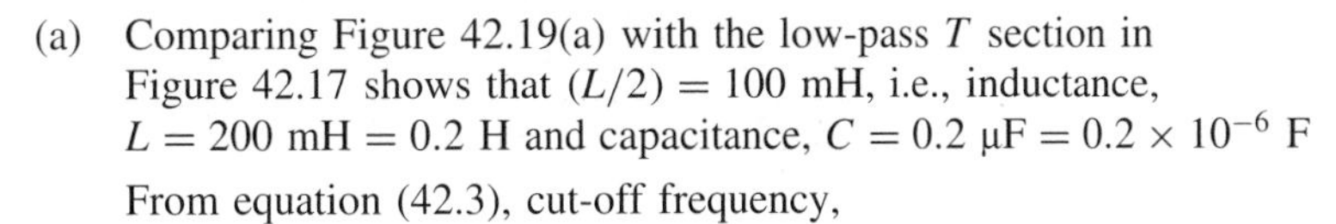

Problem 1. Determine the cut-off frequency and the nominal impedance of each of the low-pass filter sections shown in Figure 42.19.

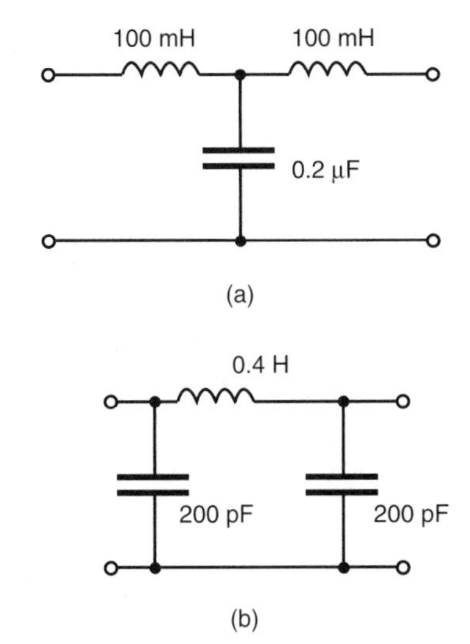

Figure 42.19

(a) Comparing Figure 42.19(a) with the low-pass T section in Figure 42.17 shows that $(L/2) = 100$ mH, i.e., inductance, $L = 200$ mH $= 0.2$ H and capacitance, $C = 0.2\ \mu\text{F} = 0.2 \times 10^{-6}$ F

From equation (42.3), cut-off frequency,

$$f_c = \frac{1}{\pi\sqrt{(LC)}} = \frac{1}{\pi\sqrt{(0.2 \times 0.2 \times 10^{-6})}} = \frac{10^3}{\pi(0.2)}$$

i.e., $f_c =$ **1592 Hz** or **1.592 kHz**

From equation (42.5), **nominal impedance,**

$$R_0 = \sqrt{\left(\frac{L}{C}\right)} = \sqrt{\left(\frac{0.2}{0.2 \times 10^{-6}}\right)} = \mathbf{1000\ \Omega} \text{ or } \mathbf{1\ k\Omega}$$

(b) Comparing Figure 42.19(b) with the low-pass π section shown in Figure 42.18 shows that $(C/2) = 200$ pF, i.e., capacitance, $C = 400$ pF $= 400 \times 10^{-12}$ F and inductance, $L = 0.4$ H,

From equation (42.3), **cut-off frequency,**

$$f_c = \frac{1}{\pi\sqrt{(LC)}} = \frac{1}{\pi\sqrt{(0.4 \times 400 \times 10^{-12})}} = \mathbf{25.16\ kHz}$$

From equation (42.5), **nominal impedance**,

$$\boldsymbol{R_0} = \sqrt{\left(\frac{L}{C}\right)} = \sqrt{\left(\frac{0.4}{400 \times 10^{-12}}\right)} = \mathbf{31.62\ k\Omega}$$

From equations (42.1) and (42.4) it is seen that the characteristic impedance Z_0 varies with ω, i.e., Z_0 varies with frequency. Thus if the nominal impedance is made to equal the load impedance into which the filter feeds then the matching deteriorates as the frequency increases from zero towards f_c. It is however convention to make the terminating impedance equal to the value of Z_0 well within the pass-band, i.e., to take the limiting value of Z_0 as the frequency approaches zero. This limit is obviously $\sqrt{(L/C)}$. This means that the filter is properly terminated at very low frequency but as the cut-off frequency is approached becomes increasingly mismatched. This is shown for a low-pass section in Figure 42.20 by curve (a). It is seen that an increasing loss is introduced into the pass band. Curve (b) shows the attenuation due to the same low-pass section being correctly terminated at all frequencies. A curve lying somewhere between curves (a) and (b) will usually result for each section if several sections are cascaded and terminated in R_0, or if a matching section is inserted between the low pass section and the load.

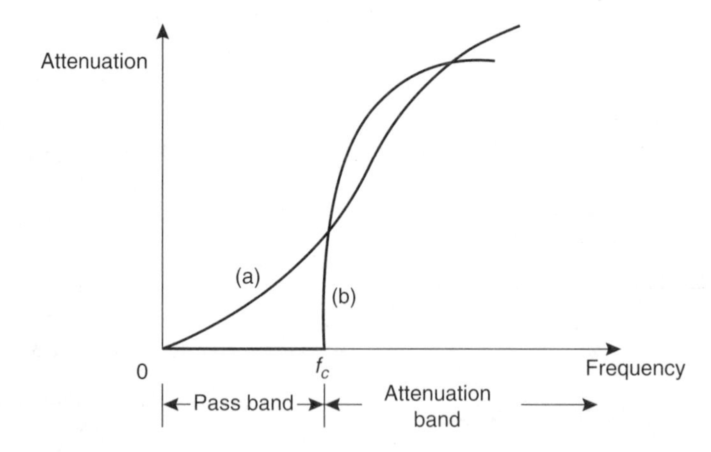

Figure 42.20

(c) To determine values of L and C given R_0 and f_c

If the values of the nominal impedance R_0 and the cut-off frequency f_c are known for a low pass T or π section it is possible to determine the values of inductance and capacitance required to form the section.

From equation (42.5), $R_0 = \sqrt{\dfrac{L}{C}} = \dfrac{\sqrt{L}}{\sqrt{C}}$ from which, $\sqrt{L} = R_0\sqrt{C}$

Substituting in equation (42.3) gives:

$$f_c = \frac{1}{\pi\sqrt{L}\sqrt{C}} = \frac{1}{\pi(R_0\sqrt{C})\sqrt{C}} = \frac{1}{\pi R_0 C}$$

from which, $$\boxed{\textbf{capacitance } C = \frac{1}{\pi R_0 f_c}} \qquad (42.6)$$

Similarly from equation (42.5), $\sqrt{C} = \dfrac{\sqrt{L}}{R_0}$

Substituting in equation (42.3) gives: $f_c = \dfrac{1}{\pi\sqrt{L}\left(\dfrac{\sqrt{L}}{R_0}\right)} = \dfrac{R_0}{\pi L}$

from which, $$\boxed{\textbf{inductance, } L = \frac{R_0}{\pi f_c}} \qquad (42.7)$$

Problem 2. A filter section is to have a characteristic impedance at zero frequency of 600 Ω and a cut-off frequency at 5 MHz. Design (a) a low-pass T section filter, and (b) a low-pass π section filter to meet these requirements.

The characteristic impedance at zero frequency is the nominal impedance R_0, i.e., $R_0 = 600\ \Omega$; cut-off frequency, $f_c = 5$ MHz $= 5 \times 10^6$ Hz. From equation (42.6),

$$\text{capacitance, } C = \frac{1}{\pi R_0 f_c} = \frac{1}{\pi(600)(5 \times 10^6)} \text{ F} = 106 \text{ pF}$$

and from equation (42.7),

$$\text{inductance, } L = \frac{R_0}{\pi f_c} = \frac{600}{\pi(5 \times 10^6)} \text{ H} = 38.2\ \mu\text{H}$$

(a) A low-pass T section filter is shown in Figure 42.21(a), where the series arm inductances are each L/2 (see Figure 42.17), i.e., $(38.2/2) = 19.1\ \mu$H

(b) A low-pass π section filter is shown in Figure 42.21(b), where the shunt arm capacitances are each $(C/2)$ (see Figure 42.18), i.e., $(106/2) = 53$ pF

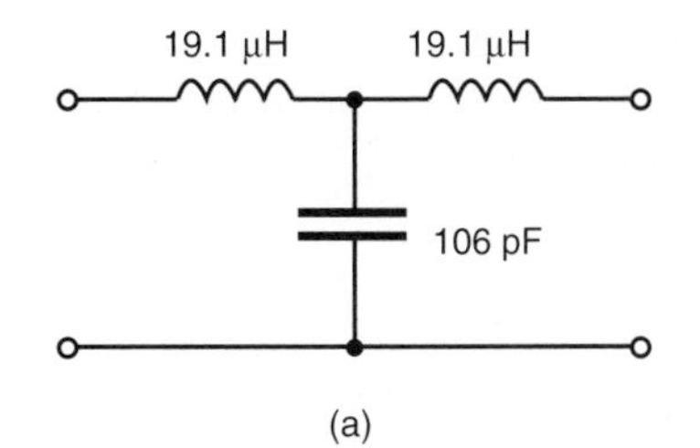

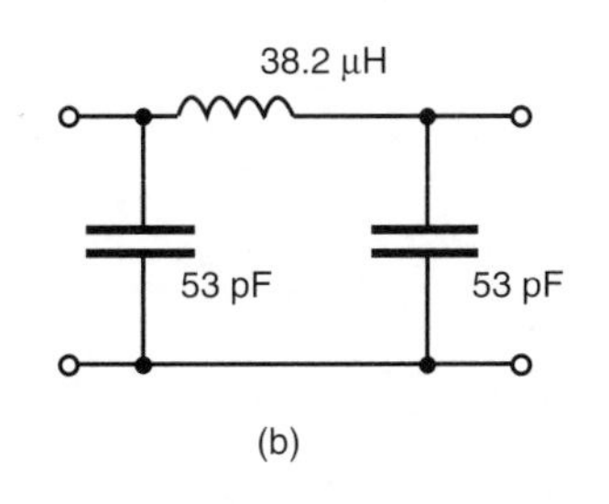

Figure 42.21

(d) 'Constant-k' prototype low-pass filter

A ladder network is shown in Figure 42.22, the elements being expressed in terms of impedances Z_1 and Z_2. The network shown in Figure 42.22(b)

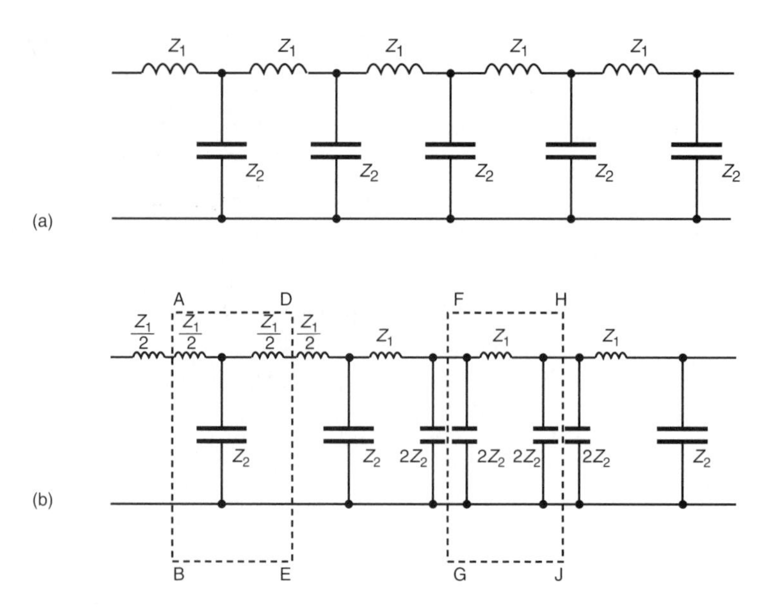

Figure 42.22

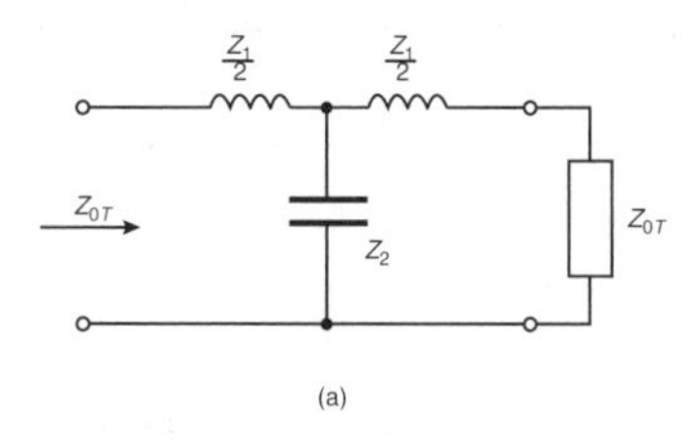

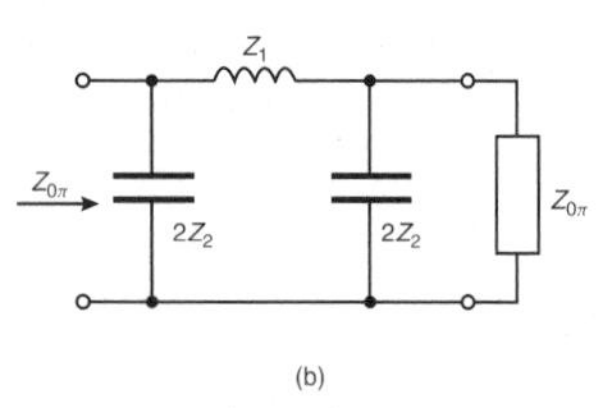

Figure 42.23

is equivalent to the network shown in Figure 42.22(a), where $(Z_1/2)$ in series with $(Z_1/2)$ equals Z_1 and $2Z_2$ in parallel with $2Z_2$ equals Z_2. Removing sections ABED and FGJH from Figure 42.22(b) gives the T section shown in Figure 42.23(a), which is terminated in its characteristic impedance Z_{OT}, and the π section shown in Figure 42.23(b), which is terminated in its characteristic impedance $Z_{0\pi}$.

From equation (41.1), page 760,

$$Z_{OT} = \sqrt{\left[\left(\frac{Z_1}{2}\right)^2 + 2\left(\frac{Z_1}{2}\right)Z_2\right]}$$

i.e., $$Z_{OT} = \sqrt{\left(\frac{Z_1^2}{4} + Z_1 Z_2\right)} \qquad (42.8)$$

From equation (41.3), page 760

$$Z_{0\pi} = \sqrt{\left[\frac{(Z_1)(2Z_2)^2}{Z_1 + 2(2Z_2)}\right]} = \sqrt{\left[\frac{Z_1(Z_1)(4Z_2^2)}{Z_1(Z_1 + 4Z_2)}\right]}$$

$$= \frac{2Z_1Z_2}{\sqrt{(Z_1^2 + 4Z_1Z_2)}} = \frac{Z_1Z_2}{\sqrt{\left(\frac{Z_1^2}{4} + Z_1Z_2\right)}}$$

i.e., $$Z_{0\pi} = \frac{Z_1Z_2}{Z_{OT}}$$ from equation (42.8)

Thus $\boxed{\mathbf{Z_{0T} Z_{0\pi} = Z_1 Z_2}}$ (42.9)

This is a general expression relating the characteristic impedances of T and π sections made up of equivalent series and shunt impedances.

From the low-pass sections shown in Figures 42.17 and 42.18,

$$Z_1 = j\omega L \text{ and } Z_2 = \frac{1}{j\omega C}.$$

Hence $\quad Z_{0T} Z_{0\pi} = (j\omega L)\left(\frac{1}{j\omega C}\right) = \frac{L}{C}$

Thus, from equation (42.5), $\boxed{\mathbf{Z_{0T} Z_{0\pi} = R_0^2}}$ (42.10)

From equations (42.9) and (42.10),

$$Z_{0T} Z_{0\pi} = Z_1 Z_2 = R_0^2 = \text{constant (k)}.$$

A ladder network composed of reactances, the series reactances being of opposite sign to the shunt reactances (as in Figure 42.23) are called **'constant-k' filter sections**. Positive (i.e., inductive) reactance is directly proportional to frequency, and negative (i.e., capacitive) reactance is inversely proportional to frequency. Thus the product of the series and shunt reactances is independent of frequency (see equations (42.9) and (42.10)). The constancy of this product has given this type of filter its name.

From equation (42.10), it is seen that Z_{0T} and $Z_{0\pi}$ will either be both real or both imaginary together (since $j^2 = -1$). Also, when Z_{0T} changes from real to imaginary at the cut-off frequency, so will $Z_{0\pi}$. The two sections shown in Figures 42.17 and 42.18 will thus have identical cut-off frequencies and thus identical pass bands. Constant-k sections of any kind of filter are known as **prototypes**.

(e) Practical low-pass filter characteristics

From equation (42.1), the characteristic impedance Z_{0T} of a low-pass T section is given by:

$$Z_{0T} = \sqrt{\left(\frac{L}{C} - \frac{\omega^2 L^2}{4}\right)}$$

Rearranging gives:

$$Z_{0T} = \sqrt{\left[\frac{L}{C}\left(1 - \frac{\omega^2 LC}{4}\right)\right]} = \sqrt{\left(\frac{L}{C}\right)}\sqrt{\left(1 - \frac{\omega^2 LC}{4}\right)}$$

$$= R_0\sqrt{\left(1 - \frac{\omega^2 LC}{4}\right)} \text{ from equation (42.5)}$$

From equation (42.2), $\omega_c^2 = \dfrac{4}{LC}$, hence $Z_{0T} = R_0\sqrt{\left(1 - \dfrac{\omega^2}{\omega_c^2}\right)}$

i.e.,
$$\boldsymbol{Z_{0T} = R_0\sqrt{\left[1 - \left(\frac{\omega}{\omega_c}\right)^2\right]}} \qquad (42.11)$$

Also, from equation (42.10), $Z_{0\pi} = \dfrac{R_0^2}{Z_{0T}} = \dfrac{R_0^2}{R_0\sqrt{\left[1 - \left(\dfrac{\omega}{\omega_c}\right)^2\right]}}$

i.e.,
$$\boldsymbol{Z_{0\pi} = \frac{R_0}{\sqrt{\left[1 - \left(\frac{\omega}{\omega_c}\right)^2\right]}}} \qquad (42.12)$$

(Alternatively, the expression for $Z_{0\pi}$ could have been obtained from equation (42.4), where

$$Z_{0\pi} = \sqrt{\left(\frac{1}{\frac{C}{L} - \frac{\omega^2 C^2}{4}}\right)} = \sqrt{\left[\frac{\frac{L}{C}}{\frac{L}{C}\left(\frac{C}{L} - \frac{\omega^2 C^2}{4}\right)}\right]}$$

$$= \frac{\sqrt{\frac{L}{C}}}{\sqrt{\left(1 - \frac{\omega^2 LC}{4}\right)}} = \frac{R_0}{\sqrt{\left[1 - \left(\frac{\omega}{\omega_c}\right)^2\right]}} \text{ as above).}$$

From equations (42.11) and (42.12), when $\omega = 0$ (i.e., when the frequency is zero),

$$Z_{0T} = Z_{0\pi} = R_0.$$

At the cut-off frequency, f_c, $\omega = \omega_c$
and from equation (42.11), Z_{0T} falls to zero,
and from equation (42.12), $Z_{0\pi}$ rises to infinity.

These results are shown graphically in Figure 42.24, where it is seen that Z_{0T} decreases from R_0 at zero frequency to zero at the cut-off frequency; $Z_{0\pi}$ rises from its initial value of R_0 to infinity at f_c.

(At a frequency, $f = 0.95 f_c$, for example, $Z_{0\pi} = \dfrac{R_0}{\sqrt{(1 - 0.95^2)}} = 3.2R_0$ from equation (42.12)).

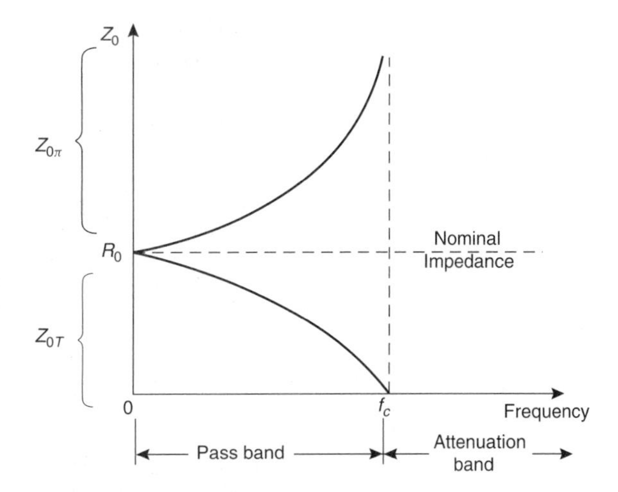

Figure 42.24

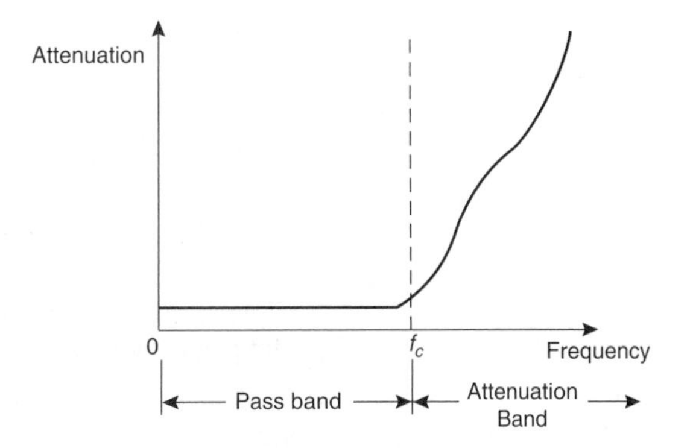

Figure 42.25

Note that since Z_0 becomes purely reactive in the attenuation band, it is not shown in this range in Figure 42.24.

Figure 42.2(a), on page 791, showed an ideal low-pass filter section characteristic. In practise, the characteristic curve of a low-pass prototype filter section looks more like that shown in Figure 42.25. The characteristic may be improved somewhat closer to the ideal by connecting two or more identical sections in cascade. This produces a much sharper cut-off characteristic, although the attenuation in the pass band is increased a little.

Problem 3. The nominal impedance of a low-pass π section filter is 500 Ω and its cut-off frequency is at 100 kHz. Determine (a) the value of the characteristic impedance of the section at a frequency of 90 kHz, and (b) the value of the characteristic impedance of the equivalent low-pass T section filter.

At zero frequency the characteristic impedance of the π and T section filters will be equal to the nominal impedance of 500 Ω.

(a) From equation (42.12), the characteristic impedance of the π section at 90 kHz is given by:

$$Z_{0\pi} = \frac{R_0}{\sqrt{\left[1 - \left(\frac{\omega}{\omega_c}\right)^2\right]}} = \frac{500}{\sqrt{\left[1 - \left(\frac{2\pi 90 \times 10^3}{2\pi 100 \times 10^3}\right)^2\right]}}$$

$$= \frac{500}{\sqrt{[1 - (0.9)^2]}} = \mathbf{1147\ \Omega}$$

(b) From equation (42.11), the characteristic impedance of the T section at 90 kHz is given by:

$$Z_{0T} = R_0\sqrt{\left[1 - \left(\frac{\omega}{\omega_c}\right)^2\right]} = 500\sqrt{[1 - (0.9)^2]} = \mathbf{218\ \Omega}$$

(Check: From equation (42.10),

$$Z_{0T}Z_{0\pi} = (218)(1147) = 250\,000 = 500^2 = R_0^2)$$

Typical low-pass characteristics of characteristic impedance against frequency are shown in Figure 42.24.

Problem 4. A low-pass π section filter has a nominal impedance of 600 Ω and a cut-off frequency of 2 MHz. Determine the frequency at which the characteristic impedance of the section is (a) 600 Ω (b) 1 kΩ (c) 10 kΩ

From equation (42.12), $Z_{0\pi} = \dfrac{R_0}{\sqrt{\left[1 - \left(\frac{\omega}{\omega_c}\right)^2\right]}}$

(a) When $Z_{0\pi} = 600\ \Omega$ and $R_0 = 600\ \Omega$, then $\omega = 0$, i.e., **the frequency is zero**

(b) When $Z_{0\pi} = 1000\ \Omega$, $R_0 = 600\ \Omega$ and $f_c = 2 \times 10^6$ Hz

then $$1000 = \frac{600}{\sqrt{\left[1 - \left(\frac{2\pi f}{2\pi 2 \times 10^6}\right)^2\right]}}$$

from which, $1 - \left(\dfrac{f}{2 \times 10^6}\right)^2 = \left(\dfrac{600}{1000}\right)^2 = 0.36$

and $\left(\dfrac{f}{2 \times 10^6}\right) = \sqrt{(1 - 0.36)} = 0.8$

Thus when $Z_{0\pi} = 1000\ \Omega$,

frequency, $f = (0.8)(2 \times 10^6) = $ **1.6 MHz**

(c) When $Z_{0\pi} = 10\ \text{k}\Omega$, then

$$10\,000 = \frac{600}{\sqrt{\left[1 - \left(\frac{f}{2}\right)^2\right]}},$$ where frequency, f is in megahertz.

Thus $1 - \left(\dfrac{f}{2}\right)^2 = \left(\dfrac{600}{10\,000}\right)^2 = (0.06)^2$

and $\dfrac{f}{2} = \sqrt{[1 - (0.06)^2]} = 0.9982$

Hence when $Z_{0\pi} = 10\ \text{k}\Omega$, **frequency** $f = (2)(0.9982)$

$= $ **1.996 MHz**

The above three results are seen to be borne out in the characteristic of $Z_{0\pi}$ against frequency shown in Figure 42.24.

Further problems on low-pass filter sections may be found in Section 42.10, problems 1 to 6, page 837.

42.6 High-pass filter sections

(a) The cut-off frequency

High-pass T and π sections are shown in Figure 42.26, (as derived in Section (42.4)), each being terminated in their characteristic impedance.

From equation (41.1), page 760, the characteristic impedance of a T section is given by:

$$Z_{0T} = \sqrt{(Z_A^2 + 2Z_A Z_B)}$$

From Figure 42.26(a), $Z_A = \dfrac{1}{j\omega 2C}$ and $Z_B = j\omega L$

Thus $$Z_{0T} = \sqrt{\left[\left(\frac{1}{j\omega 2C}\right)^2 + 2\left(\frac{1}{j\omega 2C}\right)(j\omega L)\right]}$$

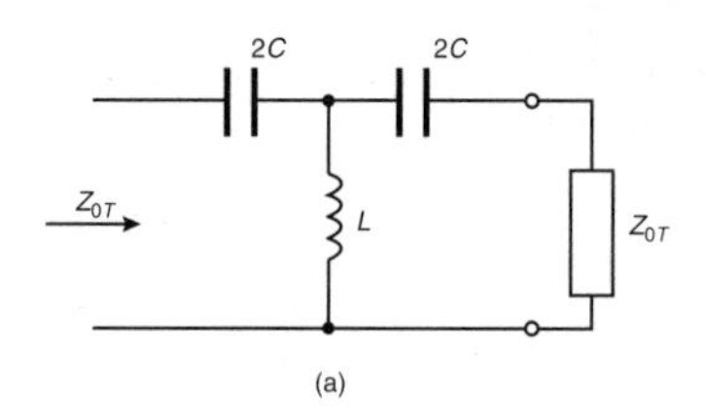

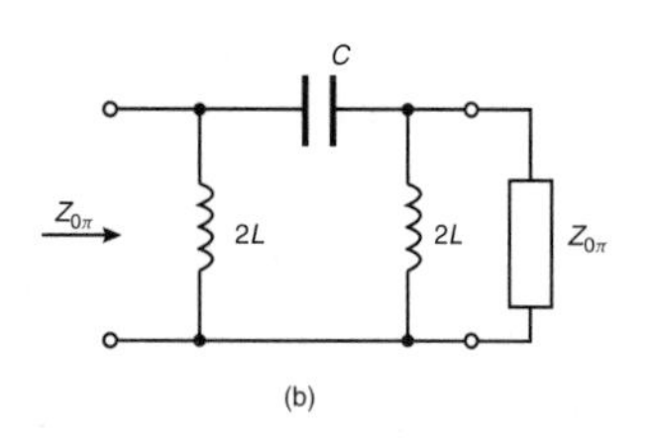

Figure 42.26

$$= \sqrt{\left[\frac{1}{-4\omega^2C^2} + \frac{L}{C}\right]}$$

i.e., $$Z_{0T} = \sqrt{\left(\frac{L}{C} - \frac{1}{4\omega^2C^2}\right)} \qquad (42.13)$$

Z_{0T} will be real when $\dfrac{L}{C} > \dfrac{1}{4\omega^2C^2}$

Thus the filter will pass all frequencies above the point

$$\text{where } \frac{L}{C} = \frac{1}{4\omega^2C^2}$$

i.e., $$\text{where } \omega_c^2 = \frac{1}{4LC} \qquad (42.14)$$

where $\omega_c = 2\pi f_c$, and f_c is the cut-off frequency.

Hence $$(2\pi f_c)^2 = \frac{1}{4LC}$$

and **the cut-off frequency**, $$\boxed{\boldsymbol{f_c = \frac{1}{4\pi\sqrt{(LC)}}}} \qquad (42.15)$$

The same equation for the cut-off frequency is obtained for the high-pass π network shown in Figure 42.26(b) as follows:

From equation (41.3), page 760, the characteristic impedance of a symmetrical π section is given by:

$$Z_{0\pi} = \sqrt{\left(\frac{Z_1Z_2^2}{Z_1 + 2Z_2}\right)}$$

From Figure 42.26(b), $Z_1 = \dfrac{1}{j\omega C}$ and $Z_2 = j2\omega L$

$$\text{Hence } Z_{0\pi} = \sqrt{\left\{\frac{\left(\dfrac{1}{j\omega C}\right)(j2\omega L)^2}{\dfrac{1}{j\omega C} + 2j2\omega L}\right\}}$$

$$= \sqrt{\left\{\frac{j4\dfrac{\omega L^2}{C}}{j\left(4\omega L - \dfrac{1}{\omega C}\right)}\right\}} = \sqrt{\left(\frac{\dfrac{4L^2}{C}}{4L - \dfrac{1}{\omega^2 C}}\right)}$$

i.e., $$Z_{0\pi} = \sqrt{\left(\frac{1}{\dfrac{C}{L} - \dfrac{1}{4\omega^2L^2}}\right)} \qquad (42.16)$$

$Z_{0\pi}$ will be real when $\dfrac{C}{L} > \dfrac{1}{4\omega^2 L^2}$ and the filter will pass all frequencies above the point where $\dfrac{C}{L} = \dfrac{1}{4\omega^2 L^2}$, i.e., where $\omega_c^2 = \dfrac{1}{4LC}$ as above.

Thus the cut-off frequency for a high-pass π network is also given by

$$\boldsymbol{f_c = \frac{1}{4\pi\sqrt{(LC)}}} \quad \text{(as in equation (42.15))} \qquad (42.15')$$

(b) Nominal impedance

When the frequency is very high, ω is a very large value and the term $(1/4\omega^2 C^2)$ in equations (42.13) and (42.16) are extremely small and may be neglected.

The characteristic impedance then becomes equal to $\sqrt{(L/C)}$, this being the nominal impedance. Thus for a high-pass filter section the nominal impedance R_0 is given by:

$$\boldsymbol{R_0 = \sqrt{\left(\frac{L}{C}\right)}} \qquad (42.17)$$

the same as for the low-pass filter sections.

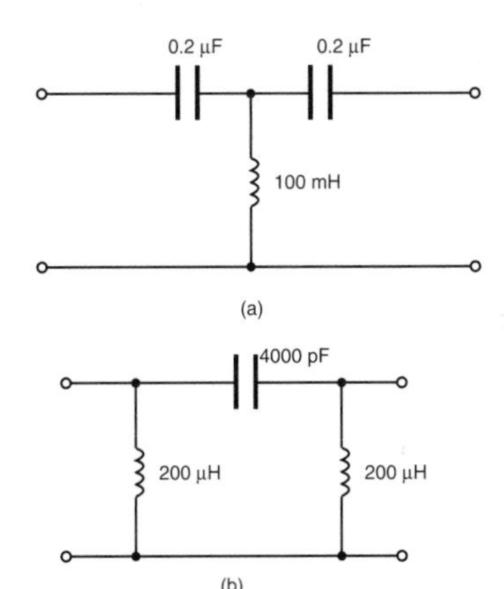

Figure 42.27

Problem 5. Determine for each of the high-pass filter sections shown in Figure 42.27 (i) the cut-off frequency, and (ii) the nominal impedance.

(a) Comparing Figure 42.27(a) with Figure 42.26(a) shows that:

$$2C = 0.2\ \mu\text{F}, \text{ i.e., capacitance, } C = 0.1\ \mu\text{F} = 0.1 \times 10^{-6}\ \text{F}$$

$$\text{and inductance, } L = 100\ \text{mH} = 0.1\ \text{H}$$

(i) From equation (42.15),

$$\text{cut-off frequency, } f_c = \frac{1}{4\pi\sqrt{(LC)}} = \frac{1}{4\pi\sqrt{[(0.1)(0.1 \times 10^{-6}]}}$$

$$\text{i.e., } \quad f_c = \frac{10^3}{4\pi(0.1)} = \mathbf{796\ Hz}$$

(ii) From equation (42.17),

$$\textbf{nominal impedance, } \boldsymbol{R_0} = \sqrt{\left(\frac{L}{C}\right)} = \sqrt{\left(\frac{0.1}{0.1 \times 10^{-6}}\right)}$$

$$= \mathbf{1000\ \Omega \text{ or } 1\ k\Omega}$$

(b) Comparing Figure 42.27(b) with Figure 42.26(b) shows that:

$2L = 200\ \mu\text{H}$, i.e., inductance, $L = 100\ \mu\text{H} = 10^{-4}$ H

and capacitance $C = 4000$ pF $= 4 \times 10^{-9}$ F

(i) From equation (42.15′),

$$\textbf{cut-off frequency, } f_c = \frac{1}{4\pi\sqrt{(LC)}}$$

$$= \frac{1}{4\pi\sqrt{[(10^{-4})(4 \times 10^{-9})]}} = \mathbf{126\ kHz}$$

(ii) From equation (42.17),

$$\textbf{nominal impedance, } R_0 = \sqrt{\left(\frac{L}{C}\right)} = \sqrt{\left(\frac{10^{-4}}{4 \times 10^{-9}}\right)}$$

$$= \sqrt{\left(\frac{10^5}{4}\right)} = \mathbf{158\ \Omega}$$

(c) To determine values of L and C given R_0 and f_c

If the values of the nominal impedance R_0 and the cut-off frequency f_c are known for a high-pass T or π section it is possible to determine the values of inductance L and capacitance C required to form the section.

From equation (42.17), $R_0 = \sqrt{\frac{L}{C}} = \frac{\sqrt{L}}{\sqrt{C}}$ from which, $\sqrt{L} = R_0\sqrt{C}$

Substituting in equation (42.15) gives:

$$f_c = \frac{1}{4\pi\sqrt{L}\sqrt{C}} = \frac{1}{4\pi(R_0\sqrt{C})\sqrt{C}} = \frac{1}{4\pi R_0 C}$$

from which,
$$\boxed{\textbf{capacitance } C = \frac{1}{4\pi R_0 f_c}} \qquad (42.18)$$

Similarly, from equation (42.17), $\sqrt{C} = \frac{\sqrt{L}}{R_0}$

Substituting in equation (42.15) gives: $f_c = \frac{1}{4\pi\sqrt{L}\left(\frac{\sqrt{L}}{R_0}\right)} = \frac{R_0}{4\pi L}$

from which,
$$\boxed{\textbf{inductance, } L = \frac{R_0}{4\pi f_c}} \qquad (42.19)$$

Problem 6. A filter is required to pass all frequencies above 25 kHz and to have a nominal impedance of 600 Ω. Design (a) a high-pass T section filter and (b) a high-pass π section filter to meet these requirements.

Cut-off frequency, $f_c = 25 \times 10^3$ Hz and nominal impedance, $R_0 = 600\ \Omega$

From equation (42.18),

$$C = \frac{1}{4\pi R_0 f_c} = \frac{1}{4\pi(600)(25 \times 10^{3})} \text{ F} = \frac{10^{12}}{4\pi(600)(25 \times 10^3)} \text{ pF}$$

i.e., $C = 5305$ pF or 5.305 nF

From equation (42.19), inductance,

$$L = \frac{R_0}{4\pi f_c} = \frac{600}{4\pi(25 \times 10^3)} \text{ H} = 1.91 \text{ mH}$$

(a) A high-pass T section filter is shown in Figure 42.28(a) where the series arm capacitances are each 2 C (see Figure 42.26(a)), i.e., $2 \times 5.305 = 10.61$ nF

(b) A high-pass π section filter is shown in Figure 42.28(b), where the shunt arm inductances are each 2 L (see Figure 42.26(b)), i.e., $2 \times 1.91 = 3.82$ mH

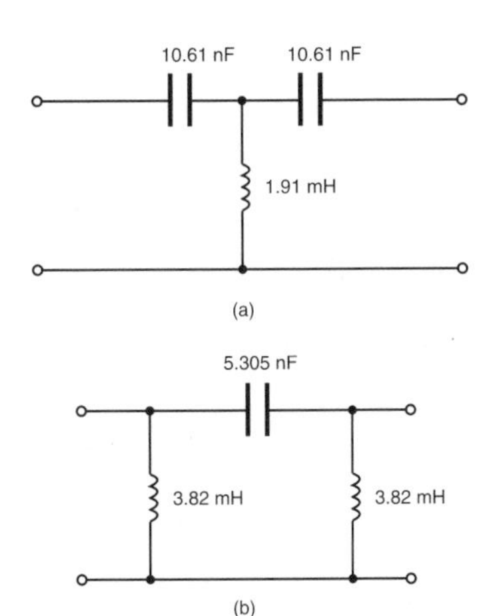

Figure 42.28

(d) 'Constant-k' prototype high-pass filter

It may be shown, in a similar way to that shown in Section 42.5(d), that for a high-pass filter section:

$$\mathbf{Z_{0T} Z_{0\pi} = Z_1 Z_2 = R_0^2}$$

where Z_1 and Z_2 are the total equivalent series and shunt arm impedances. The high-pass filter sections shown in Figure 42.26 are thus 'constant-k' prototype filter sections.

(e) Practical high-pass filter characteristics

From equation (42.13), the characteristic impedance Z_{0T} of a high-pass T section is given by:

$$Z_{0T} = \sqrt{\left(\frac{L}{C} - \frac{1}{4\omega^2 C^2}\right)}$$

Rearranging gives:

$$Z_{0T} = \sqrt{\left[\frac{L}{C}\left(1 - \frac{1}{4\omega^2 LC}\right)\right]} = \sqrt{\left(\frac{L}{C}\right)}\sqrt{\left(1 - \frac{1}{4\omega^2 LC}\right)}$$

From equation (42.14), $\omega_c^2 = \dfrac{1}{4LC}$

Thus
$$\boldsymbol{Z_{0T} = R_0\sqrt{\left[1 - \left(\frac{\omega_c}{\omega}\right)^2\right]}} \quad (42.20)$$

Also, since $\quad Z_{0T}Z_{0\pi} = R_0^2$

then
$$Z_{0\pi} = \frac{R_0^2}{Z_{0T}} = \frac{R_0^2}{R_0\sqrt{\left[1 - \left(\frac{\omega_c}{\omega}\right)^2\right]}}$$

i.e.,
$$\boldsymbol{Z_{0\pi} = \frac{R_0}{\sqrt{\left[1 - \left(\frac{\omega_c}{\omega}\right)^2\right]}}} \quad (42.21)$$

From equation (42.20),

when $\omega < \omega_c$, Z_{0T} is reactive,

when $\omega = \omega_c$, Z_{0T} is zero,

and when $\omega > \omega_c$, Z_{0T} is real, eventually increasing to R_0 when ω is very large.

Similarly, from equation (42.21),

when $\omega < \omega_c$, $Z_{0\pi}$ is reactive,

when $\omega = \omega_c$, $Z_{0\pi} = \infty$ (i.e., $\dfrac{R_0}{0} = \infty$)

and when $\omega > \omega_c$, $Z_{0\pi}$ is real, eventually decreasing to R_0 when ω is very large.

Curves of Z_{0T} and $Z_{0\pi}$ against frequency are shown in Figure 42.29.

Figure 42.4(a), on page 792, showed an ideal high-pass filter section characteristic of attenuation against frequency. In practise, the characteristic curve of a high-pass prototype filter section would look more like that shown in Figure 42.30.

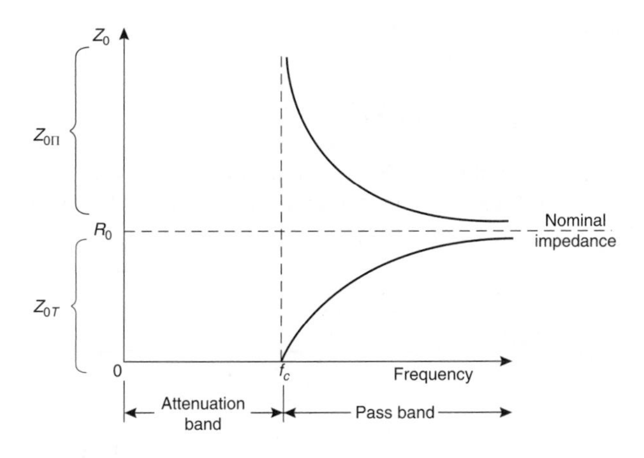

Figure 42.29

Figure 42.30

Problem 7. A low-pass T section filter having a cut-off frequency of 15 kHz is connected in series with a high-pass T section filter having a cut-off frequency of 10 kHz. The terminating impedance of the filter is 600 Ω.

(a) Determine the values of the components comprising the composite filter.

(b) Sketch the expected attenuation against frequency characteristic.

(c) State the name given to the type of filter described.

(a) **For the low-pass *T* section filter:** $f_{c_L} = 15\,000$ Hz

From equation (42.6),

$$\text{capacitance, } C = \frac{1}{\pi R_0 f_c} = \frac{1}{\pi(600)(15\,000)} \equiv 35.4 \text{ nF}$$

From equation (42.7),

$$\text{inductance, } L = \frac{R_0}{\pi f_c} = \frac{600}{\pi(15\,000)} \equiv 12.73 \text{ mH}$$

Thus from Figure 42.17, the series arm inductances are each $L/2$, i.e., $(12.73/2) = 6.37$ mH and the shunt arm capacitance is 35.4 nF.

For a high-pass *T* section filter: $f_{C_H} = 10\,000$ Hz

From equation (42.18),

$$\text{capacitance, } C = \frac{1}{4\pi R_0 f_c} = \frac{1}{4\pi(600)(10\,000)} \equiv 13.3 \text{ nF}$$

From equation (42.19),

$$\text{inductance, } L = \frac{R_0}{4\pi f_c} = \frac{600}{4\pi 10\,000} \equiv 4.77 \text{ mH}$$

Thus from Figure 42.26(a), the series arm capacitances are each $2\,C$, i.e., $2 \times 13.3 = 26.6$ nF, and the shunt arm inductance is 4.77 mH.

The composite filter is shown in Figure 42.31.

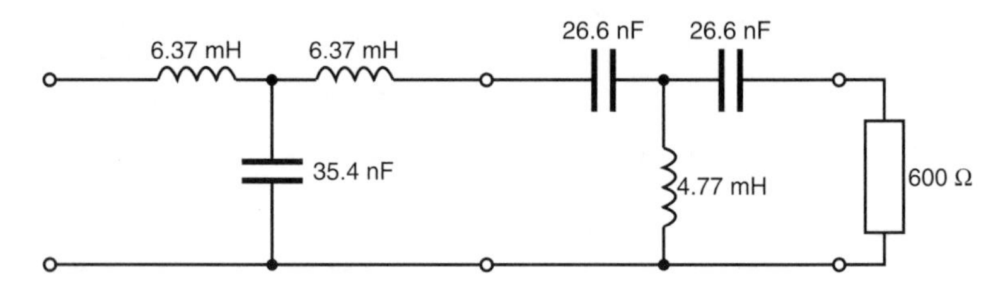

Figure 42.31

(b) A typical characteristic expected of attenuation against frequency is shown in Figure 42.32.

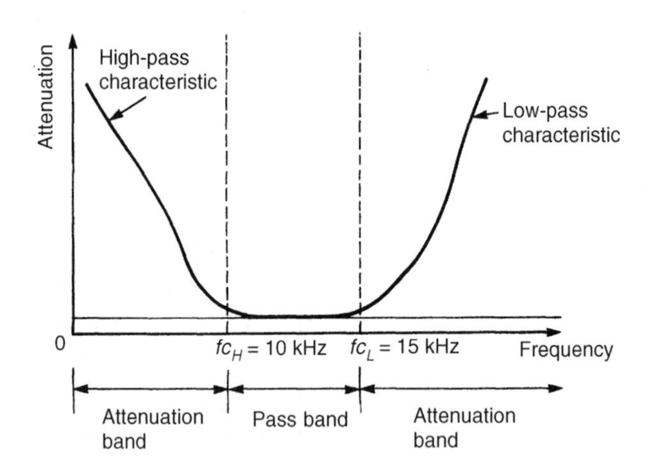

Figure 42.32

(c) The name given to the type of filter described is a **band-pass filter**. The ideal characteristic of such a filter is shown in Figure 42.5.

Problem 8. A high-pass T section filter has a cut-off frequency of 500 Hz and a nominal impedance of 600 Ω. Determine the frequency at which the characteristic impedance of the section is (a) zero, (b) 300 Ω, (c) 590 Ω.

From equation (42.20), $Z_{0T} = R_0\sqrt{\left[1 - \left(\frac{\omega_c}{\omega}\right)^2\right]}$

(a) When $Z_{0T} = 0$, then $(\omega_c/\omega) = 1$, i.e., **the frequency is 500 Hz**, the cut-off frequency.

(b) When $Z_{0T} = 300\ \Omega$, $R_0 = 600\ \Omega$ and $f_c = 500$ Hz

$$300 = 600\sqrt{\left[1 - \left(\frac{2\pi 500}{2\pi f}\right)^2\right]}$$

from which $$\left(\frac{300}{600}\right)^2 = 1 - \left(\frac{500}{f}\right)^2$$

and $$\frac{500}{f} = \sqrt{\left[1 - \left(\frac{300}{600}\right)^2\right]} = \sqrt{0.75}$$

Thus when $Z_{0T} = 300\ \Omega$, **frequency,** $\boldsymbol{f} = \frac{500}{\sqrt{0.75}} = \mathbf{577.4\ Hz}$

(c) When $Z_{0T} = 590\ \Omega$, $590 = 600\sqrt{\left[1 - \left(\frac{500}{f}\right)^2\right]}$

$$\frac{500}{f} = \sqrt{\left[1 - \left(\frac{590}{600}\right)^2\right]} = 0.1818$$

Thus when $Z_{0T} = 590\ \Omega$, **frequency,** $\boldsymbol{f} = \frac{500}{0.1818} = \mathbf{2750\ Hz}$

The above three results are seen to be borne out in the characteristic of Z_{0T} against frequency shown in Figure 42.29.

Further problems on high-pass filter sections may be found in Section 42.10, problems 7 to 12, page 837.

42.7 Propagation coefficient and time delay in filter sections

Propagation coefficient

In Figure 42.33, let A, B and C represent identical filter sections, the current ratios (I_1/I_2), (I_2/I_3) and (I_3/I_4) being equal.

Although the rate of attenuation is the same in each section (i.e., the current output of each section is one half of the current input) the amount of attenuation in each is different (section A attenuates by $\frac{1}{2}$ A, B attenuates by $\frac{1}{4}$ A and C attenuates by $\frac{1}{8}$ A). The attenuation is in fact in the

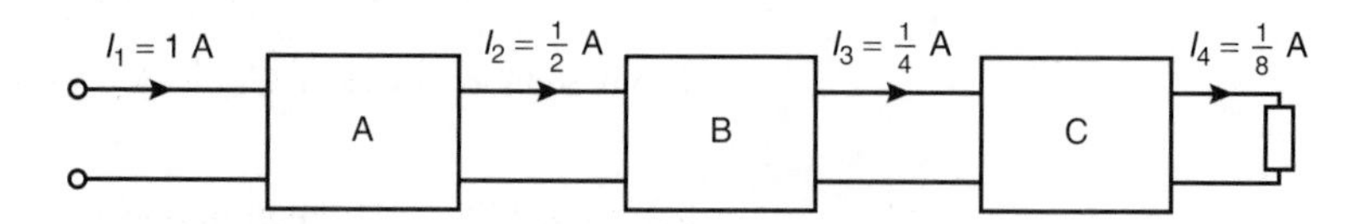

Figure 42.33

form of a logarithmic decay and

$$\frac{I_1}{I_2} = \frac{I_2}{I_3} = \frac{I_3}{I_4} = e^{\gamma} \tag{42.22}$$

where γ is called the **propagation coefficient** or the **propagation constant.**

From equation (42.22), propagation coefficient,

$$\gamma = \ln \frac{I_1}{I_2} \text{ nepers} \tag{42.23}$$

(See Section 41.3, page 761, on logarithmic units.)

Unless Sections A, B and C in Figure 42.33 are purely resistive there will be a phase change in each section. Thus the ratio of the current entering a section to that leaving it will be a phasor quantity having both modulus and argument. The propagation constant which has no units is a complex quantity given by:

$$\boxed{\gamma = \alpha + j\beta} \tag{42.24}$$

where α is called the **attenuation coefficient**, measured in nepers, and β the **phase shift coefficient**, measured in radians. β is the angle by which a current leaving a section lags behind the current entering it.

From equations (42.22) and (42.24),

$$\frac{I_1}{I_2} = e^{\gamma} = e^{\alpha + j\beta} = (e^{\alpha})(e^{j\beta})$$

Since $\quad e^x = 1 + x + \dfrac{x^2}{2!} + \dfrac{x^3}{3!} + \dfrac{x^4}{4!} + \dfrac{x^5}{5!} + \ldots\ldots$

then $\quad e^{j\beta} = 1 + (j\beta) + \dfrac{(j\beta)^2}{2!} + \dfrac{(j\beta)^3}{3!} + \dfrac{(j\beta)^4}{4!} + \dfrac{(j\beta)^5}{5!} + \ldots\ldots$

$$= 1 + j\beta - \frac{\beta^2}{2!} - j\frac{\beta^3}{3!} + \frac{\beta^4}{4!} + j\frac{\beta^5}{5!} + \ldots\ldots$$

since $j^2 = -1$, $j^3 = -j$, $j^4 = +1$, and so on.

Hence $\quad e^{j\beta} = \left(1 - \dfrac{\beta^2}{2!} + \dfrac{\beta^4}{4!} - \ldots\right) + j\left(\beta - \dfrac{\beta^3}{3!} + \dfrac{\beta^5}{5!} - \ldots\right)$

$= \cos\beta + j\sin\beta$ from the power series for $\cos\beta$ and $\sin\beta$

Thus $\frac{I_1}{I_2} = e^{\alpha}e^{j\beta} = e^{\alpha}(\cos\beta + j\sin\beta) = e^{\alpha}\angle\beta$ in abbreviated polar form,

i.e., $$\frac{\boldsymbol{I_1}}{\boldsymbol{I_2}} = \boldsymbol{e^{\alpha}\angle\beta} \qquad (42.25)$$

Now $$e^{\alpha} = \left|\frac{I_1}{I_2}\right|$$

from which

$$\textbf{attenuation coefficient, } \boldsymbol{\alpha} = \textbf{ln}\left|\frac{\boldsymbol{I_1}}{\boldsymbol{I_2}}\right| \textbf{ nepers or 20 lg }\left|\frac{\boldsymbol{I_1}}{\boldsymbol{I_2}}\right| \textbf{ dB}$$

If in Figure 42.33 current I_2 lags current I_1 by, say, 30°, i.e., $(\pi/6)$ rad, then the propagation coefficient γ of Section A is given by:

$$\gamma = \alpha + j\beta = \ln\left|\frac{1}{\frac{1}{2}}\right| + j\frac{\pi}{6}$$

i.e., $\boldsymbol{\gamma = (0.693 + j0.524)}$

If there are n identical sections connected in cascade and terminated in their characteristic impedance, then

$$\frac{I_1}{I_{n+1}} = (e^{\gamma})^n = e^{n\gamma} = e^{n(\alpha+j\beta)} = e^{n\alpha}\angle n\beta, \quad \ldots\ldots \qquad (42.26)$$

where I_{n+1} is the output current of the n'th section.

Problem 9. The propagation coefficients of two filter networks are given by

(a) $\gamma = (1.25 + j0.52)$, (b) $\gamma = 1.794\angle -39.4°$

Determine for each (i) the attenuation coefficient, and (ii) the phase shift coefficient.

(a) If $\gamma = (1.25 + j0.52)$

then (i) the attenuation coefficient, α, is given by the real part,

i.e., $\boldsymbol{\alpha = 1.25}$ **N**

and (ii) the phase shift coefficient, β, is given by the imaginary part,

i.e., $\boldsymbol{\beta = 0.52}$ **rad**

(b) $\gamma = 1.794\angle -39.4^\circ = 1.794[\cos(-39.4^\circ) + j\sin(-39.4^\circ)]$

$$= (1.386 - j1.139)$$

Hence (i) the attenuation coefficient, $\boldsymbol{\alpha} = \mathbf{1.386\ N}$

and (ii) the phase shift coefficient, $\boldsymbol{\beta} = \mathbf{-1.139\ rad}$

> Problem 10. The current input to a filter section is $24\angle 10^\circ$ mA and the current output is $8\angle -45^\circ$ mA. Determine for the section (a) the attenuation coefficient, (b) the phase shift coefficient, and (c) the propagation coefficient. (d) If five such sections are cascaded determine the output current of the fifth stage and the overall propagation constant of the network.

Let $I_1 = 24\angle 10^\circ$ mA and $I_2 = 8\angle -45^\circ$ mA, then

$$\frac{I_1}{I_2} = \frac{24\angle 10^\circ}{8\angle -45^\circ} = 3\angle 55^\circ = e^{\alpha}\angle\beta \text{ from equation (42.25).}$$

(a) Hence the attenuation constant, α, is obtained from $3 = e^{\alpha}$, i.e., $\boldsymbol{\alpha} = \mathbf{\ln 3 = 1.099\ N}$

(b) The phase shift coefficient $\beta = 55^\circ \times \dfrac{\pi}{180} = \mathbf{0.960\ rad}$

(c) The propagation coefficient $\gamma = \alpha + j\beta = \mathbf{(1.099 + j0.960)}$ or $\mathbf{1.459\angle 41.14^\circ}$

(d) If I_6 is the current output of the fifth stage, then from equation (42.26),

$$\frac{I_1}{I_6} = (e^{\gamma})^n = [3\angle 55^\circ]^5 = 243\angle 275^\circ \text{ (by De Moivre's theorem)}$$

Thus the output current of the fifth stage,

$$I_6 = \frac{I_1}{243\angle 275^\circ} = \frac{24\angle 10^\circ}{243\angle 275^\circ}$$

$$= \mathbf{0.0988\angle -265^\circ\ mA\ or\ 98.8\angle 95^\circ\ \mu A}$$

Let the overall propagation coefficient be γ'

then $\dfrac{I_1}{I_6} = 243\angle 275^\circ = e^{\gamma'} = e^{\alpha'}\angle\beta'$

The overall attenuation coefficient $\alpha' = \ln 243 = 5.49$

and the overall phase shift coefficient $\beta' = 275^\circ \times \dfrac{\pi}{180^\circ} = 4.80$ rad

Hence the overall propagation coefficient $\boldsymbol{\gamma'} = \mathbf{(5.49 + j4.80)}$ or $\mathbf{7.29\angle 41.16^\circ}$

Problem 11. For the low-pass T section filter shown in Figure 42.34 determine (a) the attenuation coefficient, (b) the phase shift coefficient and (c) the propagation coefficient γ.

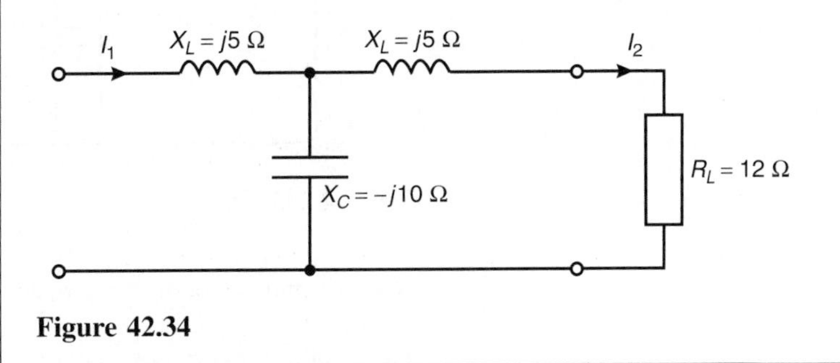

Figure 42.34

By current division in Figure 42.34, $I_2 = \left(\dfrac{X_C}{X_C + X_L + R_L}\right) I_1$

from which $\dfrac{I_1}{I_2} = \dfrac{X_C + X_L + R_L}{X_C} = \dfrac{-j10 + j5 + 12}{-j10} = \dfrac{-j5 + 12}{-j10}$

$$= \frac{-j5}{-j10} + \frac{12}{-j10}$$

$$= 0.5 + \frac{j12}{-j^2 10} = 0.5 + j1.2$$

$$= 1.3\angle 67.38^\circ \text{ or } 1.3\angle 1.176$$

From equation (42.25), $\dfrac{I_1}{I_2} = e^{\alpha}\angle\beta = 1.3\angle 1.176$

(a) The attenuation coefficient, $\alpha = \ln 1.3 = \mathbf{0.262\ N}$

(b) The phase shift coefficient, $\boldsymbol{\beta} = \mathbf{1.176\ rad}$

(c) The propagation coefficient, $\gamma = \alpha + j\beta =$ **(0.262 + j1.176)** or **1.205∠77.44°**

Variation in phase angle in the pass-band of a filter

In practise, the low and high-pass filter sections discussed in Sections 42.5 and 42.6 would possess a phase shift between the input and output voltages which varies considerably over the range of frequency comprising the pass-band.

Let the **low-pass prototype *T* section** shown in Figure 42.35 be terminated as shown in its nominal impedance R_0. The input impedance for frequencies much less than the cut-off frequency is thus also equal to R_0 and is resistive. The phasor diagram representing Figure 42.35 is shown in Figure 42.36 and is produced as follows:

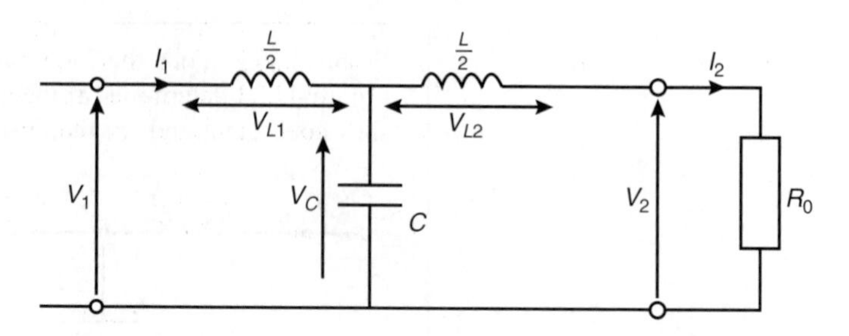

Figure 42.35

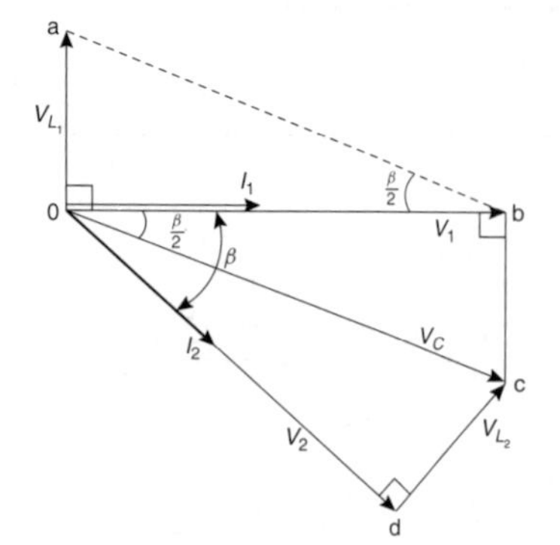

Figure 42.36

(i) V_1 and I_1 are in phase (since the input impedance is resistive).

(ii) Voltage $V_{L1} = I_1 X_L = I_1\left(\frac{\omega L}{2}\right)$, which leads I by 90°.

(iii) Voltage V_1 is the phasor sum of V_{L1} and V_C. Thus V_C is drawn as shown, completing the parallelogram oabc.

(iv) Since no power is dissipated in reactive elements $V_1 = V_2$ in magnitude.

(v) Voltage $V_{L2} = I_2\left(\frac{\omega L}{2}\right) = I_1\left(\frac{\omega L}{2}\right) = V_{L1}$

(vi) Voltage V_C is the phasor sum of V_{L2} and V_2 as shown by triangle ocd, where V_{L2} is at right angles to V_2

(vii) Current I_2 is in phase with V_2 since the output impedance is resistive. The phase lag over the section is the angle between V_1 and V_2 shown as angle β in Figure 42.36,

$$\text{where } \tan\frac{\beta}{2} = \frac{oa}{ob} = \frac{V_{L1}}{V_1} = \frac{I_1\left(\frac{\omega L}{2}\right)}{I_1 R_0} = \frac{\frac{\omega L}{2}}{R_0}$$

$$\text{From equation (42.5), } R_0 = \sqrt{\frac{L}{C}}, \text{ thus } \tan\frac{\beta}{2} = \frac{\frac{\omega L}{2}}{\sqrt{\frac{L}{C}}} = \frac{\omega\sqrt{(LC)}}{2}$$

For angles of β up to about 20°, $\tan\frac{\beta}{2} \approx \frac{\beta}{2}$ radians

Thus when $\beta < 20°$, $\frac{\beta}{2} = \frac{\omega\sqrt{(LC)}}{2}$

from which, **phase angle,** $\boldsymbol{\beta = \omega\sqrt{(LC)}}$ **radian** (42.27)

Since $\beta = 2\pi f\sqrt{(LC)} = (2\pi\sqrt{(LC)})f$ then β is proportional to f and a graph of β (vertical) against frequency (horizontal) should be a straight line of gradient $2\pi\sqrt{(LC)}$ and passing through the origin. However in practise this is only usually valid up to a frequency of about 0.7 f_c for a low-pass filter and a typical characteristic is shown in Figure 42.37. At the cut-off frequency, $\beta = \pi$ rad. For frequencies within the attenuation band, the phase shift is unimportant, since all voltages having such frequencies are suppressed.

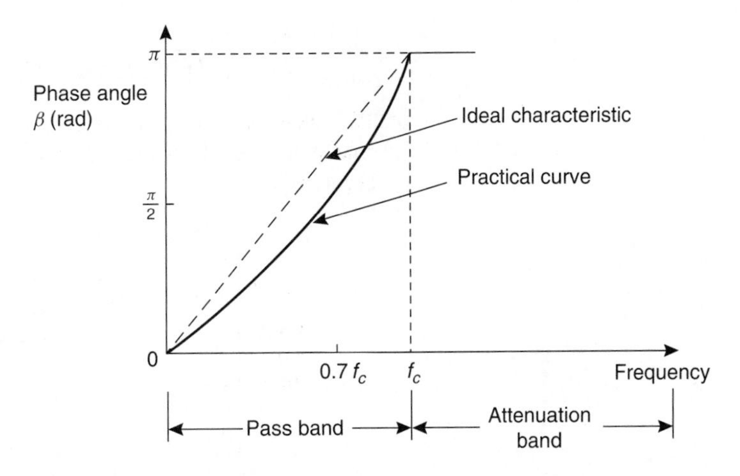

Figure 42.37

A **high-pass prototype *T* section** is shown in Figure 42.38(a) and its phasor diagram in Figure 42.38(b), the latter being produced by similar reasoning to above.

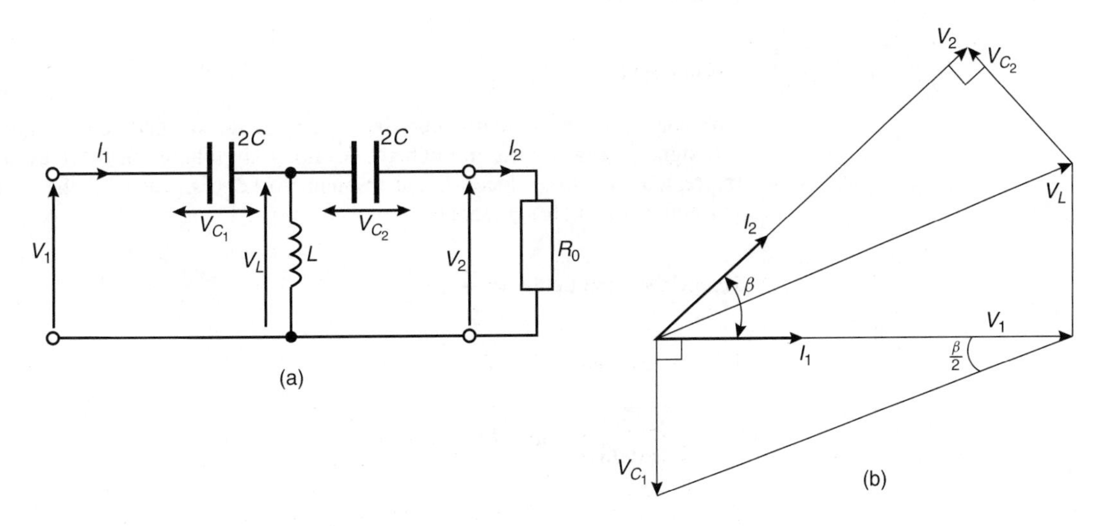

Figure 42.38

From Figure 42.38(b), $\tan\dfrac{\beta}{2} = \dfrac{V_{C1}}{V_1} = \dfrac{I_1\left(\dfrac{1}{\omega 2C}\right)}{I_1 R_0} = \dfrac{1}{2\omega C R_0}$

$$= \frac{1}{2\omega C\sqrt{\dfrac{L}{C}}} = \frac{1}{2\omega\sqrt{(LC)}}$$

i.e., $\beta = \dfrac{1}{\omega\sqrt{(LC)}} = \dfrac{1}{(2\pi\sqrt{(LC)})f}$ for small angles.

Thus the phase angle is universely proportional to frequency. The β/f characteristics of an ideal and a practical high-pass filter are shown in Figure 42.39.

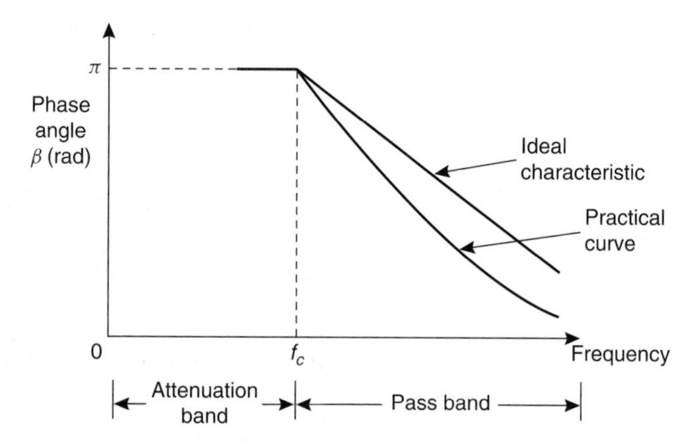

Figure 42.39

Time delay

The change of phase that occurs in a filter section depends on the time the signal takes to pass through the section. The phase shift β may be expressed as a time delay. If the frequency of the signal is f then the periodic time is $(1/f)$ seconds.

Hence the time delay $= \dfrac{\beta}{2\pi} \times \dfrac{1}{f} = \dfrac{\beta}{\omega}$.

From equation (42.27), $\beta = \omega\sqrt{(LC)}$. Thus

$$\textbf{time delay} = \frac{\omega\sqrt{(LC)}}{\omega} = \sqrt{(LC)} \qquad (42.28)$$

when angle β is small.

Equation (42.28) shows that the time delay, or **transit time**, is independant of frequency. Thus a phase shift which is proportional to frequency (equation (42.27)) results in a time delay which is independant of frequency. Hence if the input to the filter section consists of a complex wave composed of several harmonic components of differing frequency, the output will consist of a complex wave made up of the sum of corresponding components all delayed by the same amount. There will therefore be no phase distortion due to varying time delays for the separate frequency components.

In practise, however, phase shift β tends not to be constant and the increase in time delay with rising frequency causes distortion of non-sinusoidal inputs, this distortion being superimposed on that due to the attenuation of components whose frequency is higher than the cut-off frequency.

At the cut-off frequency of a prototype low-pass filter, the phase angle $\beta = \pi$ rad. Hence the time delay of a signal through such a section at the cut-off frequency is given by

$$\frac{\beta}{\omega} = \frac{\pi}{2\pi f_c} = \frac{1}{2 f_c} = \frac{1}{2\dfrac{1}{\pi\sqrt{(LC)}}} \text{ from equation (42.3),}$$

i.e., at f_c,

$$\textbf{the transit time} = \frac{\pi\sqrt{(LC)}}{2} \textbf{ seconds} \qquad (42.29)$$

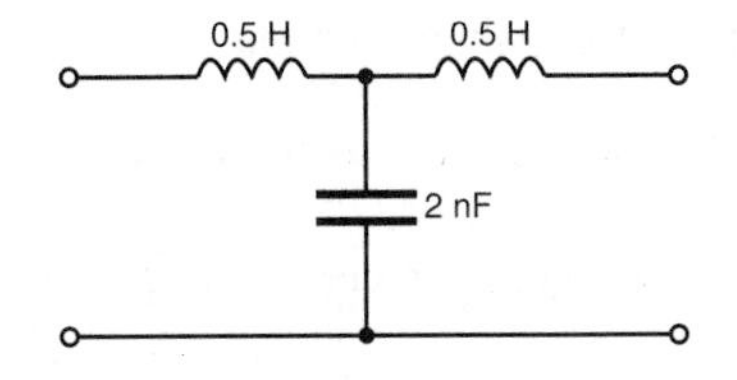

Figure 42.40

Problem 12. Determine for the filter section shown in Figure 42.40, (a) the time delay for the signal to pass through the filter, assuming the phase shift is small, and (b) the time delay for a signal to pass through the section at the cut-off frequency.

Comparing Figure 42.40 with the low-pass T section of Figure 42.13(a), shows that

$$\frac{L}{2} = 0.5 \text{ H, thus inductance } L = 1 \text{ H, and capacitance } C = 2 \text{ nF}$$

(a) From equation (42.28),

$$\text{time delay} = \sqrt{(LC)} = \sqrt{[(1)(2 \times 10^{-9})]} = \mathbf{44.7\ \mu s}$$

(b) From equation (42.29), at the cut-off frequency,

$$\text{time delay} = \frac{\pi}{2}\sqrt{(LC)} = \frac{\pi}{2}(44.7) = \mathbf{70.2\ \mu s}$$

Problem 13. A filter network comprising n identical sections passes signals of all frequencies up to 500 kHz and provides a total delay of 9.55 μs. If the nominal impedance of the circuit into which the filter is inserted is 1 kΩ, determine (a) the values of the elements in each section, and (b) the value of n.

Cut-off frequency, $f_c = 500 \times 10^3$ Hz and nominal impedance

$$R_0 = 1000\ \Omega.$$

Since the filter passes frequencies up to 500 kHz then it is a low-pass filter.

(a) From equations (42.6) and (42.7), for a low-pass filter section,

$$\text{capacitance, } C = \frac{1}{\pi R_0 f_c} = \frac{1}{\pi(1000)(500 \times 10^3)} \equiv \mathbf{636.6\ pF}$$

$$\text{and inductance, } L = \frac{R_0}{\pi f_c} = \frac{1000}{\pi(500 \times 10^3)} \equiv \mathbf{636.6\ \mu H}$$

Thus if the section is a **low-pass *T* section** then the inductance in each series arm will be $(L/2) =$ **318.3 μH** and the capacitance in the shunt arm will be **636.6 pF**.

If the section is a **low-pass π section** then the inductance in the series arm will be **636.6 μH** and the capacitance in each shunt arm will be $(C/2) =$ **318.3 pF**

(b) From equation (42.28), the time delay for a single section

$$= \sqrt{(LC)} = \sqrt{[(636.6 \times 10^{-6})(636.6 \times 10^{-12})]} = 0.6366\ \mu s$$

For a time delay of 9.55 μs therefore, the number of cascaded sections required is given by

$$\frac{9.55}{0.6366} = 15, \text{ i.e., } \boldsymbol{n = 15}$$

Problem 14. A filter network consists of 8 sections in cascade having a nominal impedance of 1 kΩ. If the total delay time is 4 μs, determine the component values for each section if the filter is (a) a low-pass T network, and (b) a high-pass π network.

Since the total delay time is 4 μs then the delay time of each of the 8 sections is $\frac{4}{8}$, i.e., 0.5 μs

From equation (42.28), time delay $= \sqrt{(LC)}$

Hence $0.5 \times 10^{-6} = \sqrt{(LC)}$ (i)

Also, from equation (42.5), $\sqrt{\frac{L}{C}} = 1000$ (ii)

From equation (ii), $\sqrt{L} = 1000\sqrt{C}$

Substituting in equation (i) gives: $0.5 \times 10^{-6} = (1000\sqrt{C})\sqrt{C} = 1000\,C$

from which, capacitance $C = \frac{0.5 \times 10^{-6}}{1000} = 0.5$ nF

From equation (ii), $\sqrt{C} = \frac{\sqrt{L}}{1000}$

Substituting in equation (i) gives: $0.5 \times 10^{-6} = (\sqrt{L})\left(\frac{\sqrt{L}}{1000}\right) = \frac{L}{1000}$

from which, inductance, $L = 500$ μH

(a) If the filter is a **low-pass *T* section** then, from Figure 42.13(a), each series arm has an inductance of $L/2$, i.e., **250 μH** and the shunt arm has a capacitance of **0.5 nF**

(b) If the filter is a **high-pass π network** then, from Figure 42.16(b), the series arm has a capacitance of **0.5 nF** and each shunt arm has an inductance of 2 L, i.e., **1000 μH or 1 mH**.

Further problems on propagation coefficient and time delay may be found in Section 42.10, problems 13 to 18, page 838

42.8 '*m*-derived' filter sections

(a) General

In a low-pass filter a clearly defined cut-off frequency followed by a high attenuation is needed; in a high-pass filter, high attenuation followed by a clearly defined cut-off frequency is needed. It is not practicable to obtain either of these conditions by wiring appropriate prototype constant-k sections in cascade. An equivalent section is therefore required having:

(i) the same cut-off frequency as the prototype but with a rapid rise in attenuation beyond cut-off for a low-pass type or a rapid decrease at cut-off from a high attenuation for the high-pass type,

(ii) the same value of nominal impedance R_0 as the prototype at all frequencies (otherwise the two forms could not be connected together without mismatch).

If the two sections, i.e., the prototype and the equivalent section, have the same value of R_0 they will have identical pass-bands.

The equivalent section is called an **'*m*-derived' filter section** (for reasons as explained below) and is one which gives a sharper cut-off at the edges of the pass band and a better impedance characteristic.

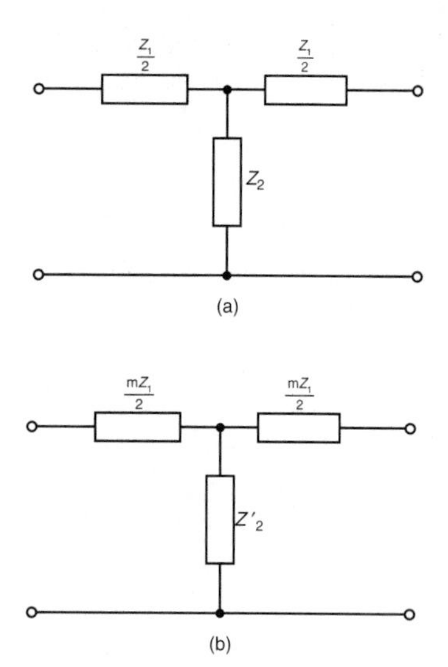

Figure 42.41

(b) *T* sections

A prototype T section is shown in Figure 42.41(a). Let a new section be constructed from this section having a series arm of the same type but of different value, say mZ_1, where m is some constant. (It is for this reason that the new equivalent section is called an 'm-derived' section.) If the characteristic impedance Z_{0T} of the two sections is to be the same then the value of the shunt arm impedance will have to be different to Z_2.

Let this be Z_2' as shown in Figure 42.41(b).

The value of Z_2' is determined as follows:

From equation (41.1), page 760, for the prototype shown in Figure 42.41(a):

$$Z_{0T} = \sqrt{\left[\left(\frac{Z_1}{2}\right)^2 + 2\left(\frac{Z_1}{2}\right)Z_2\right]}$$

i.e., $$Z_{0T} = \sqrt{\left(\frac{Z_1^2}{4} + Z_1 Z_2\right)} \qquad \text{(a)}$$

Similarly, for the new section shown in Figure 42.41(b),

$$Z_{0T} = \sqrt{\left[\left(\frac{mZ_1}{2}\right)^2 + 2\left(\frac{mZ_1}{2}\right)Z_2'\right]}$$

i.e., $$Z_{0T} = \sqrt{\left(\frac{m^2 Z_1^2}{4} + mZ_1 Z_2'\right)} \qquad \text{(b)}$$

Equations (a) and (b) will be identical if:

$$\frac{Z_1^2}{4} + Z_1 Z_2 = \frac{m^2 Z_1^2}{4} + mZ_1 Z_2'$$

Rearranging gives: $$mZ_1 Z_2' = Z_1 Z_2 + \frac{Z_1^2}{4}(1 - m^2)$$

i.e., $$Z_2' = \frac{Z_2}{m} + Z_1\left(\frac{1 - m^2}{4m}\right) \qquad (42.30)$$

Thus impedance Z_2' consists of an impedance Z_2/m in series with an impedance $Z_1((1 - m^2)/4m)$. An additional component has therefore been introduced into the shunt arm of the m-derived section. The value of m can range from 0 to 1, and when $m = 1$, the prototype and the m-derived sections are identical.

(c) π sections

A prototype π section is shown in Figure 42.42(a). Let a new section be constructed having shunt arms of the same type but of different values,

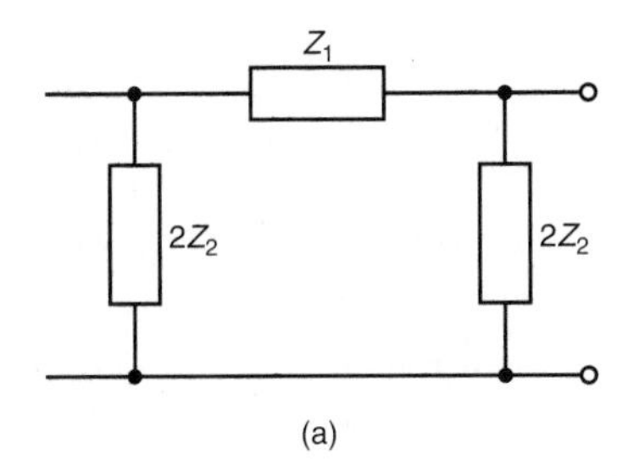

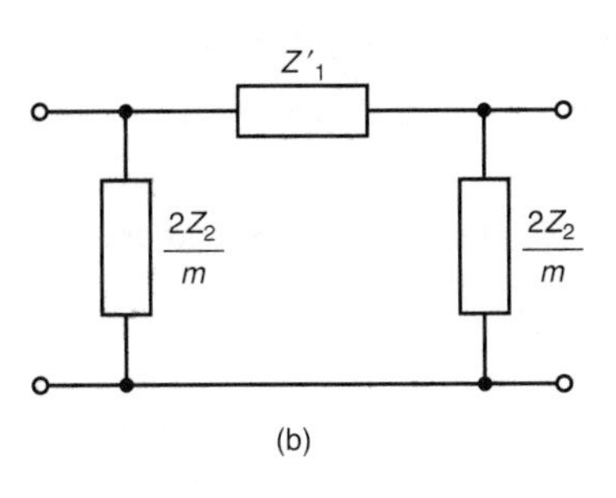

Figure 42.42

say Z_2/m, where m is some constant. If the characteristic impedance $Z_{0\pi}$ of the two sections is to be the same then the value of the series arm impedance will have to be different to Z_1.

Let this be Z'_1 as shown in Figure 42.42(b).

The value of Z'_1 is determined as follows:

From equation (42.9), $Z_{0T}Z_{0\pi} = Z_1Z_2$

Thus the characteristic impedance of the section shown in Figure 42.42(a) is given by:

$$Z_{0\pi} = \frac{Z_1Z_2}{Z_{0T}} = \frac{Z_1Z_2}{\sqrt{\left(\frac{Z_1^2}{4} + Z_1Z_2\right)}} \qquad \text{(c)}$$

from equation (a) above.

For the section shown in Figure 42.42(b),

$$Z_{0\pi} = \frac{Z'_1\frac{Z_2}{m}}{\sqrt{\left(\frac{(Z'_1)^2}{4} + Z'_1\frac{Z_2}{m}\right)}} \qquad \text{(d)}$$

Equations (c) and (d) will be identical if

$$\frac{Z_1Z_2}{\sqrt{\left(\frac{Z_1^2}{4} + Z_1Z_2\right)}} = \frac{Z'_1\frac{Z_2}{m}}{\sqrt{\left(\frac{(Z'_1)^2}{4} + Z'_1\frac{Z_2}{m}\right)}}$$

Dividing both sides by Z_2 and then squaring both sides gives:

$$\frac{Z_1^2}{\frac{Z_1^2}{4} + Z_1Z_2} = \frac{\frac{(Z'_1)^2}{m^2}}{\frac{(Z'_1)^2}{4} + \frac{Z'_1Z_2}{m}}$$

Thus $\quad Z_1^2\left(\frac{(Z'_1)^2}{4} + \frac{Z'_1Z_2}{m}\right) = \frac{(Z'_1)^2}{m^2}\left(\frac{Z_1^2}{4} + Z_1Z_2\right)$

i.e., $\quad \frac{Z_1^2(Z'_1)^2}{4} + \frac{Z_1^2Z'_1Z_2}{m} = \frac{(Z'_1)^2Z_1^2}{4m^2} + \frac{(Z'_1)^2Z_1Z_2}{m^2}$

Multiplying throughout by $4m^2$ gives:

$$m^2Z_1^2(Z'_1)^2 + 4mZ_1^2Z'_1Z_2 = (Z'_1)^2Z_1^2 + 4(Z'_1)^2Z_1Z_2$$

Dividing throughout by Z'_1 and rearranging gives:

$$4mZ_1^2Z_2 = Z'_1(Z_1^2 + 4Z_1Z_2 - m^2Z_1^2)$$

$$\text{Thus} \quad Z_1' = \frac{4mZ_1^2 Z_2}{4Z_1Z_2 + Z_1^2(1-m^2)}$$

$$\text{i.e.,} \quad \mathbf{Z_1' = \frac{4mZ_1Z_2}{4Z_2 + Z_1(1-m^2)}} \qquad (42.31)$$

An impedance mZ_1 in parallel with an impedance $(4mZ_2/1-m^2)$ gives (using (product/sum)):

$$\frac{(mZ_1)\dfrac{4mZ_2}{1-m^2}}{mZ_1 + \dfrac{4mZ_2}{1-m^2}} = \frac{(mZ_1)4mZ_2}{mZ_1(1-m^2)+4mZ_2} = \frac{4mZ_1Z_2}{4Z_2 + Z_1(1-m^2)}$$

Hence the expression for Z_1' (equation (42.31)) represents an impedance mZ_1 in parallel with an impedance $(4m/1-m^2)Z_2$

(d) Low-pass '*m*-derived' sections

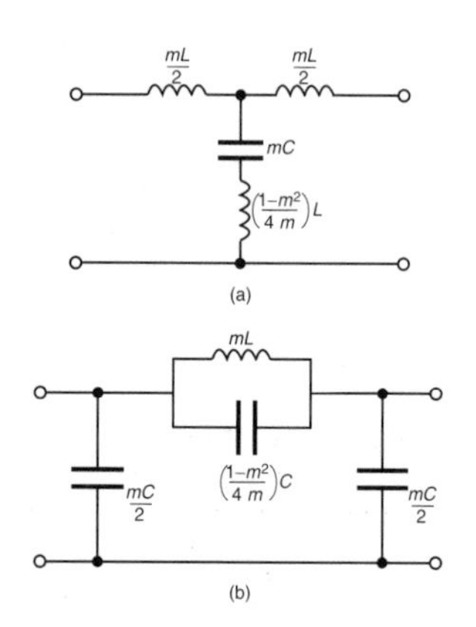

Figure 42.43

The '*m*-derived' low-pass *T* section is shown in Figure 42.43(a) and is derived from Figure 42.13(a), Figure 42.41 and equation (42.30). If Z_2 represents a pure capacitor in Figure 42.41(a), then $Z_2 = (1/\omega C)$.
A capacitance of value mC shown in Figure 42.43(a) has an impedance

$$\frac{1}{\omega mC} = \frac{1}{m}\left(\frac{1}{\omega C}\right) = \frac{Z_2}{m} \text{ as in equation (42.30).}$$

The '*m*-derived' low-pass π section is shown in Figure 42.43(b) and is derived from Figure 42.13(b), Figure 42.42 and from equation (42.31). Note that a capacitance of value $\left(\dfrac{1-m^2}{4m}\right)C$ has an impedance of

$$\frac{1}{\omega\left(\dfrac{1-m^2}{4m}\right)C} = \left(\frac{4m}{1-m^2}\right)\left(\frac{1}{\omega C}\right) = \left(\frac{4m}{1-m^2}\right)Z_2$$

where Z_2 is a pure capacitor.

In Figure 42.43(a), series resonance will occur in the shunt arm at a particular frequency—thus short-circuiting the transmission path. In the prototype, infinite attenuation is obtained only at infinite frequency (see Figure 42.25).

In the *m*-derived section of Figure 42.43(a), let the frequency of infinite attenuation be f_∞, then at resonance: $X_L = X_C$

$$\text{i.e.,} \quad \omega_\infty\left(\frac{1-m^2}{4m}\right)L = \frac{1}{\omega_\infty mC}$$

$$\text{from which,} \quad \omega_\infty^2 = \frac{1}{(mC)\left(\dfrac{1-m^2}{4m}\right)L} = \frac{4}{LC(1-m^2)}$$

From equation (42.2),

$$\frac{4}{LC} = \omega_c^2, \text{ thus } \omega_\infty^2 = \frac{\omega_c^2}{(1-m^2)},$$

where $\omega_c = 2\pi f_c$, f_c being the cut-off frequency of the prototype.

Hence $$\omega_\infty = \frac{\omega_c}{\sqrt{(1-m^2)}} \quad (42.32)$$

Rearranging gives:

$$\omega_\infty^2(1-m^2) = \omega_c^2$$

$$\omega_\infty^2 - m^2\omega_\infty^2 = \omega_c^2$$

$$m^2 = \frac{\omega_\infty^2 - \omega_c^2}{\omega_\infty^2} = 1 - \frac{\omega_c^2}{\omega_\infty^2}$$

i.e., $$\boxed{m = \sqrt{\left[1 - \left(\frac{f_c}{f_\infty}\right)^2\right]}} \quad (42.33)$$

In the m-derived π section of Figure 42.43(b), resonance occurs in the parallel arrangement comprising the series arm of the section when

$$\omega^2 = \frac{1}{mL\left(\dfrac{1-m^2}{4m}\right)C}, \text{ when } \omega^2 = \frac{4}{LC(1-m^2)}$$

as in the series resonance case (see Chapter 28).

Thus equations (42.32) and (42.33) are also applicable to the low-pass m-derived π section.

In equation (42.33), $\mathbf{0 < m < 1}$, thus $\boldsymbol{f_\infty > f_c}$.

The frequency of infinite attenuation f_∞ can be placed anywhere within the attenuation band by suitable choice of the value of m; the smaller m is made the nearer is f_∞ to the cut-off frequency, f_c.

Problem 15. A filter section is required to have a nominal impedance of 600 Ω, a cut-off frequency of 5 kHz and a frequency of infinite attenuation at 5.50 kHz. Design (a) an appropriate 'm-derived' T section, and (b) an appropriate 'm-derived' π section.

Nominal impedance $R_0 = 600\ \Omega$, cut-off frequency, $f_c = 5000$ Hz and frequency of infinite attenuation, $f_\infty = 5500$ Hz. Since $f_\infty > f_c$ the filter section is low-pass.

From equation (42.33),

$$m = \sqrt{\left[1 - \left(\frac{f_c}{f_\infty}\right)^2\right]} = \sqrt{\left[1 - \left(\frac{5000}{5500}\right)^2\right]} = 0.4166$$

For a low-pass prototype section:

$$\text{from equation (42.6), capacitance, } C = \frac{1}{\pi R_0 f_c} = \frac{1}{\pi(600)(5000)} \equiv 0.106\ \mu\text{F}$$

$$\text{and from equation (42.7), inductance, } L = \frac{R_0}{\pi f_c} = \frac{600}{\pi(5000)} \equiv 38.2\ \text{mH}$$

(a) For an 'm-derived' low-pass T section:
From Figure 42.43(a), the series arm inductances are each

$$\frac{mL}{2} = \frac{(0.4166)(38.2)}{2} = \mathbf{7.957\ mH,}$$

and the shunt arm contains a capacitor of value mC,
i.e., $(0.4166)(0.106) =$ **0.0442 μF** or **44.2 nF**, in series with an inductance of

$$\text{value } \left(\frac{1-m^2}{4m}\right) L = \left(\frac{1-0.4166^2}{4(0.4166)}\right)(38.2),$$

i.e., **18.95 mH**

The appropriate 'm-derived' T section is shown in Figure 42.44.

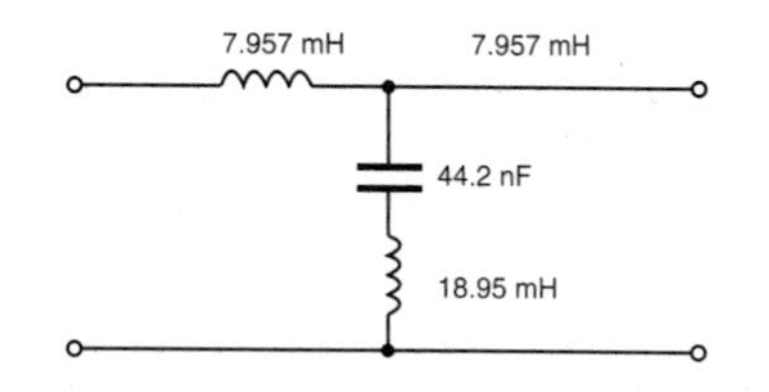

Figure 42.44

(b) For an 'm-derived' low-pass π section:
From Figure 42.43(b) the shunt arms each contain capacitances equal to $mC/2$,

$$\text{i.e., } \frac{(0.4166)(0.106)}{2} = \mathbf{0.0221\ \mu F\ or\ 22.1\ nF,}$$

and the series arm contains an inductance of value $m\,L$,
i.e., $(0.4166)(38.2) =$ **15.91 mH** in parallel with a capacitance of

$$\text{value } \left(\frac{1-m^2}{4m}\right) C = \left(\frac{1-0.4166^2}{4(0.4166)}\right)(0.106) = \mathbf{0.0526\ \mu F\ or\ 52.6\ nF}$$

The appropriate 'm-derived' π section is shown in Figure 42.45.

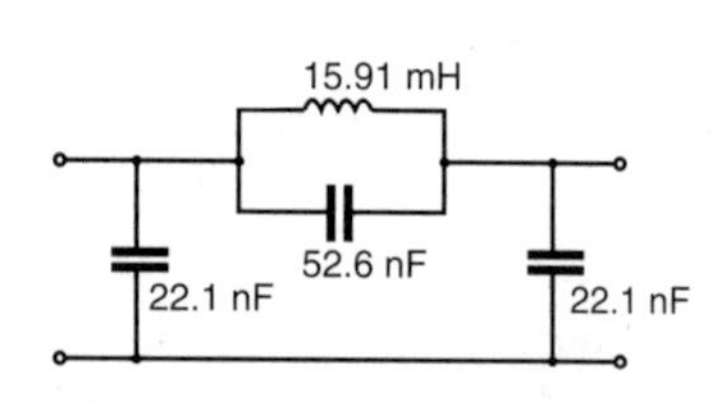

Figure 42.45

(e) High-pass '*m*-derived' sections

Figure 42.46(a) shows a high-pass prototype T section and Figure 42.46(b) shows the 'm-derived' high-pass T section which is derived from Figure 42.16(a), Figure 42.41 and equation (42.30).

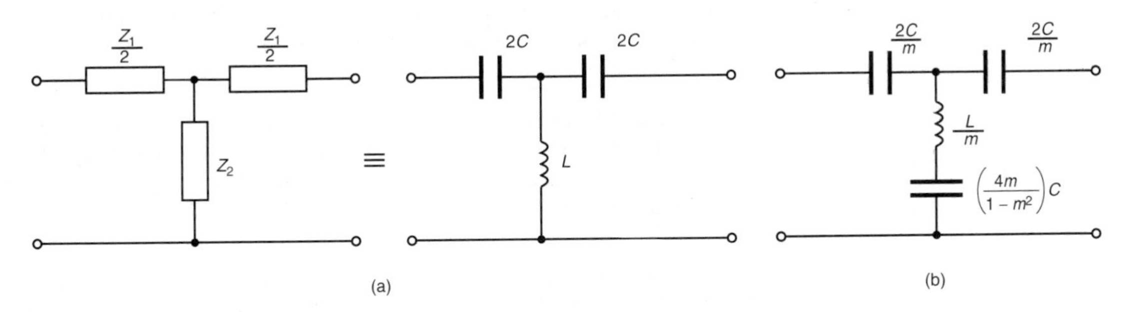

Figure 42.46

Figure 42.47(a) shows a high-pass prototype π section and Figure 42.47(b) shows the 'm-derived' high-pass π section which is derived from Figure 42.16(b), Figure 42.42 and equation (42.31). In Figure 42.46(b), resonance occurs in the shunt arm when:

$$\omega_\infty \frac{L}{m} = \frac{1}{\omega_\infty\left(\dfrac{4m}{1-m^2}\right)C}$$

i.e., when $\omega_\infty^2 = \dfrac{1-m^2}{4LC} = \omega_c^2(1-m^2)$ from equation (42.14)

i.e., $$\omega_\infty = \omega_c\sqrt{(1-m^2)} \qquad (42.34)$$

Hence $$\frac{\omega_\infty^2}{\omega_c^2} = 1 - m^2$$

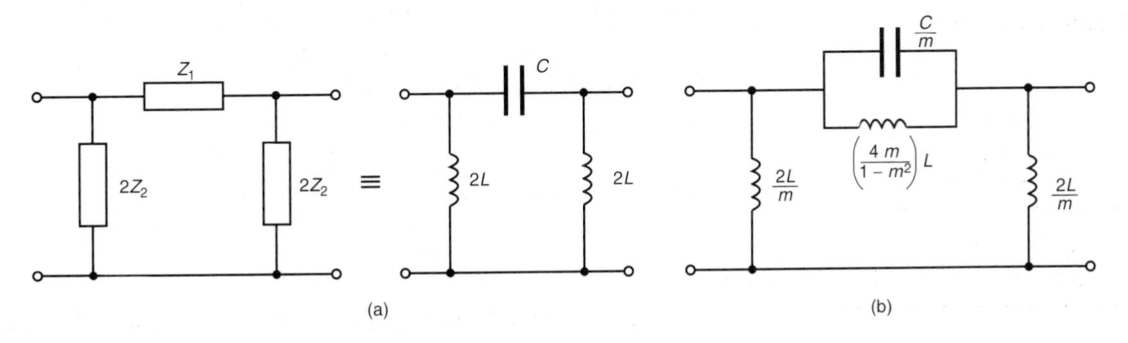

Figure 42.47

from which, $$\boldsymbol{m = \sqrt{\left[1 - \left(\frac{f_\infty}{f_c}\right)^2\right]}} \qquad (42.35)$$

For a high-pass section, $f_\infty < f_c$.

It may be shown that equations (42.34) and (42.35) also apply to the 'm-derived' π section shown in Figure 42.47(b).

> Problem 16. Design (a) a suitable 'm-derived' T section, and (b) a suitable 'm-derived' π section having a cut-off frequency of 20 kHz, a nominal impedance of 500 Ω and a frequency of infinite attenuation 16 kHz.

Nominal impedance $R_0 = 500\ \Omega$, cut-off frequency, $f_c = 20$ kHz and the frequency of infinite attenuation, $f_\infty = 16$ kHz. Since $f_\infty < f_c$ the filter is high-pass.

From equation (42.35), $$m = \sqrt{\left[1 - \left(\frac{f_\infty}{f_c}\right)^2\right]} = \sqrt{\left[1 - \left(\frac{16}{20}\right)^2\right]} = 0.60$$

For a high-pass prototype section:

From equation (42.18), capacitance,

$$C = \frac{1}{4\pi R_0 f_c} = \frac{1}{4\pi(500)(20\,000)} \equiv 7.958\ \text{nF}$$

and from equation (42.19), inductance,

$$L = \frac{R_0}{4\pi f_c} = \frac{500}{4\pi(20\,000)} \equiv 1.989\ \text{mH}$$

(a) For an 'm-derived' high-pass T section:

From Figure 42.46(b), each series arm contains a capacitance of value $2C/m$, i.e., 2(7.958)/0.60, i.e., **26.53 nF**, and the shunt arm contains an inductance of value L/m, i.e., $(1.989/0.60) =$ **3.315 mH** in series with a capacitance of value

$$\left(\frac{4m}{1 - m^2}\right) C \text{ i.e., } \left(\frac{4(0.60)}{1 - 0.60^2}\right)(7.958) = \mathbf{29.84\ nF}$$

A suitable 'm-derived' T section is shown in Figure 42.48.

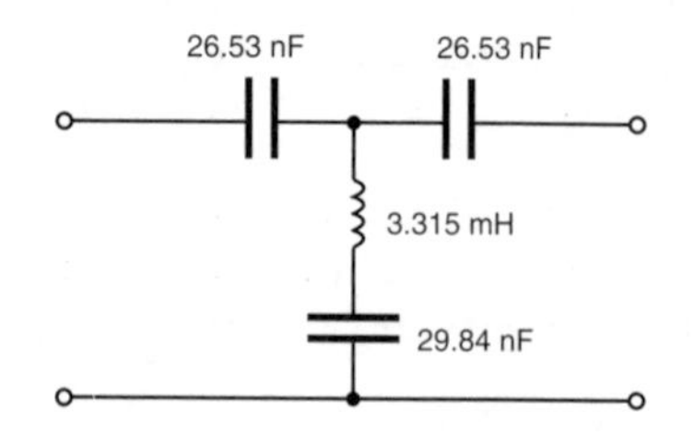

Figure 42.48

(b) For an 'm-derived' high pass π section:

From Figure 42.47(b), the shunt arms each contain inductances equal to $2L/m$, i.e., (2(1.989)/0.60), i.e., **6.63 mH** and the series arm

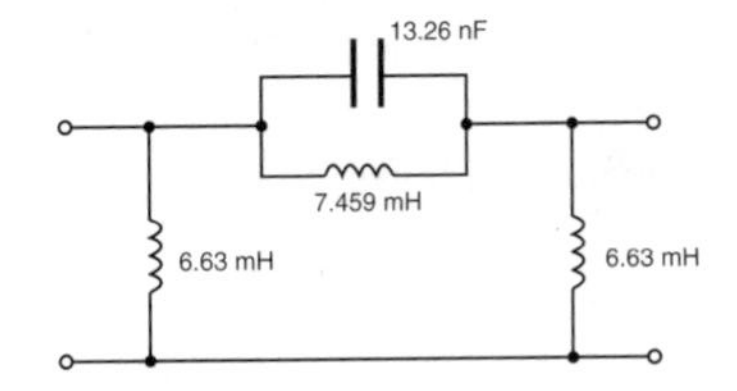

Figure 42.49

contains a capacitance of value C/m, i.e., $(7.958/0.60) = \mathbf{13.26\ nF}$ in parallel with an inductance of value $(4m/1 - m^2)L$,

i.e., $\left(\dfrac{4(0.60)}{1 - 0.60^2}\right)(1.989) \equiv \mathbf{7.459\ mH}$

A suitable 'm-derived' π section is shown in Figure 42.49.

Further problems on 'm-derived' filter sections may be found in Section 42.10, problems 19 to 22, page 839

42.9 Practical composite filters

In practise, filters to meet a given specification often have to comprise a number of basic networks. For example, a practical arrangement might consist of (i) a basic prototype, in series with (ii) an 'm-derived' section, with (iii) terminating half-sections at each end. The 'm-derived' section improves the attenuation immediately after cut-off, the prototype improves the attenuation well after cut-off, whilst the terminating half-sections are used to obtain a constant match over the pass-band.

Figure 42.50(a) shows an 'm-derived' low-pass T section, and Figure 42.50(b) shows the same section cut into two halves through AB, each of the two halves being termed a 'half-section'. The 'm-derived' half section also improves the steepness of attenuation outside the pass-band.

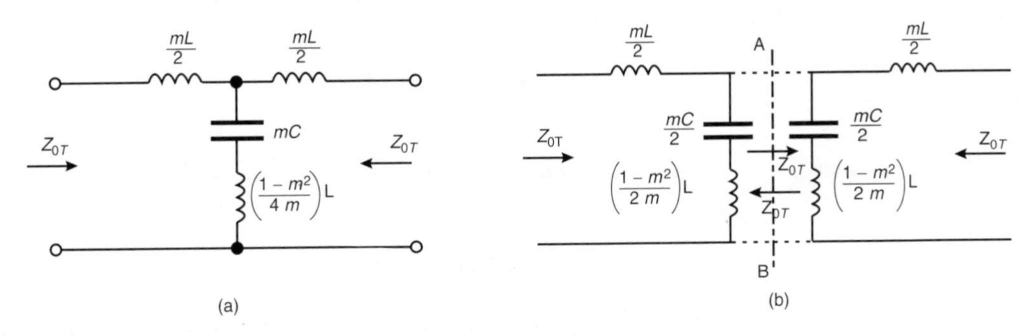

Figure 42.50

As shown in Section 42.8, the 'm-derived' filter section is based on a prototype which presents its own characteristic impedance at its terminals. Hence, for example, the prototype of a T section leads to an 'm-derived' T section and $Z_{0T} = Z_{0T(m)}$ where Z_{0T} is the characteristic impedance of the prototype and $Z_{0T(m)}$ is the characteristic impedance of the 'm-derived' section. It is shown in Figure 42.24 that Z_{0T} has a non-linear characteristic against frequency; thus $Z_{0T(m)}$ will also be non-linear.

Since from equation (42.9), $Z_{0\pi} = (Z_1Z_2/Z_{0T})$, then the characteristic impedance of the 'm-derived' π section,

$$Z_{0\pi(m)} = \frac{Z_1'Z_2'}{Z_{0T(m)}} = \frac{Z_1'Z_2'}{Z_{0T}}$$

where Z'_1 and Z'_2 are the equivalent values of impedance in the 'm-derived' section.

From Figure 42.41, $Z'_1 = mZ_1$ and from equation (42.30),

$$Z'_2 = \frac{Z_2}{m} + \left(\frac{1-m^2}{4m}\right) Z_1$$

Thus

$$Z_{0\pi(m)} = \frac{mZ_1\left[\dfrac{Z_2}{m} + \left(\dfrac{1-m^2}{4m}\right) Z_1\right]}{Z_{0T}}$$

$$= \frac{Z_1 Z_2}{Z_{0T}}\left[1 + \left(\frac{1-m^2}{4Z_2}\right) Z_1\right] \quad (42.36)$$

or

$$\mathbf{Z_{0\pi(m)} = Z_{0\pi}\left[1 + \left(\frac{1-m^2}{4Z_2}\right) Z_1\right]} \quad (42.37)$$

Thus the impedance of the 'm-derived' section is related to the impedance of the prototype by a factor of $[1 + (1 - m^2/4Z_2)Z_1]$ and will vary as m varies.

When $m = 1$, $Z_{0\pi(m)} = Z_{0\pi}$

When $m = 0$, $Z_{0\pi(m)} = \dfrac{Z_1 Z_2}{Z_{0T}}\left[1 + \dfrac{Z_1}{4Z_2}\right]$ from equation (42.36)

$$= \frac{1}{Z_{0T}}\left[Z_1 Z_2 + \frac{Z_1^2}{4}\right]$$

However from equation (42.8), $Z_1 Z_2 + .\dfrac{Z_1^2}{4} = Z_{0T}^2$

Hence, when $m = 0$, $Z_{0\pi(m)} = \dfrac{Z_{0T}^2}{Z_{0T}} = Z_{0T}$

Thus the characteristic of impedance against frequency for $m = 1$ and $m = 0$ shown in Figure 42.51 are the same as shown in Figure 42.24. Further characteristics may be drawn for values of m between 0 and 1 as shown.

It is seen from Figure 42.51 that when $m = 0.6$ the impedance is practically constant at R_0 for most of the pass-band. In a composite filter, 'm-derived' half-sections having a value of $m = 0.6$ are usually used at each end to provide a good match to a resistive source and load over the pass-band.

Figure 42.51 shows characteristics of 'm-derived' low-pass filter sections; similar curves may be constructed for m-derived high-pass filters with the two curves shown in Figure 42.29 representing the limiting values of $m = 0$ and $m = 1$.

The value of m needs to be small for the frequency of input attenuation, f_∞, to be close to the cut-off frequency, f_c. However, it is not practical

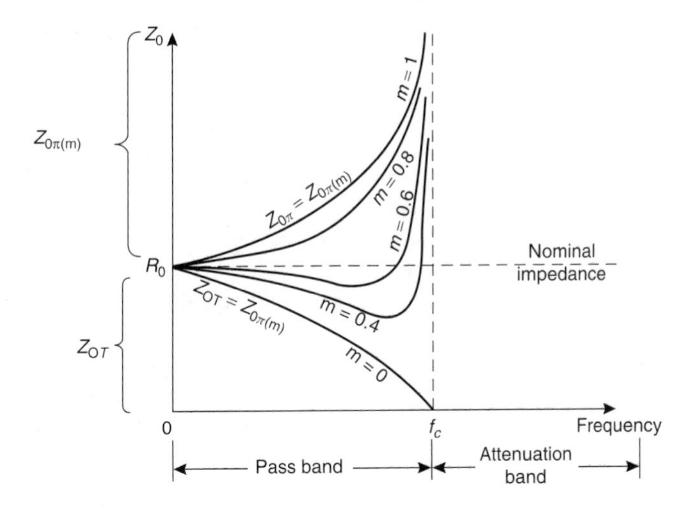

Figure 42.51

to make m very small, below 0.3 being very unusual. When $m = 0.3$, $f_\infty \approx 1.05 f_c$ (from equation (42.32)) and when $m = 0.6$, $f_\infty = 1.25 f_c$. The effect of the value of m on the frequency of infinite attenuation is shown in Figure 42.52 although the ideal curves shown would be modified a little in practise by resistance losses.

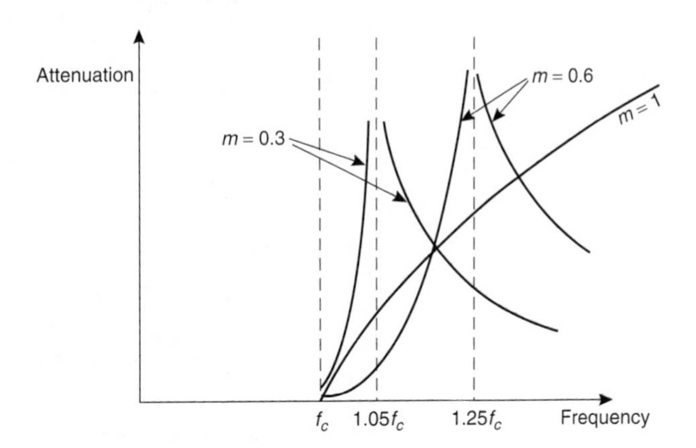

Figure 42.52

Problem 17. It is required to design a composite filter with a cut-off frequency of 10 kHz, a frequency of infinite attenuation 11.8 kHz and nominal impedance of 600 Ω. Determine the component values needed if the filter is to comprise a prototype T section, an 'm-derived' T section and two terminating 'm-derived' half-sections.

$R_0 = 600\ \Omega$, $f_c = 10$ kHz and $f_\infty = 11.8$ kHz. Since $f_c < f_\infty$ the filter is a low-pass T section.

For the prototype:

From equation (42.6), capacitance,

$$C = \frac{1}{\pi f_c R_0} = \frac{1}{\pi(10\,000)(600)} \equiv 0.0531\ \mu\text{F},$$

and from equation (42.7), inductance,

$$L = \frac{R_0}{\pi f_c} = \frac{600}{\pi(10\,000)} \equiv 19\ \text{mH}$$

Thus, from Figure 42.13(a), the series arm components are $(L/2) = (19/2) = \mathbf{9.5\ mH}$ and the shunt arm component is $\mathbf{0.0531\ \mu F}$.

For the 'm-derived' section:

From equation (42.33),

$$m = \sqrt{\left[1 - \left(\frac{f_c}{f_\infty}\right)^2\right]} = \sqrt{\left[1 - \left(\frac{10\,000}{11800}\right)^2\right]} = 0.5309$$

Thus from Figure 42.43(a), the series arm components are

$$\frac{mL}{2} = \frac{(0.5309)(19)}{2} = \mathbf{5.04\ mH}$$

and the shunt arm comprises $mC = (0.5309)(0.0531) = \mathbf{0.0282\ \mu F}$ in series with

$$\left(\frac{1 - m^2}{4m}\right) L = \left(\frac{1 - 0.5309^2}{4(0.5309)}\right)(19) = \mathbf{6.43\ mH}$$

For the half-sections a value of $m = 0.6$ is taken to obtain matching. Thus from Figure 42.50,

$$\frac{mL}{2} = \frac{(0.6)(19)}{2} = \mathbf{5.7\ mH},\ \frac{mC}{2} = \frac{(0.6)(0.0531)}{2} \equiv \mathbf{0.0159\ \mu F}$$

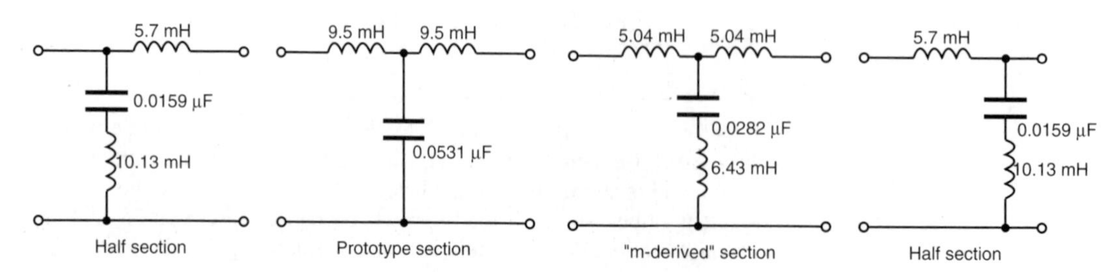

Figure 42.53

$$\text{and} \quad \left(\frac{1-m^2}{2m}\right)L = \left(\frac{1-0.6^2}{2(0.6)}\right)(19) \equiv \mathbf{10.13\ mH}$$

The complete filter is shown in Figure 42.53.

Further problems on practical composite filter sections may be found in Section 42.10 following, problems 23 and 24, page 840

42.10 Further problems on filter networks

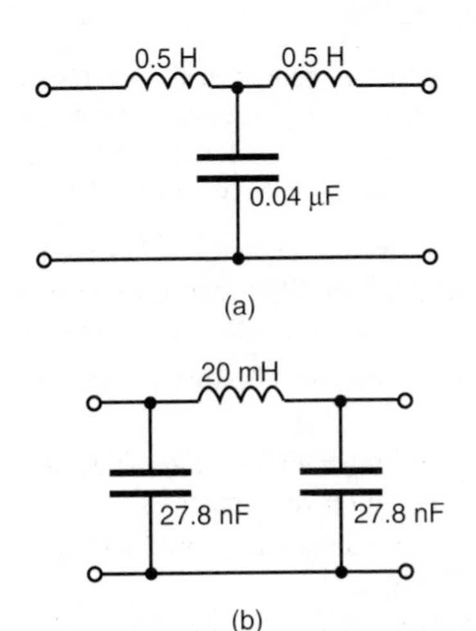

Figure 42.54

Low-pass filter sections

1 Determine the cut-off frequency and the nominal impedance of each of the low-pass filter sections shown in Figure 42.54.
[(a) 1592 Hz; 5 kΩ (b) 9545 Hz; 600 Ω]

2 A filter section is to have a characteristic impedance at zero frequency of 500 Ω and a cut-off frequency of 1 kHz. Design (a) a low-pass T section filter, and (b) a low-pass π section filter to meet these requirements.
[(a) Each series arm 79.6 mH, shunt arm 0.637 μF
(b) Series arm 159.2 mH, each shunt arm 0.318 μF]

3 Determine the value of capacitance required in the shunt arm of a low-pass T section if the inductance in each of the series arms is 40 mH and the cut-off frequency of the filter is 2.5 kHz.
[0.203 μF]

4 The nominal impedance of a low-pass π section filter is 600 Ω and its cut-off frequency is at 25 kHz. Determine (a) the value of the characteristic impedance of the section at a frequency of 20 kHz and (b) the value of the characteristic impedance of the equivalent low-pass T section filter. [(a) 1 kΩ (b) 360 Ω]

5 The nominal impedance of a low-pass π section filter is 600 Ω. If the capacitance in each of the shunt arms is 0.1 μF determine the inductance in the series arm. Make a sketch of the ideal and the practical attenuation/frequency characteristic expected for such a filter section. [72 mH]

6 A low-pass T section filter has a nominal impedance of 600 Ω and a cut-off frequency of 10 kHz. Determine the frequency at which the characteristic impedance of the section is (a) zero, (b) 300 Ω, (c) 600 Ω [(a) 10 kHz (b) 8.66 kHz (c) 0]

High-pass filter sections

7 Determine for each of the high-pass filter sections shown in Figure 42.55 (i) the cut-off frequency, and (ii) the nominal impedance.
[(a) (i) 22.51 kHz (ii) 14.14 kΩ (b) (i) 281.3 Hz (ii) 1414 Ω]

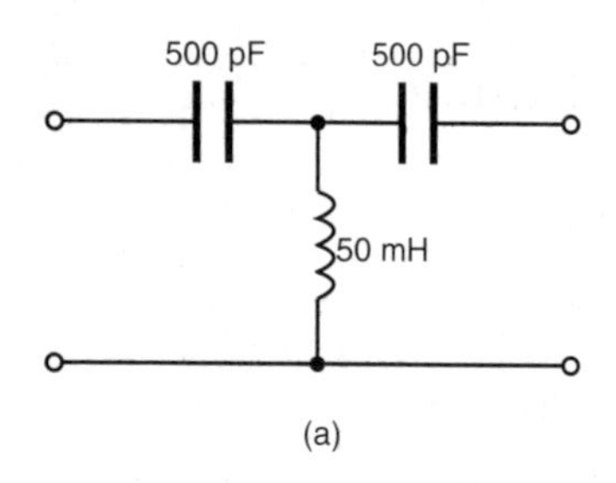

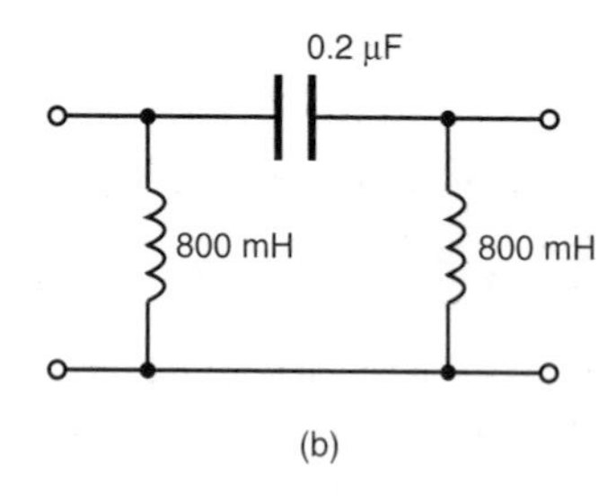

Figure 42.55

8 A filter is required to pass all frequencies above 4 kHz and to have a nominal impedance of 750 Ω. Design (a) an appropriate T section filter, and (b) an appropriate π section filter to meet these requirements.
[(a) Each series arm = 53.1 nF, shunt arm = 14.92 mH
(b) Series arm = 26.5 nF, each shunt arm = 29.84 mH]

9 The inductance in each of the shunt arms of a high-pass π section filter is 50 mH. If the nominal impedance of the section is 600 Ω, determine the value of the capacitance in the series arm.
[69.44 nF]

10 Determine the value of inductance required in the shunt arm of a high-pass T section filter if in each series arm it contains a 0.5 μF capacitor. The cut-off frequency of the filter section is 1500 Hz. Sketch the characteristic curve of characteristic impedance against frequency expected for such a filter section. [11.26 mH]

11 A high-pass π section filter has a nominal impedance of 500 Ω and a cut-off frequency of 50 kHz. Determine the frequency at which the characteristic impedance of the section is (a) 1 kΩ (b) 800 Ω (c) 520 Ω. [(a) 57.74 kHz (b) 64.05 kHz (c) 182 kHz]

12 A low-pass T section filter having a cut-off frequency of 3 kHz is connected in series with a high-pass T section filter having a cut-off frequency of 4 kHz. The terminating impedance of the filter is 600 Ω.

(a) Determine the values of the components comprising the composite filter.

(b) Sketch the expected attenuation/frequency characteristic and state the name given to the type of filter described.

[(a) Low-pass Tsection: each series arm 31.83 mH, shunt-arm 0.177 μF]
High-pass Tsection: each series arm 66.32 nF, shunt arm 11.94 mH
(b) Band-stop filter]

Propagation coefficient and time delay

13 A filter section has a propagation coefficient given by (a) $(1.79 - j0.63)$ (b) $1.378\angle 51.6°$. Determine for each (i) the attenuation coefficient and (ii) the phase angle coefficient.
[(a) (i) 1.79 N (ii) −0.63 rad (b) (i) 0.856 N (ii) 1.08 rad]

14 A filter section has a current input of $200\angle 20°$ mA and a current output of $16\angle -30°$ mA. Determine (a) the attenuation coefficient (b) the phase shift coefficient, and (c) the propagation coefficient. (d) If four such sections are cascaded determine the current output of the fourth stage and the overall propagation coefficient.
[(a) 2.526 N (b) 0.873 rad
(c) $(2.526 + j0.873)$ or $2.673\angle 19.07°$
(d) $8.19\angle -180°$ μA; $(10.103 + j3.491)$ or $10.69\angle 19.06°$]

15 Determine for the high-pass T section filter shown in Figure 42.56, (a) the attenuation coefficient, (b) the phase shift coefficient, and (c) the propagation coefficient.

[(a) 1.61 N (b) -2.50 rad
(c) $(1.61 - j2.50)$ or $2.97\angle -57.22°$]

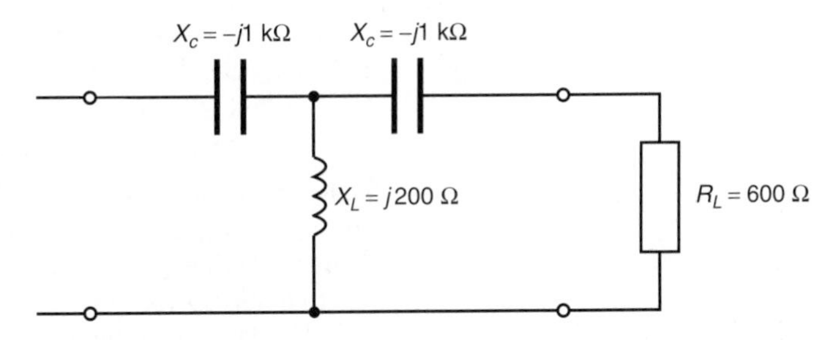

Figure 42.56

16 A low-pass T section filter has an inductance of 25 mH in each series arm and a shunt arm capacitance of 400 nF. Determine for the section (a) the time delay for the signal to pass through the filter, assuming the phase shift is small, and (b) the time delay for a signal to pass through the section at the cut-off frequency.

[(a) 141.4 μs (b) 222.1 μs]

17 A filter network comprising n identical sections passes signals of all frequencies over 8 kHz and provides a total delay of 69.63 μs. If the characteristic impedance of the circuit into which the filter is inserted is 600 Ω, determine (a) the values of the components comprising each section, and (b) the value of n.

[(a) Each series arm 33.16 nF; shunt arm 5.97 mH (b) 7]

18 A filter network consists of 15 sections in cascade having a nominal impedance of 800 Ω. If the total delay time is 30 μs determine the component value for each section if the filter is (a) a low-pass π network, (b) a high-pass T network.

[(a) Series arm 1.60 mH, each shunt arm 1.25 nF
(b) Each series arm 5 nF, shunt arm 1.60 nF]

'*m*-derived' filter sections

19 A low-pass filter section is required to have a nominal impedance of 450 Ω, a cut-off frequency of 150 kHz and a frequency of infinite attenuation at 160 kHz. Design an appropriate 'm-derived' T section filter.

[Each series arm 0.166 mH; shunt arm comprises 1.641 nF capacitor in series with 0.603 mH inductance]

20 In a filter section it is required to have a cut-off frequency of 1.2 MHz and a frequency of infinite attenuation 1.3 MHz. If the

nominal impedance of the line into which the filter is to be inserted is 600 Ω, determine suitable component values if the section is an 'm-derived' π type.

[Each shunt arm 85.1 pF; series arm contains 61.21 μH inductance in parallel with 244.9 pF capacitor]

21 Determine the component values of an 'm-derived' T section filter having a nominal impedance of 600 Ω, a cut-off frequency of 1220 Hz and a frequency of infinite attenuation of 1100 Hz.

[Each series arm 0.503 μF; series arm comprises 90.50 mH inductance in series with 0.231 μF capacitor]

22 State the advantages of an 'm-derived' filter section over its equivalent prototype.
A filter section is to have a nominal impedance of 500 Ω, a cut-off frequency of 5 kHz and a frequency of infinite attenuation of 4.5 kHz. Determine the values of components if the section is to be an 'm-derived' π filter.

[Each shunt arm 36.51 mH inductance;
series arm comprises 73.02 nF
capacitor in parallel with 17.13 mH inductance]

Composite filter sections

23 A composite filter is to have a nominal impedance of 500 Ω, a cut-off frequency of 1500 Hz and a frequency of infinite attenuation of 1800 Hz. Determine the values of components required if the filter is to comprise a prototype T section, an 'm-derived' T section and two terminating half-sections (use $m = 0.6$ for the half-sections).

[Prototype: Each series arm 53.1 mH;
shunt arm comprises 0.424 μF
'm-derived': Each series arm 29.3 mH;
shunt arm comprises 0.235 μH
capacitor in series with 33.32 mH inductance.
Half-sections: Series arm 31.8 mH;
shunt arm comprises 0.127 μF
capacitor in series with 56.59 mH inductance]

24 A filter made up of a prototype π section, an 'm-derived' π section and two terminating half-sections in cascade has a nominal impedance of 1 kΩ, a cut-off frequency of 100 kHz and a frequency of infinite attenuation of 90 kHz. Determine the values of the components comprising the composite filter and explain why such a filter is more suitable than just the prototype. (Use $m = 0.6$ for the half-sections.)

[Prototype: Series arm 795.8 pF, each shunt arm 1.592 mH
'm-derived': Each shunt arm 3.651 mH; series arm 1.826 nF
capacitor in parallel with 1.713 mH inductance.
Half-sections: Shunt arm 238.7 pF; series arm 10.61 nF
capacitor in parallel with 1.492 mH inductance]

43 Magnetically coupled circuits

At the end of this chapter you should be able to:

- define mutual inductance
- deduce that $E_2 = -M\frac{dI_1}{dt}$, $M = N_2\frac{d\Phi_2}{dI_1}$, $M = N_1\frac{d\Phi_1}{dI_2}$ and perform calculations
- show that $M = k\sqrt{(L_1L_2)}$ and perform calculations
- perform calculations on mutually coupled coils in series
- perform calculations on coupled circuits
- describe and use the dot rule in coupled circuit problems.

43.1 Introduction

When the interaction between two loops of a circuit takes place through a magnetic field instead of through common elements, the loops are said to be inductively or **magnetically coupled.** The windings of a transformer, for example, are magnetically coupled (see Chapter 20).

43.2 Self-inductance

It was shown in Chapter 9, that the e.m.f. E induced in a coil of inductance L henrys is given by:

$$\boldsymbol{E = -L\frac{di}{dt}\ \text{volts}}$$

, where $\frac{di}{dt}$ is the rate of change of current,

the magnitude of the e.m.f. induced in a coil of N turns is given by:

$$\boldsymbol{E = -N\frac{d\Phi}{dt}\ \text{volts}}$$

, where $\frac{d\Phi}{dt}$ is the rate of change of flux,

and the inductance of a coil L is given by:

$$\boldsymbol{L = \frac{N\Phi}{I}\ \text{henrys}}$$

43.3 Mutual inductance

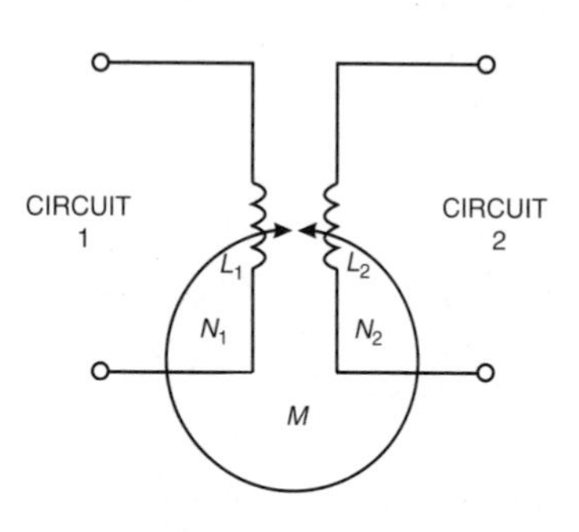

Figure 43.1

Mutual inductance is said to exist between two circuits when a changing current in one induces, by electromagnetic induction, an e.m.f. in the other. An ideal equivalent circuit of a mutual inductor is shown in Figure 43.1.

L_1 and L_2 are the self inductances of the two circuits and M the mutual inductance between them. The mutual inductance M is defined by the relationship:

$$\boxed{E_2 = -M\frac{dI_1}{dt}} \text{ or } \boxed{E_1 = -M\frac{dI_2}{dt}} \qquad (43.1)$$

where E_2 is the e.m.f. in circuit 2 due to current I_1 in circuit 1 and E_1 is the e.m.f. in circuit 1 due to the current I_2 in circuit 2.
The unit of M is the henry.

From Section 43.2, $E_2 = -N_2\frac{d\Phi_2}{dt}$ or $E_1 = -N_1\frac{d\Phi_1}{dt}$ (43.2)

Equating the E_2 terms in equations (43.1) and (43.2) gives:

$$-M\frac{dI_1}{dt} = -N_2\frac{d\Phi_2}{dt}$$

from which $\boxed{M = N_2\frac{d\Phi_2}{dI_1}}$ (43.3)

Equating the E_1 terms in equations (43.1) and (43.2) gives:

$$-M\frac{dI_2}{dt} = -N_1\frac{d\Phi_1}{dt}$$

from which $\boxed{M = N_1\frac{d\Phi_1}{dI_2}}$ (43.4)

If the coils are linked with air as the medium, the flux and current are linearly related and equations (43.3) and (43.4) become:

$$\boxed{M = \frac{N_2\Phi_2}{I_1}} \text{ and } \boxed{M = \frac{N_1\Phi_1}{I_2}} \qquad (43.5)$$

Problem 1. A and B are two coils in close proximity. A has 1200 turns and B has 1000 turns. When a current of 0.8 A flows in coil A a flux of 100 μWb links with coil A and 75% of this flux links coil B. Determine (a) the self inductance of coil A, and (b) the mutual inductance.

(a) From Section (43.2),

$$\textbf{self inductance of coil A, } L_A = \frac{N_A\Phi_A}{I_A} = \frac{(1200)(100\times10^{-6})}{0.80}$$

$$= \mathbf{0.15\ H}$$

(b) From equation (43.5),

$$\textbf{mutual inductance, } M = \frac{N_B\Phi_B}{I_A} = \frac{(1000)(0.75\times100\times10^{-6})}{0.80}$$

$$= \mathbf{0.09\,375\ H} \text{ or } \mathbf{93.75\ mH}$$

Problem 2. Two circuits have a mutual inductance of 600 mH. A current of 5 A in the primary is reversed in 200 ms. Determine the e.m.f. induced in the secondary, assuming the current changes at a uniform rate.

Secondary e.m.f., $E_2 = -M\dfrac{dI_1}{dt}$, from equation (43.5).

Since the current changes from $+5A$ to $-5A$, the change of current is 10 A.

Hence $\dfrac{dI_1}{dt} = \dfrac{10}{200\times10^{-3}} = 50$ A/s

Hence **secondary induced e.m.f.,** $E_2 = -M\dfrac{dI_1}{dt} = -(600\times10^{-3})(50)$

$$= \mathbf{-30\ volts}$$

Further problems on mutual inductance may be found in Section 43.8, problems, 1 to 4, page 864.

43.4 Coupling coefficient

The coupling coefficient k is the degree or fraction of magnetic coupling that occurs between circuits.

$$k = \frac{\text{flux linking two circuits}}{\text{total flux produced}}$$

When there is no magnetic coupling, $k = 0$. If the magnetic coupling is perfect, i.e., all the flux produced in the primary links with the secondary then $k = 1$. Coupling coefficient is used in communications engineering to denote the degree of coupling between two coils. If the coils are close together, most of the flux produced by current in one coil passes through the other, and the coils are termed **tightly coupled.** If the coils are spaced apart, only a part of the flux links with the second, and the coils are termed **loosely-coupled.**

From Section 43.2, the inductance of a coil is given by $L = \dfrac{N\Phi}{I}$

Thus for the circuit of Figure 43.1, $L_1 = \dfrac{N_1\Phi_1}{I_1}$

$$\text{from which,}\quad \Phi_1 = \frac{L_1 I_1}{N_1} \qquad (43.6)$$

From equation (43.5), $M = (N_2\Phi_2/I_1)$, but the flux that links the second circuit, $\Phi_2 = k\Phi_1$

$$\text{Thus}\quad M = \frac{N_2\Phi_2}{I_1} = \frac{N_2(k\Phi_1)}{I_1} = \frac{N_2 k}{I_1}\left(\frac{L_1 I_1}{N_1}\right) \text{ from equation (43.6)}$$

$$\text{i.e.,}\quad M = \frac{kN_2L_1}{N_1} \text{ from which, } \frac{N_2}{N_1} = \frac{M}{kL_1} \qquad (43.7)$$

Also, since the two circuits can be reversed,

$$M = \frac{kN_1L_2}{N_2} \text{ from which, } \frac{N_2}{N_1} = \frac{kL_2}{M} \qquad (43.8)$$

Thus from equations (43.7) and (43.8),

$$\frac{N_2}{N_1} = \frac{M}{kL_1} = \frac{kL_2}{M}$$

from which, $M^2 = k^2 L_1 L_2$ and $\boxed{\mathbf{M = k\sqrt{(L_1L_2)}}}$ (43.9)

or, $\boxed{\textbf{coefficient of coupling, } \mathbf{k = \dfrac{M}{\sqrt{(L_1L_2)}}}}$ (43.10)

Problem 3. Two coils have self inductances of 250 mH and 400 mH respectively. Determine the magnetic coupling coefficient of the pair of coils if their mutual inductance is 80 mH.

From equation (43.10), coupling coefficient,

$$k = \frac{M}{\sqrt{(L_1L_2)}} = \frac{80\times10^{-3}}{\sqrt{[(250\times10^{-3})(400\times10^{-3})]}} = \frac{80\times10^{-3}}{\sqrt{(0.1)}} = \mathbf{0.253}$$

Problem 4. Two coils, X and Y, having self inductances of 80 mH and 60 mH respectively, are magnetically coupled. Coil X has 200 turns and coil Y has 100 turns. When a current of 4 A is reversed in coil X the change of flux in coil Y is 5 mWb. Determine (a) the mutual inductance between the coils, and (b) the coefficient of coupling.

(a) From equation (43.3),

$$\textbf{mutual inductance, } M = N_Y \frac{d\Phi_Y}{dI_X} = \frac{(100)(5 \times 10^{-3})}{(4 - -4)}$$

$$= \mathbf{0.0625\ H} \text{ or } \mathbf{62.5\ mH}$$

(b) From equation (43.10),

$$\textbf{coefficient of coupling, } k = \frac{M}{\sqrt{(L_X L_Y)}}$$

$$= \frac{0.0625}{\sqrt{[(80 \times 10^{-3})(60 \times 10^{-3})]}} = \mathbf{0.902}$$

Further problems on coupling coefficient may be pound in Section 43.8, problems 5 and 6, page 865.

43.5 Coils connected in series

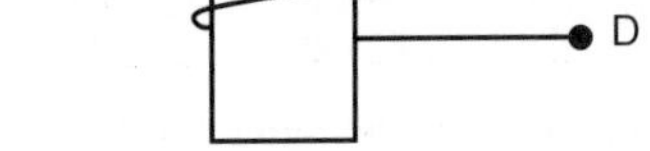

Figure 43.2

Figure 43.2 shows two coils 1 and 2 wound on an insulating core with terminals B and C joined. The fluxes in each coil produced by current i are in the same direction and the coils are termed **cumulatively coupled.**

Let the self inductance of coil 1 be L_1 and that of coil 2 be L_2 and let their mutual inductance be M.

If in dt seconds, the current increases by di amperes then the e.m.f. induced in coil 1 due to its self inductance is $L_1(di/dt)$ volts, and the e.m.f. induced in coil 2 due to its self inductance is $L_2(di/dt)$ volts. Also, the e.m.f. induced in coil 1 due to the increase of current in coil 2 is $M(di/dt)$ volts and the e.m.f. induced in coil 2 due to the increase of current in coil 1 is $M(di/dt)$.

Hence the total e.m.f. induced in coils 1 and 2 is:

$$L_1\frac{di}{dt} + L_2\frac{di}{dt} + 2\left(M\frac{di}{dt}\right) \text{ volts} = (L_1 + L_2 + 2M)\frac{di}{dt} \text{ volts}$$

If the winding between terminals A and D in Figure 43.2 are considered as a single circuit having a self inductance L_A henrys then if the same increase in dt seconds is di amperes then the e.m.f. induced in the complete circuit is $L_A(di/dt)$ volts.

$$\text{Hence} \quad L_A\frac{di}{dt} = (L_1 + L_2 + 2M)\frac{di}{dt}$$

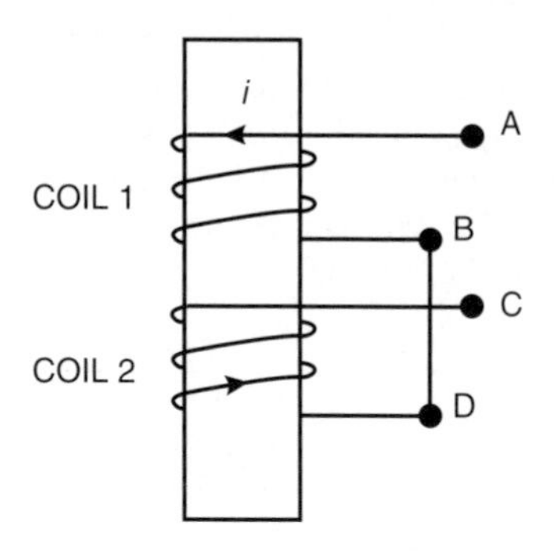

Figure 43.3

i.e., $$\boldsymbol{L_A = L_1 + L_2 + 2M} \qquad (43.11)$$

If terminals B and D are joined as shown in Figure 43.3 the direction of the current in coil 2 is reversed and the coils are termed **differentially coupled.** In this case, the total e.m.f. induced in coils 1 and 2 is:

$$L_1\frac{di}{dt} + L_2\frac{di}{dt} - 2M\frac{di}{dt}$$

The e.m.f. $M(di/dt)$ induced in coil 1 due to an increase di amperes in dt seconds in coil 2 is in the same direction as the current and is hence in opposition to the e.m.f. induced in coil 1 due to its self inductance. Similarly, the e.m.f. induced in coil 2 by mutual inductance is in opposition to that induced by the self inductance of coil 2.

If L_B is the self inductance of the whole circuit between terminals A and C in Figure 43.3 then:

$$L_B\frac{di}{dt} = L_1\frac{di}{dt} + L_2\frac{di}{dt} - 2M\frac{di}{dt}$$

i.e., $$\boldsymbol{L_B = L_1 + L_2 - 2M} \qquad (43.12)$$

Thus the total inductance L of inductively coupled circuits is given by:

$$\boxed{\boldsymbol{L = L_1 + L_2 \pm 2M}} \qquad (43.13)$$

Equation (43.11) - equation (43.12) gives:

$$L_A - L_B = (L_1 + L_2 + 2M) - (L_1 + L_2 - 2M)$$

i.e., $$L_A - L_B = 2M - (-2M) = 4M$$

from which, $$\boxed{\textbf{mutual inductance, } \boldsymbol{M = \frac{L_A - L_B}{4}}} \qquad (43.14)$$

An experimental method of determining the mutual inductance is indicated by equation (43.14), i.e., connect the coils both ways and determine the equivalent inductances L_A and L_B using an a.c. bridge. The mutual inductance is then given by a quarter of the difference between the two values of inductance.

Problem 5. Two coils connected in series have self inductance of 40 mH and 10 mH respectively. The total inductance of the circuit is found to be 60 mH. Determine (a) the mutual inductance between the two coils, and (b) the coefficient of coupling.

(a) From equation (43.13), total inductance, $L = L_1 + L_2 \pm 2M$

Hence $$60 = 40 + 10 \pm 2M$$

Since $(L_1 + L_2) < L$ then $60 = 40 + 10 + 2M$

from which $2M = 60 - 40 - 10 = 10$

and **mutual inductance,** $\boldsymbol{M} = \dfrac{10}{2} = \mathbf{5\ mH}$

(b) From equation (43.10), coefficient of coupling,

$$k = \frac{M}{\sqrt{(L_1L_2)}} = \frac{5 \times 10^{-3}}{\sqrt{[(40 \times 10^{-3})(10 \times 10^{-3})]}} = \frac{5 \times 10^{-3}}{0.02}$$

i.e., **coefficient of coupling,** $\boldsymbol{k = 0.25}$

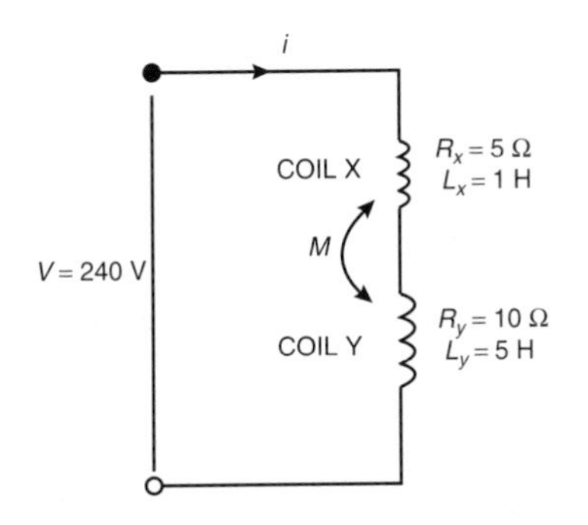

Figure 43.4

Problem 6. Two mutually coupled coils X and Y are connected in series to a 240 V d.c. supply. Coil X has a resistance of 5 Ω and an inductance of 1 H. Coil Y has a resistance of 10 Ω and an inductance of 5 H. At a certain instant after the circuit is connected, the current is 8 A and increasing at a rate of 15 A/s. Determine (a) the mutual inductance between the coils and (b) the coefficient of coupling.

The circuit is shown in Figure 43.4.

(a) From Kirchhoff's voltage law:

$$V = iR + L\frac{di}{dt}$$

i.e., $240 = 8(5 + 10) + L(15)$

i.e., $240 = 120 + 15L$

from which, $L = \dfrac{240 - 120}{15} = 8$ H

From equation (43.11),

$$L = L_X + L_Y + 2M$$

Hence $8 = 1 + 5 + 2M$

from which, **mutual inductance,** $\boldsymbol{M} = \mathbf{1\ H}$

(b) From equation (43.10),

coefficient of coupling, $\boldsymbol{k} = \dfrac{M}{\sqrt{(L_XL_Y)}} = \dfrac{1}{\sqrt{[(1)(5)]}} = \mathbf{0.447}$

Problem 7. Two coils are connected in series and their effective inductance is found to be 15 mH. When the connection to one coil is reversed, the effective inductance is found to be 10 mH. If the coefficient of coupling is 0.7, determine (a) the self inductance of each coil, and (b) the mutual inductance.

(a) From equation (43.13), total inductance, $L = L_1 + L_2 \pm 2M$

and from equation (43.9), $M = k\sqrt{(L_1 L_2)}$

hence $L = L_1 + L_2 \pm 2k\sqrt{(L_1 L_2)}$

Since in equation (43.11),

$$L_A = 15 \text{ mH}, \quad 15 = L_1 + L_2 + 2k\sqrt{(L_1 L_2)} \tag{43.15}$$

and since in equation (43.12),

$$L_B = 10 \text{ mH}, \quad 10 = L_1 + L_2 - 2k\sqrt{(L_1 L_2)} \tag{43.16}$$

Equation (43.15) + equation (43.16) gives:

$$25 = 2L_1 + 2L_2 \text{ and } 12.5 = L_1 + L_2 \tag{43.17}$$

From equation (43.17), $L_2 = 12.5 - L_1$

Substituting in equation (43.15), gives:

$$15 = L_1 + (12.5 - L_1) + 2(0.7)\sqrt{[L_1(12.5 - L_1)]}$$

i.e., $15 = 12.5 + 1.4\sqrt{(12.5L_1 - L_1^2)}$

$$\frac{15 - 12.5}{1.4} = \sqrt{(12.5L_1 - L_1^2)}$$

and $\left(\dfrac{15 - 12.5}{1.4}\right)^2 = 12.5L_1 - L_1^2$

i.e., $3.189 = 12.5L_1 - L_1^2$

from which, $L_1^2 - 12.5L_1 + 3.189 = 0$

Using the quadratic formula:

$$L_1 = \frac{-(-12.5) \pm \sqrt{[(-12.5)^2 - 4(1)(3.189)]}}{2(1)}$$

i.e., $\mathbf{L_1} = \dfrac{12.5 \pm (11.98)}{2} = \mathbf{12.24\ mH}$ or $\mathbf{0.26\ H}$

From equation (43.17):

$\mathbf{L_2} = 12.5 - L_1 = (12.5 - 12.24) = \mathbf{0.26\ mH}$

or $(12.5 - 0.26) = \mathbf{12.24\ mH}$

(b) From equation (43.14),

$$\textbf{mutual inductance, } M = \frac{L_A - L_B}{4} = \frac{15 - 10}{4} = \textbf{1.25 mH}$$

Further problems on coils in series may be found in Section 43.8, problems 7 to 11, page 865.

43.6 Coupled circuits

The magnitude of the secondary e.m.f. E_2 in Figure 43.5 is given by:

$$E_2 = M\frac{dI_1}{dt}, \text{ from equation (43.1)}$$

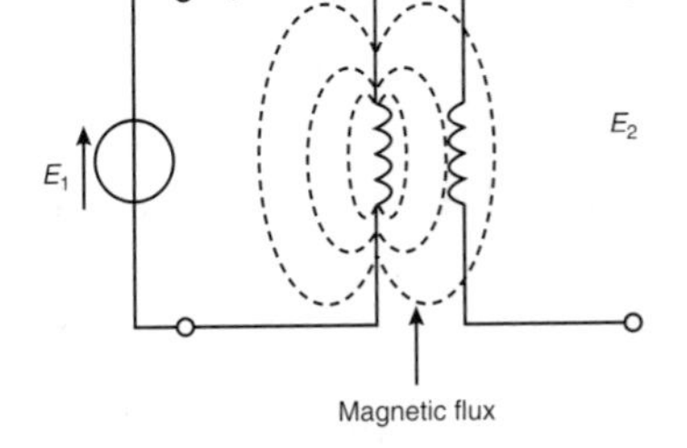

Figure 43.5

If the current I_1 is sinusoidal, i.e., $I_1 = I_{1m}\sin\omega t$

then $E_2 = M\frac{d}{dt}(I_{1m}\sin\omega t) = M\omega I_{1m}\cos\omega t$

Since $\cos\omega t = \sin(\omega t + 90°)$ then $\cos\omega t = j\sin\omega t$ in complex form.

Hence $E_2 = M\omega I_{1m}(j\sin\omega t) = j\omega M(I_{1m}\sin\omega t)$

i.e., $$\boxed{\mathbf{E_2 = j\omega M I_1}} \qquad (43.18)$$

If L_1 is the self inductance of the primary winding in Figure 43.5, there will be an e.m.f. generated equal to $j\omega L_1 I_1$ induced into the primary winding since the flux set up by the primary current also links with the primary winding.

(a) Secondary open-circuited

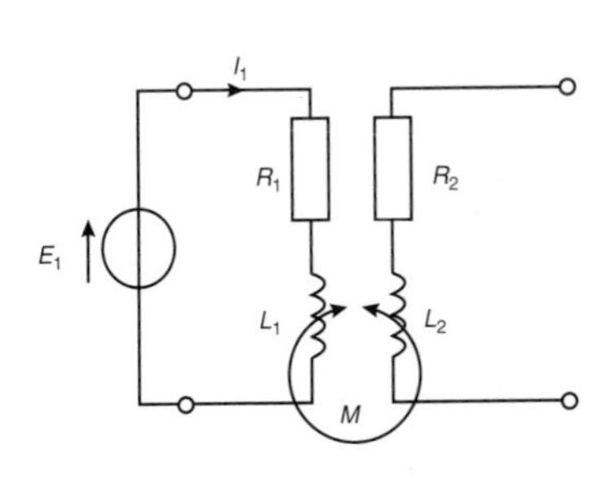

Figure 43.6

Figure 43.6 shows two coils, having self inductances of L_1 and L_2 which are inductively coupled together by a mutual inductance M. The primary winding has a voltage generator of e.m.f. E_1 connected across its terminals. The internal resistance of the source added to the primary resistance is shown as R_1 and the secondary winding which is open-circuited has a resistance of R_2.

Applying Kirchhoff's voltage law to the primary circuit gives:

$$E_1 = I_1 R_1 + L_1\frac{dI_1}{dt} \qquad (43.19)$$

If E_1 and I_1 are both sinusoidal then equation (43.19) becomes:

$$\begin{aligned} E_1 &= I_1 R_1 + L_1\frac{d}{dt}(I_{1m}\sin\omega t) \\ &= I_1 R_1 + L_1\omega I_{1m}\cos\omega t \\ &= I_1 R_1 + L_1\omega(jI_{1m}\sin\omega t) \end{aligned}$$

i.e., $E_1 = I_1R_1 + j\omega I_1L_1 = I_1(R_1 + j\omega L_1)$

i.e., $$I_1 = \frac{E_1}{R_1 + j\omega L_1} \qquad (43.20)$$

From equation (43.18), $E_2 = j\omega MI_1$

from which, $$I_1 = \frac{E_2}{j\omega M} \qquad (43.21)$$

Equating equations (43.20) and (43.21) gives: $\dfrac{E_2}{j\omega M} = \dfrac{E_1}{R_1 + j\omega L_1}$

and $$\boxed{\boldsymbol{E_2 = \frac{j\omega ME_1}{R_1 + j\omega L_1}}} \qquad (43.22)$$

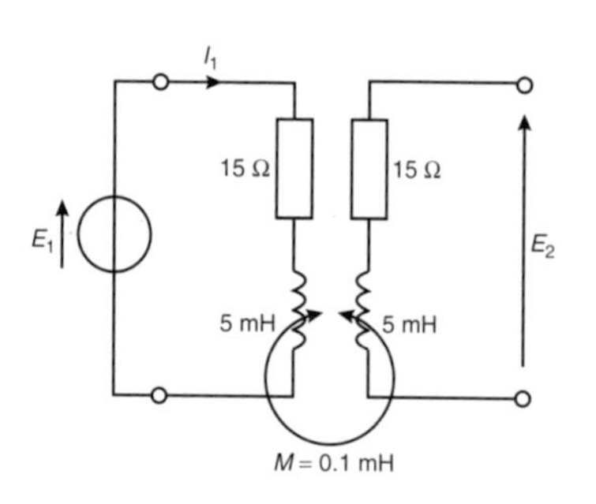

Figure 43.7

Problem 8. For the circuit shown in Figure 43.7, determine the p.d. E_2 which appears across the open-circuited secondary winding, given that $E_1 = 8\sin 2500t$ volts.

Impedance of primary, $Z_1 = R_1 + j\omega L_1 = 15 + j(2500)(5 \times 10^{-3})$

$= (15 + j12.5)\Omega$ or $19.53\angle 39.81°\ \Omega$

Primary current $I_1 = \dfrac{E_1}{Z_1} = \dfrac{8\angle 0°}{19.53\angle 39.81°}$

From equation (43.18),

$$E_2 = j\omega MI_1 = \frac{j\omega ME_1}{(R_1 + j\omega L_1)} = \frac{j(2500)(0.1 \times 10^{-3})(8\angle 0°)}{19.53\angle 39.81°}$$

$$= \frac{2\angle 90°}{19.53\angle 39.81°} = \mathbf{0.102\angle 50.19°\ V}$$

Problem 9. Two coils x and y, with negligible resistance, have self inductances of 20 mH and 80 mH respectively, and the coefficient of coupling between them is 0.75. If a sinusoidal alternating p.d. of 5 V is applied to x, determine the magnitude of the open circuit e.m.f. induced in y.

From equation (43.9), mutual inductance,

$$M = k\sqrt{(L_xL_y)} = 0.75\sqrt{[(20 \times 10^{-3})(80 \times 10^{-3})]} = 0.03\text{ H}$$

From equation (43.22), the magnitude of the open circuit e.m.f. induced in coil y,

$$|E_y| = \frac{j\omega M E_x}{R_x + j\omega L_x}$$

When $R_1 = 0$, $|\boldsymbol{E}_y| = \dfrac{j\omega M E_x}{j\omega L_x} = \dfrac{M E_x}{L_x} = \dfrac{(0.03)(5)}{20 \times 10^{-3}} = \mathbf{7.5\ V}$

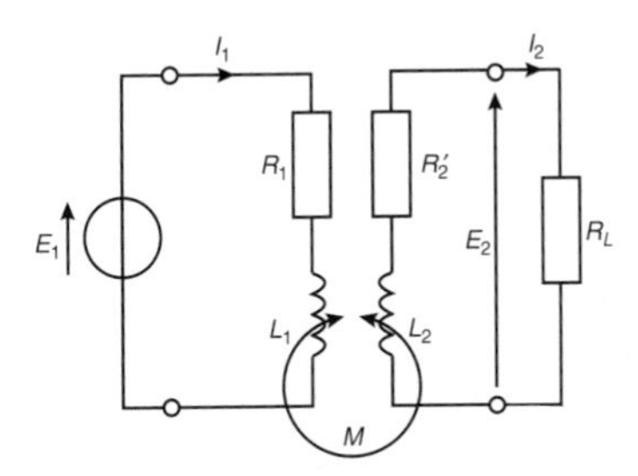

Figure 43.8

(b) Secondary terminals having load impedance

In the circuit shown in Figure 43.8 a load resistor R_L is connected across the secondary terminals. Let $R_2' + R_L = R_2$

When an e.m.f. is induced into the secondary winding a current I_2 flows and this will induce an e.m.f. into the primary winding.

Applying Kirchhoff's voltage law to the primary winding gives:

$$E_1 = I_1(R_1 + j\omega L_1) \pm j\omega M I_2 \qquad (43.23)$$

Applying Kirchhoff's voltage law to the secondary winding gives:

$$0 = I_2(R_2 + j\omega L_2) \pm j\omega M I_1 \qquad (43.24)$$

From equation (43.24), $\quad I_2 = \dfrac{\mp j\omega I_1}{(R_2 + j\omega L_2)}$

Substituting this in equation (43.23) gives:

$$E_1 = I_1(R_1 + j\omega L_1) \pm j\omega M\left(\frac{\mp j\omega M I_1}{(R_2 + j\omega L_2)}\right)$$

i.e., $\quad E_1 = I_1\left[(R_1 + j\omega L_1) + \dfrac{\omega^2 M^2}{(R_2 + j\omega L_2)}\right]$ since $j^2 = -1$

$$= I_1\left[(R_1 + j\omega L_1) + \frac{\omega^2 M^2 (R_2 - j\omega L_2)}{R_2^2 + \omega^2 L_2^2}\right]$$

$$= I_1\left[R_1 + j\omega L_1 + \frac{\omega^2 M^2 R_2}{R_2^2 + \omega^2 L_2^2} - \frac{j\omega^3 M^2 L_2}{R_2^2 + \omega^2 L_2^2}\right]$$

The effective primary impedance $Z_{1(\text{eff})}$ of the circuit is given by:

$$\boldsymbol{Z_{1(\text{eff})} = \frac{E_1}{I_1} = R_1 + \frac{\omega^2 M^2 R_2}{R_2^2 + \omega^2 L_2^2} + j\left(\omega L_1 - \frac{\omega^3 M^2 L_2}{R_2^2 + \omega^2 L_2^2}\right)} \qquad (43.25)$$

In equation (43.25), the primary impedance is $(R_1 + j\omega L_1)$. The remainder, i.e., $$\left(\frac{\omega^2 M^2 R_2}{R_2^2+\omega^2 L_2^2} - j\frac{\omega^3 M^2 L_2}{R_2^2+\omega^2 L_2^2}\right)$$

is known as the **reflected impedance** since it represents the impedance reflected back into the primary side by the presence of the secondary current.

Hence reflected impedance

$$= \frac{\omega^2 M^2 R_2}{R_2^2+\omega^2 L_2^2} - j\frac{\omega^3 M^2 L_2}{R_2^2+\omega^2 L_2^2} = \omega^2 M^2\left(\frac{R_2 - j\omega L_2}{R_2^2+\omega^2 L_2^2}\right)$$

$$= \omega^2 M^2 \frac{(R_2 - j\omega L_2)}{(R_2 + j\omega L_2)(R_2 - j\omega L_2)} = \frac{\omega^2 M^2}{R_2 + j\omega L_2}$$

i.e., $$\textbf{reflected impedance, } Z_r = \frac{\omega^2 M^2}{Z_2} \qquad (43.26)$$

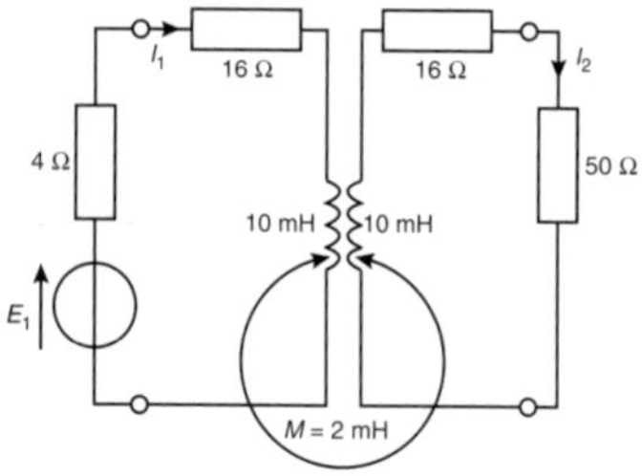

Problem 10. For the circuit shown in Figure 43.9, determine the value of the secondary current I_2 if $E_1 = 2\angle 0°$ volts and the frequency is $\frac{10^3}{\pi}$ Hz.

Figure 43.9

From equation (43.25), $R_{1(\text{eff})}$ is the real part of $Z_{1(\text{eff})}$,

i.e., $$R_{1(\text{eff})} = R_1 + \frac{\omega^2 M^2 R_2}{R_2^2+\omega^2 L_2^2}$$

$$= (4+16) + \frac{\left(2\pi\frac{10^3}{\pi}\right)^2 (2\times 10^{-3})^2(16+50)}{66^2 + \left(2\pi\frac{10^3}{\pi}\right)^2 (10\times 10^{-3})^2}$$

$$= 20 + \frac{1056}{4756} = 20.222\ \Omega$$

and $X_{1(\text{eff})}$ is the imaginary part of $Z_{1(\text{eff})}$, i.e.,

$$X_{1(\text{eff})} = \omega L_1 - \frac{\omega^3 M^2 L_2}{R_2^2+\omega^2 L_2^2}$$

$$= \left(2\pi\frac{10^3}{\pi}\right)(10\times 10^{-3}) - \frac{\left(2\pi\frac{10^3}{\pi}\right)^3 (2\times 10^{-3})^2(10\times 10^{-3})}{66^2 + \left(2\pi\frac{10^3}{\pi}\right)^2 (10\times 10^{-3})^2}$$

$$= 20 - \frac{320}{4756} = 19.933\ \Omega$$

$$\text{Hence primary current, } I_1 = \frac{E_1}{Z_{1(\text{eff})}} = \frac{2\angle 0^\circ}{(20.222 + j19.933)}$$

$$= \frac{2\angle 0^\circ}{28.395\angle 44.59^\circ} = 0.0704\angle -44.59^\circ \text{ A}$$

$$\text{From equation (43.18), } E_2 = j\omega M I_1$$

$$= j\left(2\pi \frac{10^3}{\pi}\right)(2 \times 10^{-3})(0.0704\angle -44.59^\circ)$$

$$= (4\angle 90^\circ)(0.0704\angle -44.59^\circ)$$

$$= 0.282\angle 45.41^\circ \text{ V}$$

$$\text{Hence secondary current } I_2 = \frac{E_2}{Z_2} = \frac{0.282\angle 45.41^\circ}{66 + j\left(2\pi \frac{10^3}{\pi}\right)(10 \times 10^{-3})}$$

$$= \frac{0.282\angle 45.41^\circ}{(66 + j20)}$$

$$= \frac{0.282\angle 45.41^\circ}{68.964\angle 16.86^\circ} = 4.089 \times 10^{-3}\angle 28.55^\circ \text{ A}$$

i.e., $\boldsymbol{I_2 = 4.09\angle 28.55^\circ}$ **mA**

Problem 11. For the coupled circuit shown in Figure 43.10, calculate (a) the self impedance of the primary circuit, (b) the self impedance of the secondary circuit, (c) the impedance reflected into the primary circuit, (d) the effective primary impedance, (e) the primary current, and (f) the secondary current.

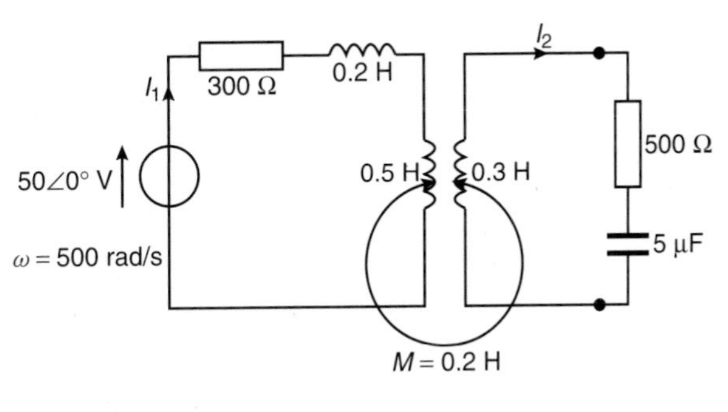

Figure 43.10

(a) Self impedance of primary circuit, $Z_1 = 300 + j(500)(0.2 + 0.5)$

i.e., $\boldsymbol{Z_1 = (300 + j350)\,\Omega}$

(b) Self impedance of secondary circuit,

$$Z_2 = 500 + j\left[(500)(0.3 - \frac{1}{(500)(5 \times 10^{-6})}\right]$$

$$= 500 + j(150 - 400)$$

i.e., $\mathbf{Z_2 = (500 - j250)\,\Omega}$

(c) From equation (43.26),

$$\textbf{reflected impedance, } Z_r = \frac{\omega^2 M^2}{Z_2} = \frac{(500)^2(0.2)^2}{(500 - j250)}$$

$$= \frac{10^4(500 + j250)}{500^2 + 250^2} = \mathbf{(16 + j8)\,\Omega}$$

(d) Effective primary impedance,

$$Z_{1(\text{eff})} = Z_1 + Z_r \text{ (note this is equivalent to equation 43.25)}$$

$$= (300 + j350) + (16 + j8)$$

i.e., $\mathbf{Z_{1(eff)} = (316 + j358)\,\Omega}$

(e) Primary current $\mathbf{I_1} = \dfrac{E_1}{Z_{1(\text{eff})}} = \dfrac{50\angle 0^\circ}{(316 + j358)}$

$$= \frac{50\angle 0^\circ}{477.51\angle 48.57^\circ}$$

$$= \mathbf{0.105\angle -48.57^\circ\ A}$$

(f) Secondary current, $I_2 = \dfrac{E_2}{Z_2}$, where

$$E_2 = j\omega M I_1 \text{ from equation (43.18)}$$

Hence $\mathbf{I_2} = \dfrac{j\omega M I_1}{Z_2} = \dfrac{j(500)(0.2)(0.105\angle -48.57^\circ)}{(500 - j250)}$

$$= \frac{(100\angle 90^\circ)(0.105\angle -48.57^\circ)}{559.02\angle -26.57^\circ}$$

$$= \mathbf{0.0188\angle 68^\circ\ A}$$

(c) Resonance by tuning capacitors

Tuning capacitors may be added to the primary and/or secondary circuits to cause it to resonate at particular frequencies. These may be connected either in series or in parallel with the windings. Figure 43.11 shows each winding tuned by series-connected capacitors C_1 and C_2. The expression for the effective primary impedance $Z_{1(\text{eff})}$, i.e., equation (43.25) applies except that ωL_1 becomes $(\omega L_1 - (1/\omega C_1))$ and ωL_2 becomes $(\omega L_2 - (1/\omega C_2))$

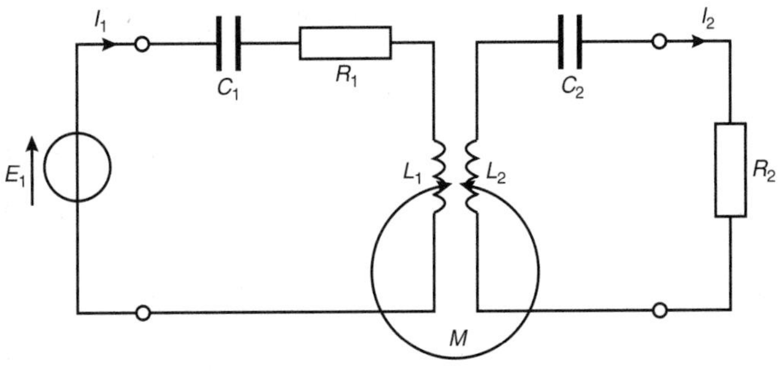

Figure 43.11

Problem 12. For the circuit shown in Figure 43.12 each winding is tuned to resonate at the same frequency. Determine (a) the resonant frequency, (b) the value of capacitor C_2, (c) the effective primary impedance, (d) the primary current, (e) the voltage across capacitor C_2 and (f) the coefficient of coupling.

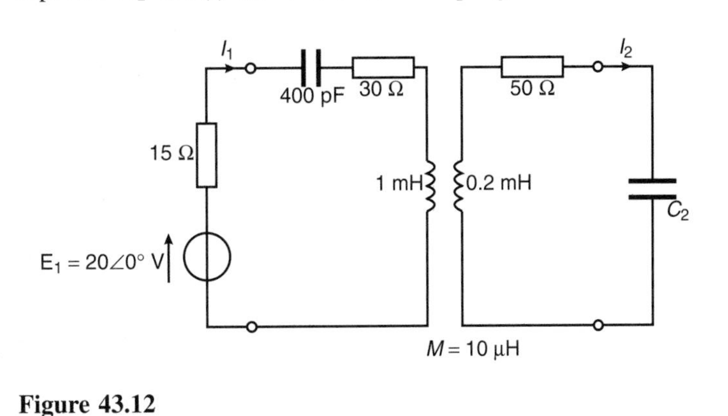

Figure 43.12

(a) For resonance in a series circuit, the resonant frequency, f_r, is given by:

$$f_r = \frac{1}{2\pi\sqrt{(LC)}} \text{ Hz}$$

$$\text{Hence} \quad f_r = \frac{1}{2\pi\sqrt{(L_1C_1)}} = \frac{1}{2\pi\sqrt{(1 \times 10^{-3})(400 \times 10^{-12})}}$$

$$= \mathbf{251.65\ kHz}$$

(b) The secondary is also tuned to a resonant frequency of 251.65 kHz.

$$\text{Hence} \quad f_r = \frac{1}{2\pi\sqrt{(L_2C_2)}} \quad \text{i.e., } (2\pi f_r)^2 = \frac{1}{L_2C_2}$$

and **capacitance**, $C_2 = \dfrac{1}{L_2(2\pi f_r)^2}$

$$= \frac{1}{(0.2 \times 10^{-3})[2\pi(251.65 \times 10^3)]^2}$$

$$= 2.0 \times 10^{-9} \text{ F or } \mathbf{2.0\ nF}$$

(Note that since $f_r = \dfrac{1}{2\pi\sqrt{(L_1C_1)}} = \dfrac{1}{2\pi\sqrt{(L_2C_2)}}$

then $L_1C_1 = L_2C_2$

and $C_2 = \dfrac{L_1C_1}{L_2} = \dfrac{(1 \times 10^{-3})(400 \times 10^{-12})}{0.2 \times 10^{-3}} = 2.0$ nF)

(c) Since both the primary and secondary circuits are resonant, the effective primary impedance $Z_{1(eff)}$, from equation (43.25) is resistive,

i.e., $$\mathbf{Z_{1(eff)}} = R_1 + \frac{\omega^2M^2R_2}{R_2^2 + \left(\omega L_1 - \dfrac{1}{\omega C_1}\right)^2} = R_1 + \frac{\omega^2M^2R_2}{R_2^2}$$

$$= R_1 + \frac{\omega^2M^2}{R_2} = (15 + 30) + \frac{[2\pi(251.65 \times 10^3)]^2(10 \times 10^{-6})^2}{50}$$

$$= 45 + 5 = \mathbf{50\ \Omega}$$

(d) **Primary current,** $\mathbf{I_1} = \dfrac{E_1}{Z_{1(eff)}} = \dfrac{20\angle 0°}{50} = \mathbf{0.40\angle 0°\ A}$

(e) From equation (43.18), secondary voltage

$$E_2 = j\omega MI_1$$

$$= j(2\pi)(251.65 \times 10^3)(10 \times 10^{-6})(0.40\angle 0°)$$

$$= 6.325\angle 90° \text{ V}$$

Secondary current, $I_2 = \dfrac{E_2}{Z_2} = \dfrac{6.325\angle 90°}{50\angle 0°} = 0.1265\angle 90°$ A

Hence **voltage across capacitor** $\mathbf{C_2}$

$$= (I_2)(X_{C_2}) = (I_2)\left(\frac{1}{\omega C_2}\right)$$

$$= (0.1265\angle 90^\circ)\left(\frac{1}{[2\pi(251.65\times 10^3)](2.0\times 10^{-9})}\angle -90^\circ\right)$$

$$= \mathbf{40\angle 0^\circ\ V}$$

(f) From equation (43.10), the

$$\textbf{coefficient of coupling, } k = \frac{M}{\sqrt{(L_1L_2)}}$$

$$= \frac{10\times 10^{-6}}{\sqrt{(1\times 10^{-3})(0.2\times 10^{-3})}} = \mathbf{0.0224}$$

Further problem on coupled circuits may be found in Section 43.8, problems 12 to 16, page 866.

43.7 Dot rule for coupled circuits

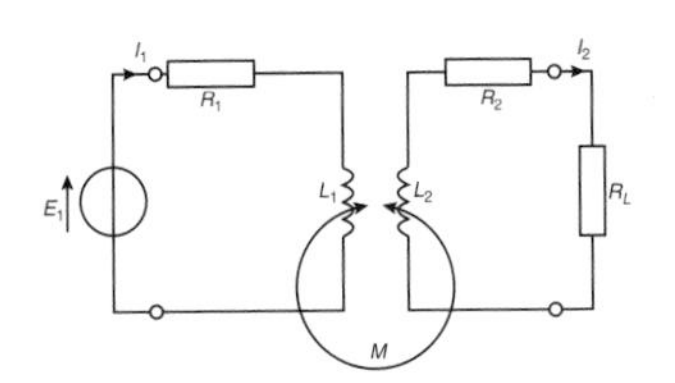

Figure 43.13

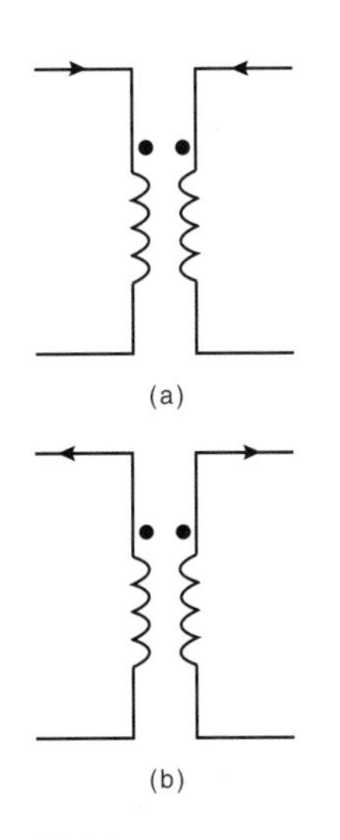

Figure 43.14

Applying Kirchhoff's voltage law to each mesh of the circuit shown in Figure 43.13 gives:

$$E_1 = I_1(R_1 + j\omega L_1) \pm j\omega M I_2$$

and $$0 = I_2(R_2 + R_L + j\omega L_2) \pm j\omega M I_1$$

In these equations the 'M' terms have been written as $\pm$ because it is not possible to state whether the magnetomotive forces due to currents I_1 and I_2 are added or subtracted. To make this clearer a dot notation is used whereby the polarity of the induced e.m.f. due to mutual inductance is identified by placing a dot on the diagram adjacent to that end of each equivalent winding which bears the same relationship to the magnetic flux.

The **dot rule** determines the sign of the voltage of mutual inductance in the Kirchhoff's law equations shown above, and states:

(i) *when both currents enter, or both currents leave, a pair of coupled coils at the dotted terminals, the signs of the 'M' terms will be the same as the signs of the 'L' terms, or*

(ii) *when one current enters at a dotted terminal and one leaves by a dotted terminal, the signs of the 'M' terms are opposite to the signs of the 'L' terms.*

Thus Figure 43.14 shows two cases in which the signs of M and L are the same, and Figure 43.15 shows two cases where the signs of M and L are opposite. In Figure 43.13, therefore, if dots had been placed at the top end of coils L_1 and L_2 then the terms $j\omega MI_2$ and $j\omega MI_1$ in the Kirchhoff's equation would be negative (since current directions are similar to Figure 43.15(a)).

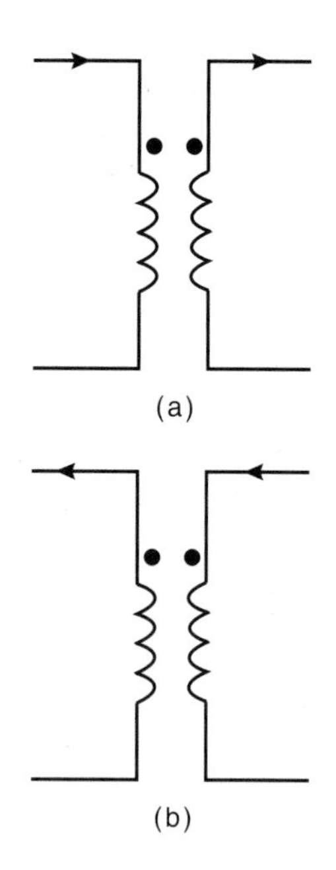

Figure 43.15

Problem 13. For the coupled circuit shown in Figure 43.16, determine the values of currents I_1 and I_2

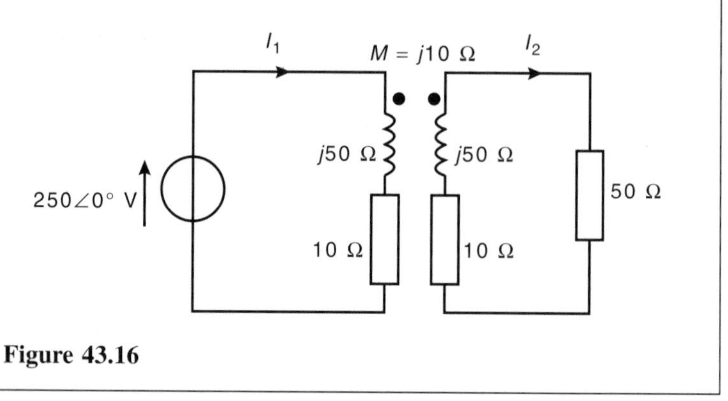

Figure 43.16

The position of the dots and the current directions correspond to Figure 43.15(a), and hence the signs of the M and L terms are opposite. Applying Kirchhoff's voltage law to the primary circuit gives:

$$250\angle 0° = (10 + j50)I_1 - j10I_2 \quad (1)$$

and applying Kirchhoff's voltage law to the secondary circuit gives:

$$0 = (10 + 50 + j50)I_2 - j10I_1 \quad (2)$$

From equation (2), $j10I_1 = (60 + j50)I_2$

and $I_1 = \dfrac{(60 + j50)I_2}{j10} = \left(\dfrac{60}{j10} + \dfrac{j50}{j10}\right) I_2 = (-j6 + 5)I_2$

i.e., $I_1 = (5 - j6)I_2 \quad (3)$

Substituting for I_1 in equation (1) gives:

$$\begin{aligned} 250\angle 0° &= (10 + j50)(5 - j6)I_2 - j10I_2 \\ &= (50 - j60 + j250 + 300 - j10)I_2 \\ &= (350 + j180)I_2 \end{aligned}$$

from which, $\mathbf{I_2} = \dfrac{250\angle 0°}{(350 + j180)} = \dfrac{250\angle 0°}{393.57\angle 27.22°}$

$= \mathbf{0.635\angle -27.22° \ A}$

From equation (3), $I_1 = (5 - j6)I_2$

$= (5 - j6)(0.635\angle -27.22°)$

$= (7.810\angle -50.19°)(0.635\angle -27.22°)$

i.e., $\mathbf{I_1 = 4.959\angle -77.41° \ A}$

i.e., $0 = I_2(53 + j60 + j15) - j15I_1 - I_1(8 + j15)$ (ii)

i.e., $0 = (53 + j75)I_2 - (8 + j30)I_1$ (2)

Hence the simultaneous equations to solve are:

$$(33 + j45)I_1 - (8 + j30)I_2 - 50\angle 0^\circ = 0 \quad (1)$$

$$-(8 + j30)I_1 + (53 + j75)I_2 = 0 \quad (2)$$

Using determinants gives:

$$\frac{I_1}{\begin{vmatrix} -(8+j30) & -50\angle 0^\circ \\ (53+j75) & 0 \end{vmatrix}} = \frac{-I_2}{\begin{vmatrix} (33+j45) & -50\angle 0^\circ \\ -(8+j30) & 0 \end{vmatrix}} = \frac{1}{\begin{vmatrix} (33+j45) & -(8+j30) \\ -(8+j30) & (53+j75) \end{vmatrix}}$$

i.e.,
$$\frac{I_1}{50(53 + j75)} = \frac{-I_2}{-50(8 + j30)} = \frac{1}{(33 + j45)(53 + j75) - (8 + j30)^2}$$

i.e.,
$$\frac{I_1}{50(91.84\angle 54.75^\circ)} = \frac{I_2}{50(31.05\angle 75.07^\circ)} = \frac{1}{(55.80\angle 53.75^\circ)(91.84\angle 54.75^\circ) - (31.05\angle 75.07^\circ)^2}$$

$$\frac{I_1}{4592\angle 54.75^\circ} = \frac{I_2}{1552.5\angle 75.07^\circ} = \frac{1}{5124.672\angle 108.50^\circ - 964.103\angle 150.14^\circ}$$

$$\frac{I_1}{4592\angle 54.75^\circ} = \frac{I_2}{1552.5\angle 75.07^\circ} = \frac{1}{-789.97 + j4379.84} = \frac{1}{4450.51\angle 100.22^\circ}$$

Hence **source current,** $I_1 = \dfrac{4592\angle 54.75^\circ}{4450.51\angle 100.22^\circ} = \mathbf{1.03\angle -45.47^\circ\ A}$

and **load current;** $I_2 = \dfrac{1552.5\angle 75.07^\circ}{4450.51\angle 100.22^\circ} = \mathbf{0.35\angle -25.15^\circ\ A}$

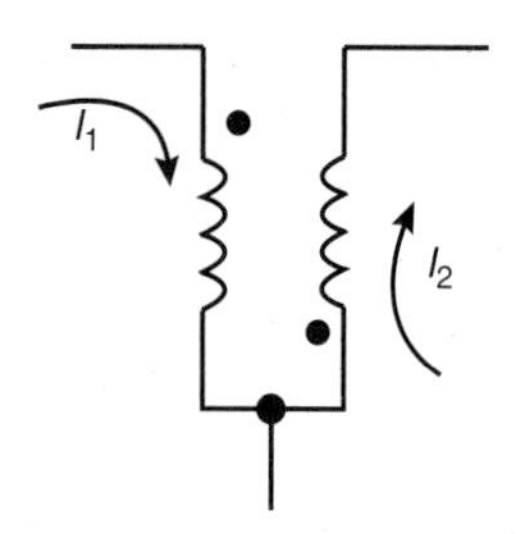

Figure 43.20

(b) When one of the windings of the mutual inductor is reversed, with, say, the dots as shown in Figure 43.20, the $j\omega MI$ terms change sign, i.e., are positive. With both currents entering the dot ends of the windings as shown, it compares with Figure 43.14(a), which indicates that the 'L' and 'M' terms are of similar sign.

Thus equations (i) and (ii) of part (a) become:

$$50\angle 0° = I_1(33 + j30 + j15) + j15I_2 - I_2(8 + j15)$$

and $$0 = I_2(53 + j60 + j15) + j15I_1 - I_1(8 + j15)$$

i.e., $$I_1(33 + j45) - I_2(8) - 50\angle 0° = 0$$

and $$-I_1(8) + I_2(53 + j75) = 0$$

Using determinants:

$$\frac{I_1}{\begin{vmatrix} -8 & -50\angle 0° \\ (53 + j75) & 0 \end{vmatrix}} = \frac{-I_2}{\begin{vmatrix} (33 + j45) & -50\angle 0° \\ -8 & 0 \end{vmatrix}}$$

$$= \frac{1}{\begin{vmatrix} (33 + j45) & -8 \\ -8 & (53 + j75) \end{vmatrix}}$$

i.e., $$\frac{I_1}{50(53 + j75)} = \frac{-I_2}{-400\angle 0°} = \frac{1}{(33 + j45)(53 + j75) - 64}$$

$$\frac{I_1}{4592\angle 54.75°} = \frac{I_2}{400\angle 0°} = \frac{1}{5124.672\angle 108.50° - 64}$$

$$= \frac{1}{-1690.08 + j4859.85} = \frac{1}{5145.34\angle 109.18°}$$

Hence **source current,** $$I_1 = \frac{4592\angle 54.75°}{5145.34\angle 109.18°}$$

$$= \mathbf{0.89\angle -54.43°\ A}$$

and **load current,** $$I_2 = \frac{400\angle 0°}{5145.34\angle 109.18°}$$

$$= \mathbf{0.078\angle -109.18°\ A}$$

Further problems on the dot rule for coupled circuits may be found in Section 43.8 following, problems 17 to 20, page 867.

43.8 Further problems on magnetically coupled circuits

Mutual inductance

1 If two coils have a mutual inductance of 500 μH, determine the magnitude of the e.m.f. induced in one coil when the current in the other coil varies at a rate of 20×10^3 A/s [10 V]

2 An e.m.f. of 15 V is induced in a coil when the current in an adjacent coil varies at a rate of 300 A/s. Calculate the value of the mutual inductance of the two coils [50 mH]

3 Two circuits have a mutual inductance of 0.2 H. A current of 3 A in the primary is reversed in 200 ms. Determine the e.m.f. induced in the secondary, assuming the current changes at a uniform rate. [−6 V]

4 A coil, x, has 1500 turns and a coil, y, situated close to x has 900 turns. When a current of 1 A flows in coil x a flux of 0.2 mWb links with x and 0.65 of this flux links coil y. Determine (a) the self inductance of coil x, and (b) the mutual inductance between the coils. [(a) 0.30 H (b) 0.117 H]

Coefficient of coupling

5 Two coils have a mutual inductance of 0.24 H. If the coils have self inductances of 0.4 H and 0.9 H respectively, determine the magnetic coefficient of coupling. [0.40]

6 Coils A and B are magnetically coupled. Coil A has a self inductance of 0.30 H and 300 turns, and coil B has a self inductance of 0.20 H and 120 turns. A change of flux of 8 mWb occurs in coil B when a current of 3 A is reversed in coil A. Determine (a) the mutual inductance between the coils, and (b) the coefficient of coupling. [(a) 0.16 H (b) 0.653]

Coils in series

7 Two coils have inductances of 50 mH and 100 mH respectively. They are placed so that their mutual inductance is 10 mH. Determine their effective inductance when the coils are (a) in series aiding (i.e., cumulatively coupled), and (b) in series opposing (i.e., differentially coupled). [(a) 170 mH (b) 130 mH]

8 The total inductance of two coils connected in series is 0.1 H. The coils have self inductance of 25 mH and 55 mH respectively. Determine (a) the mutual inductance between the two coils, and (b) the coefficient of coupling. [(a) 10 mH (b) 0.270]

9 A d.c. supply of 200 V is applied across two mutually coupled coils in series, A and B. Coil A has a resistance of 2 Ω and a self inductance of 0.5 H; coil B has a resistance of 8 Ω and a self inductance of 2 H. At a certain instant after the circuit is switched on, the current is 10 A and increasing a at rate of 25 A/s. Determine (a) the mutual inductance between the coils, and (b) the coefficient of coupling. [(a) 0.75 H (b) 0.75]

10 A ferromagnetic-cored coil is in two sections. One section has an inductance of 750 mH and the other an inductance of 148 mH. The coefficient of coupling is 0.6. Determine (a) the mutual inductance,

(b) the total inductance when the sections are connected in series aiding, and (c) the total inductance when the sections are in series opposing. [(a) 200 mH (b) 1.298 H (c) 0.498 H]

11 Two coils are connected in series and their total inductance is measured as 0.12 H, and when the connection to one coil is reversed, the total inductance is measured as 0.04 H. If the coefficient of coupling is 0.8, determine (a) the self inductance of each coil, and (b) the mutual inductance between the coils.

[(a) $L_1 = 71.22$ mH or 8.78 mH,
$L_2 = 8.78$ mH or 71.22 mH
(b) 20 mH]

Coupled circuits

12 Determine the value of voltage E_2 which appears across the open circuited secondary winding of Figure 43.21. [$0.93\angle 68.20°$ V]

13 The coefficient of coupling between two coils having self inductances of 0.5 H and 0.9 H respectively is 0.85. If a sinusoidal alternating voltage of 50 mV is applied to the 0.5 H coil, determine the magnitude of the open circuit e.m.f. induced in the 0.9 H coil. [57 mV]

14 Determine the value of (a) the primary current, I_1, and (b) the secondary current I_2, for the circuit shown in Figure 43.22.
[(a) $0.197\angle -71.91°$ A (b) $0.030\angle -48.48°$ A]

15 For the magnetically coupled circuit shown in Figure 43.23, determine (a) the self impedance of the primary circuit, (b) the self impedance of the secondary circuit, (c) the impedance reflected into the primary circuit, (d) the effective primary impedance, (e) the primary current, and (f) the secondary current.

[(a) $(100 + j200)\Omega$ (b) $(40 + j80)\Omega$
(c) $(40.5 - j81.0)\Omega$ (d) $(140.5 + j119)\Omega$
(e) $0.543\angle -40.26°$ A (f) $0.546\angle -13.69°$ A]

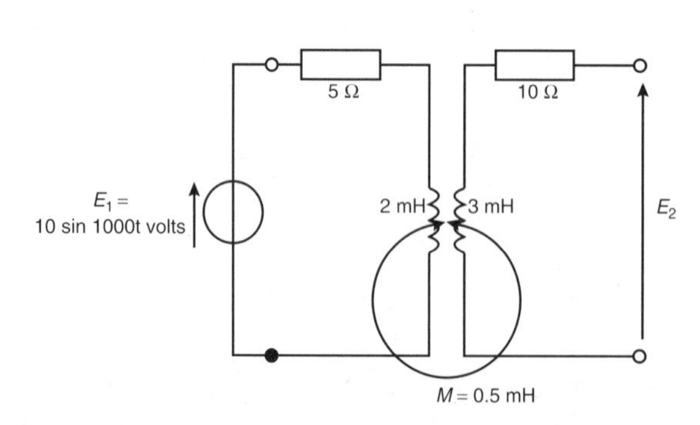

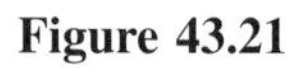
Figure 43.21

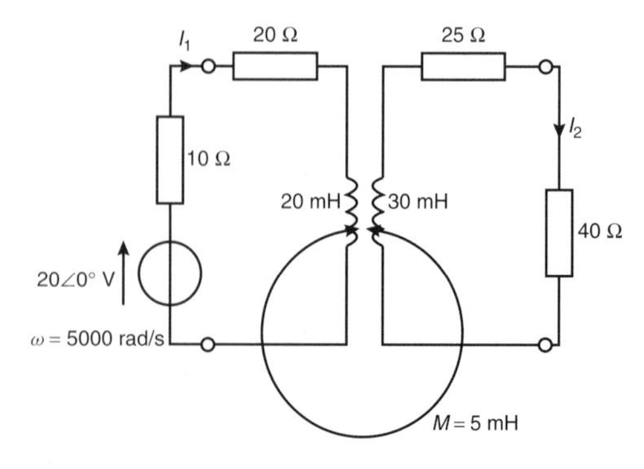

Figure 43.22

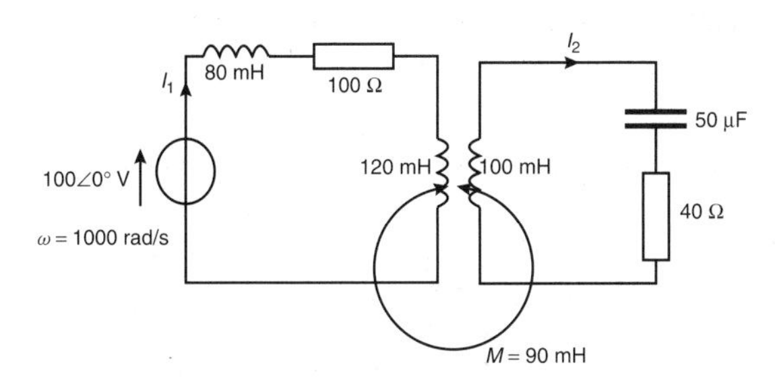

Figure 43.23

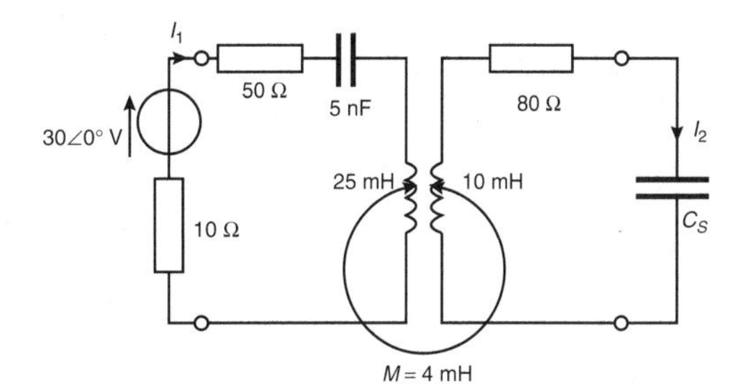

Figure 43.24

16 In the coupled circuit shown in Figure 43.24, each winding is tuned to resonance at the same frequency. Calculate (a) the resonant frequency, (b) the value of C_S, (c) the effective primary impedance, (d) the primary current, (e) the secondary current, (f) the p.d. across capacitor C_S, and (g) the coefficient of coupling.

[(a) 14.235 kHz (b) 12.5 nF (c) 1659.9 Ω (d) 18.1∠0° mA (e)80.9∠90° mA (f) 72.4∠0° V (g) 0.253]

Dot rule for coupled circuits

17 Determine the values of currents I_p and I_s in the coupled circuit shown in Figure 43.25.

[$I_p = 893.3\angle{-60.57^\circ}$ mA, $I_s = 99.88\angle 2.86^\circ$ mA]

18 The coefficient of coupling between the primary and secondary windings for the air-cored transformer shown in Figure 43.26 is 0.84. Calculate for the circuit (a) the mutual inductance M, (b) the primary current I_p, (c) the secondary current I_s, and (d) the secondary terminal p.d.

[(a) 13.28 mH(b) 1.603∠−28.98° A (c) 0.913∠17.70° A (d) 73.04∠−27.30° V]

19 A mutual inductor is used to couple a 50 Ω resistive load to a 250∠0° V generator as shown in Figure 43.27. Calculate (a) the generator current I_g and (b) the load current I_L.

[(a) $I_g = 9.653\angle{-36.03^\circ}$ A (b) $I_L = 1.084\angle 27.28^\circ$ A]

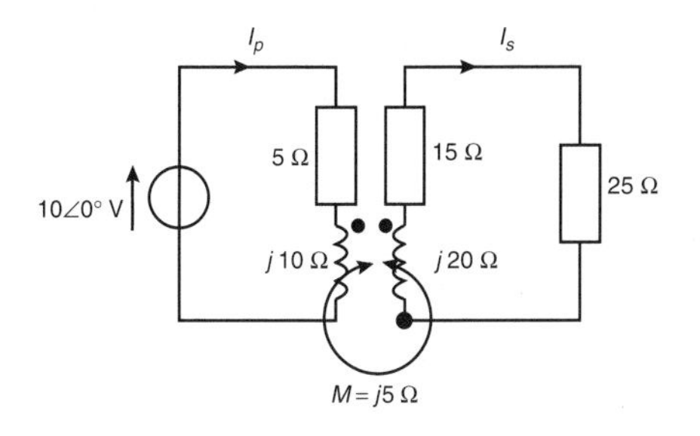

Figure 43.25

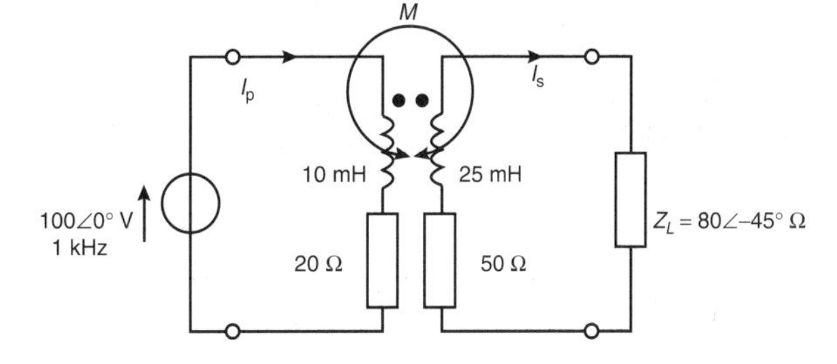

Figure 43.26

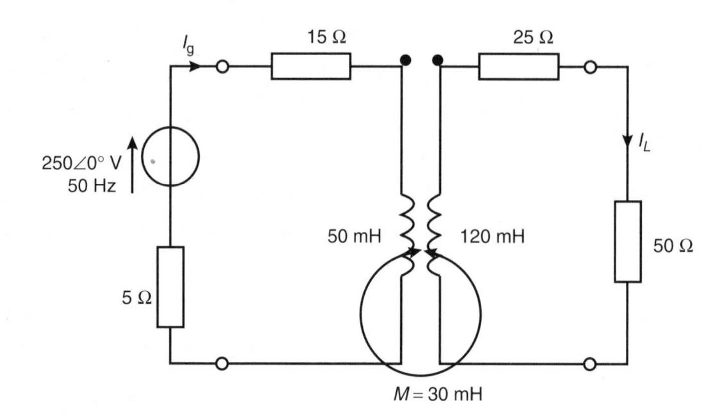

Figure 43.27

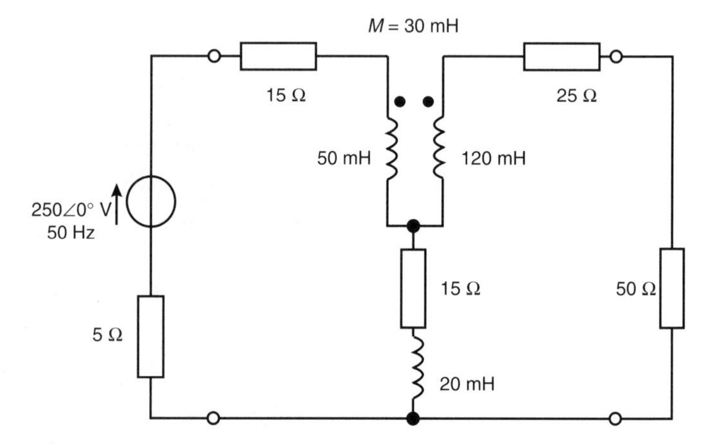

Figure 43.28

20 The mutual inductor of problem 19 is connected to the circuit as shown in Figure 43.28. Determine (a) the source current, and (b) the load current. (c) If one of the windings is reversed, determine the new value of source and load currents.

[(a) $6.658\angle -28.07°$ A (b) $1.444\angle -7.79°$ A
(c) $8.239\angle -23.09°$ A, $1.261\angle -60.96°$ A]

44 Transmission lines

At the end of this chapter you should be able to:

- appreciate the purpose of a transmission line
- define the transmission line primary constants R, L, C and G
- calculate phase delay, wavelength and velocity of propagation on a transmission line
- appreciate current and voltage relationships on a transmission line
- define the transmission line secondary line constants Z_0, γ, α and β
- calculate characteristic impedance and propagation coefficient in terms of the primary line constants
- understand and calculate distortion on transmission lines
- understand wave reflection and calculate reflection coefficient
- understand standing waves and calculate standing wave ratio

44.1 Introduction

A transmission line is a system of conductors connecting one point to another and along which electromagnetic energy can be sent. Thus telephone lines and power distribution lines are typical examples of transmission lines; in electronics, however, the term usually implies a line used for the transmission of radio-frequency (r.f.) energy such as that from a radio transmitter to the antenna.

An important feature of a transmission line is that it should guide energy from a source at the sending end to a load at the receiving end without loss by radiation. One form of construction often used consists of two similar conductors mounted close together at a constant separation. The two conductors form the two sides of a balanced circuit and any radiation from one of them is neutralized by that from the other. Such twin-wire lines are used for carrying high r.f. power, for example, at transmitters. The coaxial form of construction is commonly employed for low power use, one conductor being in the form of a cylinder which surrounds the other at its centre, and thus acts as a screen. Such cables are often used to couple f.m. and television receivers to their antennas.

At frequencies greater than 1000 MHz, transmission lines are usually in the form of a waveguide which may be regarded as coaxial lines without the centre conductor, the energy being launched into the guide or abstracted from it by probes or loops projecting into the guide.

44.2 Transmission line primary constants

Let an a.c. generator be connected to the input terminals of a pair of parallel conductors of infinite length. A sinusoidal wave will move along the line and a finite current will flow into the line. The variation of voltage with distance along the line will resemble the variation of applied voltage with time. The moving wave, sinusoidal in this case, is called a voltage **travelling wave**. As the wave moves along the line the capacitance of the line is charged up and the moving charges cause magnetic energy to be stored. Thus the propagation of such an **electromagnetic wave** constitutes a flow of energy.

After sufficient time the magnitude of the wave may be measured at any point along the line. The line does not therefore appear to the generator as an open circuit but presents a definite load Z_0. If the sending-end voltage is V_S and the sending end current is I_S then $Z_0 = V_S/I_S$. Thus all of the energy is absorbed by the line and the line behaves in a similar manner to the generator as would a single 'lumped' impedance of value Z_0 connected directly across the generator terminals.

There are **four parameters** associated with transmission lines, these being resistance, inductance, capacitance and conductance.

(i) **Resistance *R*** is given by $R = \rho l/A$, where ρ is the resistivity of the conductor material, A is the cross-sectional area of each conductor and l is the length of the conductor (for a two-wire system, l represents twice the length of the line). Resistance is stated in ohms per metre length of a line and represents the imperfection of the conductor. A resistance stated in ohms per loop metre is a little more specific since it takes into consideration the fact that there are two conductors in a particular length of line.

(ii) **Inductance *L*** is due to the magnetic field surrounding the conductors of a transmission line when a current flows through them. The inductance of an isolated twin line is considered in Section 40.7. From equation (40.23), page 748, the inductance L is given by

$$L = \frac{\mu_0 \mu_r}{\pi}\left\{\frac{1}{4} + \ln\frac{D}{a}\right\} \text{ henry/metre}$$

where D is the distance between centres of the conductor and a is the radius of each conductor. In most practical lines $\mu_r = 1$. An inductance stated in henrys per loop metre takes into consideration the fact that there are two conductors in a particular length of line.

(iii) **Capacitance *C*** exists as a result of the electric field between conductors of a transmission line. The capacitance of an isolated twin line is considered in Section 40.3. From equation (40.14), page 736, the capacitance between the two conductors is given by

$$C = \frac{\pi \varepsilon_0 \varepsilon_r}{\ln(D/a)} \text{ farads/metre}$$

In most practical lines $\varepsilon_r = 1$

(iv) **Conductance *G*** is due to the insulation of the line allowing some current to leak from one conductor to the other. Conductance is measured in siemens per metre length of line and represents the imperfection of the insulation. Another name for conductance is leakance.

Each of the four transmission line constants, R, L, C and G, known as the **primary constants**, are uniformly distributed along the line.

From Chapter 41, when a symmetrical T-network is terminated in its characteristic impedance Z_0, the input impedance of the network is also equal to Z_0. Similarly, if a number of identical T-sections are connected in cascade, the input impedance of the network will also be equal to Z_0.

A transmission line can be considered to consist of a network of a very large number of cascaded T-sections each a very short length (δl) of transmission line, as shown in Figure 44.1. This is an approximation of the uniformly distributed line; the larger the number of lumped parameter sections, the nearer it approaches the true distributed nature of the line. When the generator V_S is connected, a current I_S flows which divides between that flowing through the leakage conductance G, which is lost, and that which progressively charges each capacitor C and which sets up the voltage travelling wave moving along the transmission line. The loss or attenuation in the line is caused by both the conductance G and the series resistance R.

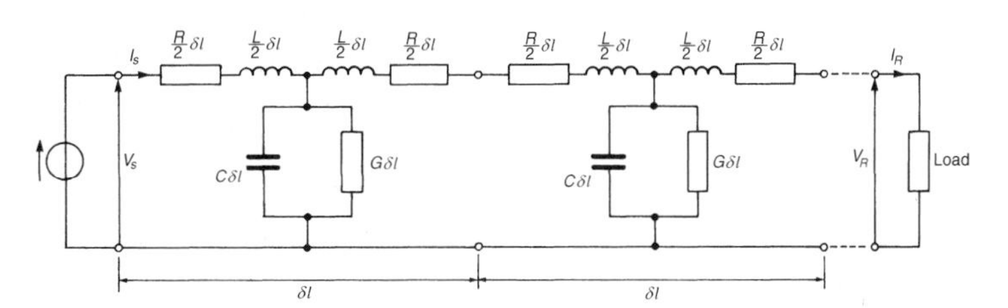

Figure 44.1

44.3 Phase delay, wavelength and velocity of propagation

Each section of that shown in Figure 44.1 is simply a low-pass filter possessing losses R and G. If losses are neglected, and R and G are removed, the circuit simplifies and the infinite line reduces to a repetitive T-section low-pass filter network as shown in Figure 44.2. Let a generator be connected to the line as shown and let the voltage be rising to a maximum positive value just at the instant when the line is connected to it. A current I_S flows through inductance L_1 into capacitor C_1. The capacitor charges and a voltage develops across it. The voltage sends a current through inductance L_1' and L_2 into capacitor C_2. The capacitor

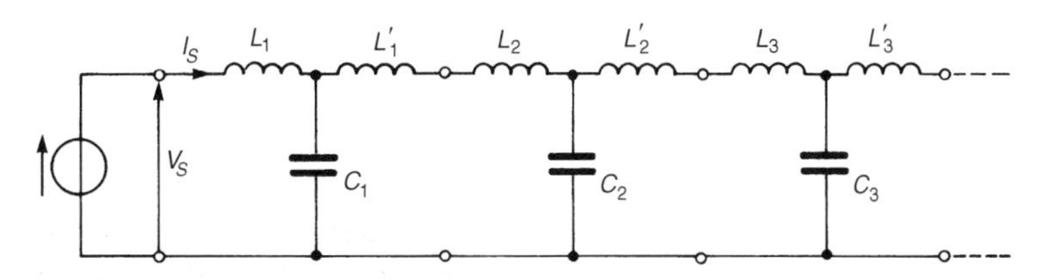

Figure 44.2

charges and the voltage developed across it sends a current through L_2' and L_3 into C_3, and so on. Thus all capacitors will in turn charge up to the maximum input voltage. When the generator voltage falls, each capacitor is charged in turn in opposite polarity, and as before the input charge is progressively passed along to the next capacitor. In this manner voltage and current waves travel along the line together and depend on each other.

The process outlined above takes time; for example, by the time capacitor C_3 has reached its maximum voltage, the generator input may be at zero or moving towards its minimum value. There will therefore be a time, and thus a phase difference between the generator input voltage and the voltage at any point on the line.

Phase delay

Since the line shown in Figure 44.2 is a ladder network of low-pass T-section filters, it is shown in equation (42.27), page 820, that the phase delay, β, is given by:

$$\boldsymbol{\beta = \omega\sqrt{(LC)}} \textbf{ radians/metre} \qquad (44.1)$$

where L and C are the inductance and capacitance per metre of the line.

Wavelength

The wavelength λ on a line is the distance between a given point and the next point along the line at which the voltage is the same phase, the initial point leading the latter point by 2π radian. Since in one wavelength a phase change of 2π radians occurs, the phase change per metre is $2\pi/\lambda$. Hence, phase change per metre, $\beta = 2\pi/\lambda$

or

$$\textbf{wavelength, } \boldsymbol{\lambda = \frac{2\pi}{\beta}} \textbf{ metres} \qquad (44.2)$$

Velocity of propagation

The velocity of propagation, u, is given by $u = f\lambda$, where f is the frequency and λ the wavelength. Hence

$$\boldsymbol{u = f\lambda = f(2\pi/\beta) = \frac{2\pi f}{\beta} = \frac{\omega}{\beta}} \qquad (44.3)$$

The velocity of propagation of free space is the same as that of light, i.e., approximately 300×10^6 m/s. The velocity of electrical energy along a line is always less than the velocity in free space. The wavelength λ of radiation in free space is given by $\lambda = c/f$ where c is the velocity of light. Since the velocity along a line is always less than c, the wavelength corresponding to any particular frequency is always shorter on the line than it would be in free space.

> Problem 1. A parallel-wire air-spaced transmission line operating at 1910 Hz has a phase shift of 0.05 rad/km. Determine (a) the wavelength on the line, and (b) the speed of transmission of a signal.

(a) From equation (44.2), wavelength $\lambda = 2\pi/\beta = 2\pi/0.05$

$$= \mathbf{125.7\ km}$$

(b) From equation (44.3), speed of transmission,

$$u = f\lambda = (1910)(125.7) = \mathbf{240 \times 10^3\ km/s} \text{ or } \mathbf{240 \times 10^6\ m/s}$$

> Problem 2. A transmission line has an inductance of 4 mH/loop km and a capacitance of 0.004 μF/km. Determine, for a frequency of operation of 1 kHz, (a) the phase delay, (b) the wavelength on the line, and (c) the velocity of propagation (in metres per second) of the signal.

(a) From equation (44.1), phase delay,

$$\beta = \omega\sqrt{(LC)} = (2\pi 1000)\sqrt{[(4 \times 10^{-3})(0.004 \times 10^{-6})]}$$
$$= \mathbf{0.025\ rad/km}$$

(b) From equation (44.2), wavelength $\lambda = 2\pi/\beta = 2\pi/0.025$

$$= \mathbf{251\ km}$$

(c) From equation (44.3), velocity of propagation,

$$u = f\lambda = (1000)(251)\text{ km/s} = \mathbf{251 \times 10^6\ m/s}$$

Further problems on phase delay, wavelength and velocity of propagation may be found in Section 44.9, problems 1 to 3, page 897.

44.4 Current and voltage relationships

Figure 44.3 shows a voltage source V_S applied to the input terminals of an infinite line, or a line terminated in its characteristic impedance, such that a current I_S flows into the line. At a point, say, 1 km down the line let the current be I_1. The current I_1 will not have the same magnitude as I_S because of line attenuation; also I_1 will lag I_S by some angle β. The ratio I_S/I_1 is therefore a phasor quantity. Let the current a further 1 km

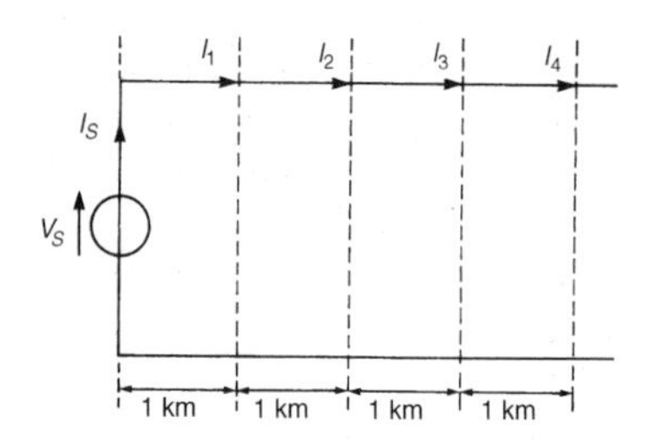

Figure 44.3

down the line be I_2, and so on, as shown in Figure 44.3. Each unit length of line can be treated as a section of a repetitive network, as explained in Section 44.2. The attenuation is in the form of a logarithmic decay and

$$\frac{I_S}{I_1} = \frac{I_1}{I_2} = \frac{I_2}{I_3} = e^{\gamma}$$

where γ is the **propagation constant**, first introduced in Section 42.7, page 815. γ has no unit.

The propagation constant is a complex quantity given by $\gamma = \alpha + j\beta$, where α is the **attenuation constant**, whose unit is the neper, and β is the **phase shift coefficient**, whose unit is the radian. For n such 1 km sections, $I_S/I_R = e^{n\gamma}$ where I_R is the current at the receiving end.

Hence $\dfrac{I_S}{I_R} = e^{n(\alpha+j\beta)} = e^{(n\alpha+jn\beta)} = e^{n\alpha}\angle n\beta$

from which, $\boxed{\boldsymbol{I_R = I_S e^{-n\gamma} = I_S e^{-n\alpha}\angle -n\beta}}$ (44.4)

In equation (44.4), the attenuation on the line is given by $n\alpha$ nepers and the phase shift is $n\beta$ radians.

At all points along an infinite line, the ratio of voltage to current is Z_0, the characteristic impedance. Thus from equation (44.4) it follows that:

receiving end voltage, $\boxed{\boldsymbol{V_R = V_S e^{-n\gamma} = V_S e^{-n\alpha}\angle -n\beta}}$ (44.5)

Z_0, γ, α, and β are referred to as the **secondary line constants** or **coefficients.**

Problem 3. When operating at a frequency of 2 kHz, a cable has an attenuation of 0.25 Np/km and a phase shift of 0.20 rad/km. If a 5 V rms signal is applied at the sending end, determine the voltage at a point 10 km down the line, assuming that the termination is equal to the characteristic impedance of the line.

Let V_R be the voltage at a point n km from the sending end, then from equation (44.5), $V_R = V_S e^{-n\gamma} = V_S e^{-n\alpha}\angle -n\beta$

Since $\alpha = 0.25$ Np/km, $\beta = 0.20$ rad/km, $V_S = 5$ V and $n = 10$ km, then

$$V_R = (5)e^{-(10)(0.25)}\angle -(10)(0.20) = 5e^{-2.5}\angle -2.0 \text{ V}$$

$$= \mathbf{0.41\angle -2.0\ V} \text{ or } \mathbf{0.41\angle -114.6^\circ\ V}$$

Thus the voltage 10 km down the line is 0.41 V rms lagging the sending end voltage of 5 V by 2.0 rad or 114.6°

Problem 4. A transmission line 5 km long has a characteristic impedance of $800\angle -25^\circ\ \Omega$. At a particular frequency, the attenuation coefficient of the line is 0.5 Np/km and the phase shift coefficient is 0.25 rad/km. Determine the magnitude and phase of the current at the receiving end, if the sending end voltage is $2.0\angle 0^\circ$ V r.m.s.

The receiving end voltage (from equation (44.5)) is given by:

$$V_R = V_S e^{-n\gamma} = V_S e^{-n\alpha}\angle -n\beta = (2.0\angle 0^\circ)e^{-(5)(0.5)}\angle -(5)(0.25)$$

$$= 2.0e^{-2.5}\angle -1.25 = 0.1642\angle -71.62^\circ \text{ V}$$

Receiving end current,

$$I_R = \frac{V_R}{Z_0} = \frac{0.1642\angle -71.62^\circ}{800\angle -25^\circ} = 2.05 \times 10^{-4}\angle(-71.62^\circ - (-25^\circ))\text{A}$$

$$= \mathbf{0.205\angle -46.62^\circ\ mA}$$

Problem 5. The voltages at the input and at the output of a transmission line properly terminated in its characteristic impedance are 8.0 V and 2.0 V rms respectively. Determine the output voltage if the length of the line is doubled.

The receiving-end voltage V_R is given by $V_R = V_S e^{-n\gamma}$.

Hence $2.0 = 8.0e^{-n\gamma}$, from which, $e^{-n\gamma} = 2.0/8.0 = 0.25$

If the line is doubled in length, then

$$V_R = 8.0e^{-2n\gamma} = 8.0(e^{-n\gamma})^2$$

$$= 8.0(0.25)^2 = \mathbf{0.50\ V}$$

Further problems on current and voltage relationships may be found in Section 44.9, problems 4 to 6, page 897.

44.5 Characteristic impedance and propagation coefficient in terms of the primary constants

Characteristic impedance

At all points along an infinite line, the ratio of voltage to current is called the characteristic impedance Z_0. The value of Z_0 is independent of the length of the line; it merely describes a property of a line that is a function of the physical construction of the line. Since a short length of line may be considered as a ladder of identical low-pass filter sections, the characteristic impedance may be determined from equation (41.2), page 760, i.e.,

$$\mathbf{Z_0 = \sqrt{(Z_{OC} Z_{SC})}} \qquad (44.6)$$

since the open-circuit impedance Z_{OC} and the short-circuit impedance Z_{SC} may be easily measured.

> Problem 6. At a frequency of 1.5 kHz the open-circuit impedance of a length of transmission line is $800\angle -50°\ \Omega$ and the short-circuit impedance is $413\angle -20°\ \Omega$. Determine the characteristic impedance of the line at this frequency.

From equation (44.6),

$$\text{characteristic impedance } Z_0 = \sqrt{(Z_{OC}Z_{SC})}$$
$$= \sqrt{[(800\angle -50°)(413\angle -20°)]}$$
$$= \sqrt{(330400\angle -70°)} = \mathbf{575\angle -35°\ \Omega}$$

by de Moivre's theorem.

The characteristic impedance of a transmission line may also be expressed in terms of the primary constants, R, L, G and C. Measurements of the primary constants may be obtained for a particular line and manufacturers usually state them for a standard length.

Let a very short length of line δl metres be as shown in Figure 44.4 comprising a single T-section. Each series arm impedance is $Z_1 = \frac{1}{2}(R + j\omega L)\delta l$ ohms, and the shunt arm impedance is

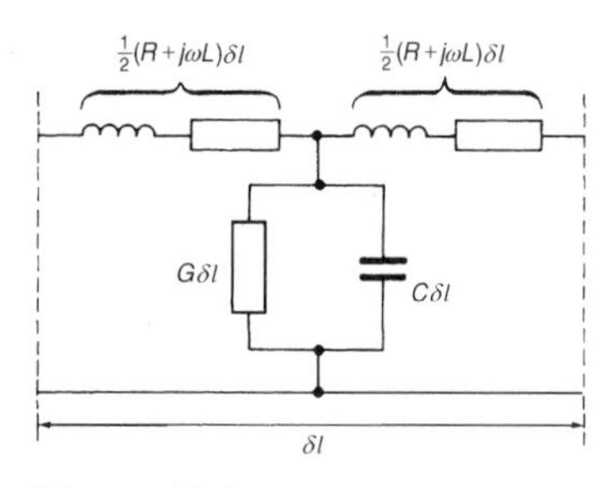

Figure 44.4

$$Z_2 = \frac{1}{Y_2} = \frac{1}{(G + j\omega C)\delta l}$$

[i.e., from Chapter 25, the total admittance Y_2 is the sum of the admittance of the two parallel arms, i.e., in this case, the sum of

$$G\delta l \text{ and } \left(\frac{1}{1/(j\omega C)}\right)\delta l]$$

From equation (41.1), page 760, the characteristic impedance Z_0 of a T-section having in each series arm an impedance Z_1 and a shunt arm impedance Z_2 is given by: $Z_0 = \sqrt{(Z_1{}^2 + 2Z_1Z_2)}$

Hence the characteristic impedance of the section shown in Figure 44.4 is

$$Z_0 = \sqrt{\left\{\left[\frac{1}{2}(R + j\omega L)\delta l\right]^2 + 2\left[\frac{1}{2}(R + j\omega L)\delta l\right]\left[\frac{1}{(G + j\omega C)\delta l}\right]\right\}}$$

The term $Z_1{}^2$ involves δl^2 and, since δl is a very short length of line, δl^2 is negligible. Hence

$$\boxed{\mathbf{Z_0 = \sqrt{\frac{R + j\omega L}{G + j\omega C}}\ \text{ohms}}} \qquad (44.7)$$

If losses R and G are neglected, then

$$\boxed{\mathbf{Z_0 = \sqrt{(L/C)}\ ohms}} \qquad (44.8)$$

Problem 7. A transmission line has the following primary constants: resistance $R = 15\ \Omega$/loop km, inductance $L = 3.4$ mH/loop km, conductance $G = 3\ \mu$S/km and capacitance $C = 10$ nF/km. Determine the characteristic impedance of the line when the frequency is 2 kHz.

From equation (44.7),

$$\text{characteristic impedance } Z_0 = \sqrt{\frac{R + j\omega L}{G + j\omega C}} \text{ ohms}$$

$$R + j\omega L = 15 + j(2\pi 2000)(3.4 \times 10^{-3})$$
$$= (15 + j42.73)\Omega = 45.29\angle 70.66^\circ\ \Omega$$
$$G + j\omega C = 3 \times 10^{-6} + j(2\pi 2000)(10 \times 10^{-9})$$
$$= (3 + j125.66)10^{-6}\ \text{S} = 125.7 \times 10^{-6}\angle 88.63^\circ\ \text{S}$$

$$\text{Hence } Z_0 = \sqrt{\frac{45.29\angle 70.66^\circ}{125.7 \times 10^{-6}\angle 88.63^\circ}} = \sqrt{[0.360 \times 10^6\angle -17.97^\circ]}\ \Omega$$

i.e., characteristic impedance, $\mathbf{Z_0 = 600\angle -8.99^\circ\ \Omega}$

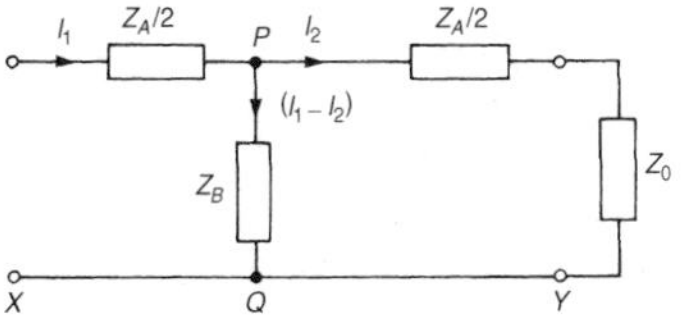

Propagation coefficient

Figure 44.5

Figure 44.5 shows a T-section with the series arm impedances each expressed as $Z_A/2$ ohms per unit length and the shunt impedance as Z_B ohms per unit length. The p.d. between points P and Q is given by:

$$V_{PQ} = (I_1 - I_2)Z_B = I_2\left(\frac{Z_A}{2} + Z_0\right)$$

i.e., $$I_1 Z_B - I_2 Z_B = \frac{I_2 Z_A}{2} + I_2 Z_0$$

Hence $$I_1 Z_B = I_2\left(Z_B + \frac{Z_A}{2} + Z_0\right)$$

from which $$\frac{I_1}{I_2} = \frac{Z_B + (Z_A/2) + Z_0}{Z_B}$$

From equation (41.1), page 760, $Z_0 = \sqrt{(Z_1{}^2 + 2Z_1Z_2)}$. In Figure 44.5, $Z_1 \equiv Z_A/2$ and $Z_2 \equiv Z_B$

Thus $$Z_0 = \sqrt{\left[\left(\frac{Z_A}{2}\right)^2 + 2\left(\frac{Z_A}{2}\right)Z_B\right]} = \sqrt{\left(\frac{Z_A{}^2}{4} + Z_AZ_B\right)}$$

Hence $$\frac{I_1}{I_2} = \frac{Z_B + (Z_A/2) + \sqrt{(Z_AZ_B + (Z_A{}^2/4))}}{Z_B}$$

$$= \frac{Z_B}{Z_B} + \frac{(Z_A/2)}{Z_B} + \frac{\sqrt{(Z_AZ_B + (Z_A{}^2/4))}}{Z_B}$$

$$= 1 + \frac{1}{2}\left(\frac{Z_A}{Z_B}\right) + \sqrt{\left(\frac{Z_AZ_B}{Z_B{}^2} + \frac{(Z_A{}^2/4)}{Z_B{}^2}\right)}$$

i.e., $$\frac{I_1}{I_2} = 1 + \frac{1}{2}\left(\frac{Z_A}{Z_B}\right) + \left[\frac{Z_A}{Z_B} + \frac{1}{4}\left(\frac{Z_A}{Z_B}\right)^2\right]^{1/2} \qquad (44.9)$$

From Section 44.4, $I_1/I_2 = e^\gamma$, where γ is the propagation coefficient. Also, from the binomial theorem:

$$(a+b)^n = a^n + na^{n-1}b + \frac{n(n-1)}{2!}a^{n-2}b^2 + \cdots$$

Thus $$\left[\frac{Z_A}{Z_B} + \frac{1}{4}\left(\frac{Z_A}{Z_B}\right)^2\right]^{1/2}$$

$$= \left(\frac{Z_A}{Z_B}\right)^{1/2} + \frac{1}{2}\left(\frac{Z_A}{Z_B}\right)^{-1/2}\frac{1}{4}\left(\frac{Z_A}{Z_B}\right)^2 + \cdots$$

Hence, from equation (44.9),

$$\frac{I_1}{I_2} = e^\gamma = 1 + \frac{1}{2}\left(\frac{Z_A}{Z_B}\right) + \left[\left(\frac{Z_A}{Z_B}\right)^{1/2} + \frac{1}{8}\left(\frac{Z_A}{Z_B}\right)^{3/2} + \cdots\right.$$

Rearranging gives: $e^\gamma = 1 + \left(\frac{Z_A}{Z_B}\right)^{1/2} + \frac{1}{2}\left(\frac{Z_A}{Z_B}\right) + \frac{1}{8}\left(\frac{Z_A}{Z_B}\right)^{3/2} + \cdots$

Let length XY in Figure 44.5 be a very short length of line δl and let impedance $Z_A = Z\delta l$, where $Z = R + j\omega L$ and $Z_B = 1/(Y\delta l)$, where $Y = G + j\omega C$

Then

$$e^{\gamma\delta l} = 1 + \left(\frac{Z\delta l}{1/Y\delta l}\right)^{1/2} + \frac{1}{2}\left(\frac{Z\delta l}{1/Y\delta l}\right) + \frac{1}{8}\left(\frac{Z\delta l}{1/Y\delta l}\right)^{3/2} + \cdots$$

$$= 1 + (ZY\delta l^2)^{1/2} + \frac{1}{2}(ZY\delta l^2) + \frac{1}{8}(ZY\delta l^2)^{3/2} + \cdots$$

$$= 1 + (ZY)^{1/2}\delta l + \frac{1}{2}(ZY)(\delta l)^2 + \frac{1}{8}(ZY)^{3/2}(\delta l)^3 + \cdots$$

$$= 1 + (ZY)^{1/2}\delta l,$$

if $(\delta l)^2$, $(\delta l)^3$ and higher powers are considered as negligible.

e^x may be expressed as a series:

$$e^x = 1 + x + \frac{x^2}{2!} + \frac{x^3}{3!} + \cdots$$

Comparison with $e^{\gamma\delta l} = 1 + (ZY)^{1/2}\delta l$ shows that $\gamma\delta l = (ZY)^{1/2}\delta l$ i.e., $\gamma = \sqrt{(ZY)}$. Thus

propagation coefficient, $$\boxed{\gamma = \sqrt{[(R + j\omega L)(G + j\omega C)]}} \qquad (44.10)$$

The unit of γ is $\sqrt{(\Omega)(S)}$, i.e., $\sqrt{[(\Omega)(1/\Omega)]}$, thus γ is dimensionless, as expected, since $I_1/I_2 = e^\gamma$, from which $\gamma = \ln(I_1/I_2)$, i.e., a ratio of two currents. For a lossless line, $R = G = 0$ and

$$\boxed{\gamma = \sqrt{(j\omega L)(j\omega C)} = j\omega\sqrt{(LC)}} \qquad (44.11)$$

Equations (44.7) and (44.10) are used to determine the characteristic impedance Z_0 and propagation coefficient γ of a transmission line in terms of the primary constants R, L, G and C. When $R = G = 0$, i.e., losses are neglected, equations (44.8) and (44.11) are used to determine Z_0 and γ.

Problem 8. A transmission line having negligible losses has primary line constants of inductance $L = 0.5$ mH/loop km and capacitance $C = 0.12$ μF/km. Determine, at an operating frequency of 400 kHz, (a) the characteristic impedance, (b) the propagation coefficient, (c) the wavelength on the line, and (d) the velocity of propagation, in metres per second, of a signal.

(a) Since the line is lossfree, from equation (44.8), the characteristic impedance Z_0 is given by

$$Z_0 = \sqrt{\frac{L}{C}} = \sqrt{\frac{0.5 \times 10^{-3}}{0.12 \times 10^{-6}}} = \mathbf{64.55\ \Omega}$$

(b) From equation (44.11), for a lossfree line, the propagation coefficient γ is given by

$$\gamma = j\omega\sqrt{(LC)} = j(2\pi 400 \times 10^3)\sqrt{[(0.5 \times 10^{-3})(0.12 \times 10^{-6})]}$$

$$= j19.47 \text{ or } \mathbf{0 + j19.47}$$

Since $\gamma = \alpha + j\beta$, the attenuation coefficient $\alpha = 0$ and the phase-shift coefficient, $\beta = 19.47$ rad/km.

(c) From equation (44.2), wavelength $\lambda = \dfrac{2\pi}{\beta} = \dfrac{2\pi}{19.47}$

$$= \mathbf{0.323\ km\ or\ 323\ m}$$

(d) From equation (44.3), velocity of propagation $u = f\lambda$

$$= (400 \times 10^3)(323) = \mathbf{129 \times 10^6\ m/s.}$$

> Problem 9. At a frequency of 1 kHz the primary constants of a transmission line are resistance $R = 25\ \Omega$/loop km, inductance $L =$ 5 mH/loop km, capacitance $C = 0.04\ \mu$F/km and conductance $G =$ 80 μS/km. Determine for the line (a) the characteristic impedance, (b) the propagation coefficient, (c) the attenuation coefficient and (d) the phase-shift coefficient.

(a) From equation (44.7),

characteristic impedance $Z_0 = \sqrt{\dfrac{R + j\omega L}{G + j\omega C}}$ ohms

$$R + j\omega L = 25 + j(2\pi 1000)(5 \times 10^{-3}) = (25 + j31.42)$$

$$= 40.15\angle 51.49^\circ\ \Omega$$

$$G + j\omega C = 80 \times 10^{-6} + j(2\pi 1000)(0.04 \times 10^{-6})$$

$$= (80 + j251.33)10^{-6} = 263.76 \times 10^{-6}\angle 72.34^\circ\ \text{S}$$

Thus characteristic impedance

$$Z_0 = \sqrt{\frac{40.15\angle 51.49^\circ}{263.76 \times 10^{-6}\angle 72.34^\circ}} = \mathbf{390.2\angle -10.43^\circ\ \Omega}$$

(b) From equation (44.10), propagation coefficient

$$\gamma = \sqrt{[(R + j\omega L)(G + j\omega C)]}$$

$$= \sqrt{[(40.15\angle 51.49^\circ)(263.76 \times 10^{-6}\angle 72.34^\circ)]}$$

$$= \sqrt{(0.01059\angle 123.83^\circ)} = \mathbf{0.1029\angle 61.92^\circ}$$

(c) $\gamma = \alpha + j\beta = 0.1029(\cos 61.92^\circ + j \sin 61.92^\circ)$,
i.e., $\gamma = 0.0484 + j0.0908$

Thus the attenuation coefficient, $\boldsymbol{\alpha} = \mathbf{0.0484}$ **nepers/km**

(d) The phase shift coefficient, $\boldsymbol{\beta} = \mathbf{0.0908}$ **rad/km**

Problem 10. An open wire line is 300 km long and is terminated in its characteristic impedance. At the sending end is a generator having an open-circuit e.m.f. of 10.0 V, an internal impedance of $(400 + j0)\Omega$ and a frequency of 1 kHz. If the line primary constants are $R = 8\ \Omega$/loop km, $L = 3$ mH/loop km, $C =$ 7500 pF/km and $G = 0.25\ \mu$S/km, determine (a) the characteristic impedance, (b) the propagation coefficient, (c) the attenuation and phase-shift coefficients, (d) the sending-end current, (e) the receiving-end current, (f) the wavelength on the line, and (g) the speed of transmission of signal.

(a) From equation (44.7),

$$\text{characteristic impedance, } Z_0 = \sqrt{\frac{R + j\omega L}{G + j\omega C}} \text{ ohms}$$

$$R + j\omega L = 8 + j(2\pi 1000)(3 \times 10^{-3})$$
$$= 8 + j6\pi = 20.48\angle 67.0^\circ\ \Omega$$

$$G + j\omega C = 0.25 \times 10^{-6} + j(2\pi 1000)(7500 \times 10^{-12})$$
$$= (0.25 + j47.12)10^{-6} = 47.12 \times 10^{-6}\angle 89.70^\circ\ \text{S}$$

Hence characteristic impedance

$$Z_0 = \sqrt{\frac{20.48\angle 67.0^\circ}{47.12 \times 10^{-6}\angle 89.70^\circ}} = \mathbf{659.3\angle{-11.35^\circ}\ \Omega}$$

(b) From equation (44.10), propagation coefficient

$$\gamma = \sqrt{[(R + j\omega L)(G + j\omega C)]} =$$
$$\sqrt{[(20.48\angle 67.0^\circ)(47.12 \times 10^{-6}\angle 89.70^\circ)]} = \mathbf{0.03106\angle 78.35^\circ}$$

(c) $\gamma = \alpha + j\beta = 0.03106(\cos 78.35^\circ + j \sin 78.35^\circ)$

$$= 0.00627 + j0.03042$$

Hence the attenuation coefficient, $\boldsymbol{\alpha = 0.00627}$ **Np/km** and the phase shift coefficient, $\boldsymbol{\beta = 0.03042}$ **rad/km**

(d) With reference to Figure 44.6, since the line is matched, i.e., terminated in its characteristic impedance, $V_S/I_S = Z_0$. Also

$$V_S = V_G - I_S Z_G = 10.0 - I_S(400 + j0)$$

Thus $I_S = \dfrac{V_S}{Z_0} = \dfrac{10.0 - 400 I_S}{Z_0}$

Rearranging gives: $I_S Z_0 = 10.0 - 400\ I_S$, from which,

$$I_S(Z_0 + 400) = 10.0$$

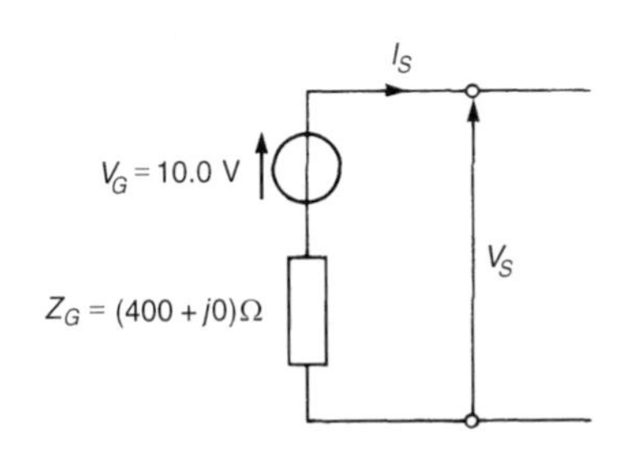

Figure 44.6

Thus the sending-end current,

$$I_S = \frac{10.0}{Z_0 + 400} = \frac{10.0}{659.3\angle -11.35^\circ + 400}$$

$$= \frac{10.0}{646.41 - j129.75 + 400} = \frac{10.0}{1054.4\angle -7.07^\circ}$$

$$= \mathbf{9.484\angle 7.07^\circ\ mA}$$

(e) From equation (44.4), the receiving-end current,

$$I_R = I_S e^{-n\gamma} = I_S e^{-n\alpha}\angle -n\beta$$

$$= (9.484\angle 7.07^\circ)e^{-(300)(0.00627)}\angle -(300)(0.03042)$$

$$= 9.484\angle 7.07^\circ e^{-1.881}\angle -9.13 \text{ rad}$$

$$= 1.446\angle -516^\circ \text{ mA} = \mathbf{1.446\angle -156^\circ\ mA}$$

(f) From equation (44.2),

$$\text{wavelength, } \lambda = \frac{2\pi}{\beta} = \frac{2\pi}{0.03042} = \mathbf{206.5\ km}$$

(g) From equation (44.3),

$$\text{speed of transmission, } u = f\lambda = (1000)(206.5)$$

$$= 206.5 \times 10^3 \text{ km/s} = \mathbf{206.5 \times 10^6\ m/s}$$

Further problems on the characteristic impedance and the propagation coefficient in terms of the primary constants may be found in Section 44.9, problems 7 to 11, page 898.

44.6 Distortion on transmission lines

If the waveform at the receiving end of a transmission line is not the same shape as the waveform at the sending end, **distortion** is said to have occurred. The three main causes of distortion on transmission lines are as follows.

(i) The characteristic impedance Z_0 of a line varies with the operating frequency, i.e., from equation (44.7),

$$Z_0 = \sqrt{\frac{R + j\omega L}{G + j\omega C}} \text{ ohms}$$

The terminating impedance of the line may not vary with frequency in the same manner.

In the above equation for Z_0, if the frequency is very low, ω is low and $Z_0 \approx \sqrt{(R/G)}$. If the frequency is very high, then $\omega L \gg R$,

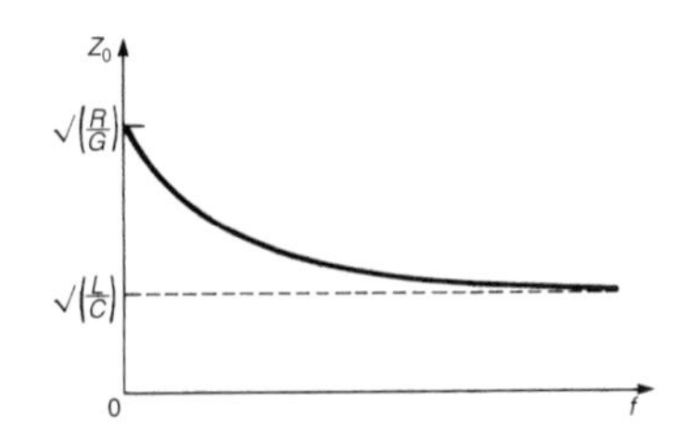

Figure 44.7

$\omega C \gg G$ and $Z_0 \approx \sqrt{(L/C)}$. A graph showing the variation of Z_0 with frequency f is shown in Figure 44.7.

If the characteristic impedance is to be constant throughout the entire operating frequency range then the following condition is required: $\sqrt{(L/C)} = \sqrt{(R/G)}$, i.e., $L/C = R/G$, from which

$$\boxed{LG = CR} \tag{44.12}$$

Thus, in a transmission line, if $LG = CR$ it is possible to provide a termination equal to the characteristic impedance Z_0 at all frequencies.

(ii) The attenuation of a line varies with the operating frequency (since $\gamma = \sqrt{[(R + j\omega L)(G + j\omega C)]}$, from equation (44.10)), thus waves of differing frequencies and component frequencies of complex waves are attenuated by different amounts.

From the above equation for the propagation coefficient:

$$\gamma^2 = (R + j\omega L)(G + j\omega C)$$
$$= RG + j\omega(LG + CR) - \omega^2 LC$$

If $LG = CR = x$, then $LG + CR = 2x$ and $LG + CR$ may be written as $2\sqrt{x^2}$, i.e., $LG + CR$ may be written as $2\sqrt{[(LG)(CR)]}$.

Thus $\gamma^2 = RG + j\omega(2\sqrt{[(LG)(CR)]}) - \omega^2 LC$

$$= [\sqrt{(RG)} + j\omega\sqrt{(LC)}]^2$$

and $\gamma = \sqrt{(RG)} + j\omega\sqrt{(LC)}$

Since

$\gamma = \alpha + j\beta$, **attenuation coefficient,** $\boxed{\boldsymbol{\alpha = \sqrt{(RG)}}}$ (44.13)

and **phase shift coefficient,** $\boxed{\boldsymbol{\beta = \omega\sqrt{(LC)}}}$ (44.14)

Thus, in a transmission line, if $LG = CR$, $\alpha = \sqrt{(RG)}$, i.e., the attenuation coefficient is independent of frequency and all frequencies are equally attenuated.

(iii) The delay time, or the time of propagation, and thus the velocity of propagation, varies with frequency and therefore waves of different frequencies arrive at the termination with differing delays. From equation (44.14), the phase-shift coefficient, $\beta = \omega\sqrt{(LC)}$ when $LG = CR$.

$$\text{Velocity of propagation, } u = \frac{\omega}{\beta} = \frac{\omega}{\omega\sqrt{(LC)}} = \frac{1}{\sqrt{(LC)}} \tag{44.15}$$

Thus, in a transmission line, if $LG = CR$, the velocity of propagation, and hence the time delay, is independent of frequency.

From the above it appears that the condition $LG = CR$ is appropriate for the design of a transmission line, since under this condition no distortion is introduced. This means that the signal at the receiving end is the same as the sending-end signal except that it is reduced in amplitude and delayed by a fixed time. Also, with no distortion, the attenuation on the line is a minimum. In practice, however, $R/L \gg G/C$. The inductance is usually low and the capacitance is large and not easily reduced. Thus if the condition $LG = CR$ is to be achieved in practice, either L or G must be increased since neither C or R can really be altered. It is undesirable to increase G since the attenuation and power losses increase. Thus the inductance L is the quantity that needs to be increased and such an artificial increase in the line inductance is called **loading**. This is achieved either by inserting inductance coils at intervals along the transmission line — this being called **'lumped loading'** — or by wrapping the conductors with a high-permeability metal tape — this being called **'continuous loading'**.

Problem 11. An underground cable has the following primary constants: resistance $R = 10\ \Omega$/loop km, inductance $L = 1.5$ mH/loop km, conductance $G = 1.2\ \mu$S/km and capacitance $C = 0.06\ \mu$F/km. Determine by how much the inductance should be increased to satisfy the condition for minimum distortion.

From equation (44.12), the condition for minimum distortion is given by $LG = CR$, from which,

$$\text{inductance } L = \frac{CR}{G} = \frac{(0.06 \times 10^{-6})(10)}{1.2 \times 10^{-6}} = 0.5 \text{ H or } 500 \text{ mH}$$

Thus the inductance should be increased by $(500 - 1.5)$ mH, i.e., **498.5 mH** per loop km, for minimum distortion.

Problem 12. A cable has the following primary constants: resistance $R = 80\ \Omega$/loop km, conductance, $G = 2\ \mu$S/km, and capacitance $C = 5$ nF/km. Determine, for minimum distortion at a frequency of 1.5 kHz (a) the value of inductance per loop kilometre required, (b) the propagation coefficient, (c) the velocity of propagation of signal, and (d) the wavelength on the line

(a) From equation (44.12), for minimum distortion, $LG = CR$, from which, inductance per loop kilometre,

$$L = \frac{CR}{G} = \frac{(5 \times 10^{-9})(80)}{(2 \times 10^{-6})} = \mathbf{0.20\ H\ or\ 200\ mH}$$

(b) From equation (44.13), attenuation coefficient,

$$\alpha = \sqrt{(RG)} = \sqrt{[(80)(2 \times 10^{-6})]} = 0.0126 \text{ Np/km}$$

and from equation (44.14), phase shift coefficient,

$$\beta = \omega\sqrt{(LC)} = (2\pi 1500)\sqrt{[(0.20)(5 \times 10^{-9})]} = 0.2980 \text{ rad/km}$$

Hence the propagation coefficient,

$$\gamma = \alpha + j\beta = \mathbf{(0.0126 + j0.2980)} \text{ or } \mathbf{0.2983\angle 87.58^\circ}$$

(c) From equation (44.15), velocity of propagation,

$$u = \frac{1}{\sqrt{(LC)}} = \frac{1}{\sqrt{[(0.2)(5 \times 10^{-9})]}}$$
$$= \mathbf{31\,620\ km/s} \text{ or } \mathbf{31.62 \times 10^6\ m/s}$$

(d) Wavelength, $\lambda = \dfrac{u}{f} = \dfrac{31.62 \times 10^6}{1500} \text{ m} = \mathbf{21.08\ km}$

Further problems on distortion on transmission lines may be found in Section 44.9, problems 12 and 13, page 899.

44.7 Wave reflection and the reflection coefficient

In earlier sections of this chapter it was assumed that the transmission line had been properly terminated in its characteristic impedance or regarded as an infinite line. In practice, of course, all lines have a definite length and often the terminating impedance does not have the same value as the characteristic impedance of the line. When this is the case, the transmission line is said to have a **'mismatched load'**.

The forward-travelling wave moving from the source to the load is called the **incident wave** or the sending-end wave. With a mismatched load the termination will absorb only a part of the energy of the incident wave, the remainder being forced to return back along the line toward the source. This latter wave is called the **reflected wave**.

Electrical energy is transmitted by a transmission line; when such energy arrives at a termination that has a value different from the characteristic impedance, it experiences a sudden change in the impedance of the medium. When this occurs, some reflection of incident energy occurs and the reflected energy is lost to the receiving load. (Reflections commonly occur in nature when a change of transmission medium occurs; for example, sound waves are reflected at a wall, which can produce echoes, and light rays are reflected by mirrors.)

If a transmission line is terminated in its characteristic impedance, no reflection occurs; if terminated in an open circuit or a short circuit, total reflection occurs, i.e., the whole of the incident wave reflects along the line. Between these extreme possibilities, all degrees of reflection are possible.

Open-circuited termination

If a length of transmission line is open-circuited at the termination, no current can flow in it and thus no power can be absorbed by the termination. This condition is achieved if a current is imagined to be reflected from the termination, the reflected current having the same magnitude as the incident wave but with a phase difference of 180°. Also, since no power is absorbed at the termination (it is all returned back along the line), the reflected voltage wave at the termination must be equal to the incident wave. Thus the voltage at the termination must be doubled by the open circuit. The resultant current (and voltage) at any point on the transmission line and at any instant of time is given by the sum of the currents (and voltages) due to the incident and reflected waves (see Section 44.8).

Short-circuit termination

If the termination of a transmission line is short-circuited, the impedance is zero, and hence the voltage developed across it must be zero. As with the open-circuit condition, no power is absorbed by the termination. To obtain zero voltage at the termination, the reflected voltage wave must be equal in amplitude but opposite in phase (i.e., 180° phase difference) to the incident wave. Since no power is absorbed, the reflected current wave at the termination must be equal to the incident current wave and thus the current at the end of the line must be doubled at the short circuit. As with the open-circuited case, the resultant voltage (and current) at any point on the line and at any instant of time is given by the sum of the voltages (and currents) due to the incident and reflected waves.

Energy associated with a travelling wave

A travelling wave on a transmission line may be thought of as being made up of electric and magnetic components. Energy is stored in the magnetic field due to the current (energy $= \frac{1}{2}LI^2$ — see page 751) and energy is stored in the electric field due to the voltage (energy $= \frac{1}{2}CV^2$ — see page 738). It is the continual interchange of energy between the magnetic and electric fields, and *vice versa*, that causes the transmission of the total electromagnetic energy along the transmission line.

When a wave reaches an open-circuited termination the magnetic field collapses since the current I is zero. Energy cannot be lost, but it can change form. In this case it is converted into electrical energy, adding to that already caused by the existing electric field. The voltage at the termination consequently doubles and this increased voltage starts the movement of a reflected wave back along the line. A magnetic field will be set up by this movement and the total energy of the reflected wave will again be shared between the magnetic and electric field components.

When a wave meets a short-circuited termination, the electric field collapses and its energy changes form to the magnetic energy. This results in a doubling of the current.

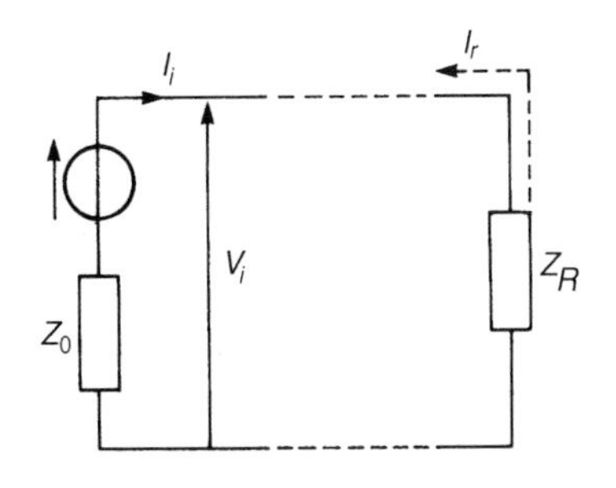

Figure 44.8

Reflection coefficient

Let a generator having impedance Z_0 (this being equal to the characteristic impedance of the line) be connected to the input terminals of a transmission line which is terminated in an impedance Z_R, where $Z_0 \neq Z_R$, as shown in Figure 44.8. The sending-end or incident current I_i flowing from the source generator flows along the line and, until it arrives at the termination Z_R behaves as though the line were infinitely long or properly terminated in its characteristic impedance, Z_0.

The incident voltage V_i shown in Figure 44.8 is given by:

$$V_i = I_i Z_0 \tag{44.12}$$

from which, $$I_i = \frac{V_i}{Z_0} \tag{44.13}$$

At the termination, the conditions must be such that:

$$Z_R = \frac{\text{total voltage}}{\text{total current}}$$

Since $Z_R \neq Z_0$, part of the incident wave will be reflected back along the line from the load to the source. Let the reflected voltage be V_r and the reflected current be I_r. Then

$$V_r = -I_r Z_0 \tag{44.14}$$

from which, $$I_r = -\frac{V_r}{Z_0} \tag{44.15}$$

(Note the minus sign, since the reflected voltage and current waveforms travel in the opposite direction to the incident waveforms.)

Thus, at the termination,

$$Z_R = \frac{\text{total voltage}}{\text{total current}} = \frac{V_i + V_r}{I_i + I_r}$$

$$= \frac{I_i Z_0 - I_r Z_0}{I_i + I_r} \text{ from equations (44.12) and (44.14)}$$

i.e., $$Z_R = \frac{Z_0(I_i - I_r)}{(I_i + I_r)}$$

Hence $$Z_R(I_i + I_r) = Z_0(I_i - I_r)$$

$$Z_R I_i + Z_R I_r = Z_0 I_i - Z_0 I_r$$

$$Z_0 I_r + Z_R I_r = Z_0 I_i - Z_R I_i$$

$$I_r(Z_0 + Z_R) = I_i(Z_0 - Z_R)$$

from which $$\frac{I_r}{I_i} = \frac{Z_0 - Z_r}{Z_0 + Z_r}$$

The ratio of the reflected current to the incident current is called the **reflection coefficient** and is often given the symbol ρ, i.e.,

$$\boxed{\frac{I_r}{I_i} = \rho = \frac{Z_0 - Z_R}{Z_0 + Z_R}} \qquad (44.16)$$

By similar reasoning to above an expression for the ratio of the reflected to the incident voltage may be obtained. From above,

$$Z_R = \frac{V_i + V_r}{I_i + I_r} = \frac{V_i + V_r}{(V_i/Z_0) - (V_r/Z_0)}$$

from equations (44.13) and (44.15),

i.e., $Z_R = \dfrac{V_i + V_r}{(V_i - V_r)/Z_0}$

Hence $\dfrac{Z_R}{Z_0}(V_i - V_r) = V_i + V_r$

from which, $\dfrac{Z_R}{Z_0}V_i - \dfrac{Z_R}{Z_0}V_r = V_i + V_r$

Then $\dfrac{Z_R}{Z_0}V_i - V_i = V_r + \dfrac{Z_R}{Z_0}V_r$

and $V_i\left(\dfrac{Z_R}{Z_0} - 1\right) = V_r\left(1 + \dfrac{Z_R}{Z_0}\right)$

Hence $V_i\left(\dfrac{Z_R - Z_0}{Z_0}\right) = V_r\left(\dfrac{Z_0 + Z_R}{Z_0}\right)$

from which $\dfrac{V_r}{V_i} = \dfrac{Z_R - Z_0}{Z_0 + Z_R} = -\left(\dfrac{Z_0 - Z_R}{Z_0 + Z_R}\right) \qquad (44.17)$

Hence $\boxed{\dfrac{V_r}{V_i} = -\dfrac{I_r}{I_i} = -\rho} \qquad (44.18)$

Thus the ratio of the reflected to the incident voltage has the same magnitude as the ratio of reflected to incident current, but is of opposite sign. From equations (44.16) and (44.17) it is seen that when $Z_R = Z_0$, $\rho = 0$ and there is no reflection.

Problem 13. A cable which has a characteristic impedance of 75 Ω is terminated in a 250 Ω resistive load. Assuming that the cable has negligible losses and the voltage measured across the terminating load is 10 V, calculate the value of (a) the reflection coefficient for the line, (b) the incident current, (c) the incident voltage, (d) the reflected current, and (e) the reflected voltage.

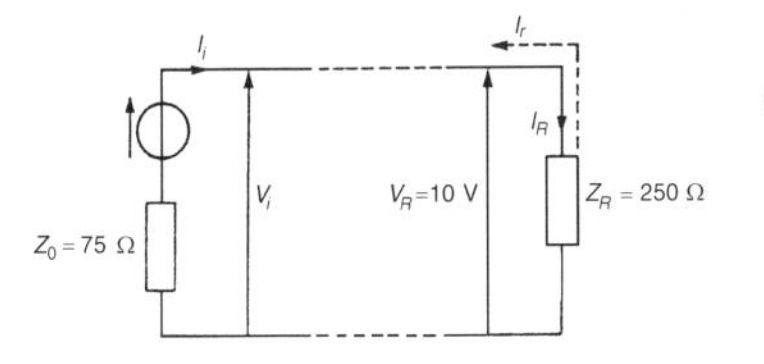

Figure 44.9

(a) From equation (44.16),

$$\textbf{reflection coefficient, } \rho = \frac{Z_0 - Z_R}{Z_0 + Z_R} = \frac{75 - 250}{75 + 250} = \frac{-175}{325} = \mathbf{-0.538}$$

(b) The circuit diagram is shown in Figure 44.9. Current flowing in the terminating load,

$$I_R = \frac{V_R}{Z_R} = \frac{10}{250} = 0.04 \text{ A}$$

However, current $I_R = I_i + I_r$. From equation (44.16), $I_r = \rho I_i$

Thus $I_R = I_i + \rho I_i = I_i(1 + \rho)$

from which **incident current,** $\boldsymbol{I_i} = \dfrac{I_R}{(1 + \rho)}$

$$= \frac{0.04}{1 + (-0.538)} = \mathbf{0.0866\ A} \text{ or } \mathbf{86.6\ mA}$$

(c) From equation (44.12),

incident voltage, $\boldsymbol{V_i} = I_i Z_0 = (0.0866)(75) = \mathbf{6.50\ V}$

(d) Since $I_R = I_i + I_r$

reflected current, $\boldsymbol{I_r} = I_R - I_i = 0.04 - 0.0866$

$$= \mathbf{-0.0466\ A} \text{ or } \mathbf{-46.6\ mA}$$

(e) From equation (44.14),

reflected voltage, $\boldsymbol{V_r} = -I_r Z_0 = -(-0.0466)(75) = \mathbf{3.50\ V}$

Problem 14. A long transmission line has a characteristic impedance of $(500 - j40)\Omega$ and is terminated in an impedance of (a) $(500 + j40)\Omega$ and (b) $(600 + j20)\Omega$. Determine the magnitude of the reflection coefficient in each case.

(a) From equation (44.16), reflection coefficient,

$$\rho = \frac{Z_0 - Z_R}{Z_0 + Z_R}$$

When $Z_0 = (500 - j40)\Omega$ and $Z_R = (500 + j40)\Omega$

$$\rho = \frac{(500 - j40) - (500 + j40)}{(500 - j40) + (500 + j40)} = \frac{-j80}{1000} = -j0.08$$

Hence the magnitude of the reflection coefficient, $|\rho| = \mathbf{0.08}$

(b) When $Z_0 = (500 - j40)\Omega$ and $Z_R = (600 + j20)\Omega$

$$\rho = \frac{(500 - j40) - (600 + j20)}{(500 - j40) + (600 + j20)} = \frac{-100 - j60}{1100 - j20}$$

$$= \frac{116.62\angle - 149.04^\circ}{1100.18\angle{-1.04^\circ}}$$

$$= 0.106\angle{-148^\circ}$$

Hence the magnitude of the reflection coefficient, $|\rho| = \mathbf{0.106}$

Problem 15. A loss-free transmission line has a characteristic impedance of $500\angle 0^\circ\ \Omega$ and is connected to an aerial of impedance $(320 + j240)\Omega$. Determine (a) the magnitude of the ratio of the reflected to the incident voltage wave, and (b) the incident voltage if the reflected voltage is $20\angle 35^\circ$ V

(a) From equation (44.17), the ratio of the reflected to the incident voltage is given by:

$$\frac{V_r}{V_i} = \frac{Z_R - Z_0}{Z_R + Z_0}$$

where Z_0 is the characteristic impedance $500\angle 0^\circ\ \Omega$ and Z_R is the terminating impedance $(320 + j240)\Omega$.

Thus $$\frac{V_r}{V_i} = \frac{(320 + j240) - 500\angle 0^\circ}{500\angle 0^\circ + (320 + j240)} = \frac{-180 + j240}{820 + j240}$$

$$= \frac{300\angle 126.87^\circ}{854.4\angle 16.31^\circ} = 0.351\angle 110.56^\circ$$

Hence the magnitude of the ratio $V_r : V_i$ is **0.351**

(b) Since $V_r/V_i = 0.351\angle 110.56^\circ$,

$$\text{incident voltage, } V_i = \frac{V_r}{0.351\angle 110.56^\circ}$$

Thus, when $V_r = 20\angle 35^\circ$ V,

$$V_i = \frac{20\angle 35^\circ}{0.351\angle 110.56^\circ} = \mathbf{57.0\angle{-75.56^\circ}\ V}$$

Further problems on the reflection coefficient may be found in Section 44.9, problems 14 to 16, page 899.

44.8 Standing waves and the standing wave ratio

Consider a lossfree transmission line **open-circuited** at its termination. An incident current waveform is completely reflected at the termination, and, as stated in Section 44.7, the reflected current is of the same magnitude as the incident current but is 180° out of phase. Figure 44.10(a) shows the incident and reflected current waveforms drawn separately (shown as

I_i moving to the right and I_r moving to the left respectively) at a time $t = 0$, with $I_i = 0$ and decreasing at the termination.

The resultant of the two waves is obtained by adding them at intervals. In this case the resultant is seen to be zero. Figures 44.10(b) and (c) show the incident and reflected waves drawn separately as times $t = T/8$ seconds and $t = T/4$, where T is the periodic time of the signal. Again, the resultant is obtained by adding the incident and reflected waveforms at intervals. Figures 44.10(d) to (h) show the incident and reflected

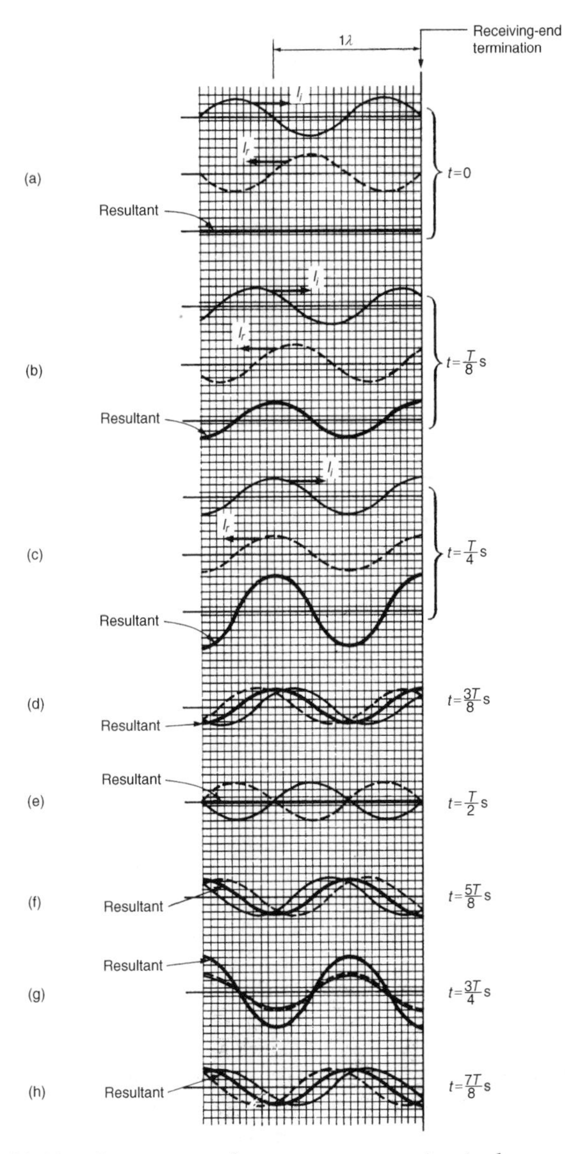

Figure 44.10 *Current waveforms on an open-circuited transmission line*

current waveforms plotted on the same axis, together with their resultant waveform, at times $t = 3T/8$ to $t = 7T/8$ at intervals of $T/8$.

If the resultant waveforms shown in Figures 44.10(a) to (g) are superimposed one upon the other, Figure 44.11 results. (Note that the scale has been increased for clarity.) The waveforms show clearly that waveform (a) moves to (b) after $T/8$, then to (c) after a further period of $T/8$, then to (d), (e), (f), (g) and (h) at intervals of $T/8$. It is noted that at any particular point the current varies sinusoidally with time, but the amplitude of oscillation is different at different points on the line.

Whenever two waves of the same frequency and amplitude travelling in opposite directions are superimposed on each other as above, interference takes place between the two waves and a **standing** or **stationary wave** is produced. The points at which the current is always zero are called **nodes** (labelled N in Figure 44.11). The standing wave does not progress to the left or right and the nodes do not oscillate. Those points on the wave that undergo maximum disturbance are called **antinodes** (labelled A in Figure 44.11). The distance between adjacent nodes or adjacent antinodes is $\lambda/2$, where λ is the wavelength. A standing wave is therefore seen to be a periodic variation in the vertical plane taking place on the transmission line without travel in either direction.

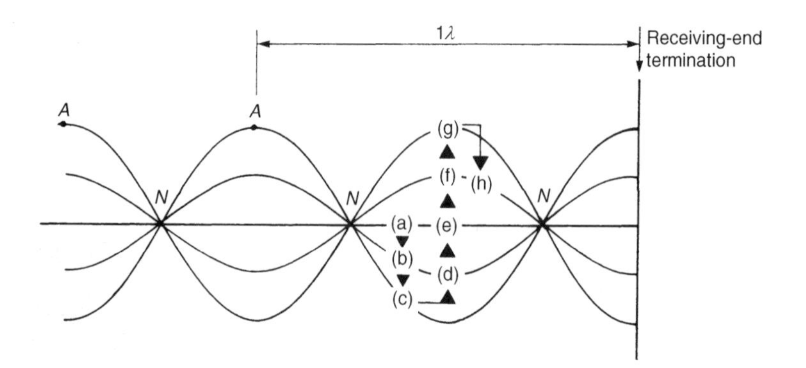

Figure 44.11

The resultant of the incident and reflected voltage for the open-circuit termination may be deduced in a similar manner to that for current. However, as stated in Section 44.7, when the incident voltage wave reaches the termination it is reflected without phase change. Figure 44.12 shows the resultant waveforms of incident and reflected voltages at intervals of $t = T/8$. Figure 44.13 shows all the resultant waveforms of Figure 44.12(a) to (h) superimposed; again, standing waves are seen to result. Nodes (labelled N) and antinodes (labelled A) are shown in Figure 44.13 and, in comparison with the current waves, are seen to occur 90° out of phase.

If the transmission line is short-circuited at the termination, it is the incident current that is reflected without phase change and the incident voltage that is reflected with a phase change of 180°. Thus the diagrams shown in Figures 44.10 and 44.11 representing current at an

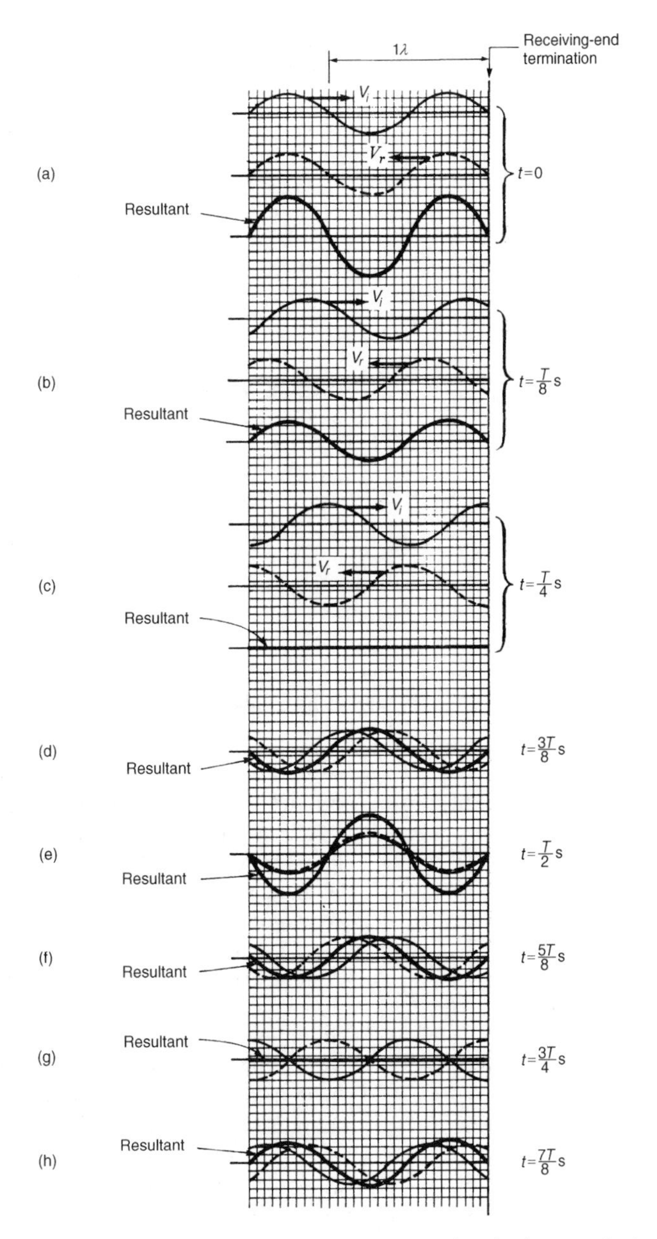

Figure 44.12 *Voltage waveforms on an open-circuited transmission line*

open-circuited termination may be used to represent voltage conditions at a short-circuited termination and the diagrams shown in Figures 44.12 and 44.13 representing voltage at an open-circuited termination may be used to represent current conditions at a short-circuited termination.

Figure 44.14 shows the rms current and voltage waveforms plotted on the same axis against distance for the case of total reflection, deduced

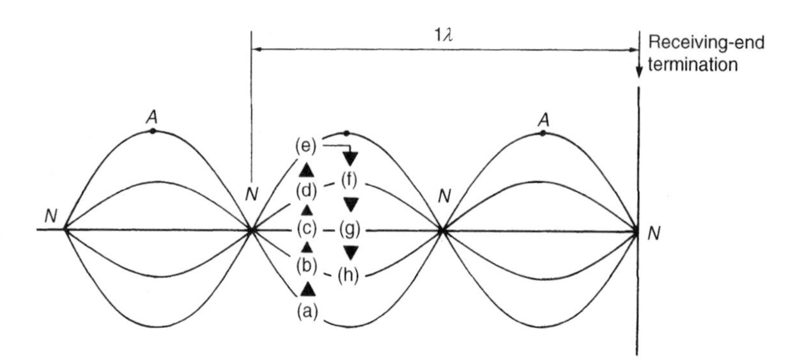

Figure 44.13

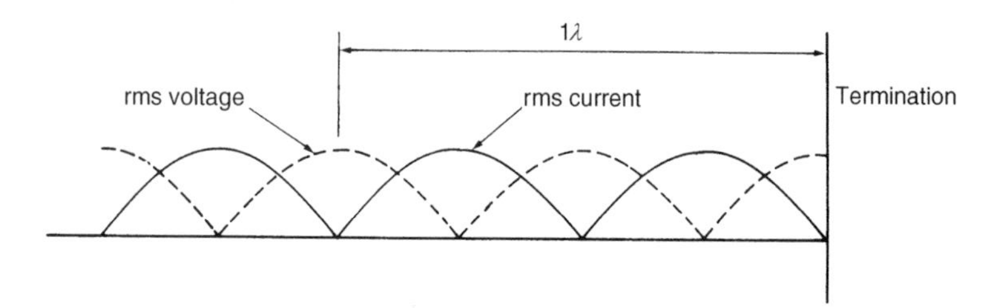

Figure 44.14

from Figures 44.11 and 44.13. The rms values are equal to the amplitudes of the waveforms shown in Figures 44.11 and 44.13, except that they are each divided by $\sqrt{2}$ (since, for a sine wave, rms value $= (1/\sqrt{2}) \times$ maximum value). With total reflection, the standing-wave patterns of rms voltage and current consist of a succession of positive sine waves with the voltage node located at the current antinode and the current node located at the voltage antinode. The termination is a current nodal point. The rms values of current and voltage may be recorded on a suitable rms instrument moving along the line. Such measurements of the maximum and minimum voltage and current can provide a reasonably accurate indication of the wavelength, and also provide information regarding the amount of reflected energy relative to the incident energy that is absorbed at the termination, as shown below.

Standing-wave ratio

Let the incident current flowing from the source of a mismatched low-loss transmission line be I_i and the current reflected at the termination be I_r. If I_{MAX} is the sum of the incident and reflected current, and I_{MIN} is their difference, then the **standing-wave ratio** (symbol **s**) on the line is defined as:

$$\boxed{s = \frac{I_{MAX}}{I_{MIN}} = \frac{I_i + I_r}{I_i - I_r}} \qquad (44.19)$$

Hence $s(I_i - I_r) = I_i + I_r$

$$sI_i - sI_r = I_i + I_r$$

$$sI_i - I_i = sI_r + I_r$$

$$I_i(s-1) = I_r(s+1)$$

i.e., $$\boldsymbol{\frac{I_r}{I_i} = \left(\frac{s-1}{s+1}\right)} \qquad (44.20)$$

The power absorbed in the termination $P_t = I_i^2 Z_0$ and the reflected power, $P_r = I_r^2 Z_0$. Thus $\dfrac{P_r}{P_t} = \dfrac{I_r^2 Z_0}{I_i^2 Z_0} = \left(\dfrac{I_r}{I_i}\right)^2$

Hence, from equation (44.20),

$$\boldsymbol{\frac{P_r}{P_t} = \left(\frac{s-1}{s+1}\right)^2} \qquad (44.21)$$

Thus the ratio of the reflected to the transmitted power may be calculated directly from the standing-wave ratio, which may be calculated from measurements of I_{MAX} and I_{MIN}. When a transmission line is properly terminated there is no reflection, i.e., $I_r = 0$, and from equation (44.19) the standing-wave ratio is 1. From equation (44.21), when $s = 1$, $P_r = 0$, i.e., there is no reflected power. In practice, the standing-wave ratio is kept as close to unity as possible.

From equation (44.16), the reflection coefficient, $\rho = I_r/I_i$ Thus, from equation (44.20), $|\rho| = \dfrac{s-1}{s+1}$

Rearranging gives: $|\rho|(s+1) = (s-1)$

$$|\rho|s + |\rho| = s - 1$$

$$1 + |\rho| = s(1 - |\rho|)$$

from which $$\boldsymbol{s = \frac{1+|\rho|}{1-|\rho|}} \qquad (44.22)$$

Equation (44.22) gives an expression for the standing-wave ratio in terms of the magnitude of the reflection coefficient.

Problem 16. A transmission line has a characteristic impedance of $600\angle 0^\circ\ \Omega$ and negligible loss. If the terminating impedance of the line is $(400 + j250)\Omega$, determine (a) the reflection coefficient and (b) the standing-wave ratio.

(a) From equation (44.16),

$$\text{reflection coefficient, } \rho = \frac{Z_0 - Z_R}{Z_0 + Z_R} = \frac{600\angle 0° - (400 + j250)}{600\angle 0° + (400 + j250)}$$

$$= \frac{200 - j250}{1000 + j250} = \frac{320.16\angle{-51.34°}}{1030.78\angle 14.04°}$$

Hence $\rho = \mathbf{0.3106\angle - 65.38°}$

(b) From above, $|\rho| = 0.3106$. Thus from equation (44.22),

$$\text{standing-wave ratio, } s = \frac{1 + |\rho|}{1 - |\rho|} = \frac{1 + 0.3106}{1 - 0.3106} = \mathbf{1.901}$$

> Problem 17. A low-loss transmission line has a mismatched load such that the reflection coefficient at the termination is $0.2\angle{-120°}$. The characteristic impedance of the line is 80 Ω. Calculate (a) the standing-wave ratio, (b) the load impedance, and (c) the incident current flowing if the reflected current is 10 mA.

(a) From equation (44.22),

$$\textbf{standing-wave ratio, } s = \frac{1 + |\rho|}{1 - |\rho|} = \frac{1 + 0.2}{1 - 0.2} = \frac{1.2}{0.8} = \mathbf{1.5}$$

(b) From equation (44.16) reflection coefficient, $\rho = \dfrac{Z_0 - Z_R}{Z_0 + Z_R}$

Rearranging gives: $\rho(Z_0 + Z_R) = Z_0 - Z_R$,

from which $\quad Z_R(\rho + 1) = Z_0(1 - \rho)$

and $$\frac{Z_R}{Z_0} = \frac{1 - \rho}{1 + \rho} = \frac{1 - 0.2\angle{-120°}}{1 + 0.2\angle{-120°}} = \frac{1 - (-0.10 - j0.173)}{1 + (-0.10 - j0.173)}$$

$$= \frac{1.10 + j0.173}{0.90 - j0.173} = \frac{1.1135\angle 8.94°}{0.9165\angle{-10.88°}}$$

$$= 1.215\angle 19.82°$$

Hence load impedance $Z_R = Z_0(1.215\angle 19.82°) = (80)(1.215\angle 19.82°)$

$$= \mathbf{97.2\angle 19.82°\ \Omega} \text{ or } \mathbf{(91.4 + j33.0)\Omega}$$

(c) From equation (44.20),

$$\frac{I_r}{I_i} = \frac{s - 1}{s + 1}$$

Hence $$\frac{10}{I_i} = \frac{1.5 - 1}{1.5 + 1} = \frac{0.5}{2.5} = 0.2$$

Thus the **incident current,** $I_i = 10/0.2 = \mathbf{50\ mA}$

Problem 18. The standing-wave ratio on a mismatched line is calculated as 1.60. If the incident power arriving at the termination is 200 mW, determine the value of the reflected power.

From equation (44.21),

$$\frac{P_r}{P_t} = \left(\frac{s-1}{s+1}\right)^2 = \left(\frac{1.60-1}{1.60+1}\right)^2 = \left(\frac{0.60}{2.60}\right)^2 = 0.0533$$

Hence the **reflected power,** $\boldsymbol{P_r} = 0.0533P_t = (0.0533)(200)$

$= \mathbf{10.66\ mW}$

Further problems on the standing wave ratio may be found in Section 44.9 following, problems 17 to 21, page 899.

44.9 Further problems on transmission lines

Phase delay, wavelength and velocity of propagation

1 A parallel-wire air-spaced line has a phase-shift of 0.03 rad/km. Determine (a) the wavelength on the line, and (b) the speed of transmission of a signal of frequency 1.2 kHz.
[(a) 209.4 km (b) 251.3×10^6 m/s]

2 A transmission line has an inductance of 5 μH/m and a capacitance of 3.49 pF/m. Determine, for an operating frequency of 5 kHz, (a) the phase delay, (b) the wavelength on the line and (c) the velocity of propagation of the signal in metres per second.
[(a) 0.131 rad/km (b) 48 km (c) 240×10^6 m/s]

3 An air-spaced transmission line has a capacitance of 6.0 pF/m and the velocity of propagation of a signal is 225×10^6 m/s. If the operating frequency is 20 kHz, determine (a) the inductance per metre, (b) the phase delay, and (c) the wavelength on the line.
[(a) 3.29 μH/m (b) 0.558×10^{-3} rad/m (c) 11.25 km]

Current and voltage relationships

4 When the working frequency of a cable is 1.35 kHz, its attenuation is 0.40 Np/km and its phase-shift is 0.25 rad/km. The sending-end voltage and current are 8.0 V rms and 10.0 mA rms. Determine the voltage and current at a point 25 km down the line, assuming that the termination is equal to the characteristic impedance of the line.
[$V_R = 0.363\angle{-6.25}$ mV or $0.363\angle 1.90^\circ$ mV
$I_R = 0.454\angle{-6.25}$ μA or $0.454\angle 1.90^\circ$ μA]

5 A transmission line 8 km long has a characteristic impedance $600\angle{-30^\circ}\ \Omega$. At a particular frequency the attenuation coefficient of the line is 0.4 Np/km and the phase-shift coefficient is 0.20 rad/km.

Determine the magnitude and phase of the current at the receiving end if the sending-end voltage is $5\angle 0°$ V rms. [$0.340\angle -61.67$ mA]

6 The voltages at the input and at the output of a transmission line properly terminated in its characteristic impedance are 10 V and 4 V rms respectively. Determine the output voltage if the length of the line is trebled. [0.64 V]

Characteristic impedance and propagation constant

7 At a frequency of 800 Hz, the open-circuit impedance of a length of transmission line is measured as $500\angle -35°$ Ω and the short-circuit impedance as $300\angle -15°$ Ω. Determine the characteristic impedance of the line at this frequency. [$387.3\angle -25°$ Ω]

8 A transmission line has the following primary constants per loop kilometre run: $R = 12$ Ω, $L = 3$ mH, $G = 4$ μS and $C = 0.02$ μF. Determine the characteristic impedance of the line when the frequency is 750 Hz. [$443.3\angle -18.95°$ Ω]

9 A transmission line having negligible losses has primary constants: inductance $L = 1.0$ mH/loop km and capacitance $C = 0.20$ μF/km. Determine, at an operating frequency of 50 kHz, (a) the characteristic impedance, (b) the propagation coefficient, (c) the attenuation and phase-shift coefficients, (d) the wavelength on the line, and (e) the velocity of propagation of signal in metres per second.

[(a) 70.71 Ω (b) $j4.443$ (c) 0; 4.443 rad/km
(d) 1.414 km (e) 70.71×10^6 m/s]

10 At a frequency of 5 kHz the primary constants of a transmission line are: resistance $R = 12$ Ω/loop km, inductance $L = 0.50$ mH/loop km, capacitance $C = 0.01$ μF/km and $G = 60$ μS/km. Determine for the line (a) the characteristic impedance, (b) the propagation coefficient, (c) the attenuation coefficient, and (d) the phase-shift coefficient.

[(a) $248.6\angle -13.29°$ Ω (b) $0.0795\angle 65.91°$
(c) 0.0324 Np/km (d) 0.0726 rad/km]

11 A transmission line is 50 km in length and is terminated in its characteristic impedance. At the sending end a signal emanates from a generator which has an open-circuit e.m.f. of 20.0 V, an internal impedance of $(250 + j0)$Ω at a frequency of 1592 Hz. If the line primary constants are $R = 30$ Ω/loop km, $L = 4.0$ mH/loop km, $G = 5.0$ μS/km, and $C = 0.01$ μF/km, determine (a) the value of the characteristic impedance, (b) the propagation coefficient, (c) the attenuation and phase-shift coefficients, (d) the sending-end current, (e) the receiving-end current, (f) the wavelength on the line, and (g) the speed of transmission of a signal, in metres per second.

[(a) $706.6\angle -17°$ Ω (b) $0.0708\angle 70.14°$
(c) 0.024 Np/km; 0.067 rad/km
(d) $21.1\angle 12.58°$ mA (e) $6.35\angle -178.21°$ mA
(f) 94.34 km (g) 150.2×10^6 m/s]

Distortion on transmission lines

12 A cable has the following primary constants: resistance $R = 90\ \Omega$/loop km, inductance $L = 2.0$ mH/loop km, capacitance $C = 0.05\ \mu$F/km and conductance $G = 3.0\ \mu$S/km. Determine the value to which the inductance should be increased to satisfy the condition for minimum distortion. [1.5 H]

13 A condition of minimum distortion is required for a cable. Its primary constants are: $R = 40\ \Omega$/loop km, $L = 2.0$ mH/loop km, $G = 2.0\ \mu$S/km and $C = 0.08\ \mu$F/km. At a frequency of 100 Hz determine (a) the increase in inductance required, (b) the propagation coefficient, (c) the speed of signal transmission and (d) the wavelength on the line.
[(a) 1.598 H (b) $(8.944 + j225)10^{-3}$
(c) 2.795×10^6 m/s (d) 27.93 km]

Reflection coefficient

14 A coaxial line has a characteristic impedance of 100 Ω and is terminated in a 400 Ω resistive load. The voltage measured across the termination is 15 V. The cable is assumed to have negligible losses. Calculate for the line the values of (a) the reflection coefficient, (b) the incident current, (c) the incident voltage, (d) the reflected current, and (e) the reflected voltage.
[(a) -0.60 (b) 93.75 mA (c) 9.375 V
(d) -56.25 mA (e) 5.625 V]

15 A long transmission line has a characteristic impedance of $(400 - j50)\Omega$ and is terminated in an impedance of (i) $(400 + j50)\Omega$, (ii) $(500 + j60)\Omega$ and (iii) $400\angle 0°\ \Omega$. Determine the magnitude of the reflection coefficient in each case.
[(i) 0.125 (ii) 0.165 (iii) 0.062]

16 A transmission line which is loss-free has a characteristic impedance of $600\angle 0°\ \Omega$ and is connected to a load of impedance $(400 + j300)\Omega$. Determine (a) the magnitude of the reflection coefficient and (b) the magnitude of the sending-end voltage if the reflected voltage is 14.60 V [(a) 0.345 (b) 42.32 V]

Standing-wave ratio

17 A transmission line has a characteristic impedance of $500\angle 0°\ \Omega$ and negligible loss. If the terminating impedance of the line is $(320 + j200)\Omega$ determine (a) the reflection coefficient and (b) the standing-wave ratio. [(a) $0.319\angle -61.72°$ (b) 1.937]

18 A low-loss transmission line has a mismatched load such that the reflection coefficient at the termination is $0.5\angle -135°$. The characteristic impedance of the line is 60 Ω. Calculate (a) the standing-wave

ratio, (b) the load impedance, and (c) the incident current flowing if the reflected current is 25 mA.

[(a) 3 (b) $113.93\angle 43.32°\ \Omega$ (c) 50 mA]

19 The standing-wave ratio on a mismatched line is calculated as 2.20. If the incident power arriving at the termination is 100 mW, determine the value of the reflected power.

[14.06 mW]

20 The termination of a coaxial cable may be represented as a 150 Ω resistance in series with a 0.20 μH inductance. If the characteristic impedance of the line is $100\angle 0°\ \Omega$ and the operating frequency is 80 MHz, determine (a) the reflection coefficient and (b) the standing-wave ratio. [(a) $0.417\angle -138.35°$ (b) 2.43]

21 A cable has a characteristic impedance of $70\angle 0°\ \Omega$. The cable is terminated by an impedance of $60\angle 30°\ \Omega$. Determine the ratio of the maximum to minimum current along the line. [1.77]

Assignment 14

This assignment covers the material contained in chapters 42 to 44.

The marks for each question are shown in brackets at the end of each question.

1 A filter section is to have a characteristic impedance at zero frequency of 720 Ω and a cut-off frequency of 2 MHz. To meet these requirements, design (a) a low-pass T section filter, and (b) a low-pass π section filter. (8)

2 A filter is required to pass all frequencies above 50 kHz and to have a nominal impedance of 620 Ω. Design (a) a high-pass T section filter, and (b) a high-pass π section filter to meet these requirements. (8)

3 Design (a) a suitable 'm-derived' T section, and (b) a suitable 'm-derived' π section having a cut-off frequency of 50 kHz, a nominal impedance of 600 Ω and a frequency of infinite attenuation 30 kHz. (14)

4 Two coils, A and B, are magnetically coupled; coil A has 400 turns and a self inductance of 20 mH and coil B has 250 turns and a self inductance of 50 mH. When a current of 10 A is reversed in coil A, the change of flux in coil B is 2 mWb. Determine (a) the mutual inductance between the coils, and (b) the coefficient of coupling. (4)

5 Two mutually coupled coils P and Q are connected in series to a 200 V d.c. supply. Coil P has an inductance of 0.8 H and resistance 2 Ω; coil Q has an inductance of 1.2 H and a resistance of 5 Ω. Determine the mutual inductance between the coils if, at a certain instant after the circuit is connected, the current is 5 A and increasing at a rate of 7.5 A/s. (5)

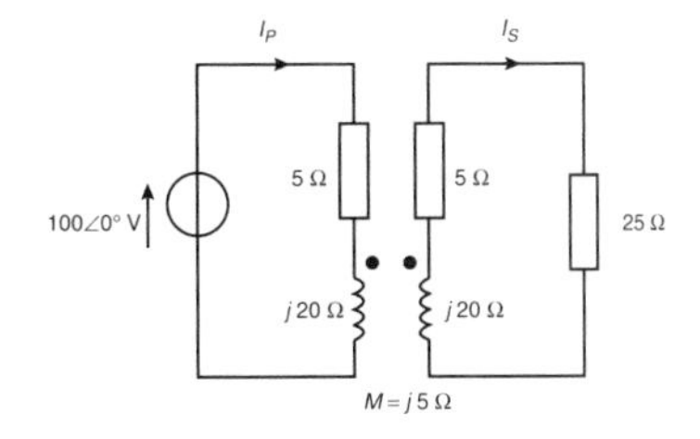

Figure A14.1

6 For the coupled circuit shown in Figure A14.1, calculate the values of currents I_P and I_S. (9)

7 A 4 km transmission line has a characteristic impedance of $600\angle -30°$ Ω. At a particular frequency, the attenuation coefficient of the line is 0.4 Np/km and the phase-shift coefficient is 0.20 rad/km. Calculate (a) the magnitude and phase of the voltage at the receiving end if the sending end voltage is $5.0\angle 0°$ V, and (b) the magnitude and phase of the receiving end current. (5)

8 The primary constants of a transmission line at a frequency of 5 kHz are: resistance, $R = 20$ Ω/loop km, inductance, $L = 3$ mH/loop km,

capacitance, $C = 50\,\text{nF/km}$, and conductance, $G = 0.4\,\text{mS/km}$. Determine for the line (a) the characteristic impedance, (b) the propagation coefficient, (c) the attenuation coefficient, (d) the phase-shift coefficient, (e) the wavelength on the line, and (f) the speed of transmission of signal. (12)

9 A loss-free transmission line has a characteristic impedance of $600\angle 0°\,\Omega$ and is connected to an aerial of impedance $(250 + j200)\,\Omega$. Determine (a) the magnitude of the ratio of the reflected to the incident voltage wave, and (b) the incident voltage if the reflected voltage is $10\angle 60°\,\Omega$. (5)

10 A low loss transmission line has a mismatched load such that the reflection coefficient at the termination is $0.5\angle -150°$. The characteristic impedance of the line is $200\,\Omega$. Determine (a) the standing wave ratio, (b) the load impedance, and (c) the incident current flowing if the reflected current is $15\,\text{mA}$. (10)

Main formulae for part 3 advanced circuit theory and technology

Complex numbers:

$z = a + jb = r(\cos\theta + j\sin\theta) = r\angle\theta$,

where $j^2 = -1$ Modulus, $r = |z| = \sqrt{(a^2 + b^2)}$

Argument, $\theta = \arg z = \arctan\dfrac{b}{a}$

Addition: $(a + jb) + (c + jb) = (a + c) + j(b + d)$

Subtraction: $(a + jb) - (c + jd) = (a - c) + j(b - d)$

Complex equations: If $a + jb = c + jd$, then $a = c$ and $b = d$

If $z_1 = r_1\angle\theta_1$ and $z_2 = r_2\angle\theta_2$ then

Multiplication: $z_1 z_2 = r_1 r_2\angle(\theta_1 + \theta_2)$ and Division: $\dfrac{z_1}{z_2} = \dfrac{r_1}{r_2}\angle(\theta_1 - \theta_2)$

De Moivre's theorem: $[r\angle\theta]^n = r^n\angle n\theta = r^n(\cos n\theta + j\sin n\theta)$

General:

$Z = \dfrac{V}{I} = R + j(X_L - X_C) = |Z|\angle\phi$,

where $|Z| = \sqrt{[R^2 + (X_L - X_C)^2]}$ and $\phi = \arctan\dfrac{X_L - X_C}{R}$

$X_L = 2\pi f L \qquad X_C = \dfrac{1}{2\pi f C} \qquad Y = \dfrac{I}{V} = \dfrac{1}{Z} = G + jB$

Series: $Z_T = Z_1 + Z_2 + Z_3 \cdots$

Parallel: $\dfrac{1}{Z_T} = \dfrac{1}{Z_1} + \dfrac{1}{Z_2} + \dfrac{1}{Z_3} + \cdots$

$P = VI\cos\phi$ or $P = I_R^2 R \qquad S = VI \qquad Q = VI\sin\phi$

Power factor $= \cos\phi = \dfrac{R}{Z}$

If $V = a + jb$ and $I = c + jd$ then $P = ac + bd$

$Q = bc - ad \qquad S = VI^* = P + jQ$

***R–L–C* series circuit:**

$$f_r = \frac{1}{2\pi\sqrt{(LC)}} \qquad Q = \frac{\omega_r L}{R} = \frac{1}{\omega_r CR} = \frac{1}{R}\sqrt{\frac{L}{C}} = \frac{V_L}{V} = \frac{V_C}{V} = \frac{f_r}{f_2 - f_1}$$

$f_r = \sqrt{(f_1 f_2)}$

***LR–C* network:**

$$f_r = \frac{1}{2\pi}\sqrt{\left(\frac{1}{LC} - \frac{R^2}{L^2}\right)} \qquad R_D = \frac{L}{CR} \qquad Q = \frac{I_C}{I_r} = \frac{\omega_r L}{R}$$

***LR–CR* network:**

$$f_r = \frac{1}{2\pi\sqrt{(LC)}}\sqrt{\left(\frac{R_L^2 - L/C}{R_C^2 - L/C}\right)}$$

Determinants:

$$\begin{vmatrix} a & b \\ c & d \end{vmatrix} = ad - bc \qquad \begin{vmatrix} a & b & c \\ d & e & f \\ g & h & j \end{vmatrix} = a\begin{vmatrix} e & f \\ h & j \end{vmatrix} - b\begin{vmatrix} d & f \\ g & j \end{vmatrix} + c\begin{vmatrix} d & e \\ g & h \end{vmatrix}$$

Delta-star:

$$Z_1 = \frac{Z_A Z_B}{Z_A + Z_B + Z_C} \text{ etc}$$

Star-delta:

$$Z_A = \frac{Z_1 Z_2 + Z_2 Z_3 + Z_3 Z_1}{Z_2} \text{ etc}$$

Impedance matching:

$$|z| = \left(\frac{N_1}{N_2}\right)^2 |Z_L|$$

Complex waveforms:

$$I = \sqrt{\left(I_0^2 + \frac{I_{1m}^2 + I_{2m}^2 + \ldots}{2}\right)}$$

$$i_{AV} = \frac{1}{\pi}\int_0^{\pi} i\,d(\omega t) \qquad \text{form factor} = \frac{\text{r.m.s}}{\text{mean}}$$

$$P = V_0 I_0 + V_1 I_1 \cos\phi_1 + V_2 I_2 \cos\phi_2 + \ldots \text{ or } P = I^2 R$$

$$\text{power factor} = \frac{P}{VI}$$

Harmonic resonance: $n\omega L = \dfrac{1}{n\omega C}$

Harmonic analysis:

$$a_0 \approx \frac{1}{p}\sum_{k=1}^{p} y_k \qquad a_n \approx \frac{2}{p}\sum_{k=1}^{p} y_k \cos nx_k$$

$$b_n \approx \frac{2}{p}\sum_{k=1}^{p} y_k \sin nx_k$$

Hysteresis and Eddy current:

Hysteresis loss/cycle $= A\alpha\beta$ J/m^3 or hysteresis loss $= k_h v f (B_m)^n$ W

Eddy current loss/cycle $= k_e (B_m)^2 f^2 t^3$ W

Dielectric loss:

Series representation: $\tan\delta = R_S\omega C_S = 1/Q$

Parallel representation: $\tan\delta = \dfrac{1}{R_p\omega C_p}$

Loss angle $\delta = (90^\circ - \phi)$

Power factor $= \cos\phi \approx \tan\delta$

Dielectric power loss $= V^2\omega C\tan\delta$

Field theory:

Coaxial cable: $C = \dfrac{2\pi\varepsilon_0\varepsilon_r}{\ln\dfrac{b}{a}}$ F/m $\quad E = \dfrac{V}{r\ln\dfrac{b}{a}}$ V/m

$$L = \frac{\mu_0\mu_r}{2\pi}\left(\frac{1}{4} + \ln\frac{b}{a}\right)\ \mathbf{H/m}$$

Twin line: $C = \dfrac{\pi\varepsilon_0\varepsilon_r}{\ln\dfrac{D}{a}}$ F/m $\quad L = \dfrac{\mu_0\mu_r}{\pi}\left(\dfrac{1}{4} + \ln\dfrac{D}{a}\right)$ H/m

Energy stored: in a capacitor, $W = \frac{1}{2}CV^2$ J; in an inductor $W = \frac{1}{2}LI^2$ J

in electric field per unit volume, $\omega_f = \frac{1}{2}DE = \frac{1}{2}\varepsilon_0\varepsilon_r E^2 = \dfrac{D^2}{2\varepsilon_0\varepsilon_r}$ J/m^3

in a non-magnetic medium, $\omega_f = \frac{1}{2}BH = \frac{1}{2}\mu_0 H^2 = \dfrac{B^2}{2\mu_0}$ J/m^3

Attenuators:

Logarithmic ratios: in decibels $= 10\ \lg\dfrac{P_2}{P_1} = 20\ \lg\dfrac{V_2}{V_1} = 20\ \lg\dfrac{I_2}{I_1}$

in nepers $= \frac{1}{2}\ln\dfrac{P_2}{P_1} = \ln\dfrac{V_2}{V_1} = \ln\dfrac{I_2}{I_1}$

Symmetrical T-attenuator: $R_0 = \sqrt{(R_1^2 + 2R_1R_2)} = \sqrt{(R_{OC}R_{SC})}$

$$R_1 = R_0\left(\frac{N-1}{N+1}\right) \qquad R_2 = R_0\left(\frac{2N}{N^2-1}\right)$$

Symmetrical π-attenuator: $R_0 = \sqrt{\left(\dfrac{R_1R_2^2}{R_1 + 2R_2}\right)} = \sqrt{(R_{OC}R_{SC})}$

$$R_1 = R_0\left(\frac{N^2-1}{2N}\right) \qquad R_2 = R_0\left(\frac{N+1}{N-1}\right)$$

L-section attenuator: $R_1 = \sqrt{[R_{OA}(R_{OA} - R_{OB})]}$

$$R_2 = \sqrt{\left(\frac{R_{OA}R_{OB}^2}{R_{OA} - R_{OB}}\right)}$$

Filter networks

Low-pass T or π: $f_C = \dfrac{1}{\pi\sqrt{(LC)}}$ $\quad R_0 = \sqrt{\dfrac{L}{C}}$

$$C = \frac{1}{\pi R_0 f_C} \qquad L = \frac{R_0}{\pi f_C}$$

$$Z_{0T} = R_0\sqrt{\left[1 - \left(\frac{\omega}{\omega_C}\right)^2\right]}$$

$$Z_{0\pi} = \frac{R_0}{\sqrt{\left[1 - \left(\frac{\omega}{\omega_C}\right)^2\right]}}$$

High-pass T or π: $f_C = \dfrac{1}{4\pi\sqrt{(LC)}}$ $\quad R_0 = \sqrt{\dfrac{L}{C}}$

$$C = \frac{1}{4\pi R_0 f_C} \qquad L = \frac{R_0}{4\pi f_C}$$

$$Z_{0T} = R_0\sqrt{\left[1 - \left(\frac{\omega_C}{\omega}\right)^2\right]}$$

$$Z_{0\pi} = \frac{R_0}{\sqrt{\left[1 - \left(\frac{\omega_C}{\omega}\right)^2\right]}}$$

Low and high-pass:

$$Z_{0T}Z_{0\pi} = Z_1Z_2 = R_0^2$$

$$\frac{I_1}{I_2} = \frac{I_2}{I_3} = \frac{I_3}{I_4} = e^{\gamma} = e^{\alpha + j\beta} = e^{\alpha}\angle\beta$$

Phase angle $\beta = \omega\sqrt{(LC)}$
time delay $= \sqrt{(LC)}$
'm-derived filter sections:

Low-pass $\quad m = \sqrt{\left[1 - \left(\dfrac{f_C}{f_\infty}\right)^2\right]}$

High pass $\quad m = \sqrt{\left[1 - \left(\dfrac{f_\infty}{f_C}\right)^2\right]}$

Magnetically coupled circuits

$$E_2 = -M\frac{dI_1}{dt} = \pm j\omega M I_1$$

$$M = N_2\frac{d\phi_2}{dI_1} = N_1\frac{d\phi_1}{dI_2} = k\sqrt{(L_1L_2)} = \frac{L_A - L_B}{4}$$

Transmission lines:

Phase delay $\beta = \omega\sqrt{(LC)}$ wavelength $\lambda = \frac{2\pi}{\beta}$

velocity of propagation $u = f\lambda = \frac{\omega}{\beta}$

$$I_R = I_S e^{-n\gamma} = I_S e^{-n\alpha}\angle -n\beta$$

$$V_R = V_S e^{-n\gamma} = V_S e^{-n\alpha}\angle -n\beta$$

$$Z_0 = \sqrt{(Z_{OC}Z_{SC})} = \sqrt{\frac{R + j\omega L}{G + j\omega C}}$$

$$\gamma = \sqrt{[(R + j\omega L)(G + j\omega C)]}$$

Reflection coefficient, $\rho = \frac{I_r}{I_i} = \frac{Z_O - Z_R}{Z_O + Z_R} = -\frac{V_r}{V_i}$

Standing-wave ratio, $s = \frac{I_{max}}{I_{min}} = \frac{I_i + I_r}{I_i - I_r} = \frac{1 + |\rho|}{1 - |\rho|}$

$$\frac{P_r}{P_t} = \left(\frac{s - 1}{s + 1}\right)^2$$

Part 4 General Reference

Standard electrical quantities—their symbols and units

QUANTITY	QUANTITY SYMBOL	UNIT	UNIT SYMBOL
Admittance	Y	siemen	S
Angular frequency	ω	radians per second	rad/s
Area	A	square metres	m^2
Attenuation coefficient (or constant)	α	neper per metre	Np/m
Capacitance	C	farad	F
Charge	Q	coulomb	C
Charge density	σ	coulomb per square metre	C/m^2
Conductance	G	siemen	S
Current	I	ampere	A
Current density	J	ampere per square metre	A/m^2
Efficiency	η	per-unit or per cent	p.u. or %
Electric field strength	E	volt per metre	V/m
Electric flux	Ψ	coulomb	C
Electric flux density	D	coulomb per square metre	C/m^2
Electromotive force	E	volt	V
Energy	W	joule	J
Field strength, electric	E	volt per metre	V/m
Field strength, magnetic	H	ampere per metre	A/m
Flux, electric	Ψ	coulomb	C
Flux, magnetic	Φ	weber	Wb
Flux density, electric	D	coulomb per square metre	C/m^2
Flux density, magnetic	B	tesla	T
Force	F	newton	N
Frequency	f	hertz	Hz
Frequency, angular	ω	radians per second	rad/s
Frequency, rotational	n	revolutions per second	rev/s
Impedance	Z	ohm	Ω
Inductance, self	L	henry	H

QUANTITY	*QUANTITY SYMBOL*	*UNIT*	*UNIT SYMBOL*
Inductance, mutual	M	henry	H
length	l	metre	m
Loss angle	δ	radian or degrees	rad or °
Magnetic field strength	H	ampere per metre	A/m
Magnetic flux	Φ	weber	Wb
Magnetic flux density	B	tesla	T
Magnetic flux linkage	Ψ	weber	Wb
Magnetising force	H	ampere per metre	A/m
Magnetomotive force	F_m	ampere	A
Mutual inductance	M	henry	H
Number of phases	m	–	–
Number of pole-pairs	p	–	–
Number of turns (of a winding)	N	–	–
Period, Periodic time	T	second	s
Permeability, absolute	μ	henry per metre	H/m
Permeability of free space	μ_0	henry per metre	H/m
Permeability, relative	μ_r	–	–
Permeance	Λ	weber per ampere or per henry	Wb/A or /H
Permittivity, absolute	ε	farad per metre	F/m
Permittivity of free space	ε_0	farad per metre	F/m
Permittivity, relative	ε_r	–	–
Phase-change coefficient	β	radian per metre	rad/m
Potential, Potential difference	V	volt	V
Power, active	P	watt	W
Power, apparent	S	volt ampere	VA
Power, reactive	Q	volt ampere reactive	var
Propagation coefficient (or constant)	γ	–	–
Quality factor, magnification	Q	–	–
Quantity of electricity	Q	coulomb	C
Reactance	X	ohm	Ω
Reflection coefficient	ρ	–	–
Relative permeability	μ_r	–	–
Relative permittivity	ε_r	–	–
Reluctance	R_m	ampere per weber or per henry	A/Wb or /H
Resistance	R	ohm	Ω

Resistance, temperature coefficient of	α	per degree Celsius or per kelvin	/°C or/K
Resistivity	ρ	ohm metre	Ωm
Slip	s	per unit or per cent	p.u. or %
Standing wave ratio	s	–	–
Susceptance	B	siemen	S
Temperature coefficient of resistance	α	per degree Celsius or per kelvin	/°C or/K
Temperature, thermodynamic	T	kelvin	K
Time	t	second	s
Torque	T	newton metre	Nm
Velocity	v	metre per second	m/s
Velocity, angular	ω	radian per second	rad/s
Volume	V	cubic metres	m^3
Wavelength	λ	metre	m

(Note that m/s may also be written as ms^{-1}, C/m^2 as Cm^{-2}, /K as K^{-1}, and so on.)

Greek alphabet

LETTER	*UPPER CASE*	*LOWER CASE*
Alpha	A	α
Beta	B	β
Gamma	Γ	γ
Delta	Δ	δ
Epsilon	E	ε
Zeta	Z	ζ
Eta	H	η
Theta	Θ	θ
Iota	I	ι
Kappa	K	κ
Lambda	Λ	λ
Mu	M	μ
Nu	N	ν
Xi	Ξ	ξ
Omicron	O	o
Pi	Π	π
Rho	P	ρ
Sigma	Σ	σ
Tau	T	τ
Upsilon	Υ	υ
Phi	Φ	ϕ
Chi	X	χ
Psi	Ψ	ψ
Omega	Ω	ω

Common prefixes

PREFIX	*NAME*	*MEANING: multiply by*
E	exa	10^{18}
P	peta	10^{15}
T	tera	10^{12}
G	giga	10^{9}
M	mega	10^{6}
k	kilo	10^{3}
h	hecto	10^{2}
da	deca	10^{1}
d	deci	10^{-1}
c	centi	10^{-2}
m	milli	10^{-3}
μ	micro	10^{-6}
n	nano	10^{-9}
p	pico	10^{-12}
f	femto	10^{-15}
a	atto	10^{-18}

Resistor colour coding and ohmic values

Colour code for fixed resistors

COLOUR	*SIGNIFICANT FIGURES*	*MULTIPLIER*	*TOLERANCE*
Silver	–	10^{-2}	± 10%
Gold	–	10^{-1}	± 5%
Black	0	1	–
Brown	1	10	± 1%
Red	2	10^2	± 2%
Orange	3	10^3	–
Yellow	4	10^4	–
Green	5	10^5	± 0.5%
Blue	6	10^6	± 0.25%
Violet	7	10^7	± 0.1%
Grey	8	10^8	–
White	9	10^9	–
None	–	–	± 20%

Thus, for a **four-band fixed resistor** (i.e. resistance values with two significant figures):

yellow-violet-orange-red indicates 47 kΩ with a tolerance of ± 2%

orange-orange-silver-brown indicates 0.33 Ω with a tolerance of ± 1%

and brown-black-brown indicates 100 Ω with a tolerance of ± 20%

(Note that the first band is the one nearest the end of the resistor).

For a five-band fixed resistor (i.e. resistance values with three significant figures):

red-yellow-white-orange-brown indicates 249 kΩ with a tolerance of ± 1%

(Note that the fifth band is 1.5 to 2 times wider than the other bands).

Letter and digit code for resistors

RESISTANCE VALUE	*MARKED AS:*
0.47 Ω	R47
1 Ω	1R0
4.7 Ω	4R7
47 Ω	47R
100 Ω	100R
1 kΩ	1K0
10 kΩ	10K
10 MΩ	10M

Tolerance is indicated as follows:

$$F = \pm 1\%,\ G = \pm 2\%,\ J = \pm 5\%,\ K = \pm 10\% \text{ and } M = \pm 20\%$$

Thus, for example, R33M = 0.33 Ω ± 20%

4R7K = 4.7 Ω ± 10%

390RJ = 390 Ω ± 5%

6K8F = 6.8 kΩ ± 1%

68KK = 68 kΩ ± 10%

4M7M = 4.7 MΩ ± 20%

Index